ML

1

R

M

RADIATIVE HEAT TRANSFER

McGraw-Hill Series in Mechanical Engineering

Consulting Editors

Jack P. Holman, Southern Methodist University
John R. Lloyd, Michigan State University

Anderson: *Modern Compressible Flow: With Historical Perspective*
Arora: *Introduction to Optimum Design*
Bray and Stanley: *Nondestructive Evaluation: A Tool for Design, Manufacturing, and Service*
Culp: *Principles of Energy Conservation*
Dally: *Packaging of Electronic Systems: A Mechanical Engineering Approach*
Dieter: *Engineering Design: A Materials and Processing Approach*
Eckert and Drake: *Analysis of Heat and Mass Transfer*
Edwards and McKee: *Fundamentals of Mechanical Component Design*
Gebhart: *Heat Conduction and Mass Diffusion*
Heywood: *Internal Combustion Engine Fundamentals*
Hinze: *Turbulence*
Howell and Buckius: *Fundamentals of Engineering Thermodynamics*
Hutton: *Applied Mechanical Vibrations*
Juvinall: *Engineering Considerations of Stress, Strain, and Strength*
Kane and Levinson: *Dynamics: Theory and Applications*
Kays and Crawford: *Convective Heat and Mass Transfer*
Kelly: *Fundamentals of Mechanical Vibrations*
Kimbrell: *Kinematics Analysis and Synthesis*
Martin: *Kinematics and Dynamics of Machines*
Modest: *Radiative Heat Transfer*
Norton: *Design of Machinery*
Phelan: *Fundamentals of Mechanical Design*
Raven: *Automatic Control Engineering*
Reddy: *Introduction to the Finite Element Method*
Rosenberg and Karnopp: *Introduction to Physics*
Schlichting: *Boundary-Layer Theory*
Shames: *Mechanics of Fluids*
Sherman: *Viscous Flow*
Shigley: *Kinematic Analysis of Mechanisms*
Shigley and Mischke: *Mechanical Engineering Design*
Shigley and Uicker: *Theory of Machines and Mechanisms*
Stiffler: *Design with Microprocessors for Mechanical Engineers*
Stoecker and Jones: *Refrigeration and Air Conditioning*
Ullman: *The Mechanical Design Process*
Vanderplaats: *Numerical Optimization: Techniques for Engineering Design, with Applications*
White: *Viscous Fluid Flow*
Zeid: *CAD/CAM Theory and Practice*

Also Available from McGraw-Hill

Schaum's Outline Series in Mechanical Engineering

Most outlines include basic theory, definitions, and hundreds of solved problems and supplementary problems with answers.

Titles on the Current List Include:

Acoustics

Basic Equations of Engineering Science

Continuum Mechanics

Engineering Economics

Engineering Mechanics, 4th edition

Engineering Thermodynamics

Fluid Dynamics, 2d edition

Fluid Mechanics & Hydraulics, 2d edition

Heat Transfer

Introduction to Engineering Calculations

Lagrangian Mechanics

Machine Design

Mathematical Handbook of Formulas
   & Tables

Mechanical Vibrations

Operations Research

Statics & Mechanics of Materials

Strength of Materials, 2d edition

Theoretical Mechanics

Thermodynamics, 2d edition

Schaum's Solved Problems Books

Each title in this series is a complete and expert source of solved problems containing thousands of problems with worked out solutions.

Related Titles on the Current List Include:

3000 Solved Problems in Calculus

2500 Solved Problems in Differential Equations

2500 Solved Problems in Fluid Mechanics and Hydraulics

1000 Solved Problems in Heat Transfer

3000 Solved Problems in Linear Algebra

2000 Solved Problems in Mechanical Engineering Thermodynamics

2000 Solved Problems in Numerical Analysis

700 Solved Problems in Vector Mechanics for Engineers: Dynamics

800 Solved Problems in Vector Mechanics for Engineers: Statics

Available at your college bookstore. A complete list of Schaum titles may be obtained by writing to: Schaum Division
                                  McGraw-Hill, Inc.
                                  Princeton Road, S-1
                                  Hightstown, NJ 08520

RADIATIVE HEAT TRANSFER

Michael F. Modest

The Pennsylvania State University

McGraw-Hill, Inc.

New York   St. Louis   San Francisco   Auckland   Bogotá
Caracas   Lisbon   London   Madrid   Mexico   Milan   Montreal
New Delhi   Paris   San Juan   Singapore   Sydney   Tokyo   Toronto

RADIATIVE HEAT TRANSFER

Acknowledgments appear on pages 803–806, and on this page by reference.

2 3 4 5 6 7 8 9 0 DOC DOC 9 0 9 8 7 6 5 4 3 2

ISBN 0–07–042675–9

This book was set in Times Roman by the author and Publication Services.
The editors were John J. Corrigan and John M. Morriss;
the production supervisor was Denise L. Puryear.
The cover was designed by Carla Bauer.
Project supervision was done by Publication Services.
R. R. Donnelley & Sons Company was printer and binder.

Library of Congress Cataloging-in-Publication Data

Modest, M. F. (Michael F.)
    Radiative heat transfer / Michael F. Modest.
        p.   cm. — (McGraw-Hill series in mechanical engineering)
    Includes bibliographical references and index.
    ISBN 0–07–042675–9
    1. Heat—Transmission.   2. Heat—Radiation and absorption.
    I. Title.   II. Series.
    QC320.M63   1993
    621.402′2—dc20   92-24671

ABOUT THE AUTHOR

Michael F. Modest was born in Berlin and spent the first half of his life in Germany. After receiving his Dipl.-Ing. degree from the Technical University in Munich, he came to the United States, and in 1972 obtained his M.S. and Ph.D. in Mechanical Engineering from the University of California at Berkeley, where he was first introduced to theory and experiment in thermal radiation. Since then, he has carried out many research projects in all areas of radiative heat transfer (measurement of surface, liquid, and gas properties; theoretical modeling for surface transport and within participating media). Since many laser beams are a form of thermal radiation, his work also encompasses the heat transfer aspects in the field of laser processing of materials.

For several years he has taught at Rensselaer Polytechnic Institute and the University of Southern California, and he is currently Professor of Mechanical Engineering at the Pennsylvania State University at University Park, PA. He is a fellow of the American Society of Mechanical Engineers and a member of the American Institute of Aeronautics and Astronautics and the Laser Institute of America. Dr. Modest resides in Boalsburg, PA, with his wife, two children, a dog, a cat, and an assortment of other animals.

To the m&m's in my life,
Monika, Mara, *and* Michelle

CONTENTS

PREFACE

Over the past 30 years the analysis of radiative heat transfer has received increasing attention. This was first due to the advent of the space age, which made it necessary to develop tools to predict heat transfer rates in such high-temperature applications as rocket nozzles and space vehicle reentry, and in vacuum applications for space craft in outer space. Following a lull during the 1970s and early 1980s, interest in radiative heat transfer has recently increased again because of the need to predict and measure heat transfer rates in ever higher temperature applications in furnaces, engines, MHD generators, and the like.

Today a course on radiative heat transfer is offered by virtually every university in the country with a graduate program in mechanical and/or aeronautical engineering; indeed, it is a required core course for most thermal science majors. A few universities, such as The Pennsylvania State University, with large programs in the fields of combustion, propulsion, etc., sometimes offer a second course on thermal radiation, covering a number of special topics.

The objectives of this book are more extensive than to provide a standard textbook for a one-semester core course on thermal radiation, since it does not appear possible to cover all important topics in the field of radiative heat transfer in a single graduate course. A number of important areas that would not be part of a "standard" one-semester course have been treated in some detail. It is anticipated that the engineer who may have used this book as his or her graduate textbook will be able to master these advanced topics through self-study. By including all important advanced topics, as well as a large number of references for further reading, the book may also be used as a reference book by the practicing engineer.

In each chapter all analytical methods are developed in substantial detail, containing a number of examples to show how the developed relations may be applied to practical problems. At the end of each chapter a number of exercises are included to give the student additional opportunity to familiarize him- or herself with the application of analytical methods developed in the preceding sections. The breadth of the description of analytical developments is such that any scientist with a satisfactory background in calculus and differential equations will be able to grasp the subject through self-study—for example, the heat transfer engineer involved in furnace calculations, the architectural engineer interested in lighting calculations, the oceanographer concerned with solar penetration into the ocean, or the meteorologist who studies atmospheric radiation problems.

The book is divided into 20 chapters, covering the four major areas in the field of radiative heat transfer. After the Introduction, there are two chapters dealing with theoretical and practical aspects of radiative properties of opaque surfaces, including a brief discussion of experimental methods. These are followed by four chapters dealing with radiative exchange between surfaces in an enclosure without a "radiatively participating" medium. The rest of the book deals with radiative transfer through absorbing, emitting, and scattering media (or "participating media"). After a detailed development of the equation of radiative transfer, radiative properties of gases, particulates, and semitransparent media are discussed, again including brief descriptions of experimental methods. Finally, over the last nine chapters the theory of radiative heat transfer through participating media is covered, separated into a number of basic problem areas and solution methods.

I have attempted to write the book in a modular fashion as much as possible. Chapter 2 is a fairly detailed (albeit concise) treatment of electromagnetic wave theory, which can (and will) be skipped by most instructors for a first course in radiative heat transfer. The chapter on opaque surface properties is self-contained and is not required reading for the rest of the book. The four chapters on surface transport (Chapters 4 through 7) are also self-contained and not required for the study of radiation in participating media. Similarly, the treatment of participating medium properties is not a prerequisite to studying the solution methods. Finally, any of the different solution aspects and methods discussed in Chapters 12 through 19 may be studied in any sequence.

I have not tried to mark those parts of the book that should be included in a one-semester course on thermal radiation, since I feel that different instructors will, and should, have different opinions on that matter. Indeed, the relative importance of different subjects may not only vary with different instructors, but also depend on student background, location, or the year of instruction. My personal opinion is that a one-semester course should touch on all four major areas (surface properties, surface transport, properties of participating media, and transfer through participating media) in a balanced way. For the average U.S. student who has had very little exposure to thermal radiation during his or her undergraduate heat transfer experience, I suggest that about half the course be devoted to Chapters 1, 3, 4, 5, and 6 (leaving out the more advanced features). The second half should be devoted to Chapters 8, 9, and 10 (again omitting less important features); some coverage of Chapter 12; and a thorough discussion of Chapter 13. If time permits (primarily, if surface properties and surface transport are treated in less detail than suggested above), I suggest to cover the P_1-approximation (which may be studied by itself, as outlined in the beginning of Chapter 14) and a portion of Chapter 17 (solution methods for nongray media). A second, special-topic course could include detailed discussions of radiative properties (Chapters 2, 3, 9, 10, and 11) and/or special solution methods (such as P_N and S_N approximations, nongray media, or zonal and Monte Carlo methods).

At this point the eager beaver is ready to embark on his or her journey through this book. For the inquisitive of mind and/or procrastinator at heart, here is a short account of *why* and *how* this book was written. Having taught a graduate course on radiative heat transfer every year for many years, I have always lamented the fact that there appeared to be no book that adequately treats all the subjects I feel belong in such a course. In particular, good discussions of radiative properties of, and heat transfer in, participating media are difficult to find. This provided the desire, but not the fire, for this book. The fire came in early 1987, when I was exposed to TEX, a mathematical typesetting language that may be familiar to a large number of readers. Although I began as just a user, TEX quickly became an obsession leading to many homemade macros (which are used throughout this book) and, finally, the development of a preprocessor that allows the typing of text and equations in near-WYSIWIG (what-you-see-is-what-you-get) fashion.

McGraw-Hill and I would like to thank the following reviewers for their many helpful comments and suggestions: D. K. Edwards, University of California–Irvine; James D. Felske, State University of New York–Buffalo; Woodrow A. Fiveland, Babcock and Wilcox; John R. Lloyd, Michigan State University; Theodore Smith, The University of Iowa; and Timothy Tong, Arizona State University.

My thanks go to Ms. Eileen Stephenson, who typed almost all of the manuscript, becoming quite a TEXpert in her own right, hammering out more equations during one day than I could proofread in one evening. Thanks also go to my colleague, Stefan Thynell, who patiently read and criticized a large part of the manuscript. Finally, last but not least, I am grateful to my family, who endured losing their husband and daddy for five long years.

Michael F. Modest

LIST OF SYMBOLS

The following is a list of symbols used frequently in this book. A number of symbols have been used for several different purposes. Alas, the Roman alphabet has only 26 lowercase and another 26 uppercase letters, and the Greek alphabet provides 34 more different ones, for a total of 86, which is, unfortunately, not nearly enough. Hopefully, the context will always make it clear which meaning of the symbols is to be used. I have used what I hope is a simple and uncluttered set of variable names. This usage, of course, comes at a price. For example, the subscript "λ" is often dropped (meaning "at a given wavelength," or "per unit wavelength"), assuming that the reader recognizes the variable as a spectral quantity from the context. Whenever applicable, units have been attached to the variables in the following table. Variables without indicated units have multiple sets of units. For example, the units for total band absorptance depend on the spectral variable used (λ, η, or ν), and on the absorption coefficient (linear, density, or pressure based), for a total of nine different possibilities.

$\quad\quad a$ semimajor axis of polarization ellipse, [N/C]

$\quad\quad a$ plane-polarized component of electric field, [N/C]

$\quad\quad a$ particle radius, [cm]

$\quad\quad a_k$ weight factors for sum-of-gray-gases, [−]

a_n, b_n Mie scattering coefficients, [−]

$\quad\quad A$ total band absorptance (or effective band width)

$\quad\quad A^*$ nondimensional band absorptance $= A/\omega$, [−]

A, A_n slab absorptivity (of n parallel sheets), [−]

A, A_p area, projected area, [cm^2]

$\quad\quad A_m$ scattering phase function coefficients, [−]

A_{ij}, B_{ij} Einstein coefficients

b line half-width

b self-broadening coefficient, $[-]$

b semiminor axis of polarization ellipse, [N/C]

B rotational constant

Bo convection-to-radiation parameter, (Boltzmann number), $[-]$

c, c_0 speed of light, (in vacuum), [cm/s]

c specific heat, [kJ/kg K]

C_1, C_2, C_3 constants for Planck function and Wien's displacement law

C_1, C_2, C_3 wide band parameters for outdated model

d line spacing

D diameter [cm]

D, D^* detectivity (normalized), [1/W] ([cm Hz$^{1/2}$/W])

$\hat{\mathbf{e}}$ unit vector into local coordinate direction, $[-]$

E, E_b emissive power, blackbody emissive power

$\mathbf{E}$ electric field vector, [N/C]

E_n exponential integral of order n, $[-]$

f_v, f_s, f_l volume, solid, liquid fractions, $[-]$

$f(n\lambda T)$ fractional blackbody emissive power, $[-]$

F_{i-j} (diffuse) view factor, $[-]$

F_{i-j}^s specular view factor, $[-]$

$\mathscr{F}_{i \to j}$ radiation exchange factor, $[-]$

g_k degeneracy, $[-]$

g nondimensional incident radiation, $[-]$

$\overline{g_i s_j}, \overline{g_i g_k}$ direct exchange areas in zonal method, [cm^2]

$\overline{\mathbf{gs}}, \overline{\mathbf{gg}}$ direct exchange area matrix, [cm^2]

G incident radiation = direction-integrated intensity

$\overline{G_i S_j}, \overline{G_i G_k}$ total exchange areas in zonal method, [cm^2]

$\overline{\mathbf{GS}}, \overline{\mathbf{GG}}$ total exchange area matrix, [cm^2]

h Planck's constant, $= 6.6262 \times 10^{-34}$ J s

h convective heat transfer coefficient, [W/cm^{2}K]

H irradiation onto a surface

H Heaviside's unit step function, $[-]$

H nondimensional heat transfer coefficient, $[-]$

$\mathscr{H}$ nondimensional irradiation onto a surface, $[-]$

$\mathbf{H}$ magnetic field vector, [C/cm s]

$\hat{\mathbf{i}}$ unit vector into the x-direction, $[-]$

I intensity of radiation

I first Stokes' parameter for polarization, [N^2/C^2]

I moment of inertia, [kg cm^2]

I_b blackbody intensity (Planck function)

I_l, I_l^m position-dependent intensity functions

I_0, I_1 modified Bessel functions, [−]

$\Im$ imaginary part of complex number,

j rotational quantum number, [−]

$\hat{\mathbf{j}}$ unit vector into the y-direction, [−]

J radiosity, [W/cm²]

$\mathcal{J}$ nondimensional radiosity, [−]

k thermal conductivity, [W/m K]

k Boltzmann's constant, $= 1.3806 \times 10^{-23}$ J/K

k absorptive index in complex index of refraction, [−]

$\hat{\mathbf{k}}$ unit vector into the z-direction, [−]

K kernel function

K luminous efficacy, [lm/W]

l, m, n direction cosines with x-, y-, z-axis, [−]

L length, [cm]

L latent heat of fusion, [J/kg]

L luminous intensity or luminance

L_e mean beam length, [cm]

L_0, L_m geometric, or average mean beam length, [cm]

m mass, [kg]

m complex index of refraction, [−]

$\dot{m}$ mass flow rate, [kg/s]

n self-broadening exponent, [−]

n refractive index, [−]

n number distribution function particle for particles, [cm⁻⁴]

$\hat{\mathbf{n}}$ unit surface normal (pointing away from surface into the medium), [−]

N conduction-to-radiation parameter, (Stark number), [−]

N_c conduction-to-radiation parameter, [−]

N_T number of particles per unit volume, [cm⁻³]

Nu Nusselt number, [−]

$\mathcal{O}$ order of magnitude, [−]

p pressure, [atm]; radiation pressure, [N/m²]

P probability function, [−]

P_l, P_l^m (associated) Legendre polynomials, [−]

Pr Prandtl number, [−]

q, $\mathbf{q}$ heat flux, heat flux vector [W/cm²]

q_{lum} luminous flux, [lm/m² = lx]

Q heat rate, [W]

Q second Stokes' parameter for polarization, [N²/C²]

$\dot{Q}'''$ heat production per unit volume, [W/cm³]

r radial coordinate, [cm]

r reflection coefficient, $[-]$

$\mathbf{r}$ position vector, [cm]

R radius, [cm]

R random number, $[-]$

R radiative resistance, $[\text{cm}^{-2}]$

R, R_n slab reflectivity (of n parallel sheets), $[-]$

$\Re$ real part of complex number

Re Reynolds number, $[-]$

s geometric path length, [cm]

$\hat{s}$ unit vector into a given direction, $[-]$

$\overline{s_i s_j}, \overline{s_i g_k}$ direct exchange areas in zonal method, $[\text{cm}^2]$

$\overline{\mathbf{ss}}, \overline{\mathbf{sg}}$ direct exchange area matrix, $[\text{cm}^2]$

S distance between two zones, or between points on enclosure surface, [cm]

S line-integrated absorption coefficient = line intensity

S radiative source function

$\mathbf{S}$ Poynting vector, $[\text{W/cm}^2]$

St Stanton number, $[-]$

Ste Stefan number, $[-]$

$\overline{S_i S_j}, \overline{S_i G_k}$ total exchange areas in zonal method, $[\text{cm}^2]$

$\overline{\mathbf{SS}}, \overline{\mathbf{SG}}$ total exchange area matrix, $[\text{cm}^2]$

t time, [s]

t transmission coefficient, $[-]$

t fin thickness, [cm]

$\hat{t}$ unit vector in tangential direction, $[-]$

T temperature, [K]

T, T_n slab transmissivity (of n parallel sheets), $[-]$

u internal energy, [kJ/kg]

u radiation energy density

u velocity, [cm/s]

u_k nondimensional transition wavenumber, $[-]$

U third Stokes' parameter for polarization, $[\text{N}^2/\text{C}^2]$

v vibrational quantum number, $[-]$

v velocity, [cm/s]

$\mathbf{v}$ velocity vector [cm/s]

V volume, $[\text{cm}^3]$

V fourth Stokes' parameter for polarization, $[\text{N}^2/\text{C}^2]$

$\mathbf{w}$ wave vector, $[\text{cm}^{-1}]$

w_i quadrature weights, $[-]$

W equivalent line width

x, y, z Cartesian coordinates, [cm]

x particle size parameter, $[-]$

x line strength parameter, $[-]$

X optical path length

X interface location, [cm]

y mole fraction, $[-]$

Y_l^m spherical harmonics, $[-]$

z nondimensional spectral variable, $[-]$

α absorptance or absorptivity, $[-]$

α band-integrated absorption coefficient $=$ band strength parameter

α opening angle, [rad]

α thermal diffusivity, [m^2/s]

β extinction coefficient

β line overlap parameter, $[-]$

γ line overlap parameter for dilute gas, $[-]$

γ complex permittivity, [C^2/N m^2]

γ azimuthal rotation angle for polarization ellipse, [rad]

γ oscillation damping factor, [Hz]

γ_E Euler's constant, $= 0.57221\ldots$

δ Dirac-delta function, $[-]$

δ polarization phase angle, [rad]

δ_{ij} Kronecker's delta, $[-]$

δ_k vibrational transition quantum step $= \Delta v$, $[-]$

ϵ emittance or emissivity, $[-]$

ϵ energy level, [J]

ϵ electrical permittivity, [C^2/N m^2]

ε complex dielectric function, or relative permittivity, $= \varepsilon' - i\varepsilon''$, $[-]$

η wavenumber, [cm^{-1}]

η direction cosine, $[-]$

η nondimensional (similarity) coordinate, $[-]$

η_{lum} luminous efficiency, $[-]$

θ polar angle, [rad]

θ nondimensional temperature, $[-]$

Θ scattering angle, [rad]

κ absorption coefficient

λ wavelength, [μm]

μ dynamic viscosity, [kg/m s]

μ magnetic permeability, [N s^2/C^2]

μ direction cosine (of polar angle), $= \cos\theta$, $[-]$

ν frequency, [Hz]

ν kinematic viscosity, [m^2/s]

ξ direction cosine, $[-]$

ξ nondimensional coordinate, [−]

ρ reflectance or reflectivity, [−]

ρ density, [g/cm^3]

ρ_f charge density, [C/m^3]

σ Stefan-Boltzmann constant, $= 5.670 \times 10^{-8}$ W/m^2 K^4

σ_s scattering coefficient

σ_e, σ_{dc} electrical conductivity, dc-value, [C^2/N m^2 s $= 1/\Omega$ m]

σ_m root-mean-square roughness, [cm]

τ transmittance or transmissivity, [−]

τ optical coordinate, optical thickness, [−]

ϕ phase angle, [rad]

Φ scattering phase function, [sr^{-1}]

Φ nondimensional medium emissive power function

Φ temperature function for line overlap β, [−]

Φ dissipation function, [J/kg m^2]

ψ azimuthal angle, [rad]

ψ stream function, [m^2/s]

Ψ temperature function for band strength α, [−]

Ψ nondimensional heat flux

ω single scattering albedo, [−]

ω angular frequency, [rad/s]

ω relaxation parameter, [−]

Ω solid angle, [sr]

Subscripts

0 reference value, or in vacuum, or at length $= 0$

1, 2 in medium, or at location, "1" or "2"

a absorbing, or apparent

av average

b blackbody value

B band integrated value

c at band center, or at cylinder, or critical value, or denoting a complex quantity,

C collision

D Doppler, or based on diameter

e effective value, or at equilibrium

g gas

i incoming, or dummy counter

j at a rotational state, or dummy counter

L at length $= L$

m modified Planck value, or medium value, or mean (bulk) value

n in normal direction

o outgoing, or from outside

p related to pressure, or polarizing value, or plasma

P Planck-mean

r reflected component

ref reference value

R Rosseland-mean, or at $r = R$

s along path s, or at surface, or at sphere, or at source

sol solar

t transmitted component

u upper limit

v at a vibrational state, or at constant volume

w wall value

W value integrated over spectral windows

x, y, z, r in a given direction

θ, ψ in a given direction

η at a given wavenumber, or per unit wavenumber

λ at a given wavelength, or per unit wavelength

ν at a given frequency, or per unit frequency

$\|$ polarization component, or situated in plane of incidence

$\perp$ polarization component, or situated in plane perpendicular to plane of incidence

Superscripts

$'$ $''$ real and imaginary parts of complex number, or directional values, or dummy variables

$\circ$ hemispherical value

$*$ complex conjugate, or obtained by P_1-approximation

$+, -$ into "positive" and "negative" directions

d diffuse

s specular

$^-$ average value

$^\sim$ complex number, or scaled value (for nonisothermal path), or Favre average

$^\wedge$ unit vector

CHAPTER
1

FUNDAMENTALS
OF THERMAL
RADIATION

1.1 INTRODUCTION

The terms *radiative heat transfer* and *thermal radiation* are commonly used to describe
the science of the heat transfer caused by electromagnetic waves. Obvious everyday
examples of thermal radiation include the heating effect of sunshine on a clear day, the
fact that—when one is standing in front of a fire—the body's side facing the fire feels
much hotter than the back, and so on. More subtle examples of thermal radiation are
that the clear sky is blue, that sunsets are red, and that, during a clear winter night,
we feel more comfortable in a room whose curtains are drawn than in a room (heated
to the same temperature) with open curtains.

All materials continuously emit and absorb electromagnetic waves, or photons,
by lowering or raising their molecular energy levels. The strength and wavelengths
of emission depend on the temperature of the emitting material. As we shall see, for
heat transfer applications wavelengths between 10^{-7}m and 10^{-3}m (ultraviolet, visible
and infrared) are of greatest importance and are, therefore, the only ones considered
here.

Before embarking on the analysis of *thermal radiation* we want briefly to compare the nature of this mode of heat transfer with the other two possible mechanisms of

transferring energy, conduction and convection. In the case of conduction in a solid, energy is carried through the atomic lattice by free electrons or by phonon-phonon interactions (i.e., excitation of vibrational energy levels for interatomic bonds). In gases and liquids, energy is transferred from molecule to molecule through collisions (i.e., the faster molecule loses some of its kinetic energy to the slower one). Heat transfer by convection is similar, but many of the molecules with raised kinetic energy are carried away by the flow and are replaced by colder fluid (low-kinetic-energy molecules), resulting in increased energy transfer rates. Thus, both conduction and convection require the presence of a medium for the transfer of energy. Thermal radiation, on the other hand, is transferred by electromagnetic waves, or photons, which may travel over a long distance without interacting with a medium. The fact that thermal radiation does not require a medium for its transfer makes it of great importance in vacuum and space applications.

Another distinguishing feature between conduction and convection on the one hand and thermal radiation on the other is the difference in their temperature dependencies. For the vast majority of conduction applications heat transfer rates are well described by *Fourier's law* as

$$q_x = -k \frac{\partial T}{\partial x}, \tag{1.1}$$

where q_x is conducted heat flux[1] in the x-direction, T is temperature and k is the thermal conductivity of the medium. Similarly, convective heat flux may usually be calculated from a correlation such as

$$q = h(T - T_\infty), \tag{1.2}$$

where h is known as the convective heat transfer coefficient, and T_∞ is a reference temperature. While k and h may depend on temperature, this dependence is usually not very strong. Thus, for most applications, conductive and convective heat transfer rates are *linearly proportional* to temperature differences. As we shall see, radiative heat transfer rates are generally proportional to differences in temperature to the fourth (or higher) power, i.e.,

$$q \propto T^4 - T_\infty^4. \tag{1.3}$$

Therefore, radiative heat transfer becomes more important with rising temperature levels and may be totally dominant over conduction and convection at very high temperatures. Thus, thermal radiation is important in combustion applications (fires, furnaces, rocket nozzles, engines, etc.), in nuclear reactions (such as in the sun, in a fusion reactor, or in nuclear bombs), during atmospheric reentry of space vehicles, etc. Other applications include solar energy collection and the greenhouse effect (both due to emission from our high-temperature sun).

---

[1]In this book we shall use the term *heat flux* to denote the flow of energy per unit time and per unit area and the term *heat rate* for the flow of energy per unit time (i.e., not per unit area).

The same reasons that make thermal radiation important in vacuum and high-temperature applications also make its analysis more difficult, or at least quite different from "conventional" analyses. Under normal conditions, conduction and convection are short-range phenomena: The average distance between molecular collisions *(mean free path for collision)* is generally very small, maybe around 10^{-10}m. If it takes, say, 10 collisions until a high-kinetic-energy molecule has a kinetic energy similar to that of the surrounding molecules, then any external influence is not directly felt over a distance larger than 10^{-9}m. Thus we are able to perform an energy balance on an "infinitesimal volume," i.e., a volume negligibly small in comparison with overall dimensions, but very large in comparison with the mean free path for collision. The principle of *conservation of energy* then leads to a *partial differential equation* to describe the temperature field and heat fluxes for both conduction and convection. This equation may have up to four independent variables (three space coordinates and time) and is *linear in temperature* for the case of constant properties. Thermal radiation, on the other hand, is generally a long-range phenomenon. The *mean free path for a photon* (i.e., the average distance a photon travels before interacting with a molecule) may be as short as 10^{-10}m (e.g., absorption in a metal), but can also be as long as 10^{+10}m or larger (e.g., the sun's rays hitting earth). Thus, conservation of energy cannot be applied over an infinitesimal volume, but must be applied over the entire volume under consideration. This leads to an *integral equation* in up to seven independent variables (the frequency of radiation, three space coordinates, two coordinates describing the direction of travel of photons, and time).

The analysis of thermal radiation is further complicated by the behavior of the radiative properties of materials. Properties relevant to conduction and convection (thermal conductivity, kinematic viscosity, density, etc.) are fairly easily measured and are generally well behaved (isotropic throughout the medium, perhaps with weak temperature dependence). Radiative properties are usually difficult to measure and often display erratic behavior. For liquids and solids the properties normally depend only on a very thin surface layer, which may vary strongly with surface preparation and often even from day to day. All radiative properties (in particular for gases) may vary strongly with wavelength, adding another dimension to the governing equation. Rarely, if ever, may this equation assumed to be linear.

Because of these difficulties inherent in the analysis of thermal radiation, a good portion of this book has been set aside to discuss radiative properties and different approximate methods to solve the governing energy equation for radiative transport.

1.2 THE NATURE OF THERMAL RADIATION

Thermal radiative energy may be viewed as consisting of *electromagnetic waves* (as predicted by *electromagnetic wave theory*) or as consisting of massless energy parcels, called *photons* (as predicted by *quantum mechanics*). Neither point of view is able to describe completely all radiative phenomena that have been observed. It is, therefore, customary to use both concepts interchangeably. In general, radiative properties of liquids and solids (including tiny particles), and of interfaces (surfaces) are more easily predicted using electromagnetic wave theory, while radiative properties of gases are more conveniently obtained from quantum mechanics.

All electromagnetic waves, or photons, are known to propagate through any medium at a high velocity. Since light is a part of the electromagnetic wave spectrum, this velocity is known as the *speed of light*, c. The speed of light depends on the medium through which it travels, and may be related to the speed of light obtained in vacuum, c_0, by the formula

$$c = \frac{c_0}{n}, \qquad c_0 = 2.998 \times 10^8 \, \text{m/s}, \tag{1.4}$$

where n is known as the *refractive index* of the medium. By definition, the refractive index of vacuum is $n \equiv 1$. For most gases the refractive index is very close to unity, for example, air at room temperature has $n = 1.00029$ over the visible spectrum. Therefore, light propagates through gases nearly as fast as through vacuum. Electromagnetic waves travel considerably slower through dielectrics (electric nonconductors), which have refractive indices between approximately 1.4 and 4, and they hardly penetrate at all into electrical conductors (metals). Each wave may be identified either by its

frequency, ν (measured in cycles/s $=\text{s}^{-1} =\text{Hz}$);
wavelength, λ (measured in $\mu\text{m} = 10^{-6}\text{m}$ or $\text{Å} = 10^{-10}\text{m}$);
wavenumber, η (measured in cm^{-1}); or
angular frequency, ω (measured in radians/s $=\text{s}^{-1}$).

All four quantities are related to one another through the formulae

$$\nu = \frac{\omega}{2\pi} = \frac{c}{\lambda} = c\eta. \tag{1.5}$$

Each wave or photon carries with it an amount of energy, ϵ, determined from quantum mechanics as

$$\epsilon = h\nu, \qquad h = 6.626 \times 10^{-34} \, \text{J s}, \tag{1.6}$$

where h is known as *Planck's constant*. The frequency of light does not change when light penetrates from one medium to another since the energy of the photon must be conserved. On the other hand, wavelength and wavenumber do, depending on the values of the refractive index for the two media. Sometimes electromagnetic waves are characterized in terms of the energy that a photon carries, $h\nu$, using the energy unit *electron volt* ($1\,\text{eV} = 1.6022 \times 10^{-19}\,\text{J}$). Thus, light with a photon energy (or "frequency") of a eV has a wavelength (in vacuum) of

$$\lambda = \frac{hc}{h\nu} = \frac{6.626 \times 10^{-34}\,\text{J s} \times 2.998 \times 10^8\,\text{m/s}}{a\,1.6022 \times 10^{-19}\,\text{J}} = \frac{1.240}{a}\,\mu\text{m}. \tag{1.7}$$

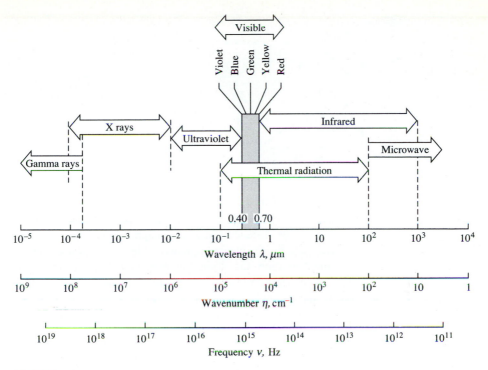

FIGURE 1-1
Electromagnetic wave spectrum.

Since electromagnetic waves of vastly different wavelengths carry vastly different amounts of energy, their behavior is often quite different. Depending on their behavior or occurrence, electromagnetic waves have been grouped into a number of different categories, as shown in Fig. 1-1. Thermal radiation may be defined to be those electromagnetic waves which are emitted by a medium due solely to its temperature [1]. As indicated earlier, this definition limits the range of wavelengths of importance for heat transfer considerations to between $0.1 \, \mu m$ (ultraviolet) and $100 \, \mu m$ (midinfrared).

1.3 BASIC LAWS OF THERMAL RADIATION

When an electromagnetic wave traveling through a medium (or vacuum) strikes the surface of another medium (solid or liquid surface, particle or bubble), the wave may be reflected (either partially or totally), and any nonreflected part will penetrate into the medium. While passing through the medium the wave may become continuously attenuated. If attenuation is complete so that no penetrating radiation reemerges, it is known as *opaque*. If a wave passes through a medium without any attenuation, it is

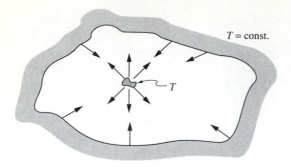

FIGURE 1-2
Kirchhoff's law.

known as *transparent*, while a body with partial attenuation is called *semitransparent*.[2] Whether a medium is transparent, semitransparent or opaque depends on the material as well as on its thickness (i.e., the distance the electromagnetic wave must travel through the medium). Metals are nearly always opaque, although it is a common high school physics experiment to show that light can penetrate through extremely thin layers of gold. Nonmetals generally require much larger thicknesses before they become opaque, and some are quite transparent over part of the spectrum (for example, window glass in the visible part of the spectrum).

An opaque surface that does not reflect *any* radiation is called a *perfect absorber* or a *black surface*: When we "see" an object, our eyes absorb electromagnetic waves from the visible part of the spectrum, which have been emitted by the sun (or artificial light) and have been reflected by the object toward our eyes. Thus, a surface that does not reflect radiation we cannot see, and it appears "black" to our eyes.[3] Since black surfaces absorb the maximum possible amount of radiative energy, they serve as a standard for the classification of all other surfaces.

It is easy to show that a black surface also emits a maximum amount of radiative energy, i.e., more than any other body at the same temperature. To show this, we use one of the many variations of *Kirchhoff's law*:[*] Consider two identical black-walled enclosures, thermally insulated to the outside, with each containing a small object—one black and the other one not—as shown in Fig. 1-2. After a long time,

---

[2] A medium that allows a fraction of light to pass through, while scattering the transmitted light into many different directions, for example, milky glass, is called *translucent*.

[3] Note that a surface appearing black to our eyes is by no means a perfect absorber at nonvisible wavelengths and vice versa; indeed, many white paints are actually quite "black" at longer wavelengths.

[*]Gustav Robert Kirchhoff (1824-1887)
German physicist. After studying in Berlin, Kirchhoff served as professor of physics at the University of Heidelberg for 21 years before
returning to Berlin as professor of mathematical physics. Together with
the chemist Robert Bunsen, he was the first to establish the theory of
spectrum analysis.

in accordance with the Second Law of Thermodynamics, both entire enclosures and the objects within them will be at a single uniform temperature. This characteristic implies that every part of the surface (of enclosure as well as objects) emits precisely as much energy as it absorbs. Both objects in the different enclosures receive exactly the same amount of radiative energy. But since the black object absorbs more energy (i.e., the maximum possible), it must also emit more energy than the nonblack object (i.e., also the maximum possible).

By the same reasoning it is easy to show that a black surface is a perfect absorber and emitter at every wavelength and for any direction (of incoming or outgoing electromagnetic waves), and that the radiation field within an isothermal black enclosure is *isotropic* (i.e., the radiative energy density is the same at any point and in any direction within the enclosure).

1.4 EMISSIVE POWER

Every medium continuously emits electromagnetic radiation randomly into all directions at a rate depending on the local temperature and on the properties of the material. The radiative heat flux emitted from a surface is called the *emissive power*, E. We distinguish between *total* and *spectral emissive power* (i.e., heat flux emitted over the entire spectrum, or at a given frequency per unit frequency interval), so that

spectral emissive power, E_ν ≡ emitted energy/time/surface area/frequency,
total emissive power, E ≡ emitted energy/time/surface area.

Here and elsewhere we use the subscripts ν, λ or η (depending on the choice of spectral variable) to express a spectral quantity whenever necessary for clarification. Thermal radiation of a single frequency or wavelength is sometimes also called *monochromatic radiation* (since, over the visible range, the human eye perceives electromagnetic waves to have the colors of the rainbow). It is clear from their definitions that the total and spectral emissive powers are related by

$$E(T) = \int_0^\infty E_\nu(T, \nu)\, d\nu. \tag{1.8}$$

Blackbody Emissive Power Spectrum

Scientists have tried for many years to predict theoretically the emission spectrum of the sun, which we know today to behave very nearly like a blackbody at approximately 5762 K. The spectral solar flux falling onto earth, or *solar irradiation,* is shown in Fig. 1-3 for extraterrestrial conditions (as measured by high-flying balloons and satellites) and for unity air mass (air mass is defined as the value of $1/\cos\theta_S$, where the *zenith angle* θ_S is the angle between the local vertical and a vector pointing toward the sun) [2, 3]. Note that solar radiation is attenuated significantly as it penetrates through the atmosphere by phenomena that will be discussed in Sections 1.11

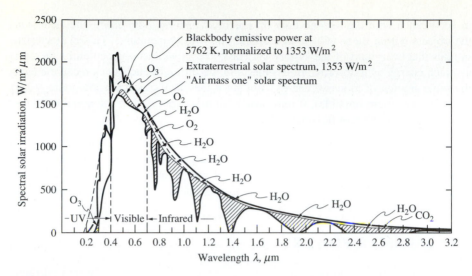

FIGURE 1-3
Solar irradiation onto earth.

and 1.13. Lord Rayleigh (1900) [4]* and Sir James Jeans (1905) [5]† independently applied the principles of classical statistics with its equipartition of energy to predict the spectrum of the sun, with dismal results. Wilhelm Wien (1896) [6]‡ used some thermodynamic arguments together with experimental data to propose a spectral dis-

---

*John William Strutt, Lord Rayleigh (1842-1919)
English physical scientist. Rayleigh obtained a mathematics degree from Cambridge, where he later served as professor of experimental physics for five years. He then became secretary, and later president, of the Royal Society. His work resulted in a number of discoveries in the fields of acoustics and optics, and he was the first to explain the blue color of the sky (cf. the Rayleigh scattering laws in Chapter 10). Rayleigh received the 1904 Nobel Prize in Physics for the isolation of argon.

† Sir James Hopwood Jeans (1877-1946)
English physicist and mathematician, whose work was primarily in the area of astrophysics. He applied mathematics to several problems in thermodynamics and electromagnetic radiation.

‡ Wilhelm Wien (1864-1928)
German physicist, who served as professor of physics at the University of Giessen and later at the University of Munich. Besides his research in the area of electromagnetic waves, his interests included other rays, such as electron beams, X-rays and α-particles. For the discovery of his displacement law he was awarded the Nobel Prize in Physics in 1911.

tribution of blackbody emissive power that was very accurate over large parts of the spectrum. Finally, in 1901 Max Planck [7]* published his work on quantum statistics: Assuming that a molecule can emit photons only at distinct energy levels, he found the spectral *blackbody emissive power* distribution, now commonly known as *Planck's law,* for a black surface bounded by a transparent medium with refractive index n, as

$$E_{b\nu}(T, \nu) = \frac{2\pi h \nu^3 n^2}{c_0^2 \left[e^{h\nu/kT} - 1 \right]}, \tag{1.9}$$

where $k = 1.3806 \times 10^{-23}$ J/K is known as *Boltzmann's constant*.[4] While frequency ν appears to be the most logical spectral variable (since it does not change when light travels from one medium into another), the spectral variables wavelength λ (primarily for surface emission and absorption) and wavenumber η (primarily for radiation in gases) are also frequently (if not more often) employed. Equation (1.9) may be readily expressed in terms of wavelength and wavenumber through the relationships

$$\nu = \frac{c_0}{n\lambda} = \frac{c_0}{n}\eta, \qquad d\nu = -\frac{c_0}{n\lambda^2}\left[1 + \frac{\lambda}{n}\frac{dn}{d\lambda}\right]d\lambda = \frac{c_0}{n}\left[1 - \frac{\eta}{n}\frac{dn}{d\eta}\right]d\eta, \quad (1.10)$$

and

$$E_b(T) = \int_0^\infty E_{b\nu}d\nu = \int_0^\infty E_{b\lambda}d\lambda = \int_0^\infty E_{b\eta}d\eta, \tag{1.11}$$

or

$$E_{b\nu}d\nu = -E_{b\lambda}d\lambda = E_{b\eta}d\eta. \tag{1.12}$$

Here λ and η are wavelength and wavenumber for the electromagnetic waves within the medium of refractive index n (while $\lambda_0 = n\lambda$ and $\eta_0 = \eta/n$ would be wavelength and wavenumber of the same wave traveling through vacuum). Equation (1.10) shows that equation (1.9) gives convenient relations for $E_{b\lambda}$ and $E_{b\eta}$ only if the refractive index is independent of frequency (or wavelength, or wavenumber). This is certainly the

---

***Max Planck (1858-1947)**
German physicist. Planck studied in Berlin with H. L. F. von Helmholtz and G. R. Kirchhoff, but obtained his doctorate at the University of Munich before returning to Berlin as professor in theoretical physics. He later became head of the Kaiser Wilhelm Society (today the Max Planck Institute). For his development of the quantum theory he was awarded the Nobel Prize in Physics in 1918.

[4]Equation (1.9) is valid for emission into a medium whose absorptive index (to be introduced in Chapter 2) is much less than the refractive index. This includes semitransparent media such as water, glass, quartz, etc., but not opaque materials. Emission into such bodies is immediately absorbed and is of no interest.

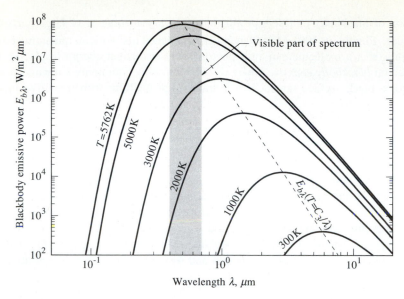

FIGURE 1-4
Blackbody emissive power spectrum.

case for vacuum ($n = 1$) and ordinary gases ($n \simeq 1$), and may be of acceptable accuracy for some semitransparent media over large parts of the spectrum (for example, for quartz $1.52 < n < 1.68$ between the wavelengths of 0.2 and 2.4 μm). Thus, with the assumption of constant refractive index,

$$E_{b\lambda}(T, \lambda) = \frac{2\pi h c_0^2}{n^2 \lambda^5 \left[e^{hc_0/n\lambda kT} - 1 \right]}, \qquad (n = \text{const}), \qquad (1.13)$$

$$E_{b\eta}(T, \eta) = \frac{2\pi h c_0^2 \eta^3}{n^2 \left[e^{hc_0\eta/nkT} - 1 \right]}, \qquad (n = \text{const}). \qquad (1.14)$$

Figure 1-4 is a graphical representation of equation (1.13) for a number of blackbody temperatures. As one can see, the overall level of emission rises with rising temperature (as dictated by the Second Law of Thermodynamics), while the wavelength of maximum emission shifts toward shorter wavelengths. The blackbody emissive power is also plotted in Fig. 1-3 for an *effective solar temperature* of 5762 K. This plot is in good agreement with extraterrestrial solar irradiation data.

It is customary to introduce the abbreviations

$$C_1 = 2\pi h c_0^2 = 3.7419 \times 10^{-16} \, \text{W}\text{m}^2,$$
$$C_2 = hc_0/k = 14{,}388 \, \mu\text{m} \, \text{K},$$

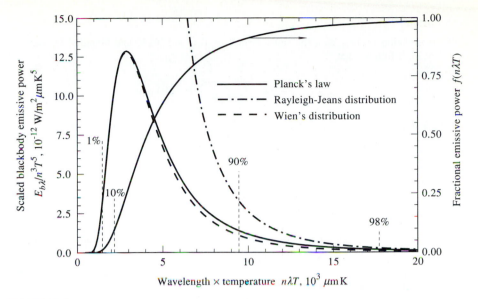

FIGURE 1-5
Normalized blackbody emissive power spectrum.

so that equation (1.13) may be recast as

$$\frac{E_{b\lambda}}{n^3T^5} = \frac{C_1}{(n\lambda T)^5[e^{C_2/(n\lambda T)} - 1]}, \qquad (n = \text{const}), \qquad (1.15)$$

which is seen to be a function of $(n\lambda T)$ only. Thus, it is possible to plot this normalized emissive power as a single line *vs.* the product of wavelength in vacuum $(n\lambda)$ and temperature (T), as shown in Fig. 1-5, and a detailed tabulation is given in Appendix C. The maximum of this curve may be determined by differentiating equation (1.15),

$$\frac{d}{d(n\lambda T)}\left(\frac{E_{b\lambda}}{n^3T^5}\right) = 0,$$

leading to a transcendental equation that may be solved numerically as

$$(n\lambda T)_{max} = C_3 = 2898\,\mu m\,K. \qquad (1.16)$$

Equation (1.16) is known as *Wien's displacement law* since it was developed independently by Wilhelm Wien [8] in 1891 (i.e., well before the publication of Plancks's emissive power law).

Example 1.1. At what wavelength has the sun its maximum emissive power? At what wavelength earth?

Solution

From equation (1.16), with the sun's surface at $T_{sun} \simeq 5762\,K$ and bounded by vacuum $(n = 1)$, it follows that

$$\lambda_{max,sun} = \frac{C_3}{T_{sun}} = \frac{2898\,\mu m\,K}{5762\,K} = 0.50\,\mu m,$$

which is near the center of the visible region. Apparently, evolution has caused our eyes to be most sensitive in that section of the electromagnetic spectrum where the maximum daylight is available. In contrast, earth's average surface temperature may be in the vicinity of $T_{earth} = 290\,K$, or

$$\lambda_{max,earth} \simeq \frac{2898\,\mu m\,K}{290\,K} = 10\,\mu m,$$

that is, earth's maximum emission occurs in the intermediate infrared, leading to infrared cameras and detectors for night "vision."

It is of interest to look at the asymptotic behavior of Planck's law for small and large wavelengths. For very small values of $hc_0/n\lambda kT$ (large wavelength, or small frequency), the exponent in equation (1.13) may be approximated by a two-term Taylor series, leading to

$$E_{b\lambda} = \frac{2\pi c_0 kT}{n\lambda^4}, \qquad \frac{hc_0}{n\lambda kT} \ll 1. \tag{1.17}$$

The same result is obtained if one lets $h \to 0$, i.e., if one allows photons of arbitrarily small energy content to be emitted, as postulated by classical statistics. Thus, equation (1.17) is identical to the one derived by *Rayleigh* and *Jeans* and bears their names. The *Rayleigh-Jeans distribution* is also included in Fig. 1-5. Obviously, this formula is accurate only for very large values of $(n\lambda T)$, where the energy of the emissive power spectrum is negligible. Thus, this formula is of little significance for engineering purposes.

For large values of $(hc_0/n\lambda kT)$ the -1 in the denominator of equation (1.13) may be neglected, leading to *Wien's distribution* (or *Wien's law*),

$$E_{b\lambda} \simeq \frac{2\pi hc_0^2}{n^2\lambda^5}e^{-hc_0/n\lambda kT} = \frac{C_1}{n^2\lambda^5}e^{-C_2/n\lambda T}, \qquad \frac{hc_0}{n\lambda kT} \gg 1, \tag{1.18}$$

since it is identical to the formula first proposed by Wien, before the advent of quantum mechanics. Examination of Wien's distribution in Fig. 1-5 shows that it is very accurate over most of the spectrum, with a total energy content of the entire spectrum approximately 8% lower than for Planck's law. Thus, Wien's distribution is frequently utilized in theoretical analyses in order to facilitate integration.

Total Blackbody Emissive Power

The total emissive power of a blackbody may be determined from equations (1.11) and (1.13) as

$$E_b(T) = \int_0^\infty E_{b\lambda}(T, \lambda)d\lambda = C_1 n^2 T^4 \int_0^\infty \frac{d(n\lambda T)}{(n\lambda T)^5 \left[e^{C_2/(n\lambda T)} - 1\right]}$$

$$= \left[\frac{C_1}{C_2^4} \int_0^\infty \frac{\xi^3 d\xi}{e^\xi - 1}\right] n^2 T^4, \quad (n = \text{const}). \tag{1.19}$$

The integral in this expression may be evaluated by complex integration, and is tabulated in many good integral tables:

$$E_b(T) = n^2 \sigma T^4, \quad \sigma = \frac{\pi^4 C_1}{15 C_2^4} = 5.670 \times 10^{-8} \frac{\text{W}}{\text{m}^2 \, \text{K}^4}, \tag{1.20}$$

where σ is known as the *Stefan-Boltzmann constant.*[*] It is often necessary to calculate the emissive power contained within a finite wavelength band, say between λ_1 and λ_2. Then

$$\int_{\lambda_1}^{\lambda_2} E_{b\lambda} d\lambda = \frac{C_1}{C_2^4} \int_{C_2/n\lambda_2 T}^{C_2/n\lambda_1 T} \frac{\xi^3 d\xi}{e^\xi - 1} n^2 T^4. \tag{1.21}$$

It is not possible to evaluate the integral in equation (1.21) analytically. Therefore, it is customary to express equation (1.21) in terms of the *fraction of blackbody emissive power* contained between 0 and $n\lambda T$,

$$f(n\lambda T) = \frac{\int_0^\lambda E_{b\lambda} d\lambda}{\int_0^\infty E_{b\lambda} d\lambda} = \int_0^{n\lambda T} \left(\frac{E_{b\lambda}}{n^3 \sigma T^5}\right) d(n\lambda T) = \frac{15}{\pi^4} \int_{C_2/n\lambda T}^\infty \frac{\xi^3 d\xi}{e^\xi - 1}, \tag{1.22}$$

so that

$$\int_{\lambda_1}^{\lambda_2} E_{b\lambda} d\lambda = [f(n\lambda_2 T) - f(n\lambda_1 T)] n^2 \sigma T^4. \tag{1.23}$$

Equation (1.22) is a function in a single variable, $n\lambda T$, and is therefore easily tabulated, as has been done in Appendix C.

---

[*]**Josef Stefan (1835-1893)**
Austrian physicist. Serving as professor at the University of Vienna, Stefan determined in 1879 that, based on his experiments, blackbody emission was proportional to temperature to the fourth power.

Ludwig Erhard Boltzmann (1844-1906)
Austrian physicist. After receiving his doctorate from the University of Vienna he held professorships in Vienna, Graz (both in Austria), Munich and Leipzig (in Germany). His greatest contributions were in the field of statistical mechanics (Boltzmann statistics). He derived the fourth-power law from thermodynamic considerations in 1889.

Example 1.2. What fraction of total solar emission falls into the visible spectrum (0.4 to 0.7 μm)?

Solution

With $n = 1$ and a solar temperature of 5762 K it follows that for $\lambda_1 = 0.4\,\mu$m, $n\lambda_1 T_{sun} = 1 \times 0.4 \times 5762 = 2304\,\mu$m K; and for $\lambda_2 = 0.7\,\mu$m, $n\lambda_2 T_{sun} = 4033\,\mu$m K. From Appendix C we find $f(n\lambda_1 T_{sun}) = 0.12011$ and $f(n\lambda_2 T_{sun}) = 0.48590$. Thus, from equations (1.20) and (1.23) the visible fraction of sunlight is

$$f(n\lambda_2 T_{sun}) - f(n\lambda_1 T_{sun}) = 0.48590 - 0.12011 = 0.36579.$$

Therefore, with a bandwidth of only 0.3 μm the human eye responds to approximately 37% of all emitted sunlight!

1.5 SOLID ANGLES

When radiative energy leaves one medium entering another (i.e., emission from a surface into another medium), this energy flux usually has different strengths in different directions. Similarly, the electromagnetic wave, or photon, flux passing through any point inside any medium may vary with direction. It is customary to describe the *direction vector* in terms of a *spherical* or *polar coordinate system*. Consider an opaque surface radiating into another medium, say air, as shown in Fig. 1-6. It is apparent that the surface can radiate into infinitely many directions, with every ray penetrating through a *hemisphere* of unit radius as indicated in the figure. The total surface area of this hemisphere, $2\pi\,1^2 = 2\pi$, is known as the *total solid angle* above the surface. An arbitrary emission direction from the surface is specified by the unit

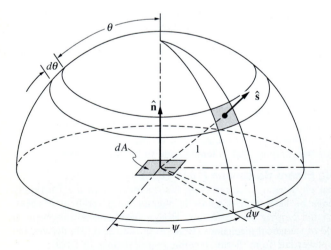

FIGURE 1-6
Emission direction and solid angles as related to a unit hemisphere.

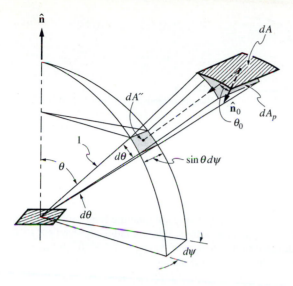

FIGURE 1-7
Definition of solid angle.

direction vector $\hat{s}$, which may be expressed in terms of the *polar angle* θ (measured from the *surface normal* $\hat{n}$) and the *azimuthal angle* ψ (measured between an arbitrary axis on the surface and the projection of $\hat{s}$ onto the surface). It is seen that, for a hemisphere, $0 \le \theta \le \pi/2$ and $0 \le \psi \le 2\pi$.

The *solid angle* with which a surface is seen from a certain point is defined as the projection of the surface onto a plane normal to the direction vector, divided by the distance squared, as shown in Fig. 1-7. If the surface is projected onto the unit hemisphere above the point, the solid angle is equal to the projected area itself, or

$$\Omega = \int_{A_p} \frac{dA_p}{S^2} = \int_A \frac{\cos\theta_0 \, dA}{S^2} = A_p''. \tag{1.24}$$

The direction vector $\hat{s}$ in Fig. 1-6 is associated with an infinitesimal solid angle equal to the infinitesimal area on the surface of the hemisphere, or

$$d\Omega = (1 \times \sin\theta \, d\psi)(1 \times d\theta) = \sin\theta \, d\theta \, d\psi. \tag{1.25}$$

Integrating over all possible directions we obtain

$$\int_{\psi=0}^{2\pi} \int_{\theta=0}^{\pi/2} \sin\theta \, d\theta \, d\psi = 2\pi, \tag{1.26}$$

for the total solid angle above the surface, as already seen earlier. While a little unfamiliar at first, the solid angle is simply a two-dimensional angular space: Similar to the way a one-dimensional angle can vary between 0 and π above a line (measured in dimensionless radians), the solid angle may vary between 0 and 2π above a surface (measured in dimensionless steradians, sr).

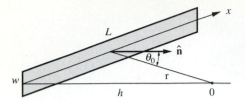

FIGURE 1-8
Solid angle subtended by a narrow strip.

Example 1.3.   Determine the solid angle with which the sun is seen from earth.

Solution

The area of the sun projected onto a plane normal to the vector pointing from earth to the sun (or, simply, the image of the sun that we see from earth) is a disk of radius $R_s \simeq 6.96 \times 10^8$ m (i.e., the radius of the sun), at a distance of approximately $S_{ES} \simeq 1.496 \times 10^{11}$ m (averaged over earth's yearly orbit). Thus the solid angle of the sun is

$$\Omega_S = \frac{(\pi R_S^2)}{S_{ES}^2} = \frac{\pi \times (6.96 \times 10^8)^2}{(1.496 \times 10^{11})^2} = 6.80 \times 10^{-5} \text{ sr.}$$

This solid angle is so small that we may generally assume that solar radiation comes from a single direction, i.e., that all the light beams are parallel.

Example 1.4.   What is the solid angle with which the narrow strip shown in Fig. 1-8 is seen from point "0"?

Solution

Since the strip is narrow we may assume that the projection angle for equation (1.24) varies only in the x-direction as indicated in Fig. 1-8, leading to

$$\Omega = w \int_0^L \frac{\cos\theta_0 \, dx}{r^2}, \qquad \cos\theta_0 = \frac{h}{r}, \qquad r^2 = h^2 + x^2,$$

and

$$\Omega = w \int_0^L \frac{h \, dx}{r^3} = wh \int_0^L \frac{dx}{(h^2 + x^2)^{3/2}} = \frac{w}{h} \left. \frac{x}{\sqrt{h^2 + x^2}} \right|_0^L = \frac{wL}{h\sqrt{h^2 + L^2}}.$$

1.6   RADIATIVE INTENSITY

While emissive power appears to be the natural choice to describe radiative heat flux leaving a surface, it is inadequate to describe the directional dependence of the radiation field, in particular inside an absorbing/emitting medium, where photons may not have originated from a surface. Therefore, very similar to the emissive power, we define the *radiative intensity I,* as radiative energy flow per unit solid angle and unit area *normal to the rays* (as opposed to surface area). Again, we distinguish between

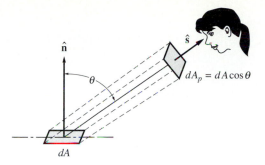

spectral and *total intensity*. Thus,

$$\text{spectral intensity, } I_\lambda \equiv \text{radiative energy flow/time/area}$$
$$\text{normal to rays/solid angle/wavelength,}$$
$$\text{total intensity, } I \equiv \text{radiative energy flow/time/area}$$
$$\text{normal to rays/solid angle.}$$

Again, spectral and total intensity are related by

$$I(\mathbf{r}, \hat{\mathbf{s}}) = \int_0^\infty I_\lambda(\mathbf{r}, \hat{\mathbf{s}}, \lambda) \, d\lambda. \tag{1.27}$$

Here, $\mathbf{r}$ is a *position vector* fixing the location of a point in space, and $\hat{\mathbf{s}}$ is a unit direction vector as defined in the previous section. While emissive power depends only on position and wavelength, the radiative intensity depends, in addition, on the direction vector $\hat{\mathbf{s}}$. The emissive power can be related to intensity by integrating over all the directions pointing away from the surface. Considering Fig. 1-9, we find that the emitted energy from dA into the direction $\hat{\mathbf{s}}$, and contained within an infinitesimal solid angle $d\Omega = \sin\theta \, d\theta \, d\psi$ is, from the definition of intensity,

$$I(\mathbf{r}, \hat{\mathbf{s}}) \, dA_p d\Omega = I(\mathbf{r}, \hat{\mathbf{s}}) \, dA \cos\theta \, \sin\theta \, d\theta \, d\psi,$$

where dA_p is the projected area of dA normal to the rays (i.e., the way dA is seen when viewed from the $-\hat{\mathbf{s}}$ direction). Thus, integrating this expression over all possible directions gives the total energy emitted from dA, or, after dividing by dA

$$E(\mathbf{r}) = \int_0^{2\pi} \int_0^{\pi/2} I(\mathbf{r}, \theta, \psi) \cos\theta \, \sin\theta \, d\theta \, d\psi = \int_{2\pi} I(\mathbf{r}, \hat{\mathbf{s}}) \hat{\mathbf{n}} \cdot \hat{\mathbf{s}} \, d\Omega. \tag{1.28}$$

This expression is, of course, also valid on a spectral basis.

The directional behavior of the radiative intensity leaving a blackbody is easily obtained from a variation of Kirchhoff's law: Consider a small, black surface suspended at the center of an isothermal spherical enclosure, as depicted in Fig. 1-10.

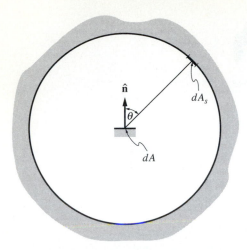

FIGURE 1-10
Kirchhoff's law for the directional behavior of
blackbody intensity.

Let us assume that the enclosure has a (hypothetical) surface coating that reflects
all incoming radiation totally and like a mirror everywhere except over a small area
dA_s, which also reflects all incoming radiation except for a small wavelength interval
between λ and $\lambda + d\lambda$. Over this small range of wavelengths dA_s behaves like a
blackbody. Now, all radiation leaving dA, traveling to the sphere (with the exception
of light of wavelength λ traveling toward dA_s), will be reflected back toward dA
where it will be absorbed (since dA is black). Thus, the net energy flow from dA to
the sphere is, recalling the definitions for intensity and solid angle

$$I_{b\lambda}(T,\theta,\psi,\lambda)(dA\cos\theta)\,d\Omega_s\,d\lambda = I_{b\lambda}(T,\theta,\psi,\lambda)(dA\cos\theta)\left(\frac{dA_s}{R^2}\right)d\lambda,$$

where $d\Omega_s$ is the solid angle with which dA_s is seen from dA. On the other hand,
also by Kirchhoff's law, the sphere does not emit any radiation (since it does not
absorb anything), except over dA_s at wavelength λ. All energy emitted from dA_s
will eventually come back to itself except for the fraction intercepted by dA. Thus,
the net energy flow from the sphere to dA is

$$I_{bn\lambda}(T,\lambda)\,dA_s\,d\Omega\,d\lambda = I_{bn\lambda}(T,\lambda)\,dA_s\left(\frac{dA\cos\theta}{R^2}\right)d\lambda,$$

where the subscript n denotes emission into the normal direction ($\theta_s = 0, \psi_s$ arbitrary), and $d\Omega$ is the solid angle with which dA is seen from dA_s. Now, from the
Second Law of Thermodynamics, these two fluxes must be equal for an isothermal
enclosure. Therefore,

$$I_{b\lambda}(T,\theta,\psi,\lambda) = I_{bn\lambda}(T,\lambda).$$

Since the direction (θ, ψ), with which dA_s is oriented, is quite arbitrary we conclude that $I_{b\lambda}$ is *independent of direction,* or

$$I_{b\lambda} = I_{b\lambda}(T, \lambda) \text{ only.} \tag{1.29}$$

Substituting this expression into equation (1.28) we obtain the following relationship between blackbody intensity and emissive power:

$$E_{b\lambda}(\mathbf{r}, \lambda) = \pi I_{b\lambda}(\mathbf{r}, \lambda). \tag{1.30}$$

This equation implies that the intensity leaving a blackbody (or any surface whose outgoing intensity is independent of direction, or *diffuse*) may be evaluated from the blackbody emissive power (or outgoing heat flux) as

$$I_{b\lambda}(\mathbf{r}, \lambda) = E_{b\lambda}(\mathbf{r}, \lambda)/\pi. \tag{1.31}$$

In the literature the spectral blackbody intensity is sometimes referred to as the *Planck function.* The directional behavior of the emission from a blackbody is found by comparing the intensity (energy flow per solid angle and *area normal to the rays*) and directional emitted flux (energy flow per solid angle and per unit surface area). The directional heat flux is sometimes called directional emissive power, and

$$E'_{b\lambda}(\mathbf{r}, \lambda, \theta, \psi)dA = I_{b\lambda}(\mathbf{r}, \lambda)dA_p,$$

or

$$E'_{b\lambda}(\mathbf{r}, \lambda, \theta, \psi) = I_{b\lambda}(\mathbf{r}, \lambda)\cos\theta, \tag{1.32}$$

that is, the directional emitted flux of a blackbody varies with the cosine of the polar angle. This is sometimes referred to as *Lambert's law*[*] or *cosine law.*

1.7   RADIATIVE HEAT FLUX

Consider the surface shown in Fig. 1-11. Let thermal radiation from an infinitesimal solid angle around the direction $\hat{\mathbf{s}}_i$ impinge onto the surface with an intensity of $I_\lambda(\hat{\mathbf{s}}_i)$. Such radiation is often called a "pencil of rays" since the infinitesimal solid angle is usually drawn looking like the tip of a sharpened pencil. Recalling the definition for intensity we see that it imparts an infinitesimal heat flow rate per wavelength on the surface in the amount of

$$dQ_\lambda = I_\lambda(\hat{\mathbf{s}}_i)d\Omega_i dA_p = I_\lambda(\hat{\mathbf{s}}_i)d\Omega_i(dA\cos\theta_i),$$

[*]**Johann Heinrich Lambert (1728-1777)**
German mathematician, astronomer, and physicist. Largely self-educated, Lambert did his work under the patronage of Frederick the Great. He made many discoveries in the areas of mathematics, heat, and light. The lambert, a measurement of diffusely reflected light intensity, is named in his honor (see Section 1.9).

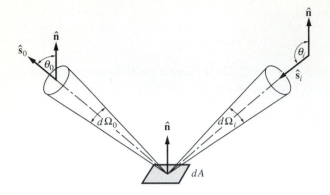

FIGURE 1-11
Radiative heat flux on an arbitrary surface.

where heat rate is taken as positive in the direction of the outward surface normal (going into the medium), so that the incoming flux going *into the surface* is negative since $\cos \theta_i < 0$. Integrating over all 2π incoming directions and dividing by the surface area gives the total incoming heat flux per unit wavelength, i.e.,

$$(q_\lambda)_{\text{in}} = \int_{\cos \theta_i < 0} I_\lambda(\hat{\mathbf{s}}_i) \cos \theta_i \, d\Omega_i. \qquad (1.33)$$

Heat loss from the surface, along a pencil of rays into the direction $\hat{\mathbf{s}}_o$, and integrated over all outgoing directions, follows as

$$(q_\lambda)_{\text{out}} = \int_{\cos \theta_o > 0} I_\lambda(\hat{\mathbf{s}}_o) \cos \theta_o \, d\Omega_o. \qquad (1.34)$$

If the surface is black ($\epsilon_\lambda = 1$), there is no energy reflected from the surface and $I_\lambda = I_{b\lambda}$, leading to $(q_\lambda)_{\text{out}} = E_{b\lambda}$. If the surface is not black, the outgoing intensity consists of contributions from emission as well as reflections. The outgoing heat flux is positive since it is going *into the medium*. The net heat flux from the surface may be calculated by adding both contributions, or

$$(q_\lambda)_{\text{net}} = (q_\lambda)_{\text{in}} + (q_\lambda)_{\text{out}} = \int_{4\pi} I_\lambda(\hat{\mathbf{s}}) \cos \theta \, d\Omega, \qquad (1.35)$$

where a single direction vector $\hat{\mathbf{s}}$ was used to describe the total range of solid angles, 4π. It is readily seen from Fig. 1-11 that $\cos \theta = \hat{\mathbf{n}} \cdot \hat{\mathbf{s}}$ and, since the net heat flux is evaluated as the flux into the positive $\hat{\mathbf{n}}$-direction, one gets

$$(q_\lambda)_{\text{net}} = \mathbf{q}_\lambda \cdot \hat{\mathbf{n}} = \int_{4\pi} I_\lambda(\hat{\mathbf{s}}) \, \hat{\mathbf{n}} \cdot \hat{\mathbf{s}} \, d\Omega. \qquad (1.36)$$

In order to obtain the *total radiative heat flux* at the surface, equation (1.36) needs to be integrated over the spectrum, and

$$\mathbf{q} \cdot \hat{\mathbf{n}} = \int_0^\infty \mathbf{q}_\lambda \cdot \hat{\mathbf{n}} \, d\lambda = \int_0^\infty \int_{4\pi} I_\lambda(\hat{\mathbf{s}}) \, \hat{\mathbf{n}} \cdot \hat{\mathbf{s}} \, d\Omega \, d\lambda. \qquad (1.37)$$

Example 1.5. A solar collector mounted on a satellite orbiting earth is directed at the sun (i.e., normal to the sun's rays). Determine the total solar heat flux incident on the collector per unit area.

Solution

The total heat rate leaving the sun is $\dot{Q}_S = 4\pi R_S^2 E_b(T_S)$, where $R_S \simeq 6.96 \times 10^8$ m is the radius of the sun. Placing an imaginary spherical shell around the sun of radius $S_{ES} = 1.496 \times 10^{11}$ m, where S_{ES} is the distance between the sun and earth, we find the heat flux going through that imaginary sphere (which includes the solar collector) as

$$q_{sol} = \frac{4\pi R_S^2 E_b(T_S)}{4\pi S_{ES}^2} = I_b(T_S)\frac{\pi R_S^2}{S_{ES}^2} = I_b(T_S)\,\Omega_S,$$

where we have replaced ths sun's emissive power by intensity, $E_b = \pi I_b$, and $\Omega_S = 6.80 \times 10^{-5}$ sr is the solid angle with which the sun is seen from earth, as determined in Example 1.3. Therefore, with $I_b(T_S) = \sigma T_S^4/\pi$ and $T_S = 5762$ K,

$$q_{in} = -(\sigma T_S^4/\pi)(\Omega_S) = -\frac{1}{\pi}5.670 \times 10^{-8} \times 5762^4 \times 6.80 \times 10^{-5} \text{ W/m}^2$$

$$= -1353 \text{ W/m}^2,$$

where we have added a minus sign to emphasize that the heat flux is going *into* the collector. The total incoming heat flux may, of course, also be determined from equation (1.33) as

$$q_{in} = \int_{\cos\theta_i < 0} I(\hat{s}_i)\cos\theta_i\,d\Omega_i.$$

Since light from the sun arrives only from the extremely small solid angle of Ω_S (over which $\cos\theta_i \simeq -1$), the integral may be written as

$$q_{in} = I_i \times (-1) \times \Omega_S.$$

Comparing this with the previous expression we see that the incoming solar intensity at the collector, I_i, is identical to the intensity leaving from the sun's surface, $I_b(T_S)$. This invariability property of radiative intensity as it travels through vacuum will be discussed in more detail in Chapter 8.

The absolute value of the heat flux determined in the previous example is known as the *solar constant* and has been measured accurately by spectral integration of the extraterrestrial solar irradiation, Fig. 1-3, as

$$q_{sol} = (1353 \pm 2.1) \text{ W/m}^2. \qquad (1.38)$$

The effective solar temperature of $T = 5762$ K has been determined by working Example 1.5 in reverse.

1.8 RADIATION PRESSURE

If we think of radiative energy as photons or (massless) particles carrying an energy of $h\nu$ and traveling at the speed of light c into a certain direction, then these particles should carry momentum in the amount of energy/speed $= h\nu/c = h\eta$ (even though they do not possess mass). Therefore, if photons hit a material surface a momentum transfer takes place, which implies that a stream of photons exerts pressure on the walls of a container, known as *radiation pressure* or *photon pressure*.

Consider a monochromatic beam of intensity $I_\lambda(\hat{s}_i)$ incident on a surface element dA from a direction $\hat{s}_i$ and over a solid angle $d\Omega_i$. The energy flow incident on dA due to this beam is $I_\lambda(\hat{s}_i)\cos\theta_i\,dA\,d\Omega_i$. Therefore, the beam carries with it momentum at a rate of

$$\frac{1}{c}I_\lambda(\hat{s}_i)\left|\cos\theta_i\right|dA\,d\Omega_i\,\hat{s}_i,$$

where the unit vector $\hat{s}_i$ has been added to emphasize that the momentum flow is a vector pointing into the direction of $\hat{s}_i$. The fraction of momentum falling onto dA in the normal direction is $\left|\cos\theta_i\right| = \left|\hat{n}\cdot\hat{s}_i\right|$. Therefore, the flow of momentum onto dA in the normal direction is, per unit area,

$$\frac{1}{c}I_\lambda(\hat{s}_i)\cos^2\theta_i\,d\Omega_i.$$

According to *Newton's Second Law* the total momentum flux normal to the surface, due to irradiation from all possible directions, must be counteracted by a pressure force $p_\lambda dA$ leading to a *spectral radiation pressure*

$$p_\lambda = \frac{1}{c}\int_0^{2\pi}\int_0^\pi I_\lambda(\hat{s})\cos^2\theta\,\sin\theta\,d\theta\,d\psi. \tag{1.39}$$

As for other spectral properties, a *total radiation pressure* is defined as

$$p = \int_0^\infty p_\lambda\,d\lambda. \tag{1.40}$$

Example 1.6.   A very light spacecraft has been entered in the Columbus Competition, scheduled for 1992, whose goal is to sail the craft to Mars, using solar radiation pressure as the only propellant. Assuming a black sail and a necessary pushing force of 100 N, determine the necessary surface area for the solar sail.

Solution

The solar pressure exerted on the sail may be calculated from equation (1.39). As in the previous example solar radiation is incident over an extremely small solid angle, $\Omega_S =$

6.8×10^{-5}sr. Thus, the solar pressure may be evaluated as

$$p = \frac{1}{c}\frac{\sigma T_S^4}{\pi}(-1)^2\Omega_S = \frac{q_{sol}}{c}$$

$$= \frac{1353 \text{ W/m}^2}{2.998 \times 10^8 \text{m/s}} = 4.5 \times 10^{-6}\text{Ws/m}^3 = 4.5 \times 10^{-6}\text{N/m}^2$$

The total force on the sail is $F = pA$, so that a force of 100 N would require a sail area of $A = 100/4.5 \times 10^{-6} = 22 \times 10^6 \text{ m}^2 = 22 \text{ km}^2$!

Note that, for a perfectly reflecting sail, the reflected intensity would be equal to the incoming intensity, but in the opposite direction with $\cos\theta = +1$. The contribution to the radiation pressure would be positive again, doubling the pressure to 9×10^{-6}N and halving the necessary sail area to 11 km^2.

This example demonstrates that radiation pressure, while measurable and nonnegligible for some applications, can certainly not compete with the pressure exerted by a molecular gas.

1.9   VISIBLE RADIATION (LUMINANCE)

Because of the great importance of visible radiation, and since much of the theory developed in this book is directly applicable to lighting design calculations, we shall very briefly discuss the nature of electromagnetic waves as perceived by the human eye. For a more detailed treatment of lighting, the reader should refer to books on the subject, such as the ones by Moon [9] and by Hopkinson and coworkers [10], or the I.E.S. (Illuminating Engineering Society) Handbook [11].

When electromagnetic waves of spectral intensity I_λ fall onto the human eye, a certain fraction of the intensity is observed as "light," known as *luminous intensity* or *luminance*,

$$L_\lambda = K_\lambda I_\lambda, \tag{1.41}$$

where the proportionality factor K_λ is called the *luminous efficacy*. Luminous intensity is measured in *lumen* per unit wavelength, per unit solid angle, and per unit area normal to the rays ($\text{lm/m}^2\,\mu\text{m sr}$). Therefore, the units of the luminous efficacy are lm/W. Luminance may also be measured using *candelas* (cd), *lux* (lx) or *lamberts* (L), where the different units are related by

$$1 \text{ L} = \frac{1}{\pi} \text{ cd/cm}^2, \quad 1 \text{ cd} = 1 \text{ lm/sr}, \quad 1 \text{ lx} = 1 \text{ lm/m}^2.$$

The average human eye responds to radiation in the wavelength interval between $\sim 0.4\,\mu\text{m}$ and $0.7\,\mu\text{m}$, and K_λ thus vanishes outside this range. A *standard luminous efficacy* to approximate the average human eye was set by the *CIE (Commission Internationale de l'Eclairage* or International Commission on Illumination) in 1924

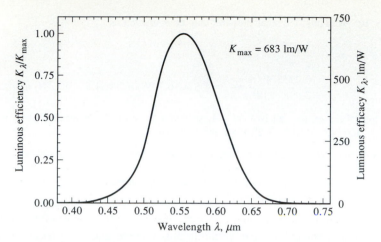

FIGURE 1-12
CIE standards for luminous efficiency and luminous efficacy.

and is shown in Fig. 1-12. Also shown is the standard *luminous efficiency*, defined as

$$\eta_{\text{lum},\lambda} = K_\lambda/K_{\max}; \quad K_{\max} = K_\lambda(\lambda = 0.555\,\mu\text{m}) = 683\,\text{lm/W}, \tag{1.42}$$

where $K_{\max}$ is the spectral maximum of K_λ, which occurs (for the standard) at $\lambda = 0.555\,\mu\text{m}$. The total luminance is calculated from

$$L = \int_0^\infty K_\lambda I_\lambda d\lambda, \tag{1.43}$$

which—if I_λ does not vary too much over the interval $0.4\,\mu\text{m} < \lambda < 0.7\,\mu\text{m}$—may be approximated as

$$L \simeq I_\lambda(\lambda = 0.555\,\mu\text{m})K_{\max} \int_0^\infty \eta_{\text{lum},\lambda} d\lambda$$

$$\simeq 86\,\frac{\text{lm}\,\mu\text{m}}{\text{W}} \times I_\lambda(\lambda = 0.555\,\mu\text{m}) \simeq 286\,\frac{\text{lm}}{\text{W}} \int_{0.4\mu\text{m}}^{0.7\mu\text{m}} I_\lambda d\lambda, \tag{1.44}$$

where $86\,\text{lm}\,\mu\text{m/W}$ is the area underneath the luminous efficiency in Fig. 1-12, and $286\,\text{lm/W}$ is an appropriate average value for K_λ.

Example 1.7.   On a clear day the strength of solar radiation has been measured as $q_{\text{sun}} = 800\,\text{W/m}^2$ (normal to the sun's rays), while total sky radiation (from all directions), falling onto a horizontal surface, has been determined as $q_{\text{sky}} = 200\,\text{W/m}^2$. Determine the luminous flux, or *illumination*, onto a horizontal surface if the sun is at a zenith angle of $60°$.

Solution

Similar to the radiative heat flux, the luminous flux onto a surface is defined as

$$q_{lum} = \int_{2\pi} L(\hat{s}) \cos \theta \, d\Omega.$$

Therefore, from equations (1.43) and (1.44)

$$q_{lum} = \int_{2\pi} 286 \frac{lm}{W} \int_{0.4\mu m}^{0.7\mu m} I_\lambda d\lambda \, \cos \theta \, d\Omega$$

$$= 286 \frac{lm}{W} \int_{0.4\mu m}^{0.7\mu m} \int_{2\pi} I_\lambda \cos \theta \, d\Omega = 286 \frac{lm}{W} \int_{0.4\mu m}^{0.7\mu m} q_\lambda d\lambda,$$

where q_λ is the total spectral irradiation from the sun and the sky. Since both contributions are due to solar emission (either by direct travel or after atmospheric scattering), we may write $q_\lambda = (q_{sun} \cos \theta_{sun} + q_{sky}) E_{b\lambda}(T_{sun}) / E_b(T_{sun})$, or

$$q_{lum} = 286 \frac{lm}{W} (q_{sun} \cos \theta_{sun} + q_{sky}) \left[f(0.7 \, \mu m \, T_{sun}) - f(0.4 \, \mu m \, T_{sun}) \right],$$

or, with $T_{sun} = 5762 \, K$,

$$q_{lum} = 286 \frac{lm}{W} \left(800 \times \frac{1}{2} + 200 \right) \frac{W}{m^2} [0.48681 - 0.12099]$$

$$= 62,775 \, lm/m^2 = 62,775 \, lx.$$

Actually, for the visible wavelengths a solar temperature of $T_{sun} = 5762 \, K$ is not very accurate (cf. Fig. 1-3,) so usually a solar temperature of 6500 K is employed for luminance calculations. With such a temperature $f(0.7 \times 6500 \, \mu m \, K) - f(0.4 \times 6500 \, \mu m \, K) = 0.57177 - 0.18311$ and

$$q_{lum} \simeq 66,694 \, lx.$$

All relationships for radiative intensity and radiative heat flux that will be developed in this book may also be used to determine luminous intensity and luminous flux simply by using units of *lumens* instead of *watts*.

1.10 INTRODUCTION TO RADIATION CHARACTERISTICS OF OPAQUE SURFACES

We have already noted that thermal radiation, unlike conduction and convection, is a long-range phenomenon. When making an energy balance for a point in space we must account for all photons that may arrive at this point, no matter from how far away. Thus, a *conservation of energy* balance must be performed on an *enclosure* bounded by *opaque walls* (i.e., a medium thick enough that no electromagnetic waves can penetrate through it). Strictly speaking, the surface of an enclosure wall can only

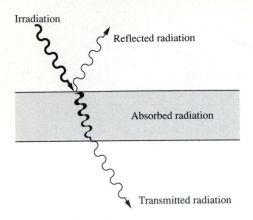

Irradiation

Reflected radiation

Absorbed radiation

Transmitted radiation

FIGURE 1-13
Absorption, reflection and transmission by a slab.

reflect radiative energy or allow a part of it to penetrate into the substrate. A surface cannot absorb or emit photons: Attenuation takes place inside the solid, as does emission of radiative energy (and some of the emitted energy escapes through the surface into the enclosure). In practical systems the thickness of the surface layer over which absorption of irradiation from inside the enclosure occurs is very small compared with the overall dimensions of an enclosure—usually a few Å for metals and a few μm for most nonmetals. The same may be said about emission from within the walls that escapes into the enclosure. Thus, in the case of opaque walls it is customary to speak of absorption by and emission from a "surface," although a thin surface layer is implied.

Consider thermal radiation impinging on a medium of finite thickness, as shown in Fig. 1-13. In general, some of the irradiation will be *reflected* away from the medium, a fraction will be *absorbed* inside the layer, and the rest will be *transmitted* through the slab. Based on this observation we define three fundamental radiative properties:

$$\textit{Reflectivity}, \quad \rho \equiv \frac{\text{reflected part of incoming radiation}}{\text{total incoming radiation}}, \qquad (1.45a)$$

$$\textit{Absorptivity}, \quad \alpha \equiv \frac{\text{absorbed part of incoming radiation}}{\text{total incoming radiation}}, \qquad (1.45b)$$

$$\textit{Transmissivity}, \quad \tau \equiv \frac{\text{transmitted part of incoming radiation}}{\text{total incoming radiation}}. \qquad (1.45c)$$

Since all radiation must be either reflected, absorbed or transmitted, we conclude

$$\rho + \alpha + \tau = 1. \qquad (1.46)$$

If the medium is sufficiently thick to be *opaque*, then $\tau = 0$ and

$$\rho + \alpha = 1. \qquad (1.47)$$

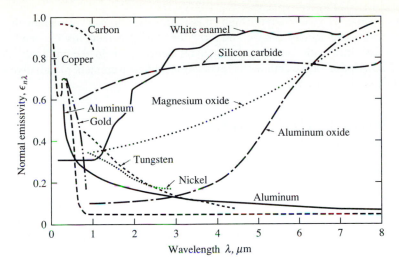

FIGURE 1-14
Normal, spectral emissivities for selected materials, from [12].

We note that all three of these properties are nondimensional and may vary in magnitude between the values 0 and 1. Since a black surface absorbs all incoming radiation it follows for such a surface that $\alpha = 1$ and $\rho = \tau = 0$.

All surfaces also emit thermal radiation (or, rather, radiative energy is emitted within the medium, some of which escapes from the surface). Since we know that, at a given temperature, the maximum possible is emitted by a black surface, we define a fourth nondimensional property;

$$Emissivity,\ \epsilon = \frac{\text{energy emitted from a surface}}{\text{energy emitted by a black surface at same temperature}}. \qquad (1.48)$$

The emissivity can, therefore, also vary between the values of 0 and 1, and for a black surface we have $\epsilon = 1$.

All four properties may be functions of temperature as well as of wavelength (or frequency). The absorptivity may be different for different directions of irradiation, while emissivity may vary with outgoing directions. Finally, the magnitude of reflectivity and transmissivity may depend on both incoming and outgoing directions. Thus, we distinguish between spectral and total properties (i.e., an average value over the spectrum), and between directional and hemispherical properties (i.e., an average value over all directions).

Typical spectral behavior of surface emissivities is shown in Fig. 1-14 for a few materials, as collected by White [12]. Shown are values for directional emissivities in the direction normal to the surface. However, the spectral behavior is the same for *hemispherical emissivities* (i.e., emissivities averaged over all directions). In general, nonmetals have relatively high emissivities, which may vary erratically across the spectrum, and metals behave similarly for short wavelengths but tend to have lower emissivities with more regular spectral dependence in the infrared. A more detailed description of the definitions, evaluation and measurement of radiative surface properties is given in Chapter 3.

1.11   INTRODUCTION TO RADIATION CHARACTERISTICS OF GASES

Like a solid medium (or the thin layer next to its surface), gases can absorb and emit radiative energy. All gas atoms or molecules carry a certain amount of energy, consisting of kinetic energy (translational energy of a molecule) and energy internal to each molecule. The internal molecular energy, in turn, consists of a number of contributions, primarily of the levels of electronic, vibrational and rotational energy states. Thus, a passing photon may be absorbed by a molecule, raising the level of one of the internal energy states. On the other hand, a molecule may spontaneously release (emit) a photon in order to lower one of its internal energy states.

Quantum mechanics postulates that only a finite number of discrete energy levels are possible, i.e., that electrons can circle around the nucleus only on a number of allowed orbits, vibration between nuclei can only occur with a number of distinct amplitudes, and the nuclei can rotate around one another only with a number of allowed rotational velocities. Therefore, changing the internal energy of a molecular gas can only destroy or generate photons with very distinct energy levels $h\nu$ and, consequently, only at distinct frequencies or wavelengths. It takes a relatively large amount of energy to change the orbit of an electron, giving rise to absorption-emission lines in the ultraviolet and visible parts of the spectrum. In the case of a monatomic gas only the electronic energy level can be altered by a photon. Changing the vibrational energy level of a molecule requires an intermediate amount of energy, resulting in spectral lines in the near to intermediate infrared. Finally, rotational energy changes take place with an even smaller amount of energy, so that rotational lines are found in the intermediate to far infrared. Usually, vibrational energy changes are accompanied by simultaneous changes in rotational energy levels, so that the vibrational lines are surrounded by many rotational lines, as illustrated in Fig. 1-15.

Unless the temperature of a gas is extremely high, the gas will essentially be free of ions and free electrons. In this case, absorption or emission of photons results in *bound-bound transitions* (i.e., there is no ionization before and after the transition), and *discrete spectral lines*, as shown in Fig. 1-15. If absorption of a photon results in ionization and the release of an electron, the transition is termed *bound-free*. Conversely, a free electron may combine with an ion in a *free-bound transition*, producing a photon. Finally, a free electron can absorb or emit photons, resulting in *free-free transitions*. Since an electron can have arbitrary kinetic energy, all transitions involving free electrons are not limited to discrete wavelengths and produce continuous spectra.

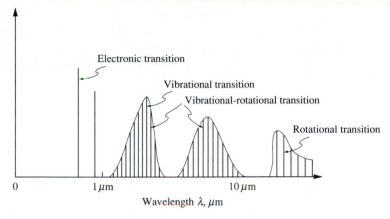

FIGURE 1-15
Spectral lines due to electronic, vibrational and rotational energy changes in a gas molecule.

In reality, not even the lines from bound-bound transitions are truly discrete, but are *broadened* slightly. The rotational lines accompanying a vibrational transition usually overlap, forming what is termed *vibration-rotation bands,* as also indicated in Fig. 1-15.

As radiative energy penetrates through a gas layer it gradually becomes attenuated by absorption. Experience (and theoretical development) shows that this absorption leads to an exponential decay of incident radiation, so that the *transmissivity* of a homogeneous isothermal gas layer may be written as

$$\tau_\eta = e^{-\kappa_\eta s}, \tag{1.49}$$

where s is the thickness of the gas layer and the proportionality constant κ_η is known as the *absorption coefficient.* In the expression for transmissivity we have used the wavenumber η as spectral variable, since the wavenumber is commonly chosen by researchers working in the field of gas radiation. Since, in the case of a gas layer, incident radiation is either transmitted or absorbed, we define the spectral absorptivity of a gas layer as

$$\alpha_\eta = 1 - \tau_\eta = 1 - e^{-\kappa_\eta s}. \tag{1.50}$$

Figure 1-16 shows a typical absorption spectrum for a nitrogen-carbon dioxide mixture, taken from the early work of Edwards [13]. The formation of vibration-rotation bands in the infrared due to bound-bound transitions, separated by spectral windows, is clearly visible in this figure. Not all gases have vibration-rotation bands. In particular, dry air (i.e., nitrogen and oxygen) at moderate temperature does not absorb or emit radiation in the infrared. This observation is in direct contrast to the high-temperature data for air, obtained by Meyerott and coworkers [14] and shown

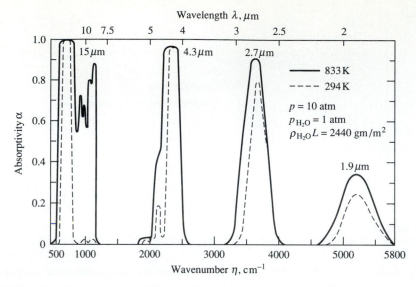

FIGURE 1-16
Spectral absorptivity of an isothermal mixture of nitrogen and carbon dioxide, from Edwards [13].

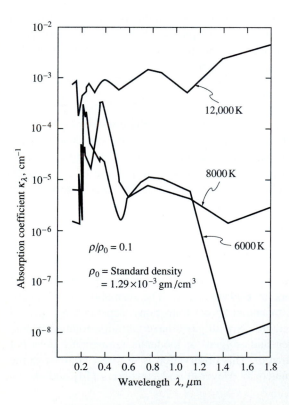

FIGURE 1-17
Spectral absorption coefficient of air at very high temperatures, from Meyerott and coworkers [14].

in Fig. 1-17. In this figure, the absorption coefficient varies gradually across the spectrum, indicating the nature of bound-free and free-free transitions in the partially ionized gas. Radiative properties of gases are described in much more detail in Chapter 9.

1.12 INTRODUCTION TO RADIATION CHARACTERISTICS OF SOLIDS AND LIQUIDS

There are a number of liquid and solid substances that absorb radiative energy only gradually, such that these materials cannot be approximated as "opaque surfaces." They are known as *semitransparent*. Typical examples for the visible part of the spectrum are water, glass, quartz, etc. Absorption and emission of photons in liquids are due to interaction with free electrons, and within solids to free electrons as well as excitation of lattice vibrations (*phonons*). Consequently, semitransparent materials are always poor electric conductors (few free electrons). The absorption behavior of these materials is qualitatively similar to that of dissociated gases (free-free transitions for the electrons), with the possibility of strong absorption bands due to *photon-phonon* interactions. As an example the absorption coefficient for window glass is shown in Fig. 1-18, taken from measurements by Neuroth [15].

A slightly more detailed discussion of the radiative properties of semitransparent media is given in Chapter 11.

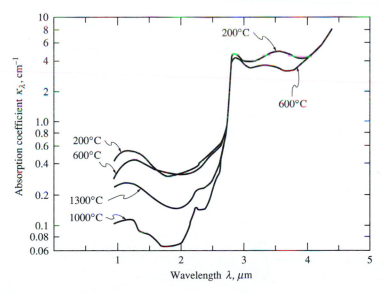

FIGURE 1-18
Spectral absorption coefficient of window glass, from Neuroth [15].

1.13   INTRODUCTION TO RADIATION CHARACTERISTICS OF PARTICLES

The interaction between photons, or electromagnetic waves, and small particles is somewhat different from that with a homogeneous gas, liquid, or solid. As for a homogeneous medium, radiation traveling through a particle cloud may be transmitted, reflected or absorbed. In addition, interaction with a particle may change the direction in which a photon travels, as schematically shown in Fig. 1-19. This can occur by one of three different mechanisms: *(i)* The path of a photon may be altered, without ever colliding with the particle, by *diffraction, (ii)* a photon may change its direction by *reflection* from the particle, and *(iii)* the photon may penetrate into the particle, changing its direction because of *refraction.* All three phenomena together are known as the *scattering* of radiation. Absorption takes place as the electromagnetic wave penetrates into the particle. Therefore, in the presence of scattering, the equation for transmissivity of a material layer, equation (1.49), must be augmented to

$$\tau_\eta = e^{-(\kappa_\eta + \sigma_{s\eta})s} = e^{-\beta_\eta s}, \tag{1.51}$$

where $\sigma_{s\eta}$ is known as the *scattering coefficient* and $\beta_\eta = \kappa_\eta + \sigma_{s\eta}$ as the *extinction coefficient.*

The nature of the interaction between electromagnetic waves and particles is determined by the relative size of the particles compared with the wavelength of the radiation. Defining a *size parameter*

$$x = \frac{2\pi a}{\lambda}, \tag{1.52}$$

where a is the effective radius of the particle, we distinguish among three different regimes:

1. $x \ll 1$, or *Rayleigh scattering,* named after Lord Rayleigh, who studied the interaction of atmospheric air (whose molecules are, in fact, very small particles) with sunlight [16]. He observed that, for very small particles, scattering is proportional to ν^4 or $1/\lambda^4$. Thus, within the visible part of the spectrum, blue light is scattered the most (accounting for blue skies) and red the least (resulting in red sunsets).

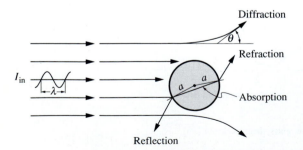

FIGURE 1-19
Interaction of electromagnetic waves with small particles.

2. $x = \mathbb{O}(1)$, or *Mie scattering,* named after Gustav Mie, who developed a comprehensive (and difficult) theory for the interaction between electromagnetic waves and particles [17].

3. $x \gg 1$. In this case the surface of the particle may be treated as a normal surface, and properties may be found through *geometric optics.*

A more detailed discussion of the radiative properties of particles is given in Chapter 10.

1.14 OUTLINE OF RADIATIVE TRANSPORT THEORY

When considering heat transfer by conduction and/or convection within a medium, we require knowledge of a number of material properties, such as thermal conductivity k, thermal diffusivity α, kinematic viscosity ν, and so on. This knowledge, together with the *law of conservation of energy,* allows us to calculate the energy field within the medium in the form of the basic variable, temperature T. Once the temperature field is determined, the local heat flux vector may be found from Fourier's law. The evaluation of radiative energy transport follows a similar pattern: Knowledge of radiative properties is required (emissivity ϵ, absorptivity α, and reflectivity ρ, in the case of surfaces, as well as absorption coefficient κ and scattering coefficient σ for semitransparent media), and the law of conservation of energy is applied to determine the energy field. Two major differences exist between conduction/convection and thermal radiation that make the analysis of radiative transport somewhat more complex: *(i)* Unlike their thermophysical counterparts, radiative properties may be functions of direction as well as of wavelength, and *(ii)* the basic variable appearing in the law of conservation of radiative energy, commonly referred to as the *equation of radiative transport,* is not temperature but radiative intensity, which is a function not only of location in space (as is temperature), but also of direction. Only after the intensity field has been determined can the local temperatures (as well as the radiative heat flux vector) be calculated.

Thermal radiation calculations are always performed by making an energy balance for an enclosure bounded by opaque walls (some of which may be artificial to account for radiation penetrating through openings in the enclosure). If the enclosure is evacuated or filled with a nonabsorbing, nonscattering medium (such as air at low to moderate temperatures), we speak of *surface radiation transport.* If the enclosure is filled with an absorbing gas or a semitransparent solid or liquid, or with absorbing and scattering particles (or bubbles), we refer to it as *radiative transport in a participating medium.* Of course, radiation in a participating medium is always accompanied by surface radiation transport.

References

1. Sparrow, E. M., and R. D. Cess: *Radiation Heat Transfer*, Hemisphere, New York, 1978.

2. Thekaekara, M. P.: "The Solar Constant and Spectral Distribution of Solar Radiant Flux," *Solar Energy*, vol. 9, no. 1, pp. 7–20, 1965.
3. Thekaekara, M. P.: "Solar Energy Outside the Earth's Atmosphere," *Solar Energy*, vol. 14, pp. 109–127, 1973.
4. Rayleigh, L.: "The Law of Complete Radiation," *Phil. Mag.*, vol. 49, pp. 539–540, 1900.
5. Jeans, J. H.: "On the Partition of Energy between Matter and the Ether," *Phil. Mag.*, vol. 10, pp. 91–97, 1905.
6. Wien, W.: "Über die Energieverteilung im Emissionsspektrum eines schwarzen Körpers," *Annalen der Physik*, vol. 58, pp. 662–669, 1896.
7. Planck, M.: "Distribution of Energy in the Spectrum," *Annalen der Physik*, vol. 4, no. 3, pp. 553–563, 1901.
8. Wien, W.: "Temperatur und Entropie der Strahlung," *Annalen der Physik*, vol. 52, pp. 132–165, 1894.
9. Moon, P.: *Scientific Basis of Illuminating Engineering*, Dover Publications, New York, 1961 (originally published by McGraw-Hill, New York, 1936).
10. Hopkinson, R. B., P. Petherbridge, and J. Longmore: *Daylighting*, Pitman Press, London, 1966.
11. Kaufman, J. E. (ed.): *IES Lighting Handbook*, Illuminating Engineering Society of North America, New York, 1981.
12. White, F. M.: *Heat Transfer*, Addison-Wesley, Reading, MA, 1984.
13. Edwards, D. K.: "Radiation Interchange in a Nongray Enclosure Containing an Isothermal CO_2–N_2 Gas Mixture," *ASME Journal of Heat Transfer*, vol. 84C, pp. 1–11, 1962.
14. Meyerott, R. E., J. Sokoloff, and R. W. Nicholls: "Absorption Coefficients of Air," in *Geophysics Research Paper No. 68*, Air Force Cambridge Research Center, 1960.
15. Neuroth, N.: "Der Einfluss der Temperatur auf die spektrale Absorption von Gläsern im Ultraroten, I (Effect of Temperature on Spectral Absorption of Glasses in the Infrared)," *Glastechnische Berichte*, vol. 25, pp. 242–249, 1952.
16. Rayleigh, L.: *Phil. Mag.*, vol. 12, 1881.
17. Mie, G. A.: "Beiträge zur Optik trüber Medien, speziell kolloidaler Metallösungen," *Annalen der Physik*, vol. 25, pp. 377–445, 1908.

Problems

1.1 Solar energy impinging on the outer layer of earth's atmosphere (usually called "solar constant") has been measured as 1353 W/m^2. What is the solar constant on mars? (Distance earth to sun $=1.496 \times 10^{11}$ m, mars to sun $=2.28 \times 10^{11}$ m).

1.2 Assuming earth to be a blackbody, what would be its average temperature if there was no internal heating from the core of earth?

1.3 Assuming earth to be a black sphere with a surface temperature of 300 K, what must earth's internal heat generation be in order to maintain that temperature (neglect radiation from the stars, but not the sun) (radius of the earth $R_E = 6.37 \times 10^6$m).

1.4 To estimate the diameter of the sun, one may use solar radiation data. The solar energy impinging onto the earth's atmosphere (called the "solar constant") has been measured as 1353 W/m^2. Assuming that the sun may be approximated to have a black surface with an effective temperature of 5762 K, estimate the diameter of the sun (distance sun to earth $S_{ES} \simeq 1.496 \times 10^{11}$ K).

1.5 Solar energy impinging on the outer layer of earth's atmosphere (usually called "solar constant") has been measured as 1353 W/m^2. Assuming the sun may be approximated as having a surface that behaves like a blackbody, estimate its effective surface temperature. (Distance sun to earth $S_{ES} \simeq 1.496 \times 10^{11}$m, radius of sun $R_S \simeq 6.96 \times 10^8$m).

1.6 A rocket in space may be approximated as a black cylinder of length $L = 20$ m and diameter $D = 2$ m. It flies past the sun at a distance of 140 million km such that the cylinder axis is perpendicular to the sun's rays. Assuming that *(i)* the sun is a blackbody at 5762 K and *(ii)* the cylinder has a high conductivity (i.e., is essentially isothermal), what is the temperature of the rocket? (Radius of sun $R_S = 696{,}000$ km; neglect radiation from earth and the stars).

1.7 A black sphere of very high conductivity (i.e., isothermal) is orbiting earth. What is its temperature? (Consider the sun but neglect radiation from the earth and the stars). What would be the temperature of the sphere if it were coated with a material that behaves like a black body for wavelengths between 0.4 μm and 3 μm, but does not absorb and emit at other wavelengths?

1.8 A 100 Watt light bulb may be considered to be an isothermal black sphere at a certain temperature. If the light flux (i.e., visible light, 0.4 μm $< \lambda <$ 0.7 μm) impinging on the floor directly (2.5 mm) below the bulb is 42.6 mW/m^2, what is the light bulb's effective temperature? What is its efficiency?

1.9 When a metallic surface is irradiated with a highly concentrated laser beam, a plume of plasma (i.e., a gas consisting of ions and free electrons) is formed above the surface that absorbs the laser's energy, often blocking it from reaching the surface. Assume that a plasma of 1 cm diameter is located 1 cm above the surface, and that the plasma behaves like a blackbody at 20,000 K. Based on these assumptions calculate the radiative heat flux and the total radiation pressure on the metal directly under the center of the plasma.

1.10 Solar energy incident on the surface of the earth may be broken into two parts: A direct component (traveling unimpeded through the atmosphere) and a sky component (reaching the surface after being scattered by the atmosphere). On a clear day the direct solar heat flux has been determined as $q_{sun} = 1000$ W/m^2 (per unit area normal to the rays), while the intensity of the sky component has been found to be diffuse (i.e., the intensity of the sky radiation hitting the surface is the same for all directions) and $I_{sky} = 70$ W/m^{2}sr. Determine the total solar irradiation onto earth's surface if the sun is located 60° above the horizon (i.e., 30° from the normal).

1.11 A window (consisting of a vertical sheet of glass) is exposed to direct sunshine at a strength of 1000 W/m^2. The window is pointing due south, while the sun is in the southwest, 30° above the horizon. Estimate the amount of solar energy that *(i)* penetrates into the building, *(ii)* is absorbed by the window, and *(iii)* is reflected by the window. The window is made of *(a)* plain glass, *(b)* tinted glass, whose radiative properties may be approximated by

$$\rho_\lambda = \quad 0.08 \quad \text{for all wavelengths (both glasses)},$$

$$\tau_\lambda = \begin{cases} 0.90 & \text{for } 0.35\ \mu\text{m} < \lambda < 2.7\ \mu\text{m} \\ 0 & \text{for all other wavelengths} \end{cases} \quad \text{(plain glass)}$$

$$\tau_\lambda = \begin{cases} 0.90 & \text{for } 0.5\ \mu\text{m} < \lambda < 1.4\ \mu\text{m} \\ 0 & \text{for all other wavelengths} \end{cases} \quad \text{(tinted glass)}.$$

(c) By what fraction is the amount of visible light (0.4 μm $< \lambda <$ 0.7 μm) reduced, if tinted rather that plain glass is used? How would you modify this statement in the light of Fig. 1-12?

1.12 On an overcast day the directional behavior of the intensity of solar radiation reaching the surface of the earth after being scattered by the atmosphere may be approximated as

$I_{sky}(\theta) = I_{sky}(\theta = 0) \cos \theta$, where θ is measured from the surface normal. For a day with $I_{sky}(0) = 100 \, \text{W}/\text{m}^2\text{sr}$ determine the solar irradiation hitting a solar collector, if the collector is (a) horizontal, (b) tilted from the horizontal by 30°. Neglect radiation from the earth's surface hitting the collector (by emission or reflection).

1.13 A 100 W light bulb is rated to have a total light output of 1750 lm. Assuming the light bulb to consist of a small, black, radiating body (the light filament) enclosed in a glass envelope (with a transmissivity $\tau_g = 0.9$ throughout the visible wavelengths), estimate the filament's temperature. If the filament has an emissivity of $\epsilon_f = 0.7$ (constant for all wavelengths and directions), how does it affect its temperature?

1.14 A *pyrometer* is a device with which the temperature of a surface may be determined remotely by measuring the radiative energy falling onto a detector. Consider a black detector of 1 mm × 1 mm area that is exposed to a 1 cm² hole in a furnace located a distance of 1 m away. The inside of the furnace is at 1500 K and the intensity escaping from the hole is essentially blackbody intensity at that temperature. (a) What is the radiative heat rate hitting the detector? (b) Assuming that the pyrometer has been calibrated for the situation in (a), what temperature would the pyrometer indicate if the nonabsorbing gas between furnace and detector were replaced by one with an (average) absorption coefficient of $\kappa = 0.1 \, \text{m}^{-1}$?

RADIATIVE PROPERTY PREDICTIONS FROM ELECTROMAGNETIC WAVE THEORY

2.1 INTRODUCTION

The basic radiative properties of surfaces forming an enclosure, i.e., emissivity, absorptivity, reflectivity and transmissivity, must be known before any radiative heat transfer calculations can be carried out. Many of these properties vary with incoming direction, outgoing direction, and wavelength, and must usually be found through experiment. However, for pure, perfectly smooth surfaces these properties may be calculated from classical electromagnetic wave theory.[1] These predictions make experimental measurements unnecessary for some cases, and help interpolating as well as extrapolating experimental data in many other situations.

---

[1]The National Institute of Standards and Technology (NIST, formerly NBS) has recommended to reserve the ending "-ivity" for radiative properties of pure, perfectly smooth materials (the ones discussed in this chapter), and "-ance" for rough and contaminated surfaces. Most real surfaces fall into the latter category, discussed in Chapter 3. However, it appears that few researchers in the field follow this convention, rather employing endings according to their own personal preference. Accordingly, we will use the ending "-ivity" throughout this book for *all* surfaces.

The first important discoveries with respect to light were made during the seventeenth century, such as the law of refraction (by Snell in 1621), the decomposition of white light into monochromatic components (by Newton in 1666), and the first determination of the speed of light (by Römer in 1675). However, the true nature of light was still unknown: The corpuscular theory (suggested by Newton) competed with a rudimentary wave theory. Not until the early nineteenth century was the wave theory finally accepted as the correct model for the description of light. Young proposed a model of purely transverse waves in 1817 (as opposed to the model prevalent until then of purely longitudinal waves), followed by Fresnel's comprehensive treatment of diffraction and other optical phenomena. In 1845 Faraday proved experimentally that there was a connection between magnetism and light. Based on these experiments, Maxwell presented in 1861 his famous set of equations for the complete description of electromagnetic waves, i.e., the interaction between electric and magnetic fields. Their success was truly remarkable, in particular because the theories of quantum mechanics and special relativity, with which electromagnetic waves are so strongly related, were not discovered until half a century later. To this day Maxwell's equations remain the basis for the study of light.[*]

2.2   THE MACROSCOPIC MAXWELL EQUATIONS

The original form of Maxwell's equations is based on electrical experiments available at the time, with their very coarse temporal and spatial resolution. Thus any of these measurements were spatial averages taken over many layers of atoms and temporal averages over many oscillations of an electromagnetic wave. For this reason the original set of equations is termed *macroscopic*. Today we know that electromagnetic waves interact with matter at the molecular level, with strong field fluctuations over each wave period. Therefore, more detailed treatises on optics and electromagnetic waves now generally start with a *microscopic* description of the wave equations, for example, the book by Stone [1]. While there is little disagreement in the literature on the microscopic equations, the macroscopic equations often differ somewhat from book to book, depending on assumptions made and constitutive relations used. Following the development of Stone [1], we may state the *macroscopic Maxwell equations* as

$$\nabla \cdot (\epsilon \mathbf{E}) = \rho_f, \qquad (2.1)$$

$$\nabla \cdot (\mu \mathbf{H}) = 0, \qquad (2.2)$$

---

[*]**James Clerk Maxwell (1831-1879)**
Scottish physicist. After attending the University of Edinburgh he obtained a mathematics degree from Trinity College in Cambridge. Following an appointment at Kings College in London he became the first Cavendish Professor of Physics at Cambridge. While best known for his electromagnetic theory, he made important contributions in many fields, such as thermodynamics, mechanics, and astronomy.

$$\nabla \times \mathbf{E} = -\mu \frac{\partial \mathbf{H}}{\partial t}, \tag{2.3}$$

$$\nabla \times \mathbf{H} = \epsilon \frac{\partial \mathbf{E}}{\partial t} + \sigma_e \mathbf{E}, \tag{2.4}$$

where $\mathbf{E}$ and $\mathbf{H}$ are the *electric field* and *magnetic field* vectors, respectively, ϵ is the electrical permittivity, μ the magnetic permeability, σ_e the electrical conductivity, and ρ_f is the charge density due to free electrons, which is generally assumed to be related to the electric field by the equation

$$\frac{\partial \rho_f}{\partial t} = -\nabla \cdot (\sigma_e \mathbf{E}). \tag{2.5}$$

The *phenomenological coefficients* σ_e, μ and ϵ depend on the medium under consideration, but may be assumed independent of the fields (for a *linear medium*) and independent of position and direction (for a *homogeneous and isotropic medium*); they may, however, depend on the wavelength of the electromagnetic waves [2].

2.3   ELECTROMAGNETIC WAVE PROPAGATION IN UNBOUNDED MEDIA

We seek a solution to the above set of equations in the form of a wave. The most general form of a *time-harmonic field* (i.e., a wave of constant frequency or wavelength) is

$$\mathbf{F} = \mathbf{A} \cos \omega t + \mathbf{B} \sin \omega t = \mathbf{A} \cos 2\pi \nu t + \mathbf{B} \sin 2\pi \nu t, \tag{2.6}$$

where ω is the *angular frequency* (in radians/s), and $\nu = \omega/2\pi$ is the frequency in cycles per second. While a little less convenient, we will use the cyclical frequency ν in the following development in order to limit the number of different spectral variables employed in this book. When it comes to the time-harmonic solution of linear partial differential equations, it is usually advantageous to introduce a *complex representation* of the real field. Thus, setting

$$\mathbf{F}_c = \overline{\mathbf{F}}_c e^{2\pi i \nu t}, \quad \overline{\mathbf{F}}_c = \mathbf{A} - i\mathbf{B}, \tag{2.7}$$

where $\overline{\mathbf{F}}_c$ is the time-average of the complex field, results in

$$\mathbf{F} = \Re\{\mathbf{F}_c\}, \tag{2.8}$$

where the symbol $\Re$ denotes that the real part of the complex vector $\mathbf{F}_c$ is to be taken. Since the Maxwell equations are linear in the fields $\mathbf{E}$ and $\mathbf{H}$, one may solve them for their complex fields, and then extract their real parts after a solution has been found. Therefore, setting

$$\mathbf{E} = \Re\{\mathbf{E}_c\} = \Re\{\overline{\mathbf{E}}_c e^{2\pi i \nu t}\}, \tag{2.9}$$

$$\mathbf{H} = \Re\{\mathbf{H}_c\} = \Re\{\overline{\mathbf{H}}_c e^{2\pi i \nu t}\}, \tag{2.10}$$

results in

$$\nabla \cdot (\gamma \mathbf{E}_c) = 0, \tag{2.11}$$

$$\nabla \cdot \mathbf{H}_c = 0, \tag{2.12}$$

$$\nabla \times \mathbf{E}_c = -2\pi i \nu \mu \mathbf{H}_c, \tag{2.13}$$

$$\nabla \times \mathbf{H}_c = 2\pi i \nu \gamma \mathbf{E}_c, \tag{2.14}$$

where

$$\gamma = \epsilon - i \frac{\sigma_e}{2\pi \nu} \tag{2.15}$$

is the *complex permittivity*. If $\gamma \neq 0$, then it can be shown that the solution to the above set of equations must be *plane waves*, i.e., the electric and magnetic fields are *transverse* to the direction of propagation (have no component in the direction of propagation). Thus, the solution of equations (2.11) through (2.14) will be of the form

$$\mathbf{E} = \Re\{\overline{\mathbf{E}}_c e^{2\pi i \nu t}\} = \Re\{\mathbf{E}_0 e^{-2\pi i (\mathbf{w} \cdot \mathbf{r} - \nu t)}\}, \tag{2.16}$$

$$\mathbf{H} = \Re\{\overline{\mathbf{H}}_c e^{2\pi i \nu t}\} = \Re\{\mathbf{H}_0 e^{-2\pi i (\mathbf{w} \cdot \mathbf{r} - \nu t)}\}, \tag{2.17}$$

where $\mathbf{r}$ is a vector pointing to an arbitrary point in space, $\mathbf{w}$ is known as the *wave vector*[2] and $\mathbf{E}_0$ and $\mathbf{H}_0$ are constant vectors. In general $\mathbf{w}$ is a complex vector,

$$\mathbf{w} = \mathbf{w}' - i\mathbf{w}'', \tag{2.18}$$

where $\mathbf{w}'$ turns out to be a vector whose magnitude is the *wavenumber*, and $\mathbf{w}''$ is known as the *attenuation vector*. Employing equation (2.18), equations (2.16) and (2.17) may be rewritten as

$$\mathbf{E}_c = \mathbf{E}_0 e^{-2\pi \mathbf{w}'' \cdot \mathbf{r}} e^{-2\pi i (\mathbf{w}' \cdot \mathbf{r} - \nu t)}, \tag{2.19}$$

$$\mathbf{H}_c = \mathbf{H}_0 e^{-2\pi \mathbf{w}'' \cdot \mathbf{r}} e^{-2\pi i (\mathbf{w}' \cdot \mathbf{r} - \nu t)}. \tag{2.20}$$

Thus, the complex electric and magnetic fields have local amplitude vectors $\mathbf{E}_0 e^{-2\pi \mathbf{w}'' \cdot \mathbf{r}}$ and $\mathbf{H}_0 e^{-2\pi \mathbf{w}'' \cdot \mathbf{r}}$ and an oscillatory part $e^{-2\pi i (\mathbf{w}' \cdot \mathbf{r} - \nu t)}$ with *phase angle* $\phi = 2\pi(\mathbf{w}' \cdot \mathbf{r} - \nu t)$. The position vector $\mathbf{r}$ may be considered to have two components: One parallel to $\mathbf{w}'$, and the other perpendicular to it. The vector product $\mathbf{w}' \cdot \mathbf{r}$ is constant for all vectors $\mathbf{r}$ that have the same component parallel to $\mathbf{w}'$, i.e., on planes normal to the vector $\mathbf{w}'$; these planes are known as *planes of equal phase*. To see how the wave travels let us look at the phase angle at two different times and locations (Fig. 2-1). First, consider the point $\mathbf{r} = 0$ at time $t = 0$ with a zero phase angle. Second, consider another point a distance z away into the direction of $\mathbf{w}'$; we see that the phase angle is zero at that point when $t = |\mathbf{w}'|z/\nu$. Thus, the phase velocity with which the wave travels from one point to the other is $c = z/t = \nu/w'$. We conclude

---

[2]The present definition of the *wave vector* differs by a factor of 2π and in name from the definition $k = 2\pi \mathbf{w}$ in most optics texts in order to conform with our definition of *wavenumber*.

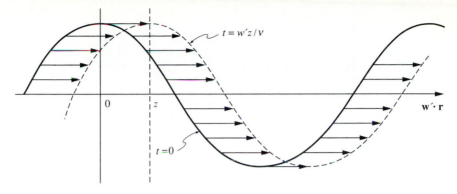

FIGURE 2-1
Phase propagation of an electromagnetic wave.

that the wave propagates into the direction of $\mathbf{w}'$, and that the vector's magnitude, w', is equal to the wavenumber η. Examining the amplitude vectors we see that $\mathbf{w}'' \cdot \mathbf{r} = \text{const}$ are *planes of equal amplitude*, and that the amplitude of the fields diminishes into the direction of $\mathbf{w}''$. If planes of equal phase and equal amplitude coincide (i.e., if $\mathbf{w}'$ and $\mathbf{w}''$ are parallel) we say the wave is *homogeneous*, otherwise the wave is said to be *inhomogeneous*. Since $\mathbf{E}_0$ and $\mathbf{w}$ are independent of position, we can substitute equation (2.19) into equation (2.11) and, assuming γ to be also invariant with space, find that

$$\nabla \cdot (\gamma \mathbf{E}_c) = \gamma \nabla \cdot \left(\mathbf{E}_0 e^{-2\pi i (\mathbf{w} \cdot \mathbf{r} - vt)} \right) = \gamma \mathbf{E}_0 \cdot \nabla \left(e^{-2\pi i (\mathbf{w} \cdot \mathbf{r} - vt)} \right)$$

$$= \gamma \mathbf{E}_0 e^{-2\pi i (\mathbf{w} \cdot \mathbf{r} - vt)} \cdot \nabla (-2\pi i \mathbf{w} \cdot \mathbf{r})$$

$$= -2\pi i \gamma \mathbf{w} \cdot \mathbf{E}_0 e^{-2\pi i (\mathbf{w} \cdot \mathbf{r} - vt)} = 0. \tag{2.21}$$

Similarly, substituting equation (2.19) into equation (2.13) results in

$$\nabla \times \mathbf{E}_c = \nabla \times \left(\mathbf{E}_0 e^{-2\pi i (\mathbf{w} \cdot \mathbf{r} - vt)} \right) = \nabla \left(e^{-2\pi i (\mathbf{w} \cdot \mathbf{r} - vt)} \right) \times \mathbf{E}_0$$

$$= -2\pi i \mathbf{w} e^{-2\pi i (\mathbf{w} \cdot \mathbf{r} - vt)} \times \mathbf{E}_0 = -2\pi i v \mu \mathbf{H}_0 e^{-2\pi i (\mathbf{w} \cdot \mathbf{r} - vt)}. \tag{2.22}$$

Thus, the partial differential equations (2.11) through (2.14) may be replaced by a set of algebraic equations,

$$\mathbf{w} \cdot \mathbf{E}_0 = 0, \tag{2.23}$$

$$\mathbf{w} \cdot \mathbf{H}_0 = 0, \tag{2.24}$$

$$\mathbf{w} \times \mathbf{E}_0 = v \mu \mathbf{H}_0, \tag{2.25}$$

$$\mathbf{w} \times \mathbf{H}_0 = -v \gamma \mathbf{E}_0. \tag{2.26}$$

It is clear from equations (2.23) and (2.24) that both $\mathbf{E}_0$ and $\mathbf{H}_0$ are perpendicular to $\mathbf{w}$, and it follows then from equations (2.25) and (2.26) that they are also perpendicular

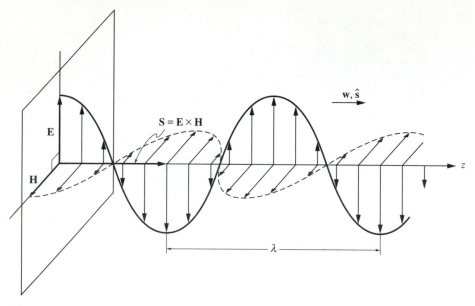

FIGURE 2-2
Electric and magnetic fields of a homogeneous wave.

to each other.[3] If the wave is homogeneous, then **w** points into the direction of wave propagation, and the electric and magnetic fields lie in planes perpendicular to this direction, as indicated in Fig. 2-2.

It remains to relate the complex wave vector **w** to the properties of the medium. Taking the vector product of equation (2.25) with **w** and recalling the vector identity derived, for example, in Wylie [3],

$$\mathbf{A} \times (\mathbf{B} \times \mathbf{C}) = \mathbf{B}(\mathbf{A} \cdot \mathbf{C}) - \mathbf{C}(\mathbf{A} \cdot \mathbf{B}), \tag{2.27}$$

lead to

$$\mathbf{w} \times (\mathbf{w} \times \mathbf{E}_0) = \mathbf{w}(\mathbf{w} \cdot \mathbf{E}_0) - \mathbf{E}_0 \mathbf{w} \cdot \mathbf{w} = \nu\mu\mathbf{w} \times \mathbf{H}_0 = -\nu^2\mu\gamma\mathbf{E}_0,$$

or

$$\mathbf{w} \cdot \mathbf{w} = \nu^2\mu\gamma. \tag{2.28}$$

If the wave travels through vacuum there can be no attenuation ($\mathbf{w}'' = 0$) and $\mu = \mu_0$, $\gamma = \epsilon_0$. We thus obtain the *speed of light in vacuum* as

$$c_0 = \nu/w' = \nu/\sqrt{\mathbf{w} \cdot \mathbf{w}} = \frac{1}{\sqrt{\epsilon_0\mu_0}}. \tag{2.29}$$

---

[3]Remember that all three vectors are complex and, therefore, the interpretation of "perpendicular" is not straightforward.

It is customary to introduce the *complex index of refraction*

$$m = n - ik \tag{2.30}$$

into equation (2.28) such that

$$\mathbf{w} \cdot \mathbf{w} = v^2 \mu \gamma = v^2 \epsilon_0 \mu_0 \left(\frac{\epsilon \mu}{\epsilon_0 \mu_0} - i \frac{\sigma_e \mu}{2\pi v \epsilon_0 \mu_0} \right) = \eta_0^2 m^2, \tag{2.31}$$

where $\eta_0 = v/c_0$ is the wavenumber of a wave with frequency v and phase velocity c_0, i.e., of a wave traveling through vacuum. This definition of m demands that

$$n^2 - k^2 = \frac{\epsilon \mu}{\epsilon_0 \mu_0} = \epsilon \mu c_0^2, \tag{2.32}$$

$$nk = \frac{\sigma_e \mu}{4\pi v \epsilon_0 \mu_0} = \frac{\sigma_e \mu \lambda_0 c_0}{4\pi}, \tag{2.33}$$

where $\lambda_0 = 1/\eta_0 = c_0/v$ is the wavelength for the wave in vacuum. Equations (2.32) and (2.33) may be solved for the *refractive index n* and the *absorptive index*[4] k as

$$n^2 = \frac{1}{2} \left[\frac{\epsilon}{\epsilon_0} + \sqrt{ \left(\frac{\epsilon}{\epsilon_0} \right)^2 + \left(\frac{\lambda_0 \sigma_e}{2\pi c_0 \epsilon_0} \right)^2 } \right], \tag{2.34}$$

$$k^2 = \frac{1}{2} \left[-\frac{\epsilon}{\epsilon_0} + \sqrt{ \left(\frac{\epsilon}{\epsilon_0} \right)^2 + \left(\frac{\lambda_0 \sigma_e}{2\pi c_0 \epsilon_0} \right)^2 } \right], \tag{2.35}$$

where we have assumed the material to be nonmagnetic, or $\mu = \mu_0$. These relations do not reveal the frequency (wavelength) dependence of the complex index of refraction, since the phenomenological coefficients ϵ and σ_e may depend on frequency. If the wave is homogeneous the wave vector may be written as $\mathbf{w} = (w' - iw'')\hat{\mathbf{s}}$, where $\hat{\mathbf{s}}$ is a unit vector in the direction of wave propagation, and it follows from equation (2.31) that $w' - iw'' = \eta_0(n - ik)$, so that the electric and magnetic fields reduce to

$$\mathbf{E}_c = \mathbf{E}_0 e^{-2\pi \eta_0 k z} e^{-2\pi i \eta_0 n(z - c_0 t/n)}, \tag{2.36}$$

$$\mathbf{H}_c = \mathbf{H}_0 e^{-2\pi \eta_0 k z} e^{-2\pi i \eta_0 n(z - c_0 t/n)}, \tag{2.37}$$

where $z = \hat{\mathbf{s}} \cdot \mathbf{r}$ is distance along the direction of propagation. For a nonvacuum, the

---

[4]The *absorptive index* is often referred to as *extinction coefficient* in the literature. Since the term *extinction coefficient* is also employed for another, related property we will always use the term *absorptive index* in this book to describe the imaginary part of the index of refraction.

phase velocity c of an electromagnetic wave is[5]

$$c = \frac{c_0}{n}. \qquad (2.38)$$

Further, the field strengths decay exponentially for nonzero values of k; thus, the absorptive index gives an indication of how quickly a wave is absorbed within the medium. Inspection of equation (2.35) shows that a large absorptive index k corresponds to a large electrical conductivity σ_e: Electromagnetic waves tend to be attenuated rapidly in good electrical conductors, such as metals, but are often transmitted with weak attenuation in media with poor electrical conductivity, or *dielectrics*, such as glass.

The magnitude and direction of the transfer of electromagnetic energy is given by the Poynting vector, i.e., a vector of magnitude EH pointing into the direction of propagation (cf. Fig. 2-2),[6]

$$\mathbf{S} = \mathbf{E} \times \mathbf{H} = \Re\{\mathbf{E}_c\} \times \Re\{\mathbf{H}_c\}. \qquad (2.39)$$

The instantaneous value for the Poynting[*] vector is a rapidly varying function of time. Of greater value to the engineer is a time-averaged value of the Poynting vector, say

$$\overline{\mathbf{S}} = \frac{1}{\delta t} \int_{t}^{t+\delta t} \mathbf{S}(t)\, dt, \qquad (2.40)$$

where δt is a very small amount of time, but significantly larger than the duration of a period, $1/\nu$; since $\mathbf{S}$ repeats itself after each period (if no attenuation occurs) a δt equal to any multiple of $1/\nu$ will give the same result for $\overline{\mathbf{S}}$, namely

$$\overline{\mathbf{S}} = \tfrac{1}{2}\Re\{\mathbf{E}_c \times \mathbf{H}_c^*\}, \qquad (2.41)$$

where $\mathbf{H}^*$ denotes the complex conjugate of $\mathbf{H}$. Thus using equation (2.25) and the

---

[5]Since there are materials that have $n < 1$ it is possible to have *phase velocities* (i.e., the velocity with which the amplitude of continuous waves penetrate through a medium), larger than c_0; these should be distinguished from the *signal velocities* (i.e., the velocity with which the energy contained in the waves travels), which can never exceed the speed of light in vacuum. The difference between the two may be grasped more easily by visualizing the movement of ocean waves: The wave crests move at a certain speed across the ocean surface (phase velocity), while the actual velocity of the water (signal velocity) is relatively slow.

[6]Note that, since the vector cross product is a nonlinear operation, the Poynting vector may **not** be calculated from $\mathbf{S} = \Re\{\mathbf{E}_c \times \mathbf{H}_c\}$.

[*]**John Henry Poynting (1852-1914)**
British physicist. He served as professor of physics at the University of Birmingham from 1880 until his death. His discovery that electromagnetic energy is proportional to the product of electric and magnetic field strength is known as Poynting's theorem.

vector identity (2.27), the Poynting vector may be expressed as

$$\overline{\mathbf{S}} = \frac{1}{2\nu\mu}\Re\{\mathbf{E}_c \times (\mathbf{w}^* \times \mathbf{E}_c^*)\} = \frac{1}{2\nu\mu}\Re\{\mathbf{w}^*(\mathbf{E}_c \cdot \mathbf{E}_c^*)\}$$

$$= \frac{n}{2c_0\mu}|\mathbf{E}_0|^2 e^{-4\pi\eta_0 k z}\hat{\mathbf{s}}. \tag{2.42}$$

The vector **S** points into the direction of propagation, and—as the wave traverses the medium—its energy content is attenuated exponentially, where the attenuation factor $4\pi\eta_0 k = \kappa$ is known as the *absorption coefficient* of the medium.

Example 2.1.   A plane homogeneous wave propagates through a perfect dielectric medium ($n = 2$) in the direction of $\hat{\mathbf{s}} = 0.8\hat{\mathbf{i}} + 0.6\hat{\mathbf{k}}$ with a wavenumber of $\eta_0 = 2500\,\text{cm}^{-1}$ and an electric field amplitude vector of $\mathbf{E}_0 = E_0[(6 + 3i)\hat{\mathbf{i}} + (2 - 5i)\hat{\mathbf{j}} - (8 + 4i)\hat{\mathbf{k}}]/\sqrt{154}$, where $E_0 = 600\,\text{N/C}$, and the $\hat{\mathbf{i}}$, $\hat{\mathbf{j}}$ and $\hat{\mathbf{k}}$ are unit vectors in the x-, y- and z-directions. Determine the magnetic field amplitude vector and the energy contained in the wave, assuming that the medium is nonmagnetic.

Solution

Since $\mathbf{w} = \mathbf{w}'$ is colinear with $\hat{\mathbf{s}}$, we find from equation (2.31) that $\mathbf{w} = w\hat{\mathbf{s}} = \eta_0 n\hat{\mathbf{s}}$ and, from equation (2.25),

$$\mathbf{H}_0 = \frac{1}{\nu\mu}\mathbf{w} \times \mathbf{E}_0 = \frac{1}{\nu\mu_0}\mathbf{w} \times \mathbf{E}_0 = \frac{n}{c_0\mu_0}\hat{\mathbf{s}} \times \mathbf{E}_0$$

$$= \frac{nE_0}{c_0\mu_0\sqrt{154}}\begin{vmatrix} \hat{\mathbf{i}} & \hat{\mathbf{j}} & \hat{\mathbf{k}} \\ 0.8 & 0.0 & 0.6 \\ 6 + 3i & 2 - 5i & -8 - 4i \end{vmatrix}$$

$$= \frac{nE_0}{c_0\mu_0 5\sqrt{154}}[(-6 + 15i)\hat{\mathbf{i}} + (50 + 25i)\hat{\mathbf{j}} + (8 - 20i)\hat{\mathbf{k}}]$$

$$= \frac{H_0}{\sqrt{3850}}[(-6 + 15i)\hat{\mathbf{i}} + (50 + 25i)\hat{\mathbf{j}} + (8 - 20i)\hat{\mathbf{k}}],$$

where

$$H_0 = \frac{nE_0}{c_0\mu_0} = \frac{2 \times 600\,\text{N/C}}{2.998 \times 10^8\,\text{m/s} \times 4\pi \times 10^{-7}\,\text{Ns}^2/\text{C}^2} = 3.185\,\text{C/ms},$$

and it is assumed that, for a nonmagnetic medium, the magnetic permeability is equal to the one in vacuum, $\mu = \mu_0$ (from Table A.1). The energy content of the wave is given by the Poynting vector, either equation (2.41) or equation (2.42). Choosing the latter, we get

$$\overline{\mathbf{S}} = \frac{n}{2c_0\mu_0}E_0^2\hat{\mathbf{s}} = \overline{S}\hat{\mathbf{s}},$$

$$\overline{S} = \frac{2 \times 600^2\,\text{N}^2/\text{C}^2}{2 \times 2.2998 \times 10^{-8}\,\text{m/s} \times 4\pi \times 10^{-7}\,\text{Ns}^2/\text{C}^2} = 955.6\,\text{W/m}^2.$$

2.4 POLARIZATION

Knowledge of the frequency, direction of propagation, and the energy content [i.e., the magnitude of the Poynting vector, equation (2.42)] does not completely describe

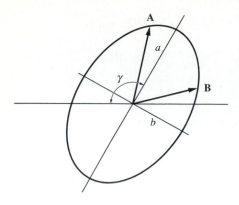

FIGURE 2-3
Vibration ellipse for a monochromatic wave.

a monochromatic (or time-harmonic) electromagnetic wave. Every train of electro-magnetic waves has a property known as the *state of polarization*. Polarization effects are generally not very important to the heat transfer engineer since emitted light generally is randomly polarized. In some applications partially or fully polarized light is employed, for example, from laser sources; and the engineer needs to know *(i)* how the reflective behavior of a surface depends on the polarization of incoming light, and *(ii)* how reflection from a surface tends to alter the state of polarization. We shall give here only a very brief introduction to polarization, based heavily on the excellent short description in Bohren and Huffman [2]. More detailed accounts on the subject may be found in the books by van de Hulst [4], Chandrasekhar [5], and others.

Consider a plane monochromatic wave with wavenumber η propagating through a nonabsorbing medium ($k \equiv 0$) in the z-direction. When describing polarization, it is customary to relate parameters to the electric field (keeping in mind that the magnetic field is simply perpendicular to it), which follows from equation (2.36) as

$$\mathbf{E} = \Re\{\mathbf{E}_c\} = \Re\{(\mathbf{A} - i\mathbf{B})e^{-2\pi i \eta n(z - ct)}\}$$
$$= \mathbf{A}\cos 2\pi\eta n(z - ct) - \mathbf{B}\sin 2\pi\eta n(z - ct), \tag{2.43}$$

where the vector $\mathbf{E}_0$ and its real components $\mathbf{A}$ and $\mathbf{B}$ are independent of position and lie, at any position z, in the plane normal to the direction of propagation. At any given location, say $z = 0$, the tip of the electric field vector traces out the curve

$$\mathbf{E}(z = 0, t) = \mathbf{A}\cos 2\pi\nu t + \mathbf{B}\sin 2\pi\nu t. \tag{2.44}$$

This curve, shown in Fig. 2-3, describes an ellipse that is known as the *vibration ellipse*. The ellipse collapses into a straight line if either $\mathbf{A}$ or $\mathbf{B}$ vanishes, in which case the wave is said to be *linearly polarized* (sometimes also called *plane polarized*). If $\mathbf{A}$ and $\mathbf{B}$ are perpendicular to one another and are of equal magnitude, the vibration ellipse becomes a circle and the wave is known as *circularly polarized*. In general, the wave in equation (2.43) is *elliptically polarized*.

At any given time, say $t = 0$, the curve described by the tip of the electric field vector is a helix (Fig. 2-4), or

$$\mathbf{E}(z, t = 0) = \mathbf{A}\cos 2\pi n\eta z - \mathbf{B}\sin 2\pi n\eta z. \tag{2.45}$$

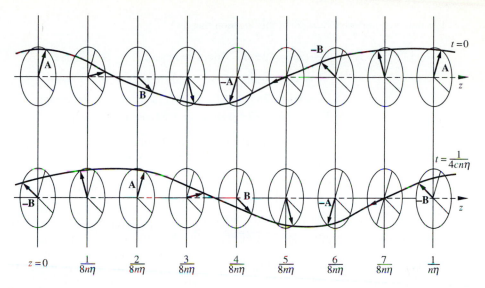

FIGURE 2-4
Space variation of electric field at fixed times.

Equation (2.45) describes the electric field at any one particular time. As time increases the helix moves into the direction of propagation, and its intersection with any plane $z = $ const describes the local vibration ellipse.

The state of polarization, which is characterized by its vibration ellipse, is defined by its *ellipticity*, b/a (the ratio of the length of its semiminor axis to that of its semimajor axis, as shown in Fig. 2-3), its *azimuth* γ (the angle between an arbitrary reference direction and its semimajor axis), and its *handedness* (i.e., the direction with which the tip of the electric field vector traverses through the vibration ellipse, clockwise or counterclockwise). These three parameters together with the magnitude of the Poynting vector are the *ellipsometric parameters* of a plane wave.

Example 2.2. Calculate the ellipsometric parameters a, b and γ for the wave considered in Example 2.1.

Solution

From equation (2.43) we find

$$\mathbf{A} = E_0(6\hat{\mathbf{i}} + 2\hat{\mathbf{j}} - 8\hat{\mathbf{k}})/\sqrt{154}, \quad \mathbf{B} = -E_0(3\hat{\mathbf{i}} - 5\hat{\mathbf{j}} - 4\hat{\mathbf{k}})/\sqrt{154},$$

and at any given location, say $z = 0$, the electric field vector may be written as

$$\mathbf{E} = E_0\big[(6\cos 2\pi\nu t - 3\sin 2\pi\nu t)\hat{\mathbf{i}} + (2\cos 2\pi\nu t + 5\sin 2\pi\nu t)\hat{\mathbf{j}}$$
$$- (8\cos 2\pi\nu t - 4\sin 2\pi\nu t)\hat{\mathbf{k}}\big]/\sqrt{154}.$$

The time-varying magnitude $|\mathbf{E}|$ at this location then is

$$
\begin{aligned}
|\mathbf{E}|^2 = \mathbf{E} \cdot \mathbf{E} = \ & \frac{E_0^2}{154}(36\cos^2 2\pi\nu t - 36\cos 2\pi\nu t \sin 2\pi\nu t + 9\sin^2 2\pi\nu t \\
& + 4\cos 2\pi\nu t + 20\cos^2 2\pi\nu t \sin 2\pi\nu t + 25\sin^2 2\pi\nu t \\
& + 64\cos 2\pi\nu t - 64\cos^2 2\pi\nu t \sin 2\pi\nu t + 16\sin^2 2\pi\nu t) \\
= \ & E_0^2(50 - 80\cos 2\pi\nu t \sin 2\pi\nu t + 54\cos^2 2\pi\nu t)/154.
\end{aligned}
$$

The maximum (a) and minimum (b) of $|\mathbf{E}|$ may be found by differentiating the last expression with respect to t and setting the result equal to zero. This operation leads to

$$
-80(\cos^2 2\pi\nu t - \sin^2 2\pi\nu t) = 108\sin 2\pi\nu t \cos 2\pi\nu t
$$
$$
-80\cos 4\pi\nu t = 54\sin 4\pi\nu t
$$

or

$$
2\pi\nu t = 0.5\tan^{-1}\left(-\frac{80}{54}\right).
$$

This function is double valued, leading to $(2\pi\nu t)_1 = -27.99°$ and $(2\pi\nu t)_2 = 62.01°$. Substituting these values into the expression for $\mathbf{E}$ gives

$$
\mathbf{E}_1 = E_0(0.5404\hat{\mathbf{i}} - 0.0468\hat{\mathbf{j}} - 0.7205\hat{\mathbf{k}}), \quad |\mathbf{E}| = a = 0.9009E_0
$$

and

$$
\mathbf{E}_2 = E_0(0.0134\hat{\mathbf{i}} + 0.4314\hat{\mathbf{j}} - 0.0179\hat{\mathbf{k}}), \quad |\mathbf{E}| = b = 0.4339E_0.
$$

The evaluation of the azimuth depends on the choice of a reference axis in the plane of the vibration ellipse. In the present problem the y-axis lies in this plane and is, therefore, the natural choice. Thus,

$$
\cos\gamma = \frac{\mathbf{E}\cdot\hat{\mathbf{j}}}{|\mathbf{E}|} = -\frac{0.0468}{0.9009} = -0.0519, \quad \gamma = 92.97°.
$$

While the ellipsometric parameters completely describe any monochromatic wave, they are difficult to measure directly (with the exception of the Poynting vector). In addition, when two or more waves of the same frequency but different polarization are superposed, only their strengths are additive: The other three ellipsometric parameters must be calculated anew. For these reasons a different but equivalent description of polarized light, known as *Stokes' parameters*, is usually preferred. The Stokes' parameters are defined by separating the wave train into two perpendicular components:

$$
\mathbf{E}_c = \mathbf{E}_0 e^{-2\pi i\eta n(z - ct)}; \quad \mathbf{E}_0 = E_\| \hat{\mathbf{e}}_\| + E_\perp \hat{\mathbf{e}}_\perp, \tag{2.46}
$$

where $\hat{\mathbf{e}}_\|$ and $\hat{\mathbf{e}}_\perp$ are *real* orthogonal unit vectors in the plane normal to wave propagation, such that $\hat{\mathbf{e}}_\|$ lies in an arbitrary reference plane that includes the wave propagation vector, and $\hat{\mathbf{e}}_\perp$ is perpendicular to it. The *parallel* ($E_\|$) and *perpendicular* ($E_\perp$) *polarization* components are generally complex and may be written as

$$
E_\| = a_\| e^{-i\delta_\|}, \quad E_\perp = a_\perp e^{-i\delta_\perp}, \tag{2.47}
$$

where a is the magnitude of the electric field and δ is the *phase angle of polarization*. Substitution into equation (2.43) leads to

$$
\begin{aligned}
\mathbf{E} &= \Re\{a_\parallel e^{-i\delta_\parallel - 2\pi i \eta n(z-ct)}\hat{\mathbf{e}}_\parallel + a_\perp e^{-i\delta_\perp - 2\pi i \eta n(z-ct)}\hat{\mathbf{e}}_\perp\} \\
&= a_\parallel \cos[\delta_\parallel + 2\pi\eta n(z-ct)]\hat{\mathbf{e}}_\parallel + a_\perp \cos[\delta_\perp + 2\pi\eta n(z-ct)]\hat{\mathbf{e}}_\perp. \qquad (2.48)
\end{aligned}
$$

Thus, the arbitrary wave given by equation (2.43) has been decomposed into two linearly polarized waves that are perpendicular to one another. The four *Stokes' parameters I, Q, U,* and *V* are defined by

$$
\begin{aligned}
I &= E_\parallel E_\parallel^* + E_\perp E_\perp^* = a_\parallel^2 + a_\perp^2, & (2.49) \\
Q &= E_\parallel E_\parallel^* - E_\perp E_\perp^* = a_\parallel^2 - a_\perp^2, & (2.50) \\
U &= E_\parallel E_\perp^* + E_\perp E_\parallel^* = 2a_\parallel a_\perp \cos(\delta_\parallel - \delta_\perp), & (2.51) \\
V &= i(E_\parallel E_\perp^* - E_\perp E_\parallel^*) = 2a_\parallel a_\perp \sin(\delta_\parallel - \delta_\perp), & (2.52)
\end{aligned}
$$

where the asterisks again denote complex conjugates. It can be shown that these four parameters may be determined through power measurements either directly (I), using a linear polarizer (arranged in the parallel and perpendicular directions for Q, rotated 45° for U), or a circular polarizer (V) (see, for example, Bohren and Huffman [2]). It is clear that only three of the Stokes' parameters are independent, since

$$
I^2 = Q^2 + U^2 + V^2. \qquad (2.53)
$$

Since the Stokes' parameters of a wave train are expressed in terms of the energy contents of its component waves [which can be seen by comparison with equation (2.42)], it follows that the Stokes' parameters for a collection of waves are additive.

The Stokes' parameters may also be related to the ellipsometric parameters by

$$
\begin{aligned}
I &= a^2 + b^2, & (2.54) \\
Q &= (a^2 - b^2)\cos 2\gamma, & (2.55) \\
U &= (a^2 - b^2)\sin 2\gamma, & (2.56) \\
V &= \pm 2ab, & (2.57)
\end{aligned}
$$

where the azimuth γ is measured from $\hat{\mathbf{e}}_\parallel$, and the sign of V specifies the handedness of the vibration ellipse. The sets of Stokes' parameters for a few special cases of polarization are shown—normalized, and written as column vectors—in Table 2.1 (from [2]). It is seen that the parameters Q and U show the degree of *linear polarization* (plus its orientation), while V is related to the degree of *circular polarization*.

The above definition of the Stokes' parameters is correct for strictly monochromatic waves as given by equation (2.46). Most natural light sources, such as the sun, light bulbs, fires, and so on, produce light whose amplitude, $\mathbf{E}_0$, is a slowly varying function of time (i.e., in comparison with a full wave period, $1/\nu$), or

$$
\mathbf{E}_0(t) = E_\parallel(t)\hat{\mathbf{e}}_\parallel + E_\perp(t)\hat{\mathbf{e}}_\perp. \qquad (2.58)
$$

TABLE 2.1
Stokes' parameters for several cases of polarized light.

Linearly Polarized

0°	90°	+45°	−45°	γ
$\leftrightarrow$	$\updownarrow$	$\searrow$	$\swarrow$	
$\begin{pmatrix} 1 \\ 1 \\ 0 \\ 0 \end{pmatrix}$	$\begin{pmatrix} 1 \\ -1 \\ 0 \\ 0 \end{pmatrix}$	$\begin{pmatrix} 1 \\ 0 \\ 1 \\ 0 \end{pmatrix}$	$\begin{pmatrix} 1 \\ 0 \\ -1 \\ 0 \end{pmatrix}$	$\begin{pmatrix} 1 \\ \cos 2\gamma \\ \sin 2\gamma \\ 0 \end{pmatrix}$

Circularly Polarized

Right	Left
$\begin{pmatrix} 1 \\ 0 \\ 0 \\ 1 \end{pmatrix}$	$\begin{pmatrix} 1 \\ 0 \\ 0 \\ -1 \end{pmatrix}$

Such waves are called *quasi-monochromatic*. If, through their slow respective variations with time, $E_\parallel$ and $E_\perp$ are *uncorrelated*, then the wave is said to be *unpolarized*. In such a case the vibration ellipse changes slowly with time, eventually tracing out ellipses of all shapes, orientations, and handedness. All waves discussed so far had a fixed relationship between $E_\parallel$ and $E_\perp$, and are known as *(completely) polarized*. If some correlation between $E_\parallel$ and $E_\perp$ exists (for example, a wave of constant handedness, ellipticity, or azimuth), then the wave is called *partially polarized*. For quasi-monochromatic waves the Stokes' parameters are defined in terms of time-averaged values, and equation (2.53) must be replaced by

$$I^2 \geq Q^2 + U^2 + V^2, \tag{2.59}$$

where the equality sign holds only for polarized light. For unpolarized light one gets $Q = U = V = 0$, while for partially polarized light the magnitudes of Q, U, and V give the following:

degree of polarization $= \sqrt{Q^2 + U^2 + V^2}/I$,
degree of linear polarization $= \sqrt{Q^2 + U^2}/I$,
degree of circular polarization $= V/I$.

Example 2.3. Reconsider the plane wave of the last two examples. Decompose the wave into two linearly-polarized waves, one in the x-z-plane, and the other perpendicular to it.

What are the Stokes' coefficients, the phase differences between the two polarizations, and the different degrees of polarization?

Solution

With $\hat{\mathbf{s}} = 0.8\hat{\mathbf{i}} + 0.6\hat{\mathbf{k}}$ and the knowledge that $\hat{\mathbf{e}}_\parallel$ must lie in the x-z-plane, i.e., $\hat{\mathbf{e}}_\parallel \cdot \hat{\mathbf{j}} = 0$, and that $\hat{\mathbf{e}}_\parallel$ must be normal to $\hat{\mathbf{s}}$, or $\hat{\mathbf{e}}_\parallel \cdot \hat{\mathbf{s}} = 0$, and finally that $\hat{\mathbf{e}}_\perp$ must be perpendicular to both of them, we get

$$\hat{\mathbf{e}}_\parallel = 0.6\hat{\mathbf{i}} - 0.8\hat{\mathbf{k}}, \quad \hat{\mathbf{e}}_\perp = \hat{\mathbf{j}},$$

where the choice of sign for both vectors is arbitrary (and we have chosen to let $\hat{\mathbf{e}}_\parallel$, $\hat{\mathbf{e}}_\perp$ and $\hat{\mathbf{s}}$ form a right-handed coordinate system). Thus, from equation (2.46) and

$$\mathbf{E}_0 = E_0[(6 + 3i)\hat{\mathbf{i}} + (2 - 5i)\hat{\mathbf{j}} - (8 + 4i)\hat{\mathbf{k}}]/\sqrt{154}$$

it follows immediately that

$$\mathbf{E}_\parallel = E_0(2 + i)(3\hat{\mathbf{i}} - 4\hat{\mathbf{k}})/\sqrt{154} = \left(5/\sqrt{154}\right)(2 + i)E_0\hat{\mathbf{e}}_\parallel$$
$$\mathbf{E}_\perp = E_0(2 - 5i)\hat{\mathbf{j}}/\sqrt{154} = \left[(2 - 5i)/\sqrt{154}\right]E_0\hat{\mathbf{e}}_\perp$$

or

$$\mathbf{E}_\parallel = \left(5/\sqrt{154}\right)(2 + i)E_0 = \sqrt{\frac{125}{154}}E_0e^{-i\delta_\parallel},$$

$$\mathbf{E}_\perp = \left[(2 - 5i)/\sqrt{154}\right]E_0 = \sqrt{\frac{29}{154}}E_0e^{-i\delta_\perp},$$

with

$$\delta_\parallel = -\tan^{-1}\left(\frac{1}{2}\right) = -26.565°,$$

$$\delta_\perp = -\tan^{-1}\left(-\frac{5}{2}\right) = 68.199°,$$

and a phase difference between the two polarizations of

$$\delta_\parallel - \delta_\perp = -94.76°,$$

(since $\tan^{-1}$ is a double-valued function, the correct value is determined by checking the signs of the real and imaginary parts of E). The Stokes' parameters can be calculated either directly from equations (2.49) through (2.52), or from equations (2.54) through (2.57) (using the ellipsometric parameters calculated in the last example). We use here the first approach so that we get

$$I = (125 + 29)E_0^2/154 = E_0^2,$$
$$Q = (125 - 29)E_0^2/154 = 48E_0^2/77,$$
$$U = 5(4 + 2i + 10i - 5 + 4 - 2i - 10i - 5)E_0^2/154 = -5E_0^2/77,$$
$$V = 5i(4 + 2i + 10i - 5 - 4 + 2i + 10i + 5)E_0^2/154 = -60E_0^2/77.$$

Finally, the degrees of polarization follow as $\sqrt{Q^2 + U^2 + V^2}/I = 100\%$ total polarization, $\sqrt{Q^2 + U^2}/I = 62.7\%$ linear polarization, and $|V|/I = 77.9\%$ of circular polarization.

2.5   REFLECTION AND TRANSMISSION

When an electromagnetic wave is incident on the interface between two homogeneous media, the wave will be partially reflected and partially transmitted into the second medium. We will limit our discussion here to plane interfaces, i.e., to cases where the local radius of curvature is much greater than the wavelength of the incoming light, λ, for which the problem may be reduced to algebraic equations. Some discussion on strongly curved surfaces in the form of small particles will be given in Chapter 10, which deals with radiative properties of particulate clouds.

In the following, after first establishing the general conditions for Maxwell's equations at the interface, we shall consider a wave traveling from one nonabsorbing medium into another nonabsorbing medium, followed by a short discussion of a wave incident from a nonabsorbing onto an absorbing medium.

Interface Conditions for Maxwell's Equations

To establish boundary conditions for $\mathbf{E}$ and $\mathbf{H}$ at an interface between two media, we shall apply the theorems of Gauss and Stokes to Maxwell's equations. Both theorems convert volume integrals to surface integrals and are discussed in detail in standard mathematical texts such as Wylie [3]. Given a vector function $\mathbf{F}$, defined within a volume V and on its boundary Γ, the theorems may be stated as

Gauss' theorem:

$$\int_V \boldsymbol{\nabla} \cdot \mathbf{F} dV = \int_\Gamma \mathbf{F} \cdot d\boldsymbol{\Gamma}, \tag{2.60}$$

Stokes' theorem:

$$\int_V \boldsymbol{\nabla} \times \mathbf{F} dV = -\int_\Gamma \mathbf{F} \times d\boldsymbol{\Gamma}, \tag{2.61}$$

where $d\boldsymbol{\Gamma} = \hat{\mathbf{n}} d\Gamma$ and $\hat{\mathbf{n}}$ is a unit surface normal pointing out of the volume.

Now consider a thin volume element $\delta V = A\delta s$ containing part of the interface as shown in Fig. 2-5. Applying Gauss' theorem to the first of Maxwell's equations, equation (2.11) yields

$$\int_{\delta V} \boldsymbol{\nabla} \cdot (\gamma \mathbf{E}_c) \, dV = \int_\Gamma \gamma \mathbf{E}_c \cdot d\boldsymbol{\Gamma} \approx \int_A [(\gamma \mathbf{E}_c)_1 \cdot (-\hat{\mathbf{n}}) + (\gamma \mathbf{E}_c)_2 \cdot \hat{\mathbf{n}}] \, dA = 0, \tag{2.62}$$

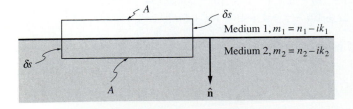

Medium 1, $m_1 = n_1 - ik_1$

Medium 2, $m_2 = n_2 - ik_2$

FIGURE 2-5
Geometry for derivation of interface conditions.

where Γ is the total surface area of δV, and contributions to the surface integral come mainly from the two sides parallel to the interface since δs is small. Also, shrinking A to an arbitrarily small area, we conclude that, everywhere along the interface,

$$m_1^2\, \mathbf{E}_{c1} \cdot \hat{\mathbf{n}} = m_2^2\, \mathbf{E}_{c2} \cdot \hat{\mathbf{n}}, \tag{2.63}$$

where equation (2.28) has been used to eliminate the complex permittivity γ. Similarly, from equation (2.12)

$$\mathbf{H}_{c1} \cdot \hat{\mathbf{n}} = \mathbf{H}_{c2} \cdot \hat{\mathbf{n}}. \tag{2.64}$$

Thus, the normal components of $m^2\mathbf{E}_c$ and $\mathbf{H}_c$ are conserved across a plane boundary. Stokes' theorem may be applied to equations (2.13) and (2.14), again for the volume element shown in Fig. 2-5. For example,

$$\int_{\delta V} \boldsymbol{\nabla} \times \mathbf{H}_c\, dV = -\int_{\Gamma} \mathbf{H}_c \times d\boldsymbol{\Gamma} \approx \int_A (\mathbf{H}_{c1} - \mathbf{H}_{c2}) \times \hat{\mathbf{n}}\, dA = \int_V 2\pi i \nu \gamma \mathbf{E}_c\, dV, \quad (2.65)$$

or, after shrinking $\delta s \to 0$ and A to a small value,

$$\mathbf{E}_{c1} \times \hat{\mathbf{n}} = \mathbf{E}_{c2} \times \hat{\mathbf{n}} \tag{2.66}$$

and

$$\mathbf{H}_{c1} \times \hat{\mathbf{n}} = \mathbf{H}_{c2} \times \hat{\mathbf{n}}. \tag{2.67}$$

Therefore, the tangential components of both $\mathbf{E}_c$ and $\mathbf{H}_c$ are conserved across a plane boundary.

Given the incident wave, it is possible to find the complete fields from Maxwell's equations and the above interface conditions. However, it is obvious that there will be a reflected wave in the medium of incidence, and a transmitted wave in the other medium. We may also assume that all waves remain plane waves. A consequence of having guessed the solution to this point is that conditions (2.66) and (2.67) are sufficient to specify the reflected and transmitted waves, and it turns out that conditions (2.63) and (2.64) are automatically satisfied (Stone [1]).

The Interface between Two Nonabsorbing Media

The reflection and transmission relationships become particularly simple if homogeneous plane waves reach the plane interface between two nonabsorbing media. For such a wave train the planes of equal phase and equal amplitude coincide and are normal to the direction of propagation, as shown in Fig. 2-6. This plane, also called the *wavefront*, moves at constant speed $c_1 = c_0/n_1$ through Medium 1, and at a constant but different speed $c_2 = c_0/n_2$ through Medium 2. If $n_2 > n_1$ then, as shown in Fig. 2-6, the wavefront will move more slowly through Medium 2, lagging behind the wavefront traveling through Medium 1. This is readily put in mathematical terms by looking at points A and B on the wavefront at a certain time t. At time $t + \Delta t$ the

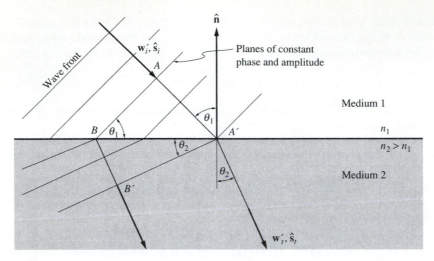

FIGURE 2-6
Transmission and reflection of a plane wave at the interface between two nonabsorbing media.

part of the wavefront initially at A will have reached point A' on the interface while the wavefront at point B, traveling a shorter distance through Medium 2, will have reached point B', where

$$\Delta t = \frac{\overline{AA'}}{c_1} = \frac{\overline{BB'}}{c_2}. \tag{2.68}$$

Using geometric relations for $\overline{AA'}$ and $\overline{BB'}$ and substituting for the phase velocities, we obtain

$$\Delta t = \frac{\overline{BA'}\sin\theta_i}{c_0/n_1} = \frac{\overline{BA'}\sin\theta_2}{c_0/n_2} = \frac{\overline{BA'}\sin\theta_r}{c_0/n_1}, \tag{2.69}$$

where the last term pertains to reflection, for which a similar relationship must exist (but which is not shown to avoid overcrowding of the figure). Thus we conclude that

$$\theta_r = \theta_i = \theta_1, \tag{2.70}$$

that is, according to electromagnetic wave theory, reflection of light is always purely specular. This is a direct consequence of a "plane" interface, i.e., a surface that is not only flat (with infinite radius of curvature) but also perfectly smooth. Equation (2.69) also gives a relationship between the directions of the incoming and transmitted waves as

$$\frac{\sin\theta_2}{\sin\theta_1} = \frac{n_1}{n_2}, \tag{2.71}$$

which is known as *Snell's law.*[*] The angles $\theta_1 = \theta_i$ and $\theta_2 = \theta_t$ are called the *angles of incidence and refraction.* The present derivation of Snell's law was based on geometric principles and is valid only for plane homogeneous waves, which limits its applicability to the interface between two nonabsorbing media, i.e., two perfect dielectrics. A more rigorous derivation of a generalized version of Snell's law is given when incidence on an absorbing medium is considered.

Besides the directions of reflection and transmission we should like to be able to determine the amounts of reflected and transmitted light. From equations (2.19) and (2.20) we can write expressions for the electric and magnetic fields in Medium 1 (consisting of incident and reflected waves) by setting $\mathbf{w}'' = 0$ for a nonabsorbing medium as

$$\mathbf{E}_{c1} = \mathbf{E}_{0i}\, e^{-2\pi i(\mathbf{w}_i' \cdot \mathbf{r} - \nu t)} + \mathbf{E}_{0r}\, e^{-2\pi i(\mathbf{w}_r' \cdot \mathbf{r} - \nu t)}, \tag{2.72}$$

$$\mathbf{H}_{c1} = \mathbf{H}_{0i}\, e^{-2\pi i(\mathbf{w}_i' \cdot \mathbf{r} - \nu t)} + \mathbf{H}_{0r}\, e^{-2\pi i(\mathbf{w}_r' \cdot \mathbf{r} - \nu t)}. \tag{2.73}$$

Similarly for Medium 2,

$$\mathbf{E}_{c2} = \mathbf{E}_{0t}\, e^{-2\pi i(\mathbf{w}_t' \cdot \mathbf{r} - \nu t)}, \tag{2.74}$$

$$\mathbf{H}_{c2} = \mathbf{H}_{0t}\, e^{-2\pi i(\mathbf{w}_t' \cdot \mathbf{r} - \nu t)}. \tag{2.75}$$

For convenience we place the coordinate origin at that point of the boundary where reflection and transmission are to be considered. Thus, at that point of the interface, with $\mathbf{r} = 0$, using boundary conditions (2.66) and (2.67),

$$(\mathbf{E}_{0i} + \mathbf{E}_{0r}) \times \hat{\mathbf{n}} = \mathbf{E}_{0t} \times \hat{\mathbf{n}}, \tag{2.76}$$

$$(\mathbf{H}_{0i} + \mathbf{H}_{0r}) \times \hat{\mathbf{n}} = \mathbf{H}_{0t} \times \hat{\mathbf{n}}. \tag{2.77}$$

To evaluate the tangential components of the electric and magnetic fields at the interface, it is advantageous to break up the fields (which, in general, may be unpolarized or elliptically polarized) into two linearly polarized waves, one parallel to the *plane of incidence* (formed by the incident wave vector $\mathbf{w}_i$ and the surface normal $\hat{\mathbf{n}}$), and the other perpendicular to it, or

$$\mathbf{E}_0 = E_\parallel \hat{\mathbf{e}}_\parallel + E_\perp \hat{\mathbf{e}}_\perp, \quad \mathbf{H}_0 = H_\parallel \hat{\mathbf{e}}_\parallel + H_\perp \hat{\mathbf{e}}_\perp. \tag{2.78}$$

This is shown schematically in Fig. 2-7. It is readily apparent from the figure that, in the plane of incidence, the unit vectors normal to the interface ($\hat{\mathbf{n}}$) and tangential to the interface ($\hat{\mathbf{t}}$) may be expressed as

$$\hat{\mathbf{n}} = \hat{\mathbf{s}}_i \cos\theta_1 - \hat{\mathbf{e}}_{i\parallel} \sin\theta_1 = -\hat{\mathbf{s}}_r \cos\theta_1 + \hat{\mathbf{e}}_{r\parallel} \sin\theta_1 = \hat{\mathbf{s}}_t \cos\theta_2 - \hat{\mathbf{e}}_{t\parallel} \sin\theta_2, \tag{2.79a}$$

$$\hat{\mathbf{t}} = \hat{\mathbf{s}}_i \sin\theta_1 + \hat{\mathbf{e}}_{i\parallel} \cos\theta_1 = \hat{\mathbf{s}}_r \sin\theta_1 + \hat{\mathbf{e}}_{r\parallel} \cos\theta_1 = \hat{\mathbf{s}}_t \sin\theta_2 + \hat{\mathbf{e}}_{t\parallel} \cos\theta_2. \tag{2.79b}$$

---

[*]**Willebord van Snel van Royen (1580-1626)**
Dutch astronomer and mathematician, who discovered Snell's law in 1621.

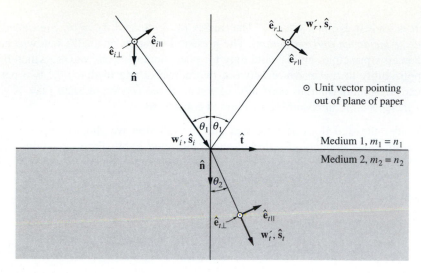

FIGURE 2-7
Orientation of wave vectors at an interface.

As defined in Fig. 2-7 the unit vectors $\hat{e}_\parallel$, $\hat{e}_\perp$ and $\hat{s}$ form right-handed coordinate systems for the incident and transmitted waves, i.e.,

$$\hat{e}_\parallel = \hat{e}_\perp \times \hat{s}, \quad \hat{e}_\perp = \hat{s} \times \hat{e}_\parallel, \quad \hat{s} = \hat{e}_\parallel \times \hat{e}_\perp, \tag{2.80}$$

and a left-handed coordinate system for the reflected wave (leading to opposite signs for the above cross-products of unit vectors).[7]

Therefore, from equation (2.79)

$$\hat{e}_\parallel \times \hat{n} = \pm\hat{e}_\parallel \times \hat{s}\cos\theta = -\hat{e}_\perp \cos\theta$$
$$\hat{e}_\perp \times \hat{n} = \pm\hat{e}_\perp \times \hat{s}\cos\theta \mp \hat{e}_\perp \times \hat{e}_\parallel \sin\theta = \hat{e}_\parallel \cos\theta + \hat{s}\sin\theta = \hat{t},$$

where the top sign applies to the incident and transmitted waves, while the lower sign applies to the reflected component. The second of these relations can also be obtained directly from Fig. 2-7. Using these relations, equations (2.76) and (2.77) may be rewritten in terms of polarized components as

$$(E_{i\parallel} + E_{r\parallel})\cos\theta_1 = E_{t\parallel}\cos\theta_2, \tag{2.81}$$
$$E_{i\perp} + E_{r\perp} = E_{t\perp}, \tag{2.82}$$
$$(H_{i\parallel} + H_{r\parallel})\cos\theta_1 = H_{t\parallel}\cos\theta_2, \tag{2.83}$$
$$H_{i\perp} + H_{r\perp} = H_{t\perp}. \tag{2.84}$$

---

[7]This is necessary for consistency, i.e., for normal incidence there should not be any difference between parallel- and perpendicular-polarized waves.

The magnetic field may be eliminated through the use of equation (2.25): With $\mathbf{w} = \eta_0 m \hat{\mathbf{s}} = (\nu/c_0) m \hat{\mathbf{s}}$ from equation (2.31) we have

$$
\begin{aligned}
\mathbf{H}_0 &= \frac{m}{c_0 \mu} \hat{\mathbf{s}} \times \mathbf{E}_0 = \pm \frac{m}{c_0 \mu \cos\theta} (\hat{\mathbf{n}} \pm \hat{\mathbf{e}}_\| \sin\theta) \times (E_\| \hat{\mathbf{e}}_\| + E_\perp \hat{\mathbf{e}}_\perp) \\
&= \pm \frac{m}{c_0 \mu \cos\theta} [E_\| \cos\theta \hat{\mathbf{e}}_\perp - E_\perp (\hat{\mathbf{t}} - \hat{\mathbf{s}} \sin\theta)] \\
&= \pm \frac{m}{c_0 \mu} (E_\| \hat{\mathbf{e}}_\perp - E_\perp \hat{\mathbf{e}}_\|).
\end{aligned}
\tag{2.85}
$$

Again, the upper sign applies to incident and transmitted waves, and the lower sign to reflected waves. The last two conditions may now be rewritten in terms of the electric field. Assuming the magnetic permeability to be the same in both media, and setting $m = n$ (nonabsorbing media), this leads to

$$
(E_{i\perp} - E_{r\perp}) n_1 \cos\theta_1 = E_{t\perp} n_2 \cos\theta_2,
\tag{2.86}
$$

$$
(E_{i\|} - E_{r\|}) n_1 = E_{t\|} n_2.
\tag{2.87}
$$

From this one may calculate the *reflection coefficient r* and the *transmission coefficient t* as

$$
r_\| = \frac{E_{r\|}}{E_{i\|}} = \frac{n_1 \cos\theta_2 - n_2 \cos\theta_1}{n_1 \cos\theta_2 + n_2 \cos\theta_1},
\tag{2.88}
$$

$$
r_\perp = \frac{E_{r\perp}}{E_{i\perp}} = \frac{n_1 \cos\theta_1 - n_2 \cos\theta_2}{n_1 \cos\theta_1 + n_2 \cos\theta_2},
\tag{2.89}
$$

$$
t_\| = \frac{E_{t\|}}{E_{i\|}} = \frac{2 n_1 \cos\theta_1}{n_1 \cos\theta_2 + n_2 \cos\theta_1},
\tag{2.90}
$$

$$
t_\perp = \frac{E_{t\perp}}{E_{i\perp}} = \frac{2 n_1 \cos\theta_1}{n_1 \cos\theta_1 + n_2 \cos\theta_2}.
\tag{2.91}
$$

For an interface between two nonabsorbing media these coefficients turn out to be real, even though the electric field amplitudes are complex. The *reflectivity* ρ is defined as the fraction of *energy* in a wave that is reflected and must, therefore, be calculated from the Poynting vector, equation (2.42), so that

$$
\rho_\| = \frac{\overline{S}_{r\|}}{\overline{S}_{i\|}} = \left(\frac{E_{r\|}}{E_{i\|}}\right)^2 = r_\|^2
\tag{2.92}
$$

gives the reflectivity of that part of the wave whose electric field vector lies in the plane of incidence (with its magnetic field normal to it), and

$$
\rho_\perp = \frac{\overline{S}_{r\perp}}{\overline{S}_{i\perp}} = \left(\frac{E_{r\perp}}{E_{i\perp}}\right)^2 = r_\perp^2
\tag{2.93}
$$

is the reflectivity for the part whose electric field vector is normal to the plane of incidence. In terms of these polarized components the overall reflectivity may be stated as "reflected energy for both polarizations, divided by the total incoming energy," or

$$\rho = \frac{E_{i\|}E_{i\|}^*\rho_\| + E_{i\perp}E_{i\perp}^*\rho_\perp}{E_{i\|}E_{i\|}^* + E_{i\perp}E_{i\perp}^*}. \tag{2.94}$$

For *unpolarized* and *circularly polarized* light $E_{i\|} = E_{i\perp}$, and the reflectivity for the entire wave train is

$$\rho = \frac{1}{2}(\rho_\| + \rho_\perp) = \frac{1}{2}\left[\left(\frac{n_1\cos\theta_2 - n_2\cos\theta_1}{n_1\cos\theta_2 + n_2\cos\theta_1}\right)^2 + \left(\frac{n_1\cos\theta_1 - n_2\cos\theta_2}{n_1\cos\theta_1 + n_2\cos\theta_2}\right)^2\right]. \tag{2.95}$$

From this relationship the refractive indices may be eliminated through Snell's law, giving

$$\rho = \frac{1}{2}\left[\frac{\tan^2(\theta_1 - \theta_2)}{\tan^2(\theta_1 + \theta_2)} + \frac{\sin^2(\theta_1 - \theta_2)}{\sin^2(\theta_1 + \theta_2)}\right], \tag{2.96}$$

which is known as *Fresnel's equation.*[*]

The overall *transmissivity* τ may similarly be evaluated from the Poynting vector, equation (2.42), but the different refractive indices and wave propagation directions in the transmitting and incident media must be considered, so that

$$\tau = \frac{n_2\cos\theta_2}{n_1\cos\theta_1}t^2 = 1 - \rho. \tag{2.97}$$

An example for the angular reflectivity at the interface between two dielectrics (with $n_2/n_1 = 1.5$) is given in Fig. 2-8. It is seen that, at an angle of incidence of $\theta_1 = \theta_p$, $r_\|$ passes through zero resulting in a zero reflectivity for the parallel component of the wave. This angle is known as the *polarizing angle* or *Brewster's angle,*[†] since light reflected from the surface—regardless of the incident polarization—will be completely polarized. Brewster's angle follows from equations (2.71) and (2.88) as

$$\tan\theta_p = \frac{n_2}{n_1}. \tag{2.98}$$

---

[*]**Augustin-Jean Fresnel (1788-1827)**
French physicist, and one of the early pioneers for the wave theory of light. Serving as an engineer for the French government he studied aberration of light and interference in polarized light. His optical theories earned him very little recognition during his lifetime.

[†]**Sir David Brewster (1781-1868)**
Scottish scientist, entered Edinburgh University at age 12 to study for the ministry. After completing his studies he turned his attention to science, particularly optics. In 1815, the year he discovered the law named after him, he was elected Fellow of the Royal Society.

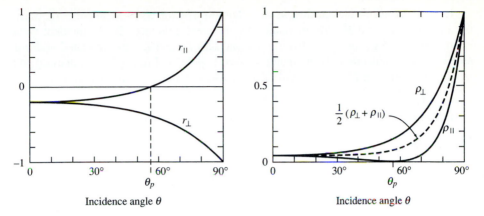

FIGURE 2-8
Reflection coefficients and reflectivities for the interface between two dielectrics ($n_2/n_1 = 1.5$).

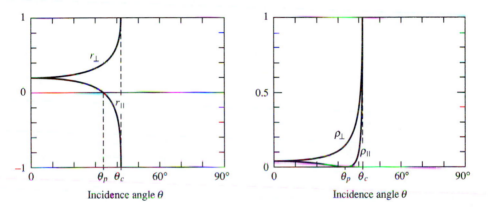

FIGURE 2-9
Reflection coefficients and reflectivities for the interface between two dielectrics ($n_1/n_2 = 1.5$).

Different behavior is observed if light travels from one dielectric into another, optically less dense medium ($n_1 > n_2$)[8], shown in Fig. 2-9. Examination of equation (2.71) shows that θ_2 reaches the value of 90° for an angle of incidence θ_c, called the *critical angle*,

$$\sin \theta_c = \frac{n_2}{n_1}. \tag{2.99}$$

It is left as an exercise for the reader to show that, for $\theta_1 > \theta_c$, light of any polarization is reflected, and nothing is transmitted into the second medium.

---

[8]The optical density of a medium is related to the number of atoms contained over a distance equal to the wavelength of the light and is proportional to the refractive index.

It is important to realize that upon reflection a wave changes its state of polarization, since $E_\parallel$ and $E_\perp$ are attenuated by different amounts. If the incident wave is unpolarized (e.g., emission from a hot surface) $E_\parallel$ and $E_\perp$ are unrelated and will remain so after reflection. If the incident wave is polarized (e.g., laser radiation), the relationship between $E_\parallel$ and $E_\perp$ will change, causing a change in polarization.

Example 2.4.   The plane homogeneous wave of the previous examples encounters the flat interface with another dielectric ($n_2 = 8/3$) that is described by the equation $z = 0$ (i.e., the x-y-plane at $z = 0$). Calculate the angles of incidence, reflection and refraction. What fraction of energy of the wave is reflected, and how much is transmitted? In addition, determine the state of polarization of the reflected wave.

Solution

Since the interface is described by $z = 0$, the surface normal (pointing into Medium 2) is simply $\hat{\mathbf{n}} = \hat{\mathbf{k}}$. From $\hat{\mathbf{s}} = 0.8\hat{\mathbf{i}} + 0.6\hat{\mathbf{k}}$ and $\hat{\mathbf{n}} \cdot \hat{\mathbf{s}} = \cos\theta_1 = 0.6$, it follows that the angle of incidence is $\theta_1 = 53.13°$ off normal, which is equal to the angle of reflection, while the angle of refraction follows from Snell's law, equation (2.71), as

$$\sin\theta_2 = \frac{n_1}{n_2}\sin\theta_1 = \frac{2}{8/3} \times 0.8 = .6, \quad \theta_2 = 36.87°.$$

It follows that $\cos\theta_2 = 0.8$ and the reflection coefficients are calculated from equations (2.88) and (2.89) as

$$r_\parallel = \frac{2 \times 0.8 - (8/3) \times 0.6}{2 \times 0.8 + (8/3) \times 0.6} = \frac{1.6 - 1.6}{3.2} = 0,$$

$$r_\perp = \frac{2 \times 0.6 - (8/3) \times 0.8}{2 \times 0.6 + (8/3) \times 0.8} = \frac{3.6 - 6.4}{10.0} = -0.28,$$

and the respective reflectivities follow as

$$\rho_\parallel = 0, \quad \text{and} \quad \rho_\perp = (-0.28)^2 = 0.0784.$$

For the present wave and interface, the wave impinges on the surface at Brewster's angle, i.e., the component of the wave that is linearly polarized in the plane of incidence is totally transmitted.

In general, to calculate the overall reflectivity, the wave must be decomposed into two linear polarized components, vibrating within the plane of incidence and perpendicular to it. Fortunately, this was already done in Example 2.3. From equation (2.94), together with the values of $E_{i\parallel} = [5(2 + i)/\sqrt{154}]E_0$ and $E_{i\perp} = [(2 - 5i)/\sqrt{154}]E_0$ from the previous example, we obtain

$$\rho = \frac{E_{i\parallel}E_{i\parallel}^*\rho_\parallel + E_{i\perp}E_{i\perp}^*\rho_\perp}{E_{i\parallel}E_{i\parallel}^* + E_{i\perp}E_{i\perp}^*} = \frac{125 \times 0 + 29 \times 0.0784}{154} = 0.0148,$$

and the overall transmissivity τ follows as

$$\tau = 1 - \rho = 0.9852.$$

To determine the polarization of the reflected beam, we first need to determine the reflected electric field amplitude vector. From the definition of the reflection coefficient we have

$$E_{r\|} = r_\| E_{i\|} = 0, \quad E_{r\perp} = r_\perp E_{i\perp} = -0.28 \times \frac{2 - 5i}{\sqrt{154}} E_0$$

and, from equations (2.49) through (2.52),

$$I = -Q = E_{r\perp} E_{r\perp}^* = \frac{0.28^2}{154} 29 E_0^2 = 0.01476 E_0^2,$$
$$U = V = 0.$$

Therefore, the wave remains 100% polarized, but the polarization is not completely linear. Indeed, any polarized radiation reflecting off a surface at Brewster's angle will become linearly polarized with only a perpendicular component.

The Interface between a Perfect Dielectric and an Absorbing Medium

The analysis of reflection and transmission at the interface between two perfect dielectrics is relatively straightforward, since an incident plane homogeneous wave remains plane and homogeneous after reflection and transmission. However, if a plane homogeneous wave is incident upon an absorbing medium, then the transmitted wave is, in general, inhomogeneous. If a beam travels from one absorbing medium into another absorbing medium, then the wave is usually inhomogeneous in both, making the analysis somewhat cumbersome. Fortunately, the interface between two absorbers is rarely important: A wave traveling through an absorbing medium is usually strongly attenuated, if not totally absorbed, before hitting a second absorber. In this section we shall consider a plane homogeneous light wave incident from a perfect dielectric on an absorbing medium.

The incident, reflected and transmitted waves are again described by equations (2.72) through (2.75), except that the wave vector for transmission, $\mathbf{w}_t$, may be complex. Thus using equations (2.66) and (2.67), the interface condition may be written as

$$\mathbf{E}_{0i} \times \hat{\mathbf{n}} \, e^{-2\pi i \mathbf{w}_i' \cdot \mathbf{r}} + \mathbf{E}_{0r} \times \hat{\mathbf{n}} \, e^{-2\pi i \mathbf{w}_r' \cdot \mathbf{r}} = \mathbf{E}_{0t} \times \hat{\mathbf{n}} \, e^{-2\pi i (\mathbf{w}_t' \cdot \mathbf{r} - i \mathbf{w}_t'' \cdot \mathbf{r})}, \quad (2.100)$$

$$\mathbf{H}_{0i} \times \hat{\mathbf{n}} \, e^{-2\pi i \mathbf{w}_i' \cdot \mathbf{r}} + \mathbf{H}_{0r} \times \hat{\mathbf{n}} \, e^{-2\pi i \mathbf{w}_r' \cdot \mathbf{r}} = \mathbf{H}_{0t} \times \hat{\mathbf{n}} \, e^{-2\pi i (\mathbf{w}_t' \cdot \mathbf{r} - i \mathbf{w}_t'' \cdot \mathbf{r})}, \quad (2.101)$$

where $\mathbf{r}$ was left arbitrary here in order to derive formally the generalized form of Snell's law although, for convenience, we still assume that the coordinate origin lies on the interface. We note that none of the amplitude vectors, $\mathbf{E}_{0i}$, $\mathbf{H}_{0i}$, etc., depends on location, and that $\mathbf{r}$ is a vector to an arbitrary point on the interface, which may be varied independently. Thus, in order for equations (2.100) and (2.101) to hold at any point on the interface, we must have

$$\mathbf{w}_i' \cdot \mathbf{r} = \mathbf{w}_r' \cdot \mathbf{r} = \mathbf{w}_t' \cdot \mathbf{r}, \quad (2.102)$$

$$0 = \mathbf{w}_t'' \cdot \mathbf{r}, \quad (2.103)$$

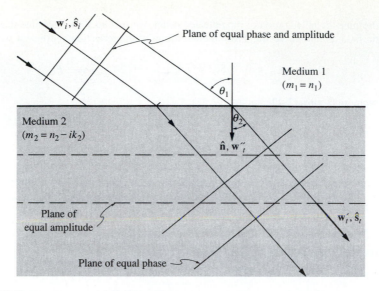

FIGURE 2-10
Transmission and reflection at the interface between a dielectric and an absorbing medium.

that is, since **r** is tangential to the interface, the tangential components of the wave vector **w**' must be continuous across the interface, while the tangential component of the attenuation vector $\mathbf{w}_t''$ must be zero, or $\mathbf{w}_t'' = w_t'' \hat{\mathbf{n}}$. Thus, within the absorbing medium, planes of equal amplitude are parallel to the interface, as indicated in Fig. 2-10. Since $\mathbf{w}_r'$ has the same tangential component as $\mathbf{w}_i'$ as well as the same magnitude [cf. equation (2.31)], it follows again that reflection must be specular, or $\theta_r = \theta_i$.

The continuity of the tangential component for the transmitted wave vector indicates that

$$w_i' \sin \theta_1 = \eta_0 n_1 \sin \theta_1 = w_t' \sin \theta_2. \tag{2.104}$$

The wave vector for transmission, w_t', may be eliminated from equation (2.104) by using equation (2.31):

$$\mathbf{w}_t \cdot \mathbf{w}_t = w_t'^2 - w_t''^2 - 2i\mathbf{w}_t' \cdot \mathbf{w}_t'' = \eta_0^2 m_2^2 = \eta_0^2 (n_2^2 - k_2^2 - 2in_2k_2), \tag{2.105}$$

or

$$w_t'^2 - w_t''^2 = \eta_0^2 (n_2^2 - k_2^2), \tag{2.106}$$

$$\mathbf{w}_t' \cdot \mathbf{w}_t'' = w_t'w_t'' \cos \theta_2 = \eta_0^2 n_2 k_2. \tag{2.107}$$

Thus, equations (2.104), (2.106) and (2.107) constitute three equations in the three unknowns θ_2, w_t' and w_t''. This system of equations may be solved to yield

$$p^2 = \left(\frac{w_t' \cos \theta_2}{\eta_0} \right)^2$$

$$= \frac{1}{2} \left[\sqrt{(n_2^2 - k_2^2 - n_1^2 \sin^2\theta_1)^2 + 4n_2^2 k_2^2} + (n_2^2 - k_2^2 - n_1^2 \sin^2\theta_1) \right], \tag{2.108}$$

$$q^2 = \left(\frac{w_t''}{\eta_0}\right)^2$$

$$= \frac{1}{2}\left[\sqrt{(n_2^2 - k_2^2 - n_1^2\sin^2\theta_1)^2 + 4n_2^2k_2^2} - (n_2^2 - k_2^2 - n_1^2\sin^2\theta_1)\right], \quad (2.109)$$

and the refraction angle θ_2 may be calculated from equation (2.104) as

$$p\tan\theta_2 = n_1\sin\theta_1. \quad (2.110)$$

Equation (2.110) together with equation (2.108) is known as the *generalized Snell's law*.

The *reflection coefficients* are calculated in the same fashion as was done for two dielectrics (left as an exercise). This leads to

$$\tilde{r}_\| = \frac{E_{r\|}}{E_{i\|}} = \frac{n_1^2(w_t'\cos\theta_2 - iw_t'') - m_2^2 w_i'\cos\theta_1}{n_1^2(w_t'\cos\theta_2 - iw_t'') + m_2^2 w_i'\cos\theta_1}, \quad (2.111)$$

$$\tilde{r}_\perp = \frac{E_{r\perp}}{E_{i\perp}} = \frac{w_i'\cos\theta_1 - (w_t'\cos\theta_2 - iw_t'')}{w_i'\cos\theta_1 + (w_t'\cos\theta_2 - iw_t'')}, \quad (2.112)$$

where the tilde has been added to indicate that the reflection coefficients are now *complex*. From equations (2.105) through (2.109) we find

$$m_2^2 = \frac{p^2}{\cos^2\theta_2} - q^2 - 2ipq = p^2(1 + \tan^2\theta_2) - q^2 - 2ipq$$

$$= p^2 - q^2 + n_1^2\sin^2\theta_1 - 2ipq. \quad (2.113)$$

Eliminating the wave vectors the reflection coefficients may be written as

$$\tilde{r}_\| = \frac{n_1(p - iq) - (p^2 - q^2 + n_1^2\sin^2\theta_1 - 2ipq)\cos\theta_1}{n_1(p - iq) + (p^2 - q^2 + n_1^2\sin^2\theta_1 - 2ipq)\cos\theta_1}, \quad (2.114)$$

$$\tilde{r}_\perp = \frac{n_1\cos\theta_1 - p + iq}{n_1\cos\theta_1 + p - iq}. \quad (2.115)$$

The expression for $\tilde{r}_\|$ may be simplified by dividing the numerator (and denominator) of $\tilde{r}_\|$ by $\cos\theta_1$ times the numerator (or denominator) of $\tilde{r}_\perp$. This operation leads to

$$\tilde{r}_\| = \frac{p - n_1\sin\theta_1\tan\theta_1 - iq}{p + n_1\sin\theta_1\tan\theta_1 - iq}\tilde{r}_\perp. \quad (2.116)$$

Finally, the reflectivities are again calculated as

$$\rho_\| = \tilde{r}_\|\tilde{r}_\|^* = \frac{(p - n_1\sin\theta_1\tan\theta_1)^2 + q^2}{(p + n_1\sin\theta_1\tan\theta_1)^2 + q^2}\rho_\perp \quad (2.117)$$

$$\rho_\perp = \tilde{r}_\perp\tilde{r}_\perp^* = \frac{(n_1\cos\theta_1 - p)^2 + q^2}{(n_1\cos\theta_1 + p)^2 + q^2}. \quad (2.118)$$

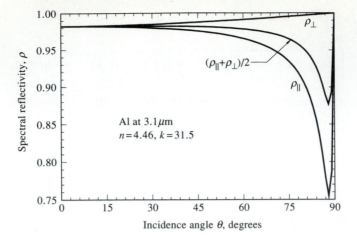

FIGURE 2-11
Directional reflectivity for a metal (aluminum at $3.1\,\mu$m with $n_2 = 4.46$, $k_2 = 31.5$) in contact with air ($n_1 = 1$).

We note that for normal incidence $\theta_1 = \theta_2 = 0$, resulting in $p = n_2$, $q = k_2$ and

$$\rho_\| = \rho_\perp = \frac{(n_1 - n_2)^2 + k_2^2}{(n_1 + n_2)^2 + k_2^2}. \tag{2.119}$$

The directional behavior of the reflectivity for a typical metal with $n_2 = 4.46$ and $k_2 = 31.5$ (corresponding to the experimental values for aluminum at $3.1\,\mu$m, [6]) exposed to air ($n_1 = 1$) is shown in Fig. 2-11.

Example 2.5. Redo Example 2.4 for a metallic interface, i.e., the plane homogeneous wave of the previous examples encounters the flat interface with a metal ($n_2 = k_2 = 90$), which again is described by the equation $z = 0$. Calculate the incidence, reflection and refraction angles. What fraction of energy of the wave is reflected, how much is transmitted?

Solution

If n_2 and k_2 are much larger than n_1 it follows from equations (2.108) and (2.109) that $p \approx n_2$ and $q \approx k_2$ and, from equation (2.104),

$$n_1 \sin\theta_1 \approx n_2 \tan\theta_2 \approx n_2 \sin\theta_2,$$

(i.e., as long as $n_2 \gg n_1$, Snell's law between dielectrics holds) and it follows that $\theta_2 = 1.02°$. With $n_2 = k_2$ equations (2.117) and (2.118) reduce to

$$\rho_\perp = \frac{(n_1 \cos\theta_1 - n_2)^2 + n_2^2}{(n_1 \cos\theta_1 + n_2)^2 + n_2^2} = \frac{(1.2 - 90)^2 + 90^2}{(1.2 + 90)^2 + 90^2} = 0.9737,$$

$$\rho_\| = \frac{(n_2 - n_1 \sin\theta_1 \tan\theta_1)^2 + n_2^2}{(n_2 + n_1 \sin\theta_1 \tan\theta_1)^2 + n_2^2}\rho_\perp = \frac{(90 - 2\times0.8^2/0.6)^2 + 90^2}{(90 + 2\times0.8^2/0.6)^2 + 90^2}\times0.9737 = 0.9286$$

and the total reflectivity is again evaluated from equation (2.94) as

$$\rho = \frac{E_{i\parallel}E_{i\parallel}^*\rho_\parallel + E_{i\perp}E_{i\perp}^*\rho_\perp}{E_{i\parallel}E_{i\parallel}^* + E_{i\perp}E_{i\perp}^*} = \frac{125 \times 0.9286 + 29 \times 0.9737}{154} = 0.9371.$$

Thus, nearly 94% of the radiation is being reflected (and even more would have been reflected if the metal was surrounded by air with $n \approx 1$), and only 6% is transmitted into the metal, where it undergoes total attenuation after a very short distance because of the large value of k_2: Equation (2.42) shows that the transmission reaches its $1/e$ value at

$$4\pi\eta_0 k_2 z = 1, \quad \text{or} \quad z = 1/(4\pi \times 2500 \times 90) = 3.5 \times 10^{-7}\text{cm} = 0.0035 \, \mu\text{m}.$$

Reflection and Transmission by a Slab

As a final topic we shall briefly consider the reflection and transmission by a slab of thickness d and complex index of refraction $m_2 = n_2 - ik_2$, embedded between two media with indices of refraction m_1 and m_3, as illustrated in Fig. 2-12. While the theory presented in this section is valid for slabs of arbitrary thickness, it is most appropriate for the study of *interference wave effects* in *thin films* or *coatings*. When an electromagnetic wave is reflected by a thin film, the waves reflected from both interfaces have different phases and *interfere* with another (i.e., they may augment each other for small phase differences, or cancel each other for phase differences of 180°). For thick slabs, such as window panes, *geometric optics* provides a much

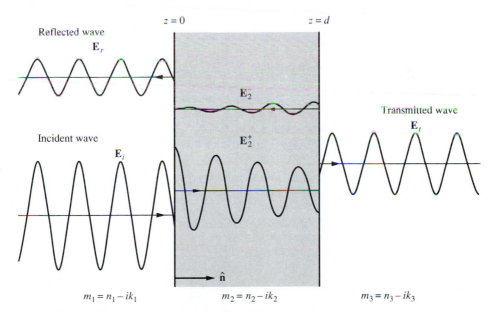

FIGURE 2-12
Reflection and transmission by a slab.

simpler vehicle to determine overall reflectivity and transmissivity. However, for an antireflective coating on a window, *thin film optics* should be considered. Since the computations become rather cumbersome, we shall limit ourselves to the simpler case of normal incidence ($\theta = 0$). For more detailed discussions, including oblique incidence angles, the reader is referred to books on the subject such as the one by Knittl [7] or to the very readable monograph by Anders [8].

Consider the slab shown in Fig. 2-12: The wave incident at the left interface is partially reflected, and partially transmitted toward the second interface. At the second interface, again, the wave is partially reflected and partially transmitted into Medium 3. The reflected part travels back to the first interface where a part is reflected back toward the second interface, and a part is transmitted into Medium 1, i.e., it is added to the reflected wave, etc. Therefore, the reflected wave $\mathbf{E}_r$ and the transmitted wave $\mathbf{E}_t$ consist of many contributions, and inside Medium 2 there are two waves $\mathbf{E}_2^+$ and $\mathbf{E}_2^-$ traveling into the directions $\hat{\mathbf{n}}$ and $-\hat{\mathbf{n}}$, respectively. Thus, the boundary conditions, equations (2.66) and (2.67), may be written for the first interface, similar to equations (2.81) through (2.84), as

$$z = \mathbf{r} \cdot \hat{\mathbf{n}} = 0: \qquad E_i + E_r = E_2^+ + E_2^-, \qquad (2.120)$$
$$H_i + H_r = H_2^+ + H_2^-, \qquad (2.121)$$

where polarization of the beam does not appear since at normal incidence $E_\parallel = E_\perp$. The magnetic field may again be eliminated using equation (2.25), as well as $\mathbf{w}_i = -\mathbf{w}_r = \eta_0 m_1 \hat{\mathbf{n}}$ and $\mathbf{w}^+ = -\mathbf{w}^- = \eta_0 m_2 \hat{\mathbf{n}}$ [from equation (2.31)], or

$$(E_i - E_r)m_1 = (E_2^+ - E_2^-)m_2. \qquad (2.122)$$

The boundary condition at the second interface follows [similar to equations (2.100) and (2.101)] as

$$z = \mathbf{r} \cdot \hat{\mathbf{n}} = d: \qquad E_2^+ e^{-2\pi i \eta_0 m_2 d} + E_2^- e^{+2\pi i \eta_0 m_2 d} = E_t e^{-2\pi i \eta_0 m_3 d} \qquad (2.123)$$
$$(E_2^+ e^{-2\pi i \eta_0 m_2 d} - E_2^- e^{+2\pi i \eta_0 m_2 d})m_2 = E_t e^{-2\pi i \eta_0 m_3 d} m_3. \qquad (2.124)$$

Equations (2.120), (2.122), (2.123) and (2.124) are four equations in the unknowns E_r, E_2^+, E_2^- and E_t, which may be solved for the *reflection and transmission coefficients of a slab*. After some algebra one obtains

$$\tilde{r}_{\text{slab}} = \frac{E_r}{E_i} = \frac{\tilde{r}_{12} + \tilde{r}_{23} e^{-4\pi i \eta_0 d m_2}}{1 + \tilde{r}_{12} \tilde{r}_{23} e^{-4\pi i \eta_0 d m_2}}, \qquad (2.125)$$

$$\tilde{t}_{\text{slab}} = \frac{E_t e^{-2\pi i \eta_0 d m_3}}{E_i} = \frac{\tilde{t}_{12} \tilde{t}_{23} e^{-2\pi i \eta_0 d m_2}}{1 + \tilde{r}_{12} \tilde{r}_{23} e^{-4\pi i \eta_0 d m_2}}, \qquad (2.126)$$

where the $\tilde{r}_{ij}$ and $\tilde{t}_{ij}$ are the complex reflection and transmission coefficients of the two interfaces,

$$\tilde{r}_{12} = \frac{m_1 - m_2}{m_1 + m_2}, \qquad \tilde{r}_{23} = \frac{m_2 - m_3}{m_2 + m_3}; \qquad (2.127a)$$

$$\tilde{t}_{12} = \frac{2m_1}{m_1 + m_2}, \qquad \tilde{t}_{23} = \frac{2m_2}{m_2 + m_3}. \tag{2.128a}$$

To evaluate the slab reflectivity and transmissivity from the complex coefficients, it is advantageous to write the coefficients in polar notation (cf., for example, Wylie [3]),

$$\tilde{r}_{ij} = r_{ij} e^{i\delta_{ij}}, \quad r_{ij} = |\tilde{r}_{ij}|, \quad \tan\delta_{ij} = \frac{\Im(\tilde{r}_{ij})}{\Re(\tilde{r}_{ij})}, \tag{2.129a}$$

$$\tilde{t}_{ij} = t_{ij} e^{i\epsilon_{ij}}, \quad t_{ij} = |\tilde{t}_{ij}|, \quad \tan\epsilon_{ij} = \frac{\Im(\tilde{t}_{ij})}{\Re(\tilde{t}_{ij})}, \tag{2.129b}$$

where r_{ij} and t_{ij} are the absolute values, and δ_{ij} and ϵ_{ij} the phase angles of the coefficients. Care must be taken in the evaluation of phase angles, since the tangent has a period of π, rather than 2π: The correct quadrant for δ_{ij} and ϵ_{ij} is found by inspecting the signs of the real and imaginary parts of $\tilde{r}_{ij}$ and $\tilde{t}_{ij}$, respectively. This calculation leads, after more algebra, to the reflectivity, R_{slab}, and transmissivity of the slab, T_{slab}, as

$$R_{slab} = \tilde{r}\tilde{r}^* = \frac{r_{12}^2 + 2r_{12}r_{23}e^{-\kappa_2 d}\cos(\delta_{12} - \delta_{23} + \zeta_2) + r_{23}^2 e^{-2\kappa_2 d}}{1 + 2r_{12}r_{23}e^{-\kappa_2 d}\cos(\delta_{12} + \delta_{23} - \zeta_2) + r_{12}^2 r_{23}^2 e^{-2\kappa_2 d}}, \tag{2.130}$$

$$T_{slab} = \frac{n_3}{n_1}\tilde{t}\tilde{t}^* = \frac{\tau_{12}\tau_{23}e^{-\kappa_2 d}}{1 + 2r_{12}r_{23}e^{-\kappa_2 d}\cos(\delta_{12} + \delta_{23} - \zeta_2) + r_{12}^2 r_{23}^2 e^{-2\kappa_2 d}}, \tag{2.131}$$

where

$$r_{ij}^2 = \rho_{ij} = \frac{(n_i - n_j)^2 + (k_i - k_j)^2}{(n_i + n_j)^2 + (k_i + k_j)^2}, \tag{2.132a}$$

$$\frac{n_j}{n_i}t_{ij}^2 = \tau_{ij} = \frac{n_i}{n_j}\frac{4(n_i^2 + k_i^2)}{(n_i + n_j)^2 + (k_i + k_j)^2}, \tag{2.132b}$$

$$\tan\delta_{ij} = \frac{2(n_i k_j - n_j k_i)}{n_i^2 + k_i^2 - (n_j^2 + k_j^2)}, \tag{2.132c}$$

$$\kappa_i = 4\pi\eta_0 k_i, \quad \zeta_i = 4\pi\eta_0 n_i d. \tag{2.132d}$$

The correct quadrant for δ_{ij} is found by checking the sign of both numerator and denominator in equation (2.132c) (which, while different from the real and imaginary parts of $\tilde{r}_{ij}$, carry their signs). If both adjacent media, i and j, are dielectrics then $\tilde{r}_{ij} = r_{ij}$ is real. In that case we set $\delta_{ij} = 0$ and let r_{ij} carry a sign. The definition of the slab transmissivity includes the factor (n_3/n_1), since it is the magnitude of the transmitted and incoming *Poynting vector*, equation (2.42), that must be compared. Knittl [7] has shown that equations (2.130) and (2.131) remain valid for each polarization for oblique incidence if the interface reflectivities, ρ_{ij}, and transmissivities, τ_{ij}, are replaced by their directional values; for example, equations (2.117) and (2.118).

Example 2.6. Determine the reflectivity and transmissivity of a $5\,\mu m$ thick manganese sulfide (MnS) crystal ($n = 2.68$, $k \ll 1$), suspended in air, for the wavelength range between $1\,\mu m$ and $1.25\,\mu m$.

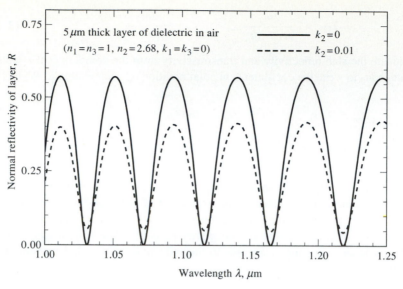

FIGURE 2-13
Normal reflectivity of a thin film with interference effects.

Solution

Assuming $n_1 = n_3 = 1$, $k_1 = k_2 = k_3 = 0$ and $n_2 = 2.68$ and substituting these into equation (2.132) leads to

$$r_{12} = r_{23} = \frac{n_2 - 1}{n_2 + 1}; \quad t_{12} = \frac{2}{n_2 + 1}, \quad t_{23} = \frac{2n_2}{n_2 + 1};$$

$$\tan \delta_{12} = \frac{0}{1 - n_2^2} = 0; \quad \tan \delta_{23} = \frac{0}{n_2^2 - 1} = 0.$$

Since the real part of $\tilde{r}_{12}$ is negative, i.e., $1 - n_2^2 < 0$, it follows that $\delta_{12} = \pi$. By similar reasoning $\delta_{23} = 0$. Alternatively, since all media are dielectrics, we could have set $\delta_{12} = \delta_{23} = 0$ and $r_{12} = -r_{23}$. Thus, with $\kappa_2 = 0$, the reflectivity and transmissivity of a dielectric slab follow as

$$R_{\text{slab}} = \frac{2\rho_{12}(1 - \cos \zeta_2)}{1 - 2\rho_{12} \cos \zeta_2 + \rho_{12}^2}, \qquad (2.133)$$

$$T_{\text{slab}} = \frac{\tau_{12}^2}{1 - 2\rho_{12} \cos \zeta_2 + \rho_{12}^2}. \qquad (2.134)$$

It is a simple matter to show that $\tau_{12} = \tau_{23} = 1 - \rho_{12}$ and, therefore, $R_{\text{slab}} + T_{\text{slab}} = 1$ for a dielectric medium. Substituting numbers for MnS gives $\rho_{12} = 0.2084$ and

$$R_{\text{slab}} = \frac{0.3995(1 - \cos \zeta_2)}{1 - 0.3995 \cos \zeta_2}, \quad T_{\text{slab}} = \frac{0.6005}{1 - 0.3995 \cos \zeta_2},$$

with $\zeta_2 = 4\pi n_2 d\eta_0 = 168.4 \, \mu\text{m} \, \eta_0 = 168.4 \, \mu\text{m}/\lambda_0$. R_{slab} and T_{slab} are periodic with a period of $\Delta\eta_0 = 2\pi/168.4 \, \mu\text{m} = 0.0373 \, \mu\text{m}^{-1}$. At $\lambda_0 = 1 \, \mu\text{m}$ this fact implies

$\Delta\lambda_0 = \lambda_0^2 \Delta\eta_0 = 0.0373 \ \mu\text{m}$. The slab reflectivity of the dielectric film in Fig. 2-13 shows a periodic reflectivity with maxima of 0.5709 (at $\zeta_2 = \pi, 3\pi, \ldots$). For values of $\zeta_2 = 2\pi, 4\pi, \ldots$, the reflectivity of the layer vanishes altogether. Also shown is the case of a slightly absorbing film, with $k_2 = 0.01$. Maximum and minimum reflectivity (as well as transmissivity) decrease and increase somewhat, respectively. This effect is less pronounced at larger wavelengths, i.e., wherever the absorption coefficient κ_2 is smaller [cf. equation (2.132d)].

While equations (2.130) through (2.132) are valid for arbitrary slab thicknesses, their application to thick slabs becomes problematic as well as unnecessary. Problematic because *(i)* for $d \gg \lambda_0$ the period of reflectivity oscillations corresponds to smaller values of $\Delta\lambda_0$ between extrema than can be measured, and *(ii)* for $d \gg \lambda_0$ it becomes rather unlikely that the distance d remains constant within a fraction of λ_0 over an extended area. Thick slab reflectivities and transmissivities may be obtained by averaging equations (2.130) and (2.131) over a period through integration, which results in

$$R_{\text{slab}} = \rho_{12} + \frac{\rho_{23}(1 - \rho_{12})^2 e^{-2\kappa_2 d}}{1 - \rho_{12}\rho_{23} e^{-2\kappa_2 d}}, \tag{2.135}$$

$$T_{\text{slab}} = \frac{(1 - \rho_{12})(1 - \rho_{23}) e^{-\kappa_2 d}}{1 - \rho_{12}\rho_{23} e^{-2\kappa_2 d}}, \tag{2.136}$$

where for T_{slab} use has been made of the fact that k_1 and k_2 must be very small, if an appreciable amount of energy is to reach Medium 3. The same relations for thick sheets without wave interference will be developed in the following chapter through geometric optics.

2.6 THEORIES FOR OPTICAL CONSTANTS

If the radiative properties of a surface—absorptivity, emissivity, and reflectivity—are to be theoretically evaluated from electromagnetic wave theory, the complex index of refraction, m, must be known over the spectral range of interest. A number of classical and quantum mechanical *dispersion theories* have been developed to predict the phenomenological coefficients ϵ (electrical permittivity) and σ_e (electrical conductivity) as functions of the frequency (or wavelength) of incident electromagnetic waves for a number of different interaction phenomena and types of surfaces. While the complex index of refraction, $m = n - ik$, is most convenient for the treatment of wave propagation, the *complex dielectric function* (or *relative permittivity*), $\varepsilon = \varepsilon' - i\varepsilon''$, is more appropriate when the microscopic mechanisms are considered that determine the magnitude of the phenomenological coefficients. The two sets of parameters are related by the expression

$$\varepsilon = \varepsilon' - i\varepsilon'' = \frac{\epsilon}{\epsilon_0} - i\frac{\sigma_e}{2\pi\nu\epsilon_0} = m^2, \tag{2.137}$$

[compare equations (2.31) through (2.35)] and, therefore,

$$\varepsilon' = \frac{\epsilon}{\epsilon_0} = n^2 - k^2, \tag{2.138a}$$

$$\varepsilon'' = \frac{\sigma_e}{2\pi\nu\epsilon_0} = 2nk, \tag{2.138b}$$

$$n^2 = \frac{1}{2}\left(\varepsilon' + \sqrt{\varepsilon'^2 + \varepsilon''^2}\right), \tag{2.139a}$$

$$k^2 = \frac{1}{2}\left(-\varepsilon' + \sqrt{\varepsilon'^2 + \varepsilon''^2}\right), \tag{2.139b}$$

where we have again assumed the medium to be nonmagnetic ($\mu = \mu_0$).

Any material may absorb or emit radiative energy at many different wavelengths as a result of impurities (presence of foreign atoms) and imperfections in the ionic crystal lattice. However, a number of phenomena tend to dominate the optical behavior of a substance. In the frequency range of interest to the heat transfer engineer (ultraviolet to midinfrared), electromagnetic waves are primarily absorbed by *free* and *bound electrons* or by change in the energy level of *lattice vibration* (converting a *photon* into a *phonon*, i.e., a quantum of lattice vibration). Since electricity is conducted by free electrons, and since free electrons are a major contributor to a solid's ability to absorb radiative energy, there are distinct optical differences between *conductors* and *nonconductors* of electricity. Every solid has a large number of electrons, resulting in a near-continuum of possible energy states (and, therefore, a near-continuum of photon frequencies that can be absorbed). However, these allowed energy states occur in *bands*. Between the bands of allowed energy states may be *band gaps*, i.e., energy states that the solid cannot attain. This is schematically shown in Fig. 2-14. If a material has a band gap between completely filled and completely empty energy bands, the material is a *nonconductor*, i.e., an *insulator* (wide band gap), or a *semiconductor* (narrow band gap). If a band of electron energy states is incompletely filled or overlaps another, empty band, electrons can be excited into adjacent energy states resulting in an electric current, and the material is called a *conductor*. Electronic absorption by nonconductors is likely only for photons with energies greater than the band gap, although sometimes two or more photons may combine to bridge the band gap. An *intraband transition* occurs when an electron changes its energy level, but stays within the same band (which can only occur in a conductor); if an electron moves into a different band (i.e., overcomes the band gap) the movement is termed an *interband transition* (and can occur in both conductors and nonconductors). This difference between conductors and nonconductors causes substantially different optical behavior: Insulators tend to be transparent and weakly reflecting for photons with energies less than the band gap, while metals tend to be highly absorbing and reflecting between the visible and infrared wavelengths [2].

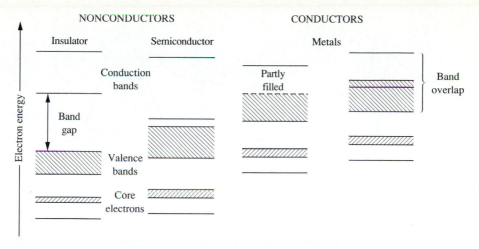

FIGURE 2-14
Electron energy bands and band gaps in a solid (shading indicates amount of electrons filling the bands) [2].

During the beginning of the century Lorentz [9]* developed a classical theory for the evaluation of the dielectric function by assuming electrons and ions are *harmonic oscillators* (i.e., springs) subjected to forces from interacting electromagnetic waves. His result was equivalent to the subsequent quantum mechanical development, and may be stated, as described by Bohren and Huffman [2], as

$$\varepsilon(\nu) = 1 + \sum_j \frac{\nu_{pj}^2}{\nu_j^2 - \nu^2 - i\gamma_j\nu},$$

(2.140)

where the summation is over different types of oscillators, ν_{pj} is known as the *plasma frequency* (and ν_{pj}^2 is proportional to the number of oscillators of type j), ν_j is the *resonance frequency*, and γ_j is the damping factor of the oscillators. Thus, the dielectric function may have a number of bands centered at ν_j, which may or may not overlap one another. Inspecting equation (2.140), we see that for $\nu \gg \nu_j$ the contribution of band j to ε vanishes, while for $\nu \ll \nu_j$ it goes to the constant value of $(\nu_{pj}/\nu_j)^2$.

Hendrik Anton Lorentz (1853-1928)

Dutch physicist. Lorentz studied at Leiden University, where he subsequently served as professor of mathematical physics for the rest of his life. His major work lay in refining the electromagnetic theory of Maxwell. For his theory that the oscillations of charged particles inside atoms were the source of light, he and his student Pieter Zeeman received the 1902 Nobel Prize in Physics. Lorentz is also famous for his Lorentz transformations, which describe the increase of mass of a moving body. These laid the foundation for Einstein's special theory of relativity.

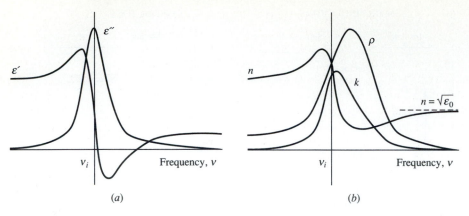

FIGURE 2-15
Lorentz model for *(a)* the dielectric function, *(b)* the index of refraction and normal, spectral reflectivity.

Therefore, for any nonoverlapping band i, we may rewrite equation (2.140) as

$$\varepsilon(\nu) = \varepsilon_0 + \frac{\nu_{pi}^2}{\nu_i^2 - \nu^2 - i\gamma_i \nu},\qquad(2.141)$$

where ε_0 incorporates the contributions from all bands with $\nu_j > \nu_i$. Equation (2.141) may be separated into its real and imaginary components, or

$$\varepsilon' = \varepsilon_0 + \frac{\nu_{pi}^2(\nu_i^2 - \nu^2)}{(\nu_i^2 - \nu^2)^2 + \gamma_i^2 \nu^2},\qquad(2.142a)$$

$$\varepsilon'' = \frac{\nu_{pi}^2 \gamma_i \nu}{(\nu_i^2 - \nu^2)^2 + \gamma_i^2 \nu^2}.\qquad(2.142b)$$

The frequency dependence of the real and imaginary parts of the dielectric function for a single oscillating band is shown qualitatively in Fig. 2-15; also shown are the corresponding curves for the real and imaginary parts of the complex index of refraction as evaluated from equation (2.139), along with the qualitative behavior of the normal, spectral reflectivity of a surface from equation (2.119). A strong band with $k \gg 0$ results in a region with strong absorption around the resonance frequency and an associated region of high reflection: Incoming photons are mostly reflected, and those few that penetrate into the medium are rapidly attenuated. On either side outside the band the refractive index n increases with increasing frequency (or decreasing wavelength); this is called *normal dispersion*. However, close to the resonance frequency, n decreases with increasing frequency; this decrease is known as *anomalous dispersion*. Note that ε' may become negative, resulting in spectral regions with $n < 1$.

All solids and liquids may absorb photons whose energy content matches the energy difference between filled and empty electron energy levels on separate bands.

Since such transitions require a substantial amount of energy, they generally occur in the ultraviolet (i.e., at high frequency). A near-continuum of electron energy levels results in an extensive region of strong absorption (and often many overlapping bands). It takes considerably less energy to excite the vibrational modes of a crystal lattice, resulting in absorption bands in the midinfrared (around $10 \, \mu$m). Since generally few different vibrational modes exist in an isotropic lattice, such transitions can often be modeled by equation (2.140) with a single band. In the case of electrical conductors photons may also be absorbed to raise the energy levels of free electrons and of bound electrons within partially filled or partially overlapping electron bands. The former, because of the nearly arbitrary energy levels that a free electron may assume, results in a single large band in the far infrared; the latter causes narrower bands in the ultraviolet to infrared.

References

1. Stone, J. M.: *Radiation and Optics*, McGraw-Hill, New York, 1963.
2. Bohren, C. F., and D. R. Huffman: *Absorption and Scattering of Light by Small Particles*, John Wiley & Sons, New York, 1983.
3. Wylie, C. R.: *Advanced Engineering Mathematics*, 5th ed., McGraw-Hill, New York, 1982.
4. van de Hulst, H. C.: *Light Scattering by Small Particles*, John Wiley & Sons, New York, 1957 (also Dover Publications, New York, 1981).
5. Chandrasekhar, S.: *Radiative Transfer*, Dover Publications, New York, 1960 (originally published by Oxford University Press, London, 1950).
6. Weast, R. C. (ed.): *CRC Handbook of Chemistry and Physics*, 68th ed., Chemical Rubber Company, Cleveland, OH, 1988.
7. Knittl, Z.: *Optics of Thin Films*, John Wiley & Sons, New York, 1976.
8. Anders, H.: *Thin Films in Optics*, The Focal Press, New York, London, 1967.
9. Lorentz, H. A.: *Collected Papers*, vol. 8, Martinus Nijhoff, The Hague, 1935.

Problems

2.1 Show that for an electromagnetic wave traveling through a dielectric ($m_1 = n_1$), impinging on the interface with another, optically less dense dielectric ($n_2 < n_1$), light of any polarization is totally reflected for incidence angles larger than $\theta_c = \sin^{-1}(n_2/n_1)$. Hint: Use equations (2.104) through (2.107) with $k_2 = 0$.

2.2 Derive equations (2.111) and (2.112) using the same approach as in the development of equations (2.88) through (2.91).
Hint: Remember that within the absorbing medium, $\mathbf{w} = \mathbf{w}' - i\mathbf{w}'' = w'\hat{\mathbf{s}} - iw''\hat{\mathbf{n}}$; this implies that $\mathbf{E}_0$ is *not* a vector normal to $\hat{\mathbf{s}}$. It is best to assume $\mathbf{E}_0 = E_\parallel \hat{\mathbf{e}}_\parallel + E_\perp \hat{\mathbf{e}}_\perp + E_s \hat{\mathbf{s}}$.

2.3 Find the normal spectral reflectivity at the interface between two absorbing media. [Hint: Use an approach similar to the one that led to equations (2.88) and (2.89), keeping in mind that all wave vectors will be complex, but that the wave will be homogeneous in both media, i.e., all components of the wave vectors are colinear with the surface normal].

2.4 A circularly polarized wave in air is incident upon a smooth dielectric surface ($n = 1.5$) with a direction of 45° off normal. What are the normalized Stokes' parameters before and after the reflection, and what are the degrees of polarization?

2.5 A circularly polarized wave in air traveling along the z-axis is incident upon a dielectric surface ($n = 1.5$). How must the dielectric-air interface be oriented so that the reflected wave is a linearly polarized wave in the y-z-plane?

2.6 A polished platinum surface is coated with a 1 μm thick layer of MgO.

   (a) Determine the material's reflectivity in the vicinity of $\lambda = 2\,\mu$m (for platinum at $2\,\mu$m $m_{Pt} = 5.29 - 6.71\,i$, for MgO $m_{MgO} = 1.65 - 0.0001\,i$).

   (b) Estimate the thickness of MgO required to reduce the average reflectivity in the vicinity of $2\,\mu$m to 0.4. What happens to the interference effects for this case?

CHAPTER
3

RADIATIVE
PROPERTIES OF
REAL SURFACES

3.1 INTRODUCTION

Ideally, electromagnetic wave theory may be used to predict all radiative properties of any material (reflectivity and transmissivity at an interface, absorption and emission within a medium). For a variety of reasons, however, the usefulness of the electromagnetic wave theory is extremely limited in practice. For one, the theory incorporates a large number of assumptions that are not necessarily good for all materials. Most importantly, electromagnetic wave theory neglects the effects of surface conditions on the radiative properties of these surfaces, instead assuming optically smooth interfaces of precisely the same (homogeneous) material as the bulk material—conditions that are very rarely met in practice. In the real world surfaces of materials are generally coated to varying degree with contaminants, oxide layers, and the like, and they usually have a certain degree of roughness (which is rarely even known on a quantitative basis). Thus, the greatest usefulness of the electromagnetic wave theory is that it provides the engineer with a tool to augment sparse experimental data through intelligent interpolation and extrapolation. Still, it is important to realize that radiative properties of opaque materials depend exclusively on the makeup of a very thin surface layer and, thus, may, for the same material, change from batch to batch and, indeed, overnight. This behavior is in contrast to most other thermophysical properties, such as thermal conductivity, which are bulk properties and as such are insensitive to surface contamination, roughness, and so on. The National Institute of Standards and Technology (NIST, formerly NBS) has recommended to reserve the ending "-ivity" for radiative properties of pure, perfectly smooth materials (the ones discussed in the previous chapter), and "-ance" for rough and contaminated surfaces. Most real surfaces fall into the latter category, discussed in the present chapter. However, few researchers in the field appear to follow this convention, rather employing endings according to their own personal preference. Accordingly, we shall use the ending "-ivity" throughout this book for *all* surfaces.

In the present chapter we shall first develop definitions of all radiative properties that are relevant for real opaque surfaces. We then apply electromagnetic wave theory to predict trends of radiative properties for metals and for dielectrics (electrical nonconductors). These theoretical results are compared with a limited number of experimental data. This is followed by a brief discussion of phenomena that cannot be predicted by electromagnetic wave theory, such as the effects of surface roughness, of surface oxidation and contamination, and of the preparation of "special surfaces," (i.e., surfaces whose properties are customized through surface coatings and/or controlled roughness).

Most experimental data available today were taken in the 1950s and 1960s during NASA's "Golden Age," when considerable resources were directed toward sending a man to the moon. Interest waned, together with NASA's funding, during the 1970s and early 1980s. More recently, because of the development of high-temperature ceramics and high-temperature applications, there has been renewed interest in the measurement of radiative surface properties.

No attempt is made here to present a complete set of experimental data for radiative surface properties. Extensive data sets of such properties have been collected in a number of references, such as [1–8], although all of these surveys are somewhat outdated.

3.2 DEFINITIONS

Emissivity

The most basic radiative property for emission from an opaque surface is its *spectral, directional emissivity*, defined as

$$\epsilon_\lambda'(T, \lambda, \hat{\mathbf{s}}_o) \equiv \frac{I_\lambda(T, \lambda, \hat{\mathbf{s}}_o) \cos \theta_o d\Omega_o}{I_{b\lambda}(T, \lambda) \cos \theta_o d\Omega_o} = \frac{I_\lambda(T, \lambda, \hat{\mathbf{s}}_o)}{I_{b\lambda}(T, \lambda)}, \tag{3.1}$$

which compares the actual spectral, directional emissive power with that of a black surface at the same conditions. We have added a prime to the letter ϵ to distinguish the directional emissivity from the hemispherical (i.e., directionally averaged) value, and the subscript λ to distinguish the spectral emissivity from the total (i.e., spectrally averaged) value. The direction vector is denoted by $\hat{s}_o$ to emphasize that, for emission, we are considering directions *away* from a surface (outgoing). Finally, we have chosen wavelength λ as the spectral variable, since this is the preferred variable by most authors in the field of surface radiation phenomena. Expressions identical to equation (3.1) hold if frequency ν or wavenumber η is employed.

Some typical trends for experimentally determined directional emissivities for actual materials are shown in Fig. 3-1a,b, as given by Schmidt and Eckert [9] (all emissivities in these figures have been averaged over the entire spectrum; see the definition of the *total, directional emissivity* below). For nonmetals the directional emissivity varies little over a large range of polar angles but decreases rapidly at grazing angles until a value of zero is reached at $\theta = \pi/2$. Similar trends hold for metals,

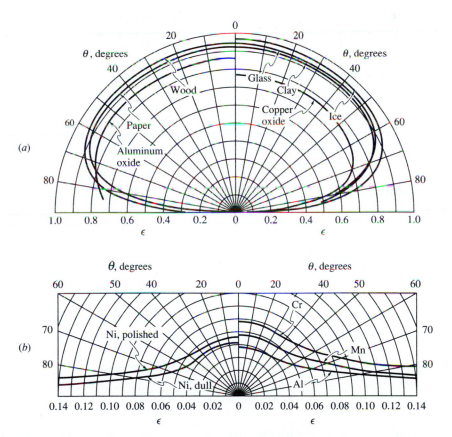

FIGURE 3-1
Directional variation of surface emissivities (*a*) for several nonmetals and (*b*) for several metals [9].

except that, at grazing angles, the emissivity first increases sharply before dropping back to zero (not shown). Note that emissivity levels are considerably higher for nonmetals.

A spectral surface whose emissivity is the same for *all* directions is called a *diffuse emitter*, or a *Lambert surface* [since it obeys Lambert's law, equation (1.32)]. No real surface can be a diffuse emitter since *electromagnetic wave theory* predicts a zero emissivity at $\theta = \pi/2$ for all materials. However, little energy is emitted into grazing directions, as seen from equation (1.28), so that the assumption of diffuse emission is often a good one.

The *spectral, hemispherical emissivity*, defined as

$$\epsilon_\lambda(T, \lambda) \equiv \frac{E_\lambda(T, \lambda)}{E_{b\lambda}(T, \lambda)}, \tag{3.2}$$

compares the actual *spectral emissive power* (i.e., emission into all directions above the surface) with that of a black surface. The spectral, hemispherical emissivity may be related to the directional one through equations (1.28) and (1.30),

$$
\begin{aligned}
\epsilon_\lambda(T, \lambda) &= \frac{\int_0^{2\pi} \int_0^{\pi/2} I_\lambda(T, \lambda, \theta, \psi) \cos\theta \, \sin\theta \, d\theta \, d\psi}{\pi I_{b\lambda}(T, \lambda)} \\
&= \frac{\int_0^{2\pi} \int_0^{\pi/2} \epsilon_\lambda'(T, \lambda, \theta, \psi) I_{b\lambda}(T, \lambda) \cos\theta \, \sin\theta \, d\theta \, d\psi}{\pi I_{b\lambda}(T, \lambda)},
\end{aligned}
\tag{3.3}
$$

which may be simplified to

$$\epsilon_\lambda(T, \lambda) = \frac{1}{\pi} \int_0^{2\pi} \int_0^{\pi/2} \epsilon_\lambda'(T, \lambda, \theta, \psi) \cos\theta \, \sin\theta \, d\theta \, d\psi, \tag{3.4}$$

since $I_{b\lambda}$ does not depend on direction. For an *isotropic surface*, i.e., a surface that has no different structure, composition or behavior for different directions on the surface (azimuthal angle), equation (3.4) reduces to

$$\epsilon_\lambda(T, \lambda) = 2 \int_0^{\pi/2} \epsilon_\lambda'(T, \lambda, \theta) \cos\theta \, \sin\theta \, d\theta. \tag{3.5}$$

We note that the hemispherical emissivity is an average over all solid angles subject to the weight factor $\cos\theta$ (arising from the directional variation of emissive power). For a *diffuse* surface, ϵ_λ' does not depend on direction and we find

$$\epsilon_\lambda(T, \lambda) = \epsilon_\lambda'(T, \lambda). \tag{3.6}$$

The *total, directional emissivity* is a spectral average of ϵ_λ', defined by

$$\epsilon'(T, \hat{s}) = \frac{I(T, \hat{s}) \cos\theta \, d\Omega}{I_b(T, \hat{s}) \cos\theta \, d\Omega} = \frac{I(T, \hat{s})}{I_b(T, \hat{s})}, \tag{3.7}$$

or, from equations (1.27) and (1.31),

$$\epsilon'(T, \hat{s}) = \frac{1}{I_b} \int_0^\infty I_\lambda \, d\lambda = \frac{1}{I_b} \int_0^\infty \epsilon_\lambda' I_{b\lambda} d\lambda$$

$$= \frac{1}{n^2 \sigma T^4} \int_0^\infty \epsilon_\lambda'(T, \lambda, \hat{s}) E_{b\lambda}(T, \lambda) \, d\lambda. \qquad (3.8)$$

Finally, the *total, hemispherical emissivity* is defined as

$$\epsilon(T) = \frac{E(T)}{E_b(T)}, \qquad (3.9)$$

and may be related to the spectral, hemispherical emissivity through

$$\epsilon(T) = \frac{\int_0^\infty E_\lambda(T, \lambda) \, d\lambda}{E_b(T)} = \frac{1}{n^2 \sigma T^4} \int_0^\infty \epsilon_\lambda(T, \lambda) E_{b\lambda}(T, \lambda) \, d\lambda. \qquad (3.10)$$

It is apparent that the total emissivity is a spectral average with the spectral emissive power as weight factor. If the spectral emissivity is the same for all wavelengths then equation (3.10) reduces to

$$\epsilon(T) = \epsilon_\lambda(T). \qquad (3.11)$$

Such surfaces are termed *gray*. If we have the very special case of a *gray, diffuse surface*, this implies

$$\epsilon(T) = \epsilon_\lambda = \epsilon' = \epsilon_\lambda'. \qquad (3.12)$$

While no real surface is truly gray, it often happens that ϵ_λ is relatively constant over that part of the spectrum where $E_{b\lambda}$ is substantial, making the simplifying assumption of a gray surface warranted.

Example 3.1.   A certain surface material has the following spectral, directional emissivity when exposed to air:

$$\epsilon_\lambda'(\lambda, \theta) = \begin{cases} 0.9 \cos \theta, & 0 < \lambda < 2 \, \mu m, \\ 0.3, & 2 \, \mu m < \lambda < \infty. \end{cases}$$

Determine the total hemispherical emissivity for a surface temperature of $T = 500 \, K$.

Solution

We first determine the hemispherical, spectral emissivity from equation (3.5) as

$$\epsilon_\lambda(\lambda) = \begin{cases} 2 \times 0.9 \int_0^{\pi/2} \cos^2 \theta \, \sin \theta \, d\theta = 0.6, & 0 < \lambda < 2 \, \mu m, \\ 2 \times 0.3 \int_0^{\pi/2} \cos \theta \, \sin \theta \, d\theta = 0.3, & 2 \, \mu m < \lambda < \infty. \end{cases}$$

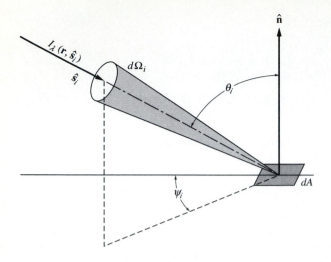

FIGURE 3-2
Directional irradiation onto a surface.

The total, hemispherical emissivity follows from equation (3.10) as

$$\epsilon(T) = \frac{1}{n^2 \sigma T^4} \left(0.6 \int_0^{2\mu\text{m}} E_{b\lambda} \, d\lambda + 0.3 \int_{2\mu\text{m}}^{\infty} E_{b\lambda} \, d\lambda \right)$$

$$= 0.3 + \frac{0.6 - 0.3}{n^2 \sigma T^4} \int_0^{2\mu\text{m}} E_{b\lambda} \, d\lambda$$

$$= 0.3 \, [1 + f(1 \times 2\mu\text{m} \times 500\,\text{K})] = 0.3 \times (1 + 0.00032) \simeq 0.3,$$

where the fractional blackbody emissive power $f(n\lambda T)$ is as defined in equation (1.22). For a temperature of 500 K the spectrum below 2 μm is unimportant, and the surface is essentially gray and diffuse.

Absorptivity

Unlike emissivity, absorptivity (as well as reflectivity and transmissivity) is not truly a surface property, since it depends on the external radiation field, as seen from its definition, equation (1.45). As for emissivity we distinguish between directional and hemispherical, as well as spectral and total absorptivities.

The radiative heat transfer rate per unit wavelength impinging onto an infinitesimal area dA, from the direction of $\hat{\mathbf{s}}_i$ over a solid angle of $d\Omega_i$ is, as depicted in Fig. 3-2,

$$I_\lambda(\mathbf{r}, \lambda, \hat{\mathbf{s}}_i)(\cos \theta_i \, dA) d\Omega_i,$$

where we have used the definition of intensity as radiative heat transfer rate per unit area normal to the rays, and per unit solid angle. I_λ is the local radiative intensity at location $\mathbf{r}$ (just above the surface). This incoming heat transfer rate, when evaluated

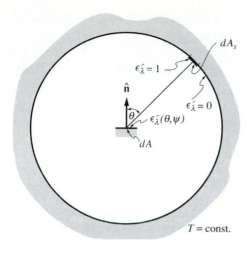

FIGURE 3-3
Kirchhoff's law for the spectral, directional absorptivity.

$T = \text{const.}$

per unit surface area dA and per unit incoming solid angle $d\Omega_i$, is known as *spectral, directional irradiation*,

$$H'_\lambda(\mathbf{r}, \lambda, \hat{\mathbf{s}}_i) = I_\lambda(\mathbf{r}, \lambda, \hat{\mathbf{s}}_i)\cos\theta_i. \tag{3.13}$$

Irradiation is a heat flux *always* pointing *into* the surface. Thus, there is no need to attach a sign to its value, and it is evaluated as an absolute value (in contrast to the definition of net heat flux in Chapter 1). The *spectral, directional absorptivity* at surface location $\mathbf{r}$ is then defined as

$$\alpha'_\lambda(\mathbf{r}, \lambda, \hat{\mathbf{s}}_i) \equiv \frac{H'_{\lambda,\text{abs}}}{H'_\lambda}, \tag{3.14}$$

where $H'_{\lambda,\text{abs}}$ is that part of H'_λ that is absorbed by dA. If local thermodynamic equilibrium prevails, the fraction α'_λ will not change if H'_λ increases or decreases. Under this condition we find that the spectral, directional absorptivity does not depend on the external radiation field and is a surface property that depends on local temperature, wavelength and incoming direction. To determine its magnitude, we consider an isothermal spherical enclosure shown in Fig. 3-3, similar to the one used in Section 1.6 to establish the directional isotropy of blackbody intensity. The enclosure coating is again perfectly reflecting except for a small area dA_s, which is also perfectly reflecting except over the wavelength interval between λ and $\lambda + d\lambda$, over which it is black. However, the small surface dA suspended at the center is now nonblack. Following the same arguments as for the development of equation (1.29), augmenting the emitted flux by ϵ'_λ and the absorbed flux by α'_λ, we find immediately

$$\alpha'_\lambda(T, \lambda, \theta, \psi) = \epsilon'_\lambda(T, \lambda, \theta, \psi). \tag{3.15}$$

Therefore, if local thermodynamic equilibrium prevails, the spectral, directional absorptivity is a true surface property and is equal to the spectral, directional emissivity.

The spectral radiative heat flux incident on a surface per unit wavelength from all directions, i.e., from the hemisphere above dA, is

$$H_\lambda(\mathbf{r}, \lambda) = \int_{2\pi} H_\lambda'(\mathbf{r}, \lambda, \hat{\mathbf{s}}_i)\, d\Omega_i = \int_{2\pi} I_\lambda(\mathbf{r}, \lambda, \hat{\mathbf{s}}_i) \cos\theta_i\, d\Omega_i. \qquad (3.16)$$

Of this the amount absorbed is, from equation (3.14),

$$\int_{2\pi} \alpha_\lambda'(T, \lambda, \hat{\mathbf{s}}_i) I_\lambda(\mathbf{r}, \lambda, \hat{\mathbf{s}}_i) \cos\theta_i\, d\Omega_i.$$

Thus, we define a *spectral, hemispherical absorptivity*

$$\alpha_\lambda(\mathbf{r}, \lambda) \equiv \frac{H_{\lambda,\text{abs}}}{H_\lambda} = \frac{\int_{2\pi} \alpha_\lambda'(T, \lambda, \hat{\mathbf{s}}_i) I_\lambda(\mathbf{r}, \lambda, \hat{\mathbf{s}}_i) \cos\theta_i\, d\Omega_i}{\int_{2\pi} I_\lambda(\mathbf{r}, \lambda, \hat{\mathbf{s}}_i) \cos\theta_i\, d\Omega_i}. \qquad (3.17)$$

Since the incoming radiation, I_λ, depends on the radiation field of the surrounding enclosure, the spectral, hemispherical absorptivity normally depends on the entire temperature field and is not a surface property. However, if the incoming radiation is approximately diffuse (i.e., if I_λ is independent of $\hat{\mathbf{s}}_i$), then the I_λ may be moved outside the integrals in equation (3.17) and canceled. Then

$$\alpha_\lambda(T, \lambda) = \frac{1}{\pi} \int_0^{2\pi} \int_0^{\pi/2} \alpha_\lambda'(T, \lambda, \theta_i, \psi_i) \cos\theta_i \sin\theta_i\, d\theta_i\, d\psi_i, \qquad (3.18)$$

or, using equations (3.4) and (3.15),

$$\alpha_\lambda(T, \lambda) = \epsilon_\lambda(T, \lambda). \qquad (3.19)$$

Therefore, spectral directional absorptivities and emissivities are equal if (and only if) the irradiation is *diffuse* (i.e., does not depend on incoming direction).

The total irradiation per unit area and per unit solid angle, but over all wavelengths, is

$$H'(\mathbf{r}) = \int_0^\infty I_\lambda(\mathbf{r}, \lambda, \hat{\mathbf{s}}_i) \cos\theta_i\, d\lambda. \qquad (3.20)$$

Thus, we may define a *total, directional absorptivity* as

$$\alpha'(\mathbf{r}, \hat{\mathbf{s}}_i) \equiv \frac{\int_0^\infty \alpha_\lambda'(T, \lambda, \hat{\mathbf{s}}_i) I_\lambda(\mathbf{r}, \hat{\mathbf{s}}_i)\, d\lambda}{\int_0^\infty I_\lambda(\mathbf{r}, \hat{\mathbf{s}}_i)\, d\lambda}, \qquad (3.21)$$

where the factor $\cos\theta_i$ has canceled out since it does not depend on wavelength. Again, α' is not normally a surface property but depends on the entire radiation field. However, if the irradiation may be written as

$$I_\lambda(\mathbf{r}, \lambda, \hat{\mathbf{s}}_i) = C(\hat{\mathbf{s}}_i) I_{b\lambda}(T, \lambda), \qquad (3.22)$$

where $C(\hat{\mathbf{s}})$ is an otherwise arbitrary function that does not depend on wavelength, i.e., if the incoming radiation is gray (based on the local surface temperature T), then, from equations (3.8) and (3.15),

$$\alpha'(T, \theta, \psi) = \epsilon'(T, \theta, \psi). \tag{3.23}$$

Finally, the total irradiation per unit area from all directions and over the entire spectrum is

$$H(\mathbf{r}) = \int_0^\infty \int_{2\pi} I_\lambda(\mathbf{r}, \lambda, \hat{\mathbf{s}}_i) \cos \theta_i \, d\Omega_i \, d\lambda. \tag{3.24}$$

Therefore, the *total, hemispherical absorptivity* is defined as

$$\alpha(\mathbf{r}) \equiv \frac{H_{\text{abs}}}{H} = \frac{\int_0^\infty \alpha_\lambda(\mathbf{r}, \lambda) H_\lambda(\mathbf{r}, \lambda) \, d\lambda}{\int_0^\infty H_\lambda(\mathbf{r}, \lambda) \, d\lambda}$$

$$= \frac{\int_0^\infty \int_{2\pi} \alpha_\lambda'(T, \lambda, \hat{\mathbf{s}}_i) I_\lambda(\mathbf{r}, \lambda, \hat{\mathbf{s}}_i) \cos \theta_i \, d\Omega_i \, d\lambda}{\int_0^\infty \int_{2\pi} I_\lambda(\mathbf{r}, \lambda, \hat{\mathbf{s}}_i) \cos \theta_i \, d\Omega_i \, d\lambda}. \tag{3.25}$$

This absorptivity is related to the total hemispherical emissivity only for the very special case of *diffuse and gray irradiation*, i.e., if

$$I_\lambda(\mathbf{r}, \lambda, \hat{\mathbf{s}}_i) = C I_{b\lambda}(T, \lambda), \tag{3.26}$$

where T is the temperature of the surface and C is a constant. Under those conditions we find, again using equation (3.15),

$$\alpha(T) = \epsilon(T). \tag{3.27}$$

Example 3.2. Let the surface considered in the previous example be irradiated by the sun from a 30° off-normal direction (i.e., a vector pointing to the sun from the surface forms a 30° angle with the outward surface normal). Determine the relevant surface absorptivity.

Solution

Since the sun irradiates the surface from only one direction, but over the entire spectrum, we need to find the *total, directional absorptivity*. From the last example, with $\theta_i = 30°$, we have

$$\alpha_\lambda'\left(\lambda, \theta_i = \frac{\pi}{6}\right) = \begin{cases} 0.45\sqrt{3}, & 0 < \lambda < 2\,\mu\text{m}, \\ 0.3, & 2\,\mu\text{m} < \lambda < \infty. \end{cases}$$

Since we know that the sun behaves like a blackbody at a temperature of $T_{\text{sun}} = 5762\,\text{K}$, we also know the spectral behavior of the sunshine falling onto our surface, or

$$I_\lambda(\lambda, \theta_i) = C I_{b\lambda}(T_{\text{sun}}, \lambda), \tag{3.28}$$

where C is a proportionality constant independent of wavelength.[1] Substituting this into equation (3.21) leads to

$$
\begin{aligned}
\alpha'\left(\theta_i = \frac{\pi}{6}\right) &= \frac{\int_0^\infty \epsilon_\lambda'(\lambda, \theta_i) I_{b\lambda}(T_{\text{sun}}, \lambda)\, d\lambda}{\int_0^\infty I_{b\lambda}(T_{\text{sun}}, \lambda)\, d\lambda} \\[2mm]
&= \frac{1}{n^2 \sigma T_{\text{sun}}^4} \left[0.45\sqrt{3} \int_0^{2\mu\text{m}} E_{b\lambda}(T_{\text{sun}}, \lambda)\, d\lambda \right. \\[2mm]
&\qquad\qquad \left. + 0.3 \int_{2\mu\text{m}}^{\infty} E_{b\lambda}(T_{\text{sun}}, \lambda)\, d\lambda \right] \\[2mm]
&= 0.3 + (0.45\sqrt{3} - 0.3) f(1 \times 2 \times 5762) \\[1mm]
&= 0.3 + (0.779 - 0.3) \times 0.93921 = 0.750.
\end{aligned}
$$

In contrast to the previous example we find that at a temperature of 5762 K the spectrum *above* 2 μm is of very little importance, and the surface is again essentially gray.

We realize from this example that *(i)* if a surface is irradiated from a gray source at temperature T_{source}, and *(ii)* if the spectral, directional emissivity of the surface is independent of temperature (as it is for most surfaces with good degree of accuracy), then the total absorptivity is equal to its total emissivity evaluated at the source temperature, or

$$
\alpha = \epsilon(T_{\text{source}}). \tag{3.29}
$$

This relation holds on a directional basis, and also for hemispherical values if the irradiation is diffuse.

Reflectivity

The reflectivity of a surface depends on *two* directions: The direction of the incoming radiation, $\hat{s}_i$, and the direction into which the reflected energy travels, $\hat{s}_r$. Therefore, we distinguish between total and spectral values, and between a number of directional reflectivities.

The heat flux per unit wavelength impinging on an area dA from a direction of $\hat{s}_i$ over a solid angle of $d\Omega_i$ was given by equation (3.13) as

$$
H_\lambda' d\Omega_i = I_\lambda(\mathbf{r}, \lambda, \hat{s}_i) \cos\theta_i\, d\Omega_i. \tag{3.30}
$$

Of this, the finite fraction α_λ' will be absorbed by the surface (assuming it to be opaque), and the rest will be reflected into all possible directions (total solid angle 2π). Therefore, in general, only an infinitesimal fraction will be reflected into an infinitesimal cone of solid angle $d\Omega_r$ around direction $\hat{s}_r$, as shown in Fig. 3-4. Denoting this fraction by $\rho_\lambda''(\mathbf{r}, \lambda, \hat{s}_i, \hat{s}_r) d\Omega_r$ we obtain the reflected energy within the

---

[1] As we have seen in Section 1.7, this constant is equal to unity.

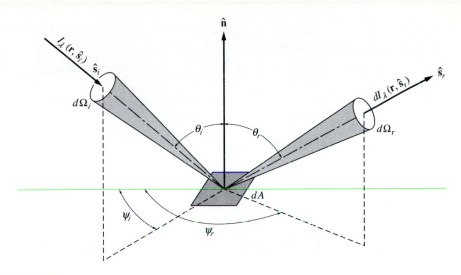

FIGURE 3-4
The bidirectional reflection function.

cone $d\Omega_r$ as

$$dI_\lambda(\mathbf{r}, \lambda, \hat{\mathbf{s}}_i, \hat{\mathbf{s}}_r)d\Omega_r = (H'_\lambda d\Omega_i)\rho''_\lambda(\mathbf{r}, \lambda, \hat{\mathbf{s}}_i, \hat{\mathbf{s}}_r)d\Omega_r. \tag{3.31}$$

The *spectral, bidirectional reflection function*[2] $\rho''_\lambda(\mathbf{r}, \lambda, \hat{\mathbf{s}}_i, \hat{\mathbf{s}}_r)$ is directly proportional to the magnitude of reflected light that travels into the direction of $\hat{\mathbf{s}}_r$,

$$\rho''_\lambda(\mathbf{r}, \lambda, \hat{\mathbf{s}}_i, \hat{\mathbf{s}}_r) = \frac{dI_\lambda(\mathbf{r}, \lambda, \hat{\mathbf{s}}_i, \hat{\mathbf{s}}_r)}{I_\lambda(\mathbf{r}, \lambda, \hat{\mathbf{s}}_i)\cos\theta_i \, d\Omega_i}, \tag{3.32}$$

Equation (3.32) is the most basic of all radiation properties: All other radiation properties of an opaque surface can be related to it. However, experimental determination of this function for all materials, temperatures, wavelengths, incoming directions and outgoing directions would be a truly Herculean task, limiting its practicality.

One may readily show that the *law of reciprocity* holds for the spectral, bidirectional reflection function, (cf. McNicholas [10] or Siegel and Howell [11]),

$$\rho''_\lambda(\mathbf{r}, \lambda, \hat{\mathbf{s}}_i, \hat{\mathbf{s}}_r) = \rho''_\lambda(\mathbf{r}, \lambda, -\hat{\mathbf{s}}_r, -\hat{\mathbf{s}}_i), \tag{3.33a}$$

or

$$\rho''_\lambda(\mathbf{r}, \lambda, \theta_i, \psi_i, \theta_r, \psi_r) = \rho''_\lambda(\mathbf{r}, \lambda, \theta_r, \psi_r, \theta_i, \psi_i). \tag{3.33b}$$

This is done with another variation of *Kirchhoff's law* by placing a surface element into an isothermal black enclosure and evaluating the net heat transfer rate—which

---

[2]ρ''_λ is sometimes referred to as a *bidirectional reflectivity*; we avoid this nomenclature since the bidirectional reflectivity function is not a fraction (i.e., constrained to values between 0 and 1), but may be larger than unity.

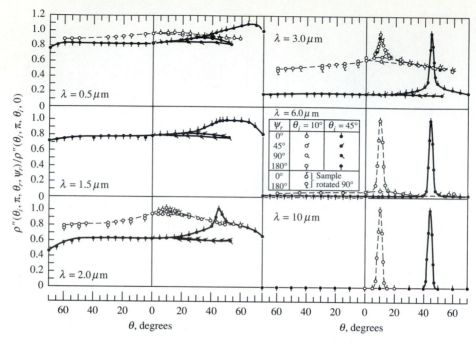

FIGURE 3-5
Normalized bidirectional reflection function for magnesium oxide [12].

must be zero—between two arbitrary, infinitesimal surface elements on the enclosure wall). The sign change on the right-hand side of equation (3.33) emphasizes that $\hat{s}_i$ points *into* the surface, while $\hat{s}_r$ points away from it. Examination of equation (3.32) shows that $0 \leq \rho_\lambda'' < \infty$. Reaching the limit of $\rho_\lambda'' \to \infty$ implies that a *finite* fraction of H_λ' is reflected into an *infinitesimal* cone of solid angle $d\Omega_r$. Such ideal behavior is achieved by an *optically smooth surface*, resulting in *specular reflection* (perfect mirror). For a specular reflector we have $\rho_\lambda'' = 0$ for all $\hat{s}_r$ except the specular direction $\theta_r = \theta_i$, $\psi_r = \psi_i + \pi$, for which $\rho_\lambda'' \to \infty$ (see Fig. 3-4).

Some measurements by Torrance and Sparrow [12] for the bidirectional reflection function are shown in Fig. 3-5 for magnesium oxide, a material widely used in radiation experiments because of its diffuse reflectivity, as defined in equation (3.36) below, in the near infrared (discussed in the last part of this chapter). The data in Fig. 3-5 are for an average surface roughness of 1 μm and are normalized with respect to the value in the specular direction. It is apparent that the material reflects rather diffusely at shorter wavelengths, but displays strong specular peaks for $\lambda > 2\ \mu$m.

A property of greater practical importance is the *spectral, directional-hemispherical reflectivity,* which is defined as the total reflected heat flux leaving dA into all directions due to the spectral, directional irradiation H_λ'. With the reflected intensity (i.e., reflected energy per unit area normal to $\hat{s}_r$) given by equation (3.31), we have, after multiplying with $\cos \theta_r$,

$$\rho_\lambda'^{\ominus}(\mathbf{r}, \lambda, \hat{\mathbf{s}}_i) \equiv \frac{\int_{2\pi} dI_\lambda(\mathbf{r}, \lambda, \hat{\mathbf{s}}_i, \hat{\mathbf{s}}_r) \cos \theta_r \, d\Omega_r}{H_\lambda'(\mathbf{r}, \lambda, \hat{\mathbf{s}}_i) \, d\Omega_i}, \tag{3.34}$$

or

$$\rho_\lambda'^{\ominus}(\mathbf{r}, \lambda, \hat{\mathbf{s}}_i) = \int_{2\pi} \rho_\lambda''(\mathbf{r}, \lambda, \hat{\mathbf{s}}_i, \hat{\mathbf{s}}_r) \cos \theta_r \, d\Omega_r, \tag{3.35}$$

where the $(H_\lambda' d\Omega_i)$ cancels out since it does not depend on outgoing direction $\hat{\mathbf{s}}_r$. Here we have temporarily added the superscript "$\ominus$" to distinguish the directional-hemispherical reflectivity ($\rho'^{\ominus}$) from the hemispherical-directional reflectivity ($\rho^{\ominus'}$, defined below). If the reflection function is independent of both $\hat{\mathbf{s}}_i$ and $\hat{\mathbf{s}}_r$, then the surface reflects equal amounts into all directions, regardless of incoming direction, and

$$\rho_\lambda'^{\ominus}(\mathbf{r}, \lambda) = \pi \rho_\lambda''(\mathbf{r}, \lambda). \tag{3.36}$$

Such a surface is called a *diffuse reflector*.

Comparing the definition of the *spectral, directional-hemispherical reflectivity* with that of the *spectral, directional absorptivity*, equation (3.14), we also find, for an opaque surface,

$$\rho_\lambda'^{\ominus}(\mathbf{r}, \lambda, \hat{\mathbf{s}}_i) = 1 - \alpha_\lambda'(\mathbf{r}, \lambda, \hat{\mathbf{s}}_i). \tag{3.37}$$

Sometimes it is of interest to determine the amount of energy reflected into a certain direction, coming from all possible incoming directions. Equation (3.31) gives the reflected intensity due to a single incoming direction. Integrating this expression over the entire hemisphere of incoming directions leads to

$$\begin{aligned} I_\lambda(\mathbf{r}, \lambda, \hat{\mathbf{s}}_r) &= \int_{2\pi} \rho_\lambda''(\mathbf{r}, \lambda, \hat{\mathbf{s}}_i, \hat{\mathbf{s}}_r) H_\lambda'(\mathbf{r}, \lambda, \hat{\mathbf{s}}_i) \, d\Omega_i \\ &= \int_{2\pi} \rho_\lambda''(\mathbf{r}, \lambda, \hat{\mathbf{s}}_i, \hat{\mathbf{s}}_r) I_\lambda(\mathbf{r}, \lambda, \hat{\mathbf{s}}_i) \cos \theta_i \, d\Omega_i. \end{aligned} \tag{3.38}$$

On the other hand, the spectral, hemispherical irradiation is

$$H_\lambda(\mathbf{r}, \lambda) = \int_{2\pi} I_\lambda(\mathbf{r}, \lambda, \hat{\mathbf{s}}_i) \cos \theta_i \, d\Omega_i. \tag{3.39}$$

If the surface were a perfect reflector, it would reflect all of H_λ, and it would reflect it equally into all outgoing directions. Thus, for the ideal case, the outgoing intensity would be, from equation (1.31), H_λ/π. Consequently, the *spectral, hemispherical-directional reflectivity* is defined as

$$\rho_\lambda^{\ominus'}(\mathbf{r}, \lambda, \hat{\mathbf{s}}_r) \equiv \frac{I_\lambda(\mathbf{r}, \lambda, \hat{\mathbf{s}}_r)}{H_\lambda(\mathbf{r}, \lambda)/\pi} = \frac{\int_{2\pi} \rho_\lambda''(\mathbf{r}, \lambda, \hat{\mathbf{s}}_i, \hat{\mathbf{s}}_r) I_\lambda(\mathbf{r}, \lambda, \hat{\mathbf{s}}_i) \cos \theta_i \, d\Omega_i}{\frac{1}{\pi} \int_{2\pi} I_\lambda(\mathbf{r}, \lambda, \hat{\mathbf{s}}_i) \cos \theta_i \, d\Omega_i}. \tag{3.40}$$

For the special case of *diffuse irradiation* (i.e., the incoming intensity does not depend on $\hat{s}_i$) equation (3.40) reduces to

$$\rho_\lambda^{\circ'}(\mathbf{r}, \lambda, \hat{s}_r) = \int_{2\pi} \rho_\lambda''(\mathbf{r}, \lambda, \hat{s}_i, \hat{s}_r) \cos \theta_i \, d\Omega_i, \tag{3.41}$$

which is identical to equation (3.35) if the reciprocity of the bidirectional reflection function, equation (3.33), is invoked. Thus, for diffuse irradiation,

$$\rho_\lambda^{\circ'}(\mathbf{r}, \lambda, \hat{s}_r) = \rho_\lambda^{'\circ}(\mathbf{r}, \lambda, \hat{s}_i), \quad \hat{s}_i = -\hat{s}_r, \tag{3.42a}$$

or

$$\rho_\lambda^{\circ'}(\mathbf{r}, \lambda, \theta_r, \psi_r) = \rho_\lambda^{'\circ}(\mathbf{r}, \lambda, \theta_i = \theta_r, \psi_i = \psi_r), \tag{3.42b}$$

that is, reciprocity exists between the spectral directional-hemispherical and hemispherical-directional reflectivities for any given irradiation/reflection direction. Use of this fact is often made in experimental measurements: While the directional-hemispherical reflectivity is of great practical importance, it is very difficult to measure; the hemispherical-directional reflectivity, on the other hand, is not very important but readily measured (see Section 3.10).

Finally, we define a *spectral, hemispherical reflectivity* as the fraction of the total irradiation from all directions reflected into all directions. From equation (3.34) we have the heat flux reflected into all directions for a single direction of incidence, $\hat{s}_i$, as

$$\rho_\lambda^{'\circ}(\mathbf{r}, \lambda, \hat{s}_i) H_\lambda'(\mathbf{r}, \lambda, \hat{s}_i) \, d\Omega_i.$$

Integrating this expression as well as H_λ' itself over all incidence angles gives

$$
\begin{aligned}
\rho_\lambda(\mathbf{r}, \lambda) &= \frac{\int_{2\pi} \rho_\lambda^{'\circ}(\mathbf{r}, \lambda, \hat{s}_i) H_\lambda'(\mathbf{r}, \lambda, \hat{s}_i) \, d\Omega_i}{\int_{2\pi} H_\lambda'(\mathbf{r}, \lambda, \hat{s}_i) \, d\Omega_i} \\
&= \frac{\int_{2\pi} \rho_\lambda^{'\circ}(\mathbf{r}, \lambda, \hat{s}_i) I_\lambda(\mathbf{r}, \lambda, \hat{s}_i) \cos \theta_i \, d\Omega_i}{\int_{2\pi} I_\lambda(\mathbf{r}, \lambda, \hat{s}_i) \cos \theta_i \, d\Omega_i}.
\end{aligned} \tag{3.43}
$$

If the incident intensity is independent of direction (diffuse irradiation), then equation (3.43) may be simplified again, and

$$\rho_\lambda(\mathbf{r}, \lambda) = \frac{1}{\pi} \int_{2\pi} \rho_\lambda^{'\circ}(\mathbf{r}, \lambda, \hat{s}_i) \cos \theta_i \, d\Omega_i. \tag{3.44}$$

Also, comparing the definitions of spectral, hemispherical absorptivity and reflectivity, we obtain, for an opaque surface,

$$\rho_\lambda(\mathbf{r}, \lambda) = 1 - \alpha_\lambda(\mathbf{r}, \lambda). \tag{3.45}$$

Finally, as for emissivity and absorptivity we need to introduce spectrally-integrated or "total" reflectivities. This is done by integrating numerator and denominator independently over the full spectrum for each of the spectral reflectivities, leading to the following relations:

Total, bidirectional reflection function

$$\rho''(\mathbf{r}, \hat{\mathbf{s}}_i, \hat{\mathbf{s}}_r) = \frac{\int_0^\infty \rho_\lambda''(\mathbf{r}, \lambda, \hat{\mathbf{s}}_i, \hat{\mathbf{s}}_r) I_\lambda(\mathbf{r}, \lambda, \hat{\mathbf{s}}_i) \, d\lambda}{\int_0^\infty I_\lambda(\mathbf{r}, \lambda, \hat{\mathbf{s}}_i) \, d\lambda} ; \tag{3.46}$$

Total, directional-hemispherical reflectivity

$$\rho'^\cap(\mathbf{r}, \hat{\mathbf{s}}_i) = \frac{\int_0^\infty \rho_\lambda'^\cap(\mathbf{r}, \lambda, \hat{\mathbf{s}}_i) I_\lambda(\mathbf{r}, \lambda, \hat{\mathbf{s}}_i) \, d\lambda}{\int_0^\infty I_\lambda(\mathbf{r}, \lambda, \hat{\mathbf{s}}_i) \, d\lambda} ; \tag{3.47}$$

Total, hemispherical-directional reflectivity

$$\rho^\cap{}'(\mathbf{r}, \hat{\mathbf{s}}_r) = \frac{\int_0^\infty \rho_\lambda^\cap{}'(\mathbf{r}, \lambda, \hat{\mathbf{s}}_r) \int_{2\pi} I_\lambda(\mathbf{r}, \lambda, \hat{\mathbf{s}}_i) \cos \theta_i \, d\Omega_i \, d\lambda}{\int_0^\infty \int_{2\pi} I_\lambda(\mathbf{r}, \lambda, \hat{\mathbf{s}}_i) \cos \theta_i \, d\Omega_i \, d\lambda} ; \tag{3.48}$$

Total, hemispherical reflectivity

$$\rho(\mathbf{r}, \hat{\mathbf{s}}_r) = \frac{\int_0^\infty \rho_\lambda(\mathbf{r}, \lambda) \int_{2\pi} I_\lambda(\mathbf{r}, \lambda, \hat{\mathbf{s}}_i) \cos \theta_i \, d\Omega_i \, d\lambda}{\int_0^\infty \int_{2\pi} I_\lambda(\mathbf{r}, \lambda, \hat{\mathbf{s}}_i) \cos \theta_i \, d\Omega_i \, d\lambda} . \tag{3.49}$$

Reciprocity relations equations (3.33) and (3.42) hold also for total reflectivities (subject to the same restrictions), as do the relations between reflectivity and absorptivity, equations (3.37) and (3.45).

3.3   PREDICTIONS FROM ELECTROMAGNETIC WAVE THEORY

In Chapter 2 we developed in some detail how the spectral, directional-hemispherical reflectivity of an optically smooth interface (specular reflector) can be predicted by the *electromagnetic wave* and *dispersion theories*. Before comparing such predictions with experimental data, we shall briefly summarize the results of Chapter 2.

Consider an electromagnetic wave traveling through air (refractive index = 1), hitting the surface of a *conducting medium* (complex index of refraction $m = n - ik$) at an angle of θ_1 with the surface normal (cf. Fig. 3-6). *Fresnel's equations* predict the reflectivities for parallel- and perpendicular-polarized light from equations (2.108) through (2.118),[3] as

$$\rho_\parallel = \frac{(p - \sin\theta_1 \tan\theta_1)^2 + q^2}{(p + \sin\theta_1 \tan\theta_1)^2 + q^2} \rho_\perp, \tag{3.50}$$

---

[3]For simplicity of notation we shall drop the superscripts $'^\cap$ for the directional-hemispherical reflectivity whenever there is no possibility of confusion.

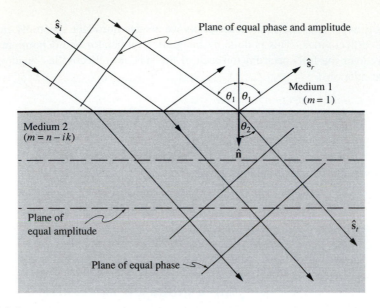

FIGURE 3-6
Transmission and reflection at an interface between air and an absorbing medium.

$$\rho_\perp = \frac{(\cos\theta_1 - p)^2 + q^2}{(\cos\theta_1 + p)^2 + q^2},\qquad(3.51)$$

where

$$p^2 = \frac{1}{2}\left[\sqrt{(n^2 - k^2 - \sin^2\theta_1)^2 + 4n^2k^2} + (n^2 - k^2 - \sin^2\theta_1)\right],\qquad(3.52)$$

$$q^2 = \frac{1}{2}\left[\sqrt{(n^2 - k^2 - \sin^2\theta_1)^2 + 4n^2k^2} - (n^2 - k^2 - \sin^2\theta_1)\right].\qquad(3.53)$$

Nonreflected light is refracted into the medium, traveling on at an angle of θ_2 with the surface normal, as predicted by the *generalized Snell's law*, from equation (2.110),

$$p\tan\theta_2 = \sin\theta_1.\qquad(3.54)$$

For normal incidence $\theta_1 = \theta_2 = 0$, and equations (3.50) through (3.53) simplify to $p = n$, $q = k$, and

$$\rho_{n\lambda} = \rho_\| = \rho_\perp = \frac{(n-1)^2 + k^2}{(n+1)^2 + k^2}.\qquad(3.55)$$

If the incident radiation is *unpolarized*, the reflectivity may be calculated as an average, i.e.,

$$\rho = \tfrac{1}{2}(\rho_\| + \rho_\perp).\qquad(3.56)$$

For a *dielectric medium* ($k = 0$), $p^2 = n^2 - \sin^2\theta_1$, and Snell's law becomes

$$n \sin\theta_2 = \sin\theta_1. \tag{3.57}$$

Therefore, $p = n \cos\theta_2$ and, with $q = 0$, Fresnel's equations reduce to

$$\rho_{\parallel} = \left(\frac{\cos\theta_2 - n\cos\theta_1}{\cos\theta_2 + n\cos\theta_1}\right)^2, \tag{3.58a}$$

$$\rho_{\perp} = \left(\frac{\cos\theta_1 - n\cos\theta_2}{\cos\theta_1 + n\cos\theta_2}\right)^2. \tag{3.58b}$$

Except for the section on semitransparent sheets, in this chapter we shall be dealing with opaque media. For such media $\rho + \alpha = 1$ and, from Kirchhoff's law,

$$\epsilon'_\lambda = \alpha'_\lambda = 1 - \rho'_\lambda. \tag{3.59}$$

To predict radiative properties from electromagnetic wave theory, the complex index of refraction, m, must be known, either from direct measurements or from dispersion theory predictions. In the dispersion theory the complex dielectric function, $\varepsilon = \varepsilon' - i\varepsilon''$, is predicted by assuming that the surface material consists of harmonic oscillators interacting with electromagnetic waves. The complex dielectric function is related to the complex index of refraction by $\varepsilon = m^2$, or

$$n^2 = \tfrac{1}{2}\left(\varepsilon' + \sqrt{\varepsilon'^2 + \varepsilon''^2}\right), \tag{3.60a}$$

$$k^2 = \tfrac{1}{2}\left(-\varepsilon' + \sqrt{\varepsilon'^2 + \varepsilon''^2}\right), \tag{3.60b}$$

where

$$\varepsilon' = \frac{\epsilon}{\epsilon_0}, \quad \varepsilon'' = \frac{\sigma_e}{2\pi\nu\epsilon_0};$$

ϵ is the electrical permittivity, ϵ_0 its value in vacuum, and σ_e is the medium's electrical conductivity. Both ϵ and σ_e are functions of the frequency of the electromagnetic wave ν. For an isolated oscillator (nonoverlapping band) ε is predicted by the Lorentz model, equation (2.142), as

$$\varepsilon' = \varepsilon_0 + \frac{\nu_{pi}^2(\nu_i^2 - \nu^2)}{(\nu_i^2 - \nu^2)^2 + \gamma_i^2\nu^2}, \tag{3.61a}$$

$$\varepsilon'' = \frac{\nu_{pi}^2\gamma_i\nu}{(\nu_i^2 - \nu^2)^2 + \gamma_i^2\nu^2}, \tag{3.61b}$$

where ε_0 is the contribution to ε' from bands at shorter wavelengths, ν_i is the resonance frequency, ν_{pi} is called the plasma frequency and γ_i is an oscillation damping factor. If these three constants can be determined or measured, then n and k can be predicted for all frequencies (or wavelengths) from equation (3.60), and the radiative properties can be calculated for all frequencies (or wavelengths) and all directions from equations (3.50) through (3.53).

3.4 RADIATIVE PROPERTIES OF METALS

In this section we shall briefly discuss how the radiative properties of metals (i.e., electrical conductors) can be predicted from electromagnetic wave theory and dispersion theory, and how these predictions compare with experimental data. The variation of the spectral, normal reflectivity with wavelength and total, normal properties will be examined, followed by a discussion of the directional dependence of radiative properties and the evaluation of hemispherical reflectivities (and emissivities). Finally, we will look at the temperature dependence of spectral as well as total properties.

Wavelength Dependence of Spectral, Normal Properties

Metals are in general excellent electrical conductors because of an abundance of free electrons. Drude [13] developed an early theory to predict the dielectric function for free electrons that is essentially a special case of the Lorentz model: Since free electrons do not oscillate but propagate freely, they may be modeled as a "spring" with a vanishing spring constant leading to a resonance frequency of $\nu_i = 0$. Thus the *Drude theory* for the dielectric function for free electrons follows from equation (3.61) as

$$\varepsilon'(\nu) = \varepsilon_0 - \frac{\nu_p^2}{\nu^2 + \gamma^2}, \tag{3.62a}$$

$$\varepsilon''(\nu) = \frac{\nu_p^2 \gamma}{\nu(\nu^2 + \gamma^2)}. \tag{3.62b}$$

Figure 3-7 shows the spectral, normal reflectivity of three metals—aluminum, copper, and silver. The theoretical lines are from Ehrenreich and coworkers [14] (aluminum) and Ehrenreich and Phillip [15] (copper and silver), who semiempirically determined the values of the unknowns ε_0, ν_p and γ in equation (3.62). The experimental data are taken from Shiles and coworkers [16] (aluminum) and Hagemann and coworkers [17] (copper and silver). The agreement between experiment and theory in the infrared is very good. For wavelengths $\lambda > 1\,\mu$m the Drude theory has been shown to represent the reflectivity of many metals accurately, if samples are prepared with great care. Discrepancies are due to surface preparation methods and the limits of experimental accuracy. Aluminum has a dip in reflectivity centered at $\sim 0.8\,\mu$m; this is due to bound electron transitions that are not considered by the Drude model. Since $\gamma \ll \nu_p$ always, there exists for each metal a frequency in the vicinity of the plasma frequency, $\nu \simeq \nu_p$, where $\varepsilon' = 1$ and $\varepsilon'' \ll 1$ or $n \simeq 1$, $k \ll 1$: This fact implies that many metals neither reflect nor absorb radiation in the ultraviolet near ν_p, but are highly transparent!

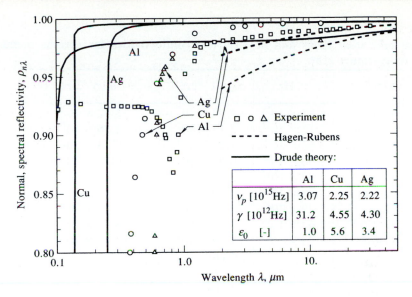

FIGURE 3-7
Spectral, normal reflectivity at room temperature for aluminum, copper, and silver.

For extremely long wavelengths (very small frequency ν), we find from equations (3.62) and (2.137) that

$$\varepsilon'' = \frac{\nu_p^2}{\nu\gamma} = \frac{\sigma_e}{2\pi\nu\epsilon_0}, \qquad \nu \ll \gamma, \tag{3.63}$$

where σ_e is the (in general, frequency-dependent) electrical conductivity, and

$$\sigma_e = 2\pi\epsilon_0\nu_p^2/\gamma = \text{const} = \sigma_{\text{dc}}. \tag{3.64}$$

Note that at the long-wavelength limit the electrical conductivity becomes independent of wavelength and is known as the *dc-conductivity*. Since the *dc-conductivity* is easily measured it is advantageous to recast equation (3.62) as

$$\varepsilon(\nu) = \varepsilon_0 - \frac{\sigma_{\text{dc}}\gamma/2\pi\epsilon_0}{\nu(\nu + i\gamma)}, \tag{3.65a}$$

$$\varepsilon' = \varepsilon_0 - \frac{\sigma_{\text{dc}}\gamma/2\pi\epsilon_0}{\nu^2 + \gamma^2}, \tag{3.65b}$$

$$\varepsilon'' = \frac{\sigma_{\text{dc}}\gamma^2/2\pi\epsilon_0}{\nu(\nu^2 + \gamma^2)}. \tag{3.65c}$$

Room temperature values for electrical resistivity, $1/\sigma_{\text{dc}}$, and for electron relaxation time, $1/2\pi\gamma$, have been given by Parker and Abbott [18] for a number of metals. They have been converted and are reproduced in Table 3.1. Note that these values

TABLE 3.1

Inverse relaxation times and dc electrical conductivities for various metals at room temperature [18].

Metal	γ, Hz	σ_{dc}, $\Omega^{-1}cm^{-1}$	$\nu_p^2 = \sigma_{dc}\gamma/2\pi\epsilon_0$, Hz^2
Lithium	1.85×10^{13}	1.09×10^5	3.62×10^{30}
Sodium	5.13×10^{12}	2.13×10^5	1.96×10^{30}
Potassium	3.62×10^{12}	1.52×10^5	9.88×10^{29}
Cesium	7.56×10^{12}	0.50×10^5	6.78×10^{29}
Copper	5.89×10^{12}	5.81×10^5	6.14×10^{30}
Silver	3.88×10^{12}	6.29×10^5	4.38×10^{30}
Gold	5.49×10^{12}	4.10×10^5	4.04×10^{30}
Nickel	1.62×10^{13}	1.28×10^5	3.72×10^{30}
Cobalt	1.73×10^{13}	1.02×10^5	3.17×10^{30}
Iron	6.63×10^{12}	1.00×10^5	1.19×10^{30}
Palladium	1.73×10^{13}	0.91×10^5	2.83×10^{30}
Platinum	1.77×10^{13}	1.00×10^5	3.18×10^{30}

differ appreciably from those given in Fig. 3-7. No values for ε_0 are given; however, the influence of ε_0 is generally negligible in the infrared. Extensive sets of spectral data for a large number of metals have been collected by Ordal and coworkers [19] (for a smaller number of metals they also give the Drude parameters, which are also conflicting somewhat with the data of Table 3.1).

For long wavelengths equation (3.60) may be simplified considerably, since for such case, $\varepsilon'' \gg |\varepsilon'|$, and it follows that

$$n^2 \approx k^2 \approx \varepsilon''/2 = \frac{\sigma_{dc}}{4\pi\nu\epsilon_0} = \frac{\sigma_{dc}\lambda_0}{4\pi c_0\epsilon_0} \gg 1, \tag{3.66}$$

where λ_0 is the wavelength in vacuum. Substituting values for the universal constants c_0 and ϵ_0, equation (3.66) becomes

$$n \approx k \approx \sqrt{30\lambda_0\sigma_{dc}}, \quad \lambda_0 \text{ in cm}, \quad \sigma_{dc} \text{ in } \Omega^{-1}cm^{-1}, \tag{3.67}$$

which is known as the *Hagen-Rubens relation* [20]. For comparison, results from equation (3.67) are also included in Fig. 3-7. It is commonly assumed that the Hagen-Rubens relation may be used for $\lambda_0 > 6\,\mu m$, although this assumption can lead to serious errors, in particular as far as evaluation of the index of refraction is concerned. While equation (3.67) is valid for the metal being adjacent to an arbitrary material, we will—for notational simplicity—assume for the rest of this discussion that the adjacent material has a refractive index of unity (vacuum or gas), that is, $\lambda_0 = \lambda$.

Substituting equation (3.67) into equation (3.55) leads to

$$\rho_{n\lambda} = \frac{2n^2 - 2n + 1}{2n^2 + 2n + 1}, \tag{3.68}$$

$$\epsilon_{n\lambda} = 1 - \rho_{n\lambda} = \frac{4n}{2n^2 + 2n + 1}. \tag{3.69}$$

Since $n \gg 1$ equation (3.69) may be further simplified to

$$\epsilon_{n\lambda} = \frac{2}{n} - \frac{2}{n^2} + \cdots, \tag{3.70a}$$

and, with equation (3.67), to

$$\epsilon_{n\lambda} \simeq \frac{2}{\sqrt{30\lambda\,\sigma_{dc}}} - \frac{1}{15\lambda\,\sigma_{dc}}, \quad \lambda \text{ in cm}, \quad \sigma_{dc} \text{ in } \Omega^{-1}\,\text{cm}^{-1}. \tag{3.70b}$$

This $1/\sqrt{\lambda}$ dependence is not predicted by the Drude theory (except for the far infrared), nor is it observed with optically smooth surfaces. However, it often approximates the behavior of polished (i.e., not entirely smooth) surfaces.

Example 3.3.   Using the constants given in Fig. 3-7 calculate the complex index of refraction and the normal, spectral reflectivity of silver at $\lambda = 6.2\,\mu$m, using (a) the Drude theory, and (b) the Hagen-Rubens relation.

Solution

(a) From Fig. 3-7 we have for silver $\varepsilon_0 = 3.4$, $\nu_p = 2.22 \times 10^{15}$ Hz and $\gamma = 4.30 \times 10^{12}$ Hz. Substituting these into equation (3.62) with $\nu = c_0/\lambda = 2.998 \times 10^8$ m/s $\times (10^6\,\mu\text{m/m})/6.2\,\mu\text{m} = 4.84 \times 10^{13}$ Hz, we obtain

$$\varepsilon' = 3.4 - \frac{(2.22 \times 10^{15})^2}{(4.84 \times 10^{13})^2 + (4.30 \times 10^{12})^2} = 3.4 - 2087 = -2084$$

$$\varepsilon'' = 2087 \times 4.30 \times 10^{12}/4.84 \times 10^{13} = 185.1.$$

The complex index of refraction follows from equation (3.60) as

$$n^2 = \frac{1}{2}\left(-2084 + \sqrt{2084^2 + 185.1^2}\right) = 4.102,$$

$$k^2 = \frac{1}{2}\left(2084 + \sqrt{2084^2 + 185.1^2}\right) = 2088,$$

or $n = 2.03$ and $k = 45.7$. Finally, the normal reflectivity follows from equation (3.55) as

$$\rho_{n\lambda} = \frac{(1 - 2.03)^2 + 45.7^2}{(1 + 2.03)^2 + 45.7^2} = 0.996.$$

(b) Using the Hagen-Rubens relation we find, from equation (3.64), that

$$\sigma_{dc} = 2\pi \times 8.8542 \times 10^{-12} \frac{C^2}{N\,m^2} \times \left(2.22 \times 10^{15}\,Hz\right)^2 / 4.30 \times 10^{12}\,Hz$$

$$= 6.376 \times 10^7 \frac{C^2}{N\,m^2\,s} = 6.376 \times 10^7 \Omega^{-1} m^{-1} = 6.376 \times 10^5 \Omega^{-1} cm^{-1}.$$

Substituting this value into equation (3.67) yields

$$n = k = \sqrt{30 \times 6.2 \times 10^{-4} \times 6.376 \times 10^5} = 108.9,$$

and

$$\rho_{n\lambda} = 1 - \epsilon_{n\lambda} = 1 - \frac{2}{n} + \frac{2}{n^2} = 1 - \frac{2}{108.9} + \frac{2}{108.9^2} = 0.982.$$

The two sets of results may be compared with experimental results of $n = 2.84$, $k = 45.7$ and $\rho_{n\lambda} = 0.995$ [17]. At first glance the Hagen-Rubens prediction for $\rho_{n\lambda}$ appears very good because, for any $k \gg 1$, $\rho_{n\lambda} \approx 1$. The values for n and k show that the Hagen-Rubens relation is in serious error even at a relatively long wavelength of $\lambda = 6.2\,\mu$m.

Total Properties for Normal Incidence

The total, normal reflectivity and emissivity may be evaluated from equation (3.8), with spectral, normal properties evaluated from the Drude theory or from the simple Hagen-Rubens relation. While the Hagen-Rubens relation is not very accurate, it does predict the emissivity *trends* correctly in the infrared, and it does allow an explicit evaluation of total, normal emissivity. Substituting equation (3.70) into equation (3.8) leads to an integral that may be evaluated in a similar fashion as for the total emissive power, equation (1.19), and, retaining the first three terms of the series expansion

$$\epsilon_n = 0.578\,(T/\sigma_{dc})^{1/2} - 0.178\,(T/\sigma_{dc}) + 0.0584\,(T/\sigma_{dc})^{3/2},$$

$$T \text{ in } K, \quad \sigma_{dc} \text{ in } \Omega^{-1}cm^{-1}. \tag{3.71}$$

Of course, equation (3.71) is only valid for small values of (T/σ_{dc}), i.e., the temperature of the surface must be such that only a small fraction of the blackbody emissive power comes from short wavelengths (where the Hagen-Rubens relation is not applicable). For pure metals, to a good approximation, the dc-conductivity is inversely proportional to absolute temperature, or

$$\sigma_{dc} = \sigma_{ref} \frac{T_{ref}}{T}. \tag{3.72}$$

Therefore, for low enough temperatures, the total, normal emissivity of a pure metal should be approximately linearly proportional to temperature. Comparison with experiment (Fig. 3-8) shows that this nearly linear relationship holds for many metals up to surprisingly high temperatures; for example, for platinum $(T/\sigma_{dc})^{1/2} = 0.5$ corresponds to a temperature of 2700 K. It is interesting to note that spectral integration

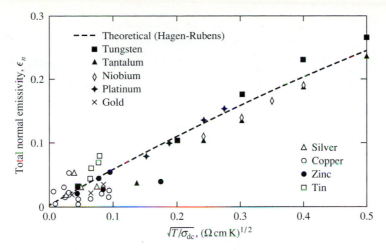

FIGURE 3-8
Total, normal emissivity of various polished metals as function of temperature [18].

of the Drude model results in 30% to 70% lower total emissivities for all metals and, thus, fails to follow experimental trends. Such integration was carried out by Parker and Abbott [18] in an approximate fashion. They attributed the discrepancy to imperfections in the molecular lattice induced by surface preparation and to the *anomalous skin effect* [21], both of which lower the electrical conductivity in the surface layer.

Directional Dependence of Radiative Properties

The spectral, directional reflectivity at the interface between an absorber and a nonabsorber is given by Fresnel's equations, (3.50) through (3.53). Since, in the infrared, n and k are generally fairly large for metals one may with little error neglect the $\sin^2 \theta_1$ in equations (3.52) and (3.53), leading to $p \simeq n$ and $q \simeq k$. Then, from equations (3.50) and (3.51) the reflectivities for parallel- and perpendicular-polarized light are evaluated from

$$\rho_\parallel = \frac{(n \cos\theta - 1)^2 + (k \cos\theta)^2}{(n \cos\theta + 1)^2 + (k \cos\theta)^2}, \tag{3.73a}$$

$$\rho_\perp = \frac{(n - \cos\theta)^2 + k^2}{(n + \cos\theta)^2 + k^2}. \tag{3.73b}$$

[The simple form for $\rho_\parallel$ is best obtained from the reflection coefficient given by equation (2.114) by neglecting $\sin^2 \theta_1$ and cancelling $m = n - ik$ from both numerator and denominator.] The directional, spectral emissivity (unpolarized) follows as

$$\epsilon'_\lambda = 1 - \tfrac{1}{2}(\rho_\parallel + \rho_\perp), \tag{3.74}$$

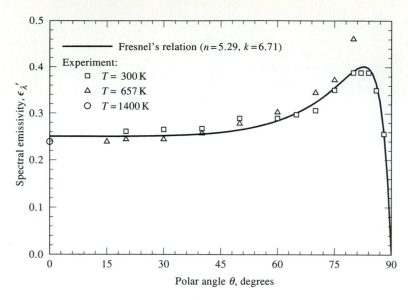

FIGURE 3-9
Spectral, directional emissivity of platinum at $\lambda = 2\,\mu$m.

and is shown in Fig. 3-9 for platinum at $\lambda = 2\,\mu$m. The theoretical line for room temperature has been calculated with $n = 5.29$, $k = 6.71$ from [22]. Comparison with experimental results of Brandenberg [23], Brandenberg and Clausen [24], and Price [25] demonstrate the validity of Fresnel's equations.[4]

Equation (3.73) may be integrated analytically over all directions to obtain the spectral, hemispherical emissivity from equation (3.5). This was done by Dunkle [26] for the two different polarizations, resulting in

$$\epsilon_\| = \frac{8n}{n^2+k^2}\left(1 - \frac{n}{n^2+k^2}\ln[(n+1)^2+k^2] + \frac{(n^2-k^2)}{k(n^2+k^2)}\tan^{-1}\frac{k}{n+1}\right), \quad (3.75a)$$

$$\epsilon_\perp = 8n\left(1 - n\ln\frac{(n+1)^2+k^2}{n^2+k^2} + \frac{(n^2-k^2)}{k}\tan^{-1}\frac{k}{n(n+1)+k^2}\right), \quad (3.75b)$$

$$\epsilon_\lambda = \frac{1}{2}\left(\epsilon_\| + \epsilon_\perp\right). \quad (3.75c)$$

Figure 3-10, from Dunkle [27], is a plot of the ratio of the hemispherical and normal emissivities, $\epsilon_\lambda/\epsilon_{n\lambda}$. For the case of $k/n = 1$ the dashed line represents results from equation (3.75), while the solid lines were obtained by numerically integrating equations (3.50) through (3.53). For $k/n > 1$ the two lines become indistinguishable. Hering and Smith [28] reported that equation (3.75) is accurate to within 1–2% for

---

[4]In the original figure of Brandenberg and Clausen [24] older values for n and k were used that gave considerably worse agreement with experiment.

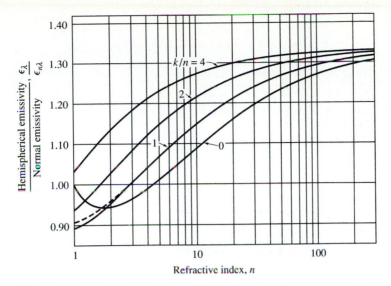

FIGURE 3-10
Ratio of hemispherical and normal spectral emissivity for electrical conductors as a function of n and k [27].

values of $n^2 + k^2$ larger than 40 and 3.25, respectively. In view of the large values that n and, in particular, k assume for metals, equation (3.75) is virtually always accurate to better than 2% for metals in the visible and infrared wavelengths.

Example 3.4.   Determine the spectral, hemispherical emissivity for room-temperature nickel at a wavelength of $\lambda = 10\ \mu$m, using (a) the Drude theory, (b) the Hagen-Rubens relation.

Solution

We first need to determine the optical constants n and k from either theory, then calculate the hemispherical emissivity from equation (3.75) or read it from Fig. 3-10.

(a) Using values for nickel from Table 3.1 in equation (3.62), we find with $\nu = c_0/\lambda = 2.998 \times 10^8\ \text{m/s}/10^{-5}\ \text{m} = 2.998 \times 10^{13}$ Hz,

$$\varepsilon' = 1.0 - 3.72 \times 10^{30} / \left[(2.998 \times 10^{13})^2 + (1.62 \times 10^{13})^2\right]$$
$$= 1 - 3204 = -3203,$$
$$\varepsilon'' = 3204 \times 1.62 \times 10^{13} / 2.998 \times 10^{13} = 1731,$$
$$n^2 = 0.5 \times \left(-3208 + \sqrt{3208^2 + 1731^2}\right) = 219,$$
$$k^2 = 0.5 \times \left(3208 + \sqrt{3208^2 + 1731^2}\right) = 3422,$$

and

$$n = 14.8, \quad k = 58.5, \quad k/n = 58.5/14.8 = 3.95.$$

To use Fig. 3-10 we first determine $\rho_{n\lambda}$ as

$$\rho_{n\lambda} = \frac{13.8^2 + 58.5^2}{15.8^2 + 58.5^2} = 0.984,$$

and

$$\epsilon_{n\lambda} = 1 - \rho_{n\lambda} = 0.016.$$

From Fig. 3-10 $\epsilon_\lambda/\epsilon_{n\lambda} \simeq 1.29$ and, therefore, $\epsilon_\lambda \simeq 0.021$.

(b) Using the Hagen-Rubens relation we find, from equation (3.70),

$$\epsilon_{n\lambda} = \frac{2}{\sqrt{30 \times 10^{-3} \times 1.28 \times 10^5}} - \frac{1}{15 \times 10^{-3} \times 1.28 \times 10^5} = 0.032.$$

Further, with $n \simeq k \simeq \sqrt{30 \times 10^{-3} \times 1.28 \times 10^5} = 62.0$, we obtain from Fig. 3-10 $\epsilon_\lambda/\epsilon_{n\lambda} \simeq 1.275$ and $\epsilon_\lambda \simeq 0.041$.

The answers from both models differ by a factor of ~ 2. This agrees with the trends shown in Fig. 3-7.

Total, directional emissivities are obtained by (numerical) integration of equations (3.73) and (3.74) over the entire spectrum. The directional behavior of total emissivities is similar to that of spectral emissivities, as shown by the early measurements of Schmidt and Eckert [9], as depicted in Fig. 3-1b in a polar diagram (as opposed to the Cartesian representation of Fig. 3-9). The emissivities were determined from total radiation measurements from samples heated to a few hundred degrees Celsius.

Total, Hemispherical Emissivity

Equation (3.75) may be integrated over the spectrum using equation (3.10), to obtain the total, hemispherical emissivity of a metal. Several approximate relations, using the Hagen-Rubens limit, have been proposed, notably the ones by Davisson and Weeks [29] and by Schmidt and Eckert [9]. Expanding equation (3.75) into a series of powers of $1/n$ (with $n = k \gg 1$), Parker and Abbott [18] were able to integrate equation (3.75) analytically, leading to

$$\epsilon(T) = 0.766(T/\sigma_{dc})^{1/2} - [0.309 - 0.0889 \ln(T/\sigma_{dc})](T/\sigma_{dc})$$
$$-0.0175(T/\sigma_{dc})^{3/2}, \quad T \text{ in K}, \quad \sigma_{dc} \text{ in } \Omega^{-1}cm^{-1}. \quad (3.76)$$

As for the total, normal emissivity the total, hemispherical emissivity is seen to be approximately linearly proportional to temperature (since $\sigma_{dc} \propto 1/T$) as long as the

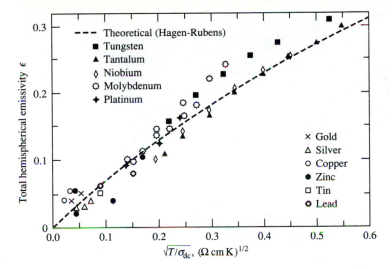

FIGURE 3-11
Total, hemispherical emissivity of various polished metals as function of temperature [18].

surface temperature is relatively low (so that only long wavelengths are of importance, for which the Hagen-Rubens relation gives reasonable results). Emissivities calculated from equation (3.76) are compared with experimental data in Fig. 3-11. Parker and Abbott also integrated the series expansion of equation (3.75) with n and k evaluated from the Drude theory. As for normal emissivities, the Drude model predicts values 30%–70% lower than the Hagen-Ruben relations, contrary to experimental evidence shown in Fig. 3-11. Again, the discrepancy was attributed to lattice imperfections and to the anomalous skin effect.

Effects of Surface Temperature

The Hagen-Rubens relation, equation (3.70), predicts that the spectral, normal emissivity of a metal should be proportional to $1/\sqrt{\sigma_{dc}}$. Since the electrical conductivity is approximately inversely proportional to temperature, the spectral emissivity should, therefore, be proportional to the square root of absolute temperature for long enough wavelengths. This trend should also hold for the spectral, hemispherical emissivity. Experiments have shown that this is indeed true for many metals. A typical example is given in Fig. 3-12, showing the spectral dependence of the hemispherical emissivity for tungsten for a number of temperatures [22]. Note that the emissivity for tungsten tends to increase with temperature beyond a *crossover wavelength* of approximately 1.3 μm, while the temperature dependence is reversed for shorter wavelengths. Similar trends of a single crossover wavelength have been observed for many metals.

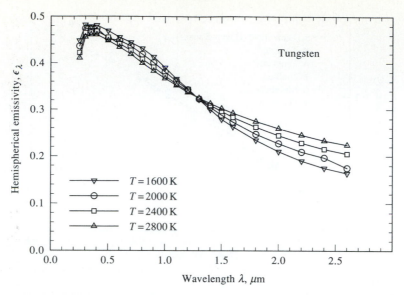

FIGURE 3-12
Temperature dependence of the spectral, hemispherical emissivity of tungsten [22].

The total, normal or hemispherical emissivities are calculated by integrating spectral values over all wavelengths, with the blackbody emissive power as weight function. Since the peak of the blackbody emissive power shifts toward shorter wavelengths with increasing temperature, we infer that hotter surfaces emit a higher fraction of energy at shorter wavelengths, where the spectral emissivity is higher, resulting in an increase in total emissivity as demonstrated in Figs. 3-8 and 3-11. Since the crossover wavelength is fairly short for many metals, the Hagen-Rubens temperature relation often holds for surprisingly high temperatures.

3.5 RADIATIVE PROPERTIES OF NONCONDUCTORS

Electrical nonconductors have few free electrons and, thus, do not display the high reflectivity/opaqueness behavior across the infrared as do metals. The radiative properties of pure nonconductors are dominated in the infrared by photon-phonon interaction, i.e., by the photon excitation of the vibrational energy levels of the solid's crystal lattice. Outside the spectral region of strong absorption by vibrational transitions there is generally a region of fairly high transparency (and low reflectivity), where absorption is dominated by impurities and imperfections in the crystal lattice. As such, these spectral regions often show irregular and erratic behavior.

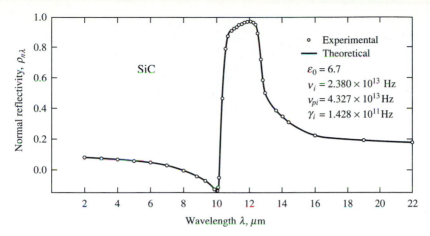

FIGURE 3-13
Spectral, normal reflectivity of α-SiC at room temperature [30].

Wavelength Dependence of Spectral, Normal Properties

The spectral behavior of pure, crystalline nonconductors is often well described by the single oscillator Lorentz model of equation (3.61). One such material is α-SiC (silicon carbide), a high-temperature ceramic of ever increasing importance. The spectral, normal reflectivity of pure, smooth α-SiC at room temperature is shown in Fig. 3-13, as given by Spitzer and coworkers [30]. The theoretical reflectivity in Fig. 3-13 is evaluated from equations (3.61), (3.60) and (3.55) with $\varepsilon_0 = 6.7$, $\nu_{pi} = 4.327 \times 10^{13}$ Hz, $\nu_i = 2.380 \times 10^{13}$ Hz, $\gamma_i = 1.428 \times 10^{11}$ Hz. Agreement between theory and experiment is superb for the entire range between $2\,\mu$m and $22\,\mu$m. Inspection of equations (3.61) and (3.60) shows that outside the spectral range $10\,\mu$m $< \lambda < 13\,\mu$m (or 2.5×10^{13} Hz $> \nu > 1.9 \times 10^{13}$ Hz), α-SiC is essentially transparent (absorptive index $k \ll 1$) and weakly reflecting. Within the range of $10\,\mu$m $< \lambda < 13\,\mu$m α-SiC is not only highly reflecting but also opaque (i.e., any radiation not reflected is absorbed within a very thin surface layer, since $k > 1$). The reflectivity drops off sharply on both sides of the absorption band. For this reason materials such as α-SiC are sometimes used as *band-pass filters*: If electromagnetic radiation is reflected several times by an α-SiC mirror, the emerging light will nearly exclusively lie in the spectral band $10\,\mu$m $< \lambda < 13\,\mu$m. This effect has led to the term *Reststrahlen band* (German for "remaining rays") for absorption bands due to crystal vibrational transitions.

   Not all crystals are well described by the single oscillator model since two or more different vibrational transitions may be possible and can result in overlapping bands. Magnesium oxide (MgO) is an example of material that can be described by a two-oscillator model (two overlapping bands), as Jasperse and coworkers [31] have

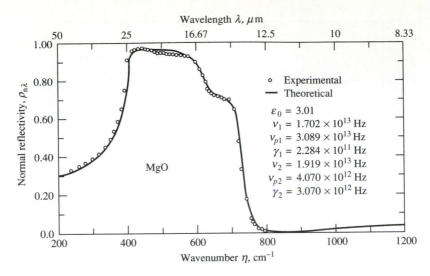

FIGURE 3-14
Spectral, normal reflectivity of MgO at room temperature [31].

shown (Fig. 3-14). The theoretical reflectivities are obtained with the parameters for the evaluation of equation (3.61) given in the figure. Note that for the calculation of ε' and ε'', equation (3.61) needs to be summed over both bands, $i = 1$ and 2. From a quantum viewpoint, the second, weaker oscillator is interpreted as the excitation of two phonons by a single photon [32].

Since the radiative properties outside a Reststrahlen band depend strongly on defects and impurities they may vary appreciably from specimen to specimen and even between different points on the same sample. For example, the spectral, normal reflectivity of silicon at room temperature is shown in Fig. 3-15 (redrawn from data collected by Touloukian and deWitt [7]). Strong influence of different types and levels of impurities is clearly evident. Therefore, looking up properties for a given material in published tables is problematical unless a detailed description of surface and material preparation is given.

Equation (3.61) demonstrates that—outside a Reststrahlen band—ε'' and, therefore, the absorptive index k of a nonconductor are very small; typically $k < 10^{-6}$ for a pure substance. While impurities and lattice defects can increase the value of k, it is very unlikely to find values of $k > 10^{-2}$ for a nonconductor outside Reststrahlen bands. At first glance it might appear, therefore, that all nonconductors must be highly transparent in the near infrared (and the visible). That this is not the case is readily seen from equation (1.49), which relates transmissivity to absorption coefficient. This, in turn, is related to the absorptive index through equation (2.42):

$$\tau = e^{-\kappa s} = e^{-4\pi k s/\lambda_0}. \tag{3.77}$$

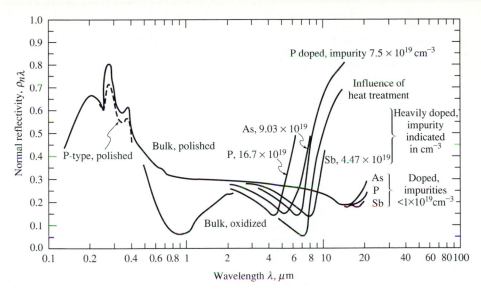

FIGURE 3-15
Spectral, normal reflectivity of silicon at room temperature [7].

For a 1 mm thick layer of a material with $k = 10^{-3}$ at a wavelength (in vacuum) of $\lambda_0 = 2\,\mu\text{m}$, equation (3.77) translates into a transmissivity of $\tau = \exp(-4\pi \times 10^{-3} \times 1/2 \times 10^{-3}) = 0.002$, i.e., the layer is essentially opaque. Still, the low values of k allow us to simplify Fresnel's equations considerably for the reflectivity of an interface. With $k^2 \ll (n-1)^2$ the nonconductor essentially behaves like a perfect dielectric and, from equation (3.55), the spectral, normal reflectivity may be evaluated as

$$\rho_{n\lambda} = \left(\frac{n-1}{n+1}\right)^2, \qquad k^2 \ll n^2. \tag{3.78}$$

Therefore, for optically smooth nonconductors the radiative properties may be calculated from refractive index data. Refractive indices for a number of semitransparent materials at room temperature are displayed in Fig. 3-16 as a function of wavelength [33]. All these crystalline materials show a similar spectral behavior: The refractive index drops rapidly in the visible region, then is nearly constant (declining very gradually) until the midinfrared, where n again starts to drop rapidly. This behavior is explained by the fact that crystalline solids tend to have an absorption band, due to electronic transitions, near the visible, and a Reststrahlen band in the infrared: The first drop in n is due to the tail end of the electronic band, as illustrated in Fig. 2-15b;[5] the second drop in the midinfrared is due to the beginning of a Reststrahlen band.

---

[5]Note that the abscissa in Fig. 2-15b is frequency ν, i.e., wavelength increases to the left.

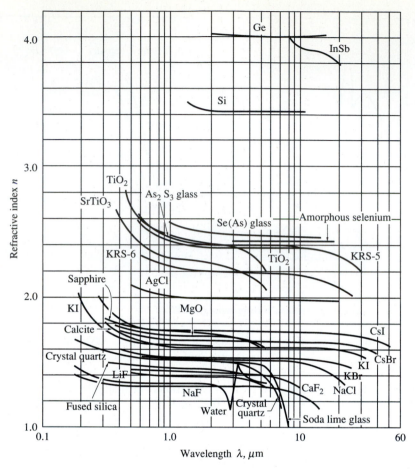

FIGURE 3-16
Refractive indices for various semitransparent materials [33].

Directional Dependence of Radiative Properties

For optically smooth nonconductors experiment has been found to follow Fresnel's equations of electromagnetic wave theory closely. Figure 3-17 shows a comparison between theory and experiment for the directional reflectivity of glass (blackened on one side to avoid multiple reflections) for polarized, monochromatic irradiation [23]. Because $k^2 \ll n^2$, the absorptive index may be eliminated from equations (3.50) and (3.51), and the relations for a perfect dielectric become valid. Thus, for unpolarized

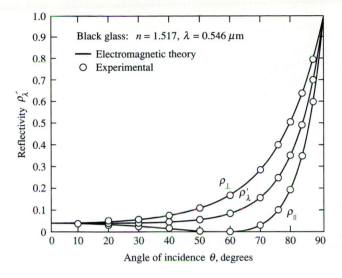

FIGURE 3-17
Spectral, directional reflectivity of glass at room temperature, for polarized light [23].

light incident from vacuum (or a gas), from equations (3.57) and (3.58)

$$\epsilon_\lambda' = 1 - \frac{1}{2}(\rho_\| + \rho_\perp) = 1 - \frac{1}{2}\left[\left(\frac{n^2\cos\theta - \sqrt{n^2 - \sin^2\theta}}{n^2\cos\theta + \sqrt{n^2 - \sin^2\theta}}\right)^2\right.$$

$$\left. + \left(\frac{\cos\theta - \sqrt{n^2 - \sin^2\theta}}{\cos\theta + \sqrt{n^2 - \sin^2\theta}}\right)^2\right]. \qquad (3.79)$$

The directional variation of the spectral emissivity of dielectrics is shown in Fig. 3-18. Comparison with Fig. 3-1 demonstrates that experiment agrees well with electromagnetic wave theory for a large number of nonconductors, even for total (rather than spectral) directional emissivities.

The spectral, hemispherical emissivity of a nonconductor may be obtained by integrating equation (3.79) with equation (3.5). While tedious, such an integration is possible, as shown by Dunkle [27]:

$$\epsilon_\| = \frac{4(2n + 1)}{3(n + 1)^2}, \qquad (3.80a)$$

$$\epsilon_\perp = \frac{4n^3(n^2 + 2n - 1)}{(n^2 + 1)(n^4 - 1)} + \frac{2n^2(n^2 - 1)^2}{(n^2 + 1)^3}\ln\left(\frac{n+1}{n-1}\right) - \frac{16n^4(n^4 + 1)\ln n}{(n^2 + 1)(n^4 - 1)^2}, \qquad (3.80b)$$

$$\epsilon_\lambda = \frac{1}{2}(\epsilon_\| + \epsilon_\perp). \qquad (3.80c)$$

The variation of normal and hemispherical emissivities with refractive index is shown in Fig. 3-19. While for metals the hemispherical emissivity is generally larger than the normal emissivity (cf., Fig. 3-10), the opposite is true for nonconductors. The reason

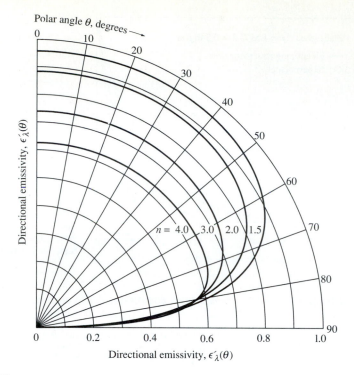

FIGURE 3-18
Directional emissivities of nonconductors as predicted by electromagnetic wave theory.

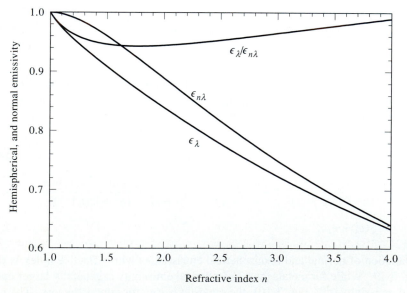

FIGURE 3-19
Normal and hemispherical emissivities for nonconductors as function of refractive index.

for this behavior is obvious from Fig. 3-1: Metals have a relatively low emissivity over most directions, but display a sharp increase for grazing angles before dropping back to zero. Nonconductors, on the other hand, have a (relatively high) emissivity for most directions, which gradually drops to zero at grazing angles (without a peak).

Example 3.5.   The directional reflectivity of silicon carbide at $\lambda = 2\,\mu$m and an incidence angle of $\theta = 10°$ has been measured as $\rho'_\lambda = 0.20$ (cf. Fig. 3-13). What is the hemispherical emissivity of SiC at $2\,\mu$m?

Solution

Since at $\theta = 10°$ the directional reflectivity does not deviate substantially from the normal reflectivity (cf. Fig. 3-18), we have $\epsilon_{n\lambda} = 1 - \rho_{n\lambda} \simeq 1 - 0.20 = 0.80$. Then, from Fig. 3-19, $n \simeq 2.6$ and $\epsilon_\lambda \simeq 0.76$.

Effects of Surface Temperature

The temperature dependence of the radiative properties of nonconductors is considerably more difficult to quantify than for metals. Infrared absorption bands in ionic solids due to excitation of lattice vibrations (*Reststrahlen bands*) generally increase in width and decrease in strength with temperature, and the wavelength of peak reflection/absorption shifts toward higher values. Figure 3-20 shows the behavior of the MgO Reststrahlen band [31]; similar results have been obtained for SiC [34]. The reflectivity for shorter wavelengths largely depends on the material's impurities. Often the behavior is similar to that of metals, i.e., the emissivity increases with temperature for the near infrared, while it decreases with shorter wavelengths. As an example, Fig. 3-21 shows the normal emissivity for zirconium carbide [35]. On the other hand, the emissivity of amorphous solids (i.e., solids without a crystal lattice) tends to be independent of temperature [36].

3.6   EFFECTS OF SURFACE ROUGHNESS

Up to this point, our discussion of radiative properties has assumed that the material surfaces are optically smooth, i.e., that the average length scale of *surface roughness* is much less than the wavelength of the electromagnetic wave. Therefore, a surface that appears rough in visible light ($\lambda \simeq 0.5\,\mu$m) may well be optically smooth in the intermediate infrared ($\lambda \simeq 50\,\mu$m). This difference is the primary reason why the electromagnetic wave theory ceases to be valid for very short wavelengths.

In this section we shall very briefly discuss some fundamental aspects of how surface roughness affects the radiative properties of opaque surfaces. Detailed discussions have been given in the books by Beckmann and Spizzichino [37] and Bass and Fuks [38], and in a review article by Ogilvy [39].

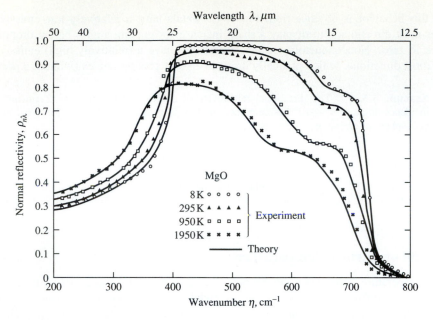

FIGURE 3-20
Variation of the spectral, normal reflectivity of MgO with temperature [31].

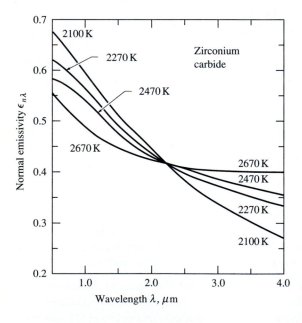

FIGURE 3-21
Temperature dependence of the spectral, normal emissivity of zirconium carbide [35].

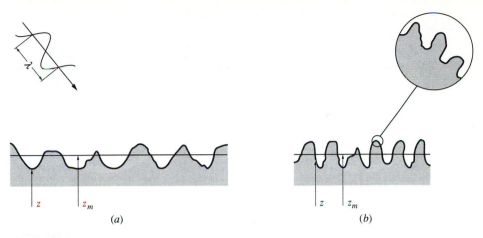

FIGURE 3-22
Topography of a rough surface: *(a)* Roughness with gradual slopes, *(b)* roughness with steep slopes. Both surfaces have similar root-mean-square roughness.

The character of roughness may be very different from surface to surface, depending on the material, method of manufacture, surface preparation, and so on, and classification of this character is difficult. A common measure of surface roughness is given by the *root-mean-square roughness* σ_m, defined as (cf. Fig. 3-22)

$$\sigma_m = \left[\frac{1}{A} \int_A (z - z_{av})^2 \, dA \right]^{1/2} , \tag{3.81}$$

where A is the surface to be examined, and $|z - z_{av}|$ is the local height deviation from the mean. The root-mean-square roughness can be readily measured with a *profilometer* (a sharp stylus that traverses the surface, recording the height fluctuations). Unfortunately, σ_m alone is woefully inadequate to describe the roughness of a surface as seen by comparing Fig. 3-22a and *b*. Surfaces of identical σ_m may have vastly different frequencies of roughness peaks, as well as different peak-to-valley Δz; in addition, σ_m gives no information on second order (or higher) roughness superimposed onto the fundamental roughness.

A first published attempt at modeling was made by Davies [40], who applied diffraction theory to a perfectly reflecting surface with roughness distributed according to a Gaussian probability distribution. The method neglects shading from adjacent peaks and, therefore, does poorly for grazing angles and for roughness with steep slopes (Fig. 3-22b). Comparison with experiments of Bennett [41] shows that, for small incidence angles, Davies' model predicts the decay of specular peaks rather well (e.g., Fig. 3-14 for MgO).

Davies' model predicts a sharp peak in the bidirectional reflection function, ρ_λ'', for the specular reflection direction, as has been found to be true experimentally for most

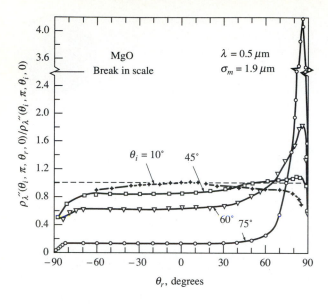

FIGURE 3-23
Normalized bidirectional reflection function (in plane of incidence) for magnesium oxide ceramic; $\sigma_m = 1.9\,\mu\text{m}$, $\lambda = 0.5\,\mu\text{m}$ [43].

cases as long as the incidence angle was not too large (e.g., Fig. 3-5). For large off-normal angles of incidence, experiment has shown that the bidirectional reflectivity function has its peak at polar angles greater than the specular direction. An example is given in Fig. 3-23 for magnesium oxide with a roughness of $\sigma_m = 1.9\,\mu\text{m}$, illuminated by radiation with a wavelength of $\lambda = 0.5\,\mu\text{m}$. Shown is the bidirectional reflection function (normalized with its value in the specular direction) for the plane of incidence (the plane formed by the surface normal and the direction of the incoming radiation). We see that for small incidence angles ($\theta_i = 10°$) the reflection function is relatively diffuse, with a small peak in the specular direction. For comparison, diffuse reflection with a direction-independent reflection function is indicated by the dashed line. For larger incidence angles the reflection function displays stronger and stronger *off-specular peaks*. For example, for an incidence angle of $\theta_i = 45°$, the off-specular peak lies in the region of $\theta = 80°$ to $85°$. Apparently, these off-specular peaks are due to *shadowing* of parts of the surface by adjacent peaks. The effects of shadowing have been incorporated into the model by Beckmann [42] and Torrance and Sparrow [43]. With the appropriate choice for two unknown constants, Torrance and Sparrow found their model agreed very well with their experimental data (Fig. 3-23).

The above models assumed that the surfaces have a certain root-mean-square roughness, but that they were otherwise random—no attempt was made to classify roughness slopes, secondary roughness, etc. Berry and coworkers [44, 45] considered diffraction of radiation from *fractal surfaces*. The behavior of fractal surfaces is such that the enlarged images appear very similar to the original surface when the surface roughness is repeatedly magnified (Fig. 3-22b). Majumdar and colleagues [46, 47] carried out roughness measurements on a variety of surfaces and found that both processed and unprocessed surfaces are generally fractal. Majumdar and Tien

[48] extended Davies' theory to include fractal surfaces, resulting in good agreement for experiments with different types of metallic surfaces [49, 50]. However, since shadowing effects have not been considered, the model is again limited to near-normal incidence.

The problem of theoretically predicting the influence of surface roughness on radiative properties is a difficult one. The state of the models is such that the bidirectional reflection function can be predicted qualitatively for many surfaces. Further experiments are necessary to ascertain whether the fractal dimensions of a surface constitute a sufficient description to make the model of Majumdar and Tien [48] universally applicable (once it is extended to include shadowing effects).

3.7 EFFECTS OF SURFACE DAMAGE AND OXIDE FILMS

Even optically smooth surfaces have a surface structure that is different from the bulk material, due to either surface damage or the presence of thin layers of foreign materials. Surface damage is usually caused by the machining process, particularly for metals and semiconductors, which distorts or damages the crystal lattice near the surface. Thin foreign coats may be formed by chemical reaction (mostly oxidation), adsorption (e.g., coats of grease or water), or electrostatics (e.g., dust particles). All of these effects may have a severe impact on the radiation properties of metals, and may cause considerable changes in the properties of semiconductors. Other materials are usually less affected, because metals have large absorptive indices, k, and thus high reflectivities. A thin, nonmetallic layer with small k can significantly decrease the composite's reflectivity (and raise its emissivity). Dielectric materials, on the other hand, have small k's and their relatively strong emission and absorption take place over a very thick surface layer. The addition of a thin, different dielectric layer cannot significantly alter their radiative properties.

A minimum amount of surface damage is introduced during sample preparation if *(i)* the technique of electropolishing is used [41], *(ii)* the surface is evaporated onto a substrate within an ultra-high vacuum environment [51], or *(iii)* the metal is evaporated onto a smooth sheet of transparent material and the reflectivity is measured at the transparent medium–metal interface [52]. Figure 3-24 shows the spectral, normal emissivity of aluminum for a surface prepared by the ultra-high vacuum method [51], and for several other aluminum surface finishes [53]. While ultra-high vacuum aluminum follows the Drude theory for $\lambda > 1\,\mu$m (cf. Fig. 3-7), polished aluminum (clean and optically smooth for large wavelengths) has a much higher emissivity over the entire spectrum. Still, the overall level of emissivity remains very low, and the reflectivity remains rather specular. Similar results have been obtained by Bennett [41], who compared electropolished and mechanically polished copper samples. As Fig. 3-24 shows, the emissivity is much larger still when off-the-shelf commercial aluminum is tested, probably due to a combination of roughness, contamination, and slight atmospheric oxidation. Bennett and colleagues [54] have shown that deposition

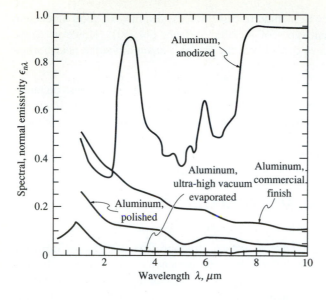

FIGURE 3-24
Spectral, normal emissivity for aluminum with different surface finishes [51, 53].

of a thin oxide layer on aluminum (up to 100 Å) appreciably increases the emissivity only for wavelengths less than 1.5 μm. This statement clearly is not true for thick oxide layers, as evidenced by Fig. 3-24: Anodized aluminum (i.e., electrolytically oxidized material with a thick layer of alumina, Al_2O_3) no longer displays the typical trends of a metal, but rather shows the behavior of the dielectric alumina. The effects of thin and thick oxide layers have been measured for many metals, with similar results. A good collection of such measurements has been given by Wood and coworkers [3]. As a rule of thumb, clean metal exposed to air at room temperature grows oxide films so thin that infrared emissivities are not affected appreciably. On the other hand, metal surfaces exposed to high-temperature oxidizing environments (furnaces, etc.) generally have radiative properties similar to those of their oxide layer.

While most severe for metallic surfaces the problem of surface modification is not unknown for nonmetals. For example, it is well known that silicon carbide, when exposed to air at high temperature, forms a silica (SiO_2) layer on its surface, resulting in a reflection band around 9 μm [55]. Nonoxidizing chemical reactions can also significantly change the radiative properties of dielectrics. For example, the strong ultraviolet radiation in outer space (from the sun) as well as gamma rays (from inside the earth's van Allen belt) can damage the surface of spacecraft protective coatings like white acrylic paint [56] or titanium dioxide/epoxy coating [57], as shown in Fig. 3-25.

In summary, radiative properties for opaque surfaces, when obtained from figures in this chapter, from the tables given in Appendix B, or from other tabulations and figures of [1–8, 58, 59], should be taken with a grain of salt. Unless detailed de-

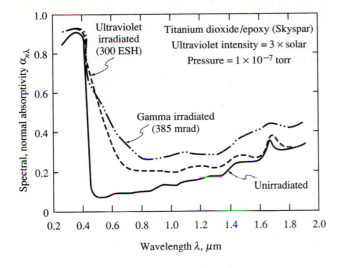

FIGURE 3-25
Effects of ultraviolet and gamma ray irradiation on a titanium dioxide/epoxy coating [57].

scriptions of surface purity, preparation, treatment, etc., are available, the data may not give any more than an order-of-magnitude estimate. One should also keep in mind that the properties of a surface may change during a process or overnight (by oxidation and/or contamination).

3.8 RADIATIVE PROPERTIES OF SEMITRANSPARENT SHEETS

The properties of radiatively participating media will be discussed in Chapters 9 through 11; i.e., semitransparent media that absorb and emit in depth and whose temperature distribution is, thus, strongly affected by thermal radiation. There are, however, important applications where thermal radiation enters an enclosure through semitransparent sheets, and where the temperature distribution within the sheet is unimportant or not significantly affected by thermal radiation. Applications include solar collector cover plates, windows in connection with light level calculations within interior spaces, and so forth. We shall, therefore, briefly present here the radiative properties of window glass, for single and multiple pane windows with and without surface coatings.

Properties of Single Pane Glasses

For an optically smooth window pane of a thickness d substantially larger than the wavelength of incident light, $d \gg \lambda$, the radiative properties are readily determined through *geometric optics* and *ray tracing*. Consider the sheet of semitransparent material depicted in Fig. 3-26. The sheet has a complex index of refraction $m_2 = n_2 - i k_2$

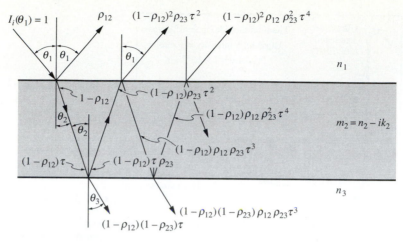

FIGURE 3-26
Reflectivity and transmissivity of a thick semitransparent sheet.

with $k_2 \ll 1$, so that the transmission through the sheet (not counting surface reflections),

$$\tau = e^{-\kappa_2 d / \cos \theta_2} = e^{-4\pi k_2 d / \lambda_0 \cos \theta_2}, \tag{3.82}$$

is appreciable, [cf. equation (2.42)]. Here $\kappa_2 = 4\pi k_2/\lambda_0$ is the absorption coefficient, λ_0 is the wavelength of the incident light in vacuum, and $d / \cos \theta_2$ is the distance a light beam of oblique incidence travels through Medium 2 in a single pass. The semitransparent sheet is surrounded by two dielectric materials with refractive indices n_1 and n_3. To calculate the reflectivity at the interfaces 1–2 and 2–3 it is sufficient to use Fresnel's equations for dielectric media, since $k_2 \ll 1$. Interchanging n_1 and n_2, as well as θ_1 and θ_2, in equation (2.95) shows that the reflectivity at the 1–2 interface is the same, regardless of whether radiation is incident from Medium 1 or Medium 2, i.e., $\rho_{12} = \rho_{21}$, and $\rho_{23} = \rho_{32}$. Now consider radiation of unit strength to be incident upon the sheet from Medium 1 in the direction of θ_1. As indicated in Fig. 3-26 the fraction ρ_{12} is reflected at the first interface, while the fraction $(1 - \rho_{12})$ is refracted into Medium 2, according to Snell's law. After traveling a distance $d / \cos \theta_2$ through Medium 2 the attenuated fraction $(1 - \rho_{12})\tau$ arrives at the 2–3 interface. Here the amount $(1 - \rho_{12})\tau\rho_{23}$ is reflected back to the 1–2 interface, while the fraction $(1 - \rho_{12})\tau(1 - \rho_{23})$ leaves the sheet and penetrates into Medium 3 in a direction of θ_3. The internally reflected fraction keeps bouncing back and forth between the interfaces, as indicated in the figure, until all energy is depleted by reflection back into Medium 1, by absorption within Medium 2, and by transmission into Medium 3. Therefore, the *slab reflectivity*, R_{slab}, may be calculated by summing over all contributions, or

$$R_{\text{slab}} = \rho_{12} + \rho_{23}(1 - \rho_{12})^2\tau^2 \left[1 + \rho_{12}\rho_{23}\tau^2 + (\rho_{12}\rho_{23}\tau^2)^2 + \cdots\right].$$

Since $\rho_{12}\rho_{23}\tau^2 < 1$ the series is readily evaluated [60], and

$$R_{\text{slab}} = \rho_{12} + \frac{\rho_{23}(1 - \rho_{12})^2\tau^2}{1 - \rho_{12}\rho_{23}\tau^2} = \frac{\rho_{12} + (1 - 2\rho_{12})\rho_{23}\tau^2}{1 - \rho_{12}\rho_{23}\tau^2}. \tag{3.83}$$

Similarly, the *slab transmissivity*, T_{slab}, follows as

$$\begin{aligned}
T_{\text{slab}} &= (1 - \rho_{12})(1 - \rho_{23})\tau\left[1 + \rho_{12}\rho_{23}\tau^2 + (\rho_{12}\rho_{23}\tau^2)^2 + \ldots\right] \\
&= \frac{(1 - \rho_{12})(1 - \rho_{23})\tau}{1 - \rho_{12}\rho_{23}\tau^2}. \tag{3.84}
\end{aligned}$$

These relations are the same as the ones evaluated for thick sheets by the electromagnetic wave theory, equations (2.135) and (2.136). From conservation of energy $A_{\text{slab}} + R_{\text{slab}} + T_{\text{slab}} = 1$, and the *slab absorptivity* follows as

$$A_{\text{slab}} = \frac{(1 - \rho_{12})(1 + \rho_{23}\tau)(1 - \tau)}{1 - \rho_{12}\rho_{23}\tau^2}. \tag{3.85}$$

If Media 1 and 3 are identical (say, air), then $\rho_{12} = \rho_{23} = \rho$ and equations (3.83) through (3.85) reduce to

$$R_{\text{slab}} = \rho\left[1 + \frac{(1 - \rho)^2\tau^2}{1 - \rho^2\tau^2}\right], \tag{3.86}$$

$$T_{\text{slab}} = \frac{(1 - \rho)^2\tau}{1 - \rho^2\tau^2}, \tag{3.87}$$

$$A_{\text{slab}} = \frac{(1 - \rho)(1 - \tau)}{1 - \rho\tau}. \tag{3.88}$$

Figure 3-27 shows typical slab transmissivities and reflectivities of several different types of glasses for normal incidence and for a pane thickness of 12.7 mm. Most glasses have fairly constant and low slab reflectivity in the spectral range from $0.1\,\mu\text{m}$ up to about $9\,\mu\text{m}$ (relatively constant refractive index n, small absorptive index k). Beyond $9\,\mu\text{m}$ the reflectivity increases because of two Reststrahlen bands [61] (not shown). Glass transmissivity tends to be very high between $0.4\,\mu\text{m}$ and $2.5\,\mu\text{m}$. Beyond $2.5\,\mu\text{m}$ the transmissivity of window glass diminishes rapidly, making windows opaque to infrared radiation. This gives rise to the so-called "greenhouse" effect: Since the sun behaves much like a blackbody at 5762 K, most of its energy ($\approx 95\%$) falling onto earth lies in the spectral range of high glass transmissivities. Therefore, solar energy falling onto a window passes readily into the space behind it. The spectral variation of solar irradiation, for extraterrestrial and unity air mass conditions, was given in Fig. 1-3. On the other hand, if the space behind the window is at low to moderate temperatures (300 to 400 K), emission from such surfaces is at fairly long wavelengths, which is absorbed by the glass and, thus, cannot escape.

The influence of pane thickness on reflectivity and transmissivity is shown in Fig. 3-28 for the case of soda-lime glass (i.e., ordinary window glass). As the pane

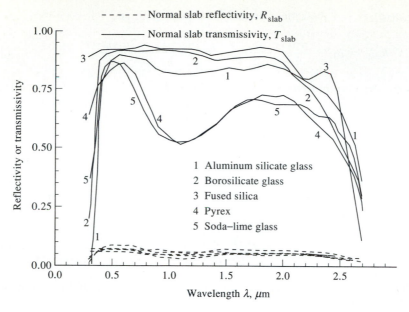

FIGURE 3-27
Spectral, normal slab transmissivity and reflectivity for panes of five different types of glasses at room temperature; data from [7].

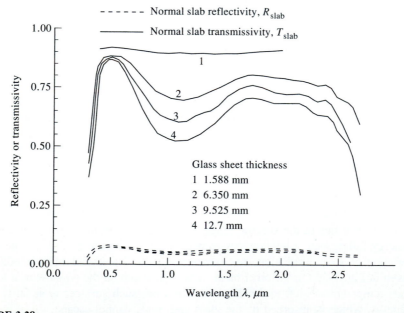

FIGURE 3-28
Spectral, normal slab transmissivity and reflectivity of soda lime glass at room temperature, for a number of pane thicknesses; data from [7].

thickness increases, transmissivity decreases due to the increasing absorption. Since the absorption coefficient is small for $\lambda < 2.7\,\mu$m (see Fig. 1-18), the effect is rather minor (and even less so for the other glasses shown in Fig. 3-27).

In some high-temperature applications the emission from hot glass surfaces becomes important (e.g., in the manufacture of glass). Gardon [62] has calculated the spectral, hemispherical and total, hemispherical emissivity of soda-lime glass sheets at 1000°C based on the data of Neuroth [63]. Spectral emissivities beyond $2.7\,\mu$m do not depend strongly on temperature since the absorption coefficient is relatively temperature-independent (see Fig. 1-18). For all but the thinnest glass sheets the material becomes totally opaque, and the hemispherical emissivity is evaluated as $\epsilon_\lambda = 1 - \rho_\lambda \simeq 0.91.$[6]

Coatings

Glass sheets and other transparent solids often have coatings on them for a variety of reasons: To eliminate transmission of ultraviolet radiation, to decrease or increase transmission over certain spectral regions, and the like. We distinguish between thick coatings ($d \gg \lambda$, no interference effects) and thin film coatings ($d = \mathbb{O}(\lambda)$), with wave interference, as discussed in Chapter 2). The effects of a thick dielectric layer (with refractive index n_2, and absorptive index $k_2 \simeq 0$) on the reflectivity of a thick sheet of glass (n_3 and $k_3 \simeq 0$) is readily analyzed with the two-interface formula given by equation (3.83). With $\tau \simeq 1$ and, for normal incidence,

$$\rho_{12} = \left(\frac{n_1 - n_2}{n_1 + n_2}\right)^2, \quad \text{and} \quad \rho_{23} = \left(\frac{n_2 - n_3}{n_2 + n_3}\right)^2,$$

the coating reflectivity becomes

$$R_{\text{coat}} = \frac{\rho_{12} + \rho_{23} - 2\rho_{12}\rho_{23}}{1 - \rho_{12}\rho_{23}} = 1 - \frac{(1 - \rho_{12})(1 - \rho_{23})}{1 - \rho_{12}\rho_{23}}$$

$$= 1 - \frac{(4n_1 n_2)(4n_2 n_3)}{(n_1 + n_2)^2(n_2 + n_3)^2 - (n_1 - n_2)^2(n_2 - n_3)^2},$$

which is readily simplified to

$$R_{\text{coat}} = 1 - \frac{4n_1 n_2 n_3}{(n_2^2 + n_1 n_3)(n_1 + n_3)}. \tag{3.89}$$

If the aim is to minimize the overall reflectivity of the semitransparent sheet, then a value for the refractive index of the coating must be chosen to make R_{coat} a minimum. Thus, setting $dR_{\text{coat}}/dn_2 = 0$ leads to

$$n_{2,\text{min}} = \sqrt{n_1 n_3}. \tag{3.90}$$

---

[6]The *hemispherical* emissivity is evaluated by first evaluating $\rho_{n\lambda}$: With $n \simeq 1.5$ (for $\lambda > 2.7\,\mu$m) from Fig. 3-16, $\rho_{n\lambda} = 0.04$ and $\epsilon_{n\lambda} = 0.96$; finally, from Fig. 3-19 $\epsilon_\lambda \simeq 0.91$.

Substituting equation (3.90) into (3.89) results in a minimum coated-surface reflectivity of

$$R_{coat,min} = 1 - \frac{2\sqrt{n_1 n_3}}{n_1 + n_3}. \tag{3.91}$$

The slab reflectivity for a thin dielectric coating on a dielectric substrate, $d = \mathbb{O}(\lambda)$, is subject to wave interference effects and has been evaluated in Chapter 2, from equation (2.130), with $\delta_{12} = \pi$ and $\delta_{23} = 0$ (cf. Example 2.6), as

$$R_{coat} = \frac{r_{12}^2 + 2r_{12}r_{23}\cos\zeta + r_{23}^2}{1 + 2r_{12}r_{23}\cos\zeta + r_{12}^2 r_{23}^2}, \tag{3.92a}$$

$$r_{12} = \frac{n_1 - n_2}{n_1 + n_2}, \quad r_{23} = \frac{n_2 - n_3}{n_2 + n_3}, \quad \zeta = \frac{4\pi n_2 d}{\lambda}. \tag{3.92b}$$

Equation (3.92) has an interference minimum when $\zeta = \pi$ (i.e., if the film thickness is a quarter of the wavelength inside the film, $d = 0.25\lambda/n_2$). For this interference minimum the reflectivity of the coated surface becomes

$$R_{coat} = \left(\frac{r_{12} - r_{23}}{1 - r_{12}r_{23}}\right)^2. \tag{3.93}$$

Clearly, this equation results in a minimum (or zero) reflectivity if $r_{12} = r_{23}$, or $n_{2,min} = \sqrt{n_1 n_3}$, which is the same as for thick films, equation (3.90). To obtain minimum reflectivities for glass ($n_3 \simeq 1.5$) facing air ($n_1 \simeq 1$) would require a dielectric film with $n_2 \simeq 1.22$. Dielectric films of such low refractive index do not appear possible. However, Yoldas and Partlow [64] showed that a porous film (pore size $\ll \lambda$) can effectively lower the refractive index, and they obtained glass transmissivities greater than 99% throughout the visible.

In other applications a strong reflectivity is desired. An example of experimentally determined reflectivity and transmissivity of a coated dielectric is given in Fig. 3-29 for a 0.35 μm thick layer of Sn-doped In_2O_3 film on glass [65]. The oscillating properties clearly demonstrate the effects of wave interference at shorter wavelengths. At wavelengths $\lambda > 1.5\,\mu$m the material has a strong absorption band, making it highly reflective and opaque. Thus, this coated glass makes a better solar collector coverplate than ordinary glass, since internally emitted infrared radiation is *reflected back* into the collector (rather than being absorbed), keeping the cover glass cool and reducing losses. Similar behavior was obtained by Yoldas and O'Keefe [66], who deposited thin (200 to 500 Å) triple-layer films (titanium dioxide–silver–titanium dioxide) on soda-lime glass.

Multiple Parallel Sheets

To minimize convection losses, two or more parallel sheets of windows are often employed, as illustrated in Fig. 3-30a. To find the total reflectivity and transmissivity

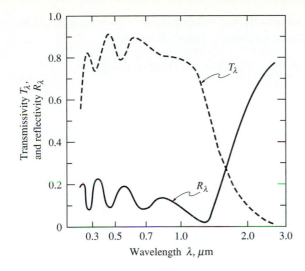

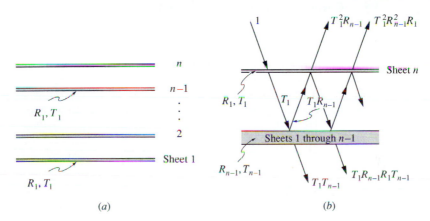

FIGURE 3-29
Spectral, normal reflectivity and transmissivity of a 0.35 μm thick Sn-doped In$_2$O$_3$ film deposited on Corning 7059 glass [65].

FIGURE 3-30
Reflectivity and transmissivity of multiple sheets: (a) Geometric arrangement, (b) ray tracing for interaction between a single layer and the remainder of the sheets.

of n layers, we break the system up into a single layer and the remaining $(n-1)$ layers. Then ray tracing (see Fig. 3-30b) results in

$$R_n = R_1 + T_1^2 R_{n-1}\left[1 + R_1 R_{n-1} + (R_1 R_{n-1})^2 + \ldots\right]$$

$$= R_1 + \frac{T_1^2 R_{n-1}}{1 - R_1 R_{n-1}}, \tag{3.94}$$

and, similarly,

$$T_n = \frac{T_1 T_{n-1}}{1 - R_1 R_{n-1}}, \tag{3.95}$$

where R_{n-1} and T_{n-1} are the net reflectivity and transmissivity of $(n-1)$ layers. The net absorptivity of the n layers can be calculated directly either from

$$A_n = A_1 + A_1 T_1 R_{n-1} (1 + R_1 R_{n-1} + \ldots) + A_{n-1} T_1 (1 + R_1 R_{n-1} + \ldots)$$

$$= A_1 + \frac{T_1 (A_1 R_{n-1} + A_{n-1})}{1 - R_1 R_{n-1}}, \qquad (3.96)$$

or from conservation of energy, i.e., $A_n + R_n + T_n = 1$. In the development of equation (3.94) we have assumed that R_1 is the same for light shining onto the top or the bottom of the sheet ($\rho_{12} = \rho_{23}$), in other words, that equation (3.86) is valid. The above recursion formulae were first derived by Edwards [67] without the restriction of $\rho_{12} = \rho_{23}$. In a later paper Edwards [68] expanded the method to include wave interference effects for stacked thin films. Multiple sheets subject to mixed diffuse and collimated irradiation, but without interference effects, were analyzed by Mitts and Smith [69].

Example 3.6.   Determine the normal transmissivity of a triple-glazed window for visible wavelengths. The window panes are thin sheets of soda-lime glass, separated by layers of air.

Solution

The reflectivity R_1 and transmissivity T_1 of a single sheet are readily calculated from equations (3.86) and (3.87). For thin sheets (e.g., curve 1 in Fig. 3-28) we have $\tau \simeq 1$, and with $n \simeq 1.5$ (cf. Fig. 3-16), $\rho = [(1.5 - 1)/(1.5 + 1)]^2 = 0.04$. Therefore,

$$R_1 = \rho \left[1 + \frac{(1 - \rho)^2}{1 - \rho^2} \right] = \frac{2\rho}{1 + \rho} = \frac{2 \times 0.04}{1 + 0.04} = 0.0769,$$

$$T_1 = \frac{(1 - \rho)^2}{1 - \rho^2} = \frac{1 - \rho}{1 + \rho} = 1 - R_1 = 0.9231$$

(and $A_1 = 0$, since we assumed $\tau \simeq 0$). For two panes, from equations (3.94) and (3.95) with $n = 2$,

$$R_2 = R_1 + \frac{T_1^2 R_1}{1 - R_1^2} = 0.0769 \left(1 + \frac{0.9231^2}{1 - 0.0769^2} \right) = 0.1429,$$

$$T_2 = \frac{T_1^2}{1 - R_1^2} = 0.8571$$

(and, again $A_2 = 0$). Finally, for three panes

$$R_3 = R_1 + \frac{T_1^2 R_2}{1 - R_1 R_2} = 0.0769 + \frac{0.9231^2 \times 0.1429}{1 - 0.0769 \times 0.1429} = 0.2000,$$

$$T_3 = \frac{T_1 T_2}{1 - R_1 R_2} = \frac{0.9231 \times 0.8571}{1 - 0.0769 \times 0.1429} = 0.8000.$$

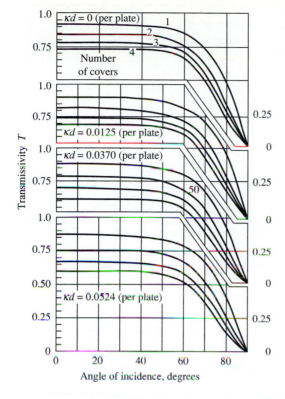

FIGURE 3-31
Transmissivities of 1, 2, 3 and 4 sheets of glass ($n = 1.526$) for different optical thicknesses per sheet, κd [71].

Assuming negligible absorption within the glass, 80% of visible radiation is transmitted through the triple pane window (at normal incidence), while 20% is reflected back.

Although they are valid, equations (3.83) and (3.84) are quite cumbersome for oblique incidence, in particular, if absorption cannot be neglected. Some calculations for nonabsorbing (for $n = 1.5$ [70] and for $n = 1.526$ [71]) and absorbing [71] ($n = 1.526$) multiple sheets of window glass have been carried out. Note that, for oblique incidence, the overall reflectivity and transmissivity are different for parallel- and perpendicular-polarized light. Even for unpolarized light the polarized components must be determined before averaging, as

$$R_n = \frac{1}{2}(R_{n\perp} + R_{n\parallel}), \quad T_n = \frac{1}{2}(T_{n\perp} + T_{n\parallel}). \tag{3.97}$$

The results of the calculations by Duffie and Beckman [71] are given in graphical form in Fig. 3-31.

3.9   SPECIAL SURFACES

For many engineering applications it would be desirable to have a surface material available with very specific radiative property characteristics. For example, the net

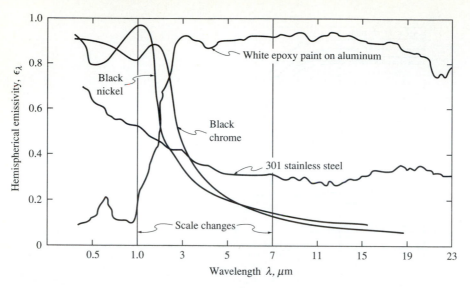

FIGURE 3-32
Spectral, hemispherical reflectivities of several spectrally selective surfaces [77].

radiative heat gain of a solar collector is the difference between absorbed solar energy and radiation losses due to emission by the collector surface. While a black absorber plate would absorb all solar irradiation, it unfortunately would also lose a maximum amount of energy due to surface emission. An ideal solar collector surface has a maximum emissivity for those wavelengths and directions over which solar energy falls onto the surface, and a minimum emissivity for all other wavelengths and directions. On the other hand, a radiative heat rejector, such as the ones used by the U.S. Space Shuttle to reject excess heat into outer space, should have a high emissivity at longer wavelengths, and a high reflectivity for those wavelengths and directions with which sunshine falls onto the heat rejector.

To a certain degree the radiative properties of a surface can be tailored toward desired characteristics. Surfaces that absorb and emit strongly over one wavelength range, and reflect strongly over the rest of the spectrum are called *spectrally selective*, while surfaces with tailored directional properties are known as *directionally selective*.

An ideal, spectrally selective surface would be black ($\alpha_\lambda = \epsilon_\lambda = 1$) over the wavelength range over which maximum absorption (or emission) is *desired*, and would be totally reflective ($\alpha_\lambda = \epsilon_\lambda = 0$) beyond a certain *cutoff wavelength* λ_c, where *undesirable* emission (or absorption) would occur. Of course, in practice such behavior can only be approximated. Such an ideal surface is indicated by the long-dash line in Fig. 3-32.

The performance of a selective surface is usually measured by the "α/ϵ-ratio," where α is the total, directional absorptivity of the material for solar irradiation,

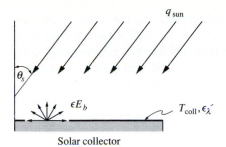

Solar collector

FIGURE 3-33
Solar irradiation on and emission from a solar collector plate

while ϵ is the total, hemispherical emissivity for infrared surface emission. Consider a solar collector plate (Fig. 3-33), irradiated by the sun at an off-normal angle of θ_s. Making an energy balance (per unit area of the collector), we find

$$q_{net} = \epsilon \sigma T_{coll}^4 - \alpha q_{sun} \cos \theta_s, \qquad (3.98)$$

where the factor $\cos \theta_s$ appears since q_{sun} is solar heat flux per unit area *normal to the sun's rays*. The total, hemispherical emissivity may be related to spectral, hemispherical values through equation (3.10), while the total, directional absorptivity is found from equation (3.21). Thus

$$\epsilon = \frac{1}{\sigma T_{coll}^4} \int_0^\infty \epsilon_\lambda(T_{coll}, \lambda) E_{b\lambda}(T_{coll}, \lambda) \, d\lambda, \qquad (3.99a)$$

$$\alpha = \frac{1}{q_{sun}} \int_0^\infty \alpha_\lambda(T_{coll}, \lambda, \theta_s) q_{sun,\lambda} \, d\lambda$$

$$= \frac{1}{\sigma T_{sun}^4} \int_0^\infty \alpha_\lambda(T_{coll}, \lambda, \theta_s) E_{b\lambda}(T_{sun}, \lambda) \, d\lambda, \qquad (3.99b)$$

where we have made use of the fact that the spectral distribution of q_{sun} is the same as the blackbody emission from the sun's surface. Clearly, for optimum performance of a collector the solar absorptivity should be maximum, while the infrared emissivity should be minimum. Therefore, a large α/ϵ-ratio indicates a better performance for a solar collector. On the other hand, for radiative heat rejectors a minimum value for α/ϵ is desirable.

Most selective absorbers are manufactured by coating a thin nonmetallic film onto a metal. Over most wavelengths the nonmetallic film is very transmissive and incoming radiation passes straight through to the metal interface with its very high reflectivity. However, many nonconductors have spectral regions over which they do absorb appreciably without being strongly reflective (usually due to lattice defects or contaminants). The result is a material that acts like a strongly reflecting metal over most of the spectrum, but like a strongly absorbing nonconductor for selected wavelength ranges. A few examples of such selective surfaces are also given in Fig. 3-32.

Black chrome (chrome-oxide coating) and black nickel (nickel-oxide coating) are popular solar collector materials, while epoxy paint may be used as an efficient solar energy rejector. If the coatings are extremely thin, interference effects can also be exploited to improve selectivity. For example, Martin and Bell [72] showed that a three-layer coating of of SiO_2-Al-SiO_2 on metallic substrates has a solar absorptivity greater than 90%, but an infrared emissivity of $< 10\%$. Fan and Bachner [65] produced a coating for glass that raised its reflectivity to $> 80\%$ for infrared wavelengths, without appreciably affecting solar transmissivity, Fig. 3-29.

The advantages of spectrally selective surface properties were first recognized by Hottel and Woertz [73]. With the growing interest in solar energy collection during the 1950s and 1960s, a number of selective coatings were developed, and the subject was discussed by Gier and Dunkle [74], and Tabor and coworkers [75, 76]. There are several compilations for radiative properties of selective absorbers [3, 8, 77]. A somewhat more detailed discussion about spectrally selective surface properties has been given by Duffie and Beckman [71].

Example 3.7. Let us assume that it is possible to manufacture a diffusely absorbing/emitting selective absorber with a spectral emissivity $\epsilon_\lambda = \epsilon_s = 0.05$ for $0 < \lambda < \lambda_c$ and $\epsilon_\lambda = \epsilon_c = 0.95$ for $\lambda > \lambda_c$, where the cutoff wavelength can be varied through manufacturing methods. Determine the optimum cutoff wavelength for a solar collector with an absorber plate at 350 K that is exposed to solar irradiation of $q_{sun} = 1000 \, \text{W}/\text{m}^2$ at an angle of $\theta_s = 30°$ off-normal. What is the net radiative energy gain for such a collector?

Solution

A simple energy balance on the surface, using equations (3.9) and (3.14) leads to

$$q_{net} = E - H'_{abs}(\theta_s) = \epsilon E_b - \alpha'(\theta_s) H'(\theta_s)$$

where $q_{net} > 0$ if a net amount of energy leaves the surface, $q_{net} < 0$ if energy is collected. Total hemispherical emissivity follows from equation (3.10) while total directional absorptivity is determined from equation (3.21). For our diffuse absorber we have $\alpha'_\lambda(\lambda, \theta) = \epsilon_\lambda(\lambda)$ and

$$\epsilon = \frac{1}{\sigma T_{coll}^4} \left[\epsilon_s \int_0^{\lambda_c} E_{b\lambda}(T_{coll}, \lambda) \, d\lambda + \epsilon_c \int_{\lambda_c}^\infty E_{b\lambda}(T_{coll}, \lambda) \, d\lambda \right]$$

$$= \epsilon_s + \frac{(\epsilon_c - \epsilon_s)}{\sigma T_{coll}^4} \int_{\lambda_c}^\infty E_{b\lambda}(T_{coll}, \lambda) \, d\lambda,$$

$$\alpha = \frac{1}{\sigma T_{sun}^4} \left[\epsilon_s \int_0^{\lambda_c} E_{b\lambda}(T_{sun}, \lambda) \, d\lambda + \epsilon_c \int_{\lambda_c}^\infty E_{b\lambda}(T_{sun}, \lambda) \, d\lambda \right]$$

$$= \epsilon_s + \frac{(\epsilon_c - \epsilon_s)}{\sigma T_{sun}^4} \int_{\lambda_c}^\infty E_{b\lambda}(T_{sun}, \lambda) \, d\lambda.$$

Substituting these expressions into our energy balance leads to

$$q_{net} = \epsilon_s(\sigma T_{coll}^4 - q_{sun}\cos\theta_s)$$
$$+ (\epsilon_c - \epsilon_s)\int_{\lambda_c}^{\infty}\left[E_{b\lambda}(T_{coll}, \lambda) - \frac{q_{sun}\cos\theta_s}{\sigma T_{sun}^4}E_{b\lambda}(T_{sun}, \lambda)\right]d\lambda.$$

Optimizing the value of λ_c implies finding a maximum for q_{net}. Therefore, from *Leibnitz' rule* (see, e.g., [60]), which states that

$$\frac{d}{dx}\int_{a(x)}^{b(x)}f(x, y)\,dy = \frac{db}{dx}f(x, b) - \frac{da}{dx}f(x, a) + \int_a^b\frac{df}{dx}(x, y)\,dy, \tag{3.100}$$

we find

$$\frac{dq_{net}}{d\lambda_c} = -(\epsilon_c - \epsilon_s)\left[E_{b\lambda}(T_{coll}, \lambda_c) - \frac{q_{sun}\cos\theta_s}{\sigma T_{sun}^4}E_{b\lambda}(T_{sun}, \lambda_c)\right] = 0,$$

or

$$E_{b\lambda}(T_{coll}, \lambda_c) = \frac{q_{sun}\cos\theta_s}{\sigma T_{sun}^4}E_{b\lambda}(T_{sun}, \lambda_c).$$

Note that the cutoff wavelength does not depend on the values for ϵ_c and ϵ_s. Using Planck's law, equation (1.13), with $n = 1$ (surroundings are air), the last expression reduces to

$$\exp(C_2/\lambda_c T_{coll}) - 1 = \frac{\sigma T_{sun}^4}{q_{sun}\cos\theta_s}\left[\exp(C_2/\lambda_c T_{sun}) - 1\right].$$

This transcendental equation needs to be solved by iteration. As a first guess one may employ Wien's distribution, equation (1.18), (dropping two '-1' terms),

$$\exp(C_2/\lambda_c T_{coll}) \simeq \frac{\sigma T_{sun}^4}{q_{sun}\cos\theta_s}\exp(C_2/\lambda_c T_{sun})$$

or

$$\exp\left[\frac{C_2}{\lambda_c}\left(\frac{1}{T_{coll}} - \frac{1}{T_{sun}}\right)\right] \simeq \frac{\sigma T_{sun}^4}{q_{sun}\cos\theta_s},$$

$$\lambda_c \simeq C_2\left(\frac{1}{T_{coll}} - \frac{1}{T_{sun}}\right)\Big/\ln\frac{\sigma T_{sun}^4}{q_{sun}\cos\theta_s}$$

$$= 14{,}388\left(\frac{1}{350} - \frac{1}{5762}\right)\mu m\Big/\ln\frac{5.670\times10^{-8}\times5762^4}{1000\times\cos30°} = 3.45\,\mu m.$$

Iterating the full Planck's law leads to a cutoff wavelength of $\lambda_c = 3.69\,\mu m$. Substituting these values into the expressions for emissivity and absorptivity,

$$\epsilon = \epsilon_s + (\epsilon_c - \epsilon_s)[1 - f(\lambda_c T_{coll})] = 0.95 - 0.90 + 0.90f(3.69\times350)$$
$$= 0.05 + 0.90\times0.00413 = 0.054,$$

$$\alpha = \epsilon_s + (\epsilon_c - \epsilon_s)[1 - f(\lambda_c T_{sun})] = 0.05 + 0.90\times f(3.69\times5762)$$
$$= 0.05 + 0.90\times0.98775 = 0.939.$$

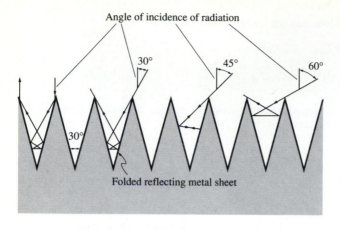

Angle of incidence of radiation

Folded reflecting metal sheet

FIGURE 3-34
Directional absorption and reflection of irradiation by a V-grooved surface [79].

The net heat flux follows then as

$$q_{net} = 0.054 \times 5.760 \times 10^{-8} \times 350^4 - 0.939 \times 1000 \times \cos 30° = -767\,\mathrm{W/m^2}.$$

Actually, neither $f(\lambda_c T_{coll}) \simeq 0$ nor $f(\lambda_c T_{sun}) \simeq 1$ is particularly sensitive to the exact value of λ_c, because there is very little spectral overlap between solar radiation (95% of which is in the wavelength range[7] of $\lambda < 2.2\,\mu m$) and blackbody emission at 350 K (95% of which is at $\lambda > 5.4\,\mu m$).

Surfaces can be made directionally selective by mechanically altering the surface finish on a microscale (microgrooves) or macroscale. For example, large V-grooves (large compared with the wavelengths of radiation) tend to reflect incoming radiation several times for near-normal incidence, as indicated in Fig. 3-34 (from Trombe and coworkers [79]) for an opening angle of $\gamma = 30°$, each time absorbing a fraction of the beam. The number of reflections decreases with increasing incidence angle, down to a single reflection for incidence angles $\theta > 90° - \gamma$ (or 60° in the case of Fig. 3-34). Hollands [80] has shown that this type of surface has a significantly higher normal emissivity, which is important for collection of solar irradiation, than hemispherical emissivity, which governs emission losses. A similarly shaped material, with flat black bottoms, was theoretically analyzed by Perlmutter and Howell [81]. Their analytical values for directional emissivity were experimentally confirmed by Brandenberg and Clausen [24], as illustrated in Fig. 3-35.

3.10 EXPERIMENTAL METHODS

It is quite apparent from the discussion in the preceding sections that, although electromagnetic wave theory can be used to *augment* experimental data, it cannot replace

---

[7]Based on a blackbody at 5762 K. This number remains essentially unchanged for true, extraterrestrial solar irradiation [78], while the 95% fraction moves to even shorter wavelengths if atmospheric absorption is taken into account (cf. Fig. 1-3).

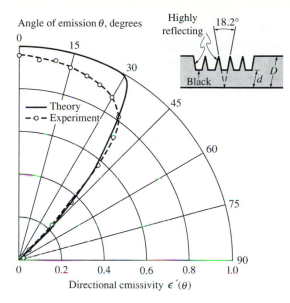

FIGURE 3-35
Directional emissivity of a grooved surface with highly reflective, specular sidewalls and near-black base. Results are for plane perpendicular to groove length. Theory ($\rho_{\text{sides}} = \epsilon_{\text{base}} = 1$) from [81], experiment [taken at $\lambda = 8\,\mu$m with aluminum sidewalls and black paint base with $\epsilon_\lambda(8\,\mu$m$) = 0.95$] from [24].

them. While the spectral, bidirectional reflection function, equation (3.32), is the most basic radiation property of an opaque surface, to which all other properties can be related, it is rarely measured. Obtaining the bidirectional reflection function is difficult because of the low achievable signal strength. It is also impractical since it is a function of both incoming and outgoing directions and of wavelength and temperature. A complete description of the surface requires enormous amounts of data. In addition, the use of the bidirectional reflection function complicates the analysis to such a point that it is rarely attempted.

If bidirectional data are not required it is sufficient, for an opaque material, to measure one of the following, from which all other ones may be inferred: Absorptivity, emissivity, directional-hemispherical reflectivity and hemispherical-directional reflectivity. Various different measurement techniques have been developed, which may be separated into three loosely-defined groups: *Calorimetric emission measurements*, *radiometric emission measurements*, and *reflection measurements*. The interest in experimental methods was at its peak during the 1960s as a result of the advent of the space age. Compilations covering the literature of that period have been given in two NASA publications [82, 83]. Interest waned during the 1970s and 1980s but has recently picked up again because of the development of better and newer materials operating at higher temperatures. Sacadura [84] has given an updated review of experimental methods.

While measurement techniques vary widely from method to method, most of them employ similar optical components, such as light sources, monochromators, and detectors. Therefore, we shall begin our discussion of experimental methods with a short description of important optical components.

Instrumentation

Radiative property measurements generally require a light source, a monochromator, a detector, and the components of the optical path, such as mirrors, lenses, beam splitters, optical windows, and so on. Depending on the nature of the experiment and/or detector, other accessories, such as optical choppers, may also be necessary.

LIGHT SOURCES.   Light sources are required for the measurement of absorption by, or reflection from, an opaque surface, as well as for the alignment of optical components in any spectroscopic system. In addition, light sources are needed for transmission and scattering measurements of absorbing/scattering media, such as gases, particles and semitransparent solids and liquids (to be discussed in later chapters). We distinguish between monochromatic and polychromatic light sources.

Monochromatic sources.   These types of sources operate through *stimulated emission*, producing light over an extremely narrow wavelength range. Their monochromaticity, low beam divergence, coherence, and high power concentration make *lasers* particularly attractive as light sources. While only invented some 30 years ago, there are today literally dozens of solid-state and gas lasers covering the spectrum between the ultraviolet and the far infrared. Although lasers are generally monochromatic, there are a number of gas lasers that can be tuned over a part of the spectrum by stimulating different transitions. For example, *dye lasers* (using large organic dye molecules as the lasing medium) may be operated at a large number of wavelengths in the range $0.2\,\mu m < \lambda < 1\,\mu m$, while the common CO_2 laser (usually operating at $10.6\,\mu m$) may be equipped with a movable grating, allowing it to lase at a large number of wavelengths in the range $9\,\mu m < \lambda < 11\,\mu m$. Even solid-state lasers can be operated at several wavelengths through frequency-doubling. For example, the Nd-YAG laser, the most common solid-state laser, can be used at $1.064\,\mu m$, $0.532\,\mu m$, $0.355\,\mu m$, and $0.266\,\mu m$. Of particular importance for radiative property measurements is the helium-neon laser because of its low price and small size and because it operates in the visible at $0.633\,\mu m$ (making it useful for optical alignment).

A different kind of monochromatic source is the *low pressure gas discharge lamp*, in which a low-density electric current passes through a low-pressure gas. Gas atoms and molecules become ionized and conduct the current. Electrons bound to the gas atoms become excited to higher energy levels, from which they fall again, emitting radiation over a number of narrow spectral lines whose wavelengths are characteristic of the gas used, such as zinc, mercury, and so on.

Polychromatic sources.   These usually incandescent light sources emit radiation by *spontaneous emission* due to the thermal excitation of source atoms and molecules, resulting in a continuous spectrum. The spectral distribution and total radiated power depend on the temperature, area, and emissivity of the surface. Incandescent sources may be of the filament type (similar to an ordinary light bulb) or of the bare-element type. The *quartz-tungsten-halogen* lamp has a doped tungsten filament inside a quartz envelope, which is filled with a rare gas and a small amount of a halogen. Operating

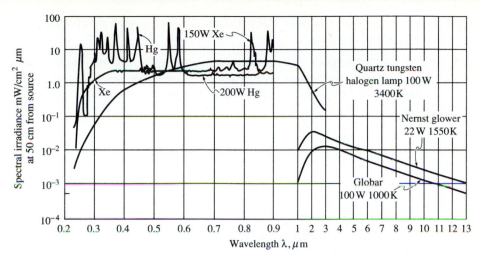

FIGURE 3-36
Spectral irradiation on a distant surface from various incandescent light sources.

at a filament temperature greater than 3000 K, this lamp produces a near-blackbody spectrum with maximum emission below 1 μm. However, because of the transmission characteristics of quartz (which is the same as fused silica, Fig. 3-27), there is no appreciable emission beyond 3 μm. Bare-element sources are either rods of silicon carbide, called *globars*, or heating wires embedded in refractory oxides, called *Nernst glowers*. Globars operate at a temperature of 1000 K and produce an almost-gray spectrum with a maximum around 2.9 μm. Nernst glowers operate at temperatures up to 1500 K, with a somewhat less ideal spectral distribution. The irradiation onto a distant surface from different incandescent sources is shown in Fig. 3-36. None of the light sources shown in Fig. 3-36 has a truly "black" spectral distribution, since their output is influenced by their spectral emissivity. In most experiments this is of little importance since, in general, sample and reference signals (coming from the same spectral source) are compared. If a true blackbody source is required (primarily for calibration of instruments) *blackbody cavity sources* are available from a number of manufacturers. In these sources a cylindrical and/or conical cavity, made of a high-temperature, high-emissivity material (such as silicon carbide) is heated to a desired temperature. Radiation leaving the cavity, also commonly called *Hohlraum* (German for "hollow space"), is essentially black (cf. Table 5.1).

The brightest conventional source of optical radiation is the *high pressure gas discharge lamp*, which combines the characteristics of spontaneous and stimulated emission. The lamp is similar to a low-pressure gas discharge source, but with high current density and gas pressure. This configuration results in an arc with highly excited atoms and molecules forming a plasma. While the hot plasma emits as an incandescent source, ionized atoms emit over substantially broadened spectral lines, resulting in a mixed spectrum (Fig. 3-36). Commonly used gases for such arc sources are xenon, mercury, and deuterium.

SPECTRAL SEPARATORS.   Spectral radiative properties can be measured over part of the spectrum in one of two ways: *(i)* Measurements are made using a variety of monochromatic light sources, which adequately represent the desired part of the spectrum, or *(ii)* a polychromatic source is used together with a device that allows light of only a few select wavelengths to reach the detector. Such devices may consist of simple *optical filters,* manually driven or motorized *monochromators,* or highly sophisticated *FTIR* (Fourier Transform InfraRed) spectrometers.

Optical filters.   These are multilayer thin-film devices that selectively transmit radiation only over desired ranges of wavelengths. *Bandpass filters* transmit light only over a finite, usually narrow, wavelength region, while *edge filters* transmit only above or below certain *cutoff* or *edge wavelengths.* Bandpass filters consist of a series of thin dielectric films that, at each interface, partially reflect and partially transmit radiation (cf. Fig. 2-13). The spacing between layers is such that beams of the desired wavelength are, after multiple reflections within the layers, in phase with the transmitted beam (constructive interference). Other wavelengths are rejected because they destructively interfere with one another. Bandpass filters for any conceivable wavelength between the ultraviolet and the midinfrared are routinely manufactured. Edge filters operate on the same principle, but are more complex in design.

Monochromators.   These devices separate an incoming polychromatic beam into its spectral components. They generally consist of an entrance slit, a prism or grating that spreads the incoming light according to its wavelengths, and an exit slit, which allows only light of desired wavelengths to escape. If a prism is used, it is made of a highly-transparent material with a refractive index that varies slightly across the spectrum (cf. Fig. 3-16). As shown in Fig. 3-37a, the incoming radiant energy is separated into its constituent wavelengths since, by Snell's law, the prism bends different wavelengths (with different refractive index) by different amounts. Rotating the prism around an axis allows different wavelengths to escape through the exit slit. Instead of a prism one can use a *diffraction grating* to separate the wavelengths of incoming light, employing the principle of constructive and destructive interference [85], as schematically indicated in Fig. 3-37b. Until a few years ago all monochromators employed salt prisms, while today almost all systems employ diffraction gratings, since they are considerably cheaper and simpler to handle (salt prisms tend to be hygroscopic, i.e., they are attacked by the water vapor in the surrounding air). However, diffraction gratings have the disadvantages that their spectral range is more limited (necessitating devices with multiple gratings), and they may give erroneous readings due to higher-order signals (frequency-doubling).

FTIR spectrometers.   These instruments collect the entire radiant energy (i.e., comprising all FTIR spectrometer wavelengths) after reflection from a moving mirror. The measured intensity depends on the position of the moving mirror owing to con-

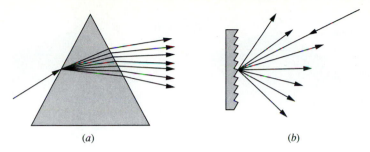

(a)   (b)

FIGURE 3-37
Schematic of spectral separation with (a) a transparent prism, (b) a diffraction grating.

structive and destructive interference. This signal is converted by a computer through an inverse *Fast Fourier Transform* into a power vs. wavelength plot. The spectral range of FTIRs is limited only by the choice of beam splitters and detectors, and is comparable to that of prism monochromators. However, while monochromators generally require several minutes to collect data over their entire spectral range, the FTIR is able to do this in a fraction of a second. Detailed descriptions of the operation of FTIRs may be found in books on the subject, such as the one by Griffiths and de Haseth [86].

DETECTORS.   In a typical spectroscopic experiment the detector measures the intensity of incoming radiation due to transmission through, emission from, or reflection by, a sample. This irradiation may be relatively monochromatic (i.e., covers a very narrow wavelength range after having passed through a filter or monochromator), or may be polychromatic (for total emissivity measurements, or if an FTIR is used). In either case, the detector converts the beam's power into an electrical signal, which is amplified and recorded. The performance of detectors is measured by certain criteria, which are generally functions of several operating conditions, such as wavelength, temperature, modulating frequency, bias voltage, and gain of any internal amplifier. The *response time* (τ) is the time for a detector's output to reach $1 - 1/e = 63\%$ of its final value, after suddenly being subjected to constant irradiation. The *linearity range* of a detector is the range of input power over which the output signal is a linear function of the input. The *noise equivalent power (NEP)* is the radiant energy rate in Watts that is necessary to give an output signal equal to the *rms* noise output from the detector. More widely used is the reciprocal of *NEP*, the *detectivity (D)*. The detectivity is known to vary inversely with the square root of the detector area, A_D, while the signal noise is proportional to the square root of the amplifier's noise-equivalent bandwidth Δf (in Hz). Thus, a *normalized detectivity (D*)* is defined to allow comparison between different types of detectors regardless of their detector areas and amplifier bandwidths as

$$D^* = (A_D \, \Delta f)^{1/2} \, D. \tag{3.101}$$

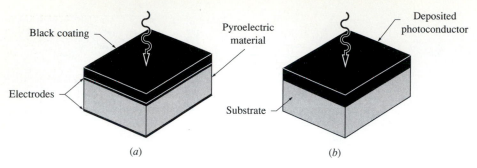

FIGURE 3-38
Schematic of *(a)* a pyroelectric detector, *(b)* a photoconductive detector.

Depending on how the incoming radiation interacts with the detector material, detectors are grouped into *thermal* and *photon* (or *quantum*) *detectors*.

Thermal detectors.   These devices convert incident radiation into a temperature rise. This temperature change is measured either through one or more thermocouples, or by using the pyroelectric effect. A single, usually blackened (to increase absorptivity) *thermocouple* is the simplest and cheapest of all thermal detectors. However, it suffers from high amplifier noise and, therefore, limited detectivity. One way to increase output voltage and detectivity is to connect a number of thermocouples in series (typically 20 to 120), constituting a *thermopile*. Thermopiles can be manufactured economically through thin-film processes. *Pyroelectric detectors* are made of crystalline materials that have permanent electric polarization. When heated by irradiation, the material expands and changes its polarization, which causes a current to flow in a circuit that connects the detector's top and bottom surfaces, as shown in the schematic of Fig. 3-38*a*. Since the change in temperature produces the current, pyroelectric detectors respond only to pulsed or chopped irradiation. They respond to changes in irradiation much more rapidly than thermocouples and thermopiles, and are not affected by steady backgound radiation.

Photon detectors.   These absorb the energy of incident radiation with their electrons, producing free charge carriers (*photoconductive* and *photovoltaic detectors*) or even ejecting electrons from the material (*photoemissive detectors*). In photoconductive and photovoltaic detectors the production of free electrons increases the electrical conductivity of the material. In the photoconductive mode an applied voltage, or *reverse bias*, causes a current that is proportional to the strength of irradiation to flow, as schematically shown in Fig. 3-38*b*. In the photovoltaic mode no bias is applied and, closing the electric circuit, a current flows as a result of the excitation of electrons (as in the operation of photovoltaic, or solar, cells). Photovoltaic detectors have greater detectivity, while photoconductive detectors exhibit extremely fast response times. For optimum performance each mode requires slightly different design, although a single

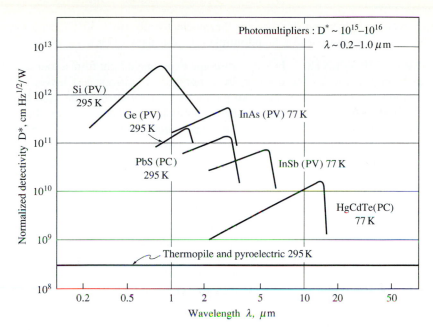

FIGURE 3-39
Typical spectral ranges and normalized detectivities for various detectors.

device may be operated in either mode. Typical semiconductor materials used for photovoltaic and photoconductive detectors are silicon (Si), germanium (Ge), indium antimonide (InSb), mercury cadmium telluride (HgCdTe),[8] lead sulfide and selenide (PbS and PbSe), and cadmium sulfide (CdS). The most basic photoemissive device is a *photodiode*, in which high-energy photons (ultraviolet to near infrared) cause emission of electrons from photocathode surfaces placed in a vacuum. Applying a voltage causes a current that is proportional to the intensity of incident radiation to flow. The signal of a vacuum photodiode is amplified in a *photomultiplier* by fitting it with a series of anodes (called *dynodes*), which produce secondary emission electrons and a current. The latter is an order-of-magnitude higher than the original photocurrent.

Thermal detectors generally respond evenly across the entire spectrum, while photon detectors have limited spectral response but higher detectivity and faster response times. The normalized detectivity of several detectors is compared in Fig. 3-39. The spectral response of photon detectors can be tailored to a degree by varying the relative amounts of detector material components. The response time of thermal detectors is relatively slow, normally in the order of milliseconds, while the response time of photon detectors ranges from microseconds to a few nanoseconds. The detectivity is

---

[8]Mercury-Cadmium-Telluride detectors are also commonly referred to as *MCT* detectors.

often increased by cooling the detector thermoelectrically (to $-30°C$), with dry ice (195 K), or by attaching it to a liquid-nitrogen Dewar flask (77 K).

OTHER COMPONENTS.   In a spectroscopic experiment light from a source and/or sample is guided toward the detector by a number of mirrors and lenses. *Plane mirrors* are employed to bend the beam path while *curved mirrors* are used to focus an otherwise diverging beam onto a sample, the monochromator entrance slit, or the detector. Today's optical mirrors provide extremely high reflectivities ($> 99.5\%$) over the entire spectrum of interest. While focusing mirrors are generally preferable for a number of reasons, sometimes *lenses* need to be used for focusing. The most important drawbacks of lenses are that they tend to have relatively large reflection losses and their spectral range (with high transmissivity) is limited. While antireflection coatings can be applied, these coatings are generally only effective over narrow spectral ranges as a result of interference effects. Common lens materials for the infrared are zinc selenide (ZnSe), calcium fluoride (CaF_2), germanium (Ge), and others. Sometimes it is necessary to split a beam into two portions (e.g., to create a reference beam that does not pass over the sample) using a *beam splitter*. Beam splitters are made of the same material as lenses, exploiting their reflecting *and* transmitting tendencies. It is also common to chop the beam using a *mechanical chopper*, which consists of a rotating blade with one or more holes or slits. Chopping may be done for a variety of reasons, such as to provide an alternating signal for a pyroelectric detector, to separate background radiation from desired radiation, to decrease electronic noise by using a lock-in amplifier tuned into the chopper frequency, and so on.

Calorimetric Emission Measurement Methods

If only knowledge of the total, hemispherical emissivity of a surface is required, this is most commonly determined by measuring the net radiative heat loss or gain of an isolated specimen [87–103]. Figure 3-40 shows a typical experimental setup, which was used by Funai [88]. The specimen is suspended inside an evacuated test chamber, the walls of which are coated with a near-black material. The chamber walls are cooled, while the specimen is heated electrically, directly (metallic samples), through a metal substrate (nonconducting samples), or by some other means. Temperatures of specimen and chamber wall are monitored by thermocouples. The emissivity of the sample can be determined from steady-state [87–95] or transient measurements [89, 96–103].

In the steady-state method the sample is heated to, and kept at, a desired temperature by passing the appropriate current through the heating element. The total, hemispherical emissivity may then be calculated by equating electric heat input to the specimen with the radiative heat loss from the specimen to the surroundings, or

$$\epsilon(T) = \frac{I^2 R}{A_s \sigma (T_s^4 - T_w)},\tag{3.102}$$

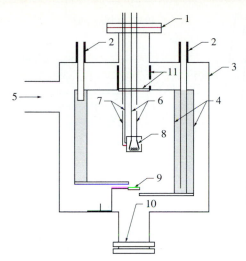

1. Vacuum feed-through flange
2. Coolant fill and vent tubes
3. Stainless steel vacuum jacket
4. Copper chamber walls
5. Vacuum inlet
6. Power leads
7. Thermocouple leads
8. Sample and heating element
9. Magnet-operated shutter
10. Sample viewing port
11. Radiation shields

FIGURE 3-40
Typical setup for calorimetric emission measurements [88].

where I^2R is the dissipated electrical power, A_s is the exposed surface area of the specimen, and T_s and T_w are the temperatures of specimen and chamber walls, respectively. As will be discussed in Chapter 5, equation (3.102) assumes that the surface area of the chamber is much larger than A_s and/or that the emissivity of the chamber wall is near unity [cf. equation (5.36)].

In the transient calorimetric technique the current is switched off when the desired temperature has been reached, and the rates of loss of internal energy and radiative heat loss are equated, or

$$\epsilon(T) = -\frac{m_s c_s \, dT_s/dt}{A_s \sigma (T_s^4 - T_w^4)},$$

(3.103)

where m_s and c_s are mass and specific heat of the sample, respectively.

Radiometric Emission Measurement Methods

High-temperature, spectral, directional surface emissivities are most often determined by comparing the emission from a sample with that from a blackbody at the same temperature and wavelength, both viewed by the same detector over an identical or equivalent optical path. Under those conditions the signal from both measurements will be proportional to emitted intensity (with the same proportionality constant), and the spectral, directional emissivity is found by taking the ratio of the two signals, or

$$\epsilon_\lambda'(T, \lambda, \theta, \psi) = \frac{I_\lambda(T, \lambda, \theta, \psi)}{I_{b\lambda}(T, \lambda)}.$$

(3.104)

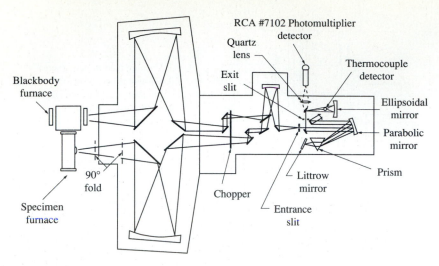

FIGURE 3-41
Emissometer with separate reference blackbody and two optical paths [104].

The comparison blackbody may be a separate blackbody kept at the same temperature, or it may be an integral part of the sample chamber. The latter is generally preferred at high temperatures, where temperature control is difficult, and for short wavelengths, where small deviations in temperatures can cause large inaccuracies.

Separate reference blackbody.   In this method a blackbody, usually a long, cylindrical, isothermal cavity with an *L/D*-ratio larger than 4, is kept separate from the sample chamber, while both are heated to the same temperature. Radiation coming from this *Hohlraum* is essentially black (cf. Table 5.1). The control system keeps sample and blackbody at the same temperature by monitoring temperature differences with a differential thermocouple and taking corrective action whenever necessary. To monitor sample and blackbody emission via an identical optical path, either two identical paths have to be constructed, or sample and blackbody must be alternately placed into the single optical path. In the former method, identical paths are formed either through two sets of optics [104], or by moving optical components back and forth [35]. Figure 3-41 shows an example of a system with two different optical paths [104], while Fig. 3-42 is an example of a linearly actuated blackbody/sample arrangement [105].

Integrated reference blackbody. At high temperatures it is preferable to incorporate the reference blackbody into the design of the sample furnace. If the sample rests at the bottom of a deep isothermal, cylindrical cavity, the radiation leaving the sample (by emission and reflection) corresponds to that of a black surface. If the hot side wall is removed or replaced by a cold one, radiation leaving the sample is due to emission only. Taking the ratio of the two signals then allows the determination of the spectral, directional emissivity from equation (3.104). Removing the reflection component

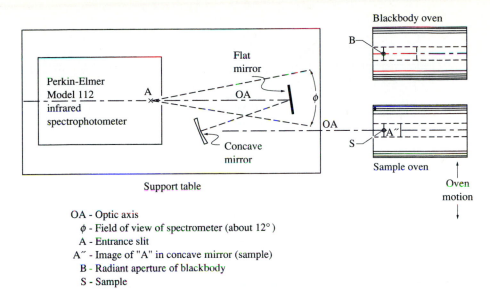

OA - Optic axis
ϕ - Field of view of spectrometer (about 12°)
A - Entrance slit
A″ - Image of "A" in concave mirror (sample)
B - Radiant aperture of blackbody
S - Sample

FIGURE 3-42
Emissometer with separate reference blackbody and linearly actuated sample/blackbody arrangement [105].

from the signal may be achieved in one of two ways. Several researchers have used a tubular furnace with the sample mounted on a movable rod [106–108]. When the sample is deep inside the furnace the signal corresponds to a blackbody. The sample is then rapidly moved to the exit of the furnace and the signal is due to emission alone. Disadvantages of the method are *(i)* maintaining isothermal conditions up to close to the end of the tube, *(ii)* keeping the sample at the same temperature after displacement, and *(iii)* stress on the high-temperature sample due to the rapid movement. In the approach of Vader and coworkers [109] and Sikka [110], reflection from the sample is suppressed by freely dropping a cold tube into the blackbody cavity. A schematic of the apparatus of Sikka is shown in Fig. 3-43. Once the cold tube has been dropped, measurements must be taken rapidly (in a few seconds' time), before substantial heating of the drop-tube (and cooling of the sample). Vader and coworkers obtained spectral measurements by placing various filters in front of their detector, performing a number of drops for each sample temperature. Sikka employed an FTIR spectrometer, allowing him to measure the entire spectral range from 1 μm to 9 μm in a single drop.

Reflection Measurements

Reflection measurements are carried out to determine the bidirectional reflection function, the directional-hemispherical reflectivity, and the hemispherical-directional

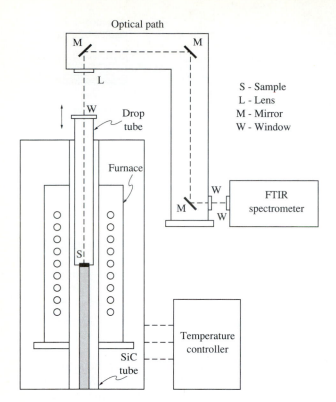

S - Sample
L - Lens
M - Mirror
W - Window

FIGURE 3-43
Schematic of a drop-tube
emissometer [110].

reflectivity. The latter two provide indirect means to determine the directional absorptivity and emissivity of opaque specimen, in particular, if sample temperatures are too low for emission measurements.

BIDIRECTIONAL REFLECTION MEASUREMENTS.   If the bidirectional reflection behavior of a surface is of interest, the bidirectional reflection function, ρ''_λ, must be measured directly, by irradiating the sample with a collimated beam from one direction and collecting the reflected intensity over various small solid angles. A sketch of an early apparatus used by Birkebak and Eckert [111] and Torrance and Sparrow [112] is shown in Fig. 3-44. Radiation from a globar A travels through a diaphragm B to a spherical mirror $M1$, which focuses it onto the test sample TS. A pencil of radiation reflected from the sample into the desired direction is collected by spherical mirror $M2$ and focused onto the entrance slit of the monochromator, in which the wavelengths are separated by the rock salt prism P, and the signal is recorded by the thermopile T. The test sample is mounted on a multiple-yoke apparatus, which allows independent rotation around three perpendicular axes. The resulting measurements are relative (i.e., absolute values can only be obtained by calibrating the apparatus with a known standard in place of the test sample). Example measurements for magnesium oxide are shown in Fig. 3-5 [12]. The main problem with bidirectional reflection measurements is the low level of reflected radi-

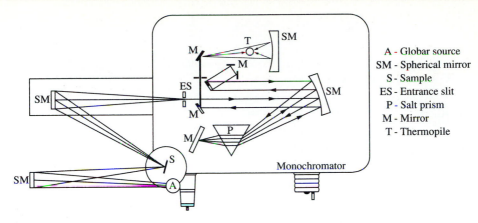

FIGURE 3-44
Schematic of the bidirectional reflection measurement apparatus of Birkebak and Eckert [111].

ation, particularly in off-specular directions, that must be detected. More recently, a number of designs have employed strong monochromatic laser sources to overcome this problem, for example, Hsia and Richmond [113], Greffet [114], and Al Hamwi and Sacadura [115].

An overview of the different methods to determine directional-hemispherical and hemispherical-directional reflectivities has been given by Touloukian and deWitt [6]. The different types of experiments may be grouped into three categories, *heated cavity reflectometers, integrating sphere reflectometers,* and *integrating mirror reflectometers,* each having their own ranges of applicability, advantages, and shortcomings.

HEATED CAVITY REFLECTOMETERS. The *heated cavity reflectometer* [6, 116–118] (sometimes known as the *Gier-Dunkle reflectometer* after its inventors [118]) consists of a uniformly heated enclosure fitted with a water-cooled sample holder and a viewport, as schematically shown in Fig. 3-45. Since the sample is situated within a more or less closed isothermal enclosure, the intensity striking it from any direction is essentially equal to the blackbody intensity $I_{b\lambda}(T_w)$ (evaluated at cavity-wall temperature, T_w). Images of the sample and a spot on the cavity wall are alternately focused onto the entrance slit of a monochromator. The signal from the specimen corresponds to emission (at the sample's temperature, T_s) plus reflection of the cavity-wall's blackbody intensity, $I_{b\lambda}(T_w)$. Since the signal from the cavity wall is proportional to $I_{b\lambda}(T_w)$, the ratio of the two signals corresponds to

$$\frac{I_s}{I_w} = \frac{\rho_\lambda^{\cap\prime}(\hat{s}) I_{b\lambda}(T_w) + \epsilon_\lambda'(\hat{s}) I_{b\lambda}(T_s)}{I_{b\lambda}(T_w)}. \tag{3.105}$$

If the sample is relatively cold ($T_s \ll T_w$), emission may be neglected and the device

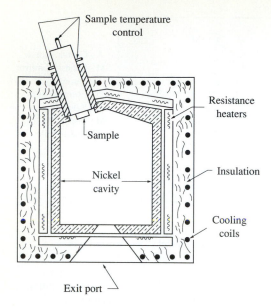

FIGURE 3-45
Schematic of a heated cavity reflectometer [116].

simply measures the hemispherical-directional reflectivity. For higher specimen temperatures, and for an opaque surface with diffuse irradiation, from equations (3.40), (3.37) and (3.42),

$$\rho_\lambda^{\cap\prime}(\hat{s}) \;=\; \rho_\lambda^{\prime\cap}(\hat{s}) \;=\; 1 - \alpha_\lambda^\prime(\hat{s}) \;=\; 1 - \epsilon_\lambda^\prime(\hat{s}), \tag{3.106}$$

and

$$\frac{I_s}{I_w} \;=\; 1 - \epsilon_\lambda^\prime(\hat{s})\left[1 - \frac{I_{b\lambda}(T_s)}{I_{b\lambda}(T_w)}\right]. \tag{3.107}$$

The principal source of error in this method is the difficulty in making the entire cavity reasonably isothermal and (as a consequence) making the reference signal proportional to a blackbody at the cavity-wall temperature. To make these errors less severe the method is generally only used for low sample temperatures.

INTEGRATING SPHERE REFLECTOMETERS.   These devices are most commonly employed for reflectivity measurements [117, 119–129] and are available commercially in a variety of forms, either as separate instruments or already incorporated into spectrophotometers.   A good early discussion of different designs was given by Edwards and coworkers [124].   The integrating sphere may be used to measure hemispherical-directional or directional-hemispherical reflectivity, depending on whether it is used in *indirect* or *direct mode*.   Schematics of integrating spheres operating in the two modes are shown in Fig. 3-46.   The ideal device is coated on its inside with a material of high and perfectly diffuse reflectivity.   The most common material in use is smoked magnesium oxide, which reflects strongly and very diffusely up to $\lambda \simeq 2.6\,\mu\mathrm{m}$ (cf. Fig. 3-5).   More recently, new materials such as "diffuse gold" [125–128] have been used to overcome the wavelength limitations.

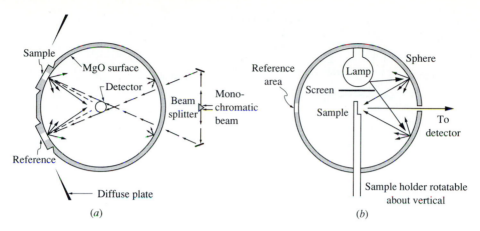

FIGURE 3-46
Typical integrating sphere reflectometers, *(a)* direct mode, *(b)* indirect mode.

The strong, diffuse reflectivity, together with the spherical geometry, assures that any external radiation hitting the surface of the sphere is converted into a perfectly diffuse intensity field due to many diffuse reflections. In the direct method the sample is illuminated directly by an external source, as shown in Fig. 3-46a. All of the reflected radiation is collected by the sphere and converted into a diffuse intensity field, which is measured by a detector. Similar readings are then taken on a comparison standard of known reflectivity, under the same conditions. The sample may be removed and replaced by the standard (*substitution method*); or there may be separate sample and standard holders, which are alternately irradiated by the external source (*comparison method*), the latter being generally preferred. In the indirect method a spot on the sphere surface is irradiated while the detector measures the intensity reflected by the sample (or the comparison standard) directly.

Errors in integrating sphere measurements are primarily caused by imperfections of the surface coating (imperfectly diffuse reflectivity), losses out of apertures, and unwanted irradiation onto the detector (direct reflection from the sample in the direct mode, direct reflection from the externally-irradiated spot on the sphere in the indirect mode). Because of temperature sensitivity of the diffuse coatings, integrating-sphere measurements have mostly been limited to moderate temperature levels to date.

INTEGRATING MIRROR REFLECTOMETERS.   An alternative to the integrating sphere is a similar design utilizing an integrating mirror. Mirrors in general have high reflectivities in the infrared and are much more efficient than integrating spheres and, hence, are highly desirable in the infrared where the energy of the light source is low. On the other hand, it is difficult to collect the radiant energy, reflected by the sample into the hemisphere above it, into a parallel beam of small cross section. For this reason, an integrating mirror reflectometer requires a large detector area. There are three types of integrating mirrors: *Hemispherical* [130],

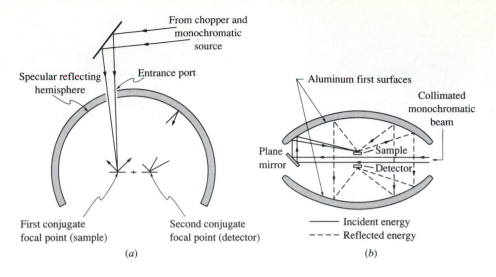

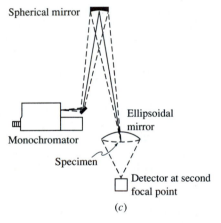

FIGURE 3-47
Design schematics of several integrating mirror reflectometers, using *(a)* a hemispherical, *(b)* a paraboloidal, *(c)* an ellipsoidal mirror.

paraboloidal [116, 131] and *ellipsoidal* [132–137]. Schematics of the three different types are shown in Fig. 3-47. The principle of operation of all three is the same, only the shape of the mirror is different. Each of these mirrors has two conjugate focal points, i.e., if a point source of light is placed at one focal point, all radiation will, after reflection off the mirror, fall onto the second focal point. Thus, in the integrating mirror technique an external beam is focused onto the sample, which is located at one of the focal points, through a small opening in the mirror. Radiation reflected from the sample into any direction will be reflected by the integrating mirror and is then collected by the detector located at the other focal point. This technique yields the directional-hemispherical reflectivity of the sample, after comparison with a reference signal. Sources for error in the integrating mirror method are absorption by the mirror, energy lost through the entrance port, nonuniform angular response of detectors, and energy missing the detector owing to mirror aberrations. To minimize aberrations, ellipsoids are preferable over hemispheres. The method has generally been limited to relatively large wavelengths, $> 2.5 \, \mu$m (because of mirror limitations), and to moderate temperatures. A design allowing sample temperatures up to $1000°$C has recently been reported by Battuello and coworkers [136]. In general, integrating mirrors are somewhat less popular than integrating spheres because mirrors are more sensitive to flux losses and misalignment errors.

References

1. Goldsmith, A., and T. E. Waterman: "Thermophysical Properties of Solid Materials," Technical Report WADC TR 58-476, Armour Research Foundation, 1959.
2. Gubareff, G. G., J. E. Janssen, and R. H. Torborg: "Thermal Radiation Properties Survey," 2d ed., Honeywell Research Center, Minneapolis, MI, 1960.
3. Wood, W. D., H. W. Deem, and C. F. Lucks: *Thermal Radiative Properties*, Plenum Publishing Company, New York, 1964.
4. Svet, D. I.: *Thermal Radiation: Metals, Semiconductors, Ceramics, Partly Transparent Bodies, and Films*, Plenum Publishing Company, New York, 1965.
5. Edwards, D. K., and I. Catton: "Radiation Characteristics of Rough and Oxidized Metals," in *Adv. Thermophys. Properties Extreme Temp. Pressures*, ed. S. Gratch, ASME, pp. 189–199, 1965.
6. Touloukian, Y. S., and D. P. DeWitt (eds.): *Thermal Radiative Properties: Metallic Elements and Alloys*, vol. 7 of *Thermophysical Properties of Matter*, Plenum Press, New York, 1970.
7. Touloukian, Y. S., and D. P. DeWitt (eds.): *Thermal Radiative Properties: Nonmetallic Solids*, vol. 8 of *Thermophysical Properties of Matter*, Plenum Press, New York, 1972.
8. Touloukian, Y. S., D. P. DeWitt, and R. S. Hernicz (eds.): *Thermal Radiative Properties: Coatings*, vol. 9 of *Thermophysical Properties of Matter*, Plenum Press, New York, 1973.
9. Schmidt, E., and E. R. G. Eckert: "Über die Richtungsverteilung der Wärmestrahlung von Oberflächen," *Forschung auf dem Gebiete des Ingenieurwesens*, vol. 7, p. 175, 1935.
10. McNicholas, H. J.: "Absolute Methods of Reflectometry," *J. Res. Natl. Bur. Std*, vol. 1, pp. 29–72, 1928.
11. Siegel, R., and J. R. Howell: *Thermal Radiation Heat Transfer*, 2nd ed., Hemisphere, New York, 1981.
12. Torrance, K. E., and E. M. Sparrow: "Biangular Reflectance of an Electric Nonconductor as a Function of Wavelength and Surface Roughness," *ASME Journal of Heat Transfer*, vol. 87, pp. 283–292, 1965.
13. Drude, P.: *Annalen der Physik*, vol. 39, p. 530, 1890.
14. Ehrenreich, H., H. R. Phillip, and B. Segall: "Optical Properties of Aluminum," *Physical Review*, vol. 132, no. 5, pp. 1918–1928, 1963.
15. Ehrenreich, H., and H. R. Phillip: "Optical Properties of Ag and Cu," *Physical Review*, vol. 128, no. 1, pp. 1622–1629, 1962.
16. Shiles, E., T. Sasaki, M. Inokuti, and D. Y. Smith: "Self-Consistency and Sum-Rule Tests in the Kramers-Kronig Analysis of Optical Data: Applications to Aluminum," *Physical Review B*, vol. 22, pp. 1612–1628, 1980.
17. Hagemann, H. J., W. Gudat, and C. Kunz: "Optical Constants from the Far Infrared to the X-ray Region: Mg, Al, Cu, Ag, Au, Bi, C and Al_2O_3," *Journal of the Optical Society of America*, vol. 65, pp. 742–744, 1975.
18. Parker, W. J., and G. L. Abbott: "Theoretical and Experimental Studies of the Total Emittance of Metals," in *Symposium on Thermal Radiation of Solids*, ed. S. Katzoff, NASA SP-55, pp. 11–28, 1965.

19. Ordal, M. A., L. L. Long, R. J. Bell, S. E. Bell, R. W. Alexander, and C. A. Ward: "Optical Properties of the Metals Al, Co, Cu, Au, Fe, Pb, Ni, Pd, Pt, Ag, Ti, and W in the Infrared and Far Infrared," *Applied Optics*, vol. 22, no. 7, pp. 1099–1119, 1983.

20. Hagen, E., and H. Rubens: "Metallic Reflection," *Annalen der Physik*, vol. 1, no. 2, pp. 352–375, 1900.

21. Reuter, G. E. H., and E. H. Sondheimer: "The Theory of the Anomalous Skin Effect in Metals," *Proc. Roy. Soc. (London) Series A*, vol. 195, no. 1042, pp. 336–364, 1948.

22. Weast, R. C. (ed.): *CRC Handbook of Chemistry and Physics*, 68th ed., Chemical Rubber Company, Cleveland, OH, 1988.

23. Brandenberg, W. M.: "The Reflectivity of Solids at Grazing Angles," in *Measurement of Thermal Radiation Properties of Solids*, ed. J. C. Richmond, NASA SP-31, pp. 75–82, 1963.

24. Brandenberg, W. M., and O. W. Clausen: "The Directional Spectral Emittance of Surfaces Between 200 and 600 C," in *Symposium on Thermal Radiation of Solids*, ed. S. Katzoff, NASA SP-55, pp. 313–319, 1965.

25. Price, D. J.: "The Emissivity of Hot Metals in the Infrared," *Proceedings of the Physical Society*, vol. 59, no. 331, pp. 118–131, 1947.

26. Dunkle, R. V.: "Thermal Radiation Characteristics of Surfaces," in *Theory and Fundamental Research in Heat Transfer*, ed. J. A. Clark, Pergamon Press, New York, pp. 1–31, 1963.

27. Dunkle, R. V.: "Emissivity and Inter-Reflection Relationships for Infinite Parallel Specular Surfaces," in *Symposium on Thermal Radiation of Solids*, ed. S. Katzoff, NASA SP-55, pp. 39–44, 1965.

28. Hering, R. G., and T. F. Smith: "Surface Radiation Properties from Electromagnetic Theory," *International Journal of Heat and Mass Transfer*, vol. 11, pp. 1567–1571, 1968.

29. Davisson, C., and J. R. Weeks: "The Relation Between the Total Thermal Emissive Power of a Metal and Its Electrical Resistivity," *Journal of the Optical Society of America*, vol. 8, no. 5, pp. 581–605, 1924.

30. Spitzer, W. G., D. A. Kleinman, C. J. Frosch, and D. J. Walsh: "Optical Properties of Silicon Carbide," in *Silicon Carbide — A High Temperature Semiconductor*, eds. J. R. O'Connor and J. Smiltens, Proceedings of the 1959 Conference on Silicon Carbide, Boston, Massachusetts, Pergamon Press, pp. 347–365, 1960.

31. Jasperse, J. R., A. Kahan, J. N. Plendl, and S. S. Mitra: "Temperature Dependence of Infrared Dispersion in Ionic Crystals LiF and MgO," *Physical Review*, vol. 140, no. 2, pp. 526–542, 1966.

32. Bohren, C. F., and D. R. Huffman: *Absorption and Scattering of Light by Small Particles*, John Wiley & Sons, New York, 1983.

33. *American Institute of Physics Handbook*, 3d ed., ch. 6, McGraw-Hill, New York, 1972.

34. Roy, S., S. Y. Bang, M. F. Modest, and V. S. Stubican: "Measurement of Spectral, Directional Reflectivities of Solids at High Temperatures Between 9 and 11 μm," in *Proceedings of the Third ASME/JSME Thermal Engineering Joint Conference*, vol. 4, pp. 19–26, 1991.

35. Riethof, T. R., and V. J. DeSantis: "Techniques of Measuring Normal Spectral Emissivity of Conductive Refractory Compounds at High Temperatures," in *Measurement of Thermal Radiation Properties of Solids*, ed. J. C. Richmond, NASA SP-31, pp. 565–584, 1963.

36. Mitra, S. S., and S. Nudelman: *Far-Infrared Properties of Solids*, Plenum Press, New York, 1970.

37. Beckmann, P., and A. Spizzichino: *The Scattering of Electromagnetic Waves from Rough Surfaces*, Macmillan, New York, 1963.

38. Bass, F. G., and I. M. Fuks: *Wave Scattering from Statistically Rough Surfaces*, Pergamon Press, Oxford, 1979.

39. Ogilvy, J. A.: "Wave Scattering from Rough Surfaces," *Reports on Progress in Physics*, vol. 50, pp. 1553–1608, 1987.

40. Davies, H.: "The Reflection of Electromagnetic Waves from a Rough Surface," *Proceedings of IEEE*, vol. 101, part IV, pp. 209–214, 1954.

41. Bennett, H. E.: "Specular Reflection of Aluminized Ground Glass and the Height Distribution of Surface Irregularities," *Journal of the Optical Society of America*, vol. 53, pp. 1389–1394, 1963.

42. Beckmann, P.: "Shadowing of Random Rough Surfaces," *IEEE Transactions on Antennas and Propagation*, vol. AP-13, pp. 384–388, 1965.

43. Torrance, K. E., and E. M. Sparrow: "Theory for Off-Specular Reflection from Roughened Surfaces," *Journal of the Optical Society of America*, vol. 57, no. 9, pp. 1105–1114, 1967.
44. Berry, M. V.: "Diffractals," *Journal of Physics A: Mathematical and General*, vol. 12, pp. 781–797, 1979.
45. Berry, M. V., and T. M. Blackwell: "Diffractal Echoes," *Journal of Physics A: Mathematical and General*, vol. 14, pp. 3101–3110, 1981.
46. Majumdar, A., and C. L. Tien: "Reflection of Radiation by Rough Fractal Surfaces," in *Radiation Heat Transfer: Fundamentals and Applications*, vol. HTD–137, ASME, pp. 27–35, June 1990.
47. Majumdar, A., and B. Bhushan: "Role of Fractal Geometry in Roughness Characterization and Contact Mechanics of Surfaces," *ASME Journal of Tribology*, vol. 112, pp. 205–216, 1990.
48. Majumdar, A., and C. L. Tien: "Fractal Characterization and Simulation of Rough Surfaces," in *Wear*, vol. 136, pp. 313–327, 1990.
49. Houchens, A. F., and R. G. Hering: "Bidirectional Reflectance of Rough Metal Surfaces," in *Thermophysics of Spacecrafts and Planetary Bodies*, ed. G. B. Heller, vol. 20, AIAA, Progress in Astronautics and Aeronautics, MIT Press, pp. 65–89, 1967.
50. Smith, T. F., and R. G. Hering: "Comparison of Bidirectional Reflectance Measurements and Model for Rough Metallic Surfaces," in *Proceedings of the Fifth Symposium on Thermophysical Properties*, Boston, ASME, pp. 429–435, 1970.
51. Bennett, H. E., M. Silver, and E. J. Ashley: "Infrared Reflectance of Aluminum Evaporated in Ultra-High Vacuum," *Journal of the Optical Society of America*, vol. 53, no. 9, pp. 1089–1095, 1963.
52. Schulz, L. G.: "The Experimental Studies of the Optical Properties of Metals and the Relation of the Results to the Drude Free Electron Theory," *Suppl. Phil. Mag.*, vol. 6, pp. 102–144, 1957.
53. Dunkle, R. V., and J. T. Gier: "Snow Characteristics Project Progress Report," Technical report, University of California at Berkeley, June 1953.
54. Bennett, H. E., J. M. Bennett, and E. J. Ashley: "Infrared Reflectance of Evaporated Aluminum Films," *Journal of the Optical Society of America*, vol. 52, pp. 1245–1250, 1962.
55. Spitzer, W. G., D. A. Kleinman, and D. J. Walsh: "Infrared Properties of Hexagonal Silicon Carbide," *Physical Review*, vol. 113, no. 1, pp. 127–132, January 1959.
56. Gaumer, R. E., E. R. Streed, and T. F. Vajta: "Methods for Experimental Determination of the Extra-Terrestrial Solar Absorptance of Spacecraft Materials," in *Measurement of Thermal Radiation Properties of Solids*, ed. J. C. Richmond, NASA SP-31, pp. 135–146, 1963.
57. Pezdirtz, G. F., and R. A. Jewell: "A Study of the Photodegradation of Selected Thermal Control Surfaces," in *Symposium on Thermal Radiation of Solids*, ed. S. Katzoff, NASA SP-55, pp. 433–441, 1965.
58. Hottel, H. C.: "Radiant Heat Transmission," in *Heat Transmission*, ed. W. H. McAdams, 3d ed., ch. 4, McGraw-Hill, New York, 1954.
59. Edwards, D. K., A. F. Mills, and V. E. Denny: *Transfer Processes*, 2nd ed., Hemisphere/McGraw-Hill, New York, 1979.
60. Wylie, C. R.: *Advanced Engineering Mathematics*, 5th ed., McGraw-Hill, New York, 1982.
61. Hsieh, C. K., and K. C. Su: "Thermal Radiative Properties of Glass from 0.32 to 206 μm," *Solar Energy*, vol. 22, pp. 37–43, 1979.
62. Gardon, R.: "The Emissivity of Transparent Materials," *Journal of The American Ceramic Society*, vol. 39, no. 8, pp. 278–287, 1956.
63. Neuroth, N.: "Der Einfluss der Temperatur auf die spektrale Absorption von Gläsern im Ultraroten, I (Effect of Temperature on Spectral Absorption of Glasses in the Infrared)," *Glastechnische Berichte*, vol. 25, pp. 242–249, 1952.
64. Yoldas, B. E., and D. P. Partlow: "Wide Spectrum Antireflective Coating for Fused Silica and Other Glasses," *Applied Optics*, vol. 23, no. 9, pp. 1418–1424, 1984.
65. Fan, J. C. C., and F. J. Bachner: "Transparent Heat Mirrors for Solar-Energy Applications," *Applied Optics*, vol. 15, no. 4, pp. 1012–1017, 1976.
66. Yoldas, B. E., and T. O'Keefe: "Deposition of Optically Transparent IR Reflective Coatings on Glass," *Applied Optics*, vol. 23, no. 20, pp. 3638–43, 1984.
67. Edwards, D. K.: "Solar Absorption by Each Element in an Absorber-Coverglass Array," *Solar Energy*, vol. 19, pp. 401–402, 1977.

68. Edwards, D. K.: "Finite Element Embedding with Optical Interference," *Presentation of the Twentieth Joint ASME/AIChE National Heat Transfer Conference*, vol. 81-HT-65, 1981.

69. Mitts, S. J., and T. F. Smith: "Solar Energy Transfer Through Semitransparent Systems," *Journal of Thermophysics and Heat Transfer*, vol. 1, no. 4, pp. 307–312, 1987.

70. Shurcliff, W. A.: "Transmittance and Reflection Loss of Multi-Plate Planar Window of a Solar-Radiation Collector: Formulas and Tabulations of Results for the Case of n =1.5," *Solar Energy*, vol. 16, pp. 149–154, 1974.

71. Duffie, J. A., and W. A. Beckman: *Solar Energy Thermal Processes*, John Wiley & Sons, New York, 1974.

72. Martin, D. C., and R. J. Bell: "The Use of Optical Interference to Obtain Selective Energy Absorption," in *Proceedings of the Conference on Coatings for the Aerospace Environment*, vol. WADD-TR-60-TB, 1960.

73. Hottel, H. C., and B. B. Woertz: "The Performance of Flat-Plate Solar-Heat Collectors," *Transactions of ASME, Journal of Heat Transfer*, vol. 64, pp. 91–104, 1942.

74. Gier, J. T., and R. V. Dunkle: "Selective Spectral Characteristics as an Important Factor in the Efficiency of Solar Collectors," in *Transactions of the Conference on the Use of Solar Energy*, vol. 2, Tucson, AZ, University of Arizona Press, p. 41, 1958.

75. Tabor, H., J. Harris, H. Weinberger, and B. Doron: "Further Studies on Selective Black Coatings," *Proceedings of the UN Conference on New Sources of Energy*, vol. 4, p. 618, 1964.

76. Tabor, H.: "Selective Surfaces for Solar Collectors," in *Low Temperature Engineering Applications of Solar Energy*, ASHRAE, 1967.

77. Edwards, D. K., K. E. Nelson, R. D. Roddick, and J. T. Gier: "Basic Studies on the Use and Control of Solar Energy," Technical Report 60-93, The University of California, Los Angeles, CA, 1960.

78. Thekaekara, M. P.: "Solar Energy Outside the Earth's Atmosphere," *Solar Energy*, vol. 14, pp. 109–127, 1973.

79. Trombe, F., M. Foex, and V. LePhat: "Research on Selective Surfaces for Air Conditioning Dwellings," *Proceedings of the UN Conference on New Sources of Energy*, vol. 4, pp. 625–638, 1964.

80. Hollands, K. G. T.: "Directional Selectivity, Emittance, and Absorptance Properties of Vee Corrugated Specular Surfaces," *Solar Energy*, vol. 7, no. 3, pp. 108–116, 1963.

81. Perlmutter, M., and J. R. Howell: "A Strongly Directional Emitting and Absorbing Surface," *ASME Journal of Heat Transfer*, vol. 85, no. 3, pp. 282–283, 1963.

82. Richmond, J. C. (ed.): *Measurement of Thermal Radiation Properties of Solids*, NASA SP-31, 1963.

83. Katzoff, S. (ed.): *Symposium on Thermal Radiation Properties of Solids*, NASA SP-55, 1964.

84. Sacadura, J. F.: "Measurement Techniques for Thermal Radiation Properties," in *Proceedings of the Ninth International Heat Transfer Conference*, Washington, D.C., Hemisphere, pp. 207–222, 1990.

85. Hutley, M. C.: *Diffraction Gratings*, Academic Press, New York, 1982.

86. Griffiths, P. R., and J. A. de Haseth: *Fourier Transform Infrared Spectrometry*, vol. 83 of *Chemical Analysis*, John Wiley & Sons, New York, 1986.

87. Sadler, R., L. Hemmerdinger, and I. Rando: "A Device for Measuring Total Hemispherical Emittance," in *Measurement of Thermal Radiation Properties of Solids*, ed. J. C. Richmond, NASA SP-31, pp. 217–223, 1963.

88. Funai, A. I.: "A Multichamber Calorimeter for High-Temperature Emittance Studies," in *Measurement of Thermal Radiation Properties of Solids*, ed. J. C. Richmond, NASA SP-31, pp. 317–327, 1963.

89. McElroy, D. L., and T. G. Kollie: "The Total Hemispherical Emittance of Platinum, Columbium-1%, Zirconium, and Polished and Oxidized Iron-8 in the Range 100°C to 1200°C," in *Measurement of Thermal Radiation Properties of Solids*, ed. J. C. Richmond, NASA SP-31, pp. 365–379, 1963.

90. Moore, V. S., A. R. Stetson, and A. G. Metcalfe: "Emittance Measurements of Refractory Oxide Coatings up to 2900°C," in *Measurement of Thermal Radiation Properties of Solids*, ed. J. C. Richmond, NASA SP-31, pp. 527–533, 1963.

91. Nyland, T. W.: "Apparatus for the Measurement of Hemispherical Emittance and Solar Absorptance from 270°C to 650°C," in *Measurement of Thermal Radiation Properties of Solids*, ed. J. C. Richmond, NASA SP-31, pp. 393–401, 1963.

92. Zerlaut, G. A.: "An Apparatus for the Measurement of the Total Normal Emittance of Surfaces at Satellite Temperatures," in *Measurement of Thermal Radiation Properties of Solids*, ed. J. C. Richmond, NASA SP-31, pp. 275–285, 1963.

93. Chen, S. H. P., and S. C. Saxena: "Experimental Determination of Hemispherical Total Emittance of Metals as a Function of Temperature," *Ind. Eng. Chem. Fundam.*, vol. 12, no. 2, pp. 220–224, 1973.

94. Jody, B. J., and S. C. Saxena: "Radiative Heat Transfer from Metal Wires: Hemispherical Total Emittance of Platinum," *Journal of Physics E: Scientific Instruments*, vol. 9, pp. 359–362, 1976.

95. Taylor, R. E.: "Determination of Thermophysical Properties by Direct Electrical Heating," *High Temperatures - High Pressures*, vol. 13, pp. 9–22, 1981.

96. Gordon, G. D., and A. London: "Emittance Measurements at Satellite Temperatures," in *Measurement of Thermal Radiation Properties of Solids*, ed. J. C. Richmond, NASA SP-31, pp. 147–151, 1963.

97. Rudkin, R. L.: "Measurement of Thermal Properties of Metals at Elevated Temperatures," in *Temperature, Its Measurement and Control in Science and Industry*, vol. 3, part 2, Reinhold Publishing Corp., New York, pp. 523–534, 1962.

98. Gaumer, R. E., and J. V. Stewart: "Calorimetric Determination of Infrared Emittance and the α/ϵ Ratio," in *Measurement of Thermal Radiation Properties of Solids*, ed. J. C. Richmond, NASA SP-31, pp. 127–133, 1963.

99. Butler, C. P., and R. J. Jenkins: "Space Chamber Emittance Measurements," in *Measurement of Thermal Radiation Properties of Solids*, ed. J. C. Richmond, NASA SP-31, pp. 39–43, 1963.

100. Butler, C. P., and E. C. Y. Inn: "A Method for Measuring Total Hemispherical and Emissivity of Metals," in *First Symposium - Surface Effects on Spacecraft Materials*, New York, John Wiley & Sons, pp. 117–137, 1960.

101. Smalley, R., and A. J. Sievers: "The Total Hemispherical Emissivity of Copper," *Journal of the Optical Society of America*, vol. 68, pp. 1516–1518, 1978.

102. Ramanathan, K. G., and S. H. Yen: "High-Temperature Emissivities of Copper, Aluminum and Silver," *Journal of the Optical Society of America*, vol. 67, pp. 32–38, 1977.

103. Masuda, H., and M. Higano: "Measurement of Total, Hemispherical Emissivities of Metal Wires by Using Transient Calorimetric Techniques," *ASME Journal of Heat Transfer*, vol. 110, pp. 166–172, 1988.

104. Limperis, T., D. M. Szeles, and W. L. Wolfe: "The Measurement of Total Normal Emittance of Three Nuclear Reactor Materials," in *Measurement of Thermal Radiation Properties of Solids*, ed. J. C. Richmond, NASA SP-31, pp. 357–364, 1963.

105. Fussell, W. B., and F. Stair: "Preliminary Studies Toward the Determination of Spectral Absorption Coefficients of Homogeneous Dielectric Material in the Infrared at Elevated Temperatures," in *Symposium on Thermal Radiation of Solids*, ed. S. Katzoff, NASA SP-55, pp. 287–292, 1965.

106. Knopken, S., and R. Klemm: "Evaluation of Thermal Radiation at High Temperatures," in *Measurement of Thermal Radiation Properties of Solids*, ed. J. C. Richmond, NASA SP-31, pp. 505–514, 1963.

107. Bennethum, W. H.: "Thin Film Sensors and Radiation Sensing Techniques for Measurement of Surface Temperature of Ceramic Components," in *HITEMP Review, Advanced High Temperature Engine Materials Technology Program*, NASA CP-10039, 1989.

108. Atkinson, W. H., and M. A. Cyr: "Sensors for Temperature Measurement for Ceramic Materials," in *HITEMP Review, Advanced High Temperaure Engine Materials Technology Program*, NASA CP-10039, pp. 287–292, 1989.

109. Vader, D. T., R. Viskanta, and F. P. Incropera: "Design and Testing of a High-Temperature Emissometer for Porous and Particulate Dielectrics," *Rev. Sci. Instrum.*, vol. 57, no. 1, pp. 87–93, 1986.

110. Sikka, K. K.: "High Temperature Normal Spectral Emittance of Silicon Carbide Based Materials," M.S. thesis, The Pennsylvania State University, University Park, PA, 1991.

111. Birkebak, R. C., and E. R. G. Eckert: "Effect of Roughness of Metal Surfaces on Angular Distribution of Monochromatic Reflected Radiation," *ASME Journal of Heat Transfer*, vol. 87, pp. 85–94, 1965.

112. Torrance, K. E., and E. M. Sparrow: "Off-Specular Peaks in the Directional Distribution of Reflected Thermal Radiation," *ASME Journal of Heat Transfer*, vol. 88, pp. 223–230, 1966.

113. Hsia, J. J., and J. C. Richmond: "A High Resolution Laser Bidirectional Reflectometer," *Journal Research of N.B.S.*, vol. 80A, no. 2, pp. 189–220, 1976.

114. Greffet, J. J.: "Design of a Fully Automated Bidirectional Laser Reflectometer," in *Applications to Emissivity Measurement, Proceedings of SPIE on Stray Light and Contamination in Optical Systems*, pp. 184–191, 1988.

115. Al Hamwi, M., and J. F. Sacadura: "Méthode de détermination des propriétes radiatives spectrales et directionnelles, dans le proche et moyen I.R., de surfaces opaques métalliques et non-métalliques," *Proceedings of JITH '89*, pp. 126–136, November 1989.

116. Dunkle, R. V.: "Spectral Reflection Measurements," in *First Symposium - Surface Effects on Spacecraft Materials*, New York, John Wiley & Sons, pp. 117–137, 1960.

117. Hembach, R. J., L. Hemmerdinger, and A. J. Katz: "Heated Cavity Reflectometer Modifications," in *Measurement of Thermal Radiation Properties of Solids*, ed. J. C. Richmond, NASA SP-31, pp. 153–167, 1963.

118. Gier, J. T., R. V. Dunkle, and J. T. Bevans: "Measurement of Absolute Spectral Reflectivity from 1.0 to 15 Microns," *Journal of the Optical Society of America*, vol. 44, pp. 558–562, 1954.

119. Fussell, W. B., J. J. Triolo, and F. A. Jerozal: "Portable Integrating Sphere for Monitoring Reflectance of Spacecraft Coatings," in *Measurement of Thermal Radiation Properties of Solids*, ed. J. C. Richmond, NASA SP-31, pp. 103–116, 1963.

120. Drummeter, L. F., and E. Goldstein: "Vanguard Emittance Studies at NRL," in *First Symposium - Surface Effects on Spacecraft Materials*, New York, John Wiley & Sons, pp. 152–163, 1960.

121. Snail, K. A., and L. M. Hangsen: "Integrating Sphere Designs with Isotropic Throughput," *Applied Optics*, vol. 28, pp. 1793–1799, May 1989.

122. Egan, W. G., and T. Hilgeman: "Integrating Spheres for Measurements between 0.185 μm and 12 μm," *Applied Optics*, vol. 14, pp. 1137–1142, May 1975.

123. Kneissl, G. J., and J. C. Richmond: "A Laser Source Integrating Sphere Reflectometer," Technical Report NBS-TN-439, National Bureau of Standards, 1968.

124. Edwards, D. K., J. T. Gier, K. E. Nelson, and R. D. Roddick: "Integrating Sphere for Imperfectly Diffuse Samples," *Applied Optics*, vol. 51, pp. 1279–1288, 1961.

125. Willey, R. R.: "Fourier Transform Infrared Spectrophotometer for Transmittance and Diffuse Reflectance Measurements," *Applied Spectroscopy*, vol. 30, pp. 593–601, 1976.

126. Richter, W.: "Fourier Transform Reflectance Spectrometry Between 8000 cm^{-1} (1.25 μm) and 800 cm^{-1} (12.5μm) Using an Integrating Sphere," *Applied Spectroscopy*, vol. 37, pp. 32–38, 1983.

127. Gindele, K., M. Köhl, and M. Mast: "Spectral Reflectance Measurements Using an Integrating Sphere in the Infrared," *Applied Optics*, vol. 24, pp. 1757–1760, 1985.

128. Richter, W., and W. Erb: "Accurate Diffuse Reflection Measurements in the Infrared Spectral Range," *Applied Optics*, vol. 26, no. 21, pp. 4620–4624, November 1987.

129. Sheffer, D., U. P. Oppenheim, D. Clement, and A. D. Devir: "Absolute Reflectometer for the 0.8–2.5μm Region," *Applied Optics*, vol. 26, no. 3, pp. 583–586, 1987.

130. Janssen, J. E., and R. H. Torborg: "Measurement of Spectral Reflectance Using an Integrating Hemisphere," in *Measurement of Thermal Radiation Properties of Solids*, ed. J. C. Richmond, NASA SP-31, pp. 169–182, 1963.

131. Neher, R. T., and D. K. Edwards: "Far Infrared Reflectometer for Imperfectly Diffuse Specimens," *Applied Optics*, vol. 4, pp. 775–780, 1965.

132. Neu, J. T.: "Design, Fabrication and Performance of an Ellipsoidal Spectroreflectometer," *NASA CR 73193*, 1968.

133. Dunn, S. T., J. C. Richmond, and J. F. Panner: "Survey of Infrared Measurement Techniques and Computational Methods in Radiant Heat Transfer," *Journal of Spacecraft and Rockets*, vol. 3, pp. 961–975, July 1966.

134. Heinisch, R. P., F. J. Bradar, and D. B. Perlick: "On the Fabrication and Evaluation of an Integrating Hemi-Ellipsoid," *Applied Optics*, vol. 9, no. 2, pp. 483–489, 1970.

135. Wood, B. E., P. G. Pipes, A. M. Smith, and J. A. Roux: "Hemi-Ellipsoidal Mirror Infrared Reflectometer: Development and Operation," *Applied Optics*, vol. 15, no. 4, pp. 940–950, 1976.
136. Battuello, M., F. Lanza, and T. Ricolfi: "Infrared Ellipsoidal Mirror Reflectometer for Measurements Between Room Temperature and 1000°C," *High Temperature*, vol. 18, pp. 683–688, 1986.
137. Snail, K. A.: "Reflectometer Design Using Nonimaging Optics," *Applied Optics*, vol. 26, no. 24, pp. 5326–5332, 1987.
138. Hale, G. M., and M. R. Querry: "Optical Constants of Water in the 200nm to 200 μm Wavelength Region," *Applied Optics*, vol. 12, pp. 555–563, 1973.

Problems

3.1 A diffusely emitting surface at 500 K has a spectral, directional emissivity that can be approximated by 0.5 in the range $0 < \lambda < 5\,\mu\text{m}$ and 0.3 for $\lambda > 5\,\mu\text{m}$. What is the total, hemispherical emissivity of this surface surrounded by (a) air and (b) a dielectric medium of refractive index $n = 2$?

3.2 A certain material at 600 K has the following spectral, directional emissivity:

$$\epsilon_\lambda' = \begin{cases} 0.9\cos\theta, & \lambda < 1\,\mu\text{m}, \\ 0.2, & \lambda > 1\,\mu\text{m}, \end{cases}$$

(a) What is the total, hemispherical emissivity of the material?

(b) If the sun irradiates this surface at an angle of $\theta = 60°$ off-normal, what is the relevant total absorptivity?

(c) What is the net radiative energy gain or loss of this surface (per unit time and area)?

3.3 A long, cylindrical antenna of 1 cm radius on an earth-orbiting satellite is coated with a material whose emissivity is

$$\epsilon_\lambda' = \begin{cases} 0, & \lambda < 1\,\mu\text{m}, \\ \cos\theta, & \lambda \geq 1\,\mu\text{m}. \end{cases}$$

Find the absorbed energy per meter length. (Assume irradiation is from the sun only; neglect the earth and stars.)

3.4 The spectral, hemispherical emissivity of a (hypothetical) metal may be approximated by the relationship

$$\epsilon_\lambda = \begin{cases} 0.5, & \lambda < \lambda_c = 0.5\,\mu\text{m}, \\ 0.5\lambda_c/\lambda, & \lambda > \lambda_c, \end{cases}$$

(independent of temperature). Determine the total, hemispherical emissivity of this material using (a) Planck's law, (b) Wien's distribution, for a surface temperature of (i) 300 K, (ii) 1000 K. How accurate is the prediction using Wien's distribution?

3.5 A treated metallic surface is used as a solar collector material; its spectral, directional emissivity may be approximated by

$$\epsilon_\lambda' = \begin{cases} 0.5\,\mu\text{m}/\lambda, & \theta < 45°, \\ 0, & \theta > 45°. \end{cases}$$

What is the relevant α/ϵ-ratio for near normal solar incidence if $T_{\text{coll}} \simeq 600\,\text{K}$?

3.6 A surface sample with $\epsilon'_\lambda = 0.9 \cos \theta$ for $\lambda < 2\,\mu$m, and $\epsilon'_\lambda = 0.2$ for $\lambda > 2\,\mu$m is irradiated by three tungsten lights as shown. The tungsten lights may be approximated by black spheres at $T = 2000\,$K fitted with mirrors to produce parallel light beams aimed at the sample. Neglecting background radiation, determine the absorptivity of the sample.

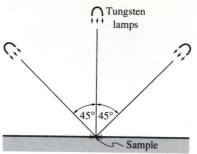

3.7 A metal ($m_2 = 50 - 50\,i$) is coated with a dielectric ($m_1 = 2 - 0\,i$), which is exposed to vacuum.

   (a) What is the range of possible directions from which radiation can impinge on the metal?

   (b) What is the normal reflectivity of the dielectric-metal interface?

   (c) What is the (approximate) relevant hemispherical reflectivity for the dielectric-metal interface?

3.8 For a certain material, temperature and wavelength the spectral, hemispherical emissivity has been measured as ϵ_λ. Estimate the refractive index of the material under these conditions, assuming the material to be (a) a dielectric with $\epsilon_\lambda = 0.8$, (b) a metal in the infrared with $\epsilon_\lambda = 0.2$ (the Hagen-Rubens relation being valid).

3.9 It can be derived from electromagnetic wave theory that

$$\frac{\epsilon_\lambda}{\epsilon_{n\lambda}} \simeq \frac{4}{3} - \frac{1}{4}\epsilon_{n\lambda} \quad \text{for} \quad \epsilon_{n\lambda} \ll 1.$$

Determine ϵ_λ for metals with $\epsilon_{n\lambda} \ll 1$ as a function of wavelength and temperature.

3.10 A solar collector surface with emissivity $\epsilon'_\lambda = 0.9 \cos \theta$ for $\lambda < 2\,\mu$m and $\epsilon'_\lambda = 0.2$ for $\lambda > 2\,\mu$m is to be kept at $T_c = 500\,$K. For $q_{\text{sol}} = 1300\,$W/m^2, what is the range of possible sun positions with respect to the surface for which at least 50% of the maximum net radiative energy is collected? Neglect conduction and convection losses from the surface.

3.11 Determine the total, normal emissivity of copper, silver and gold for a temperature of 1500 K. Check your results by comparing with Fig. 3-8.

3.12 Determine the total, hemispherical emissivity of copper, silver and gold for a temperature of 1500 K. Check your results by comparing with Fig. 3-11.

3.13 A polished platinum sphere is heated until it is glowing red. An observer is stationed a distance away, from where the sphere appears as a red disk. Using the various aspects of electromagnetic wave theory and/or Fig. 3-9 and Table 3.1, explain how the brightness of emitted radiation would vary across the disk, if observed with (a) the human eye, (b) an infrared camera.

3.14 Two aluminum plates, one covered with a layer of white enamel paint, the other polished, are directly facing the sun, which is irradiating the plates with 1000 W/m^2. Assuming that convection/conduction losses of the plates to the environment at 300 K can be calculated by using a heat transfer coefficient of 10 W/m^{2}K, and that the back sides of the plates are insulated, estimate the equilibrium temperature of each plate.

3.15 Consider a metallic surface coated with a dielectric layer.

(a) Show that the fraction of energy reflected at the vacuum-dielectric interface is negligible ($n_1 = 1.2$; $k_1 = 0$).

(b) Develop an expression for the normal, spectral emissivity for the metal substrate, similar to the Hagen-Rubens relationship.

(c) Develop an approximate relation for the directional, spectral emissivity of the metal substrate for large wavelengths and moderate incidence angles, say $\theta < 75°$.

3.16 A plate of metal with $n_2 = k_2 = 100$ is covered with a dielectric as shown. The dielectric has an absorption band such that $n_1 = 2$, and $k_1 = 1$ for $0.2\,\mu m < \lambda < 2\,\mu m$ and $k_1 = 0$ elsewhere. The dielectric is thick enough, such that any light traveling through it of wavelengths $0.2\,\mu m < \lambda < 2\,\mu m$ is entirely absorbed before it reaches the metal.

vacuum, $n_0 = 1$

_____

dielectric, n_1, k_1

//

metal, n_2, k_2

(a) What is the total, normal emissivity of the composite if its temperature is 400 K?

(b) What is the total, normal absorptivity if sun shines perpendicularly onto the composite?

3.17 Estimate the total, normal emissivity of α-SiC for a temperature of (i) 300 K, (ii) 1000 K. You may assume the spectral, normal emissivity to be independent of temperature.

3.18 Estimate the total, hemispherical emissivity of a thick slab of pure silicon at room temperature.

3.19 Estimate and compare the total, normal emissivity of room temperature aluminum for the surface finishes given in Fig. 3-24.

3.20 A satellite orbiting earth has part of its (flat) surface coated with spectrally selective "black nickel," which is a diffuse emitter and whose spectral emissivity may be approximated by

$$\epsilon_\lambda = \begin{cases} 0.9, & \lambda < 2\,\mu m, \\ 0.25, & \lambda > 2\,\mu m. \end{cases}$$

Assuming the back of the surface to be insulated, and the front exposed to solar irradiation of $q_{sol} = 1353\,W/m^2$ (normal to the surface), determine the relevant α/ϵ-ratio for the surface. What is its equilibrium temperature? What would be its equilibrium temperature if the surface is turned away from the sun, such that the sun's rays strike it at a polar angle of $\theta = 60°$?

3.21 Repeat Problem 3.20 for white paint on aluminum, whose diffuse emissivity may be approximated by

$$\epsilon_\lambda = \begin{cases} 0.1, & \lambda < 2\,\mu m, \\ 0.9, & \lambda > 2\,\mu m. \end{cases}$$

3.22 Estimate the spectral, hemispherical emissivity of the grooved materials shown in Fig. 3-35. Repeat Problem 3.20 for these materials, assuming them to be gray.

3.23 Repeat Problem 1.7 for a sphere covered with the grooved material of Fig. 3-35, whose directional, spectral emissivity may be approximated by

$$\epsilon_\lambda' = \begin{cases} 0.9, & 0 \le \theta < 40°, \\ 0.0, & 40° < \theta < 90°. \end{cases}$$

Assume the material to be gray.

3.24 A solar collector consists of a metal plate coated with "black nickel." The collector is irradiated by the sun with a strength of $q_{sol} = 1000 \, \mathrm{W/m^2}$ from a direction which is $\theta = 30°$ from the surface normal. On its top the surface loses heat by radiation and by free convection (heat transfer coefficient $h_1 = 10 \, \mathrm{W/m^2 K}$), both to an atmosphere at $T_{amb} = 20°C$. The bottom surface delivers heat to the collector fluid ($h_2 = 50 \, \mathrm{W/m^2 K}$), which flows past the surface at $T_{fluid} = 20°C$. What is the equilibrium temperature of the collector plate? How much energy (per unit area) is collected (i.e., carried away by the fluid)? Discuss the performance of this collector. Assume black nickel to be a diffuse emitter.

3.25 Make a qualitative plot of temperature vs. the total hemispherical emissivity of

   (a) a 3 mm thick sheet of window glass,
   (b) polished aluminum, and
   (c) an ideal metal, which obeys the Hagen-Rubens relation.

3.26 A horizontal sheet of 5 mm thick glass is covered with a 2 mm thick layer of water. If solar radiation is incident normal to the sheet, what are the transmissivity and reflectivity of the water/glass layer at $\lambda_1 = 0.6 \, \mu m$ and $\lambda_2 = 2 \, \mu m$? For water $m_{H_2O}(0.6 \, \mu m) = 1.332 - 1.09 \times 10^{-8} \, i$, $m_{H_2O}(2 \, \mu m) = 1.306 - 1.1 \times 10^{-3} \, i$ [138]; for glass $m_{glass}(0.6 \, \mu m) = 1.517 - 6.04 \times 10^{-7} \, i$, $m_{glass}(2 \, \mu m) = 1.497 - 5.89 \times 10^{-5} \, i$ [61].

3.27 A solar collector plate of spectral absorptivity $\alpha_{coll} = 0.90$ is fitted with two sheets of 5 mm thick glass as shown in the adjacent sketch. What fraction of normally incident solar radiation is absorbed by the collector plate at a wavelength of $0.6 \, \mu m$? At $0.6 \, \mu m \, m_{glass} = 1.517 - 6.04 \times 10^{-7} \, i$ [61].

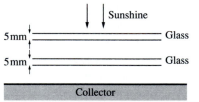

CHAPTER
4

VIEW FACTORS

4.1 INTRODUCTION

In many engineering applications the exchange of radiative energy between surfaces is virtually unaffected by the medium that separates them. Such (radiatively) *non-participating media* include vacuum as well as monatomic and most diatomic gases (including air) at low to moderate temperature levels (i.e., before ionization and dissociation occurs). Examples include spacecraft heat rejection systems, solar collector systems, radiative space heaters, illumination problems, and so on.

In the following four chapters we shall consider the analysis of *surface radiation transport,* i.e., radiative heat transfer in the absence of a participating medium, for different levels of complexity. It is common practice to simplify the analysis by making the assumption of an *idealized enclosure* and/or of *ideal surface properties.*

The greatest simplification arises if all surfaces are black: For such a situation no reflected radiation needs to be accounted for, and all emitted radiation is diffuse (i.e., the intensity leaving a surface does not depend on direction). The next level of difficulty arises if surfaces are assumed to be gray, diffuse emitters (and, thus, absorbers) as well as gray, diffuse reflectors. The vast majority of engineering calculations are limited to such ideal surfaces, which are the topic of Chapter 5.

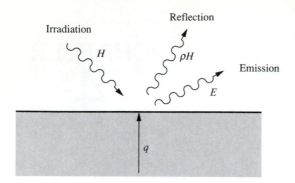

FIGURE 4-1
Surface energy balance.

If the reflective behavior of a surface deviates strongly from a diffuse reflector (e.g., a polished metal, which reflects almost like a mirror) one may often approximate the reflectivity to consist of a purely diffuse and a purely specular component. This situation is discussed in Chapter 6. However, if greater accuracy is desired, if the reflectivity cannot be approximated by purely diffuse and specular components, or if the assumption of a gray surface is not acceptable, a more general approach must be taken. A few such methods are briefly outlined in Chapter 7.

As discussed in Chapter 1 thermal radiation is generally a long-range phenomenon. This is *always* the case in the absence of a participating medium, since photons will travel unimpeded from surface to surface. Therefore, performing a thermal radiation analysis for one surface implies that all surfaces, no matter how far removed, that can exchange radiative energy with one another must be considered simultaneously. How much energy any two surfaces exchange depends in part on their size, separation distance, and orientation, leading to geometric functions known as *view factors*. In the present chapter these view factors are developed for gray, diffusely radiating (i.e., emitting and reflecting) surfaces. However, the view factor is a very basic function that will also be employed in the analysis of specular reflectors as well as for the analysis for surfaces with arbitrary emission and reflection properties.

Making an energy balance on a surface element, as shown in Fig. 4-1, we find

$$q = q_{\text{emission}} - q_{\text{absorption}} = E - \alpha H. \tag{4.1}$$

In this relation q_{emission} and $q_{\text{absorption}}$ are absolute values with directions as given by Fig. 4-1, while q is the net heat flux supplied to the surface, as defined in Chapter 1 by equation (1.35). According to this definition q is positive if the heat is coming from inside the wall material, by conduction or other means ($q > 0$), and negative if going from the enclosure into the wall ($q < 0$). Alternatively, the heat flux may be expressed as

$$q = q_{\text{out}} - q_{\text{in}} = (q_{\text{emission}} + q_{\text{reflection}}) - q_{\text{irradiation}}$$
$$= (E + \rho H) - H, \tag{4.2}$$

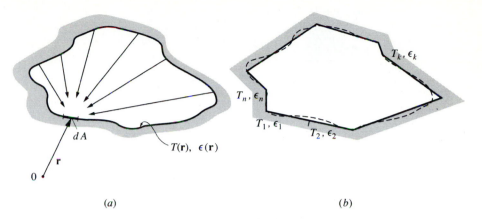

FIGURE 4-2
(a) Irradiation from different locations in an enclosure. (b) Real and ideal enclosures for radiative transfer calculations.

which is, of course, the same as equation (4.1) since, for opaque surfaces, $\rho = 1 - \alpha$. The irradiation H depends, in general, on the level of emission from surfaces far removed from the point under consideration, as schematically indicated in Fig. 4-2a. Thus, in order to make a radiative energy balance we always need to consider an entire *enclosure* rather than an infinitesimal control volume (as is normally done for other modes of heat transfer, i.e., conduction or convection). The enclosure must be *closed* so that irradiation from all possible directions can be accounted for, and the enclosure surfaces must be *opaque* so that all irradiation is accounted for, for each direction. In practice, an incomplete enclosure may be closed by introducing artificial surfaces. An enclosure may be idealized in two ways, as indicated in Fig. 4-2b: By replacing a complex geometrical shape with a few simple surfaces, and by assuming surfaces to be isothermal with constant (i.e., average) heat flux values across them. Obviously, the idealized enclosure approaches the real enclosure for sufficiently small isothermal subsurfaces.

4.2 DEFINITION OF VIEW FACTORS

To make an energy balance on a surface element, equation (4.1), the irradiation H must be evaluated. In a general enclosure the irradiation will have contributions from all visible parts of the enclosure surface. Therefore, we need to determine how much energy leaves an arbitrary surface element dA' that travels toward dA. The geometric relations governing this process for "diffuse" surfaces (for surfaces that absorb and emit diffusely, and also reflect radiative energy diffusely) are known as *view factors*. Other names used in the literature are *configuration factor*, *angle factor*, and *shape factor*, and sometimes the term *diffuse view factor* is used (to distinguish

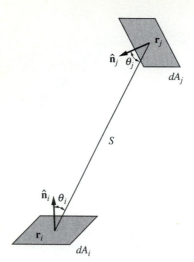

FIGURE 4-3
Radiative exchange between two infinitesimal surface elements.

from *specular view factors* for specularly reflecting surfaces; see Chapter 6). The view factor between two infinitesimal surface elements dA_i and dA_j, as shown in Fig. 4-3, is defined as

$$dF_{dA_i - dA_j} \equiv \frac{\text{diffuse energy leaving } dA_i \text{ directly toward and intercepted by } dA_j}{\text{total diffuse energy leaving } dA_i}, \quad (4.3)$$

where the word "directly" is meant to imply "on a straight path, without intervening reflections." This view factor is infinitesimal since only an infinitesimal *fraction* can be intercepted by an infinitesimal area. From the definition of intensity and Fig. 4-3 we may determine the heat transfer rate from dA_i to dA_j as

$$I(\mathbf{r}_i)(dA_i \cos\theta_i)\, d\Omega_j = I(\mathbf{r}_i)\cos\theta_i \cos\theta_j\, dA_i\, dA_j / S^2, \quad (4.4)$$

where θ_i (or θ_j) is the angle between the surface normal $\hat{\mathbf{n}}_i$ (or $\hat{\mathbf{n}}_j$) and the line connecting dA_i and dA_j (of length S). The total radiative energy leaving dA_i into the hemisphere above it is $J = E + \rho H$, where J is called the *radiosity*. Since the surface emits and reflects diffusely both E and ρH obey equation (1.30), and the outgoing flux may be related to intensity by

$$J(\mathbf{r}_i)\, dA_i = [E(\mathbf{r}_i) + \rho(\mathbf{r}_i)H(\mathbf{r}_i)]\, dA_i = \pi I(\mathbf{r}_i)dA_i.$$

Note that the radiative intensity away from dA_i, due to emission and/or reflection, does not depend on direction. Therefore, the view factor between two infinitesimal areas is

$$dF_{dA_i - dA_j} = \frac{\cos\theta_i \cos\theta_j}{\pi S^2}\, dA_j. \quad (4.5)$$

By introducing the abbreviation $\mathbf{s}_{ij} = \mathbf{r}_j - \mathbf{r}_i$, and noting that $\cos \theta_i = \hat{\mathbf{n}}_i \cdot \mathbf{s}_{ij}/|\mathbf{s}_{ij}|$, the view factor may be recast in vector form as

$$dF_{dA_i - dA_j} = \frac{(\hat{\mathbf{n}}_i \cdot \mathbf{s}_{ij})(\hat{\mathbf{n}}_j \cdot \mathbf{s}_{ji})}{\pi S^4} \, dA_j. \tag{4.6}$$

Switching subscripts i and j in equation (4.5) immediately leads to the important *law of reciprocity,*

$$dA_i \, dF_{dA_i - dA_j} = dA_j \, dF_{dA_j - dA_i}. \tag{4.7}$$

Often, enclosures are idealized to consist of a number of finite isothermal subsurfaces, as indicated in Fig. 4-2b. Therefore, we should like to expand the definition of the view factor to include radiative exchange between one infinitesimal and one finite area, and between two finite areas. Consider first the exchange between an infinitesimal dA_i and a finite A_j, as shown in Fig. 4-4. The total energy leaving dA_i toward all of A_j is, from equation (4.4),

$$I(\mathbf{r}_i) \, dA_i \int_{A_j} \frac{\cos \theta_i \cos \theta_j}{S^2} \, dA_j,$$

while the total energy leaving the dA_i into all directions remains unchanged. Thus, we find

$$F_{dA_i - A_j} = \int_{A_j} \frac{\cos \theta_i \cos \theta_j}{\pi S^2} \, dA_j, \tag{4.8}$$

which is now finite since the intercepting surface, A_j, is finite.

Next we consider the view factor from A_j to the infinitesimal dA_i. The amount of radiation leaving all of A_j toward dA_i is, from equation (4.4) (after switching subscripts i and j),

$$dA_i \int_{A_j} I(\mathbf{r}_j) \frac{\cos \theta_i \cos \theta_j}{S^2} \, dA_j,$$

and the total amount leaving A_j into all directions is

$$\pi \int_{A_j} I(\mathbf{r}_j) \, dA_j.$$

Thus, we find the view factor between surfaces A_j and dA_i is

$$dF_{A_j - dA_i} = \int_{A_j} I(\mathbf{r}_j) \frac{\cos \theta_i \cos \theta_j}{S^2} \, dA_j \, dA_i \bigg/ \pi \int_{A_j} I(\mathbf{r}_j) \, dA_j, \tag{4.9}$$

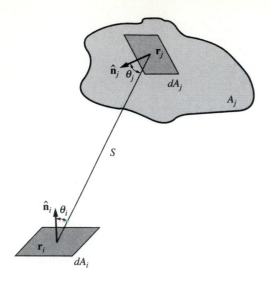

FIGURE 4-4
Radiative exchange between one infinitesimal
and one finite surface element.

which is infinitesimal since the intercepting surface, dA_i, is infinitesimal. The view factor in equation (4.9)—unlike equations (4.5) and (4.8)—is not a purely geometric parameter since it depends on the radiation field $I(\mathbf{r}_j)$. However, for an ideal enclosure as shown in Fig. 4-2b, it is usually assumed that the intensity leaving any surface is not only diffuse but also does not vary across the surface, i.e., $I(\mathbf{r}_j) = I_j = \text{const}$. With this assumption equation (4.9) becomes

$$dF_{A_j-dA_i} = \frac{1}{A_j} \int_{A_j} \frac{\cos\theta_i \cos\theta_j}{\pi S^2} dA_j \, dA_i. \tag{4.10}$$

Comparing this with equation (4.8) we find another law of reciprocity, with

$$A_j \, dF_{A_j-dA_i} = dA_i F_{dA_i-A_j}, \tag{4.11}$$

subject to the restriction that the intensity leaving A_j does not vary across the surface.

Finally, we consider radiative exchange between two finite areas A_i and A_j as depicted in Fig. 4-5. The total energy leaving A_i toward A_j is, from equation (4.4),

$$\int_{A_i} \int_{A_j} I(\mathbf{r}_i) \frac{\cos\theta_i \cos\theta_j}{S^2} dA_j \, dA_i,$$

and the view factor follows as

$$F_{A_i-A_j} = \int_{A_i} \int_{A_j} I(\mathbf{r}_i) \frac{\cos\theta_i \cos\theta_j}{S^2} dA_j \, dA_i \bigg/ \pi \int_{A_i} I(\mathbf{r}_i) \, dA_i. \tag{4.12}$$

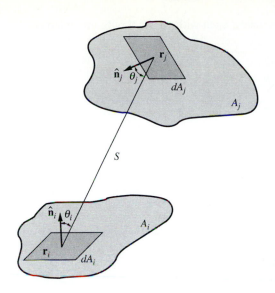

FIGURE 4-5
Radiative exchange between two finite surfaces.

If we assume again that the intensity leaving A_i does not vary across the surface, the view factor reduces to

$$F_{A_i-A_j} = \frac{1}{A_i} \int_{A_i} \int_{A_j} \frac{\cos\theta_i \cos\theta_j}{S^2} \, dA_j \, dA_i. \tag{4.13}$$

The *law of reciprocity* follows readily as

$$A_i F_{A_i-A_j} = A_j F_{A_j-A_i}, \tag{4.14}$$

which is now subject to the condition that the radiation intensities leaving A_i and A_j must both be constant across their respective surfaces.

In a somewhat more compact notation, the law of reciprocity may be summarized as

$$dA_i \, dF_{di-dj} = dA_j \, dF_{dj-di}, \tag{4.15a}$$

$$dA_i F_{di-j} = A_j \, dF_{j-di}, \qquad (I_j = \text{const}), \tag{4.15b}$$

$$A_i F_{i-j} = A_j F_{j-i}, \qquad (I_i, I_j = \text{const}). \tag{4.15c}$$

The different levels of view factors may be related to one another by

$$F_{di-j} = \int_{A_j} dF_{di-dj}, \tag{4.16a}$$

$$F_{i-j} = \frac{1}{A_i} \int_{A_i} F_{di-j} \, dA_i. \tag{4.16b}$$

Finally, an enclosure consisting of N surfaces, each with constant outgoing intensities, obeys the *summation relation,*

$$\sum_{j=1}^{N} F_{di-j} = \sum_{j=1}^{N} F_{i-j} = 1. \tag{4.17}$$

This follows directly from the definition of the view factor (i.e., the sum of all fractions must add up to unity). Note that equation (4.17) includes the view factor F_{i-i}. If surface A_i is flat or convex, no radiation leaving it will strike itself directly, and F_{i-i} simply vanishes. However, if A_i is concave, part of the radiation leaving it will be intercepted by itself and $F_{i-i} > 0$.

4.3   METHODS FOR THE EVALUATION OF VIEW FACTORS

The calculation of a radiative view factor between any two finite surfaces requires the solution to a double area integral, or a fourth-order integration. Such integrals are exceedingly difficult to evaluate analytically except for very simple geometries. Even numerical quadrature may often be problematic because of singularities in the integrand, and because of excessive CPU time requirements. Therefore, considerable effort has been directed towards tabulation and the development of evaluation methods for view factors. Early tables and charts for simple configurations were given by Hamilton and Morgan [1], Leuenberger and Pearson [2], and Kreith [3]. Fairly extensive tabulations were given in the books by Sparrow and Cess [4] and Siegel and Howell [5]. Siegel and Howell also give an exhaustive listing of sources for more involved view factors. The most complete tabulation is given in a catalogue by Howell [6]. Some experimental methods have been discussed by Jakob [7] and Liu and Howell [8]. Within the present book Appendix D gives view factor formulae for an extensive set of geometries, as well as a graphical representation for three very basic configurations.

Radiation view factors may be determined by a variety of methods. One possible grouping of different approaches could be:

1. Direct integration:

   (*i*) analytical or numerical integration of the relations given in the previous section (surface integration);

   (*ii*) conversion of the relations to contour integrals, followed by analytical or numerical integration (contour integration);

2. Statistical determination: View factors may be determined through statistical sampling with the *Monte Carlo method;*

3. Special methods: For many simple shapes integration can be avoided by employing one of the following special methods:

(*i*) view factor algebra, i.e., repeated application of the *rules of reciprocity* and the *summation relationship;*

(*ii*) crossed-strings method: A simple method for evaluation of view factors in two-dimensional geometries;

(*iii*) unit sphere method: A powerful method for view factors between one infinitesimal and one finite area;

(*iv*) inside sphere method: A simple method for a few special shapes.

All of the above methods will be discussed in the following pages, except for the *Monte Carlo method,* which is treated in considerable detail in Chapter 19.

4.4   AREA INTEGRATION

To evaluate equation (4.5) or to carry out the integrations in equations (4.8) and (4.13) the integrand (i.e., $\cos\theta_i$, $\cos\theta_j$ and S) must be known in terms of a local coordinate system that describes the geometry of the two surfaces. While the evaluation of the integrand may be straightforward for some simple configurations, it is desirable to have a more generally applicable formula at one's disposal. Using an arbitrary coordinate origin, a vector pointing from the origin to a point on a surface may be written as

$$\mathbf{r} = x\hat{\mathbf{i}} + y\hat{\mathbf{j}} + z\hat{\mathbf{k}}, \tag{4.18}$$

where $\hat{\mathbf{i}}$, $\hat{\mathbf{j}}$ and $\hat{\mathbf{k}}$ are *unit vectors* pointing into the x-, y-, and z-directions, respectively. Thus the vector from dA_i going to dA_j is determined (see Fig. 4-5) as

$$\mathbf{s}_{ij} = -\mathbf{s}_{ji} = \mathbf{r}_j - \mathbf{r}_i = (x_j - x_i)\hat{\mathbf{i}} + (y_j - y_i)\hat{\mathbf{j}} + (z_j - z_i)\hat{\mathbf{k}}. \tag{4.19}$$

The length of this vector is determined as

$$|\mathbf{s}_{ij}|^2 = |\mathbf{s}_{ji}|^2 = S^2 = (x_j - x_i)^2 + (y_j - y_i)^2 + (z_j - z_i)^2. \tag{4.20}$$

We will now assume that the local surface normals are also known in terms of the unit vectors $\hat{\mathbf{i}}$, $\hat{\mathbf{j}}$ and $\hat{\mathbf{k}}$, or, from Fig. 4-6,

$$\hat{\mathbf{n}} = l\hat{\mathbf{i}} + m\hat{\mathbf{j}} + n\hat{\mathbf{k}}, \tag{4.21}$$

where l, m, and n are the *direction cosines* for the unit vector $\hat{\mathbf{n}}$, i.e., $l = \hat{\mathbf{n}} \cdot \hat{\mathbf{i}} = \cos\theta_x$ is the cosine of the angle θ_x between $\hat{\mathbf{n}}$ and the x-axis, etc. We may now evaluate $\cos\theta_i$ and $\cos\theta_j$ as

$$\cos\theta_i = \frac{\hat{\mathbf{n}}_i \cdot \mathbf{s}_{ij}}{S} = \frac{1}{S}[(x_j - x_i)l_i + (y_j - y_i)m_i + (z_j - z_i)n_i], \tag{4.22a}$$

$$\cos\theta_j = \frac{\hat{\mathbf{n}}_j \cdot \mathbf{s}_{ji}}{S} = \frac{1}{S}[(x_i - x_j)l_j + (y_i - y_j)m_j + (z_i - z_j)n_j]. \tag{4.22b}$$

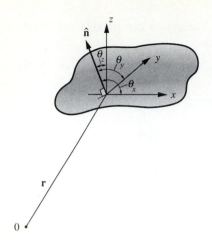

FIGURE 4-6
Unit normal and direction cosines for a surface element.

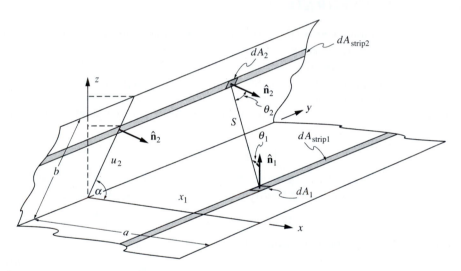

FIGURE 4-7
View factor for strips on infinitely long groove.

Example 4.1.   Consider the infinitely long $(-\infty < y < +\infty)$ wedge-shaped groove as shown in Fig. 4-7. The groove has sides of widths a and b and an opening angle α. Determine the view factor between the narrow strips shown in the figure.

Solution

After placing the coordinate system as shown in the figure, we find $z_1 = 0$, $x_2 = u_2 \cos \alpha$ and $z_2 = u_2 \sin \alpha$, leading to

$$
\begin{aligned}
S^2 &= (x_1 - u_2 \cos \alpha)^2 + (y_1 - y_2)^2 + u_2^2 \sin^2 \alpha \\
&= (x_1^2 - 2x_1 u_2 \cos \alpha + u_2^2) + (y_1 - y_2)^2 = S_0^2 + (y_1 - y_2)^2,
\end{aligned}
$$

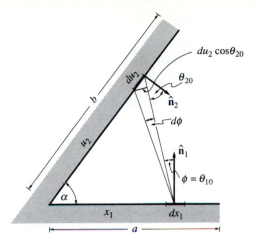

FIGURE 4-8
Two-dimensional wedge-shaped groove with projected distances.

where S_0 is the projection of S in the x-z-plane and is constant in the present problem. The two surface normals are readily determined as

$$\hat{\mathbf{n}}_1 = \hat{\mathbf{k}}, \quad \text{or} \quad l_1 = m_1 = 0, \quad n_1 = 1,$$
$$\hat{\mathbf{n}}_2 = \hat{\mathbf{i}} \sin\alpha - \hat{\mathbf{k}} \cos\alpha, \quad \text{or} \quad l_2 = \sin\alpha, \; m_2 = 0, \; n_2 = -\cos\alpha,$$

leading to

$$\cos\theta_1 = u_2 \sin\alpha / S,$$
$$\cos\theta_2 = [(x_1 - u_2 \cos\alpha)\sin\alpha + u_2 \sin\alpha \cos\alpha]/S = x_1 \sin\alpha/S.$$

For illustrative purposes we will first calculate $dF_{d1-\text{strip}2}$ from equation (4.8), and then determine $dF_{\text{strip}1-\text{strip}2}$ from equation (4.16). Thus

$$dF_{d1-\text{strip}2} = \int_{dA_{\text{strip}2}} \frac{\cos\theta_1 \cos\theta_2}{\pi S^2} dA_2 = \frac{du_2}{\pi} \int_{-\infty}^{+\infty} \frac{x_1 u_2 \sin^2\alpha \, dy_2}{[S_0^2 + (y_1 - y_2)^2]^2}$$

$$= \frac{x_1 u_2 \sin^2\alpha \, du_2}{\pi} \left[\frac{y_2 - y_1}{2S_0^2 [S_0^2 + (y_1 - y_2)^2]} + \frac{1}{2S_0^3} \tan^{-1} \frac{y_2 - y_1}{S_0} \right]_{-\infty}^{+\infty}$$

$$= \frac{x_1 u_2 \sin^2\alpha \, du_2}{2S_0^3} = \frac{1}{2} \frac{u_2 \sin\alpha}{S_0} \frac{x_1 \sin\alpha}{S_0} \frac{du_2}{S_0} = \frac{1}{2} \cos\theta_{10} \cos\theta_{20} \frac{du_2}{S_0},$$

where θ_{10} and θ_{20} are the projections of θ_1 and θ_2 in the x-z-plane. Looking at Fig. 4-8 this may be rewritten as

$$dF_{d1-\text{strip}2} = \tfrac{1}{2} \cos\phi \, d\phi,$$

where $\phi = \theta_{10}$ is the off-normal angle at which $dA_{\text{strip}2}$ is oriented from $dA_{\text{strip}1}$. We note that $dF_{d1-\text{strip}2}$ does not depend on y_1. No matter where on *strip 1* an observer is standing, he sees the same *strip 2* extending from $-\infty$ to $+\infty$. It remains to calculate $dF_{\text{strip}1-\text{strip}2}$ from equation (4.16). Since equation (4.16) simply takes an average, and since $dF_{d1-\text{strip}2}$

does not vary along $dA_{strip\,1}$, it follows immediately that

$$dF_{strip\,1-strip\,2} = \tfrac{1}{2}\cos\phi\,d\phi = \frac{x_1\sin^2\alpha\,u_2 du_2}{2S_0^3}.$$

Example 4.2. Determine the view factor F_{1-2} for the infinitely long groove shown in Fig. 4-8.

Solution

Since we already know the view factor between two infinite strips, we can write

$$F_{strip\,1-2} = \int_0^b dF_{strip\,1-strip\,2},$$

$$F_{1-2} = \frac{1}{a}\int_0^a F_{strip\,1-2}\,dx_1.$$

Therefore, from Example 4.1,

$$F_{strip\,1-2} = \frac{x_1\sin^2\alpha}{2}\int_0^b \frac{u_2\,du_2}{(x_1^2 - 2x_1 u_2\cos\alpha + u_2^2)^{3/2}}$$

$$= \frac{x_1\sin^2\alpha}{2}\left.\frac{x_1\cos\alpha\,u_2 - x_1^2}{x_1^2\sin^2\alpha\,\sqrt{x_1^2 - 2x_1 u_2\cos\alpha + u_2^2}}\right|_0^b$$

$$= \frac{1}{2}\left(1 + \frac{b\cos\alpha - x_1}{\sqrt{x_1^2 - 2bx_1\cos\alpha + b^2}}\right).$$

Finally, carrying out the second integration we obtain

$$F_{1-2} = \frac{1}{a}\int_0^a F_{strip\,1-2}\,dx_1 = \frac{1}{2}\left(1 - \frac{1}{a}\left.\sqrt{x_1^2 - 2bx_1\cos\alpha + b^2}\right|_0^a\right)$$

$$= \frac{1}{2}\left(1 + \frac{b}{a} - \sqrt{1 - 2\frac{b}{a}\cos\alpha + \left(\frac{b}{a}\right)^2}\right).$$

Example 4.3. As a final example for area integration we shall consider the view factor between two parallel, coaxial disks of radius R_1 and R_2, respectively, as shown in Fig. 4-9.

Solution

Placing x-, y-, and z-axes as shown in the figure, and making a coordinate transformation to cylindrical coordinates, we find

$$x_1 = r_1\cos\psi_1,\ y_1 = r_1\sin\psi_1,\ z_1 = 0;\quad dA_1 = r_1\,dr_1\,d\psi_1;$$
$$x_2 = r_2\cos\psi_2,\ y_2 = r_2\sin\psi_2,\ z_2 = h;\quad dA_2 = r_2\,dr_2\,d\psi_2;$$

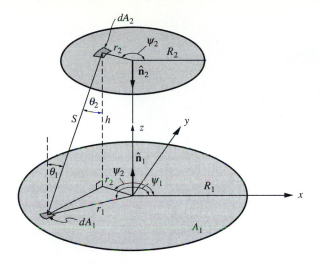

FIGURE 4-9
Coordinate systems for the view factor between parallel, coaxial disks.

$$S^2 = (r_1 \cos \psi_1 - r_2 \cos \psi_2)^2 + (r_1 \sin \psi_1 - r_2 \sin \psi_2)^2 + h^2$$
$$= h^2 + r_1^2 + r_2^2 - 2r_1 r_2 \cos(\psi_1 - \psi_2).$$

Since $\hat{\mathbf{n}}_1 = \hat{\mathbf{k}}$ and $\hat{\mathbf{n}}_2 = -\hat{\mathbf{k}}$, we also find $l_1 = l_2 = m_1 = m_2 = 0$, $n_1 = -n_2 = 1$, and from equation (4.22) $\cos \theta_1 = \cos \theta_2 = h/S$. Thus, from equation (4.13)

$$F_{1-2} = \frac{1}{(\pi R_1^2)\pi} \int_{r_1=0}^{R_1} \int_{r_2=0}^{R_2} \int_{\psi_1=0}^{2\pi} \int_{\psi_2=0}^{2\pi} \frac{h^2 r_1 r_2 \, d\psi_2 \, d\psi_1 \, dr_2 \, dr_1}{[h^2 + r_1^2 + r_2^2 - 2r_1 r_2 \cos(\psi_1 - \psi_2)]^2}.$$

Changing the dummy variable ψ_2 to $\psi = \psi_1 - \psi_2$ makes the integrand independent of ψ_1 (integrating from $\psi_1 - 2\pi$ to ψ_1 is the same as integrating from 0 to 2π, since integration is over a full period), so that the ψ_1-integration may be carried out immediately:

$$F_{1-2} = \frac{2h^2}{\pi R_1^2} \int_{r_1=0}^{R_1} \int_{r_2=0}^{R_2} \int_{\psi=0}^{2\pi} \frac{r_1 r_2 \, d\psi \, dr_2 \, dr_1}{(h^2 + r_1^2 + r_2^2 - 2r_1 r_2 \cos \psi)^2}.$$

This result can also be obtained by physical argument, since the view factor from any pie-slice of A_1 must be the same (and equal to the one from the entire disk). While a second integraton (over r_1, r_2, or ψ) can be carried out, analytical evaluation of the remaining two integrals appears bleak. We shall abandon the problem here in the hope of finding another method with which we can evaluate F_{1-2} more easily.

4.5 CONTOUR INTEGRATION

According to *Stokes' theorem*, as developed in standard mathematics texts such as Wylie [9], a surface integral may be converted to an equivalent contour integral (see Fig. 4-10) through

$$\oint_{\Gamma} \mathbf{f} \cdot d\mathbf{s} = \int_A (\nabla \times \mathbf{f}) \cdot \hat{\mathbf{n}} \, dA, \qquad (4.23)$$

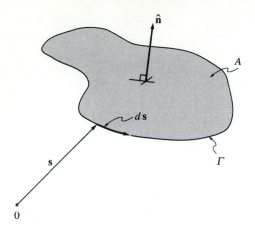

FIGURE 4-10
Conversion between surface and contour integral;
Stokes' theorem.

where $\mathbf{f}$ is a vector function defined everywhere on the surface A, including its boundary Γ, $\hat{\mathbf{n}}$ is the unit surface normal, and $\mathbf{s}$ is the position vector for a point on the boundary of A ($d\mathbf{s}$, therefore, is the vector describing the boundary contour of A). By convention the contour integration in equation (4.23) is carried out in the counterclockwise sense for an observer standing atop the surface (i.e., on the side from which the normal points up).

If a vector function $\mathbf{f}$ that makes the integrand of equation (4.23) equivalent to the one of equation (4.8) can be identified, then the area (or double) integral of equation (4.8) can be reduced to a contour (or single) integral. Applying Stokes' theorem twice, the double area integration of equation (4.13) could be converted to a double line integral. Contour integration was first applied to radiative view factor calculations (in the field of illumination engineering) by Moon [10]. The earliest applications to radiative heat transfer appear to have been by de Bastos [11] and Sparrow [12].

View Factors from Differential Elements to Finite Areas

For this case the vector function $\mathbf{f}$ may be identified as

$$\mathbf{f} = \frac{1}{2\pi} \frac{\mathbf{s}_{12} \times \hat{\mathbf{n}}_1}{S^2}, \tag{4.24}$$

leading to

$$F_{d1-2} = \frac{1}{2\pi} \oint_{\Gamma_2} \frac{(\mathbf{s}_{12} \times \hat{\mathbf{n}}_1) \cdot d\mathbf{s}_2}{S^2}, \tag{4.25}$$

where $\mathbf{s}_{12}$ is the vector pointing from dA_1 to a point on the contour of A_2 (described by vector $\mathbf{s}_2$), while $d\mathbf{s}_2$ points along the contour of A_2.

For the interested reader with some background in vector calculus we shall briefly prove that equation (4.25) is equivalent to equation (4.8). Using the identity (given, e.g., by Wylie [9]),

$$\nabla \times (\varphi \mathbf{a}) = \varphi \nabla \times \mathbf{a} - \mathbf{a} \times \nabla \varphi, \tag{4.26}$$

we may write[1]

$$2\pi \nabla_2 \times \mathbf{f} = \nabla_2 \times \left(\frac{\mathbf{s}_{12} \times \hat{\mathbf{n}}_1}{S^2} \right) = \frac{1}{S^2} \nabla_2 \times (\mathbf{s}_{12} \times \hat{\mathbf{n}}_1) - (\mathbf{s}_{12} \times \hat{\mathbf{n}}_1) \times \nabla_2 \left(\frac{1}{S^2} \right). \tag{4.27}$$

From equations (4.19) and (4.20) it follows that

$$\nabla_2 \left(\frac{1}{S^2} \right) = -\frac{2}{S^3} \nabla_2 S = -\frac{2}{S^3} \frac{\mathbf{s}_{12}}{S} = -\frac{2\mathbf{s}_{12}}{S^4}.$$

We also find, using standard vector identities,

$$(\mathbf{s}_{12} \times \hat{\mathbf{n}}_1) \times \mathbf{s}_{12} = \hat{\mathbf{n}}_1(\mathbf{s}_{12} \cdot \mathbf{s}_{12}) - \mathbf{s}_{12}(\mathbf{s}_{12} \cdot \hat{\mathbf{n}}_1) = S^2 \hat{\mathbf{n}}_1 - \mathbf{s}_{12}(\mathbf{s}_{12} \cdot \hat{\mathbf{n}}_1), \tag{4.28a}$$
$$\nabla_2 \times (\mathbf{s}_{12} \times \hat{\mathbf{n}}_1) = \hat{\mathbf{n}}_1 \cdot \nabla_2 \mathbf{s}_{12} - \mathbf{s}_{12} \cdot \nabla_2 \hat{\mathbf{n}}_1 + \mathbf{s}_{12} \nabla_2 \cdot \hat{\mathbf{n}}_1 - \hat{\mathbf{n}}_1 \nabla_2 \cdot \mathbf{s}_{12}. \tag{4.28b}$$

In the last expression the terms $\nabla_2 \hat{\mathbf{n}}_1$ and $\nabla_2 \cdot \hat{\mathbf{n}}_1$ drop out since $\hat{\mathbf{n}}_1$ is independent of surface A_2. Also, from equation (4.19) we find

$$\nabla_2 \cdot \mathbf{s}_{12} = 3, \quad \nabla_2 \mathbf{s}_{12} = \hat{\mathbf{i}}\hat{\mathbf{i}} + \hat{\mathbf{j}}\hat{\mathbf{j}} + \hat{\mathbf{k}}\hat{\mathbf{k}} = \boldsymbol{\delta}, \tag{4.29}$$

where $\boldsymbol{\delta}$ is the unit tensor whose diagonal elements are unity and whose nondiagonal elements are zero:

$$\boldsymbol{\delta} = \begin{pmatrix} 1 & 0 & 0 \\ 0 & 1 & 0 \\ 0 & 0 & 1 \end{pmatrix}. \tag{4.30}$$

With $\hat{\mathbf{n}}_1 \cdot \boldsymbol{\delta} = \hat{\mathbf{n}}_1$ equation (4.28b) reduces to

$$\nabla_2 \times (\mathbf{s}_{12} \times \hat{\mathbf{n}}_1) = \hat{\mathbf{n}}_1 - 3\hat{\mathbf{n}}_1 = -2\hat{\mathbf{n}}_1.$$

Substituting all this into equation (4.27), we obtain

$$2\pi \nabla_2 \times \mathbf{f} = -\frac{2\hat{\mathbf{n}}_1}{S^2} + \frac{2}{S^4} \left[S^2 \hat{\mathbf{n}}_1 - \mathbf{s}_{12}(\mathbf{s}_{12} \cdot \hat{\mathbf{n}}_1) \right] = -\frac{2}{S^4} \mathbf{s}_{12}(\mathbf{s}_{12} \cdot \hat{\mathbf{n}}_1),$$

and

$$(\nabla_2 \times \mathbf{f}) \cdot \hat{\mathbf{n}}_2 = -\frac{(\mathbf{s}_{12} \cdot \hat{\mathbf{n}}_1)(\mathbf{s}_{12} \cdot \hat{\mathbf{n}}_2)}{\pi S^4} = \frac{\cos \theta_1 \cos \theta_2}{\pi S^2}. \tag{4.31}$$

---

[1] We add the subscript 2 to all operators to make clear that differentiation is with respect to position coordinates on A_2, for example, x_2, y_2, and z_2 if a Cartesian coordinate system is employed.

Together with Stokes' theorem this completes the proof that equation (4.25) is equivalent to an area integral over the function given by equation (4.31).

For a Cartesian coordinate system, using equations (4.18) through (4.21), we have

$$d\mathbf{s}_2 = dx_2\hat{\mathbf{i}} + dy_2\hat{\mathbf{j}} + dz_2\hat{\mathbf{k}},$$

and equation (4.25) becomes

$$
\begin{aligned}
F_{d1-2} &= \frac{l_1}{2\pi} \oint_{\Gamma_2} \frac{(z_2-z_1)\,dy_2 - (y_2-y_1)\,dz_2}{S^2} \\
&+ \frac{m_1}{2\pi} \oint_{\Gamma_2} \frac{(x_2-x_1)\,dz_2 - (z_2-z_1)\,dx_2}{S^2} \\
&+ \frac{n_1}{2\pi} \oint_{\Gamma_2} \frac{(y_2-y_1)\,dx_2 - (x_2-x_1)\,dy_2}{S^2}.
\end{aligned}
\tag{4.32}
$$

Example 4.4.   Determine the view factor F_{d1-2} for the configuration shown in Fig. 4-11.

Solution

With the coordinate system as shown in the figure we have

$$S = \sqrt{x^2 + y^2 + c^2},$$

and, with $\hat{\mathbf{n}}_1 = -\hat{\mathbf{k}}$, or $l_1 = m_1 = 0$ and $n_1 = -1$, it follows that equation (4.32) reduces to

$$
\begin{aligned}
F_{d1-2} &= -\frac{1}{2\pi} \oint_{\Gamma_2} \frac{y\,dx - x\,dy}{S^2} \\
&= -\frac{1}{2\pi} \left\{ \left[\int_{x=0}^{x=b} \frac{y}{S^2}\,dx \right]_{y=0} + \left[\int_{y=0}^{y=a} \frac{(-x)}{S^2}\,dy \right]_{x=b} \right. \\
&\quad \left. + \left[\int_{x=b}^{x=0} \frac{y}{S^2}\,dx \right]_{y=a} + \left[\int_{y=a}^{y=0} \frac{(-x)}{S^2}\,dy \right]_{x=0} \right\} \\
&= \frac{1}{2\pi} \left(\int_{y=0}^{a} \frac{b\,dy}{b^2+y^2+c^2} + \int_{x=0}^{b} \frac{a\,dx}{x^2+a^2+c^2} \right) \\
&= \frac{1}{2\pi} \left(\frac{b}{\sqrt{b^2+c^2}} \tan^{-1} \frac{y}{\sqrt{b^2+c^2}} \Big|_0^a + \frac{a}{\sqrt{a^2+c^2}} \tan^{-1} \frac{x}{\sqrt{a^2+c^2}} \Big|_0^b \right) \\
F_{d1-2} &= \frac{1}{2\pi} \left(\frac{b}{\sqrt{b^2+c^2}} \tan^{-1} \frac{a}{\sqrt{b^2+c^2}} + \frac{a}{\sqrt{b^2+c^2}} \tan^{-1} \frac{b}{\sqrt{a^2+c^2}} \right).
\end{aligned}
$$

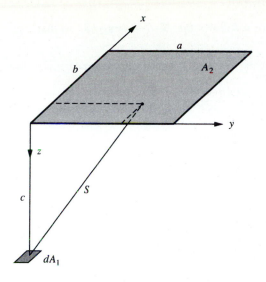

FIGURE 4-11
View factor to a rectangular plate from a parallel infinitesimal area element located opposite a corner.

View Factors between Finite Areas

To reduce the order of integration for the determination of the view factor between two finite surfaces A_1 and A_2, Stokes' theorem may be applied twice, leading to

$$A_1 F_{1-2} = \frac{1}{2\pi} \oint_{\Gamma_1} \oint_{\Gamma_2} \ln S \, d\mathbf{s}_2 \cdot d\mathbf{s}_1, \tag{4.33}$$

where the contours of the two surfaces are described by the two vectors $\mathbf{s}_1$ and $\mathbf{s}_2$. To prove that equation (4.33) is equivalent to equation (4.13) we get, comparing with equation (4.23) (for surface A_1),

$$\mathbf{f} = \frac{1}{2\pi} \oint_{\Gamma_2} \ln S \, d\mathbf{s}_2. \tag{4.34}$$

Taking the curl leads, by means of equation (4.26), to

$$2\pi \nabla_1 \times \mathbf{f} = \oint_{\Gamma_2} \nabla_1 \times (\ln S \, d\mathbf{s}_2) = \oint_{\Gamma_2} \nabla_1 (\ln S) \times d\mathbf{s}_2$$

$$= \oint_{\Gamma_2} \frac{1}{S} \nabla_1 S \times d\mathbf{s}_2, \tag{4.35}$$

where differentiation is with respect to the coordinates of surface A_1 (for which Stokes' theorem has been applied). Forming the dot product with $\hat{\mathbf{n}}_1$ then results in

$$\hat{\mathbf{n}}_1 \cdot (\nabla_1 \times \mathbf{f}) = \oint_{\Gamma_2} \frac{1}{2\pi S} \hat{\mathbf{n}}_1 \cdot (\nabla_1 S \times d\mathbf{s}_2) = \oint_{\Gamma_2} \frac{\hat{\mathbf{n}}_1 \times \nabla_1 S}{2\pi S} \cdot d\mathbf{s}_2, \tag{4.36}$$

where use has been made of the vector relationship

$$\mathbf{u} \cdot (\mathbf{v} \times \mathbf{w}) = (\mathbf{u} \times \mathbf{v}) \cdot \mathbf{w}. \tag{4.37}$$

Again, from equations (4.19) and (4.20) it follows that $\nabla_1 S = -\mathbf{s}_{12}/S$, so that

$$
\hat{\mathbf{n}}_1 \cdot (\nabla_1 \times \mathbf{f}) = -\oint_{\Gamma_2} \frac{\hat{\mathbf{n}}_1 \times \mathbf{s}_{12}}{2\pi S^2} \cdot d\mathbf{s}_2 = \oint_{\Gamma_2} \frac{\mathbf{s}_{12} \times \hat{\mathbf{n}}_1}{2\pi S^2} \cdot d\mathbf{s}_2
$$

$$
= F_{d1-2} = \int_{A_2} \frac{\cos\theta_1 \cos\theta_2}{\pi S^2} \, dA_2,
$$

where equation (4.25) has been employed. Finally,

$$
A_1 F_{1-2} = \int_{A_1} \hat{\mathbf{n}}_1 \cdot (\nabla_1 \times \mathbf{f}) \, dA_1 = \int_{A_1}\!\!\int_{A_2} \frac{\cos\theta_1 \cos\theta_2}{\pi S^2} \, dA_2 \, dA_1, \tag{4.38}
$$

which is, of course, identical to equation (4.13).

For Cartesian coordinates, with $\mathbf{s}_1$ and $\mathbf{s}_2$ from equation (4.18), equation (4.33) becomes

$$
A_1 F_{1-2} = \frac{1}{2\pi} \oint_{\Gamma_1} \oint_{\Gamma_2} \ln S \, (dx_2 \, dx_1 + dy_2 \, dy_1 + dz_2 \, dz_1). \tag{4.39}
$$

Example 4.5. Determine the view factor between two parallel, coaxial disks, Example 4.3, by contour integration.

Solution

With $d\mathbf{s} = dx\,\hat{\mathbf{i}} + dy\,\hat{\mathbf{j}} + dz\,\hat{\mathbf{k}}$ it follows immediately from the coordinates given in Example 4.3 that

$$
d\mathbf{s}_1 = R_1 \, d\psi_1(-\sin\psi_1\hat{\mathbf{i}} + \cos\psi_1\hat{\mathbf{j}}),
$$

$$
d\mathbf{s}_2 = R_2 \, d\psi_2(-\sin\psi_2\hat{\mathbf{i}} + \cos\psi_2\hat{\mathbf{j}}),
$$

$$
d\mathbf{s}_1 \cdot d\mathbf{s}_2 = R_1 R_2 \, d\psi_1 d\psi_2(\sin\psi_1 \sin\psi_2 + \cos\psi_1 \cos\psi_2)
$$

$$
= R_1 R_2 \cos(\psi_1 - \psi_2) \, d\psi_1 d\psi_2,
$$

where, it should be remembered, $d\mathbf{s}$ is along the periphery of a disk, i.e., at $r = R$. Substituting the last expression into equation (4.33) leads to

$$
F_{1-2} = \frac{R_1 R_2}{2\pi(\pi R_1^2)} \int_{\psi_1=0}^{2\pi} \int_{\psi_2=0}^{-2\pi} \ln\left[h^2 + R_1^2 + R_2^2 - 2R_1 R_2 \cos(\psi_1 - \psi_2)\right]^{1/2}
$$

$$
\times \cos(\psi_1 - \psi_2) \, d\psi_2 \, d\psi_1,
$$

where the integration for ψ_2 is from 0 to -2π since, for an observer standing on top of A_2, the integration must be in a counterclockwise sense. Just like in Example 4.3, we can eliminate one of the integrations immediately since the angles appear only as differences, i.e., $\psi_1 - \psi_2$:

$$
F_{1-2} = -\frac{1}{\pi}\frac{R_2}{R_1} \int_0^{2\pi} \ln\left(h^2 + R_1^2 + R_2^2 - 2R_1 R_2 \cos\psi\right)^{1/2} \cos\psi \, d\psi.
$$

Integrating by parts we obtain:

$$
\begin{aligned}
F_{1-2} &= -\frac{1}{\pi}\frac{R_2}{R_1}\left[\sin\psi\,\ln\left(h^2+R_1^2+R_2^2-2R_1R_2\cos\psi\right)^{1/2}\Bigg|_0^{2\pi}\right.\\
&\qquad\left.-R_1R_2\int_0^{2\pi}\frac{\sin^2\psi\,d\psi}{h^2+R_1^2+R_2^2-2R_1R_2\cos\psi}\right]\\
&= \frac{R_2/R_1}{2\pi}\int_0^{2\pi}\frac{\sin^2\psi\,d\psi}{X-\cos\psi},
\end{aligned}
$$

where we have introduced the abbreviation

$$
X = \frac{h^2+R_1^2+R_2^2}{2R_1R_2}.
$$

The integral can be found in better integral tables, or may be converted to a simpler form through trigonometric relations, leading to

$$
F_{1-2} = \frac{R_2/R_1}{2\pi}2\pi\left(X-\sqrt{X^2-1}\right) = \frac{R_2}{R_1}\left(X-\sqrt{X^2-1}\right).
$$

4.6 VIEW FACTOR ALGEBRA

Many view factors for fairly complex configurations may be calculated without any integration by simply using the rules of reciprocity and summation, and perhaps the known view factor for a more basic geometry. We shall illustrate the usefulness of this *view factor algebra* through a few simple examples.

Example 4.6. Suppose we have been given the view factor for the configuration shown in Fig. 4-11, that is, $F_{d1-2} = F(a, b, c)$ as determined in Example 4.4. Determine the view factor F_{d1-3} for the configuration shown in Fig. 4-12.

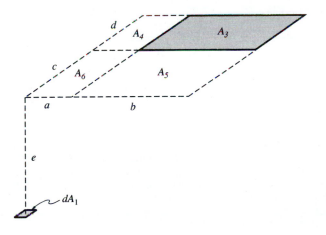

FIGURE 4-12
View factor configuration for Example 4.6.

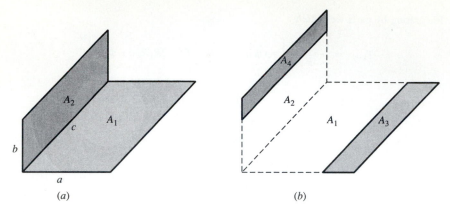

FIGURE 4-13
Configuration for Example 4.7, (a) full corner piece, (b) strips on a corner piece.

Solution

To express F_{d1-3} in terms of known view factors $F(a, b, c)$ (with the differential area opposite one of the corners of the large plate), we fill the plane of A_3 with hypothetical surfaces A_4, A_5 and A_6 as indicated in Fig. 4-12. From the definition of view factors, or equation (4.13), it follows that

$$F_{d1-(3+4+5+6)} = F_{d1-3} + F_{d1-4} + F_{d1-(5+6)},$$
$$F_{d1-4} = F_{d1-(4+6)} - F_{d1-6}.$$

Thus,

$$F_{d1-3} = F_{d1-(3+4+5+6)} - F_{d1-(4+6)} + F_{d1-6} - F_{d1-(5+6)}.$$

All four of these are of the type discussed in Example 4.4. Therefore,

$$F_{d1-3} = F(a+b, c+d, e) - F(a, c+d, e) + F(a, c, e) - F(a+b, c, e).$$

We have successfully converted the present complex view factor to a summation of four known, more basic ones.

Example 4.7. Assuming the view factor for a finite corner, as shown in Fig. 4-13a, is known as $F_{1-2} = f(a, b, c)$, where f is a known function of the dimensions of the corner pieces (as given in Appendix D), determine the view factor F_{3-4}, between the two perpendicular strips as shown in Fig. 4-13b.

Solution

From the definition of the view factor, and since the energy traveling to A_4 is the energy going to A_2 and A_4 minus the one going to A_2, it follows that

$$F_{3-4} = F_{3-(2+4)} - F_{3-2},$$

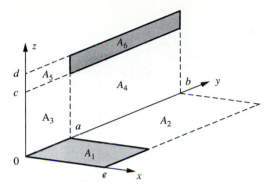

FIGURE 4-14
Configuration for Example 4.8.

and, using reciprocity,

$$F_{3-4} = \frac{1}{A_3}\left[(A_2 + A_4)F_{(2+4)-3} - A_2F_{2-3}\right].$$

Similarly, we find

$$F_{3-4} = \frac{A_2 + A_4}{A_3}\left(F_{(2+4)-(1+3)} - F_{(2+4)-1}\right) - \frac{A_2}{A_3}\left(F_{2-(1+3)} - F_{2-1}\right).$$

All view factors on the right-hand side are corner pieces and are, thus, known by evaluating the function f with appropriate dimensions.

Example 4.8. Again, assuming the view factor known for the configuration in Fig. 4-13a, determine F_{1-6} as shown in Fig. 4-14.

Solution

Examining Fig. 4-14, and employing reciprocity, we find

$$\begin{aligned}
(A_5 + A_6)F_{(5+6)-(1+2)} &= (A_5 + A_6)\left(F_{(5+6)-1} + F_{(5+6)-2}\right)\\
&= A_1\left(F_{1-5} + F_{1-6}\right) + A_2\left(F_{2-5} + F_{2-6}\right)\\
&= A_1\left(F_{1-(3+5)} - F_{1-3}\right) + A_2\left(F_{2-(4+6)} - F_{2-4}\right)\\
&\quad + A_1 F_{1-6} + A_2 F_{2-5}.
\end{aligned}$$

On the other hand, we also have

$$(A_5 + A_6)\,F_{(5+6)-(1+2)} = (A_1 + A_2)\left(F_{(1+2)-(3+4+5+6)} - F_{(1+2)-(3+4)}\right).$$

In both expressions all view factors, with the exceptions of F_{1-6} and F_{2-5}, are of the type given in Fig. 4-13a.

These last two view factors may be related to one another, as is easily seen from their integral forms. From equation (4.13) we have

$$A_2 F_{2-5} = \int_{A_2}\int_{A_5} \frac{\cos\theta_2\,\cos\theta_5}{\pi S^2}\,dA_5\,dA_2.$$

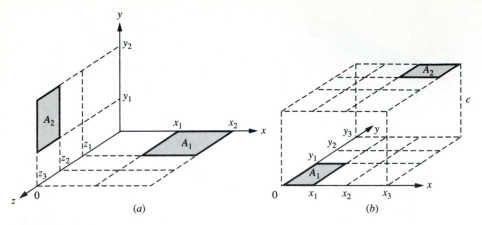

FIGURE 4-15
View factors between generalized rectangles, (a) surfaces are on perpendicular planes, (b) surfaces are on parallel planes.

With a coordinate system as shown in Fig. 4-14, we get from equations (4.20) and (4.22) $S^2 = x_2^2 + (y_2 - y_5)^2 + z_5^2$, $\cos\theta_2 = z_5/S$, $\cos\theta_5 = x_2/S$, or

$$A_2 F_{2-5} = \int_{x_2=0}^{e} \int_{y_2=a}^{b} \int_{y_5=0}^{a} \int_{z_5=c}^{d} \frac{x_2 z_5 \, dz_5 \, dy_5 \, dy_2 \, dx_2}{\pi [x_2^2 + (y_2 - y_5)^2 + z_5^2]^2}.$$

Similarly, we obtain for F_{1-6}

$$A_1 F_{1-6} = \int_{x_1=0}^{e} \int_{y_1=0}^{a} \int_{y_6=a}^{b} \int_{z_6=c}^{d} \frac{x_1 z_6 \, dz_6 \, dy_6 \, dy_1 \, dx_1}{\pi [x_1^2 + (y_1 - y_6)^2 + z_6^2]^2}.$$

Switching the names for dummy integration variables, it is obvious that

$$A_2 F_{2-5} = A_1 F_{1-6},$$

which may be called the *law of reciprocity for diagonally opposed pairs* of perpendicular rectangular plates.

Finally, solving for F_{1-6} we obtain

$$F_{1-6} = \frac{A_1 + A_2}{2A_1} \left(F_{(1+2)-(3+4+5+6)} - F_{(1+2)-(3+4)} \right)$$

$$- \frac{1}{2} \left(F_{1-(3+5)} - F_{1-3} \right) - \frac{A_2}{2A_1} \left(F_{2-(4+6)} - F_{2-4} \right).$$

Using similar arguments, one may also determine the view factor between two arbitrarily oriented rectangular plates lying in perpendicular planes (Fig. 4-15a) or in parallel planes (Fig. 4-15b). After considerable algebra, one finds [1]:

Perpendicular plates (Fig. 4-15a):

$$
\begin{aligned}
2A_1 F_{1-2} &= f(x_2, y_2, z_3) - f(x_2, y_1, z_3) - f(x_1, y_2, z_3) + f(x_1, y_1, z_3) \\
&+ f(x_1, y_2, z_2) - f(x_1, y_1, z_2) - f(x_2, y_2, z_2) + f(x_2, y_1, z_2) \\
&- f(x_2, y_2, z_3 - z_1) + f(x_2, y_1, z_3 - z_1) + f(x_1, y_2, z_3 - z_1) - f(x_1, y_1, z_3 - z_1) \\
&+ f(x_2, y_2, z_2 - z_1) - f(x_2, y_1, z_2 - z_1) - f(x_1, y_2, z_2 - z_1) + f(x_1, y_1, z_2 - z_1),
\end{aligned}
$$
$$(4.40)$$

where $f(w, h, l) = A_1 F_{1-2}$ is the product of area and view factor between two perpendicular rectangles with a common edge as given by Configuration 39 in Appendix D.

Parallel plates (Fig. 4-15b):

$$
\begin{aligned}
4A_1 F_{1-2} &= f(x_3, y_3) - f(x_3, y_2) - f(x_3, y_3 - y_1) + f(x_3, y_2 - y_1) \\
&- [f(x_2, y_3) - f(x_2, y_2) - f(x_2, y_3 - y_1) + f(x_2, y_2 - y_1)] \\
&- [f(x_3 - x_1, y_3) - f(x_3 - x_1, y_2) - f(x_3 - x_1, y_3 - y_1) + f(x_3 - x_1, y_2 - y_1)] \\
&+ f(x_2 - x_1, y_3) - f(x_2 - x_1, y_2) - f(x_2 - x_1, y_3 - y_1) + f(x_2 - x_1, y_2 - y_1),
\end{aligned}
$$
$$(4.41)$$

where $f(a, b) = A_1 F_{1-2}$ is the product of area and view factor between two directly opposed, parallel rectangles, as given by Configuration 38 in Appendix D.

Equations (4.40) and (4.41) are not restricted to $x_3 > x_2 > x_1$, and so on, but hold for arbitrary values, for example, they are valid for partially overlapping surfaces.

Example 4.9.   Show that equation (4.41) reduces to the correct expression for directly opposing rectangles.

Solution

For directly opposing rectangles, we have $x_1 = x_3 = a$, $y_1 = y_3 = b$ and $x_2 = y_2 = 0$. We note that the formula for $A_1 F_{1-2}$ for Configuration 38 in Appendix D is such that $f(a, b) = f(-a, b) = f(a, -b) = f(-a, -b)$, i.e., the view factor between "negative" areas is negative for a single negative dimension, and positive if both a and b are negative. Also, if either a or b is zero (zero area), then $f(a, b) = 0$. Thus,

$$
\begin{aligned}
4A_1 F_{1-2} = 4f(a, b) &= f(a, b) - 0 - 0 + f(a, -b) - [0 - 0 - 0 + 0] \\
&- [0 - 0 - 0 + 0] + f(-a, b) - 0 - 0 + f(-a, -b) \\
&= 4f(a, b).
\end{aligned}
$$

Many other view factors for a multitude of configurations may be obtained through view factor algebra. A few more examples will be given in this and the following chapters (when radiative exchange between black, gray-diffuse and gray-specular surfaces is discussed).

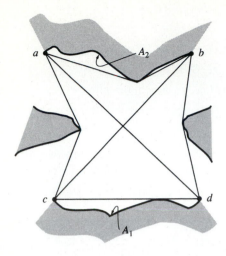

FIGURE 4-16
The crossed-strings method for arbitrary
two-dimensional configurations.

4.7 THE CROSSED-STRINGS METHOD

View factor algebra may be used to determine all view factors in long enclosures with
constant cross section. The method was first discovered by Hottel [13],[*] and is called
the crossed-strings method since the view factors can be determined experimentally
by a person armed with four pins, a roll of string, and a yardstick. Consider the configuration in Fig. 4-16, which shows the cross section of an infinitely long enclosure,
continuing into and out of the plane of the figure: We would like to determine F_{1-2}.
Obviously, the surfaces shown are rather irregular (partly convex, partly concave),
and the view between them may be obstructed. We shudder at the thought of having
to carry out the view factor determination by integration, and plant our four pins at
the two ends of each surface, as indicated by the labels a, b, c, and d. We now
connect points a and c and b and d with tight strings, making sure that no visual
obstruction remains between the two strings. Similarly, we place tight strings ab
and cd across the surfaces, and ad and bc diagonally between them, as shown in
Fig. 4-16. Now assuming the strings to be imaginary surfaces A_{ab}, A_{ac}, and A_{bc}, we
apply the summation rule to the "triangle" abc:

$$A_{ab} F_{ab-ac} + A_{ab} F_{ab-bc} = A_{ab}, \qquad (4.42a)$$

$$A_{ac} F_{ac-ab} + A_{ac} F_{ac-bc} = A_{ac}, \qquad (4.42b)$$

*Hoyte Clark Hottel (b. 1903)
American engineer. Obtained his M.S. from the Massachusetts Institute
of Technology in 1924, and has been on the Chemical Engineering
faculty at M.I.T. since 1927. Hottel's major contributions have been
his pioneering work on radiative heat transfer in furnaces, particularly
his study of the radiative properties of molecular gases (Chapter 9) and
his development of the zonal method (Chapter 18).

$$A_{bc} F_{bc-ac} + A_{bc} F_{bc-ab} = A_{bc}, \tag{4.42c}$$

where $F_{ab-ab} = F_{ac-ac} = F_{bc-bc} = 0$ since a tightened string will *always* form a convex surface. Equations (4.42) are three equations in six unknown view factors, which may be solved by applying reciprocity to three of them:

$$A_{ab} F_{ab-ac} + A_{ab} F_{ab-bc} = A_{ab}, \tag{4.43a}$$

$$A_{ab} F_{ab-ac} + A_{ac} F_{ac-bc} = A_{ac}, \tag{4.43b}$$

$$A_{ac} F_{ac-bc} + A_{ab} F_{ab-bc} = A_{bc}. \tag{4.43c}$$

Adding the first two equations and subtracting the last leads to the *view factor for an arbitrarily shaped triangle with convex surfaces*,

$$F_{ab-ac} = \frac{A_{ab} + A_{ac} - A_{bc}}{2A_{ab}}, \tag{4.44}$$

which states that the view factor between two surfaces in an arbitrary "triangle" is equal to the area of the originating surface, plus the area of the receiving surface, minus the area of the third surface, divided by twice the originating surface.

Applying equation (4.44) to triangle abd we find immediately

$$F_{ab-bd} = \frac{A_{ab} + A_{bd} - A_{ad}}{2A_{ab}}. \tag{4.45}$$

But, from the summation rule,

$$F_{ab-ac} + F_{ab-bd} + F_{ab-cd} = 1. \tag{4.46}$$

Thus

$$
\begin{aligned}
F_{ab-cd} &= 1 - \frac{A_{ab} + A_{ac} - A_{bc}}{2A_{ab}} - \frac{A_{ab} + A_{bd} - A_{ad}}{2A_{ab}} \\
&= \frac{(A_{bc} + A_{ad}) - (A_{ac} + A_{bd})}{2A_{ab}}.
\end{aligned}
\tag{4.47}
$$

Inspection of Fig. 4-16 shows that all radiation leaving A_{ab} traveling to A_{cd} will hit surface A_1. At the same time all radiation from A_{ab} going to A_1 must pass through A_{cd}. Therefore,

$$F_{ab-cd} = F_{ab-1}.$$

Using reciprocity and repeating the argument for surfaces A_{ab} and A_2, we find

$$F_{ab-cd} = F_{ab-1} = \frac{A_1}{A_{ab}} F_{1-ab} = \frac{A_1}{A_{ab}} F_{1-2},$$

and, finally,

$$F_{1-2} = \frac{(A_{bc} + A_{ad}) - (A_{ac} + A_{bd})}{2A_1}. \tag{4.48}$$

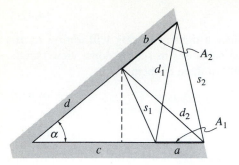

FIGURE 4-17
Infinitely long wedge-shaped groove for Examples 4.10 and 4.11.

This formula is easily memorized by looking at the configuration between any two surfaces as a generalized "rectangle," consisting of A_1, A_2, and the two sides A_{ac} and A_{bd}. Then

$$F_{1-2} = \frac{\text{diagonals} - \text{sides}}{2 \times \text{originating area}}. \tag{4.49}$$

Example 4.10.   Calculate F_{1-2} for the configuration shown in Fig. 4-17.

Solution

From the figure it is obvious that

$$s_1^2 = (c - d \cos\alpha)^2 + d^2 \sin^2\alpha = c^2 + d^2 - 2cd \cos\alpha.$$

Similarly, we have

$$
\begin{aligned}
s_2^2 &= (a + c)^2 + (b + d)^2 - 2(a + c)(b + d)\cos\alpha \\
d_1^2 &= (a + c)^2 + d^2 - 2(a + c)d \cos\alpha \\
d_2^2 &= c^2 + (b + d)^2 - 2c(b + d)\cos\alpha,
\end{aligned}
$$

and

$$F_{1-2} = \frac{d_1 + d_2 - (s_1 + s_2)}{2a}.$$

For $c = d = 0$, this reduces to the result of Example 4.2, or

$$F_{1-2} = \frac{a + b - \sqrt{a^2 + b^2 - 2ab\cos\alpha}}{2a}.$$

Example 4.11.   Find the view factor F_{d1-2} of Fig. 4-17 for the case that A_1 is an infinitesimal strip of width dx. Use the crossed-strings method.

Solution

We can obtain the result right away by replacing a by dx in the previous example. Throwing out differentials of second and higher order, we find that s_1 and d_2 remain unchanged, and

$$
\begin{aligned}
d_1 &= \sqrt{(c + dx)^2 + d^2 - 2(c + dx)d \cos \alpha} \\
&\simeq \sqrt{c^2 + d^2 - 2cd \cos \alpha + 2(c - d \cos \alpha)dx} \\
&\simeq \sqrt{c^2 + d^2 - 2cd \cos \alpha} \left[1 + \frac{(c - d \cos \alpha)dx}{c^2 + d^2 - 2cd \cos \alpha} \right] = s_1 + \frac{dx}{s_1}(c - d \cos \alpha)
\end{aligned}
$$

$$
\begin{aligned}
s_2 &= \sqrt{(c + dx)^2 + (b + d)^2 - 2(c + dx)(b + d)\cos \alpha} \\
&\simeq d_2 + \frac{dx}{d_2}[c - (b + d)\cos \alpha].
\end{aligned}
$$

Substituting this into equation (4.49), we obtain

$$
\begin{aligned}
F_{d1-2} &= \frac{s_1 + (c - d \cos \alpha)dx/s_1 + d_2 - s_1 - d_2 - [c - (b+d)\cos \alpha]dx/d_2}{2dx} \\
&= \frac{1}{2} \left[\frac{c - d \cos \alpha}{\sqrt{c^2 + d^2 - 2cd \cos \alpha}} - \frac{c - (b+d)\cos \alpha}{\sqrt{c^2 + (b+d)^2 - 2c(b+d)\cos \alpha}} \right].
\end{aligned}
$$

The same result could also have been obtained by letting

$$
F_{d1-2} = \lim_{a \to 0} F_{1-2},
$$

where F_{1-2} is the view factor from the previous example. Using de l'Hopital's rule to determine the value of the resulting expression leads to

$$
F_{d1-2} = \frac{1}{2} \left(\frac{\partial d_1}{\partial a} - \frac{\partial s_2}{\partial a} \right) \Bigg|_{a=0},
$$

and the above result.

Thus, the crossed-strings method may also be applied to strips. Example 4.1 could also have been solved this way; since the result is infinitesimal this computation would require retaining differentials up to second order. However, integration becomes simpler for strips of differential widths, while application of the crossed-strings method becomes more involved.

We shall present one final example to show how view factors for curved surfaces and for configurations with floating obstructions can be determined by the crossed-strings method.

Example 4.12. Determine the view factor F_{1-2} for the configuration shown in Fig. 4-18.

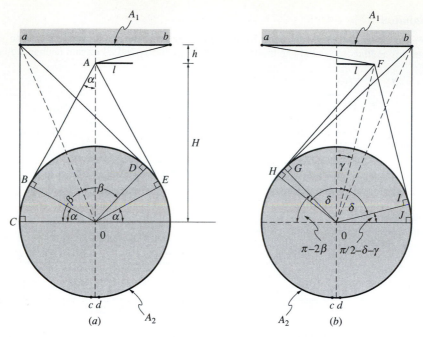

FIGURE 4-18
Configuration for view factor calculation of Example 4.12; string placement (a) for F_{1-2}^l, (b) for F_{1-2}^r.

Solution

In the figure the end points of A_1 and A_2 (pin points) have been labeled a, b, c, and d, and other strategic points have been labeled with capital letters. A closed-contour surface such as a cylinder may be modeled by placing two pins right next to each other, with surface A_2 being a strongly bulging convex surface between the pins. While the location of the two pins on the cylinder is arbitrary, it is usually more convenient to pick a location out of sight of A_1. Since A_1 can see A_2 from both sides of the obstruction, F_{1-2} cannot be determined with a single set of strings. Using view factor algebra, we can state that

$$F_{1-2} = F_{1-2}^l + F_{1-2}^r,$$

where F_{1-2}^l and F_{1-2}^r are the view factors between A_1 and A_2 when considering only light paths on the left or right of the obstruction, respectively. The placement of strings for F_{1-2}^l is given in Fig. 4-18a, and for F_{1-2}^r in Fig. 4-18b.

Considering first F_{1-2}^l, the diagonals and sides may be determined from

$$d_1 = aD + DE + Ed, \quad d_2 = bA + AB + BC + Cc,$$

$$s_1 = aC + Cc, \quad s_2 = bA + AE + Ed.$$

Substituting these expressions into equation (4.49) and canceling those terms that appear in a diagonal as well as in a side $(Ed, bA, \text{ and } Cc)$, we obtain

$$F^l_{1-2} = \frac{aD + DE + AB + BC - (aC + AE)}{2ab}.$$

Looking at Fig. 4-18a we also notice that $aC = aD$ and $AB = AE$, so that

$$F^l_{1-2} = \frac{BC + DE}{2ab} = \frac{\alpha R + (\pi - 2\beta - \alpha)R}{2 \times 2R} = \frac{1}{2}\left(\frac{\pi}{2} - \beta\right).$$

But $\cot\beta = \tan(\pi/2 - \beta) = R/(h + H)$. Thus,

$$F^l_{1-2} = \frac{1}{2}\tan^{-1}\frac{R}{h + H}.$$

Similarly, we find from Fig. 4-18b for F^r_{1-2},

$$d_1 = aF + FI + IJ + Jd, \quad d_2 = bG + GH + Hc,$$

$$s_1 = aF + FH + Hc, \quad s_2 = bJ + Jd,$$

$$F^r_{1-2} = \frac{FI + IJ + bG + GH - (FH + bJ)}{2ab}.$$

By inspection $bG = bJ$ and $FI = FH$, leading to

$$F^r_{1-2} = \frac{IJ + GH}{2ab} = \frac{\left(\frac{\pi}{2} - \delta - \gamma\right)R + \left(\pi - 2\beta + \delta - \frac{\pi}{2} - \gamma\right)R}{2 \times 2R}$$

$$= \frac{1}{2}\left(\frac{\pi}{2} - \beta - \gamma\right) = \frac{1}{2}\left(\tan^{-1}\frac{R}{h + H} - \tan^{-1}\frac{l}{h}\right),$$

Note that this formula only holds as long as $GH > 0$ (i.e., as long as the cylinder is seen without obstruction from point b). Finally, adding the left and right contributions to the view factor,

$$F_{1-2} = \tan^{-1}\frac{R}{h + H} - \frac{1}{2}\tan^{-1}\frac{l}{h}.$$

4.8 THE INSIDE-SPHERE METHOD

Consider two surfaces A_1 and A_2 that are both parts of the surface of one and the same sphere, as shown in Fig. 4-19. We note that, for this type of configuration, $\theta_1 = \theta_2 = \theta$ and $S = 2R\cos\theta$. Therefore,

$$F_{d1-2} = \int_{A_2}\frac{\cos\theta_1\cos\theta_2}{\pi S^2}dA_2 = \int_{A_2}\frac{\cos^2\theta}{\pi(2R\cos\theta)^2}dA_2$$

$$= \frac{1}{4\pi R^2}\int_{A_2}dA_2 = \frac{A_2}{A_s}, \tag{4.50}$$

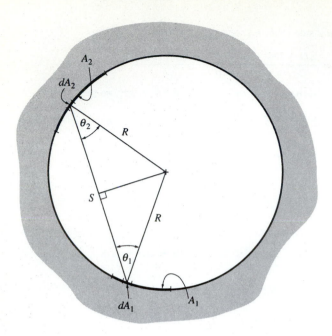

FIGURE 4-19
The inside-sphere method.

where $A_s = 4\pi R^2$ is the surface area of the entire sphere. Similarly, from equation (4.16),

$$F_{1-2} = F_{d1-2} = \frac{A_2}{A_s},\tag{4.51}$$

since F_{d1-2} does not depend on the position of dA_1. Therefore, because of the unique geometry of a sphere, the view factor between two surfaces on the same sphere only depends on *size* of the receiving surface, and not on the *location* of either one.

The inside-sphere method is primarily used in conjunction with view factor algebra, to determine the view factor between two surfaces that may not necessarily lie on a sphere.

Example 4.13.   Find the view factor between two parallel, coaxial disks of radius R_1 and R_2 using the inside-sphere method.

Solution

Inspecting Fig. 4-20 we see that it is possible to place the parallel disks inside a sphere of radius R in such a way that the entire peripheries of both disks lie on the surface of the sphere.

Since all radiation from A_1 to A_2 travels on to the spherical cap $A_{2'}$, (in the absence of A_2), and since all radiation from A_1 to $A_{2'}$ must pass through A_2, we have

$$F_{1-2} = F_{1-2'}.$$

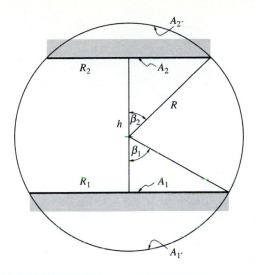

FIGURE 4-20
View factor between coaxial parallel disks.

Using reciprocity and applying a similar argument for A_1 and spherical cap $A_{1'}$, we find

$$F_{1-2} = F_{1-2'} = \frac{A_{2'}}{A_1}F_{2'-1} = \frac{A_{2'}}{A_1}F_{2'-1'} = \frac{A_{1'}A_{2'}}{A_1 A_s}.$$

The areas of the spherical caps are readily calculated as

$$A_{i'} = 2\pi R^2 \int_0^{\beta_i} \sin\beta \, d\beta = 2\pi R^2 (1 - \cos\beta_i), \quad i = 1, 2.$$

Thus, with $A_1 = \pi R_1^2$ and $A_s = 4\pi R^2$, this results in

$$F_{1-2} = \frac{(2\pi R^2)^2 (1 - \cos\beta_1)(1 - \cos\beta_2)}{\pi R_1^2 \, 4\pi R^2}.$$

From Fig. 4-20 one finds $\cos\beta_i = \sqrt{R^2 - R_i^2}/R$, and

$$F_{1-2} = \frac{1}{R_1^2}\left(R - \sqrt{R^2 - R_1^2}\right)\left(R - \sqrt{R^2 - R_2^2}\right).$$

It remains to find the radius of the sphere R, since only the distance between disks, h, is known. From Fig. 4-20

$$h = \sqrt{R^2 - R_1^2} + \sqrt{R^2 - R_2^2},$$

which may be solved (by squaring twice), to give

$$R^2 = (X^2 - 1)\left(\frac{R_1 R_2}{h}\right)^2, \quad X = \frac{h^2 + R_1^2 + R_2^2}{2R_1 R_2}.$$

This result is, of course, identical to the one given in Example 4.5, although it is not trivial to show this.

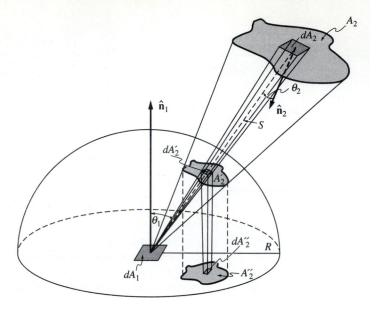

FIGURE 4-21
Surface projection for the unit sphere method.

4.9   THE UNIT SPHERE METHOD

The unit sphere method is a powerful tool to calculate view factors between one infinitesimal and one finite area. It is particularly useful for the experimental determination of such view factors, as first stated by Nusselt [14]. An experimental implementation of the method through optical projection has been discussed by Farrell [15].

To determine the view factor F_{d1-2} between dA_1 and A_2 we place a hemisphere[2] of radius R on top of A_1, centered over dA_1, as shown in Fig. 4-21. From equations (4.4) and (4.8) we may write

$$F_{d1-2} = \int_{A_2} \frac{\cos\theta_1 \cos\theta_2}{\pi S^2} dA_2 = \int_{\Omega_2} \frac{\cos\theta_1}{\pi} d\Omega_2. \qquad (4.52)$$

The solid angle $d\Omega_2$ may also be expressed in terms of area dA_2' (dA_2 projected onto the hemisphere) as $d\Omega_2 = dA_2'/R^2$. Further, the area dA_2' may be projected along the z-axis onto the plane of A_1 as $dA_2'' = \cos\theta_1 dA_2'$. Thus,

$$F_{d1-2} = \int_{A_2'} \frac{\cos\theta_1}{\pi} \frac{dA_2'}{R^2} = \int_{A_2''} \frac{dA_2''}{\pi R^2} = \frac{A_2''}{\pi R^2}, \qquad (4.53)$$

---

[2]The name *unit sphere method* originated with Nusselt, who used a sphere of unit radius; however, a sphere of arbitrary radius may be used.

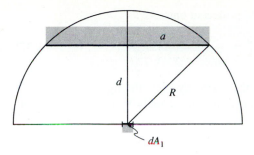

FIGURE 4-22
Geometry for the view factor in Example 4.14.

that is, F_{d1-2} is the fraction of the disk πR^2 that is occupied by the double projection of A_2. Experimentally this can be measured, for example, by placing an opaque area A_2 within a hemisphere, made of a translucent material, and which has a light source at the center (at dA_1). Looking down onto the translucent hemisphere in the negative z-direction, $A_{2'}$ will appear as a shadow. A photograph of the shadow (and the bright disk) can be taken, showing the double projection of A_2, and F_{d1-2} can be measured.

Example 4.14.   Determine the view factor for F_{d1-2} between an infinitesimal area and a parallel disk as shown in Fig. 4-22.

Solution

While a hemisphere of arbitrary radius could be employed, we shall choose here for convenience a radius of $R = \sqrt{a^2 + d^2}$, i.e., a hemisphere that includes the periphery of the disk on its surface. Then $A_2'' = A_2 = \pi a^2$, and the view factor follows as

$$F_{d1-2} = \frac{\pi a^2}{\pi R^2} = \frac{a^2}{a^2 + d^2}.$$

Obviously, only a few configurations will allow such simple calculation of view factors. For a more general case it would be desirable to have some "cookbook formula" for the application of the method. This is readily achieved by looking at the vector representation of the surfaces. Any point on the periphery of A_2 may be expressed as a vector

$$\mathbf{s}_{12} = x\hat{\mathbf{i}} + y\hat{\mathbf{j}} + z\hat{\mathbf{k}}. \tag{4.54}$$

The corresponding point on A_2' may be expressed as

$$\mathbf{s}_{12}' = x'\hat{\mathbf{i}} + y'\hat{\mathbf{j}} + z'\hat{\mathbf{k}} = \frac{R}{\sqrt{x^2 + y^2 + z^2}}\mathbf{s}_{12}, \tag{4.55}$$

and on A_2'' as

$$\mathbf{s}_{12}'' = x''\hat{\mathbf{i}} + y''\hat{\mathbf{j}} = x'\hat{\mathbf{i}} + y'\hat{\mathbf{j}}. \tag{4.56}$$

Thus, any point (x, y, z) on A_2 is double-projected onto A_2'' as

$$x_2'' = \frac{x}{\sqrt{x^2 + y^2 + z^2}}R, \quad y_2'' = \frac{y}{\sqrt{x^2 + y^2 + z^2}}R. \tag{4.57}$$

Only the area formed by the projection of the periphery of A_2 through equation (4.57) needs to be found. This integration is generally considerably less involved than the one in equation (4.8).

References

1. Hamilton, D. C., and W. R. Morgan: "Radiant Interchange Configuration Factors," *NASA TN 2836*, 1952.
2. Leuenberger, H., and R. A. Pearson: "Compilation of Radiant Shape Factors for Cylindrical Assemblies," ASME paper no. 56-A-144, 1956.
3. Kreith, F.: *Radiation Heat Transfer for Spacecraft and Solar Power Design*, International Textbook Company, Scranton, PA, 1962.
4. Sparrow, E. M., and R. D. Cess: *Radiation Heat Transfer*, Hemisphere, New York, 1978.
5. Siegel, R., and J. R. Howell: *Thermal Radiation Heat Transfer*, 2nd ed., Hemisphere, New York, 1981.
6. Howell, J. R.: *A Catalog of Radiation Configuration Factors*, McGraw Hill, New York, 1982.
7. Jakob, M.: *Heat Transfer*, vol. 2, John Wiley & Sons, New York, 1957.
8. Liu, H. P., and J. R. Howell: "Measurement of Radiation Exchange Factors," *ASME Journal of Heat Transfer*, vol. 109, no. 2, pp. 470–477, 1956.
9. Wylie, C. R.: *Advanced Engineering Mathematics*, 5th ed., McGraw-Hill, New York, 1982.
10. Moon, P.: *Scientific Basis of Illuminating Engineering*, Dover Publications, New York, 1961 (originally published by McGraw-Hill, New York, 1936).
11. de Bastos, R.: "Computation of Radiation Configuration Factors by Contour Integration," M.S. thesis, Oklahoma State University, 1961.
12. Sparrow, E. M.: "A New and Simpler Formulation for Radiative Angle Factors," *ASME Journal of Heat Transfer*, vol. 85C, pp. 73–81, 1963.
13. Hottel, H. C.: "Radiant Heat Transmission," in *Heat Transmission*, ed. W. H. McAdams, 3d ed., ch. 4, McGraw-Hill, New York, 1954.
14. Nusselt, W.: "Graphische Bestimming des Winkelverhältnisses bei der Wärmestrahlung," *VDIZ*, vol. 72, p. 673, 1928.
15. Farrell, R.: "Determination of Configuration Factors of Irregular Shape," *ASME Journal of Heat Transfer*, vol. 98, no. 2, pp. 311–313, 1976.

Problems

4.1 For Configuration 11, Appendix D, find F_{d1-2} by

  (a) Area integration

  (b) Contour integration

Compare the effort involved.

4.2 Using the results of Problem 4.1, find F_{1-2} for Configuration 33 in Appendix D.

4.3 Find F_{1-2} for Configuration 32, Appendix D, by area integration.

4.4 Evaluate F_{d1-2} for Configuration 13 in Appendix D by

  (a) area integration

(*b*) contour integration.

Compare the effort involved.

4.5 Using the result from the Problem 4.4, calculate F_{1-2} for Configuration 40, Appendix D.

4.6 Find the view factor F_{d1-2} for Configuration 11 in Appendix D, with dA_1 tilted towards A_2 by an angle ϕ.

4.7 Find F_{d1-2} for the surfaces shown in the figure, using

   (*a*) area integration;
   (*b*) view factor algebra and Configuration 11 in Appendix D.

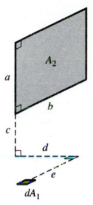

4.8 For the infinite half-cylinder depicted in the figure, find F_{1-2}.

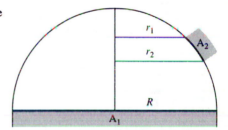

4.9 Find F_{d1-2} for the surfaces shown in the figure.

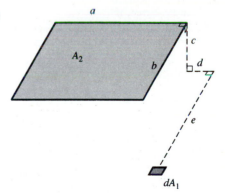

4.10 To reduce heat transfer between two infinite concentric cylinders a third cylinder is placed between them as shown in the figure. The center cylinder has an opening of half-angle θ. Calculate F_{4-2}.

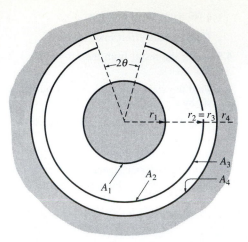

4.11 Calculate the view factor F_{1-2} for surfaces on a cone as shown in the figure.

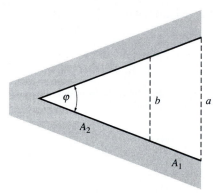

4.12 Determine the view factor F_{1-2} for the configuration shown in the figure, if

    (*a*) the bodies are two-dimensional (i.e., infinitely long perpendicular to the paper);

    (*b*) the bodies are axisymmetric (cones).

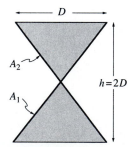

4.13 Find F_{1-2} for the configuration shown in the figure (infinitely long perpendicular to paper).

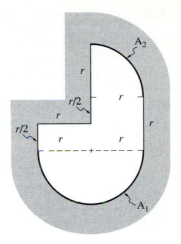

4.14 Calculate the view factor between two infinitely long cylinders as shown in the figure. If a radiation shield is placed between them to obstruct partially the view (dashed line), how does the view factor change?

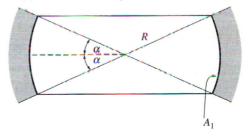

4.15 Determine the view factor for Configuration 51 in Appendix D, using (*a*) other, more basic view factors given in Appendix D, (*b*) the crossed strings rule.

4.16 Find the view factor of the spherical ring, shown in the figure, to itself F_{1-1}, using the inside-sphere method.

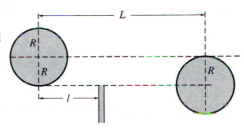

4.17 Find the view factor between spherical caps as shown in the figure, for the case of

$$H \geq \frac{R_1^2}{\sqrt{R_1^2 - a_1^2}} + \frac{R_2^2}{\sqrt{R_2^2 - a_2^2}},$$

where H = distance between sphere centers, R = sphere radius, a = radius of cap base. Why is this restriction necessary?

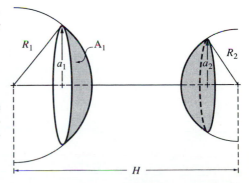

4.18 Consider the axisymmetric configuration shown in the figure. Calculate the view factor F_{1-3}.

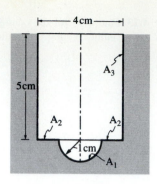

4.19 Find F_{d1-2} from the infinitesimal area to the disk as shown in the figure, with $0 \leq \beta \leq \pi$.

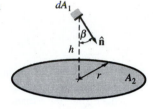

4.20 Determine the view factor for Configuration 18 in Appendix D, using the inside sphere method.

CHAPTER
5

RADIATIVE EXCHANGE BETWEEN GRAY, DIFFUSE SURFACES

5.1 INTRODUCTION

In this chapter we shall begin our analysis of radiative heat transfer rates within enclosures without a participating medium, making use of the view factors developed in the preceding chapter. We shall first deal with the simplest case of a black enclosure, that is, an enclosure where all surfaces are black.

Such simple analysis may often be sufficient, for example, for furnace applications with soot-covered walls. This will be followed by expanding the analysis to enclosures with gray, diffuse surfaces, whose radiative properties do not depend on wavelength, and which emit as well as reflect energy diffusely. Considerable experimental evidence demonstrates that most surfaces emit (and, therefore, absorb) diffusely except for grazing angles ($\theta > 60°$), which are unimportant for heat transfer calculations (for example, Fig. 3-1). Most surfaces tend to be fairly rough and, therefore, reflect in

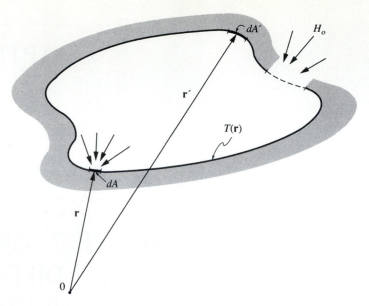

FIGURE 5-1
A black enclosure of arbitrary geometry.

a relatively diffuse fashion. Finally, if the surface properties vary little across that part of the spectrum over which the blackbody emissive powers of the surfaces are appreciable, then the simplification of gray properties may be acceptable.

In both cases—black enclosures as well as enclosures with gray, diffuse surfaces— we shall first derive the governing integral equation for arbitrary enclosures, which is then reduced to a set of algebraic equations by applying it to idealized enclosures. At the end of the chapter solution methods to the general integral equations are briefly discussed.

5.2   RADIATIVE EXCHANGE BETWEEN BLACK SURFACES

Consider a black-walled enclosure of arbitrary geometry and with arbitrary temperature distribution as shown in Fig. 5-1. An energy balance for dA yields, from equation (4.1),

$$q(\mathbf{r}) = E_b(\mathbf{r}) - H(\mathbf{r}), \tag{5.1}$$

where H is the irradiation onto dA. From the definition of the view factor, the rate with which energy leaves dA' and is intercepted by dA is $(E_b(\mathbf{r}')dA')\,dF_{dA'-dA}$. Therefore, the total rate of incoming heat transfer onto dA from the entire enclosure and from outside (for enclosures with some semitransparent surfaces and/or holes) is

$$H(\mathbf{r})\,dA = \int_A E_b(\mathbf{r}')F_{dA'-dA}dA' + H_o(\mathbf{r})\,dA, \qquad (5.2)$$

where $H_o(\mathbf{r})$ is the external contribution to the irradiation, i.e., any part not due to emission from the enclosure surface. Using reciprocity, this may be stated as

$$
\begin{aligned}
H(\mathbf{r}) &= \int_A E_b(\mathbf{r}')\,dF_{dA-dA'} + H_o(\mathbf{r}) \\
&= \int_A E_b(\mathbf{r}')\frac{\cos\theta\cos\theta'}{\pi S^2}(\mathbf{r},\mathbf{r}')\,dA' + H_o(\mathbf{r}). \qquad (5.3)
\end{aligned}
$$

For an enclosure with known surface temperature distribution, the local heat flux is readily calculated as[1]

$$\mathbf{q}(\mathbf{r}) = E_b(\mathbf{r}) - \int_A E_b(\mathbf{r}')\,dF_{dA-dA'} - H_o(\mathbf{r}). \qquad (5.4)$$

To simplify the problem it is customary to break up the enclosure into N isothermal subsurfaces, as shown in Fig. 4-2. Then equation (5.4) becomes

$$q_i(\mathbf{r}_i) = E_{bi} - \sum_{j=1}^{N} E_{bj}\int_{A_j} dF_{dA_i-dA_j} - H_{oi}(\mathbf{r}_i), \qquad (5.5)$$

or, from equation (4.16),

$$q_i(\mathbf{r}_i) = E_{bi} - \sum_{j=1}^{N} E_{bj}F_{di-j}(\mathbf{r}_i) - H_{oi}(\mathbf{r}_i). \qquad (5.6)$$

Even though the temperature may be constant across A_i, the heat flux is usually not since (*i*) the local view factor F_{di-j} nearly always varies across A_i, and (*ii*) the external irradiation H_{oi} may not be uniform. We may calculate an *average heat flux* by averaging equation (5.6) over A_i. With $\int_{A_i} F_{di-j}\,dA_i = A_iF_{i-j}$ this leads to

$$q_i = \frac{1}{A_i}\int_{A_i} q_i(\mathbf{r}_i)\,dA_i = E_{bi} - \sum_{j=1}^{N} E_{bj}F_{i-j} - H_{oi}, \quad i = 1,2,\ldots,N, \qquad (5.7)$$

where q_i and H_{oi} are now understood to be average values.

Employing equation (4.17) we rewrite E_{bi} as $\sum_{j=1}^{N} E_{bi}F_{i-j}$, or

$$q_i = \sum_{j=1}^{N} F_{i-j}(E_{bi} - E_{bj}) - H_{oi}, \quad i = 1,2,\ldots,N. \qquad (5.8)$$

---

[1]When looking at equation (5.4) one is often tempted by intuition to replace $dF_{dA-dA'}$ by $dF_{dA'-dA}$. It should always be remembered that we have used reciprocity, since $dF_{dA'-dA}$ is per unit area at $\mathbf{r}'$, while equation (5.4) is per unit area at $\mathbf{r}$.

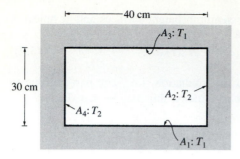

FIGURE 5-2
Two dimensional black duct for Example 5.1.

In this equation the heat flux is expressed in terms of the net radiative energy exchange between surfaces A_i and A_j,

$$Q_{i-j} = q_{i-j}A_i = A_iF_{i-j}(E_{bi} - E_{bj}) = -Q_{j-i}. \tag{5.9}$$

Example 5.1.   Consider a very long duct as shown in Fig. 5-2. The duct is $30\,\text{cm} \times 40\,\text{cm}$ in cross section, and all surfaces are black. The top and bottom walls are at temperature $T_1 = 1000\,\text{K}$, while the side walls are at temperature $T_2 = 600\,\text{K}$. Determine the net radiative heat transfer rate (per unit duct length) on each surface.

Solution

We may use either equation (5.7) or (5.8). We shall use the latter here since it takes better advantage of the symmetry of the problem (i.e., it uses the fact that the net radiative exchange between two surfaces at the same temperature must be zero). Thus, with no external irradiation, and using symmetry (e.g., $E_{b1} = E_{b3}$, $F_{1-2} = F_{1-4}$, etc.),

$$
\begin{aligned}
q_1 &= F_{1-2}(E_{b1} - E_{b2}) + F_{1-3}(E_{b1} - E_{b3}) + F_{1-4}(E_{b1} - E_{b4}) \\
&= 2F_{1-2}(E_{b1} - E_{b2}) = q_3, \\
q_2 &= q_4 = 2F_{2-1}(E_{b2} - E_{b1}).
\end{aligned}
$$

Only the view factors F_{1-2} and F_{2-1} are required, which are readily determined from the crossed-strings method as

$$F_{1-2} = \frac{30 + 40 - \left(\sqrt{30^2 + 40^2} + 0\right)}{2 \times 40} = \frac{1}{4},$$

$$F_{2-1} = \frac{A_1}{A_2}F_{1-2} = \frac{40}{30} \times \frac{1}{4} = \frac{1}{3}.$$

Therefore (using a prime to indicate "per unit duct length"),

$$
\begin{aligned}
Q_1' = Q_3' &= 2A_1'F_{1-2}\sigma(T_1^4 - T_2^4) \\
&= 2 \times 0.4\,\text{m} \times 0.25 \times 5.670 \times 10^{-8}\,\frac{\text{W}}{\text{m}^2\,\text{K}^4}(1000^4 - 600^4)\,\text{K}^4 = 9870\,\text{W/m} \\
Q_2' = Q_4' &= 2A_2'F_{2-1}\sigma(T_2^4 - T_1^4) = -9870\,\text{W/m}
\end{aligned}
$$

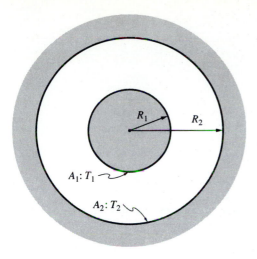

FIGURE 5-3
Concentric black spheres for Example 5.2.

It is apparent from this example that the sum of all surface heat transfer rates must vanish. This follows immediately from *conservation of energy*: The total heat transfer rate into the enclosure (i.e., the heat transfer rates summed over all surfaces) must be equal to the rate of change of radiative energy within the enclosure. Since radiation travels at the speed of light, steady state is reached almost instantaneously, so that the rate of change of radiative energy may nearly always be neglected. Mathematically, we may multiply equation (5.7) by A_i and sum over all areas:

$$\sum_{i=1}^{N}(Q_i + A_i H_{oi}) = \sum_{i=1}^{N} A_i E_{bi} - \sum_{i=1}^{N} A_i \sum_{j=1}^{N} E_{bj} F_{i-j}$$

$$= \sum_{i=1}^{N} A_i E_{bi} - \sum_{j=1}^{N} A_j E_{bj} \sum_{i=1}^{N} F_{j-i} = 0. \qquad (5.10)$$

This relationship is most useful to check the correctness of one's calculations, or their accuracy (for computer calculations).

Example 5.2. Consider two concentric, isothermal, black spheres with radii R_1 and R_2, and temperatures T_1 and T_2, respectively, as shown in Fig. 5-3. Show how the temperature of the inner sphere can be deduced, if temperature and heat flux of the outer sphere are measured.

Solution

We have only two surfaces, and equation (5.8) becomes

$$q_1 = F_{1-2}(E_{b1} - E_{b2}); \quad q_2 = F_{2-1}(E_{b2} - E_{b1}).$$

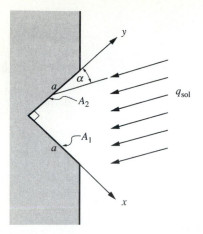

FIGURE 5-4
Right-angled groove exposed to solar irradiation, Example 5.3.

Since all radiation from Sphere 1 travels to 2, we have $F_{1-2} = 1$ and, by reciprocity, $F_{2-1} = A_1/A_2$. Thus,

$$Q_1 = -Q_2 = A_1\sigma(T_1^4 - T_2^4).$$

Solving this for T_1 we get, with $A_i = 4\pi R_i^2$,

$$T_1^4 = T_2^4 - \left(\frac{R_2}{R_1}\right)^2 \frac{q_2}{\sigma}.$$

Whenever T_1 is larger than T_2, q_2 is negative, and vice versa.

Example 5.3. A right-angled groove, consisting of two long black surfaces of width a, is exposed to solar radiation q_{sol} (Fig. 5-4). The entire groove surface is kept isothermal at temperature T. Determine the net radiative heat transfer rate from the groove.

Solution

Again, we may employ either equation (5.7) or (5.8). However, this time the enclosure is not closed; and we must close it artificially. We note that any radiation leaving the cavity will not come back (barring any reflection from other surfaces nearby). Thus, our artificial surface should be black. We also assume that, with the exception of the (parallel) solar irradiation, no external radiation enters the cavity. Since the solar irradiation is best treated separately through the external irradiation term H_o, our artificial surface is nonemitting. Both criteria are satisfied by covering the groove with a black surface at 0 K. Even though we now have three surfaces, the last one does not really appear in equation (5.7) (since $E_{b3} = 0$), but it does appear in equation (5.8). Using equation (5.7) we find

$$q_1 = E_{b1} - F_{1-2}E_{b2} - H_{o1} = \sigma T^4(1 - F_{1-2}) - q_{sol}\cos\alpha,$$
$$q_2 = E_{b2} - F_{2-1}E_{b1} - H_{o2} = \sigma T^4(1 - F_{2-1}) - q_{sol}\sin\alpha.$$

From Configuration 33 in Appendix D we find, with $H = 1$,

$$F_{1-2} = \tfrac{1}{2}\left(2 - \sqrt{2}\right) = 0.293 = F_{2-1},$$

and

$$Q' = a(q_1 + q_2) = a\left[\sqrt{2}a\sigma T^4 - q_{sol}(\cos\alpha + \sin\alpha)\right].$$

These examples demonstrate that equation (5.8) is generally more convenient to use for closed configurations, since it takes advantage of the fact that the net exchange between two surfaces at the same temperature (or with itself) is zero. Equation (5.7), on the other hand, is more convenient for open configurations, since the hypothetical surfaces employed to close the configuration do not contribute (because of their zero emissive power): With this equation the hypothetical closing surfaces may be completely ignored!

Equation (5.7) may be written in a third form that is most convenient for computer calculations. Using *Kronecker's delta function*, defined as

$$\delta_{ij} = \begin{cases} 1, & i = 1, \\ 0, & i \neq 1, \end{cases} \tag{5.11}$$

we find $\sum_{j=1}^{N} \delta_{ij} = 1$ and $\sum_{j=1}^{N} E_{bj}\delta_{ij} = E_{bi}$. Thus,

$$q_i = \sum_{j=1}^{N}(\delta_{ij} - F_{i-j})E_{bj} - H_{oi}, \quad i = 1, 2, \ldots, N. \tag{5.12}$$

Let us suppose that for surfaces $i = 1, 2, \ldots, n$ the heat fluxes are prescribed (and temperatures are unknown), while for surfaces $i = n + 1, \ldots, N$ the temperatures are prescribed (heat fluxes unknown). Unlike for the heat fluxes, no explicit relations for the unknown temperatures exist. Placing all unknown temperatures on one side of equation (5.12), we may write

$$\sum_{j=1}^{n}(\delta_{ij} - F_{i-j})E_{bj} = q_i + H_{oi} + \sum_{j=n+1}^{N} F_{i-j}E_{bj}, \quad i = 1, 2, \ldots, n, \tag{5.13}$$

where everything on the right-hand side of the equation is known. In matrix form this is written[2] as

$$\mathbf{A} \cdot \mathbf{e}_b = \mathbf{b}, \tag{5.14}$$

---

[2]For easy readability of matrix manipulations we shall follow here the convention that a two-dimensional matrix is denoted by a bold capitalized letter, while a vector is written as a bold lower-case letter.

where

$$\mathbf{A} = \begin{pmatrix} 1 - F_{1-1} & -F_{1-2} & \cdots & -F_{1-n} \\ -F_{2-1} & 1 - F_{2-2} & \cdots & -F_{2-n} \\ \vdots & & \ddots & \vdots \\ -F_{n-1} & -F_{n-2} & \cdots & 1 - F_{n-n} \end{pmatrix}, \tag{5.15}$$

$$\mathbf{e}_b = \begin{pmatrix} E_{b1} \\ E_{b2} \\ \vdots \\ E_{bn} \end{pmatrix}, \quad \mathbf{b} = \begin{pmatrix} q_1 + H_{o1} + \sum_{j=n+1}^{N} F_{1-j} E_{bj} \\ q_2 + H_{o2} + \sum_{j=n+1}^{N} F_{2-j} E_{bj} \\ \vdots \\ q_n + H_{on} + \sum_{j=n+1}^{N} F_{n-j} E_{bj} \end{pmatrix}. \tag{5.16}$$

The $n \times n$ matrix $\mathbf{A}$ is readily inverted on a computer (generally with the aid of a software library subroutine), and the unknown temperatures are calculated as

$$\mathbf{e}_b = \mathbf{A}^{-1} \cdot \mathbf{b}. \tag{5.17}$$

5.3 RADIATIVE EXCHANGE BETWEEN GRAY, DIFFUSE SURFACES

We shall now assume that all surfaces are gray, that they are diffuse emitters, absorbers and reflectors. Under these conditions $\epsilon = \epsilon'_\lambda = \alpha'_\lambda = \alpha = 1 - \rho$. The total heat flux leaving a surface at location $\mathbf{r}$ is, from Fig. 4-1,

$$J(\mathbf{r}) = \epsilon(\mathbf{r})E_b(\mathbf{r}) + \rho(\mathbf{r})H(\mathbf{r}), \tag{5.18}$$

which is called the *surface radiosity J* at location $\mathbf{r}$. Since both emission and reflection are diffuse, so is the resulting intensity leaving the surface:

$$I(\mathbf{r}, \hat{\mathbf{s}}) = I(\mathbf{r}) = J(\mathbf{r})/\pi. \tag{5.19}$$

Therefore, an observer at a different location is unable to distinguish emitted and reflected radiation on the basis of *directional* behavior. However, the observer may be able to distinguish the two as a result of their different *spectral* behavior. Consider Example 5.2 for the case of a black outer sphere but a gray, diffuse inner sphere. On the inner sphere the emitted radiation has the spectral distribution of a blackbody at temperature T_1, while the reflected radiation—which was originally emitted at the outer sphere—has the spectral distribution of a blackbody at temperature T_2. Thus,

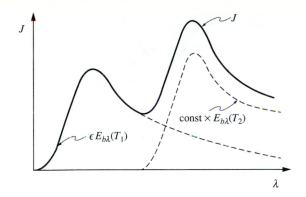

FIGURE 5-5
Qualitative spectral behavior of radiosity for irradiation from an isothermal source.

the spectral radiosity will behave as shown qualitatively in Fig. 5-5. An observer will be able to distinguish between emitted and reflected radiation if he has the ability to distinguish between radiation at different wavelengths. A gray surface does not have this ability, since it behaves in the same fashion toward *all* incoming radiation at *any* wavelength, i.e., it is "color blind." Consequently, a gray surface does not "know" whether its irradiation comes from a gray, diffuse surface or from a black surface with an effective emissive power J. This fact simplifies the analysis considerably since it allows us to calculate radiative heat transfer rates between surfaces by balancing the net outgoing radiation (i.e., emission and reflection) traveling directly from surface to surface (as opposed to emitted radiation traveling to another surface directly or after any number of reflections). For this reason the following analysis is often referred to as the *net radiation method*.

Making an energy balance on a surface dA in the enclosure shown in Fig. 5-6 we obtain from equation (4.2)

$$q(\mathbf{r}) = \epsilon(\mathbf{r})E_b(\mathbf{r}) - \alpha(\mathbf{r})H(\mathbf{r}) = J(\mathbf{r}) - H(\mathbf{r}). \qquad (5.20)$$

The irradiation $H(\mathbf{r})$ is again found by determining the contribution from a differential area $dA'(\mathbf{r}')$, followed by integrating over the entire surface. From the definition of the view factor the heat transfer rate leaving dA' intercepted by dA is $(J(\mathbf{r}')dA')dF_{dA'-dA}$. Thus, similar to the black-surfaces case,

$$H(\mathbf{r})\, dA = \int_A J(\mathbf{r}')\, dF_{dA'-dA} dA' + H_o(\mathbf{r})\, dA, \qquad (5.21)$$

where $H_o(\mathbf{r})$ is again any external radiation arriving at dA. Using reciprocity this equation reduces to

$$H(\mathbf{r}) = \int_A J(\mathbf{r}')\, dF_{dA-dA'} + H_o(\mathbf{r}). \qquad (5.22)$$

Substitution into equation (5.20) yields

$$q(\mathbf{r}) = \epsilon(\mathbf{r})E_b(\mathbf{r}) - \alpha(\mathbf{r})\left[\int_A J(\mathbf{r}')\, dF_{dA-dA'} + H_o(\mathbf{r})\right]. \qquad (5.23)$$

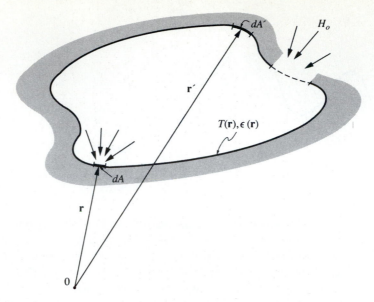

FIGURE 5-6
Radiative exchange in a gray, diffuse enclosure.

Thus, the unknown heat flux (or temperature) could be calculated if the radiosity field had been known. A governing integral equation for radiosity is readily established by solving equation (5.20) for J:

$$J(\mathbf{r}) = \epsilon(\mathbf{r})E_b(\mathbf{r}) + \rho(\mathbf{r})\left[\int_A J(\mathbf{r}')\, dF_{dA-dA'} + H_o(\mathbf{r})\right], \tag{5.24}$$

for those surface locations where the temperature is known, or

$$J(\mathbf{r}) = q(\mathbf{r}) + \int_A J(\mathbf{r}')\, dF_{dA-dA'} + H_o(\mathbf{r}), \tag{5.25}$$

for those parts of the surface where the local heat flux is specified. However, in problems without participating media there is rarely a need to determine radiosity, and it is usually best to eliminate radiosity from equation (5.23). Expressing radiosity in terms of local temperature and heat flux and eliminating irradiation H from equation (5.20) we have

$$q - \alpha q = (\epsilon E_b - \alpha H) - \alpha(J - H) = \epsilon E_b - \alpha J.$$

Up to this point we have differentiated between emissivity and absorptivity, to keep the relations as general as possible (i.e., to accommodate nongray surface properties if necessary). We shall now invoke the assumption of gray, diffuse surfaces, or $\alpha = \epsilon$. Then

$$q(\mathbf{r}) = \frac{\epsilon(\mathbf{r})}{1 - \epsilon(\mathbf{r})}[E_b(\mathbf{r}) - J(\mathbf{r})]. \tag{5.26}$$

Solving for radiosity, we get

$$J(\mathbf{r}) = E_b(\mathbf{r}) - \left(\frac{1}{\epsilon(\mathbf{r})} - 1\right)q(\mathbf{r}). \tag{5.27}$$

Substituting this into equation (5.23), we obtain an integral equation relating temperature T and heat flux q:

$$\frac{q(\mathbf{r})}{\epsilon(\mathbf{r})} - \int_A \left(\frac{1}{\epsilon(\mathbf{r'})} - 1\right)q(\mathbf{r'})\,dF_{dA-dA'} + H_o(\mathbf{r})$$

$$= E_b(\mathbf{r}) - \int_A E_b(\mathbf{r'})\,dF_{dA-dA'}. \tag{5.28}$$

Note that equation (5.28) reduces to equation (5.4) for a black enclosure. However, for a black enclosure with known temperature field the local heat flux can be determined with a simple integration over emissive power. For a gray enclosure an *integral equation* must be solved, i.e., an equation where the unknown dependent variable $q(\mathbf{r})$ appears inside an integral. This requirement makes the solution considerably more difficult.

As for a black enclosure it is customary to break up a gray enclosure into N subsurfaces, over each of which the *radiosity* is assumed constant. Then equation (5.23) becomes

$$\frac{q_i(\mathbf{r}_i)}{\epsilon_i(\mathbf{r}_i)} = E_{bi}(\mathbf{r}_i) - \sum_{j=1}^{N} J_j F_{di-j}(\mathbf{r}_i) - H_{oi}(\mathbf{r}_i), \quad i = 1, 2, \ldots, N, \tag{5.29}$$

and, taking an average over subsurface A_i,

$$\frac{q_i}{\epsilon_i} = E_{bi} - \sum_{j=1}^{N} J_j F_{i-j} - H_{oi}, \quad i = 1, 2, \ldots, N. \tag{5.30}$$

Taking a similar average for equation (5.26) gives

$$q_i = \frac{\epsilon_i}{1 - \epsilon_i}[E_{bi} - J_i]. \tag{5.31}$$

Solving for J and substituting into equation (5.30) then leads to

$$\frac{q_i}{\epsilon_i} - \sum_{j=1}^{N} \left(\frac{1}{\epsilon_j} - 1\right)F_{i-j}q_j + H_{oi} = E_{bi} - \sum_{j=1}^{N} F_{i-j}E_{bj}, \quad i = 1, 2, \ldots, N. \tag{5.32}$$

This relation also follows directly from equation (5.28) if both $(1/\epsilon - 1)q$ and E_b (the components of J) are assumed constant across the subsurfaces. Recalling that $\sum_{j=1}^{N} F_{i-j} = 1$, we may also write equation (5.32) as an interchange between surfaces,

$$\frac{q_i}{\epsilon_i} - \sum_{j=1}^{N} \left(\frac{1}{\epsilon_j} - 1\right)F_{i-j}\,q_j + H_{oi} = \sum_{j=1}^{N} F_{i-j}(E_{bi} - E_{bj}), \quad i = 1, 2, \ldots, N. \tag{5.33}$$

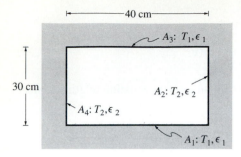

FIGURE 5-7
Two-dimensional gray, diffuse duct for Example 5.4.

Either one of these equations, of course, reduces to equation (5.8) for a black enclosure. Equation (5.32) is preferred for open configurations, since it allows one to ignore hypothetical closing surfaces; and equation (5.33) is preferred for closed enclosures because it eliminates transfer between surfaces at the same temperature.

Sometimes one wishes to determine the radiosity of a surface, for example, in the field of pyrometry (relating surface temperature to radiative intensity leaving a surface). Depending on which of the two is unknown, elimination of q_i or E_{bi} from equation (5.30) with the help of equation (5.31) leads to

$$J_i = \epsilon_i E_{bi} + (1 - \epsilon_i) \left(\sum_{i=1}^{N} J_j F_{i-j} + H_{oi} \right) \tag{5.34a}$$

$$= q_i + \sum_{i=1}^{N} J_j F_{i-j} + H_{oi}, \quad i = 1, 2, \dots, N. \tag{5.34b}$$

These two relations simply repeat the definition of radiosity, the first stating that radiosity consists of emitted and reflected heat fluxes and the second that radiosity, or outgoing heat flux, is equal to net heat flux (with negative q_{in}) plus the absolute value of q_{in}.

Example 5.4.   Reconsider Example 5.1 for a gray, diffuse surface material. Top and bottom walls are at $T_1 = T_3 = 1000\,\text{K}$ with $\epsilon_1 = \epsilon_2 = 0.3$, while the side walls are at $T_2 = T_4 = 600\,\text{K}$ with $\epsilon_2 = \epsilon_4 = 0.8$ as shown in Fig. 5-7. Determine the net radiative heat transfer rates for each surface.

Solution

Using equation (5.33) for $i = 1$ and $i = 2$, and recalling that $F_{1-2} = F_{1-4}$ and $F_{2-1} = F_{2-3}$,

$$i = 1: \quad \frac{q_1}{\epsilon_1} - 2\left(\frac{1}{\epsilon_2} - 1\right) F_{1-2}\, q_2 - \left(\frac{1}{\epsilon_1} - 1\right) F_{1-3}\, q_1 = 2 F_{1-2}(E_{b1} - E_{b2})$$

$$i = 2: \quad \frac{q_2}{\epsilon_2} - 2\left(\frac{1}{\epsilon_1} - 1\right) F_{2-1}\, q_1 - \left(\frac{1}{\epsilon_2} - 1\right) F_{2-4}\, q_2 = 2 F_{2-1}(E_{b2} - E_{b1})$$

We have already evaluated $F_{1-2} = \frac{1}{4}$ and $F_{2-1} = \frac{1}{3}$ in Example 5.1. From the summation rule $F_{1-3} = 1 - 2F_{1-2} = \frac{1}{2}$ and $F_{2-4} = 1 - 2F_{2-1} = \frac{1}{3}$. Substituting these, as well as emissivity values, into the relations reduces them to the simpler form of

$$\left[\frac{1}{3} - \left(\frac{1}{0.3} - 1\right)\frac{1}{2}\right]q_1 - 2\left(\frac{1}{0.8} - 1\right)\frac{1}{4}q_2 = 2 \times \frac{1}{4}(E_{b1} - E_{b2}),$$

$$-2\left(\frac{1}{0.3} - 1\right)\frac{1}{3}q_1 + \left[\frac{1}{0.8} - \left(\frac{1}{0.8} - 1\right)\frac{1}{3}\right]q_2 = 2 \times \frac{1}{3}(E_{b1} - E_{b2}),$$

or

$$\frac{13}{6}q_1 - \frac{1}{8}q_2 = \frac{1}{2}(E_{b1} - E_{b2}),$$

$$-\frac{14}{9}q_1 + \frac{7}{6}q_2 = -\frac{2}{3}(E_{b1} - E_{b2}).$$

Thus,

$$\left(\frac{13}{6} \times \frac{7}{6} - \frac{14}{9} \times \frac{1}{8}\right)q_1 = \left(\frac{1}{2} \times \frac{7}{6} - \frac{2}{3} \times \frac{1}{8}\right)(E_{b1} - E_{b2}),$$

$$q_1 = \frac{3}{7} \times \frac{1}{2}(E_{b1} - E_{b2}) = \frac{3}{14}\sigma(T_1^4 - T_2^4),$$

and

$$\left(-\frac{1}{8} \times \frac{14}{9} + \frac{7}{6} \times \frac{13}{6}\right)q_2 = \left(\frac{1}{2} \times \frac{14}{9} - \frac{2}{3} \times \frac{13}{6}\right)(E_{b1} - E_{b2}),$$

$$q_2 = \frac{3}{7} \times \frac{2}{3}(E_{b1} - E_{b2}) = -\frac{2}{7}\sigma(T_1^4 - T_2^4).$$

Finally, substituting values for temperatures,

$$Q_1' = 0.4\,\mathrm{m} \times \tfrac{3}{14} \times 5.670 \times 10^{-8}\,\frac{\mathrm{W}}{\mathrm{m^2\,K^4}}(1000^4 - 600^4)\,\mathrm{K^4} = 4230\,\mathrm{W/m},$$

$$Q_2' = -0.3\,\mathrm{m} \times \tfrac{2}{7} \times 5.670 \times 10^{-8}\,\frac{\mathrm{W}}{\mathrm{m^2\,K^4}}(1000^4 - 600^4)\,\mathrm{K^4} = -4230\,\mathrm{W/m}.$$

Of course, both heat transfer rates must again add up to zero. We observe that these rates are less than half the ones for the black duct.

Example 5.5. Determine the radiative heat flux between two isothermal gray concentric spheres with radii R_1 and R_2, temperatures T_1 and T_2, and emissivities ϵ_1 and ϵ_2, respectively, as shown in Fig. 5-8a.

Solution

Again applying equation (5.33) for $i = 1$ (inner sphere) and $i = 2$ (outer sphere), we obtain:

$$i = 1: \quad \frac{q_1}{\epsilon_1} - \left(\frac{1}{\epsilon_1} - 1\right)F_{1-1}\,q_1 - \left(\frac{1}{\epsilon_2} - 1\right)F_{1-2}\,q_2 = F_{1-2}(E_{b1} - E_{b2}),$$

$$i = 2: \quad \frac{q_2}{\epsilon_2} - \left(\frac{1}{\epsilon_1} - 1\right)F_{2-1}\,q_1 - \left(\frac{1}{\epsilon_2} - 1\right)F_{2-2}\,q_2 = F_{2-1}(E_{b2} - E_{b1}).$$

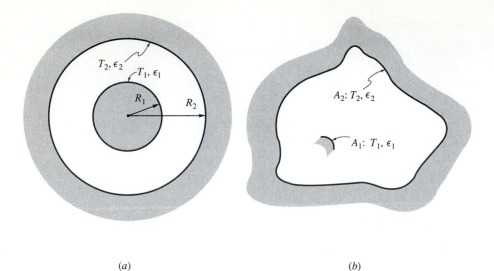

(a) (b)

FIGURE 5-8
Radiative transfer between (a) two concentric spheres, (b) a convex surface and a large isothermal enclosure.

With $F_{1-1} = 0$, $F_{1-2} = 1$, $F_{2-1} = A_1/A_2$, and $F_{2-2} = 1 - F_{2-1} = 1 - A_1/A_2$, these two equations reduce to

$$\frac{1}{\epsilon_1} q_1 - \left(\frac{1}{\epsilon_2} - 1\right) q_2 = \sigma(T_1^4 - T_2^4),$$

$$\left(\frac{1}{\epsilon_1} - 1\right)\frac{A_1}{A_2} q_1 + \left[\frac{1}{\epsilon_2} - \left(\frac{1}{\epsilon_2} - 1\right)\left(1 - \frac{A_1}{A_2}\right)\right] q_2 = -\frac{A_1}{A_2}\sigma(T_1^4 - T_2^4).$$

This may be solved for q_1 by eliminating q_2 (or using conservation of energy, i.e., $A_1 q_1 + A_2 q_2 = 0$), or

$$q_1 = \frac{\sigma(T_1^4 - T_2^4)}{\dfrac{1}{\epsilon_1} + \dfrac{A_1}{A_2}\left(\dfrac{1}{\epsilon_2} - 1\right)}. \tag{5.35}$$

We note that equation (5.35) is not just limited to concentric spheres, but holds for any convex surface A_1 (i.e., with $F_{1-1} = 0$) that radiates only to A_2 (i.e., $F_{1-2} = 1$) as indicated in Fig. 5-8b. This is often convenient for a convex surface A_i placed into a large, isothermal environment ($A_a \gg A_i$) at temperature T_a, leading to

$$q_i = \epsilon_i \sigma(T_i^4 - T_a^4). \tag{5.36}$$

Surface A_i may also be a hypothetical one, closing an open configuration contained within a large environment.

Example 5.6.   Repeat Example 5.3 for a groove whose surface is gray and diffuse, with emissivity ϵ, rather than black.

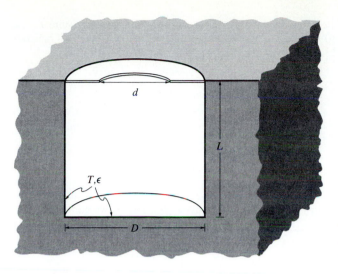

FIGURE 5-9
Cylindrical cavity with partial cover plate, Example 5.7.

Solution

Using equation (5.32) for the open configuration we obtain

$$i = 1: \qquad \frac{q_1}{\epsilon} - \left(\frac{1}{\epsilon} - 1\right) F_{1-2}\, q_2 + H_{o1} = \sigma T^4 (1 - F_{1-2}),$$

$$i = 2: \qquad \frac{q_2}{\epsilon} - \left(\frac{1}{\epsilon} - 1\right) F_{2-1}\, q_1 + H_{o2} = \sigma T^4 (1 - F_{2-1}),$$

where we have made use of the fact that $E_{b1} = E_{b2} = \sigma T^4$ and $\epsilon_1 = \epsilon_2 = \epsilon$. As in Example 5.3 we have $F_{1-2} = F_{2-1} = 1 - \sqrt{2}/2$ and $H_{o1} = q_{\text{sol}} \cos \alpha$, $H_{o2} = q_{\text{sol}} \sin \alpha$. Since we are only interested in the total heat loss we add the two equations, leading to

$$\left[\frac{1}{\epsilon} - \left(\frac{1}{\epsilon} - 1\right) F_{1-2}\right] (q_1 + q_2) = \sqrt{2}\sigma T^4 - q_{\text{sol}}(\cos \alpha + \sin \alpha),$$

and

$$Q' = a(q_1 + q_2) = \frac{a\left[\sqrt{2}\sigma T^4 - q_{\text{sol}}(\cos \alpha + \sin \alpha)\right]}{1 + \left(\frac{1}{\epsilon} - 1\right)/\sqrt{2}}.$$

Comparing this result with that of Example 5.3, we see that the heat loss due to emission is decreased (less emission, but more effective heat loss of emitted energy due to reflection from the opposing surface), as is the solar heat gain (since some of the irradiation is reflected back out of the cavity).

Example 5.7. Consider the cavity shown in Fig. 5-9, which consists of a cylindrical hole of diameter D and length L. The top of the cavity is covered with a disk, which has a hole of diameter d. The entire inside of the cavity is isothermal at temperature T, and is covered with a gray, diffuse material of emissivity ϵ. Determine the amount of radiation escaping from the cavity.

Solution

For simplicity, since the entire surface is isothermal and has the same emissivity, we use a single zone A_1, which comprises the entire groove surface (sides, bottom and top). Therefore, equation (5.32) reduces to

$$\left[\frac{1}{\epsilon_1} - \left(\frac{1}{\epsilon_1} - 1\right)F_{1-1}\right]q_1 = (1 - F_{1-1})E_{b1}.$$

Since the total radiative energy rate leaving the cavity is $Q_1 = A_1q_1$, we get

$$Q_1 = \frac{1 - F_{1-1}}{\dfrac{1}{\epsilon_1} - \left(\dfrac{1}{\epsilon_1} - 1\right)F_{1-1}}A_1E_{b1}.$$

The view factor F_{1-1} is easily determined by recognizing that $F_{o-1} = 1$ (and A_o is the opening at the top) and, by reciprocity,

$$F_{1-1} = 1 - F_{1-o} = 1 - \frac{A_o}{A_1}F_{o-1} = 1 - \frac{A_o}{A_1}.$$

Therefore, the radiative heat flux leaving the cavity, per unit area of opening, is

$$\frac{Q_1}{A_o} = \frac{\left(1 - 1 + \dfrac{A_o}{A_1}\right)\dfrac{A_1}{A_o}E_{b1}}{\dfrac{1}{\epsilon_1} - \left(\dfrac{1}{\epsilon_1} - 1\right)\left(1 - \dfrac{A_o}{A_1}\right)} = \frac{E_{b1}}{1 + \left(\dfrac{1}{\epsilon_1} - 1\right)\dfrac{A_o}{A_1}}.$$

Thus, if $A_o/A_1 \ll 1$, the opening of the cavity behaves like a blackbody with emissive power E_{b1}. Such cavities are commonly used in experimental methods in which blackbodies are needed for comparison. For example, a cavity with $d/D = 1/2$ and $L/D = 2$ has

$$\frac{A_o}{A_1} = \frac{\pi d^2/4}{2\pi D^2/4 - \pi d^2/4 + \pi DL} = \frac{d^2}{2D^2 - d^2 + 4DL}$$

$$= \frac{(d/D)^2}{2 - (d/D)^2 + 4(L/D)} = \frac{1/4}{2 - 1/4 + 4\times 2} = \frac{1}{39}.$$

For $\epsilon_1 = 0.5$ this results in an *apparent emissivity* of

$$\epsilon_a = \frac{Q_1}{A_oE_{b1}} = \frac{1}{1 + \left(\dfrac{1}{\epsilon_1} - 1\right)\dfrac{A_o}{A_1}} = \frac{1}{1 + \left(\dfrac{1}{0.5} - 1\right)\dfrac{1}{39}} = \frac{39}{40} = 0.975.$$

For computer calculations the Kronecker delta is introduced into equation (5.32), as was done for a black enclosure, leading to

$$\sum_{j=1}^{N}\left[\frac{\delta_{ij}}{\epsilon_j} - \left(\frac{1}{\epsilon_j} - 1\right)F_{i-j}\right]q_j = \sum_{j=1}^{N}[\delta_{ij} - F_{i-j}]E_{bj} - H_{oi}. \qquad (5.37)$$

If all the temperatures are known and the radiative heat fluxes are to be determined, equation (5.37) may be cast in matrix form as

$$\mathbf{C} \cdot \mathbf{q} = \mathbf{A} \cdot \mathbf{e_b} - \mathbf{h_o}, \tag{5.38}$$

where $\mathbf{C}$ and $\mathbf{A}$ are matrices with elements

$$C_{ij} = \frac{\delta_{ij}}{\epsilon_j} - \left(\frac{1}{\epsilon_j} - 1\right)F_{i-j},$$

$$A_{ij} = \delta_{ij} - F_{i-j},$$

and $\mathbf{q}, \mathbf{e_b}$ and $\mathbf{h_o}$ are vectors of the unknown heat fluxes q_j and the known emissive powers E_{bj} and external irradiations H_{oj}. Equation (5.38) is solved by matrix inversion as

$$\mathbf{q} = \mathbf{C}^{-1} \cdot [\mathbf{A} \cdot \mathbf{e_b} - \mathbf{h_o}]. \tag{5.39}$$

If the emissive power is known over only some of the surfaces, and the heat fluxes are specified elsewhere, equation (5.38) may be rearranged into a similar equation for the vector containing all the unknowns.

5.4 ELECTRICAL NETWORK ANALOGY

While equation (5.37) represents the most convenient set of governing equations for numerical calculations on today's digital computers, some people prefer to get a physical feeling for the radiative exchange problem by representing it through an analogous electrical network, a method more suitable for analog computers—now nearly extinct. For completeness, we shall briefly present this electrical network method, which was first introduced by Oppenheim [1].

From equation (5.20) we have

$$q_i = J_i - H_i, \quad i = 1, 2, \ldots, N, \tag{5.40}$$

or, with equations (5.30) and (5.31),

$$q_i = J_i - \sum_{j=1}^{N} J_j F_{i-j} - H_{oi}, \tag{5.41}$$

$$= \sum_{j=1}^{N} (J_i - J_j)F_{i-j} - H_{oi}, \quad i = 1, 2, \ldots, N. \tag{5.42}$$

We shall first consider the simple case of two infinite parallel plates without external irradiation. Thus, $N = 2$, $H_{oi} = 0$, and

$$Q_1 = A_1 q_1 = \frac{J_1 - J_2}{\dfrac{1}{A_1 F_{1-2}}} = -Q_2. \tag{5.43}$$

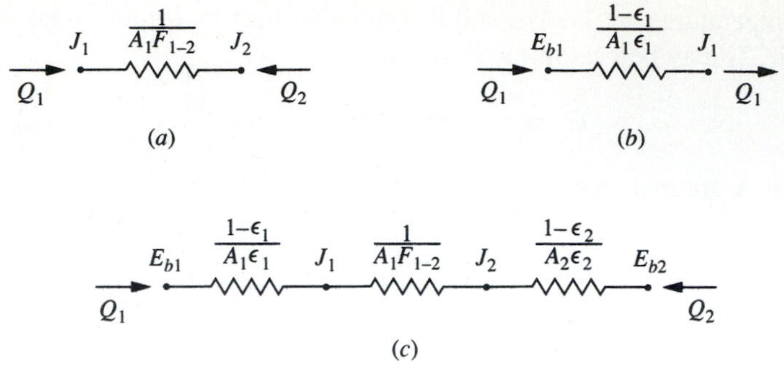

FIGURE 5-10
Electric network analogy for infinite parallel plates: (a) Space resistance, (b) surface resistance, (c) total resistance.

As written, equation (5.43) may be interpreted as follows: If the radiosities are considered *potentials*, $1/A_1F_{1-2}$ is a *radiative resistance between surfaces*, or a *space resistance*, and Q is a radiative heat flow "current," then equation (5.43) is identical to the one governing an electrical current flowing across a resistor due to a voltage potential, as indicated in Fig. 5-10a. The space resistance is a measure of how easily a radiative heat flux flows from one surface to another: The larger F_{1-2}, the more easily heat can travel from A_1 to A_2, resulting in a smaller resistance. The same heat flux is also given by equation (5.31) as

$$Q_1 = \frac{E_{b1} - J_1}{\dfrac{1 - \epsilon_1}{A_1\epsilon_1}} = \frac{J_2 - E_{b2}}{\dfrac{1 - \epsilon_2}{A_2\epsilon_2}} = -Q_2, \tag{5.44}$$

where $(1 - \epsilon_i)/A_i\epsilon_i$ are *radiative surface resistances*. This situation is shown in Fig. 5-10b. The surface resistance describes a surface's ability to radiate. For the maximum radiator, a black surface, the resistance is zero. This fact implies that, for a finite heat flux, the potential drop across a zero resistance must be zero, i.e., $J_i = E_{bi}$. Of course, the radiosities may be eliminated from equations (5.43) and (5.44), and

$$Q_1 = \frac{E_{b1} - E_{b2}}{\dfrac{1 - \epsilon_1}{A_1\epsilon_1} + \dfrac{1}{A_1F_{1-2}} + \dfrac{1 - \epsilon_2}{A_2\epsilon_2}} = -Q_2, \tag{5.45}$$

where the denominator is the *total radiative resistance* between surfaces A_1 and A_2. Since the three resistances are in series they simply add up as electrical resistances do; see Fig. 5-10c.

This network analogy is readily extended to more complicated situations by rewrit-

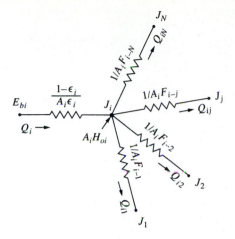

FIGURE 5-11
Network representation for radiative heat flux between surface A_i and all other surfaces.

ing equation (5.42) as

$$Q_i = \frac{E_{bi} - J_i}{\dfrac{1 - \epsilon_i}{A_i \epsilon_i}} = \sum_{j=1}^{N} \frac{J_i - J_j}{\dfrac{1}{A_i F_{i-j}}} - A_i H_{oi} = \sum_{j=1}^{N} Q_{i-j} - A_i H_{oi}. \tag{5.46}$$

Thus, the total heat flux at surface i is the net radiative exchange between A_i and all the other surfaces in the enclosure. The electrical analog is shown in Fig. 5-11, where the current flowing from E_{bi} to J_i is divided into N parallel lines, each with a different potential difference and with different resistors.

Example 5.8.   Consider a solar collector shown in Fig. 5-12a. The collector consists of a glass cover plate, a collector plate, and side walls. We shall assume that the glass is totally transparent to solar irradiation, which penetrates through the glass and hits the absorber plate with a strength of $1000 \, \text{W/m}^2$. The absorber plate is black and is kept at a constant temperature $T_1 = 77°C$ by heating water flowing underneath it. The side walls are insulated and made of a material with emissivity $\epsilon_2 = 0.5$. The glass cover may be considered opaque to thermal (i.e., infrared) radiation with an emissivity $\epsilon_3 = 0.9$. The collector is $1 \, \text{m} \times 1 \, \text{m} \times 10 \, \text{cm}$ in dimension and is reasonably evacuated to suppress free convection between absorber plate and glass cover. The convective heat transfer coefficient at the top of the glass cover is known to be $h = 5.0 \, \text{W/m}^2 \, \text{K}$, and the temperature of the ambient is $T_a = 17°C$. Estimate the collected energy for normal solar incidence.

Solution

We may construct an equivalent network (Fig. 5-12b), leading to

$$Q_1 = \frac{\sigma(T_1^4 - T_a^4)}{R_{13} + \dfrac{1 - \epsilon_3}{A_3 \epsilon_3} + R_{3a}} - A_1 q_s,$$

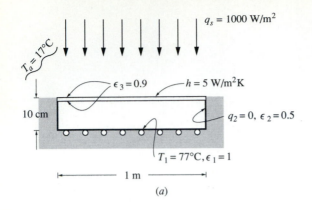

$q_s = 1000 \text{ W/m}^2$

$T_a = 17°C$

$\epsilon_3 = 0.9$ $h = 5 \text{ W/m}^2\text{K}$

10 cm

$q_2 = 0, \epsilon_2 = 0.5$

$T_1 = 77°C, \epsilon_1 = 1$

1 m

(a)

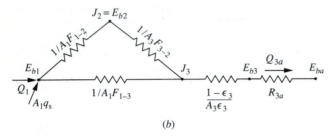

$J_2 = E_{b2}$

$1/A_1F_{1-2}$ $1/A_3F_{3-2}$

E_{b1} J_3 E_{b3} Q_{3a} E_{ba}

Q_1 $1/A_1F_{1-3}$ $\dfrac{1-\epsilon_3}{A_3\epsilon_3}$ R_{3a}

$A_1 q_s$

(b)

FIGURE 5-12
Schematics for Example 5.8
(a) geometry, (b) network.

where R_{13} is the total resistance between surfaces A_1 and A_3, and R_{3a} is the resistance, by radiation as well as free convection, between glass cover and environment. We note that, since A_2 is insulated, there is no heat flux entering/leaving at E_{b2} and, from equation (5.44), $J_2 = E_{b2}$. Thus, the total resistance between A_1 and A_3 comes from two parallel circuits, one with resistance $1/(A_1F_{1-3})$ and the other with two resistances in series, $1/(A_1F_{1-2})$ and $1/(A_3F_{3-2})$, or

$$\frac{1}{R_{13}} = \frac{1}{1/(A_1F_{1-3})} + \frac{1}{1/(A_1F_{1-2}) + 1/(A_3F_{3-2})}$$
$$= A_1F_{1-3} + \tfrac{1}{2}A_1F_{1-2} = A_1\left(F_{1-3} + \tfrac{1}{2}F_{1-2}\right),$$

where we have used the fact that $A_1F_{1-2} = A_3F_{3-2}$ by symmetry. From Configuration 38 or Fig. D-2 in Appendix D we obtain, with $X = Y = 10$, $F_{1-3} = 0.827$ and $F_{1-2} = 1 - F_{1-3} = 0.173$, and

$$R_{13} = 1/\left[1\text{ m}^2 \times (0.827 + 0.5 \times 0.173)\right] = 1.095\text{ m}^{-2}.$$

The resistance between glass cover and ambient is a little more complicated. The total heat loss from the cover plate, by free convection and radiation, is

$$Q_{3a} = \epsilon_3 A_3 \sigma(T_3^4 - T_a^4) + hA_3(T_3 - T_a),$$

where we have assumed that the environment (sky) radiates to the collector with the ambient temperature T_a. To convert this to the correct form we rewrite it as

$$Q_{3a} = \sigma(T_3^4 - T_a^4)A_3 \left[\epsilon_3 + \frac{h(T_3 - T_a)}{\sigma(T_3^4 - T_a^4)} \right],$$

or

$$\frac{1}{R_{3a}} = A_3 \left[\epsilon_3 + \frac{h}{\sigma} \frac{T_3 - T_a}{T_3^4 - T_a^4} \right] = A_3 \left[\epsilon_3 + \frac{h}{\sigma} \frac{1}{T_3^3 + T_3^2 T_a + T_3 T_a^2 + T_a^3} \right].$$

As a first approximation, if T_3 is not too different from T_a,

$$\frac{1}{R_{3a}} \simeq A_3 \left(\epsilon_3 + \frac{h}{4\sigma T_a^3} \right).$$

$$= 1\,\mathrm{m^2} \left(0.9 + \frac{5\,\mathrm{W/m^2\,K}}{4 \times 5.670 \times 10^{-8}\,\mathrm{W/m^2\,K^4} \times (273 + 17)^3\,\mathrm{K^3}} \right) = \frac{1}{0.554}\,\mathrm{m^2}.$$

Finally, substituting the resistances into the expression for Q_1 we get

$$Q_1 = \frac{5.670 \times 10^{-8}\,\mathrm{W/m^2\,K^4}\,[(273 + 77)^4 - (273 + 17)^4]\,\mathrm{K^4}}{1.095\,\mathrm{m^{-2}} + \dfrac{1 - 0.9}{0.9\,\mathrm{m^2}} + 0.554\,\mathrm{m^{-2}}} - 1\,\mathrm{m^2} \times 1000\,\mathrm{W/m^2}$$

$$= -744\,\mathrm{W}.$$

Since the system could collect a theoretical maximum of $-1000\,\mathrm{W}$, the collector efficiency is

$$\eta_{\mathrm{collector}} = \frac{Q_1}{A_1 q_s} = \frac{744}{1000} = 0.744 = 74.4\%.$$

This efficiency should be compared with an uncovered black collector plate, whose net heat flux would be

$$Q_1 = A_1 \left[\sigma(T_1^4 - T_a^4) + h(T_1 - T_a) - q_s \right]$$

$$= 1\,\mathrm{m^2} \left[5.670 \times 10^{-8} \times (350^4 - 290^4) + 5 \times (350 - 290) - 1000 \right]\,\mathrm{W/m^2}$$

$$= -250\,\mathrm{W}.$$

Thus, an unprotected collector at that temperature would have an efficiency of only 25%.

The electrical network analogy is a very simple and physically appealing approach for simple two- and three-surface enclosures, such as the one of the previous example. However, in more complicated enclosures with multiple surfaces the method quickly becomes tedious and intractable.

5.5 SOLUTION METHODS FOR THE GOVERNING INTEGRAL EQUATIONS

The usefulness of the method described in the previous sections is limited by the fact that it requires the radiosity to be constant over each subsurface. This is rarely the case if the subsurfaces of the enclosure are relatively large (as compared with typical distances between surfaces). Today, with the advent of powerful digital computers, more accurate solutions are usually obtained by increasing the number of subsurfaces, N, in equation (5.37), which then become simply a finite-difference solution to the integral equation (5.28). Still, there are times when more accurate methods for the solution of equation (5.28) are desired (for computational efficiency), or when exact or approximate solutions are sought in explicit form. Therefore, we shall give here a very brief outline of such solution methods.

If radiosity J is to be determined, the governing equation that needs to be solved is either equation (5.24), if the surface temperature is given, or equation (5.25), if surface heat flux is specified. If unknown temperatures or heat fluxes are to be determined directly, equation (5.28) must be solved. In all cases the governing equation may be written as a *Fredholm integral equation of the second kind,*

$$\phi(\mathbf{r}) = f(\mathbf{r}) + \int_A K(\mathbf{r}, \mathbf{r}') \phi(\mathbf{r}') \, dA', \tag{5.47}$$

where $K(\mathbf{r}, \mathbf{r}')$ is called the *kernel* of the integral equation, $f(\mathbf{r})$ is a known function, and $\phi(\mathbf{r})$ is the function to be determined (e.g., radiosity or heat flux). Comprehensive discussions for the treatment of such integral equations are given in mathematical texts such as Courant and Hilbert [2] or Hildebrand [3]. A number of radiative heat transfer examples have been discussed by Özişik [4].

Numerical solutions to equation (5.47) may be found in a number of ways. In the *method of successive approximation* a first guess of $\phi(\mathbf{r}) = f(\mathbf{r})$ is made with which the integral in equation (5.47) is evaluated (analytically in some simple situations, but more often through numerical quadrature). This leads to an improved value for $\phi(\mathbf{r})$, which is substituted back into the integral, and so on. This scheme is known to converge for all surface radiation problems. Another possible solution method is *reduction to algebraic equations* by using numerical quadrature for the integral, i.e., replacing it by a series of quadrature coefficients and nodal values. This leads to a set of equations similar to equation (5.37), but of higher accuracy. A third method of solution has been given by Sparrow and Haji-Sheikh [5], who demonstrated that the method of *variational calculus* may be applied to general problems governed by a Fredholm integral equation.

Most early numerical solutions in the literature dealt with two very basic systems. The problem of two-dimensional parallel plates of finite width was studied in some detail by Sparrow and coworkers [5–7], using the variational method. The majority of studies have concentrated on radiation from cylindrical holes because of the importance of this geometry for cylindrical tube flow, as well as for the preparation of a blackbody for calibrating radiative property measurements. The problem of an infinitely long isothermal hole radiating from its opening was first studied by Buckley [8] and by Eckert [9]. Buckley's work appears to be the first employing the kernel approximation method. Much later, the same problem was solved exactly through the method of

TABLE 5.1

Apparent emissivity, $\epsilon_a = J/\sigma T^4$, at the bottom center of an isothermal partially covered cylindrical cavity [17, 18].

			ϵ_a	
ϵ	R_i/R	$(L/R = 2)$	$(L/R = 4)$	$(L/R = 8)$
0.25	0.4	0.916	0.968	0.990
	0.6	0.829	0.931	0.981
	0.8	0.732	0.888	0.969
	1.0	0.640	0.844	0.965
0.50	0.4	0.968	0.990	0.998
	0.6	0.932	0.979	0.995
	0.8	0.887	0.964	0.992
	1.0	0.839	0.946	0.989
0.75	0.4	0.988	0.997	0.999
	0.6	0.975	0.997	0.998
	0.8	0.958	0.988	0.997
	1.0	0.939	0.982	0.996

successive approximation (with numerical quadrature) by Sparrow and Albers [10]. A finite hole, but with both ends open, was studied by a number of investigators. Usiskin and Siegel [11] considered the constant wall heat flux case, using the kernel approximation as well as a variational approach. The constant wall temperature case was studied by Lin and Sparrow [12], and combined convection/surface radiation was investigated by Perlmutter and Siegel [13, 14]. Of greater importance for the manufacture of a blackbody is the isothermal cylindrical cavity of finite depth, which was studied by Sparrow and coworkers [15, 16] using successive approximations. If part of the opening is covered by a flat ring with a smaller hole, such a cavity behaves like a blackbody for very small L/R ratios. This problem was studied by Alfano [17] and Alfano and Sarno [18]. Because of their importance for the manufacture of blackbody cavities these results are summarized in Table 5.1. A detector removed from the cavity will sense a signal proportional to the intensity leaving the bottom center of the cavity in the normal direction. Thus the effectiveness of the blackbody is measured by how close to unity the ratio $I_n/I_b(T)$ is. For perfectly diffuse reflectors, $I_n = J/\pi$, and with $I_b = \sigma T/\pi$ an apparent emissivity is defined as

$$\epsilon_a = I_n/I_b(T) = J/\sigma T^4. \tag{5.48}$$

To give an outline how the different methods may be applied we shall, over the following few pages, solve the same simple example by three different methods, the first two being "exact," and the third being the kernel approximation.

Example 5.9.   Consider two long parallel plates of width w as shown in Fig. 5-13. Both plates are isothermal at the (same) temperature T, and both have a gray, diffuse emissivity of ϵ. The plates are separated by a distance h and are placed in a large, cold environment. Determine the local radiative heat fluxes along the plate using the method of successive approximation.

Solution

From equation (5.24) we find, with $dF_{di-di} = 0$,

$$J_1(x_1) = \epsilon \sigma T^4 + (1 - \epsilon) \int_0^w J_2(x_2)\, dF_{d1-d2},$$

$$J_2(x_2) = \epsilon \sigma T^4 + (1 - \epsilon) \int_0^w J_1(x_1)\, dF_{d2-d1},$$

and, from Configuration 1 in Appendix D, with $s_{12} = h/\cos\phi$, $s_{12}d\phi = dx_2\cos\phi$, and $\cos\phi = h/\sqrt{h^2 + (x_2 - x_1)^2}$,

$$dx_1 dF_{d1-d2} = dx_2 dF_{d2-d1} = \frac{1}{2}\cos\phi\, d\phi\, dx_1 = \frac{\cos^3\phi}{2h} dx_1 dx_2$$

$$= \frac{1}{2}\frac{h^2 dx_1 dx_2}{[h^2 + (x_1 - x_2)^2]^{3/2}}.$$

Introducing nondimensional variables $W = w/h$, $\xi = x/h$, $\mathcal{J}(x) = J(x)/\sigma T^4$ and realizing that, as a result of symmetry, $J_1 = J_2$ (and $q_1 = q_2$), we may simplify the governing integral equation to

$$\mathcal{J}(\xi) = \epsilon + \frac{1}{2}(1 - \epsilon)\int_0^W \mathcal{J}(\xi')\frac{d\xi'}{[1 + (\xi' - \xi)^2]^{3/2}}. \tag{5.49}$$

Making a first guess of $\mathcal{J}^{(1)} = \epsilon$ we obtain a second guess by substitution,

$$\mathcal{J}^{(2)}(\xi) = \epsilon\left\{1 + \frac{1}{2}(1 - \epsilon)\int_0^W \frac{d\xi'}{[1 + (\xi' - \xi)^2]^{3/2}}\right\}$$

$$= \epsilon\left\{1 + \frac{1}{2}(1 - \epsilon)\left[\frac{W - \xi}{\sqrt{1 + (W - \xi)^2}} + \frac{\xi}{\sqrt{1 + \xi^2}}\right]\right\}.$$

Repeating the procedure we get

$$\mathcal{J}^{(3)}(\xi) = \epsilon\left\{1 + \frac{1}{2}(1 - \epsilon)\left[\frac{W - \xi}{\sqrt{1 + (W - \xi)^2}} + \frac{\xi}{\sqrt{1 + \xi^2}}\right]\right.$$

$$\left. + \frac{1}{4}(1 - \epsilon)^2\int_0^W\left[\frac{W - \xi'}{\sqrt{1 + (W - \xi')^2}} + \frac{\xi'}{\sqrt{1 + \xi'^2}}\right]\frac{d\xi'}{[1 + (\xi' - \xi)^2]^{3/2}}\right\},$$

where the last integral becomes quite involved. We shall stop at this point since further successive integrations would have to be carried out numerically. It is clear from the above

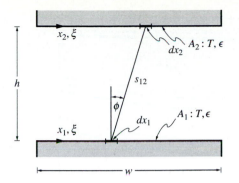

FIGURE 5-13
Radiative exchange between two long isothermal parallel plates.

expression that the terms in the series diminish as $\epsilon\,[(1-\epsilon)W]^n$, i.e., few successive iterations are necessary for surfaces with low reflectivities and/or w/h ratios. Once the radiosity has been determined the local heat flux follows from equation (5.26). Limiting ourselves to $\mathcal{J}^{(2)}$ (single successive approximation), this yields

$$\Psi(\xi) \;=\; \frac{q(\xi)}{\sigma T^4} \;=\; \frac{\epsilon}{1-\epsilon}[1-\mathcal{J}(\xi)]$$

$$=\; \epsilon - \frac{\epsilon^2}{2}\left[\frac{W-\xi}{\sqrt{1+(W-\xi)^2}} + \frac{\xi}{\sqrt{1+\xi^2}}\right] - \mathcal{O}\!\left(\epsilon^2(1-\epsilon)W^2\right),$$

where $\mathcal{O}(z)$ is shorthand for "order of magnitude z." Some results are shown in Fig. 5-14 and compared with other solution methods for the case of $W = w/h = 1$ and three values of the emissivity. Observe that the heat loss is a minimum at the center of the plate, since this location receives maximum irradiation from the other plate (i.e., the view factor from this location to the opposing plate is maximum). For decreasing ϵ the heat loss increases, of course, since more is emitted; however, this increase is less than linear since also more energy is coming in, of which a larger fraction is absorbed. The first successive approximation does very well for small and large ϵ as expected from the order of magnitude of the neglected terms.

Example 5.10. Repeat Example 5.9 using numerical quadrature.

Solution

The governing equation is, of course, again equation (5.49). We shall approximate the integral on the right-hand-side by a series obtained through numerical integration, or *quadrature.* In this method an integral is approximated by a weighted series of the integrand evaluated at a number of nodal points; or

$$\int_a^b f(\xi,\xi')\,d\xi' \simeq (b-a)\sum_{j=1}^{J} c_j f(\xi,\xi_j), \qquad \sum_{j=1}^{J} c_j = 1. \tag{5.50}$$

Here the ξ_j represent J locations between a and b, and the c_j are weight coefficients. The nodal points ξ_j may be equally spaced for easy presentation of results *(Newton-Cotes*

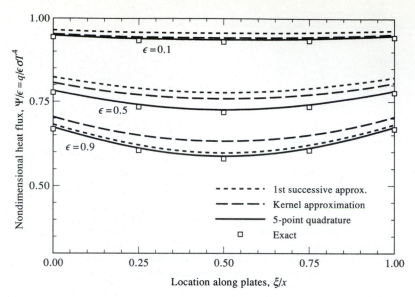

FIGURE 5-14
Local radiative heat flux on long, isothermal parallel plates, determined by various methods.

quadrature), or their location may be optimized for increased accuracy *(Gaussian quadrature);* for a detailed treatment of quadrature see, for example, the book by Fröberg [19].

Using equation (5.50) in equation (5.49) we obtain

$$\mathcal{J}_i = \epsilon + (1 - \epsilon)W \sum_{j=1}^{J} c_j \mathcal{J}_j f_{ij}, \quad i = 1, 2, \ldots, J,$$

where

$$f_{ij} = \frac{1}{2} \Big/ \big[1 + (\xi_j - \xi_i)^2\big]^{3/2}.$$

This system of equations may be further simplified by utilizing the symmetry of the problem, i.e., $\mathcal{J}(\xi) = \mathcal{J}(W-\xi)$. Assuming that nodes are placed symmetrically about the centerline, $\xi_{J+1-j} = \xi_j$, leads to $c_{J+1-j} = c_j$ and $\mathcal{J}_{J+1-j} = \mathcal{J}_j$, or

$$J \text{ odd:} \quad \mathcal{J}_i = \epsilon + (1 - \epsilon)W \left\{ \sum_{j=1}^{(J-1)/2} c_j \mathcal{J}_j [f_{ij} + f_{i,J+1-j}] \right.$$

$$\left. + c_{(J+1)/2} \mathcal{J}_{(J+1)/2} f_{i,(J+1)/2} \right\}, \quad i = 1, 2, \ldots, \frac{J+1}{2},$$

$$J \text{ even:} \quad \mathcal{J}_i = \epsilon + (1 - \epsilon)W \sum_{j=1}^{J/2} c_j \mathcal{J}_j (f_{ij} + f_{i,J+1-j}), \quad i = 1, 2, \ldots, \frac{J}{2}.$$

The values of the radiosities may be determined by successive approximation, or by direct matrix inversion. In Fig. 5-14 the simple case of $J = 5$ (resulting in three simultaneous equations) is included, using Newton-Cotes quadrature with $\xi_j = W(j-1)/4$ and $c_1 = c_5 = 7/90$, $c_2 = c_4 = 32/90$, and $c_3 = 12/90$ [19].

Exact analytical solutions that yield explicit relations for the unknown radiosity are rare and limited to a few special geometries. However, approximate analytical solutions may be found for many geometries through the *kernel approximation method*. In this method the kernel $K(x, x')$ is approximated by a linear series of special functions such as $e^{-ax'}$, $\cos ax'$, $\cosh ax'$, and so on (i.e., functions that, after one or two differentiations with respect to x', turn back into the original function except for a constant factor). It is then often possible to convert integral equation (5.47) into a differential equation that may be solved explicitly. The method is best illustrated through an example.

Example 5.11. Repeat Example 5.10 using the kernel approximation method.

Solution

We again need to solve equation (5.49), this time by approximating the kernel. For convenience we shall choose a simple exponential form,

$$K(\xi, \xi') = \frac{1}{[1 + (\xi' - \xi)^2]^{3/2}} \simeq a\, e^{-b|\xi'-\xi|}.$$

We shall determine "optimum" parameters a and b by letting the approximation satisfy the *0th* and *1st moments*. This implies multiplying the expression by $|\xi' - \xi|$ raised to the *0th* and *1st* powers, followed by integration over the entire domain for $|\xi' - \xi|$, i.e., from 0 to ∞ (since W could be arbitrarily large).[3] Thus,

$$\text{0th moment:} \quad \int_0^\infty \frac{dx}{(1 + x^2)^{3/2}} = 1 = \int_0^\infty a\, e^{-bx}\, dx = \frac{a}{b},$$

$$\text{1st moment:} \quad \int_0^\infty \frac{x\, dx}{(1 + x^2)^{3/2}} = 1 = \int_0^x a\, e^{-bx} x\, dx = \frac{a}{b^2},$$

leading to $a = b = 1$ and

$$K(\xi, \xi') \simeq e^{-|\xi'-\xi|}.$$

Substituting this expression into equation (5.49) leads to

$$\mathcal{J}(\xi) \simeq \epsilon + \frac{1}{2}(1 - \epsilon)\left[\int_0^\xi \mathcal{J}(\xi')e^{-(\xi-\xi')}d\xi' + \int_\xi^W \mathcal{J}(\xi')e^{-(\xi'-\xi)}d\xi'\right].$$

We shall now differentiate this expression twice with respect to ξ, for which we need to employ *Leibnitz' rule*, equation (3.100). Therefore,

$$\frac{d\mathcal{J}}{d\xi} = \frac{1}{2}(1 - \epsilon)\left[\mathcal{J}(\xi) - \int_0^\xi \mathcal{J}(\xi')e^{-(\xi-\xi')}d\xi' - \mathcal{J}(\xi) + \int_\xi^W \mathcal{J}(\xi')e^{-(\xi'-\xi)}d\xi'\right],$$

---

[3]Using the actual W at hand will result in a better approximation, but new values for a and b must be determined if W is changed; in addition, the mathematics become considerably more involved.

$$\frac{d^2\mathcal{I}}{d\xi^2} = \frac{1}{2}(1-\epsilon)\left[-\mathcal{I}(\xi) + \int_0^\xi \mathcal{I}(\xi')e^{-(\xi-\xi')}d\xi' - \mathcal{I}(\xi) + \int_\xi^W \mathcal{I}(\xi')e^{-(\xi'-\xi)}d\xi'\right],$$

or, by comparison with the expression for $\mathcal{I}(\xi)$,

$$\frac{d^2\mathcal{I}}{d\xi^2} = \mathcal{I} - \epsilon - (1-\epsilon)\mathcal{I} = \epsilon(\mathcal{I}-1).$$

Thus, the governing integral equation has been converted into a second order ordinary differential equation, which is readily solved as

$$\mathcal{I}(\xi) = 1 + C_1 e^{-\sqrt{\epsilon}\xi} + C_2 e^{+\sqrt{\epsilon}\xi}.$$

While an integral equation does not require any boundary conditions, we have converted the governing equation into a differential equation that requires two boundary conditions in order to determine C_1 and C_2. The dilemma is overcome by substituting the general solution back into the governing integral equation (with approximated kernel). This calculation can be done for variable values of ξ by comparing coefficients of independent functions of ξ, or simply for two arbitrarily selected values for ξ. The first method gives the engineer proof that his analysis is without mistake, but is usually considerably more tedious. Often it is also possible to employ symmetry, as is the case here, since $\mathcal{I}(\xi) = \mathcal{I}(W-\xi)$ or

$$C_1\left[e^{-\sqrt{\epsilon}\xi} - e^{-\sqrt{\epsilon}(W-\xi)}\right] = -C_2\left[e^{\sqrt{\epsilon}\xi} - e^{\sqrt{\epsilon}(W-\xi)}\right]$$
$$= C_2 e^{\sqrt{\epsilon}W}\left[e^{-\sqrt{\epsilon}\xi} - e^{-\sqrt{\epsilon}(W-\xi)}\right],$$

or

$$C_1 = C_2 e^{\sqrt{\epsilon}W}.$$

Consequently,

$$\mathcal{I}(\xi) = 1 + C_1\left[e^{-\sqrt{\epsilon}\xi} + e^{-\sqrt{\epsilon}(W-\xi)}\right],$$

and substituting this expression into the governing equation at $\xi = 0$ gives

$$\mathcal{I}(0) = 1 + C_1\left(1 + e^{-\sqrt{\epsilon}W}\right) = \epsilon + \frac{1}{2}(1-\epsilon)\int_0^W \left\{1 + C_1\left[e^{-\sqrt{\epsilon}\xi'} + e^{-\sqrt{\epsilon}(W-\xi')}\right]\right\}e^{-\xi'}d\xi'$$

$$= \epsilon + \frac{1}{2}(1-\epsilon)\int_0^W \left\{e^{-\xi'} + C_1\left[e^{-(1+\sqrt{\epsilon})\xi'} + e^{-\xi'-\sqrt{\epsilon}(W-\xi')}\right]\right\}d\xi'$$

$$= \epsilon - \frac{1}{2}(1-\epsilon)\left\{e^{-\xi'} + C_1\left[\frac{e^{-(1+\sqrt{\epsilon})\xi'}}{1+\sqrt{\epsilon}} + \frac{e^{-\xi'-\sqrt{\epsilon}(W-\xi')}}{1-\sqrt{\epsilon}}\right]\right\}\Bigg|_0^W$$

$$= \epsilon + \frac{1}{2}(1-\epsilon)\left\{1 - e^{-W} + C_1\left[\frac{1-e^{-(1+\sqrt{\epsilon})W}}{1+\sqrt{\epsilon}} + \frac{e^{-\sqrt{\epsilon}W} - e^{-W}}{1-\sqrt{\epsilon}}\right]\right\}.$$

Solving this for C_1 gives

$$1-\epsilon-\tfrac{1}{2}(1-\epsilon)(1-e^{-W}) \;=\; C_1\Big[\tfrac{1}{2}(1-\sqrt{\epsilon})\Big(1-e^{-(1+\sqrt{\epsilon})W}\Big)$$
$$+\tfrac{1}{2}(1+\sqrt{\epsilon})\Big(e^{-\sqrt{\epsilon}W}-e^{-W}\Big)-\Big(1-e^{-\sqrt{\epsilon}W}\Big)\Big]$$

$$\tfrac{1}{2}(1-\epsilon)\Big(1+e^{-W}\Big) \;=\; C_1\Big\{\tfrac{1}{2}(1-\sqrt{\epsilon})+\tfrac{1}{2}(1+\sqrt{\epsilon})e^{-\sqrt{\epsilon}W}-1-e^{-\sqrt{\epsilon}W}$$
$$-\Big[\tfrac{1}{2}(1-\sqrt{\epsilon})e^{-\sqrt{\epsilon}W}+\tfrac{1}{2}(1+\sqrt{\epsilon})\Big]e^{-W}\Big\}\,,$$

or

$$C_1 \;=\; -\frac{1-\epsilon}{(1+\sqrt{\epsilon})+(1-\sqrt{\epsilon})e^{-\sqrt{\epsilon}W}}$$

and

$$\mathcal{J}(\xi) \;=\; 1-(1-\epsilon)\frac{e^{-\sqrt{\epsilon}\xi}+e^{-\sqrt{\epsilon}(W-\xi)}}{(1+\sqrt{\epsilon})+(1-\sqrt{\epsilon})e^{-\sqrt{\epsilon}W}}\,.$$

Finally, the nondimensional heat flux follows as

$$\Psi(\xi) \;=\; \frac{\epsilon}{1-\epsilon}[1-\mathcal{J}(\xi)] \;=\; \frac{\epsilon\Big[e^{-\sqrt{\epsilon}\xi}+e^{-\sqrt{\epsilon}(W-\xi)}\Big]}{(1+\sqrt{\epsilon})+(1-\sqrt{\epsilon})e^{-\sqrt{\epsilon}W}}\,,$$

which is also included in Fig. 5-14.

Note that $e^{-|\xi'-\xi|}$ is not a particularly good approximation for the kernel, since the actual kernel has a zero first derivative at $\xi'=\xi$. A better approximation can be obtained by using

$$K(\xi,\xi') \simeq a_1\,e^{-b_1|\xi'-\xi|}+a_2\,e^{-b_2|\xi'-\xi|}$$

(with $a_1>1$ and $a_2<0$). If W is relatively small, say $<\tfrac{1}{2}$, a good approximation may be obtained using

$$K(\xi,\xi') \simeq \cos a(\xi'-\xi)$$

(since the kernel has an inflection point at $|\xi'-\xi|=\tfrac{1}{2}$).

We shall conclude this chapter with two examples that demonstrate that exact analytical solutions are possible for a few simple geometries for which the view factors between area elements attain certain special forms.

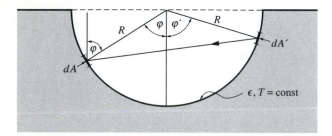

FIGURE 5-15
Isothermal hemispherical cavity irradiated normally by the sun, Example 5.12.

Example 5.12.   Consider a hemispherical cavity irradiated by the sun as shown in Fig. 5-15. The surface of the cavity is kept isothermal at temperature T and is coated with a gray, diffuse material with emissivity ϵ. Assuming that the cavity is, aside from the solar irradiation, exposed to cold surroundings, determine the local heat flux rates that are necessary to maintain the cavity surface at constant temperature.

Solution

From equation (5.24) the local radiosity at position (φ, ψ) is determined as

$$J(\varphi) = \epsilon \sigma T^4 + (1 - \epsilon) H(\varphi)$$

$$= \epsilon \sigma T^4 + (1 - \epsilon) \left[\int_A J(\varphi') dF_{dA-dA'} + H_o(\varphi) \right],$$

where we have already stated that radiosity is a function of φ only, i.e., there is no dependence on azimuthal angle ψ. The view factor between infinitesimal areas on a sphere is known from the *inside sphere method*, equation (4.51), as

$$dF_{dA-dA'} = \frac{dA'}{4\pi R^2} = \frac{R^2 \sin\varphi' \, d\varphi' \, d\psi'}{4\pi R^2}.$$

The external irradiation at dA is readily determined as $H_o(\varphi) = q_{sun} \cos\varphi$, and the expression for radiosity becomes

$$J(\varphi) = \epsilon \sigma T^4 + (1 - \epsilon) \left[\int_0^{2\pi} \int_0^{\pi/2} J(\varphi') \frac{\sin\varphi' \, d\varphi' \, d\psi'}{4\pi} + q_{sun} \cos\varphi \right]$$

$$= \epsilon \sigma T^4 + \frac{1 - \epsilon}{2} \int_0^{\pi/2} J(\varphi') \sin\varphi' \, d\varphi' + (1 - \epsilon) q_{sun} \cos\varphi.$$

Because of the unique behavior of view factors between sphere surface elements we note that the irradiation at location φ that arrives from other parts of the sphere, H_s, *does not depend on* φ. Thus,

$$H_s = \frac{1}{2} \int_0^{\pi/2} J(\varphi') \sin \varphi' \, d\varphi' = \text{const},$$

and

$$J(\varphi) = \epsilon \sigma T^4 + (1 - \epsilon) H_s + (1 - \epsilon) q_{\text{sun}} \cos \varphi.$$

Substituting this equation into the expression for H_s leads to

$$H_s = \frac{1}{2} \int_0^{\pi/2} \left[\epsilon \sigma T^4 + (1 - \epsilon) H_s + (1 - \epsilon) q_{\text{sun}} \cos \varphi' \right] \sin \varphi' \, d\varphi'$$

$$= \tfrac{1}{2} \epsilon \sigma T^4 + \tfrac{1}{2}(1 - \epsilon) H_s + \tfrac{1}{4}(1 - \epsilon) q_{\text{sun}},$$

or

$$H_s = \frac{\epsilon}{1 + \epsilon} \sigma T^4 + \frac{1 - \epsilon}{2(1 + \epsilon)} q_{\text{sun}}.$$

An energy balance at dA gives

$$q(\varphi) = \epsilon \sigma T^4 - \epsilon H(\varphi) = \epsilon (\sigma T^4 - H_s - q_{\text{sun}} \cos \varphi)$$

or

$$q(\varphi) = \epsilon \left[\frac{\sigma T^4}{1 + \epsilon} - \left(\frac{1 - \epsilon}{2(1 + \epsilon)} + \cos \varphi \right) q_{\text{sun}} \right].$$

We observe from this example that in problems where all radiating surfaces are part of a sphere, none of the view factors involved depends on the location of the originating surface, and an exact analytical solution can always be found in a similar fashion. Apparently, this was first recognized by Jensen [20] and reported in the book by Jakob [21].

Exact analytical solutions are also possible for such configurations where all relevant view factors have repeating derivatives (as in the kernel approximation).

Example 5.13.   A long thin radiating wire is to be employed as an infrared light source. To maximize the output of infrared energy into the desired direction, the wire is fitted with an insulated, highly reflective sheath as shown in Fig. 5-16. The sheath is cylindrical with radius R (which is much larger than the diameter of the wire), and has a cutout of half-angle φ to let the concentrated infrared light escape. Assuming that the wire is heated with a power of Q' W/m length of wire, and that the sheath can lose heat only by radiation and only from its inside surface, determine the temperature distribution across the sheath.

Solution

From an energy balance on a surface element dA it follows from equation (5.20) that, with $q(\theta) = 0$,

$$\sigma T^4(\theta) = J(\theta) = H(\theta),$$

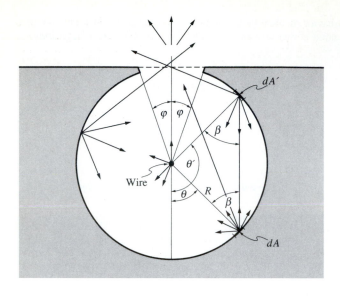

FIGURE 5-16
Thin radiating wire with radiating sheath, Example 5.13.

and

$$H(\theta) = \int_A J(\theta') \, dF_{dA-dA'} + H_o(\theta).$$

We may treat the energy emitted from the wire as external radiation (neglecting absorption by the wire since it is so small). Since the total released energy will spread equally into all directions, we find

$$H_o(\theta) = Q'/2\pi R = \text{const}.$$

The view factor $dF_{dA-dA'}$ between two infinitely long strips on the cylinder surface is given by Configuration 1 in Appendix D as

$$F_{dA-dA'} = \tfrac{1}{2} \cos \beta \, d\beta,$$

where the angle β is indicated in Fig. 5-16 and may be related to θ through

$$2\beta + |\theta' - \theta| = \pi.$$

Differentiating β with respect to θ' we obtain $d\beta = \pm d\theta'/2$, depending on whether θ' is larger or less than θ. Substituting for β in the view factor, this becomes

$$F_{dA-dA'} = \frac{1}{2} \cos \left(\frac{\pi}{2} - \left| \frac{\theta' - \theta}{2} \right| \right) \frac{1}{2} d\theta' = \frac{1}{4} \sin \left| \frac{\theta' - \theta}{2} \right| d\theta',$$

where the $\pm$ has been omitted since the view factor is always positive (i.e., $|d\beta|$ is to be used). Substituting this into the above relationship for radiosity we obtain

$$J(\theta) = \frac{1}{4} \int_{-\pi+\varphi}^{\pi-\varphi} J(\theta') \sin \left| \frac{\theta' - \theta}{2} \right| d\theta' + H_o$$

$$= \frac{1}{4} \int_{-\pi+\varphi}^{\theta} J(\theta') \sin \frac{\theta - \theta'}{2} d\theta' + \frac{1}{4} \int_{\theta}^{\pi-\varphi} J(\theta') \sin \frac{\theta' - \theta}{2} d\theta' + H_o.$$

Since the view factor in the integrand has repetitive derivatives we may convert this integral equation into a second-order differential equation, as was done in the kernel approximation method. Differentiating twice, we have

$$\frac{dJ}{d\theta} = \frac{1}{8}\int_{-\pi+\varphi}^{\theta} J(\theta')\cos\frac{\theta-\theta'}{2}\,d\theta' - \frac{1}{8}\int_{\theta}^{\pi-\varphi} J(\theta')\cos\frac{\theta'-\theta}{2}\,d\theta'$$

$$\frac{d^2J}{d\theta^2} = \frac{1}{8}J(\theta) - \frac{1}{16}\int_{-\pi+\varphi}^{\theta} J(\theta')\sin\frac{\theta-\theta'}{2}\,d\theta'$$
$$+\frac{1}{8}J(\theta) - \frac{1}{16}\int_{\theta}^{\pi-\varphi} J(\theta')\sin\frac{\theta'-\theta}{2}\,d\theta'.$$

Comparing this result with the above integral equation for $J(\theta)$ we find

$$\frac{d^2J}{d\theta^2} = \tfrac{1}{4}J(\theta) - \tfrac{1}{4}[J(\theta) - H_o] = \tfrac{1}{4}H_o.$$

This equation is readily solved as

$$J(\theta) = \tfrac{1}{8}H_o\theta^2 + C_1\theta + C_2.$$

The two integration constants must now be determined by substituting the solution back into the governing integral equation. However, C_1 may be determined from symmetry since, for this problem, $J(\theta) = J(-\theta)$ and $C_1 = 0$. To determine C_2 we evaluate J at $\theta = 0$:

$$J(0) = C_2 = \frac{1}{4}\int_{-\pi+\varphi}^{0} J(\theta')\sin\left(-\frac{\theta'}{2}\right)d\theta' + \frac{1}{4}\int_{0}^{\pi-\varphi} J(\theta')\sin\frac{\theta'}{2}\,d\theta' + H_o$$

$$= \frac{1}{2}\int_{0}^{\pi-\varphi} J(\theta')\sin\frac{\theta'}{2}\,d\theta' + H_o$$

$$= \frac{1}{2}\int_{0}^{\pi-\varphi}\left(C_2 + \frac{H_o}{8}\theta'^2\right)\sin\frac{\theta'}{2}\,d\theta' + H_o.$$

Integrating twice by parts we obtain

$$C_2 = H_o - \left(C_2 + \frac{H_o}{8}\theta'^2\right)\cos\frac{\theta'}{2}\Bigg|_0^{\pi-\varphi} + \frac{H_o}{4}\int_0^{\pi-\varphi}\theta'\cos\frac{\theta'}{2}\,d\theta'$$

$$= H_o - \left[C_2 + \frac{H_o}{8}(\pi-\varphi)^2\right]\cos\left(\frac{\pi}{2}-\frac{\varphi}{2}\right) + C_2$$

$$+\frac{H_o}{2}\left(\theta'\sin\frac{\theta'}{2}\Bigg|_0^{\pi-\varphi} - \int_0^{\pi-\varphi}\sin\frac{\theta'}{2}\,d\theta'\right)$$

$$= H_o + C_2 - \left[C_2 + \frac{H_o}{8}(\pi-\varphi)^2\right]\sin\frac{\varphi}{2}$$

$$+\frac{H_o}{2}\left((\pi-\varphi)\sin\left(\frac{\pi}{2}-\frac{\varphi}{2}\right) + 2\cos\frac{\theta'}{2}\Bigg|_0^{\pi-\varphi}\right)$$

$$= H_o + C_2 - \left[C_2 + \frac{H_o}{8}(\pi-\varphi^2)\right]\sin\frac{\varphi}{2} + \frac{H_o}{2}(\pi-\varphi)\cos\frac{\varphi}{2} + H_o\sin\frac{\varphi}{2} - H_o.$$

Solving this equation for C_2 we get

$$C_2 = H_o \left[1 + \frac{\pi - \varphi}{2} \cos \frac{\varphi}{2} - \frac{1}{8}(\pi - \varphi)^2 \right].$$

Therefore,

$$T^4(\theta) = \frac{J}{\sigma} = \frac{Q'}{2\pi R\sigma} \left\{ 1 + \frac{\pi - \varphi}{2} \cos \frac{\varphi}{2} - \frac{1}{8} \left[(\pi - \varphi)^2 - \theta^2 \right] \right\}.$$

We find that the temperature has a minimum at $\theta = 0$, since around that location the view factor to the opening is maximum, resulting in a maximum of escaping energy. The temperature level increases as φ decreases (since less energy can escape) and reaches $T \to \infty$ as $\varphi = 0$ (since this produces an insulated closed enclosure with internal heat production).

The fact that long cylindrical surfaces lend themselves to exact analysis was apparently first recognized by Sparrow [22]. The preceding two examples have shown that exact solutions may be found for a number of special geometries, namely, (*i*) enclosures whose surfaces all lie on a single sphere, and (*ii*) enclosures for which view factors between surface elements have repetitive derivatives. For other still fairly simple geometries an approximate analytical solution may be determined from the *kernel approximation method*. However, the vast majority of radiative heat transfer problems in enclosures without a participating medium must be solved by numerical methods. A large majority of these are solved using the *net radiation method* described in the first few sections of this chapter. If greater accuracy or better numerical efficiency is desired, one of the numerical methods briefly described in this section needs to be used, such as *numerical quadrature* leading to a set of linear algebraic equations (as in the net radiation method).

References

1. Oppenheim, A. K.: "Radiation Analysis by the Network Method," *Transactions of ASME, Journal of Heat Transfer*, vol. 78, pp. 725–735, 1956.
2. Courant, R., and D. Hilbert: *Methods of Mathematical Physics*, Interscience Publishers, New York, 1953.
3. Hildebrand, F. B.: *Methods of Applied Mathematics*, Prentice Hall, Englewood Cliffs, NJ, 1952.
4. Özişik, M. N.: *Radiative Transfer and Interactions With Conduction and Convection*, John Wiley & Sons, New York, 1973.
5. Sparrow, E. M., and A. Haji-Sheikh: "A Generalized Variational Method for Calculating Radiant Interchange Between Surfaces," *ASME Journal of Heat Transfer*, vol. 87C, pp. 103–109, 1965.
6. Sparrow, E. M.: "Application of Variational Methods to Radiation Heat Transfer Calculations," *ASME Journal of Heat Transfer*, vol. 82C, pp. 375–380, 1960.
7. Sparrow, E. M., J. L. Gregg, J. V. Szel, and P. Manos: "Analysis, Results, and Interpretation for Radiation Between Simply Arranged Gray Surfaces," *ASME Journal of Heat Transfer*, vol. 83C, pp. 207–214, 1961.
8. Buckley, H.: "On the Radiation From the Inside of a Circular Cylinder," *Phil. Mag.*, vol. 4, no. 23, pp. 753–762, 1927.
9. Eckert, E. R. G.: "Das Strahlungsverhältnis von Flächen mit Einbuchtungen und von zylindrischen Bohrungen," *Arch. Wärmewirtsch.*, vol. 16, pp. 135–138, 1935.

10. Sparrow, E. M., and L. U. Albers: "Apparent Emissivity and Heat Transfer in a Long Cylindrical Hole," *ASME Journal of Heat Transfer*, vol. 82C, pp. 253–255, 1960.

11. Usiskin, C. M., and R. Siegel: "Thermal Radiation from a Cylindrical Enclosure with Specified Wall Heat Flux," *ASME Journal of Heat Transfer*, vol. 82C, pp. 369–374, 1960.

12. Lin, S. H., and E. M. Sparrow: "Radiant Interchange Among Curved Specularly Reflecting Surfaces, Application to Cylindrical and Conical Cavities," *ASME Journal of Heat Transfer*, vol. 87C, pp. 299–307, 1965.

13. Perlmutter, M., and R. Siegel: "Effect of Specularly Reflecting Gray Surface on Thermal Radiation Through a Tube and from Its Heated Wall," *ASME Journal of Heat Transfer*, vol. 85, pp. 55–62, 1963.

14. Siegel, R., and M. Perlmutter: "Convective and Radiant Heat Transfer for Flow of a Transparent Gas in a Tube with a Gray Wall," *International Journal of Heat and Mass Transfer*, vol. 5, pp. 639–660, 1962.

15. Sparrow, E. M., L. U. Albers, and E. R. G. Eckert: "Thermal Radiation Characteristics of Cylindrical Enclosures," *ASME Journal of Heat Transfer*, vol. 84C, pp. 73–81, 1962.

16. Sparrow, E. M., and R. P. Heinisch: "The Normal Emittance of Circular Cylindrical Cavities," *Applied Optics*, vol. 9, pp. 2569–2572, 1970.

17. Alfano, G.: "Apparent Thermal Emittance of Cylindrical Enclosures with and without Diaphragms," *International Journal of Heat and Mass Transfer*, vol. 15, no. 12, pp. 2671–2674, 1972.

18. Alfano, G., and A. Sarno: "Normal and Hemispherical Thermal Emittances of Cylindrical Cavities," *ASME Journal of Heat Transfer*, vol. 97, no. 3, pp. 387–390, 1975.

19. Fröberg, C. E.: *Introduction to Numerical Analysis*, Addison-Wesley, Reading, Massachusetts, 1969.

20. Jensen, H. H.: "Some Notes on Heat Transfer by Radiation," *Kgl. Danske Videnskab. Selskab. Mat.-Fys. Medd.*, vol. 24, no. 8, pp. 1–26, 1948.

21. Jakob, M.: *Heat Transfer*, vol. II, John Wiley & Sons., New York, 1957.

22. Sparrow, E. M.: "Radiant Absorption Characteristics of Concave Cylindrical Surfaces," *ASME Journal of Heat Transfer*, vol. 84C, pp. 283–293, 1962.

Problems

5.1 A firefighter (approximated by a two-sided black surface at 310 K 180 cm long and 40 cm wide) is facing a large fire at a distance of 10 m (approximated by a semi-infinite black surface at 1500 K). Ground and sky are at 0°C (and may also be approximated as black). What are the net radiative heat fluxes on the front and back of the firefighter? Compare these with heat rates by free convection ($h = 10 \, \text{W/m}^2\text{K}$, $T_{amb} = 0°\text{C}$).

5.2 A small star has a radius of 100,000 km. Suppose that the star is originally at a uniform temperature of 1,000,000 K before it "dies," i.e., before nuclear fusion stops supplying heat. If it is assumed that the star has a constant heat capacity of $\rho c_p = 1 \, \text{kJ/m}^3\text{K}$, and that it remains isothermal during cool-down, estimate the time required until the star has cooled to 10,000 K. Note: A body of such proportions radiates like a blackbody (Why?).

5.3 A small furnace consists of a cylindrical, blackwalled enclosure, 20 cm long and with a diameter of 10 cm. The bottom surface is electrically heated to 1500 K, while the cylindrical sidewall is insulated. The top plate is exposed to the environment, such that its temperature is 500 K. Estimate the heating requirements for the bottom wall, and the temperature of the cylindrical sidewall, by treating the sidewall as (*a*) a single zone, (*b*) two equal rings of 10 cm height each.

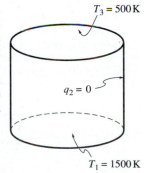

$T_3 = 500 \, \text{K}$

$q_2 = 0$

$T_1 = 1500 \, \text{K}$

5.4 Repeat Problem 5.3 for a 20 cm high furnace of quadratic (10 cm × 10 cm) cross section.

5.5 For the configuration shown in the figure, determine the temperature of Surface 2 with the following data:

Surface 1 : T_1 = 1000 K,

   q_1 = −1 W/cm²,

   ϵ_1 = 0.6;

Surface 2 : ϵ_2 = 0.2;

Surface 3 : ϵ_3 = 0.3, perfectly insulated.

All configurations are gray and diffuse.

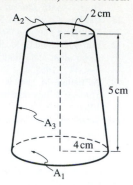

5.6 During launch the heat rejector radiative panels of the Space Shuttle are folded against the inside of the Shuttle doors. During orbit the doors are opened and the panels are rotated out by an angle φ as shown in the figure. Assuming door and panel can be approximated by infinitely long, isothermal quarter-cylinders of radius a and emissivity ϵ = 0.8, calculate the necessary rotation angle φ so that half the total energy emitted by panel (2) and door (1) escapes through the opening. At what opening angle will a maximum amount of energy be rejected? How much and why?

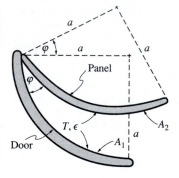

5.7 A cubical enclosure has gray, diffuse walls which interchange energy. Four of the walls are isothermal at T_s with emissivity ϵ_s, the other two are isothermal at T_t with emissivity ϵ_t. Calculate the heat flux rates per unit time and area.

5.8 A row of equally-spaced, cylindrical heating elements (s = 2d) is used to heat the inside of a furnace as shown. Assuming that the outer wall is made of firebrick with ϵ_3 = 0.3 and is perfectly insulated, that the heating rods are made of silicon carbide (ϵ_1 = 0.8), and that the inner wall has an emissivity of ϵ_2 = 0.6, what must be the operating temperature of the rods to supply a net heat flux of 300 kW/m² to the furnace, if the inner wall is at a temperature of 1300 K?

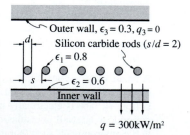

5.9 Two identical circular disks are connected at one point of their periphery by a hinge. The configuration is then opened by an angle ϕ as shown in the figure. Assuming the opening angle to be $\phi = 60°$, $d = 1\,\text{m}$, calculate the average equilibrium temperature for each of the two disks, with solar radiation entering the configuration parallel to Disk 2 with a strength of $q_{sun} = 1000\,\text{W/m}^2$. Disk 1 is gray and diffuse with $\alpha = \epsilon = 0.5$, Disk 2 is black. Both disks are insulated.

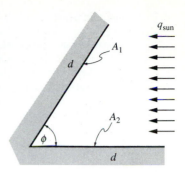

5.10 A long, black V-groove is irradiated by the sun as shown. Assuming the groove to be perfectly insulated, and radiation to be the only mode of heat transfer, determine the average groove temperature as function of solar incidence angle θ (give values for $\theta = 0°$, $15°$, $30°$, $60°$, $90°$). For simplicity the V-groove wall may be taken as a single zone.

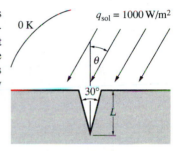

5.11 A (simplified) radiation heat flux meter consists of a conical cavity coated with a gray, diffuse material, as shown in the figure. To measure the radiative heat flux, the cavity is perfectly insulated.

(a) Develop an expression that relates the flux, H_o, to the cavity temperature, T.

(b) If the cavity is turned away from the incoming flux by an angle α, what happens to the cavity temperature?

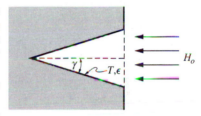

5.12 A very long solar collector plate is to collect energy at a temperature of $T_1 = 350\,\text{K}$. To improve its performance for off-normal solar incidence, a highly reflective surface is placed next to the collector as shown in the adjacent figure. How much energy (per unit length) does the collector plate collect for a solar incidence angle of $30°$? For simplicity you may make the following assumptions: The collector is isothermal and gray-diffuse with emissivity $\epsilon_1 = 0.8$; the reflector is gray-diffuse with $\epsilon_2 = 0.1$, and heat losses from the reflector by convection as well as all losses from the collector ends may be neglected.

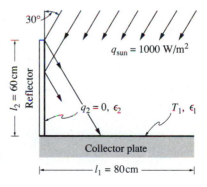

5.13 A long thin black heating wire radiates 300 W/cm length of wire and is used to heat a flat surface by thermal radiation. To increase its efficiency the wire is surrounded by an insulated half-cylinder as shown in the figure. Both surfaces are gray and diffuse with emissivities ϵ_2 and ϵ_3, respectively. What is the net heat flux at Surface 3? How does this compare with the case without cylinder?

Hint: You may either treat the heating wire as a thin cylinder whose radius you eventually shrink to zero, or treat radiation from the wire as external radiation (the second approach being somewhat simpler).

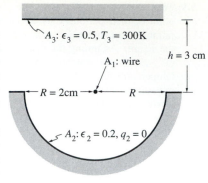

5.14 Determine F_{1-2} for the rotationally symmetric configuration shown in the figure (i.e., a big sphere, $R = 13$ cm, with a circular hole, $r = 5$ cm, and a hemispherical cavity, $r = 5$ cm). Assuming Surface 2 to be gray and diffuse ($\epsilon = 0.5$) and insulated and Surface 1 to be black and also insulated, what is the average temperature of the black cavity if collimated irradiation of $1000 \, \text{W/m}^2$ is penetrating through the hole as shown?

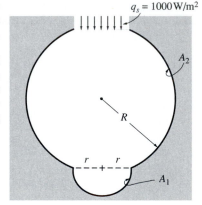

5.15 An integrating sphere (a device to measure surface properties) is 10 cm in radius. It contains on its inside wall a 1 cm² black detector, a 1×2 cm entrance port and a 1×1 cm sample as shown. The remaining portion of the sphere is smoked with magnesium oxide having a short-wavelength reflectance of 0.98, which is almost perfectly diffuse. A collimated beam of radiant energy (i.e., all energy is contained within a very small cone of solid angles) enters the sphere through the entrance port, falls onto the sample, and then is reflected and interreflected,

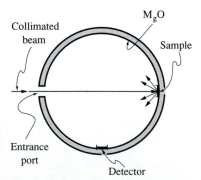

giving rise to a sphere wall radiosity and irradiation. Radiation emitted from the walls is not detected because the source radiation is chopped, and the detector-amplifier system responds only to the chopped radiation. Find the fractions of the chopped incoming radiation that are

    (a) lost out the entrance port,

    (b) absorbed by the MgO-smoked wall, and

    (c) absorbed by the detector.

[Item (c) is called the "sphere efficiency."]

5.16 The side wall of a flask holding liquid helium may be approximated as a long double-walled cylinder as shown in the adjacent sketch. The container walls are made of 1 mm thick stainless steel ($k = 15\,\text{W/m K}$, $\epsilon = 0.2$), and have outer radii of $R_2 = 10\,\text{cm}$ and $R_4 = 11\,\text{cm}$. The space between walls is evacuated, and the outside is exposed to free convection with the ambient at $T_{\text{amb}} = 20°\text{C}$ and a heat transfer coefficient of $h_o = 10\,\text{W/m}^2\text{K}$ (for the combined effects of free convection and radiation). It is reasonable to assume that the temperature of the inner wall is at liquid helium temperature, or $T(R_2) = 4\,\text{K}$.

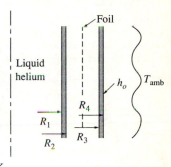

    (a) Determine the heat gain by the helium, per unit length of flask.

    (b) To reduce the heat gain a thin silver foil ($\epsilon = 0.02$) is placed midway between the two walls. How does this affect the heat flux?

For the sake of the problem, you may assume both steel and silver to be diffuse reflectors.

5.17 Consider Configuration 33 in Appendix D with $h = w$. The bottom wall is at constant temperature T_1 and has emissivity ϵ_1; the side wall is at $T_2 = \text{const}$ and ϵ_2. Find the exact expression for $q_1(x)$ if $\epsilon_2 = 1$.

5.18 An infinitely long half-cylinder is irradiated by the sun as shown in the figure. The inside of the cylinder is gray and diffuse, the outside is insulated. There is no radiation from the background. What is the equilibrium temperature distribution along the cylinder periphery,

    (a) using four isothermal zones of 45° each,

    (b) using the exact relations. (Hint: Use differentiation as in the kernel approximation method).

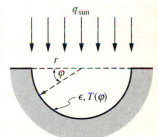

5.19 A large isothermal surface (exposed to vacuum, temperature T_w, diffuse-gray emissivity ϵ_w) is irradiated by the sun. To reduce the heat gain/loss from the surface, a thin copper shield (emissivity ϵ_c and initially at temperature T_{c0}) is placed between surface and sun as shown in the figure.

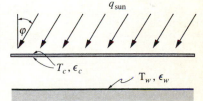

(a) Determine the relationship between T_c and time t (it is sufficient to leave the answer in implicit form with an unsolved integral).

(b) Give the steady state temperature for T_c (i.e., for $t \to \infty$).

(c) Briefly discuss qualitatively the following effects:

(i) The shield is replaced by a moderately thick slab of styrofoam coated on both sides with a very thin layer of copper.

(ii) The surfaces are finite in size.

5.20 Consider a 90° pipe elbow as shown in the figure (pipe diameter $=D = 1$ m; inner elbow radius $= 0$, outer elbow radius $= D$). The elbow is isothermal at temperature $T = 1000$ K, has a gray diffuse emissivity $\epsilon = 0.4$, and is placed in a cool environment. What is the total heat loss from the isothermal elbow (inside and outside)?

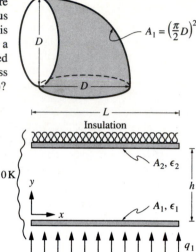

5.21 Consider two infinitely long, parallel, black plates of width L as shown. The bottom plate is uniformly heated electrically with a heat flux of $q_1 = $ const, while the top plate is insulated. The entire configuration is placed into a large cold environment.

(a) Determine the governing equations for the temperature variation across the plates.

(b) Find the solution by the kernel substitution method. To avoid tedious algebra, you may leave the final result in terms of *two* constants to be determined, as long as you outline carefully how these constants may be found.

(c) If the plates are gray and diffuse with emissivities ϵ_1 and ϵ_2, how can the temperature distribution be determined, using the solution from part *(b)*?

CHAPTER
6

RADIATIVE
EXCHANGE
BETWEEN
PARTIALLY-SPECULAR,
GRAY SURFACES

6.1 INTRODUCTION

In the previous two chapters it was assumed that all surfaces constituting the enclosure are—besides being gray—diffuse emitters as well as diffuse reflectors of radiant energy. Diffuse emission is nearly always an acceptable simplification. The assumption of diffuse reflection, on the other hand, often leads to considerable error, since many surfaces deviate substantially from this behavior. Electromagnetic wave theory predicts reflection to be specular for optically smooth surfaces, i.e., to reflect light like a mirror. All clean metals, many nonmetals such as glassy materials, and most polished materials display strong specular reflection peaks. Nevertheless, they all, to some extent, reflect somewhat into other directions as a result of their surface roughness. Surfaces may appear dull (i.e., diffusely reflecting) to the eye, but are rather specular in the infrared, since the ratio of every surface's root-mean-square roughness to wavelength decreases with increasing wavelength.

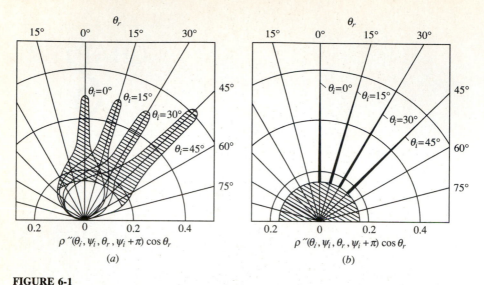

FIGURE 6-1
(*a*) Subdivision of the reflectivity of oxidized brass (shown for plane of incidence) into specular (shaded) and diffuse components (unshaded), from [1], (*b*) Equivalent idealized reflectivity.

For a surface with diffuse reflectivity the reflected radiation has the same (diffuse) directional distribution as the emitted energy, as discussed in the beginning of Section 5.3. Therefore, the radiation field within the enclosure is completely specified in terms of the radiosity, which is a function of location along the enclosure walls (but *not* a function of *direction* as well). If reflection is nondiffuse, then the radiation intensities leaving any surface are functions of direction as well as surface location, and the analysis becomes immensely more complicated.[1] To make the analysis tractable, one may make the idealization that the reflectivity, while not diffuse, can be adequately represented by a combination of a diffuse and a specular component, as illustrated in Fig. 6-1 for oxidized brass [1]. Thus, for the present chapter, we assume the radiative properties to be of the form

$$\rho = \rho^s + \rho^d = 1 - \alpha = 1 - \epsilon = 1 - \epsilon'_\lambda, \tag{6.1}$$

where ρ^s and ρ^d are the specular and diffuse components of the reflectivity, respectively. Since the surfaces are assumed to be gray, diffuse emitters ($\epsilon = \epsilon'_\lambda$), it follows that neither α nor ρ may depend on wavelength or on *incoming* direction (i.e., the *magnitude* of ρ does not depend on incoming direction); how ρ is distributed over *outgoing* directions depends on incoming direction through ρ^s. With this approximation, the separate reflection components may be found analytically by splitting the

---

[1]In addition, if the irradiation is polarized (e.g., owing to irradiation from a laser source), specular reflections will change the state of polarization (because of the different values for $\rho_\parallel$ and $\rho_\perp$, as discussed in Chapter 2). We shall only consider unpolarized radiation.

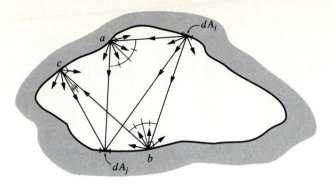

FIGURE 6-2
Radiative exchange in an enclosure with specular reflectors.

bidirectional reflection function into two parts,

$$\rho''(\mathbf{r}, \hat{\mathbf{s}}_i, \hat{\mathbf{s}}_r) = \rho''^s(\mathbf{r}, \hat{\mathbf{s}}_i, \hat{\mathbf{s}}_r) + \rho''^d(\mathbf{r}, \hat{\mathbf{s}}_i, \hat{\mathbf{s}}_r). \qquad (6.2)$$

Substituting this expression into equation (3.41) and equation (3.44) then leads to ρ^s and ρ^d. Values of ρ^s and ρ^d may also be determined directly from experiment, as reported by Birkebak and coworkers [2], making detailed measurements of the bidirectional reflection function unnecessary.

Within an enclosure consisting of surfaces with purely diffuse and purely specular reflection components, the complexity of the problem may be reduced considerably by realizing that any specularly reflected beam may be traced back to a point on the enclosure surface from which it emanated diffusely (i.e., any beam was part of an energy stream leaving the surface after emission or diffuse reflection), as illustrated in Fig. 6-2. Therefore, by redefining the view factors to include specular reflection paths in addition to direct view, the radiation field may again be described by a diffuse energy function that is a function of surface location but not of direction.

6.2 SPECULAR VIEW FACTORS

To accommodate surfaces with reflectivities described by equation (6.1), we define a *specular view factor* as

$$dF^s_{dA_i - dA_j} \equiv \frac{\begin{array}{c}\text{diffuse energy leaving } dA_i \text{ intercepted by } dA_j, \text{ by}\\ \text{direct travel or any number of specular reflections}\end{array}}{\text{total diffuse energy leaving } dA_i}. \qquad (6.3)$$

The concept of the specular view factor is illustrated in Figs. 6-2 and 6-3. Diffuse radiation leaving dA_i (by emission or diffuse reflection) can reach dA_j either directly or after one or more reflections. Usually only a finite number of specular reflection paths such as $dA_i - a - dA_j$ or $dA_i - b - c - dA_j$ (and others not indicated in the figure) will be possible. The surface at points a, b, and c behaves like a perfect mirror as far as the specular part of the reflection is concerned. Therefore, if an observer

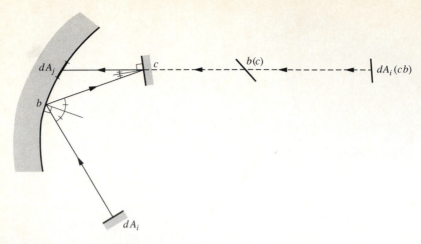

FIGURE 6-3
Specular view factor between infinitesimal surface elements; formation of images.

stood on top of dA_j looking toward c, it would appear as if point b as well as dA_i were situated *behind* point c as indicated in Fig. 6-3; the point labeled $b(c)$ is the *image* of point b as *mirrored* by the surface at c, and $dA_i(cb)$ is the image of dA_i as mirrored by the surfaces at c and b. Therefore, as we examine Figs. 6-2 and 6-3, we may formally evaluate the specular view factor between two infinitesimal areas as

$$dF^s_{dA_i-dA_j} = dF_{dA_i-dA_j} + \rho^s_a dF_{dA_i(a)-dA_j} + \rho^s_b \rho^s_c dF_{dA_i(cb)-dA_j}$$
$$+ \text{ other possible reflection paths.} \qquad (6.4)$$

Thus, the specular view factor may be expressed as a sum of diffuse view factors, with one contribution for each possible direct or reflection path. Note that, for images, the diffuse view factors must be multiplied by the specular reflectivities of the mirroring surfaces, since radiation traveling from dA_i to dA_j is attenuated by every reflection.

If all specularly reflecting parts of the enclosure are flat, then all images of dA_i have the same shape and size as dA_i itself. However, curved surfaces tend to distort the images (focusing and defocusing effects). In the case of only flat, specularly reflecting surfaces we may multiply equation (6.4) by dA_i and, invoking the law of reciprocity for diffuse view factors, equation (4.7), we obtain

$$dA_i dF^s_{dA_i-dA_j} = dA_j dF_{dA_j-dA_i} + \rho^s_a dA_j dF_{dA_j-dA_i(a)}$$
$$+ \rho^s_b \rho^s_c dA_j dF_{dA_j-dA_i(bc)}$$
$$= dA_j dF_{dA_j-dA_i} + \rho^s_a dA_j dF_{dA_j(a)-dA_i}$$
$$+ \rho^s_b \rho^s_c dA_j dF_{dA_j(bc)-dA_i}$$
$$= dA_j dF^s_{dA_j-dA_i}, \qquad (6.5)$$

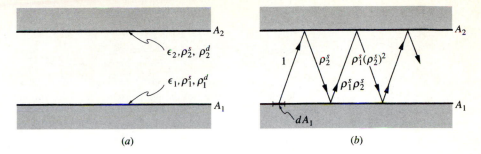

FIGURE 6-4
(a) Geometry for Example 6.1, (b) Ray tracing for the evaluation of F_{1-1}^s and F_{1-2}^s.

that is, the *law of reciprocity* holds for specular view factors as long as all specularly reflecting surfaces are flat. Although considerably more complicated, it is possible to show that the law of reciprocity also holds for curved specular reflectors. If we also assume that the diffuse energy leaving A_i and A_j is constant across each respective area, we have the equivalent to equation (4.15),

$$dA_i \, dF_{di-dj}^s = dA_j \, dF_{dj-di}^s, \tag{6.6a}$$

$$dA_i F_{di-j}^s = A_j \, dF_{j-di}^s, \qquad (J_j = \text{const}), \tag{6.6b}$$

$$A_i F_{i-j}^s = A_j F_{j-i}^s, \qquad (J_i, J_j = \text{const}), \tag{6.6c}$$

where we have adopted the compact notation first introduced in Chapter 4, and J_i is the total diffuse energy (per unit area) leaving surface A_i (again called the *radiosity*).

Example 6.1.   Evaluate the specular view factors F_{1-1}^s and F_{1-2}^s for the parallel plate geometry shown in Fig. 6-4a.

Solution

We note that, because of the one-dimensionality of the problem, F_{d1-2}^s must be the same for any dA_1 on surface A_1. Since F_{1-2}^s is nothing but a surface average of F_{d1-2}^s, we conclude that $F_{d1-2}^s = F_{1-2}^s$. It is sufficient to consider energy leaving from an infinitesimal area (rather than all of A_1). Examining Fig. 6-4b we see that every beam (assumed to have unity strength) leaving dA_1, regardless of its direction, must travel to surface A_2 (a beam of strength "1" is intercepted). After reflection at A_2 a beam of strength ρ_2^s returns to A_1, where it is reflected again and a beam of strength $\rho_2^s \rho_1^s$ returns to A_2. After one more reflection a beam of strength $(\rho_2^s \rho_1^s)\rho_2^s$ returns to A_1, and so on. Thus, the specular view factor may be evaluated as

$$F_{d1-2}^s = F_{1-2}^s = 1 + \rho_1^s \rho_2^s + (\rho_1^s \rho_2^s)^2 + (\rho_1^s \rho_2^s)^3 + \ldots. \tag{6.7}$$

Since $\rho_1^s \rho_2^s < 1$ the sum in this equation is readily evaluated by the methods given in

Wylie [3], and

$$F_{1-2}^s = \frac{1}{1 - \rho_1^s \rho_2^s} = F_{2-1}^s. \tag{6.8}$$

The last part of this relation is found by switching subscripts or by invoking reciprocity (and $A_1 = A_2$). We notice that specular view factors are not limited to values between zero and one, but are often greater than unity because much of the radiative energy leaving a surface is accounted for more than once. All energy from A_1 is intercepted by A_2 after direct travel, but only the fraction $(1 - \rho_2^s)$ is removed (by absorption and/or diffuse reflection) from the specular reflection path. The fraction ρ_2^s travels on specularly and is, therefore, counted a second time, etc. Thus, it is $(1 - \rho_2^s)F_{1-2}^s$ that must have a value between zero and one, and the *summation relation*, equation (4.17), must be replaced by

$$\sum_{j=1}^N (1 - \rho_j^s)F_{i-j}^s = 1. \tag{6.9}$$

Equation (6.9), formed here through intuition, will be developed rigorously in the next section.

F_{1-1}^s may be found similarly as

$$F_{1-1}^s = \rho_2^s + (\rho_1^s \rho_2^s)\rho_2^s + (\rho_1^s \rho_2^s)^2 \rho_2^s + \ldots = \frac{\rho_2^s}{1 - \rho_1^s \rho_2^s}.$$

We note in passing that

$$(1 - \rho_1^s)F_{1-1}^s + (1 - \rho_2^s)F_{1-2}^s = \frac{(1 - \rho_1^s)\rho_2^s + 1 - \rho_2^s}{1 - \rho_1^s \rho_2^s} = 1,$$

as postulated by equation (6.9).

Example 6.2.   Evaluate all specular view factors for two concentric cylinders or spheres.

Solution

Possible beam paths with specular reflections from inner to outer cylinders (or spheres) and *vice versa* are shown in Fig. 6-5a. As in the previous example a beam leaving A_1 in any direction must hit surface A_2 (with strength "1"). Because of the circular geometry, after specular reflection the beam (now of strength ρ_2^s) *must* return to A_1 (i.e., it cannot hit A_2 again before hitting A_1). After renewed reflections the beam keeps bouncing back and forth between A_1 and A_2. Thus, as for parallel plates,

$$F_{1-2}^s = 1 + \rho_1^s \rho_2^s + (\rho_1^s \rho_2^s)^2 + \ldots = \frac{1}{1 - \rho_1^s \rho_2^s}.$$

Similarly, we have

$$F_{1-1}^s = \rho_2^s + (\rho_1^s \rho_2^s)\rho_2^s + \ldots = \frac{\rho_2^s}{1 - \rho_1^s \rho_2^s}.$$

A beam emanating from A_2 will first hit either A_1, and then keep bouncing back and forth between A_1 and A_2 (cf. Fig. 6-5a), or A_2, and then keep bouncing along A_2 without ever

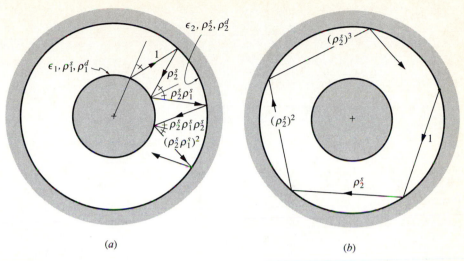

FIGURE 6-5

(a) Geometry for Example 6.2, (b) repeated reflections along outer surface.

hitting A_1 (cf. Fig. 6-5b). Thus, since the fraction F_{2-1} of the diffuse energy leaving A_2 hits A_1 after direct travel, we have

$$F_{2-1}^s = F_{2-1}\left[1 + \rho_1^s \rho_2^s + (\rho_1^s \rho_2^s)^2 + \ldots\right] = \frac{A_1/A_2}{1 - \rho_1^s \rho_2^s},$$

$$F_{2-2}^s = F_{2-2}\left[1 + \rho_2^s + (\rho_2^s)^2 + (\rho_2^s)^3 + \ldots\right]$$
$$+ F_{2-1}\left[\rho_1^s + \rho_1^s(\rho_1^s \rho_2^s) + \ldots\right]$$
$$= \frac{1 - A_1/A_2}{1 - \rho_2^s} + \frac{\rho_1^s A_1/A_2}{1 - \rho_1^s \rho_2^s},$$

where the simple diffuse view factors F_{2-1} and F_{2-2} have been evaluated in terms of A_1 and A_2. Of course, F_{2-1}^s could have been found from F_{1-2}^s by reciprocity, and F_{2-2}^s with the aid of equation (6.9).

A few more examples of specular view factor determinations will be given once the appropriate heat transfer relations have been developed.

6.3 ENCLOSURES WITH PARTIALLY-SPECULAR SURFACES

Consider an enclosure of arbitrary geometry as shown in Fig. 6-2. All surfaces are gray, diffuse emitters and gray reflectors with purely diffuse and purely specular components, i.e., their radiative properties obey equation (6.1). Under these conditions the net heat flux at a surface at location $\mathbf{r}$ is, from Fig. 6-6,

$$q(\mathbf{r}) = q_{\text{emission}} - q_{\text{absorption}} = \epsilon(\mathbf{r})[E_b(\mathbf{r}) - H(\mathbf{r})]$$
$$= q_{\text{out}} - q_{\text{in}} = \epsilon(\mathbf{r})E_b(\mathbf{r}) + \rho^d(\mathbf{r})H(\mathbf{r}) + \rho^s(\mathbf{r})H(\mathbf{r}) - H(\mathbf{r}). \qquad (6.10)$$

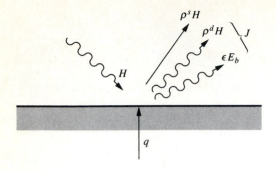

FIGURE 6-6
Energy balance for surfaces with partially specular reflection.

The first two terms on the last right-hand side of equation (6.10) or the part of the outgoing heat flux that leaves diffusely, we will again call the *surface radiosity*,

$$J(\mathbf{r}) = \epsilon(\mathbf{r})E_b(\mathbf{r}) + \rho^d(\mathbf{r})H(\mathbf{r}), \tag{6.11}$$

so that

$$q(\mathbf{r}) = J(\mathbf{r}) - [1 - \rho^s(\mathbf{r})]H(\mathbf{r}). \tag{6.12}$$

Eliminating the irradiation $H(\mathbf{r})$ from equations (6.10) and (6.12) leads to

$$q(\mathbf{r}) = \frac{\epsilon(\mathbf{r})}{\rho^d(\mathbf{r})}\left[[1 - \rho^s(\mathbf{r})]E_b(\mathbf{r}) - J(\mathbf{r})\right], \tag{6.13}$$

which, of course, reduces to equation (5.26) for a diffusely reflecting surface if $\rho^s = 0$ and $\rho^d = 1 - \epsilon$. For a purely specular reflecting surface ($\rho^d = 0$) equation (6.13) is indeterminate since the radiosity consists only of emission, or $J = \epsilon E_b$.

As in Chapter 5 the irradiation $H(\mathbf{r})$ is found by determining the contribution to H from a differential area $dA'(\mathbf{r}')$, followed by integration over the entire enclosure surface. A subtle difference is that we do not track the total energy leaving dA' (multiplied by a suitable direct-travel view factor); rather, the contribution from specular reflections is subtracted and attributed to the surface from which it leaves diffusely. The more complicated path of such energy is then accounted for by the definition of the specular view factor. Thus, similar to equation (5.21),

$$H(\mathbf{r})dA = \int_A J(\mathbf{r}')\,dF^s_{dA'-dA}dA' + H^s_o(\mathbf{r})\,dA, \tag{6.14}$$

where $H^s_o(\mathbf{r})$ is any external irradiation arriving at dA (through openings or semitransparent walls). Similar to the specular view factors, the H^s_o includes external radiation hitting dA *directly* or *after any number of specular reflections*. Using reciprocity, equation (6.14) becomes

$$H(\mathbf{r}) = \int_A J(\mathbf{r}') \, dF^s_{dA-dA'} + H^s_o(\mathbf{r}), \tag{6.15}$$

and, after substitution into equation (6.11), an integral equation for the unknown radiosity is obtained as

$$J(\mathbf{r}) = \epsilon(\mathbf{r})E_b(\mathbf{r}) + \rho^d(\mathbf{r}) \left[\int_A J(\mathbf{r}') \, dF^s_{dA-dA'} + H^s_o(\mathbf{r}) \right]. \tag{6.16}$$

For surface locations for which heat flux $q(\mathbf{r})$ is given rather than $E_b(\mathbf{r})$, equation (6.12) should be used rather than equation (6.11). It is usually more desirable to eliminate the radiosity, to obtain a single relationship between surface blackbody emissive powers and heat fluxes. Solving equation (6.13) for J gives

$$J(\mathbf{r}) = [1 - \rho^s(\mathbf{r})] E_b(\mathbf{r}) - \frac{\rho^d(\mathbf{r})}{\epsilon(\mathbf{r})} q(\mathbf{r}), \tag{6.17}$$

and substituting this expression into equation (6.16) leads to

$$(1 - \rho^s)E_b - \frac{\rho^d}{\epsilon} q = (1 - \rho^s - \rho^d)E_b + \rho^d \left[\int_A (1 - \rho^s) \, E_b \, dF^s_{dA-dA'} \right.$$
$$\left. - \int_A \frac{\rho^d}{\epsilon} q \, dF^s_{dA-dA'} + H^s_o \right],$$

or

$$E_b(\mathbf{r}) - \int_A [1 - \rho^s(\mathbf{r}')] E_b(\mathbf{r}') \, dF^s_{dA-dA'}$$
$$= \frac{q(\mathbf{r})}{\epsilon(\mathbf{r})} - \int_A \frac{\rho^d(\mathbf{r}')}{\epsilon(\mathbf{r}')} q(\mathbf{r}') \, dF^s_{dA-dA'} + H^s_o(\mathbf{r}). \tag{6.18}$$

We note that, for diffusely reflecting surfaces with $\rho^s = 0$, $\rho^d = 1 - \epsilon$, $F^s_{i-j} = F_{i-j}$ and $H^s_o = H_o$, equation (6.18) reduces to equation (5.28). If the specular view factors can be calculated (and that is often a big "if"), then equation (6.18) is not any more difficult to solve than equation (5.28). Indeed, if part or all of the surface is purely specular ($\rho^d = 0$), equation (6.18) becomes considerably simpler.

As for black and gray-diffuse enclosures, it is customary to simplify the analysis by using an idealized enclosure, consisting of N relatively simple subsurfaces, over each of which the radiosity is assumed constant. Then

$$\int_A J(\mathbf{r}') \, dF^s_{dA-dA'} \simeq \sum_{j=1}^N J_j \int_{A_j} dF^s_{dA-dA_j} = \sum_{j=1}^N J_j F^s_{dA-A_j},$$

and, after averaging over a subsurface A_i on which dA is situated, equation (6.16) simplifies to

$$J_i = \epsilon_i E_{bi} + \rho_i^d \left(\sum_{j=1}^{N} J_j F_{i-j}^s + H_{oi}^s \right), \quad i = 1, 2, \ldots, N. \tag{6.19}$$

Eliminating radiosity through equation (6.17) then simplifies equation (6.18) to

$$E_{bi} - \sum_{j=1}^{N} (1 - \rho_j^s) F_{i-j}^s E_{bj} = \frac{q_i}{\epsilon_i} - \sum_{j=1}^{N} \frac{\rho_j^d}{\epsilon_j} F_{i-j}^s q_j + H_{oi}^s, \quad i = 1, 2, \ldots, N. \tag{6.20}$$

The *summation relation*, equation (6.9), is easily obtained from equation (6.20) by considering a special case: In an isothermal enclosure ($E_{b1} = E_{b2} = \cdots = E_{bN}$) without external irradiation ($H_{o1}^s = H_{o2}^s = \cdots = 0$), according to the Second Law of Thermodynamics, all heat fluxes must vanish ($q_1 = q_2 = \cdots = 0$). Thus, canceling emissive powers,

$$\sum_{j=1}^{N} (1 - \rho_j^s) F_{i-j}^s = 1, \quad i = 1, 2, \ldots, N. \tag{6.21}$$

Since the F_{i-j}^s are *geometric* factors and do not depend on temperature distribution, equation (6.21) is valid for arbitrary emissive power values.

Finally, for computer calculations it may be advantageous to write the emissive power and heat fluxes in matrix form. Introducing Kronecker's delta equation (6.20) becomes

$$\sum_{j=1}^{N} \left[\delta_{ij} - (1 - \rho_j^s) F_{i-j}^s \right] E_{bj} = \sum_{j=1}^{N} \left(\frac{\delta_{ij}}{\epsilon_j} - \frac{\rho_j^d}{\epsilon_j} F_{i-j}^s \right) q_j + H_{oi}^s,$$

$$i = 1, 2, \ldots, N, \tag{6.22}$$

or[2]

$$\mathbf{A} \cdot \mathbf{e}_b = \mathbf{C} \cdot \mathbf{q} + \mathbf{h}_o^s, \tag{6.23}$$

where $\mathbf{C}$ and $\mathbf{A}$ are matrices with elements

$$A_{ij} = \delta_{ij} - (1 - \rho_j^s) F_{i-j}^s,$$

$$C_{ij} = \frac{\delta_{ij}}{\epsilon_j} - \frac{\rho_j^d}{\epsilon_j} F_{i-j}^s,$$

and $\mathbf{q}$, $\mathbf{e}_b$ and $\mathbf{h}_o^s$ are vectors for the surface heat fluxes, emissive powers, and external

---

[2]Again, for easy readability of matrix manipulations we shall follow here the convention that a two-dimensional matrix is denoted by a bold capitalized letter, while a vector is written as a bold lower-case letter.

irradiations, respectively. If all temperatures and external irradiations are known, the unknown heat fluxes are readily found by matrix inversion as

$$\mathbf{q} = \mathbf{C}^{-1} \cdot [\mathbf{A} \cdot \mathbf{e}_b - \mathbf{h}_o^s]. \tag{6.24}$$

If the emissive power is only known over some of the surfaces, and the heat fluxes are specified elsewhere, equation (6.23) may be rearranged into a similar equation for the vector containing all the unknowns.

Example 6.3.　Two large parallel plates are separated by a nonparticipating medium as shown in Fig. 6-4a. The bottom surface is isothermal at T_1, with emissivity ϵ_1 and a partially specular, partially diffuse reflectivity $\rho_1 = \rho_1^d + \rho_1^s$. Similarly, the top surface is isothermal at T_2 with ϵ_2 and $\rho_2 = \rho_2^d + \rho_2^s$. Determine the radiative heat flux between the surfaces.

Solution

From equation (6.20) we have, for $i = 1$, with $H_{o1}^s = 0$,

$$E_{b1} - (1 - \rho_1^s)F_{1-1}^s E_{b1} - (1 - \rho_2^s)F_{1-2}^s E_{b2} = \frac{q_1}{\epsilon_1} - \frac{\rho_1^d}{\epsilon_1}F_{1-1}^s q_1 - \frac{\rho_2^d}{\epsilon_2}F_{1-2}^s q_2.$$

While we could apply $i = 2$ to equation (6.20) to obtain a second equation for q_1 and q_2, it is simpler here to use *overall conservation of energy*, or $q_2 = -q_1$. Thus,

$$q_1 = \frac{[1 - (1 - \rho_1^s)F_{1-1}^s]E_{b1} - (1 - \rho_2^s)F_{1-2}^s E_{b2}}{\dfrac{1}{\epsilon_1} - \dfrac{1 - \epsilon_1 - \rho_1^s}{\epsilon_1}F_{1-1}^s + \dfrac{1 - \epsilon_2 - \rho_2^s}{\epsilon_2}F_{1-2}^s}.$$

Using the results from Example 6.1 and/or equation (6.21), we obtain

$$q_1 = \frac{(1 - \rho_2^s)F_{1-2}^s(E_{b1} - E_{b2})}{\left(\dfrac{1}{\epsilon_1} + \dfrac{1}{\epsilon_2}\right)(1 - \rho_2^s)F_{1-2}^s + F_{1-1}^s - F_{1-2}^s}$$

$$= \frac{(1 - \rho_2^s)(1)(E_{b1} - E_{b2})}{\left(\dfrac{1}{\epsilon_1} + \dfrac{1}{\epsilon_2}\right)(1 - \rho_2^s)(1) + \rho_2^s - 1} = \frac{E_{b1} - E_{b2}}{\dfrac{1}{\epsilon_1} + \dfrac{1}{\epsilon_2} - 1}, \tag{6.25}$$

which produces the same result whether we have diffusely or specularly reflecting surfaces. Indeed, equation (6.25) is valid for the radiative transfer between two isothermal parallel plates, regardless of the directional behavior of the reflectivity (i.e., it is not limited to the idealized reflectivities considered in this chapter). *Any* beam leaving A_1 *must* hit surface A_2 and vice versa, regardless of whether the reflectivity is diffuse, specular or neither of the two; the surface locations will be different but the directional variation of reflectivity has no influence on the heat transfer rate since the surfaces are isothermal.

Example 6.4.　Repeat the previous example for concentric spheres and cylinders.

Solution

Again, from equation (6.20) with $i = 1$ and $H_{oi}^s = 0$, we obtain

$$E_{b1} - (1 - \rho_1^s)F_{1-1}^s E_{b1} - (1 - \rho_2^s)F_{1-2}^s E_{b2} = \frac{q_1}{\epsilon_1} - \frac{\rho_1^d}{\epsilon_1}F_{1-1}^s q_1 - \frac{\rho_2^d}{\epsilon_2}F_{1-2}^s q_2.$$

In this case conservation of energy demands $q_2 A_2 = -q_1 A_1$, and

$$q_1 = \frac{[1 - (1 - \rho_1^s)F_{1-1}^s]E_{b1} - (1 - \rho_2^s)F_{1-2}^s E_{b2}}{\dfrac{1}{\epsilon_1} - \dfrac{1 - \epsilon_1 - \rho_1^s}{\epsilon_1}F_{1-1}^s + \dfrac{1 - \epsilon_2 - \rho_2^s}{\epsilon_2}\dfrac{A_1}{A_2}F_{1-2}^s}$$

$$= \frac{(1 - \rho_2^s)F_{1-2}^s(E_{b1} - E_{b2})}{\left(\dfrac{1}{\epsilon_1} + \dfrac{1}{\epsilon_2}\dfrac{A_1}{A_2}\right)(1 - \rho_2^s)F_{1-2}^s + F_{1-1}^s - \dfrac{A_1}{A_2}F_{1-2}^s}.$$

The specular view factors F_{1-1}^s and F_{1-2}^s are the same as in the previous example (cf. Example 6.2), leading to

$$q_1 = \frac{E_{b1} - E_{b2}}{\dfrac{1}{\epsilon_1} + \dfrac{1}{\epsilon_2}\dfrac{A_1}{A_2} - \dfrac{A_1/A_2 - \rho_2^s}{1 - \rho_2^s}}. \tag{6.26}$$

We note that equation (6.26) does not depend on ρ_1^s: Again, any radiation reflected off surface A_1 *must* return to surface A_2, regardless of the directional behavior of its reflectivity. If surface A_2 is purely specular ($\rho_2^s = 1 - \epsilon_2$), all radiation from A_1 bounces back and forth between A_1 and A_2, and equation (6.26) reduces to equation (6.25), i.e., the heat flux between these concentric spheres or cylinders is the same as between parallel plates. On the other hand, if A_2 is diffuse ($\rho_2^s = 0$) equation (6.26) reduces to the purely diffuse case since the directional behavior of ρ_1 is irrelevant.

Example 6.5.   A very long solar collector plate is to collect energy at a temperature of $T_1 = 350\,\text{K}$. To improve its performance for off-normal solar incidence, a highly reflective surface is placed next to the collector as shown in Fig. 6-7. For simplicity you may make the following assumptions: The collector is isothermal and gray-diffuse with emissivity $\epsilon_1 = 1 - \rho_1^d = 0.8$; the mirror is gray and specular with $\epsilon_2 = 1 - \rho_2^s = 0.1$, and heat losses from the mirror by convection as well as all losses from the collector ends may be neglected. How much energy (per unit length) does the collector plate collect for solar irradiation of $q_{\text{sun}} = 1000\,\text{W/m}^2$ at an incidence angle of $30°$?

Solution

Applying equation (6.22) to absorber plate ($i = 1$) as well as mirror ($i = 2$) we obtain

$$[1 - (1 - \rho_1^s)F_{1-1}^s]E_{b1} - (1 - \rho_2^s)F_{1-2}^s E_{b2}$$

$$= \left[\frac{1}{\epsilon_1} - \frac{\rho_1^d}{\epsilon_1}F_{1-1}^s\right]q_1 - \frac{\rho_2^d}{\epsilon_2}F_{1-2}^s q_2 + H_{o1}^s,$$

$$-(1 - \rho_1^s)F_{2-1}^s E_{b1} + [1 - (1 - \rho_2^s)F_{2-2}^s]E_{b2}$$

$$= -\frac{\rho_1^d}{\epsilon_1}F_{2-1}^s q_1 + \left[\frac{1}{\epsilon_2} - \frac{\rho_2^d}{\epsilon_2}F_{2-2}^s\right]q_2 + H_{o2}^s.$$

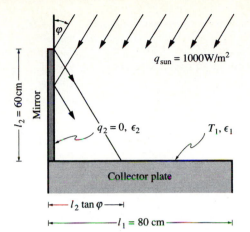

$q_{sun} = 1000 \text{W/m}^2$

$q_2 = 0, \ \epsilon_2$ T_1, ϵ_1

Mirror

$l_2 = 60 \text{cm}$

Collector plate

$l_2 \tan \varphi$

$l_1 = 80 \text{ cm}$

FIGURE 6-7
Geometry for Example 6.5.

Since $\rho_1^s = 0$, it follows that $F_{1-1}^s = F_{2-2}^s = 0$ and also $F_{1-2}^s = F_{1-2}$, $F_{2-1}^s = F_{2-1}$. For this configuration no specular reflections from one surface to another surface are possible (radiation leaving the absorber plate, after specular reflection from the mirror, always leaves the open enclosure). Thus, with $q_2 = 0$,

$$E_{b1} - \epsilon_2 F_{1-2} E_{b2} = \frac{q_1}{\epsilon_1} + H_{o1}^s,$$

$$-F_{2-1} E_{b1} + E_{b2} = -\left(\frac{1}{\epsilon_1} - 1\right) F_{2-1} q_1 + H_{o2}^s.$$

Eliminating E_{b2}, by multiplying the second equation by $\epsilon_2 F_{1-2}$ and adding, leads to

$$(1 - \epsilon_2 F_{1-2} F_{2-1}) E_{b1} = \left[\frac{1}{\epsilon_1} - \left(\frac{1}{\epsilon_1} - 1\right) \epsilon_2 F_{1-2} F_{2-1}\right] q_1 + H_{o1}^s + \epsilon_2 F_{1-2} H_{o2}^s.$$

The external fluxes are evaluated as follows: The mirror receives solar flux only directly (no specular reflection off the absorber plate is possible), i.e., $H_{o2}^s = q_{sun} \sin \varphi$. The absorber plate receives a direct contribution, $q_{sun} \cos \varphi$, and a second contribution after specular reflection off the mirror. This second contribution has the strength of $\rho_2^s q_{sun} \cos \varphi$ per unit area. However, only part of the collector plate ($l_2 \tan \varphi$) receives this secondary contribution, which, for our crude two-node description, must be averaged over l_1. Thus,

$$H_{o1}^s = q_{sun} \cos \varphi + \rho_2^s q_{sun} \cos \varphi \frac{l_2 \tan \varphi}{l_1} = q_{sun} \left[\cos \varphi + (1 - \epsilon_2) \frac{l_2}{l_1} \sin \varphi\right].$$

Therefore,

$$q_1 = \frac{(1 - \epsilon_2 F_{1-2} F_{2-1}) E_{b1} - [\cos \varphi + (1 - \epsilon_2) \sin \varphi (l_2/l_1) + \epsilon_2 F_{1-2} \sin \varphi] q_{sun}}{\frac{1}{\epsilon_1} - \epsilon_2 \left(\frac{1}{\epsilon_1} - 1\right) F_{1-2} F_{2-1}}.$$

The view factors are readily evaluated by the crossed-strings method as $F_{1-2} = (80 + 60 - 100)/(2 \times 80) = \frac{1}{4}$ and $F_{2-1} = 80 \times \frac{1}{4}/60 = \frac{1}{3}$. Substituting numbers, we obtain

$$q_1 = \frac{\left(1 - 0.1 \times \frac{1}{4} \times \frac{1}{3}\right)5.670 \times 10^{-8} \times 350^4 - \left(\frac{\sqrt{3}}{2} + 0.9 \times \frac{1}{2} \times \frac{60}{80} + 0.1 \times \frac{1}{4} \times \frac{1}{2}\right)1000}{\frac{1}{0.8} - 0.1\left(\frac{1}{0.8} - 1\right) \times \frac{1}{4} \times \frac{1}{3}}$$

$$= -298 \, \text{W/m}^2.$$

Under these conditions, therefore, the collector is about 30% efficient. This result should be compared with a collector without mirror ($l_2 = 0$ and $F_{1-2} = 0$), for which we get

$$q_{1,\text{no mirror}} = \frac{E_{b1} - q_{\text{sun}} \cos \varphi}{1/\epsilon_1}$$

$$= 0.8 \times \left(5.670 \times 10^{-8} \times 350^4 - 1000 \times \frac{\sqrt{3}}{2}\right) = -12 \, \text{W/m}^2.$$

This absorber plate collects hardly any energy at all (indeed, after accounting for convection losses, it would experience a net energy loss). If the mirror had been a diffuse reflector the heat gain would have been $q_{1,\text{diffuse mirror}} = -172 \, \text{W/m}^2$, which is significantly less than for the specular mirror (cf. Problem 5.12).

We conclude from this example that (*i*) mirrors can significantly improve collector performance, and (*ii*) infrared re-radiation losses from near-black collectors are very substantial. Of course, re-radiation losses may be significantly reduced by using *selective surfaces* or glass-covered collectors (cf. Chapter 3).

We shall conclude this section with three more examples designed to clarify certain aspects of evaluating the specular view factors in enclosures comprised of only simple planar elements.

Example 6.6.   Consider the triangular enclosures shown in Figs. 6-8*a* and *b*. Surfaces A_1 and A_2 are isothermal at T_1 and T_2, respectively, and are purely diffuse reflectors with $\epsilon_1 = 1 - \rho_1^d$ and $\epsilon_2 = 1 - \rho_2^d$. Surface A_3 is isothermal at T_3 and is a purely specular reflector with $\epsilon_3 = 1 - \rho_3^s$. Set up the system of equations for the unknown surface heat fluxes.

Solution

Since there is only a single (and flat) specular surface, no multiple specular reflections are possible. While F_{1-1}^s and F_{2-2}^s are nonzero, it is clear that $F_{3-3}^s = 0$. Thus, from equation (6.22), with $H_{oi}^s = 0$,

$$(1 - F_{1-1}^s)E_{b1} - F_{1-2}^s E_{b2} - \epsilon_3 F_{1-3}^s E_{b3}$$

$$= \left[\frac{1}{\epsilon_1} - \left(\frac{1}{\epsilon_1} - 1\right)F_{1-1}^s\right]q_1 - \left(\frac{1}{\epsilon_2} - 1\right)F_{1-2}^s q_2,$$

$$-F_{2-1}^s E_{b1} + (1 - F_{2-2}^s)E_{b2} - \epsilon_3 F_{2-3}^s E_{b3}$$

$$= -\left(\frac{1}{\epsilon_1} - 1\right)F_{2-1}^s q_1 + \left[\frac{1}{\epsilon_2} - \left(\frac{1}{\epsilon_2} - 1\right)F_{2-2}^s\right]q_2,$$

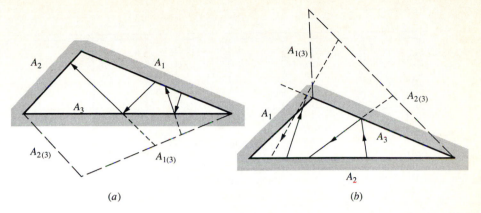

FIGURE 6-8
Triangular enclosure with a single specularly reflecting surface, with a few possible beam paths indicated, (*a*) without obstructions, (*b*) with partial obstructions.

$$-F_{3-1}^s E_{b1} - F_{3-2}^s E_{b2} + E_{b3}$$
$$= -\left(\frac{1}{\epsilon_1} - 1\right) F_{3-1}^s q_1 - \left(\frac{1}{\epsilon_2} - 1\right) F_{3-2}^s q_2 + \frac{q_3}{\epsilon_3}.$$

We note that q_3 only enters the last equation, so we only have two simultaneous equations to solve (i.e., as many as we have surfaces with *diffuse* reflection components). We shall need to determine the specular view factors F_{1-1}^s, F_{1-2}^s, and F_{2-2}^s, while the rest can be evaluated through reciprocity and the summation rule. Considering the first case of Fig. 6-8*a*, we find

$$F_{1-1}^s = \rho_3^s F_{1(3)-1}$$
$$F_{1-2}^s = F_{1-2} + \rho_3^s F_{1(3)-2}, \quad \epsilon_3 F_{1-3}^s = 1 - F_{1-1}^s - F_{1-2}^s,$$
$$F_{2-1}^s = A_1 F_{1-2}^s / A_2$$
$$F_{2-2}^s = \rho_3^s F_{2(3)-2}, \quad \epsilon_3 F_{2-3}^s = 1 - F_{2-1}^s - F_{2-2}^s,$$
$$F_{3-1}^s = A_1 F_{1-3}^s / A_3, \quad F_{3-2}^s = A_2 F_{2-3}^s / A_3,$$

where all view factors on the right hand sides are readily evaluated through standard diffuse view factor analysis. The problem becomes slightly more difficult in the configuration shown in Fig. 6-8*b*, where the specular surface is attached to another surface with an opening angle of $> 90°$. Standing in the left corner on surface A_2, one obviously cannot see all of the image $A_{2(3)}$ from there by looking through "mirror" A_3. Care must be taken that these visual obstructions are not overlooked. If the enclosure is two-dimensional, such partially obstructed view factors are no problem for the crossed-strings method, but may pose great difficulty for an analytical solution otherwise.

The effects of partial shading become somewhat more obvious when configurations with two or more adjacent specular surfaces are considered.

Example 6.7. Consider the rectangular enclosure shown in Fig. 6-9. Surfaces A_1 and A_2 are purely specular, surfaces A_3 and A_4 are purely diffuse reflectors. Top and bottom

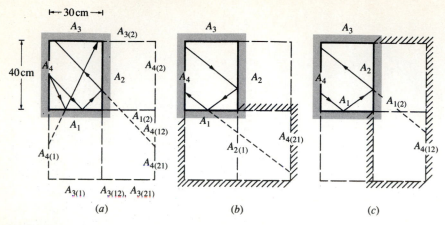

FIGURE 6-9
Rectangular enclosure with two adjacent specular reflectors, with some possible beam paths indicated: (a) Evaluation of F^s_{4-3}, (b) evaluation of $A_{4(21)}$ contribution to F^s_{4-4}, (c) evaluation of $A_{4(12)}$ contribution to F^s_{4-4}.

walls are at $T_1 = T_3 = 1000\,\text{K}$, with $\epsilon_1 = 1 - \rho^s_1 = \epsilon_3 = 1 - \rho^d_3 = 0.3$; the side walls are at $T_2 = T_4 = 600\,\text{K}$ with emissivities $\epsilon_2 = 1 - \rho^s_2 = \epsilon_4 = 1 - \rho^d_4 = 0.8$. Determine the net radiative heat flux for each surface.

Solution

Looking at Fig. 6-9a, one sees that $F^s_{1-1} = F^s_{2-2} = 0$, while all other specular view factors are nonzero. Again, with $H^s_{oi} = 0$, we have from equation (6.22)

$$E_{b1} - \epsilon_2 F^s_{1-2} E_{b2} - F^s_{1-3} E_{b3} - F^s_{1-4} E_{b4}$$
$$= \frac{q_1}{\epsilon_1} - \left(\frac{1}{\epsilon_3} - 1\right) F^s_{1-3} q_3 - \left(\frac{1}{\epsilon_4} - 1\right) F^s_{1-4} q_4,$$

$$-\epsilon_1 F^s_{2-1} E_{b1} + E_{b2} - F^s_{2-3} E_{b3} - F^s_{2-4} E_{b4}$$
$$= \frac{q_2}{\epsilon_2} - \left(\frac{1}{\epsilon_3} - 1\right) F^s_{2-3} q_3 - \left(\frac{1}{\epsilon_4} - 1\right) F^s_{2-4} q_4,$$

$$-\epsilon_1 F^s_{3-1} E_{b1} - \epsilon_2 F^s_{3-2} E_{b2} + (1 - F^s_{3-3}) E_{b3} - F^s_{3-4} E_{b4}$$
$$= \left[\frac{1}{\epsilon_3} - \left(\frac{1}{\epsilon_3} - 1\right) F^s_{3-3}\right] q_3 - \left(\frac{1}{\epsilon_4} - 1\right) F^s_{3-4} q_4,$$

$$-\epsilon_1 F^s_{4-1} E_{b1} - \epsilon_2 F^s_{4-2} E_{b2} - F^s_{4-3} E_{b3} + (1 - F^s_{4-4}) E_{b4}$$
$$= -\left(\frac{1}{\epsilon_3} - 1\right) F^s_{4-3} q_3 + \left[\frac{1}{\epsilon_4} - \left(\frac{1}{\epsilon_4} - 1\right) F^s_{4-4}\right] q_4.$$

Again, we have only two simultaneous equations to solve for the two (diffuse) heat fluxes q_3 and q_4: The first two equations are explicit expressions for q_1 and q_2, respectively (once q_3 and q_4 have been determined). Checking the various images in Fig. 6-9a, we find that the specular view factors for surface A_1 are

$$F_{1-1}^s = 0,$$
$$F_{1-2}^s = F_{1-2},$$
$$F_{1-3}^s = F_{1-3} + \rho_2^s F_{1(2)-3},$$
$$F_{1-4}^s = F_{1-4} + \rho_2^s F_{1(2)-4}.$$

Checking the summation rule, we find

$$(1 - \rho_1^s)F_{1-1}^s + (1 - \rho_2^s)F_{1-2}^s + F_{1-3}^s + F_{1-4}^s$$
$$= 0 + F_{1-2} + F_{1-3} + F_{1-4} - \rho_2^s(F_{1-2} - F_{1(2)-3} - F_{1(2)-4}) = 1$$

or

$$F_{1(2)-3} + F_{1(2)-4} = F_{1-2}.$$

Indeed, by checking Fig. 6-9a, we find

$$F_{1(2)-3} + F_{1(2)-4} = F_{1(2)-(3+4)} = F_{1(2)-2} = F_{1-2}.$$

Similarly, we have

$$F_{2-2}^s = 0,$$
$$F_{2-1}^s = F_{2-1},$$
$$F_{2-3}^s = F_{2-3} + \rho_1^s F_{2(1)-3},$$
$$F_{2-4}^s = F_{2-4} + \rho_1^s F_{2(1)-4}.$$

For surfaces A_3 and A_4 dual specular reflections are possible:

$$F_{3-1}^s = F_{3-1} + \rho_2^s F_{3(2)-1},$$
$$F_{3-2}^s = F_{3-2} + \rho_1^s F_{3(1)-2},$$
$$F_{3-3}^s = \rho_1^s F_{3(1)-3} + \rho_1^s \rho_2^s F_{3(12)-3} + \rho_2^s \rho_1^s F_{3(21)-3},$$
$$F_{3-4}^s = F_{3-4} + \rho_1^s F_{3(1)-4} + \rho_2^s F_{3(2)-4} + \rho_1^s \rho_2^s F_{3(12)-4} + \rho_2^s \rho_1^s F_{3(21)-4},$$

$$F_{4-1}^s = F_{4-1} + \rho_2^s F_{4(2)-1},$$
$$F_{4-2}^s = F_{4-2} + \rho_1^s F_{4(1)-2},$$
$$F_{4-3}^s = F_{4-3} + \rho_1^s F_{4(1)-3} + \rho_2^s F_{4(2)-3} + \rho_1^s \rho_2^s F_{4(12)-3} + \rho_2^s \rho_1^s F_{4(21)-3},$$
$$F_{4-4}^s = \rho_2^s F_{4(2)-4} + \rho_1^s \rho_2^s F_{4(12)-4} + \rho_2^s \rho_1^s F_{4(21)-4}.$$

It is tempting to assume that $F_{4(12)-4} = F_{4(21)-4}$, etc. Closer inspection of Figs. 6-9b and c reveals, however, that these view factors are partially obstructed: For example, for $F_{4(21)-4}$ all rays from $A_{4(21)}$ to A_4 must pass through the image $A_{2(1)}$ as well as A_1, i.e., all rays must stay *below* the corner between A_1 and A_2 (center point of Fig. 6-9b). On the other hand, for $F_{4(12)-4}$ all rays from $A_{4(12)}$ must stay *above* the corner between A_1 and A_2, and *both together* add up to the *unobstructed* view factor from the image to A_4. The same is true for $F_{3(12)-3} + F_{3(21)-3}$. However, the geometry is such that $F_{4(21)-3} = 0$, while $F_{4(12)-3}$ is unobstructed (thus, still adding up to the unobstructed view factor). Similarly, $F_{3(12)-4} = 0$, while $F_{3(21)-4}$ is unobstructed.

Simplifications for partially obstructed view factor were found for this particular simple geometry. Care must be taken before extrapolating these results to other configurations.

Before actually evaluating view factors one should take advantage of the fact that there are only two different surface temperatures, i.e., $E_{b3} = E_{b1}$ and $E_{b4} = E_{b2}$, and only two emissivities, $\epsilon_3 = \epsilon_1$ and $\epsilon_4 = \epsilon_2$:

$$(1 - F_{1-3}^s)E_{b1} - (\epsilon_2 F_{1-2}^s + F_{1-4}^s)E_{b2}$$
$$= \frac{q_1}{\epsilon_1} - \left(\frac{1}{\epsilon_1} - 1\right)F_{1-3}^s q_3 - \left(\frac{1}{\epsilon_2} - 1\right)F_{1-4}^s q_4,$$

$$-(\epsilon_1 F_{2-1}^s + F_{2-3}^s)E_{b1} + (1 - F_{2-4}^s)E_{b2}$$
$$= \frac{q_2}{\epsilon_2} - \left(\frac{1}{\epsilon_1} - 1\right)F_{2-3}^s q_3 - \left(\frac{1}{\epsilon_2} - 1\right)F_{2-4}^s q_4,$$

$$(1 - \epsilon_1 F_{3-1}^s - F_{3-3}^s)E_{b1} - (\epsilon_2 F_{3-2}^s + F_{3-4}^s)E_{b2}$$
$$= \left[\frac{1}{\epsilon_1} - \left(\frac{1}{\epsilon_1} - 1\right)F_{3-3}^s\right]q_3 - \left(\frac{1}{\epsilon_2} - 1\right)F_{3-4}^s q_4,$$

$$-(\epsilon_1 F_{4-1}^s + F_{4-3}^s)E_{b1} + (1 - \epsilon_2 F_{4-2}^s - F_{4-4}^s)E_{b2}$$
$$= -\left(\frac{1}{\epsilon_1} - 1\right)F_{4-3}^s q_3 + \left[\frac{1}{\epsilon_2} - \left(\frac{1}{\epsilon_2} - 1\right)F_{4-4}^s\right]q_4.$$

The necessary view factors are readily found from the crossed-strings method, [equation (4.49)], reciprocity, and the summation rule [equation (6.21)], as well as from Example 5.1 for the diffuse view factors:

$$F_{1-2}^s = F_{1-2} = 0.25;$$
$$F_{1-3} = 0.5, \quad F_{1(2)-3} = (\sqrt{64 + 9} + 3 - 2 \times 5)/2 \times 4 = 0.1930 :$$
$$F_{1-3}^s = 0.5 + 0.2 \times 0.1930 = 0.5386;$$
$$F_{1-4} = 0.25, \quad F_{1(2)-4} = (5 + 8 - 4 - \sqrt{73})/8 = 0.0570 :$$
$$F_{1-4}^s = 0.25 + 0.2 \times 0.0570 = 0.2614;$$

$$F_{2-1}^s = F_{2-1} = 0.3333;$$
$$F_{2-3} = 0.3333, \quad F_{2(1)-3} = (5 + 6 - \sqrt{52} - 3)/6 = 0.1315 :$$
$$F_{2-3}^s = 0.3333 + 0.7 \times 0.1315 = 0.4254;$$
$$F_{2-4} = 0.3333, \quad F_{2(1)-4} = (\sqrt{52} + 4 - 2 \times 5)/6 = 0.2019 :$$
$$F_{2-4}^s = 0.3333 + 0.7 \times 0.2019 = 0.4746;$$

$$F_{3-1}^s = F_{1-3}^s = 0.5386;$$
$$F_{3-2}^s = A_2 F_{2-3}^s/A_3 = 0.75 \times 0.4254 = 0.3191;$$
$$F_{3(1)-3} = (\sqrt{52} - 6)/4 = 0.3028,$$
$$F_{3(12)-3} + F_{3(21)-3} = (10 + 6 - 2\sqrt{52})/8 = 0.1972 :$$
$$F_{3-3}^s = 0.7 \times 0.3028 + 0.2 \times 0.7 \times 0.1972 = 0.2396;$$
$$F_{3-4}^s = 1 - \epsilon_1 F_{3-1}^s - \epsilon_2 F_{3-2}^s - F_{3-3}^s$$
$$= 1 - 0.3 \times 0.5386 - 0.8 \times 0.3191 - 0.2396 = 0.3436;$$

$$F_{4-1}^s = A_1 F_{1-4}^s/A_4 = 0.2614/0.75 = 0.3485;$$
$$F_{4-2}^s = F_{2-4}^s = 0.4746;$$

$$F_{4-3}^s = A_3 F_{3-4}^s / A_4 = 0.3436/0.75 = 0.4581;$$
$$F_{4-4}^s = 1 - \epsilon_1 F_{4-1}^s - \epsilon_2 F_{4-2}^s - F_{4-3}^s$$
$$= 1 - 0.3 \times 0.3485 - 0.8 \times 0.4746 - 0.4581 = 0.0576.$$

Substituting these values into the heat flux equations and realizing, from the summation rule, that the two coefficients in front of E_{b1} and E_{b2} are the same for each equation, we obtain

$$(1 - 0.5386)(E_{b1} - E_{b2}) = \frac{q_1}{0.3} - \left(\frac{1}{0.3} - 1\right)0.5386q_3 - \left(\frac{1}{0.8} - 1\right)0.2614q_4,$$

$$-(1 - 0.4746)(E_{b1} - E_{b2}) = \frac{q_2}{0.8} - \left(\frac{1}{0.3} - 1\right)0.4254q_3 - \left(\frac{1}{0.8} - 1\right)0.4746q_4,$$

$$(0.8 \times 0.3191 + 0.3436)(E_{b1} - E_{b2}) =$$
$$\left[\frac{1}{0.3} - \left(\frac{1}{0.3} - 1\right)0.2396\right]q_3 - \left(\frac{1}{0.8} - 1\right)0.3436q_4,$$

$$-(0.3 \times 0.3485 + 0.4581)(E_{b1} - E_{b2}) =$$
$$-\left(\frac{1}{0.3} - 1\right)0.4581q_3 + \left[\frac{1}{0.8} - \left(\frac{1}{0.8} - 1\right)0.0576\right]q_4.$$

After a little cleaning up these equations become

$$2.7743q_3 - 0.0859q_4 = 0.5989(E_{b1} - E_{b2}),$$
$$-1.0689q_3 + 1.2356q_4 = -0.5627(E_{b1} - E_{b2}),$$
$$q_1 = 0.3770q_3 + 0.0196q_4 + 0.1384(E_{b1} - E_{b2}),$$
$$q_2 = 0.7941q_3 + 0.0949q_4 - 0.4203(E_{b1} - E_{b2}) \quad .$$

Solving the first two equations leads to

$$q_3 = \frac{0.5989 \times 1.2356 - 0.5627 \times 0.0859}{2.7743 \times 1.2356 - 1.0689 \times 0.0859}(E_{b1} - E_{b2}) = 0.2073(E_{b1} - E_{b2}),$$

$$q_4 = \frac{0.5989 \times 1.0689 - 0.5627 \times 2.7743}{2.7743 \times 1.2356 - 1.0689 \times 0.0859}(E_{b1} - E_{b2}) = -0.2761(E_{b1} - E_{b2}),$$

and

$$q_1 = [0.3770 \times 0.2073 + 0.0196 \times (-0.2761) + 0.1384](E_{b1} - E_{b2})$$
$$= 0.2111(E_{b1} - E_{b2}),$$
$$q_2 = [0.7941 \times 0.2073 + 0.0949 \times (-0.2761) - 0.4203](E_{b1} - E_{b2})$$
$$= -0.2819(E_{b1} - E_{b2}).$$

To determine the net surface heat fluxes we evaluate

$$E_{b1} - E_{b2} = \sigma(T_1^4 - T_2^4) = 5.670 \times 10^{-8}(1000^4 - 600^4) \, \text{W/m}^2 = 4.935 \, \text{W/cm}^2$$

and multiply by the respective surface areas. Thus,

$$Q_1' = 40 \, \text{cm} \times 0.2111 \times 4.935 \, \text{W/cm}^2 = 41.7 \, \text{W/cm},$$
$$Q_2' = 30 \, \text{cm} \times (-0.2819) \times 4.935 \, \text{W/cm}^2 = -41.7 \, \text{W/cm},$$
$$Q_3' = 40 \, \text{cm} \times 0.2073 \times 4.935 \, \text{W/cm}^2 = 40.9 \, \text{W/cm},$$
$$Q_4' = 30 \, \text{cm} \times (-0.2761) \times 4.935 \, \text{W/cm}^2 = -40.9 \, \text{W/cm}.$$

Checking our results, we note that the four heat fluxes add up to zero as they should.

The results of the present example—an enclosure with two adjacent specular reflectors—should be compared with those of Example 5.4, dealing with the identical problem except that all four surfaces were perfectly diffuse reflectors. For Example 5.4, we had found $Q_1' = -Q_2' = Q_3' = -Q_4' = 42.3\,\text{W/cm}$. For the present configuration the heat fluxes of the specular surfaces are reduced by 1%, while the heat fluxes of the diffuse surfaces are reduced a little more, by approximately 3%. Overall, the effects of specularity are found to be rather minor.

In the last two examples only two simultaneous equations had to be solved, even though there were three and four unknown surface heat fluxes, respectively, because for any *purely specular surface with known temperature* the radiosity is not unknown, but is given as $J = \epsilon E_b$. Thus, for an enclosure consisting of N surfaces, of which n are purely specular with known temperature, only $N - n$ simultaneous equations need to be solved. While this fact simplifies specular enclosure analysis as compared with diffuse enclosures, one should remember that, in general, specular view factors are considerably more difficult to evaluate.

As a final example for configurations with flat surfaces we shall consider a case where many specular reflections are possible.

Example 6.8.   Since solar energy strikes the absorbing plate of a strategically oriented solar collector only over a narrow band of incidence directions (varying somewhat during the day, as well as during the year), the ideal collector material would be directionally selective: The emissivity should be high for directions of solar incidence (to maximize energy collection), and low for all other directions (to minimize re-radiation losses). One such material is a V-corrugated specular surface shown in Fig. 6-10a. Assuming that the V-corrugated groove, with opening angle 2γ, is coated with a purely specular reflecting material, with emissivity $\epsilon = 1 - \rho^s$, what is the apparent hemispherical emissivity of such a surface (i.e., what is its heat loss compared with a flat black plate at the same temperature)?

Solution

Calling the two surfaces in a single "V" A_1 and A_2, as indicated in Fig. 6-10b, with $E_{b1} = E_{b2} = E_b$, $\epsilon_1 = \epsilon_2 = \epsilon$, and $H_{o1}^s = H_{o2}^s = 0$ we obtain from equation (6.22) (for $i = 1$)

$$[1 - \epsilon\,(F_{1-1}^s + F_{1-2}^s)]\,E_b = \frac{q}{\epsilon}.$$

Total heat lost from both surfaces of the groove is $Q = q \times 2L = qd/\sin\gamma$; on the other hand, heat lost from a black surface covering the opening would be $Q_b = E_b d$. Thus, the apparent emissivity is

$$\epsilon_a = \frac{Q}{Q_b} = \frac{q}{E_b \sin\gamma} = \frac{\epsilon\,[1 - \epsilon\,(F_{1-1}^s + F_{1-2}^s)]}{\sin\gamma}.$$

This expression could be further simplified, using summation rule and reciprocity, to $\epsilon_a = \epsilon\,F_{1-3}^s/\sin\gamma = 2\epsilon\,F_{3-1}^s$, where A_3 is the open top of the V (and of width d). However,

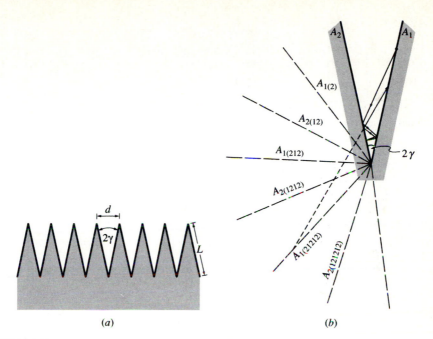

FIGURE 6-10
Geometry for Example 6.8: (*a*) V-corrugated surface, (*b*) images for a single V for the evaluation of F^s_{1-1}.

F^s_{1-1} and F^s_{1-2} are somewhat simpler to evaluate, and we shall do so here: A beam leaving surface A_1 can return to A_1 (*i*) after a single reflection off surface A_2 [appearing to come from the image $A_{1(2)}$, as indicated in Fig. 6-10*b*], or (*ii*) after hitting A_2, traveling back to A_1, returning one more time to A_2, and hitting A_1 a second time [i.e., a beam that appears to come from image $A_{1(212)}$], and so on. Thus,

$$F^s_{1-1} = \rho F_{1(2)-1} + \rho^3 F_{1(212)-1} + \rho^5 F_{1(21212)-1} + \ldots.$$

F^s_{1-2} may be similarly evaluated. We shall here determine $F^s_{2-1} = F^s_{1-2}$ instead, since this expression allows us to employ the images shown in Fig. 6-10*b*: Energy may travel directly from A_2 to A_1, or go from A_2 to A_1, get reflected back to A_2, and reflected back to A_1 again [appearing to come from image $A_{2(12)}$], and so forth. Therefore,

$$F^s_{1-2} = F^s_{2-1} = F_{2-1} + \rho^2 F_{2(12)-1} + \rho^4 F_{2(1212)-1} + \ldots.$$

Adding both together and using reciprocity (with all areas being the same), we obtain

$$F^s_{1-1} + F^s_{1-2} = F_{1-2} + \rho F_{1-1(2)} + \rho^2 F_{1-2(12)} + \rho^3 F_{1-1(212)} + \ldots.$$

Each one of these view factors F_{i-j} is subject to the restriction that all beams from A_1 to the image A_j must pass through *all* the images between A_1 and A_j; however, in this geometry no partial obstruction occurs as seen from Fig. 6-10*b*. The series above ends as soon as the image can no longer be seen from A_1, i.e., when the opening angle between

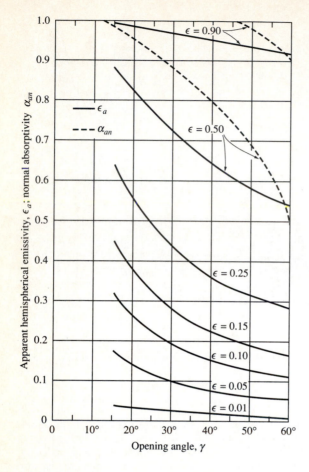

FIGURE 6-11
Apparent normal and hemispherical emissivities for specularly reflecting V-corrugated surfaces [6].

A_1 and the image exceeds $180°$. The view factor for a V-groove with opening angle 2ϕ from Configuration 34 in Appendix D, is $F_{2\phi} = 1 - \sin\phi$. Thus,

$$F^s_{1-1} + F^s_{1-2} = 1 - \sin\gamma + \rho(1 - \sin 2\gamma) + \rho^2(1 - \sin 3\gamma)$$
$$+ \ldots + \rho^{n-1}(1 - \sin n\gamma), \qquad n\gamma < \pi/2.$$

Finally, the apparent hemispherical emissivity of the V-corrugated surface is

$$\epsilon_a = \frac{\epsilon}{\sin\gamma}\left[1 - \epsilon\sum_{k=1}^{n}\rho^{k-1}(1 - \sin k\gamma)\right], \qquad n < \pi/2\gamma.$$

Figure 6-11 shows the apparent hemispherical emissivity of V-corrugated surfaces as a function of opening angle for a number of flat-surface emissivities. Also shown in the figure is the normal emissivity (or absorptivity), which may also be calculated from equation (6.22) (left as an exercise). For example, for $\epsilon = 0.5$ and a groove opening angle of

$\gamma = 30°$, the apparent hemispherical emissivity (important for re-radiation losses) is 0.72, and the normal emissivity (important for solar energy collection) is 0.88. While the difference between these two values is not huge, the corrugated groove (*i*) helps to make the absorber plate more black, and (*ii*) substantially reduces the re-radiation losses (by $\simeq 20\%$ for the $\epsilon = 0.5$, $\gamma = 30°$ surface). More detail about the radiative properties of V-corrugated grooves may be found in the papers by Eckert and Sparrow [4], Sparrow and Lin [5], and Hollands [6], and the book by Sparrow and Cess [7].

Curved Surfaces with Specular Reflection Components

In all our examples we have only considered *idealized enclosures* consisting of *flat surfaces*, for which the mirror images necessary for specular view factor calculations are relatively easily determined. If some or all of the reflecting surfaces are *curved* then equations (6.18) and (6.20) remain valid, but the specular view factors tend to be much more difficult to obtain. Analytical solutions can be found only for relatively simple geometries, such as axisymmetric surfaces, but even then they tend to get very involved. The very simple case of cylindrical cavities (with and without specularly reflecting end plate) has been studied by Sparrow and coworkers [8–10] and by Perlmutter and Siegel [11]. The more involved case of conical cavities has been treated by Sparrow and colleagues [9, 10, 12] as well as Polgar and Howell [13], while spherical cavities have been addressed by Tsai and coworkers [14, 15] and Sparrow and Jonsson [16, 17]. Somewhat more generalized discussions on the determination of specular view factors for curved surfaces have been given by Plamondon and Horton [18] and by Burkhard and coworkers [19]. In view of the complexity involved in these evaluations, specular view factors for curved surfaces are probably most conveniently calculated by a statistical method, such as the Monte Carlo method, which will be discussed in detail in Chapter 19. A considerably more detailed discussion of thermal radiation from and within grooves and cavities is given in the book by Sparrow and Cess [7].

6.4 ELECTRICAL NETWORK ANALOGY

The electrical network analogy, first introduced in Section 5.4, may be readily extended to allow for partially specular reflectors. This possibility was first demonstrated by Ziering and Sarofim [20]. Expressing equations (6.12) and (6.15) for an idealized enclosure [i.e., an enclosure with finite surfaces of constant radiosity, exactly as was done in equation (6.19)], we can evaluate the nodal heat fluxes as

$$q_i = J_i - (1 - \rho_i^s) \left[\sum_{j=1}^{N} J_j \, F_{i-j}^s + H_{oi}^s \right], \quad i = 1, 2, \ldots, N. \tag{6.27}$$

Using the summation rule, equation (6.21), this relation may also be written as the

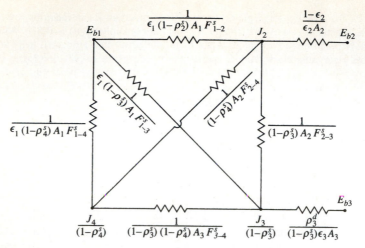

FIGURE 6-12

Electrical network equivalent for a four-surface enclosure (A_1 = specular, A_2 = diffuse, A_3 = partially diffuse and specular, A_4 = insulated, partially specular).

sum of net radiative interchange between any two surfaces,

$$
\begin{aligned}
q_i &= \sum_{j=1}^{N}\left[(1-\rho_j^s)J_i - (1-\rho_i^s)J_j\right]F_{i-j}^s - (1-\rho_i^s)H_{oi}^s \\
&= \sum_{j=1}^{N}\left[\frac{J_i}{1-\rho_i^s} - \frac{J_j}{1-\rho_j^s}\right](1-\rho_i^s)(1-\rho_j^s)F_{i-j}^s - (1-\rho_i^s)H_{oi}^s. \quad (6.28)
\end{aligned}
$$

Similarly, from equation (6.13),

$$
q_i = \frac{(1-\rho_i^s)\epsilon_i}{\rho_i^d}\left(E_{bi} - \frac{J_i}{1-\rho_i^s}\right). \quad (6.29)
$$

After multiplication with A_i these relations may be combined and written in terms of potentials $[E_{bi}$ and $J_i/(1-\rho_i^s)]$ and resistances as

$$
Q_i = \frac{E_{bi} - \dfrac{J}{1-\rho_i^s}}{\dfrac{\rho_i^d}{(1-\rho_i^s)\epsilon_i A_i}} = \sum_{j=1}^{N}\frac{\dfrac{J_i}{1-\rho_i^s} - \dfrac{J_j}{1-\rho_j^s}}{\dfrac{1}{(1-\rho_i^s)(1-\rho_j^s)A_i F_{i-j}^s}} - (1-\rho_i^s)A_i H_{oi}^s. \quad (6.30)
$$

Of course, this relation reduces to equation (5.46) for the case of purely diffuse surfaces ($\rho_i^s = 0$, $i = 1, 2, \ldots, N$). As an example, Fig. 6-12 shows the equivalent electrical network for an enclosure consisting of four surfaces: Surface A_1 is a specular reflector ($\rho_1^d = 0$), surface A_2 is a diffuse reflector ($\rho_2^s = 0$), surface A_3 has specular and diffuse reflectivity components, and surface A_4 (also partially specular) is insulated. Note that, unlike diffuse reflectivity, the specular reflectivity is *not* irrelevant for insulated surfaces.

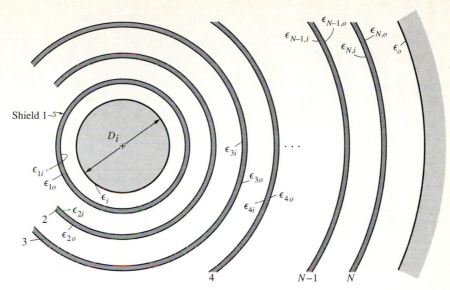

FIGURE 6-13
Concentric cylinders (or spheres) with N radiation shields between them.

6.5 RADIATION SHIELDS

In high-performance insulating materials it is common to suppress conductive and convective heat transfer by evacuating the space between two surfaces. This leaves thermal radiation as the dominant heat loss mode even for low-temperature applications such as insulation in cryogenic storage tanks. The radiation loss may be minimized by placing a multitude of closely spaced, parallel, highly reflective *radiation shields* between the surfaces. The radiation shields are generally made of thin metallic foils or, to reduce conductive losses further, of dielectric foils coated with metallic films. In either case radiation shields tend to be very specular reflectors.

A typical arrangement for N radiation shields between two concentric cylinders (or spheres concentric) is shown in Fig. 6-13. This geometry includes the case of parallel plates for large (and nearly equal) radii. Let the inner cylinder have temperature T_i, surface area A_i, and emissivity $\epsilon_i = 1 - \rho_i^s - \rho_i^d$. Similarly, each shield has temperature T_n (unknown), A_n, $\epsilon_{ni} = 1 - \rho_{ni}^s - \rho_{ni}^d$ (on its inner surface), and $\epsilon_{no} = 1 - \rho_{no}^s - \rho_{no}^d$ (on its outer surface). The last shield, A_N, faces the outer cylinder with T_o, A_o and $\epsilon_o = 1 - \rho_o^s - \rho_o^d$. The net radiative heat flux leaving A_i is, of course, equal to the heat flux going through each shield and to the one arriving at A_o. This net heat flux may be readily determined from the electrical network analogy, or by repeated application of the enclosure relations, equation (6.20). Since the case of concentric cylinders was already evaluated in Example 6.4 we prefer here the latter.

The net heat flux between any two of the concentric surfaces is then

$$Q = \frac{E_{bj} - E_{bk}}{R_{j-k}}, \quad R_{j-k} = \frac{1}{\epsilon_j A_j} + \frac{1}{\epsilon_k A_k} - \frac{1}{1-\rho_k^s}\left(\frac{1}{A_k} - \frac{\rho_k^s}{A_j}\right). \tag{6.31}$$

Therefore, we may write

$$Q R_{i-1i} = E_{bi} - E_{b1},$$
$$Q R_{1o-2i} = E_{b1} - E_{b2},$$
$$\vdots$$
$$Q R_{No-o} = E_{bN} - E_{bo}.$$

Adding all these equations eliminates all the unknown shield temperatures, and, after solving for the heat flux, we obtain

$$Q = \frac{E_{bi} - E_{bo}}{R_{i-1i} + \sum_{n=1}^{N-1} R_{no-n+1,i} + R_{No-o}}. \tag{6.32}$$

The resistances given in equation (6.31) may be simplified somewhat if surface A_k is either a purely diffuse reflector ($\rho_k^s = 0$), or a purely specular reflector ($1 - \rho_k^s = \epsilon_k$):

$$A_k \quad \text{diffuse}: \quad R_{j-k} = \frac{1}{\epsilon_j A_j} + \left(\frac{1}{\epsilon_k} - 1\right)\frac{1}{A_k}, \tag{6.33a}$$

$$A_k \quad \text{specular}: \quad R_{j-k} = \left(\frac{1}{\epsilon_j} + \frac{1}{\epsilon_k} - 1\right)\frac{1}{A_j}. \tag{6.33b}$$

Example 6.9.   A Dewar holding 4 liters of liquid helium at 4.2 K consists essentially of two concentric stainless steel ($\epsilon = 0.3$) cylinders of 50 cm length, and inner and outer diameters of $D_i = 10$ cm and $D_o = 20$ cm, respectively. The space between the cylinders is evacuated to a high vacuum to eliminate conductive/convective heat losses. Radiation shields are to be placed between the Dewar walls to reduce radiative losses to the point that it takes 24 hours for the 4-liter filling to evaporate if the Dewar is placed into an environment at 298 K. For the purpose of this example it may be assumed: (i) End losses as well as conduction/convection losses are negligible, (ii) the wall temperatures are at $T_i = 4.2$ K and $T_o = 298$ K, respectively, and (iii) radiation is one-dimensional. Thin plastic sheets coated on both sides with aluminum ($\epsilon = 0.05$) are available as shield material. Estimate the number of shields required. The heat of evaporation for helium at atmospheric pressure is $h_{fg} = 20.94$ J/g (which is a very low value compared with other liquids), the liquid density is $\rho_{He} = 0.125$ g/cm^3 [21].

Solution

The total heat required to evaporate 4 liters of liquid helium is

$$Q = \rho_{He} V_{He} h_{fg} = 0.125 \, \frac{\text{g}}{\text{cm}^3} \times 4 \, \text{liters} \times \frac{10^3 \, \text{cm}^3}{\text{liter}} \times 20.94 \, \frac{\text{J}}{\text{g}} = 10.47 \, \text{kJ.}$$

If all of this energy is supplied through radial radiation over a time period of 24 h, one infers that the heat flux in equation (6.32) must be held at or below $\dot{Q} = Q/24\,\text{h} = 10{,}470\,\text{J}/24\,\text{h} \times (1\,\text{h}/3600\,\text{s}) = 0.1212\,\text{W}$, or $q_1 = \dot{Q}/A_1 = 0.1212\,\text{W}/(\pi \times 10\,\text{cm} \times 50\,\text{cm}) = 7.71 \times 10^{-5}\,\text{W}/\text{cm}^2$. Therefore, the total resistance must, from equation (6.32), be a minimum of

$$A_1 R_{\text{tot}} = |E_{bi} - E_{bo}|/q_1 = 5.670 \times 10^{-12} \times |4.2^4 - 298^4|/7.71 \times 10^{-5}$$
$$= 580.0.$$

Since the shields are aluminum-coated plastic films they will certainly be specular reflectors. We note from equation (6.33) that the resistances are inversely proportional to shield area. Therefore, it is best to place the shields as close to the inner cylinder as possible. We will assume that the shields can be so closely spaced that $A_1 \simeq A_2 \simeq \ldots A_N = A_s = \pi D_s L$, with $D_s = 11\,\text{cm}$. Evaluating the total resistance from equations (6.32) and (6.33), we find

$$A_1 R_{\text{tot}} = \frac{1}{\epsilon_w} + \frac{1}{\epsilon_s} - 1 + \sum_{n=1}^{N-1}\left(\frac{2}{\epsilon_s} - 1\right)\frac{A_1}{A_s} + \frac{1}{\epsilon_s}\frac{A_1}{A_s} + \left(\frac{1}{\epsilon_w} - 1\right)\frac{A_1}{A^*},$$

where $\epsilon_w = 0.3$ is the emissivity of the (stainless steel) walls, $\epsilon_s = 0.05$ is the emissivity of the (aluminized) shields, and A^* depends on the nature of the stainless steel: If the steel is specular $A^* = A_s$, if it is diffuse $A^* = A_o$. We shall investigate both possibilities to see whether specularity is an important factor in this arrangement. Since the elements of the series in the last equation do not depend on n, we may solve for N as

$$N = \frac{A_1 R_{\text{tot}} - \dfrac{1}{\epsilon_w} - \left(\dfrac{1}{\epsilon_w} - 1\right)\dfrac{A_1}{A^*} - \left(\dfrac{1}{\epsilon_s} - 1\right)\left(1 - \dfrac{A_1}{A_s}\right)}{\left(\dfrac{2}{\epsilon_s} - 1\right)\dfrac{A_1}{A_s}}$$

$$= \frac{580.0 - \dfrac{1}{0.3} - \left(\dfrac{1}{0.3} - 1\right)\dfrac{10}{\{11 \text{ or } 20\}} - \left(\dfrac{1}{0.05} - 1\right)\left(1 - \dfrac{10}{11}\right)}{\left(\dfrac{2}{0.05} - 1\right)\dfrac{10}{11}}$$

$$= \begin{cases} 16.18, & A_o \text{ diffuse}, \\ 16.16, & A_o \text{ specular}. \end{cases}$$

Therefore, a minimum of 17 radiation shields would be required. While the performance of a specular outer cylinder is better than a diffuse one, the improvement is marginal. Note from equation (6.26) with $A^* = A_i$ (A_o specular) or $A^* = A_o$ (A_o diffuse) that, without radiation shields,

$$q_1 = \frac{|E_{bi} - E_{bo}|}{\dfrac{1}{\epsilon_w} + \left(\dfrac{1}{\epsilon_w} - 1\right)\dfrac{A_i}{A^*}} = \frac{5.670 \times 10^{-12}|4.2^4 - 298^4|}{\dfrac{1}{0.3} + \left(\dfrac{1}{0.3} - 1\right) \times \{1 \text{ or } \tfrac{1}{2}\}}$$

$$= \begin{cases} 9.94 \times 10^{-3}\,\text{W}/\text{cm}^2, & A_o \text{ diffuse}, \\ 7.89 \times 10^{-3}\,\text{W}/\text{cm}^2, & A_o \text{ specular}, \end{cases}$$

that is, the heat loss is approximately 100 times larger! Also, without shields the aspect ratio $A_i/A_o = 1/2$ deviates considerably from unity, making the differences between specular and diffuse cylinders more apparent.

6.6 SEMITRANSPARENT SHEETS (WINDOWS)

When we developed the governing relations for radiative heat transfer in an enclosure bounded by diffusely reflecting surfaces (Chapter 5) or by partially diffuse/partially specular reflectors (this chapter), we made allowance for external radiation to penetrate into the enclosure through holes and/or semitransparent surfaces (windows). While we have investigated some examples with external radiation entering through holes, only one (Example 5.8) has dealt with a simple semitransparent surface.

Radiative heat transfer in enclosures with semitransparent windows occurs in a number of important applications, such as solar collectors, externally irradiated specimens kept in a controlled atmosphere, furnaces with sight windows, and so on. We shall briefly outline in this section how such enclosures may be analyzed with equation (6.18) or (6.22). To this purpose we shall assume that properties of the semitransparent window are wavelength-independent (gray), that equation (6.1) describes the reflectivity (facing the inside of the enclosure), and that the transmissivity of the window also has specular (light is transmitted without change of direction) and diffuse (light leaving the window is perfectly diffuse) components.[3] Thus,

$$\rho + \tau + \alpha = \rho^s + \rho^d + \tau^s + \tau^d + \alpha = 1, \quad \epsilon = \alpha. \tag{6.34}$$

Further, we shall assume that radiation hitting the outside of the window has a collimated component q_{oc} (i.e., parallel rays coming from a single direction, such as sunshine) and a diffuse component q_{od} (such as sky radiation coming in from all directions with equal intensity). Making an energy balance for the *net radiative heat flux* from the semitransparent window into the enclosure leads to (cf. Fig. 6-14):

$$\begin{aligned} q(\mathbf{r}) &= q_{em} + q_{tr,in} - q_{abs} - q_{tr,out} \\ &= \epsilon(\mathbf{r})E_b(\mathbf{r}) + \tau^d(\mathbf{r})q_{oc}(\mathbf{r}) + \tau(\mathbf{r})q_{od}(\mathbf{r}) - \alpha(\mathbf{r})H(\mathbf{r}) - \tau(\mathbf{r})H(\mathbf{r}), \end{aligned} \tag{6.35}$$

where the specularly transmitted fraction of the collimated external radiation, $\tau^s q_{oc}$, has not been accounted for since it enters the enclosure in a nondiffuse fashion; it is accounted for in $H_o^s(\mathbf{r}')$ as part of the irradiation at another enclosure location $\mathbf{r}'$ (traveling there directly, or after any number of specular reflections). Using equation (6.34), equation (6.35) may also be written as

$$q(\mathbf{r}) = q_{out} - q_{in} = \left(\epsilon E_b + \tau^d q_{oc} + \tau q_{od} + \rho^d H + \rho^s H\right) - H, \tag{6.36}$$

where q_{in} is the energy falling onto the inside of the window coming from within the enclosure. The first four terms of q_{out} are diffuse and may be combined to form the radiosity

$$J(\mathbf{r}) = \epsilon E_b + \tau^d q_{oc} + \tau q_{od} + \rho^d H. \tag{6.37}$$

---

[3]It is unlikely that a realistic window has both specular and diffuse transmissivity components; rather its transmissivity will either be specular (clear windows) or diffuse (milky windows, glass blocks, etc.). We simply use the more general expression to make it valid for all types of windows.

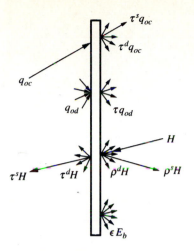

FIGURE 6-14
Energy balance for a semitransparent window.

Examination of equations (6.34) through (6.37) shows that they may be reduced to equations (6.10) through (6.12) if we introduce an apparent emissivity ϵ_a and an apparent blackbody emissive power $E_{b,a}$ as

$$\epsilon_a(\mathbf{r}) = \epsilon + \tau = 1 - \rho, \qquad (6.38a)$$

$$\epsilon_a E_{b,a}(\mathbf{r}) = \epsilon E_b + \tau^d q_{oc} + \tau q_{od}, \qquad (6.38b)$$

Thus, the semitransparent window is equivalent to an opaque surface with apparent emissivity ϵ_a and apparent emissive power $E_{b,a}$ (if the radiative properties are *gray*). Therefore, equations (6.18) and (6.22) remain valid as long as the emissivity and blackbody emissive powers of semitransparent surfaces are understood to be apparent values.

Example 6.10. A long hallway 3m wide by 4m high is lighted with a skylight that covers the entire ceiling. The skylight is double-glazed with an optical thickness of $\kappa d = 0.037$ per window plate. The floor and sides of the hallway may be assumed to be gray and diffuse with $\epsilon = 0.2$. The outside of the skylight is exposed to a clear sky, so that diffuse visible light in the amount of $q_{sky} = 20,000 \, \text{lm/m}^2$ is incident on the skylight. Direct sunshine also falls on the skylight in the amount of $q_{sun} = 80,000 \, \text{lm/m}^2$ (normal to the rays). For simplicity assume that the sun angle is $\theta_s = 36.87°$ as indicated in Fig. 6-15. Determine the amount of light incident on a point in the lower right-hand corner (also indicated in the figure) if (a) the skylight is clear, (b) the skylight is diffusing (with the same transmissivity and reflectivity).

Solution

From Fig. 3-31 for double glazing and $\kappa d = 0.037$ we find a hemispherical transmissivity (i.e., directionally averaged) of $\tau \simeq 0.70$, while for solar incidence with $\theta = 36.87°$ we have $\tau_\theta \simeq 0.75$. The hemispherical reflectivity of the skylight may be estimated by assuming that the reflectivity is the same as the one of a nonabsorbing glass. Then, from Fig. 3-30 $\rho_1 = \rho_1^s = 1 - \tau(\kappa d = 0) \simeq 1 - 0.75 = 0.25$. From equation (6.38) we find $\epsilon_{1,a} = 1 - \rho_1 = 0.75$ and, for a clear skylight, $\epsilon_{1,a} E_{b1,a} = 0 + 0 + \tau q_{sky}$ since $\tau^d = 0$,

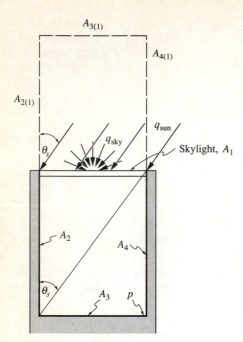

FIGURE 6-15
Geometry for a skylit hallway (Example 6.10).

and since there is no luminous emission from the window (or from any of the other walls, for that matter). Because of the special sun angle, direct sunshine falls only onto surface A_2, filling the entire wall, i.e., $H^s_{o2} = \tau_\theta q_{sun} \sin \theta_s$.

To determine the illumination at the point in the corner, we need to calculate the local irradiation H (in terms of *lumens*). This calculation, in turn, requires knowledge of the radiosity for all the surfaces of the hallway (for the skylight it is already known as $J_1 = \epsilon_{1,a} E_{b1,a} = \tau q_{sky}$, since $\rho^d_1 = 0$). To this purpose we shall approximate the hallway as a four-surface enclosure for which we shall calculate the average radiosities. Based on these radiosities we may then calculate the local irradiation for a point from equation (6.15). While equation (6.22) is most suitable for heat transfer calculations, we shall use equation (6.19) for this example since radiosities are more useful in lighting calculations.[4] Therefore, for $i = 2, 3$, and 4,

$$J_2 = \rho_2 (J_1 F^s_{2-1} + J_2 F^s_{2-2} + J_3 F^s_{2-3} + J_4 F^s_{2-4}) + H^s_{o2},$$
$$J_3 = \rho_3 (J_1 F^s_{3-1} + J_2 F^s_{3-2} + J_3 F^s_{3-3} + J_4 F^s_{3-4}),$$
$$J_4 = \rho_4 (J_1 F^s_{4-1} + J_2 F^s_{4-2} + J_3 F^s_{4-3} + J_4 F^s_{4-4}).$$

The necessary view factors are readily calculated from the crossed-strings method:

$$F^s_{2-1} = F_{2-1} = \frac{3 + 4 - 5}{2 \times 4} = 0.25, \qquad F^s_{2-2} = 0,$$

---

[4]If equation (6.22) is used the resulting heat fluxes are converted to radiosities using equation (6.13), or $J = -\rho^d q/\epsilon$ (since $E_b = 0$).

$$F_{2-3}^s = F_{2-3} + \rho_1 F_{2(1)-3} = 0.25 + 0.25 \times \frac{8+5-(4+\sqrt{73})}{2 \times 4}$$
$$= 0.25(1+0.05700) = 0.26425,$$

$$F_{2-4}^s = F_{2-4} + \rho_1 F_{2(1)-4} = 0.5 + 0.25 \times \frac{3+\sqrt{73}-2 \times 5}{2 \times 4}$$
$$= 0.5 + 0.25 \times 0.19300 = 0.54825,$$

$$F_{3-1}^s = F_{3-1} = \frac{2 \times 5 - 2 \times 4}{2 \times 3} = 0.33333,$$

$$F_{3-2}^s = \frac{A_2}{A_3} F_{2-3}^s = \frac{4}{3} \times 0.26425 = 0.35233,$$

$$F_{3-3}^s = \rho_1 F_{3(1)-3} = 0.25 \times \frac{2 \times \sqrt{73} - 2 \times 8}{2 \times 3} = 0.25 \times 0.18133 = 0.04533,$$

$$F_{3-4}^s = F_{3-2}^s = 0.35233,$$

$$F_{4-1}^s = F_{2-1}^s = 0.2500, \quad F_{4-2}^s = F_{2-4}^s = 0.54825$$

$$F_{4-3}^s = F_{2-3}^s = 0.26425, \quad F_{4-4}^s = 0.$$

Therefore, after normalization with $\mathscr{J}_i = J_i/J_1$ and $\mathscr{H} = H_{o2}^s/J_1$, and with $\rho_2 = \rho_3 = \rho_4 = 1 - 0.2 = 0.8$.

$$\mathscr{J}_2 = 0.8(0.25 + 0 + 0.26425\mathscr{J}_3 + 0.54825\mathscr{J}_4) + \mathscr{H},$$
$$\mathscr{J}_3 = 0.8(0.33333 + 0.35233\mathscr{J}_2 + 0.04533\mathscr{J}_3 + 0.35233\mathscr{J}_4),$$
$$\mathscr{J}_4 = 0.8(0.25 + 0.54825\mathscr{J}_2 + 0.26425\mathscr{J}_3 + 0),$$

or

$$\mathscr{J}_2 - 0.21140\mathscr{J}_3 - 0.43860\mathscr{J}_4 = \mathscr{H} + 0.2,$$
$$-0.28186\mathscr{J}_2 + 0.96374\mathscr{J}_3 - 0.28186\mathscr{J}_4 = 0.26667,$$
$$-0.43860\mathscr{J}_2 - 0.21140\mathscr{J}_3 + \mathscr{J}_4 = 0.2.$$

Omitting the details of solving these three simultaneous equations, we find

$$\mathscr{J}_2 = 1.48978\mathscr{H} + 0.59051,$$
$$\mathscr{J}_3 = 0.66812\mathscr{H} + 0.62211,$$
$$\mathscr{J}_4 = 0.79466\mathscr{H} + 0.59051.$$

The irradiation onto the corner point is, from equation (6.15)

$$H_p = \sum_{j=1}^{4} J_j F_{p-j}^s = J_1\left(F_{p-1}^s + \mathscr{J}_2 F_{p-2}^s + \mathscr{J}_3 F_{p-3}^s + \mathscr{J}_4 F_{p-4}^s\right),$$

where the view factors may be determined from Configurations 10 and 11 in Appendix D (with $b \to \infty$, and multiplying by 2 since the strip tends to infinity in both directions):

$$F_{p-1}^s = F_{p-1} = \frac{1}{2}\frac{a}{\sqrt{a^2+c^2}} = \frac{1}{2} \times \frac{3}{5} = 0.3,$$

$$F_{p-2}^s = F_{p-2} + \rho_1 F_{p(1)-2} = F_{p-2} + \rho_1\left[F_{p(1)-2+2(1)} - F_{p(1)-2(1)}\right],$$

$$F_{p-2} = \frac{1}{2}\left(1 - \frac{c}{\sqrt{a^2 + c^2}}\right) = \frac{1}{2}\left(1 - \frac{3}{5}\right) = 0.2,$$

$$F_{p(1)-2(1)} = F_{p-2} = 0.2, \quad F_{p(1)-2+2(1)} = \frac{1}{2}\left(1 - \frac{3}{\sqrt{73}}\right) = 0.32444,$$

$$F_{p-2}^s = 0.2 + 0.25 \times (0.32444 - 0.2) = 0.23111,$$

$$F_{p-3}^s = \rho_1 F_{p(1)-3} = 0.25 \times \frac{1}{2} \times \frac{3}{\sqrt{73}} = 0.04389, \quad F_{p-4}^s = 0.5.$$

Therefore,

$$\mathcal{H}_p = \frac{H_p}{J_1} = 0.3 + 0.23111 \times (1.48978\mathcal{H} + 0.59051)$$

$$+ 0.04389 \times (0.66812\mathcal{H} + 0.62211) + 0.5 \times (0.79466\mathcal{H} + 0.59051)$$

$$= 0.77096\mathcal{H} + 0.75903.$$

Finally, for a clear window, $J_1 = \tau_1 q_{sky} = 0.7 \times 20,000 = 14,000\,\text{lx}$, and $H_{o2}^s = \tau_\theta q_{sun} \sin 36.87° = 0.75 \times 80,000 \times 0.6 = 36,000\,\text{lx}$, and

$$H_p = 0.77096 \times 36,000 + 0.75903 \times 14,000 = 38,381\,\text{lx}.$$

On the other hand, if the window has a diffusing transmissivity $\tau = \tau^d = 0.7$, then $H_{o2}^s = 0$ and, from equation (6.37), $J_1 = \tau(q_{sky} + q_{sun} \cos 36.87°) = 0.7 \times (20,000 + 80,000 \times 0.8) = 58,800\,\text{lx}$. This results in

$$H_p = 0.75903 \times 58,800 = 44,631\,\text{lx}.$$

For a diffusing window the light is more evenly distributed throughout the hallway, resulting in higher illumination at point p.

6.7 SOLUTION OF THE GOVERNING INTEGRAL EQUATION

As in the case for diffusely reflecting surfaces the methods of the previous sections require the radiosity to be constant over each subsurface, a condition rarely met in practice. More accurate results may be obtained by solving the governing integral equation, either equation (6.16) (to determine radiosity J) or equation (6.18) (to determine the unknown heat flux and/or surface temperature directly), by any of the methods outlined in Chapter 5. This is best illustrated by repeating Examples 5.9 to 5.11.

Example 6.11.   Consider two long parallel plates of width w as shown in Fig. 6-16. Both plates are isothermal at the (same) temperature T, and both have a gray, diffuse emissivity of ϵ. The reflectivity of the material is partly diffuse, partly specular, so that $\epsilon = 1 - \rho^s - \rho^d$. The plates are separated by a distance h and are placed in a large, cold environment. Determine the local radiative heat fluxes along the plate using numerical quadrature.

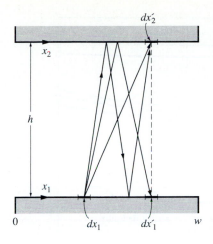

FIGURE 6-16
Radiative exchange between two long isothermal plates with specular reflection components.

Solution

From equation (6.18) we find, for location x_1 on the lower plate,

$$E_b - (1 - \rho^s)E_b \left[\int_0^w dF^s_{dx_1-dx_1'} + \int_0^w dF^s_{dx_1-dx_2'} \right]$$

$$= \frac{q(x_1)}{\epsilon} - \frac{\rho^d}{\epsilon} \left[\int_0^w q(x_1') \, dF^s_{dx_1-dx_1'} + \int_0^w q(x_2') \, dF^s_{dx_1-dx_2'} \right].$$

The necessary specular view factors are readily found from

$$dF^s_{dx_1-dx_1'} = \rho^s \, dF_{dx_1(2)-dx_1'} + (\rho^s)^3 \, dF_{dx_1(212)-dx_1'} + \cdots$$

$$dF^s_{dx_1-dx_2'} = dF_{dx_1-dx_2'} + (\rho^s)^2 \, dF_{dx_1(21)-dx_2'} + \cdots$$

The view factor between two infinitely long parallel strips of infinitesimal width and separated by a distance kh ($k = 1, 2, \ldots$) is given by Example 5.9 as

$$\frac{1}{2} \frac{(kh)^2 \, dx'}{[(kh)^2 + (x - x')^2]^{3/2}}.$$

Thus,

$$dF^s_{dx_1-dx_1'} + dF^s_{dx_1-dx_2'} = dF_{dx_1-dx_2'} + \rho^s \, dF_{dx_1(2)-dx_1'}$$

$$+ (\rho^s)^2 \, dF_{dx_1(21)-dx_2'} + \cdots$$

$$= \frac{1}{2} \sum_{k=1}^{\infty} (\rho^s)^{k-1} \frac{(kh)^2 \, dx'}{[(kh)^2 + (x_1 - x')^2]^{3/2}},$$

where we have made use of $x_1' = x_2' = x'$. This expression may be substituted into the governing integral equation. Realizing that, by symmetry, $q(x_1') = q(x_2') = q(x')$ and

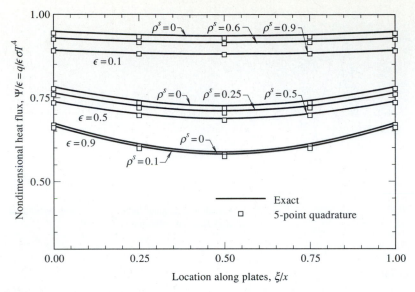

FIGURE 6-17
Local radiative heat flux on isothermal, parallel plates with diffuse and specular reflection components.

nondimensionalizing with $\xi = x/h$, $W = w/h$ and $\Psi = q(\xi)/E_b$, lead to

$$1 - (1 - \rho^s) \int_0^W \frac{1}{2} \sum_{k=1}^{\infty} \frac{(\rho^s)^{k-1} k^2 \, d\xi'}{[k^2 + (\xi - \xi')^2]^{3/2}}$$

$$= \frac{\Psi(\xi)}{\epsilon} - \frac{\rho^d}{\epsilon} \left[\int_0^W \Psi(\xi') \frac{1}{2} \sum_{k=1}^{\infty} \frac{(\rho^s)^{k-1} k^2 \, d\xi'}{[k^2 + (\xi - \xi')^2]^{3/2}} \right].$$

As in Example 5.10 this equation may be solved by numerical quadrature as

$$\Psi_i - \rho^d W \sum_{j=1}^{J} c_j \Psi_j f_{ij} = \epsilon \left[1 - (1 - \rho^s) W \sum_{j=1}^{J} c_j f_{ij} \right],$$

where Ψ_i is evaluated at J nodal positions ξ_i, $i = 1, 2, \ldots, J$, and the c_j are weight coefficients for the numerical integration. The f_{ij} are an abbreviation for the integration kernel,

$$f_{ij} = \frac{1}{2} \sum_{k=1}^{\infty} \frac{k^2 (\rho^s)^{k-1}}{[k^2 + (\xi_i - \xi_j)^2]^{3/2}}.$$

They must be evaluated by summing as many terms as necessary (decreasing as $(\rho^s)^{k-1}/k$ for large k). Results for the same simple $J = 5$ quadrature of Example 5.10 are given in Fig. 6-17, together with "exact" solutions (high-order quadrature). The results show that, for $W = w/h = 1$, the heat loss from the plates decreases if reflection is specular: Specular reflection traps emitted radiation somewhat more through repeated reflections between the plates.

Note that, if both surfaces are purely specular, the heat flux may be calculated directly (i.e., no solution of an integral equation is necessary). This calculation was first done for the parallel-plate case by Eckert and Sparrow [4]. In general, equation (6.18) is actually easier to solve than its diffuse-reflection counterpart if some or all of the surfaces are purely specular. However, the necessary specular view factors are generally much more difficult—if not impossible—to evaluate. Such a case arises, for example, for curved surfaces with multiple specular reflections. Since the specular view factors for such problems are most easily found from statistical methods, such as the Monte Carlo method (Chapter 19), it is usually best to solve the entire heat transfer problem using the Monte Carlo method.

6.8 CONCLUDING REMARKS

Before leaving the topic of specularly reflecting surfaces we want to discuss briefly under what circumstances the assumption of a partly diffuse, partly specular reflector is appropriate. The analysis for such surfaces is generally considerably more involved than for diffusely reflecting surfaces, as a result of the more difficult evaluation of specular view factors. On the other hand, the analysis is substantially less involved than for surfaces with more irregular reflection behavior (as will be discussed in the following chapter).

Examples 6.3 and 6.7 have shown that in fully closed configurations (without external irradiation) the heat fluxes show very little dependence on specularity. This is true for all closed configurations as long as there are no long and narrow channels separating surfaces of widely different temperatures (cf. Problems 6.1 and 6.21). Therefore, for most practical enclosures it should be sufficient to evaluate heat fluxes assuming purely diffuse reflectors—even though a number of surfaces may be decidedly specular. On the other hand, in open configurations, in long and narrow channels, in configurations with collimated irradiation—whenever there is a possibility of *beam channeling*—the influence of specularity can be very substantial and must be accounted for.

It is tempting to think of diffuse and specular reflection as not only *extreme* but also *limiting* cases: This leads to the thought that—if heat fluxes have been determined for purely diffuse reflection, and again for purely specular reflection—the heat flux for a surface with more irregular reflection behavior must always lie between these two limiting values. This consideration is true in most cases, in particular since most real surfaces tend to have a reflectivity maximum near the specular direction. However, there are cases when the actual heat flux is *not* bracketed by the diffuse and specular reflection models, particularly for directionally selective surfaces. As an example consider the local radiative heat flux from an isothermal groove, such as the one given by Fig. 6-10. Toor [23] has investigated this problem for diffuse reflectors, for specular reflectors, and for three different types of surface roughnesses analyzed with the Monte Carlo method, and his results are shown in Fig. 6-18. It is quite apparent that, near the vertex of the groove, diffuse and specular reflectors both seriously overpredict the heat loss. The reason is that, at grazing angles, rough surfaces tend to reflect strongly back into the direction of incidence.

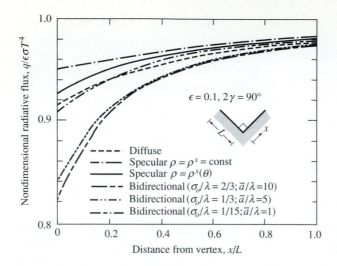

FIGURE 6-18
Local radiative heat flux from the surface of an isothermal V-groove for different reflection behavior; for all surfaces $2\gamma = 90°$ and $\epsilon = 0.1$; σ_o is root-mean-square optical roughness, $\bar{a}$ is a measure [22] for average distance between roughness peaks [23].

References

1. Sarofim, A. F., and H. C. Hottel: "Radiation Exchange Among Non-Lambert Surfaces," *ASME Journal of Heat Transfer*, vol. 88C, pp. 37–44, 1966.
2. Birkebak, R. C., E. M. Sparrow, E. R. G. Eckert, and J. W. Ramsey: "Effect of Surface Roughness on the Total and Specular Reflectance of Metallic Surfaces," *ASME Journal of Heat Transfer*, vol. 86, pp. 193–199, 1964.
3. Wylie, C. R.: *Advanced Engineering Mathematics*, 5th ed., McGraw-Hill, New York, 1982.
4. Eckert, E. R. G., and E. M. Sparrow: "Radiative Heat Exchange Between Surfaces with Specular Reflection," *International Journal of Heat and Mass Transfer*, vol. 3, pp. 42–54, 1961.
5. Sparrow, E. M., and S. L. Lin: "Absorption of Thermal Radiation in V-Groove Cavities," *International Journal of Heat and Mass Transfer*, vol. 5, pp. 1111–1115, 1962.
6. Hollands, K. G. T.: "Directional Selectivity, Emittance, and Absorptance Properties of Vee Corrugated Specular Surfaces," *Solar Energy*, vol. 7, no. 3, pp. 108–116, 1963.
7. Sparrow, E. M., and R. D. Cess: *Radiation Heat Transfer*, Hemisphere, New York, 1978.
8. Sparrow, E. M., L. U. Albers, and E. R. G. Eckert: "Thermal Radiation Characteristics of Cylindrical Enclosures," *ASME Journal of Heat Transfer*, vol. 84C, pp. 73–81, 1962.
9. Lin, S. H., and E. M. Sparrow: "Radiant Interchange Among Curved Specularly Reflecting Surfaces, Application to Cylindrical and Conical Cavities," *ASME Journal of Heat Transfer*, vol. 87C, pp. 299–307, 1965.
10. Sparrow, E. M., and S. L. Lin: "Radiation Heat Transfer at a Surface Having Both Specular and Diffuse Reflectance Components," *International Journal of Heat and Mass Transfer*, vol. 8, pp. 769–779, 1965.
11. Perlmutter, M., and R. Siegel: "Effect of Specularly Reflecting Gray Surface on Thermal Radiation Through a Tube and from Its Heated Wall," *ASME Journal of Heat Transfer*, vol. 85, pp. 55–62, 1963.
12. Sparrow, E. M., and V. K. Jonsson: "Radiant Emission Characteristics of Diffuse Conical Cavities," *Journal of the Optical Society of America*, vol. 53, pp. 816–821, 1963.
13. Polgar, L. G., and J. R. Howell: "Directional Thermal-Radiative Properties of Conical Cavities," *NASA TN D-2904*, 1965.
14. Tsai, D. S., F. G. Ho, and W. Strieder: "Specular Reflection in Radiant Heat Transport Across a Spherical Void," *Chemical Engineering Science–Genie Chimique*, vol. 39, pp. 775–779, 1984.
15. Tsai, D. S., and W. Strieder: "Radiation across a Spherical Cavity Having Both Specular and Diffuse Reflectance Components," *Chemical Engineering and Science*, vol. 40, no. 1, p. 170, 1985.

16. Sparrow, E. M., and V. K. Jonsson: "Absorption and Emission Characteristics of Diffuse Spherical Enclosures," *NASA TN D-1289*, 1962.

17. Sparrow, E. M., and V. K. Jonsson: "Absorption and Emission Characteristics of Diffuse Spherical Enclosures," *ASME Journal of Heat Transfer*, vol. 84, pp. 188–189, 1962.

18. Plamondon, J. A., and T. E. Horton: "On the Determination of the View Function to the Images of a Surface in a Nonplanar Specular Reflector," *International Journal of Heat and Mass Transfer*, vol. 10, no. 5, pp. 665–679, 1967.

19. Burkhard, D. G., D. L. Shealy, and R. U. Sexl: "Specular Reflection of Heat Radiation From an Arbitrary Reflector Surface to an Arbitrary Receiver Surface," *International Journal of Heat and Mass Transfer*, vol. 16, pp. 271–280, 1973.

20. Ziering, M. B., and A. F. Sarofim: "The Electrical Network Analog to Radiative Transfer: Allowance for Specular Reflection," *ASME Journal of Heat Transfer*, vol. 88, pp. 341–342, 1966.

21. Kropschot, R. H., B. W. Birmingham, and D. B. Mann (eds.): *Technology of Liquid Helium*, National Bureau of Standards, Monograph 111, Washington, D.C., 1968.

22. Beckmann, P., and A. Spizzichino: *The Scattering of Electromagnetic Waves from Rough Surfaces*, Macmillan, New York, 1963.

23. Toor, J. S.: "Radiant Heat Transfer Analysis Among Surfaces Having Direction Dependent Properties by the Monte Carlo Method," M.S. thesis, Purdue University, Lafayette, IN, 1967.

Problems

6.1 Two infinitely long black plates of width D are separated by a long, narrow channel, as indicated in the adjacent sketch. One plate is isothermal at T_1, the other is isothermal at T_2. The emissivity of the insulated channel wall is

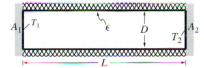

ϵ. Determine the radiative heat flux between the plates if the channel wall is (*a*) specular, (*b*) diffuse. For simplicity you may treat the channel wall as a single node. The diffuse case approximates the behavior of a *light guide*, a device used to pipe daylight into interior, windowless spaces.

6.2 Two infinitely long parallel plates of width w are spaced $h = 2w$ apart. Surface 1 has $\epsilon_1 = 0.2$ and $T_1 = 1000\,\text{K}$, Surface 2 has $\epsilon_2 = 0.5$ and $T_2 = 2000\,\text{K}$. Calculate the heat transfer on these plates if

(*a*) the surfaces are diffuse reflectors,

(*b*) the surfaces are specular.

6.3 A long duct has the cross-section of an equilateral triangle with side lengths $L = 1\,\text{m}$. Surface 1 is a diffuse reflector to which an external heat flux at the rate of $Q_1' = 1\,\text{kW/m}$ length of duct is supplied. Surfaces 2 and 3 are isothermal at $T_2 = 1000\,\text{K}$ and $T_3 = 500\,\text{K}$, respectively, and are purely specular reflectors with $\epsilon_1 = \epsilon_2 = \epsilon_3 = 0.5$.

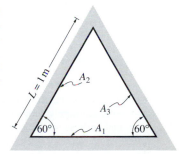

(a) Determine the average temperature of Surface 1, and the heat fluxes for Surfaces 2 and 3.

(b) How would the results change if Surfaces 2 and 3 were also diffusely reflecting?

6.4 Determine the temperature of surface A_2 in the axisymmetric configuration shown in the adjacent sketch, with the following data:

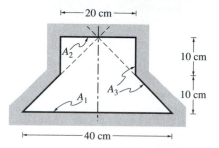

A_1 : $T_1 = 1000\,\text{K}$, $q_1 = -1\,\text{W/cm}^2$,

$\epsilon_1 = 0.6$ (diffuse reflector);

A_2 : $\epsilon_2 = 0.2$ (specular reflector);

A_3 : $q_3 = 0.0$ (perfectly insulated),

$\epsilon_3 = 0.3$ (diffuse reflector).

All surfaces are gray and emit diffusely.
Note: Some view factors may have to be approximated if integration is to be avoided

6.5 Consider the infinite groove cavity shown. The entire surface of the groove is isothermal at T and coated with a gray, diffusely emitting material with emissivity ϵ.

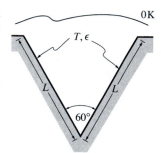

(a) Assuming the coating is a diffuse reflector, what is the total heat loss (per unit length) of the cavity?

(b) If the coating is a specular reflector, what is the total heat loss for the cavity?

6.6 Consider the infinite groove cavity shown in the adjacent sketch. The entire surface ($L = 2\,\text{cm}$) is isothermal at $T = 1000\,\text{K}$ and is coated with a gray material whose reflectivity may be idealized to consist of purely diffuse and specular components such that $\epsilon = \rho^d = \rho^s = \frac{1}{3}$. What is the total heat loss from the cavity? What is its apparent emissivity, defined by

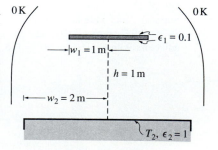

$$\epsilon_a = \frac{\text{total flux leaving cavity}}{\text{area of groove opening} \times E_b} ?$$

6.7 To calculate the net heat loss from a part of a spacecraft, this part may be approximated by an infinitely long black plate at temperature $T_2 = 600\,\text{K}$, as shown. Parallel to this plate is another (infinitely long) thin plate that is gray and reflects specularly with the same emissivity ϵ_1 on both sides. You may assume the surroundings to be black at $0\,\text{K}$. Calculate the net heat loss from the black plate.

6.8 A long isothermal plate (at T_1) is a gray, diffuse emitter (ϵ_1) and purely specular reflector, and is used to reject heat into space. To regulate the heat flux the plate is shielded by another (black) plate, which is perfectly insulated as illustrated in the adjacent sketch. Give an expression for heat loss as function of shield opening angle (neglect variations along plates). At what opening angle $0 \le \phi \le 180°$ does maximum heat loss occur?

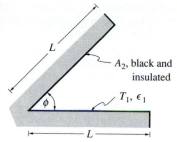

6.9 Reconsider the configuration of Problem 6.8, but assume the entire configuration to be isothermal at temperature T, and covered with a partially-diffuse, partially-specular material, $\epsilon = 1 - \rho^s - \rho^d$. Determine an expression for the heat lost from the cavity.

6.10 An infinitely long cylinder with a gray, diffuse surface ($\epsilon_1 = 0.8$) at $T_1 = 2000\,\text{K}$ is situated with its axis parallel to an infinite plane with $\epsilon_2 = 0.2$ at $T_2 = 1000\,\text{K}$ in a vacuum environment with a background temperature of $0\,\text{K}$. The axis of the cylinder is two diameters from the plane. Specify the heat loss from the cylinder when the plate surface is (*a*) gray and diffuse, or (*b*) gray and specular.

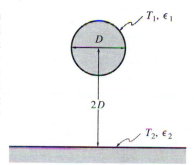

6.11 A typical space radiator may have a shape as shown in the adjacent sketch, i.e., a small tube to which are attached a number of flat plate fins, spaced at equal angle intervals. Assume that the central tube is negligibly small, and that a fixed amount of specularly-reflecting fin material is available ($\epsilon = \rho^s = 0.5$), to give (per unit length of tube) a total, one-sided fin area of $A' = N \times L$. Also assume the whole structure to be isothermal. Develop an expression for the total heat loss from the radiator as a function of the number of fins (each fin having length $L = A'/N$). Does an

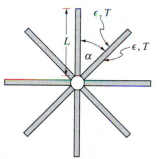

optimum exist? Qualitatively discuss the more realistic case of supplying a fixed amount of heat to the bases of the fins (rather than assuming isothermal fins).

6.12 An infinitely long corner of characteristic length $w =$ 1 m is a gray, diffuse emitter and purely specular reflector with $\epsilon = \rho^s = \frac{1}{2}$. The entire corner is kept at a constant temperature $T = 500$ K, and is irradiated externally by a line source of strength $S' = 20$ kW/m, located a distance w away from both sides of the corner, as shown in the sketch. What is the total heat flux Q' (per m length) to be supplied or extracted from the corner to keep the temperature at 500 K?

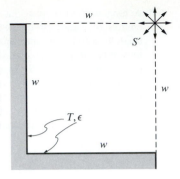

6.13 A long, thin heating wire, radiating energy in the amount of $S' = 300$ W/cm (per cm length of wire), is located between two long, parallel plates as shown in the adjacent sketch. The bottom plate is insulated and specularly reflecting with $\epsilon_2 = 1 - \rho_2^s = 0.2$, while the top plate is isothermal at $T_1 = 300$ K and diffusely reflecting with $\epsilon_1 = 1 - \rho_2^d = 0.5$. Determine the net radiative heat flux on the top plate.

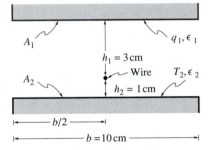

6.14 A long groove has diffuse walls that are insulated. All surfaces are gray with $\epsilon = 0.5$. A parallel beam of radiation, $q_0 = 1$ W/cm² enters the open end of the cavity in the center line direction, flooding the cavity opening completely.

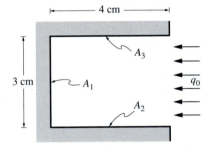

    (a) What is the apparent reflectivity of the groove (i.e., how much radiative energy is leaving it), and what is the temperature of surface A_1?

    (b) What are these values if surface A_1 is a specular reflector instead of diffuse?

6.15 Evaluate the normal emissivity for the V-corrugated surface shown in Fig. 6-10a. Hint: This is most easily calculated by determining the normal absorptivity, or the net heat flux on a cold groove irradiated by parallel light from the normal direction.

6.16 Redo 6.15 for an arbitrary off-normal direction $0 < \theta < \pi/2$ in a two-dimensional sense (i.e., what is the off-normal absorptivity for parallel incoming light whose propagation vector is in the same plane as all the surface normal, namely the plane of the paper in Fig. 6-10).

6.17 Consider the solar collector shown. The collector plate is gray and diffuse, while the insulated guard plates are gray and specularly reflecting. Sun strikes the cavity at an angle α ($\alpha < 45°$). How much heat is collected? Compare with a collector without guard plates. For what values of α is your theory valid?

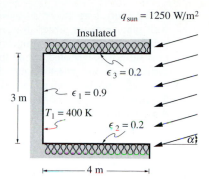

6.18 Reconsider the spacecraft of Problem 6.7. To decrease the heat loss from Surface 1 a specularly reflecting shield, of the same dimensions as the black surface and with emissivity $\epsilon_3 = 0.1$, is placed between the two plates. Determine the net heat loss from the black plate as a function of shield location. Where would you place the shield?

6.19 Repeat Problem 5.16, but assume steel and silver to be specular reflectors.

6.20 A rectangular cavity as shown is irradiated by a parallel-light source of strength $q_s = 1000$ W/m². The entire cavity is held at constant temperature $T = 300$ K and is coated with a gray material whose reflectivity may be idealized to consist of purely diffuse and specular components, such that $\epsilon = \rho^d = \rho^s = \frac{1}{3}$. How must the cavity be oriented toward the light source (i.e., what is ϕ?) so that there is no net heat flux on surface A_1?

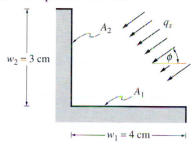

6.21 Two circular black plates of diameter D are separated by a long, narrow tubular channel, as indicated in the sketch next to Problem 6.1. One disk is isothermal at T_1, the other is isothermal at T_2. The channel wall is a perfect reflector, i.e., $\epsilon = 0$. Determine the radiative heat flux between the disks if the channel wall is (a) specular, (b) diffuse. For simplicity, you may treat the channel wall as a single node. If the channel is made of a transparent material, the specular arrangement approximates the behavior of an optical fiber; if the channel is filled with air, the diffuse case approximates the behavior of a *light guide,* a device used to pipe daylight into interior, windowless spaces.

6.22 An infinitely long corner piece as shown is coated with a material of (diffuse and gray) emissivity ϵ, and purely specular reflectivity. Calculate the variation of heat flux along the surfaces per unit area. Both surfaces are isothermal at T_1 and T_2, respectivley.

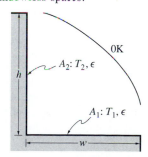

6.23 An infinitely long cavity as shown is coated with gray, specular materials ϵ_1 and ϵ_2 (but the materials are diffuse emitters). The vertical surface is insulated, while the horizontal surface is at constant temperature T_1. The surroundings may be assumed to be black at $0\,\mathrm{K}$. Specify the variation of the temperature along the vertical plate.

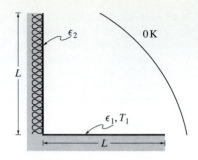

6.24 Consider the corner for Problem 6.20, which is irradiated by sunshine at an angle ϕ. Both plates are gray and specularly reflecting (emissivity $\epsilon = 1 - \rho^s$) and isothermal at T. Develop an expression for the local heat fluxes as a function of ϵ, T, x, y, q_s, and ϕ.

CHAPTER
7

RADIATIVE EXCHANGE BETWEEN NONIDEAL SURFACES

7.1 INTRODUCTION

In Chapter 6 we saw that, in certain situations, the directional nature of the reflectivity of surfaces can strongly influence radiative heat transfer rates. This effect occurs particularly in open configurations, in enclosures with long channels, or in applications with collimated irradiation. Since real surfaces are neither diffuse nor specular reflectors, the actual directional behavior may have substantial impact, as we saw from the data in Fig. 6-18. We also noted that solar collectors did not appear to perform very well because, in our gray analysis, the reradiation losses were rather large. However, experience has shown that reradiation losses can be reduced substantially if *selective surfaces* (i.e., strongly nongray) surfaces are used for the collector plates. Apparently, there are a substantial number of applications for which our idealized treatment (gray, diffuse—i.e., direction-independent—absorptivity and emissivity, gray and diffuse or specular reflectivity) is not sufficiently accurate. Actual surface properties deviate from our idealized treatment in a number of ways:

1. As seen from the discussion in Chapter 3, radiative properties can vary appreciably across the spectrum.

2. Spectral properties and, in particular, spectrally averaged properties may depend on the local surface temperature.

3. Absorptivity and reflectivity of a surface may depend on the direction of the *incoming* radiation.

4. Emissivity and reflectivity of a surface may depend on the direction of the *outgoing* radiation.

5. The components of polarization of incident radiation are reflected differently by a surface. Even for unpolarized radiation this difference can cause errors if many consecutive specular reflections take place. In the case of polarized laser irradiation this effect will always be important.

In this chapter we shall briefly discuss how nongray effects may be incorporated into the analyses of the previous chapters. We shall also develop the governing equation for the intensity leaving the surface of an enclosure with arbitrary radiative properties (spectrally and directionally), from which heat transfer rates may be calculated. This expression will be applied to a simple geometry to show how directionally irregular surface properties may be incorporated in the analysis.

7.2 RADIATIVE EXCHANGE BETWEEN NONGRAY SURFACES

In this section we shall consider radiative exchange between nongray surfaces that are directionally ideal: Their absorptivities and emissivities are independent of direction, while their reflectivity is idealized to consist of purely diffuse and/or specular components. For such a situation equation (6.22) becomes, on a spectral basis,

$$\sum_{j=1}^{N} \left[\delta_{ij} - (1 - \rho_{\lambda j}^s) F_{\lambda, i-j}^s \right] E_{b\lambda j} = \sum_{j=1}^{N} \left(\frac{\delta_{ij}}{\epsilon_{\lambda j}} - \frac{\rho_{\lambda j}^d}{\epsilon_{\lambda j}} F_{\lambda, i-j}^s \right) q_{\lambda j} + H_{o\lambda i}^s,$$

$$i = 1, 2, \ldots, N. \tag{7.1}$$

While diffuse view factors are purely geometric quantities and, therefore, never depend on wavelength, the specular view factors depend on the spectral variation of specular reflectivities. In principle, equation (7.1) may be solved for all the unknown $q_{\lambda j}$ and/or $E_{b\lambda j}$. This operation is followed by integrating the results over the entire spectrum, leading to

$$q_j = \int_0^\infty q_{\lambda j} \, d\lambda, \qquad E_{bj} = \int_0^\infty E_{b\lambda j} \, d\lambda. \tag{7.2}$$

In matrix form this may be written, similar to equation (6.23), as

$$\mathbf{A}_\lambda \cdot \mathbf{e}_{b\lambda} = \mathbf{C}_\lambda \cdot \mathbf{q}_\lambda + \mathbf{h}_{o\lambda}^s, \tag{7.3}$$

where $\mathbf{A}_\lambda$, $\mathbf{e}_{b\lambda}$, $\mathbf{C}_\lambda$, $\mathbf{q}_\lambda$ and $\mathbf{h}_{o\lambda}^s$ are defined as in Chapter 6, but on a spectral basis. Assuming that all the q_j are unknown (and all temperatures are known), equation (7.3) may be solved and integrated as

$$\mathbf{q} = \int_0^\infty \mathbf{q}_\lambda \, d\lambda = \int_0^\infty \mathbf{C}_\lambda^{-1} \cdot [\mathbf{A}_\lambda \cdot \mathbf{e}_{b\lambda} - \mathbf{h}_{o\lambda}^s] \, d\lambda. \tag{7.4}$$

A similar expression may be found if the heat flux is specified over some of the surfaces (with temperatures unknown). Branstetter [1] carried out integration of equation (7.4) for two infinite, parallel plates with platinum surfaces. In practice, accurate numerical evaluation of equation (7.4) is considered too complicated for most applications: For every wavelength used in the numerical integration (or *quadrature*) the matrix $\mathbf{C}$ needs to be inverted, which—for large numbers of nodes—is generally done by iteration. In addition, if one or more of the surfaces are specular reflectors, the specular view factors need to be recalculated for each wavelength (though not the diffuse view factors of which they are composed). Therefore, nongray effects are usually addressed by simplified models such as the *semigray approximation* or the *band approximation*.

Semigray Approximation

In some applications there is a natural division of the radiative energy within an enclosure into two or more distinct spectral regions. For example, in a solar collector the incoming energy comes from a high-temperature source with most of its energy below 3 μm, while radiation losses for typical collector temperatures are at wavelengths above 3 μm. In the case of laser heating and processing the incoming energy is monochromatic (at the laser wavelength), while reradiation takes place over the entire near- to midinfrared (depending on the workpiece temperature), etc. In such a situation equation (6.22) may be split into two sets of N equations each, one set for each spectral range, and with different radiative properties for each set. For example, consider an enclosure subject to external irradiation, which is confined to a certain spectral range "(1)". The surfaces in the enclosure, owing to their temperature, emit over spectral range "(2)".[1] Then from equation (6.22),

$$\sum_{j=1}^N \left[\frac{\delta_{ij}}{\epsilon_j^{(1)}} - \frac{\rho_j^{d(1)}}{\epsilon_j^{(1)}} F_{i-j}^{s(1)} \right] q_j^{(1)} + H_{oi}^s = 0, \tag{7.5a}$$

$$\sum_{j=1}^N \left[\frac{\delta_{ij}}{\epsilon_j^{(2)}} - \frac{\rho_j^{d(2)}}{\epsilon_j^{(2)}} F_{i-j}^{s(2)} \right] q_j^{(2)} = \sum_{j=1}^N \left[\delta_{ij} - (1 - \rho_j^{s(2)}) F_{i-j}^{s(2)} \right] E_{bj}, \tag{7.5b}$$

$$q_i = q_i^{(1)} + q_i^{(2)}, \qquad i = 1, 2, \ldots, N, \tag{7.5c}$$

where $\epsilon_j^{(1)}$ is the average emissivity for surface j over spectral interval (1), and so on.

---

[1] Note that spectral ranges "(1)" and "(2)" do not need to cover the entire spectrum and, indeed, they may overlap.

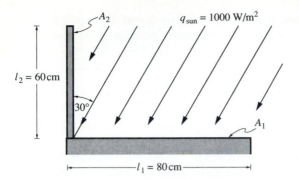

FIGURE 7-1
Solar collector geometry for Example 7.1.

Example 7.1.   A very long solar collector plate is to collect energy at a temperature of $T_1 = 350\,\mathrm{K}$. To improve its performance for off-normal solar incidence, a surface, which is highly reflective at short wavelengths, is placed next to the collector as shown in Fig. 7-1. For simplicity you may make the following assumptions: (*i*) The collector A_1 is isothermal and a diffuse reflector; (*ii*) the mirror A_2 is a specular reflector; (*iii*) the spectral properties of collector and mirror may be approximated as

$$\epsilon_1 = 1 - \rho_1^d = \begin{cases} 0.8, & \lambda < \lambda_c = 4\,\mu\mathrm{m}, \\ 0.1, & \lambda > \lambda_c, \end{cases}$$

$$\epsilon_2 = 1 - \rho_2^s = \begin{cases} 0.1, & \lambda < \lambda_c, \\ 0.8, & \lambda > \lambda_c, \end{cases}$$

and (*iv*) heat losses from the mirror by convection as well as all losses from the collector ends may be neglected. How much energy (per unit length) does the collector plate collect for a solar incidence angle of $30°$?

Solution

From equation (7.5) we find, with $F_{1-2}^s = F_{1-2}$, $F_{2-1}^s = F_{2-1}$, and $F_{1-1}^s = F_{2-2}^s = 0$, for range (1),

$$\frac{q_1^{(1)}}{\epsilon_1^{(1)}} + H_{o1}^s = 0,$$

$$-\left(\frac{1}{\epsilon_1^{(1)}} - 1\right)F_{2-1}q_1^{(1)} + \frac{q_2^{(1)}}{\epsilon_2^{(1)}} + H_{o2}^s = 0,$$

and for range (2),

$$\frac{q_1^{(2)}}{\epsilon_1^{(2)}} = E_{b1} - \epsilon_2^{(2)}F_{1-2}E_{b2},$$

$$-\left(\frac{1}{\epsilon_1^{(2)}} - 1\right)F_{2-1}q_1^{(2)} + \frac{q_2^{(2)}}{\epsilon_2^{(2)}} = -F_{2-1}E_{b1} + E_{b2}.$$

Eliminating E_{b2} from the last two equations, we find

$$\left[\frac{1}{\epsilon_1^{(2)}} - \left(\frac{1}{\epsilon_1^{(2)}} - 1\right)\epsilon_2^{(2)}F_{1-2}F_{2-1}\right]q_1^{(2)} + F_{1-2}q_2^{(2)} = (1 - \epsilon_2^{(2)}F_{1-2}F_{2-1})E_{b1}.$$

Multiplying the second equation for range (1) by $\epsilon_2^{(1)}F_{1-2}$ results in

$$-\left(\frac{1}{\epsilon_1^{(1)}}-1\right)\epsilon_2^{(1)}F_{1-2}F_{2-1}q_1^{(1)}+F_{1-2}q_2^{(1)}=-\epsilon_2^{(1)}F_{1-2}H_{o2}^s.$$

Adding the last two equations and using $q_2 = q_2^{(1)}+q_2^{(2)}=0$ then leads to

$$\left[\frac{1}{\epsilon_1^{(2)}}-\left(\frac{1}{\epsilon_1^{(2)}}-1\right)\epsilon_2^{(2)}F_{1-2}F_{2-1}\right]q_1^{(2)}=\left(\frac{1}{\epsilon_1^{(1)}}-1\right)\epsilon_2^{(1)}F_{1-2}F_{2-1}q_1^{(1)}$$

$$-\epsilon_2^{(1)}F_{1-2}H_{o2}^s+\left(1-\epsilon_2^{(2)}F_{1-2}F_{2-1}\right)E_{b1},$$

or, with $q_1^{(1)}=-\epsilon_1^{(1)}H_{o1}^s$,

$$q_1 = q_1^{(1)}+q_1^{(2)}$$

$$= \frac{(1-\epsilon_2^{(2)}F_{1-2}F_{2-1})E_{b1}-(1-\epsilon_1^{(1)})\epsilon_2^{(1)}F_{1-2}F_{2-1}H_{o1}^s-\epsilon_2^{(1)}F_{1-2}H_{o2}^s}{1/\epsilon_1^{(2)}-\left(1/\epsilon_1^{(2)}-1\right)\epsilon_2^{(2)}F_{1-2}F_{2-1}}-\epsilon_1^{(1)}H_{o1}^s.$$

From Example 6.5 we have

$$H_{o2}^s = q_{sun}\sin\varphi = 1000\times\sin 30° = 800\,\text{W/m}^2,$$

$$H_{o1}^s = q_{sun}\left[\cos\varphi+\rho_2^{s(1)}\sin\varphi\,(l_2/l_1)\right]$$

$$= 1000\left[\cos 30°+0.9\times\sin 30°(60/80)\right]=1204\,\text{W/m}^2.$$

With $F_{1-2}=\frac{1}{4}$, $F_{2-1}=\frac{1}{3}$, $F_{1-2}F_{2-1}=\frac{1}{12}$, and $E_{b1}=5.670\times 10^{-8}\times 350^4 = 851\,\text{W/m}^2$, q_1 may now be evaluated as

$$q_1 = \frac{\left(1-\frac{0.8}{12}\right)\times 851-\frac{0.2\times 0.1}{12}\times 1204-\frac{0.1}{4}\times 500}{\frac{1}{0.1}-\left(\frac{1}{0.1}-1\right)\times\frac{0.8}{12}}-0.8\times 1204$$

$$= 83-963 = -880\,\text{W/m}^2,$$

or a collection efficiency of 88%! In addition, surface A_2 remains much cooler than for the gray case (Example 6.5); from the first equation for region (2)

$$E_{b2} = \left(E_{b1}-\frac{q_1^{(2)}}{\epsilon_1^{(2)}}\right)\bigg/\epsilon_2^{(2)}F_{1-2}=\left(851-\frac{83}{0.1}\right)\bigg/\frac{0.8}{4}=114\,\text{W/m}^2,$$

or

$$T_2 = (E_{b2}/\sigma)^{1/4}=\left[114/5.670\times 10^{-8}\right]^{1/4}=209\,\text{K}.$$

Obviously, surface A_2 would heat up by convection from the surroundings. Surface emission from A_2 would then further improve the collection efficiency.

Thus, selective surfaces can have enormous impact on radiative heat fluxes in configurations with irradiation from high-temperature sources.

The semigray approximation is not limited to two distinct spectral regions. Each surface of the enclosure may be given a set of absorptivities and reflectivities, one value for each different surface temperature (with its different emission spectra).

Armaly and Tien [2] have indicated how such absorptivities may be determined. However, while simple and straightforward, the method can never become "exact," no matter how many different values of absorptivity and reflectivity are chosen for each surface.

Bobco and coworkers [3] have given a general discussion of the semigray approximation. The method has been applied to solar irradiation falling into a V-groove cavity with a spectrally selective, diffusely reflecting surface by Plamondon and Landram [4]. Comparison with exact (i.e., spectrally integrated) results proved the method to be very accurate. Shimoji [5] used the semigray approximation to model solar irradiation onto conical and V-groove cavities whose reflectivities had purely diffuse and specular components.

Band Approximation

Another commonly used method of solving equation (7.1) is the *band approximation*. In this method the spectrum is broken up into M bands, over which the radiative properties of *all* surfaces in the enclosure are constant. Therefore,

$$\sum_{j=1}^{N}\left[\delta_{ij} - (1 - \rho_j^{s(m)})F_{i-j}^{s(m)}\right]E_{bj}^{(m)} = \sum_{j=1}^{N}\left[\frac{\delta_{ij}}{\epsilon_j^{(m)}} - \frac{\rho_j^{d(m)}}{\epsilon_j^{(m)}}F_{i-j}^{s(m)}\right]q_j^{(m)} + H_{oi}^{s(m)},$$

$$i = 1, 2, \ldots, N, \quad m = 1, 2, \ldots, M; \quad (7.6a)$$

$$E_{bj} = \sum_{m=1}^{M}E_{bj}^{(m)}, \quad q_j = \sum_{m=1}^{M}q_j^{(m)}, \quad H_{oi}^s = \sum_{m=1}^{M}H_{oi}^{s(m)}. \quad (7.6b)$$

Equation (7.6) is, of course, nothing but a simple numerical integration of equation (7.1), using the trapezoidal rule with varying steps. This method has the advantage that the widths of the bands can be tailored to the spectral variation of properties, resulting in good accuracy with relatively few bands. For very few bands the accuracy of this method is similar to that of the semigray approximation, but is a little more cumbersome to apply. On the other hand, the *band approximation* can achieve any desired accuracy by using many bands.

Example 7.2.   Repeat Example 7.1 using the band approximation.

Solution

Since the emissivities in this example have been idealized to have constant values across the spectrum with the exception of a step at $\lambda = 4\,\mu$m, a two-band approximation ($\lambda < \lambda_c = 4\,\mu$m and $\lambda > 4\,\mu$m) will produce the "exact" solution (within the framework of the net radiation method). From equation (7.6)

$$E_{b1}^{(m)} - \epsilon_2^{(m)}F_{1-2}E_{b2}^{(m)} = \frac{q_1^{(m)}}{\epsilon_1^{(m)}} + H_{o1}^{s(m)},$$

$$-F_{2-1}E_{b1}^{(m)} + E_{b2}^{(m)} = -\left(\frac{1}{\epsilon_1^{(m)}} - 1\right)F_{2-1}q_1^{(m)} + \frac{q_2^{(m)}}{\epsilon_2^{(m)}} + H_{o2}^{s(m)}, \quad m = 1, 2,$$

where $E_{bi}^{(1)} = \int_0^{\lambda_c} E_{b\lambda i}\, d\lambda = f(\lambda_c T_i)E_{bi}$, $E_{bi}^{(2)} = [1 - f(\lambda_c T_i)]E_{bi}$, etc. These are four equations in the six unknowns $q_1^{(m)}$, $q_2^{(m)}$, $E_{b2}^{(m)}$, $m = 1, 2$. Two more conditions are obtained from $q_2 = q_2^{(1)} + q_2^{(2)} = 0$ and $E_{b2}^{(1)} + E_{b2}^{(2)} = E_{b2} = \sigma T_2^4$. The problem is that $E_{b2}^{(m)}$ are nonlinear relations in T_2, making it impossible to find explicit relations for the desired $q_1 = q_1^{(1)} + q_1^{(2)}$. The system is solved by iteration, by solving for $q_i^{(m)}$:

$$q_1^{(m)} = \epsilon_1^{(m)}\left(E_{b1}^{(m)} - \epsilon_2^{(m)} F_{1-2} E_{b2}^{(m)} - H_{o1}^{s(m)}\right),$$

$$q_2^{(m)} = \epsilon_2^{(m)}\left[\left(\frac{1}{\epsilon_1^{(m)}} - 1\right)F_{2-1} q_1^{(m)} - F_{2-1} E_{b1}^{(m)} + E_{b2}^{(m)} - H_{o2}^{s(m)}\right],$$

$$m = 1, 2.$$

First, T_2 is guessed, from which the $E_{b2}^{(m)}$ may be evaluated. This computation is followed by determining the $q_1^{(m)}$, after which the $q_2^{(m)}$ can be calculated. If $q_2 > 0$, surface A_2 is too hot and its temperature is reduced and vice versa until the correct temperature is obtained. This calculation may be done by writing a simple computer code, resulting in $T_2 = 212\,\mathrm{K}$ and $q_1 = -867\,\mathrm{W/m}^2$. As expected, for the present example the band approximation offers little improvement while complicating the analysis. However, the band approximation is the method of choice if no distinct spectral regions are obvious and/or the spectral behavior of properties is more involved.

Dunkle and Bevans [6] applied the band approximation to the same problem as Branstetter [1] (infinite, parallel, tungsten plates) as well as to some other configurations, showing that the band approximation generally achieves accuracies of 2% and better with very few bands, while a gray analysis may result in errors of 30% or more.

7.3   DIRECTIONALLY NONIDEAL SURFACES

In the vast majority of applications the assumption of "directionally ideal" surfaces gives results of sufficient accuracy, i.e., surfaces may be assumed to be diffusely emitting and absorbing and to be diffusely and/or specularly reflecting (with the magnitude of reflectivity independent of incoming direction). However, that these results are not always accurate and that heat fluxes are not necessarily bracketed by the diffuse- and specular-reflection cases have been shown in Fig. 6-18 for V-grooves. There will be situations where (*i*) the directional properties, (*ii*) the geometrical considerations and/or (*iii*) the accuracy requirements are such that the directional behavior of radiation properties must be addressed.

If radiative properties with arbitrary directional behavior are to be accounted for, it is no longer possible to reduce the governing equation to an integral equation in a single quantity (the radiosity) that is a function of surface location only (but not of direction). Rather, applying conservation of energy to this problem produces an equation governing the directional intensity leaving a surface that is a function of both location on the enclosure surface and direction.

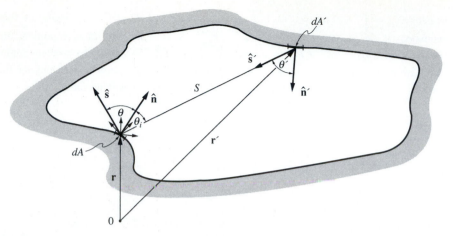

FIGURE 7-2
Radiative exchange in an enclosure with arbitrary surface properties.

The Governing Equation for Intensity

Consider the arbitrary enclosure shown in Fig. 7-2. The spectral radiative heat flux leaving an infinitesimal surface element dA' into the direction of $\hat{\mathbf{s}}'$ and arriving at surface element dA is

$$I_\lambda(\mathbf{r}', \lambda, \hat{\mathbf{s}}') \, dA'_p \, d\Omega = I_\lambda(\mathbf{r}', \lambda, \hat{\mathbf{s}}')(dA' \cos \theta') \frac{dA \cos \theta_i}{S^2}, \tag{7.7}$$

where $S = |\mathbf{r}' - \mathbf{r}|$ is the distance between dA' and dA, $\cos \theta' = \hat{\mathbf{s}}' \cdot \hat{\mathbf{n}}'$ is the cosine of the angle between the unit direction vector $\hat{\mathbf{s}}' = (\mathbf{r} - \mathbf{r}')/S$ and the outward surface normal $\hat{\mathbf{n}}'$ at dA' and, similarly, $\cos \theta_i = (-\hat{\mathbf{s}}') \cdot \hat{\mathbf{n}}$ at dA. This irradiation at dA coming from dA' may also be expressed, from equation (3.30), as

$$H'_\lambda(\mathbf{r}, \lambda, \hat{\mathbf{s}}') \, dA \, d\Omega_i = I_\lambda(\mathbf{r}, \lambda, \hat{\mathbf{s}}') \, dA \cos \theta_i \, \frac{dA' \cos \theta'}{S^2}. \tag{7.8}$$

Equating these two expressions, we find

$$I_\lambda(\mathbf{r}, \lambda, \hat{\mathbf{s}}') = I_\lambda(\mathbf{r}', \lambda, \hat{\mathbf{s}}'), \tag{7.9}$$

that is, the radiative intensity remains unchanged as it travels from dA' to dA.

The outgoing intensity at dA into the direction of $\hat{\mathbf{s}}$ consists of two contributions: Locally emitted intensity and reflected intensity. The locally emitted intensity is, from equation (3.1),

$$\epsilon'_\lambda(\mathbf{r}, \lambda, \hat{\mathbf{s}}) I_{b\lambda}(\mathbf{r}, \lambda).$$

The amount of irradiation at dA coming from dA' [equation (7.8)] that is reflected into a solid angle $d\Omega_o$ around the direction $\hat{\mathbf{s}}$ is, from the definition of the bidirectional reflection function, equation (3.31)

$$dI_\lambda(\mathbf{r}, \lambda, \hat{\mathbf{s}}) d\Omega_o = \rho_\lambda''(\mathbf{r}, \lambda, \hat{\mathbf{s}}', \hat{\mathbf{s}}) \left(H_\lambda'(\mathbf{r}, \lambda, \hat{\mathbf{s}}') d\Omega_i \right) d\Omega_o,$$

or

$$dI_\lambda(\mathbf{r}, \lambda, \hat{\mathbf{s}}) = \rho_\lambda''(\mathbf{r}, \lambda, \hat{\mathbf{s}}', \hat{\mathbf{s}}) I_\lambda(\mathbf{r}, \lambda, \hat{\mathbf{s}}') \cos\theta_i \, d\Omega_i$$

$$= \rho_\lambda''(\mathbf{r}, \lambda, \hat{\mathbf{s}}', \hat{\mathbf{s}}) I_\lambda(\mathbf{r}, \lambda, \hat{\mathbf{s}}') \frac{\cos\theta_i \cos\theta'}{S^2} dA'.$$

Integrating the reflected intensity over all incoming directions (or over the entire enclosure surface), and adding the locally emitted intensity, we find an expression for the outgoing intensity at dA as

$$I_\lambda(\mathbf{r}, \lambda, \hat{\mathbf{s}}) = \epsilon_\lambda'(\mathbf{r}, \lambda, \hat{\mathbf{s}}) I_{b\lambda}(\mathbf{r}, \lambda) + \int_{2\pi} \rho_\lambda''(\mathbf{r}, \lambda, \hat{\mathbf{s}}', \hat{\mathbf{s}}) I_\lambda(\mathbf{r}', \lambda, \hat{\mathbf{s}}') \cos\theta_i \, d\Omega_i$$

$$= \epsilon_\lambda'(\mathbf{r}, \lambda, \hat{\mathbf{s}}) I_{b\lambda}(\mathbf{r}, \lambda) + \int_A \rho_\lambda''(\mathbf{r}, \lambda, \hat{\mathbf{s}}', \hat{\mathbf{s}}) I_\lambda(\mathbf{r}', \lambda, \hat{\mathbf{s}}') \frac{\cos\theta_i \cos\theta'}{S^2} dA'. \quad (7.10)$$

Equation (7.10) is an integral equation for outgoing intensity ($\hat{\mathbf{n}} \cdot \hat{\mathbf{s}} > 0$) anywhere on the surface enclosure. Once a solution to equation (7.10) has been obtained (analytically, numerically or statistically; approximately or "exactly"), the net radiative heat flux is determined from

$$q_\lambda(\mathbf{r}, \lambda) = q_{\text{out}} - q_{\text{in}}$$

$$= \int_{\hat{\mathbf{n}} \cdot \hat{\mathbf{s}} > 0} I_\lambda(\mathbf{r}, \lambda, \hat{\mathbf{s}}) \cos\theta \, d\Omega - \int_{\hat{\mathbf{n}} \cdot \hat{\mathbf{s}} < 0} I_\lambda(\mathbf{r}, \lambda, \hat{\mathbf{s}}') \cos\theta_i \, d\Omega_i$$

$$= \int_{\hat{\mathbf{n}} \cdot \hat{\mathbf{s}} > 0} I_\lambda(\mathbf{r}, \lambda, \hat{\mathbf{s}}) \cos\theta \, d\Omega - \int_A I_\lambda(\mathbf{r}', \lambda, \hat{\mathbf{s}}') \frac{\cos\theta_i \cos\theta'}{S^2} dA', \quad (7.11)$$

or, equivalently, from

$$q_\lambda(\mathbf{r}, \lambda) = q_{\text{em}} - q_{\text{abs}} = \epsilon_\lambda E_{b\lambda} - \alpha_\lambda H_\lambda$$

$$= \int_{\hat{\mathbf{n}} \cdot \hat{\mathbf{s}} > 0} \epsilon_\lambda'(\mathbf{r}, \lambda, \hat{\mathbf{s}}) \cos\theta \, d\Omega \, I_{b\lambda}(\mathbf{r}, \lambda)$$

$$- \int_A \alpha_\lambda'(\mathbf{r}, \lambda, \hat{\mathbf{s}}') I_\lambda(\mathbf{r}', \lambda, \hat{\mathbf{s}}') \frac{\cos\theta_i \cos\theta'}{S^2} dA'. \quad (7.12)$$

Both forms of equation (7.10) (solid angle and area integration) may be employed, depending on the problem at hand. For example, if dA is a diffuse emitter and reflector then, from equation (3.36), $\rho_\lambda''(\mathbf{r}, \lambda, \hat{\mathbf{s}}', \hat{\mathbf{s}}) = \rho_\lambda'(\mathbf{r}, \lambda)/\pi$ and, from equation (5.19), $I_\lambda(\mathbf{r}, \lambda, \hat{\mathbf{s}}) = J_\lambda(\mathbf{r}, \lambda)/\pi$. If dA' is also diffuse, we obtain from the second form of equation (7.10)

$$J_\lambda(\mathbf{r}, \lambda) = \epsilon_\lambda(\mathbf{r}, \lambda) E_{b\lambda}(\mathbf{r}, \lambda) + \rho_\lambda(\mathbf{r}, \lambda) \int_A J_\lambda(\mathbf{r}', \lambda) \, dF_{dA-dA'}, \quad (7.13)$$

which is nothing but the spectral form of equation (5.24) without external irradiation.[2] Similarly, equation (7.11) reduces to

$$q_\lambda(\mathbf{r}, \lambda) = J_\lambda(\mathbf{r}, \lambda) - \int_A J_\lambda(\mathbf{r}', \lambda) \, dF_{dA-dA'}, \qquad (7.14)$$

the spectral form of equation (5.25).

On the other hand, if dA is a specular reflector the first form of equation (7.10) becomes more convenient: For a specular surface we have $\rho_\lambda'' = 0$ for all $\hat{\mathbf{s}}'$ except for $\hat{\mathbf{s}}' = \hat{\mathbf{s}}_s$, where $\hat{\mathbf{s}}_s$ is the "specular direction" from which a beam must originate in order to travel on into the direction of $\hat{\mathbf{s}}$ after specular reflection. For that direction $\rho_\lambda'' \to \infty$, and it is clear that the integrand of the integral in equation (7.10) will be nonzero only in the immediate vicinity of $\hat{\mathbf{s}}' = \hat{\mathbf{s}}_s$. In that vicinity $I_\lambda(\mathbf{r}', \lambda, \hat{\mathbf{s}}')$ varies very little and we may remove it from the integral. From the definition of the spectral, directional-hemispherical reflectivity, equation (3.35), and the law of reciprocity for the bidirectional reflectivity function, equation (3.33), we obtain

$$\int_{2\pi} \rho_\lambda''(\mathbf{r}, \lambda, \hat{\mathbf{s}}', \hat{\mathbf{s}}) I_\lambda(\mathbf{r}, \lambda, \hat{\mathbf{s}}') \cos\theta_i \, d\Omega_i = I_\lambda(\mathbf{r}', \lambda, \hat{\mathbf{s}}_s) \int_{2\pi} \rho_\lambda''(\mathbf{r}, \lambda, \hat{\mathbf{s}}', \hat{\mathbf{s}}) \cos\theta_i \, d\Omega_i$$

$$= I_\lambda(\mathbf{r}', \lambda, \hat{\mathbf{s}}_s) \int_{2\pi} \rho_\lambda''(\mathbf{r}, \lambda, -\hat{\mathbf{s}}, -\hat{\mathbf{s}}') \cos\theta_i \, d\Omega_i$$

$$= I_\lambda(\mathbf{r}', \lambda, \hat{\mathbf{s}}_s) \rho_\lambda'(\mathbf{r}, \lambda, -\hat{\mathbf{s}})$$

where $-\hat{\mathbf{s}}$ denotes an incoming direction, pointing toward dA, and $\rho_\lambda'(\mathbf{r}, \lambda, -\hat{\mathbf{s}})$ is the directional-hemispherical reflectivity. From the same Kirchhoff's law used to establish equation (3.33), it follows that $\rho_\lambda'(\mathbf{r}, \lambda, -\hat{\mathbf{s}}) = \rho_\lambda'(\mathbf{r}, \lambda, \hat{\mathbf{s}}_s)$ and

$$I_\lambda(\mathbf{r}, \lambda, \hat{\mathbf{s}}) = \epsilon_\lambda'(\mathbf{r}, \lambda, \hat{\mathbf{s}}) I_{b\lambda}(\mathbf{r}, \lambda) + \rho_\lambda'(\mathbf{r}, \lambda, \hat{\mathbf{s}}_s) I_\lambda(\mathbf{r}', \lambda, \hat{\mathbf{s}}_s). \qquad (7.15)$$

Example 7.3.   Consider a very long V-groove with an opening angle of $2\gamma = 90°$ and with optically smooth metallic surfaces with index of refraction $m = n - ik = 23.452(1 - i)$, i.e., the surfaces are specularly reflecting and their directional dependence obeys Fresnel's equations. The groove is isothermal at temperature T, and no external irradiation is entering the configuration. Calculate the local net radiative heat loss as a function of the distance from the vertex of the groove.

Solution

This is one of the problems studied by Toor [7], using the Monte Carlo method (the solid line in Fig. 6-18). The directional emissivity may be calculated from Fresnel's equations for a metal, equations (3.73) and (3.74), as

$$\epsilon'(\theta) = 1 - \rho'(\theta) = \frac{2n\cos\theta}{(n + \cos\theta)^2 + k^2} + \frac{2n\cos\theta}{(n\cos\theta + 1)^2 + (k\cos\theta)^2},$$

---

[2]External irradiation is readily included in equations (7.10) and (7.11) by replacing I_λ with $I_\lambda + I_{o\lambda}$ inside the integrals.

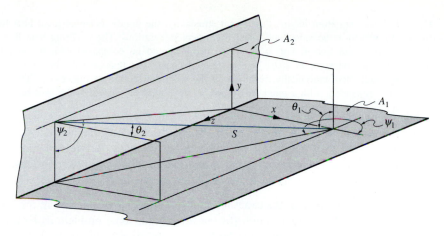

FIGURE 7-3
Isothermal V-groove with specularly reflecting, directionally dependent reflectivity (Example 7.3).

while the hemispherical emissivity follows from equation (3.75) or Fig. 3-10 as $\epsilon = 0.1$.

The present problem is particularly simple since the surfaces are specular reflectors and since the opening angle of the groove is 90° (cf. Fig. 7-3). Any radiation leaving surface A_1 traveling toward A_2 will be absorbed by A_2 or reflected out of the groove; none can be reflected back to A_1. This fact implies that all radiation arriving at A_1 is due to emission from A_2, which is a known quantity. Therefore, for those azimuthal angles ψ_2 pointing toward A_1 we have

$$-\frac{\pi}{2} < \psi_2 < \frac{\pi}{2}: \qquad I_2(\theta_2) = \epsilon'(\theta_2)I_b,$$

and the local heat flux follows from equation (7.12) as

$$q(x) = \epsilon E_b - \int_{2\pi} \epsilon'(\theta_1)I_2(\theta_2)\cos\theta_1 \, d\Omega_1$$

$$= \epsilon E_b - 2\int_{\psi_1=0}^{\pi/2}\int_{\theta_1=\theta_{1\min}(\psi_1)}^{\pi/2} \epsilon'(\theta_1)\epsilon'(\theta_2)I_b\cos\theta_1 \sin\theta_1 \, d\theta_1 \, d\psi_1$$

or

$$\frac{q(x)}{\epsilon E_b} = 1 - \frac{2}{\pi\epsilon}\int_{\psi_1=0}^{\pi/2}\int_{\theta_1=\theta_{1\min}(\psi_1)}^{\pi/2} \epsilon'(\theta_1)\epsilon'(\theta_2)\cos\theta_1 \sin\theta_1 \, d\theta_1 \, d\psi_1.$$

Here the limits on the integral express the fact that the solid angle, with which A_2 is seen from A_1, is limited. It remains to express $\theta_{1\min}$ as well as θ_2 in terms of θ_1 and ψ_1. From Fig. 7-3 it follows that

$$\cos\theta_1 = \frac{y}{S}, \qquad \cos\theta_2 = \frac{x}{S}, \qquad S\sin\theta_1 = \frac{x}{\cos\psi_1}.$$

From these three relations and the fact that the minimum value of θ_1 occurs when $y - L$, we find

$$\cos\theta_2 = \sin\theta_1 \cos\psi_1, \quad\text{and}\quad \theta_{1\min}(\psi_1) = \tan^{-1}\frac{x}{L\cos\psi_1}.$$

Using Fresnel's equation for the directional emissivity, the nondimensional local heat flux $q(x)/\epsilon E_b$ may now be calculated using numerical integration. The resulting heat flux is shown as the solid line in Fig. 6-18. This result should be compared with the simpler case of diffuse emission, or $\epsilon'(\theta) = \epsilon = 0.1 = $ const. For that case the integral above is readily integrated analytically, resulting in the dash-dotted line of Fig. 6-18. The two results are very close, with a maximum error of $\simeq 2\%$ near the vertex of the groove.

While the evaluation of the "exact" heat flux, using Fresnel's equations, was quite straightforward in this very simple problem, these calculations are normally much, much more involved than the diffuse-emission approximation. Before embarking on such extensive calculations it is important to ask oneself whether employing Fresnel's equations will lead to substantially different results for the problem at hand.

Few numerical solutions of the exact integral equations have appeared in the literature. For example, Hering and Smith [8] considered the same problem as Example 7.3, but for varying opening angles and for rough surface materials (with the bidirectional reflection function as given in an earlier paper [9]). Lack of detailed knowledge of bidirectional reflection distributions, as well as the enormous complexity involved in the solution of the integral equation (7.10), makes it necessary in practice to make additional simplifying assumptions or to employ a different approach, such as the Monte Carlo method (to be discussed in Chapter 19).

Net Radiation Method

It is possible to apply the net radiation method to surfaces with directionally non-ideal properties, although its application is considerably more difficult and restrictive. Breaking up the enclosure into N subsurfaces we may write equation (7.10), for $\mathbf{r}$ pointing to a location on subsurface A_i, as

$$I(\mathbf{r}, \lambda, \hat{\mathbf{s}}) = \epsilon'(\mathbf{r}, \lambda, \hat{\mathbf{s}})I_b(\mathbf{r}, \lambda) + \pi \sum_{j=1}^{N} \rho_j''(\mathbf{r}, \lambda, \hat{\mathbf{s}})I_j(\mathbf{r}, \lambda)F_{di-j}(\mathbf{r}), \tag{7.16}$$

where we have dropped the subscript λ for simplicity of notation, and where ρ_j'' and I_j are "suitable" average values between point $\mathbf{r}$ and surface A_j. Averaging equation (7.16) over A_i leads to

$$I_i(\lambda, \hat{\mathbf{s}}) = \epsilon_i'(\lambda, \hat{\mathbf{s}})I_{bi}(\lambda) + \pi \sum_{j=1}^{N} \rho_{ji}'(\lambda, \hat{\mathbf{s}})I_{ji}(\lambda)F_{i-j}, \quad i = 1, 2, \ldots, N. \tag{7.17}$$

Here I_{ji} is an average value of the intensity leaving surface A_j traveling toward A_i, and ρ_{ji}' is a corresponding value for the bidirectional reflection function. If we assume that the enclosure temperature and surface properties are known everywhere, then equation (7.17) has N unknown intensities I_{ji} ($j = 1, 2, \ldots, N$) for each subsurface A_i. Thus, if equation (7.17) is averaged over all the solid angles with which subsurface A_k is seen from A_i, it becomes a set of $N \times N$ equations in the N^2

unknown I_{ik}:

$$I_{ik}(\lambda) = \epsilon_{ik}(\lambda)I_{bi}(\lambda) + \pi \sum_{j=1}^{N} \rho_{jik}(\lambda)I_{ji}(\lambda)F_{i-j}, \quad i,k = 1,2,\ldots,N. \quad (7.18)$$

Here ρ_{jik} is an average value of the bidirectional reflection function for radiation traveling from A_j to A_k via reflection at A_i. For a diffusely emitting, absorbing and reflecting enclosure we have $\epsilon_{ik} = \epsilon_i$, $\pi\rho_{jik} = \rho_i$, and equation (7.18) becomes, with $I_{ji} = I_j = J_j/\pi$,

$$J_i = \epsilon_i E_{bi} + \rho_i \sum_{j=1}^{N} J_j F_{i-j}, \quad i = 1,2,\ldots,N, \quad (7.19)$$

which is identical to equations (5.30) and (5.31) (without external irradiation).

If the N subsurfaces are relatively small (as compared with the distance-squared between them), average properties ϵ_{ik} and ρ_{jik} may be obtained simply by evaluating ϵ' and ρ'' at the directions given by connecting the centerpoints of surface A_i with A_j and A_k. For larger subsurfaces a more elaborate averaging may be desirable. A discussion on that subject has been given by Bevans and Edwards [10].

Once the N^2 unknown I_{ik} have been determined, the average heat flux on A_i may be calculated from equations (7.18) and (7.11) or (7.12) as

$$q_i(\lambda) = \pi \sum_{k=1}^{N} I_{ik}(\lambda)F_{i-k} - \pi \sum_{j=1}^{N} I_{ji}(\lambda)F_{i-j} = \pi \sum_{j=1}^{N}(I_{ij} - I_{ji})F_{i-j} \quad (7.20a)$$

$$= \epsilon_i(\lambda)E_{bi}(\lambda) - \pi \sum_{j=1}^{N} \alpha_{ij}(\lambda)I_{ji}(\lambda)F_{i-j}, \quad i = 1,2,\ldots,N, \quad (7.20b)$$

where ϵ_i is the hemispherical emissivity of A_i and α_{ij} is the average absorptivity of subsurface A_i for radiation coming from A_j.

It is apparent from equations (7.10) and (7.18) that the net radiation method for directionally nonideal surfaces is valid (*i*) if each I_{bi} varies little over each subsurface A_i, (*ii*) if each I_{ik} varies little between any two positions on A_i and A_k, and (*iii*) if similar restrictions apply to ϵ_{ik}, α_{ij} and ρ_{jik}. Restrictions (*ii*) and (*iii*) are likely to be easily violated unless the surfaces are near-diffuse reflectors or are very small (as compared with the distance between them).

Equations (7.10) and (7.18) are valid for an enclosure with gray surface properties, or on a spectral basis. For nongray surface properties the governing equations are readily integrated over the spectrum using the methods outlined in the previous section.

To illustrate the difficulties associated with directionally nonideal surfaces, we shall consider one particularly simple example.

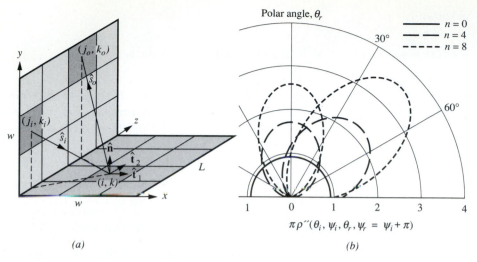

FIGURE 7-4
(a) Geometry for Example 7.4, (b) Bidirectional reflection function in plane of incidence for $\theta_i = 0°$ and $\theta_i = 45°$, for the material of Example 7.4.

Example 7.4. Consider the isothermal corner of finite length as depicted in Fig. 7-4a. The surface material is similar to the one of the infinitely long corner of the previous example, i.e., the absorptivity and emissivity obey Fresnel's equations with $m = n - ik = 23.452(1 - i)$, and a hemispherical emissivity of $\epsilon = 0.1$. However, in the present example we assume that the material is reflecting in a nonspecular fashion with a bidirectional reflection function of

$$\rho''(\hat{s}_i, \hat{s}_r) = \frac{\rho'(\hat{s}_i)}{\pi C_n(\hat{s}_i)}(1 + \hat{s}_s \cdot \hat{s}_r)^n,$$

where $\hat{s}_i$ is the direction of incoming radiation, $\hat{s}_s$ is the specular reflection direction (i.e., $\theta_s = \theta_i$, $\psi_s = \psi_i + \pi$), and $\hat{s}_r$ is the actual direction of reflection. This form of the bidirectional reflection function describes a surface that has a reflectivity maximum in the specular direction, and whose reflectivity drops off equally in all directions away from the specular direction (i.e., with changing polar angle and/or azimuthal angle.). Since the directional-hemispherical reflectivity must obey $\rho'(\hat{s}_i) = 1 - \epsilon'(\hat{s}_i)$, the function $C_n(\hat{s}_i)$ follows from equation (3.35) as

$$C_n(\hat{s}_i) = \frac{1}{\pi} \int_{2\pi} (1 + \hat{s}_s \cdot \hat{s}_r)^n \cos\theta_r \, d\Omega_r.$$

Determine the local radiative heat loss rates from the plates for the case that both plates are isothermal at the same temperature.

Solution

The direction vectors $\hat{s}$ may be expressed in terms of polar angle θ and azimuthal angle ψ, or $\hat{s} = \sin\theta(\cos\psi\hat{t}_1 + \sin\psi\hat{t}_2) + \cos\theta\hat{n}$, where $\hat{n}$ is the unit surface normal and $\hat{t}_1$ and $\hat{t}_2$ are two perpendicular unit vectors tangential to the surface. Therefore, the bidirectional reflection function may be written as

$$\rho''(\theta_i, \psi_i, \theta_r, \psi_r) = \frac{\rho'(\theta_i, \psi_i)}{\pi C_n(\theta_i)} [1 + \cos\theta_i \cos\theta_r - \sin\theta_i \sin\theta_r \cos(\psi_i - \psi_r)]^n, \quad (7.21a)$$

$$C_n(\theta_i) = \frac{1}{\pi} \int_0^{2\pi} \int_0^{\pi/2} (1 + \cos\theta_i \cos\theta$$

$$+ \sin\theta_i \sin\theta \cos\psi)^n \cos\theta \sin\theta \, d\theta \, d\psi. \qquad (7.21b)$$

The bidirectional reflection function within the plane of incidence ($\psi_r = \psi_i$ or $\psi_i + \pi$) is shown in Fig. 7-4b for two different incidence directions and three different values of n. Obviously, for $n = 0$ the surface reflects diffusely (but the *amount* of reflection, as well as absorption and emission, depends on direction through Fresnel's equation). As n grows, the surface becomes more specular, and purely specular reflection would be reached with $n \to \infty$. For this configuration and surface material we should like to determine the heat lost from the plates using the net radiation method.

As indicated in Fig. 7-4a we shall apply the net radiation method, equations (7.18) and (7.20), by breaking up each surface into $M \times N$ subsurfaces (M divisions in the x- and y-directions, N in the z-direction). Considering the intensity at node (i, k) on the bottom surface directed toward node (j_o, k_o) on the vertical wall, we find that equation (7.18) becomes, after division by I_b,

$$\Phi_{i,k \to j_o,k_o} = \frac{I_{i,k \to j_o,k_o}}{I_b}$$

$$= \epsilon_{i,k \to j_o,k_o} + \sum_{j_i=1}^{M} \sum_{k_i=1}^{N} \pi \rho_{j_i,k_i \to i,k \to j_o,k_o} F_{i,k \to j_i,k_i} \Phi_{j_i,k_i \to i,k}. \qquad (7.22)$$

In this relation we have made use of the fact that a node on the bottom surface can only see nodes on the side wall and vice versa. Also, by symmetry we have

$$\Phi_{i,k \to j_o,k_o} = \Phi_{j,k \to i_o,k_o} \quad \text{if} \quad j = i \quad \text{and} \quad i_o = j_o,$$

and

$$\Phi_{i,k \to j_o,k_o} = \Phi_{i,N+1-k \to j_o,N+1-k_o},$$

that is, the intensity must be symmetric to the two planes $x = y$ and $z = L/2$. We, therefore, have a total of $M \times (N/2)$ unknowns (assuming N to be even) and need to apply equation (7.22) for $i = 1, 2, \ldots, M$ and $k = 1, 2, \ldots, N/2$. To calculate the necessary ϵ' and ρ'' values, one must establish a number of polar and azimuthal angles. From Fig. 7-4a it follows that

$$(\cos\theta_i)_{i,k \to j_i,k_i} = \frac{y_{j_i}}{\sqrt{x_i^2 + y_{j_i}^2 + (z_k - z_{k_i})^2}},$$

$$(\cos\theta_r)_{i,k \to j_o,k_o} = \frac{y_{j_o}}{\sqrt{x_i^2 + y_{j_o}^2 + (z_k - z_{k_o})^2}}.$$

Using the values for $(\cos\theta_r)_{i,k \to j_o,k_o}$ one can readily calculate the directional emissivities $\epsilon_{i,k \to j_o,k_o} = 1 - \rho'(\cos\theta_r)$ from Fresnel's equation as given in Example 7.3. Similarly, $\rho'(\cos\theta_i)$ and $C_n(\cos\theta_i)$ are determined from Fresnel's equation and equation (7.21b),[3] respectively; and all values of $\rho_{j_i,k_i \to i,k \to j_o,k_o}$ follow from equation (7.21b). All necessary view factors may be calculated from equation (4.40), for arbitrarily oriented perpendicular

---

[3] For integer values of n the integration may be carried out analytically, either by hand or on a computer using a symbolic mathematics analyzer (the latter having been used here).

plates. For all view factors the opposing surfaces are of identical and constant size with $x_2 - x_1 = y_2 - y_1 = w/M$ and $z_1 = z_3 - z_2 = L/N$. Offsets x_1 and y_1 may vary between 0 and $(M - 1)w/M$ and z_2 between 0 and $(N - 1)L/N$. Thus, using symmetry and reciprocity, one must evaluate a total of $(M/2) \times M \times N$ view factors. In many of today's workstations and computers all different values of directional emissivity, the factor ρ'/C_n in the bidirectional reflection function, and all view factors may be calculated—once and for all—and stored (requiring memory allocation for often millions of numbers). The bidirectional reflection function itself depends on surface locations and on all possible incoming as well as all possible outgoing directions. Even after employing symmetry and reciprocity (for the bidirectional reflection function), this would require storing $[M \times (N/2)] \times [M \times N]^2/2 = (MN)^3/4$ numbers. Unless relatively few subdivisions are used (say $M, N < 10$), it will be impossible to precalculate and store values of the bidirectional reflection function; rather, part of it must be recalculated every time it is required.

The nondimensional intensities are now easily found from equation (7.22) by successive approximation: A first guess for the intensity field is made by setting $\Phi_{i,k \to j_o,k_o} = \epsilon_{i,k \to j_o,k_o}$. Improved values for $\Phi_{i,k \to j_o,k_o}$ are found by evaluating equation (7.22) again and again until the intensities have converged to within specified error bounds. The local net radiative heat flux may then be determined from equation (7.20b) as

$$\Psi_{i,k} = \frac{q_{i,k}}{\epsilon E_b} = 1 - \frac{1}{\epsilon} \sum_{j_i=1}^{M} \sum_{k_i=1}^{N} \epsilon_{i,k \to j_i,k_i} F_{i,k \to j_i,k_i} \Phi_{j_i,k_i \to i,k}.$$

Some representative results for the local radiative heat flux near $z = L/2$ (i.e., for $k = N/2$) are shown in Fig. 7-5 for the case of $w = L$ (square plates). Clearly, taking into consideration substantially different reflective properties has rather small effects on the local heat transfer rates. Obviously, as the surface becomes more specular (increasing n) the heat loss rates increase (since less radiation will be reflected back to the emitting surface), but the increases are very minor except for the region close to the vertex (and even there, they are less than 4%).

The directional distribution of the emissivity is just as important as that of the bidirectional reflection function: The curve labeled "diffuse" shows the case of diffuse emission and reflection, i.e., $\epsilon'(\hat{s}) = \alpha'(\hat{s}) = \epsilon = 0.1$ and $\pi\rho''(\hat{s}_i, \hat{s}_r) = \rho' = 1 - \epsilon = 0.9$. In contrast, the curve labeled "Fresnel, $n = 0$" corresponds to the case of $\epsilon'(\hat{s}) = \alpha'(\hat{s}) = 1 - \rho'(\hat{s})$ evaluated from Fresnel's equation and $\pi\rho(\hat{s}_i, \hat{s}_r) = \rho'(\hat{s}_i)$. All lines in Fig. 7-5 have been calculated by breaking up each surface into 20×20 subsurfaces. Also included are the data points for results obtained by breaking up each surface into only 2×2 (solid symbols) and 4×4 surfaces (open symbols). Local heat fluxes are predicted accurately with few subsurfaces, even for strongly nondiffuse reflection. Total heat loss is predicted even more accurately, with maximum errors of $< 0.6\%$ (2×2 subsurfaces) and $< 0.3\%$ (4×4 subsurfaces), respectively.

The results should be compared with those of Toor [7] for $w/L \to 0$, as shown in Fig. 6-18: The "diffuse" case of Fig. 7-5 virtually coincides with the corresponding case in Fig. 6-18, while the $n = 8$ case falls very close to the specular case with Fresnel-varying reflectivity of Toor (solid line in Fig. 6-18).

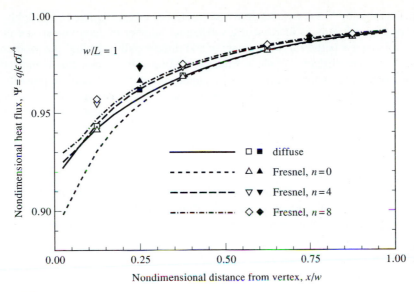

FIGURE 7-5
Nondimensional, local heat fluxes for the corner geometry of Example 7.4, for $w/L = 1$. Solid symbols: Surfaces are broken up into 2×2 subsurfaces; open symbols: 4×4 subsurfaces; lines: 20×20 subsurfaces.

For the present example at least, taking into account the directional behavior of emissivity and reflectivity is rarely justifiable in view of the additional complexity and computational effort required. Only if the radiative properties are known with great accuracy, and if heat fluxes need to be determined with similar accuracy, should this type of analysis be attempted. Similar statements may be made for most other configurations. For example, if Example 7.4 is recalculated for directly opposed parallel quadratic plates, the effects of Fresnel's equation and the bidirectional reflection function are even less: Heat fluxes for diffuse reflection—whether Fresnel's equation is used or not—differ by less than 0.6%, while differences due to the value of n in the bidirectional reflection function never exceed 0.2%. Only in configurations with collimated irradiation and/or strong beam-channeling possibilities should one expect substantial impact as a result of the directional variations of surface properties.

7.4   ANALYSIS FOR ARBITRARY SURFACE CHARACTERISTICS

The discussion in the previous two sections has demonstrated that the evaluation of radiative transfer rates in enclosures with nonideal surface properties, while relatively straightforward to formulate, is considerably more complex and time-consuming. If one considers nongray surface properties, the computational effort increases roughly by a factor of M if M spectral bands (band approximation) or M sets of property values (semigray approximation) are employed. In an analysis with directional properties for an enclosure with N subsurfaces, the computational effort is increased roughly by a factor of N (an enormous increase if a substantial number of subdivisions are made). If the radiative properties are both nongray and directionally varying, the problem becomes even more difficult. While it is relatively simple to combine the methods of the previous two sections for the analysis of an enclosure with such surface properties, to the author's knowledge this has not yet been done in any reported work. Few analytical solutions for such problems can be found (for the very simplest of geometries), and even standard numerical techniques may fail for nontrivial geometries; because of the four-dimensional character, huge matrices would have to be inverted. Therefore, such calculations are normally carried out with statistical methods such as the Monte Carlo method (to be discussed in detail in Chapter 19). For example, Toor [7] has studied the radiative interchange between simply arranged flat surfaces having theoretically determined directional surface properties; Modest and Poon [11] and Modest [12] evaluated the heat rejection and solar absorption rates of the U.S. Space Shuttle's heat rejector panels, using nongray and directional properties determined from experimental data. The validity and accuracy of several directional models have been tested and verified experimentally by Toor and Viskanta [13, 14]. They studied radiative transfer among three simply arranged parallel rectangles, comparing experimental results with a simple analysis employing (*i*) the semigray model, (*ii*) Fresnel's equation for the evaluation of directional properties, and (*iii*) reflectivities consisting of purely diffuse and specular parts. They found good agreement with experiment and concluded that, for the gold surfaces studied, (*i*) directional effects are more pronounced than nongray effects, (*ii*) in the presence of one or more diffusely reflecting surfaces the effects of specularity of other surfaces become unimportant.

Employing a combination of band approximation and the net radiation method has the disadvantage that (*i*) either large amounts of directional properties and/or view factors must be calculated repeatedly in the iterative solution process (making the method numerically inefficient), or (*ii*) large amounts of precalculated properties and/or view factors must be stored (requiring enormous amounts of computer storage). In addition, this method tends to have a voracious appetite for computer CPU time. On the other hand, it avoids the statistical scatter that is always present in Monte Carlo solutions. In light of today's rapid development in the computer field, with many small workstations boasting internal storage capabilities of 128 Mbyte and more, as well as enormous number-crunching capabilities, it appears that the methods discussed in this chapter are quickly becoming attractive alternatives to the Monte Carlo method.

References

1. Branstetter, J. R.: "Radiant Heat Transfer between Nongray Parallel Plates of Tungsten," *NASA TN D-1088*, 1961.
2. Armaly, B. F., and C. L. Tien: "A Note on the Radiative Interchange Among Nongray Surfaces," *ASME Journal of Heat Transfer*, vol. 92, pp. 178–179, 1970.
3. Bobco, R. P., G. E. Allen, and P. W. Othmer: "Local Radiation Equilibrium Temperatures in Semigray Enclosures," *Journal of Spacecraft and Rockets*, vol. 4, no. 8, pp. 1076–1082, 1967.
4. Plamondon, J. A., and C. S. Landram: "Radiant Heat Transfer from Nongray Surfaces with External Radiation. Thermophysics and Temperature Control of Spacecraft and Entry Vehicles," *Progress in Astronautics and Aeronautics*, vol. 18, pp. 173–197, 1966.

5. Shimoji, S.: "Local Temperatures in Semigray Nondiffuse Cones and V-grooves," *AIAA Journal*, vol. 15, no. 3, pp. 289–290, 1977.

6. Dunkle, R. V., and J. T. Bevans: "Part 3, A Method for Solving Multinode Networks and a Comparison of the Band Energy and Gray Radiation Approximations," *ASME Journal of Heat Transfer*, vol. 82, no. 1, pp. 14–19, 1960.

7. Toor, J. S.: "Radiant Heat Transfer Analysis Among Surfaces Having Direction Dependent Properties by the Monte Carlo Method," M.S. thesis, Purdue University, Lafayette, IN, 1967.

8. Hering, R. G., and T. F. Smith: "Surface Roughness Effects on Radiant Energy Interchange," *ASME Journal of Heat Transfer*, vol. 93, no. 1, pp. 88–96, 1971.

9. Hering, R. G., and T. F. Smith: "Apparent Radiation Properties of a Rough Surface," AIAA paper no. 69-622, 1969.

10. Bevans, J. T., and D. K. Edwards: "Radiation Exchange in an Enclosure with Directional Wall Properties," *ASME Journal of Heat Transfer*, vol. 87, no. 3, pp. 388–396, 1965.

11. Modest, M. F., and S. C. Poon: "Determination of Three-Dimensional Radiative Exchange Factors for the Space Shuttle by Monte Carlo," ASME paper no. 77-HT-49, 1977.

12. Modest, M. F.: "Determination of Radiative Exchange Factors for Three Dimensional Geometries with Nonideal Surface Properties," *Numerical Heat Transfer*, vol. 1, pp. 403–416, 1978.

13. Toor, J. S., and R. Viskanta: "A Critical Examination of the Validity of Simplified Models for Radiant Heat Transfer Analysis," *International Journal of Heat and Mass Transfer*, vol. 15, pp. 1553–1567, 1972.

14. Toor, J. S., and R. Viskanta: "Experiment and Analysis of Directional Effects on Radiant Heat Transfer," *ASME Journal of Heat Transfer*, vol. 94, pp. 459–466, November 1972.

Problems

7.1 Two identical circular disks of diameter $D = 1$ m are connected at one point of their periphery by a hinge. The configuration is then opened by an angle ϕ. Surface 1 is a diffuse reflector, but emits and absorbs according to

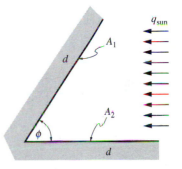

$$\epsilon'_\lambda = \begin{cases} 0.95\cos\theta, & \lambda \le 3\ \mu\text{m}, \\ 0.5, & \lambda > 3\ \mu\text{m}. \end{cases}$$

Disk 2 is black. Both disks are insulated. Assuming the opening angle to be $\phi = 60°$, calculate the average equilibrium temperature for each of the two disks, with solar radiation entering the configuration parallel to Disk 2 with a strength of $q_{\text{sun}} = 1000\ \text{W/m}^2$.

7.2 Repeat Problem 5.9 using the semigray approximation. Disk 1 is covered with a diffuse coating of black chrome (Fig. 3-32).

7.3 Repeat Example 5.8 for an absorber plate made of black chrome (Fig. 3-32) and a glass cover made of soda-lime glass (Fig. 3-27). Use the semigray or the band approximation.

7.4 Repeat Problem 5.19 for the case that the top of the copper shield is coated with white epoxy paint (Fig. 3-32).

7.5 A cubical enclosure has five of its surfaces maintained at 300 K, while the sixth is isothermal at 1200 K. The entire enclosure is coated with a material that emits and reflects diffusely with

$$\epsilon_\lambda = \begin{cases} 0.2, & 0 \le \lambda < 4\ \mu\text{m}, \\ 0.8, & 4\ \mu\text{m} < \lambda < \infty. \end{cases}$$

Determine the net radiative heat fluxes on the surfaces.

7.6 Repeat Problem 6.10 for the case that Surface 1 is coated with a material described in Problem 7.5.

7.7 Repeat Problem 6.12 for the case that the corner is coated with a diffusely emitting, specularly reflecting layer whose spectral behavior may be approximated by

$$\epsilon_\lambda = \begin{cases} 0.8, & 0 \leq \lambda < 3\,\mu\text{m}, \\ 0.2, & 3\,\mu\text{m} < \lambda < \infty. \end{cases}$$

The line source consists of a long filament at 2500 K inside a quartz tube, i.e., the source behaves like a gray body for $\lambda < 2.5\,\mu$m but has no emission beyond $2.5\,\mu$m.

7.8 Repeat Problem 6.12 for the case that the corner is cold (i.e., has negligible emission), and that the surface is gray and specularly reflecting with $\epsilon = \rho^s = 0.5$, but has a directional emissivity/absorptivity of

$$\epsilon'(\theta) = \epsilon_n \cos\theta.$$

Determine local and total absorbed radiative heat fluxes.

7.9 Consider two infinitely long, parallel plates of width $w = 1\,$m, spaced a distance $h = 0.5\,$m apart (see Configuration 32 in Appendix D). Both plates are isothermal at 1000 K and are coated with a gray material with a directional emissivity of

$$\epsilon'(\theta_i) = \alpha'(\theta_i) = 1 - \rho'(\theta_i) = \epsilon_n \cos\theta_i,$$

and a hemispherical emissivity of $\epsilon = 0.5$. Reflection is neither diffuse nor specular, but the bidirectional reflection function of the material is

$$\rho''(\theta_i, \theta_r) = \frac{3}{2\pi}\rho'(\theta_i)\cos\theta_r.$$

Write a small computer program to determine the total heat lost (per unit length) from each plate. Compare with the case for a diffusely emitting/reflecting surface.

CHAPTER
8

THE EQUATION OF RADIATIVE TRANSFER IN PARTICIPATING MEDIA

8.1 INTRODUCTION

In previous chapters we have looked at radiative transfer between surfaces that were separated by vacuum or by a transparent ("radiatively nonparticipating") medium. However, in many engineering applications the interaction of thermal radiation with an absorbing, emitting, and scattering ("radiatively participating") medium must be accounted for. Examples in the heat transfer area are the burning of any fuel (be it gaseous, liquid or solid; be it for power production, within fires, within explosions, etc.), rocket propulsion, hypersonic shock layers, ablation systems on reentry vehicles, nuclear explosions, plasmas in fusion reactors, and many more.

In the present chapter we shall develop the general relationships that govern the behavior of radiative heat transfer in the presence of an absorbing, emitting, and/or scattering medium. We shall begin by making a radiative energy balance, known

295

as the *Equation of Radiative Transfer*, which describes the radiative intensity field within the enclosure as a function of location (fixed by location vector **r**), direction (fixed by unit direction vector ŝ) and spectral variable (wavenumber η).[1] To obtain the net radiative heat flux crossing a surface element, we must sum the contributions of radiative energy irradiating the surface from all possible directions and for all possible wavenumbers. Therefore, integrating the equation of transfer over all directions and wavenumber leads to a *conservation of radiative energy* statement applied to an infinitesimal volume. Finally, this will be combined with a balance for all types of energy (including conduction and convection), leading to the *Overall Conservation of Energy* equation.

In the following three chapters we shall deal with the radiation properties of participating media, i.e., with how a substance can absorb, emit, and scatter thermal radiation. In Chapter 9 we discuss how a molecular gas can absorb and emit photons by changing its energy states, how to predict the radiation properties, and how to measure them experimentally. Chapter 10 is concerned with how small particles interact with electromagnetic waves—how they absorb, emit, and scatter radiative energy. Again, theoretical as well as experimental methods are covered. Finally, in Chapter 11 a very brief account is given of the radiation properties of solids and liquids that allow electromagnetic waves of certain wavelengths to penetrate into them for appreciable distances, known as semitransparent media.

8.2 RADIATIVE INTENSITY IN VACUUM

Before we discuss how radiative intensity is affected by absorption, emission, and scattering, it is important to understand how intensity penetrates through a vacuum. When discussing surface radiation we noticed that the concept of intensity had one advantage over emissive power, namely, that the emitted intensity from a black surface did not vary with direction. Within a medium, the definition of an emissive power is not possible since there is no surface to which to relate it. The intensity, defined as radiative energy transferred per unit time, solid angle, spectral variable, and area normal to the pencil of rays, is the most appropriate variable to describe the radiative transfer within a medium.

Consider radiative intensity penetrating at normal angle through a (fictitious) infinitesimal area dA_1, at location s_1 and time t_1, as shown in the sketch of Fig. 8-1. Based on the definition of intensity, we see that the amount of energy passing through dA_1 over a duration dt and spectral range $d\eta$ that will—a little later—fall onto the infinitesimal surface dA_2, is

$$I_\eta(s_1, t_1)\, dt\, d\Omega_{1\to 2}\, d\eta\, dA_1 = I_\eta(s_1, t_1)\, dt\, \frac{dA_2}{(s_2 - s_1)^2}\, d\eta\, dA_1,$$

---

[1] In our discussion of surface radiative transport we have used wavelength λ as the spectral variable throughout, largely to conform with the majority of other publications. However, for gases, frequency ν or wavenumber η are considerably more convenient to use. Again, to conform with the majority of the literature, we shall use wavenumber throughout this part.

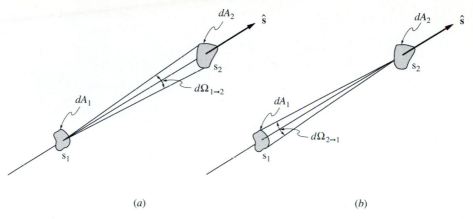

(a)   (b)

FIGURE 8-1
Radiative intensity in vacuum.

where $d\Omega_{1\to2}$ is the solid angle with which dA_2 is seen from an observer on dA_1. Since it takes the radiation until the time $t_2 = t_1 + (s_2 - s_1)/c$ to travel from s_1 to s_2, we can say that the energy going through dA_2 that is coming from dA_1 is

$$I_\eta(s_2, t_2)\, dt\, d\Omega_{2\to1}\, d\eta\, dA_2 = I_\eta(s_2, t_2)\, dt\, \frac{dA_1}{(s_2 - s_1)^2}\, d\eta\, dA_2.$$

Since both energies must be equal, we conclude that

$$I_\eta(s_2, t_1 + (s_2 - s_1)/c) = I_\eta(s_1, t_1). \tag{8.1}$$

Since the speed of light is so large in comparison with nearly every time scale in engineering problems, we may almost always assume that the radiative energy arrives "instantaneously" everywhere in the medium,[2] or

$$I_\eta(s_2) = I_\eta(s_1), \tag{8.2}$$

or

$$I_\eta(\hat{s}) = \text{const.} \tag{8.3}$$

Therefore, within a radiatively nonparticipating medium, the radiative intensity in any given direction is constant along its path. This property of the intensity makes it a most suitable quantity for the description of absorption, emission and scattering of energy within a medium, because any changes in intensity along any given path must be due to one or more of these phenomena.

---

[2]Using slightly different arguments this relation had already been established during the discussion on surface radiation between nonideal surfaces, equation (7.9).

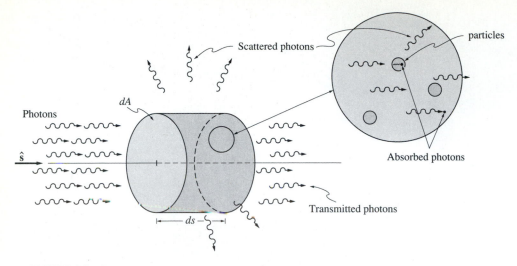

FIGURE 8-2
Attenuation of radiative intensity by absorption and scattering.

8.3   ATTENUATION BY ABSORPTION AND SCATTERING

If the medium through which radiative energy travels is "participating," then any incident beam will be attenuated by absorption and scattering while it travels through the medium, as schematically shown in Fig. 8-2. In the following we shall develop expressions for this attenuation for a light beam which travels within a pencil of rays into the direction $\hat{s}$. The present discussion will be limited to media with constant refractive index, i.e., media through which electromagnetic waves travel along straight lines [while a varying refractive index will bend the ray, as shown by Snell's law, equation (2.71), for an abrupt change].

Absorption

The absolute amount of absorption has been observed to be directly proportional to the magnitude of the incident energy as well as the distance the beam travels through the medium. Thus, we may write,

$$(dI_\eta)_{\text{abs}} = -\kappa_\eta I_\eta \, ds, \tag{8.4}$$

where the proportionality constant κ_η is known as the *(linear) absorption coefficient*, and the negative sign has been introduced since the intensity decreases. As will be discussed in the following chapter, the absorption of radiation in molecular gases depends also on the number of receptive molecules per unit volume, so that some researchers use a *mass absorption coefficient* or a *pressure absorption coefficient*, defined by

$$(dI_\eta)_{abs} = -\kappa_{\rho\eta} I_\eta \rho\, ds = -\kappa_{p\eta} I_\eta\, p\, ds. \tag{8.5}$$

The subscripts ρ and p are used here only to demonstrate the differences between the coefficients. The reader of scientific literature often must rely on the physical units to determine the coefficient used.

Integration of equation (8.4) over a geometric path s results in

$$I_\eta(s) = I_\eta(0) \exp\left(-\int_0^s \kappa_\eta ds\right) = I_\eta(0) e^{-\tau_\eta}, \tag{8.6}$$

where

$$\tau_\eta = \int_0^s \kappa_\eta\, ds \tag{8.7}$$

is the optical thickness (for absorption) through which the beam has traveled and $I_\eta(0)$ is the intensity entering the medium at $s = 0$. Note that the (linear) absorption coefficient is the inverse of the mean free path for a photon until it undergoes absorption. One may also define an *absorptivity* for the participating medium (for a given path within the medium) as

$$\alpha_\eta \equiv \frac{I_\eta(0) - I_\eta(s)}{I_\eta(0)} = 1 - e^{-\tau_\eta}. \tag{8.8}$$

Scattering

Attenuation by scattering, or "out-scattering" (away from the direction under consideration), is very similar to absorption, i.e., a part of the incoming intensity is removed from the direction of propagation, $\hat{s}$. The only difference between the two phenomena is that absorbed energy is converted into internal energy, while scattered energy is simply redirected and appears as augmentation along another direction (discussed in the next section), also known as "in-scattering." Thus, we may write

$$(dI_\eta)_{sca} = -\sigma_{s\eta} I_\eta\, ds, \tag{8.9}$$

where the proportionality constant $\sigma_{s\eta}$ is the *(linear) scattering coefficient* for scattering from the pencil of rays under consideration into all other directions. Again, scattering coefficients based on density or pressure may be defined. It is also possible to define an optical thickness for scattering, where the scattering coefficient is the inverse of the mean free path for scattering.

Total Attenuation

The total attenuation of the intensity in a pencil of rays by both absorption and scattering is known as *extinction*. Thus, an *extinction coefficient* is defined[3] as

$$\beta_\eta = \kappa_\eta + \sigma_{s\eta}. \tag{8.10}$$

The optical distance based on extinction is defined as

$$\tau_\eta = \int_0^s \beta_\eta \, ds. \tag{8.11}$$

As for absorption and scattering, the extinction coefficient is sometimes based on density or pressure.

8.4   AUGMENTATION BY EMISSION AND SCATTERING

A light beam traveling through a participating medium in the direction of $\hat{s}$ loses energy by absorption and by scattering away from the direction of travel. But at the same time it also gains energy by emission, as well as by scattering from other directions into the direction of travel $\hat{s}$.

Emission

The rate of emission from a volume element will be proportional to the magnitude of the volume. Therefore, the emitted intensity (which is the rate of emitted energy per unit area) along any path again must be proportional to the length of the path, and it must be proportional to the local energy content in the medium. Since, at thermodynamic equilibrium, the intensity everywhere must be equal to the blackbody intensity, it will be shown in Chapter 9, equation (9.15), that

$$(dI_\eta)_{em} = \kappa_\eta I_{b\eta} \, ds, \tag{8.12}$$

that is, the proportionality constant for emission is the same as for absorption. Similar to absorptivity, one may also define an *emissivity of an isothermal medium* as the amount of energy emitted over a certain path s that escapes into a given direction (without having been absorbed between point of emission and point of exit), as compared to the maximum possible. Combining equations (8.4) and (8.12) gives the complete equation of transfer for an absorbing-emitting (but not scattering) medium as

$$\frac{dI_\eta}{ds} = \kappa_\eta (I_{b\eta} - I_\eta), \tag{8.13}$$

---

[3]Care must be taken to distinguish the dimensional extinction coefficient β_η from the absorptive index, i.e., the imaginary part of the index of refraction complex k (sometimes referred to in the literature as the "extinction coefficient").

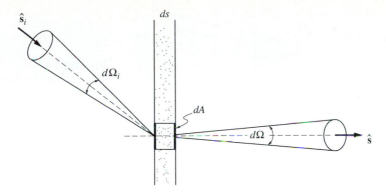

FIGURE 8-3
Redirection of radiative intensity by scattering.

where the first term of the right-hand side is augmentation due to emission and the second term is attenuation due to absorption. The solution to the equation of transfer for an isothermal gas layer of thickness s is

$$I_\eta(s) = I_\eta(0)e^{-\tau_\eta} + I_{b\eta}\left(1 - e^{-\tau_\eta}\right), \tag{8.14}$$

where the optical distance has been defined in equation (8.7). If only emission is considered, $I_\eta(0) = 0$, and the emissivity is defined as

$$\epsilon_\eta = I_\eta(s)/I_{b\eta} = 1 - e^{-\tau_\eta}, \tag{8.15}$$

which, as is the case with surface radiation, is identical to the expression for absorptivity.

Scattering

Augmentation due to scattering, or "in-scattering," has contributions from all directions and, therefore, must be calculated by integration over all solid angles. Consider the radiative heat flux impinging on a volume element $dV = dA\,ds$, from an infinitesimal pencil of rays in the direction $\hat{s}_i$ as depicted in Fig. 8-3. Recalling the definition for radiative intensity as energy flux per unit area normal to the rays, per unit solid angle, and per unit wavenumber interval, one may calculate the total spectral radiative heat flux impinging on dA from within the solid angle $d\Omega_i$ as

$$I_\eta(\hat{s}_i)(dA\,\hat{s}_i \cdot \hat{s})\,d\Omega_i\,d\eta.$$

This flux travels through dV for a distance $ds/\hat{s}_i \cdot \hat{s}$. Therefore, the total amount of energy scattered away from $\hat{s}_i$ is, according to equation (8.9),

$$\sigma_{s\eta}\left(I_\eta(\hat{s}_i)(dA\,\hat{s}_i \cdot \hat{s})\,d\Omega_i\,d\eta\right)\left(\frac{ds}{\hat{s}_i \cdot \hat{s}}\right) = \sigma_{s\eta}I_\eta(\hat{s}_i)\,dA\,d\Omega_i\,d\eta\,ds. \tag{8.16}$$

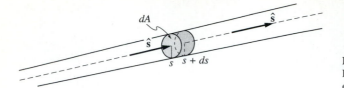

FIGURE 8-4
Pencil of rays for radiative energy balance.

Of this amount, the fraction $\Phi_\eta(\hat{s}_i, \hat{s}) \, d\Omega/4\pi$ is scattered into the cone $d\Omega$ around the direction $\hat{s}$. The function Φ_η is called the *scattering phase function* and describes the probability that a ray from one direction, $\hat{s}_i$, will be scattered into a certain other direction, $\hat{s}$. The constant 4π is arbitrary and is included for convenience [see equation (8.19) below].

The amount of energy flux from the cone $d\Omega_i$ scattered into the cone $d\Omega$ is then

$$\sigma_{s\eta} I_\eta(\hat{s}_i) \, dA \, d\Omega_i \, d\eta \, ds \frac{\Phi_\eta(\hat{s}_i, \hat{s})}{4\pi} \, d\Omega. \tag{8.17}$$

We can now calculate the energy flux scattered into the direction $\hat{s}$ from *all* incoming directions $\hat{s}_i$ by integrating:

$$(dI_\eta)_{\text{sca}}(\hat{s}) \, dA \, d\Omega \, d\eta = \int_{4\pi} \sigma_{s\eta} I_\eta(\hat{s}_i) \, dA \, d\Omega_i \, d\eta \, ds \, \Phi_\eta(\hat{s}_i, \hat{s}) \frac{d\Omega}{4\pi},$$

or

$$(dI_\eta)_{\text{sca}}(\hat{s}) = ds \frac{\sigma_{s\eta}}{4\pi} \int_{4\pi} I_\eta(\hat{s}_i) \, \Phi_\eta(\hat{s}_i, \hat{s}) \, d\Omega_i. \tag{8.18}$$

Returning to equation (8.17), we find that the amount of energy flux scattered from $d\Omega_i$ into all directions is

$$\sigma_{s\eta} I_\eta(\hat{s}_i) \, dA \, d\Omega_i \, d\eta \, ds \frac{1}{4\pi} \int_{4\pi} \Phi_\eta(\hat{s}_i, \hat{s}) \, d\Omega,$$

which must be equal to the amount in equation (8.16). We conclude that

$$\frac{1}{4\pi} \int_{4\pi} \Phi_\eta(\hat{s}_i, \hat{s}) \, d\Omega \equiv 1. \tag{8.19}$$

Therefore, if $\Phi_\eta = \text{const}$, i.e., if equal amounts of energy are scattered into all directions (called *isotropic scattering*), then $\Phi_\eta \equiv 1$. This is the reason for the inclusion of the factor 4π.

8.5   THE EQUATION OF TRANSFER

We can now make an energy balance on the radiative energy traveling in the direction of $\hat{s}$ within a small pencil of rays as shown in Fig. 8-4. The change in intensity is found by summing the contributions from emission, absorption, scattering away from

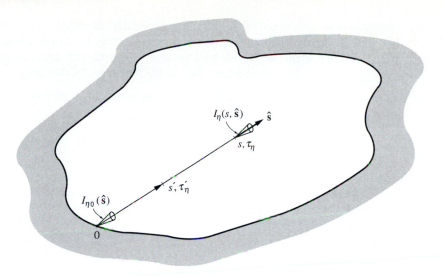

FIGURE 8-5
Enclosure for derivation of equation of transfer.

the direction $\hat{s}$, and scattering into the direction of $\hat{s}$, from equations (8.4), (8.9) (8.12), and (8.18) as

$$I_\eta(s+ds, \hat{s}, t+dt) - I_\eta(s, \hat{s}, t) = \kappa_\eta I_{b\eta}(s, t)\, ds - \kappa_\eta I_\eta(s, \hat{s}, t)\, ds$$

$$- \sigma_{s\eta} I_\eta(s, \hat{s}, t)\, ds + \frac{\sigma_{s\eta}}{4\pi} \int_{4\pi} I_\eta(\hat{s}_i)\, \Phi_\eta(\hat{s}_i, \hat{s})\, d\Omega_i\, ds. \quad (8.20)$$

The outgoing intensity may be developed into a truncated Taylor series, or

$$I_\eta(s+ds, \hat{s}, t+dt) = I_\eta(s, \hat{s}, t) + dt\frac{\partial I_\eta}{\partial t} + ds\frac{\partial I_\eta}{\partial s}, \quad (8.21)$$

so that equation (8.20) may be simplified to

$$\frac{1}{c}\frac{\partial I_\eta}{\partial t} + \frac{\partial I_\eta}{\partial s} = \kappa_\eta I_{b\eta} - \kappa_\eta I_\eta - \sigma_{s\eta} I_\eta + \frac{\sigma_{s\eta}}{4\pi} \int_{4\pi} I_\eta(\hat{s}_i)\, \Phi_\eta(\hat{s}_i, \hat{s})\, d\Omega_i, \quad (8.22)$$

where $c = ds/dt$ is the speed with which the radiation intensity propagates. In this equation all quantities may vary with location in space, time and wavenumber, while the intensity and the phase function also depend on direction $\hat{s}$ (and $\hat{s}_i$). Only the directional dependence, and only whenever necessary, has been explicitly indicated in this and the following equations, to simplify notation. Equation (8.22) is valid anywhere inside an arbitrary enclosure. Its solution requires knowledge of the intensity for each direction at some location s, usually the intensity entering the medium through or from the enclosure boundary into the direction of $\hat{s}$, as indicated in Fig. 8-5. We have not yet brought equation (8.22) into its most compact form so that the four different contributions to the change of intensity may be clearly identified. Again we can state that the time dependence of the radiative intensity may be almost always neglected in heat transfer applications. We have presented here the full equation for completeness, but will omit the transient term during the remainder of this book.

After introducing the extinction coefficient defined in equation (8.10), one may restate equation (8.22) as

$$\frac{dI_\eta}{ds} = \hat{\mathbf{s}} \cdot \boldsymbol{\nabla} I_\eta = \kappa_\eta I_{b\eta} - \beta_\eta I_\eta + \frac{\sigma_{s\eta}}{4\pi} \int_{4\pi} I_\eta(\hat{\mathbf{s}}_i) \Phi_\eta(\hat{\mathbf{s}}_i, \hat{\mathbf{s}}) \, d\Omega_i, \qquad (8.23)$$

where the intensity gradient has been converted into a total derivative since we assume the process to be quasi-steady. The equation of transfer is often rewritten in terms of nondimensional optical coordinates (see Fig. 8-5),

$$\tau_\eta = \int_0^s (\kappa_\eta + \sigma_{s\eta}) \, ds = \int_0^s \beta_\eta \, ds, \qquad (8.24)$$

and the *single scattering albedo*, defined as

$$\omega_\eta \equiv \frac{\sigma_{s\eta}}{\kappa_\eta + \sigma_{s\eta}} = \frac{\sigma_{s\eta}}{\beta_\eta}, \qquad (8.25)$$

leading to

$$\frac{dI_\eta}{d\tau_\eta} = -I_\eta + (1 - \omega_\eta)I_{b\eta} + \frac{\omega_\eta}{4\pi} \int_{4\pi} I_\eta(\hat{\mathbf{s}}_i) \Phi_\eta(\hat{\mathbf{s}}_i, \hat{\mathbf{s}}) \, d\Omega_i. \qquad (8.26)$$

The last two terms in equation (8.26) are often combined and are then known as the *source function* for radiative intensity,

$$S_\eta(\tau_\eta, \hat{\mathbf{s}}) = (1 - \omega_\eta)I_{b\eta} + \frac{\omega_\eta}{4\pi} \int_{4\pi} I_\eta(\hat{\mathbf{s}}_i) \Phi_\eta(\hat{\mathbf{s}}_i, \hat{\mathbf{s}}) \, d\Omega_i. \qquad (8.27)$$

Equation (8.26) then assumes the deceptively simple form of

$$\frac{dI_\eta}{d\tau_\eta} + I_\eta = S_\eta(\tau_\eta, \hat{\mathbf{s}}), \qquad (8.28)$$

which is, of course, an integro-differential equation (in space, and in two directional coordinates with local origin). Furthermore, the Planck function $I_{b\eta}$ is generally not known and must be found by considering the overall energy equation (adding derivatives in the three space coordinates and integrations over two more directional coordinates and the wavenumber spectrum).

8.6 FORMAL SOLUTION TO THE EQUATION OF TRANSFER

If the source function is known (or assumed known), equation (8.28) can be formally integrated by the use of an integrating factor. Thus, multiplying through by e^{τ_η} results in

$$\frac{d}{d\tau_\eta}\left(I_\eta e^{\tau_\eta}\right) = S_\eta(\tau_\eta, \hat{\mathbf{s}})\, e^{\tau_\eta}, \tag{8.29}$$

which may be integrated from a point $s' = 0$ at the wall to a point $s' = s$ inside the medium (see Fig. 8-5), so that

$$I_\eta(\tau_\eta) = I_\eta(0)e^{-\tau_\eta} + \int_0^{\tau_\eta} S_\eta(\tau'_\eta, \hat{\mathbf{s}})\, e^{-(\tau_\eta - \tau'_\eta)}\, d\tau'_\eta, \tag{8.30}$$

where τ'_η is the optical coordinate at $s = s'$.

Physically, one can readily appreciate that the first term on the right-hand side of equation (8.30) is the contribution to the local intensity by the intensity entering the enclosure at $s = 0$, which decays exponentially due to extinction over the optical distance τ_η. The integrand of the second term, $S_\eta(\tau'_\eta)d\tau'_\eta$, on the other hand, is the contribution from the local emission at τ'_η, attenuated exponentially by selfextinction over the optical distance between the emission point and the point under consideration, $\tau_\eta - \tau'_\eta$. The integral, finally, sums all the contributions over the entire emission path.

Equation (8.30) is a third-order integral equation in intensity I_η. The integral over the source function must be carried out over the optical coordinate (for all directions), while the source function itself is also an integral over a set of direction coordinates (with varying local origin) containing the unknown intensity. Furthermore, usually the temperature and, therefore, the blackbody intensity are not known and must be found in conjunction with overall conservation of energy. There are, however, a few cases for which the equation of transfer becomes considerably simplified.

Nonscattering Medium

If the medium only absorbs and emits, the source function reduces to the local blackbody intensity, and

$$I_\eta(\tau_\eta) = I_\eta(0)e^{-\tau_\eta} + \int_0^{\tau_\eta} I_{b\eta}(\tau'_\eta)\, e^{-(\tau_\eta - \tau'_\eta)}d\tau'_\eta. \tag{8.31}$$

This equation is an explicit expression for the radiation intensity if the temperature field is known. However, generally the temperature is not known and must be found in conjunction with overall conservation of energy.

Example 8.1.   What is the spectral intensity emanating from an isothermal sphere bounded by vacuum or a cold black wall?

Solution

Because of the symmetry in this problem, the intensity emanating from the sphere surface is only a function of the exit angle. Examining Fig. 8-6, we see that equation (8.31)

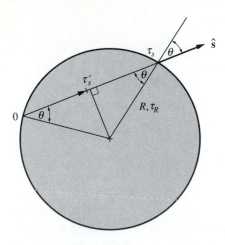

FIGURE 8-6
Isothermal sphere for Example 8.1.

reduces to

$$I_\eta(\tau_R, \theta) = \int_0^{\tau_s} I_{b\eta}(\tau_s') \, e^{-(\tau_s - \tau_s')} \, d\tau_s'.$$

But for a sphere

$$\tau_s = 2\tau_R \cos\theta,$$

regardless of the azimuthal angle. Therefore, with $I_{b\eta}(\tau_s') = I_{b\eta} = \text{const}$, the desired intensity turns out to be

$$I_\eta(\tau_R, \theta) = I_{b\eta} e^{-(2\tau_R \cos\theta - \tau_s')} \Big|_0^{2\tau_R \cos\theta} = I_{b\eta} \left(1 - e^{-2\tau_R \cos\theta}\right).$$

Thus, for $\tau_R \gg 1$ the isothermal sphere emits equally into all directions, like a black surface at the same temperature.

The Cold Medium

If the temperature of the medium is so low that the blackbody intensity at that temperature is small as compared with incident intensity, then the equation of transfer is decoupled from other modes of heat transfer. However, the governing equation remains a third-order integral equation, namely,

$$I_\eta(\tau_\eta, \hat{s}) = I_\eta(0)e^{-\tau_\eta} + \int_0^{\tau_\eta} \frac{\omega_\eta}{4\pi} \int_{4\pi} I_\eta(\tau_\eta', \hat{s}_i) \Phi_\eta(\hat{s}_i, \hat{s}) \, d\Omega_i \, e^{-(\tau_\eta - \tau_\eta')} \, d\tau_\eta'. \quad (8.32)$$

If the scattering is isotropic, or $\Phi \equiv 1$, the directional integration in equation (8.32) may be carried out, so that

$$I_\eta(\tau_\eta, \hat{s}) = I_\eta(0)e^{-\tau_\eta} + \frac{1}{4\pi} \int_0^{\tau_\eta} \omega_\eta \, G_\eta(\tau_\eta') \, e^{-(\tau_\eta - \tau_\eta')} \, d\tau_\eta', \quad (8.33)$$

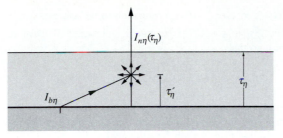

FIGURE 8-7
Geometry for Example 8.2.

where

$$G_\eta(\tau) \equiv \int_{4\pi} I_\eta(\tau_\eta', \hat{s}_i) \, d\Omega_i \tag{8.34}$$

is known as the *incident radiation function* (since it is the total intensity impinging on a point from all sides). The problem is then much simplified since it is only necessary to find a solution for G [by direction-integrating equation (8.33)] rather than determining the direction-dependent intensity.

Purely Scattering Medium

If the medium scatters radiation, but does not absorb or emit, then the radiative transfer is again decoupled from other heat transfer modes. In this case $\omega_\eta \equiv 1$, and the equation of transfer reduces to a form essentially identical to equation (8.32), i.e.,

$$I_\eta(\tau_\eta, \hat{s}) = I_\eta(0) \, e^{-\tau_\eta} + \frac{1}{4\pi} \int_0^{\tau_\eta} \int_{4\pi} I_\eta(\tau_\eta', \hat{s}_i) \, \Phi_\eta(\hat{s}_i, \hat{s}) \, d\Omega_i \, e^{-(\tau_\eta - \tau_\eta')} \, d\tau_\eta'. \tag{8.35}$$

Again, for isotropic scattering, this equation may be simplified by introducing the incident radiation, so that

$$I_\eta(\tau_\eta, \hat{s}) = I_\eta(0) \, e^{-\tau_\eta} + \frac{1}{4\pi} \int_0^{\tau_\eta} G_\eta(\tau_\eta', \hat{s}) \, e^{-(\tau_\eta - \tau_\eta')} \, d\tau_\eta'. \tag{8.36}$$

Example 8.2.   A large isothermal black plate is covered with a thin layer of isotropically scattering, nonabsorbing (and, therefore, nonemitting) material with unity index of refraction. Assuming that the layer is so thin that any ray emitted from the plate is scattered at most once before leaving the scattering layer, estimate the radiative intensity above the layer in the direction normal to the plate.

Solution

The exiting intensity in the normal direction (see Fig. 8-7) may be calculated from equation (8.36) by retaining only terms of order τ_η or higher (since $\tau_\eta \ll 1$). This process leads to $e^{-\tau_\eta} = 1 - \tau_\eta + \mathcal{O}(\tau_\eta^2)$, $G(\tau_\eta') = G(\tau_\eta) + \mathcal{O}(\tau_\eta)$ (radiation to be scattered arrives unattenuated at a point), and $e^{-(\tau_\eta - \tau_\eta')} = 1 - \mathcal{O}(\tau_\eta)$ (scattered radiation will leave the medium without further attenuation), so that

$$I_{n\eta} = I_{b\eta}(1 - \tau_\eta) + \frac{1}{4\pi}G_\eta\,\tau_\eta + \mathcal{O}(\tau_\eta^2),$$

where the intensity emanating from the plate is known since the plate is black. The incident radiation at any point is due to unattenuated emission from the bottom plate arriving from the lower 2π solid angles, and nothing coming from the top 2π solid angles, i.e., $G_\eta \approx 2\pi I_{b\eta}$ and

$$I_{n\eta} = I_{b\eta}(1 - \tau_\eta) + \frac{1}{2}I_{b\eta}\,\tau_\eta + \mathcal{O}(\tau_\eta^2) = I_{b\eta}\left(1 - \frac{\tau_\eta}{2}\right) + \mathcal{O}(\tau_\eta^2).$$

Physically this result tells us that the emission into the normal direction is attenuated by the fraction τ_η (scattered away from normal direction), and augmented by the fraction $\tau_\eta/2$ (scattered into the normal direction): Since scattering is isotropic, exactly half of the attenuation is scattered upward and half downward; the latter is then absorbed by the emitting plate. Thus, the scattering layer acts as a heat shield for the hot plate.

8.7 BOUNDARY CONDITIONS FOR THE EQUATION OF TRANSFER

The equation of transfer in its quasi-steady form, equation (8.23), is a first-order differential equation in intensity (for a fixed direction $\hat{s}$). As such, the equation requires knowledge of the radiative intensity at a single point in space, into the direction of $\hat{s}$. Generally, the point where the intensity can be specified independently lies on the surface of an enclosure surrounding the participating medium, as indicated by the formal solution in equation (8.30). This intensity, leaving a wall into a specified direction, may be determined by the methods given in Chapter 5 (diffusely emitting and reflecting surfaces), Chapter 6 (diffusely emitting and specularly reflecting surfaces) and Chapter 7 (surfaces with arbitrary characteristics).

Diffusely Emitting and Reflecting Opaque Surfaces

For a surface that emits and reflects diffusely, the exiting intensity is independent of direction. Therefore, at a point $\mathbf{r}_w$ on the surface, from equations (5.18) and (5.19),

$$I(\mathbf{r}_w, \hat{s}) = I(\mathbf{r}_w) = J(\mathbf{r}_w)/\pi = \epsilon(\mathbf{r}_w)\,I_b(\mathbf{r}_w) + \rho(\mathbf{r}_w)\,H(\mathbf{r}_w)/\pi, \tag{8.37}$$

where $H(\mathbf{r}_w)$ is the hemispherical irradiation (i.e., incoming radiative heat flux) defined by equation (3.39), leading to

$$I(\mathbf{r}_w, \hat{s}) = \epsilon(\mathbf{r}_w)\,I_b(\mathbf{r}_w) + \frac{\rho(\mathbf{r}_w)}{\pi}\int_{\hat{n}\cdot\hat{s}'<0} I(\mathbf{r}_w, \hat{s}')\,|\hat{n}\cdot\hat{s}'|\,d\Omega', \tag{8.38}$$

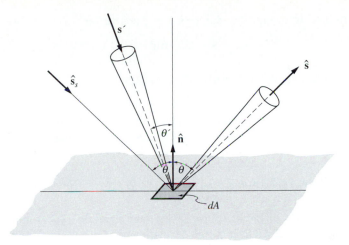

FIGURE 8-8
Radiative intensity reflected from a surface.

where $\hat{\mathbf{n}}$ is the local outward surface normal and $\hat{\mathbf{n}} \cdot \hat{\mathbf{s}}' = \cos \theta'$ is the cosine of the angle between any incoming direction $\hat{\mathbf{s}}'$ and the surface normal, as indicated in Fig. 8-8. Therefore, the outgoing intensity is not generally known explicitly, but is related to the incoming intensity. An exception is the black surface, for which (with $\rho = 0$),

$$I(\mathbf{r}_w, \hat{\mathbf{s}}) = I_b(\mathbf{r}_w). \tag{8.39}$$

Diffusely Emitting, Specularly Reflecting, Opaque Surfaces

If the reflectivity of the surface has a specular as well as a diffuse component, i.e., the reflectivity obeys equation (6.1), then the outgoing intensity also consists of two components. One part of the outgoing intensity is due to diffuse emission as well as the diffuse fraction of reflected energy, as described by equation (8.38). In addition, the outgoing intensity has a specularly reflected component,[4] so that

$$I(\mathbf{r}_w, \hat{\mathbf{s}}) = \epsilon(\mathbf{r}_w) I_b(\mathbf{r}_w) + \frac{\rho^d(\mathbf{r}_w)}{\pi} \int_{\hat{\mathbf{n}} \cdot \hat{\mathbf{s}}' < 0} I(\mathbf{r}_w, \hat{\mathbf{s}}') |\hat{\mathbf{n}} \cdot \hat{\mathbf{s}}'| \, d\Omega'$$
$$+ \rho^s(\mathbf{r}_w) I(\mathbf{r}_w, \hat{\mathbf{s}}_s), \tag{8.40}$$

where $\hat{\mathbf{s}}_s$ is the "specular direction," defined as the direction from which a light beam must hit the surface in order to travel into the direction of $\hat{\mathbf{s}}$ after a specular reflection.

---

[4]Note that the specularly reflected component cannot be "assigned" to the surface where it leaves in diffuse fashion, as was done for surface transport in Chapter 6. The reason is that the intensity changes while radiation travels from surface to surface within a participating medium.

This direction is, from Fig. 8-8, $\hat{s} + (-\hat{s}_s) = 2\cos\theta\hat{n}$, or

$$\hat{s}_s = \hat{s} - 2(\hat{s} \cdot \hat{n})\hat{n}. \tag{8.41}$$

Opaque Surfaces with Arbitrary Surface Properties

Reflection from a surface with nonideal radiative properties is governed by the bidirectional reflection function, as discussed in Chapter 7. From equation (7.10) it follows immediately that

$$I(\mathbf{r}_w, \hat{s}) = \epsilon(\mathbf{r}_w, \hat{s}) I_b(\mathbf{r}_w) + \int_{\hat{n}\cdot\hat{s}<0} \rho''(\mathbf{r}_w, \hat{s}', \hat{s}) I(\mathbf{r}_w, \hat{s}') |\hat{n}\cdot\hat{s}'| \, d\Omega'. \tag{8.42}$$

If the surface reflects diffusely, $\rho'' = \rho^d/\pi$ and equation (8.42) reduces to equation (8.38). For specular reflection the development of equation (7.15) shows that it reduces to equation (8.40).

Semitransparent Boundaries

If the boundary is a semitransparent wall, external radiation may penetrate into the enclosure and must be added to equations (8.38), (8.40), and (8.42) as $I_o(\mathbf{r}_w, \hat{s})$. The emissivity ϵ in these boundary conditions is then an effective value for the internal emission from the entire semitransparent wall thickness. If the bounding surface is totally transparent (or simply an opening), then there is no emission from the boundary and $\epsilon = 0$. This type of boundary condition was discussed in some detail in Section 6.6.

8.8 RADIATION ENERGY DENSITY

A volume element inside an enclosure is irradiated from all directions and, at any instant in time t, contains a certain amount of radiative energy in the form of photons. Consider, for example, an element $dV = dA\,ds$ irradiated perpendicularly to dA with intensity $I_\eta(\hat{s})$ as shown in Fig. 8-4. Therefore, per unit time radiative energy in the amount of $I_\eta(\hat{s})\,d\Omega\,dA$ enters dV. From equation (8.1) we see that this energy remains inside dV for a duration of $dt = ds/c$, before exiting at the other side. Thus, due to irradiation from a single direction, the volume contains the amount of radiative energy $I_\eta(\hat{s})\,d\Omega\,dA\,ds/c = I_\eta(\hat{s})\,d\Omega\,dV/c$ at any instant in time. Adding the contributions from all possible directions, we find the total radiative energy stored within dV is $u_\eta\,dV$, where u_η is the *spectral radiation energy density*

$$u_\eta \equiv \frac{1}{c}\int_{4\pi} I_\eta(\hat{s})\,d\Omega. \tag{8.43}$$

Integration over the spectrum gives the *total radiation energy density*,

$$u = \int_0^\infty u_\eta\,d\eta = \frac{1}{c}\int_{4\pi}\int_0^\infty I_\eta(\hat{s})\,d\eta\,d\Omega = \frac{1}{c}\int_{4\pi} I(\hat{s})\,d\Omega. \tag{8.44}$$

Although the radiation energy density is a very basic quantity akin to internal energy for energy stored within matter, it is not widely used by heat transfer engineers. Instead, it is common practice to employ the *incident radiation G_η*, which is related to the energy density through

$$G_\eta \equiv \int_{4\pi} I_\eta(\hat{s}) \, d\Omega = c \, u_\eta; \quad G = cu. \tag{8.45}$$

8.9 RADIATIVE HEAT FLUX

The spectral radiative heat flux onto a surface element has been expressed in terms of incident and outgoing intensity in equation (1.36) as

$$\mathbf{q}_\eta \cdot \hat{\mathbf{n}} = \int_{4\pi} I_\eta \, \hat{\mathbf{n}} \cdot \hat{\mathbf{s}} \, d\Omega. \tag{8.46}$$

This relationship also holds, of course, for a hypothetical (i.e., totally transmissive) surface element placed arbitrarily inside an enclosure. Removing the surface normal from equation (1.36), we obtain the definition for the *spectral, radiative heat flux vector* inside a participating medium. To obtain the *total radiative heat flux*, equation (8.46) needs to integrated over the spectrum, and

$$\mathbf{q} = \int_0^\infty \mathbf{q}_\eta \, d\eta = \int_0^\infty \int_{4\pi} I_\eta(\hat{s}) \, \hat{s} \, d\Omega \, d\eta. \tag{8.47}$$

Depending on the coordinate system used, or the surface being described, the radiative heat flux vector may be separated into its coordinate components, for example q_x, q_y, and q_z (for a Cartesian coordinate system), or into components normal and tangential to a surface, and so on.

Example 8.3. Evaluate the total heat loss from an isothermal spherical medium bounded by vacuum, assuming that $\kappa_\eta = $ const (i.e., does not vary with location, temperature or wavenumber).

Solution

Here we are dealing with a spherical coordinate system, and we are interested in the radial component of the radiative heat flux (the other two being equal to zero by symmetry). We saw in the last example that the intensity emanating from the sphere is

$$I_\eta(\tau_R, \theta) = I_{b\eta} \left(1 - e^{-2\tau_R \cos\theta} \right), \quad 0 \le \theta \le \frac{\pi}{2},$$

where θ is measured from the surface normal pointing away from the sphere (Fig. 8-6). Since the sphere is bounded by vacuum, there is no incoming radiation and

$$I_\eta(\tau_R, \theta) = 0, \quad \frac{\pi}{2} \le \theta \le \pi.$$

Therefore, from equation (8.47),

$$
\begin{aligned}
q(\tau_R) &= \int_0^\infty \int_0^{2\pi} \int_0^\pi I_\eta(\tau_R, \theta) \cos\theta \, \sin\theta \, d\theta \, d\psi \, d\eta \\
&= 2\pi \int_0^\infty \int_0^{\pi/2} I_{b\eta}\left(1 - e^{-2\tau_R \cos\theta}\right) \cos\theta \, \sin\theta \, d\theta \, d\eta \\
&= \pi I_b \left\{ 1 - \frac{1}{2\tau_R^2}\left[1 - (1 + 2\tau_R)e^{-2\tau_R}\right] \right\} \\
&= n^2 \sigma T^4 \left\{ 1 - \frac{1}{2\tau_R^2}\left[1 - (1 + 2\tau_R)e^{-2\tau_R}\right] \right\},
\end{aligned}
$$

where n is the refractive index of the medium (usually $n \approx 1$ for gases, but $n > 1$ for semitransparent liquids and solids). As discussed in the previous example, if $\tau_R \to \infty$ the heat flux approaches the same value as the one from a black surface.

If the sphere in the last example is optically thin $\tau_R \ll 1$ (i.e., the medium *emits* radiative energy, but does not *absorb* any of the emitted energy), then the total heat loss (total emission) from the sphere is

$$
Q = 4\pi R^2 q = 4\pi R^2 \times \tfrac{4}{3}\tau_R n^2 \sigma T^4 = 4\kappa n^2 \sigma T^4 V. \tag{8.48}
$$

This result may be generalized to govern emission from any isothermal volume V *without selfabsorption*, or

$$
Q_{\text{emission}} = 4\kappa n^2 \sigma T^4 V. \tag{8.49}
$$

8.10   DIVERGENCE OF THE RADIATIVE HEAT FLUX

While the heat transfer engineer is interested in the radiative heat flux, this interest usually holds true only for fluxes at physical boundaries. Inside the medium, on the other hand, we need to know how much net radiative energy is deposited into (or withdrawn from) each volume element. Thus, making a radiative energy balance on an infinitesimal volume $dV = dx \, dy \, dz$ as shown in Fig. 8-9, we have

$$
\left(\begin{array}{c} \text{radiative energy} \\ \text{stored in } dV \\ \text{per unit time} \end{array}\right) - \left(\begin{array}{c} \text{rad. energy generated} \\ \text{(emitted) by } dV \\ \text{per unit time} \end{array}\right) + \left(\begin{array}{c} \text{rad. energy destroyed} \\ \text{(absorbed) by } dV \\ \text{per unit time} \end{array}\right)
$$

$$
= \left(\begin{array}{c} \text{flux in at } x \ - \ \text{flux out at } x + dx \\ + \text{ flux in at } y \ - \ \text{flux out at } y + dy \\ + \text{ flux in at } z \ - \ \text{flux out at } z + dz \end{array}\right).
$$

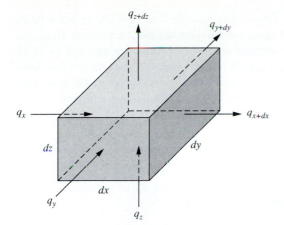

FIGURE 8-9
Control volume for derivation of divergence
of radiative heat flux.

The right-hand side may be written in mathematical form as

$$\left.\begin{array}{r} q(x)\,dy\,dz - q(x+dx)\,dy\,dz \\ +q(y)\,dx\,dz - q(y+dy)\,dx\,dz \\ +q(z)\,dx\,dy - q(z+dz)\,dx\,dy \end{array}\right\} = -\left(\frac{\partial q}{\partial x} + \frac{\partial q}{\partial y} + \frac{\partial q}{\partial z}\right)dx\,dy\,dz = -\boldsymbol{\nabla}\cdot\mathbf{q}\,dV.$$

Thus, within the overall energy equation, it is the divergence of the radiative heat
flux that is of interest inside the participating medium.[5]

We have already established an energy balance for thermal radiation, the equation
of transfer [for example, equation (8.23)],

$$\frac{dI_\eta}{ds} = \hat{\mathbf{s}}\cdot\boldsymbol{\nabla}I_\eta = \kappa_\eta I_{b\eta} - \beta_\eta I_\eta(\hat{\mathbf{s}}) + \frac{\sigma_{s\eta}}{4\pi}\int_{4\pi} I_\eta(\hat{\mathbf{s}}_i)\,\Phi_\eta(\hat{\mathbf{s}}_i,\hat{\mathbf{s}})\,d\Omega_i, \qquad (8.50)$$

which is a radiation balance for an infinitesimal pencil of rays. Thus, in order to get
a volume balance, we integrate this equation over all solid angles, or

$$\int_{4\pi} \hat{\mathbf{s}}\cdot\boldsymbol{\nabla}I_\eta\,d\Omega = \int_{4\pi} \kappa_\eta I_{b\eta}\,d\Omega - \int_{4\pi} \beta_\eta I_\eta(\hat{\mathbf{s}})\,d\Omega$$

$$+ \int_{4\pi} \frac{\sigma_{s\eta}}{4\pi}\int_{4\pi} I_\eta(\hat{\mathbf{s}}_i)\,\Phi_\eta(\hat{\mathbf{s}}_i,\hat{\mathbf{s}})\,d\Omega_i\,d\Omega, \qquad (8.51)$$

and

$$\boldsymbol{\nabla}\cdot\int_{4\pi} I_\eta\hat{\mathbf{s}}\,d\Omega = 4\pi\kappa_\eta I_{b\eta} - \int_{4\pi} \beta_\eta I_\eta(\hat{\mathbf{s}})\,d\Omega$$

$$+ \frac{\sigma_{s\eta}}{4\pi}\int_{4\pi} I_\eta(\hat{\mathbf{s}}_i)\left(\int_{4\pi} \Phi_\eta(\hat{\mathbf{s}}_i,\hat{\mathbf{s}})\,d\Omega\right)d\Omega_i. \qquad (8.52)$$

---

[5]For simplicity, this equation was derived for a Cartesian coordinate system but the result holds, of
course, for any arbitrary coordinate system.

On the left side of equation (8.52) the integral and the direction vector were taken into the gradient since direction and space coordinates are all independent from one another.[6] The expression inside the operator is now, of course, the spectral radiative heat flux. On the right side of equation (8.52) the order of integration has been changed, applying the Ω-integration to the only part depending on it, the scattering phase function Φ_η. This last integration can be carried out using equation (8.19), leading to

$$\nabla \cdot \mathbf{q}_\eta = 4\pi\kappa_\eta I_{b\eta} - \beta_\eta \int_{4\pi} I_\eta(\hat{s})\, d\Omega + \sigma_{s\eta} \int_{4\pi} I_\eta(\hat{s}_i)\, d\Omega_i. \tag{8.53}$$

Since Ω and Ω_i are dummy arguments for integration over all solid angles, the last two terms can be pulled together, using $\kappa_\eta = \beta_\eta - \sigma_{s\eta}$:

$$\nabla \cdot \mathbf{q}_\eta = \kappa_\eta \left(4\pi I_{b\eta} - \int_{4\pi} I_\eta\, d\Omega\right) = \kappa_\eta (4\pi I_{b\eta} - G_\eta). \tag{8.54}$$

Equation (8.54) states that physically the net loss of radiative energy from a control volume is equal to emitted energy minus absorbed irradiation. This direction-integrated form of the equation of transfer no longer contains the scattering coefficient. This fact is not surprising since scattering only redirects the stream of photons; it does not affect the energy content of any given unit volume.

Equation (8.54) is a spectral relationship, i.e., it gives the heat flux per unit wavenumber at a certain spectral position. If the divergence of the total heat flux is desired, the integration over the spectrum is carried out to give

$$\nabla \cdot \mathbf{q} = \nabla \cdot \int_0^\infty \mathbf{q}_\eta\, d\eta = \int_0^\infty \kappa_\eta \left(4\pi I_{b\eta} - \int_{4\pi} I_\eta d\Omega\right) d\eta$$

$$= \int_0^\infty \kappa_\eta (4\pi I_{b\eta} - G_\eta)\, d\eta. \tag{8.55}$$

Equation (8.55) is a statement of the *conservation of radiative energy*. For the special case of a gray medium ($\kappa_\eta = \kappa = $ constant) this may be simplified to

$$\nabla \cdot \mathbf{q} = \kappa \left(4\sigma T^4 - \int_{4\pi} I\, d\Omega\right) = \kappa (4\sigma T^4 - G). \tag{8.56}$$

Example 8.4. Calculate the divergence of the total radiative heat flux at the center and at the surface of the gray, isothermal spherical medium in the previous example.

---

[6]While this statement is always true, care must be taken in non-Cartesian coordinate systems: Although the direction *vector* is independent from space coordinates, the *three components* may be tied to locally defined unit vectors. For example, in a cylindrical coordinate system the direction vector is usually defined in terms of $\hat{e}_r$ and $\hat{e}_\theta$, which vary with r and θ.

Solution

We already know the intensity at the surface of the sphere and, therefore,

$$G_\eta(\tau_R) = 2\pi \int_0^\pi \sin\theta\, I_\eta\, d\theta = 2\pi I_{b\eta} \int_0^{\pi/2} \left(1 - e^{-2\tau_R \cos\theta}\right)\sin\theta\, d\theta$$

$$= 2\pi I_{b\eta}\left(1 - \frac{e^{-2\tau_R \cos\theta}}{2\tau_R}\bigg|_0^{\pi/2}\right) = \frac{\pi I_{b\eta}}{\tau_R}\left(2\tau_R - 1 + e^{-2\tau_R}\right),$$

and

$$\mathbf{\nabla}\cdot\mathbf{q}(\tau_R) = \kappa\,(4\pi I_b - G) = \frac{\sigma T^4}{R}\left(2\tau_R + 1 - e^{-2\tau_R}\right). \tag{8.57}$$

At the center of the sphere the intensity is easily evaluated as

$$I_\eta(0) = I_{b\eta}\left(1 - e^{-\tau_R}\right),$$

and

$$G_\eta(0) = 4\pi I_{b\eta}\left(1 - e^{-\tau_R}\right),$$

so that

$$\mathbf{\nabla}\cdot\mathbf{q}(0) = \kappa 4\sigma T^4\, e^{-\tau_R}. \tag{8.58}$$

The right-hand sides of equations (8.57) and (8.58) are radiative heat losses per unit time and volume, which must be made up for by a volumetric heat source if the sphere is to stay isothermal.

8.11 OVERALL ENERGY CONSERVATION

Thermal radiation is only one mode of transferring heat which, in general, must compete with conductive and convective heat transfer. Therefore, the temperature field must be determined through an energy conservation equation that incorporates all three modes of heat transfer. The radiation intensity, through emission and temperature-dependent properties, depends on the temperature field and, therefore, cannot be decoupled from the overall energy equation.

The general form of the energy conservation equation for a moving compressible fluid may be stated as

$$\rho\frac{Du}{Dt} = \rho\left(\frac{\partial u}{\partial t} + \mathbf{v}\cdot\mathbf{\nabla}u\right) = -\mathbf{\nabla}\cdot\mathbf{q} - p\mathbf{\nabla}\cdot\mathbf{v} + \mu\Phi + \dot{Q}''', \tag{8.59}$$

where u is internal energy, $\mathbf{v}$ is the velocity vector, $\mathbf{q}$ is the total heat flux vector, Φ is the dissipation function, and $\dot{Q}'''$ is heat generated within the medium (such as energy release due to chemical reactions). For a detailed derivation of equation (8.59), the reader is referred to standard textbooks such as [1, 2]. If the medium is radiatively participating through emission, absorption, and scattering, then the conservation equations for momentum and energy are altered by three effects [3]:

1. The heat flux term in equation (8.59), which without radiation is in most applications due only to molecular diffusion (heat conduction), now has a second component, the radiative heat flux, due to radiative energy interacting with the medium within the control volume.

2. The internal energy now contains a radiative contribution [the incident radiation G, due to the first term in equation (8.22)].

3. The radiation pressure tensor must be added to the traditional fluid dynamics pressure tensor.

We have already seen that the second effect is almost always negligible, and the same is true for the augmentation of the pressure tensor. Under these conditions the energy conservation equation can be simplified. If we assume that $du = c_v dT$ and Fourier's law for heat conduction to hold,

$$\mathbf{q} = \mathbf{q}_C + \mathbf{q}_R = -k\nabla T + \mathbf{q}_R, \tag{8.60}$$

equation (8.59) becomes

$$\rho c_v \frac{DT}{Dt} = \rho c_v \left(\frac{\partial T}{\partial t} + \mathbf{v} \cdot \nabla T \right)$$

$$= \nabla \cdot (k\nabla T) - p\nabla \cdot \mathbf{v} + \mu \Phi + \dot{Q}''' - \nabla \cdot \mathbf{q}_R. \tag{8.61}$$

This is an integro-differential equation for the calculation of the temperature field, since the evaluation of the divergence of the radiative heat flux must come from (8.54), which is an integral equation in temperature. Obviously, a complete solution of this equation, even with the recent advent of supercomputers, is a truly formidable task.

Example 8.5. State the equation of transfer and its boundary conditions for the case of combined steady-state conduction and radiation within a one-dimensional, planar, gray and nonscattering medium, bounded by isothermal black walls.

Solution

Since the problem is steady-state and there is no movement in the medium, the left side of equation (8.61) vanishes, and only the first (conduction) and last (radiation) terms on the right side remain. For a one-dimensional planar medium this reduces to[7]

$$\frac{d}{dz}\left(k\frac{dT}{dz} - q_R \right) = 0, \tag{8.62}$$

and the divergence of radiative heat flux is related to temperature and incident radiation through equation (8.54),

$$\frac{dq_R}{dz} = \kappa(4\sigma T^4 - G), \tag{8.63}$$

---

[7]While in the science of conduction the variable x is usually employed for one-dimensional planar problems, for thermal radiation problems the variable z is more convenient. The reason for this is that, by convention, the polar angle for the direction vector is measured from the z-axis.

where the spectral integration for the gray medium has been carried out by simply dropping the subscript η. Finally, the incident radiation is found from direction-integrating equation (8.31) (not a trivial task). The necessary boundary conditions are $T = T_i$, $i = 1, 2$ at the two walls (for conduction) and $I(0, \hat{s}) = \sigma T_i^4 / \pi$ (for radiation) needed in equation (8.31). Solution of this seemingly simple problem is by no means trivial, and can only be achieved through relatively involved numerical analysis.

Radiative Equilibrium

Much attention in the following chapters will be given to the situation in which radiation is the dominant mode of heat transfer, meaning that when conduction and convection are negligible. This situation is referred to as *radiative equilibrium*, meaning that thermodynamic equilibrium within the medium is achieved by virtue of thermal radiation alone. As is commonly done in the discussion of "pure" conduction or convection, we allow volumetric heat sources throughout the medium. Thus, we may write

$$\mathbf{\nabla} \cdot \mathbf{q}_R = \dot{Q}''', \tag{8.64}$$

which is identical in form to the basic steady-state heat conduction equation (before substitution of Fourier's law).[8] Radiative equilibrium is often a good assumption in applications with extremely high temperatures, such as plasmas, nuclear explosions, and such. The inclusion of a volumetric heat source allows the treatment of conduction and convection "through the back door:" A guess is made for the temperature field and the nonradiation terms in equation (8.61) are calculated to give $\dot{Q}'''$ for the radiation calculations. This process is then repeated until a convergence criterion is met.

8.12 SOLUTION METHODS FOR THE EQUATION OF TRANSFER

Exact analytical solutions to the equation of transfer [equation (8.23)] are exceedingly difficult, and explicit solutions are impossible for all but the very simplest situations. Therefore, research on radiative heat transfer in participating media has generally proceeded in two directions: (*i*) Exact (analytical and numerical) solutions of highly idealized situations, and (*ii*) approximate solution methods for more involved scenarios. Phenomena that make a radiative heat transfer problem difficult may be placed into four different categories:

Geometry: The problem may be one-dimensional, two-dimensional or three-dimensional. Most investigations to date have dealt with one-dimensional geometries, and the vast majority of these dealt with the simplest case of a one-dimensional plane-parallel slab.

---

[8] In fact, we should have an unsteady equation had we not dropped the unsteady term in equation (8.22).

Temperature Field: The least difficult situation arises if the temperature profile within the medium is known, making equation (8.23) a relatively "simple" integral equation. Consequently, the most basic case of an isothermal medium has been studied extensively. Alternatively, if radiative equilibrium prevails, the temperature field is unknown but uncoupled from conduction and convection, and must be found from directional and spectral integration of the equation of transfer. In the most complicated scenario, radiative heat transfer is combined with conduction and/or convection, resulting in a highly nonlinear integro-differential equation.

Scattering: The solution to a radiation problem is greatly simplified if the medium does not scatter. In that case the equation of transfer reduces to a simple first-order differential equation if the temperature field is known, and a relatively simpler integral equation if radiative equilibrium prevails. If scattering must be considered, isotropic scattering is often assumed. Relatively few investigations have dealt with the case of anisotropic scattering, and most of those are limited to the case of linear-anisotropic scattering (see Section 10.8).

Properties: Although most participating media display strong nongray character, as discussed in the previous three chapters, the vast majority of investigations to date have centered on the study of gray media. In addition, while radiative properties also generally depend strongly on temperature, concentration, etc., most calculations are limited to situations with constant properties.

Most "exact" solutions are limited to gray media with constant properties in one-dimensional, mainly plane-parallel geometries. The media are isothermal or at radiative equilibrium, and if they scatter, the scattering is usually isotropic. Since the usefulness of such one-dimensional solutions in heat transfer applications is rather limited, they are only briefly discussed in Chapter 12.

Seven chapters are devoted to the various approximate methods that have been devised for the solution of the equation of transfer. Still, these seven chapters by no means cover all the different methods that have been and still are used by investigators in the field. A number of approximate methods for one-dimensional problems are discussed in Chapter 13. The *optically thin* and *diffusion* (or optically thick) approximations have historically been developed for a one-dimensional plane-parallel medium, but can readily be applied to more complicated geometries. Similarly, the *Schuster-Schwarzschild* or *two-flux approximation* [4, 5] is a forerunner to the multi-dimensional *discrete ordinates method*. In this method the intensity is assumed to be constant over discrete parts of the total solid angle of 4π. Several other flux methods exist, but they are usually tailored toward special geometries, and cannot easily be applied to other scenarios, for example, the *six-flux methods* of Chu and Churchill [6] and Shih and coworkers [7, 8]. Another early one-dimensional model was the *moment method* or *Eddington approximation* [9]. In this model the directional dependence is expressed by a truncated series representation (rather than discretized). In general geometries this expansion is usually achieved through the use of spherical harmonics, leading to the *spherical harmonics method*. Several variations to the moment method that are tailored toward specific geometries have been proposed [10, 11], but these are of limited general utility. Finally, the exponential kernel approximation, already discussed in Chapter 5 for surface radiation problems, may be used as a tool for many one-dimensional problems. However, its extension to multidimensional geometries is problematic.

A survey of the literature over the past several years demonstrates that some solution methods have been used frequently, while others that appeared promising at one time are no longer employed on a regular basis. Apparently, some methods have been found to be more readily adapted to more difficult situations than others (such as multidimensionality, variable properties, anisotropic scattering and/or nongray effects). The majority of radiative heat transfer analyses today appear to use one of four methods: (*i*) The *spherical harmonics method* or a variation of it, (*ii*) the *discrete ordinates method*, (*iii*) the *zonal method*, and (*iv*) the *Monte Carlo method*. The first two of these have already been discussed briefly above with the one-dimensional approximations. The zonal method was developed by Hottel [12] in his pioneering work on furnace heat transfer. Unlike the spherical harmonics and discrete ordinates methods, the zonal method approximates spatial, rather than directional, behavior by breaking up an enclosure into finite, isothermal subvolumes. On the other hand, the Monte Carlo method [13] is a statistical method, in which the history of bundles of photons is traced as they travel through the enclosure. While the statistical nature of the Monte Carlo method makes it difficult to match it with other calculations, it is the only method that can satisfactorily deal with effects of irregular radiative properties (nonideal directional and/or nongray behavior).

Because of their importance, an entire chapter is devoted to each of these four solution methods. Several other methods can be found in the literature that are not covered in this book. For example, the *discrete transfer method*, recently proposed by Shah [14] and Lockwood and Shah [15], combines features of the discrete ordinates, zonal and Monte Carlo methods. Another hybrid proposed by Edwards [16] combines elements of the Monte Carlo and zonal methods.

References

1. Rohsenow, W. M., and H. Y. Choi: *Heat, Mass and Momentum Transfer*, Prentice Hall, Englewood Cliffs, NJ, 1961.
2. Kays, W. M., and M. E. Crawford: *Convective Heat and Mass Transfer*, McGraw-Hill, 1980.
3. Sparrow, E. M., and R. D. Cess: *Radiation Heat Transfer*, Hemisphere, New York, 1978.
4. Schuster, A.: "Radiation through a Foggy Atmosphere," *Astrophysical Journal*, vol. 21, pp. 1–22, 1905.
5. Schwarzschild, K.: "Über das Gleichgewicht der Sonnenatmosphären, (Equilibrium of the Sun's Atmosphere)," *Akad. Wiss. Göttingen, Math.-Phys. Kl. Nachr.*, vol. 195, pp. 41–53, 1906.
6. Chu, C. M., and S. W. Churchill: "Numerical Solution of Problems in Multiple Scattering of Electromagnetic Radiation," *Journal of Physical Chemistry*, vol. 59, pp. 855–863, 1960.
7. Shih, T.-M., and Y. N. Chen: "A Discretized-Intensity Method Proposed for Two-Dimensional Systems Enclosing Radiative and Conductive Media," *Numerical Heat Transfer*, vol. 6, pp. 117–134, 1983.

8. Shih, T.-M., and A. L. Ren: "Combined Radiative and Convective Recirculating Flows in Enclosures," *Numerical Heat Transfer*, vol. 8, no. 2, pp. 149–167, 1985.

9. Eddington, A. S.: *The Internal Constitution of the Stars*, Dover Publications, New York, 1959.

10. Chou, Y. S., and C. L. Tien: "A Modified Moment Method for Radiative Transfer in Non-Planar Systems," *Journal of Quantitative Spectroscopy and Radiative Transfer*, vol. 8, pp. 719–733, 1968.

11. Hunt, G. E.: "The Transport Equation of Radiative Transfer with Axial Symmetry," *SIAM J. Appl. Math.*, vol. 16, no. 1, pp. 228–237, 1968.

12. Hottel, H. C., and E. S. Cohen: "Radiant Heat Exchange in a Gas-Filled Enclosure: Allowance for Nonuniformity of Gas Temperature," *AIChE Journal*, vol. 4, pp. 3–14, 1958.

13. Howell, J. R.: "Application of Monte Carlo to Heat Transfer Problems," in *Advances in Heat Transfer*, eds. J. P. Hartnett and T. F. Irvine, vol. 5, Academic Press, New York, 1968.

14. Shah, N. G.: "New Method of Computation of Radiation Heat Transfer in Combustion Chambers," Ph.D. thesis, Imperial College of Science and Technology, London, England, 1979.

15. Lockwood, F. C., and N. G. Shah: "A New Radiation Solution Method for Incorporation in General Combustion Prediction Procedures," in *Eighteenth Symposium (International) on Combustion*, The Combustion Institute, pp. 1405–1409, 1981.

16. Edwards, D. K.: "Hybrid Monte-Carlo Matrix-Inversion Formulation of Radiation Heat Transfer with Volume Scattering," in *Heat Transfer in Fire and Combustion Systems*, vol. HTD–45, ASME, pp. 273–278, 1985.

Problems

8.1 A semi-infinite medium $0 \le z < \infty$ consists of a gray, absorbing-emitting gas that does not scatter, bounded by vacuum at the interface $z = 0$. The gas is isothermal at $1000\,\text{K}$, and the absorption coefficient is $\kappa = 1\,\text{m}^{-1}$. The interface is nonreflecting; conduction and convection may be neglected.

   (*a*) What is the local heat generation that is necessary to keep the gas at $1000\,\text{K}$?

   (*b*) What is the intensity distribution at the interface, that is, $I(z = 0, \theta, \psi)$, for all θ and ψ?

   (*c*) What is the total heat flux leaving the semi-infinite medium?

8.2 Reconsider the semi-infinite medium of Problem 8.1 for a temperature distribution of $T = T_0\, e^{-z/L}$, $T_0 = 1000\,\text{K}$, $L = 1\,\text{m}$. What are exiting intensity and heat flux for this case? Discuss how the answer would change if κ would vary between 0 and ∞.

8.3 Repeat Problem 8.1 for a medium of thickness $L = 1\,\text{m}$. Discuss how the answer would change if κ would vary between 0 and ∞.

8.4 A semi-infinite, gray, nonscattering medium ($n = 2$, $\kappa = 1\,\text{m}^{-1}$) is irradiated by the sun normal to its surface at a rate of $q_{\text{sun}} = 1000\,\text{W/m}^2$. Neglecting emission from the relatively cold medium, determine the local heat generation rate due to absorption of solar energy.

Hint: the solar radiation may be thought of as being due to a radiative intensity which has a large value I_o over a very small cone of solid angles $\delta\Omega$, and is zero elsewhere, i.e.,

$$I(\hat{s}) = \begin{cases} I_o & \text{over } \delta\Omega \text{ along } \hat{n}, \\ 0 & \text{elsewhere,} \end{cases}$$

and

$$q_{\text{sun}} = \int_{4\pi} I(\hat{s})\hat{n} \cdot \hat{s}\, d\Omega = I_o\, \delta\Omega.$$

8.5 A 1 m thick slab of an absorbing/emitting gas has an approximately linear temperature distribution as shown in the sketch. On both sides the medium is bounded by vacuum with nonreflecting boundaries.

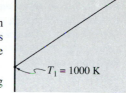

(a) If the medium has a constant and gray absorption coefficient of $\kappa = 1\,\mathrm{m}^{-1}$, what is the intensity (as function of direction) leaving the hot side of the slab?

(b) Give an expression for the radiative heat flux leaving the hot side.

8.6 A relatively cold sphere with a radius of $R_o = 1\,\mathrm{m}$ consists of a nonscattering gray medium that absorbs with an absorption coefficient of $\kappa = 0.1\,\mathrm{cm}^{-1}$ and has a refractive index $n = 2$. At the center of the sphere is a small black sphere with radius $R_i = 1\,\mathrm{cm}$ at a temperature of 1000 K. On the outside, the sphere is bounded by vaccum. What is the total heat flux leaving the sphere? Explain what happens as κ is increased from zero to a large value.

8.7 A semitransparent sphere of radius $R = 10\,\mathrm{cm}$ has a parabolic temperature profile $T = T_c(1 - r^2/R^2)$, $T_c = 2000\,\mathrm{K}$. The sphere is gray with $\kappa = 0.1\,\mathrm{cm}^{-1}$, $n = 1.0$, does not scatter, and has nonreflective boundaries. Outline how to calculate the total heat loss from the sphere (i.e., there is no need actually to carry out cumbersome integrations).

8.8 Repeat Problem 8.7, but assume that the temperature is uniform at 2000 K. What must be the local production of heat if the sphere is to remain at 2000 K everywhere? Note: The answer may be left in integral form (which must be solved numerically). Carry out the integration for $r = 0$ and $r = R$.

8.9 Repeat Problem 8.7, but assume that the temperature is uniform at 2000 K. Also, there is no heat production, meaning that the sphere cools down. How long will it take for the sphere to cool down to 500 K (the heat capacity of the medium is $\rho c = 1000\,\mathrm{kJ/m^3\,K}$ and the conductivity is very large, i.e., the sphere is isothermal at all times).

8.10 A laser beam is directed onto the atmosphere of a (hypothetical) planet. The planet's atmosphere contains 0.01% by volume of an absorbing gas. The absorbing gas has a molecular weight of 20 and, at the laser wavelength, an absorption coefficient $\kappa_\eta = 10^{-4}\,\mathrm{cm}^{-1}/(\mathrm{g/m^3})$. It is known that pressure and temperature distribution of the atmosphere can be approximated by $p = p_0 e^{-2z/L}$ and $T = T_0 e^{-z/L}$, where $p_0 = 0.75\,\mathrm{atm}$, $T_0 = 400\,\mathrm{K}$ are values at the planet surface $z = 0$, and $L = 2\,\mathrm{km}$ is a characteristic length. What fraction of the laser energy arrives at the planet's surface?

8.11 A CO_2 laser with a total power output of $Q = 10\,\mathrm{W}$ is directed (at right angle) onto a 10 cm thick, isothermal, absorbing/emitting (but not scattering) medium at 1000 K. It is known that the laser beam is essentially monochromatic at a wavelength of 10.6 μm with a Gaussian power distribution. Thus, the intensity falling onto the medium is

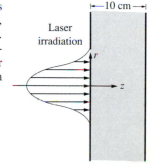

$$I(0) \propto e^{-(r/R)^2}/(\delta\Omega\,\delta\eta), \quad 0 \le r \le \infty;$$

$$Q = \int_A I(0)\,dA\,\delta\Omega\,\delta\eta,$$

where r is distance from beam center, $R = 100\,\mu m$ is the "effective radius" of the laser beam, $\delta\Omega = 5 \times 10^{-3}$ sr is the range of solid angles over which the laser beam outputs intensity (assumed uniform over $\delta\Omega$), and $\delta\eta$ is the range of wavenumbers over which the intensity is distributed (also assumed uniform). At $10.6\,\mu m$ the medium is known to have an absorption coefficient $\kappa_\eta = 0.15\,\text{cm}^{-1}$. Assuming that the medium has nonreflecting boundaries, determine the exiting total intensity in the normal direction (transmitted laser radiation plus emission, assuming the medium to be gray). Is the emission contribution important? How thick would the medium have to be to make transmission and emission equally important?

8.12 Repeat Problem 8.11 for a medium with refractive index $n = 2$, bounded by vacuum (i.e., a slab with reflecting surfaces). Hint: (1) Part of the laser beam will be reflected when first hitting the slab, part will penetrate into the slab. Part of this energy will be absorbed by the layer, part will hit the rear face, where again a fraction will be reflected back into the slab, and the rest will emerge from the slab, etc. Similar multiple internal reflections will take place with the emitted energy before emerging from the slab. (2) To calculate the slab–surroundings reflectivity, show that the value of the absorptive index is negligible.

8.13 A thin column of gas of cross section δA and length L contains a uniform suspension of small particles that absorb and scatter radiation. The scattering is according to the phase function (a) $\Phi = 1$ (isotropic scattering), (b) $\Phi = 1 + A_1 \cos\Theta$ (linear anisotropic scattering, A_1 is a constant), and (c) $\Phi = \frac{3}{4}(1 + \cos^2\Theta)$ (Rayleigh scattering), where Θ is the angle between incoming and scattered directions. A laser beam hits the column normal to δA. What is the transmitted fraction of the laser power? What fraction of the laser flux goes through an infinite plane at L normal to the gas column? What fraction goes back through a plane at 0? What happens to the rest?

8.14 Repeat Example 8.2 for (a) $\Phi = 1 + A_1 \cos\Theta$ (linear anisotropic scattering, A_1 is a constant), and (b) $\Phi = \frac{3}{4}(1 + \cos^2\Theta)$ (Rayleigh scattering), where Θ is the angle between incoming and scattered directions.

CHAPTER
9

RADIATIVE PROPERTIES OF MOLECULAR GASES

9.1   FUNDAMENTAL PRINCIPLES

Radiative transfer characteristics of an opaque wall can often be described with good accuracy by the very simple model of gray and diffuse emission, absorption, and reflection. The radiative properties of a molecular gas, on the other hand, vary so strongly and rapidly across the spectrum that the assumption of a "gray" gas is almost never a good one [1]. In the present chapter a short development of the radiative properties of molecular gases is given. A more elaborate discussion can be found, for example, in the monograph by Tien [2].

Most of the earlier work was not in the area of heat transfer but rather was carried out by astronomers, who had to deal with light absorption within earth's atmosphere, and by astrophysicists, who studied the spectra of stars. The study of atmospheric radiation was apparently initiated by Lord Rayleigh [3] and Langley [4] in the late nineteenth century. The radiation spectra of stars started to receive attention in the early twentieth century, for example by Eddington [5] and Chandrasekhar [6, 7].

The earliest measurements of radiation from hot gases were reported by Paschen, a physicist, in 1894 [8], but his work was apparently ignored by heat transfer engineers for many years [9].

The last few decades have seen much progress in the understanding of molecular gas radiation, in particular the radiation from water vapor and carbon dioxide, which is of great importance in the combustion of hydrocarbon fuels. Much of the pioneering work since the late 1920s was done by Hottel and coworkers [10–17] (measurements and practical calculations) and by Penner [18] and Plass [19, 20] (theoretical basis).

When a photon (or an electromagnetic wave) interacts with a gas molecule, it may be either absorbed, raising the molecule's energy level, or scattered, changing the direction of travel of the photon. Conversely, a gas molecule may spontaneously lower its energy level by the emission of an appropriate photon. As will be seen in the next chapter on particle properties (since every molecule is, of course, a very small particle), the scattering of photons by molecules is always negligible for heat transfer applications. There are three different types of radiative transitions that lead to a change of molecular energy level by emission or absorption of a photon: (*i*) Transition between nondissociated ("bound") atomic or molecular states, called *bound-bound transitions*, (*ii*) transitions from a "bound" state to a "free" (dissociated) one (absorption) or from "free" to "bound" (emission), called *bound-free transitions*, and (*iii*) transitions between two different "free" states, *free-free transitions*.

The internal energy of every atom and molecule depends on a number of factors, primarily on the energies associated with electrons spinning at varying distances around the nucleus, atoms within a molecule spinning around one another, and atoms within a molecule vibrating against each other. Quantum mechanics postulates that the energy levels for atomic or molecular electron orbit as well as the energy levels for molecular rotation and vibration are quantized; i.e., electron orbits and rotational and vibrational frequencies can only change by certain discrete amounts. Since the energy contained in a photon or electromagnetic wave is directly proportional to frequency, quantization means that, in *bound-bound* transitions, photons must have a certain frequency (or wavelength) in order to be captured or released, resulting in discrete spectral lines for absorption and emission. Since, according to Heisenberg's *uncertainty principle*, the energy level of an atom or molecule cannot be fixed precisely, this phenomenon (and, as we shall see, some others as well) results in a slight broadening of these spectral lines.

Changing the orbit of an electron requires a relatively large amount of energy, or a high-frequency photon, resulting in absorption-emission lines at short wavelengths between the ultraviolet and the near-infrared (between 10^{-2} μm and 1.5 μm). Vibrational energy level changes require somewhat less energy, so that their spectral lines are found in the infrared (between 1.5 μm and 20 μm), while changes in rotational energy levels call for the least amount of energy and, thus, rotational lines are found in the far infrared (beyond 20 μm). Changes in vibrational energy levels may (and often must) be accompanied by rotational transitions, leading to closely spaced groups of spectral lines that, as a result of line broadening, may partly overlap and lead to so-called *vibration-rotation bands* in the infrared.

If the initial energy level of a molecule is very high (e.g., in very high-temperature gases), then the absorption of a photon may cause the breaking-away of an electron or the breakup of the entire molecule because of too strong vibration, i.e., a *bound-free* transition. The postabsorption energy level of the molecule depends on the kinetic energy of the separated part, which is essentially not quantized. Therefore, *bound-free* transitions result in a continuous absorption spectrum over all wavelengths or frequencies for which the photon energy exceeds the required ionization or dissociation energy. The same is true for the reverse process, emission of a photon in a *free-bound* transition (often called *radiative combination*).

In an ionized gas free electrons can interact with the electric field of ions resulting in a *free-free* transition (also known as *Bremsstrahlung*, which is German for *brake radiation*); i.e., the release of a photon lowers the kinetic energy of the electron (decelerates it), or the capture of a photon accelerates it. Since kinetic energy levels of electrons are essentially not quantized, these photons may have any frequency or wavelength.

Bound-free and *free-free* transitions generally occur at very high temperatures (when dissociation and ionization become substantial). The continuum radiation associated with them is usually found at short wavelengths (ultraviolet to visible). Therefore, these effects are of importance only in extremely high-temperature situations. Most engineering applications occur at moderate temperature levels, with little ionization and dissociation, making *bound-bound* transitions most important. At combustion temperatures the emissive power has its maximum in the infrared (between 1 μm and 6 μm), giving special importance to vibration-rotation bands.

9.2 EMISSION AND ABSORPTION PROBABILITIES

There are three different processes leading to the release or capture of a photon, namely, *spontaneous emission*, *induced emission* (or *negative absorption*,) and *induced absorption*. The absorption and emission coefficients associated with these transitions may, at least theoretically, be calculated from quantum mechanics. Complete descriptions of the microscopic phenomena may be found in books on statistical mechanics [21, 22] or spectroscopy [23, 24]. An informative (rather than precise) synopsis has been given by Tien [2] that we shall essentially follow here.

Let there be n_i atoms or molecules (per unit volume) at energy state i and n_j at (lower) energy state j. The difference of energy between the two states is $h\nu$. The number of transitions from state i to state j by release of a photon with energy $h\nu$ (spontaneous emission) must be proportional to the number of atoms or molecules at that level. Thus

$$\left(\frac{dn_i}{dt}\right)_{i \to j} = -A_{ij}n_i, \tag{9.1}$$

where the proportionality constant A_{ij} is known as the *Einstein coefficient for spontaneous emission.* Spontaneous emission is isotropic, meaning that the direction of the emitted photon is random, resulting in equal emission intensity in all directions. Quantum mechanics postulates that, in addition to spontaneous emission, incoming radiative intensity (or photon streams) with the appropriate frequency may induce the molecule to emit photons into the same direction as the incoming intensity (induced emission). Therefore, the total number of transitions from state i to state j may be written as

$$\left(\frac{dn_i}{dt}\right)_{i \to j} = -n_i \left(A_{ij} + B_{ij} \int_{4\pi} I_\nu d\Omega\right), \tag{9.2}$$

where I_ν is the incoming intensity which must be integrated over all directions to account for all possible transitions, and B_{ij} is the *Einstein coefficient for induced emission.* Finally, part of the incoming radiative intensity may be absorbed by molecules at energy state j. Obviously, the absorption rate will be proportional to the strength of incoming radiation as well as the number of molecules that are at energy state j, leading to

$$\left(\frac{dn_j}{dt}\right)_{j \to i} = n_j B_{ji} \int_{4\pi} I_\nu d\Omega, \tag{9.3}$$

where B_{ji} is the *Einstein coefficient for absorption.* The three Einstein coefficients may be related to one another by considering the special case of equilibrium radiation. Equilibrium radiation occurs in an isothermal black enclosure, where the radiative intensity is everywhere equal to the blackbody intensity $I_{b\nu}$ and where the average number of molecules at any given energy level is constant at any given time, i.e., the number of transitions from energy level i to j is equal to the ones from j to i, or

$$\left(\frac{dn_i}{dt}\right)_{i \to j} + \left(\frac{dn_j}{dt}\right)_{j \to i} = -n_i \left(A_{ij} + B_{ij} \int_{4\pi} I_{b\nu} d\Omega\right)$$

$$+ n_j B_{ji} \int_{4\pi} I_{b\nu} d\Omega = 0. \tag{9.4}$$

At equilibrium the number of particles at any energy level is governed by Boltzmann's distribution law [21], leading to

$$n_j / n_i = e^{-\epsilon_j/kT} \Big/ e^{-\epsilon_i/kT} = e^{h\nu/kT}, \tag{9.5}$$

where ϵ_i and ϵ_j are the energy levels associated with states i and j, respectively. Thus, the blackbody intensity may be evaluated from equation (9.4) as

$$I_{b\nu} = \frac{1}{4\pi} \frac{A_{ij}/B_{ij}}{(B_{ji}/B_{ij}) e^{h\nu/kT} - 1}. \tag{9.6}$$

Comparison with Planck's law, equation (1.9), shows that all three Einstein coefficients are dependent upon another, namely,

$$A_{ij} = \frac{8\pi h \nu^3}{c_0^2} B_{ij}, \quad B_{ij} = B_{ji}. \tag{9.7}$$

The one remaining independent Einstein coefficient is clearly an indicator of how strongly a gas is able to absorb radiation. This is most easily seen by examining the number of induced transitions (by emission and absorption) in a single direction (or within a thin pencil of rays). If

$$\frac{d}{d\Omega}\left(\frac{dn}{dt}\right)_{j \to i} = (n_j B_{ji} - n_i B_{ij}) I_\nu \tag{9.8}$$

is the net number of photons removed from the pencil of rays per unit time and per unit volume, then—since each photon carries the energy $h\nu$—the change of radiative energy per unit time, per unit area and distance, and per unit solid angle is

$$\frac{dI_\nu}{ds} = -h\nu \frac{d}{d\Omega}\left(\frac{dn}{dt}\right)_{i \to j} = -(n_j B_{ji} - n_i B_{ij}) h\nu I_\nu, \tag{9.9}$$

where s is distance along the pencil of rays. The *(linear) spectral absorption coefficient* κ_ν of a gas is defined as

$$\kappa_\nu = (n_j B_{ji} - n_i B_{ij}) h\nu. \tag{9.10}$$

so that

$$\frac{dI_\nu}{ds} = -\kappa_\nu I_\nu, \tag{9.11}$$

which is, of course, identical to equation (8.4). The absorption coefficient as defined here is often termed the *effective absorption coefficient* since it incorporates induced emission (or negative absorption). Sometimes a *true absorption coefficient* is defined as

$$\kappa_\nu = n_j B_{ji} h\nu . \tag{9.12}$$

Since induced emission and induced absorption always occur together and cannot be separated, it is general practice to incorporate induced emission into the absorption coefficient, so that only the effective absorption coefficient needs to be considered.[1] Examination of equation (9.10) shows that the absorption coefficient is proportional to molecular number density. Therefore, as mentioned earlier, a number of researchers take the number density out of the definition for κ_ν either in the form of density or

---

[1] Since it is experimentally impossible to distinguish induced emission from induced absorption, its existence had initially been questioned. Equation (9.6) is generally accepted as proof that induced emission does indeed exist: Without it $B_{ij} \to 0$ and the blackbody intensity would be governed by Wien's distribution, equation (1.18), which is known to be incorrect.

pressure, by defining a *mass absorption coefficient* or a *pressure absorption coefficient*, respectively, as

$$\kappa_{\rho\nu} \equiv \frac{\kappa_\nu}{\rho}, \quad \kappa_{p\nu} \equiv \frac{\kappa_\nu}{p}. \tag{9.13}$$

If a mass or pressure absorption coefficient is used, then a ρ or p must, of course, be added to equation (9.11).[2]

The negative of equation (9.1) gives the rate at which molecules emit photons of strength $h\nu$ randomly into all directions (into a solid angle of 4π) and per unit volume. Thus, multiplying this equation by $-h\nu$ and dividing by 4π gives isotropic energy emitted per unit time, per unit solid angle, per unit area and distance along a pencil of rays or, in short, the change of intensity per unit distance due to spontaneous emission:

$$\frac{dI_\nu}{ds} = -h\nu \frac{d}{d\Omega}\left(\frac{dn}{dt}\right)_{i \to j} = n_i A_{ij} h\nu/4\pi. \tag{9.14}$$

Using equations (9.5), (9.7) and (9.10) in equation (9.14) leads to

$$\frac{dI_\nu}{ds} = \kappa_\nu \frac{2h\nu^3}{c_0^2}\left(\frac{n_i}{n_j - n_i}\right) = \kappa_\nu I_{b\nu}, \tag{9.15}$$

which represents the augmentation of directional intensity due to spontaneous emission.

9.3 ATOMIC AND MOLECULAR SPECTRA

We have already seen that the emission or absorption of a photon goes hand in hand with the change of rotational and/or vibrational energy levels in molecules, or with the change of electron orbits (in atoms and molecules). This change, in turn, causes a change in radiative intensity resulting in *spectral lines*. In this section we discuss briefly how the position of spectral lines within a vibration-rotation band can be calculated, since it is these bands that are of great importance to the heat transfer engineer. More detailed information as well as discussion of electronic spectra, bound-free and free-free transitions may be found in more specialized books on quantum mechanics [22, 23, 25] or spectroscopy [24, 26–28], or in the detailed monograph on gas radiation properties by Tien [2].

Since every particle moves in three-dimensional space, it has three degrees of freedom: It can move in the forward-backward, left-right and/or upward-downward directions. If two or more particles are connected with each other (diatomic and

---

[2]Thus, depending on what spectral variable is employed (wavelength λ, wavenumber η, or frequency ν), a spectrally integrated absorption coefficient may appear in nine different variations. Often the only way to determine which definition has been used is to check the units given carefully.

polyatomic molecules), then each of the atoms making up the molecule has three degrees of freedom. However, it is more convenient to say that a molecule consisting of N atoms has three degrees of freedom for translation, and $3N-3$ degrees of freedom for relative motion between atoms. These $3N-3$ degrees of internal freedom may be further separated into rotational and vibrational degrees of freedom. This fact is illustrated in Fig. 9-1 for a diatomic molecule and for linear and nonlinear triatomic molecules. The diatomic molecule has three internal degrees of freedom. Obviously, it can rotate around its center of gravity within the plane of the paper or, similarly, perpendicularly to the paper (with the rotation axis lying in the paper). It could also rotate around its own axis; however, neither one of the atoms would move (except for rotating around itself). Thus, the last degree of freedom must be used for vibrational motion between the two atoms as indicated in the figure. The situation gets rapidly more complicated for molecules with increasing number of atoms. For linear triatomic molecules there are, again, only two rotational modes. Since there are six internal degrees of freedom, there are four vibrational modes, as indicated in Fig. 9-1. In contrast, a nonlinear triatomic molecule has three rotational modes: In this case rotation around the horizontal axis in the plane of the paper is legitimate, so that there are only three vibrational degrees of freedom.

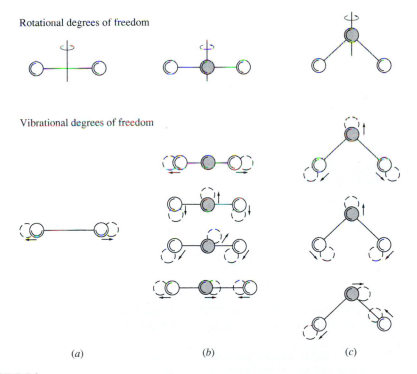

(a)   (b)   (c)

FIGURE 9-1
Rotational and vibrational degrees of freedom for (a) diatomic, (b) linear triatomic, and (c) nonlinear triatomic molecules.

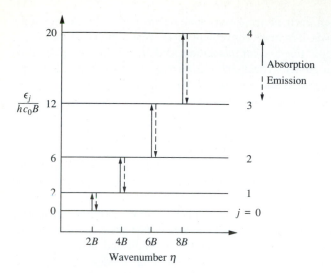

FIGURE 9-2
Spectral position and energy levels for a rigid rotator.

Rotational Transitions

To calculate the allowed rotational energy level from quantum mechanics using *Schrödinger's wave equation* (see, for example, [21, 22]), we generally assume that the molecule consists of point masses connected by rigid massless rods, the so-called *rigid rotator model*. The solution to this wave equation dictates that possible energy levels for a linear molecule are limited to

$$\epsilon_j = \frac{\hbar^2}{2I}j(j+1) = hc_0 Bj(j+1), \quad j = 0, 1, 2, \ldots \; (j \text{ integer}), \qquad (9.16)$$

where $\hbar = h/2\pi$ is the modified Planck's constant, I is the moment of inertia of the molecule, j is the rotational quantum number, and the abbreviation B has been introduced for later convenience. Allowed transitions are $\Delta j = \pm 1$ and 0 (the latter being of importance for a simultaneous vibrational transition); this expression is known as the *selection rule*. In the case of the absorption of a photon ($j \rightarrow j + 1$ transition) the wavenumbers of the resulting spectral lines can then be determined[3] as

$$\eta = (\epsilon_{j+1} - \epsilon_j)/hc_0 = B(j+1)(j+2) - Bj(j+1)$$
$$= 2B(j+1), \qquad\qquad j = 0, 1, 2, \ldots . \quad (9.17)$$

The results of this equation produce a number of equidistant spectral lines (in units of wavenumber or frequency), as shown in the sketch of Fig. 9-2.

---

[3]In our discussion of surface radiative transport we have used wavelength λ as the spectral variable throughout, largely to conform with the majority of other publications. However, for gases frequency ν or wavenumber η are considerably more convenient to use [see, for example, equation (9.17)]. Again, to conform with the majority of the literature, we shall use wavenumber throughout this part.

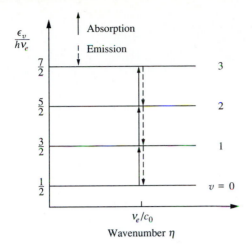

FIGURE 9-3
Spectral position and energy levels for harmonic oscillator.

The rigid rotator model turns out to be surprisingly accurate, although for high rotation rates ($j \gg 0$) a small correction factor due to the centrifugal contribution (stretching of the "rod") may be considered. It should also be noted that not all linear molecules exhibit rotational lines, since an electric dipole moment is required for a transition to occur. Thus, diatomic molecules such as O_2 and N_2 never undergo rotational transitions, while symmetric molecules such CO_2 show a rotational spectrum only if accompanied by a vibrational transition [2]. Evaluation of the spectral lines of nonlinear polyatomic molecules are always rather complicated and the reader is referred to specialized treatises such as the one by Herzberg [27].

Vibrational Transitions

The simplest model of a vibrating diatomic molecule assumes two point masses connected by a perfectly elastic massless spring. Such a model leads to a harmonic oscillation and is, therefore, called the *harmonic oscillator*. For this case the solution to Schrödinger's wave equation for the determination of possible vibrational energy levels is readily found to be

$$\epsilon_v = h\nu_e(v + \tfrac{1}{2}), \quad v = 0, 1, 2, \ldots \ (v \text{ integer}), \tag{9.18}$$

where ν_e is the equilibrium frequency of harmonic oscillation or *eigenfrequency*, and v is the vibrational quantum number. The selection rule for a harmonic oscillator is $\Delta v = \pm 1$ and, thus, one would expect a single spectral line at the same frequency as the harmonic oscillation, or at a wavenumber

$$\eta = (\epsilon_{v+1} - \epsilon_v)/hc_0 = (\nu_e/c_0)(v + 1 - v) = \nu_e/c_0, \tag{9.19}$$

as indicated in Fig. 9-3. Unfortunately, the assumption of a harmonic oscillator leads to considerably less accurate results than the one of a rigid rotator. This fact is easily appreciated by looking at Fig. 9-4, which depicts molecular energy level of a

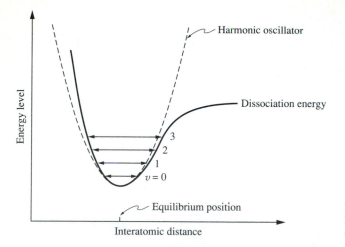

FIGURE 9-4
Energy level vs. interatomic distance.

diatomic molecule vs. interatomic distance: When atoms move toward each other repulsive forces grow more and more rapidly, while the opposite is true when the atoms move apart. The heavy line in Fig. 9-4 shows the minimum and maximum distances between atoms for any given vibrational energy state (showing also that the molecule may dissociate if the energy level becomes too high). In a perfectly elastic spring, force increases linearly with displacement, leading to a symmetric quadratic polynomial for the displacement limits as also indicated in the figure. If a more complicated spring constant is included in the analysis, this results in additional terms in equation (9.18); and the selection rule changes to $\Delta v = \pm 1, \pm 2, \pm 3, \ldots$, producing several approximately equally spaced spectral lines. The transition corresponding to $\Delta v = \pm 1$ is called the *fundamental*, or the *first harmonic*, and usually is by far the strongest one. The transition corresponding to $\Delta v = \pm 2$ is called the *first overtone* or *second harmonic*, and so on. For example, CO has a strong fundamental band at $\eta_0 = 2143\,\text{cm}^{-1}$ and a much weaker first overtone band at $\eta_0 = 4260\,\text{cm}^{-1}$ (see the data in Table 9.3 in Section 9.6).

Combined Transitions

Since the energy required to change the vibrational energy is so much larger than that needed for rotational changes, and since both transitions can (and indeed often must) occur simultaneously, this requirement leads to many closely spaced lines, also called a vibration-rotation band, centered around the wavenumber $\eta = \nu_e/c_0$, which is known as the *band origin* or *band center*.

For the simplest model of a rigid rotator combined with a harmonic oscillator, assuming both modes to be independent, the combined energy level at quantum numbers j, v is given by

$$\epsilon_{vj} = h\nu_e(v + \tfrac{1}{2}) + B_v j(j + 1), \quad v, j = 0, 1, 2, \ldots . \tag{9.20}$$

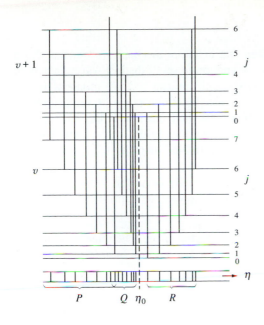

FIGURE 9-5
Typical spectrum of vibration-rotation bands.

Since the small error due to the assumption of a totally rigid rotator can result in appreciable total error when a large collection of simultaneous vibration-rotation transition is considered, allowance has been made in the above expression for the fact that B_v (or the molecular moment of inertia) may depend on the vibrational energy level. The allowed transitions ($\Delta v = \pm 1$ combined with $\Delta j = \pm 1, 0$) lead to three separate branches of the band, namely P ($\Delta j = -1$), Q ($\Delta j = 0$) and R ($\Delta j = +1$) branches, with spectral lines at wavenumbers

$$\eta_P = \eta_0 - (B_{v+1} + B_v)j + (B_{v+1} - B_v)j^2, \quad j = 1, 2, 3, \ldots \quad (9.21a)$$

$$\eta_Q = \eta_0 + (B_{v+1} - B_v)j + (B_{v+1} - B_v)j^2, \quad j = 1, 2, 3, \ldots \quad (9.21b)$$

$$\eta_R = \eta_0 + 2B_{v+1} + (3B_{v+1} - B_v)j$$
$$+ (B_{v+1} - B_v)j^2, \quad j = 0, 1, 2, \ldots \quad (9.21c)$$

where j is the rotational state before the transition. It is seen that there is no line at the band origin. If $B_{v+1} = B_v = \text{const}$, then the Q-branch vanishes and the two remaining branches yield equally spaced lines on both sides of the band center. If $B_{v+1} \neq B_v$, then the R-branch will, for sufficiently large j, fold back toward and beyond the band origin. In that case all lines within the band are on one side of a limiting wavenumber. Those bands, where this occurs close to the band center (i.e, for small j where the line intensity is strong), are known as *bands with a head*. A sketch of a typical vibration-rotation band spectrum is shown in Fig. 9-5. Note that in linear molecules the Q-branch often does not occur as a result of forbidden transitions [2].

9.4 LINE RADIATION

In the previous two sections we have seen that quantum mechanics postulates that a molecular gas can emit or absorb photons at an infinite set of distinct wavenumbers or frequencies. We already observed that no spectral line can be truly monochromatic; rather, absorption or emission occurs over a tiny but finite range of wavenumbers. The results are *broadened spectral lines* that have their maxima at the wavenumber predicted by quantum mechanics. In this section we look briefly at the causes of line broadening and at *line shapes*, i.e., the variation of line intensity with wavenumber for an isolated line. More detailed accounts may be found in more specialized works [2, 18, 24]. The effects of line overlap, which usually occurs in vibration-rotation bands in the infrared, will be discussed in the following section on "Narrow Band Models."

Numerous phenomena cause broadening of spectral lines. The three most important ones are *collision broadening*, *natural line broadening*, and *Doppler broadening*.

Collision Broadening

As the name indicates, *collision broadening* of spectral lines is attributable to the frequency of collisions between gas molecules. The shape of such a line can be calculated from the electron theory of Lorentz[*] (and is, therefore, often called *Lorentz broadening*) or from quantum mechanics [29, 30] as

$$\kappa_\eta = \frac{S}{\pi} \frac{b_C}{(\eta - \eta_0)^2 + b_C^2}, \quad S \equiv \int_{\Delta\eta} \kappa_\eta d\eta, \tag{9.22}$$

where S is the line-integrated absorption coefficient or line strength, b_C is the so-called line half-width in units of wavenumber (half the line width at half the maximum absorption coefficient), and η_0 is the wavenumber at the line center. The shape of a collision-broadened line is shown in Fig. 9-6 (together with the shape of a Doppler-broadened line). Since molecular collisions are proportional to the number density of molecules ($n \propto \rho \propto p/T$) and to the average molecular speed ($v_{av} \propto \sqrt{T}$), it is not surprising that the half-width can be calculated from kinetic theory [30] as

$$b_C = \frac{2}{\sqrt{\pi}} \frac{D^2 p}{c_0 \sqrt{mkT}} = b_{C0} \left(\frac{p}{p_0}\right)\sqrt{\frac{T_0}{T}}, \tag{9.23}$$

where D is the effective diameter of the molecule, m is its mass, p is total gas pressure, T is absolute temperature, and the subscript 0 denotes a reference state. If the absorbing-emitting gas is part of a mixture, the fact that collisions involving only nonradiating gas do not cause broadening must be accounted for by using an *effective*

---

[*]A biographical footnote for Hendrik A. Lorentz may be found in Section 2.6.

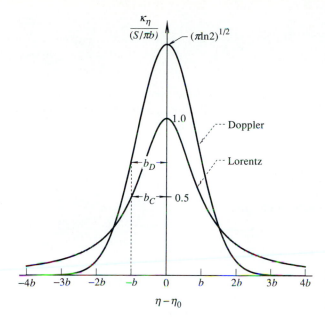

FIGURE 9-6
Spectral line shape for collision and Doppler broadening (for equal line intensity).

pressure [31]

$$p_e = F_a p_a + \sum_k F_k p_k, \tag{9.24}$$

where F_a and p_a are the self-broadening coefficient and partial pressure of the radiating gas, while the F_k and p_k are broadening coefficients and partial pressures of the kth nonradiating gas, respectively.

Natural Line Broadening

Heisenberg's uncertainty principle states that no energy (and, therefore, no spectral position caused by an energy transition) can be measured precisely, thus causing spectral lines to be broadened. The line shape for natural line broadening is identical to the one of collision broadening and, therefore, the line half-width b_N is usually simply added to the b_C obtained from collision broadening in equation (9.22).

Doppler Broadening

According to the Doppler effect a wave traveling toward an observer appears slightly compressed (shorter wavelength or higher frequency) if the emitter is also moving toward the observer, and slightly expanded (longer wavelength or lower frequency) if the emitter is moving away. This is true whether the wave is a sound wave (for example, the pitch of a whistle of a train passing an observer) or an electromagnetic wave. Thus,

$$\eta_{obs} = \eta_{em}\left(1 + \frac{\mathbf{v} \cdot \hat{\mathbf{s}}}{c}\right),\tag{9.25}$$

where $\mathbf{v}$ is the velocity of the emitter and $\hat{\mathbf{s}}$ is a unit vector pointing from the emitter to the observer. Assuming local thermodynamic equilibrium so that Maxwell's velocity distribution applies, one can calculate the line profile as [18]

$$\kappa_\eta = \sqrt{\frac{\ln 2}{\pi}}\left(\frac{S}{b_D}\right)\exp\left[-(\ln 2)\left(\frac{\eta - \eta_0}{b_D}\right)^2\right],\tag{9.26}$$

where b_D is the Doppler line half-width, given by

$$b_D = \frac{\eta_0}{c_0}\sqrt{\frac{2kT}{m}}\ln 2,\tag{9.27}$$

and m is the mass of the radiating molecule. Note that, unlike during collision and natural line broadening, the Doppler line width depends on its spectral position. The different line shapes were compared in Fig. 9-6. For equal overall strength, the Doppler line is much more concentrated near the line center.

Combined Effects

In most engineering applications, collision broadening, which is proportional to $p/\sqrt{T}$, is by far the most important broadening mechanism. Only at very high temperatures (when, owing to the distribution of the Planck function, transitions at large η are most important) may Doppler broadening, with its proportionality to $\eta\sqrt{T}$, become dominant. Thus, while an absorption coefficient for combined effects has been given [18], the use of the collision broadening profile is usually sufficient.

Example 9.1. The half-width of a certain spectral line of a certain gas has been measured to be 0.05 cm^{-1} at room temperature (300 K) and 1 atm. When the line-half width is measured at 1 atm and 3000 K, it turns out that the width has remained unchanged. Estimate the contributions of Doppler and collision broadening in both cases.

Solution

As a first approximation we assume that the widths of both contributions may be added to give the total line half-width (this is a fairly good approximation if one makes a substantially larger contribution than the other). Therefore, we may estimate

$$b_{C1} + b_{D1} \approx b_1 = b_2 \approx b_{C2} + b_{D2}$$

and, from equations (9.23) and (9.27),

$$\frac{b_{C2}}{b_{C1}} = \sqrt{\frac{T_1}{T_2}} = \frac{1}{\sqrt{10}}, \qquad \frac{b_{D2}}{b_{D1}} = \sqrt{\frac{T_2}{T_1}} = \sqrt{10}.$$

Eliminating the Doppler widths from these equations we obtain

$$\frac{b_{C1}}{b_1} = \frac{\sqrt{10}}{9}\left(\sqrt{10} - \frac{b_2}{b_1}\right) = 0.76,$$

and

$$\frac{b_{C2}}{b_2} = \frac{1}{9}\left(\sqrt{10}\frac{b_1}{b_2} - 1\right) = 0.24.$$

We see that at room temperature, collision broadening is about three times stronger than Doppler broadening, while exactly the reverse is true at $3000\,\mathrm{K}$.

Radiation from Isolated Lines

Combining equations (9.11) and (9.15) gives the complete equation of transfer for an absorbing-emitting (but not scattering) medium,

$$\frac{dI_\eta}{ds} = \kappa_\eta(I_{b\eta} - I_\eta), \tag{9.28}$$

where the first term of the right-hand side represents augmentation due to emission and the second term is attenuation due to absorption. Let us assume we have a layer of an isothermal and homogeneous gas of thickness L. Then neither $I_{b\eta}$ nor κ_η is a function of location and the solution to the equation of transfer is

$$I_\eta(X) = I_\eta(0)e^{-\kappa_\eta X} + I_{b\eta}\left(1 - e^{-\kappa_\eta X}\right), \tag{9.29}$$

where the *optical path length* X is equal to L if a linear absorption coefficient is used (*geometric path length*), or equal to L multiplied by partial density (*density path length*) or pressure (*pressure path length*) of the radiating gas if either mass or pressure absorption coefficient is used. Thus, the difference between entering and exiting intensity, integrated over the entire spectral line, is

$$I(X) - I(0) = \int_{\Delta\eta} [I_\eta(X) - I_\eta(0)]d\eta$$

$$\approx [I_{b\eta 0} - I_{\eta 0}(0)] \int_{\Delta\eta} \left(1 - e^{-\kappa_\eta X}\right)d\eta, \tag{9.30}$$

where the assumption has been used that neither incoming nor blackbody intensity can vary appreciably over the width of a single spectral line. The integrand of the factor

$$W = \int_{\Delta\eta} \left(1 - e^{-\kappa_\eta X}\right)d\eta \tag{9.31}$$

is the fraction of incoming radiation absorbed by the gas layer at any given wavenumber, and it is also the fraction of the total emitted radiation that escapes from the layer (not undergoing self-absorption). W is commonly called the *equivalent line width* since a line of width W with infinite absorption coefficient would have the identical effect on absorption and emission. The equivalent line width for a Lorentz line may be evaluated by substituting equation (9.22) into equation (9.31) to yield

$$W = SXe^{-x}[I_0(x) + I_1(x)], \tag{9.32}$$

where

$$b_L \equiv b_C + b_N, \qquad x \equiv SX/2\pi b_L, \tag{9.33}$$

and the I_0 and I_1 are modified Bessel functions. For simpler evaluation, equation (9.32) may be approximated as reported by Tien [2] as

$$W = SX\sqrt{\frac{2}{\pi x}}\sqrt{1 - \exp(-\pi x/2)}. \tag{9.34}$$

Asymptotic values for W are easily obtained as

$$W = \quad SX, \qquad x \ll 1, \tag{9.35}$$

$$W = \ 2\sqrt{SXb_L}, \qquad x \gg 1. \tag{9.36}$$

Comparing equation (9.33) with equation (9.22), evaluated at the line center, shows that x is the nondimensional *optical thickness* of the gas layer $\kappa_\eta X$ at the line center. Therefore, the parameter x gives an indication of the strength of the line. For a weak line ($x \ll 1$) little absorption takes place so that every position in the gas layer receives the full irradiation, resulting in a linear absorption rate (with distance). In the case of a strong line ($x \gg 1$) the radiation intensity has been appreciably weakened before exiting the gas layer, resulting in locally lesser absorption and causing the square-root dependence of equation (9.36).

9.5   NARROW BAND MODELS

A single spectral line at a certain spectral position was fully characterized by its strength (the intensity, or integrated absorption coefficient) and its line half-width (plus knowledge of the broadening mechanism, i.e., collision or Doppler broadening). However, a vibration-rotation band has many closely spaced spectral lines that may overlap considerably. While the absorption coefficients for individual lines may simply be added to give the absorption coefficient of an entire band at any spectral position,

$$\kappa_\eta = \sum_j \kappa_{\eta j}, \tag{9.37}$$

the resulting function tends to gyrate violently across the band unless the lines overlap very strongly. This tendency makes heat transfer calculations extremely difficult to carry out, if the exact relationship is to be used in the spectral integration for total intensity [equation (8.30)], total radiative heat flux [equation (8.47)], or the divergence of the heat flux [equation (8.54)]. Careful examination of the governing equations shows, however, that all spectral integrations may be reduced to two cases, namely,

$$\int_0^\infty \kappa_\eta I_{(b)\eta} \, d\eta \quad \text{and} \quad \int_0^\infty I_{(b)\eta} \left[1 - \exp\left(- \int_0^s \kappa_\eta \, ds\right) \right] d\eta, \tag{9.38}$$

where $I_{(b)\eta}$ denotes that either $I_{b\eta}$ or I_η can occur. It is clear from inspection of Fig. 1-5 that the Planck function will never vary appreciably over the spectral range of a few lines, considering that adjacent lines are very closely spaced (measured in fractions of cm^{-1}). But the local radiation intensity can also be expected to be relatively smooth, since it is due to emission from all locations in the medium (with their varying temperatures) and the bounding walls (which will emit at wavenumbers where the gas is very transparent), and is further smoothed by absorption and scattering. Therefore, we may simplify expressions (9.38), with extremely good accuracy, to

$$\int_0^\infty I_{(b)\eta} \left\{ \frac{1}{\Delta\eta} \int_{\eta-\Delta\eta/2}^{\eta+\Delta\eta/2} \kappa_\eta \, d\eta' \right\} d\eta \tag{9.39a}$$

and

$$\int_0^\infty I_{(b)\eta} \left\{ \frac{1}{\Delta\eta} \int_{\eta-\Delta\eta/2}^{\eta+\Delta\eta/2} \left[1 - \exp\left(- \int_0^s \kappa_\eta \, ds\right) \right] d\eta' \right\} d\eta. \tag{9.39b}$$

The expressions within the large braces are local averages of the spectral absorption coefficient and of the spectral emissivity, respectively, indicated by an overbar:[4]

$$\bar{\kappa}_\eta(\eta) = \frac{1}{\Delta\eta} \int_{\eta-\Delta\eta/2}^{\eta+\Delta\eta/2} \kappa_\eta \, d\eta', \tag{9.40}$$

$$\bar{\epsilon}_\eta(\eta) = \frac{1}{\Delta\eta} \int_{\eta-\Delta\eta/2}^{\eta+\Delta\eta/2} \left[1 - \exp\left(- \int_0^s \kappa_\eta \, ds\right) \right] d\eta'. \tag{9.41}$$

One can expect that the spectral variation of $\bar{\kappa}$ and $\bar{\epsilon}$ is relatively smooth over the band, making spectral integration of radiative heat fluxes feasible.

To find spectrally averaged or "narrow band" values of the absorption coefficient and the emissivity, some information must be available on the spacing of individual lines within the group and on their relative strengths. A number of models have been proposed to this purpose, of which the two extreme ones are the *Elsasser model*, in which equally spaced lines of equal intensity are considered, and the *statistical model*, in which the spectral lines are assumed to have random spacing and/or intensity. A typical spectral line arrangement for these two extreme models is shown in Fig. 9-7. The main distinction between the two models is the difference in line overlap. Both models will predict the same narrow band parameters for optically thin situations or

---

[4]It should be understood that the definition of $\bar{\kappa}$ in equation (9.40) is not sufficient since $\bar{\epsilon} \neq 1 - \exp(-\bar{\kappa}s)$. This fact will be demonstrated in Example 9.2.

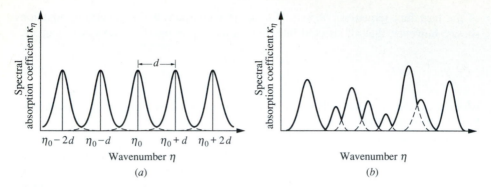

FIGURE 9-7
Typical spectral line arrangement for (a) Elsasser and (b) statistical model.

nonoverlap conditions (since overlap has no effect), as well as for optically very strong situations (since no beam can penetrate through the gas, regardless of the overlapping characteristics). Under intermediate conditions the Elsasser model will always predict a higher emissivity/absorptivity than the statistical model, since regular spacing always results in less overlap (for the same average absorption coefficient) [2]. The deviation between the two models is never more than 20%. Thus, if we consider (*i*) the poor accuracy of measurements and of correlations of radiative properties (see, for example, Section 9.8), and (*ii*) the difficulty of obtaining accurate analytical solutions, there appears to be little need for other, more complicated models tailored toward the behavior of particular gases.

The Elsasser Model

We saw earlier in this chapter that diatomic molecules and linear polyatomic molecules have only two identical rotational modes, resulting in a single set of lines (consisting of two or three branches, as shown in Figs. 9-2 and 9-5). For these gases one may expect spectral lines with nearly constant spacing and slowly varying intensity, in particular if the Q-branch is unimportant (or "forbidden") and if the folding-back of the R-branch gives also only a small contribution.

Since in most engineering applications, collision broadening is the dominant broadening mechanism, we shall limit ourselves here to lines of the Lorentz profile. Summing up the contributions from infinitely many lines on both sides of an arbitrary line with center at η_0, we get

$$\kappa_\eta = \sum_{i=-\infty}^{\infty} \frac{S}{\pi} \frac{b_L}{(\eta - \eta_0 - id)^2 + b_L^2}, \qquad (9.42)$$

where d is the (constant) spacing between spectral lines.[5] This series may be evaluated

---

[5] Since we are using wavenumber here, the value for d is measured in units of wavenumbers, cm^{-1}. If we were to use frequency or wavelength, the definition and units of d would correspondingly change.

in closed form, as was first done by Elsasser, resulting in

$$\kappa_\eta = \frac{S}{d} \frac{\sinh 2\beta}{\cosh 2\beta - \cos(z - z_0)} ,$$ (9.43)

where

$$\beta \equiv \pi b_L/d, \qquad z \equiv 2\pi\eta/d.$$ (9.44)

From equation (9.40), the average absorption coefficient is simply

$$\overline{\kappa}_\eta = \frac{S}{d}.$$ (9.45)

This expression is the equivalent of forcing each line to stay within "its" box of width d, over which it is then averaged. The spectrally averaged emissivity may now be evaluated from equation (9.41) as

$$\overline{\epsilon}_\eta = 1 - \frac{1}{2\pi} \int_{-\pi}^{\pi} \exp\left(-\frac{2\beta x \sinh 2\beta}{\cosh 2\beta - \cos z}\right) dz,$$ (9.46)

where, since the absorption coefficient is a periodic function, one full period was chosen for the averaging wavenumber range and, thus, the arbitrary location z_0 could be eliminated. As one may see from its definition, equation (9.44), β is the *line overlap parameter*; β gives an indication of how much the individual lines overlap each other, and x, already defined in equation (9.33), is the *line strength parameter*. At this point we may also define another nondimensional parameter, the *narrow band optical thickness* $\tau = \overline{\kappa}X$, so that we now have three characterizing parameters, namely,

$$x = \frac{SX}{2\pi b_L}, \qquad \beta = \pi \frac{b_L}{d}, \qquad \tau = \frac{S}{d}X = 2\beta x.$$ (9.47)

Equation (9.46) cannot be solved in closed form, but an accurate approximate expression, known as the Godson approximation, has been given [30]:

$$\overline{\epsilon}_\eta = \text{erf}\left(\frac{\sqrt{\pi}}{2}\frac{W}{d}\right) = \text{erf}\left(\frac{\sqrt{\pi}}{2}\frac{S}{d}Xe^{-x}[I_0(x) + I_1(x)]\right),$$ (9.48)

where erf is the *error function* and is tabulated in standard mathematical texts [32]. While one continuous expression is very useful in machine calculations, it is often desirable to have simpler expressions for hand calculations. We can distinguish among three different limiting regimes:

$$\text{weak lines } (x \ll 1): \ \overline{\epsilon}_\eta = 1 - \exp\left(-\frac{S}{d}X\right) = 1 - e^{-\tau},$$ (9.49)

$$\text{strong lines } (x \gg 1): \ \overline{\epsilon}_\eta = \text{erf}\left(\sqrt{\pi \frac{S}{d} \frac{b_L}{d} X}\right) = \text{erf}\left(\sqrt{\tau\beta}\right),$$ (9.50)

$$\text{no overlap } (\beta \ll 1): \ \overline{\epsilon}_\eta = \frac{W}{d} \approx \sqrt{\frac{4}{\pi}\tau\beta \left(1 - e^{-\pi\tau/4\beta}\right)},$$ (9.51)

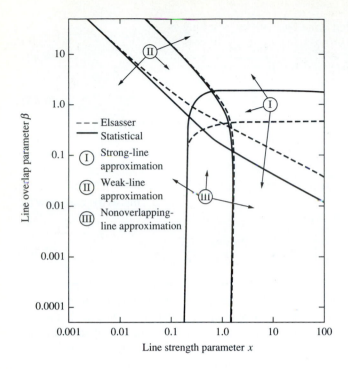

FIGURE 9-8

Regions of validity for Elsasser and statistical band model regimes [2].

where the W/d in equation (9.51) can possibly be further simplified using equations (9.35) and (9.36). These relations are summarized in Table 9.1 and their ranges of validity are indicated in Fig. 9-8 [2].

The Statistical Model

In the statistical model it is assumed that the spectral lines are not equally spaced but, rather, are randomly distributed across the narrow band. This assumption can be expected to be an accurate representation for complex molecules for which lines from different rotational modes overlap in an irregular fashion. A statistical analysis has been carried out for two different intensity distribution models [18, 19, 30]. In the *uniform statistical model* the line intensity is assumed to be constant (as in the Elsasser model), leading to

$$\bar{\epsilon}_\eta = 1 - \exp\left(-\frac{W}{d}\right),$$
(9.52)

where W/d may again be approximated by equations (9.35) and (9.36) for weak and strong lines, respectively, and $W/d \ll 1$ for nonoverlapping lines.

The statistical model has also been evaluated for exponentially decaying line

TABLE 9.1
Summary of effective line widths and narrow band emissivities for Lorentz lines

	Weak line	Strong line	No overlap	All regimes
	$x \ll 1$	$x \gg 1$	$\beta \ll 1$	
Single line, W	SX	$2\sqrt{SXb_L}$		$SXe^{-x}[I_0(x) + I_1(x)]$
$\dfrac{W}{d}$	τ	$2\sqrt{\dfrac{\tau\beta}{\pi}}$		$2\sqrt{\dfrac{\tau\beta}{\pi}}\left(1 - e^{-\pi\tau/4\beta}\right)$
Elsasser Model, $\bar{\epsilon}_\eta$				
$S = \text{const}$ $\quad$ $d = \text{const}$	$1 - e^{-\tau}$	$\text{erf}\left(\sqrt{\tau\beta}\right)$	$\dfrac{W}{d}$	$\text{erf}\left(\dfrac{\sqrt{\pi}}{2}\dfrac{W}{d}\right)$
Statistical Model, $\bar{\epsilon}_\eta$				
$S = \text{const}$ $\quad$ d random	$1 - e^{-\tau}$	$1 - \exp\left(-2\sqrt{\dfrac{\tau\beta}{\pi}}\right)$	$\dfrac{W}{d}$	$1 - \exp\left(-\dfrac{W}{d}\right)$
Statistical Model, $\bar{\epsilon}_\eta$				
S exponential $\quad$ d random	$1 - e^{-\tau}$	$1 - \exp\left(-\sqrt{\tau\beta}\right)$	$\dfrac{\tau}{\sqrt{1 + \tau/\beta}}$	$1 - \exp\left(-\dfrac{\tau}{\sqrt{1 + \tau/\beta}}\right)$

$$\text{Definitions:} \quad x = \frac{SX}{2\pi b_L}, \quad \beta = \pi\frac{b_L}{d}, \quad \tau = \frac{S}{d}X = 2\beta x$$

intensities (*general statistical model*), resulting in

$$\bar{\epsilon}_\eta = 1 - \exp\left(-\frac{(S/d)X}{\sqrt{1 + SX/(\pi b_L)}}\right) = 1 - e^{-\tau/\sqrt{1+\tau/\beta}}. \qquad (9.53)$$

The results from the statistical models have also been summarized in Table 9.1 and Fig. 9-8. The narrow band emissivities from all three models are compared in Fig. 9-9 as a function of the optical path (i.e., average absorption coefficient S/d multiplied by distance X). Note that all predictions are relatively close to each other, although the statistical models may predict up to 20% lower emissivities for optically thick situations. Both statistical models more or less coincide except for optically thin situations where the uniform statistical model coincides with the Elsasser model.

Example 9.2. The following data are known at a certain spectral location for a pure gas at 300 K and 0.75 atm: The mean line spacing is $0.6\,\text{cm}^{-1}$, the mean line half-width is $0.03\,\text{cm}^{-1}$, and the mean line intensity (or integrated absorption coefficient) is $0.08\,\text{cm}^{-2}\,\text{atm}^{-1}$. What is the mean spectral emissivity for geometric path lengths of 1 cm and 1 m, if the gas is diatomic (such as CO), or if the gas is polyatomic (such as water vapor)?

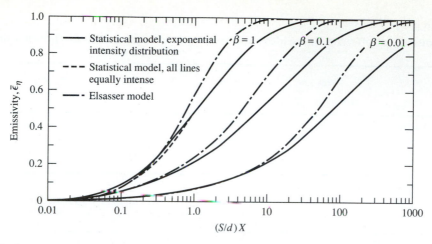

FIGURE 9-9
Mean spectral emissivities as function of optical path $(S/d)X$ [19].

Solution

Since the units of the given line intensity tell us that a pressure absorption coefficient has been used, we need to employ a pressure path length $X = ps$. For a path length of 1 cm we get $X = 0.75\,\text{atm} \times 1\,\text{cm} = 0.75\,\text{cm atm}$ and $x = SX/2\pi b = 0.08\,\text{cm}^{-2}\,\text{atm}^{-1} \times 0.75\,\text{cm atm}/(2\pi\,0.03\,\text{cm}^{-1}) = 1/\pi$, while the overlapping parameter turns out to be $\beta = \pi b/d = \pi \times 0.03\,\text{cm}^{-1}/0.6\,\text{cm}^{-1} = \pi/20$, and $\tau = 2\beta x = 2(\pi/20)(1/\pi) = 0.1$. Checking Fig. 9-8, we see that these values of x and β lie in the "nonoverlapping line" range. Thus, for a diatomic gas for which the Elsasser model should be more accurate, we can use either equation (9.32) or (9.34). The latter leads to $\bar{\epsilon}_\eta = 2\sqrt{(1/10)(\pi/20)(1/\pi)}\sqrt{1 - \exp[-(\pi/10)/(4\pi/20)]} = 0.0887$ or 8.9%. If the gas is polyatomic we may want to use equation (9.53), leading to $\bar{\epsilon}_\eta = 0.0752$. If the path length is a full meter, we have $X = 75\,\text{cm atm}$ and $x = 100/\pi$ while still $\beta = \pi/20$ and now $\tau = 10$. Thus we are in the strong-line region. For the diatomic gas, from equation (9.50) $\bar{\epsilon}_\eta = \text{erf}[\sqrt{10(\pi/20)}] = \text{erf}(1.2533) = 0.924$. For the polyatomic gas, again using equation (9.53), we get $\bar{\epsilon}_\eta = 0.712$.

In the first two cases, using the simple relation $\bar{\epsilon} = 1 - \exp(-\bar{\kappa}s)$ actually would have given fairly good results (0.095) because the gas is optically thin resulting in linear absorption at every wavenumber. For the larger path we would have gotten $1 - e^{-10} \approx 1$. Thus, using an average value for the absorption coefficient makes the gas opaque at *all* wavenumbers rather than only near the line centers.

Example 9.3. For a certain polyatomic gas the line-width-to-spacing ratio and the average absorption coefficient for a vibration-rotation band in the infrared are known as

$$\left(\frac{S}{d}\right)_\eta \approx \left(\frac{S}{d}\right)_0 e^{-2|\eta - \eta_0|/\omega}, \qquad \left(\frac{S}{d}\right)_0 = 10\,\text{cm}^{-1}, \tag{9.54}$$

$$\omega = 50\,\text{cm}^{-1}, \qquad \frac{b}{d} \approx 0.1 \approx \text{const.}$$

Find an expression for the averaged spectral emissivity and for the total band absorptance, defined by

$$A \equiv \int_{band} \epsilon_\eta d\eta = \int_0^\infty \left(1 - e^{-\kappa_\eta X}\right) d\eta$$

for a path length of 20 cm.

Solution

Calculating the optical thickness $\tau_0 = (S/d)_0 X = 10 \times 20 = 200$, the overlap parameter $\beta = \pi/10$, and the line strength $x_0 = \tau_0/2\beta = 1000/\pi$, we find that this band falls into the "strong-line" regime everywhere except in the (unimportant) far band wings. Since we have a polyatomic molecule with exponential decay of intensity, the general statistical model, equation (9.53), should provide the best answer, or

$$\bar{\epsilon}_\eta = 1 - e^{-\tau/\sqrt{1+\tau/\beta}} \approx 1 - e^{-\sqrt{\tau\beta}},$$

since $\tau/\beta \gg 1$. Substituting the calculated values into equation (9.53) yields the spectral emissivity,

$$\bar{\epsilon}_\eta = 1 - \exp\left(-\sqrt{\tau_0\beta}\, e^{-|\eta-\eta_0|/\omega}\right).$$

Integrating this equation over the entire band gives the total band absorptance,

$$A = \int_0^\infty \left[1 - \exp\left(-\sqrt{\tau_0\beta}\, e^{-|\eta-\eta_0|/\omega}\right)\right] d\eta.$$

Realizing that this integral has two symmetric parts and with $\ln z = -(\eta - \eta_0)/\omega$, we have

$$A = 2\omega \int_0^1 \left[1 - \exp\left(-\sqrt{\tau_0\beta}\, z\right)\right] \frac{dz}{z}.$$

This integral may be solved in terms of exponential integrals[6] as given, for example, in Abramowitz and Stegun [32]. This leads to

$$A = 2\omega \left(E_1(\sqrt{\tau_0\beta}) + \ln(\sqrt{\tau_0\beta}) + \gamma_E\right) = 264.7\,\text{cm}^{-1},$$

where $\gamma_E = 0.57721\ldots$ is Euler's constant.

Nonisothermal Gases

Up to this point in calculating narrow band emissivities we have tacitly assumed that the gas is isothermal, i.e., we replaced the integral $\int_0^X \kappa\, dX$ in equation (9.41) by

---

[6]Exponential integrals are discussed in some detail in Appendix E.

κX. We now want to expand our results to include nonisothermal gases. Clearly, the solution to equation (9.46) is too complex to make a straightforward solution with temperature-dependent properties possible. Instead we start by looking at extreme situations: If the lines overlap very strongly ($\beta \to \infty$), then the spectral absorption coefficient is a smooth function and $\kappa = S/d$, so that equation (9.53) reduces to equation (9.49) and we may replace them by

$$\beta \gg 1 : \bar{\epsilon}_\eta = 1 - e^{-\bar{\tau}}, \quad \bar{\tau} = \int_0^X \frac{S}{d} \, dX, \tag{9.55}$$

where $\bar{\tau}$ is now a path-averaged optical depth. As another extreme we look at the case of very strong, narrow lines that overlap very little. Then we get from equation (9.53)

$$\frac{1}{\Delta\eta} \int_{\Delta\eta} \exp\left(-\int_0^X \kappa_\eta \, dX\right) d\eta \approx \exp\left(-\sqrt{\bar{\tau}\tilde{\beta}}\right),$$

and, thus, we may define a suitable $\tilde{\beta}$ as

$$\tilde{\beta} \equiv \frac{1}{\bar{\tau}} \int_0^X \frac{S}{d} \beta \, dX. \tag{9.56}$$

One may now patch these two definitions together, following equation (9.53), to find

$$\bar{\epsilon}_\eta = 1 - e^{-\bar{\tau}/\sqrt{1+\bar{\tau}/\tilde{\beta}}}, \tag{9.57}$$

where the $\bar{\tau}$ and $\tilde{\beta}$ are determined from their definitions in equations (9.55) and (9.56). This equation, known as the Curtis-Godson two-parameter narrow band scaling approximation [1, 30], has been successfully applied to nonisothermal gases.

9.6 WIDE BAND MODELS

The heat transfer engineer is usually only interested in obtaining heat fluxes or divergences of heat fluxes integrated over the entire spectrum. Therefore, it is desirable to have models that can more readily predict the total absorption or emission from a band than was done in Example 9.3. These models are known as *wide band models* since they treat the spectral range of the entire band.

Strength of Spectral Lines within a Band

In equation (9.10) we related the spectral absorption coefficient to the Einstein coefficients B_{ji} and B_{ij} before knowing how such a transition takes place. Since equation (9.9) describes the entire absorption for a transition from one allowed energy level to another, we must replace κ_ν by the line-integrated absorption coefficient S to account for line broadening. Thus, for a combined vibrational (from vibrational quantum number v to v') and rotational (from rotational quantum number j to j')

transition, equation (9.10) may be rewritten as

$$S = B_{(vj)(v'j')}(n_{vj} - n_{v'j'})hc\eta, \tag{9.58}$$

where η is the associated transition wavenumber from equations (9.21a) through (9.21c). The molecular number density n_{vj} and Einstein coefficient $B_{(vj)(v'j')}$ may be calculated from quantum mechanics through very lengthy and complex calculations. For example, much of Penner's book [18] is devoted to this subject. For a rigid rotator–harmonic oscillator, with the additional assumptions that the bandwidth is small compared with the wavenumber at the band center and that only the P and R branches are important, equation (9.58) may be restated as

$$S_{Pj} = \alpha j \frac{e^{-hcB_v j(j+1)}}{\sum\limits_{k=0}^{\infty}(2k+1)e^{-hcB_v k(k+1)}}, \qquad j = 1, 2, 3, \ldots \tag{9.59}$$

$$S_{Rj} = \alpha(j+1) \frac{e^{-hcB_v j(j+1)}}{\sum\limits_{k=0}^{\infty}(2k+1)e^{-hcB_v k(k+1)}}, \qquad j = 0, 1, 2, \ldots \tag{9.60}$$

where

$$\alpha = \sum\limits_{j=1}^{\infty} S_{Pj} + \sum\limits_{j=0}^{\infty} S_{Rj} \tag{9.61}$$

is the combined strength of all lines. If the lines are closely spaced, the sums may be replaced by an integral (with $\Delta j \to d\eta/d$), that is,

$$\alpha = \int_0^{\infty}\left(\frac{S}{d}\right)_{\eta} d\eta. \tag{9.62}$$

Indeed, the *band strength parameter* or *band intensity* α is defined by

$$\alpha \equiv \int_0^{\infty} \kappa_\eta \, d\eta = \int_0^{\infty}\left(\frac{S}{d}\right)_{\eta} d\eta. \tag{9.63}$$

Examination of quantum mechanical results also shows that S_η and α (when based on a linear absorption coefficient) are proportional to density of the absorbing/emitting gas and are a (rather involved) function of temperature. Thus, we find the very important result that band intensity and the spectrally averaged absorption coefficient—when based on a mass absorption coefficient—are functions of temperature only (no pressure dependence)![7]

Equations (9.59) and (9.60) may be used for accurate wide band calculations. These have been attempted by Greif and coworkers [33, 34] in a series of papers.

---

[7]See also equation (9.69) in the exponential wide band model.

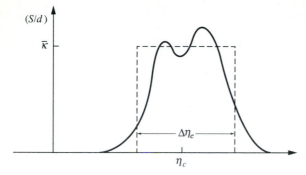

FIGURE 9-10
The box model for the approximation of total band absorptance.

While such calculations are more accurate, they tend to be too involved, so simpler methods are sought for practical applications.

The Box Model

In this very simple model the band is approximated by a rectangular box of width $\Delta\eta_e$ (the effective band width) and height $\overline{\kappa}$ as shown in Fig. 9-10. With these assumptions we can calculate the total band absorptance as

$$A \equiv \int_{\text{band}} \epsilon_\eta \, d\eta = \int_0^\infty \left(1 - e^{-\kappa_\eta X}\right) d\eta = \Delta\eta_e \left(1 - e^{-\overline{\kappa} X}\right), \tag{9.64}$$

where both $\Delta\eta_e$ and $\overline{\kappa}$ may be functions of temperature and pressure. The box model was developed by Penner [18] and successfully applied to diatomic gases. However, the determination of the effective band width is something of a "black art." Once $\Delta\eta_e$ has been found (by using the somewhat arbitrary criterion given by Penner [18] or some other means), $\overline{\kappa}$ may be related to the *band intensity* α by

$$\overline{\kappa} = \alpha/\Delta\eta_e. \tag{9.65}$$

If the molecular gas layer forms a radiation barrier between two surfaces of unequal temperature, then a suitable choice for the effective band width can give quite reasonable results. However, if emission from a hot gas is considered, then the results become very sensitive to the correct choice of $\Delta\eta_e$. Nevertheless, the box model—because of its great simplicity—enjoys considerable popularity for use in heat transfer models (see Chapter 17).

Example 9.4.   Calculate the effective band width $\Delta\eta_e$ for which the box model predicts the correct total band absorptance for Example 9.3.

Solution

Integrating equation (9.54) over the entire band gives $\alpha = (S/d)_0 \times \omega = 500 \, \text{cm}^{-2}$ and $\overline{\kappa} X = \alpha X/\Delta\eta_e = 10{,}000 \, \text{cm}^{-1}/\Delta\eta_e$. Equation (9.64) then gives, with $A = 264.7 \, \text{cm}^{-1}$, by trial and error $\Delta\eta_e = 264.7 \, \text{cm}^{-1}$. $\Delta\eta_e$ is seen to be substantially larger than ω and essentially equal to A, because the band in this example is optically very thick. Even in

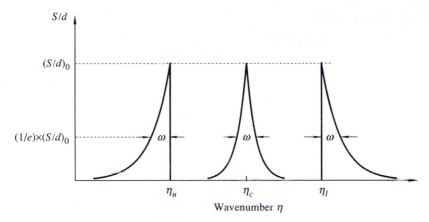

FIGURE 9-11
Band shapes for exponential wide band model.

the band wings far away from the band center the band is optically opaque ($\tau \gg 1$). This result must be accounted for in the choice of $\Delta \eta_e$. For optically thick gases finding the correct $\Delta \eta_e$ is equivalent to finding A itself. Drawing a box seemingly best approximating the actual band shape can lead to large errors!

The Exponential Wide Band Model

The exponential wide band model, first developed by Edwards and Menard [35], is by far the most successful of the wide band models. The original model has been further developed in a series of papers by Edwards and coworkers [36–39]. The word "successful" here implies that the model is able to correlate experimental data for band absorptances with an average error of approximately ±20% (but with maximum errors as high as 50% to 80%). We present here the latest, most accurate version, together with its terminology. For a more exhaustive discussion the reader may want to consult Edwards' monograph on gas radiation [1].

Since it is known from quantum mechanics that the line intensity decreases exponentially in the band wings far away from the band center,[8] Edwards assumed that the smoothed absorption coefficient S/d has one of the following three shapes, as shown in Fig. 9-11:

$$\text{with upper limit head} \quad \frac{S}{d} = \frac{\alpha}{\omega} e^{-(\eta_u - \eta)/\omega}, \tag{9.66a}$$

$$\text{symmetric band} \quad \frac{S}{d} = \frac{\alpha}{\omega} e^{-2|\eta_c - \eta|/\omega}, \tag{9.66b}$$

$$\text{with lower limit head} \quad \frac{S}{d} = \frac{\alpha}{\omega} e^{-(\eta - \eta_l)/\omega}, \tag{9.66c}$$

[8]This fact is easily seen by letting $j \gg 1$ in equations (9.21a) and (9.59) for the P-branch, and in equations (9.21c) and (9.60) for the R-branch.

where, as defined earlier,

$$\alpha \equiv \int_0^\infty \kappa_\eta \, d\eta = \int_0^\infty \left(\frac{S}{d}\right)_\eta d\eta \tag{9.67}$$

is the integrated absorption coefficient or the *band strength parameter* (or area under the curves in Fig. 9-11) and ω is the *band width parameter*,[9] giving the width of the band at $1/e$ of maximum intensity. The band can be expected to be fairly symmetric if, during rotational energy changes, the B does not change too much [recall equations (9.21a) through (9.21c)]. η_c is then the wavenumber connected with the vibrational transition. On the other hand, if the change in B is substantial, then either the R or the P branch may fold back, leading to bands with upper or lower head. Thus, the wavenumbers η_u and η_l are the wavenumbers where this folding back occurs, and not the band center. The sharp exponential apex is, of course, not very realistic. The rationale is that, if the band center is optically thick, then it is opaque no matter what the shape, while if it is thin, then only the total α is of importance. Edwards and Menard [35] proceeded to evaluate the band absorptance using the general statistical model by substituting expressions (9.66) into equation (9.53) and carrying out the integration in an approximate fashion. Since equation (9.53) contains the line overlap parameter β and the optical thickness τ, the authors were able to describe the total band absorptance as a function of three parameters, namely,

$$A^* = A/\omega = A^*(\alpha, \beta, \tau_0), \tag{9.68}$$

where τ_0 is the optical thickness at the band center (symmetric band) or the band head. Their results are summarized in Table 9.2.[10]

Example 9.5. Determine the total band absorptance of the previous two examples by the exponential wide band model.

Solution

From Example 9.3 we have $\tau_0 = 200$ and $\beta = \pi/10$. Thus, since $\tau_0 > 1/\beta$, we find from Table 9.2 $A^* = \ln(\tau_0\beta) + 2 - \beta = \ln(200 \times \pi/10) + 2 - \pi/10 = 5.826$ and $A = A^*\omega = 5.826 \times 50 = 291.3 \, \text{cm}^{-1}$. The difference between the two results is primarily due to the fact that in Example 9.3 we treated the optically thin band wings as optically thick.

The parameters α, β, and ω are functions of temperature and must be determined experimentally. Values for the most important combustion gases—CO_2, H_2O, CO,

---

[9]The band width parameter ω, as used here, applies only to the wide band correlation. If equations (9.66) are used for spectral (i.e., narrow band) calculations, Edwards [1] suggests increasing the value of ω by 20% for better agreement between wide band model and band-integrated narrow band model calculations.

[10]In the original version the parameters $C_1 = \alpha$, $C_3 = \omega$, and $C_2 = \sqrt{4C_1C_3\gamma}$ were used, where γ is the value of β for a gas mixture at a total pressure of 1 atm with zero partial pressure of the absorbing gas. Also, limits between regimes were slightly different, using A itself rather than τ_0.

TABLE 9.2
Exponential wide band correlation for an isothermal gas

$\beta \leq 1$	$0 \leq \tau_0 \leq \beta$	$A^* = \tau_0$	Linear regime
	$\beta \leq \tau_0 \leq 1/\beta$	$A^* = 2\sqrt{\tau_0\beta} - \beta$	Square root regime
	$1/\beta \leq \tau_0 < \infty$	$A^* = \ln(\tau_0\beta) + 2 - \beta$	Logarithmic regime
$\beta \geq 1$	$0 \leq \tau_0 \leq 1$	$A^* = \tau_0$	Linear regime
	$1 \leq \tau_0 < \infty$	$A^* = \ln \tau_0 + 1$	Logarithmic regime

α, β, and ω from Table 9.3 and equations (9.69) through (9.76), $\tau_0 = \alpha X/\omega$.

CH_4, NO, and SO_2—for a reference temperature of $T_0 = 100\,\mathrm{K}$ are given in Table 9.3. Most of these correlation data are based on work by Edwards and coworkers and are summarized in [1]. Data for the purely rotational band of H_2O have been taken from the more modern work of Modak [40]. Values for other bands and other gases may be found in the literature, e.g., for H_2O, CO_2, and CH_4 [1, 36, 39, 41–44], for CO [1, 36, 39, 45–47], for SO_2 [1, 39, 48], for NH_3 [49], for NO [50], for N_2O [51], and for C_2H_2 [52] (in the older of these references the parameters for the slightly different original model are given; in a number of papers a pressure path length has been used instead of a density path length). The temperature dependence of the band correlation parameters for vibration-rotation bands is given by Edwards [1] as

$$\alpha(T) = \alpha_0 \frac{\Psi^*(T)}{\Psi^*(T_0)} = \alpha_0 \frac{\left\{ 1 - \exp\left(-\sum_{k=1}^{m} u_k \delta_k\right)\right\} \Psi(T)}{\left\{ 1 - \exp\left(-\sum_{k=1}^{m} u_{0,k} \delta_k\right)\right\} \Psi(T_0)}, \tag{9.69}$$

$$\beta(T) = \gamma P_e = \gamma_0 \sqrt{\frac{T_0}{T}} \frac{\Phi(T)}{\Phi(T_0)} P_e, \tag{9.70}$$

$$\omega(T) = \omega_0 \sqrt{\frac{T}{T_0}} \tag{9.71}$$

where

$$\Psi(T) = \frac{\prod_{k=1}^{m} \sum_{v_k = v_{0,k}}^{\infty} \frac{(v_k + g_k + \delta_k - 1)!}{(g_k - 1)! v_k!} e^{-u_k v_k}}{\prod_{k=1}^{m} \sum_{v_k = 0}^{\infty} \frac{(v_k + g_k - 1)!}{(g_k - 1)! v_k!} e^{-u_k v_k}}, \tag{9.72}$$

TABLE 9.3

Wide band model correlation parameters for various gases

Band Location		Pressure Parameters		Correlation Parameters		
λ	(δ_k)	n	b	α_0	γ_0	ω_0
$[\mu\mathrm{m}]$				$[\mathrm{cm}^{-1}/(\mathrm{g/m}^2)]$		$[\mathrm{cm}^{-1}]$

H₂O $m = 3$, $\eta_1 = 3652\,\mathrm{cm}^{-1}$, $\eta_2 = 1595\,\mathrm{cm}^{-1}$, $\eta_3 = 3756\,\mathrm{cm}^{-1}$, $g_k = (1, 1, 1)$

$71\,\mu\mathrm{m}^a$	$\eta_c = 140\,\mathrm{cm}^{-1}$ (Rotational) (0,0,0)	1	$8.6\sqrt{\frac{T_0}{T}}+0.5$	44,205	0.14311	69.3
$6.3\,\mu\mathrm{m}$	$\eta_c = 1600\,\mathrm{cm}^{-1}$ (0,1,0)	1	$8.6\sqrt{\frac{T_0}{T}}+0.5$	41.2	0.09427	56.4
$2.7\,\mu\mathrm{m}$	$\eta_c = 3760\,\mathrm{cm}^{-1}$ (0,2,0)			0.2		
	(1,0,0) (0,0,1)	1	$8.6\sqrt{\frac{T_0}{T}}+0.5$	2.3 23.4	$0.13219^{b,c}$	60.0^b
$1.87\,\mu\mathrm{m}$	$\eta_c = 5350\,\mathrm{cm}^{-1}$ (0,1,1)	1	$8.6\sqrt{\frac{T_0}{T}}+1.5$	3.0	0.08169	43.1
$1.38\,\mu\mathrm{m}$	$\eta_c = 7250\,\mathrm{cm}^{-1}$ (1,0,1)	1	$8.6\sqrt{\frac{T_0}{T}}+1.5$	2.5	0.11628	32.0

CO₂ $m = 3$, $\eta_1 = 1351\,\mathrm{cm}^{-1}$, $\eta_2 = 666\,\mathrm{cm}^{-1}$, $\eta_3 = 2396\,\mathrm{cm}^{-1}$, $g_k = (1, 2, 1)$

$15\,\mu\mathrm{m}$	$\eta_c = 667\,\mathrm{cm}^{-1}$ (0,1,0)	0.7	1.3	19.0	0.06157	12.7
$10.4\,\mu\mathrm{m}$	$\eta_c = 960\,\mathrm{cm}^{-1}$ (−1,0,1)	0.8	1.3	2.47×10^{-9}	0.04017	13.4
$9.4\,\mu\mathrm{m}$	$\eta_c = 1060\,\mathrm{cm}^{-1}$ (0,−2,1)	0.8	1.3	2.48×10^{-9}	0.11888	10.1
$4.3\,\mu\mathrm{m}$	$\eta_u = 2410\,\mathrm{cm}^{-1}$ (0,0,1)	0.8	1.3	110.0	0.24723	11.2
$2.7\,\mu\mathrm{m}$	$\eta_c = 3660\,\mathrm{cm}^{-1}$ (1,0,1)	0.65	1.3	4.0	0.13341	23.5
$2.0\,\mu\mathrm{m}$	$\eta_c = 5200\,\mathrm{cm}^{-1}$ (2,0,1)	0.65	1.3	0.060	0.39305	34.5

[a] For the rotational band $\alpha = \alpha_0 \exp\left(-9\sqrt{T_0/T}\right)$, $\gamma = \gamma_0\sqrt{T_0/T}$.

[b] Combination of three bands, all but weak (0,2,0) band are fundamental bands, $\alpha_0 = 25.9\,\mathrm{cm}^{-1}$.

[c] Line overlap for overlapping bands from equation (9.80).

$$\alpha = \alpha_0\frac{\Psi^*}{\Psi_0^*}, \quad \omega = \omega_0\sqrt{\frac{T}{T_0}}, \quad \beta = \gamma P_e = \gamma_0\sqrt{\frac{T_0}{T}}\frac{\Phi}{\Phi_0}P_e, \quad P_e = \left[\frac{p}{p_0}\left(1+(b-1)\frac{p_a}{p}\right)\right]^n$$

Ψ^* from equations (9.69) and (9.72), Φ from equation (9.73), $T_0 = 100\,\mathrm{K}$, $p_0 = 1\,\mathrm{atm}$

TABLE 9.3

Wide band model correlation parameters for various gases (cont'd)

Band Location		Pressure Parameters		Correlation Parameters		
λ	(δ_k)	n	b	α_0	γ_0	ω_0
$[\mu m]$				$[cm^{-1}/(g/m^2)]$		$[cm^{-1}]$
CO $\quad m = 1,\ \eta_1 = 2143\,cm^{-1},\ g_1 = 1$						
$4.7\,\mu m$	$\eta_c = 2143\,cm^{-1}$ (1)	0.8	1.1	20.9	0.07506	25.5
$2.35\,\mu m$	$\eta_c = 4260\,cm^{-1}$ (2)	0.8	1.0	0.14	0.16758	20.0
CH₄ $\quad m = 4,\ \eta_1 = 2914\,cm^{-1},\ \eta_2 = 1526\,cm^{-1},\ \eta_3 = 3020\,cm^{-1},\ \eta_4 = 1306\,cm^{-1},$ $g_k = (1, 2, 3, 3)$						
$7.7\,\mu m$	$\eta_c = 1310\,cm^{-1}$ (0,0,0,1)	0.8	1.3	28.0	0.08698	21.0
$3.3\,\mu m$	$\eta_c = 3020\,cm^{-1}$ (0,0,1,0)	0.8	1.3	46.0	0.06973	56.0
$2.4\,\mu m$	$\eta_c = 4220\,cm^{-1}$ (1,0,0,1)	0.8	1.3	2.9	0.35429	60.0
$1.7\,\mu m$	$\eta_c = 5861\,cm^{-1}$ (1,1,0,1)	0.8	1.3	0.42	0.68598	45.0
NO $\quad m = 1,\ \eta_1 = 1876\,cm^{-1},\ g_1 = 1$						
$5.3\,\mu m$	$\eta_c = 1876\,cm^{-1}$ (1)	0.65	1.0	9.0	0.18050	20.0
SO₂ $\quad m = 3,\ \eta_1 = 1151\,cm^{-1},\ \eta_2 = 519\,cm^{-1},\ \eta_3 = 1361\,cm^{-1},\ g_k = (1, 1, 1)$						
$19.3\,\mu m$	$\eta_c = 519\,cm^{-1}$ (0,1,0)	0.7	1.28	4.22	0.05291	33.1
$8.7\,\mu m$	$\eta_c = 1151\,cm^{-1}$ (1,0,0)	0.7	1.28	3.67	0.05952	24.8
$7.3\,\mu m$	$\eta_c = 1361\,cm^{-1}$ (0,0,1)	0.65	1.28	29.97	0.49299	8.8
$4.3\,\mu m$	$\eta_c = 2350\,cm^{-1}$ (2,0,0)	0.6	1.28	0.423	0.47513	16.5
$4.0\,\mu m$	$\eta_c = 2512\,cm^{-1}$ (1,0,1)	0.6	1.28	0.346	0.58937	10.9

$$\alpha = \alpha_0 \frac{\Psi^*}{\Psi_0^*}, \quad \omega = \omega_0 \sqrt{\frac{T}{T_0}}, \quad \beta = \gamma P_e = \gamma_0 \sqrt{\frac{T_0}{T}} \frac{\Phi}{\Phi_0} P_e, \quad P_e = \left[\frac{p}{p_0} \left(1 + (b-1) \frac{p_a}{p} \right) \right]^n$$

Ψ^* from equations (9.69) and (9.72), Φ from equation (9.73), $T_0 = 100\,K$, $p_0 = 1\,atm$

$$\Phi(T) = \frac{\left\{ \prod\limits_{k=1}^{m} \sum\limits_{v_k=v_{0,k}}^{\infty} \sqrt{\frac{(v_k + g_k + \delta_k - 1)!}{(g_k - 1)!\, v_k!}}\, e^{-u_k v_k} \right\}^2}{\prod\limits_{k=1}^{m} \sum\limits_{v_k=v_{0,k}}^{\infty} \frac{(v_k + g_k + \delta_k - 1)!}{(g_k - 1)!\, v_k!}\, e^{-u_k v_k}}, \tag{9.73}$$

and

$$u_k = hc\eta_k/kT, \quad u_{0,k} = hc\eta_k/kT_0 \quad (T_0 = 100\,\text{K}), \tag{9.74}$$

$$v_{0,k} = \begin{cases} 0 & \text{for } \delta_k \geq 0 \\ |\delta_k| & \text{for } \delta_k \leq 0 \end{cases}, \tag{9.75}$$

$$P_e = \left[\frac{p}{p_0} \left(1 + (b - 1)\frac{p_a}{p} \right) \right]^n, \quad (p_0 = 1\,\text{atm}). \tag{9.76}$$

In these rather complicated expressions the v_k is the vibrational quantum number, δ_k is the change in vibrational quantum number during transition (± 1 for a fundamental band, etc.), and the g_k are statistical weights for the transition (degeneracy =number of ways the transition can take place). Values for the η_k, δ_k, and g_k are given in Table 9.3. The effective pressure P_e gives the pressure dependence of line broadening due to collisions of absorbing molecules with other absorbing molecules and with nonabsorbing molecules that may be present (for example, nitrogen and other inert gases contained in a mixture). Note that the definition for P_e is slightly different here from equation (9.24) (this was done for empirical reasons, to achieve better agreement with experimental data). For the case of nonnegative δ_k or $v_{0,k} = 0$ (the majority of gas bands listed in Table 9.3), the series in the expression for Ψ and the denominator of Φ may be simplified [32] to

$$\sum_{v_k=0}^{\infty} \frac{(v_k + g_k + \delta_k - 1)!}{(g_k - 1)!}\, e^{-u_k v_k} = \frac{(g_k + \delta_k - 1)!}{(g_k - 1)!} \left(1 - e^{-u_k} \right)^{-g_k - \delta_k}. \tag{9.77}$$

If $v_{0,k} \neq 0$, then $v_{0,k}$ terms need to be subtracted from the above result.

Because of the low reference temperature of $T_0 = 100\,\text{K}$, the values for $u_{0,k}$ are relatively large, so both Φ_0 and Ψ_0 are very simple to evaluate as

$$\Psi_0 \approx \prod_{k=1}^{m} \frac{(g_k + \delta_k - 1)!}{(g_k - 1)!}, \quad \Phi_0 \approx 1. \tag{9.78}$$

If only one of the vibrational modes undergoes a transition (only one $\delta_k \neq 0$), then all other modes cancel out of the expression for Ψ; and if the transition results in a fundamental band (single transition with $\delta_k = 1$), then $\Psi^* \equiv 1$. This implies that, for a fundamental band, $\alpha(T) = \alpha_0 = \text{const}$. Unfortunately, the temperature dependence of the broadening mechanism is always more complicated, and Φ must generally be evaluated from equation (9.73).

If several bands overlap each other (e.g., the three H_2O-bands situated around 2.7 μm), then also the individual lines overlap lines from other bands, resulting in an effective overlap parameter β that is larger than for any of the individual bands. The band strength and overlap parameter for overlapping bands are calculated [1] from

$$\alpha = \sum_{j=1}^{J} \alpha_j, \tag{9.79}$$

$$\beta = \frac{1}{\alpha} \left[\sum_{j=1}^{J} \sqrt{\alpha_j \beta_j} \right]^2, \tag{9.80}$$

where J is the number of overlapping bands.

When the exponential wide band model was first presented by Edwards and Menard, the temperature dependence for the broadening parameter was not calculated by quantum statistics but was rather correlated from experimental data that, because of their scatter, generally resulted in fairly simple formulae; but extrapolation to higher temperatures tended to be very inaccurate. Most of the bands listed in Table 9.3 are fundamental bands, not because calculations for these bands are simpler, but because fundamental bands tend to be much stronger than overtones or combined-mode bands, often making them the only important ones for heat transfer calculations.

To facilitate hand calculations, the temperature dependence of band strength parameters α (for nonfundamental bands) and overlap parameters γ are shown in graphical form in Fig. 9-12 for water vapor. A similar plot is given in Fig. 9-13 for the important bands of carbon dioxide, and Fig. 9-14 shows the temperature dependence of the line overlap parameter for the fundamental bands of methane and carbon monoxide (with $\alpha = \alpha_0 = $ const).

Example 9.6.   Consider a water vapor–air mixture at 3 atm and 600 K, with 5% water vapor by volume. What is the most important H_2O band and what is its total band absorptance for a path of 10 cm?

Solution

At 600 K the Planck function has its maximum around 5 μm. Since total emission will depend on the blackbody intensity [see equation (9.38)], we seek a band with large α in the vicinity of 5 μm. Inspection of Table 9.3 shows that the strongest vibration-rotation band for water vapor lies at 6.3 μm and is, therefore, the band we are interested in. From the table we find $\alpha = \alpha_0 = 41.2 \, cm^{-1}/(g/m^2)$, $\beta = \gamma_0 \sqrt{T_0/T}(\Phi/\Phi_0)P_e$ with $\gamma_0 = 0.09427$, and $\omega = \omega_0 \sqrt{T/T_0} = 56.4 \sqrt{600/100} = 138.15 \, cm^{-1}$. To evaluate the effective broadening pressure we find $n = 1$ and $b = 8.6 \sqrt{100/600} + 0.5 = 4.01$ and with a volume fraction $y = p_a/p$ the effective pressure becomes $P_e = (p/1 \, atm)[1 + (b-1)y]^n = 3[1 + 3.01 \times 0.05] = 3.4515$. Estimating the temperature dependence of the line overlap parameter

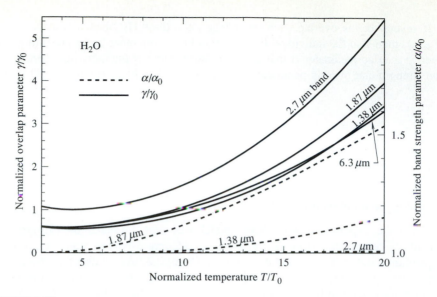

FIGURE 9-12
Temperature dependence of the line overlap parameter, γ, and band strength parameter, α, for water vapor.

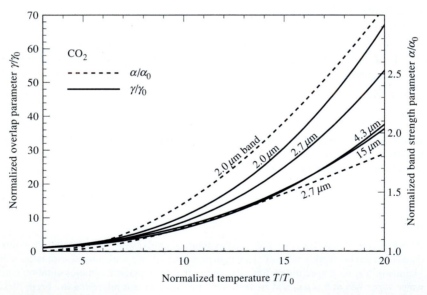

FIGURE 9-13
Temperature dependence of line overlap parameter, γ, and band strength parameter, α, for carbon dioxide.

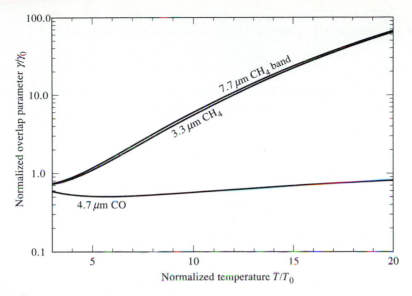

FIGURE 9-14
Temperature dependence of the line overlap parameter, γ, for the fundamental bands of methane and carbon monoxide.

from Fig. 9-12 leads to $\gamma/\gamma_0 \simeq 0.65$ and $\beta = 0.09427 \times 0.65 \times 3.4515 = 0.211$. Since all values for α in Table 9.3 are based on a mass absorption coefficient, we must calculate X as $X = \rho_a s$, where ρ_a is the partial density of the absorbing gas (**not** the density of the gas mixture). We know that every gas at STP ($p_{STP} = 1$ atm and $T_{STP} = 273$ K) occupies 22.4 liter per mol. For our water vapor with a partial pressure of $0.05 \times 3 = 0.15$ atm and a molecular weight of 18 g/mol, we get

$$\rho_a = \frac{18 \text{ g/mol} \times 1000 \, \ell/\text{m}^3}{22.4 \, \ell/\text{mol}} \frac{273}{600} \frac{0.15}{1} = 54.84 \text{ g/m}^3$$

and $X = 54.84 \times 0.1 = 5.48 \text{ g/m}^2$. Finally, from $\tau_0 = \alpha X/\omega$ we get $\tau_0 = 41.2 \times 5.48/138.15 = 1.634$. Since the value of τ_0 lies between the values of β and $1/\beta$ we are in the square-root regime and

$$A^* = 2\sqrt{\tau_0 \beta} - \beta = 2\sqrt{1.634 \times 0.211} - 0.211 = 0.964$$

or $A = 0.964 \times 138.15 = 133 \text{ cm}^{-1}$.

The calculation of exact values for Φ and Ψ for nonfundamental bands is rather tedious and is best left to computer calculations. For illustrative purposes we shall look here at an example for the particularly simple case of CO, a diatomic gas with only a single vibrational mode.

Example 9.7.   Consider pure CO at 800 K and 1 atm. What is the total band absorptance of a 2 m thick layer of isothermal CO for the strong 4.7 μm band?

Solution

Since values in Table 9.3 are based on the mass absorption coefficient, we need to calculate the density of the CO, as we did in the last example, that is,

$$\rho_a = \frac{28 \text{ g/mol} \times 1000 \text{ } \ell/\text{m}^3}{22.4 \text{ } \ell/\text{mol}} \frac{273}{800} \frac{1}{1} = 42.66 \text{ g/m}^3$$

and $X = \rho_a s = 85.32 \text{ g/m}^2$. We also find from Table 9.3 $n = 0.8$ and $b = 1.1$, so that $P_e = 1.1^{0.8} = 1.079$ and $\gamma_0 P_e = 0.07506 \times 1.079 = 0.08101$. Further we find $\alpha = 20.9 \text{ cm}^{-1}/(\text{g/m}^2)$, $\omega = 25.5 \sqrt{800/100} = 72.125 \text{ cm}^{-1}$, and $\tau_0 = \alpha X/\omega = 20.9 \times 85.32/72.125 = 24.72$. To demonstrate the analytical determination of the temperature dependence of β we need to calculate Φ from equation (9.73), which, for CO with $m = 1$ and $g = \delta = 1$ reduces to

$$\Phi = \left(\sum_{v=0}^{\infty} \sqrt{(v+1)} e^{-vu} \right)^2 \bigg/ \sum_{v=0}^{\infty} (v+1) e^{-vu} .$$

The series in the denominator may be evaluated since

$$\sum_{v=0}^{\infty} e^{-vu} = \sum_{v=0}^{\infty} (e^{-u})^v = 1/(1 - e^{-u}),$$

and

$$\sum_{v=0}^{\infty} v e^{-vu} = -\frac{d}{du} \left(\sum_{v=0}^{\infty} e^{-vu} \right) = \frac{e^{-u}}{(1 - e^{-u})^2} .$$

Thus,

$$\Phi = (1 - e^{-u})^2 \left(\sum_{v=0}^{\infty} \sqrt{(v+1)} e^{-vu} \right)^2 .$$

We may evaluate u as

$$u = hc\eta_1/kT = C_2\eta_1/T = 1.4388 \text{ cm K} \times 2143 \text{ cm}^{-1}/800 \text{ K} = 3.854.$$

Substituting gives

$$\Phi = 0.9788^2 \left(1 + \sqrt{2} e^{-1.927} + \sqrt{3} e^{-3.854} + \sqrt{4} e^{-5.781} + \ldots \right)^2 = 1.500.$$

Similarly, $\Phi_0 \approx 1.000$. We are finally in a position to calculate the temperature-adjusted value for β as $\gamma/\gamma_0 = \sqrt{T_0/T} \Phi/\Phi_0 = \sqrt{1/8} \times 1.500/1.000 = 0.5303$ and $\beta = (\gamma/\gamma_0)\gamma_0 P_e = 0.5303 \times 0.08101 = 0.0430$. Thus, $\tau_0 > 1/\beta$ and we are in the logarithmic regime, and

$$A^* = \ln(\tau_0 \beta) + 2 - \beta = 2.018 \quad \text{and} \quad A = 145.6 \text{ cm}^{-1}.$$

Had we instead used Fig. 9-14, doing so would have led to $\gamma/\gamma_0 \simeq 0.52$ and $\beta = 0.08101 \times 0.52 = 0.0421$, and

$$A^* = \ln(\tau_0 \beta) + 2 - \beta = 1.999 \quad \text{and} \quad A = 144.1 \, \text{cm}^{-1}.$$

While the correlation in Table 9.2 is simple and straightforward (aside from the temperature dependence of α and β), it is often preferable to have a single continuous correlation formula. A simple analytical expression can be obtained for the high pressure limit, i.e., when the lines become very wide from broadening resulting in very strong overlap, or $\beta \to \infty$, leading to

$$A^* = E_1(\tau_0) + \ln \tau_0 + \gamma_E, \qquad \beta \to \infty, \tag{9.81}$$

where $E_1(\tau)$ is known as an *exponential integral function*, which is discussed in some detail in Appendix E. Felske and Tien [53] have given a formula for all ranges of β, based on results from the numerical quadrature of equation (9.53):

$$A^* = 2E_1\left(\sqrt{\frac{\tau_0 \beta}{1 + \beta/\tau_0}}\right) + E_1\left(\frac{1}{2}\sqrt{\frac{\tau_0/\beta}{1 + \beta/\tau_0}}\right) - E_1\left(\frac{1 + 2\beta}{2}\sqrt{\frac{\tau_0/\beta}{1 + \beta/\tau_0}}\right)$$

$$+ \ln\left(\frac{\tau_0}{(1 + \beta/\tau_0)(1 + 2\beta)}\right) + 2\gamma_E. \tag{9.82}$$

A previous, somewhat simpler expression was given by Tien and Lowder [54] as

$$A^* = \ln\left(1 + \tau_0 f(\beta) \frac{\tau_0 + 2}{\tau_0 + 2f(\beta)}\right) \tag{9.83a}$$

where

$$f(\beta) = 2.94\left[1 - e^{-2.60\beta}\right]. \tag{9.83b}$$

This correlation is known today to be seriously in error for small values of β [53, 55]; however, it is simple and accurate whenever the band absorptance is appreciable, as seen from Fig. 9-15, in which the correlation of Edwards and Menard is compared with that of Tien and Lowder and the one of Felske and Tien. A considerable number of other band correlations are available in the literature, based on numerous variations of the Elsasser and statistical models. An exhaustive discussion of these correlations and their accuracies (as compared with numerical quadrature results based on the plain Elsasser and the general statistical models) has been given by Tiwari [56].

Wide Band Model for Nonisothermal Gases

As indicated in the previous section on narrow band models, the spectral emissivity for a nonisothermal path [cf. equation (9.41)] is

$$\epsilon_\eta = 1 - \exp\left(-\int_0^X \kappa_\eta \, dX\right), \tag{9.84}$$

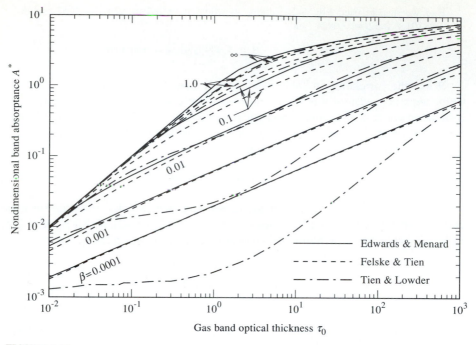

FIGURE 9-15
Comparison of various band absorptance correlations.

from which we may calculate the total band absorptance as

$$A = \int_0^\infty \epsilon_\eta \, d\eta = \int_0^\infty \left[1 - \exp\left(-\int_0^X \kappa_\eta \, dX\right)\right] d\eta. \qquad (9.85)$$

Here we have replaced the geometric path s by X in case a linear absorption coefficient is not used, but rather one based on density (as was done for the correlation parameters in Table 9.3) or pressure. Since we would still like to use the simple wide band model, appropriate path-averaged values for the correlation parameters α, β, and ω must be found. Attempts at such scaling were made by Chan and Tien [57], Cess and Wang [58], and Edwards and Morizumi [59], and are summarized by Edwards [1]. The average value for α follows readily from the weak line limit (linear regime in Table 9.2) as

$$\tilde{\alpha} \equiv \frac{1}{X} \int_0^X \int_0^\infty \kappa_\eta \, d\eta \, dX = \frac{1}{X} \int_0^X \int_0^\infty \left(\frac{S}{d}\right)_\eta d\eta \, dX = \frac{1}{X} \int_0^X \alpha \, dX. \qquad (9.86)$$

The definition of an average value for ω is

$$\tilde{\omega} \equiv \frac{1}{\tilde{\alpha} X} \int_0^X \omega \alpha \, dX, \qquad (9.87)$$

while the averaged value for β is found by comparison with the square root regime in Table 9.2 as

$$\tilde{\beta} \equiv \frac{1}{\tilde{\omega}\tilde{\alpha}X} \int_0^X \beta \omega \alpha \, dX. \tag{9.88}$$

There is little theoretical justification for the choice of $\tilde{\omega}$ and $\tilde{\beta}$,[11] but comparison with spectral calculations using equations (9.55), (9.56), and (9.57) showed that they give excellent results [59].

Example 9.8. Reconsider Example 9.6, but assume that the water vapor–air mixture temperature varies linearly between 400 K and 800 K over its path of 10 cm. How does this affect the total band absorptance for the 6.3 μm band?

Solution

We may express the temperature variation as $T = 400\,\text{K}(1 + s'/s)$, where s' is distance along path s, and the density variation as

$$\rho_a = 6\rho_{600}\frac{T_0}{T} = \frac{\frac{3}{2}\rho_{600}}{1 + s'/s} = 6\rho_{600}\frac{T_0}{T}.$$

Thus,

$$X = \int_0^s \rho_a \, ds' = 6\rho_{600}\int_0^s \left(\frac{T_0}{T}\right) ds' = \frac{3}{2}\rho_{600}s\int_0^1 \frac{d\xi}{1 + \xi}$$
$$= \frac{3}{2}X_{600}\ln 2 = 1.040\,X_{600} = 5.702\,\text{g/m}^2.$$

The path-averaged band strength becomes

$$\tilde{\alpha} = \frac{1}{X}\int_0^s \alpha\,\rho_a\,ds = \frac{1}{X}\alpha_0 X = \alpha_0 = 41.2\,\text{cm}^{-1}/(\text{g/m}^2),$$

since the 6.3 μm band is a fundamental band and α is independent of temperature. For the averaged $\tilde{\omega}$ we get, from $\omega = \omega_0\sqrt{T/T_0} = \omega_0\sqrt{4}\sqrt{1 + s'/s}$,

$$\tilde{\omega} = \frac{1}{\tilde{\alpha}X}\int_0^s \omega\alpha\,\rho_a\,ds' = \frac{6\omega_0\rho_{600}}{X}\int_0^s \sqrt{\frac{T}{T_0}\frac{T_0}{T}}\,ds' = \frac{3\omega_0\rho_{600}s}{X}\int_0^1 \frac{d\xi}{\sqrt{1+\xi}}$$
$$= \frac{3\omega_0 X_{600}}{X} \times 2\sqrt{1+\xi}\Big|_0^1 = 6(\sqrt{2}-1)\frac{X_{600}}{X}\,\omega_0 = \frac{6(\sqrt{2}-1)}{\frac{3}{2}\ln 2} \times 56.4\,\text{cm}^{-1}$$
$$= 134.8\,\text{cm}^{-1}.$$

---

[11] Note that there are two different definitions for $\tilde{\beta}$, one for narrow band calculations and the present one for the wide band model.

And, finally, the overlap parameter is obtained from

$$\tilde{\beta} = \frac{1}{\tilde{\omega}\tilde{\alpha}X} \int_0^s \beta\omega\alpha\rho_a \, ds' = \frac{6\rho_{600}}{\tilde{\omega}X} \int_0^s \left(\gamma_0 P_e \frac{\gamma}{\gamma_0}\right)\left(\omega_0 \sqrt{\frac{T}{T_0}}\right)\left(\frac{T_0}{T}\right) ds'$$

$$= 6\gamma_0 P_e \frac{\omega_0}{\tilde{\omega}} \frac{X_{600}}{X} \int_0^1 \frac{\gamma}{\gamma_0} \sqrt{\frac{T_0}{T}} \, d\xi'.$$

Inspection of Fig. 9-12 reveals that the integrand varies between $0.59/\sqrt{4} \simeq 0.30$ (at 400 K), to $0.66/\sqrt{6} \simeq 0.27$ (at 600 K), back to $0.80/\sqrt{8} \simeq 0.29$ (at 500 K); i.e., the integrand is relatively constant. Keeping in mind the inherent inaccuracies of the wide band model, the integral may be approximated by using an average value of 0.28. Then

$$\tilde{\beta} \simeq 0.28 \times 6\gamma_0 P_e \frac{\omega_0}{\tilde{\omega}} \frac{X_{600}}{X} = \frac{0.28 \times 6\gamma_0 P_e}{6(\sqrt{2}-1)} = \frac{0.28 \times 0.09427 \times 3.4515}{\sqrt{2}-1} = 0.220.$$

The effective optical thickness at the band center is now

$$\tau_0 = \tilde{\alpha}X/\tilde{\omega} = 41.2 \times 5.702/134.8 = 1.743.$$

Again we are in the square root regime and

$$A^* = 2\sqrt{\tau_0\tilde{\beta}} - \tilde{\beta} = 2\sqrt{1.743 \times 0.220} - 0.220 = 1.018 \quad \text{and} \quad A = 137 \, cm^{-1}.$$

Thus, although the temperature varied considerably over the path (by a factor of two) values for α, β, and ω changed only slightly, and the final value for the band absorptance changed by less than 3%. In view of the accuracy of the wide band correlation, the assumption of an isothermal gas can often lead to satisfactory results. This has been corroborated by Felske and Tien [60], who suggested a linear average for temperature, and a second independent linear average for density (as opposed to density evaluated at average temperature). They found negligible discrepancy for a large number of nonisothermal examples.

9.7 TOTAL EMISSIVITY AND MEAN ABSORPTION COEFFICIENT

Total Emissivity

We shall see in later chapters that, for accurate heat transfer calculations that take into account the interdependence between temperature and radiation, the total band absorptance is the basic property needed in the analysis. In less sophisticated, more practical engineering treatment it is usually sufficient to evaluate the emission from a hot gas (usually considered isothermal) that reaches a wall. The total emissivity is defined as the portion of total emitted radiation over a path X that is not attenuated by self-absorption divided by the maximum possible emission or, from equation (9.29) and considering only emission within the gas,

$$
\epsilon \equiv \frac{\int_0^\infty I_{b\eta}\epsilon_\eta\, d\eta}{\int_0^\infty I_{b\eta}\, d\eta} = \frac{\int_0^\infty I_{b\eta}\left(1 - e^{-\kappa_\eta X}\right) d\eta}{\int_0^\infty I_{b\eta}\, d\eta}
$$

$$
= \sum_{i=1}^{N} \left(\frac{\pi I_{b\eta 0}}{\sigma T^4}\right)_i \int_{\Delta\eta_{\text{band}}} \left(1 - e^{-\kappa_\eta X}\right) d\eta = \sum_{i=1}^{N} \left(\frac{\pi I_{b\eta 0}}{\sigma T^4}\right)_i A_i, \tag{9.89}
$$

where two simplifying assumptions have been made: (*i*) The spectral width of each of the N bands is so narrow that the Planck function varies only negligibly over this range, and (*ii*) the bands do not overlap. While the first assumption is generally very good (with the exception of pure rotational bands such as the one for water vapor listed in Table 9.3), bands do sometimes overlap (for example, the 2.7 μm bands in a water vapor–CO_2 mixture).

If two or more bands of the species contained in a gas mixture overlap, the emission from the mixture will be smaller than the sum of the individual contributions (because of increased self-absorption). This problem has been dealt with, in an approximate fashion, by Hottel and Sarofim [9]. They argued that the transmissivities of species a and b over the overlapping region $\Delta\eta$ are independent from another, that is,

$$
\bar{\tau}_{a+b} = \frac{1}{\Delta\eta}\int_{\Delta\eta} e^{-\kappa_{\eta a}X} e^{-\kappa_{\eta b}X}\, d\eta
$$

$$
\approx \left[\frac{1}{\Delta\eta}\int_{\Delta\eta} e^{-\kappa_{\eta a}X}\, d\eta\right]\left[\frac{1}{\Delta\eta}\int_{\Delta\eta} e^{-\kappa_{\eta b}X}\, d\eta\right] = \bar{\tau}_a\,\bar{\tau}_b. \tag{9.90}
$$

If we define a total emissivity for a single band as

$$
\epsilon_i \equiv \left(\frac{\pi I_{b\eta 0}}{\sigma T^4}\right)_i A_i, \tag{9.91}
$$

then this expression leads to the total emissivity of two overlapping bands, or

$$
\epsilon_{a+b} = \epsilon_a + \epsilon_b - \epsilon_a\epsilon_b. \tag{9.92}
$$

This equation is only accurate if both bands fully overlap. If the overlap is only partial, then the correction term, $\epsilon_a\epsilon_b$, should be calculated based on the fractions of band emissivity that pertain to the overlap region (i.e., a quantity that is not available from wide band correlations). An approximate way of dealing with this problem has been suggested by Felske and Tien [60].

A total absorptivity for the gas may be defined in the same way as equation (9.89). However, as for surfaces, in the absorptivity the absorption coefficient must be evaluated at the temperature of the gas, while the Planck function is based on the blackbody temperature of the radiation source.

It is clear from equation (9.89) that the total emissivity is equal to the sum of band absorptances multiplied by the weight factor $(\pi I_{b\eta 0}/\sigma T^4)$. Since the band absorptance is roughly proportional to the band strength parameter α (exactly proportional for small values of optical path X), comparison of the factors $[\alpha(\pi I_{b\eta 0}/\sigma T^4)]_i$ shows which bands need to be considered for the calculation of the total emissivity.

Example 9.9. What is the total emissivity of a 1 m thick layer of pure CO at 800 K and 1 atm?

Solution

For these conditions CO has a single important absorption band in the infrared. Comparing $\alpha I_{b\eta 0}$ for the 4.7 μm and 2.35 μm bands (see Table 9.3) we find with $(\eta_0/T)_{4.3} = 2143$ cm$^{-1}/800$ K $= 2.6788$ cm$^{-1}/$K and $(\eta_0/T)_{2.35} = 4260$ cm$^{-1}/800$ K $= 5.325$ cm$^{-1}/$K,

$$\left(\frac{\alpha E_{b\eta 0}}{T^3}\right)_{4.7} \bigg/ \left(\frac{\alpha E_{b\eta 0}}{T^3}\right)_{2.35} = \frac{20.9 \times 1.5563}{0.14 \times 0.2659} = 874.$$

Therefore, since the 4.7 μm band is much stronger $(\alpha_{4.7}/\alpha_{2.35} \simeq 150)$ and located in a more important part of the spectrum $(E_{b\eta 4.7}/E_{b\eta 2.35} \simeq 6)$, the influence of the 2.35 μm band can be neglected. The band absorptance for the 4.7 μm band has already been evaluated in Example 9.7 as $A = 145.6$ cm^{-1} for the present conditions. Thus,

$$\epsilon_{CO}(800\,\text{K, 1 atm}) = \left(\frac{\pi I_{b\eta 0}}{\sigma T^4}\right)_{\eta_0 = 2143\,\text{cm}^{-1}} \times 145.6\,\text{cm}^{-1},$$

$$= \left(\frac{E_{b\eta 0}}{T^3}\right)_{\eta_0 = 2143\,\text{cm}^{-1}} \times \frac{145.6\,\text{cm}^{-1}}{\sigma T},$$

$$= 1.5563 \times 10^{-8}\,\frac{\text{W}}{\text{m}^2\,\text{cm}^{-1}\,\text{K}^3} \times \frac{145.6\,\text{cm}^{-1}}{5.670 \times 10^{-8} \times 800\,\text{W}/\text{m}^2\,\text{K}^3},$$

$$= 0.0500.$$

If only total emissivities are desired, it would be very convenient to have correlations, tables or charts from which the total emissivity can be read directly, rather than having to go through the algebra of the wide band correlations plus equation (9.89). A number of investigators have included total emissivity charts with their wide band correlation data; for example, Brosmer and Tien [44, 52] compiled data on CH$_4$ and C$_2$H$_2$, and Tien and coworkers [51] did the same for N$_2$O. However, by far the most monumental work has been collected by Hottel [16] and Hottel and Sarofim [9]. They considered primarily combustion gases, but they also presented charts for a number of other gases. Their data for total emissivity and absorptivity are presented in the form

$$\epsilon = \epsilon(p_a L, p, T_g), \tag{9.93}$$

$$\alpha = \alpha(p_a L, p, T_g, T_s) \approx \left(\frac{T_g}{T_s}\right)^{1/2} \epsilon\left(p_a L \frac{T_g}{T_s}, p, T_s\right), \tag{9.94}$$

where T_g is the gas temperature and T_s is the temperature of an external blackbody (or gray) source such as a hot surface. Originally, the power for T_g/T_s recommended by Hottel was 0.65 for CO$_2$ and 0.45 for water vapor, but with greater theoretical understanding the single value of 0.5 has become accepted [9]. In equation (9.94)

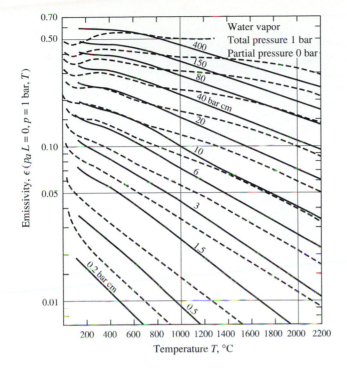

FIGURE 9-16
Total emissivity of water vapor at a total gas pressure of 1 bar and zero partial pressure, from Hottel [16] (solid lines) and Leckner [61] (dashed lines).

p_a is the partial pressure of the absorbing gas and p is the total pressure. (Hottel and Sarofim preferred a pressure path length over the density path length used by Edwards). The emissivities were given in chart form vs. temperature, with pressure path length as parameter, and for an overall pressure of 1 atmosphere. More recent work by Leckner [61], Ludwig and coworkers [62, 63], Sarofim and coworkers [64] and others has shown that the original charts by Hottel [9, 16], while accurate for many conditions (in particular, over the ranges covered by experimental data of the times), are seriously in error for some conditions (primarily those based on extrapolation of experimental data). New charts, based on the integration of spectral data, have been prepared by Leckner [61] and Ludwig and coworkers [62, 63], and show good agreement among each other. Emissivity charts, comparing the newly calculated data by Leckner [61] with Hottel's [16], are shown in Fig. 9-16 for water vapor and in Fig. 9-17 for carbon dioxide. These charts give the emissivities for the limiting case of vanishing partial pressure of the absorbing gas ($p_a \rightarrow 0$).

The original charts by Hottel also included pressure correction charts for the evaluation of cases with $p_a \neq 0$ and $p \neq 1$ bar, as well as charts for the overlap parameter $\Delta\epsilon$. Again, these factors were found to be somewhat inaccurate under extreme conditions and have been improved upon in more recent work. Particularly useful for calculations are the correlation expressions given by Leckner [61], which (for temperatures above 400 K) have a maximum error of 5% for water vapor and 10% for CO_2, respectively, as compared with his spectrally integrated emissivities. In his correlation

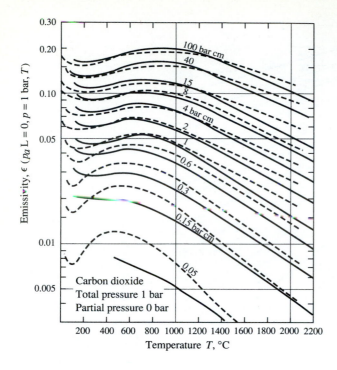

FIGURE 9-17
Total emissivity of carbon dioxide at a total gas pressure of 1 bar and zero partial pressure, from Hottel [16] (solid lines) and Leckner [61] (dashed lines).

the zero-partial-pressure emissivity is given by

$$\epsilon_0(p_a, p = 1 \text{ bar}, T_g) = \exp\left[\sum_{i=0}^{M}\sum_{j=0}^{N} c_{ji}\left(\frac{T}{T_0}\right)^j \left(\log_{10}\frac{p_a L}{(p_a L)_0}\right)^i\right],$$

$$T_0 = 1000 \text{ K}, \quad (p_a L)_0 = 1 \text{ bar cm}, \tag{9.95}$$

and the c_{ji} are correlation constants given in Table 9.4 for water vapor and in Table 9.5 for carbon dioxide. The emissivity for different pressure conditions is then found from

$$\frac{\epsilon(p_a L, p, T_g)}{\epsilon_0(p_a L, 1 \text{ bar}, T_g)} = 1 - \frac{(a-1)(1-P_E)}{a+b-1+P_E} \exp\left(-c\left[\log_{10}\frac{(p_a L)_m}{p_a L}\right]^2\right) \tag{9.96}$$

where P_E is an effective pressure, and a, b, c, and $(p_a L)_m$ are correlation parameters, also given in Tables 9.4 and 9.5.

As noted before, in a mixture that contains both carbon dioxide and water vapor, the bands partially overlap and another correction factor must be introduced, which is found from

$$\Delta\epsilon = \left[\frac{\zeta}{10.7 + 101\zeta} - 0.0089\zeta^{10.4}\right]\left(\log_{10}\frac{(p_{H_2O} + p_{CO_2})L}{(p_a L)_0}\right)^{2.76}, \tag{9.97}$$

TABLE 9.4

Correlation constants for the determination of the total emissivity for water vapor [61]

M, N		2,2	
c_{00} $\quad \ldots \quad$ c_{0M} $\vdots \quad \ddots \quad \vdots$ c_{N0} $\quad \ldots \quad$ c_{NM}	-2.2118 0.85667 -0.10838	-1.1987 0.93048 -0.17156	0.035596 -0.14391 0.045915
P_E	$(p + 2.56 p_a / \sqrt{t}) / p_0$		
$(p_a L)_m / (p_a L)_0$	$13.2 t^2$		
a	$2.479, \qquad\qquad t < 0.75$ $1.888 - 2.053 \log_{10} t, \quad t > 0.75$		
b	$1.10 / t^{1.4}$		
c	0.5		
$T_0 = 1000\,\mathrm{K}, \qquad p_0 = 1\,\mathrm{bar}, \; t = T/T_0, \; (p_a L)_0 = 1\,\mathrm{bar\,cm}$			

with

$$\zeta = \frac{p_{H_2O}}{p_{H_2O} + p_{CO_2}}. \qquad (9.98)$$

This factor is directly applicable to emissivity and absorptivity.

To summarize, the total emissivity and absorptivity of gases containing CO_2, water vapor, or both, may be calculated from:

$$\epsilon_i = \epsilon_{0i}(p_i L, 1\,\mathrm{bar}, T_g)\left(\frac{\epsilon}{\epsilon_0}\right)_i, \qquad i = CO_2 \text{ or } H_2O, \qquad (9.99a)$$

$$\alpha_i = \left(\frac{T_g}{T_s}\right)^{1/2} \epsilon_{0i}\left(p_i L \frac{T_g}{T_s}, 1\,\mathrm{bar}, T_s\right)\left(\frac{\epsilon}{\epsilon_0}\right)_i, \quad i = CO_2 \text{ or } H_2O, \qquad (9.99b)$$

$$\epsilon_{CO_2+H_2O} = \epsilon_{CO_2} + \epsilon_{H_2O} - \Delta\epsilon(T_g), \qquad (9.99c)$$

$$\alpha_{CO_2+H_2O} = \alpha_{CO_2} + \alpha_{H_2O} - \Delta\epsilon(T_s). \qquad (9.99d)$$

Example 9.10. Consider a 1 m thick layer of a gas mixture at 1000 K and 5 bar that consists of 10% carbon dioxide, 20% water vapor, and 70% nitrogen. What is the total normal intensity escaping from this layer?

TABLE 9.5

Correlation constants for the determination of the total emissivity for carbon dioxide [61]

M, N	2,3
c_{00} ... c_{0M} $\vdots$ $\ddots$ $\vdots$ c_{N0} ... c_{NM}	$\begin{array}{rrrr} -3.9893 & 2.7669 & -2.1081 & 0.39163 \\ 1.2710 & -1.1090 & 1.0195 & -0.21897 \\ -0.23678 & 0.19731 & -0.19544 & 0.044644 \end{array}$
P_E	$(p + 0.28\,p_a)/p_0$
$(p_a L)_m / (p_a L)_0$	$\begin{array}{ll} 0.054/t^2, & t < 0.7 \\ 0.225 t^2, & t > 0.7 \end{array}$
a	$1 + 0.1/t^{1.45}$
b	0.23
c	1.47
$T_0 = 1000\,\text{K}, \quad p_0 = 1\,\text{bar}, \quad t = T/T_0, \quad (p_a L)_0 = 1\,\text{bar cm}$	

Solution

From equations (9.29) and (9.89) we see that the exiting total intensity is

$$I = \int_0^\infty I_{b\eta}\left(1 - e^{-\kappa_\eta X}\right) d\eta = \int_0^\infty I_{b\eta}\epsilon_\eta \, d\eta = \frac{\epsilon\sigma T^4}{\pi},$$

where the ϵ is the total emissivity of the water vapor–carbon dioxide mixture. First we calculate the emissivity of CO_2 at a total pressure of 1 bar from Table 9.5: With a $p_{CO_2}L = 0.1 \times 5\,\text{m bar} = 50\,\text{bar cm}$ and $T_g = 1000\,\text{K}$ we find $\epsilon_{CO_2,0}(1\,\text{bar}) = 0.157$ (which may also be estimated from Fig. 9-17); for a total pressure of 5 bar we find from Table 9.4 the effective pressure is $P_E = 5.14$, $a = 1.1$, $b = 0.23$, $c = 1.47$ and $(p_a L)_m = 0.225\,\text{bar cm}$. Thus, from equation (9.96)

$$\left(\frac{\epsilon}{\epsilon_0}\right)_{CO_2} = 1 - \frac{0.1 \times (-4.14)}{0.33 + 5.14} \exp\left[-1.47 \times \left(\log_{10}\frac{0.225}{50}\right)^2\right] \approx 1.00,$$

and

$$\epsilon_{CO_2} \approx 0.157.$$

Similarly, for water vapor with $p_{H_2O}L = 0.2 \times 5\,\text{m bar} = 100\,\text{bar cm}$ we find $\epsilon_{H_2O,0}(1\,\text{bar}) \approx 0.359$ and the pressure correction factor becomes, with $P_E = 7.56$, $a = 1.88$, $b = 1.1$,

$c = 0.5$ and $(p_aL)_m = 13.2\ \text{bar cm}$,

$$\left(\frac{\epsilon}{\epsilon_0}\right)_{H_2O} = 1 - \frac{0.888 \times (-6.56)}{1.988 + 7.56}\exp\left[-0.5 \times \left(\log_{10}\frac{13.2}{100}\right)^2\right] = 1.414,$$

and

$$\epsilon_{H_2O} \approx 0.359 \times 1.414 = 0.508.$$

Finally, since we have a mixture of carbon dioxide and water vapor, we need to deduct for the band overlaps: From equation (9.97), with $\zeta = \frac{2}{3}$, $\Delta\epsilon = 0.072$. Thus, the total emissivity is $\epsilon = 0.157 + 0.508 - 0.072 = 0.593$. The total normal intensity is then

$$I = 0.593 \times 5.670 \times 10^{-8}\ \text{W/(m}^2\,\text{K}^4) \times (500\ \text{K})^4/\pi\ \text{sr} = 669\ \text{W/m}^2\ \text{sr}.$$

It is obvious from this example that the calculation of total emissivities is far from an exact science and carries a good deal of uncertainty. Carrying along three digits in the above calculations is optimistic at best. The reader should understand that accurate emissivity values are difficult to measure, and that too many parameters are involved to make simple and accurate correlations possible.

Total Absorption Coefficients

We noted in the previous chapter that the emission term in the equation of transfer, equation (8.23), and in the divergence of the radiative heat flux, equation (8.54), is proportional to $\kappa_\eta I_{b\eta}$. Thus, for the evaluation of total intensity or heat flux divergence it is convenient to define the following total absorption coefficient, known as the *Planck-mean absorption coefficient*:

$$\kappa_P \equiv \frac{\int_0^\infty I_{b\eta}\kappa_\eta\,d\eta}{\int_0^\infty I_{b\eta}\,d\eta} = \frac{\pi}{\sigma T^4}\int_0^\infty I_{b\eta}\kappa_\eta\,d\eta. \tag{9.100}$$

Using narrow band averaged values for the absorption coefficient, and making again the assumption that the Planck function varies little across each band, equation (9.100) may be restated as

$$\kappa_P = \sum_{i=1}^N \left(\frac{\pi I_{b\eta 0}}{\sigma T^4}\right)_i \int_{\Delta\eta_{\text{band}}}\left(\frac{S}{d}\right)d\eta = \sum_{i=1}^N \left(\frac{\pi I_{b\eta 0}}{\sigma T^4}\right)_i \alpha_i, \tag{9.101}$$

where the sum is over all N bands. It is interesting to note that the Planck-mean absorption coefficient depends only on the band strength parameter α and, therefore, on temperature (but not on pressure). Values for α have been measured and tabulated by a number of investigators for various gases and, using them, Planck-mean absorption coefficients have been presented by Tien [2], as shown in Figs. 9-18 and 9-19. Sometimes the Planck-mean absorption coefficient is required for absorption (rather than emission), for example, when gas and radiation source are at different temperatures. This expression is known as the *modified Planck-mean absorption coefficient*,

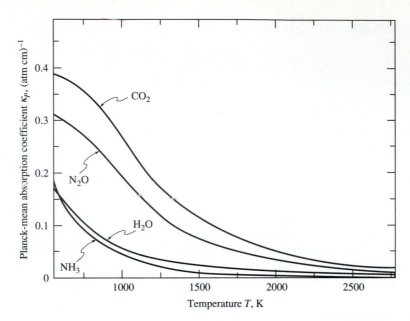

FIGURE 9-18
Planck-mean absorption coefficients for carbon dioxide, water vapor, ammonia, and nitrous oxide [2].

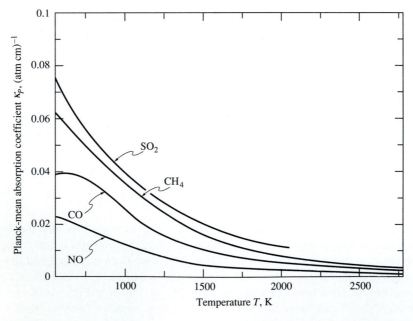

FIGURE 9-19
Planck-mean absorption coefficients for carbon monoxide, nitric oxide, methane, and sulfur dioxide [2].

and is defined as

$$\kappa_m(T, T_s) \equiv \frac{\int_0^\infty I_{b\eta}(T_s)\kappa_\eta(T)\, d\eta}{\int_0^\infty I_{b\eta}(T_s)\, d\eta} \tag{9.102}$$

An approximate expression relating κ_m to κ_P has been given by Cess and Mighdoll [65] as

$$\kappa_m(T, T_s) = \kappa_P(T_s)\left(\frac{T_s}{T}\right). \tag{9.103}$$

In later chapters we shall see that in optically thick situations the radiative heat flux becomes proportional to

$$\frac{1}{\kappa_\eta}\boldsymbol{\nabla} I_{b\eta} = \frac{1}{\kappa_\eta}\frac{dI_{b\eta}}{dT}\boldsymbol{\nabla} T.$$

This has led to the definition of an optically thick or *Rosseland-mean absorption coefficient* as

$$\frac{1}{\kappa_R} \equiv \int_0^\infty \frac{1}{\kappa_\eta}\frac{dI_{b\eta}}{dT}\, d\eta \bigg/ \int_0^\infty \frac{dI_{b\eta}}{dT}\, d\eta = \frac{\pi}{4\sigma T^3}\int_0^\infty \frac{1}{\kappa_\eta}\frac{dI_{b\eta}}{dT}\, d\eta. \tag{9.104}$$

Even though they noted the difficulty of integrating equation (9.104) over the entire spectrum (with zero absorption coefficient between bands), Abu-Romia and Tien [47] and Tien [2] attempted to evaluate the Rosseland-mean absorption coefficient for pure gases. Since the results are, at least by this author, regarded as very dubious they will not be reproduced here. We shall return to the Rosseland absorption coefficient when its use is warranted, i.e., when a medium is optically thick over the entire spectrum (for example, an optically thick particle background with or without molecular gases).

9.8   EXPERIMENTAL METHODS

Before going on to employ the above concepts of radiation properties of molecular gases in the solution of the radiative equation of transfer and the calculation of radiative heat fluxes, we want briefly to look at some of the more common experimental methods of determining these properties. While light sources, monochromators, detectors, and optical components are similar to the ones used for surface property measurements, as discussed in Section 3.10, gas property measurements result in transmission studies (as opposed to reflection measurements for surfaces).

All transmission measurements resemble one another to a certain extent: They consist of a light source, a monochromator (unless, for measurements over a narrow spectral range, a tunable laser is used as source), a chopper, a test cell with the (approximately isothermal) gas whose properties are to be measured, a detector, associated optics and an amplifier-recorder device. The chopper often serves two pur-

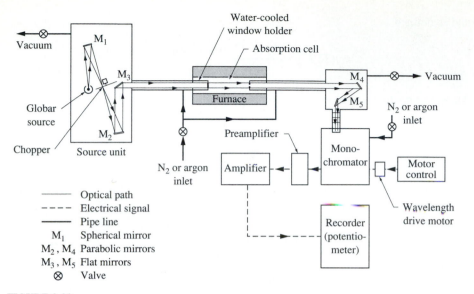

FIGURE 9-20
General set-up of gas radiation measurement apparatus [66].

poses: (*i*) A pyroelectric detector cannot measure radiative intensity, rather, it measures *changes* in intensity; and (*ii*) if the beam is chopped before going through the sample gas then, by measuring the difference in intensity between chopper open and closed conditions, indeed only transmission of the incident light beam is measured. That is, any emission from the (possibly very hot) test gas and/or stray radiation will not be part of the signal. A typical set-up is shown in Fig. 9-20, depicting an apparatus used by Tien and Giedt [66].

Measurements of radiative properties of gases may be characterized by the nature of the test gas containment and by the spectral width of the measurements. As indicated by Edwards [1], we distinguish among (1) hot window cell, (2) cold window cell, (3) nozzle seal cell, and (4) free jet devices; these may be used to make (*a*) narrow band measurements, (*b*) total band absorptance measurements, or (*c*) total emissivity/absorptivity measurements.

The hot-window cell uses an isothermal gas within a container that is closed off at both ends by windows that are kept at the same temperature as the gas. While this set-up is the most nearly ideal situation for measurements, it is generally very difficult to find window material that (*i*) can withstand the high temperatures at which gas properties are often measured, (*ii*) are transparent in the spectral regions where measurements are desired (usually near-infrared to infrared) and do not experience "thermal runaway" (strong increase in absorptivity at a certain temperature level), and (*iii*) do not succumb to chemical attack from the test gas and other gases. Such cells have been used, for example, by Penner [18], Goldstein [67] and Oppenheim and Goldman [68].

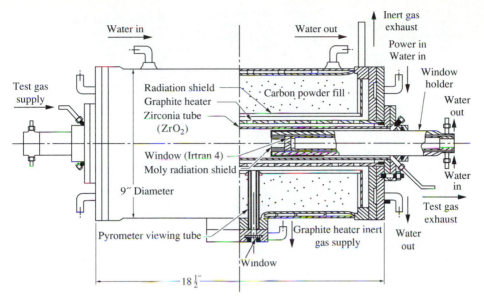

FIGURE 9-21
Schematic of the high-temperature gas furnace used by Tien and Giedt [66].

The cold-window cell, as the name implies, lets the probing beam enter and exit the test cell through water-cooled windows. This method has the advantage that the problems in a hot-window cell are nearly nonexistent. However, if the geometric path of the gas is relatively short, this method introduces serious temperature and density variations along the path. Tien and Giedt [66] designed a high-temperature furnace, consisting of a zirconia tube surrounded by a graphite heater, that allowed temperatures up to 2000 K. The furnace was fitted with water-cooled, movable zinc selenide windows, which are transmissive between $0.5\ \mu$m and $20\ \mu$m and stay inert to reactions with water vapor and carbon oxides for temperatures below 550 K. A schematic of their furnace is shown in Fig. 9-21. This apparatus was used by Tien and coworkers to measure the properties of various gases [46, 48–51, 69].

Nozzle seal cells are open flow cells in which the absorbing gas is contained within the cell by layers on each end of inert gases such as argon or nitrogen. This system eliminates some of the problems with windows, but may also cause density and temperature gradients near the seal; in addition, some scattering may be introduced by the turbulent eddies of the mixing flows [70]. This type of apparatus has been used by Hottel and Mangelsdorf [11] and Eckert [71] for total emissivity measurements of water vapor and carbon dioxide. Most of the measurements made by Edwards and coworkers also used nozzle seal cells [36, 37, 39, 45, 70, 72, 73]. A schematic of the apparatus used by Bevans and coworkers [72] is shown in Fig. 9-22.

Using a burner and jet for gas radiation measurements eliminates the window problems, and is in many ways similar to the nozzle seal cell. Free jet devices can be

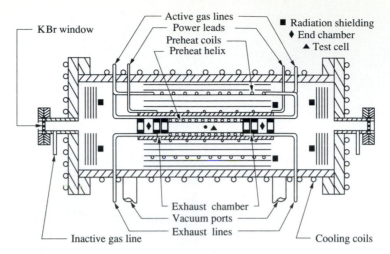

FIGURE 9-22
Schematic of nozzle seal gas containment system by Bevans and coworkers [72].

used for extremely high temperatures, but they also introduce considerable uncertainty with respect to gas temperature and density distribution and to path length. Ferriso and Ludwig [74] used such a device for spectral measurements of the 2.7 μm water vapor band.

Data Correlation

The half-width of a typical spectral line in the infrared is on the order of $0.1\,\text{cm}^{-1}$. To get a strong enough signal, any spectral measurement is by experimental necessity an average over several wavenumbers and, therefore, dozens or even hundreds of lines, unless an extremely monochromatic laser beam is employed. Thus, the measured transmissivity or (after subtracting from unity) absorptivity/emissivity is of the narrow band average type. A correlation for the average absorption coefficient may be found by inverting equation (9.48) or equation (9.53), depending on whether the Elsasser or the general statistical model is to be used, in either case yielding

$$\frac{S}{d} = \frac{S}{d}(\bar{\epsilon}_\eta, X, b/d), \tag{9.105}$$

where the $\bar{\epsilon}_\eta$ and X (density or pressure path length) are measured quantities, and the width-to-spacing ratio must be determined independently. If the common assumption of a constant b/d for the entire band is used, the width-to-spacing ratio can be obtained in a number of ways: (*i*) Direct prediction of b and d, (*ii*) using an independently determined band intensity, α, as the closing parameter, or (*iii*) finding a best fit for β (which is directly related to b/d) in the exponential wide band model. A typical correlation for the average absorption coefficient of the 7.8 μm nitrous oxide band is shown in Fig. 9-23 (from Tien, Modest and McCreight [51]). Note that S/d depends

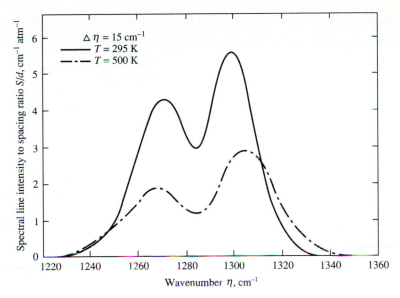

FIGURE 9-23
Narrow band correlation for the $7.8\,\mu\text{m}$ band of nitrous oxide [51] (monochromator slit opening is such that measurements are a narrow band average over $\Delta\eta = 15\,\text{cm}^{-1}$).

only on wavenumber and temperature and, thus, collapses the data points of any arbitrary optical path lengths.

Measured spectral absorptivities may be integrated to determine total band absorptances. Plotting those band absorptances that fall into the logarithmic regime vs. XP_e on semilog paper gives a straight line whose slope is the band width parameter (cf. Table 9.2). Preparing a linear plot of A/P_e vs. $\sqrt{X/P_e}$ for data in the square root regime gives again a straight line, this time with $\sqrt{\alpha\omega\gamma}$ as the slope (where $\gamma = \beta/P_e = \pi b/d$ is the width-to-spacing ratio for a dilute mixture, cf. Tables 9.2 and 9.3).

Finally, total emissivity values may be calculated by substituting the measured total band absorptances into equation (9.89).

Experimental Errors

No matter which method is used to measure gaseous radiation properties, the data will be subject to considerable errors. Edwards [70] lists as the major sources:

1. Reflection and scattering by optical windows;

2. Scattering by mixing zones in nozzle seals and free jets;

3. Inhomogeneity and uncertainty in the values of temperature, pressure, and composition;

4. Deterioration of the window material due to adsorption or "thermal runaway;"

5. Improper correlation model.

References

1. Edwards, D. K.: "Molecular Gas Band Radiation," in *Advances in Heat Transfer*, vol. 12, Academic Press, New York, pp. 115–193, 1976.
2. Tien, C. L.: "Thermal Radiation Properties of Gases," in *Advances in Heat Transfer*, vol. 5, Academic Press, New York, pp. 253–324, 1968.
3. Rayleigh, L.: "On the Light from the Sky, Its Polarization and Colour," *Philos. Mag.*, vol. 41, pp. 107–120, 274–279, 1871 (reprinted in *Scientific Papers by Lord Rayleigh*, vol. I: 1869–1881, no. 8, Dover, New York, 1964).
4. Langley, S. P.: "Experimental Determination of Wave-Lengths in the Invisible Prismatic Spectrum," *Mem. Natl. Acad. Sci.*, vol. 2, pp. 147–162, 1883.
5. Eddington, A. S.: *The Internal Constitution of the Stars*, Cambridge University Press, England, 1926 (also Dover Publications, New York, 1959).
6. Chandrasekhar, S.: *An Introduction to the Study of Stellar Structure*, University of Chicago Press, 1939.
7. Chandrasekhar, S.: *Radiative Transfer*, Dover Publications, New York, 1960 (originally published by Oxford University Press, London, 1950).
8. Paschen, F.: *Annalen der Physik und Chemie*, vol. 53, p. 334, 1894.
9. Hottel, H. C., and A. F. Sarofim: *Radiative Transfer*, McGraw-Hill, New York, 1967.
10. Hottel, H. C.: "Heat Transmission by Radiation from Non-Luminous Gases," *Transactions of AIChE*, vol. 19, pp. 173–205, 1927.
11. Hottel, H. C., and H. G. Mangelsdorf: "Heat Transmission by Radiation from Non-Luminous Gases II. Experimental Study of Carbon Dioxide and Water Vapor," *Transactions of AIChE*, vol. 31, pp. 517–549, 1935.
12. Hottel, H. C., and V. C. Smith: "Radiation from Non-Lumunious Flames," *Transactions of ASME, Journal of Heat Transfer*, vol. 57, pp. 463–470, 1935.
13. Hottel, H. C., and I. M. Stewart: "Space Requirement for the Combustion of Pulverized Coal," *Industrial Engineering Chemistry*, vol. 32, pp. 719–730, 1940.
14. Hottel, H. C., and R. B. Egbert: "The Radiation of Furnace Gases," *Transactions of ASME, Journal of Heat Transfer*, vol. 63, pp. 297–307, 1941.
15. Hottel, H. C., and R. B. Egbert: "Radiant Heat Transmission from Water Vapor," *Transactions of AIChE*, vol. 38, pp. 531–565, 1942.
16. Hottel, H. C.: "Radiant Heat Transmission," in *Heat Transmission*, ed. W. H. McAdams, 3d ed., ch. 4, McGraw-Hill, New York, 1954.
17. Hottel, H. C., and E. S. Cohen: "Radiant Heat Exchange in a Gas-Filled Enclosure: Allowance for Nonuniformity of Gas Temperature," *AIChE Journal*, vol. 4, pp. 3–14, 1958.
18. Penner, S. S.: *Quantitative Molecular Spectroscopy and Gas Emissivities*, Addison Wesley, Reading, Massachusetts, 1959.
19. Plass, G. N.: "Models for Spectral Band Absorption," *Journal of the Optical Society of America*, vol. 48, no. 10, pp. 690–703, 1958.
20. Plass, G. N.: "Spectral Emissivity of Carbon Dioxide From 1800–2500 cm^{-1}," *Journal of the Optical Society of America*, vol. 49, pp. 821–828, 1959.
21. Tien, C. L., and J. H. Lienhard: *Statistical Thermodynamics*, Holt, Rinehart & Winston, New York, 1971.
22. Davidson, N.: *Statistical Mechanics*, McGraw-Hill, New York, 1962.
23. Heitler, W.: *The Quantum Theory of Radiation*, Oxford University Press, Oxford and New York, 1954.
24. Griem, H. R.: *Plasma Spectroscopy*, McGraw-Hill, New York, 1964.
25. Bethe, H. A., and E. E. Salpeter: *Quantum Mechanics of One- and Two-Electron Systems*, Academic Press, New York, 1957.

26. Herzberg, G.: *Atomic Spectra and Atomic Structure*, 2d ed., Dover Publications, New York, 1944.

27. Herzberg, G.: *Molecular Spectra and Molecular Structure, Vol. II: Infrared and Raman Spectra of Polyatomic Molecules*, Van Nostrand, Princeton, New Jersey, 1945.

28. Herzberg, G.: *Molecular Spectra and Molecular Structure, Vol. I: Spectra of Diatomic Molecules*, 2d ed., Van Nostrand, Princeton, New Jersey, 1950.

29. Breene, R. G.: *The Shift and Shape of Spectral Lines*, Pergamon Press, Oxford, 1961.

30. Goody, R. J.: *Atmospheric Radiation*, Oxford University Press, Oxford and New York, 1964.

31. Burch, D. E., E. B. Singleton, and D. Williams: "Absorption Line Broadening in the Infrared," *Applied Optics*, vol. 1, pp. 359–363, 1962.

32. Abramowitz, M., and I. A. Stegun (eds.): *Handbook of Mathematical Functions*, Dover Publications, New York, 1965.

33. Hsieh, T. C., and R. Greif: "Theoretical Determination of the Absorption Coefficient and the Total Band Absorptance Including a Specific Application to Carbon Monoxide," *International Journal of Heat and Mass Transfer*, vol. 15, pp. 1477–1487, 1972.

34. Chu, K. H., and R. Greif: "Theoretical Determination of Band Absorption for Nonrigid Rotation with Applications to CO, NO, N_2O, and CO_2," *ASME Journal of Heat Transfer*, vol. 100, pp. 230–234, 1978.

35. Edwards, D. K., and W. A. Menard: "Comparison of Models for Correlation of Total Band Absorption," *Applied Optics*, vol. 3, pp. 621–625, 1964.

36. Edwards, D. K., L. K. Glassen, W. C. Hauser, and J. S. Tuchscher: "Radiation Heat Transfer in Nonisothermal Nongray Gases," *ASME Journal of Heat Transfer*, vol. 89, pp. 219–229, 1967.

37. Weiner, M. M., and D. K. Edwards: "Non-Isothermal Gas Radiation in Superposed Vibration-Rotation Bands," *Journal of Quantitative Spectroscopy and Radiative Transfer*, vol. 8, pp. 1171–1183, 1968.

38. Edwards, D. K.: "Radiative Transfer Characteristics of Materials," *ASME Journal of Heat Transfer*, vol. 91, pp. 1–15, 1969.

39. Edwards, D. K., and A. Balakrishnan: "Thermal Radiation by Combustion Gases," *International Journal of Heat and Mass Transfer*, vol. 16, pp. 25–40, 1973.

40. Modak, A. T.: "Exponential Wide Band Parameters for the Pure Rotational Band of Water Vapor," *Journal of Quantitative Spectroscopy and Radiative Transfer*, vol. 21, pp. 131–142, 1979.

41. Edwards, D. K., and W. A. Menard: "Correlations for Absorption by Methane and Carbon Dioxide Gases," *Applied Optics*, vol. 3, pp. 847–852, 1964.

42. Edwards, D. K., and W. A. Menard: "Correlation of Absorption by Water Vapor at Temperatures From 300 K to 1100 K," *Applied Optics*, vol. 4, pp. 715–721, 1965.

43. Edwards, D. K., B. J. Flornes, L. K. Glassen, and W. Sun: "Comparison of Models for Correlation of Total Band Absorption," *Applied Optics*, vol. 3, pp. 621–625, 1964.

44. Brosmer, M. A., and C. L. Tien: "Infrared Radiation Properties of Methane at Elevated Temperatures," *Journal of Quantitative Spectroscopy and Radiative Transfer*, vol. 33, no. 5, pp. 521–532, 1985.

45. Edwards, D. K.: "Absorption of Radiation by Carbon Monoxide Gas According to the Exponential Wide-Band Model," *Applied Optics*, vol. 4, no. 10, pp. 1352–1353, 1965.

46. Abu-Romia, M. M., and C. L. Tien: "Measurements and Correlations of Infrared Radiation of Carbon Monoxide at Elevated Temperatures," *Journal of Quantitative Spectroscopy and Radiative Transfer*, vol. 6, pp. 143–167, 1966.

47. Abu-Romia, M. M., and C. L. Tien: "Appropriate Mean Absorption Coefficients for Infrared Radiation of Gases," *ASME Journal of Heat Transfer*, vol. 89C, pp. 321–327, 1967.

48. Chan, S. H., and C. L. Tien: "Infrared Radiation Properties of Sulfur Dioxide," *ASME Journal of Heat Transfer*, vol. 93, pp. 172–177, 1971.

49. Tien, C. L.: "Band and Total Emissivity of Ammonia," *International Journal of Heat and Mass Transfer*, vol. 16, pp. 856–857, 1973.

50. Green, R. M., and C. L. Tien: "Infrared Radiation Properties of Nitric Oxide at Elevated Temperatures," *Journal of Quantitative Spectroscopy and Radiative Transfer*, vol. 10, pp. 805–817, 1970.

51. Tien, C. L., M. F. Modest, and C. R. McCreight: "Infrared Radiation Properties of Nitrous Oxide," *Journal of Quantitative Spectroscopy and Radiative Transfer*, vol. 12, pp. 267–277, 1972.

52. Brosmer, M. A., and C. L. Tien: "Thermal Radiation Properties of Acetylene," *ASME Journal of Heat Transfer*, vol. 107, pp. 943–948, 1985.

53. Felske, J. D., and C. L. Tien: "A Theoretical Closed Form Expression for the Total Band Absorptance of Infrared-Radiating Gases," *ASME Journal of Heat Transfer*, vol. 96, pp. 155–158, 1974.

54. Tien, C. L., and J. E. Lowder: "A Correlation for Total Band Absorption of Radiating Gases," *International Journal of Heat and Mass Transfer*, vol. 9, pp. 698–701, 1966.

55. Cess, R. D., and S. N. Tiwari: "Infrared Radiative Energy Transfer in Gases," in *Advances in Heat Transfer*, vol. 8, Academic Press, New York, pp. 229–283, 1972.

56. Tiwari, S. N.: "Models for Infrared Atmospheric Radiation," in *Advances in Geophysics*, vol. 20, Academic Press, New York, 1978.

57. Chan, S. H., and C. L. Tien: "Total Band Absorptance of Non-Isothermal Infrared-Radiating Gases," *Journal of Quantitative Spectroscopy and Radiative Transfer*, vol. 9, pp. 1261–1271, 1969.

58. Cess, R. D., and L. S. Wang: "A Band Absorptance Formulation for Non-Isothermal Gaseous Radiation," *International Journal of Heat and Mass Transfer*, vol. 13, pp. 547–555, 1970.

59. Edwards, D. K., and S. J. Morizumi: "Scaling Vibration-Rotation Band Parameters for Nonhomogeneous Gas Radiation," *Journal of Quantitative Spectroscopy and Radiative Transfer*, vol. 10, pp. 175–188, 1970.

60. Felske, J. D., and C. L. Tien: "Infrared Radiation from Non-Homogeneous Gas Mixtures Having Overlapping Bands," *Journal of Quantitative Spectroscopy and Radiative Transfer*, vol. 14, pp. 35–48, 1974.

61. Leckner, B.: "Spectral and Total Emissivity of Water Vapor and Carbon Dioxide," *Combustion and Flame*, vol. 19, pp. 33–48, 1972.

62. Boynton, F. P., and C. B. Ludwig: "Total Emissivity of Hot Water Vapor – II, Semi-Empirical Charts Deduced from Long-Path Spectral Data," *International Journal of Heat and Mass Transfer*, vol. 14, pp. 963–973, 1971.

63. Ludwig, C. B., W. Malkmus, J. E. Reardon, and J. A. L. Thomson: "Handbook of Infrared Radiation from Combustion Gases," Technical Report SP-3080, NASA, 1973.

64. Sarofim, A. F., I. H. Farag, and H. C. Hottel: "Radiative Heat Transmission from Nonluminous Gases. Computational Study of the Emissivities of Carbon Dioxide," ASME paper no. 78-HT-55, 1978.

65. Cess, R. D., and P. Mighdoll: "Modified Planck Mean Coefficients for Optically Thin Gaseous Radiation," *International Journal of Heat and Mass Transfer*, vol. 10, pp. 1291–1292, 1967.

66. Tien, C. L., and W. H. Giedt: "Experimental Determination of Infrared Absorption of High-Temperature Gases," in *Advances in Thermophysical Properties at Extreme Temperatures and Pressures*, ASME, pp. 167–173, 1965.

67. Goldstein, R. J.: "Measurements of Infrared Absorption by Water Vapor at Temperatures to 1000 K," *Journal of Quantitative Spectroscopy and Radiative Transfer*, vol. 4, pp. 343–352, 1964.

68. Oppenheim, U. P., and A. Goldman: "Spectral Emissivity of Water Vapor at 1200 K," in *Tenth Symposium (International) on Combustion*, The Combustion Institute, pp. 185–188, 1965.

69. Abu-Romia, M. M., and C. L. Tien: "Spectral and Integrated Intensity of CO Fundamental Band at Elevated Temperatures," *International Journal of Heat and Mass Transfer*, vol. 10, pp. 1779–1784, 1967.

70. Edwards, D. K.: "Thermal Radiation Measurements," in *Measurements in Heat Transfer*, eds. E. R. G. Eckert and R. J. Goldstein, ch. 10, Hemisphere, Washington, DC, 1976.

71. Eckert, E. R. G.: "Messung der Gesamtstrahlung von Wasserdampf und Kohlensäure in Mischung mit nichtstrahlenden Gasen bei Temperaturen bis 1300°C," *VDI Forschungshefte*, vol. 387, pp. 1–20, 1937.

72. Bevans, J. T., R. V. Dunkle, D. K. Edwards, J. T. Gier, L. L. Levenson, and A. K. Oppenheim: "Apparatus for the Determination of the Band Absorption of Gases at Elevated Pressures and Temperatures," *Journal of the Optical Society of America*, vol. 50, pp. 130–136, 1960.

73. Edwards, D. K.: "Absorption by Infrared Bands of Carbon Dioxide Gas at Elevated Pressures and Temperatures," *Journal of the Optical Society of America*, vol. 50, pp. 617–626, 1960.

74. Ferriso, C. C., and C. B. Ludwig: "Spectral Emissivities and Integrated Intensities of the 2.7 μm H_2O Band between 530 and 2200 K," *Journal of Quantitative Spectroscopy and Radiative Transfer*, vol. 4, pp. 215–227, 1964.

Problems

9.1 Estimate the eigenfrequency for vibration, ν_e, for a CO molecule.

9.2 A certain gas at 1 bar pressure has a molecular mass of $m = 10^{-22}$ g and a diameter of $D = 5 \times 10^{-8}$ cm. At what temperature would Doppler and collision broadening result in identical broadening widths for a line at a wavenumber of 4000 cm^{-1}?

9.3 Water vapor is known to have spectral lines in the vicinity of $\lambda = 1.38\,\mu$m. Consider a single, broadened spectral line centered at $\lambda_0 = 1.33\,\mu$m. If the water vapor is at a pressure of 0.1 atm and a temperature of 1000 K, what would you expect to be the main cause for broadening? Over what range of wavenumbers would you expect the line to be appreciable, i.e., over what range is the absorption coefficient at least 1% of its value at the line center?

9.4 Compute the half-width for a spectral line of CO_2 at 2.8 μm for both Doppler and collision broadening as a function of pressure and temperature. Find the temperature as function of pressure for which both broadening phenomena result in the same half-width. (Note: The effective diameter of the CO_2 molecule is 4.0×10^{-8} cm).

9.5 Repeat Problem 9.4 for CO at a spectral location of 4.8 μm (Note: The effective diameter of the CO molecule is 3.4×10^{-8} cm).

9.6 A certain gas has two important vibration-rotation bands centered at 4 μm and 10 μm. Measurements of spectral lines in the 4 μm band (taken at 300 K and 1 bar $= 10^5$ N/m^2) indicate a half-width of $b_\eta = 0.5$ cm^{-1}. Predict the half-width in the 10 μm band for the gas at 500 K, 3 bar. (The diameter of the gas molecules is known to be between 5 Å $< D < 40$ Å).

9.7 A polyatomic gas has an absorption band in the infrared. For a certain small wavelength range to the following is known:
Average line-half width: 0.04 cm^{-1},
Average integrated absorption coefficient: 0.20 cm^{-2}/(g/cm^3),
Average line spacing: 0.25 cm^{-1},
The density of the gas at STP is 3×10^{-3} g/cm^3.
For a 50 cm thick gas layer at 500 K and 1 atm calculate the local mean spectral emissivity using

    (a) the Elsasser model,
    (b) the statistical model.

Which result can be expected to be more accurate?

9.8 Consider a gas for which the semistatistical model is applicable, i.e., $\bar{\epsilon}_\eta = 1 - \exp(-W_\eta/d)$. To predict $\bar{\epsilon}_\eta$ for arbitrary situations, a band-averaged (or constant) value for b_η/d must be known. Experimentally available are values for $\alpha = \int_{\Delta\eta}(S_\eta/d)\,d\eta$ and $\bar{\epsilon}_\eta = \bar{\epsilon}_\eta(\eta)$ (for optically thick situations) for given p_e and T. It is also known that

$$\frac{b_\eta}{d} \simeq \left(\frac{b_\eta}{d}\right)_0 p_e \left(\frac{T_0}{T}\right)^{1/2}.$$

Outline how an average value for $(b_\eta/d)_0$ can be found.

9.9 The following is known for a gas mixture at 600 K and 2 atm total pressure and in the vicinity of a certain spectral position: The gas consists of 80% (by volume) N_2 and 20%

of a diatomic absorbing gas with a molecular weight of $20 \, \text{g/mol}$, a mean line half-width $b = 0.01 \, \text{cm}^{-1}$, a mean line spacing of $d = 0.1 \, \text{cm}^{-1}$, and a mean line intensity of $S = 8 \times 10^{-5} \, \text{cm}^{-2}/(\text{g/m}^3)$. (a) For a gas column $10 \, \text{cm}$ thick determine the mean spectral emissivity of the gas. (b) What happens if the pressure is increased to $20 \, \text{atm}$? (Since no broadening parameters are known you may assume the effective broadening pressure to be equal to the total pressure).

9.10 One kg of a gas mixture at $2000 \, \text{K}$ and $1 \, \text{atm}$ occupies a container of $1 \, \text{m}$ height. The gas consists of 70% nitrogen (by volume) and of 30% an absorbing species. It is known that, at a certain spectral location, the line half-width is $b = 300 \, \text{MHz}$, the mean line spacing is $d = 2000 \, \text{MHz}$, and the line intensity is $S = 100 \, \text{cm}^{-1} \, \text{MHz}$. (a) Calculate the mean spectral emissivity under these conditions. (b) What will happen to the emissivity if the sealed container is cooled to $300 \, \text{K}$?

9.11 A $50 \, \text{cm}$ thick layer of a pure gas is maintained at $1000 \, \text{K}$ and $1 \, \text{atm}$. It is known that, at a certain spectral location, the mean line half-width is $b = 0.1 \, \text{nm}$, the mean line spacing is $d = 2 \, \text{nm}$, and the mean line intensity is $S = 0.002 \, \text{cm}^{-1} \, \text{nm} \, \text{atm}^{-1} = 2 \times 10^{-10} \, \text{atm}^{-1}$. What is the mean spectral emissivity under these conditions? ($1 \, \text{nm} = 10^{-9} \, \text{m}$.)

9.12 The following data for a diatomic gas at $300 \, \text{K}$ and $1 \, \text{atm}$ are known: The mean line spacing is $0.6 \, \text{cm}^{-1}$ and the mean line half-width is $0.03 \, \text{cm}^{-1}$; the mean line intensity ($=$integrated absorption coefficient) is $0.8 \, \text{cm}^{-2} \, \text{atm}^{-1}$ (based on a pressure absorption coefficient). Calculate the mean spectral emissivity for a path length of $1 \, \text{cm}$. In what band approximation is the optical condition?

9.13 Consider again the gas of Problem 9.12, but replace the line intensity by

$$S_{p\eta} = S_{p\eta_0} e^{-|\eta - \eta_0|/\omega}, \quad S_{p\eta_0} = 0.02 \, \text{cm}^{-2} \, \text{atm}^{-1}, \quad \omega = 20 \, \text{cm}^{-1}.$$

In what regime is this optical condition? What is the total band absorptance?

9.14 A mixture of nitrogen and sulfur dioxide (with 0.05 volume-% SO_2) is at $1 \, \text{atm}$ and $300 \, \text{K}$. For the strong ν_3-band of SO_2, centered at $\eta_3 = 1361 \, \text{cm}^{-1}$, it is known that

$$\frac{b_\eta}{d} \simeq 0.06, \quad \frac{S_\eta}{d} = 17 \, \text{cm}^{-1} \, \text{atm}^{-1} \exp\left[-\left(\frac{\eta - \eta_3}{25 \, \text{cm}^{-1}}\right)^2\right].$$

For a $10 \, \text{cm}$ thick gas-mixture layer:

(a) Develop an expression for the average spectral emissivity. What regime(s) apply?
(b) Calculate the total band absorptance.
(c) For comparison, calculate the total band absorptance from the wide band model. (Hint: For $300 \, \text{K} \, \gamma/\gamma_0 > 1$).

Note: Under these conditions collision broadening is the predominant broadening mechanism.

9.15 A certain gas is known to behave almost according to the rigid-rotor/harmonic-oscillator model, resulting in gradually changing line strengths (with wavenumber) and somewhat irregular line spacing. Calculate the local mean emissivity for a $1 \, \text{m}$ thick layer of the gas at $0.1 \, \text{atm}$ pressure. In the wavelength range of interest, it is known that the integrated absorption coefficient is equal to $0.80 \, \text{cm}^{-2} \text{atm}^{-1}$, the line half-width is $0.04 \, \text{cm}^{-1}$ and the average line spacing is $0.40 \, \text{cm}^{-1}$.

9.16 A gas mixture at 1500 K and 1 atm is known to contain a small amount of CO_2. To remotely determine the partial pressure of CO_2 the band absorptance of the CO_2 4.3 μm band is measured and is found to be 100 cm^{-1} for a path length of 1 m. Assuming that the gas may be treated as a nitrogen–CO_2 mixture, determine the partial pressure of the CO_2.

9.17 It is desired to predict the fraction of the sunlight absorbed by the nitrous oxide (N_2O) contained in the atmosphere. You may assume that the atmosphere is a 20 km high layer of N_2–N_2O mixture, that the atmosphere is isothermal with a linear pressure variation from 1 atm at the ground to zero at the top of the atmosphere, and that N_2O makes up 10^{-4}% by volume of the mixture everywhere.
N_2O has two vibration-rotation bands with the following wide band coefficients (at the temperature of the atmosphere):

	α	γ	ω	n	b
4.5 μm band	2035 cm^{-2} atm^{-1}	0.145	22 cm^{-1}	0.6	1.12
7.8 μm band	161 cm^{-2} atm^{-1}	0.377	18.5 cm^{-1}	0.6	1.12

(a) Show that the influence of the 7.8 μm band is negligible.

(b) Calculate the total absorptivity for this atmosphere assuming a constant average pressure of 0.5 atm.

(c) Show that the absorptivity is the same as in (b) if the linear pressure variation is taken into account. Hint: You may assume that the pressure-absorption coefficient, $\kappa_{p\eta}$, is independent of pressure.

9.18 A 1 m thick layer of a mixture of nitrogen and methane (CH_4) at T $= 300$ K and $p = 1$ atm has a measured total emissivity of $\epsilon = 0.010$. Estimate the partial pressure of the methane ($R_{CH_4} = 5.128 \times 10^{-6}$ atm m^3/g K). It is known that $p_{CH_4} \ll p$.

9.19 A mixture of water vapor and nitrogen at a total pressure of 1 atm and a temperature of 300 K is found to have a total band absorptance of 100 cm^{-1} for the 6.3 μm band for a geometric path length of 50 cm. Determine the partial pressure of the water vapor.

9.20 One method of measuring the temperature of a high-temperature gas is to determine the total band absorptance of a vibration-rotation band of the gas under the prevailing conditions. Consider a 20 cm thick layer of pure methane, CH_4, at 1 atm pressure. If the total band absorptance of the 3.3 μm band is 587 cm^{-1}, what is the temperature of the CH_4?
Note: Such instruments are generally used only for $T > 1000\,°C$.

9.21 Estimate the total band absorptance of the 2.7 μm CO_2 band at 833 K, a total pressure of 10 atm, a partial pressure of 1 atm, and a mass-path length of $\rho_{CO_2}L = 2440$ g/m^2, from Fig. 1-16. Compare with the result from the exponential wide band model.

9.22 Using the exponential wide band model, evaluate the total emissivity of a 1 m thick layer of a nitrogen–water vapor mixture at 2 atm and 400 K if the water vapor content by volume is (a) 0.01%, (b) 1%, or (c) 100%.

9.23 Repeat Problem 9.22 for a CO_2–nitrogen mixture at 600 K and a total pressure of 0.75 atm.

9.24 Evaluate the Planck-mean absorption coefficients for the two gases in Problems 9.22 and 9.23, based on the data given in Table 9.3. Compare the results with Fig. 9-18.

9.25 Write a small computer program that calculates the total emissivity of a CO_2–inert gas mixture as a function of temperature, pressure, CO_2 volume fraction, and path length.

For a given set of pressure, volume fraction and length, plot the emissivity as a function of temperature.

9.26 Repeat problem 9.22 for a path with a temperature profile given by $T = 300\,\text{K}[1 + 4s(L - s)/L^2]$, where s is distance across the gas layer.

9.27 A mixture of nitrogen and sulfur dioxide (with 5% SO_2 by volume) is at 1 atm total pressure. To measure the temperature of the mixture in a furnace environment ($T > 1000\,\text{K}$), an instrument is used that measures total band absorptance for the strong SO_2-band centered at $\eta_c = 1361\,\text{cm}^{-1}$. For that band it is known that $\alpha = 2340\,(T_0/T)\,\text{cm}^{-2}\,\text{atm}^{-1}$, $\beta = 0.357\,\sqrt{T/T_0}P_e$, $\omega = 8.8\,\sqrt{T/T_0}\,\text{cm}^{-1}$, $b = 1.28$ and $n = 0.65$. What is the temperature of the mixture if the total band absorptance has been measured as $142\,\text{cm}^{-1}$ for a 1 m thick gas layer?

9.28 To determine the average atmospheric temperature on a distant planet, the total band absorptance for the $3.3\,\mu\text{m}$ CH_4 band has been measured as $A_{3.3} = 100\,\text{cm}^{-1}$. It is known from other measurements that methane is a trace element in the atmosphere (which contains mostly nitrogen and whose total pressure is 2 atm), and that the absorption path length for methane on that planet, for which $A_{3.3}$ was measured, is $4.14\,\text{g/m}^2$. What is the temperature?

9.29 Develop a simple box model for the evaluation of the effective band width, i.e., $A = \int_0^\infty \epsilon_\eta\,d\eta = \bar{\epsilon}_\eta\,\Delta\eta$, based on an average emissivity (rather than absorption coefficient). You may assume that the line spacing and line intensity are constant across the band. Calculate the total band absorptance of water vapor at 0.1 atm and 400 K for path lengths of 1 mm and 1 m, assuming that $\Delta\eta \approx \omega$, where ω is the band width parameter from the exponential wide band model. Compare with results from that model.

9.30 Two black plates at 500 K are separated by 1 m of a mixture consisting of 10% carbon dioxide, 10% water vapor, and 80% nitrogen at a temperature of 1200 K and a total pressure of 1 atm. Estimate the heat gain of each plate (you may carry out a numerical quadrature by looking at, say, three different angles).

CHAPTER
10

RADIATIVE PROPERTIES OF PARTICULATE MEDIA

10.1 INTRODUCTION

When an electromagnetic wave or a photon interacts with a medium containing small particles, the radiative intensity may be changed by absorption and/or scattering. Common examples of this interaction are sunlight being absorbed by a cloud of smoke (which is nothing but a multitude of fine particles suspended in air), scattering of sunshine by the atmosphere (the atmosphere consisting of molecules which are, in fact, tiny particles) resulting in blue skies and red sunsets, and the colors of the rainbow. Radiation scattering by particles was first dealt with by astrophysicists, who were interested in the scattering of starlight by interstellar dust. Scientists from many other disciplines are concerned with the scattering of electromagnetic waves: Meteorologists are concerned with scattering within the earth's atmosphere (scattering of sunlight as well as scattering of radar waves for observation of precipitation); electrical engineers and physicists deal with the propagation of radio waves through the atmosphere; physicists, chemists, and engineers today use light scattering as diagnostic tools for nonintrusive and nondestructive measurements in gases, liquids, and solids.

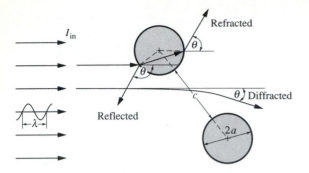

FIGURE 10-1
Interaction between electromagnetic waves and spherical particles.

A review of thermal radiation phenomena in particulate media has recently been given by Tien and Drolen [1].

How much and into which direction a particle scatters an electromagnetic wave passing through its vicinity depends on (*i*) the shape of the particle, (*ii*) the material of the particle (i.e., the complex index of refraction, $m = n - ik$), (*iii*) its relative size, and (*iv*) the clearance between particles. In radiative analyses the shape of particles is usually assumed to be spherical (for spherical and irregularly shaped objects) or cylindrical (for long fibrous materials). These simplifying assumptions give generally excellent results, since averaging over many millions of irregular shapes tends to smoothen the irregularities [1]. In the following discussion we shall limit ourselves to absorption and scattering from spherical particles, as shown in Fig. 10-1.

An electromagnetic wave or photon passing through the immediate vicinity of spherical particles will be absorbed or scattered. The scattering is due to three separate phenomena, namely, (*i*) diffraction (waves never come into contact with the particle, but their direction of propagation is altered by the presence of the particle), (*ii*) reflection by a particle (waves reflected from the surface of the sphere), and (*iii*) refraction in a particle (waves that penetrate into the sphere and, after partial absorption, reemerge traveling into a different direction). The vast majority of photons are scattered elastically, i.e., their wavelength (and energy) remain unchanged. A tiny fraction undergo *inelastic* or *Raman scattering* (the photons reemerge with a different wavelength). While very important for optical diagnostics, the *Raman effect* is unimportant for the evaluation of radiative heat transfer rates, and we shall treat only elastic scattering in this book. If scattering by one particle is not affected by the presence of surrounding particles, we speak of *independent scattering*, otherwise we have *dependent scattering*. Thus, the radiative properties of a cloud of spherical particles of radius a, interacting with an electromagnetic wave of wavelength λ, are governed by three independent nondimensional parameters:

$$\text{complex index of refraction:} \quad m = n - ik, \tag{10.1}$$

$$\text{size parameter:} \quad x = 2\pi a/\lambda, \tag{10.2}$$

$$\text{clearance-to-wavelength ratio:} \quad c/\lambda. \tag{10.3}$$

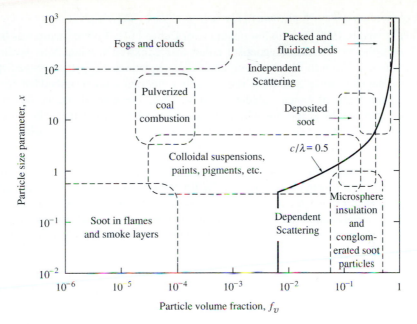

FIGURE 10-2
Scattering regime map for independent and dependent scattering [1].

If scattering is independent ($c/\lambda \gg 1$), then only the first two parameters are needed. For the classification of dependent scattering, the clearance-to-wavelength ratio is often replaced by a purely geometric parameter, c/a, which in turn may be related to the volume fraction of particles, f_v. While in earlier works, for example that by van de Hulst [2], it was assumed that dependent effects were a function of particle separation only, it is now known that wavelength effects also play a role. This fact was first recognized by Hottel and coworkers [3]. Since then, a number of investigators, notably Tien and coworkers [1, 4–8], have established limits for when dependent effects must be considered. Their results, summarized in Fig. 10-2, show that dependent scattering effects may be ignored as long as $f_v < 0.006$ or $c/\lambda > 0.5$. Since these values include nearly all heat transfer applications, only independent scattering is discussed in the present chapter. The reader interested in the prediction of dependent scattering properties should consult the monograph by Tien and Drolen [1].

10.2 ABSORPTION AND SCATTERING FROM A SINGLE SPHERE

The scattering and absorption of radiation by single spheres was first discussed during the later part of the nineteenth century by Lord Rayleigh [9, 10], who obtained a simple solution for spheres whose diameters are much smaller than the wavelength of radiation (small size parameter, $x \ll 1$). This work was followed in the 1890s by the work of Lorenz, a Danish physicist [11, 12], in 1908 by the classical paper of Gustav Mie* [13], and in 1909 by a similar treatment of Debye [14]. Lorenz' work was based on his own theory of electromagnetism rather than Maxwell's, while Mie developed an equivalent solution to Maxwell's equations [cf. equations (2.1) through (2.5)] for an electromagnetic wave train traveling through a medium with an imbedded sphere. Although the work of Lorenz predates that of Mie, the general theory describing radiative scattering by absorbing spheres is generally referred to as the "Mie Theory." An exhaustive review of the history of the development of particle scattering theory has been given by Kerker [15].

The complicated *Mie scattering theory* must generally be used if the size of the sphere is such that it is too large to apply the Rayleigh theory, but too small to employ geometric optics (which requires $x \gg 1$ as well as $kx \gg 1$). We shall give here a very brief discussion of the Mie theory and some representative results. Detailed derivations may be found in the books on the subject by van de Hulst [2], Kerker [15], Deirmendjian [16], and Bohren and Huffman [17].

The amount of scattering and absorption by a particle is usually expressed in terms of the *scattering cross section, C_{sca}* and *absorption cross section C_{abs}*. The total amount of absorption and scattering, or extinction, is expressed in terms of the *extinction cross section,*

$$C_{ext} = C_{abs} + C_{sca}. \tag{10.4}$$

Often *efficiency factors Q* are used instead of cross sections, being nondimensionalized with the projected surface area of the sphere, or

$$\text{absorption efficiency factor:} \quad Q_{abs} = \frac{C_{abs}}{\pi a^2}, \tag{10.5}$$

$$\text{scattering efficiency factor:} \quad Q_{sca} = \frac{C_{sca}}{\pi a^2}, \tag{10.6}$$

$$\text{extinction efficiency factor:} \quad Q_{ext} = \frac{C_{ext}}{\pi a^2}, \tag{10.7}$$

and

$$Q_{ext} = Q_{abs} + Q_{sca}. \tag{10.8}$$

Radiation interacting with a spherical particle may be scattered away from its original direction by an angle Θ, i.e., the propagation vector of the electric and magnetic fields may be redirected by the scattering angle (Fig. 10-1). This deflection from the

---

*__*Gustav Mie (1868–1957)__
German physicist. After studying at the universities of Rostock and Heidelberg, he served as professor of physics at various German universities.

incident direction is described by the angle Θ alone because, for a spherical particle, there can be no azimuthal variation. The intensity of the wave scattered by the angle Θ [i.e., the magnitude of the Poynting vector, equation (2.42)] is proportional to two complex *amplitude functions* $S_1(\Theta)$ and $S_2(\Theta)$, where the subscripts denote two perpendicular polarizations. Once these amplitude functions have been determined, the intensity of radiation I_{sca}, scattered by an angle Θ from the incident unpolarized beam of strength I_{in} may be calculated [2, 15, 16] from

$$\frac{I_{\text{sca}}(\Theta)}{I_{\text{in}}} = \frac{1}{2}\frac{i_1 + i_2}{x^2}, \tag{10.9}$$

where i_1 and i_2 are the nondimensional polarized intensities calculated from

$$i_1(x, m, \Theta) = |S_1|^2, \quad i_2(x, m, \Theta) = |S_2|^2. \tag{10.10}$$

From equation (10.9) it follows that the total amount of energy scattered by one sphere into all directions [2] is

$$Q_{\text{sca}} = \frac{C_{\text{sca}}}{\pi a^2} = \frac{a^2}{\pi a^2} \int_{4\pi} \frac{I_{\text{sca}}}{I_{\text{in}}} \, d\Omega = \frac{1}{x^2} \int_0^\pi (i_1 + i_2) \sin \Theta \, d\Theta. \tag{10.11}$$

The fraction of this energy that is scattered into any given direction is denoted by the *scattering phase function* $\Phi(\Theta)$, which is normalized such that

$$\frac{1}{4\pi} \int_{4\pi} \Phi(\hat{s}_i, \hat{s}) \, d\Omega \equiv 1. \tag{10.12}$$

Thus, together with equation (10.9), the scattering phase function may be expressed as

$$\Phi(\Theta) = \frac{i_1 + i_2}{\dfrac{1}{4\pi} \displaystyle\int_{4\pi} (i_1 + i_2) \, d\Omega} = 2\frac{i_1 + i_2}{x^2 Q_{\text{sca}}}. \tag{10.13}$$

Finally, total extinction by a single particle (absorption within the particle, plus scattering into all directions) is related to the amplitude functions by

$$Q_{\text{ext}} = \frac{4}{x^2} \Re\{S(0)\}, \tag{10.14}$$

where the amplitude function S is without a subscript because $S_1(0) = S_2(0)$.

The major difficulty in the evaluation of scattering properties lies in the calculation of the complex amplitude functions $S_1(\Theta)$ and $S_2(\Theta)$. For the general case of arbitrary values for the complex index of refraction m and the size parameter x, the full Mie equations, as expressed by van de Hulst [2], must be employed,

$$S_1(\Theta) = \sum_{n=1}^\infty \frac{2n+1}{n(n+1)} [a_n \pi_n(\cos \Theta) + b_n \tau_n(\cos \Theta)], \tag{10.15}$$

$$S_2(\Theta) = \sum_{n=1}^\infty \frac{2n+1}{n(n+1)} [b_n \pi_n(\cos \Theta) + a_n \tau_n(\cos \Theta)], \tag{10.16}$$

where the direction-dependent functions π_n and τ_n are related to Legendre polynomials P_n (for a description of these polynomials, see, e.g., Wylie [18]) by

$$\pi_n(\cos\Theta) = \frac{dP_n(\cos\Theta)}{d\cos\Theta}, \tag{10.17}$$

$$\tau_n(\cos\Theta) = \cos\Theta\,\pi_n(\cos\Theta) - \sin^2\Theta\,\frac{d\pi_n(\cos\Theta)}{d\cos\Theta}, \tag{10.18}$$

and the Mie scattering coefficients a_n and b_n are complex functions of x and $y = mx$,

$$a_n = \frac{\psi_n'(y)\psi_n(x) - m\psi_n(y)\psi_n'(x)}{\psi_n'(y)\zeta_n(x) - m\psi_n(y)\zeta_n'(x)}, \tag{10.19}$$

$$b_n = \frac{m\psi_n'(y)\psi_n(x) - \psi_n(y)\psi_n'(x)}{m\psi_n'(y)\zeta_n(x) - \psi_n(y)\zeta_n'(x)}. \tag{10.20}$$

The functions ψ_n and ζ_n are known as *Riccati-Bessel functions*, and are related to Bessel and Hankel functions [18, 19] by

$$\psi_n(z) = \left(\frac{\pi z}{2}\right)^{1/2} J_{n+1/2}(z), \quad \zeta_n(z) = \left(\frac{\pi z}{2}\right)^{1/2} H_{n+1/2}(z). \tag{10.21}$$

Equations (10.15) and (10.16) may be substituted into equations (10.11) and (10.14). Using the fact that—like Legendre polynomials—the functions π_n and τ_n constitute sets of orthogonal functions leads to

$$Q_{\text{sca}} = \frac{2}{x^2}\sum_{n=1}^{\infty}(2n+1)(|a_n|^2 + |b_n|^2), \tag{10.22}$$

$$Q_{\text{ext}} = \frac{2}{x^2}\sum_{n=1}^{\infty}(2n+1)\Re\{a_n + b_n\}. \tag{10.23}$$

Once all Mie scattering coefficients a_n and b_n have been determined, the phase function Φ may also be evaluated from equation (10.13), but this calculation tends to be extremely tedious because of the nature of equation (10.10), and because the calculations must be carried out anew for every scattering angle Θ. To facilitate the calculations Chu and Churchill [20] and Clark, Chu and Churchill [21] expressed the scattering phase function as a series in Legendre polynomials,

$$\Phi(\Theta) = 1 + \sum_{n=1}^{\infty} A_n P_n(\cos\Theta), \tag{10.24}$$

where the coefficients A_n are directly related to the Mie scattering coefficients a_n and b_n through some rather complicated formulae not reproduced here. The great advantage of this formulation is that, once the A_n have been determined, the value of the phase function Φ is determined quickly for any or all scattering directions.

In many applications the use of the complicated scattering phase function described by equation (10.24) is too involved. For a simpler analysis the directional scattering behavior may be described by the average cosine of the scattering angle, known as *asymmetry factor*, and related to the phase function by

$$g = \overline{\cos \Theta} = \frac{1}{4\pi} \int_{4\pi} \Phi(\Theta) \cos \Theta \, d\Omega. \tag{10.25}$$

For the case of *isotropic scattering* (i.e., equal amounts are scattered into all directions, and $\Phi \equiv 1$) the asymmetry factor vanishes; g also vanishes if scattering is symmetrical about the plane perpendicular to beam propagation. If the particle scatters more radiation into the forward directions ($\Theta < \pi/2$), g is positive; if more radiation is scattered into the backward direction ($\Theta > \pi/2$), g is negative. For spherical particles the asymmetry factor is readily calculated [17] as

$$g = \overline{\cos \Theta} = \frac{4}{x^2 Q_{\text{sca}}} \sum_{n=1}^{\infty} \left[\frac{n(n+2)}{n+1} \Re\{a_n a_{n+1}^* + b_n b_{n+1}^*\} \right.$$

$$\left. + \frac{2n+1}{n(n+1)} \Re\{a_n b_n^*\} \right]. \tag{10.26}$$

The calculation of the scattering Mie coefficients a_n and b_n is no trivial matter even in these days of supercomputers: The relationships leading to their determination are involved and require the frequent evaluation of complicated functions with complex arguments. For large size parameters x many terms need to be calculated ($n_{\max} \approx 2x$). Recursion formulae for the functions π_n, τ_n, ψ_n and ζ_n have been given by Deirmendjian [16] and others, which evaluate these functions for increasing values of n in terms of previously calculated functions. Deirmendjian observed that the accuracy of calculations decreases for increasing n, causing complete failure of the calculations for large values of the size parameter x (for which many terms are required in the series for the amplitude functions), even if double-precision arithmetic is employed. This problem was overcome by Kattawar and Plass [22] who showed that all four functions may be reduced to functions each belonging to one of two sets: One set has stable recursion formulae for increasing n (i.e., round-off error *decreases* with growing n), and the other set is stable for decreasing values of n (setting the function to zero for a larger n than required in the series results in very accurate values for slightly smaller n). Wiscombe [23] compared the accuracy and stability of several Mie scattering computer solution routines and discussed the efficiency of different calculation methods (whether to use upward or downward recursion, what recursion formulae to use, etc.). Some representative results of Mie calculations are shown in Figs. 10-3 through 10-6. Figure 10-3 shows typical behavior of efficiency factors, demonstrated with the extinction efficiency of a dielectric ($k \equiv 0$) for a number of different refractive indices n. Observe that there is a primary oscillation in the variation of Q_{ext} with size parameter, upon which secondary oscillations are superimposed (stronger for larger refractive indices). Note also that the oscillations become smaller for larger size parameters, and $Q_{\text{ext}} \to 2$ as $x \to \infty$ (for dielectrics as well

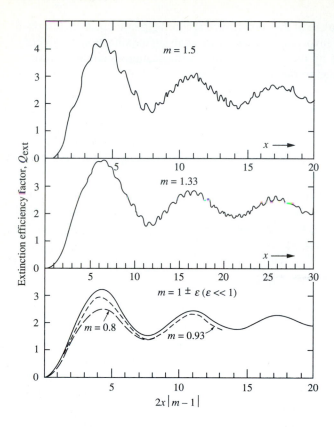

FIGURE 10-3
Extinction efficiency factors for dielectric spheres for several refractive indices [2].

as metals). Figures 10-4 and 10-5 show the qualitative behavior of efficiency factors for absorption, Q_{abs}, and scattering, Q_{sca}, respectively, for a fixed value of the size parameter ($x = 1$), as a function of absorptive index k. The absorption efficiency factors may vary by many orders of magnitude over the range of absorptive index k, while the scattering efficiency remains constant over great changes of k. Finally, Fig. 10-6 shows some representative scattering phase functions, $\Phi(\Theta)$. Figure 10-6a shows the scattering behavior of very small particles (known as *Rayleigh scattering*): The scattering is symmetric to the plane perpendicular to the incident beam, and is nearly isotropic with slight forward and backward scattering peaks and somewhat lesser scattering to the sides. Figure 10-6b demonstrates the behavior of particles with refractive indices close to unity (known as *Rayleigh-Gans scattering*): Nearly all of the scattered energy is scattered into forward directions with some scattering into a few preferred other directions. This behavior becomes more extreme as the size parameter increases. Figure 10-6c shows the phase function of a typical dielectric: The scattering has a strong forward component; otherwise the scattering behavior demonstrates rapid maxima and minima at varying scattering angles, with much stronger amplitudes than for *Rayleigh-Gans scattering* (note the change in scale). The variations are not quite so extreme as in Fig. 10-6b owing to the large value for n. The

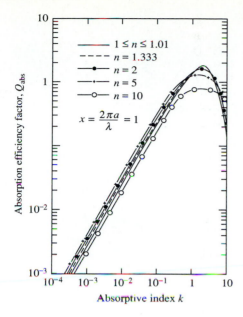

FIGURE 10-4
The absorption efficiency factor as function of complex index of refraction, $m = n - ik$, for a size parameter of $x = 1$ [22].

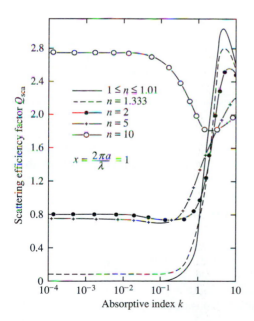

FIGURE 10-5
The scattering efficiency factor as function of complex index of refraction, $m = n - ik$, for a size parameter of $x = 1$ [22].

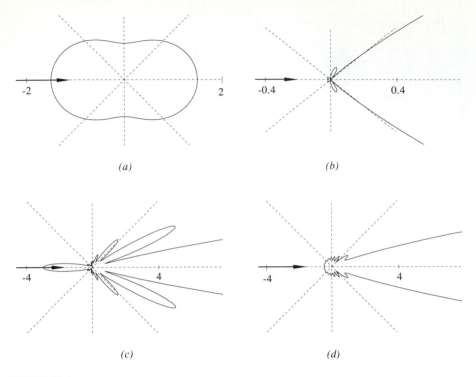

FIGURE 10-6
Polar plot of scattering phase functions for single spherical particles, (a) small sphere with $x = 0.001$, (b) dielectric with $x = 5$ and $m = 1.0001$, (c) dielectric sphere with $x = 10$ and $m = 2$, (d) metallic sphere (aluminum) with $x = 10$ and $m = 4.46 - 31.5i$.

behavior of a typical metal (aluminum at $3.1\ \mu$m) is shown in Fig. 10-6d: Besides a strong forward-scattering peak these particles display lesser-degree oscillations than dielectrics.

10.3   RADIATIVE PROPERTIES OF A PARTICLE CLOUD

In all problems of radiative heat transfer with particulate scattering and absorption, we have to deal with a large collection of particles. If the scattering is independent, as is assumed in this chapter, then the effects of large numbers of particles are simply additive. For simplicity, it is often assumed that particle clouds consist of spheres that are all equally large. More accurate analyses take into account that particles of many different sizes may occur within a single cloud, and that these sizes often vary by orders of magnitude. We shall briefly describe both approaches in the following paragraphs.

Clouds of Uniform Size Particles

The fraction of energy scattered by all particles per unit length along the direction of the incoming beam is called the *scattering coefficient* [as defined by equation (8.9)] and is equal to the scattering cross section summed over all particles. If N_T is the number of particles per unit volume, all of uniform radius a, then

$$\sigma_{s\lambda} = N_T C_{sca} = \pi a^2 N_T Q_{sca}, \tag{10.27}$$

and, similarly, for absorption and extinction,

$$\kappa_\lambda = N_T C_{abs} = \pi a^2 N_T Q_{abs}, \tag{10.28}$$
$$\beta_\lambda = \kappa_\lambda + \sigma_{s\lambda} = N_T C_{ext} = \pi a^2 N_T Q_{ext}. \tag{10.29}$$

Since the scattering phase function (or the directional distribution of scattered energy) in a cloud of uniform particles is the same for each particle, it is also the same for the particle cloud, or

$$\Phi_{T\lambda}(\Theta) = \Phi(\Theta), \tag{10.30}$$

and similarly for the asymmetry factor,

$$g_{T\lambda} = \overline{(\cos \Theta)}_{T\lambda} = \overline{\cos \Theta}. \tag{10.31}$$

In both cases we have temporarily added the subscript T (to distinguish the total cloud of particles from a single particle) and λ to emphasize the fact that both quantities are spectral quantities that may vary with wavelength.

If total (i.e., spectrally integrated) properties are desired, equations (10.27) through (10.29) may be integrated to obtain Planck-mean or Rosseland-mean coefficients (for absorption, scattering, and/or extinction), as defined by equations (9.100) and (9.104), or

$$y_P = \frac{\pi}{\sigma T^4} \int_0^\infty I_{b\lambda} y_\lambda \, d\eta, \qquad y = \kappa, \sigma_s, \text{ or } \beta, \tag{10.32}$$

$$\frac{1}{y_R} = \frac{\pi}{4\sigma T^3} \int_0^\infty \frac{1}{y_\lambda} \frac{dI_{b\lambda}}{dT} d\lambda, \qquad y = \kappa, \sigma_s, \text{ or } \beta. \tag{10.33}$$

Similarly, total emissivities and absorptivities may be obtained from equation (9.89). Since the efficiency factors Q may vary rapidly across the spectrum, these integrations generally need to be done numerically.

Clouds of Nonuniform Size Particles

For clouds of particles of nonuniform size it is customary to describe the number of particles as a function of radius in the form of a *particle distribution function*. A number of different forms for the distribution function have been used by various researchers. We introduce here the so-called *modified gamma distribution* [16],

$$n(a) = Aa^\gamma \exp(-Ba^\delta), \quad 0 \le a < \infty, \tag{10.34}$$

which vanishes at $a = 0$ and $a \to \infty$. This distribution function reduces to the *gamma distribution* if $\delta = 1$. The four constants A, B, γ, and δ are positive and real, and γ and δ are usually chosen to be integers. They must be determined from measurable quantities such as total number of particles (per unit volume),

$$N_T = \int_0^\infty n(a)\,da = A \int_0^\infty a^\gamma \exp(-Ba^\delta)\,da = \frac{A\Gamma\left(\frac{\gamma+1}{\delta}\right)}{\delta B^{(\gamma+1)/\delta}}. \tag{10.35}$$

Here Γ is the *gamma function*,

$$\Gamma(z) = \int_0^\infty e^{-t}\,t^{z-1}\,dt, \tag{10.36}$$

and has been tabulated, e.g., by Abramawitz and Stegun [19]. Equation (10.35) shows that the constant A is essentially given by N_T. The total volume of particles per unit volume, or *volume fraction*, is given by

$$f_v = \int_0^\infty \tfrac{4}{3}\pi a^3\, n(a)\,da = \frac{4\pi A\Gamma\left(\frac{\gamma+4}{\delta}\right)}{3\delta B^{(\gamma+4)/\delta}}. \tag{10.37}$$

Assuming that all particles have the same optical properties, we may again determine the scattering coefficient for a particle cloud by adding the scattering cross section over all particles but, because of the particle size distribution, this is now an integral rather than a simple sum,

$$\sigma_{s\lambda} = \int_0^\infty C_{\text{sca}}\, n(a)\,da = \pi \int_0^\infty Q_{\text{sca}}\, a^2 n(a)\,da, \tag{10.38}$$

and, similarly, for absorption and extinction,

$$\kappa_\lambda = \int_0^\infty C_{\text{abs}}\, n(a)\,da = \pi \int_0^\infty Q_{\text{abs}}\, a^2 n(a)\,da, \tag{10.39}$$

$$\beta_\lambda = \int_0^\infty C_{\text{ext}}\, n(a)\,da = \pi \int_0^\infty Q_{\text{ext}}\, a^2 n(a)\,da. \tag{10.40}$$

For nonuniform particles the scattering phase function is not the same for all particles. From the definition of the phase function, it follows that the scattered energies into a given direction must be summed over all particles and then normalized, or

$$\Phi_{T\lambda}(\Theta) = \frac{\displaystyle\int_0^\infty (i_1 + i_2)\, n(a)\,da}{\dfrac{1}{4\pi}\displaystyle\int_{4\pi}\left[\int_0^\infty (i_1 + i_2)\, n(a)\,da\right] d\Omega} = \frac{\displaystyle\int_0^\infty C_{\text{sca}}(a)\,\Phi(a,\Theta)\, n(a)\,da}{\displaystyle\int_0^\infty C_{\text{sca}}(a)\, n(a)\,da}$$

$$= \frac{1}{\sigma_{s\lambda}} \int_0^\infty C_{\text{sca}}(a)\,\Phi(a,\Theta)\, n(a)\,da, \tag{10.41}$$

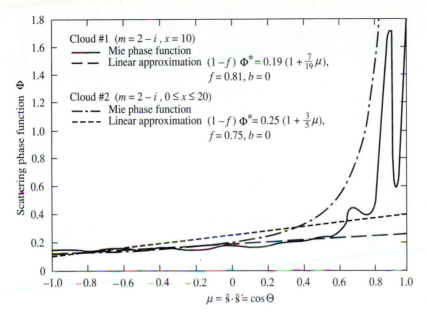

FIGURE 10-7
Mie scattering phase function for clouds of absorbing particles [40].

and, similarly,

$$g_{T\lambda} = \overline{(\cos\Theta)}_{T\lambda} = \frac{1}{\sigma_{s\lambda}} \int_0^\infty C_{\text{sca}}(a)\, g(a)\, n(a)\, da. \tag{10.42}$$

Again, if total properties are needed, equations (10.38) through (10.40) may be integrated over the entire spectrum.

Figures 10-7 and 10-8 show a few typical scattering phase functions for absorbing and nonabsorbing particle clouds. Two types of particles are considered, one nonabsorbing with an index of refraction $m = 2$, the other one absorbing with $m = 2 - i$. The particles are either in clouds of constant radius $a = 5\,\mu$m, or in clouds with a distribution function

$$n(a) = 27{,}230 a^2 \exp(-1.7594a), \tag{10.43}$$

which has its maximum at $a = 5\,\mu$m. All the particle clouds have a number density of 10^4 particles/cm^3, and the Mie calculations have been carried out for a typical wavelength of $\lambda = 3.1416\,\mu$m, resulting in a size parameter of $x = 2\pi a/\lambda = 10$ for the constant-radius clouds, and a range of significant size parameters of $0 < x \leq 20$ for clouds with particle size distribution. The radiative properties for the four different particle clouds are summarized in Table 10.1. Absorption and scattering coefficients for constant-radius and particle distribution clouds differ considerably,

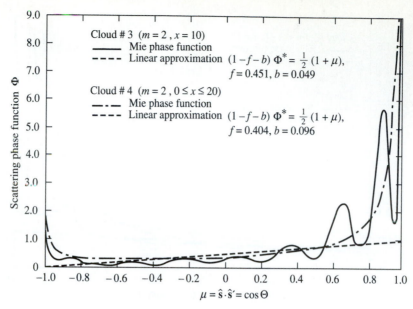

FIGURE 10-8
Mie scattering phase function for clouds of dielectric particles [40].

TABLE 10.1
Radiative properties of typical particle clouds $(N_T = 10^4/\text{cm}^3, \lambda = 3.1416\mu\text{m})$

	Cloud #1 Const. Radius $a = 5\,\mu\text{m}$ $m = 2 - i$	Cloud#2 Size Distr. $n(a)$ $m = 2 - i$	Cloud#3 Const. Radius $a = 5\,\mu\text{m}$ $m = 2$	Cloud#4 Size Distr. $n(a)$ $m = 2$
Absorption coefficient $\kappa\ [\text{cm}^{-1}]$	8.307×10^{-3}	1.524×10^{-3}	0	0
Scattering coefficient $\sigma_s\ [\text{cm}^{-1}]$	1.073×10^{-2}	1.674×10^{-3}	6.420×10^{-2}	3.363×10^{-3}
Extinction coefficient $\beta\ [\text{cm}^{-1}]$	1.904×10^{-2}	3.198×10^{-3}	6.420×10^{-2}	3.363×10^{-3}
Scattering albedo	0.5634	0.5235	1	1
Terms needed for phase function	26	35	27	33

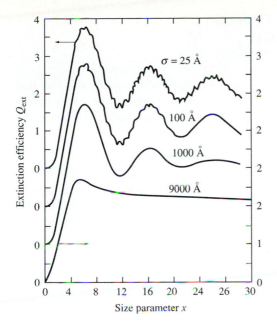

FIGURE 10-9
The effect of size dispersion on the extinction efficiency for water droplets and visible light (σ =standard deviation in Gaussian distribution function) [17].

primarily because the average particle size in equation (10.43) is less than 5μm, being 2.33μm for the volume- or mass-averaged radius, and 1.52μm for the number-averaged radius. Observe that the phase functions for uniform particle size clouds display strong oscillations due to diffraction peaks, because the phase function is identical to the one of single particles (cf. Fig. 10-6). Since the diffraction peaks shift slightly with changing size parameters, these peaks and valleys are smoothed out for clouds with varying particle sizes. For these types of clouds the phase function becomes very smooth with only a strong forward-scattering peak remaining (plus a weaker backward-scattering peak for dielectric particles). Thus, the analysis of scattering phenomena may actually be simpler if there is a particle size distribution!

Bohren and Huffman [17] have shown that this smoothing effect occurs for the efficiency factors as well as for the phase function, and that the effect requires only a small deviation from uniform-size particles. Figure 10-9 shows the extinction efficiency for clouds of water droplets, which are assumed to have a Gaussian distribution function centered around a mean particle size with standard deviation σ. Small deviations from uniform size blur out the high-frequency variation (called the *ripple structure*), while slightly larger deviations also dampen out the low-frequency variations of the extinction efficiency (called the *interference structure*). Similar smoothing effects occur in a cloud of uniform-size particles of irregular shape as shown by Hodkinson [24] for aqueous suspensions of irregular quartz particles.

10.4 RADIATIVE PROPERTIES OF SMALL SPHERES (RAYLEIGH SCATTERING)

Radiative scattering by spheres that are small compared with wavelength was first described by Lord Rayleigh [9, 10] long before the development of Mie's theory [13]. However, results for small particles are here most easily obtained by taking the appropriate limits in the general solution to Mie's equations.

If the scattering particles are extremely small, then the size parameter $x = 2\pi a/\lambda$ becomes very small. Such behavior is primarily observed with gas molecules (which are, in fact, very tiny particles). There are, however, also some multimolecule solid particles that fall into the Rayleigh scattering regime, e.g., soot particles (whose diameters are often smaller than $0.1\,\mu$m and which, in combustion applications, are irradiated by light of approximately $3\,\mu$m, resulting in $x \approx 0.1$).

In the limit of $x \to 0$ it is relatively straightforward to show that only the a_1 in equations (10.19) and (10.20) is nonzero, or

$$S_2(\Theta) = S_1(\Theta)\cos\Theta = i\frac{m^2 - 1}{m^2 + 2}x^3\cos\Theta, \tag{10.44}$$

that is, the amplitude function for one polarization is independent of scattering angle Θ. Substitution into equations (10.11) and (10.14) then gives the efficiency factors as

$$Q_{sca} = \frac{8}{3}\left|\frac{m^2 - 1}{m^2 + 2}\right|^2 x^4, \tag{10.45}$$

$$Q_{ext} = -4\Im\left\{\frac{m^2 - 1}{m^2 + 2}\right\}x \approx Q_{abs}, \tag{10.46}$$

where the last equality in equation (10.46) is due to the fact that $x^4 \ll x$, so that scattering may be neglected as compared with absorption. We observe the wavelength dependence of the scattering efficiency to be

$$Q_{sca} \propto \frac{1}{\lambda^4} \propto \nu^4. \tag{10.47}$$

We note in passing that this fact explains the colors of the sky: During most of the day, when the sun's rays travel a relatively short distance through earth's atmosphere (cf. Fig. 10-10), only the shortest wavelengths are scattered away in any appreciable amounts from the sun's direct path, scattered again and again by the molecules in the atmosphere, providing us with a blue sky (blue light having the shortest wavelength within the visible spectrum). Close to sunset, however, the sun's rays travel at a grazing angle through the atmosphere to the observer, so that all but the very longest wavelengths (of the visible spectrum) have been scattered away from the direct path,

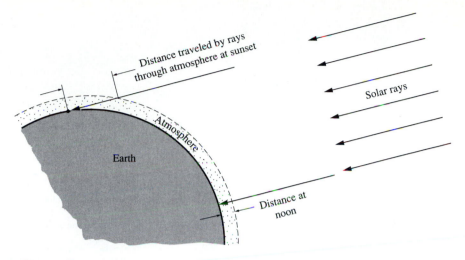

FIGURE 10-10
Distance traveled through earth's atmosphere by solar rays.

giving the sun a red appearance. Without the atmosphere the sky would appear black to us, as witnessed by the astronauts visiting the (atmosphere-less) moon.

The wavelength dependence of the absorption efficiency, on the other hand, is

$$Q_{abs} \propto \frac{1}{\lambda} \propto \nu, \tag{10.48}$$

which describes the spectral behavior of small particles such as soot reasonably well.

The phase function for Rayleigh scattering follows immediately from equations (10.44) and (10.13) as

$$\Phi(\Theta) = \tfrac{3}{4}(1 + \cos^2 \Theta), \tag{10.49}$$

where the two terms are the contributions from the two perpendicular polarizations, as shown in Fig. 10-11. It is observed that the phase function is symmetric as far as forward and backward scattering is concerned, and does not deviate too strongly from isotropic scattering.

The absorption coefficient for a cloud of nonuniform-size small particles follows from equations (10.39) and (10.46) as

$$\kappa_\lambda = \pi \int_0^\infty Q_{abs} \, a^2 n(a) \, da = -4\Im\left\{\frac{m^2 - 1}{m^2 + 2}\right\} \int_0^\infty \left(\frac{2\pi a}{\lambda}\right) \pi \, a^2 n(a) \, da. \tag{10.50}$$

The integral in this equation may be related to the *volume fraction* f_v,

$$f_v = \int_0^\infty \left(\frac{4}{3}\pi a^3\right) n(a) \, da, \tag{10.51}$$

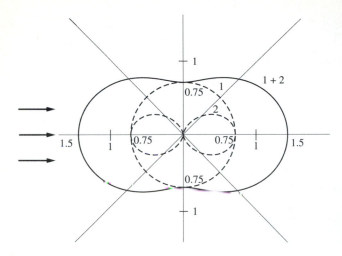

FIGURE 10-11
Polar diagram of Rayleigh phase function: 1, polarized with electric vector in plane perpendicular to paper, 2, polarized in plane of paper; 1+2, unpolarized.

so that the absorption coefficient for small particles reduces to

$$\kappa_\lambda = -\Im\left\{\frac{m^2 - 1}{m^2 + 2}\right\}\frac{6\pi f_v}{\lambda}. \tag{10.52}$$

Therefore, for particles small enough that Rayleigh scattering holds, the absorption coefficient does not depend on particle size distribution, but only on the total volume occupied by all particles (per unit system volume).

Example 10.1. During the burning of propane it is observed that the products contain a volume fraction of $10^{-4}\%$ of soot with complex index of refraction $m = 2.21 - 1.23i$ (measured at a wavelength of $3\,\mu$m). Assuming a mean particle diameter of $0.05\,\mu$m, determine the absorption and scattering efficiency of this soot cloud as well as its absorption coefficient, all at a wavelength of $3\,\mu$m.

Solution

For the given diameter and wavelength the particle size parameter is $x = \pi \times 0.05\,\mu\text{m}/3\,\mu\text{m}$ $= 0.0524 \ll 1$ and we assume Rayleigh scattering to hold for all particles. For all three properties we need to evaluate the complex ratio $(m^2 - 1)/(m^2 + 2)$:

$$\frac{m^2 - 1}{m^2 + 2} = \frac{2.21^2 - 2 \times 2.21 \times 1.23i - 1.23^2 - 1}{2.21^2 - 2 \times 2.21 \times 1.23i - 1.23^2 + 2}$$

$$= \frac{2.3712 - 5.4366i}{5.3712 - 5.4366i} \times \frac{5.3712 + 5.4366i}{5.3712 + 5.4366i}$$

$$= \frac{42.2928 - 16.3098i}{58.4064} = 0.7241 - 0.2792i.$$

Thus, the efficiencies can be evaluated as

$$Q_{sca} = \frac{8}{3}|0.7241 - 0.2792i|^2 \times (0.0524)^4 = 1.21 \times 10^{-5},$$

and

$$Q_{abs} = -4 \times (-0.2792) \times 0.0524 = 5.85 \times 10^{-2},$$

showing that scattering may indeed be neglected compared with absorption. The absorption coefficient follows from equation (10.52) as

$$\kappa_\lambda = -(-0.2792) \times \frac{6\pi \times 10^{-4}/100}{3 \times 10^{-4}\text{cm}} = 0.01754\,\text{cm}^{-1},$$

that is, any radiation (at $3\,\mu$m) penetrating into such a soot cloud would be attenuated to $1/e$ of its original intensity over a distance of $1/\kappa_\lambda = 57\,$cm.

10.5 RAYLEIGH-GANS SCATTERING

A near-dielectric sphere with $k \approx 0$ and with a refractive index close to unity, i.e., $|m - 1| \ll 1$, has negligible reflectivity and, thus, lets light pass into the sphere unattenuated and unrefracted. If also $x|m - 1| \ll 1$, then the light will exit the sphere again essentially unattenuated. However, since the phase velocity of light is slightly less inside the particle, light traveling through the sphere will display a small phase lag as opposed to the incident light. This phenomenon is known as *Rayleigh-Gans scattering*.

As described by van de Hulst [2], taking the appropriate limits reduces equations (10.15) and (10.16) to

$$S_2(\Theta) = S_1(\Theta)\cos\Theta = ix^3(m - 1)G(u)\cos\Theta, \tag{10.53}$$

where

$$G(u) = \frac{2}{u^3}(\sin u - u\cos u), \qquad u = 2x\sin\tfrac{1}{2}\Theta. \tag{10.54}$$

The absorption and extinction efficiencies are identical to the ones for Rayleigh scattering, that is,

$$Q_{ext} = -4\Im\left\{\frac{m^2 - 1}{m^2 + 2}\right\}x \approx Q_{abs}, \tag{10.55}$$

while the scattering efficiency turns out to be

$$Q_{sca} = |m - 1|^2 x^4 \int_0^\pi G^2(u)(1 + \cos^2\Theta)\sin\Theta\,d\Theta. \tag{10.56}$$

Finally, the phase function for Rayleigh-Gans scattering is now easily determined as

$$\Phi(\Theta) = \frac{2G^2(u)(1 + \cos^2\Theta)}{\displaystyle\int_0^\pi G^2(u)(1 + \cos^2\Theta)\sin\Theta\,d\Theta}. \tag{10.57}$$

An example of this phase function is included in Fig. 10-6b for $x = 5$ and $m = 1.0001$. The phase function displays a strong forward-scattering peak (which increases with increasing size parameter), with very rapid oscillations of varying amplitude into the other directions.

10.6 RADIATIVE PROPERTIES OF LARGE SPHERES

If the spheres are very large ($x \gg 1$), very many terms are required in the evaluation of equations (10.15) and (10.16). However, in this case it is sufficient to resort to geometric optics, and one may separate diffraction from reflection and refraction. For very large spheres it is always true that

$$Q_{\text{ext}} = 2. \tag{10.58}$$

This relationship is sometimes called the *extinction paradox* since it states that a large particle removes exactly *twice* the amount of light from the beam as it can intercept, and has been discussed by van de Hulst [2]. Since, for geometric optics, the projected area of a particle for reflection and absorption is πa^2, this means that half of the extinction efficiency is due to diffraction. How much of the rest is due to absorption, and how much due to reflection, depend on the value of the complex index of refraction m, or the reflectivity of the sphere's surface.

In the following we shall determine the scattering properties of *large opaque spheres*, i.e., such spheres for which any ray refracted into the particle will be totally absorbed within, without exiting the sphere at another location. This requires the additional assumption that $kx \gg 1$ (say, 2 or 3). Thus, k may be fairly small as long as $x \gg 1$. A consequence of this is that, for a metal, "large particle" may mean $x > 10$, while for a near-dielectric it may mean $x > 10,000$.

While electromagnetic wave theory always assumes optically smooth surfaces, resulting in specular reflection, very large spheres (as compared with wavelength) may have roughness levels at the sphere's surface that are also large as compared with wavelength, resulting in nonspecular reflection. Treatment of very irregular directional behavior for the reflectivity is, of course, extremely difficult (as it was for surface transport, cf. Chapter 7). However, the extreme case of perfectly diffuse reflection lends itself to straightforward analysis (similar to the treatment of surface transport in Chapter 5), and is, therefore, also included in the present section.

Diffraction from Large Spheres

The diffraction pattern of light passing through the vicinity of a large sphere is, by Babinet's principle, equal to that of a circular hole with the same diameter [2]. As a consequence the directional behavior of the diffracted light consists of alternating bright and dark rings. The amplitude functions for diffraction have been given by

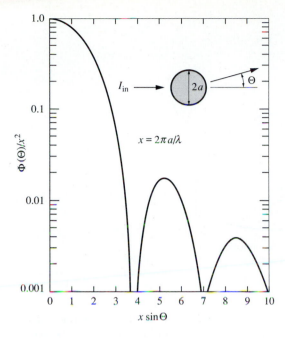

FIGURE 10-12
Phase function for diffraction over a large sphere.

van de Hulst [2] as

$$S_1(\Theta) = S_2(\Theta) = x\frac{J_1(x \sin\Theta)}{\sin\Theta}, \tag{10.59}$$

where J_1 is a Bessel function [18]. Therefore, the phase function for diffraction over a large sphere follows from equation (10.13) (noting that $Q_{\text{sca}} = 1$ for diffraction) as

$$\Phi(\Theta) = 2\frac{i_1 + i_2}{x^2} = 4\frac{J_1^2(x \sin\Theta)}{\sin^2\Theta}. \tag{10.60}$$

This phase function, depicted in Fig. 10-12, demonstrates that almost all energy is scattered forward within a narrow cone of $\Theta < (150/x)°$ from the direction of transmission. Thus, in heat transfer applications we may usually neglect diffraction and treat it as transmission. Then, for large particles without diffraction,

$$Q_{\text{ext}} = 1. \tag{10.61}$$

Large Specularly Reflecting Spheres

Consider a specularly reflecting opaque sphere irradiated by an intensity I_i distributed over a thin pencil of rays of solid angle $d\Omega_i$ as shown in Fig. 10-13. Under these conditions the infinitesimal band at an angle β from the incident direction (indicated by shading in the figure) receives radiation from a direction which is off-normal (from

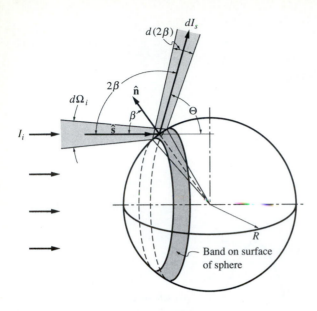

FIGURE 10-13
Scattering of incident radiation by a
large specularly reflecting sphere.

its surface) by an angle β. Recalling the definition of intensity as "heat rate per unit
area normal to the rays, per unit solid angle, and per unit wavelength," the energy
intercepted by the band over a wavelength range of $d\lambda$ is

$$d^2Q_i = I_i \, d\Omega_i \, d\lambda \, (dA_{\text{band}} \cos \beta) = I_i \, d\Omega_i \, d\lambda \, 2\pi a \sin \beta \, a \, d\beta \, \cos \beta. \qquad (10.62)$$

Of that, the fraction $\rho^s(\beta)$ is reflected into the direction 2β as measured from the
incoming pencil of rays. The total heat rate intercepted by the sphere is

$$dQ_i = \int_0^{\pi/2} I_i \, d\Omega_i \, d\lambda \, 2\pi a^2 \sin \beta \, \cos \beta \, d\beta = I_i \, d\Omega_i \, d\lambda \, \pi a^2, \qquad (10.63)$$

while the total reflected (or scattered) heat rate is

$$dQ_s = \int_0^{\pi/2} \rho^s(\beta) \, I_i \, d\Omega_i \, d\lambda \, 2\pi a^2 \sin \beta \, \cos \beta \, d\beta$$

$$= I_i \, d\Omega_i \, d\lambda \, \pi a^2 \, 2 \int_0^{\pi/2} \rho^s(\beta) \sin \beta \, \cos \beta \, d\beta = \rho^s \, I_i \, d\Omega_i \, d\lambda \, \pi a^2, \qquad (10.64)$$

where ρ^s is the hemispherical reflectivity, averaged over all incoming directions
[cf. equation (3.5)]:

$$\rho^s = 2 \int_0^{\pi/2} \rho^s(\beta) \sin \beta \, \cos \beta \, d\beta. \qquad (10.65)$$

Thus, the scattering efficiency for a large, opaque, specularly reflecting particle is
simply

$$Q_{\text{sca}} = \frac{dQ_s}{dQ_i} = \rho^s, \qquad (10.66)$$

and the absorption efficiency follows as

$$Q_{abs} = Q_{ext} - Q_{sca} = 1 - \rho^s = \alpha, \tag{10.67}$$

that is, the hemispherical absorptivity.

To evaluate the scattering phase function we consider the amount of energy scattered into any given direction Θ, where Θ is measured from the transmission direction $\hat{s}$, as also indicated in Fig. 10-13. It is clear that, for a homogeneous sphere, the scattered intensity can only vary with the polar angle Θ (and not azimuthally). Furthermore, for a specularly reflecting sphere the outgoing intensity in a certain direction Θ can only come from a single position on the sphere's surface. For example, radiation scattered into the direction $\Theta = \pi - 2\beta$ comes from the shaded band in Fig. 10-13. Recalling that the scattering phase function is defined as $4\pi \times$ scattered intensity/total scattered heat flux [cf. equation (8.17)] we get, for $\Theta = \pi - 2\beta$ or $\beta = (\pi - \Theta)/2$,

$$\begin{aligned}
\Phi(\Theta) &= 4\pi \frac{\rho^s(\beta) d^2 Q_i / d\Omega_r}{dQ_s} \\
&= 4\pi \rho^s \left(\frac{\pi - \Theta}{2}\right) \frac{I_i d\Omega_i \, d\lambda \, 2\pi a^2 \sin\beta \cos\beta \, d\beta / d\Omega_r}{\rho^s \, I_i d\Omega_i \, d\lambda \, \pi a^2}.
\end{aligned} \tag{10.68}$$

The solid angle for the reflection is best visualized by letting the reflected intensity fall upon a concentric (and very large) sphere of radius R. The solid angle is then the area of the illuminated band divided by R^2, or

$$d\Omega_r = 2\pi \sin 2\beta \, d(2\beta), \tag{10.69}$$

leading to

$$\Phi(\Theta) = \rho^s \left(\frac{\pi - \Theta}{2}\right) / \rho^s. \tag{10.70}$$

Alternatively, we could use the fact that the scattering phase function is proportional to intensity into any given direction, and then normalize the resulting expression with equation (8.19). The actual directional scattering behavior (or the behavior of the phase function) depends on the material of which the particles are made. Figure 10-14 shows a comparison of the phase function between a "typical" metal (aluminum at 3.1 μm with an index of refraction of $m = 4.46 - 31.5i$) and a "typical" dielectric ($m = 2$).[1] These two phase functions should be compared with Fig. 10-6c,d, which are for identical materials but for a smaller size parameter (and are shown in a polar rather than Cartesian plot). Since the size parameter in Fig. 10-6c,d is fairly large ($x = 10$), the major difference between Fig. 10-6c,d and Fig. 10-14 lies in the omission of diffraction in Fig. 10-14. For large particles all materials have their maximum scattering into the forward direction, $\Theta = 0$, since $\rho^s(\pi/2) = 1$

---

[1] Note that in the case of a dielectric the absorptive index k is assumed negligible as compared with the refractive index n, but k is assumed large enough to make the spheres opaque.

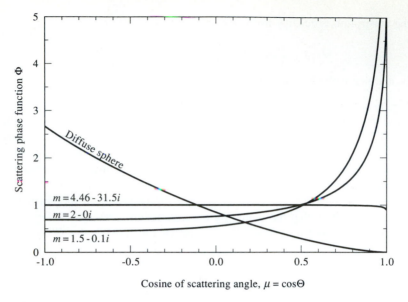

FIGURE 10-14
Scattering phase functions for large spheres of various materials.

always. However, this peak is considerably more pronounced for dielectrics, and is hardly noticeable for the metal because of the dip in reflectivity at near-grazing angles (compare also Fig. 2-11, which shows the directional variation of the reflectivity of silver). Because of their relatively high reflectivity at all directions, large metallic particles tend to be almost isotropic scatterers.

Example 10.2. Consider glass particles with a complex index of refraction $m = 1.5 - 0.1i$ and a density of $\rho_{glass} = 2\,g/cm^3$, suspended in an inert gas, with a particle loading ratio of $1\,kg$ of particles per m^3 of suspension volume. Particle sizes range between $100\,\mu m$ and $1000\,\mu m$, with an equal distribution over all sizes by weight-%. Determine the absorption coefficient, the scattering coefficient, and the phase function for the infrared ($3\,\mu m < \lambda < 10\,\mu m$).

Solution

First, we need to determine the particle distribution function by number (rather than mass). Since the mass distribution function is a constant we get with

$$m(a) = \frac{1\,kg/m^3}{(1000 - 100)\,\mu m} = \tfrac{4}{3}\pi a^3 \rho_{glass} n(a), \quad 100\,\mu m \leq a \leq 1000\,\mu m,$$

$$n(a) = \frac{3m(a)}{4\pi a^3 \rho_{glass}} = 1.3226 \times 10^{-7}\,\mu m^{-1}/a^3, \quad 100\,\mu m \leq a \leq 1000\,\mu m.$$

Next we need to determine the range of the size parameter x to see whether Rayleigh scattering, Mie scattering or large-particle scattering must be considered. The minimum value for x will occur for the smallest particle at the longest wavelength, or

$$x_{min} = \frac{2\pi a_{min}}{\lambda_{max}} = \frac{2\pi 100}{10} = 62.83 \gg 1,$$

$$(kx)_{min} = 6.283 \gg 1,$$

that is, the large-particle assumption will be acceptable for all conditions encountered in this example. Thus, the absorption and scattering coefficients may be related to the hemispherical emissivity of the glass. Since for this glass $k \ll n$, the material behaves essentially like a dielectric, and the hemispherical emissivity may be found from Fig. 3-19 or equation (3.80). Either method leads to $\epsilon = \alpha = 1 - \rho = 0.91$. The absorption and scattering coefficients may then be calculated from equations (10.39) and (10.38) as

$$\kappa_\lambda = \pi \int_0^\infty \alpha a^2 n(a)\, da = \pi \alpha \int_{100\,\mu m}^{1000\,\mu m} a^2 \frac{1.3226 \times 10^{-7}}{\mu m\, a^3}\, da$$

$$= 1.3226 \times 10^{-7}\, \mu m^{-1} \pi \alpha \ln \frac{1000}{100} = 9.60 \times 10^{-3} \alpha\, cm^{-1}$$

$$= 8.74 \times 10^{-3} cm^{-1},$$

$$\sigma_{s\lambda} = 9.60 \times 10^{-3} \rho\, cm^{-1} = 0.86 \times 10^{-3} cm^{-1}. \tag{10.71}$$

The scattering phase function must be evaluated from equation (10.70) and is also included in Fig. 10-14. Because of the small value for k, the directional behavior is very similar to that of the perfect dielectric ($m = 2$), but the forward scattering peak is more pronounced because of the smaller refractive index.

Large Diffusely Reflecting Spheres

In equations (10.62) through (10.67) the directional characteristics of the sphere reflectivity did not enter the development. Thus, for a diffusely reflecting sphere the amount of incident radiation on a surface element, as well as the expression for the heat flux reflected into all directions, is the same as for a specularly reflecting sphere. Therefore, equations (10.62) through (10.67) also hold for the diffusely reflecting sphere, or

$$Q_{abs} = \alpha, \tag{10.72}$$

$$Q_{sca} = \rho. \tag{10.73}$$

However, while for a specularly reflecting sphere the energy scattered into any given direction resulted from reflection from a single location on the sphere's surface, this is not true for a diffusely reflecting sphere. This complicates the development for the scattering phase function a bit. Consider Fig. 10-15: Incident radiation traveling into the direction of the unit vector $\hat{s}_i$ illuminates one half of the diffusely reflecting sphere. An observer, located far away from the sphere in the direction of $\hat{s}_o$, sees a different half of the sphere, part of which is illuminated by the incident radiation (shown by shadowing), part of which is in the shade. This illuminated region seen by the observer has the shape of a circular wedge similar to a slice of lemon. To describe the surface in polar coordinates it is most convenient to define the plane formed by

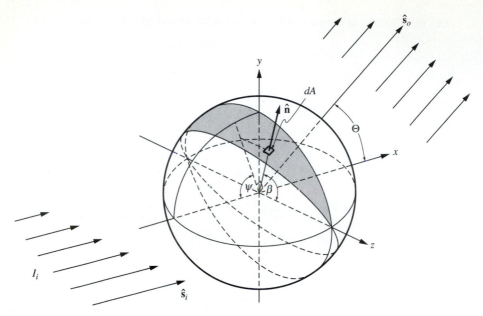

FIGURE 10-15
Scattering of incident radiation by a large diffusely reflecting sphere.

the two unit vectors $\hat{\mathbf{s}}_i$ and $\hat{\mathbf{s}}_o$ to be the x-y-plane with polar angle β measured from the z-axis and the azimuthal angle ψ measured from the negative x-axis as indicated in Fig. 10-15. With this coordinate system the normal to a surface element in the illuminated region may be expressed as

$$\hat{\mathbf{n}}(\beta, \psi) = -\sin\beta\,\cos\psi\,\hat{\mathbf{i}} + \sin\beta\,\sin\psi\,\hat{\mathbf{j}} + \cos\beta\,\hat{\mathbf{k}}, \qquad (10.74)$$

and also

$$\hat{\mathbf{s}}_i = \hat{\mathbf{i}}, \qquad \hat{\mathbf{s}}_o = \cos\Theta\,\hat{\mathbf{i}} + \sin\Theta\,\hat{\mathbf{j}}. \qquad (10.75)$$

The energy reflected from an infinitesimal surface area is, as developed by equation (10.62),

$$d^2Q_s = \rho I_i\,d\Omega_i\,d\lambda\,[dA(-\hat{\mathbf{n}}\cdot\hat{\mathbf{s}}_i)], \qquad (10.76)$$

where dA is two-dimensionally infinitesimal as indicated in Fig. 10-15 (i.e., not a ring as in the previous section, Fig. 10-13). Thus, the *radiosity* at that location, because of diffuse reflection of incident radiation, is

$$dJ = \rho I_i\,d\Omega_i\,d\lambda\,(-\hat{\mathbf{n}}\cdot\hat{\mathbf{s}}_i). \qquad (10.77)$$

Some of the reflected radiation will travel toward the observer into the direction of $\hat{\mathbf{s}}_o$. If we assume the observer stands on a large sphere with radius $R \gg a$, then the heat flux through a surface element dA_R on the large sphere due to reflection from the small sphere is

$$dI_s \, d\Omega = \int_{A_{shaded}} d J \, dF_{dA-dA_R} \, dA, \tag{10.78}$$

where

$$dF_{dA-dA_R} = \frac{\hat{\mathbf{n}} \cdot \hat{\mathbf{s}}_o \, dA_R}{\pi R^2} = \frac{1}{\pi} \hat{\mathbf{n}} \cdot \hat{\mathbf{s}}_o \, d\Omega \tag{10.79}$$

is the view factor between dA and dA_R, $\hat{\mathbf{n}} \cdot \hat{\mathbf{s}}_o$ is the cosine of the angle between surface normal at dA and the line to dA_R, while the surface normal at dA_R points directly to the particle. Thus,

$$dI_s = \frac{1}{\pi} \int_{A_{shaded}} d J \, \hat{\mathbf{n}} \cdot \hat{\mathbf{s}}_o \, dA, \tag{10.80}$$

and, again recalling that the scattering phase function is equal to $4\pi \times$ scattered intensity/total scattered heat flux, we get

$$\Phi(\hat{\mathbf{s}}_i, \hat{\mathbf{s}}_o) = 4\pi \, dI_s/dQ_s = 4 \int_{A_{shaded}} (\rho I_i \, d\Omega_i \, d\lambda)(-\hat{\mathbf{n}} \cdot \hat{\mathbf{s}}_i)(\hat{\mathbf{n}} \cdot \hat{\mathbf{s}}_o) \, dA / \rho I_i d\Omega_i \, d\lambda \, \pi a^2$$

$$= \frac{4}{\pi a^2} \int_{A_{shaded}} (-\hat{\mathbf{n}} \cdot \hat{\mathbf{s}}_i)(\hat{\mathbf{n}} \cdot \hat{\mathbf{s}}_o) \, dA$$

$$= \frac{4}{\pi a^2} \int_{\frac{\pi}{2}-\Theta}^{\frac{\pi}{2}} \int_0^{\pi} \sin \beta \cos \psi \, \sin \beta (\sin \psi \sin \Theta - \cos \psi \cos \Theta) \, a^2 \sin \beta \, d\beta \, d\psi,$$

which may readily be integrated to yield

$$\Phi(\Theta) = \frac{8}{3\pi} (\sin \Theta - \Theta \cos \Theta). \tag{10.81}$$

The phase function for diffuse spheres, equation (10.81), is also depicted in Fig. 10-14. Unlike for specularly reflecting spheres, the phase function for diffusely reflecting spheres displays a strong backward-scattering peak, and it is independent of the reflectivity (or the complex index of refraction) of the materials.

10.7 EXPERIMENTAL DETERMINATION OF RADIATIVE PROPERTIES OF PARTICLES

Experimental measurements of radiative properties of particles and clouds of particles are useful to verify the Mie theory, to ascertain the applicability of the Mie theory (for nonspherical particles, for nonisotropic particles, for closely-spaced particles, etc.), or simply to determine the radiative properties of particles for which no

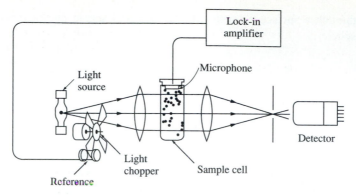

FIGURE 10-16
Schematic for measurement of extinction coefficient and absorption coefficient [17].

theory exists. Properties that can be measured are *extinction coefficient, absorption coefficient*, and *scattered intensity*. The easiest property to measure is the *extinction coefficient*. In principle, a standard spectrometer can be used for this measurement. The results, however, may be unreliable unless the detector is modified to eliminate forward-scattered light, which may account for the majority of total extinction [17] (in particular for large particle sizes, cf. Figs. 10-7 and 10-8). A schematic of such an apparatus is shown in Fig. 10-16. Light from a point source is collimated by a lens, transmitted through the sample cell (with its suspension of particles), and then focused onto a detector by a second lens. In order to reject forward scattered light, the detector is covered by a guard plate with a small pinhole located at the focal point of the second lens. The diameter of the pinhole must be carefully optimized: If the hole is too small then the signal from the transmitted light may become too weak, while a hole too large will admit an unacceptable amount of forward scattered energy to the detector. Normally the light beam is chopped by a rotating blade since most detectors only respond to *changes* in irradiation.

To distinguish between absorption and scattering, either the absorption coefficient or total scattering must be measured independently. To measure scattering over all (forward and backward) directions is very difficult, requiring a spectrometer capable of collecting radiation going into all directions (usually accomplished with an integrating sphere technique described in Chapter 3; cf., for example, Bryant *et al.* [25]. Absorption can also be detected fairly easily with a method usually referred to as *photoacoustic* [17]. Particles irradiated by a chopped beam are heated periodically, causing periodic changes in the particle temperature, which in turn cause slight pressure oscillations that may be detected by a sensitive microphone. These signals are then amplified by a lock-in amplifier synchronized with the light chopper. Since only absorbed light causes a temperature change in the particles, the acoustic signal must be proportional to the absorption coefficient of the suspension. Details may be found in the papers by Roessler and Faxvog [26] and Faxvog and Roessler [27], who measured the absorption coefficients of acetylene smoke and diesel emissions using this method. An ingenious way to separate transmitted and scattered radiation in the visible has been developed by Härd and Nilsson [28], who utilized the Doppler effect that occurs when an electromagnetic wave is scattered by a moving particle.

Angular scattering measurements are carried out with a *scattering photometer* (sometimes called a *nephelometer*). We distinguish between measurements with *single scattering* (i.e., the cell contains a dilute particle mixture which is optically thin, $\sigma_s L \ll 1$, so that every light beam is scattered at most once before exiting the particle layer) and *multiple scattering*, between *monodisperse suspensions* (i.e., all particles are exactly the same size) and *polydisperse suspensions* (i.e., the particle sizes obey a certain distribution function), between *near-forward scattering* (to measure the strong forward-scattering peak, but separating it from transmission) and scattering into all directions. Angular light scattering measurements are sometimes classified as either *absolute* or *relative*. In an absolute measurement the ratio between intercepted and scattered radiation, $\delta I_s(\Theta)/I_i$, is measured directly, while in a relative measurement the scattered intensity is related to intensity scattered into a reference direction, $\delta I_s(\Theta)/\delta I_s(\Theta_{\text{ref}})$. Thus neither measurement is truly "absolute;" in both cases a relative (i.e., nondimensional) intensity is recorded [17]. Since relative measurements are considerably easier to make, this method is employed by most experimentalists.

Single scattering experiments have been carried out primarily to verify the Mie theory, or to assess the accuracy of a device to be used for other scattering measurements. Hottel and coworkers [29] described such an experiment, in which they measured the nondimensional polarized intensities given by equation (10.10) for monodisperse polystyrene latex spheres (it appears that polystyrene spheres are favored by most experimenters, since it is relatively easy to manufacture spheres of constant diameter and of known index of refraction in the visible, $m = 1.60$, i.e., the spheres scatter but do not absorb). Their equipment consisted of a mercury arc and optics to produce an unpolarized near-parallel beam, a polarizer, a test cell manufactured from parallel microscope slides, and optics to confine the received beam to a small divergence angle. Their results for single scattering of small ($2a = 0.106 \, \mu$m) spheres are shown in Fig. 10-17. It is seen that the agreement between experiment and Mie theory is excellent. Hottel and coworkers attribute the small discrepancies primarily to the unavoidable spread in particle sizes. A more modern device to measure the scattering phase function for (almost) single scattering, as well as extinction and scattering coefficients, has been reported by Menart and colleagues [30]. Their apparatus employed a globar light source and an open gas-particle column, both mounted on a rotatable table, together with collection optics and a highly sensitive dual element (InSb–HgCdTe) detector. Measurements taken for soda-lime glass beads and aluminum oxide particles in the wavelength range between 2.5 and 11 μm showed good agreement with Mie theory.

Multiple scattering experiments were reported by Woodward [31, 32], also on polystyrene spheres, using dispersions with narrow size distribution. Woodward found

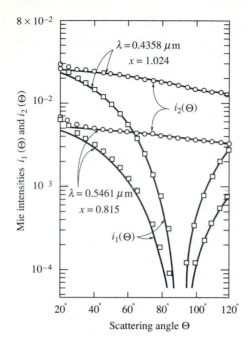

FIGURE 10-17
Mie scattering intensities i_1 and i_2 for 0.106μm diameter spheres; data points experimental, solid lines theoretical [29].

good agreement between his data and the multiple-scattering theory of Hartel [33], a somewhat dated approximate solution of the equation of transfer for a purely scattering medium. However, as Smart and coworkers [34] pointed out, Woodward did not correct for the reflection of the emergent beam at the water-glass-air interfaces of the test cell, nor did he account for the change in scattering path length for larger angles. To compensate for these errors, Smart and coworkers [34] devised an apparatus whose schematic is shown in Fig. 10-18. Employing a standard Brice-Phoenix spectrometer, they placed the simple parallel-microscope slide test section inside a special cell filled with Nujol (a liquid paraffin). Nujol has the same refractive index as the glass bounding the test section as well as the outer jacket of the cell. Thus, the Nujol serves two purposes: Reflections at the interfaces are almost eliminated and—from Snell's law—all scattering angles up to 90° are contained in experimental angles below 65°, making otherwise impossible-to-measure scattering angles measurable.

Some representative results of their multiple-scattering measurements for varying optical thicknesses of the particle suspension are shown in Figs. 10-19 and 10-20. Agreement between experiment and theory is excellent except for very small and very large optical thicknesses. For small thicknesses the experiment could not re-generate the maxima and minima, probably as a result of uncertainty in the parti-cle size distribution. Disagreement for large thicknesses stems from the fact that Smart and coworkers also used Hartel's approximate theory. Orchard [35] pointed out that using Hartel's approximation leads to a transmissivity of 0.5 for a medium of

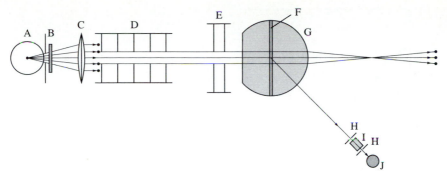

FIGURE 10-18
Schematic for angular scattering experiment [34]: A, Hg arc; B, monochromatic filter; C, lens; D,E,H, light stops; F, test section; G, jacket with Nujol; I, analyzer; J, photomultiplier.

infinite optical thickness (rather than the correct value of zero). Therefore, Hottel and coworkers [29] used the method of discrete ordinates, which may be made arbitrarily accurate for sufficient numbers of "ordinates,"[2] to calculate bidirectional reflectivity and transmissivity for a particle layer. Some representative data in Fig. 10-21, show the excellent agreement between theory and experiment for optical thicknesses up to 775. Very similar experiments, also using the method of discrete ordinates for theoretical calculations, were carried out by Brewster and Tien [4] and Yamada, Cartigny and Tien [5] for large polydivinyl spheres in air, resulting in equally good agreement between experiment and theory.

A different approach to avoiding reflection and refraction losses, and to measuring scattering intensities at oblique angles, was taken by Daniel and coworkers [36], who measured the phase function for aqueous suspensions of unicellular algae. They used a rotatable fiberoptic detector immersed *inside* the large dish filled with a dilute algae suspension.

The measurement of scattered intensity into the near-forward direction poses a unique set of problems, because the signal may vary by several orders of magnitude over a few degrees of scattering angle, and because separating the transmitted radiation from forward scattering is difficult. Although sometimes employed for particle sizing and the determination of the index of refraction, near-forward scattering is of importance primarily in applications with large geometric paths, such as atmospheric scattering, scattering effects on visibility in the seas, astrophysical applications, and so on. In heat transfer applications forward scattering is generally of small importance, since treating it as transmitted radiation usually results in negligible errors. The reader interested in such experiments is referred to the papers by Spinrad and coworkers [37].

---

[2]This method for the solution of the radiative transport equation is described in detail in Chapter 15.

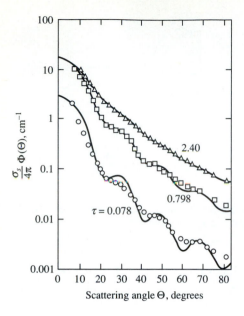

FIGURE 10-19
Relative scattered intensity vs. angle of observation for relatively low concentration; data points experimental, solid lines theoretical [34].

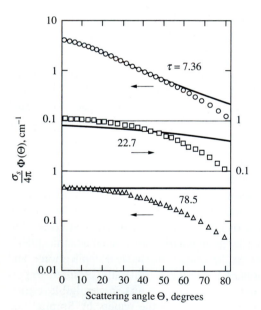

FIGURE 10-20
Relative scattered intensity vs. angle of observation for relatively high concentration; data points experimental, solid lines theoretical [34].

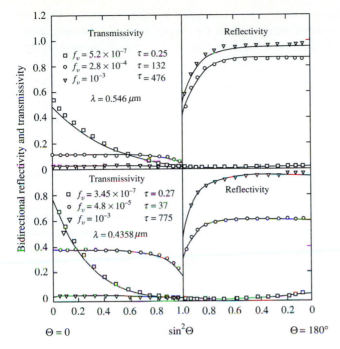

FIGURE 10-21
Bidirectional reflectivity and transmissivity for an aqueous solution of $0.530\mu m$ polystyrene spheres, for varying concentrations; data points experimental, solid lines theoretical [29].

10.8 APPROXIMATE SCATTERING PHASE FUNCTIONS

It is clear from Figs. 10-3 and 10-9 that radiative properties of particles may display strong oscillatory behavior with size parameter and, therefore, wavelength, particularly for the case of large, monodisperse, dielectric particles. Even more bothersome is the fact that the scattering phase function may undergo strong angular oscillations at any given single wavelength, again particularly for the case of large, monodisperse, dielectric particles (cf. Figs. 10-6, 10-7, 10-8). Since radiative calculations for media with spectrally varying properties are generally carried out on a spectral basis with subsequent integration over all relevant wavelengths, this fact means that these spectral oscillations are somewhat inconvenient, but they do not make the analysis intractable. Strong angular oscillations in the scattering phase function, on the other hand, will enormously complicate the analysis for any given wavelength. Indeed, most solution methods described in the following chapters cannot accept highly oscillatory phase functions, or else they must be carried to unacceptably high orders or node numbers. It is, therefore, common practice to approximate oscillatory phase functions by simpler expressions with more regular behavior.

It is observed that large particles generally have strong forward-scattering peaks (due to diffraction, cf. Figs. 10-7 and 10-8). Indeed, if $x \to \infty$, half of the total extinction is due to diffraction into near-forward directions, as described in Section 10.6. Since diffraction was neglected (i.e., treated as transmission) in that section, the phase functions for large particles are in fact simplified. If either geometric optics cannot be used or diffraction effects must be retained for other reasons, then the approximate phase function must accommodate the strong forward-scattering peak. To this purpose many investigators have used the *Henyey-Greenstein phase function*,

$$\Phi_{HG}(\Theta) = \frac{1 - g^2}{[1 + g^2 - 2g \cos \Theta]^{3/2}}, \tag{10.82}$$

where g is the asymmetry factor. Sometimes the Henyey-Greenstein function is written in the form of a Legendre polynomial series, or

$$\Phi_{HG}(\Theta) = 1 + \sum_{n=1}^{\infty} (2n + 1)g^n P_n(\cos \Theta). \tag{10.83}$$

Thus, this expression is equivalent to equation (10.24) with approximate values for the A_n being related to the asymmetry factor. A representative comparison between Mie and Henyey-Greenstein phase functions is given in Fig. 10-22 for a dielectric with index of refraction $m = n = 1.33$ and size parameter $x = 300$ (water droplets). Both van de Hulst [38] and Hansen [39] have shown that the Henyey-Greenstein formulation gives very accurate results for radiative heat fluxes as long as the particles are nondielectric: Dielectric particles may have a relatively strong backward scattering peak besides a strong forward scattering peak. This situation cannot be described by the aysmmetry factor alone, and the Henyey-Greenstein formulation must fail. That neglect of backward-scattering peaks can cause considerable error in heat flux calculations has been shown by Modest and Azad [40].

For many calculations the Henyey-Greenstein phase function is still too complicated. As mentioned earlier, in heat transfer applications forward scattering may usually be treated as transmission. This fact has led a number of researchers to the use of so-called *Dirac-delta* or *Delta-Eddington approximations*, where the forward scattering peak is separated from the rest of the scattering phase function by

$$\Phi(\Theta) \approx 2f \, \delta(1 - \cos \Theta) + (1 - f)\Phi^*(\Theta), \tag{10.84}$$

where Φ^* is the new approximate phase function, f is a forward scattering fraction to be determined, and δ is the *Dirac-delta function* defined by

$$\delta(x) = \lim_{\delta\epsilon \to 0} \begin{cases} 0, & |x| > \delta\epsilon, \\ \dfrac{1}{\delta\epsilon}, & |x| < \delta\epsilon, \end{cases} \tag{10.85}$$

or

$$\int_{-\infty}^{\infty} \delta(x) \, dx = 1. \tag{10.86}$$

Substitution of equation (10.84) into equation (8.19) shows that the approximate phase function is properly normalized, that is,

$$\frac{1}{4\pi} \int_{4\pi} \Phi^*(\Theta) \, d\Omega = 1. \tag{10.87}$$

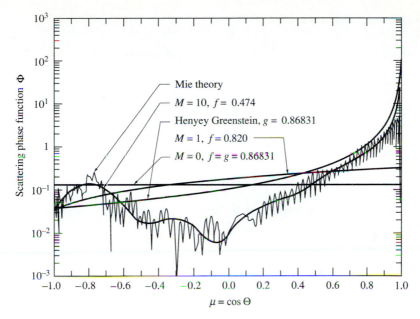

FIGURE 10-22
Comparison of Mie, Henyey-Greenstein, linear-anisotropic and isotropic phase functions for water droplets
($m = 1.33$, $x = 100$).

Different authors have used different approaches to define f and Φ^*. Potter [41] was one of the first to use the following scheme for his work on atmospheric scattering. He truncated the peak by extrapolating the phase function from directions outside the peak into the forward direction; otherwise he left the phase function unchanged. Not surprisingly, his method produced excellent results, but it still leaves the approximate phase function in a rather complex form.

It appears more promising to express the approximate phase function as a truncated Legendre series,

$$\Phi^*(\Theta) = 1 + \sum_{n=1}^{M} A_n^* P_n(\cos \Theta), \tag{10.88}$$

where the constant M is the chosen order of approximation, mostly taken as $M = 1$ (linear-anisotropic scattering) [40, 42–44], $M = 0$ (isotropic scattering) [43], while higher order approximations have been carried out by Crosbie and Davidson [43]. There is considerable disagreement among authors about the criteria to be used to determine the forward fraction f as well as the coefficients A_n^*. Both Joseph and coworkers [42] and Crosbie and Davidson [43] agreed that at least one of the *moments* of equation (10.84) should be satisfied: Multiplying equation (10.84) by $P_m(\Theta)$ and integrating over all Θ results in

$$\int_0^\pi \Phi(\Theta) P_m(\cos \Theta)\, d\Theta = \int_0^\pi 2f\delta(1-\cos \Theta) P_m(\cos \Theta)\, d\Theta$$

$$+(1-f)\sum_{n=1}^M \int_0^\pi A_n^* P_n(\cos \Theta) P_m(\cos \Theta)\, d\Theta, \qquad (10.89)$$

or, using the fact that Legendre polynomials are orthogonal functions over the interval $(0, \pi)$ [18],

$$(1-f)A_m^* = A_m - (2m+1)f, \qquad m = 1, 2, \ldots \qquad (10.90)$$

If the approximate phase function is to be *isotropic*, equation (10.90) yields, with $A_1^* = 0$,

$$f = \frac{A_1}{3} = g, \qquad (10.91)$$

and

$$\Phi(\Theta) \approx 2g\,\delta(1-\cos \Theta) + (1-g). \qquad (10.92)$$

Joseph and colleagues [42] developed an approximate *linear-anisotropic* phase function. They employed equation (10.90) for the first two moments to find f and A_1^*, using an approximate value of $A_2 \approx 5g^2$ (from the Henyey-Greenstein phase function). However, their approximate phase function may turn out to be negative for some backscattering directions, which is physically impossible. Crosbie and Davidson [43] overcame this difficulty by applying the second moment only conditionally. From the first moment it follows that

$$A_1^* = 3\frac{g-f}{1-f}. \qquad (10.93)$$

Requiring the phase function to be positive for all angles is equivalent to $|A_1^*| \leq 1$, or

$$\frac{1}{2}(3g-1) \leq f \leq g. \qquad (10.94)$$

Instead of using the second moment directly, i.e., $f = A_2/5$, they require $|f - A_2/5|$ to be a minimum without violating equation (10.94). Obviously, this method is readily extended to arbitrarily high orders. Their linear-anisotropic and order-10 phase function approximations are also included in Fig. 10-22 for water droplets. It should be noted that this method will work only for *positive* asymmetry factors. In the case of $g < 0$ the method breaks down and $f = 0$ should be used. Even then one may find $A_1^* < -1$, in which case one has to force $A_1^* = -1$ to avoid negative forward scattering. Obviously the method will break down completely for strong backward-scattering peaks.

None of the above approximations allows for simultaneous forward and backward scattering peaks. Modest and Azad [40] have shown that neglecting the backward scattering peaks that may appear in dielectrics may cause considerable error in heat flux calculations. Thus, they proposed a *double Dirac-delta phase function* approximation. In this model, both the forward and backward scattered peaks are removed by letting

$$\Phi(\Theta) \approx 2f\,\delta(1 - \cos\Theta) + 2b\,\delta(1 + \cos\Theta) + (1 - f - b)\,\Phi^*(\Theta), \qquad (10.95)$$

where b is the backward scattering peak to be removed. Noting that the approximate phase function of Joseph and coworkers [42] often becomes negative over backward angles, Modest and Azad did not choose to evaluate f, b and A_1^* through moments. Rather, they selected visually (*i*) the slope A_1^* of the approximate linear-anisotropic phase function Φ^*, (*ii*) a cutoff angle for forward scattering, Θ_f, and (*iii*) a cutoff angle for backward scattering, Θ_b. [In practice, rather than by selecting a forward angle Θ_f, the forward fraction f is determined by the choice of $(1 - f - b)\,\Phi^*(\Theta = \pi/2) = 1 - f - b$.] The peak fractions and/or forward-scattering cutoff angle are determined from

$$f = \frac{1}{2}\int_0^{\Theta_f} \left[\Phi(\Theta) - (1 - f - b)\,\Phi^*(\Theta)\right]\sin\Theta\,d\Theta,$$

$$b = \frac{1}{2}\int_{\Theta_b}^{\pi} \left[\Phi(\Theta) - (1 - f - b)\,\Phi^*(\Theta)\right]\sin\Theta\,d\Theta. \qquad (10.96)$$

This subjective method also makes it possible to keep the phase function approximation as simple as possible if backward (or forward) scattering peaks are not very pronounced, by setting b (or f) equal to zero. Examples for their choice of phase function approximation are included in Figs. 10-7 and 10-8. If a mathematical rather than visual determination of the factors f and A_1^* is desired, a least-mean-square-error fit may be used over the midrange of the phase function, or

$$\int_{\Theta_f}^{\Theta_b} \left[\Phi(\Theta) - (1 - f - b)\,\Phi^*(\Theta)\right]^2 \sin\Theta\,d\Theta = \text{minimum}, \qquad (10.97)$$

or

$$\int_{\Theta_f}^{\Theta_b} \left[\Phi(\Theta) - (1 - f - b)\,\Phi^*(\Theta)\right]\cos^n\Theta\,\sin\Theta\,d\Theta = 0, \quad n = 0, 1. \qquad (10.98)$$

While this method still necessitates a subjective choice for Θ_f and Θ_b, the results for f and A_1^* are rather insensitive to this choice as long as the range defined by Θ_f and Θ_b is outside of any strong forward and/or backward scattering peaks.

Substitution of equation (10.95) into the *equation of transfer*, equation (8.23), yields, for the terms describing attenuation and augmentation due to scattering,

$$-\sigma_s I(\hat{s}) + \frac{\sigma_s}{4\pi}\int_{4\pi} I(\hat{s}')\,\Phi(\hat{s}\cdot\hat{s}')\,d\Omega'$$

$$\approx -(1 - f)\sigma_s I(\hat{s}) + b\sigma_s I(-\hat{s}) + (1 - f - b)\frac{\sigma_s}{4\pi}\int_{4\pi} I(\hat{s}')\,\Phi^*(\hat{s}\cdot\hat{s}')\,d\Omega'. \qquad (10.99)$$

If there is no appreciable backscattering ($b \approx 0$), then the fraction f of the scattered energy is treated as transmitted, and the original equation of transfer may be used with an effective scattering coefficent of $\sigma_s^* = (1 - f)\sigma_s$ and a linear-anisotropic phase function. If there is an appreciable backward scattering peak ($b \neq 0$), the equation of transfer will become a little more complicated and may indeed become too complex for some of the solution methods discussed in the following chapters.

Example 10.3.   Calculate approximate phase functions for monodisperse suspensions of large specular dielectric spheres ($m = 2$) and diffusely reflecting spheres, using the Henyey-Greenstein function, the Crosbie and Davidson model, and the Modest and Azad model.

Solution

The Henyey-Greenstein function requires the calculation of the asymmetry factor

$$g = \frac{A_1}{3} = \frac{1}{2} \int_{-1}^{+1} \Phi(\mu)\mu \, d\mu,$$

where μ is the cosine of the scattering angle. The Crosbie and Davidson approximation requires the calculation of g as well as the calculation of

$$\frac{A_2}{5} = \frac{1}{2} \int_{-1}^{+1} \Phi(\mu) P_2(\mu) \, d\mu.$$

For the Modest and Azad approximation the parameters are normally found by eye-fitting plots of the phase functions, in this case from Fig. 10-14.

Numerical integration of the phase function yields for the specular dielectric spheres $g = 0.229$ and $A_2/5 = 0.138$. Since $A_2/5 < g$ it follows for the Crosbie and Davidson model that $f = A_2/5 = 0.138$ and, from equation (10.93), $A_1^* = 0.315$. For the Modest and Azad model we fit a straight line in Fig. 10-14 that best approximates the dielectric phase function for, say, $\mu < 0.6$. The value at $\mu = 0$ corresponds to $1 - f$, while the slope of the line corresponds to A_1^*; this approach yields $f = 0.2$ and $A_1^* = 0.15$. All three approximate phase functions are shown in Fig. 10-23 together with the exact expression. The Henyey-Greenstein function does not try to remove the forward scattering peak, but is unable to follow the sharp peak for large μ. The Crosbie-Davidson and Modest-Azad models follow the actual function well, except for the forward peak that has been removed. They would actually more or less coincide if a $\mu_f \approx 0.7$ had been chosen for the Modest-Azad model.

The integration for the diffuse-sphere phase function could be carried out analytically but is rather tedious. Numerical integration of the phase function yields for the diffuse spheres $g = -0.444$ and $A_2/5 = 0.062$. Since $g < 0$, the scattering is predominantly backward and the Crosbie and Davidson model cannot be applied. Thus, for this model, we force $f = 0$ and, from equation (10.93), $A_1^* = -1.333$; since this would result in negative values for forward directions we also force $A_1^* = -1$. For the Modest and Azad model we again fit a straight line in Fig. 10-14 that best approximates the diffuse phase function for, say, $\mu > -0.6$. Two different fits are shown in Fig. 10-23: (*i*) To avoid the extra complication of a back-scattered fraction b we set $b = 0$ and $A_1^* = -1$ (to force zero forward scattering), which is the same as the "forced" Crosbie-Davidson function;

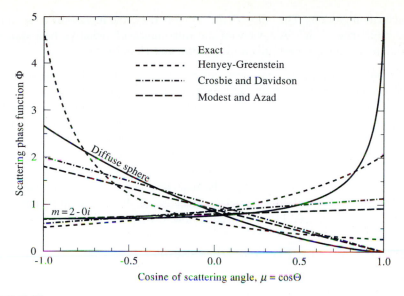

FIGURE 10-23

Scattering phase function approximations for Example 10.3.

(*ii*) we set $b = 0.1$ and again $A_1^* = -1$. It is seen that the Henyey-Greenstein function does not work very well for back scattering, while the other two models give acceptable results. Note that the Modest-Azad model is easier to use, but leaves one with the "pain" of having to make a subjective choice (unless one wants to employ least-mean-square-error fits). Note also that this phase function has no distinct backward scattering peak so that the Crosbie-Davidson model could be forced to give reasonable results. In the case of strong backward scattering peak (with or without simultaneous forward peak) only the Modest-Azad model will give acceptable results.

10.9 RADIATION PROPERTIES OF COMBUSTION PARTICLES

Undoubtedly, some of the most important engineering applications of thermal radiation are in the areas of the combustion of gaseous, liquid (usually in droplet form) or solid (often pulverized) fuels, be it for power production or for propulsion. During combustion thermal radiation will carry energy directly from the combustion products to the burner walls, often at rates higher than for convection. In the case of liquid and solid fuels thermal radiation also plays an important role in the preheating of the fuel and its ignition. Nearly all flames are visible to the human eye and are, therefore, called *luminous* (sending out light). Apparently, there is some radiative emission from within the flame at wavelengths where there are no vibration-rotation bands for any combustion gases. This luminous emission is today known to come from tiny *char* (almost pure carbon) particles, called *soot*, which are generated during the combustion process. The "dirtier" the flame is (i.e., the higher the soot content), the more luminous it is. A review of the importance of radiative heat transfer in combustion systems has been given by Sarofim and Hottel [45].

All combustion processes are very complicated. Uusually there are many intermediate chemical reactions in sequence and/or parallel, intermittent generation of a variety of intermediate species, generation of soot, agglomeration of soot particles, and subsequent partial burning of the soot. Since thermal radiation contributes strongly to the heat transfer mechanism of the combustion, any understanding and modeling of the process must include knowledge of the radiation properties of the combustion gases as well as any particulates that are present. The most important particles during the combustion of pulverized coal are the relatively large coal and fly ash particles as well as the very small soot particles. Because of their great importance, these suspensions will be treated in some detail below.

Pulverized Coal and Fly Ash Dispersions

To calculate the radiative properties of arbitrary size distributions of coal and ash particles, one must have knowledge of their complex index of refraction as a function of wavelength and temperature. Data for carbon and different types of coal indicate that its real part, n, varies little over the infrared and is relatively insensitive to the type of coal (e.g., anthracite, lignite, bituminous), while the absorptive index, k, may vary strongly over the spectrum and from coal to coal [46–48]. The composition of fly ash and, therefore, its optical properties may vary greatly from coal to coal. The few data in the literature [49–51], report consistent values for the refractive index ($n \approx 1.5$) and widely varying values for the absorptive index. Recently, Wall and coworkers [51] calculated the absorptive index for a number of Australian coals (based on their ash composition), and found that k varied between 0.008 and 0.020. Nothing at all appears to be known about the temperature dependence of these optical properties. A summary of representative values for the optical constants of coals and ashes has been reported by Viskanta and colleagues [52] and is reproduced in Table 10.2.

TABLE 10.2
Representative values for the complex index of refraction in the near infrared for different coals and ashes [52]

particle type	$m = n - ik$
carbon	$2.20 - 1.12i$
anthracite	$2.05 - 0.54i$
bituminous	$1.85 - 0.22i$
lignite	$1.70 - .066i$
fly ash	$1.50 - .020i$

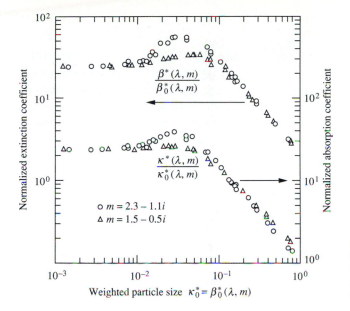

FIGURE 10-24

Index of refraction effects on the extinction and absorption properties of pulverized coal [54].

A first attempt to establish formulae for extinction by carbon particles was made by Tien and coworkers [53], who looked at a single index of refraction ($m = 1.5 - 0.5i$) for a gamma size distribution of particles [cf. equation (10.34)]. They found a relatively simple (but not very accurate) smooth correlation for the extinction coefficient β. Buckius and Hwang [54] carried out a large number of Mie calculations for a variety of complex indices of refraction (simulating different coals) and a variety of different particle distribution functions [gamma distributions and "rectangular" distributions, i.e., $n(a) = $ const over a certain range of radii]. They found that, when normalized with the Rayleigh small-particle limit, absorption coefficient and extinction coefficient as well as asymmetry factor are virtually independent of the particle size distribution function, and only depend on a mean particle diameter. Employing the range for m given by Foster and Howarth [46] for different coals, they found a similar insensitivity to the index of refraction, at least in the limits of small and large particles; in the intermediate size range, deviations of up to nearly $\pm 50\%$ were reported as shown in a sample of their calculations, Fig. 10-24. The spectral results were also wavelength-integrated to yield Planck-mean and Rosseland-mean absorption and extinction coefficients. Considering a temperature range of 750 K to 2500 K they found that their data could be correlated to within 30% for the different coals. Based on their numerical data for different types of coals they developed correlations for a number of nondimensional radiation properties. Spectral properties correlated were absorption and extinction coefficients and the asymmetry factor, with nondimensional κ and β defined by

$$\kappa^*(\lambda, m) = \kappa(\lambda, m, N_T)/f_A, \quad \beta^*(\lambda, m) = \beta(\lambda, m, N_T)/f_A, \quad (10.100)$$

where

$$f_A = \int_0^\infty \pi a^2 \, n(a) \, da \tag{10.101}$$

is the total projected area of the particles per unit volume. Thus, these nondimensional values are essentially size-averaged absorption and extinction efficiencies [cf. equations (10.39) and (10.40)]. For extremely small particles $\kappa^* \approx \beta^*$ may be calculated from Rayleigh scattering theory, equation (10.52), as

$$\kappa_0^*(\lambda, m) = \beta_0^*(\lambda, m) = -\Im\left\{ \frac{m^2 - 1}{m^2 + 2} \right\} \frac{6\pi f_v}{\lambda f_A}$$

$$= -4\bar{x} \, \Im\left\{ \frac{m^2 - 1}{m^2 + 2} \right\}, \tag{10.102}$$

where $\bar{x}$ is a mean size parameter based on a mean particle radius defined by

$$\bar{r} = \frac{3 f_v}{4 f_A} = \frac{\int_0^\infty a^3 \, n(a) \, da}{\int_0^\infty a^2 \, n(a) \, da}. \tag{10.103}$$

Since β_0^* is linear in $\bar{x}$ it may also be regarded as a weighted (by a function of m) size parameter. The asymmetry factor for Rayleigh scattering is zero (because of its symmetric phase function) and g_0 for the small particle limit must be found from a higher order expansion given by [54], which may be simplified to

$$g_0(\lambda, m) = \frac{1}{15} \Re\left\{ \frac{(m^2 + 2)(m^2 + 3)}{2m^2 + 3} \right\} \left(\frac{2\pi}{\lambda} \right)^2 \frac{\int_0^\infty a^8 \, n(a) \, da}{\int_0^\infty a^6 \, n(a) \, da}. \tag{10.104}$$

In a similar fashion, they defined nondimensional Planck-mean and Rosseland-mean absorption and extinction coefficients, all normalized by f_A. All correlations obey the same basic formula,

$$\frac{1}{y^z} = \frac{1}{y_0^z} + \frac{1}{y_\infty^z}, \tag{10.105}$$

where y stands for one of the above nondimensional properties, y_0 is that property for small average particle sizes, and y_∞ the one for large average particle sizes. The correlation parameters y_0, y_∞ and z for the various properties are summarized in Table 10.3.

The results of Buckius and Hwang were essentially corroborated by Viskanta and coworkers [52]. They too found that variations with particle distribution functions are relatively minor, and that the different indices of refraction made a difference only for midsized particles. However, they felt that these differences were too large to use a single correlation and presented individual graphs for different coals. Table 10.3

TABLE 10.3

Correlation parameters for the prediction of nondimensional coal properties from
$y^{-z} = y_0^{-z} + y_\infty^{-z}$ **[54].**

y	y_0	y_∞	z
$\beta^*(\lambda, m)$	$\beta_0^*(1 + 6.78\beta_0^{*2})$	$3.09/\beta_0^{*0.1}$	1.2
$\kappa^*(\lambda, m)$	$\beta_0^*(1 + 2.30\beta_0^{*2})$	$1.66/\beta_0^{*0.16}$	1.6
$g(\lambda, m)$	g_0	0.9	1.0
β_P^*	$0.0032\phi[1 + (\phi/355)^{1.9}]$	$10.99/\phi^{0.02}$	1.2
β_R^*	$0.0032\phi[1 + (\phi/485)^{1.75}]$	$10.99/\phi^{0.02}$	1.2
κ_P^*	$0.0032\phi[1 + (\phi/725)^{1.65}]$	$13.75/\phi^{0.13}$	1.5
κ_R^*	$0.0032\phi[1 + (\phi/650)^{2.3}]$	$15.65/\phi^{0.143}$	1.15

$\phi = \bar{r}T/1\,\mu\text{m K}$, β and κ nondimensionalized by f_A from eq. (10.101);
β_0^* from eq. (10.102), g_0 from eq. (10.104), $\bar{r}$ from eq. (10.103).

indicates that—according to Buckius and Hwang [54]—Planck-mean and Rosseland-mean coefficients do not depend on the optical properties of the coal and are very close to one another. Again, this observation was corroborated by Viskanta and coworkers [52] for carbon, anthracite and bituminous coal, as well as for lignite at high temperature (above 1000 K). For fly ash and for lower temperature lignite mean absorption coefficients were considerably lower due to the significantly lower absorptive indices of these materials. Thus, Table 10.3 should be regarded as a relatively crude approximation, which should be replaced when more accurate data for different coals and ashes become available (optical properties varying with wavelength and temperature, particle size distributions).

Radiative Properties of Soot

Soot particles are produced in fuel-rich flames, or fuel-rich parts of flames, as a result of incomplete combustion of hydrocarbon fuels. As shown by electron microscopy, soot particles are generally small and spherical, ranging in size between approximately 50 Å and 800 Å (0.005 μm to 0.08 μm), and up to about 3000 Å in extreme cases [55, 56]. While mostly spherical in shape, soot particles may also appear in agglomerated chunks and even as long agglomerated filaments. It has been determined experimentally in typical diffusion flames of hydrocarbon fuels that the volume percentage of soot generally lies in the range between $10^{-4}\%$ to $10^{-6}\%$ [45, 57, 58].

Since soot particles are very small, they are generally at the same temperature as the flame and, therefore, strongly emit thermal radiation in a continuous spectrum over the infrared region. Experiments have shown that soot emission often is considerably stronger than the emission from the combustion gases. Still, even today the mechanisms leading to soot production and destruction as well as the chemical makeup of soot particles are not yet well understood. In order to predict the radiative properties of a soot cloud, it is necessary to determine the amount, shape and distribution of soot particles, as well as their optical properties, which depend on chemical composition and particle porosity.

Early work on soot radiation properties concentrated on predicting the absorption coefficient κ_λ for a given flame as a function of wavelength. For all but the largest soot particles the size parameter $x = 2\pi a/\lambda$ is very small for all but the shortest wavelengths in the infrared, so one may expect that Rayleigh's theory for small particles will, at least approximately, hold. This condition would, according to equation (10.52), lead to negligible scattering and an absorption coefficient of

$$\kappa_\lambda = \beta_\lambda = -\Im\left\{\frac{m^2 - 1}{m^2 + 2}\right\}\frac{6\pi f_v}{\lambda}, \tag{10.106}$$

or, expanding the complex index of refraction, $m = n - ik$,

$$\kappa_\lambda = \beta_\lambda = \frac{36\pi n k}{(n^2 - k^2 + 2)^2 + 4n^2 k^2}\frac{f_v}{\lambda}. \tag{10.107}$$

Experiments have confirmed that scattering may indeed be neglected [59]. Equation (10.107) would lead one to expect that the absorption coefficient should vary with wavelength as $1/\lambda$. However, this assumption is only approximately correct and it is customary to write

$$\kappa_\lambda = \frac{C f_v}{\lambda^a}, \tag{10.108}$$

where C and a are empirical constants. Many different values for the *dispersion exponent a* have been measured by investigators for many different flame conditions, ranging from as low as 0.7 to as high as 2.2. Earlier theories explained this deviation from Rayleigh theory to be a consequence of particle size. While it is true that Mie theory predicts a growing value for a for increasing particle size, it is easy to show that this alone cannot explain the large values for the dispersion exponent in some flames. Rather, this increase in a must be due to spectral variations of the effective complex index of refraction, resulting from the chemical composition and the porosity of the soot particles. Millikan [60, 61] investigated the dependence between dispersion exponent and chemical composition. While for many years soot was assumed to be amorphous carbon, he found the particles contained considerable amounts of hydrogen (up to 40 atom-%), and he determined that a was approximately directly proportional to the hydrogen-carbon ratio of the soot material as shown in Fig. 10-25. He further showed that the radiative properties of the soot were the same for in situ flame measurements as for soot collected from the flame, suggesting that the optical properties are fairly independent of temperature. Unfortunately, his experimental setup did not allow for the determination of the constant C in equation (10.108), so that quantitative evaluation of the extinction coefficient is not possible.

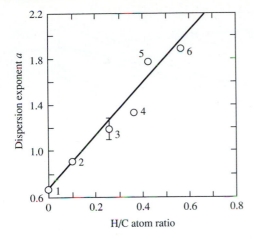

FIGURE 10-25

Dispersion exponent a of soot deposits vs. hydrogen-to-carbon ratio: 1, pure carbon (arc evaporated); 2, acetylene/oxygen flame; 3, ethylene/oxygen flame; 4,5,6, ethylene/air flames [60].

The optical properties of soot material, i.e., the complex index of refraction m, have received a very considerable amount of attention during the last twenty years, using different forms of carbon and various experimental methods. Foster and Howarth [46] were the first to report experimental measurements for the complex index of refraction of hydrocarbon soot, based on various carbon black powders. This work was followed shortly thereafter with measurements by Dalzell and Sarofim [62] on soot collected on cooled brass plates from laminar diffusion flames burning either acetylene or propane. In both cases pellets with very smooth, quasi-specular surfaces were formed by compressing small soot samples between optically flat surfaces with pressures up to 2760 bar. The index of refraction was then deduced from reflectivity measurements employing Fresnel's relations for specular reflectors. They found the optical properties of the two different soots to be fairly similar, with values for acetylene soot somewhat higher than for propane soot, apparently because of the higher H/C ratio in propane soot. Comparing their results with values reported by Stull and Plass [63] (based on amorphous carbon) and by Howarth, Foster and Thring [64] (based on pyrographite) they note that optical properties of amorphous or graphitic carbon are not equal to those of soot, primarily because of the different H/C ratios.

The data of Dalzell and Sarofim [62] have been employed in many subsequent studies (and continue to be used today). For example, Hubbard and Tien [65] used them to evaluate Planck-mean and Rosseland-mean absorption coefficients for soot clouds and soot-gas mixtures. However, the accuracy of Dalzell and Sarofim's data has been questioned by a number of researchers. All ex situ measurements suffer from the fact that during the analysis the soot is not in the same state as in the flame. The soot particles are at a different temperature, and they may have different morphologies because of agglomeration during the sampling process. The severest criticism concerns the pellet-reflection technique. Medalia and Richards [66], Graham [67] and Janzen [68] have pointed out that the pellets must contain a considerable amount of void (33% even after compression to 2760 bar, according to Medalia and Richards [66]),

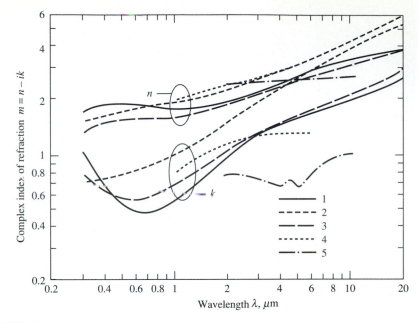

FIGURE 10-26
Complex index of refraction for soot based on different studies: 1, Lee and Tien [69] (polystyrene and Plexiglas soot); 2, Stull and Plass [63] (amorphous carbon); 3, Dalzell and Sarofim [62] (propane soot); 4, Howarth and coworkers [64] (pyrographite at 300 K); 5, Felske and coworkers [73] (propane soot).

since the sample is made by compressing a powder. This technique leads to two serious sources for errors: (*i*) Since the pellets are actually a two-phase dispersion of soot and air, the inferred index of refraction is the one of the dispersion and not the one of the soot particles themselves, and (*ii*) at least at short wavelengths the pellet cannot assumed to be optically smooth and Fresnel's relations become invalid.

These problems prompted Lee and Tien [69] to obtain soot optical properties from in situ flame transmission data together with application of the *dispersion theory* [17, 70] (i.e., the theory that predicts the wavenumber dependence of the optical constants n and k by relating them to bound- and free-electron densities). Their results for polystyrene and Plexiglas flame soot, based on data by Buckius and Tien [71] and Bard and Pagni [72], are shown in Fig. 10-26 together with the data of Stull and Plass [63], Howarth and coworkers [64], and the propane soot results of Dalzell and Sarofim [62]. Lee and Tien's data agree fairly well with those of Dalzell and Sarofim, except for the visible where the pellet-reflection technique is particularly suspect. In contrast to Dalzell and Sarofim as well as Millikan [60, 61], Tien and Lee noted that the optical properties varied little from flame to flame despite their different fuel (not necessarily soot) H/C ratios. Conceivably the *soot* of their different flames had similar H/C ratios. They also applied the dispersion theory to determine the temperature dependence of the optical properties, observing that $m = n - ik$ is very insensitive to temperature changes at high temperature levels. This would imply negligible effect of spatial temperature variation on soot properties, as is commonly assumed. It should be noted that, like the pellet-reflection technique, the spectral transmission technique has its own set of difficulties: For its data reduction, a scattering theory and a theory describing the spectral variation of the refractive index (the dispersion theory) must be used. Usually the *Mie scattering theory* based on monodisperse spherical soot particles is employed. Thus, only when the particles are spherical with a single diameter can these results be used with confidence.

In a later paper Felske and coworkers [73] returned to the pellet-reflection technique, arguing that—for a carefully prepared pellet—the data in the infrared do obey Fresnel's relations. They also measured the void fraction over the first few layers of particles (where all absorption occurs) and found that the proportion of voids in these layers is significantly lower (18%) than in the bulk of the material (33%). For their data evaluation they first determined the applicability of Fresnel's relations by measuring the *specularity index* defined as

$$ s = \rho_\perp^2 \left(\frac{\pi}{4} \right) / \rho_\parallel \left(\frac{\pi}{4} \right), \tag{10.109} $$

where $\rho_\perp$ and $\rho_\parallel$ are the perpendicular and parallel polarized components of the pellet reflectivity, respectively. For a surface obeying Fresnel's relations $s \equiv 1$ always, regardless of the complex index of refraction of the material [cf. equations (3.50) and (3.51)]. They determined that their surfaces could be considered specular reflectors for wavelengths $\lambda \geq 2.0\ \mu$m. They then proceeded to correct their data for the measured void fraction using a number of different models. Their data for the refractive index of propane soot are also included in Fig. 10-26. It is seen that their data, even after correction for voidage (which raises the value for n by approximately 0.3, and for k by approximately 0.15), differ significantly from those of other investigations and depend only weakly on wavelength.

If there is evidence that the soot particles have agglomerated into large chunks or even long chains, use of the spherical-particle assumption becomes very questionable. A number of textbooks have considered scattering by nonspherical particles [2, 15, 17]. Approximating chunks of soot as prolate spheroids, Jones [74] found their absorption behavior to be considerably different from that of spheres of identical volumes. Lee and Tien [75] investigated the extreme case of long chains approximated by infinite cylinders. They found that the extinction coefficient for spheres drops off in the infrared much faster than the one for cylinders of the same radius, as shown in Fig. 10-27. However, the wavelength-integrated extinction coefficient is rather insensitive to particle shape at elevated temperatures, say $T > 1000$ K (i.e., at flame temperatures where soot emission may be important) [75]. Similar results were found by Mackowski and coworkers [76], who looked at infinite soot cylinders also using Lee and Tien's optical properties. Investigating the behavior of polydisperse cylindrical soot particles, they found the behavior to be similar to that observed by Buckius and

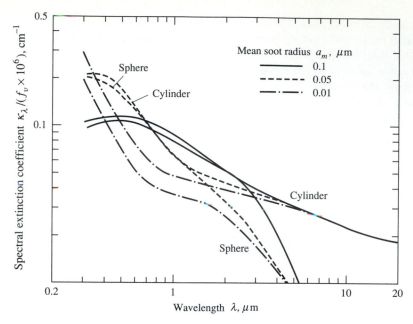

FIGURE 10-27
Spectral extinction coefficients for cylindrical and spherical soot particles [75].

Hwang [54] for polydisperse coal particles. While they generated correlations for absorption and extinction coefficients according to equation (10.105), unfortunately their correlation is rather cumbersome to use since different sets of parameters apply to each of a large number of wavelengths.

Little is known about the size distribution of soot particles in any given flame. Some measurements for specific conditions have been carried out using electron microscopy, e.g., Chippett and Gray [77], or dynamic light scattering, e.g., Charalmpopoulos and Felske [78], but no general results have been reported that would allow the a priori prediction of the soot distribution function. On the other hand, results by Lee and Tien [69] have shown that application of the small-particle limit, equation (10.107), may lead to large errors. Figure 10-28 shows the variation of the soot extinction coefficient with mean particle radius using a gamma distribution function [equation (10.34) with $\gamma = 3$ and $\delta = 1$] and a volume fraction of $f_v = 3.3 \times 10^{-6}$.

For a simplified heat transfer analysis it is generally desirable to use suitably defined mean absorption and extinction coefficients such as the Planck-mean and Rosseland-mean. If the soot particles are very small so that the Rayleigh theory applies for all particles and relevant wavelengths, then the extinction coefficient is described by equation (10.107). By choosing appropriate spectral average values for the refractive index n and absorptive index k one may approximate the extinction

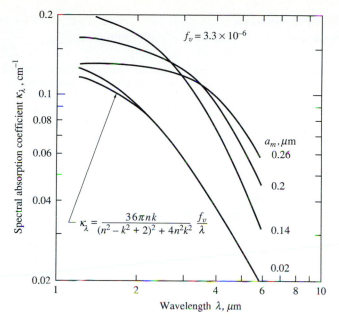

FIGURE 10-28
Variation of the spectral extinction coefficient of soot with particle size distribution [69]. Soot particles have a gamma distribution function [equation (10.34) with $\gamma = 3$ and $\delta = 1$] and a volume fraction of $f_v = 3.3 \times 10^{-6}$.

coefficient by

$$\kappa_\lambda = \beta_\lambda = C_0 \frac{f_v}{\lambda}, \quad C_0 = \frac{36\pi n k}{(n^2 - k^2 + 2)^2 + 4n^2 k^2}, \tag{10.110}$$

where C_0 is now a constant depending only on the soot index of refraction. With this simple $1/\lambda$ wavelength dependence, Planck-mean and Rosseland-mean extinction coefficients are readily calculated as

$$\kappa_P = \beta_P = 3.83 f_v C_0 T / C_2, \quad \kappa_R = \beta_R = 3.60 f_v C_0 T / C_2, \tag{10.111}$$

where $C_2 = 1.4388 \, \text{cm K}$ is the second Planck function constant. It is interesting to note that the Planck-mean coefficient (appropriate for optically thin situations) differs by only 6% from the Rosseland-mean (appropriate for optically thick situations). Thus, Felske and Tien [79] have suggested using an average value of

$$\kappa_m = \beta_m = 3.72 f_v C_0 T / C_2 \tag{10.112}$$

for all optical regimes. It is important to keep in mind that the above formulae apply only to very small soot particles, and that the extinction coefficient will increase for larger particle sizes as indicated in Figs. 10-27 and 10-28. However, it appears that no correlations for mean extinction coefficients have yet been published for polydisperse spherical soot particles not obeying the Rayleigh limit.

Example 10.4. Propane is burned with air under fuel-rich conditions, resulting in a volume fraction of soot of $f_v = 10^{-5}$. Determine the extinction coefficient for very small particles at a wavelength of 3 μm using the refractive index data of (*i*) Lee and Tien, (*ii*) Stull and Plass, (*iii*) Dalzell and Sarofim, and (*iv*) Felske and coworkers. If the soot consisted of long cylinders with radius 0.1 μm, how would the extinction coefficient change? If the soot was spherical (but of same mean radius of 0.1 μm) what would be the extinction coefficient?

Solution

To determine the extinction coefficient for small spherical soot particles we use equation (10.107) together with optical property data from Fig. 10-26:

Lee and Tien:	$n = 2.24$,	$k = 1.26$,	$\kappa_\lambda = 0.1734\,\mathrm{cm}^{-1}$;
Stull and Plass:	$n = 2.63$,	$k = 1.95$,	$\kappa_\lambda = 0.1472\,\mathrm{cm}^{-1}$;
Dalzell and Sarofim:	$n = 2.19$,	$k = 1.30$,	$\kappa_\lambda = 0.1835\,\mathrm{cm}^{-1}$;
Felske and coworkers:	$n = 2.31$,	$k = 0.71$,	$\kappa_\lambda = 0.1077\,\mathrm{cm}^{-1}$.

Thus, the values found from the data of Lee and Tien and from Dalzell and Sarofim are fairly consistent, while the absorptive index based on Stull and Plass' data is considerably higher, probably because amorphous carbon simply does not represent soot well. The absorptive index based on the data of Felske and coworkers is by far the lowest. The discrepancy remains unresolved with the present state of knowledge of soot characteristics. For soot of 0.1 μm mean radius we find from Fig. 10-27 $\kappa_\lambda = 0.29\,\mathrm{cm}^{-1}$ for spheres and $\kappa_\lambda = 0.40\,\mathrm{cm}^{-1}$ for long cylinders. These results must be compared with the Rayleigh limit based on Lee and Tien's data, $\kappa_\lambda = 0.1734\,\mathrm{cm}^{-1}$. Clearly, for particles this size the Rayleigh limit strongly underpredicts the extinction coefficient, and the extinction coefficient for cylindrical soot is still larger. The effect of particle size can also be inferred from Fig. 10-28: For a wavelength of $\lambda = 3\,\mu$m and a volume fraction of $f_v = 3.3 \times 10^{-6}$ the Rayleigh-limit extinction coefficient is approximately $0.058\,\mathrm{cm}^{-1}$; by interpolation the extinction coefficient for 0.1 μm particles is approximately $0.12\,\mathrm{cm}^{-1}$. Since the extinction coefficient is directly proportional to volume fraction, these values may be corrected by multiplying each by $10/3.3$ resulting in $\kappa_\lambda = 0.176\,\mathrm{cm}^{-1}$ for the Rayleigh limit and $\kappa_\lambda = 0.36\,\mathrm{cm}^{-1}$ for the larger sizes, which are consistent with the values obtained from Figs. 10-26 and 10-27.

In summary, as of today little is known about the nature of the production and the destruction of soot particles in a flame. Nor is much known of the chemical composition of these particles (and, consequently, their optical properties), of their shape and size distribution, and so on. Therefore, any prediction of radiative properties as described in this section carries a good amount of uncertainty. Results from this section may be used for fairly crude approximations, while accurate predictions would have to depend on in situ measurements for the flame under consideration.

References

1. Tien, C. L., and B. L. Drolen: "Thermal Radiation in Particulate Media with Dependent and Independent Scattering," in *Annual Review of Numerical Fluid Mechanics and Heat Transfer*, vol. 1, Hemisphere, New York, pp. 1–32, 1987.

2. van de Hulst, H. C.: *Light Scattering by Small Particles*, John Wiley & Sons, New York, 1957 (also Dover Publications, New York, 1981).

3. Hottel, H. C., A. F. Sarofim, W. H. Dalzell, and I. A. Vasalos: "Optical Properties of Coatings. Effect of Pigment Concentration," *AIAA Journal*, vol. 9, pp. 1895–1898, 1971.

4. Brewster, M. Q., and C. L. Tien: "Radiative Transfer in Packed/Fluidized Beds: Dependent vs. Independent Scattering," *ASME Journal of Heat Transfer*, vol. 104, pp. 573–579, 1982.

5. Yamada, Y., J. D. Cartigny, and C. L. Tien: "Radiative Transfer with Dependent Scattering by Particles, Part 2: Experimental Investigation," *ASME Journal of Heat Transfer*, vol. 108, pp. 614–618, 1986.

6. Cartigny, J. D., Y. Yamada, and C. L. Tien: "Radiative Transfer with Dependent Scattering by Particles, Part 1: Theoretical Investigation," *ASME Journal of Heat Transfer*, vol. 108, pp. 608–613, 1986.

7. Drolen, B. L., and C. L. Tien: "Independent and Dependent Scattering in Packed-Sphere Systems," *Journal of Thermophysics and Heat Transfer*, vol. 1, pp. 63–68, 1987.

8. Drolen, B. L., K. Kumar, and C. L. Tien: "Experiments on Dependent Scattering of Radiation," AIAA paper no. TP-87-210, 1987.

9. Rayleigh, L.: "On the Light from the Sky, Its Polarization and Colour," *Philos. Mag.*, vol. 41, pp. 107–120, 274–279, 1871 (reprinted in *Scientific Papers by Lord Rayleigh*, vol. I: 1869–1881, no. 8, Dover, New York, 1964).

10. Rayleigh, L.: *Phil. Mag.*, vol. 12, 1881.

11. Lorenz, L.: in *Videnskab Selskab Skrifter*, vol. 6, Copenhagen, Denmark, 1890.

12. Lorenz, L.: in *Oeuvres Scientifiques*, vol. I, Copenhagen, Denmark, p. 405, 1898.

13. Mie, G. A.: "Beiträge zur Optik trüber Medien, speziell kolloidaler Metallösungen," *Annalen der Physik*, vol. 25, pp. 377–445, 1908.

14. Debye, P.: *Annalen der Physik*, vol. 30, no. 4, p. 57, 1909.

15. Kerker, M.: *The Scattering of Light and Other Electromagnetic Radiation*, Academic Press, New York, 1969.

16. Deirmendjian, D.: *Electromagnetic Scattering on Spherical Polydispersions*, Elsevier, New York, 1969.

17. Bohren, C. F., and D. R. Huffman: *Absorption and Scattering of Light by Small Particles*, John Wiley & Sons, New York, 1983.

18. Wylie, C. R.: *Advanced Engineering Mathematics*, 5th ed., McGraw-Hill, New York, 1982.

19. Abramowitz, M., and I. A. Stegun (eds.): *Handbook of Mathematical Functions*, Dover Publications, New York, 1965.

20. Chu, C. M., and S. W. Churchill: "Representation of the Angular Distribution of Radiation Scattered by a Spherical Particle," *Journal of the Optical Society of America*, vol. 45, no. 11, pp. 958–962, 1955.

21. Clark, G. C., C. M. Chu, and S. W. Churchill: "Angular Distribution Coefficients for Radiation Scattered by a Spherical Particle," *Journal of the Optical Society of America*, vol. 47, pp. 81–84, 1957.

22. Kattawar, G. W., and G. N. Plass: "Electromagnetic Scattering from Absorbing Spheres," *Applied Optics*, vol. 6, no. 8, pp. 1377–1383, 1967.

23. Wiscombe, W. J.: "Improved Mie Scattering Algorithms," *Applied Optics*, vol. 19, pp. 1505–1509, 1980.

24. Hodkinson, J. R.: "Light Scattering and Extinction by Irregular Particles Larger than the Wavelength," in *Electromagnetic Scattering*, ed. M. Kerker, Macmillan, New York, pp. 87–100, 1963.

25. Bryant, F. D., B. A. Sieber, and P. Latimer: "Absolute Optical Cross Sections of Cells and Chloroplasts," *Arch. Biochem. Biophys.*, vol. 135, pp. 97–108, 1969.

26. Roessler, D. M., and F. R. Faxvog: "Optoacoustic Measurement of Optical Absorption in Acetylene Smoke," *Journal of the Optical Society of America*, vol. 69, pp. 1699–1704, 1979.

27. Faxvog, F. R., and D. M. Roessler: "Optoacoustic Measurement of Diesel Particulate Emission," *Journal of Applied Physics*, vol. 50, pp. 7880–7882, 1979.

28. Härd, S., and O. Nilsson: "Laser Heterodyne Apparatus for Measuring Small Angle Scattering from Particles," *Applied Optics*, vol. 18, pp. 3018–3026, 1979.

29. Hottel, H. C., A. F. Sarofim, I. A. Vasalos, and W. H. Dalzell: "Multiple Scatter: Comparison of Theory with Experiment," *ASME Journal of Heat Transfer*, vol. 92, pp. 285–291, 1970.

30. Menart, J. A., H. S. Lee, and R. O. Buckius: "Experimental Determination of Radiative Properties for Scattering Particulates," *Experimental Heat Transfer*, vol. 2, no. 4, p. 309, 1989.

31. Woodward, D. H.: "He-Ne Laser as Source for Light Scattering Measurements," *Applied Optics*, vol. 2, pp. 1205–1207, 1963.

32. Woodward, D. H.: "Multiple Light Scattering by Spherical Dielectric Particles," *Journal of the Optical Society of America*, vol. 54, pp. 1325–1331, 1964.

33. Hartel, W.: "Zur Theorie der Lichstreuung durch trübe Schichten, besonders Trübgläser," *Licht*, vol. 10, pp. 141–143, 232–234, 1940.

34. Smart, C., R. Jacobsen, M. Kerker, P. Kratohvil, and E. Matijevic: "Experimental Study of Multiple Light Scattering," *Journal of the Optical Society of America*, vol. 55, no. 8, pp. 947–955, 1965.

35. Orchard, S. E.: "Multiple Scattering by Spherical Dielectric Particles," *Journal of the Optical Society of America*, vol. 55, pp. 737–738, 1965.

36. Daniel, K. J., N. M. Laurendeau, and F. P. Incropera: "Optical Property Measurements for Suspensions of Unicellular Algae," ASME paper no. 78-HT-14, 1978.

37. Spinrad, R. W., J. R. V. Zaneveld, and H. Pak: "Volume Scattering Function of Suspended Particulate Matter at Near-Forward Angles: A Comparison of Experimental and Theoretical Values," *Applied Optics*, vol. 17, no. 7, pp. 1125–1130, 1978.

38. van de Hulst, H. C.: "Asymptotic Fitting, A Method for Solving Anisotropic Transfer Problems in Thick Layers," *J. of Comput. Phys.*, vol. 3, pp. 291–306, 1968.

39. Hansen, J. E.: "Exact and Approximate Solutions for Multiple Scattering by Cloudy and Hazy Planetary Atmospheres," *Journal of the Atmospheric Sciences*, vol. 26, pp. 478–487, 1969.

40. Modest, M. F., and F. H. Azad: "The Influence and Treatment of Mie-Anisotropic Scattering in Radiative Heat Transfer," *ASME Journal of Heat Transfer*, vol. 102, pp. 92–98, 1980.

41. Potter, J. F.: "The Delta Function Approximation in Radiative Transfer Theory," *Journal of the Atmospheric Sciences*, vol. 27, pp. 943–949, 1970.

42. Joseph, J. H., W. J. Wiscombe, and J. A. Weinman: "The Delta-Eddington Approximation for Radiative Flux Transfer," *Journal of the Atmospheric Sciences*, vol. 33, pp. 2452–2459, 1976.

43. Crosbie, A. L., and G. W. Davidson: "Dirac-Delta Function Approximations to the Scattering Phase Function," *Journal of Quantitative Spectroscopy and Radiative Transfer*, vol. 33, no. 4, pp. 391–409, 1985.

44. Davies, R.: "Fast Azimuthally Dependent Model of the Reflection of Solar Radiation by Plane-Parallel Clouds," *Applied Optics*, vol. 19, pp. 250–255, 1980.

45. Sarofim, A. F., and H. C. Hottel: "Radiative Transfer in Combustion Chambers: Influence of Alternative Fuels," in *Proceedings of the Sixth International Heat Transfer Conference*, vol. 6, Washington, D.C., Hemisphere, pp. 199–217, 1978.

46. Foster, P. J., and C. R. Howarth: "Optical Constants of Carbons and Coals in the Infrared," *Carbon*, vol. 6, pp. 719–729, 1968.

47. Blokh, A. G.: "The Problem of Flame as a Disperse System," in *Heat Transfer in Flames*, eds. N. F. Afghan and J. M. Beer, Scripta Book Co., Washington, pp. 111–130, 1974.

48. Blokh, A. G., and L. D. Burak: "Primary Radiation Characteristics of Solid Fuels," *Thermal Engineering*, vol. 20, no. 8, pp. 65–70, 1973.

49. Lowe, A., I. M. Stewart, and T. F. Wall: "The Measurement and Interpretation of Radiation from Fly-Ash Particles in Large Pulverized Coal Flames," in *Seventeenth Symposium (International) on Combustion*, The Combustion Institute, pp. 105–114, 1979.

50. Blokh, A. G., Sagadeev, and V. D. Vyushin: "Experimental and Theoretical Investigation of Radiation Properties of Flame with Coal Burning in Powerful Boiler Furnaces," in *Heat and Mass Transfer-VI*, vol. VIII, Soviet Academy of Sciences (in Russian), Minsk, pp. 70–73, 1980.

51. Wall, T. F., A. Lowe, L. J. Wibberley, T. Mai-Viet, and R. P. Gupta: "Fly-Ash Characteristics and Radiative Heat Transfer in Pulverized-Coal-Fired Furnaces," *Combustion Science and Technology*, vol. 26, pp. 107–121, 1981.

52. Viskanta, R., A. Ugnan, and M. P. Mengüç: "Predictions of Radiative Properties of Pulverized Coal and Fly-Ash Polydispersions," ASME paper no. 81-HT-24, 1981.

53. Tien, C. L., D. G. Doornink, and D. A. Rafferty: "Attenuation of Visible Radiation by Carbon Smokes," *Combustion Science and Technology*, vol. 6, pp. 55–59, 1972.

54. Buckius, R. O., and D. C. Hwang: "Radiation Properties for Polydispersions: Application to Coal," *ASME Journal of Heat Transfer*, vol. 102, pp. 99–103, 1980.

55. Singer, J. M., and J. Grumer: "Carbon Formation in Very Rich Hydrocarbon–Air Flames—I. Studies of Chemical Content, Temperature, Ionization and Particulate Matter," in *Seventh Symposium (International) on Combustion*, The Combustion Institute, pp. 559–572, 1959.

56. Wersborg, B. L., J. B. Howard, and G. C. Williams: "Physical Mechanisms in Carbon Formation in Flames," in *Fourteenth Symposium (International) on Combustion*, The Combustion Institute, pp. 929–940, 1972.

57. Kunugi, M., and H. Jinno: "Determination of Size and Concentration of Soot Particles in Diffusion Flames by a Light-Scattering Technique," in *Eleventh Symposium (International) on Combustion*, The Combustion Institute, pp. 257–266, 1966.

58. Sato, T., T. Kunitomo, S. Yoshi, and T. Hashimoto: "On the Monochromatic Distribution of the Radiation from the Luminous Flame," *Bulletin of Japan Society of Mechanical Engineers*, vol. 12, pp. 1135–1143, 1969.

59. Becker, A.: *Annalen der Physik*, vol. 28, p. 1017, 1909.

60. Millikan, R. C.: "Optical Properties of Soot," *Journal of the Optical Society of America*, vol. 51, pp. 698–699, 1961.

61. Millikan, R. C.: "Sizes, Optical Properties and Temperatures of Soot Particles," in *The Fourth Symposium on Temperature, Its Measurement and Control in Science and Industry*, vol. 3, pp. 497–507, 1961.

62. Dalzell, W. H., and A. F. Sarofim: "Optical Constants of Soot and Their Application to Heat-Flux Calculations," *ASME Journal of Heat Transfer*, vol. 91, no. 1, pp. 100–104, 1969.

63. Stull, V. R., and G. N. Plass: "Emissivity of Dispersed Carbon Particles," *Journal of the Optical Society of America*, vol. 50, no. 2, pp. 121–129, 1960.

64. Howarth, C. R., P. J. Foster, and M. W. Thring: "The Effect of Temperature on the Extinction of Radiation by Soot Particles," in *Proceedings of the Third International Heat Transfer Conference*, vol. 5, Washington, D.C., Hemisphere, pp. 122–128, 1966.

65. Hubbard, G. L., and C. L. Tien: "Infrared Mean Absorption Coefficients of Luminous Flames and Smoke," *ASME Journal of Heat Transfer*, vol. 100, pp. 235–239, 1978.

66. Medalia, A. I., and L. W. Richards: "Tinting Strength of Carbon Black," *Journal of Colloid and Interface Science*, vol. 40, pp. 233–252, 1972.

67. Graham, S. C.: "The Refractive Indices of Isolated and of Aggregated Soot Particles," *Combustion Science and Technology*, vol. 9, pp. 159–163, 1974.

68. Janzen, J.: "The Refractive Index of Colloidal Carbon," *Journal of Colloid and Interface Science*, vol. 69, 1979.

69. Lee, S. C., and C. L. Tien: "Optical Constants of Soot in Hydrocarbon Flames," in *Eighteenth Symposium (International) on Combustion*, The Combustion Institute, pp. 1159–1166, 1980.

70. Moss, T. S., G. J. Burrell, and B. Ellis: *Semiconductor Opto-Electronics*, John Wiley & Sons, New York, 1972.

71. Buckius, R. O., and C. L. Tien: "Infrared Flame Radiation," *International Journal of Heat and Mass Transfer*, vol. 20, pp. 93–106, 1977.

72. Bard, S., and P. J. Pagni: "Carbon Particulate in Small Pool Fire Flames," *ASME Journal of Heat Transfer*, vol. 103, pp. 357–362, 1981.

73. Felske, J. D., T. T. Charalmpopoulos, and H. S. Hura: "Determination of Refractive Indices of Soot Particles from the Reflectivities of Compressed Soot Particles," *Combustion Science and Technology*, vol. 37, pp. 263–284, 1984.

74. Jones, A. R.: "An Estimate of the Possible Effects of Particle Agglomeration on the Emissivity of Sooty Flames," in *Combustion Institute European Symposium*, pp. 376–381, 1973.

75. Lee, S. C., and C. L. Tien: "Effect of Soot Shape on Soot Radiation," *Journal of Quantitative Spectroscopy and Radiative Transfer*, vol. 29, pp. 259–265, 1983.

76. Mackowski, D. W., R. A. Altenkirch, and M. P. Mengüç: "Extinction and Absorption Coefficients of Cylindrically-Shaped Soot Particles," *Combustion Science and Technology*, vol. 40, pp. 399–410, 1987.

77. Chippett, S., and W. A. Gray: "The Size and Optical Properties of Soot Particles," *Combustion and Flame*, vol. 31, pp. 149–159, 1978.
78. Charalmpopoulos, T. T., and J. D. Felske: "Refractive Indices of Soot Particles Deduced from In-Situ Laser Light Scattering Measurements," *Combustion and Flame*, vol. 68, pp. 57–67, 1987.
79. Felske, J. D., and C. L. Tien: "The Use of the Milne-Eddington Absorption Coefficient for Radiative Heat Transfer in Combustion Systems," *Transactions of ASME, Journal of Heat Transfer*, vol. 99, no. 3, pp. 458–465, 1977.

Problems

10.1 One way to determine the number of particles in a gas is to measure the absorption coefficient for the cloud. For a cloud of large, diffuse particles ($x \gg 1$, $\epsilon_\lambda = 0.4$), the particle distribution function is known to be of the form

$$n(a) = \begin{cases} C = \text{const}, & 100\,\mu\text{m} < a < 500\,\mu\text{m}, \\ 0, & \text{elsewhere}. \end{cases}$$

If κ_λ is measured as $1\,\text{cm}^{-1}$, determine C and the total number of particles per cm^3.

10.2 Consider a particle cloud of fixed-size particles (radius a) contained between parallel plates $0 \leq x \leq L = 1\,\text{m}$. The volume fraction of particles is $f_v(x) = f_0 + \Delta f\,(x/L)$, and their temperature is $T(x) = T_0 + \Delta T\,(x/L)$, where $\Delta f/f_0 = \Delta T/T_0 = 1$, $f_0 = 1\%$, $T_0 = 500\,\text{K}$. Assuming the particle size to be $a = 500\,\mu\text{m}$, and made of a material with a gray hemispherical emissivity of $\epsilon_\lambda = 0.7$, show that the large-particle approximation may be used for the infrared. Calculate the local, spectral absorption and scattering coefficients. Determine the local Planck-mean extinction coefficient as well as the total optical thickness of the slab (based on the Planck-mean).

10.3 Redo Problem 10.2 for propane soot with a single mean radius of $a_m = 0.1\,\mu\text{m}$ in a flame with $f_0 = 10^{-6}$ and $T_0 = 1500\,\text{K}$. Show that the small particle limit is appropriate for, say, $\lambda > 3\,\mu\text{m}$. For hand calculations you may approximate the index of refraction by a single average value (say, at $3\,\mu\text{m}$), and the emissive power by Wien's law.

10.4 Redo Problem 10.3 for the case that the soot has agglomerated into long cylindrical chains.

10.5 The distribution function of a particle cloud may be approximated by an exponential function such as

$$n(a) = C_1 a^2 \exp(-C_2\,a),$$

where C_1 and C_2 are constants. It is proposed to determine the distribution function of a set of particles by suspending a measured mass of particles between parallel plates, followed by measuring extinction across the particle layer. Given that $m'' = 0.05\,\text{g}/\text{cm}^2$ of particles are present between the plates, which are $10\,\text{cm}$ apart, and that the optical thickness based on extinction has been measured as $\tau_0 = 2$,

    (*a*) Determine the distribution function above (i.e., C_1 and C_2).

    (*b*) If a single particle size were to be used to achieve the same extinction with the same mass of particles, what would the particle radius be?

You may assume all particles to be "large" and diffuse spheres with an emissivity of 0.7 and a density of $\rho = 2\,\text{g}/\text{cm}^3$.

10.6 Consider a particle cloud with a distribution function of $n(a) = C a^2 e^{-ba}$, where a is particle radius and b and C are constants. The particles are soot ($m \simeq 1.5 - 0.5i$), and measurements show the soot occupies a volume fraction of 1%, while the number density has been measured as $N_T = 8\pi^2 \times 10^9/\text{cm}^3$. Calculate the extinction, absorption, and scattering coefficients of the cloud for the wavelength range $1\,\mu\text{m} < \lambda < 4\,\mu\text{m}$.

10.7 In a coal-burning plant, pulverized coal is used that is known to have a particle size distribution function of

$$n(a) \propto a^2 e^{-Aa^6}, \quad A = 3 \times 10^{-11}\,\mu\text{m}^{-6}.$$

The coal may be approximated as diffuse spheres with a gray emissivity of $\epsilon = 0.3$. What is the effective minimum size parameter, $x_{\min}$, (i.e., 90% by weight of all particles have a size parameter larger than that)? You may assume a combustion temperature of $\approx 2000\,\text{K}$, i.e., the relevant wavelengths range from about $1\,\mu\text{m}$ to about $10\,\mu\text{m}$. If the furnace is loaded with 10 kg coal particles per cubic meter, what are the spectral absorption and scattering coefficients? (Density of the coal $= 2000\,\text{kg/m}^3$).

10.8 Consider nitrogen mixed with spherical particles at a rate of 10^8 particles/m^3. The particles have a radius of $300\,\mu\text{m}$ and are diffuse-gray with $\epsilon = 0.5$.

(a) Determine absorption and scattering coefficients, and the scattering phase function.
(b) Show how the phase function can be approximated by a Henyey-Greenstein function.
(c) Show qualitatively how the phase function may be approximated with the Modest and Azad model by
(i) $A_1 = -1, \quad f = b = 0$;
(ii) $A_1 = -1, \quad f = 0, \quad b = \frac{1}{4}$.
(d) Can the Crosbie-Davidson model be used for this mixture?
(e) Compare the different versions of the phase function in a Φ vs. $\cos\Theta$ plot.

CHAPTER
11

RADIATIVE
PROPERTIES OF
SEMITRANSPARENT
MEDIA

11.1   INTRODUCTION

Any solid or liquid that allows electromagnetic waves to penetrate an appreciable distance into it is known as a semitransparent medium. What constitutes an "appreciable distance" depends, of course, on the physical system at hand. If a thick film on top of a substrate allows a substantial amount of photons to propagate, say, $100\,\mu$m into it, the film material would be considered semitransparent. On the other hand, if heat transfer within a large vat of liquid glass is of interest, the glass cannot be considered semitransparent for those wavelengths that cannot penetrate several centimeters through the glass.

Pure solids with perfect crystalline or very regular amorphous structures, as well as pure liquids, gradually absorb radiation as it travels through the medium, but they do not scatter it appreciably within that part of the spectrum that is of interest to the heat transfer engineer. If a solid crystal has defects, or if a solid or liquid contains inclusions (foreign molecules or particles, bubbles, etc.), the material may scatter as well as absorb. In some instances semitransparent media are inhomogeneous and tend

438

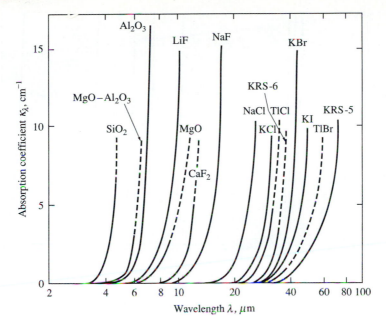

FIGURE 11-1
Spectral absorption coefficients of several ionic crystals at room temperature [2].

to scatter radiation as a result of their inhomogeneities. An example of such material is aerogel [1] a highly transparent, low heat-loss window material made of tiny hollow glass spheres pressed together.

A number of theoretical models exist to predict the absorption and scattering characteristics of semitransparent media. As for opaque surfaces, the applicability of theories is limited, and they must be used in conjunction with experimental data. In this chapter we shall limit ourselves to absorption within semitransparent media. The models describing scattering behavior are the same as the ones presented in the previous chapter and will not be further discussed here.

11.2 ABSORPTION BY SEMITRANSPARENT SOLIDS

The absorption behavior of ionic crystals can be rather successfully modeled by the Lorentz model, which was discussed in some detail in Chapters 2 and 3. The Lorentz theory predicts that an ionic crystal has one or more Reststrahlen bands in the mid-infrared ($\lambda > \simeq 5\,\mu$m) (photon excitation of lattice vibrations). The wavelength at which strong absorption commences because of Reststrahlen bands is often called the *long-wavelength absorption edge*. The spectral absorption coefficients and their long-wavelength absorption edges are shown for a number of ionic crystals in Fig. 11-1.

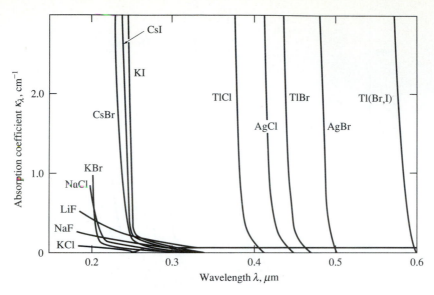

FIGURE 11-2
Spectral absorption coefficient of several halides at room temperature [2].

Note that these crystals are essentially transparent over much of the near infrared, and become very rapidly opaque at the onset of Reststrahlen bands.

The Lorentz model also predicts that the excitation of valence band electrons, across the band gap into the conduction band, results in several absorption bands at short wavelengths (usually around the ultraviolet). Figure 11-2 shows the absorption coefficient and *short-wavelength absorption edge* for several halides: Materials that are essentially opaque in the ultraviolet become highly transparent in the visible and beyond.

Pure solids are generally highly transparent between the two absorption edges. If large amounts of localized lattice defects and/or dopants (foreign-material molecules called *color centers*) are present, electronic excitations may occur at other wavelengths in between. A number of models predict the absorption characteristics of such defects, some sophisticated, some simple and semiempirical. For example, Bhattacharyya and Streetman [3] and Blomberg and coworkers [4] developed models predicting the effect of dopants on the absorption coefficient of silicon. Figure 11-3 shows a comparison of the model by Blomberg and coworkers with experimental data of Siregar and colleagues [5] and Boyd and coworkers [6] for phosporus-doped silicon at $10.6\,\mu m$ (a wavelength of great importance for materials processing with CO_2 lasers). The absorption coefficient increases strongly with dopant concentration and with temperature. According to both models, the rise with temperature is due to increases in the number of free electrons and to their individual contributions.

The absorption behavior of amorphous, i.e., noncrystalline solids is much more

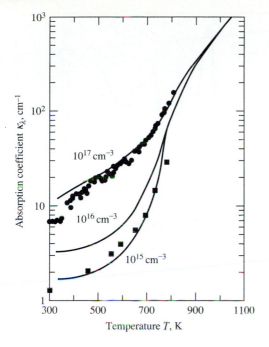

FIGURE 11-3
Spectral absorption coefficient of phosphorus-doped Si at $10.6\,\mu$m; solid lines: Model of Blomberg and coworkers [4]; square symbols (■): Data of Boyd and coworkers [6] (dopant concentration of $1.1 \times 10^{15}\,\text{cm}^{-3}$); circular symbols (●): Data of Siregar and coworkers [5] (dopant concentration unknown).

difficult to predict, although the general trends are quite similar. By far the most important semitransparent amorphous solid is soda-lime glass (ordinary window glass, as opposed to quartz or silicon dioxide crystals depicted in Fig. 11-1). A number of investigators measured the absorption behavior of window glass, notably Genzel [7], Neuroth [8, 9], Grove and Jellyman [10] and Bagley and coworkers [11]. Figure 1-18 shows the behavior of the spectral absorption coefficient of window glass for a number of different temperatures. As expected from the data for the transmissivity of window panes (Figs. 3-27 and 3-28), glass is fairly transparent for wavelengths $\lambda < 2.5\,\mu$m; beyond that it tends to become rather opaque.

The temperature dependence for quartz has been observed to be similar to that of silicon by Beder and coworkers [12], who reported a four-fold increase of the absorption coefficient between room temperature and 1500°C.

11.3 ABSORPTION BY SEMITRANSPARENT LIQUIDS

The absorption properties of semitransparent liquids are quite similar to those of solids, while they also display some behavior similar to molecular gases. Remnants of intermolecular vibrations (Reststrahlen bands) are observed in many liquids, as are remnants of electronic band gap transitions in the ultraviolet. In the wavelengths in between, molecular vibration bands are observed for molecules with permanent dipole moments, similar to the vibration-rotation bands of gases.

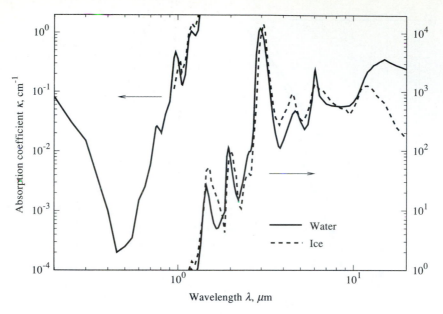

FIGURE 11-4
Spectral absorption coefficient of clear water (at room temperature) and clear ice (at $-10°C$) [13].

Because of its abundance in the world around us (and, indeed, inside our own bodies) the absorption properties of water (and its solid form as ice) are by far the most important and, therefore, have been studied extensively, indeed for centuries. The data of many investigators for clear water and clear ice have been collected and interpreted by Irvine and Pollack [13] and by Ray [14]. Another review, limited to pure water but including some more recent data, has been given by Hale and Querry [15]. The spectral absorption coefficient of clear water (at room temperature) and of clear ice (at $-10°C$) is shown in Fig. 11-4, based on the tabulations of Irvine and Pollack [13]. Note the similarity between solid ice and liquid water. The lowest points of the absorption spectra of water and ice lie in the visible, making them virtually transparent over short distances. The minimum point lies in the blue part of the visible ($\lambda \simeq 0.45 \, \mu m$): Large bodies of water (or clear ice) transmit blue light the most, giving them a bluish hue. In the near- to midinfrared water and ice display several absorption bands (at 1.45, 1.94, 2.95, 4.7 and 6.05 μm in water, somewhat shifted for ice). These bands are very similar to the water vapor bands at 1.38, 1.87, 2.7 and 6.3 μm (see Table 9.3). The temperature dependence of the absorption coefficient of water has been investigated by Goldstein and Penner [16] (up to 209°C) and by Hale and coworkers [17] (up to 70°C) and was found to be fairly weak. As temperature increases, water becomes somewhat more transparent in relatively transparent regions and somewhat more opaque in absorbing regions. A rather detailed discussion of the absorption behavior of clean water and ice has been given by Bohren and Huffman [18]. Natural waters and ice generally contain

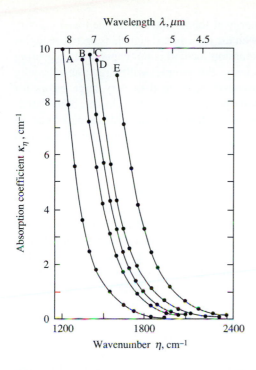

FIGURE 11-5
Spectral absorption coefficient of LiF for various temperatures; A: 300 K; B: 705 K; C: 835 K; D: 975 K; E: 1160 K. Melting point of LiF is 1115 K [19].

significant amounts of particulates (small organisms, detritus) and gas bubbles, which tend to increase the absorption rate as well as to scatter radiation. While a number of measurements have been made on varieties of natural waters and ice, the results are difficult to correlate since the composition of natural waters varies greatly.

The similarity of absorption behavior between the solid and liquid states of a substance is not limited to water. Barker [19] has measured the absorption coefficient of three alkali halides (KBr, NaCl, and LiF) for several temperatures between 300 K and temperatures above the melting point. Since Reststrahlen bands tend to widen with increasing temperature (see Section 3.5), the long-wavelength absorption edge moves towards shorter wavelengths. No distinct discontinuity in absorption coefficient was observed as the material changed phase from solid to liquid. As an example, the behavior of lithium fluoride (LiF) is depicted in Fig. 11-5. Semiempirical models for the absorption coefficient of alkali halide crystals, resulting in simple formulae, have been given by Skettrup [20] and Woodruff [21], while a similar formula for alkali halide melts has been developed by Senatore and coworkers [22].

11.4   EXPERIMENTAL METHODS

The spectral absorption coefficient of a semitransparent solid or liquid can be measured in several ways. The simplest and most common method is to measure the transmissivity of a sample of known thickness, as described in Section 10.7 for particulate clouds. Since solids and liquids reflect energy at the air interfaces, the transmissivity is often determined by forming a ratio betwen the transmitted signals from two samples of different thickness. However, the transmission method is not capable of measuring very small or very large absorption coefficients: For samples with large transmissivity small errors in the determination of transmissivity, τ, lead to very large errors for the absorption coefficient, κ (since κ is proportional to $\ln \tau$). On the other hand, for a material with large κ sufficient energy for transmission measurements can be passed only through extremely thin samples. Such samples are usually prepared as vacuum-deposited thin films, which do not have the same properties as the parent material [23].

The absorption coefficient may also be determined through a number of different reflection techniques. The reflectivity of an optically-smooth interface of a semitransparent medium depends, through the complex index of refraction, on the refractive index n as well as the absorptive index k. In turn, k is related to the absorption coefficient through equation (3.77) as $\kappa = 4\pi\eta k/n$, where $\eta = 1/\lambda$ is the wavenumber of the radiation inside the medium. Thus, two data points are necessary to determine n and k. Noting the directional dependence of reflectivity on $m = n - ik$, some researchers have measured the specular reflectivity at two different angles. Leupacher and Penzkofer [24] showed that this can lead to very substantial errors. Other researchers have measured the reflectivity at a single angle, using parallel- and perpendicular-polarized light (known as *ellipsometric technique*). However, this may also lead to large errors [24]. A new method overcoming these problems has been proposed by Lu and Penzkofer [25]. Using parallel-polarized light they vary the incidence angle until the point of minimum reflectivity at Brewster's angle is found (cf. Figs. 2-8 and 2-11).

Another reflection technique exploits the fact that a *causal relationship* exists between n and k, i.e., they are not independent of one another. This causal relationship is known as the *Kramers-Kronig relation*, which may be expressed as

$$\delta(\eta) = \frac{\eta}{\pi} \int_0^\infty \frac{\ln \rho_n(\eta')}{\eta^2 - \eta'^2} \, d\eta', \tag{11.1}$$

where $\rho_n(\eta)$ is the spectral, normal reflectivity of the sample surface [cf. equation (2.119)], and $\delta(\eta)$ is the phase angle of the complex reflection coefficient, equation (2.115),

$$\tilde{r}_n = \sqrt{\rho_n} \, e^{i\delta} = \frac{n - ik - 1}{n - ik + 1}. \tag{11.2}$$

Thus, if ρ_n is measured for a large part of the spectrum, the phase angle δ may be determined from equation (11.1) for wavenumbers well inside the measured spectrum; n and k are then readily found from equation (11.2). The method is particularly well suited to experiments employing an FTIR (Fourier transform infrared) spectrometer, which can take broad spectrum measurements over very short times, and which

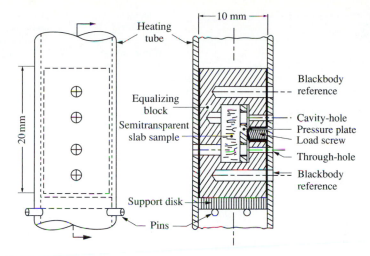

FIGURE 11-6
Sample and holder, mounted within heating tube, for device to determine the optical properties of small, semitransparent solid samples [27].

often have a built-in Kramers-Kronig analysis capability. More detailed discussions on the various Kramers-Kronig relations may be found, for example, in the books by Wooten [26] and Bohren and Huffman [18]. A description of the numerical evaluation of equation (11.1) has been given by Wooten [26].

Measurement of physical properties at high temperatures is always difficult, but particularly so for semitransparent media since *two* properties need to be measured (absorption coefficient as well as interface reflectivity, or equivalently, n and k). Myers and coworkers [27] have given a good review of such methods for solid samples. They also developed a new method to determine the optical properties of small, semi-transparent, solid samples. Their device is essentially a compact arrangement of that employed by Stierwalt [28], which takes three different radiance measurements in rapid succession. A front and cross-sectional view of their sample heating arrangement is shown in Fig. 11-6. The slab-shaped sample is mounted within an equalizing nickel block, which is coupled radiatively to the electrically-heated tube. The nickel block has four cavities and holes serving as radiance targets. A water-cooled graphite block (not shown) is positioned behind the heating tube to provide a room-temperature background for the through-hole as well as a reference for the detector. Three radiance measurements are made and compared with the reference: (*i*) The slab sample positioned in front of the blackbody (cavity-hole), (*ii*) the freely radiating sample (through-hole), and (*iii*) the blackbody reference. With the relations given in Section 3.8 one can use these measurements to deduce the optical properties (n, k, and κ). The method has the advantages that measurements at high temperatures ($\simeq 1000°C$) can be taken, that only a single sample is necessary, and that no optically smooth surfaces are required. On the other hand, the method suffers from the standard weak-

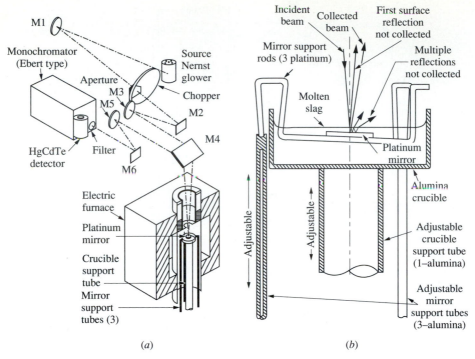

(a) (b)

FIGURE 11-7
Measurement of absorption coefficients of high-temperature liquids. (a) Schematic of apparatus of Ebert and Self [31], (b) schematic of their submerged reflector arrangement.

nesses of transmission methods (see discussion at the beginning of this section), and is restricted to high temperatures (to produce a strong enough emission signal).

Measurements of the optical properties of a high-temperature liquid are even more challenging. It is more difficult to confine a liquid in a sample holder (which must be horizontal), and more difficult to measure the thickness of the liquid layer. In addition, the layer thickness may be nonuniform because of (often unknown) surface tension effects. Furthermore, high-temperature liquids are often highly reactive, making a sealed chamber necessary. If the vapor pressure becomes substantial at high temperatures, the windows of the sealed chamber will be attacked. Shvarev and coworkers [29] have measured the optical properties of liquid silicon in the wavelength range of 0.4-1.0 μm with such a sealed-chamber furnace apparatus, using an ellipsometric technique. Barker [19, 30] designed an apparatus to measure the optical properties of semitransparent solid slabs and corrosive melts. To isolate the specimen he relied on a windowless chamber with continuous inert-gas purging. His data evaluation required independent measurements of the interface reflectivity, the reflectivity of a platinum mirror, the sample overall reflectivity, and the thickness of the sample.

In addition, the reflectivity of the platinum-liquid interface must be estimated. As such, Barker's method appears to be very vulnerable to experimental error.

A more accurate device, limited to absorption coefficients of liquids, has been reported by Ebert and Self [31]. A schematic of their apparatus is shown in Fig. 11-7a. The aperture of a blackbody source at 1700°C is imaged (by the spherical mirror M3) onto the platinum mirror located in an alumina crucible inside the furnace. The reflected signal is focused onto the monochromator and detector via another spherical mirror (M5). The beam is chopped to eliminate emission as well as background radiation from the signal. The transmissivity of the liquid is measured by what they called a "submerged reflector method," illustrated in Fig. 11-7b: A platinum mirror, which may be adjusted via three support rods, is submerged below the surface of the liquid filling the crucible. The platinum mirror is tilted slightly from the horizontal to allow the first surface reflection and multiple internal reflections to be rejected from the collection optics. The thickness of the liquid layer is adjusted by raising and lowering the crucible (leaving the platinum mirror in place). As in the transmission technique, signals for two different layer thicknesses (d_1 and d_2) are ratioed, giving the transmissivity for a layer of thickness $2(d_2 - d_1)$. By rejecting the first reflection, and by being able to produce and measure very thin liquid layers, they were able to measure absorption coefficients an order of magnitude higher than Barker, reporting values as high as $70\,\mathrm{cm}^{-1}$ for synthetic molten slags [31].

References

1. Caps, R., and J. Fricke: "Infrared Radiative Heat Transfer in Highly Transparent Silica Aerogel," *Solar Energy*, vol. 36, no. 4, pp. 361–364, 1986.
2. Smakula, A.: "Synthetic Crystals and Polarizing Materials," *Optica Acta*, vol. 9, pp. 205–222, 1962.
3. Bhattacharyya, A., and B. G. Streetman: "Theoretical Considerations Regarding Pulsed CO_2 Laser Annealing of Silicon," *Solid State Communications*, vol. 36, pp. 671–675, 1980.
4. Blomberg, M., K. Naukkarinen, T. Tuomi, V. M. Airaksinen, M. Luomajarvi, and E. Rauhala: "Substrate Heating Effects in CO_2 Laser Annealing of Ion-Implanted Silicon," *Journal of Applied Physics*, vol. 54, no. 5, pp. 2327–2328, 1983.
5. Siregar, M. R. T., W. Lüthy, and K. Affolter: "Dynamics of CO_2 Laser Heating in the Processing of Silicon," *Applied Physics Letters*, vol. 36, pp. 787–788, 1980.
6. Boyd, I. W., J. I. Binnie, B. Wilson, and M. J. Colles: "Absorption of Infrared Radiation in Silicon," *Journal of Applied Physics*, vol. 55, no. 8, pp. 3061–3063, 1984.
7. Genzel, L.: "Messung der Ultrarot-Absorption von Glas zwischen 20°C and 1360°C (Measurement of Infrared Absorption of Glass Between 20°C and 1360°C)," *Glastechnische Berichte*, vol. 24, no. 3, pp. 55–63, 1951.
8. Neuroth, N.: "Der Einfluss der Temperatur auf die spektrale Absorption von Gläsern im Ultraroten, I (Effect of Temperature on Spectral Absorption of Glasses in the Infrared)," *Glastechnische Berichte*, vol. 25, pp. 242–249, 1952.
9. Neuroth, N.: "Der Einfluss der Temperatur auf die spektrale Absorption von Gläsern im Ultraroten, II (Effect of Temperature on Spectral Absorption of Glasses in the Infrared II)," *Glastechnische Berichte*, vol. 26, pp. 66–69, 1953.
10. Grove, F. J., and P. E. Jellyman: "The Infrared Transmission of Glass in the Range from Room Temperature to 1400°C," *Journal of the Society of Glass Technology*, vol. 39, no. 186, pp. 3–15, 1955.

11. Bagley, B. G., E. M. Vogel, W. G. French, G. A. Pasteur, J. N. Gan, and J. Tauc: "The Optical Properties of Soda-Lime-Silica Glass in the Region from 0.006 to 22 eV," *Journal of Non-Crystalline Solids*, vol. 22, pp. 423–436, 1976.

12. Beder, E. C., C. D. Bass, and W. L. Shackleford: "Transmissivity and Absorption of Fused Quartz Between 0.2 μm and 3.5 μm From Room Temperature to 1500° C," *Applied Optics*, vol. 10, pp. 2263–2268, 1971.

13. Irvine, W. M., and J. B. Pollack: "Infrared Optical Properties of Water and Ice Spheres," *ICARUS*, vol. 8, pp. 324–360, 1968.

14. Ray, P. S.: "Broadband Complex Refractive Indices of Ice and Water," *Applied Optics*, vol. 11, pp. 1836–1844, 1972.

15. Hale, G. M., and M. R. Querry: "Optical Constants of Water in the 200nm to 200μm Wavelength Region," *Applied Optics*, vol. 12, pp. 555–563, 1973.

16. Goldstein, R. J., and S. S. Penner: "The Near-Infrared Absorption of Liquid Water at Temperatures between 27 and 209°C," *Journal of Quantitative Spectroscopy and Radiative Transfer*, vol. 4, pp. 441–451, 1964.

17. Hale, G. M., M. R. Querry, A. N. Rusk, and D. Williams: "Influence of Temperature on the Spectrum of Water," *Journal of the Optical Society of America*, vol. 62, pp. 1103–1108, 1972.

18. Bohren, C. F., and D. R. Huffman: *Absorption and Scattering of Light by Small Particles*, John Wiley & Sons, New York, 1983.

19. Barker, A. J.: "The Effect of Melting on the Multiphonon Infrared Absorption Spectra of KBr, NaCl, and LiF," *Journal of Physics C: Solid State Physics*, vol. 5, pp. 2276–2282, 1972.

20. Skettrup, T.: "Urbach's Rule and Phase Fluctuations of the Transmitted Light," *Physica Status Solidi (b)*, vol. 103, pp. 613–621, 1981.

21. Woodruff, T. O.: "Empirically Derived Formula for the Energies of the First Ultraviolet Absorption Maximum of 20 Alkali-Halide Crystals," *Solid State Communications*, vol. 46, pp. 139–142, 1983.

22. Senatore, G., M. P. Tosi, and T. O. Woodruff: "A Simple Formula for the Fundamental Optical Absorption of Alkali Halide Melts," *Solid State Communications*, vol. 52, no. 2, pp. 173–176, 1984.

23. Viskanta, R., and E. E. Anderson: "Heat Transfer in Semi-Transparent Solids," in *Advances in Heat Transfer*, vol. 11, Academic Press, New York, pp. 317–441, 1975.

24. Leupacher, W., and A. Penzkofer: "Refractive-Index Measurement of Absorbing Condensed Media," *Applied Optics*, vol. 23, pp. 1554–1558, 1984.

25. Lu, Y., and A. Penzkofer: "Optical Constants Measurements of Strongly Absorbing Media," *Applied Optics*, vol. 25, no. 1, pp. 221–225, 1986.

26. Wooten, F.: *Optical Properties of Solids*, Academic Press, New York, 1972.

27. Myers, V. H., A. Ono, and D. P. DeWitt: "A Method for Measuring Optical Properties of Semi-transparent Materials at High Temperatures," *AIAA Journal*, vol. 24, no. 2, pp. 321–326, 1986.

28. Stierwalt, D. L.: "Infrared Spectral Emittance of Optical Materials," *Applied Optics*, vol. 5, no. 12, pp. 1911–1915, 1966.

29. Shvarev, K. M., B. A. Baum, and P. V. Gel'd: "Optical Properties of Liquid Silicon," *Sov. Phys. Solid State*, vol. 16, no. 11, pp. 2111–2112, May 1975.

30. Barker, A. J.: "A Compact, Windowless Reflectance Furnace for Infrared Studies of Corrosive Melts," *Journal of Physics E: Scientific Instruments*, vol. 6, pp. 241–244, 1973.

31. Ebert, J. L., and S. A. Self: "The Optical Properties of Molten Coal Slag," in *Heat Transfer Phenomena in Radiation, Combustion and Fires*, vol. HTD–106, ASME, pp. 123–126, 1989.

Problems

11.1 The absorption coefficient of a liquid, confined between two parallel and transparent windows, is to be measured by the transmission method. The detector signals from transmission measurements with varying liquid thickness are to be used.

(a) Using transmission measurements for two thicknesses, show how the absorption coefficient κ may be deduced. Determine how errors in the transmissivity value and the liquid layer thickness affect the accuracy of κ.

(b) If transmission measurements are made for many thicknesses, can you devise a method that measures small absorption coefficients more accurately?

11.2 Show how the optical properties (n, k, and κ) of a semitransparent solid may be deduced from the three measurements taken with the apparatus of Myers and coworkers [27], as depicted in Fig. 11-6.

CHAPTER
12

EXACT
SOLUTIONS FOR
ONE-DIMENSIONAL
GRAY MEDIA

12.1 INTRODUCTION

The governing equation for radiative transfer of absorbing, emitting and scattering media was developed in Chapter 8, resulting in an integro-differential equation for radiative intensity in five independent variables (three space coordinates and two direction coordinates). The problem becomes even more complicated if the medium is nongray (which introduces an additional variable, such as wavelength or frequency, and makes the equation nonlinear) and/or if other modes of heat transfer are present (which make it necessary to solve simultaneously for overall conservation of energy, to which intensity is related in a nonlinear way). Consequently, exact analytical solutions exist for only a few extremely simple situations. The simplest case arises when one considers thermal radiation in a one-dimensional plane-parallel gray medium that is either at radiative equilibrium (i.e., radiation is the only mode of heat transfer) or whose temperature field is known. Analytical solutions for such simple problems have been studied extensively, partly because of the great importance of one-dimensional plane-parallel media, partly because the simplicity of such solutions allows testing of more general solution methods, and partly because such a solution can give qualitative indications for more difficult situations.

In the present chapter we develop some analytical solutions for one-dimensional plane-parallel media and also include a few solutions for one-dimensional cylindrical and spherical media (without development). In general, we shall assume the medium to be gray, and all radiative intensity-related quantities are total, i.e., frequency-integrated quantities, for example, $I_b = \int_0^\infty I_{b\nu} d\nu = n^2 \sigma T^4 / \pi$. Most relations also hold, on a spectral basis, for nongray media, except for those that utilize the statement of radiative equilibrium, $\nabla \cdot \mathbf{q} = 0$ (since this relation does not hold on a spectral basis).

12.2   GENERAL FORMULATION FOR A PLANE-PARALLEL MEDIUM

The governing equation for the intensity field in an absorbing, emitting, and scattering medium is, from equation (8.26),

$$\hat{\mathbf{s}} \cdot \nabla I = \kappa I_b - \beta I + \frac{\sigma_s}{4\pi} \int_{4\pi} I(\hat{\mathbf{s}}_i) \, \Phi(\hat{\mathbf{s}}_i, \hat{\mathbf{s}}) \, d\Omega_i, \tag{12.1}$$

which describes the change of radiative intensity along a path in the direction of $\hat{\mathbf{s}}$. The formal solution to equation (12.1) is given by equation (8.30) as

$$I(\mathbf{r}, \hat{\mathbf{s}}) = I_w(\hat{\mathbf{s}}) e^{-\tau_s} + \int_0^{\tau_s} S(\tau_s', \hat{\mathbf{s}}) \, e^{-(\tau_s - \tau_s')} d\tau_s', \tag{12.2}$$

where S is the radiative source term, equation (8.27),

$$S(\tau_s', \hat{\mathbf{s}}) = (1 - \omega) I_b(\tau_s') + \frac{\omega}{4\pi} \int_{4\pi} I(\tau_s', \hat{\mathbf{s}}_i) \, \Phi(\hat{\mathbf{s}}, \hat{\mathbf{s}}_i) \, d\Omega_i, \tag{12.3}$$

and $\tau_s = \int_0^s \beta(s) ds$ is *optical thickness* or *optical depth* based on extinction coefficient[1] measured from a point on the wall ($\tau_s' = 0$) toward the point under consideration ($\tau_s' = \tau_s$), in the direction of $\hat{\mathbf{s}}$. For a plane-parallel medium the change of intensity is illustrated in Fig. 12-1a, measuring polar angle θ from the direction perpendicular to the plates (z-direction), and azimuthal angle ψ in a plane parallel to the plates (x-y-plane): Radiative intensity of strength $I_w(\hat{\mathbf{s}}) = I_w(\theta, \psi)$ leaves the point on the bottom surface into the direction of θ, ψ, toward the point under consideration, P. This intensity is augmented by the radiative source (by emission and by in-scattering, i.e., scattering of intensity from other directions into the direction of P). The amount of energy $S(\tau_s', \theta, \psi) d\tau_s'$ is released over the infinitesimal optical depth $d\tau_s'$ and travels toward P. Since this energy also undergoes absorption and out-scattering along its

---

[1] We use here the notation τ_s to describe optical depth along s so that we will be able to use the simpler τ for optical depth perpendicular to the plates, i.e., $\tau = \int_0^z \beta \, dz$.

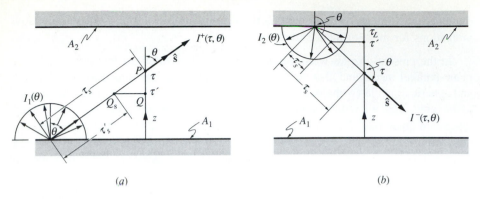

FIGURE 12-1
Coordinates for radiative intensities in a one-dimensional plane-parallel medium; (a) upward directions, (b) downward directions.

path from τ_s' to τ_s, only the fraction $e^{-(\tau_s - \tau_s')}$ actually arrives at P. In general, the intensity leaving the bottom wall may vary across the bottom surface, and radiative source and medium properties may vary throughout the medium, i.e., in the directions parallel to the plates as well as normal to them.

We shall now assume that both plates are isothermal and isotropic, i.e., neither temperature nor radiative properties vary across each plate and properties may show a directional dependence on polar angle θ, but not on azimuthal angle ψ. Thus, the intensity leaving the bottom plate at a certain location is the same for all azimuthal angles and, indeed, for all positions on that plate; it is a function of polar angle θ alone. We also assume that the temperature field and radiative properties of the medium vary only in the direction perpendicular to the plates. This assumption implies that the radiative source at position Q, $S(\tau', \theta)$, is identical to the one at position Q_s, $S(\tau_s', \theta)$, or any horizontal position with identical z-coordinate $\tau' = \int_0^z \beta\,dz$ (based on extinction coefficient). Therefore, radiative source, $S(\tau, \theta)$, and radiative intensity, $I(\tau, \theta)$, both depend only on a single space coordinate plus a single direction coordinate. The radiative source term may be simplified for the one-dimensional case to

$$S(\tau', \theta) = (1 - \omega)I_b(\tau')$$
$$+ \frac{\omega}{4\pi} \int_{\psi_i = 0}^{2\pi} \int_{\theta_i = 0}^{\pi} I(\tau', \theta_i)\,\Phi\,(\theta, \psi, \theta_i, \psi_i)\sin\theta_i\,d\theta_i\,d\psi_i. \qquad (12.4)$$

For *isotropic scattering*, $\Phi \equiv 1$, and we find immediately from the definition for incident radiation, G, [equation (8.34)], that

$$S(\tau') = (1 - \omega)I_b(\tau') + \frac{\omega}{4\pi}G(\tau'). \qquad (12.5)$$

In other words, the source term does not depend on direction, that is, the radiative source due to isotropic emission and isotropic in-scattering is also isotropic.

If the scattering is *anisotropic*, we may write, from equation (10.88)[2]

$$\Phi(\hat{\mathbf{s}} \cdot \hat{\mathbf{s}}_i) = 1 + \sum_{m=1}^{M} A_m P_m(\hat{\mathbf{s}} \cdot \hat{\mathbf{s}}_i), \tag{12.6}$$

where it is assumed that the series may be truncated after M terms. Measuring the polar angle from the z-axis and the azimuthal angle from the x-axis (in the x-y-plane) for both $\hat{\mathbf{s}}$ and $\hat{\mathbf{s}}_i$, we get

$$\hat{\mathbf{s}} = \sin\theta(\cos\psi\hat{\mathbf{i}} + \sin\psi\hat{\mathbf{j}}) + \cos\theta\hat{\mathbf{k}}, \tag{12.7}$$

$$\hat{\mathbf{s}}_i = \sin\theta_i(\cos\psi_i\hat{\mathbf{i}} + \sin\psi_i\hat{\mathbf{j}}) + \cos\theta_i\hat{\mathbf{k}}, \tag{12.8}$$

and

$$\Phi(\theta, \psi, \theta_i, \psi_i) = 1 + \sum_{m=1}^{M} A_m P_m[\cos\theta\cos\theta_i + \sin\theta\sin\theta_i\cos(\psi - \psi_i)]. \tag{12.9}$$

Using a relationship between Legendre polynomials [1], one may separate the directional dependence in the last relationship by

$$P_m[\cos\theta\cos\theta_i + \sin\theta\sin\theta_i\cos(\psi - \psi_i)] = P_m(\cos\theta)P_m(\cos\theta_i)$$

$$+ 2\sum_{n=1}^{m} \frac{(m-n)!}{(m+n)!} P_n^m(\cos\theta)P_n^m(\cos\theta_i)\cos m(\psi - \psi_i), \tag{12.10}$$

where the P_n^m are *associated Legendre polynomials*. Thus, the scattering phase function may be rewritten as

$$\Phi(\theta, \psi, \theta_i, \psi_i) = 1 + \sum_{m=1}^{M} P_m(\cos\theta)P_m(\cos\theta_i)$$

$$+ 2\sum_{m=1}^{M}\sum_{n=1}^{m} A_m \frac{(m-n)!}{(m+n)!} P_n^m(\cos\theta)P_n^m(\cos\theta_i)\cos m(\psi - \psi_i). \tag{12.11}$$

For a one-dimensional plane-parallel geometry, the intensity does not depend on azimuthal angle, and we may carry out the ψ_i-integration in equation (12.4). This integration leads to a one-dimensional scattering phase function of

$$\Phi(\theta, \theta_i) = \frac{1}{2\pi}\int_0^{2\pi} \Phi(\hat{\mathbf{s}} \cdot \hat{\mathbf{s}}_i)\,d\psi_i = 1 + \sum_{m=1}^{M} A_m P_m(\cos\theta)P_m(\cos\theta_i), \tag{12.12}$$

since $\int_0^{2\pi} \cos m(\psi - \psi_i)\,d\psi_i = 0$. The radiative source then becomes

$$S(\tau', \theta) = (1 - \omega)I_b(\tau') + \frac{\omega}{2}\int_0^{\pi} I(\tau', \theta_i)\Phi(\theta, \theta_i)\sin\theta_i\,d\theta_i. \tag{12.13}$$

---

[2] In Chapter 10 we used Θ to denote the angle between incoming and scattered ray and, therefore, $\cos\Theta = \hat{\mathbf{s}} \cdot \hat{\mathbf{s}}_i$.

For *linear-anisotropic scattering*, with

$$\Phi(\hat{s} \cdot \hat{s}_i) = 1 + A_1 P_1(\hat{s} \cdot \hat{s}_i) = 1 + A_1 \hat{s} \cdot \hat{s}_i, \quad M = 1, \qquad (12.14)$$

and, using the definitions for incident radiation and radiative heat flux, equations (8.34) and (8.47), respectively, equation (12.13) reduces to

$$S(\tau', \theta) = (1 - \omega)I_b(\tau') + \frac{\omega}{4\pi}\left[G(\tau') + A_1 q(\tau') \cos \theta\right]. \qquad (12.15)$$

We may now simplify the equation of radiative transfer, equation (12.1), using the geometric relations $\tau_s = \tau / \cos \theta$ and $\tau_s' = \tau' / \cos \theta$ (see Fig. 12-1a),

$$\frac{1}{\beta} \frac{dI}{ds} = \frac{dI}{d\tau_s}$$

$$= \cos \theta \frac{dI}{d\tau} = (1 - \omega)I_b - I + \frac{\omega}{2} \int_0^\pi I(\tau, \theta_i) \Phi(\theta, \theta_i) \sin \theta_i \, d\theta_i. \quad (12.16)$$

Similarly, the expression for intensity, equation (12.2), may be simplified to

$$I^+(\tau, \theta) = I_1(\theta) e^{-\tau/\cos \theta} + \int_0^\tau S(\tau', \theta) e^{-(\tau-\tau')/\cos \theta} \frac{d\tau'}{\cos \theta}, \quad 0 < \theta < \frac{\pi}{2}, \quad (12.17)$$

where the intensity is denoted by I^+ since equation (12.17) is limited to directions with wall intensities emanating from the lower wall, at $\tau = 0$ ("positive" directions). Here the radiative source $S(\tau', \theta)$ is given by equation (12.5) for *isotropic scattering* (or no scattering with $\omega = 0$), by equation (12.15) for *linear-anisotropic scattering*, and by equations (12.12) and (12.13) for *general anisotropic scattering*.

A similar relationship is readily developed for intensity emanating from the top wall (traveling into "negative" directions). With $\tau_s' = -(\tau_L - \tau')/\cos \theta$ and $\tau_s = -(\tau_L - \tau)/\cos \theta$ (keeping in mind that $\cos \theta < 0$ for "negative" directions, $\theta > \pi/2$) we obtain (see Fig. 12-1b)

$$I^-(\tau, \theta) = I_2(\theta) e^{(\tau_L - \tau)/\cos \theta} + \int_{\tau_L}^\tau S(\tau', \theta) e^{(\tau'-\tau)/\cos \theta} \frac{d\tau'}{\cos \theta}$$

$$= I_2(\theta) e^{(\tau_L - \tau)/\cos \theta} - \int_\tau^{\tau_L} S(\tau', \theta) e^{(\tau'-\tau)/\cos \theta} \frac{d\tau'}{\cos \theta}, \quad \frac{\pi}{2} < \theta < \pi, \quad (12.18)$$

where $I_2(\theta)$ is the intensity leaving the wall at $\tau = \tau_L$ (Wall 2). It is customary (and somewhat more compact) to rewrite equations (12.16) through (12.18) in terms of the direction cosine $\mu = \cos \theta$, or

$$\mu \frac{dI}{d\tau} + I = (1 - \omega)I_b + \frac{\omega}{2} \int_{-1}^1 I(\tau, \mu_i) \Phi(\mu, \mu_i) \, d\mu_i = S(\tau, \mu), \qquad (12.19)$$

$$I^+(\tau, \mu) = I_1(\mu)e^{-\tau/\mu} + \int_0^\tau S(\tau', \mu) e^{-(\tau-\tau')/\mu} \frac{d\tau'}{\mu}, \qquad 0 < \mu < 1, \quad (12.20a)$$

$$I^-(\tau, \mu) = I_2(\mu)e^{(\tau_L - \tau)/\mu} - \int_\tau^{\tau_L} S(\tau', \mu) \, e^{(\tau' - \tau)/\mu} \frac{d\tau'}{\mu}, \quad -1 < \mu < 0. \quad (12.20b)$$

For heat transfer purposes the incident radiation, G, and radiative heat flux, q, are of interest. From the definition of incident radiation, equation (8.34), it follows that

$$
\begin{aligned}
G(\tau) &= \int_0^{2\pi} \int_0^\pi I(\tau, \theta) \sin\theta \, d\theta \, d\psi = 2\pi \int_{-1}^{+1} I(\tau, \mu) \, d\mu \\
&= 2\pi \left[\int_{-1}^0 I^-(\tau, \mu) \, d\mu + \int_0^{+1} I^+(\tau, \mu) \, d\mu \right] \\
&= 2\pi \left[\int_0^1 I^-(\tau, -\mu) \, d\mu + \int_0^1 I^+(\tau, \mu) \, d\mu \right] \\
&= 2\pi \left\{ \int_0^1 I_1(\mu)e^{-\tau/\mu} \, d\mu + \int_0^1 I_2(-\mu)e^{-(\tau_L - \tau)/\mu} \, d\mu \right. \\
&\quad \left. + \int_0^1 \left[\int_0^\tau S(\tau', \mu)e^{-(\tau - \tau')/\mu} \, d\tau' + \int_\tau^{\tau_L} S(\tau', -\mu)e^{-(\tau' - \tau)/\mu} \, d\tau' \right] \frac{d\mu}{\mu} \right\}. \quad (12.21)
\end{aligned}
$$

Similarly, for the radiative heat flux for a plane-parallel medium, equation (8.47),

$$
\begin{aligned}
q(\tau) &= \int_0^{2\pi} \int_0^\pi I(\tau, \theta) \cos\theta \sin\theta \, d\theta \, d\psi = 2\pi \int_{-1}^{+1} I(\tau, \mu)\mu \, d\mu \\
&= 2\pi \left\{ \int_0^1 I_1(\mu)e^{-\tau/\mu} \mu \, d\mu - \int_0^1 I_2(-\mu)e^{-(\tau_L - \tau)/\mu} \mu \, d\mu \right. \\
&\quad \left. + \int_0^1 \left[\int_0^\tau S(\tau', \mu)e^{-(\tau - \tau')/\mu} \, d\tau' - \int_\tau^{\tau_L} S(\tau', -\mu)e^{-(\tau' - \tau)/\mu} \, d\tau' \right] d\mu \right\}. \quad (12.22)
\end{aligned}
$$

During a large part of this chapter we shall study the solution to equations (12.21) and (12.22) for a number of different situations. We shall assume either that the temperature across the medium and, therefore, $I_b(\tau)$ is known or that radiative equilibrium prevails, $dq/d\tau = 0$. In either case we are interested in the direction-integrated form of the equation of transfer, equation (12.1), which has been given by equation (8.54) as

$$\nabla \cdot \mathbf{q} = \kappa(4\pi I_b - G), \quad (12.23)$$

or, for the present one-dimensional case after division by extinction coefficient β (and remembering that $\kappa/\beta = 1 - \sigma_s/\beta$),

$$\frac{dq}{d\tau} = (1 - \omega)(4\pi I_b - G). \quad (12.24)$$

We note in passing that, up to this point, all relations, and in particular equations (12.21), (12.22), and (12.24), hold on a total basis for a gray medium and on a spectral basis for any medium. If radiative equilibrium prevails, then $dq/d\tau = 0$

or, in the presence of a heat source,[3]

$$\frac{dq}{d\tau} = \frac{\dot{Q}'''}{\beta}(\tau),\tag{12.25}$$

where $\dot{Q}'''$ is local heat generation per unit time and volume. Equation (12.25) is valid only for total radiative heat flux and may, therefore, in this form be applied only to gray media. For such a case we see that the incident radiation is closely related to the blackbody intensity (and, therefore, temperature) by

$$4\pi I_b(\tau) = G(\tau) + \frac{\dot{Q}'''}{\kappa}(\tau).\tag{12.26}$$

12.3 RADIATIVE EQUILIBRIUM OF A NONSCATTERING MEDIUM

Enclosure with Black Bounding Surfaces

Since this is the most basic of cases, we shall rederive the relationships for this simple problem. From equation (12.3), with $\omega = 0$, it follows that $S(\tau', \hat{s}) = I_b(\tau')$; for black bounding surfaces, the intensity leaving the lower plate is $I_1(\theta) = I_{b1}$ and the intensity leaving the top plate is $I_2(\theta) = I_{b2}$. Thus, for this simple case, neither radiative source nor boundary intensities are direction-dependent. Equations (12.17) and (12.18) may then be rewritten as

$$I^+(\tau', \theta) = I_{b1}e^{-\tau/\cos\theta} + \int_0^\tau I_b(\tau')e^{-(\tau-\tau')/\cos\theta}\frac{d\tau'}{\cos\theta}, \quad 0 < \theta < \frac{\pi}{2},\tag{12.27a}$$

$$I^-(\tau', \theta) = I_{b2}e^{(\tau_L-\tau)/\cos\theta} - \int_\tau^{\tau_L} I_b(\tau')e^{(\tau'-\tau)/\cos\theta}\frac{d\tau'}{\cos\theta}, \quad \frac{\pi}{2} < \theta < \pi.\tag{12.27b}$$

Making the substitution $\mu = \cos\theta$ transforms this to

$$I^+(\tau', \mu) = I_{b1}e^{-\tau/\mu} + \frac{1}{\mu}\int_0^\tau I_b(\tau')e^{-(\tau-\tau')/\mu}\,d\tau', \quad 0 < \mu < 1,\tag{12.28a}$$

$$I^-(\tau', \mu) = I_{b2}e^{(\tau_L-\tau)/\mu} - \frac{1}{\mu}\int_\tau^{\tau_L} I_b(\tau')e^{(\tau'-\tau)/\mu}\,d\tau', \quad -1 < \mu < 0.\tag{12.28b}$$

From the definition for incident radiation it follows, from equation (12.21), that

$$\begin{aligned}
G(\tau) &= 2\pi\left[\int_{-1}^0 I^-(\tau, \mu)\,d\mu + \int_0^1 I^+(\tau, \mu)\,d\mu\right]\\
&= 2\pi\left[I_{b1}\int_0^1 e^{-\tau/\mu}\,d\mu + I_{b2}\int_0^1 e^{-(\tau_L-\tau)/\mu}\,d\mu\right.\\
&\quad \left. +\int_0^\tau I_b(\tau')\int_0^1 e^{-(\tau-\tau')/\mu}\frac{d\mu}{\mu}\,d\tau' + \int_\tau^{\tau_L} I_b(\tau')\int_0^1 e^{-(\tau'-\tau)/\mu}\frac{d\mu}{\mu}\,d\tau'\right].
\end{aligned}\tag{12.29}$$

---

[3]Such heat sources are often used to couple the radiation problem with overall energy conservation.

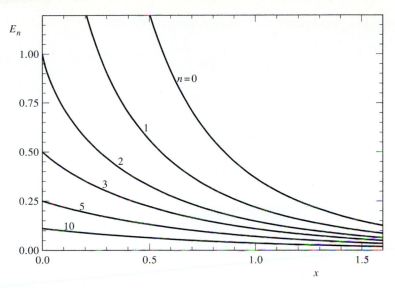

FIGURE 12-2
General behavior of exponential integrals $E_n(x)$.

Taking advantage of the fact that wall intensities and radiative sources do not depend on direction, we have taken these terms out of the direction integrals and reversed the order of integration for the terms describing medium emission.

A similar relationship may be established for radiative heat flux, from equation (12.22), as

$$q(\tau) = 2\pi \left[I_{b1} \int_0^1 e^{-\tau/\mu} \mu \, d\mu - I_{b2} \int_0^1 e^{-(\tau_L - \tau)/\mu} \mu \, d\mu \right.$$
$$\left. + \int_0^\tau I_b(\tau') \int_0^1 e^{-(\tau - \tau')/\mu} d\mu \, d\tau' - \int_\tau^{\tau_L} I_b(\tau') \int_0^1 e^{-(\tau' - \tau)/\mu} d\mu \, d\tau' \right]. \quad (12.30)$$

We see that none of the important parameters G, q, and I_b depends on direction, and that direction μ enters equations (12.29) and (12.30) only as a dummy integration variable. We may write these equations in more compact form by introducing the *exponential integral of order n*,

$$E_n(x) = \int_1^\infty e^{-xt} \frac{dt}{t^n} = \int_0^1 \mu^{n-2} e^{-x/\mu} d\mu. \quad (12.31)$$

Since exponential integrals are of great importance in radiative transfer, a sketch of them is shown in Fig. 12-2, and a somewhat more detailed discussion is given in Appendix E. For our present purposes, we note that exponential integrals behave somewhat like "generalized negative exponentials" and that

$$E_n(0) = \int_1^\infty \frac{dt}{t^n} = \frac{1}{n-1},$$ (12.32)

$$\frac{d}{dx}E_n(x) = -E_{n-1}(x); \quad \text{or} \quad E_n(x) = \int_x^\infty E_{n-1}(x)\,dx.$$ (12.33)

Substituting equation (12.31) into equations (12.29) and (12.30) then leads to

$$G(\tau) = 2\pi\left[I_{b1}E_2(\tau) + I_{b2}E_2(\tau_L - \tau)\right.$$
$$\left. + \int_0^\tau I_b(\tau')E_1(\tau - \tau')\,d\tau' + \int_\tau^{\tau_L} I_b(\tau')E_1(\tau' - \tau)\,d\tau'\right],$$ (12.34)

$$q(\tau) = 2\pi\left[I_{b1}E_3(\tau) - I_{b2}E_3(\tau_L - \tau)\right.$$
$$\left. + \int_0^\tau I_b(\tau')E_2(\tau - \tau')\,d\tau' - \int_\tau^{\tau_L} I_b(\tau')E_2(\tau' - \tau)\,d\tau'\right].$$ (12.35)

We shall now assume that the medium is gray [i.e., equations (12.34) and (12.35) deal with total properties, or $I_b = n^2\sigma T^4/\pi$] and that radiative equilibrium prevails, $dq/d\tau = 0$. Thus, we find $q(\tau) = \text{const}$ and, from equation (12.26), $G = 4\pi I_b = 4n^2\sigma T^4$. Equation (12.34) now becomes an integral equation governing the temperature distribution within the medium, or

$$T^4(\tau) = \frac{1}{2}\left[T_1^4 E_2(\tau) + T_2^4 E_2(\tau_L - \tau) + \int_0^{\tau_L} T^4(\tau')E_1(|\tau' - \tau|)\,d\tau'\right].$$ (12.36)

Since the heat flux,

$$q(\tau) = 2n^2\sigma T_1^4 E_3(\tau) - 2n^2\sigma T_2^4 E_3(\tau_L - \tau)$$
$$+2\int_0^\tau n^2\sigma T^4(\tau')E_2(\tau - \tau')\,d\tau' - 2\int_\tau^{\tau_L} n^2\sigma T^4(\tau')E_2(\tau' - \tau)\,d\tau',$$ (12.37)

does not vary across the medium, it may be evaluated at any location, conveniently chosen as $\tau = 0$:

$$q = n^2\sigma T_1^4 - 2n^2\sigma T_2^4 E_3(\tau_L) - 2\int_0^{\tau_L} n^2\sigma T^4(\tau')E_2(\tau')\,d\tau'.$$ (12.38)

The difference between equations (12.38) and (12.37) is that equation (12.38) is only valid for radiative equilibrium, and equation (12.37) is valid for the more general case of any gray medium between black plates. For an overall solution the temperature field is found first by solving the integral equation (12.36), after which knowledge of the temperature field is used to determine radiative heat flux from equation (12.37). Unfortunately, no closed-form solution exists to integral equations such as (12.36); a solution has to be found by numerical and/or approximate means. Before proceeding to a solution it is advantageous to reduce the number of parameters in equations (12.36) and (12.37) to a minimum. We introduce a nondimensional emissive power or temperature

$$\Phi_b(\tau) = \frac{T^4(\tau) - T_2^4}{T_1^4 - T_2^4}, \tag{12.39}$$

and a nondimensional radiative heat flux

$$\Psi_b = \frac{q}{n^2\sigma(T_1^4 - T_2^4)}. \tag{12.40}$$

If we substitute these expressions into equations (12.36) and (12.37) and use equations (12.32) and (12.33), we find that

$$\Phi_b(\tau) = \frac{1}{2}\left[E_2(\tau) + \int_0^{\tau_L} \Phi_b(\tau')E_1(|\tau - \tau'|)\,d\tau'\right], \tag{12.41}$$

$$\Psi_b(\tau) = 2\left[E_3(\tau) + \int_0^{\tau} \Phi_b(\tau')E_2(\tau - \tau')\,d\tau' - \int_\tau^{\tau_L} \Phi_b(\tau')E_2(\tau' - \tau)\,d\tau'\right], \tag{12.42}$$

or

$$\Psi_b = 1 - 2\int_0^{\tau_L} \Phi_b(\tau')E_2(\tau')\,d\tau'. \tag{12.43}$$

Besides the independent variable τ, only one parameter, the medium's optical thickness τ_L, appears in the governing equations for Φ_b and Ψ_b: Once Φ_b has been determined for a given τ_L, the temperature field and radiative heat flux may be determined for any combination of surface temperatures. Equation (12.41) is a *Fredholm integral equation* and is readily solved by any of the methods described in Section 5.5. The numerical solution to equations (12.41) and (12.43) was first given by Heaslet and Warming [2]. Figure 12-3 shows the nondimensional temperature field for a range of optical thicknesses. Some representative nondimensional fluxes are given in Table 12.1.

Examination of Fig. 12-3 shows that, for radiative equilibrium, there may be a temperature discontinuity at the walls.[4] In the limiting case of a transparent medium, $\tau_L \to 0$, we have $\Phi_b = \frac{1}{2}$ or $T^4 \to (T_1^4 + T_2^4)/2 =$ const, with corresponding temperature jumps at the boundaries (strictly speaking, a transparent or nonparticipating medium, $\tau_L = 0$, could have any temperature distribution since it would not enter the calculations). The temperature slip decreases as the optical thickness increases until

---

[4]This discontinuity must, of course, vanish if heat is transferred by conduction and/or convection in addition to radiation.

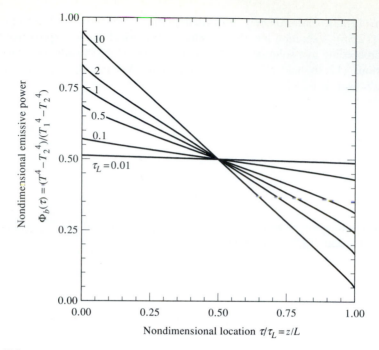

FIGURE 12-3
Nondimensional temperature distribution for a gray medium at radiative equilibrium between isothermal plates.

TABLE 12.1
Nondimensional radiative heat flux for radiative equilibrium between parallel black plates, $\Psi_b = q/n^2\sigma(T_1^4 - T_2^4)$, from Heaslet and Warming [2]

Optical thickness, τ_L	Ψ_b	Optical thickness, τ_L	Ψ_b
0.0	1.0000	0.8	0.6046
0.1	0.9157	1.0	0.5532
0.2	0.8491	1.5	0.4572
0.3	0.7934	2.0	0.3900
0.4	0.7458	2.5	0.3401
0.5	0.7040	3.0	0.3016
0.6	0.6672	5.0	0.2077

$$\text{For } \tau_L \gg 1, \quad \Psi_b = \frac{4/3}{1.42089 + \tau_L}$$

it vanishes for $\tau_L \to \infty$. In that optically thick limit the nondimensional emissive power profile becomes linear. The situation is not unlike conduction in a rarefied gas: When the mean free path for collision (absorption) is very large, molecules (photons) travel between plates without interference with an average energy equal to the average of surface temperatures (emissive powers). If the mean free path becomes very small compared with physical dimensions, the conductive flux obeys Fourier's law and the diffusion limit is reached.

Gray, Diffuse Boundaries

If the walls are not black, but are gray, diffuse emitters and reflectors, the entire development of this section still holds, except that the fluxes leaving the bottom and top plates are no longer πI_{b1} and πI_{b2}, but must be replaced by the radiosities J_1 and J_2, respectively. The radiosities, accounting for emission as well as diffuse reflection, may be related to the Planck function through equation (5.26) as

$$\mathbf{q}_w \cdot \hat{\mathbf{n}} = \frac{\epsilon_w}{1 - \epsilon_w}(\pi I_{bw} - J_w), \tag{12.44}$$

or

$$\tau = 0: \qquad q = \frac{\epsilon_1}{1 - \epsilon_1}(n^2\sigma T_1^4 - J_1), \tag{12.45a}$$

$$\tau = \tau_L: \qquad -q = \frac{\epsilon_2}{1 - \epsilon_2}(n^2\sigma T_2^4 - J_2). \tag{12.45b}$$

Replacing $\pi I_{bi} = n^2\sigma T_i^4$ by J_i in equations (12.39) and (12.40) transforms equation (12.43) into

$$\frac{q}{J_1 - J_2} = \Psi_b = 1 - 2\int_0^{\tau_L} \Phi_b(\tau')E_2(\tau')\,d\tau',$$

and from equation (12.45)

$$J_1 - J_2 = n^2\sigma(T_1^4 - T_2^4) - \left(\frac{1}{\epsilon_1} + \frac{1}{\epsilon_2} - 2\right)q.$$

Thus,

$$q = \Psi_b(J_1 - J_2) = \Psi_b\left[n^2\sigma(T_1^4 - T_2^4) - \left(\frac{1}{\epsilon_1} + \frac{1}{\epsilon_2} - 2\right)q\right],$$

or

$$\Psi = \frac{q}{n^2\sigma(T_1^4 - T_2^4)} = \frac{\Psi_b}{1 + \Psi_b\left(\dfrac{1}{\epsilon_1} + \dfrac{1}{\epsilon_2} - 2\right)}. \tag{12.46}$$

Similarly, for the nondimensional temperature distribution one obtains

$$\Phi(\tau) = \frac{T^4(\tau) - T_2^4}{T_1^4 - T_2^4} = \frac{\Phi_b(\tau) + \left(\frac{1}{\epsilon_2} - 1\right)\Psi_b}{1 + \Psi_b\left(\frac{1}{\epsilon_1} + \frac{1}{\epsilon_2} - 2\right)}. \tag{12.47}$$

Example 12.1.   A gray, nonscattering medium with refractive index $n = 1$ and an absorption coefficient $\kappa = 0.1\,\mathrm{cm}^{-1}$ is contained between two isothermal cylinders. The inner cylinder is hot ($T_1 = 2000\,\mathrm{K}$) and highly reflective ($\epsilon_1 = 0.1$); the outer cylinder is a strong absorber ($\alpha_2 = \epsilon_2 = 0.9$), and it must be kept relatively cool ($T_2 \leq 400\,\mathrm{K}$). The gap between the two cylinders is 25 cm. Assuming that conductive and convective heat transfer can be neglected as compared to radiation, and assuming that the cylinders have large diameters ($D_1 \gg 25\,\mathrm{cm}$), determine the necessary cooling rate for the outer cylinder to avoid overheating.

Solution

Since the thickness of the medium is small as compared with the diameters of the cylinders, we may model the gap as a one-dimensional plane-parallel slab of optical thickness $\tau_L = 0.1\,\mathrm{cm}^{-1} \times 25\,\mathrm{cm} = 2.5$. Thus, from Table 12.1 $\Psi_b = 0.3401$ and from equation (12.46)

$$\Psi = \frac{q}{\sigma(T_1^4 - T_2^4)} = \frac{0.3401}{1 + 0.3401\left(\frac{1}{0.1} + \frac{1}{0.9} - 2\right)} = 0.0830,$$

and

$$q_{min} = \Psi\,\sigma\left(T_1^4 - T_{2,max}^4\right)$$
$$= 0.0830 \times 5.670 \times 10^{-12}(2000^4 - 400^4)\,\mathrm{W/cm}^2 = 7.52\,\mathrm{W/cm}^2.$$

12.4   RADIATIVE EQUILIBRIUM OF A SCATTERING MEDIUM

Isotropic Scattering

For isotropic scattering the source function is found from equations (12.5) and (12.26) (assuming no internal heat generation takes place) as

$$S(\tau) = (1 - \omega)I_b(\tau) + \frac{\omega}{4\pi}G(\tau) = I_b(\tau). \tag{12.48}$$

Thus, all relations developed in the previous section are equally valid for isotropically scattering media with optical thickness $\tau = \int_0^z \beta\,dz$ based on extinction coefficient rather than absorption coefficient. For a gray medium at radiative equilibrium there is no distinction between absorption and isotropic scattering: Any energy absorbed at τ must be reemitted isotropically at the same location, although at different wavelengths;

any isotropically scattered energy is simply redirected isotropically (without change of wavelength). Since a gray medium is "colorblind" it cannot distinguish between emission and isotropic scattering. However, for the purely scattering case, $\omega \rightarrow 1$, there is no emission and, therefore, I_b no longer enters the calculations. For this extreme case the $T^4(\tau)$ in equations (12.39) and (12.47) should be replaced by $G(\tau)/4n^2\sigma$.

Anisotropic Scattering

For *linear-anisotropic scattering* the source function is given by equation (12.15), which, for radiative equilibrium in a one-dimensional plane-parallel slab, reduces to

$$S(\tau, \mu) = I_b(\tau) + \frac{A_1\omega}{4\pi} q\mu. \qquad (12.49)$$

Therefore, equations (12.21) and (12.22) become

$$\frac{G(\tau)}{4\pi} = I_b(\tau) = \frac{J_1}{2\pi} E_2(\tau) + \frac{J_2}{2\pi} E_2(\tau_L - \tau)$$

$$+ \frac{1}{2} \int_0^\tau I_b(\tau') E_1(\tau - \tau')\,d\tau' + \frac{1}{2} \int_\tau^{\tau_L} I_b(\tau') E_1(\tau' - \tau)\,d\tau'$$

$$+ \frac{A_1\omega}{8\pi} q\,[E_3(\tau_L - \tau) - E_3(\tau)], \qquad (12.50)$$

$$q(\tau) = q = 2J_1 E_3(\tau) - 2J_2 E_3(\tau_L - \tau)$$

$$+ 2\pi \left[\int_0^\tau I_b(\tau') E_2(\tau - \tau')\,d\tau' - \int_\tau^{\tau_L} I_b(\tau') E_2(\tau' - \tau)\,d\tau' \right]$$

$$+ \frac{A_1\omega}{2} q \left[\frac{2}{3} - E_4(\tau) - E_4(\tau_L - \tau) \right]. \qquad (12.51)$$

In nondimensional form these relations reduce to

$$\Phi_b(\tau) = \frac{\pi I_b(\tau) - J_2}{J_1 - J_2} = \frac{1}{2} \left\{ E_2(\tau) + \int_0^{\tau_L} \Phi_b(\tau') E_1(|\tau - \tau'|)\,d\tau' \right.$$

$$\left. + \frac{A_1\omega}{4} \Psi_b\,[E_3(\tau_L - \tau) - E_3(\tau)] \right\}, \qquad (12.52)$$

$$\Psi_b = \frac{q}{J_1 - J_2} = 2 \left\{ E_3(\tau) + \int_0^\tau \Phi_b(\tau') E_2(\tau - \tau')d\tau' \right.$$

$$\left. - \int_\tau^{\tau_L} \Phi_b(\tau') E_2(\tau' - \tau)\,d\tau' + \frac{A_1\omega}{4} \Psi_b \left[\frac{2}{3} - E_4(\tau) - E_4(\tau_L - \tau) \right] \right\}. \qquad (12.53)$$

The problem of radiative equilibrium in a one-dimensional, plane-parallel, anisotropically scattering medium has been solved by Modest and Azad [3]. They considered full Mie-anisotropic scattering for a number of particulate clouds, whose relevant

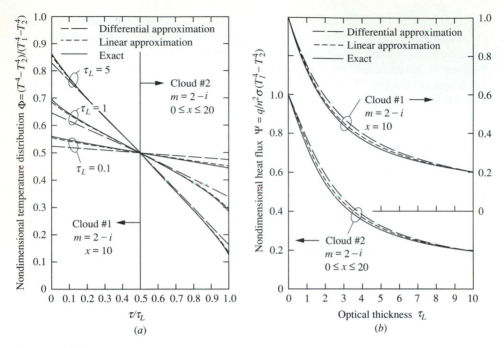

FIGURE 12-4
(*a*) Nondimensional temperature profiles, and (*b*) Nondimensional heat flux rates; for a slab at radiative equilibrium.

parameters have been given in Chapter 10, Table 10.1. Figure 12-4 shows representative results for radiative heat fluxes and temperature distributions in Clouds 1 and 2. Also included are results for linear-anisotropic scattering (approximating the phase functions as indicated in Fig. 10-7), using the exact relations, equations (12.52) and (12.53), as well as the differential approximation (to be discussed in the following two chapters). It was observed that approximating a complicated phase function by a linear-anisotropic one (after removing forward- and backward-scattering peaks) always lead to accurate results for heat transfer applications.

12.5 PLANE MEDIUM WITH SPECIFIED TEMPERATURE FIELD

If the medium is not at radiative equilibrium (i.e., radiative heat transfer is not so dominant that conduction and/or convection can be neglected), the problem of finding the temperature distribution and heat fluxes is always nonlinear. For the simplest case of a gray medium with constant properties, the incident radiation, as calculated from equation (12.21), is proportional to temperature to the fourth power, while the conductive and/or convective terms are proportional to temperature itself. Therefore, the temperature field must always be determined through an iterative procedure. In general, this involves guessing a temperature field, which is then used to determine incident radiation G [from equation (12.21)] and divergence of radiative heat flux $\nabla \cdot \mathbf{q}$ [from equation (12.23)]. This radiative source term is then substituted into the equation for overall conservation of energy, equation (8.61), from which an improved temperature field is determined. This process is then repeated until the temperature field has converged to within specified criteria. The treatment of combined radiation together with conduction and/or convection is discussed in more detail in Chapter 20. There are also some important industrial applications where outright knowledge of the temperature field may be assumed, for example, swirling combustion chambers that are essentially isothermal as a result of very strong convection.

We start with equations (12.21) and (12.22), which are fairly general expressions for incident radiation G and radiative heat flux q [the only restrictions are of being limited to (i) one-dimensional plane-parallel media, and (ii) gray media or spectral calculations]. Limiting ourselves to diffuse surfaces ($I_w = J_w/\pi$) and isotropic scattering (i.e., a direction-independent source function S), we may rewrite these two equations using the definition of the exponential integrals, as

$$G(\tau) = 2J_1 E_2(\tau) + 2J_2 E_2(\tau_L - \tau)$$
$$+ 2\pi \int_0^\tau S(\tau') E_1(\tau - \tau') \, d\tau' + 2\pi \int_\tau^{\tau_L} S(\tau') E_1(\tau' - \tau) \, d\tau', \qquad (12.54)$$

$$q(\tau) = 2J_1 E_3(\tau) - 2J_2 E_3(\tau_L - \tau)$$
$$+ 2\pi \int_0^\tau S(\tau') E_2(\tau - \tau') \, d\tau' - 2\pi \int_\tau^{\tau_L} S(\tau') E_2(\tau' - \tau) \, d\tau'. \qquad (12.55)$$

Again limiting ourselves to isotropic scattering, we find the radiative source is, from equation (12.15),

$$S(\tau) = (1 - \omega) I_b(\tau) + \frac{\omega}{4\pi} G(\tau). \qquad (12.56)$$

Since radiative equilibrium *may not be assumed* if the temperature field is given, the radiative heat flux is *not constant* across the plane layer. In fact, if the temperature field is given, usually the divergence of the radiative heat flux is desired, equation (12.24),

$$\frac{dq}{d\tau} = (1 - \omega)(4\pi I_b - G). \qquad (12.57)$$

In the absence of scattering we have $S(\tau) = I_b(\tau)$, and the integrals in equations (12.54) and (12.55) are readily evaluated; heat flux q, incident radiation G and, therefore, divergence of radiative heat flux $dq/d\tau$ may be determined explicitly. On the other hand, if the medium scatters isotropically, equation (12.54) becomes an integral equation for the unknown $G(\tau)$. Defining a general nondimensional function similar to equation (12.39),

$$\Phi(\tau) = \frac{\pi S(\tau) - J_2}{J_1 - J_2}, \qquad (12.58)$$

equation (12.54) may be simplified to

$$\Phi(\tau) = (1 - \omega)\frac{\pi I_b(\tau) - J_2}{J_1 - J_2} + \frac{\omega}{2}\left[E_2(\tau) + \int_0^{\tau_L} \Phi(\tau)E_1(|\tau' - \tau|)\,d\tau'\right]. \qquad (12.59)$$

Equation (12.59) reduces to equation (12.41) for the case of $\omega \to 1$ (purely scattering medium). For such a medium thermal radiation is decoupled from the temperature field (since there is no emission), and radiative equilibrium prevails regardless of the temperature distribution. This behavior is also seen from equation (12.24), which states that $dq/d\tau = 0$ if $\omega = 1$, regardless of $I_b(\tau)$.

Similarly, equation (12.55) may be nondimensionalized as

$$\begin{aligned}
\Psi(\tau) &= \frac{q(\tau)}{J_1 - J_2} \\
&= 2\left[E_3(\tau) + \int_0^\tau \Phi(\tau')E_2(\tau - \tau')\,d\tau' - \int_\tau^{\tau_L} \Phi(\tau')E_2(\tau' - \tau)\,d\tau'\right]. \qquad (12.60)
\end{aligned}$$

Example 12.2.   A gray, nonscattering medium with refractive index $n = 1$ is contained between two parallel, gray plates. The medium is isothermal at temperature T_m, with constant absorption coefficient κ. The two plates are both isothermal at temperature T_w, have the same gray-diffuse emissivity ϵ, and are spaced a distance L apart. Determine the radiative heat flux between the plates as well as its divergence.

Solution

For a nonscattering medium ($\sigma_s = 0$ or $\omega = 0$), equation (12.59) becomes superfluous and the flux may be determined directly from equations (12.55) or (12.60). From equation (12.56) $\pi S(\tau) = \pi I_{bm} = \sigma T_m^4 = \text{const.}$ Therefore, from equation (12.55) with $\tau = \kappa z$,

$$\begin{aligned}
q(\tau) &= 2J_w E_3(\tau) - 2J_w E_3(\tau_L - \tau) \\
&\quad + 2\sigma T_m^4 \int_0^\tau E_2(\tau - \tau')\,d\tau' - 2\sigma T_m^4 \int_\tau^{\tau_L} E_2(\tau' - \tau)\,d\tau' \\
&= 2J_w E_3(\tau) - 2J_w E_3(\tau_L - \tau) \\
&\quad + 2\sigma T_m^4\left[E_3(\tau - \tau')\big|_0^\tau + E_3(\tau' - \tau)\big|_\tau^{\tau_L}\right] \\
&= (J_w - \sigma T_m^4)\,2\,[E_3(\tau) - E_3(\tau_L - \tau)],
\end{aligned}$$

where we have made use of the symmetry of the problem, i.e., $J_1 = J_2 = J_w$. The necessary relationship between surface flux, temperature, and radiosity has been given by equation (12.44), or for the plane slab at $\tau = 0$,

$$q(0) = \frac{\epsilon}{1 - \epsilon}(\sigma T_w^4 - J_w) = (J_w - \sigma T_m^4)\,[1 - 2E_3(\tau_L)].$$

Solving for J_w, we find

$$J_w = \frac{\sigma T_w^4 + (1/\epsilon - 1)\,[1 - 2E_3(\tau_L)]\,\sigma T_m^4}{1 + (1/\epsilon - 1)[1 - 2E_3(\tau_L)]},$$

and

$$\frac{q(\tau)}{\sigma(T_w^4 - T_m^4)} = \frac{2\left[E_3(\tau) - E_3(\tau_L - \tau)\right]}{1 + (1/\epsilon - 1)[1 - 2E_3(\tau_L)]}.$$

The divergence of the flux may be evaluated by first calculating the incident radiation from equation (12.54) and then using equation (12.57). While this method is preferable for numerical and/or multidimensional calculations, it is more convenient here simply to differentiate the above expression for the heat flux. Thus,

$$\frac{dq}{d\tau}(\tau) \bigg/ \sigma(T_w^4 - T_m^4) = -\frac{2[E_2(\tau) + E_2(\tau_L - \tau)]}{1 + (1/\epsilon - 1)[1 - 2E_3(\tau_L)]}.$$

If $T_w > T_m$, then $dq/d\tau$ is always negative: The flux is positive at $\tau = 0$ (going into the medium), zero at the midplane, and turning more and more negative as the $\tau = \tau_L$ plate is approached.

12.6 RADIATIVE TRANSFER IN SPHERICAL MEDIA

In a plane-parallel medium, if the temperature field (as well as any radiative property) varies only in the direction normal to the plates, the problem is one-dimensional. If the polar angle θ is measured from the direction normal to the plates, the radiative intensity depends only on the spatial coordinate z and polar angle θ (but not on azimuthal angle ψ, because of symmetry). A similar situation exists in a one-dimensional spherical medium. Let the temperature field vary only in the radial direction r, but not with polar angle θ_s or azimuthal angle ψ_s (where the subscripts s have been added to emphasize that these angles specify *position* in a spherical coordinate system, and are independent of angles θ and ψ, which are employed to describe *direction*). If the polar direction angle θ is measured from the radial position vector as shown in Fig. 12-5, then the radiative intensity depends only on polar angle θ and—owing to symmetry— not on azimuthal angle ψ. However, unlike in the plane layer of Fig. 12-1, in spherical symmetry the polar angle changes as a beam travels in the direction of $\hat{s}$ through the medium (with θ steadily decreasing with increasing pathlength s). Therefore, with intensity depending on radial location r and direction angle θ, the left-hand side of equation (12.1) must be expressed[5] as

$$\hat{s} \cdot \nabla I = \frac{dI}{ds} = \frac{\partial I}{\partial r}\frac{dr}{ds} + \frac{\partial I}{\partial \theta}\frac{d\theta}{ds}. \tag{12.61}$$

From inspection of Fig. 12-5, we find that $\cos\theta = dr/ds$. Also, from the law of sines,

$$\frac{\sin\theta}{r'\,d\theta_s} = \frac{\sin\theta'}{r\,d\theta_s} \quad \text{or} \quad r\sin\theta = r'\sin\theta' = \text{const} \tag{12.62}$$

---

[5]While this expression is also valid for a one-dimensional plane layer (with r replaced by z), the polar angle does not change along a path through the slab, or $d\theta/ds = 0$.

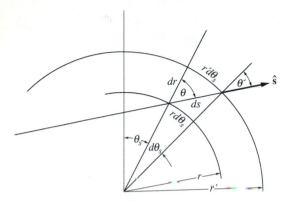

FIGURE 12-5
Coordinates for a one-dimensional spherical medium.

along s. Differentiating this relation gives

$$dr \sin \theta + r \cos \theta \, d\theta = 0, \quad \text{or} \quad \frac{d\theta}{dr} = -\frac{\tan \theta}{r}. \tag{12.63}$$

Substituting both relations into equation (12.61) leads to

$$\frac{dI}{ds} = \cos \theta \frac{\partial I}{\partial r}(r, \theta) - \frac{\sin \theta}{r} \frac{\partial I}{\partial \theta}(r, \theta), \tag{12.64}$$

or, if the shorthand $\mu = \cos \theta$ is preferred,

$$\frac{dI}{ds} = \mu \frac{\partial I}{\partial r}(r, \mu) + \frac{1 - \mu^2}{r} \frac{\partial I}{\partial \mu}(r, \mu), \tag{12.65}$$

where we have used $d\mu = -\sin \theta \, d\theta$. Substituting equation (12.65) into equations (12.1) and (12.3), we obtain, with $\tau = \int_0^r \beta \, dr$,

$$\mu \frac{\partial I}{\partial \tau}(\tau, \mu) + \frac{1 - \mu^2}{\tau} \frac{\partial I}{\partial \mu}(\tau, \mu,) = S(\tau, \mu) - I(\tau, \mu), \tag{12.66}$$

where the radiative source for linear-anisotropic scattering has been given by equation (12.15).

The problem of one-dimensional heat transfer through a spherical medium was first considered by Sparrow and coworkers [4], who investigated radiative equilibrium in a gray nonscattering medium contained between concentric black spheres. They assumed that there was uniform heat generation within the medium and that both surfaces had identical and constant temperatures. Ryhming [5] considered the same problem, but without heat generation and with the two surfaces at different temperatures T_1 and T_2. The condition of black walls was relaxed by Viskanta and Crosbie [6], who considered a nonscattering, gray, heat-generating medium between two gray, isothermal spheres of radius R_1 and R_2, respectively (and at temperatures T_1

and T_2, and with gray diffuse emissivities ϵ_1 and ϵ_2). They found that the temperature field, in terms of emissive power E_b, may be calculated from

$$E_b(\tau) = J_1 + (J_2 - J_1)\Phi(\tau) + \frac{\dot{Q}'''}{\kappa}\Phi_s(\tau). \tag{12.67}$$

Here J_1 and J_2 are the radiosities of the two spherical surfaces, and $\dot{Q}'''$ is the (uniform) heat generation within the medium. $\Phi(\tau)$ is the nondimensional emissive power for a medium without heat generation, determined from

$$\Phi(\tau) = \frac{E_b(\tau) - J_1}{J_2 - J_1} = \frac{1}{2\tau}\left[g(\tau) + \int_{\tau_1}^{\tau_2} K(\tau, t)\Phi(t)\,dt\right], \tag{12.68}$$

$$g(\tau) = \tau_2 E_2(\tau_2 - \tau) - \sqrt{\tau_2^2 - \tau_1^2}\,E_2\left(\sqrt{\tau_2^2 - \tau_1^2} + \sqrt{\tau^2 - \tau_1^2}\right)$$
$$+ E_3(\tau_2 - \tau) - E_3\left(\sqrt{\tau_2^2 - \tau_1^2} + \sqrt{\tau^2 - \tau_1^2}\right), \tag{12.69}$$

$$K(\tau, t) = \left[E_1(|\tau - t|) - E_1\left(\sqrt{\tau_2^2 - \tau_1^2} + \sqrt{\tau^2 - \tau_1^2}\right)\right]t. \tag{12.70}$$

$\Phi_s(\tau)$ is the nondimensional emissive power for a medium with uniform heat generation, but with both surfaces having the same radiosity, J_1, obtained from

$$\Phi_s(\tau) = \frac{E_b - J_1}{\dot{Q}'''/\kappa} = \frac{1}{4} + \frac{1}{2\tau}\int_{\tau_1}^{\tau_2} K(\tau, t)\Phi_s(t)\,dt. \tag{12.71}$$

Once the functions $\Phi(\tau)$ and/or $\Phi_s(\tau)$ have been determined, the radiative heat fluxes can be calculated from

$$\tau^2 q(\tau) = (J_1 - J_2)\tau_1^2\,\Psi(\tau) + \frac{\dot{Q}'''}{\kappa}\left[\frac{\tau^3}{3} - \tau_1^2\,\Psi_s(\tau)\right], \tag{12.72}$$

where

$$\Psi(\tau) = \frac{2}{\tau_1^2}\left[h(\tau) + \int_{\tau_1}^{\tau_2} H(\tau, t)\Phi(t)\,dt\right], \tag{12.73}$$

$$\Psi_s(\tau) = \frac{1}{\tau_1^2}\left[\frac{\tau^3}{3} - 2\int_{\tau_1}^{\tau_2} H(\tau, t)\Phi_s(t)\,dt\right], \tag{12.74}$$

$$h(\tau) = -\tau\tau_2 E_3(\tau_2 - \tau) - \sqrt{(\tau_2^2 - \tau_1^2)(\tau^2 - \tau_1^2)}\,E_3\left(\sqrt{\tau_2^2 - \tau_1^2} + \sqrt{\tau^2 - \tau_1^2}\right)$$
$$+ (\tau_2 - \tau)E_4(\tau_2 - \tau) - \left(\sqrt{\tau_2^2 - \tau_1^2} + \sqrt{\tau^2 - \tau_1^2}\right)E_4\left(\sqrt{\tau_2^2 - \tau_1^2} + \sqrt{\tau^2 - \tau_1^2}\right)$$
$$+ E_5(\tau_2 - \tau) - E_5\left(\sqrt{\tau_2^2 - \tau_1^2} + \sqrt{\tau^2 - \tau_1^2}\right), \tag{12.75}$$

TABLE 12.2

Values of nondimensional flux functions for radiative equilibrium between concentric spheres, from Viskanta and Crosbie [6]

τ_2	Ψ			Ψ_s
	$R_1/R_2 = 0.1$	$R_1/R_2 = 0.5$	$R_1/R_2 = 0.9$	$R_1/R_2 = 0.5$
0	1.0000	1.0000	1.0000	0.0000
0.1	0.9970	0.9900	0.9946	0.0321
0.5	0.9844	0.9488	0.9728	0.1678
1.0	0.9680	0.8976	0.9459	0.3525
2.0		0.8006	0.8944	0.7619
5.0	0.8316	0.5797	0.7625	2.1552
10.0	0.6839	0.3834	0.6077	
20.0		0.2250	0.4312	

$$H(\tau, t) = \left[\tau \, \text{sgn} \, (\tau - t) E_2(|\tau - t|) - \sqrt{\tau^2 - \tau_1^2} \, E_2 \left(\sqrt{\tau^2 - \tau_1^2} + \sqrt{t^2 - \tau_1^2} \right) \right.$$

$$\left. + E_3(|\tau - t|) - E_3 \left(\sqrt{\tau^2 - \tau_1^2} + \sqrt{t^2 - \tau_1^2} \right) \right] t, \tag{12.76}$$

where $\text{sgn}(t) = t/|t| = \pm 1$, depending on the sign of t. Solutions to Φ, Ψ, Φ_s, and Ψ_s have been tabulated by Viskanta and Crosbie [6] for a number of radius ratios R_1/R_2 and optical thicknesses τ_2. Their results for the nondimensional flux functions Ψ and Ψ_s are given in Table 12.2. As for the plane slab, the statement of radiative equilibrium,

$$\frac{1}{\tau^2} \frac{d}{d\tau} (\tau^2 q) = \frac{\dot{Q}'''}{\kappa}, \tag{12.77}$$

implies that Ψ and Ψ_s are constants, and equations (12.73) and (12.74) may be evaluated for any arbitrary value of τ.

It remains to eliminate the radiosities J_1 and J_2 from equation (12.72) for the case of nonblack boundaries. Similar to the development for parallel plates, equations (12.44) through (12.46), we have

$$\tau = \tau_1 : \qquad q_1 = \frac{\epsilon_1}{1 - \epsilon_1} \left(n^2 \sigma T_1^4 - J_1 \right), \tag{12.78a}$$

$$\tau = \tau_2 : \qquad -q_2 = \frac{\epsilon_2}{1 - \epsilon_2} \left(n^2 \sigma T_2^4 - J_2 \right). \tag{12.78b}$$

Performing an energy balance (i.e., stating that energy coming in at Sphere 1, plus energy generated in the volume between spheres, equals energy going out at

Sphere 2), we obtain

$$4\pi R_1^2 q_1 + \dot{Q}'''\frac{4}{3}\pi(R_2^3 - R_1^3) = 4\pi R_2^2 q_2,$$

or

$$q_2 = \left(\frac{\tau_1}{\tau_2}\right)^2 q_1 + \frac{1}{3}\frac{\dot{Q}'''}{\kappa}\frac{\tau_2^3 - \tau_1^3}{\tau_2^2}. \tag{12.79}$$

Substituting equation (12.79) into (12.78) leads to

$$\left(\frac{1}{\epsilon_1} - 1\right)q_1 + \left(\frac{1}{\epsilon_2} - 1\right)\left[\left(\frac{\tau_1}{\tau_2}\right)^2 q_1 + \frac{1}{3}\frac{\dot{Q}'''}{\kappa}\frac{\tau_2^3 - \tau_1^3}{\tau_2^2}\right]$$
$$= n^2\sigma(T_1^4 - T_2^4) - (J_1 - J_2)$$

or

$$(J_1 - J_2)\tau_1^2 = n^2\sigma(T_1^4 - T_2^4)\tau_1^2 - \left[\frac{1}{\epsilon_1} - 1 + \left(\frac{\tau_1}{\tau_2}\right)^2\left(\frac{1}{\epsilon_2} - 1\right)\right]\tau^2 q$$
$$- \frac{1}{3}\frac{\dot{Q}'''}{\kappa}\left(\frac{1}{\epsilon_2} - 1\right)\left(\frac{\tau_1}{\tau_2}\right)^2(\tau_2^3 - \tau_1^3), \tag{12.80}$$

which may be employed to eliminate the radiosities from equation (12.72).

More recently, a few investigators have considered somewhat more involved situations. The governing integral equations for an isotropically scattering spherical medium were first stated by Pomraning and Siewert [7]. These equations were used by Thynell and Özişik [8] to investigate the gray isotropically scattering solid sphere with gray, diffusely reflecting boundary. Their analysis applied to any given temperature fields or variable internal heat generation. Finally, the problem of nondiffuse reflectivity at the outer face, obeying Fresnel's laws, was investigated by Wu and Wang [9] for an isothermal, isotropically scattering, solid sphere.

Example 12.3.   A gray, nonscattering medium with refractive inex $n = 1$ and and absorption coefficient $\kappa = 0.1\,\text{cm}^{-1}$ is contained between two concentric isothermal spheres with radii $R_1 = 25\,\text{cm}$ and $R_2 = 50\,\text{cm}$. The inner sphere is hot ($T_1 = 2000\,\text{K}$) and highly reflective ($\epsilon_1 = 0.1$); the outer sphere is a strong absorber ($\alpha_2 = \epsilon_2 = 0.9$), which must be kept relatively cool ($T_2 = 400\,\text{K}$). Assuming that conductive and convective heat transfer can be neglected as compared with radiation, determine the necessary cooling rate.

Solution

We have the situation of one-dimensional radiative equilibrium between concentric spheres, and equation (12.72) applies with $\dot{Q}''' = 0$. With $R_1/R_2 = 25/50 = 0.5$ and $\tau_2 = \kappa R_2 = 0.1 \times 50 = 5$ we obtain, from Table 12.2, $\Psi = 0.5797$. The radiosities are eliminated with equation (12.80) so that

$$\tau^2 q = \sigma(T_1^4 - T_2^4)\tau_1^2\Psi - \left[\frac{1}{\epsilon_1} - 1 + \left(\frac{\tau_1}{\tau_2}\right)^2\left(\frac{1}{\epsilon_2} - 1\right)\right]\tau^2 q\,\Psi,$$

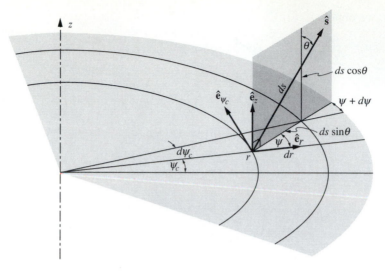

FIGURE 12-6
Coordinates for a one-dimensional cylindrical medium.

or

$$\frac{\tau^2 q}{\tau_1^2 \sigma(T_1^4 - T_2^4)} = \frac{q_1}{\sigma(T_1^4 - T_2^4)} = \frac{\Psi}{1 + \left[\frac{1}{\epsilon_1} - 1 + \left(\frac{\tau_1}{\tau_2}\right)^2 \left(\frac{1}{\epsilon_2} - 1\right)\right]\Psi}$$

$$= \frac{0.5797}{1 + \left[\frac{1}{0.1} - 1 + 0.5^2 \left(\frac{1}{0.9} - 1\right)\right]0.5797} = 0.0930.$$

This result should be compared with the value of 0.0830 found in Example 12.1 for the identical situation between parallel plates. The flux density at the inner sphere then turns out to be

$$q_1 = 0.0930\sigma(T_1^4 - T_2^4)$$
$$= 0.0930 \times 5.670 \times 10^{-12} (2000^4 - 400^4)\,\text{W/cm}^2 = 8.42\,\text{W/cm}^2.$$

12.7   RADIATIVE TRANSFER IN CYLINDRICAL MEDIA

We shall now briefly consider the case of a one-dimensional cylindrical medium, with temperature and radiative properties varying only in the radial direction r, but not with axial position z or azimuthal angle ψ_c (where we have again added the subscript c to emphasize that this angle specifies *position* in the cylindrical coordinate system, and is independent of azimuthal *direction* angle ψ). For this geometry it is advantageous to place the direction coordinate system such that polar angle θ is measured from the positive z-axis, while the azimuthal angle ψ is measured in the

r-ψ_c-plane perpendicular to it, as shown in Fig. 12-6. Measuring the azimuthal angle from the radial coordinate as indicated in the figure, we recognize that radiative intensity may vary with radial position r and *both* direction angles θ and ψ. Therefore, similar to equation (12.61), we have

$$\hat{s} \cdot \nabla I = \frac{dI}{ds} = \frac{\partial I}{\partial r}\frac{dr}{ds} + \frac{\partial I}{\partial \theta}\frac{d\theta}{ds} + \frac{\partial I}{\partial \psi}\frac{d\psi}{ds}. \tag{12.81}$$

From symmetry, it follows that $I(r, \theta, \psi) = I(r, \theta, -\psi) = I(r, \pi - \theta, \psi)$. Inspecting Fig. 12-6, we find

$$\cos \psi = \frac{dr}{ds \sin \theta}, \quad \text{or} \quad \frac{dr}{ds} = \sin \theta \cos \psi, \tag{12.82}$$

where $ds \sin \theta$ is the projection of ds into the r-ψ_c-plane. Traveling along a beam in the direction of $\hat{s}$ we see that, similar to the spherical case, the azimuthal angle ψ steadily decreases (instead of θ in the spherical case, see Fig. 12-5). Therefore, replacing θ by ψ in equation (12.63), we find

$$\frac{d\psi}{dr} = -\frac{\tan \psi}{r}, \quad \text{or} \quad \frac{d\psi}{ds} = \frac{d\psi}{dr}\frac{dr}{ds} = -\frac{\sin \theta \sin \psi}{r}. \tag{12.83}$$

On the other hand, traveling along $\hat{s}$, we see that the angle with the z-axis remains unchanged, or $d\theta/ds = 0$. Sticking these relations into equation (12.81) yields

$$\frac{dI}{ds} = \sin \theta \left[\cos \psi \frac{\partial I}{\partial r}(r, \theta, \psi) - \frac{\sin \psi}{r}\frac{\partial I}{\partial \psi}(r, \theta, \psi) \right]. \tag{12.84}$$

The equation of transfer appropriate for the one-dimensional cylindrical medium follows then from equations (12.1) and (12.3) as

$$\sin \theta \left[\cos \psi \frac{\partial I}{\partial \tau}(\tau, \theta, \psi) - \frac{\sin \psi}{\tau}\frac{\partial I}{\partial \psi}(\tau, \theta, \psi) \right] = S(\tau, \theta, \psi) - I(\tau, \theta, \psi), \tag{12.85}$$

where again $\tau = \int_0^r \beta \, dr$. This relationship can also be found from the left-hand side of equation (12.81) by recognizing (cf. Fig. 12-6) that

$$\hat{s} = \sin \theta \cos \psi \, \hat{e}_r + \sin \theta \sin \psi \, \hat{e}_{\psi_c} + \cos \theta \, \hat{e}_z, \tag{12.86a}$$

$$\psi + \psi_c = \text{const along } \hat{s}, \tag{12.86b}$$

and using, for cylindrical coordinates,

$$\nabla = \frac{\partial}{\partial r}\hat{e}_r + \frac{1}{r}\frac{\partial}{\partial \psi_c}\hat{e}_{\psi_c} + \frac{\partial}{\partial z}\hat{e}_z.$$

with $d\psi_c = -d\psi$ and $\partial I/\partial z = 0$. The general form for the radiative source function is given by equation (12.3); for linear-anisotropic scattering, equation (12.15) remains valid with $\hat{e}_r \cdot \hat{s} = \cos\theta$ (valid for slab and sphere, cf. Figs. 12-1 and 12-5) replaced by $\hat{e}_r \cdot \hat{s} = \sin\theta \cos\psi$ (valid for the cylinder, cf. Fig. 12-6), or

$$S(\tau, \theta, \psi) = (1 - \omega) I_b(\tau) + \frac{\omega}{4\pi} [G(\tau) + A_1 q(\tau) \sin\theta \cos\psi]. \tag{12.87}$$

The problem of one-dimensional heat transfer through a cylindrical medium was first considered by Heaslet and Warming [10]. They investigated the case of an isotropically scattering medium contained within an isothermal black cylindrical container. Two cases were treated: (*i*) Radiative equilibrium with uniform heat generation within the medium, and (*ii*) an isothermal medium. While they displayed a few graphical results, no tabulated results were given.

The solution for a one-dimensional, gray, linear-anisotropically scattering cylinder with arbitrary, but specified, temperature distribution has been given by Azad and Modest [11], using a different approach. We list here their solution for the simple case of no scattering ($\omega = 0$), for which

$$q(\tau) = [n^2\sigma T^4(\tau_R) - J_w] F(\tau, \tau_R)$$
$$- \int_0^\tau \frac{d}{d\tau'} (n^2\sigma T^4) \frac{\tau'}{\tau} F(\tau', \tau) d\tau' - \int_\tau^{\tau_R} \frac{d}{d\tau'} (n^2\sigma T^4) F(\tau, \tau') d\tau, \tag{12.88}$$

with

$$F(\tau, \tau') = -\frac{4}{\pi} \int_0^\pi \int_0^{\pi/2} \exp\left[-\frac{\tau}{\sin\theta}\left(\cos\psi + \sqrt{\left(\frac{\tau'}{\tau}\right)^2 - \sin^2\psi}\right)\right]$$
$$\times \sin^2\theta \cos\psi \, d\theta \, d\psi. \tag{12.89}$$

Equation (12.88) is very similar to the equivalent expression for the one-dimensional slab, equation (12.37) (after integration by parts): Instead of the relatively simple (and widely tabulated) exponential integral $E_3(\tau' - \tau)$ we have another, somewhat more complicated geometric function, $F(\tau, \tau')$. For an isothermal cylinder equation (12.88) may be evaluated in explicit form at $\tau = \tau_R$, with $F(\tau_R, \tau_R)$ expressed in terms of modified Bessel functions. Some representative results of equation (12.88) for $\tau = \tau_R$ have been tabulated in Table 12.3.

Thermal radiation between concentric cylinders was first treated by Kesten [12], who considered a nonscattering, gray gas with known temperature distribution. The case of a gray, isotropically scattering medium at radiative equilibrium between concentric cylinders has been studied by Pandey and Cogley [13] and Loyalka [14]. The governing equations become rather involved and will not be reproduced here. Because of this complexity, both solutions are not quite exact: Pandey and Cogley used some approximate geometric functions, while Loyalka used a simple variational approach to solve the governing integral equation. Comparison with the "exact" Monte Carlo

TABLE 12.3

Nondimensional heat loss from a gray, nonscattering, isothermal cylinder, $\Psi = q(\tau_R)/(n^2\sigma T^4 - J_w)$

τ_R	Ψ	τ_R	Ψ
0.1	0.1770	1.0	0.8143
0.2	0.3172	1.5	0.9047
0.3	0.4299	2.0	0.9458
0.4	0.5213	2.5	0.9662
0.5	0.5960	3.0	0.9772
0.6	0.6573	3.5	0.9836
0.7	0.7080	4.0	0.9877
0.8	0.7500	4.5	0.9904
0.9	0.7850	5.0	0.9923

solution[6] by Perlmutter and Howell [15] shows that Loyalka's results may essentially be taken as exact. Some representative results for the nondimensional radiative heat flux in terms of surface radiosities,

$$\Psi = \frac{q(\tau_1)}{J_1 - J_2}, \qquad (12.90)$$

are given in Table 12.4.

For nonblack walls the radiosities may be eliminated from equation (12.90) in precisely the same fashion as was done for concentric spheres. From equation (12.80), with $A_1/A_2 = R_1/R_2 = \tau_1/\tau_2$ and $\dot{Q}''' = 0$,

$$J_1 - J_2 = n^2\sigma(T_1^2 - T_2^4) - \left[\frac{1}{\epsilon_1} - 1 + \frac{\tau_1}{\tau_2}\left(\frac{1}{\epsilon_2} - 1\right)\right]q_1,$$

and

$$\frac{q_1}{n^2\sigma(T_1^2 - T_2^4)} = \frac{\Psi}{1 + \left[\frac{1}{\epsilon_1} - 1 + \frac{\tau_1}{\tau_2}\left(\frac{1}{\epsilon_2} - 1\right)\right]\Psi}. \qquad (12.91)$$

Example 12.4. Repeat Example 12.3 for the case of concentric cylinders.

Solution

With $\tau_2 = 5$ and $\tau_1 = 2.5$ we have $\tau_2 - \tau_1 = 2.5$ and, after interpolating (somewhat nonlinearly) between values from Table 12.4 for $\tau_2 - \tau_1 = 2.0$ and $\tau_2 - \tau_1 = 3.0$, $\Psi \approx 0.48$. Thus,

$$\frac{q_1}{\sigma(T_1^4 - T_2^4)} = \frac{0.48}{1 + \left[\frac{1}{0.1} - 1 + 0.5\left(\frac{1}{0.9} - 1\right)\right]0.48} = 0.0898$$

---

[6]For a discussion of the Monte Carlo method, see Chapter 19.

TABLE 12.4

Nondimensional radiative heat transfer between concentric cylinders at radiative equilibrium, $\Psi = q(\tau_1)/(J_1 - J_2)$

Optical Thickness	Ψ Radius Ratio R_1/R_2		
$\tau_2 - \tau_1$	0.1	0.5	0.9
0.1	0.9893	0.9677	0.9462
0.5	0.9464	0.8476	0.7688
1.0	0.8937	0.7225	0.6167
2.0	0.7956	0.5446	0.4371
3.0	0.7105	0.4313	0.3367
4.0	0.6377	0.3549	0.2727
5.0	0.5763	0.3010	0.2291
6.0	0.5250	0.2615	0.1976
7.0	0.4810	0.2308	0.1738
8.0	0.4429	0.2060	0.1549
9.0	0.4102	0.1864	0.1398
10.0	0.3821	0.1703	0.1278

and

$$q_1 = 0.0898 \times 5.670 \times 10^{-12} (2000^4 - 400^4) = 8.13 \, \text{W/cm}^2.$$

We observe that, for identical conditions, the heat loss is greatest between concentric spheres, followed by concentric cylinders, and finally by parallel plates. Also, from Tables 12.2 and 12.4 we see that heat loss increases with decreasing radius ratio R_1/R_2. This observation may be explained by the fact that, per unit area, the surface of an (inner) sphere exchanges heat with a larger area on the (outer) sphere $(A_2/A_1 = R_2^2/R_1^2)$ than is the case for concentric cylinders $(A_2/A_1 = R_2/R_1)$ or parallel plates $(A_2/A_1 = 1)$. The same argument applies to decreasing radius ratios.

12.8 NUMERICAL SOLUTION OF THE GOVERNING INTEGRAL EQUATIONS

The governing integral equations may be solved with several analytical and/or numerical techniques, which will not be discussed in this text in any detail.

An example of analytical techniques is the use of *Chandrasekhar's X- and Y-functions*, based on the *principle of invariance*, which is described in some detail in Chandrasekhar's book [16]. For example, Heaslet and Warming [2] expressed the nondimensional temperature Φ_b in equation (12.41) in terms of moments of these X- and Y-functions. The moments are determined by numerical quadrature, using tabulated values for Chadrasekhar's X- and Y-functions. *Case's normal-mode expansion technique* [17, 18] is to linear integral equations what separation-of-variables is to partial differential equations. The technique was originally developed for neutron transport theory, and has been applied to radiative heat transfer by Ferziger and Simmons [19, 20] and Siewert and Özişik [21–27]. A detailed account may be found in the book by Özişik [28]. Siewert and coworkers [29, 30] have further developed Case's normal-mode approach, by finding solutions in terms of power series. This approach, known as the F_N-method, was also originally applied to neutron transport, and only later to thermal radiation [31].

Numerical solutions to the governing integral equations may be found by a variety of methods. The simplest such method is the standard numerical quadrature as discussed in Section 5.5. Since integrands often contain singularities [cf. equation (12.41)], these must be removed before quadrature can be applied [2]. The problem of singularities may also be overcome by approximating the unknown variable in functional form, which allows the analytical evaluation of such integrals. For example, Özişik and coworkers [32–35] solved the governing integral equation for several plane-parallel problems with the *Galerkin method* [36]. In this method, the unknown dependent variable [say, $\Phi(\tau)$ in equation (12.41)] is approximated by a series of independent functions $\varphi_i(\tau)$, that is,

$$\Phi(\tau) = C_1 \varphi_1(\tau) + C_2 \varphi_2(\tau) + \cdots = \sum_{i=1}^{N} C_i \varphi_i(\tau), \tag{12.92}$$

where the C_i are unknown constants. These are determined by multiplying the governing equation by each of the functions $\varphi_i(\tau)$, followed by integration over the entire domain, resulting in N simultaneous algebraic equations for the unknown constants. Most often the independent functions in equation (12.92) are chosen to be powers in the independent variable, for example, $\varphi_i(\tau) = \tau^{i-1}$ ($i = 1, 2, \ldots$), although Legendre polynomials have also been used [37], in order to exploit the orthogonality properties of such polynomials. The Galerkin method offers results of great accuracy even for series truncated after very few terms, albeit at the price of very tedious analytical or numerical integrations. For more general geometries the use of the related *finite element method* becomes more practical, as applied by Reddy and Murty [38].

A somewhat simpler method is the *point collocation method* together with approximating the unknown variable by piecewise-continuous splines. In this method the unknown dependent variable, say $\Phi(\tau)$, is approximated by a spline function involving $N + 1$ nodal values $\Phi_i = \Phi(\tau_i)$, $i = 0, 1, 2, \ldots N$. For example, if standard cubic splines are employed [39],

$$\Phi = \Phi_i + B_i (\tau - \tau_i) + C_i (\tau - \tau_i)^2 + D_i (\tau - \tau_i)^3,$$
$$\tau_i \leq \tau \leq \tau_{i+1}, \quad i = 0, 1, \ldots, N - 1, \tag{12.93}$$

where the constants B_i, C_i, and D_i depend on all values of Φ_i and are readily found through standard software packages resident on most computers. More sophisticated splines, such as *B*-splines [40–42] or Chebyshev polynomials [43], result in more complicated expressions. Equation (12.93) is now substituted into the governing integral equation, and the piece-wise integrals are evaluated analytically. Applying the governing equation to the $N + 1$ nodal points (point collocation) results in $N + 1$ simultaneous, linear algebraic equations for the unknown Φ_i.

References

1. MacRobert, T. M.: *Spherical Harmonics*, 3rd ed., Pergamon Press, New York, 1967.
2. Heaslet, M. A., and R. F. Warming: "Radiative Transport and Wall Temperature Slip in an Absorbing Planar Medium," *International Journal of Heat and Mass Transfer*, vol. 8, pp. 979–994, 1965.
3. Modest, M. F., and F. H. Azad: "The Influence and Treatment of Mie-Anisotropic Scattering in Radiative Heat Transfer," *ASME Journal of Heat Transfer*, vol. 102, pp. 92–98, 1980.
4. Sparrow, E. M., C. M. Usiskin, and H. A. Hubbard: "Radiation Heat Transfer in a Spherical Enclosure Containing a Participating, Heat-Generating Gas," *ASME Journal of Heat Transfer*, vol. 83, pp. 199–206, 1961.
5. Ryhming, I. L.: "Radiative Transfer Between Two Concentric Spheres Separated by an Absorbing and Emitting Gas," *International Journal of Heat and Mass Transfer*, vol. 9, pp. 315–324, 1966.
6. Viskanta, R., and A. L. Crosbie: "Radiative Transfer Through a Spherical Shell of an Absorbing-Emitting Gray Medium," *Journal of Quantitative Spectroscopy and Radiative Transfer*, vol. 7, pp. 871–889, 1967.
7. Pomraning, G. C., and C. E. Siewert: "On the Integral Form of the Equation of Transfer for a Homogeneous Sphere," *Journal of Quantitative Spectroscopy and Radiative Transfer*, vol. 28, no. 6, pp. 503–506, 1982.
8. Thynell, S. T., and M. N. Özişik: "Radiation Transfer in an Isotropically Scattering Homogeneous Solid Sphere," *Journal of Quantitative Spectroscopy and Radiative Transfer*, vol. 33, no. 4, pp. 319–330, 1985.
9. Wu, C.-Y., and C.-J. Wang: "Emittance of a Finite Spherical Scattering Medium with Fresnel Boundary," *Journal of Thermophysics and Heat Transfer*, vol. 4, no. 2, pp. 250–251, 1990.
10. Heaslet, M. A., and R. F. Warming: "Theoretical Predictions of Radiative Transfer in Homogeneous Cylindrical Medium," *Journal of Quantitative Spectroscopy and Radiative Transfer*, vol. 6, pp. 751–774, 1966.
11. Azad, F. H., and M. F. Modest: "Evaluation of Radiative Heat Fluxes in Absorbing, Emitting and Anisotropically Scattering Cylindrical Media," *ASME Journal of Heat Transfer*, vol. 103, pp. 350–356, 1981.
12. Kesten, A. S.: "Radiant Heat Flux Distribution in a Cylindrically-Symmetric Nonisothermal Gas with Temperature-Dependent Absorption Coefficient," *Journal of Quantitative Spectroscopy and Radiative Transfer*, vol. 8, pp. 419–434, 1968.
13. Pandey, D. K., and A. C. Cogley: "An Integral Solution Procedure for Radiative Transfer in Concentric Cylindrical Media," ASME paper no. 83-WA/HT-78, 1983.
14. Loyalka, S. K.: "Radiative Heat Transfer Between Parallel Plates and Concentric Cylinders," *International Journal of Heat and Mass Transfer*, vol. 12, pp. 1513–1517, 1969.
15. Perlmutter, M., and J. R. Howell: "Radiant Transfer through a Gray Gas Between Concentric Cylinders Using Monte Carlo," *ASME Journal of Heat Transfer*, vol. 86, no. 2, pp. 169–179, 1964.
16. Chandrasekhar, S.: *Radiative Transfer*, Oxford University Press, London, 1950.
17. Case, K. M.: "Elementary Solutions of the Transport Equation and Their Applications," *Annals of Physics*, vol. 9, pp. 1–23, 1960.
18. Case, K. M., and P. F. Zweifel: *Linear Transport Theory*, Addison-Wesley, Reading, MA, 1967.
19. Ferziger, J. H., and G. M. Simmons: "Application of Case's Method to Plane-Parallel Radiative Transfer," *International Journal of Heat and Mass Transfer*, vol. 9, pp. 987–992, 1966.
20. Simmons, G. M., and J. H. Ferziger: "Non-Grey Radiative Heat Transfer Between Parallel Plates," *International Journal of Heat and Mass Transfer*, vol. 11, pp. 1611–1620, 1968.

21. Siewert, C. E., and P. F. Zweifel: "An Exact Solution of Equations of Radiative Transfer for Local Thermodynamic Equilibrium in the Non-Gray Case: Picket Fence Approximation," *Ann. Phys. (N.Y.)*, vol. 36, pp. 61–85, 1966.

22. Siewert, C. E., and P. F. Zweifel: "Radiative Transfer II," *J. Math Phys.*, vol. 7, pp. 2092–2102, 1966.

23. Siewert, C. E., and M. N. Özişik: "An Exact Solution in the Theory of Line Formation," *Monthly Notices Royal Astronomical Society*, vol. 146, pp. 351–360, 1969.

24. Kriese, J. T., and C. E. Siewert: "Radiative Transfer in a Conservative Finite Slab with an Internal Source," *International Journal of Heat and Mass Transfer*, vol. 9, pp. 987–992, 1966.

25. Özişik, M. N., and C. E. Siewert: "On the Normal-Mode Expansion Technique for Radiative Transfer in a Scattering, Absorbing and Emitting Slab with Specularly Reflecting Boundaries," *International Journal of Heat and Mass Transfer*, vol. 12, pp. 611–620, 1969.

26. Beach, H. L., M. N. Özişik, and C. E. Siewert: "Radiative Transfer in Linearly Anisotropic-Scattering, Conservative and Non-Conservative Slabs with Reflective Boundaries," *International Journal of Heat and Mass Transfer*, vol. 14, pp. 1551–1565, 1971.

27. Reith, R. J., C. E. Siewert, and M. N. Özişik: "Non-Grey Radiative Heat Transfer in Conservative Plane-Parallel Media with Reflecting Boundaries," *Journal of Quantitative Spectroscopy and Radiative Transfer*, vol. 11, pp. 1441–1462, 1971.

28. Özişik, M. N.: *Radiative Transfer and Interactions With Conduction and Convection*, John Wiley & Sons, New York, 1973.

29. Siewert, C. E., and P. Benoist: "The F_N Method in Neutron-Transport Theory. Part I: Theory and Applications," *Nuclear Science and Engineering*, vol. 69, pp. 156–160, 1979.

30. Grandjean, P., and C. E. Siewert: "The F_N Method in Neutron-Transport Theory. Part II: Applications and Numerical Results," *Nuclear Science and Engineering*, vol. 69, pp. 161–168, 1979.

31. Siewert, C. E., J. R. Maiorino, and M. N. Özişik: "The Use of the F_N Method for Radiative Transfer Problems with Reflective Boundary Conditions," *Journal of Quantitative Spectroscopy and Radiative Transfer*, vol. 23, pp. 565–573, 1980.

32. Özişik, M. N., and Y. Yener: "The Galerkin Method for Solving Radiation Transfer in Plane-Parallel Media," *ASME Journal of Heat Transfer*, vol. 104, pp. 351–354, 1982.

33. Cengel, Y. A., M. N. Özişik, and Y. Yener: "Determination of Angular Distribution of Radiation in an Isotropically Scattering Slab," *ASME Journal of Heat Transfer*, vol. 106, pp. 248–252, 1984.

34. Cengel, Y. A., and M. N. Özişik: "The Use of Galerkin Method for Radiation Transfer in an Anisotropically Scattering Slab with Reflecting Boundaries," *Journal of Quantitative Spectroscopy and Radiative Transfer*, vol. 32, pp. 225–234, 1982.

35. Cengel, Y. A., and M. N. Özişik: "Integrals Involving Legendre Polynomials that Arise in the Solution of Radiation Transfer," *Journal of Quantitative Spectroscopy and Radiative Transfer*, vol. 31, pp. 215–219, 1982.

36. Kantorovich, L. V., and V. I. Krylov: *Approximate Methods of Higher Analysis*, John Wiley & Sons, New York, 1964.

37. Condiff, D.: "Anisotropic Scattering in Three Dimensional Differential Approximation of Radiation Heat Transfer," in *Fundamentals and Applications of Radiation Heat Transfer*, vol. HTD–72, ASME, pp. 19–29, 1987.

38. Reddy, J. N., and V. D. Murty: "Finite-Element Solution of Integral Equations Arising in Radiative Heat Transfer and Laminar Boundary-Layer Theory," *Numerical Heat Transfer*, vol. 1, pp. 389–401, 1978.

39. Modest, M. F.: "Oblique Collimated Irradiation of an Absorbing, Scattering Plane-Parallel Layer," *Journal of Quantitative Spectroscopy and Radiative Transfer*, vol. 45, pp. 309–312, May 1991.

40. Chawla, T. C., G. Leaf, and W. Chen: "A Collocation Method Using B-Splines for One-Dimensional Heat or Mass-Transfer-Controlled Moving Boundary Problems," *Nuclear Engineering Design*, vol. 35, pp. 163–180, 1975.

41. Chawla, T. C., and S. H. Chan: "Solution of Radiation-Conduction Problems with Collocation Method Using B-Splines as Approximating Functions," *International Journal of Heat and Mass Transfer*, vol. 22, no. 12, pp. 1657–1667, 1979.

42. Chawla, T. C., W. J. Minkowycz, and G. Leaf: "Spline-Collocation Solution of Integral Equations Occurring in Radiative Transfer and Laminar Boundary-Layer Problems," *Numerical Heat Transfer*, vol. 3, pp. 133–148, 1980.

43. Kamiuto, K.: "Chebyshev Collocation Method for Solving the Radiative Transfer Equation," *Journal of Quantitative Spectroscopy and Radiative Transfer*, vol. 35, no. 4, pp. 329–336, 1986.

Problems

12.1 The gap between two parallel black plates at T_1 and T_2, respectively, is filled with a particle-laden gas. Radiative equilibrium prevails, the particle loading is a fixed volume fraction, with particles manufactured from two different materials (one a specular reflector, the other a diffuse reflector, both having the same ϵ). Sketch the nondimensional heat flux $\Psi = q/\upsilon(T_1^4 \quad T_2^4)$ vs. particle size (but keeping volume-fraction constant).

12.2 Consider a space enclosed by infinite, diffuse-gray parallel plates filled with a gray nonscattering medium. The surfaces are isothermal (both at T_w), and there is uniform and constant heat generation within the medium per unit volume, $\dot{Q}'''$. Conduction and convection are negligible so that $\nabla \cdot \mathbf{q} = \dot{Q}'''$. Set up the integral equations describing temperature and heat flux distribution in the enclosure, i.e., show that

$$\Phi(\tau) = \frac{\sigma T^4 - J_w}{\dot{Q}'''/4\kappa} = 1 + \frac{1}{2}\int_0^{\tau_L} \Phi(\tau')E_1\left(|\tau - \tau'|\right)d\tau',$$

$$\Psi(\tau) = \frac{q}{\dot{Q}'''/\kappa} = \tau - \frac{\tau_L}{2}.$$

12.3 A semi-infinite, gray, isotropically scattering medium, originally at a temperature of $0\,\mathrm{K}$, is subjected to collimated irradiation with a constant heat flux q_0 normal to its nonreflecting surface. Set up the integral relationships governing steady-state temperature and radiative heat flux within the medium, assuming radiative equilibrium. Hint: Collimated irradiation with heat flux q_0 has the radiative intensity

$$I_0(\theta, \psi) = \begin{cases} \dfrac{q_0}{2\pi \sin\theta\, \delta\theta}, & 0 \le \theta < \delta\theta,\ 0 \le \psi \le 2\pi, \\ 0, & \text{elsewhere.} \end{cases}$$

12.4 An infinite, black, isothermal plate bounds a semi-infinite space filled with black spheres. At any given distance, z, away from the plate the particle number density is identical, namely $N_T = 6.3662 \times 10^8\,\mathrm{m}^{-3}$. However, the radius of the suspended spheres diminishes monotonically away from the surface as

$$a = a_0\, e^{-z/L}; \quad a_0 = 10^{-4}\mathrm{m}, \quad L = 1\,\mathrm{m}.$$

(a) Determine the absorption coefficient as a function of z (you may make the large-particle assumption).

(b) Determine the optical coordinate as a function of z. What is the total optical thickness of the semi-infinite space?

(c) Assuming that radiative equilibrium prevails, determine the heat loss from the plate.

12.5 The radiative heat transfer between two isothermal, black plates at temperatures T_1 and T_2 and separated by a nonparticipating gas is to be minimized. Enough of a black material is available to place a 1 mm thick radiation shield between the plates. Alternatively, the same amount of material could be used in the form of small spheres of 0.1 mm radius to be suspended between the plates. Which possibility results in lower heat flux, assuming conduction and convection to be negligible?

12.6 Two infinite, isothermal plates at temperatures T_1 and T_2 are separated by a cold, gray medium of optical thickness $\tau_L = \kappa L$ (no scattering).

(a) Calculate the radiative heat flux at the bottom plate, the top plate, and the net radiative energy going into the gray medium, assuming that both plates are black.

(b) Repeat (a), but assume that both plates have the same temperature T, and that both plates are gray with equal emissivity ϵ (diffuse emission and reflection).

12.7 A semi-infinite, absorbing-emitting, nonscattering medium at uniform temperature is in contact with a gray-diffuse wall at T_w and with emissivity ϵ_w.

(a) The medium is gray, and has a constant absorption coefficient. Determine the net radiative heat flux at the wall.

(b) Let the medium be nongray with nonconstant absorption coefficient κ_λ, and the wall be nongray and nondiffuse with spectral, directional emissivity ϵ_λ'. How would this affect the wall heat flux?

12.8 A 1 m thick, isothermal slab bounded by two cold black plates has a temperature of 3000 K, and a nongray absorption coefficient that can be approximated by

$$\kappa_\lambda = \begin{cases} 0, & \lambda < 2\,\mu\text{m}, \\ 0.20\,\text{cm}^{-1}, & \lambda > 2\,\mu\text{m}. \end{cases}$$

Calculate the total heat loss by radiation from the slab (in W/cm^2).

12.9 Consider (a) two parallel plates, (b) two concentric spheres, (c) two concentric cylinders. The bottom/inner surface needs to dissipate a heat flux of $30\,\text{W/cm}^2$ and has a gray-diffuse emissivity $\epsilon_1 = 0.5$. The top/outer surface is at $T_2 = 1000\,\text{K}$ with $\epsilon_2 = 0.8$. The medium in between the surfaces is gray and nonscattering ($\kappa = 0.1\,\text{cm}^{-1}$), has a thickness of $L = 5\,\text{cm}$ and is at radiative equilibrium. Determine the temperature at the bottom/inner surface necessary to dissipate the supplied heat for the three different cases (the radii of the inner cylinder and sphere are $R_1 = 5\,\text{cm}$). Discuss the results.

12.10 Consider a very hot sphere of a nongray gas of radius $R = 1\,\text{m}$ in 0 K surroundings that have been evacuated. The gas has a single absorption-emission band in the infrared, with an absorption coefficient

$$\kappa_\eta = \begin{cases} 0, & \eta < 3000\,\text{cm}^{-1} = \eta_0, \\ \kappa_0\, e^{-(\eta - \eta_0)/\omega}, & \eta > 3000\,\text{cm}^{-1}, \end{cases}$$

where $\kappa_0 = 1\,\text{cm}^{-1}$, $\omega = 200\,\text{cm}^{-1}$. During cool-down the sphere is always isothermal, and remains of constant size (i.e., constant density $\rho = 1000\,\text{g/m}^3$). The heat capacity of the gas is $c_p = 1\,\text{kJ/kg K}$. Determine the time required to cool the gas from $T_i = 6000\,\text{K}$ to $T_e = 1000\,\text{K}$. Sketch qualitatively the behavior of $\Psi = q/\sigma T^4$ vs. T. Hint: To make an analytical solution possible, you may make the following assumptions:

(a) $\mathrm{Ein}(x) = \int_0^1 (1 - e^{-x\xi})d\xi/\xi = E_1(x) + \ln x + \gamma_E \simeq \ln x + \gamma_E$ (for sufficiently large x; see also Appendix E).

(b) Wien's distribution may be used.

12.11 It is proposed to construct a high-temperature heating element by guiding hot combustion gases through a silicon carbide tube. The outside of the SiC tube then radiates heat toward the load. Such devices are known as "radiant tubes." For the design of such a radiant tube you may make the following assumptions:

(i) the combustion gas inside the radiant tube is essentially gray and isothermal with $\kappa = 0.2\,\mathrm{cm}^{-1}$ and $T_{\mathrm{gas}} = 2000\,\mathrm{K}$;

(ii) the silicon carbide tube wall is essentially isothermal and of negligible thickness, with a gray-diffuse emissivity of 0.8 on both sides,

(iii) the long tube is contained in a large furnace with a background temperature of 1000 K.

Determine the necessary tube diameter to achieve a radiant heating rate of 100 kW per m length of tube.

CHAPTER
13

APPROXIMATE
SOLUTION
METHODS FOR
ONE-DIMENSIONAL
MEDIA

Because of their importance and their relative simplicity, the cases of gray (or spectral), plane-parallel, or other one-dimensional media bounded by isothermal, gray, diffusely emitting and reflecting walls have been studied extensively. Even for the simplest case the exact solution can only be cast implicitly in the form of an integral equation, thus prompting the development of numerous approximate solution techniques. These approximate methods may roughly be classified into those applicable for limiting conditions (cold medium approximation, optically thin approximation, optically thick approximation) and those making approximations for the directional distribution of intensity (two-flux approximation, moment method). In the following we shall discuss a few of these methods as applied to the one-dimensional case of a medium contained between isothermal gray surfaces. In principle, all of these methods could also be applied to more complicated geometries, although such an extension is not obvious for all of them (and may, indeed, be very tedious). We shall assume that the medium is gray and, therefore, not carry along any spectral subscripts on intensity and other quantities; however, all of the methods discussed here are equally valid on a spectral basis.

13.1 THE OPTICALLY THIN APPROXIMATION

The exact integral equations describing incident radiation G and radiative heat flux q for a gray (or on a spectral basis) medium confined between two isothermal, gray-diffuse, and parallel plates were developed with equations (12.21) and (12.22) as

$$G(\tau) = 2J_1 E_2(\tau) + 2J_2 E_2(\tau_L - \tau)$$
$$+2\pi \int_0^\tau S(\tau') E_1(\tau - \tau') \, d\tau' + 2\pi \int_\tau^{\tau_L} S(\tau') E_1(\tau' - \tau) \, d\tau', \qquad (13.1)$$

$$q(\tau) = 2J_1 E_3(\tau) - 2J_2 E_3(\tau_L - \tau)$$
$$+2\pi \int_0^\tau S(\tau') E_2(\tau - \tau') \, d\tau' - 2\pi \int_\tau^{\tau_L} S(\tau') E_2(\tau' - \tau) \, d\tau', \qquad (13.2)$$

where J_1 and J_2 are the radiosities at the two surfaces and the radiative source $S(\tau)$ has been assumed to be independent of direction [limiting us to isotropic scattering, see equation (12.5)].

We shall now assume that the medium is optically thin, i.e., $\tau_L \ll 1$. If we want to evaluate q accurately up to $\mathcal{O}(\tau)$ [i.e., neglecting terms of $\mathcal{O}(\tau^2)$ or smaller], we must evaluate the E_3 in equation (13.2) accurate up to $\mathcal{O}(\tau)$; it is sufficient to evaluate E_2 (and S) accurate up to $\mathcal{O}(1)$ [since the integration itself is $\mathcal{O}(\tau)$]. Thus, with

$$E_2(x) = 1 + \mathcal{O}(x), \qquad E_3(x) = \tfrac{1}{2} - x + \mathcal{O}(x^2),$$

we get

$$q(\tau) \simeq J_1(1-2\tau) - J_2(1-2\tau_L+2\tau) + 2\pi \left[\int_0^\tau S(\tau') \, d\tau' - \int_\tau^{\tau_L} S(\tau') \, d\tau' \right]. \qquad (13.3)$$

Evaluating the radiative source, equation (12.5),

$$S(\tau) = (1-\omega) I_b(\tau) + \frac{\omega}{4\pi} G(\tau), \qquad (13.4)$$

implies evaluating $G(\tau)$ accurate up to $\mathcal{O}(1)$, i.e., from equation (13.1),

$$G(\tau) = 2J_1 + 2J_2 + \mathcal{O}(\tau) \qquad (13.5)$$

[keeping in mind that, while $\lim_{x \to 0} E_1(x) \to \infty$, $\lim_{x \to 0} x E_1(x) \to 0$]. Either the blackbody intensity, $I_b(\tau)$, is "known" (by considering other modes of heat transfer), or we have $S = I_b = G/4\pi = (J_1 + J_2)/2\pi$ (radiative equilibrium in a gray medium). Thus, the radiative heat flux for an optically thin slab is:

$I_b(\tau)$ specified:

$$q = J_1[1 - 2(1-\omega)\tau - \omega\tau_L] - J_2[1 + 2(1-\omega)\tau - (2-\omega)\tau_L]$$
$$+2\pi(1-\omega) \left[\int_0^\tau I_b(\tau') \, d\tau' - \int_\tau^{\tau_L} I_b(\tau') \, d\tau' \right]; \qquad (13.6)$$

Radiative equilibrium:

$$q = (J_1 - J_2)(1 - \tau_L) = \text{const.} \tag{13.7}$$

If the temperature of the medium is specified, we are usually interested in the divergence of the radiative heat flux, or dq/dz (the radiative source within the overall energy equation). From equation (12.24)

$$\frac{dq}{d\tau} = (1 - \omega)(4\pi I_b - G). \tag{13.8}$$

In order to predict dq/dz accurate to $\mathbb{O}(\tau)$, we must specify G accurate to $\mathbb{O}(1)$ [since we are using τ rather than z as the independent variable in equation (13.8)]. Thus, from equation (13.5)

$$\frac{dq}{d\tau} = 2(1 - \omega)(2\pi I_b - J_1 - J_2). \tag{13.9}$$

Equation (13.6) may be interpreted physically: Intensity leaving the surfaces (radiosities J_1 and J_2) is attenuated by absorption and scattering, but the attenuation is linear since the strength of the radiosities is diminished very little (i.e., every point within the medium has essentially the same incident radiation and, therefore, attenuation rate). The integral terms describe emission within the medium, traveling up $(+)$ and down $(-)$. Emission is virtually unattenuated by self-absorption since the extinction coefficient is too low. Equation (13.7) shows no dependence on scattering albedo ω: In a gray medium at radiative equilibrium it is not possible to distinguish between absorption and isotropic scattering. If a photon is absorbed at a certain location, the same amount of energy must immediately be reemitted isotropically. Since the (different) wavelength of emission cannot be detected for a gray medium, this process is equivalent to isotropic scattering.

13.2 THE OPTICALLY THICK APPROXIMATION (DIFFUSION APPROXIMATION)

A particularly simple expression for the radiative heat flux can be obtained if the slab is optically thick, or $\tau_L \gg 1$. We note from equation (13.2) that the radiative source S is accompanied by an exponential integral that acts as a weight function. If the medium is optically thick, this exponential integral decays very rapidly over a short (geometrical) distance away from $\tau' = \tau$. To exploit this fact we first rewrite equation (13.2) by changing the integration variable τ' to $\tau'' = |\tau - \tau'|$:

$$\begin{aligned} q(\tau) = \ &2J_1 E_3(\tau) - 2J_2 E_3(\tau_L - \tau) \\ &+ 2\pi \int_0^\tau S(\tau - \tau'')E_2(\tau'')\,d\tau'' - 2\pi \int_0^{\tau_L - \tau} S(\tau + \tau'')E_2(\tau'')\,d\tau''. \end{aligned} \tag{13.10}$$

We shall now assume that we are a large optical distance away from either of the surfaces, i.e., $\tau \gg 1$ and $\tau_L - \tau \gg 1$. Under these conditions the influence of the boundaries (J_1 and J_2) becomes negligible and the integration limit may be replaced by infinity [since $E_2(\tau'') \simeq 0$ beyond the actual limits]. Thus,

$$q(\tau) \simeq 2\pi \int_0^\infty S(\tau - \tau'')E_2(\tau'')\, d\tau'' - 2\pi \int_0^\infty S(\tau + \tau'')E_2(\tau'')\, d\tau'', \qquad (13.11)$$

where the arguments in q and S simply denote a physical location between the two plates. Since $E_2(\tau'')$ is expected to vanish a very short geometrical distance away from $\tau'' = 0$ (or $\tau' = \tau$), the radiative source can vary only slightly over this distance. Therefore, we may expand S into a Taylor series as

$$S(\tau \pm \tau'') = S(\tau) \pm \tau'' \left(\frac{dS}{d\tau}\right)_\tau + \frac{(\tau'')^2}{2}\left(\frac{d^2S}{d\tau^2}\right)_\tau \pm \cdots.$$

Substituting this expression into equation (13.11) leads to

$$\frac{q(\tau)}{2\pi} = S(\tau)\int_0^\infty E_2(\tau'')\, d\tau'' - \frac{dS}{d\tau}\int_0^\infty \tau'' E_2(\tau'')\, d\tau''$$

$$+ \frac{1}{2}\frac{d^2S}{d\tau^2}\int_0^\infty (\tau'')^2 E_2(\tau'')\, d\tau'' - + \cdots$$

$$-S(\tau)\int_0^\infty E_2(\tau'')\, d\tau'' - \frac{dS}{d\tau}\int_0^\infty \tau'' E_2(\tau'')\, d\tau''$$

$$- \frac{1}{2}\frac{d^2S}{d\tau^2}\int_0^\infty (\tau'')^2 E_2(\tau'')\, d\tau'' - - \cdots$$

$$= -2\frac{dS}{d\tau}\int_0^\infty x E_2(x)\, dx + O\left(\frac{1}{\tau^3}\right).$$

Evaluating the integral we get, using the relations for exponential integrals given in Appendix E,

$$\int_0^\infty x E_2(x)\, dx = -x E_3(x)\Big|_0^\infty + \int_0^\infty E_3(x)\, dx = -E_4(x)\Big|_0^\infty = \frac{1}{3},$$

and

$$q(\tau) = -\frac{4\pi}{3}\frac{dS}{d\tau}. \qquad (13.12)$$

For a nonscattering medium, or a gray medium at radiative equilibrium, $S = I_b$, and equation (13.12) reduces to

$$q(\tau) = -\frac{4\pi}{3}\frac{dI_b}{d\tau}. \qquad (13.13)$$

For the general case of an isotropically scattering medium the radiative source must be determined from equation (13.4) by first determining the incident radiation $G(\tau)$ from equation (13.1). Following the same procedure as for $q(\tau)$, we get

$$
\frac{G(\tau)}{2\pi} = S(\tau) \int_0^\infty E_1(\tau'') \, d\tau'' - \frac{dS}{d\tau} \int_0^\infty \tau'' E_1(\tau'') \, d\tau''
$$

$$
+ \frac{1}{2} \frac{d^2 S}{d\tau^2} \int_0^\infty (\tau'')^2 E_1(\tau'') \, d\tau'' - + \cdots
$$

$$
+ S(\tau) \int_0^\infty E_1(\tau'') \, d\tau'' + \frac{dS}{d\tau} \int_0^\infty \tau'' E_1(\tau'') \, d\tau''
$$

$$
+ \frac{1}{2} \frac{d^2 S}{d\tau^2} \int_0^\infty (\tau'')^2 E_1(\tau'') \, d\tau'' + + \cdots
$$

$$
= 2 S(\tau) \int_0^\infty E_1(\tau'') \, d\tau'' + \mathcal{O}\left(\frac{1}{\tau^2}\right) = 2 S(\tau),
$$

or

$$
\frac{G}{4\pi} = S = (1 - \omega) I_b + \omega \frac{G}{4\pi}
$$

and

$$
S(\tau) = \frac{G}{4\pi}(\tau) = I_b(\tau). \tag{13.14}
$$

Thus, for an optically thick, isotropically scattering medium equation (13.13) holds, whether the medium is at radiative equilibrium or not, and the heat flux, on a spectral basis, is determined from

$$
q_\eta = -\frac{4\pi}{3\beta_\eta} \frac{dI_{b\eta}}{dz}, \tag{13.15}
$$

or, after integration over wavenumbers, the total heat flux from

$$
q = -\frac{4\sigma}{3\beta_R} \frac{d(n^2 T^4)}{dz}, \tag{13.16}
$$

where β_R is the *Rosseland-mean extinction coefficient* as defined in equation (9.104). Equations (13.15) and (13.16) are commonly known as the *Rosseland approximation*, since they were originally derived by Rosseland [1], or the *diffusion approximation*, since equation (13.16) is of the same type as Fourier's law of heat diffusion and Fick's law of mass diffusion.

The diffusion approximation is extremely convenient to use. One may even define a "radiative conductivity"

$$
k_R = \frac{16 n^2 \sigma T^3}{3\beta_R}, \tag{13.17}
$$

so that

$$q = -k_R \frac{dT}{dz},\tag{13.18}$$

and the radiation problem reduces to a simple conduction problem with strongly temperature-dependent conductivity. Similar to Fourier's law, equation (13.15) may be extended to three-dimensional geometries by writing

$$\mathbf{q}_\eta = -\frac{4\pi}{3\beta_\eta}\nabla I_{b\eta},\tag{13.19}$$

and

$$\mathbf{q} = -\frac{4\sigma}{3\beta_R}\nabla(n^2 T^4) = -k_R\nabla T.\tag{13.20}$$

However, it is important to keep in mind that the diffusion approximation is not valid near a boundary, where it often fails quite miserably. In practice, the method is useful only in optically extremely thick situations (for example, heat transfer through hot glass and other semitransparent materials).

Example 13.1.   Consider a gray, isothermal medium at temperature T, confined between two parallel, black, isothermal plates both at temperature T_w. Determine the radiative heat flux as function of distance across the layer, using the diffusion approximation.

Solution

From equation (13.16) we find $q \equiv 0$ everywhere inside the medium, while $q \to \infty$ at both surfaces (because of the temperature jump there). The diffusion approximation is clearly inadequate in the optical vicinity of walls, in particular if temperature discontinuities are present.

Example 13.2.   Now consider a gray medium contained between two black, isothermal cylinders. The inner cylinder has a radius of R_1 and is at temperature T_1. The outer cylinder has a radius of R_2 and is at temperature T_2. The medium absorbs and emits but does not scatter radiation, and radiative equilibrium prevails. Determine temperature profile and heat flux across the medium.

Solution

From the condition of radiative equilibrium,

$$\nabla \cdot \mathbf{q} = \frac{1}{r}\frac{d}{dr}(rq) = 0,$$

we find $rq = C_1' = $ const, or $q = C_1'/r = C_1/\tau$ if the optical coordinate $\tau = \kappa r$ is used. Thus, from equation (13.13)

$$q = \frac{C_1}{\tau} = -\frac{4\sigma}{3}\frac{dT^4}{d\tau}$$

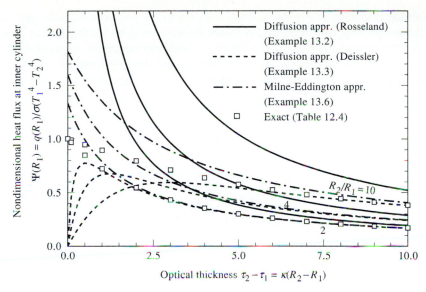

FIGURE 13-1
Nondimensional radiative heat flux in a medium at radiative equilibrium, confined between isothermal black cylinders.

(assuming a refractive index of unity). We may integrate this expression to find

$$\sigma T^4 = -\frac{3}{4} C_1 \ln \tau + C_2, \tag{13.21}$$

or, using the boundary conditions $T = T_1$ at $r = R_1$ and $T = T_2$ at $r = R_2$,

$$\frac{T^4 - T_1^4}{T_2^4 - T_1^4} = \frac{\ln(\tau/\tau_1)}{\ln(\tau_2/\tau_1)},$$

and

$$\Psi = \frac{q}{\sigma(T_1^4 - T_2^4)} = \frac{4}{3\tau \ln(\tau_2/\tau_1)}. \tag{13.22}$$

As seen by comparison with the exact and other approximate solutions in Fig. 13-1, equation (13.22) does reasonably well for optically thick situations, but fails for the optically thin limit as $\kappa \to 0$, where $\Psi \to \infty$ [as opposed to the correct limit, $\Psi(R_1) = 1$].

Deissler's Jump Boundary Conditions

When we derived equation (13.13) we assumed that the medium was optically thick ($\tau_L \gg 1$) and that we were far removed from a boundary ($\tau \gg 0$, $\tau_L - \tau \gg 0$). But in the examples we assumed that equation (13.13) holds also at the walls and found the temperature profile by using the boundary temperatures (Example 13.2). The examples showed that this assumption is not very good. Deissler [2] argued that, while flux must be conserved [and, therefore, equation (13.13) must hold at the surface if it holds inside the medium adjacent to it], no *radiative* principle states that the temperature of surface and adjacent medium must be continuous (as already known from the exact solution).[1] To develop a boundary condition, he used the same principles that were applied to equation (13.10) and the development following it, and applied them to a point at the boundary. For $\tau = 0$ equation (13.10) becomes (with $\tau_L \gg 1$)

$$
\begin{aligned}
q(0) = J_1 &- 2\pi \int_0^\infty S(\tau'') E_2(\tau'')\, d\tau'' \\
&- J_1 - 2\pi \left[S(0) \int_0^\infty E_2(\tau'')\, d\tau'' + \frac{dS}{d\tau}(0) \int_0^\infty \tau'' E_2(\tau'')\, d\tau'' \right. \\
&\left. + \frac{1}{2}\frac{d^2S}{d\tau^2}(0) \int_0^\infty (\tau'')^2 E_2(\tau'')\, d\tau'' + \mathcal{O}\left(\frac{1}{\tau^3}\right) \right].
\end{aligned}
$$

Using equation (13.14) this expression becomes

$$
q(0) = J_1 - \pi I_b(0) - \frac{2\pi}{3}\frac{dI_b}{d\tau}(0) - \frac{\pi}{2}\frac{d^2 I_b}{d\tau^2}(0) + \mathcal{O}\left(\frac{1}{\tau^3}\right). \tag{13.23}
$$

Deissler truncated the series after the second derivative since doing so gives the same level of approximation as equation (13.13). Substituting equation (13.13) into equation (13.23) leads to

$$
J_1 = \pi I_b(0) - \frac{2\pi}{3}\frac{dI_b}{d\tau}(0) + \frac{\pi}{2}\frac{d^2 I_b}{d\tau^2}(0). \tag{13.24}
$$

For radiative equilibrium of a one-dimensional slab this further simplifies to

$$
J_1 - \pi I_b(0) = -\frac{2\pi}{3}\frac{dI_b}{d\tau}(0) = \frac{1}{2}q(0) = \frac{1}{2}q, \tag{13.25}
$$

since $q = \text{const}$ and, therefore, $d^2 I_b/d\tau^2 = 0$.

The jump boundary condition may be generalized to multidimensional geometries [2].

$$
J_w(\mathbf{r}_w) = \pi I_b(\mathbf{r}_w) - \frac{2\pi}{3}\frac{\partial I_b}{\partial \tau_z}(\mathbf{r}_w) + \frac{\pi}{4}\left(2\frac{\partial^2 I_b}{\partial \tau_z^2} + \frac{\partial^2 I_b}{\partial \tau_x^2} + \frac{\partial^2 I_b}{\partial \tau_y^2}\right)(\mathbf{r}_w), \tag{13.26}
$$

where τ_z is, as before, an optical coordinate measured in the direction of the outward surface normal, and τ_x and τ_y are optical coordinates tangential to the surface.

Example 13.3.   Repeat Example 13.2 using Deissler's jump boundary conditions.

---

[1]Of course, in the presence of conduction and/or convection, temperature continuity is forced by those other heat transfer modes.

Solution

For a cylindrical coordinate system, equation (13.26) becomes[2]

$$\tau = \tau_1: \quad \sigma T_1^4 = \sigma T^4 - \frac{2}{3}\sigma\frac{dT^4}{d\tau} + \frac{\sigma}{4}\left(\frac{1}{\tau}\frac{dT^4}{d\tau} + 2\frac{d^2T^4}{d\tau^2}\right)$$

$$\tau = \tau_2: \quad \sigma T_2^4 = \sigma T^4 + \frac{2}{3}\sigma\frac{dT^4}{d\tau} + \frac{\sigma}{4}\left(\frac{1}{\tau}\frac{dT^4}{d\tau} + 2\frac{d^2T^4}{d\tau^2}\right),$$

[the change of sign in the second boundary condition is due to the fact that in equation (13.24) τ is measured *away* from the surface, while at τ_2 it is measured *toward* the surface]. Utilizing the general diffusion solution for a one-dimensional cylinder at radiative equilibrium, equation (13.21), we find

$$\tau = \tau_1: \quad \sigma T_1^4 = C_2 - \frac{3}{4}C_1\ln\tau_1 + \frac{C_1}{2\tau_1} + \frac{3C_1}{16\tau_1^2},$$

$$\tau = \tau_2: \quad \sigma T_2^4 = C_2 - \frac{3}{4}C_1\ln\tau_2 - \frac{C_1}{2\tau_2} + \frac{3C_1}{16\tau_2^2},$$

From these two equations we obtain

$$C_1 = \frac{\sigma(T_1^4 - T_2^4)}{\frac{3}{4}\ln\frac{\tau_2}{\tau_1} + \frac{1}{2}\left(\frac{1}{\tau_1} + \frac{1}{\tau_2}\right) + \frac{3}{16}\left(\frac{1}{\tau_1^2} - \frac{1}{\tau_2^2}\right)},$$

and, with $q_1 = C_1/\tau_1$,

$$\Psi = \frac{q_1}{\sigma(T_1^4 - T_2^4)} = 1\bigg/\left[\frac{3\tau_1}{4}\ln\frac{\tau_2}{\tau_1} + \frac{1}{2}\left(1 + \frac{\tau_1}{\tau_2}\right) + \frac{3}{16\tau_1}\left(1 - \frac{\tau_1^2}{\tau_2^2}\right)\right].$$

A plot of this nondimensional flux is also included in Fig. 13-1.

13.3 THE SCHUSTER-SCHWARZSCHILD APPROXIMATION

A very simple solution method for a one-dimensional, plane-parallel slab was given independently by Schuster [3] and Schwarzschild [4]. While they limited their derivation to nonscattering media, the method is readily extended to include isotropic scattering, which we will consider here. The equation of transfer for a one-dimensional, plane-parallel, isotropically scattering, gray (or on a spectral basis) medium is, setting $\Phi \equiv 1$ in equation (12.19),

$$\mu\frac{dI}{d\tau} = (1 - \omega)I_b - I + \frac{\omega}{2}\int_{-1}^{+1} I\,d\mu, \quad -1 < \mu < +1. \tag{13.27}$$

---

[2]The development of these boundary conditions is left as an exercise; see Problem 13.1.

Schuster and Schwarzschild assumed the radiative intensity to be isotropic, but different, over the upper and lower hemisphere, that is,

$$I(\tau, \mu) = \begin{cases} I^-(\tau), & -1 < \mu < 0, \\ I^+(\tau), & 0 < \mu < +1. \end{cases} \tag{13.28}$$

Substituting this expression into equation (13.27) leads to

$$\mu \frac{dI}{d\tau} = (1 - \omega)I_b - I + \frac{\omega}{2}(I^- + I^+). \tag{13.29}$$

Because of the approximation made for I, equation (13.29) can, of course, only be solved in an approximate way. Since intensity has been reduced to two unknown functions of space only, equation (13.29) must be reduced to two space-dependent equations. Integrating equation (13.29) over the upper and lower hemispheres, respectively, achieves this goal and results in

$$\frac{1}{2}\frac{dI^+}{d\tau} = (1 - \omega)I_b - I^+ + \frac{\omega}{2}(I^- + I^+), \tag{13.30a}$$

$$-\frac{1}{2}\frac{dI^-}{d\tau} = (1 - \omega)I_b - I^- + \frac{\omega}{2}(I^- + I^+), \tag{13.30b}$$

subject to the boundary conditions

$$\tau = 0: \quad I^+ = J_1/\pi, \tag{13.31a}$$

$$\tau = \tau_L: \quad I^- = J_2/\pi, \tag{13.31b}$$

where J_1 and J_2 are the radiosities of the bounding plates. From the definitions for incident radiation and radiative heat flux we find

$$G = 2\pi \int_{-1}^{1} I \, d\mu = 2\pi(I^+ + I^-), \tag{13.32}$$

and

$$q = 2\pi \int_{-1}^{1} I\mu \, d\mu = \pi(I^+ - I^-). \tag{13.33}$$

Thus, adding and subtracting equations (13.30) gives

$$\frac{dq}{d\tau} = (1 - \omega)(4\pi I_b - G), \tag{13.34}$$

$$\frac{dG}{d\tau} = -4q, \tag{13.35}$$

$$\tau = 0: \quad G + 2q = 4J_1, \tag{13.36a}$$

$$\tau = \tau_L: \quad G - 2q = 4J_2. \tag{13.36b}$$

Example 13.4. Find an expression for the heat flux within a gray, nonscattering isothermal medium (temperature T) confined between two isothermal, parallel, black plates at (the same) temperature T_w. Use the Schuster-Schwarzschild approximation.

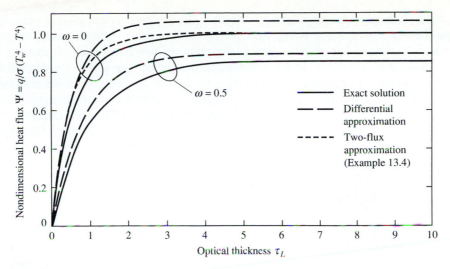

FIGURE 13-2
Radiative heat flux through a gray, isothermal, isotropically scattering medium bounded by black plates.

Solution

Differentiating equation (13.34) and using equation (13.35) gives

$$\frac{d^2 q}{d\tau^2} = -\frac{dG}{d\tau} = 4q,$$

or

$$q = C_1 e^{2\tau} + C_2 e^{-2\tau},$$

subject to the boundary conditions

$$\tau = 0: \qquad 4\sigma T^4 - \frac{dq}{d\tau} + 2q = 4\sigma T_w^4,$$

$$\tau = \tau_L: \qquad 4\sigma T^4 - \frac{dq}{d\tau} - 2q = 4\sigma T_w^4,$$

in which G has been eliminated using equation (13.34) and $J_1 = J_2 = \sigma T_w^4$. Substituting the expression for heat flux into the boundary condition gives

$$(2 - 2)C_1 + (2 + 2)C_2 = 4\sigma(T_w^4 - T^4),$$
$$-(2 + 2)C_1 e^{2\tau_L} - (2 - 2)C_2 e^{-2\tau_L} = 4\sigma(T_w^4 - T^4),$$

or

$$C_2 = -C_1 e^{2\tau_L} = \sigma(T_w^4 - T^4)$$

and

$$\Psi = \frac{q}{\sigma(T^4 - T_w^4)} = e^{-2(\tau_L - \tau)} - e^{-2\tau}.$$

This nondimensional flux, evaluated at a wall, is plotted in Fig. 13-2, along with the exact solution and other approximate methods.

The Schuster-Schwarzschild approximation always goes to the correct optically thin limit ($\tau_L \to 0$), since in that case the treatment is exact. Since it breaks up the intensity functions into two constant components for two directions, the method is also commonly referred to as the *two-flux approximation*. Obviously, the method is easily generalized to higher order (breaking up the 4π directions in more than two components and directions, or *discrete ordinates*), as well as to multidimensional geometries. For example, six-flux methods have been used by Chin and Churchill [5] and by Shih and Chen [6]; a review of the six-flux method has been given by Chan [7]. The general discrete ordinates method will be discussed in detail in Chapter 15.

13.4   THE MILNE-EDDINGTON APPROXIMATION (MOMENT METHOD)

Another simple method for the case of a one-dimensional, plane-parallel medium has been developed independently by Milne [8] and Eddington [9]. The method is also commonly referred to as the *differential approximation*, especially when generalized to more complicated geometries. Starting from equation (13.27) they took the *zeroth* and *first moments* of the equation, i.e., they integrated equation (13.27) over all directions after multiplication with $\mu^0 = 1$ (zeroth moment) and $\mu^1 = \mu$ (first moment). Defining intensity moments as

$$I_k = 2\pi \int_{-1}^{1} I \mu^k d\mu, \quad k = 0, 1, \cdots \tag{13.37}$$

leads to

$$\frac{dI_1}{d\tau} = (1 - \omega)4\pi I_b - I_0 + \omega I_0 = (1 - \omega)(4\pi I_b - I_0), \tag{13.38}$$

$$\frac{dI_2}{d\tau} = -I_1, \tag{13.39}$$

or two equations in three unknowns, I_0, I_1 and I_2. To make the system determinate, a *closing condition* must be found, i.e., a relationship between I_0, I_1 and I_2. Milne and Eddington, like Schuster and Schwarzschild, assumed the intensity to be isotropic over both the upper and lower hemisphere. Thus,

$$I_k = 2\pi \left(I^- \int_{-1}^{0} \mu^k d\mu + I^+ \int_{0}^{1} \mu^k d\mu \right) = \frac{2\pi}{k+1} \left[(-1)^k I^- + I^+ \right], \tag{13.40}$$

or

$$I_2 = \tfrac{1}{3} I_0. \tag{13.41}$$

With $G = I_0$ and $q = I_1$ equation (13.41) transforms (13.38) and (13.39) to

$$\frac{dq}{d\tau} = (1 - \omega)(4\pi I_b - G), \tag{13.42}$$

$$\frac{dG}{d\tau} = -3q. \tag{13.43}$$

The boundary conditions are identical to those of the Schuster-Schwarzschild approximation, equation (13.31), again leading to

$$\tau = 0: \qquad G + 2q = 4J_1, \qquad (13.44a)$$
$$\tau = \tau_L: \qquad G - 2q = 4J_2. \qquad (13.44b)$$

In the case of radiative equilibrium we have $dq/d\tau = 0$ and, therefore, $G = 4\pi I_b$. For this case equation (13.43) reduces to

$$q = -\frac{4\pi}{3} \frac{dI_b}{d\tau}, \qquad (13.45)$$

which is the same as for the diffusion approximation (although the boundary conditions are different and, for large τ_L, are carried to one additional order of accuracy in the diffusion approximation).

Example 13.5.   Consider a gray medium with refractive index $n = 1$, confined between two isothermal, black, parallel plates at temperatures T_1 and T_2, respectively. As in Example 13.2 the medium is at radiative equilibrium and absorbs and emits, but does not scatter radiation. Determine the heat flux between the plates using the differential approximation.

Solution

For a gray, nonscattering medium at radiative equilibrium equations (13.42) and (13.43) reduce to

$$\frac{dq}{d\tau} = 4\sigma T^4 - G = 0, \quad \text{or} \quad q = \text{const},$$
$$\frac{dG}{d\tau} = -3q, \quad \text{or} \quad G = 4\sigma T^4 = C - 3q\tau.$$

Applying the boundary conditions we get

$$\tau = 0: \qquad C \qquad + 2q = 4\sigma T_1^4,$$
$$\tau = \tau_L: \qquad C - 3q\tau_L - 2q = 4\sigma T_2^4,$$

or

$$\Psi = \frac{q}{\sigma(T_1^4 - T_2^4)} = \frac{1}{1 + \frac{3}{4}\tau_L}$$

and

$$C = 4\sigma T_1^4 - 2q,$$

$$\Phi = \frac{T_1^4 - T^4}{T_1^4 - T_2^4} = \frac{2 + 3\tau}{4 + 3\tau_L}.$$

It is easy to show that this result is identical to the one obtained from the diffusion approximation with Deissler's jump boundary conditions (for example, by letting $\tau_2 = \tau_1 + \tau_L$ and $\tau_1 \to \infty$ in Example 13.3).

Example 13.6. Repeat Example 13.2 using the differential approximation.

Solution

For the one-dimensional case of a medium at radiative equilibrium between concentric cylinders the divergence of the radiative heat flux is, in cylindrical coordinates,

$$\frac{1}{\tau}\frac{d}{d\tau}(\tau q) = 4\sigma T^4 - G = 0,$$

or

$$q = \frac{C_1}{\tau}.$$

Substituting this expression into equation (13.45) gives

$$\sigma T^4 = -\tfrac{3}{4}C_1 \ln \tau + C_2,$$

which is, of course, the same as for the diffusion approximation (since we have radiative equilibrium). Applying the boundary conditions (with $G = 4\sigma T^4$) gives

$$\tau = \tau_1 : \qquad -\frac{3}{4}C_1 \ln \tau_1 + C_2 + \frac{C_1}{2\tau_1} = \sigma T_1^4,$$

$$\tau = \tau_2 : \qquad -\frac{3}{4}C_1 \ln \tau_2 + C_2 - \frac{C_1}{2\tau_2} = \sigma T_2^4,$$

or

$$C_1 = \frac{\sigma(T_1^4 - T_2^4)}{\dfrac{1}{2}\left(\dfrac{1}{\tau_1} + \dfrac{1}{\tau_2}\right) + \dfrac{3}{4}\ln\dfrac{\tau_2}{\tau_1}},$$

and

$$\Psi = \frac{q_1}{\sigma(T_1^4 - T_2^4)} = 1 \left/ \left[\frac{1}{2}\left(1 + \frac{\tau_1}{\tau_2}\right) + \frac{3}{4}\tau_1 \ln \frac{\tau_2}{\tau_1}\right]\right. .$$

Results from the differential approximation are also included in Fig. 13-1. Note that the diffusion approximation with jump conditions outperforms the differential approximation over a large range of optical depths (since it has a higher order boundary condition), but fails much more severely in optically thin cases.

Like the Schuster-Schwarzschild approximation, the Milne-Eddington approximation may be generalized to higher order as well as to more general geometries. It is then known as the *moment method*, in which the radiative intensity is approximated by

$$\begin{aligned}
I(\mathbf{r}, \hat{\mathbf{s}}) &= I_0(\mathbf{r}) + I_{1x}(\mathbf{r})s_x + I_{1y}(\mathbf{r})s_y + I_{1z}(\mathbf{r})s_z + I_{2xx}(\mathbf{r})s_x^2 + I_{2xy}(\mathbf{r})s_x s_y + \cdots \\
&= I_0(\mathbf{r}) + \mathbf{I}_1(\mathbf{r}) \cdot \hat{\mathbf{s}} + \mathbf{I}_2(\mathbf{r}) : \hat{\mathbf{s}}\hat{\mathbf{s}} + \cdots .
\end{aligned} \qquad (13.46)$$

Here the $s_x = \hat{\mathbf{s}} \cdot \hat{\mathbf{i}} = \sin\theta \cos\psi$, $s_y = \hat{\mathbf{s}} \cdot \hat{\mathbf{j}} = \sin\theta \sin\psi$, and $s_z = \hat{\mathbf{s}} \cdot \hat{\mathbf{k}} = \cos\theta$ are the direction cosines of the unit direction vector $\hat{\mathbf{s}}$. I_0 is a scalar to be determined (related to G), $\mathbf{I}_1$ is a vector (related to q), $\mathbf{I}_2$ is a second-rank tensor (which may be related to radiation pressure), and so on. The unknowns are determined by taking moments of the equation of transfer, i.e., by integrating it over all directions after multiplication by 1, s_x, s_y, s_z, s_x^2, $s_x s_y$, ... The method has been shown by Krook [10] to be completely equivalent to the method of spherical harmonics (using spherical harmonics, which are functions of direction cosines, and exploiting their orthogonality properties). That method will be discussed in detail in Chapter 14.

13.5   THE EXPONENTIAL KERNEL APPROXIMATION

Another popular way to solve equations (13.1) and (13.2) in an approximate way is known as the *exponential kernel approximation*. In this method the *kernels* of the integrals [E_1 in equation (13.1), and E_2 in equation (13.2)], are approximated by functions of exponentials (e^x, $\cosh x$, $\sinh x$, $\cos x$, $\sin x$). Since these functions have repetitive derivatives (except for constant factors), this fact enables us to eliminate the integrals from equations (13.1) and (13.2) and transform them into differential equations. We shall demonstrate the method here by solving equation (13.2) with a very simple approximate kernel of the form

$$E_2(x) \simeq a e^{-bx}. \tag{13.47}$$

More elaborate approximations could consist of a sum of exponentials, for example, (as long as derivatives with respect to x are repetitive). To determine the "best" values for a and b, one could choose either to satisfy equation (13.47) at two selected points, or to satisfy equation (13.47) in an integral sense. How exactly the values of a and b are determined is somewhat arbitrary and should be decided by studying the problem at hand (is the medium optically thin, thick, covering all ranges?). The most common method of finding a and b is to take the zeroth and first moments of equation (13.47):

$$\int_0^\infty E_2(x)\, dx = -E_3(x)\Big|_0^\infty = \frac{1}{2} = a \int_0^\infty e^{-bx}\, dx = -\frac{a}{b} e^{-bx}\Big|_0^\infty = \frac{a}{b},$$

$$\int_0^\infty x E_2(x)\, dx = \frac{1}{3} = a \int_0^\infty x e^{-bx}\, dx = \frac{a}{b^2},$$

or

$$a = \tfrac{3}{4}, \quad b = \tfrac{3}{2}, \quad E_2(x) \simeq \tfrac{3}{4} e^{-3x/2}.$$

While $E_1(x)$ and $E_3(x)$ could be found in a similar fashion, it is usually preferred (for consistency, and numerical simplicity) to use the recursion formulae for exponential integrals, as given in Appendix E. Thus,

$$E_1(x) = -\frac{dE_2}{dx} \simeq \tfrac{9}{8} e^{-3x/2},$$

$$E_3(x) = \int_x^\infty E_2(x)\, dx \simeq \tfrac{1}{2} e^{-3x/2}.$$

A plot of these approximations is given in Fig. 13-3. With them equation (13.2)

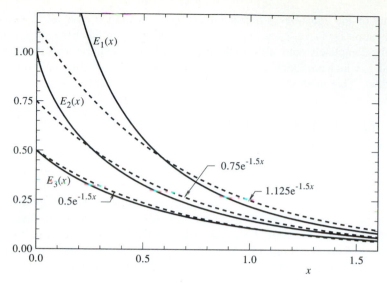

FIGURE 13-3
Approximations for exponential integrals.

may be rewritten as

$$q(\tau) = J_1 e^{-3\tau/2} - J_2 e^{-3(\tau_L - \tau)/2} + \frac{3\pi}{2} \left[\int_0^\tau S(\tau') e^{-3(\tau - \tau')/2} d\tau' \right.$$
$$\left. - \int_\tau^{\tau_L} S(\tau') e^{-3(\tau' - \tau)/2} d\tau' \right]. \qquad (13.48)$$

Differentiating this equation twice with respect to τ results in

$$\frac{d^2 q}{d\tau^2} = \frac{9}{4} J_1 e^{-3\tau/2} - \frac{9}{4} J_2 e^{-3(\tau_L - \tau)/2} + 3\pi \frac{dS}{d\tau}$$
$$+ \frac{27\pi}{8} \left[\int_0^\tau S(\tau') e^{-3(\tau - \tau')/2} d\tau' - \int_\tau^{\tau_L} S(\tau') e^{-3(\tau' - \tau)/2} d\tau' \right], \qquad (13.49)$$

or, using equation (13.48) to eliminate the integrals,

$$\frac{d^2 q}{d\tau^2} - \frac{9}{4} q = 3\pi \frac{dS}{d\tau} = 3\pi \frac{d}{d\tau} \left[(1 - \omega) I_b + \frac{\omega}{4\pi} G \right]. \qquad (13.50)$$

The source function is either known or must be determined by performing a similar procedure on equation (13.1). Equation (13.50) is a second-order differential equation and, thus, requires two boundary conditions (while an integral equation does not require any boundary conditions). The problem is overcome by substituting the solution to equation (13.50) back into equation (13.48). Since two boundary conditions are required it is sufficient to do this at two selected locations, say $\tau = 0$ and $\tau = \tau_L$.

Example 13.7. Redo Example 13.5 using the exponential kernel approximation.

Solution

For $\omega = 0$ we have $S = I_b = \sigma T^4/\pi$, and for radiative equilibrium $dq/d\tau = 0$ (and, therefore, $d^2q/d\tau^2 = 0$). Thus, from equation (13.50), we get

$$q = -\frac{4\sigma}{3}\frac{dT^4}{d\tau} = \text{const},$$

which is the same as for the diffusion approximation as well as for the differential approximation. Integration gives

$$\sigma T^4 = C - \tfrac{3}{4}q\tau.$$

Substituting this expression into equation (13.48) leads to

$$q = \sigma T_1^4 e^{-3\tau/2} - \sigma T_2^4 e^{-3(\tau_L - \tau)/2} + \frac{3}{2}\left[\int_0^\tau \left(C - \frac{3}{4}q\tau'\right)e^{-3(\tau - \tau')/2}d\tau'\right.$$

$$\left. - \int_\tau^{\tau_L} \left(C - \frac{3}{4}q\tau'\right)e^{-3(\tau' - \tau)/2}d\tau'\right]$$

$$= \left(\sigma T_1^4 - C - \frac{q}{2}\right)e^{-3\tau/2} - \left(\sigma T_2^4 - C + \frac{q}{2} + \frac{3}{4}q\tau_L\right)e^{-3(\tau_L - \tau)/2} + q.$$

Since this equation must hold for all values of τ, both expressions within the parentheses must vanish,

$$\sigma T_1^4 = C + \frac{q}{2},$$

$$\sigma T_2^4 = C - \frac{q}{2} - \frac{3}{4}\tau_L q,$$

or

$$\Psi = \frac{q}{\sigma(T_1^4 - T_2^4)} = \frac{1}{1 + \frac{3}{4}\tau_L}.$$

This result is identical to the ones from the diffusion approximation with jump condition, and from the differential approximation, Example 13.5.

References

1. Rosseland, S.: *Theoretical Astrophysics; Atomic Theory and the Analysis of Stellar Atmospheres and Envelopes*, Clarendon Press, Oxford, 1936.
2. Deissler, R. G.: "Diffusion Approximation for Thermal Radiation in Gases with Jump Boundary Conditions," *ASME Journal of Heat Transfer*, vol. 86, pp. 240–246, 1964.
3. Schuster, A.: "Radiation through a Foggy Atmosphere," *Astrophysical Journal*, vol. 21, pp. 1–22, 1905.
4. Schwarzschild, K.: "Über das Gleichgewicht der Sonnenatmosphären, (Equilibrium of the Sun's Atmosphere)," *Akad. Wiss. Göttingen, Math.-Phys. Kl. Nachr.*, vol. 195, pp. 41–53, 1906.

5. Chin, J. H., and S. W. Churchill: "Anisotropic, Multiply Scattered Radiation from an Arbitrary, Cylindrical Source in an Infinite Slab," *ASME Journal of Heat Transfer*, vol. 87, pp. 167–172, 1965.

6. Shih, T.-M., and Y. N. Chen: "A Discretized-Intensity Method Proposed for Two-Dimensional Systems Enclosing Radiative and Conductive Media," *Numerical Heat Transfer*, vol. 6, pp. 117–134, 1983.

7. Chan, S. H.: "Numerical Methods for Multidimensional Radiative Transfer Analysis in Participating Media," in *Annual Review of Numerical Fluid Mechanics and Heat Transfer*, vol. 1, Hemisphere, New York, pp. 305–350, 1987.

8. Milne, F. A.: "Thermodynamics of the Stars," in *Handbuch der Astrophysik*, Springer-Verlag, OHG, Berlin, pp. 65–255, 1930.

9. Eddington, A. S.: *The Internal Constitution of the Stars*, Dover Publications, New York, 1959.

10 Krook, M.: "On the Solution of the Equation of Transfer, I," *Astrophysical Journal*, vol. 122, pp. 488–497, 1955.

Problems

13.1 Derive the jump boundary condition for the diffusion approximation, equation (13.26), for the case of concentric cylinders. Assume the heat transfer to be one-dimensional (only radial, no azimuthal or axial dependence).

Hint: Introduce a local Cartesian coordinate system at a point at the boundary, express any other r-location within the medium in terms of x, y, z, and transform the derivatives in equation (13.26) to r-derivatives; finally let x, y, z go to zero (since the derivatives at the boundary are needed).

13.2 The gap between two parallel black plates at T_1 and T_2, respectively, is filled with a particle-laden gas. Radiative equilibrium prevails, the particle loading is a fixed volume fraction, with particles manufactured from two different materials (one a specular reflector, the other a diffuse reflector, both having the same ϵ). Sketch nondimensional heat flux $\Psi = q/\sigma(T_1^4 - T_2^4)$ vs. particle size (but keeping volume-fraction constant).

13.3 Consider radiative equilibrium of a gray, absorbing, emitting, and isotropically scattering medium contained between two isothermal, gray-diffuse, parallel plates spaced a distance L apart. Determine the nondimensional temperature variation within the medium, $\Phi = (\sigma T^4 - J_2)/(J_1 - J_2)$ for the optically thin case ($\tau_L \ll 1$).

13.4 Consider a gray, absorbing-emitting, linear-anisotropically scattering medium at radiative equilibrium. The medium is confined between two parallel, isothermal, black plates (at temperatures T_1 and T_2). Determine an expression for the radiative heat flux between the two plates using the diffusion approximation with jump boundary conditions.

13.5 Do Problem 13.4 using the Schuster-Schwarzschild (2-flux) approximation.

13.6 Do Problem 13.4 using the Milne-Eddington (differential) approximation.

13.7 Do Problem 13.4 using the exponential kernel approximation method.

13.8 Consider a space enclosed by infinite, diffuse-gray, parallel plates 1 m apart filled with a gray, nonscattering medium ($\kappa = 5\,\mathrm{m}^{-1}$). The surfaces are isothermal (both at $T_w = 500\,\mathrm{K}$ with emissivity $\epsilon_w = 0.6$), and there is uniform and constant heat generation within the medium per unit volume, $\dot{Q}''' = 10^6\,\mathrm{W/m^3}$. Conduction and convection are negligible such that $\nabla \cdot \mathbf{q} = \dot{Q}'''$. Determine the radiative heat flux to the walls as well as the maximum temperature within the medium, using the diffusion approximation with jump boundary conditions.

13.9 Do Problem 13.8 using the Schuster-Schwarzschild approximation.

13.10 Do Problem 13.8 using the Milne-Eddington approximation.

13.11 Do Problem 13.8, using the exponential-kernel approximation. Note: The necessary exact integral relations have been given in Problem 12.2.

13.12 Do Problem 12.4 using the Milne-Eddington (differential) approximation.

13.13 Do Problem 12.8 using the Milne-Eddington (differential) approximation.

13.14 A material produces an amount of heat that is constant per unit volume, i.e., $\dot{Q}''' =$ const. This heat production needs to be removed by thermal radiation. It is proposed to grind up the (fixed volume of) material into small particles, which are to be suspended evenly between two cold plates of (identical) emissivity ϵ. Since it is important to keep the overall temperature level in the particles as low as possible, should the particles be ground as fine as possible, as large as possible, or does some optimum radius exist? What is the optimum particle size, and what is the maximum temperature if this size is employed? You may assume one-dimensional parallel plates with a constant volume fraction of particles, black particles with relatively large size parameters, and you may use the Schuster-Schwarzschild approximation.

13.15 Do Problem 13.14 using the Milne-Eddington (differential) approximation.

13.16 Do Problem 13.14 using the exponential-kernel approximation.

13.17 Consider parallel, black plates, spaced 1 m apart, at constant temperatures T_1 and T_2. Due to pressure variations, the (gray) absorption coefficient is equal to

$$\kappa = \kappa_0 + \kappa_1 z; \quad \kappa_0 = 0.01 \, \text{cm}^{-1}; \quad \kappa_1 = 0.0002 \, \text{cm}^{-2},$$

where z is measured from Plate 1. The medium does not scatter radiation. Determine, for radiative equilibrium, the nondimensional heat flux $\Psi = q/\sigma(T_1^4 - T_2^4)$ by (a) the exact method, (b) the regular diffusion approximation, (c) the diffusion approximation with jump boundary conditions, (d) the two-flux method, (e) the differential approximation, and (f) the kernel approximation.

CHAPTER
14

THE METHOD OF
SPHERICAL
HARMONICS
(P_N-APPROXIMATION)

14.1 INTRODUCTION

For a gray medium (or on a spectral basis) with known temperature distribution (or for the case of radiative equilibrium), the general problem of radiative transfer entails determining the radiative intensity from an integro-differential equation in five independent variables—three space coordinates and two direction coordinates—a prohibitive task. The method of *spherical harmonics* provides a vehicle to obtain an approximate solution of arbitrarily high order (i.e., accuracy), by transforming the equation of transfer into a set of simultaneous partial differential equations. The approach was first proposed by Jeans [1] in his work on radiative transfer in stars. Further description of the method may be found in the books by Kourganoff [2], Davison [3] and Murray [4] (the latter two dealing with the closely related neutron transport theory). The *spherical harmonics method* is identical to the *moment method* described in Chapter 13, except that moments are taken in such a way as to take advantage of the orthogonality of spherical harmonics.

In this chapter we shall first develop the set of partial differential equations for the general P_N-method for one-dimensional plane-parallel media and their boundary conditions.[1] Next we deal in more detail with the most popular P_1-approximation for arbitrary geometries. Then a brief presentation of the P_3 and higher-order approximation is given.

The great advantage of the method of spherical harmonics is the conversion of the governing equation to relatively simple partial differential equations. The drawback of the method is that low-order approximations are usually only accurate in optically thick media, and accuracy improves only slowly for higher-order approximations while mathematical complexity increases extremely rapidly. Therefore, this chapter includes a discussion of a number of variations on the P_1-approximation that attempt to overcome its inaccuracy in optically thin situations, most notably the *modified differential approximation (MDA)* and the *improved differential approximation (IDA)*. Both methods separate radiation emanating from walls from the radiation emanating from within the medium, either before *(MDA)* or after *(IDA)* applying the P_1-approximation. While these methods deliver excellent accuracy, they are no longer the solution to a simple partial differential equation, but also require the evaluation of some integral correction factors.

14.2 DEVELOPMENT OF THE GENERAL P_N-APPROXIMATION

We may think of the radiative intensity field $I(\mathbf{r}, \hat{\mathbf{s}})$[2] at location $\mathbf{r}$ within the medium as the value of a scalar function on the surface of a sphere of unit radius, surrounding the point $\mathbf{r}$. Any such function may be expressed in terms of a two-dimensional generalized Fourier series as

$$I(\mathbf{r}, \hat{\mathbf{s}}) = \sum_{l=0}^{\infty} \sum_{m=-l}^{l} I_l^m(\mathbf{r})Y_l^m(\hat{\mathbf{s}}),\tag{14.1}$$

where the $I_l^m(\mathbf{r})$ are position-dependent coefficients and the $Y_l^m(\hat{\mathbf{s}})$ are *spherical harmonics*, given by

$$Y_l^m(\hat{\mathbf{s}}) = (-1)^{(m+|m|)/2}\left[\frac{(l-|m|)!}{(l+|m|)!}\right]^{1/2} e^{im\psi}P_l^{|m|}(\cos\theta),\tag{14.2}$$

that satisfy Laplace's equation in spherical coordinates. Here θ and ψ are the polar and azimuthal angles describing the direction unit vector $\hat{\mathbf{s}}$, respectively, and the P_l^m are *associated Legendre polynomials*.

---

[1] The reader only interested in the P_1-approximation may skip directly to Section 14.4 after reading the first three paragraphs of Section 14.2.

[2] All relations in this chapter are valid on a spectral basis and, for a gray medium, also for total quantities. For notational simplicity we omit any subscript used to emphasize the spectral nature of quantities.

We may substitute equation (14.1) into the general equation of radiative transfer, equation (8.23),

$$\hat{s} \cdot \nabla_\tau I + I = (1 - \omega)I_b + \frac{\omega}{4\pi} \int_{4\pi} I(\hat{s}') \Phi(\hat{s} \cdot \hat{s}') \, d\Omega', \qquad (14.3)$$

where space coordinates have been nondimensionalized using the extinction coefficient, i.e., $d\tau = \beta \, ds$ (as indicated by the subscript τ in ∇_τ). Equation (14.3) requires the outgoing intensity to be specified everywhere along the surface of the enclosure. Equation (14.3) is multiplied by Y_k^n after also expanding the scattering phase function into a series of Legendre polynomials, equation (10.88), followed by integration over all directions. Exploiting the orthogonality properties of spherical harmonics [5] leads to infinitely many coupled partial differential equations in the unknown position-dependent functions $I_l^m(\mathbf{r})$.[3] Up to this point the above representation is an exact method for the determination of the intensity field. To simplify the problem an approximation is now made by truncating the series in equation (14.1) after very few terms. By doing so, we have replaced the single unknown I (which is a function of space and direction) by $1 + 3 + \cdots + (2N + 1) = (N + 1)^2$ unknown I_l^m that are functions of space only. Therefore, we need to replace equation (14.3) (a function of space and direction) by $(N + 1)^2$ equations (which are functions of space only). This is achieved by multiplying equation (14.3) by Y_k^m and integrating over all directions.

The highest value for l retained, N, gives the method its order and its name. Most often employed is the P_N or *differential approximation* ($l = 0, 1$), while the P_3-approximation ($l = 0, 1, 2, 3$) has been used a few times. It is known from neutron transport theory that approximations of odd order are more accurate than even ones of next highest order, so that the P_2-approximation is never used. In most early developments and applications the P_N-method was derived only for the one-dimensional plane-parallel case, for example, as in Jeans [1], Kourganoff [2], and Krook [6]. A detailed derivation of the general three-dimensional case in Cartesian coordinates has been given by Cheng [7, 8]. The extension to general coordinate systems has been given by Ou and Liou [9]. Another general three-dimensional derivation has been given by Condiff [10], who expanded the intensity in terms of *polyadic Legendre polynomials* given by Brenner [11], that is, Legendre functions $P_n(\hat{s})$ whose arguments are tensors of order n (rather than scalars).

We shall now develop the general P_N-method in some detail for the one-dimensional plane-parallel medium, in order to (i) shed further light on the general method, and (ii) facilitate the difficult problem of developing a consistent set of boundary conditions. For such a simple case the intensity does not depend on azimuthal angle ψ (assuming the polar angle θ is measured from an axis perpendicular to the plates, as shown in

---

[3] Obviously, a thorough understanding of the method requires the reader to be familiar with the method of separation-of-variables and generalized Fourier series, as applied to the solution of linear partial differential equations.

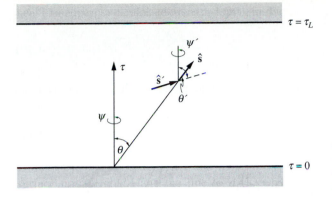

FIGURE 14-1
Coordinates for the
one-dimensional plane-parallel
medium.

Fig. 14-1), i.e., $I_l^m = 0$ for $m \neq 0$. Thus, equation (14.1) may be simplified to

$$I(\tau, \mu) \simeq \sum_{l=0}^{N} I_l(\tau) P_l(\mu), \tag{14.4}$$

where we set $\mu = \cos\theta$ and omitted the superscript "0" from I_l since it is no longer necessary. The scattering phase function for such a medium, expanded into Legendre polynomials, is [see equation (12.12)]

$$\Phi(\mu, \mu') = \sum_{m=0}^{M} A_m P_m(\mu') P_m(\mu), \tag{14.5}$$

where M is the order of approximation for the phase function; and we find

$$\int_{-1}^{1} \Phi(\mu, \mu') I(\tau, \mu') \, d\mu' = \sum_{l=0}^{N} I_l(\tau) \sum_{m=0}^{M} A_m P_m(\mu) \int_{-1}^{1} P_l(\mu') P_m(\mu') \, d\mu'. \tag{14.6}$$

We may now utilize the orthogonality of Legendre polynomials (see, for example, Abramowitz and Stegun [12]), to write

$$\int_{-1}^{1} P_l(\mu) P_m(\mu) \, d\mu = \frac{2\delta_{lm}}{2m+1} = \begin{cases} 0 & \text{for } m \neq l, \\ \dfrac{2}{2m+1} & \text{for } m = l. \end{cases} \tag{14.7}$$

Employing this orthogonality relation in equation (14.6) leads to

$$\int_{-1}^{1} \Phi(\mu, \mu') I(\tau, \mu') \, d\mu' = \sum_{l=0}^{N} \frac{2A_l}{2l+1} I_l(\tau) P_l(\mu), \tag{14.8}$$

where it is implied that $A_l = 0$ for $l > M$. (On the other hand, if $M > N$, the A_l for $l = N+1, \ldots, M$ disappear and this information about the phase function is lost in the N-th order approximation.) We may now recast the equation of transfer for the one-dimensional plane-parallel medium as

$$\mu \frac{dI}{d\tau} + I(\tau) = (1 - \omega) I_b(\tau) + \frac{\omega}{2} \int_{-1}^{1} \Phi(\mu, \mu') I(\tau, \mu') \, d\mu', \tag{14.9}$$

or

$$\sum_{l=0}^{N} \left[\frac{dI_l}{d\tau} \mu P_l(\mu) + I_l(\tau) P_l(\mu) \right] = (1 - \omega) I_b(\tau) + \omega \sum_{l=0}^{N} \frac{A_l I_l(\tau)}{2l + 1} P_l(\mu). \qquad (14.10)$$

To exploit the orthogonality of the Legendre polynomials, we shall use the recursion relation [12]

$$(2l + 1)\mu P_l(\mu) = l P_{l-1}(\mu) + (l + 1) P_{l+1}(\mu). \qquad (14.11)$$

Thus, we may recast equation (14.10) as

$$\sum_{l=0}^{N} \left\{ \frac{I_l'(\tau)}{2l + 1} [l P_{l-1}(\mu) + (l + 1) P_{l+1}(\mu)] + I_l(\tau) P_l(\mu) \right\}$$

$$= (1 - \omega) I_b(\tau) + \sum_{l=0}^{N} \frac{\omega A_l I_l(\tau)}{2l + 1} P_l(\mu), \qquad (14.12)$$

where the prime denotes differentiation with respect to τ. Since we have introduced $(N + 1)$ new variables, $I_0, I_1, \ldots I_N$, we need to convert equation (14.12) into $(N + 1)$ equations independent of direction. Thus, multiplying by $P_k(\mu)$ $(k = 0, 1, \ldots, N)$ and integrating over all μ leads to

$$\frac{k + 1}{2k + 3} I_{k+1}'(\tau) + \frac{k}{2k - 1} I_{k-1}'(\tau) + \left(1 - \frac{\omega A_k}{2k + 1}\right) I_k(\tau) = (1 - \omega) I_b(\tau) \delta_{0k},$$

$$k = 0, 1, \ldots, N, \qquad (14.13)$$

where equation (14.7) has been utilized. Equation (14.13) is a set of $(N + 1)$ simultaneous first-order ordinary differential equations for the unknown functions $I_0(\tau)$, $I_1(\tau), \ldots, I_N(\tau)$.[4] As such it requires a set of $(N + 1)$ boundary conditions for its solution.

14.3 BOUNDARY CONDITIONS FOR THE P_N-METHOD

The equation of radiative transfer, equation (14.3), is a first-order partial differential equation in intensity, requiring a boundary condition of the type

$$I(\mathbf{r} = \mathbf{r}_w, \hat{\mathbf{s}}) = I_w(\mathbf{r}_w, \hat{\mathbf{s}}) \text{ for } \hat{\mathbf{n}} \cdot \hat{\mathbf{s}} > 0 \qquad (14.14)$$

everywhere on the surface, that is, the intensity leaving a surface (described by the vector $\mathbf{r}_w$) must be prescribed in some fashion for all outgoing directions $\hat{\mathbf{n}} \cdot \hat{\mathbf{s}} > 0$ (with $\hat{\mathbf{n}}$ being the outward surface normal), as shown in Fig. 14-2.

---

[4]Remember that equation (14.4) is truncated beyond $l = N$, so that $I_{N+1}(\tau) = 0$.

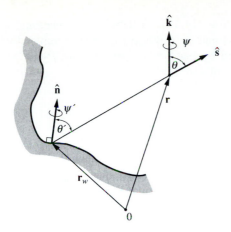

FIGURE 14-2
Prescribed boundary intensities for P_N-method.

When the P_N-approximation is applied [truncating equation (14.1) after $l = N$] this boundary condition can no longer be satisfied and must be replaced by one that either satisfies equation (14.14) at selected directions $\hat{s}_i$ or satisfies it in an integral sense. Mark [13, 14] and Marshak [15] proposed two different sets of boundary conditions for the spherical harmonics method as applied to neutron transport within a one-dimensional plane-parallel medium.

Mark's Boundary Condition

For a one-dimensional slab of optical thickness τ_L, equation (14.14) may be rewritten as

$$I(0, \mu) = I_{w1}(\mu), \qquad 0 < \mu < 1, \qquad (14.15a)$$
$$I(\tau_L, \mu) = I_{w2}(\mu), \qquad -1 < \mu < 0, \qquad (14.15b)$$

where I_{w1} and I_{w2} are the prescribed intensities at Surfaces 1 ($\tau = 0$) and 2 ($\tau = \tau_L$).[5]

The P_N-method for such a medium, equation (14.13), requires $(N + 1)$ boundary conditions, say $\frac{1}{2}(N + 1)$ each, at $\tau = 0$ and $\tau = \tau_L$ (assuming that N is odd). Noting that the equation

$$P_{N+1}(\mu) = 0 \qquad (14.16)$$

has precisely $\frac{1}{2}(N+1)$ roots μ_i with values between 0 and 1, Mark suggested replacing the boundary conditions of equation (14.15) by

$$I(0, \mu = \mu_i) = I_{w1}(\mu_i), \qquad i = 1, 2, \ldots, \tfrac{1}{2}(N + 1), \qquad (14.17a)$$
$$I(\tau_L, \mu = -\mu_i) = I_{w2}(-\mu_i), \qquad i = 1, 2, \ldots, \tfrac{1}{2}(N + 1), \qquad (14.17b)$$

---

[5] We include the subscript w here to distinguish the I_{wi} from the intensity moments I_i defined by equation (14.4).

where the μ_i are the positive roots of equation (14.16). A detailed explanation for this choice has been given by Mark [13, 14] and by Davison [3]. For example, for the P_1-approximation for a medium bounded by black walls we get with $P_2(\mu) = \frac{1}{2}(3\mu^2 - 1)$, $\mu_1 = 1/\sqrt{3}$ and, from equation (14.4),

$$I\left(0, \mu = \frac{1}{\sqrt{3}}\right) = I_0(0) + \frac{I_1(0)}{\sqrt{3}} = I_{b1}, \tag{14.18a}$$

$$I\left(\tau_L, \mu = -\frac{1}{\sqrt{3}}\right) = I_0(\tau_L) - \frac{I_1(\tau_L)}{\sqrt{3}} = I_{b2}. \tag{14.18b}$$

One serious drawback of Mark's boundary conditions is the fact that they are difficult, if not impossible, to apply to more complicated geometries.

Marshak's Boundary Conditions

An alternative set of boundary conditions for the one-dimensional plane-parallel P_N-approximation was proposed by Marshak, who suggested that equation (14.15) be satisfied in an integral sense by setting

$$\int_0^1 I(0, \mu)P_{2i-1}(\mu)\,d\mu = \int_0^1 I_{w1}(\mu)P_{2i-1}(\mu)\,d\mu,$$

$$i = 1, 2, \ldots, \tfrac{1}{2}(N + 1); \tag{14.19a}$$

$$\int_{-1}^0 I(\tau_L, \mu)P_{2i-1}(\mu)\,d\mu = \int_{-1}^0 I_{w2}(\mu)P_{2i-1}(\mu)\,d\mu,$$

$$i = 1, 2, \ldots, \tfrac{1}{2}(N + 1). \tag{14.19b}$$

Again, the reason for choosing all the Legendre polynomials of odd order has been explained in detail by Marshak [15] and Davison [3]. Since the orthogonality properties of Legendre polynomials do not hold for half-range integrals, such as the ones in equation (14.19), it is often more convenient to replace equation (14.19) by the equivalent statement

$$\int_0^1 I(0, \mu)\mu^{2i-1}\,d\mu = \int_0^1 I_{w1}(\mu)\mu^{2i-1}\,d\mu,$$

$$i = 1, 2, \ldots, \tfrac{1}{2}(N + 1); \tag{14.20a}$$

$$\int_{-1}^0 I(\tau_L, \mu)\mu^{2i-1}\,d\mu = \int_{-1}^0 I_{w2}(\mu)\mu^{2i-1}\,d\mu,$$

$$i = 1, 2, \ldots, \tfrac{1}{2}(N + 1). \tag{14.20b}$$

As an example we again consider the P_1-approximation for a medium bounded by black walls. Then

$$\int_0^1 I(0, \mu)\mu\,d\mu = \int_0^1 [I_0(0) + I_1(0)\mu]\,\mu\,d\mu = \int_0^1 I_{b1}\mu\,d\mu,$$

or

$$I_0(0) + \tfrac{2}{3}I_1(0) = I_{b1},\tag{14.21a}$$

$$I_0(\tau_L) - \tfrac{2}{3}I_1(\tau_L) = I_{b2}.\tag{14.21b}$$

We note that replacing the factor 2 in Marshak's boundary condition by a $\sqrt{3}$ converts it to Mark's boundary condition.

One advantage of Marshak's boundary condition is that it may be extended to more general problems, although not painlessly. Note that the integration in equation (14.19) is carried out over all directions above the surface (i.e., a hemisphere) with the Legendre polynomials of equation (14.4) as weight factors. Thus, it appears natural to generalize the boundary condition to (see Fig. 14-2)

$$\int_{\hat{\mathbf{n}}\cdot\hat{\mathbf{s}}>0} I(\mathbf{r}_w, \hat{\mathbf{s}})Y_{2i-1}^m(\hat{\mathbf{s}})\,d\Omega = \int_{\hat{\mathbf{n}}\cdot\hat{\mathbf{s}}>0} I_w(\hat{\mathbf{s}})Y_{2i-1}^m(\hat{\mathbf{s}})\,d\Omega,$$

$$i = 1, 2, \ldots, \tfrac{1}{2}(N+1), \quad \text{all relevant } m.\tag{14.22}$$

The statement "all relevant m" rather than $-i \le m \le +i$ appears in equation (14.22) since not all terms in equation (14.1) may be nonzero as a result of symmetry. For example, for a one-dimensional plane-parallel medium there is no azimuthal dependence, so that the only "relevant" value for m is $m = 0$. This term leads to a single boundary condition on each surface for the P_1-approximation (as already seen to be correct), two for the P_3-approximation, and so on. Unfortunately, equation (14.22) still leads to too many boundary conditions in many situations. For example, for the P_1-approximation for a general three-dimensional medium without symmetry, equation (14.22) leads to three boundary conditions everywhere ($i = 1, m = 0, \pm 1$), while only one is needed (as explained in the following section). Davison [3] has shown that the number of superfluous conditions is always at least one less than the relevant m at $i = \tfrac{1}{2}(N+1)$. Thus, on intuitive grounds it is accepted practice to satisfy equation (14.22) for all relevant m for $i = 1, 2, \ldots, \tfrac{1}{2}(N-1)$, and for as many relevant m as possible for $i = \tfrac{1}{2}(N+1)$. For $i = \tfrac{1}{2}(N+1)$, the $Y_N^m(\hat{\mathbf{s}})$ are expressed in terms of a local coordinate system, in which polar angle θ' is measured from the surface normal (i.e., $\cos\theta' = \hat{\mathbf{n}}\cdot\hat{\mathbf{s}}$), and azimuthal angle ψ' is measured on the surface. Using this local coordinate system, equation (14.22) for $i = \tfrac{1}{2}(N+1)$ is satisfied for as many m as possible, starting with the smallest $|m|$ [3].

Example 14.1. Consider the infinite quarter-space $\tau_x > 0$, $\tau_z > 0$ bounded by isothermal black surfaces at T_1 and T_2 as shown in Fig. 14-3. Develop the boundary conditions for the P_1-approximation at both surfaces (i.e., $\tau_x = 0$ and $\tau_z = 0$).

Solution

For the P_1-approximation equation (14.1) reduces to

$$I(\tau_x, \tau_z, \theta, \psi) = I_0^0(\tau_x, \tau_z) + I_1^{-1}(\tau_x, \tau_z)\frac{1}{\sqrt{2}}e^{-i\psi}P_1^1(\cos\theta)$$

$$+ I_1^0(\tau_x, \tau_z)P_1^0(\cos\theta) - I_1^1(\tau_x, \tau_z)\frac{1}{\sqrt{2}}e^{+i\psi}P_1^1(\cos\theta).$$

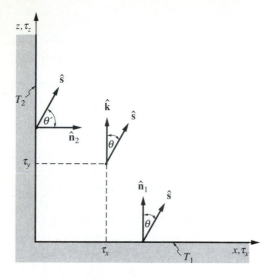

FIGURE 14-3
Geometry for Example 14.1.

For this two-dimensional problem it is convenient to measure polar angle θ from the τ_z-axis, and azimuthal angle ψ in the τ_x-τ_y-plane from the τ_x-axis. Then $I(\psi) = I(-\psi)$ and, with $P_1^0(\cos\theta) = \cos\theta$, $P_1^1(\cos\theta) = \sin\theta$, and $e^{\pm i\psi} = \cos\psi \pm i\sin\psi$,

$$
\begin{aligned}
I(\tau_x, \tau_y, \theta, \psi) &= I_0^0 + I_1^0 \cos\theta + \frac{1}{\sqrt{2}}(I_1^{-1} - I_1^1)\cos\psi\sin\theta \\
&\quad - \frac{i}{\sqrt{2}}(I_1^{-1} + I_1^1)\sin\psi\sin\theta \\
&= I_0^0 + I_1^0 \cos\theta + I_1^* \cos\psi\sin\theta,
\end{aligned}
$$

since the last term must vanish owing to symmetry. Therefore, equation (14.22) is able to provide two boundary conditions everywhere on the surface ($i = 1$ and $m = 0, 1$), while we need only one (as to be developed in the next section). Thus, following the discussion of equation (14.22), we introduce local direction coordinate systems on the surfaces and satisfy equation (14.22) only for $m = 0$. For the bottom surface, $\tau_z = 0$, the problem is simple since the surface normal is parallel to the τ_z-axis, from which the polar angle is measured. Thus,

$$
\int_{\psi=0}^{2\pi}\int_{\theta=0}^{\pi/2}\left(I_0^0 + I_1^0 \cos\theta + I_1^* \cos\psi\sin\theta\right)\cos\theta \sin\theta\, d\theta\, d\psi
$$

$$
= \int_0^{2\pi}\int_0^{\pi/2} I_{b1}\cos\theta \sin\theta\, d\theta\, d\psi,
$$

or

$$
I_0^0(\tau_x, 0) + \tfrac{2}{3}I_1^0(\tau_x, 0) = I_{b1}.
$$

At the vertical surface ($\tau_x = 0$) $P_1^0 = \cos\theta'$, where θ' is the angle between a direction vector and the surface normal $\hat{n} = \hat{i}$. Thus, with $\cos\theta' = \hat{s}\cdot\hat{i}$ and $\hat{s} = \sin\theta(\cos\psi\hat{i}+$

$\sin \psi \hat{\mathbf{j}}) + \cos \theta \hat{\mathbf{k}}$, it follows that $\cos \theta' = \sin \theta \cos \psi$ and

$$\int_{\psi=-\pi/2}^{\pi/2} \int_{\theta=0}^{\pi} \left(I_0^0 + I_1^0 \cos \theta + I_1^* \cos \psi \sin \theta \right) \sin \theta \cos \psi \sin \theta \, d\theta \, d\psi$$

$$= \pi \left(I_0^0 + \frac{2}{3} I_1^* \right) = \pi I_{b2},$$

or

$$I_0^0(0, \tau_z) + \frac{2}{3} I_1^*(0, \tau_z) = I_{b2}.$$

We shall see in the next section that I_0^0 is directly proportional to incident radiation, while I_1^0 and I_1^* are proportional to radiative heat flux into the τ_y- and τ_x-directions, respectively.

Davison [3] stated that for low-order approximations Marshak's boundary conditions would give superior results, but that for high-order approximations Mark's boundary conditions should be more accurate. However, subsequent numerical work by Pelland [16] and Schmidt and Gelbard [17] showed Marshak's boundary condition leads to more accurate results, even in high-order approximations.

14.4   THE P_1-APPROXIMATION

If the series in equation (14.1) is truncated beyond $l = 1$ (i.e., $I_l^m = 0$ for $l \geq 2$), we get the lowest-order, or P_1, approximation, or

$$I(\mathbf{r}, \hat{\mathbf{s}}) = I_0^0 Y_0^0 + I_1^{-1} Y_1^{-1} + I_1^0 Y_1^0 + I_1^1 Y_1^1. \tag{14.23}$$

From standard mathematical texts, such as MacRobert [18], we find the associated Legendre polynomials as $P_0^0 = 1, P_1^0 = \cos \theta, P_1^1 = \sin \theta$, and, using equation (14.2),

$$I(\mathbf{r}, \theta, \psi) = I_0^0 + I_1^0 \cos \theta - \frac{1}{\sqrt{2}} \left(I_1^1 e^{i\psi} - I_1^{-1} e^{-i\psi} \right) \sin \theta$$

$$= I_0^0 + I_1^0 \cos \theta + \frac{1}{\sqrt{2}} (I_1^{-1} - I_1^1) \sin \theta \cos \psi$$

$$- \frac{i}{\sqrt{2}} (I_1^{-1} + I_1^1) \sin \theta \sin \psi \tag{14.24}$$

We notice that equation (14.24) has four terms: The first term is independent of direction, the second is proportional to the z-component of the direction vector $\hat{\mathbf{s}} = \sin \theta \cos \psi \hat{\mathbf{i}} + \sin \theta \sin \psi \hat{\mathbf{j}} + \cos \theta \hat{\mathbf{k}}$, the third is proportional to s_x and the last to s_y.[6] Each term is preceded by an unknown function of the space coordinates, which are

---

[6]Provided the polar angle is measured from the z-axis, and the azimuthal angle from the x-axis.

to be determined. Equation (14.24) may be written more compactly by introducing two new functions, a (a scalar) and $\mathbf{b}$ (a vector having three components) as

$$I(\mathbf{r}, \hat{\mathbf{s}}) = a(\mathbf{r}) + \mathbf{b}(\mathbf{r}) \cdot \hat{\mathbf{s}}. \tag{14.25}$$

Obviously, solving for the four unknowns—a and the three components of $\mathbf{b}$—is equivalent to determining the I_l^m (to which they can be directly related). Substituting equation (14.25) into the definition for incident radiation yields

$$G(\mathbf{r}) = \int_{4\pi} I(\mathbf{r}, \hat{\mathbf{s}}) \, d\Omega = a(\mathbf{r}) \int_{4\pi} d\Omega + \mathbf{b}(\mathbf{r}) \cdot \int_{4\pi} \hat{\mathbf{s}} \, d\Omega = 4\pi a(\mathbf{r}), \tag{14.26}$$

since

$$\int_{4\pi} \hat{\mathbf{s}} \, d\Omega = \int_0^{2\pi} \int_0^{\pi} \begin{pmatrix} \sin\theta\cos\psi \\ \sin\theta\sin\psi \\ \cos\theta \end{pmatrix} \sin\theta \, d\theta \, d\psi = \begin{pmatrix} 0 \\ 0 \\ 0 \end{pmatrix} = \mathbf{0}. \tag{14.27}$$

Similarly, substituting equation (14.25) into the definition for the radiative heat flux gives

$$\mathbf{q}(\mathbf{r}) = \int_{4\pi} I(\mathbf{r}, \hat{\mathbf{s}}) \hat{\mathbf{s}} \, d\Omega = a(\mathbf{r}) \int_{4\pi} \hat{\mathbf{s}} \, d\Omega + \mathbf{b}(\mathbf{r}) \cdot \int_{4\pi} \hat{\mathbf{s}}\hat{\mathbf{s}} \, d\Omega = \frac{4\pi}{3} \mathbf{b}(\mathbf{r}), \tag{14.28}$$

since

$$\int_{4\pi} \hat{\mathbf{s}}\hat{\mathbf{s}} \, d\Omega = \int_0^{2\pi} \int_0^{\pi} \begin{pmatrix} \sin^2\theta\cos^2\psi & \sin^2\theta\sin\psi\cos\psi & \sin\theta\cos\theta\cos\psi \\ \sin^2\theta\sin\psi\cos\psi & \sin^2\theta\sin^2\psi & \sin\theta\cos\theta\sin\psi \\ \sin\theta\cos\theta\cos\psi & \sin\theta\cos\theta\sin\psi & \cos^2\theta \end{pmatrix}$$
$$\times \sin\theta \, d\theta \, d\psi$$

$$= \int_0^{\pi} \begin{pmatrix} \pi\sin^2\theta & 0 & 0 \\ 0 & \pi\sin^2\theta & 0 \\ 0 & 0 & 2\pi\cos^2\theta \end{pmatrix} \sin\theta \, d\theta$$

$$= \frac{4\pi}{3} \begin{pmatrix} 1 & 0 & 0 \\ 0 & 1 & 0 \\ 0 & 0 & 1 \end{pmatrix} = \frac{4\pi}{3} \boldsymbol{\delta}, \tag{14.29}$$

where $\boldsymbol{\delta}$ is the unit tensor, and $\mathbf{b} \cdot \boldsymbol{\delta} = \mathbf{b}$. Therefore, we may rewrite equation (14.25) in terms of incident radiation and radiative heat flux as

$$I(\mathbf{r}, \hat{\mathbf{s}}) = \frac{1}{4\pi} [G(\mathbf{r}) + 3\mathbf{q}(\mathbf{r}) \cdot \hat{\mathbf{s}}]. \tag{14.30}$$

The preceding development is useful to show that equation (14.30) indeed corresponds to the lowest order of the P_N-approximation, equation (14.1). Of course, equation (14.30) should have physical significance and it should be possible to derive it from physical principles. This was done by Modest [19], who treated radiation as a "photon gas" with momentum and energy, and derived the intensity field through quantum statistics. He showed that the average photon velocity (which is proportional to heat flux) is inversely proportional to optical thickness, and that equation (14.30) holds for a location a large optical distance away from any points not at thermodynamic equilibrium (sharp temperature gradients, steps in temperature, etc.).

Now, substituting equation (14.30) into equation (14.3) and assuming linear-anisotropic scattering,[7]

$$\Phi(\hat{\mathbf{s}} \cdot \hat{\mathbf{s}}') = 1 + A_1 \hat{\mathbf{s}} \cdot \hat{\mathbf{s}}', \tag{14.31}$$

leads to

$$
\begin{aligned}
\int_{4\pi} I(\hat{\mathbf{s}}') \Phi(\hat{\mathbf{s}} \cdot \hat{\mathbf{s}}') \, d\Omega' &= \frac{1}{4\pi} \int_{4\pi} (G + 3\mathbf{q} \cdot \hat{\mathbf{s}}')(1 + A_1 \hat{\mathbf{s}} \cdot \hat{\mathbf{s}}') \, d\Omega' \\
&= \frac{G}{4\pi} \left[\int_{4\pi} d\Omega' + A_1 \hat{\mathbf{s}} \cdot \int_{4\pi} \hat{\mathbf{s}}' d\Omega' \right] \\
&\quad + \frac{3\mathbf{q}}{4\pi} \cdot \left[\int_{4\pi} \hat{\mathbf{s}}' d\Omega' + A_1 \left(\int_{4\pi} \hat{\mathbf{s}}'\hat{\mathbf{s}}' d\Omega' \right) \cdot \hat{\mathbf{s}} \right] \\
&= G + A_1 \mathbf{q} \cdot \boldsymbol{\delta} \cdot \hat{\mathbf{s}} = G + A_1 \mathbf{q} \cdot \hat{\mathbf{s}}, \tag{14.32}
\end{aligned}
$$

where equations (14.27) and (14.29) have been employed (and the last step is easily verified by, say, using Cartesian coordinates and carrying out the dot product). Thus, equation (14.3) becomes

$$\frac{1}{4\pi} \nabla_\tau \cdot [\hat{\mathbf{s}}(G + 3\mathbf{q} \cdot \hat{\mathbf{s}})] + \frac{1}{4\pi}(G + 3\mathbf{q} \cdot \hat{\mathbf{s}}) \simeq (1-\omega)I_b + \frac{\omega}{4\pi}(G + A_1\mathbf{q} \cdot \hat{\mathbf{s}}), \tag{14.33}$$

where we were able to pull the direction vector $\hat{\mathbf{s}}$ inside the gradient, since direction is independent of position. Multiplying equation (14.33) by $Y_0^0 = 1$ and integrating over all solid angles gives

$$\nabla_\tau \cdot \mathbf{q} = (1 - \omega)(4\pi I_b - G), \tag{14.34}$$

where again equations (14.27) and (14.29) have been invoked. Equation (14.34) is, of course, identical to equation (8.54) since it does not depend on the functional form for intensity.

To obtain additional equations we may multiply equation (14.33) by $Y_1^m(m = -1, 0, +1)$ or equivalently, by the components of the direction vector $\hat{\mathbf{s}}$. Choosing

---

[7]Because of the orthogonality of spherical harmonics the P_1-approximation remains unchanged for nonlinear anisotropic scattering. The choice of the functional form for intensity, equation (14.30), does not allow such scattering behavior, i.e., the medium must be so optically thick that any nonlinear anisotropically scattered intensity is smoothed out in the immediate vicinity of the scattering point. In reality, this smoothing implies that a "best" linear-anisotropic scattering factor A_1^* must be determined.

the latter and integrating over all directions leads to

$$\frac{1}{4\pi}\nabla_\tau \cdot \left[G \int_{4\pi} \hat{s}\hat{s}\, d\Omega + 3\mathbf{q} \cdot \int_{4\pi} \hat{s}\hat{s}\hat{s}\, d\Omega \right] + \frac{1}{4\pi}\left[G \int_{4\pi} \hat{s}\, d\Omega + 3\mathbf{q} \cdot \int_{4\pi} \hat{s}\hat{s}\, d\Omega \right]$$

$$= (1 - \omega)I_b \int_{4\pi} \hat{s}\, d\Omega + \frac{\omega}{4\pi}\left[G \int_{4\pi} \hat{s}\, d\Omega + A_1\mathbf{q} \cdot \int_{4\pi} \hat{s}\hat{s}\, d\Omega \right]. \tag{14.35}$$

It is easy to show that $\int_{4\pi} \hat{s}\hat{s}\hat{s}\, d\Omega = 0$ (and, indeed, the integral over any odd multiple of $\hat{s}$) and, therefore, this equation reduces to

$$\frac{1}{3}\nabla_\tau \cdot (G\boldsymbol{\delta}) + \mathbf{q} \cdot \boldsymbol{\delta} = \frac{\omega A_1}{3}\mathbf{q} \cdot \boldsymbol{\delta},$$

or

$$\nabla_\tau G = -(3 - A_1\omega)\mathbf{q}. \tag{14.36}$$

Equations (14.34) and (14.36) are a complete set of one scalar and one vector equation in the unknowns G and $\mathbf{q}$, and are the governing equations for the P_1 or *differential approximation*. The heat flux may be eliminated from these equations by taking the divergence of equation (14.36) after dividing by $(1 - A_1\omega/3)$:

$$\nabla_\tau \cdot \left(\frac{1}{1 - A_1\omega/3}\nabla G \right) = -3\nabla_\tau \cdot \mathbf{q} = -3(1 - \omega)(4\pi I_b - G). \tag{14.37}$$

If $A_1\omega$ is constant (does not vary across the volume) this equation reduces to

$$\nabla_\tau^2 G - (1 - \omega)(3 - A_1\omega)G = -(1 - \omega)(3 - A_1\omega)4\pi I_b. \tag{14.38}$$

Equation (14.38) is a *Helmholtz equation*, closely related to Laplace's equation, and is *elliptic* in nature (see, for example, a standard mathematics text such as Pipes and Harvill [20]). As such, it requires a single boundary condition specified everywhere on the enclosure surface.

If radiative equilibrium prevails, then $\nabla \cdot \mathbf{q} = 0$, and

$$\nabla_\tau^2 G = 0, \tag{14.39}$$

or

$$\nabla_\tau^2 I_b = 0. \tag{14.40}$$

In either case we get the elliptic *Laplace's equation* with the same boundary condition requirements. Once the incident radiation and/or blackbody intensity has been determined, the radiative heat flux is found from equation (14.36) as

$$\mathbf{q} = -\frac{1}{3 - A_1\omega}\nabla_\tau G. \tag{14.41}$$

According to equation (14.22) the P_1-approximation is able to supply three boundary conditions while equations (14.38) or (14.39) only require a single one. Thus, following the discussion of Marshak's boundary condition, equation (14.22), we choose only the case of $m = 0$ for the weight function in equation (14.22), with polar angle measured from the surface normal. Thus,

$$Y_1^0(\hat{s}) = P_1^0(\cos\theta') = \cos\theta' = \hat{s}\cdot\hat{n},$$ (14.42)

where θ' is the polar angle of $\hat{s}$ in the local coordinate system as shown in Fig. 14-2. Physically, that is, without reference to the general P_N-approximation, this choice of boundary condition implies that the directional distribution of the outgoing intensity along the enclosure wall is satisfied in an integral sense, by requiring the normal heat flux to be continuous (from enclosure surface into the participating medium). Then the boundary condition becomes

$$\int_{\hat{n}\cdot\hat{s}>0} I_w(\hat{s})\,\hat{s}\cdot\hat{n}\,d\Omega = \frac{1}{4\pi}\int_{\hat{n}\cdot\hat{s}>0}(G + 3\mathbf{q}\cdot\hat{s})\hat{s}\cdot\hat{n}\,d\Omega$$

$$= \frac{1}{4\pi}\int_{\psi'=0}^{2\pi}\int_{\theta'=0}^{\pi/2}\Big(G + 3q_{t1}\sin\theta'\cos\psi'$$

$$+ 3q_{t2}\sin\theta'\sin\psi' + 3q_n\cos\theta'\Big)\cos\theta'\sin\theta'd\theta d\psi'$$

$$= \frac{1}{2}\int_0^{\pi/2}(G + 3q_n\cos\theta')\cos\theta'\sin\theta'd\theta' = \frac{1}{4}(G + 2q_n)$$

or

$$G + 2\mathbf{q}\cdot\hat{n} = 4\int_{\hat{n}\cdot\hat{s}>0} I_w(\hat{s})\,\hat{s}\cdot\hat{n}\,d\Omega.$$ (14.43)

Here q_{t1} and q_{t2} are the two components of the heat flux vector tangential to the surface and $q_n = \mathbf{q}\cdot\hat{n}$ is the normal component.

For an opaque surface which emits and reflects radiation diffusely, $I_w(\hat{s}) = J_w/\pi$, where J_w is the surface's radiosity. Substituting this into equation (14.43) leads to

$$G + 2\mathbf{q}\cdot\hat{n} = \frac{4}{\pi}J_w\int_0^{2\pi}\int_0^{\pi/2}\cos\theta'\sin\theta'd\theta'd\psi' = 4J_w.$$ (14.44)

Recalling equation (5.26),

$$\mathbf{q}\cdot\hat{n} = \frac{\epsilon}{1-\epsilon}(\pi I_{bw} - J_w),$$ (14.45)

equation (14.43) finally becomes

$$2\mathbf{q}\cdot\hat{n} = 4J_w - G = \frac{\epsilon}{2-\epsilon}(4\pi I_{bw} - G),$$ (14.46)

where ϵ is the local surface emissivity. Modest [19] has shown that equation (14.46) also holds if the surface reflectivity consists of purely diffuse and purely specular components, i.e., if

$$\epsilon = 1 - \rho^d - \rho^s.$$ (14.47)

Thus, within the accuracy of the P_1, or differential, approximation, the results for enclosures with diffusely and/or specularly reflecting surfaces are identical. Since equation (14.38) is a second order equation in G, it is of advantage to eliminate $\mathbf{q} \cdot \hat{\mathbf{n}}$ from the boundary condition using equation (14.41). Thus,

$$-\frac{2 - \epsilon}{\epsilon} \frac{2}{3 - A_1\omega} \hat{\mathbf{n}} \cdot \nabla_\tau G + G = 4\pi I_{bw} \tag{14.48}$$

is the correct boundary condition to go with equation (14.38). Equation (14.48) is known as a *boundary condition of the third kind* (since it incorporates both the dependent variable and its normal gradient).

Example 14.2.   Consider an isothermal, gray slab at temperature T and of optical thickness τ_L, bounded by two isothermal black surfaces at temperature T_w. The medium scatters linear-anisotropically. Determine an expression of the nondimensional heat flux as a function of the optical parameters.

Solution

For this simple case we may write equation (14.38) as

$$\frac{d^2G}{d\tau^2} - (1 - \omega)(3 - A_1\omega)G = -(1 - \omega)(3 - A_1\omega)4n^2\sigma T^4,$$

or

$$G(\tau) = C_1 \cosh \gamma\tau + C_2 \sinh \gamma\tau + 4n^2\sigma T^4,$$

where

$$\gamma = \sqrt{(1 - \omega)(3 - A_1\omega)}.$$

Because of the symmetry of the problem it is advantageous to place the origin at the center of the slab, i.e., $-\tau_L/2 \leq \tau \leq +\tau_L/2$. Then

$$\frac{dG}{d\tau}(\tau = 0) = 0 = \gamma C_1 \sinh(\gamma \times 0) + \gamma C_2 \cosh(\gamma \times 0) + 0,$$

or $C_2 = 0$. Applying equation (14.48) at $\tau = \tau_L/2$, with $\epsilon = 1$, we get

$$\frac{2}{3 - A_1\omega} \frac{dG}{d\tau}(\tau_L/2) + G(\tau_L/2) = 4n^2\sigma T_w^4,$$

or

$$\frac{2\gamma}{3 - A_1\omega} C_1 \sinh \tfrac{1}{2}\gamma\tau_L + C_1 \cosh \tfrac{1}{2}\gamma\tau_L + 4n^2\sigma T^4 = 4n^2\sigma T_w^4,$$

$$C_1 = -\frac{4n^2\sigma(T^4 - T_w^4)}{\cosh \tfrac{1}{2}\gamma\tau_L + 2\sqrt{\frac{1-\omega}{3-A_1\omega}} \sinh \tfrac{1}{2}\gamma\tau_L},$$

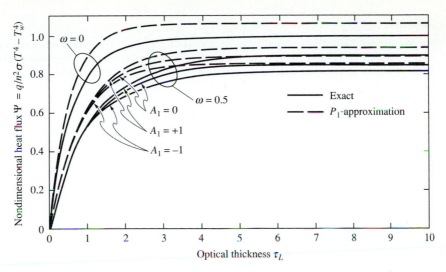

FIGURE 14-4
Nondimensional wall heat fluxes for a constant-temperature slab with linear-anisotropic scattering.

and

$$G(\tau) = 4n^2\sigma T^4 - 4n^2\sigma(T^4 - T_w^4)\frac{\cosh \gamma\tau}{\cosh \frac{1}{2}\gamma\tau_L + 2\sqrt{\frac{1-\omega}{3-A_1\omega}}\sinh \frac{1}{2}\gamma\tau_L}.$$

The heat flux is determined from equation (14.41) as

$$\Psi = \frac{q}{n^2\sigma(T^4 - T_w^4)} = -\frac{1}{n^2\sigma(T^4 - T_w^4)}\frac{1}{3 - A_1\omega}\frac{dG}{d\tau}$$

$$= \frac{2\sinh \gamma\tau}{\sinh \frac{1}{2}\gamma\tau_L + \frac{1}{2}\sqrt{\frac{3-A_1\omega}{1-\omega}}\cosh \frac{1}{2}\gamma\tau_L}.$$

Some sample results for the heat flux at the wall ($\tau = \tau_L/2$) are given in Fig. 14-4. We note that in this case the P_1-approximation goes to the correct optically thin limit (since there is no emission and, therefore, no heat flux), but not to the correct optically thick limit (since, as a result of the temperature step at the wall, there will always be an intensity discontinuity at the wall). In fact, for this problem the results of the P_1-approximation are worst (in absolute magnitude) close to that location.

Example 14.3.   Let us look at a gray medium at radiative equilibrium placed between two black concentric cylinders of radius R_1 and R_2 that are isothermal at temperatures T_1 and T_2. For simplicity, we shall assume that the medium does not scatter ($\sigma_s = 0$), and that its absorption coefficient, κ, is constant. We desire to find the heat flux from inner to outer cylinder as a function of the ratio R_1/R_2 and the optical thickness of the medium, $\tau_{12} = \tau_2 - \tau_1 = \kappa(R_2 - R_1)$.

Solution

For one-dimensional radiative equilibrium problems such as this, it is usually advantageous to apply equations (14.34) and (14.36) independently (rather than eliminating radiative heat flux). Then, from equation (14.34) we have, in cylindrical coordinates (with $\omega = 0$ and $\tau = \kappa r$),

$$\frac{1}{\tau}\frac{d}{d\tau}(\tau q) = 4n^2\sigma T^4 - G = 0.$$

If we multiply by τ and integrate, we find

$$\tau q = C_1, \quad \text{or} \quad q = \frac{C_1}{\tau}.$$

Substituting this expression into equation (14.36) gives

$$\frac{dG}{d\tau} = -3q = -\frac{3C_1}{\tau},$$

or

$$G = -3C_1 \ln \tau + C_2.$$

The boundary conditions are, from equation (14.46) with $\epsilon = 1$,

$$\tau = \tau_1 : \quad 2\mathbf{q}\cdot\hat{\mathbf{n}} = 2q = 4n^2\sigma T_1^4 - G,$$
$$\tau = \tau_2 : \quad 2\mathbf{q}\cdot\hat{\mathbf{n}} = -2q = 4n^2\sigma T_2^4 - G,$$

from which C_1 and C_2 may be determined as

$$C_1 = \frac{4n^2\sigma(T_1^4 - T_2^4)}{\dfrac{2}{\tau_1} + \dfrac{2}{\tau_2} + 3\ln\dfrac{\tau_2}{\tau_1}}, \quad C_2 = 4n^2\sigma T_2^4 + C_1\left(\frac{2}{\tau_2} + 3\ln\tau_2\right).$$

Heat flux and temperature then follow as

$$\Psi = \frac{q}{n^2\sigma(T_1^4 - T_2^4)} = \frac{2}{1 + \dfrac{\tau_2}{\tau_1} + \dfrac{3}{2}\tau_2\ln\dfrac{\tau_2}{\tau_1}}\left(\frac{\tau_2}{\tau}\right),$$

$$\Phi = \frac{T^4 - T_2^4}{T_1^4 - T_2^4} = \frac{1 + \dfrac{3}{2}\tau_2\ln\dfrac{\tau_2}{\tau}}{1 + \dfrac{\tau_2}{\tau_1} + \dfrac{3}{2}\tau_2\ln\dfrac{\tau_2}{\tau_1}}.$$

The resulting nondimensional heat flux, Ψ, evaluated at the inner cylinder, is shown in Fig. 14-5 together with exact results (Table 12.4), results from the diffusion approximation with jump boundary condition (Example 13.3) and results from the P_3-approximation given by Bayazitoğlu and Higenyi [21]. As expected, the P_1-approximation does well for optically thick media. For the optically thin case, however, as $\kappa \to 0$ the heat flux goes to

$$\Psi_1 \to \frac{2}{1 + R_2/R_1}\frac{R_2}{R_1} = 2\Big/\left(1 + \frac{R_1}{R_2}\right),$$

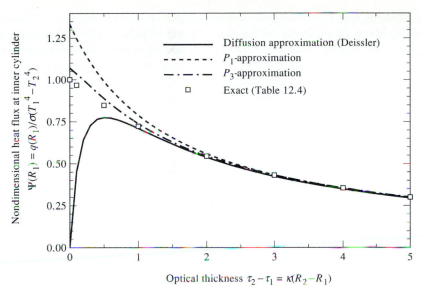

FIGURE 14-5
Nondimensional heat fluxes between concentric black cylinders at radiative equilibrium.

while the correct answer should be $\Psi_1 \to 1$, as we know from Chapter 5, equation (5.35). Therefore, for $R_1/R_2 \to 1$ the correct optically thin limit is obtained (and the gap between such cylinders becomes a plane-parallel slab), while for small inner cylinders, $R_1/R_2 \ll 1$, the error becomes larger and may be as large as 100%!

The P_1-approximation is a very popular method since it reduces the (spectral or gray) equation of transfer from a very complicated integral equation to a relatively simple partial differential equation. The P_1-approximation is powerful (allowing non-black surfaces, nonconstant properties, anisotropic scattering, etc.), and the average heat transfer engineer is much better trained in solving differential equations than integral equations. Furthermore, if overall energy conservation (also a partial differential equation) is computed, compatibility of the solution methods is virtually assured. However, it is important to remember that the P_1-approximation may be substantially in error in optically thin situations, in particular in multidimensional geometries with large aspect ratios (i.e., long and narrow configurations) and/or when surface emission dominates over medium emission. Another drawback of this method is that it is difficult to apply it to nongray media, in particular if accurate modeling of gas band wings is important.

14.5 P_3- AND HIGHER-ORDER APPROXIMATIONS

The general P_N-approximation for absorbing/emitting, anisotropically scattering, one-dimensional cylindrical media has been given by Kofink [22], and the P_3-approximation for one-dimensional slabs, concentric cylinders and concentric spheres has been developed in terms of moments by Bayazitoğlu and Higenyi [21]. The general P_N-formulation for three-dimensional Cartesian coordinate systems has been derived by Cheng [7, 8], including Marshak's boundary conditions for surfaces normal to one of the coordinates. Mengüç and Viskanta [23, 24] limited their development to the P_3-approximation in terms of moments (rather than spherical harmonics), but considered three-dimensional Cartesian coordinates [23] as well as axisymmetric cylindrical geometries [24]. Finally, the three-dimensional P_N-approximation for arbitrary coordinate systems has been derived by Ou and Liou [9]. With the exception of Cheng [7], no boundary conditions beyond a reference to equation (14.22) have been given in the literature. We present here only the final form of the governing equations, without derivation.

Expressing intensity with the relationship given by equation (14.1), Ou and Liou found the governing equations for the I_l^m as

$$\mathcal{L}_+ X_l^m + \mathcal{L}_- Y_l^m + \mathcal{L}Z_l^m + \left(1 - \frac{\omega A_l}{2l + 1}\right)I_l^m = (1 - \omega)I_b\delta_{0l},$$

$$l \le N, \quad -N \le m \le N, \quad (14.49)$$

where

$$X_l^m = -\frac{\sqrt{(l + m + 1)(l + m + 2)}}{2(2l + 3)} I_{l+1}^{m+1}$$

$$+ \frac{\sqrt{(l - m + 1)(l - m + 2)}}{2(2l - 1)} I_{l-1}^{m+1}, \quad (14.50a)$$

$$Y_l^m = +\frac{\sqrt{(l - m + 1)(l - m + 2)}}{2(2l + 3)} I_{l+1}^{m-1}$$

$$- \frac{\sqrt{(l + m - 1)(l + m)}}{2(2l - 1)} I_{l-1}^{m-1}, \quad (14.50b)$$

$$Z_l^m = +\frac{\sqrt{(l - m + 1)(l + m + 1)}}{2l + 3} I_{l+1}^m$$

$$+ \frac{\sqrt{(l - m)(l + m)}}{2l - 1} I_{l-1}^m, \quad (14.50c)$$

and the $\mathcal{L}_+$, $\mathcal{L}_-$, and $\mathcal{L}$ are linear, first-order differential operators. For the common Cartesian, cylindrical, and spherical coordinate systems these turn out to be as follows:

Cartesian:

$$\mathcal{L}_+ = \mathcal{L}_-^* = \frac{\partial}{\partial \tau_x} + i\frac{\partial}{\partial \tau_y}, \quad \mathcal{L} = \frac{\partial}{\partial \tau_z}; \quad (14.51)$$

Cylindrical:

$$\mathscr{L}_+ = \mathscr{L}_-^* = e^{i\psi_c}\left(\frac{\partial}{\partial\tau_r} + \frac{i}{\tau_r}\frac{\partial}{\partial\psi_c}\right), \quad \mathscr{L} = \frac{\partial}{\partial\tau_z}; \tag{14.52}$$

Spherical:

$$\mathscr{L}_+ = \mathscr{L}_-^* = e^{i\psi_s}\left[\sin\theta_s\frac{\partial}{\partial\tau_r} + \frac{\cos\theta_s}{\tau_r}\frac{\partial}{\partial\theta_s} + \frac{i}{\tau_r\sin\theta_s}\frac{\partial}{\partial\psi_s}\right],$$

$$\mathscr{L} = \cos\theta_s\frac{\partial}{\partial\tau_r} - \frac{\sin\theta_s}{\tau_r}\frac{\partial}{\partial\theta_s}. \tag{14.53}$$

In these formulae the asterisk denotes the complex conjugate, and subscripts have been attached to the angles θ and ψ to make certain they are recognized as belonging to the *position* coordinate system (as opposed to the angles θ and ψ describing the independent *direction* vector). For the direction vector $\hat{s}$ the polar angle θ is measured from the positive z-axis, and the azimuthal angle ψ is measured counterclockwise from the positive x-axis in the x-y-plane. Note that doing so fixes the directional coordinates θ, ψ to a Cartesian coordinate system, making equation (14.49) unsuitable for cylindrical and spherical geometries with symmetry (one- and two-dimensional problems). Finally, the boundary conditions for equations (14.49) must be found from equation (14.22).

For three-dimensional geometries, it is obvious that anything but low-order approximations quickly become extremely cumbersome to deal with. Already the P_3-approximation may result in as many as 16 simultaneous partial differential equations (depending on the symmetry), for which complicated boundary conditions need to be developed from equation (14.22). Furthermore, the nature of the boundary conditions is such that they are not easily applied to equation (14.49), that is, it is generally necessary to transform the first-order partial differential equations (14.49) analytically to elliptic ones, as was done for the P_1-approximation, equation (14.37). As a result of this complexity, very few multidimensional problems have been solved by the P_3-approximation, and apparently none by higher orders. We shall limit ourselves here to a simple example for a one-dimensional plane-parallel slab. The problem of one-dimensional concentric cylinders has been discussed in detail by Kofink [22], and results for one-dimensional slabs, concentric cylinders, and concentric spheres have been given by Bayazitoğlu and Higenyi [21].

Example 14.4.   Consider an isothermal medium at temperature T, confined between two large, parallel black plates that are isothermal at the (same) temperature T_w. The medium is gray and absorbs and emits, but does not scatter. Determine an expression for the heat transfer rates within the medium using the P_3-approximation.

Solution

For this one-dimensional problem we choose $\tau = \tau_z$ as the (nondimensional) space coordinate between the plates, as was done in Example 14.2. This choice assures that intensity does not depend on azimuthal angle ψ and, in turn, implies that all $I_l^m = 0$ for $m \neq 0$. Because of the one-dimensionality we find $\mathcal{L}_+ = \mathcal{L}_- = 0$ and $\mathcal{L} = d/d\tau_z$. Thus, equation (14.49) reduces to

$$\frac{dZ_l^0}{d\tau} + I_l^0 = I_b \, \delta_{0l}, \quad l \leq 3.$$

Now, if we use equation (14.50) (and recognize that $I_{-1}^0 = I_4^0 = 0$) and drop the superscripts on the I_l^0, since they are no longer needed, we have four simultaneous first-order differential equations:

$$l = 0: \quad \frac{1}{3}I_1' \qquad + I_0 = I_b,$$

$$l = 1: \quad \frac{2}{5}I_2' + I_0' + I_1 = 0,$$

$$l = 2: \quad \frac{3}{7}I_3' + \frac{2}{3}I_1' + I_2 = 0,$$

$$l = 3: \quad \frac{3}{5}I_2' + I_3 = 0,$$

where the primes denote differentiation with respect to τ. After some minor manipulation, which is left to the reader, these four equations may be consolidated into a single fourth-order equation,

$$\frac{3}{35}I_0^{(iv)} - \frac{6}{7}I_0'' + I_0 = I_b,$$

and

$$I_1 = -\frac{18}{7}I_0' + \frac{9}{35}I_0''',$$

$$I_2 = \frac{55}{14}(I_0 - I_b) - \frac{9}{14}I_0'',$$

$$I_3 = -\frac{33}{14}I_0' + \frac{27}{70}I_0'''.$$

The general solution to the above equation (keeping in mind that I_b =const) is

$$I_0(\tau) = I_b + (I_{bw} - I_b)[C_1 \cosh \lambda_1 \tau + C_2 \cosh \lambda_2 \tau + C_3 \sinh \lambda_1 \tau + C_4 \sinh \lambda_2 \tau],$$

where the constant factor $(I_{bw} - I_b)$ was included to make the C_i dimensionless. The λ_1 and λ_2 are the positive roots of the equation

$$\frac{3}{35}\lambda^4 - \frac{6}{7}\lambda^2 + 1 = 0,$$

or $\lambda_1 = 1.1613$ and $\lambda_2 = 2.9413$. To exploit the symmetry of the problem, we again choose the origin for τ_z to be at the midpoint between the two plates. Then $I_0'(0) = 0$ and $C_3 = C_4 = 0$. We still need two boundary conditions at one of the plates, say $\tau = -\tau_L/2$ (and τ_L is the total optical thickness of the medium). From equation (14.22) or, more directly, from equation (14.19), we obtain

$$\int_0^1 [I_0 + I_1 P_1(\mu) + I_2 P_2(\mu) + I_3 P_3(\mu)] \, \mu^{2i-1} d\mu = \int_0^1 I_{bw} \mu^{2i-1} d\mu, \quad i = 1, 2$$

and, using $P_1(\mu) = \mu$, $P_2(\mu) = \frac{1}{2}(3\mu^2 - 1)$ and $P_3(\mu) = \frac{1}{2}(5\mu^3 - 3\mu)$, as described by Abramowitz and Stegun [12],

$$i = 1: \quad \tfrac{1}{2}I_0 + \tfrac{1}{3}I_1 + \left(\tfrac{3}{8} - \tfrac{1}{4}\right)I_2 + \left(\tfrac{1}{2} - \tfrac{1}{2}\right)I_3 = \tfrac{1}{2}I_{bw}$$

$$i = 2: \quad \tfrac{1}{4}I_0 + \tfrac{1}{5}I_1 + \left(\tfrac{1}{4} - \tfrac{1}{8}\right)I_2 + \left(\tfrac{5}{14} - \tfrac{3}{10}\right)I_3 = \tfrac{1}{4}I_{bw}$$

or

$$I_{bw} = I_0 + \tfrac{2}{3}I_1 + \tfrac{1}{4}I_2,$$

$$I_{bw} = I_0 + \tfrac{4}{5}I_1 + \tfrac{1}{2}I_2 + \tfrac{8}{35}I_3.$$

Substituting for I_1, I_2 and I_3 results in

$$I_{bw} = I_0 - \tfrac{12}{7}I_0' + \tfrac{6}{35}I_0''' + \tfrac{55}{56}(I_0 - I_b) - \tfrac{9}{56}I_0'',$$

$$I_{bw} = I_0 - \tfrac{72}{35}I_0' + \tfrac{36}{175}I_0''' + \tfrac{55}{28}(I_0 - I_b) - \tfrac{9}{28}I_0'' - \tfrac{132}{245}I_0' + \tfrac{108}{1225}I_0''',$$

or

$$I_{bw} - I_b = \tfrac{111}{56}(I_0 - I_b) - \tfrac{12}{7}I_0' - \tfrac{9}{56}I_0'' + \tfrac{6}{35}I_0''',$$

$$I_{bw} - I_b = \tfrac{83}{28}(I_0 - I_b) - \tfrac{636}{245}I_0' - \tfrac{9}{28}I_0'' + \tfrac{72}{245}I_0'''.$$

Now, substituting the solution for I_0 into these boundary conditions leads to

$$1 = a_1C_1 + a_2C_2 = b_1C_1 + b_2C_2,$$

where

$$a_i = \left(\frac{111}{56} - \frac{9}{56}\lambda_i^2\right)\cosh\lambda_i\frac{\tau_L}{2} + \left(\frac{12}{7}\lambda_i - \frac{6}{35}\lambda_i^3\right)\sinh\lambda_i\frac{\tau_L}{2}, \quad i = 1, 2,$$

$$b_i = \left(\frac{83}{28} - \frac{9}{28}\lambda_i^2\right)\cosh\lambda_i\frac{\tau_L}{2} + \left(\frac{636}{245}\lambda_i - \frac{72}{245}\lambda_i^3\right)\sinh\lambda_i\frac{\tau_L}{2}, \quad i = 1, 2.$$

Finally, we get

$$C_1 = \frac{b_2 - a_2}{a_1b_2 - a_2b_1}, \qquad C_2 = \frac{a_1 - b_1}{a_1b_2 - a_2b_1}.$$

The heat flux through the medium is determined from

$$q(\tau) = 2\pi\int_{-1}^{1} I(\tau, \mu)\mu\, d\mu = \frac{4\pi}{3}I_1(\tau),$$

[since $\mu = P_1(\mu)$, and using the orthogonality property of Legendre polynomials]. Using the solutions for I_1 and I_0, the heat flux may be expressed in nondimensional form as

$$\Psi = \frac{q(\tau)}{n^2\sigma(T_w^4 - T^4)} = -\frac{12}{35}\frac{10I_0' - I_0'''}{I_{bw} - I_b}$$

$$= -\frac{12}{35}\sum_{i=1}^{2}(10\lambda_i^2 - \lambda_i^3)C_i\sinh\lambda_i\tau,$$

where, for simplicity, it was assumed that the medium is gray, or $I_b = n^2\sigma T^4/\pi$.

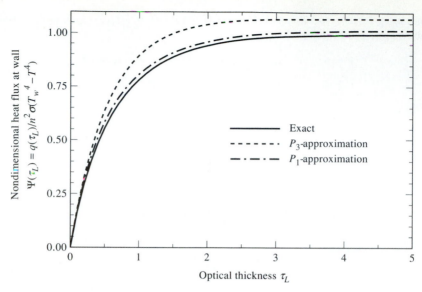

FIGURE 14-6
Nondimensional wall heat fluxes for an isothermal slab; comparison of P_1- and P_3-approximations with the exact solution.

The nondimensional heat flux at the top surface ($\tau = \tau_L/2$) is shown in Fig. 14-6, as a function of optical depth of the slab. The results are compared with those of the P_1- or *differential approximation* (Example 14.2), and with the exact result,

$$\Psi = 1 - 2E_3(\tau_L),$$

which is readily found from equation (12.35). For this particular example the P_1-approximation is very accurate (maximum error ~15%) and, as to be expected, the P_3-approximation performs even better (maximum error ~7%).

It should be clear from the above example that P_3- and higher-order P_N-approximations quickly become very tedious, even for simple geometries. However, P_3 results can be substantially more accurate than P_1 results, particularly in optically thin media and/or geometries with large aspect ratios. Another example, shown in Fig. 14-5, depicts nondimensional heat flux through a gray, nonscattering medium at radiative equilibrium, confined between infinitely long, concentric, black and isothermal cylinders. The P_3-solution of Bayazitoğlu and Higenyi [21] is compared with the P_1 solution (Example 14.3) and with results from an enhanced P_1-approximation (to be discussed in the next section). Observe that the P_3-approximation introduces roughly half the error of the P_1-method. One outstanding advantage of the P_3-method is that, once the problem has been formulated (setting up the governing equations suitable for a numerical solution), the increase in computer time required (compared with the P_1-method) is very minor. In addition, P_3-calculations are also usually very grid-compatible with conduction/convection calculations, if one must account for combined modes of heat transfer.

14.6   ENHANCEMENTS TO THE P_1-APPROXIMATION

As indicated earlier, the P_1- or *differential approximation* enjoys great popularity because of its relative simplicity and because of its compatibility with standard methods for the solution of the (overall) energy equation. The fact that the P_1-approximation may become very inaccurate in optically thin media—and thus of limited use—has prompted a number of investigators to seek enhancements or modifications to the differential approximation to make it reasonably accurate for all conditions. We shall briefly describe here two such methods, the so-called *modified differential approximation* and the *improved differential approximation*.

The Modified Differential Approximation

The directional intensity at any given point inside the medium is due to two sources: Radiation originating from a surface (due to emission and reflection), and radiation originating from within the medium (due to emission and in-scattering). The contribution due to radiation emanating from walls may display very irregular directional behavior, especially in optically thin situations (due to surface radiosities varying across the enclosure surface, causing irradiation to change rapidly over incoming directions). Intensity emanating from inside the medium generally varies very slowly with direction because emission and isotropic scattering result in an isotropic radiation source. Only for highly anisotropic scattering may the radiation source—and, therefore, at least locally also the intensity--display irregular directional behavior.

In what they termed the *modified differential approximation (MDA)* Olfe [25–28] and Glatt and Olfe [29] separated wall emission from medium emission in simple black and gray-walled enclosures with gray, nonscattering media. While very accurate, their model was limited to nonscattering media in simple, mostly one-dimensional enclosures. Wu and coworkers [30] demonstrated, for one-dimensional plane-parallel media, that the MDA may be extended to scattering media with reflecting boundaries. Finally, Modest [31] showed that the method can be applied to three-dimensional linear-anisotropically scattering media with reflecting boundaries.

Consider an arbitrary enclosure as shown in Fig. 14-7. The equation of transfer is, from equation (14.3),

$$\frac{dI}{d\tau}(\mathbf{r}, \hat{\mathbf{s}}) = \hat{\mathbf{s}} \cdot \boldsymbol{\nabla}_\tau I = S(\mathbf{r}, \hat{\mathbf{s}}) - I(\mathbf{r}, \hat{\mathbf{s}}), \tag{14.54}$$

where, for linear-anisotropic scattering with a phase function given by equation (14.31), the radiative source term is, from equation (14.32),

$$S(\mathbf{r}, \hat{\mathbf{s}}) = (1 - \omega)I_b(\mathbf{r}) + \frac{\omega}{4\pi}[G(\mathbf{r}) + A_1\mathbf{q}(\mathbf{r}) \cdot \hat{\mathbf{s}}]. \tag{14.55}$$

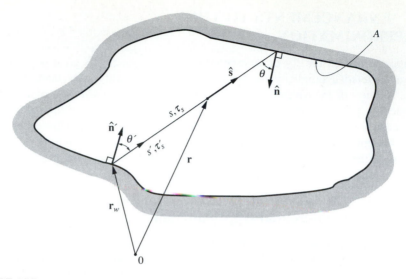

FIGURE 14-7
Radiative intensity within an arbitrary enclosure.

For diffusely reflecting walls, equations (14.54) and (14.55) are subject to the boundary condition

$$I(\mathbf{r}_w, \hat{\mathbf{s}}) = \frac{J_w}{\pi}(\mathbf{r}_w) = I_{bw}(\mathbf{r}_w) - \frac{1-\epsilon}{\pi\epsilon}\mathbf{q}\cdot\hat{\mathbf{n}}(\mathbf{r}_w), \tag{14.56}$$

where J_w is the surface radiosity related to I_{bw} and $q_w = \mathbf{q}\cdot\hat{\mathbf{n}}$ through equation (14.45).

We now break up the intensity at any point into two components: One, I_w, which may be traced back to emission from the enclosure wall (but may have been attenuated by absorption and scattering in the medium, and by reflections from the enclosure walls), and the remainder, I_m, which may be traced back to the radiative source term (i.e., radiative intensity released within the medium into a given direction by emission and scattering). Thus, we write

$$I(\mathbf{r}, \hat{\mathbf{s}}) = I_w(\mathbf{r}, \hat{\mathbf{s}}) + I_m(\mathbf{r}, \hat{\mathbf{s}}) \tag{14.57}$$

and let I_w satisfy the equation

$$\frac{dI_w}{d\tau_s}(\mathbf{r}, \hat{\mathbf{s}}) = -I_w(\mathbf{r}, \hat{\mathbf{s}}), \tag{14.58}$$

leading to

$$I_w(\mathbf{r}, \hat{\mathbf{s}}) = \frac{J_w}{\pi}(\mathbf{r}_w)\, e^{-\tau_s}, \tag{14.59}$$

as indicated in Fig. 14-7. Since for I_w no radiative source within the medium is considered, the radiosity in equation (14.59) is the one caused by wall emission only (with attenuation within the medium). The radiosity variation along the enclosure wall may be determined by invoking the definition of the radiosity as the sum of emission plus reflected irradiation, or

$$J_w(\mathbf{r}) = \epsilon \pi I_{bw}(\mathbf{r}) + (1 - \epsilon) \int_{\hat{\mathbf{s}} \cdot \hat{\mathbf{n}} < 0} I_w(\mathbf{r}, \hat{\mathbf{s}}) \left| \hat{\mathbf{s}} \cdot \hat{\mathbf{n}} \right| d\Omega$$

$$= \epsilon \pi I_{bw}(\mathbf{r}) + (1 - \epsilon) \int_A J_w(\mathbf{r}_w) \frac{\cos \theta \cos \theta'}{\pi S^2} e^{-\tau_s} dA, \qquad (14.60)$$

as also indicated in Fig. 14-7. The surface integral representation of equation (14.60) is obtained by invoking the definition of solid angle, equation (1.24), or $d\Omega = \cos \theta' \, dA / S^2$, equivalent to the definition of view factors in Chapter 4.

Equation (14.60) is the standard integral equation for the radiosity in an enclosure without a participating medium, except for the attenuation factor $e^{-\tau_s}$, and may be solved by standard methods such as breaking up the enclosure surface into N small subsurfaces of constant radiosity. Assuming that the attenuation term may be approximated by the value between node centers leads to

$$J_i = \epsilon_i \pi I_{bi} + (1 - \epsilon_i) \sum_{j=1}^{N} J_i e^{-\tau_{ij}} F_{i-j}, \qquad i = 1, 2, \ldots, N, \qquad (14.61)$$

where the F_{i-j} are the view factors between the subsurfaces.

It remains to calculate the contribution to the intensity from within the medium. Assuming that the P_1-approximation adequately represents intensity emanating from within the medium, we write, using equation (14.30),

$$I_m(\mathbf{r}, \hat{\mathbf{s}}) \simeq \frac{1}{4\pi} [G_m(\mathbf{r}) + 3\mathbf{q}_m(\mathbf{r}) \cdot \hat{\mathbf{s}}], \qquad (14.62)$$

where G_m and $\mathbf{q}_m$ are medium-related incident radiation and heat flux, respectively, defined by

$$G_m(\mathbf{r}) = \int_{4\pi} I_m(\mathbf{r}, \hat{\mathbf{s}}) \, d\Omega, \qquad (14.63)$$

$$\mathbf{q}_m(\mathbf{r}) = \int_{4\pi} I_m(\mathbf{r}, \hat{\mathbf{s}}) \hat{\mathbf{s}} \, d\Omega. \qquad (14.64)$$

Substituting equations (14.58) and (14.62) into equation (14.54) we get

$$\frac{dI_m}{d\tau_s} = \hat{\mathbf{s}} \cdot \boldsymbol{\nabla}_\tau I_m \simeq (1 - \omega) I_b + \frac{\omega}{4\pi} [G_w + G_m + A_1 (\mathbf{q}_w + \mathbf{q}_m) \cdot \hat{\mathbf{s}}] - I_m, \qquad (14.65)$$

where the wall-related incident radiation and heat flux are defined as

$$G_w(\mathbf{r}) = \int_{4\pi} I_w(\mathbf{r}, \hat{\mathbf{s}}) \, d\Omega = \frac{1}{\pi} \int_{4\pi} J_w(\mathbf{r}_w) e^{-\tau_s} \, d\Omega, \tag{14.66}$$

$$\mathbf{q}_w(\mathbf{r}) = \int_{4\pi} I_w(\mathbf{r}, \hat{\mathbf{s}}) \hat{\mathbf{s}} \, d\Omega = \frac{1}{\pi} \int_{4\pi} J_w(\mathbf{r}_w) e^{-\tau_s} \hat{\mathbf{s}} \, d\Omega. \tag{14.67}$$

After integrating equation (14.65) over all solid angles, we have

$$\boldsymbol{\nabla}_\tau \cdot \mathbf{q}_m = (1 - \omega)4\pi I_b + \omega(G_w + G_m) - G_m. \tag{14.68}$$

If equation (14.65) is multiplied by $\hat{\mathbf{s}}$ before integration over all directions, we get

$$\boldsymbol{\nabla}_\tau G_m = A_1 \omega(\mathbf{q}_w + \mathbf{q}_m) - 3\mathbf{q}_m. \tag{14.69}$$

Equations (14.68) and (14.69) are a complete set of equations for the unknowns G_m and $\mathbf{q}_m$. [For higher-order P_N-approximations, additional equations would need to be generated by multiplying equation (14.65) by successively higher order harmonics before integration.] The necessary boundary conditions for equations (14.68) and (14.69) are found by making an energy balance for medium-related radiation at a point on the surface

$$q_m \cdot \hat{\mathbf{n}} = \epsilon \int_{\hat{\mathbf{s}} \cdot \hat{\mathbf{n}} < 0} I_m(\mathbf{r}, \hat{\mathbf{s}}) \hat{\mathbf{s}} \cdot \hat{\mathbf{n}} \, d\Omega, \tag{14.70}$$

which, after substituting equation (14.62) for I_m, leads to Marshak's boundary condition for a cold surface,

$$2\left(\frac{2}{\epsilon} - 1\right) \mathbf{q}_m \cdot \hat{\mathbf{n}} + G_m = 0. \tag{14.71}$$

[For a more detailed derivation of these relations see the similar development of the P_1-approximation, equations (14.33) through (14.36)]. Equations (14.68) and (14.69), together with boundary condition (14.71), constitute a set of equations for the determination of the medium-related incident radiation and heat flux, with enclosure wall-related incident radiation and heat flux given by equations (14.66), (14.67), and (14.60). Finally, the total values for incident radiation and heat flux are given by

$$G = G_w + G_m, \tag{14.72}$$

$$\mathbf{q} = \mathbf{q}_w + \mathbf{q}_m. \tag{14.73}$$

This solution will reduce to the correct solution for the optically thin limit (where the medium-related contribution vanishes), and to the unmodified P_1-approximation for the optically thick limit. For nonscattering or isotropically scattering media, the method requires the solution to a Helmholtz equation (similar to the ordinary P_1-approximation), and the additional evaluation of a scalar surface integral for every point in the medium (G_w); whereas, for anisotropic scattering, G_w as well as a

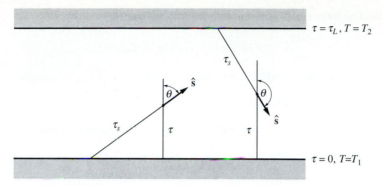

$\tau = \tau_L, T = T_2$

$\tau = 0, T=T_1$

FIGURE 14-8
Intensities for a one-dimensional plane-parallel slab.

vector surface integral ($\mathbf{q}_w$) must be evaluated. In addition, for the determination of radiosities, a surface integral equation must be solved (or view factors evaluated). If the extinction coefficient is independent of temperature, the G_w (and $\mathbf{q}_w$) integrals do not depend on medium temperature and, thus, may be evaluated once and for all (i.e., they will not enter any iterative process if the temperature field of the medium is to be determined). If the extinction coefficient is temperature dependent, a solution for G_w and $\mathbf{q}_w$, based on gross estimates for the temperature field [in order to calculate the optical distances τ_{ij} in equation (14.61)], still will result in the correct optically thin and thick limits and, therefore, can be expected to be of reasonable accuracy everywhere in between.

Example 14.5.   Consider a one-dimensional, gray, absorbing/emitting and isotropically scattering slab with refractive index $n = 1$ at radiative equilibrium, contained between two isothermal black walls at temperatures T_1 and T_2, respectively. Determine the radiative heat flux between the plates using the modified differential approximation.

Solution

Measuring optical distance $\tau = \int_0^z \beta \, dz$ perpendicular to the plates, as shown in Fig. 14-8, one may readily determine the wall-related intensity as

$$I_w(\tau, \mu) = \begin{cases} \dfrac{\sigma T_1^4}{\pi} e^{-\tau/\mu}, & 0 < \mu \leq 1, \\[2mm] \dfrac{\sigma T_2^4}{\pi} e^{(\tau_L - \tau)/\mu}, & -1 \leq \mu < 0, \end{cases}$$

leading to

$$G_w(\tau) = 2\pi \int_{-1}^{1} I_w \, d\mu = 2\sigma T_1^4 E_2(\tau) + 2\sigma T_2^4 E_2(\tau_L - \tau),$$

$$q_w(\tau) = 2\pi \int_{-1}^{1} I_w \, d\mu = 2\sigma T_1^4 E_3(\tau) - 2\sigma T_2^4 E_3(\tau_L - \tau).$$

Substituting these expressions into equation (14.69) with $A_1 = 0$ and $q_m = q - q_w$ gives

$$\frac{dG_m}{d\tau} = -3q + 6\left[\sigma T_1^4 E_3(\tau) - \sigma T_2^4 E_3(\tau_L - \tau)\right].$$

Since q =const, due to radiative equilibrium, this equation is readily integrated, and

$$G_m = C - 3q\tau - 6\left[\sigma T_1^4 E_4(\tau) + \sigma T_2^4 E_4(\tau_L - \tau)\right].$$

The constants C and q may be found from the boundary conditions, equation (14.71), or

$$\tau = 0: \quad 2q_m + G_m = 2q - 2\sigma T_1^4 + 4\sigma T_2^4 E_3(\tau_L)$$
$$+C - 2\sigma T_1^4 - 6\sigma T_2^4 E_4(\tau_L) = 0,$$
$$\tau = \tau_L: \quad -2q_m + G_m = -2q + 4\sigma T_1^4 E_3(\tau_L) - 2\sigma T_2^4$$
$$+C - 3q\tau_L - 6\sigma T_1^4 E_4(\tau_L) - 2\sigma T_2^4 = 0.$$

Subtracting the second boundary condition from the first yields the nondimensional heat flux as

$$\Psi = \frac{q}{\sigma(T_1^4 - T_2^4)} = \frac{1 + E_3(\tau_L) - \frac{3}{2}E_4(\tau_L)}{1 + 3\tau_L/4}.$$

This result is compared with exact values as well as those from the *ordinary differential approximation (ODA)* in Fig. 14-9. Note that the ODA has a maximum error of $\approx 3.3\%$, compared with $\approx 1.3\%$ for the MDA (both at $\tau_L = 1$). This comparison, however, in no way demonstrates the power of the present method, since this problem is one of the very few in which the ordinary differential approximation actually reduces to the correct optically thin limit.

Example 14.6.   Consider a one-dimensional gray absorbing/emitting nonscattering slab between two isothermal black plates, both at temperature T_w. The medium has a refractive index of $n = 1$ and is isothermal at T_m. Determine the radiative heat flux between the plates using the MDA.

Solution

The wall-related heat flux follows immediately from the previous example as

$$q_w(\tau) = 2\sigma T_w^4[E_3(\tau) - E_3(\tau_L - \tau)].$$

For a nonscattering medium equations (14.68) and (14.69) contain no wall-related terms, and

$$\frac{d^2 G_m}{d\tau^2} = -3\frac{dq_m}{d\tau} = -3(4\sigma T_m^4 - G_m),$$

or

$$G_m(\tau) = C_1 e^{-\sqrt{3}\tau} + C_2 e^{+\sqrt{3}\tau} + 4\sigma T_m^4$$

$$= C\left[e^{-\sqrt{3}\tau} + e^{-\sqrt{3}(\tau_L - \tau)}\right] + 4\sigma T_m^4,$$

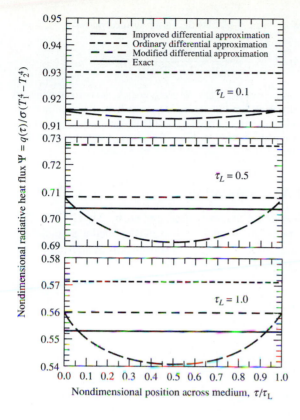

FIGURE 14-9
Normalized heat flux for a one-dimensional slab at radiative equilibrium, bounded by isothermal black plates; influence of optical depth on radiative heat flux obtained by various methods.

where, in the last equation, we have used the fact that $G_m(\tau) = G_m(\tau_L - \tau)$, as a result of symmetry. The medium-related heat flux follows as

$$q_m = -\frac{1}{3}\frac{dG_m}{d\tau} = \frac{C}{\sqrt{3}}\left[e^{-\sqrt{3}\tau} - e^{-\sqrt{3}(\tau_L-\tau)}\right].$$

Applying the boundary condition at $\tau = 0$, we obtain

$$2q_m + G_m = \frac{2}{\sqrt{3}}C\left(1 - e^{-\sqrt{3}\tau_L}\right) + C\left(1 + e^{-\sqrt{3}\tau_L}\right) + 4\sigma T_m^4 = 0,$$

and

$$q_m(\tau) = -\frac{4\sigma T_m^4\left[e^{-\sqrt{3}\tau} - e^{-\sqrt{3}(\tau_L-\tau)}\right]}{2 + \sqrt{3} - (2 - \sqrt{3})e^{-\sqrt{3}\tau_L}}.$$

The total heat flux is evaluated as $q = q_w + q_m$. At the lower surface we have

$$q(0) = [1 - 2E_3(\tau_L)]\sigma T_w^4 - \frac{4\left(1 - e^{-\sqrt{3}\tau_L}\right)}{2 + \sqrt{3} - (2 - \sqrt{3})e^{-\sqrt{3}\tau_L}}\sigma T_m^4.$$

This example illustrates the one weakness of the MDA: While q goes to the correct optically thin limit ($q \to 0$), for the optically thick limit the term due to medium emission goes to the value predicted by the ODA, which is not equipped to handle temperature jumps within optically thick media. The result is not included in Fig. 14-6 since it may lie anywhere between the exact and the P_1 results, depending on the values of T_w and T_m.

The Improved Differential Approximation

The *improved differential approximation (IDA)* was first applied by Modest [32, 33] and Modest and Stevens [34] to two-dimensional, nonscattering media at radiative equilibrium. The method has been extended to linear-anisotropically scattering three-dimensional media (at radiative equilibrium or not) by Modest [35, 36].

We start again by considering the equation of transfer, equation (14.54), together with the radiative source term, equation (14.55). Solving equation (14.54) along a pencil of rays (see Fig. 14-7) gives

$$I(\mathbf{r}, \hat{\mathbf{s}}) = \frac{J_w}{\pi}(\mathbf{r}')e^{-\tau_s} + \int_0^{\tau_s} S(\mathbf{r}' + s'\hat{\mathbf{s}}, \hat{\mathbf{s}})e^{-(\tau_s - \tau_s')}d\tau_s'. \tag{14.74}$$

The problem with equation (14.74) is, of course, that for scattering media and/or for radiative equilibrium the radiative source S is not known a priori and depends on intensity I itself. The intensity consists of two components, one related to radiosity leaving the walls, the other related to medium emission and in-scattering. As for the MDA we break up the intensity according to equation (14.57) and approximate I_m from the P_1-approximation (ODA). In the IDA these tasks are achieved by setting

$$I_m(\mathbf{r}, \hat{\mathbf{s}}) \simeq \int_0^{\tau_s} S^*(\mathbf{r}' + s'\hat{\mathbf{s}}, \hat{\mathbf{s}})e^{-(\tau_s - \tau_s')}d\tau_s', \tag{14.75}$$

where

$$S^*(\mathbf{r}, \hat{\mathbf{s}}) = (1 - \omega)I_b^*(\mathbf{r}) + \frac{\omega}{4\pi}\left[G^*(\mathbf{r}) + A_1\mathbf{q}^*(\mathbf{r}) \cdot \hat{\mathbf{s}}\right] \tag{14.76}$$

is the source term evaluated from the ODA, and the asterisks have been added to distinguish the differential approximation results from the improved values; i.e., in equation (14.76) the radiative heat flux $\mathbf{q}^*$ and incident radiation G^* (as well as Planck function I_b^* for the case of radiative equilibrium) are evaluated approximately from equations (14.34) and (14.36) together with boundary condition (14.46). This approximation goes to the correct optically thin limit ($I_m \to 0$ as $\tau_s \to 0$) as well as the correct optically thick limit (where the P_1-approximation predicts the correct radiative source, $I_m \to S^* \to S$ as $\tau_s \to \infty$).[8] Since for optically thin media the accuracy of S^* is irrelevant, and for optically thick media only $S^*(\mathbf{r})$ is of importance (rather than the integral over it), it may be justified to approximate equation (14.75) further. To this effect we shall assume that the radiative source varies linearly from the point under consideration, $\mathbf{r}$, toward the point at the wall, $\mathbf{r}'$ (i.e., into the $-\hat{\mathbf{s}}$ direction), or

$$S^*(\mathbf{r}' + s'\hat{\mathbf{s}}, \hat{\mathbf{s}}) \simeq S^*(\mathbf{r}, \hat{\mathbf{s}}) - (\tau_s - \tau_s')\frac{dS^*}{d\tau_s}(\mathbf{r}, \hat{\mathbf{s}}). \tag{14.77}$$

Substituting this expression into equation (14.75) then leads to

$$\int_0^{\tau_s} S^*(\mathbf{r}' + s'\hat{\mathbf{s}}, \hat{\mathbf{s}})\, e^{-(\tau_s - \tau_s')}d\tau_s' = S^*(\mathbf{r}, \hat{\mathbf{s}})(1 - e^{-\tau_s})$$

$$- \frac{dS^*}{d\tau_s}(\mathbf{r}, \hat{\mathbf{s}})[1 - (1 + \tau_s)e^{-\tau_s}]. \tag{14.78}$$

To eliminate $dS^*/d\tau_s$ from equation (14.78), we use the assumed linearity of the source function in reverse, i.e., we set

$$\int_0^{\tau_s} S^*(\mathbf{r}' + s'\hat{\mathbf{s}}, \hat{\mathbf{s}})e^{-(\tau_s - \tau_s')}d\tau_s' \simeq S^*(\mathbf{r} - s_0\hat{\mathbf{s}}, \hat{\mathbf{s}})(1 - e^{-\tau_s}), \tag{14.79}$$

where the geometrical distance s_0 is related to the optical distance

$$\tau_0 = \int_0^{s_0} \beta(\mathbf{r} - s''\hat{\mathbf{s}})\, ds'' = 1 - \frac{\tau_s\, e^{-\tau_s}}{1 - e^{-\tau_s}}. \tag{14.80}$$

where s'' is distance measured from $\mathbf{r}$ into the $-\hat{\mathbf{s}}$ direction. Thus, for a linearly varying source the integral is easily evaluated from equation (14.78), using a representative value for the source evaluated an optical distance τ_0 away ($0 < \tau_0 < 1$). We note that equations (14.78) and (14.79) also reduce to the correct optically thick and thin limits. Equation (14.79) is preferable over equation (14.78) since at moderate optical thicknesses it can account to a certain degree for nonlinear variation of the source term.

Substituting equation (14.79) into equation (14.74) leads to

$$I(\mathbf{r}, \hat{\mathbf{s}}) \simeq \frac{J_w}{\pi}(\mathbf{r}')e^{-\tau_s} + S^*(\mathbf{r} - s_0\hat{\mathbf{s}}, \hat{\mathbf{s}})(1 - e^{-\tau_s}). \tag{14.81}$$

The wall radiosities are found similarly to the MDA, by invoking their definition,

---

[8] It should be noted that, in the optically thick limit, the P_1-approximation will *always* predict the correct *radiative source*, but not necessarily the correct *radiative heat flux* (in the presence of temperature steps), which we found to be a minor shortcoming of the MDA.

that is,

$$J_w(\mathbf{r}) = \epsilon\pi I_{bw}(\mathbf{r}) + (1-\epsilon)\int_{\hat{\mathbf{s}}\cdot\hat{\mathbf{n}}<0} I(\mathbf{r},\hat{\mathbf{s}})|\hat{\mathbf{s}}\cdot\hat{\mathbf{n}}|\,d\Omega$$

$$= \epsilon\pi I_{bw}(\mathbf{r}) + (1-\epsilon)\int_A \left[J_w(\mathbf{r}')e^{-\tau_s} \right.$$

$$\left. + \pi S^*(\mathbf{r}-s_0\hat{\mathbf{s}},\hat{\mathbf{s}})(1-e^{-\tau_s})\right]\frac{\cos\theta\cos\theta'}{\pi S^2}\,dA. \quad (14.82)$$

Equation (14.82) may again be solved by standard surface transport methods, such as small zones (with little variation of J_w, τ_s and $\hat{\mathbf{s}}$), to give

$$J_i = \epsilon_i\pi I_{bi} + (1-\epsilon_i)\sum_{j=1}^N\left[J_j e^{-\tau_{ij}} + \pi S_{ij}^*(1-e^{-\tau_{ij}})\right]F_{i-j}, \quad i = 1,2,\ldots,N. \quad (14.83)$$

Once the radiosities are known, the improved values for incident radiation and radiative heat flux anywhere inside the medium follow from their definition as

$$G(\mathbf{r}) = \frac{1}{\pi}\int_{4\pi}\left[J_w(\mathbf{r}')e^{-\tau_s} + \pi S^*(\mathbf{r}-s_0\hat{\mathbf{s}},\hat{\mathbf{s}})(1-e^{-\tau_s})\right]d\Omega$$

$$= \int_A\left[J_w(\mathbf{r}')e^{-\tau_s} + \pi S^*(\mathbf{r}-s_0\hat{\mathbf{s}},\hat{\mathbf{s}})(1-e^{-\tau_s})\right]\frac{\cos\theta'dA}{\pi S^2}, \quad (14.84)$$

$$\mathbf{q}(\mathbf{r}) = \frac{1}{\pi}\int_{4\pi}\left[J_w(\mathbf{r}')e^{-\tau_s} + \pi S^*(\mathbf{r}-s_0\hat{\mathbf{s}},\hat{\mathbf{s}})(1-e^{-\tau_s})\right]\hat{\mathbf{s}}\,d\Omega$$

$$= \int_A\left[J_w(\mathbf{r}')e^{-\tau_s} + \pi S^*(\mathbf{r}-s_0\hat{\mathbf{s}},\hat{\mathbf{s}})(1-e^{-\tau_s})\right]\hat{\mathbf{s}}\,\frac{\cos\theta'dA}{\pi S^2}. \quad (14.85)$$

Thus, the *improved differential approximation* requires (*i*) the solution to the ordinary P_1-approximation, (*ii*) the solution to a surface integral equation (or the evaluation of view factors) for the determination of radiosities, and (*iii*) the evaluation of a surface integral for any point inside the medium for which heat flux or incident radiation is desired.

Example 14.7.   Repeat Example 14.5 using the improved differential approximation.

Solution

For radiative equilibrium in a gray medium we have $G^* = 4\pi I_b^* = 4\sigma T^{*4}$ and, therefore, from equation (14.76) for an isotropically scattering medium $S^* = I_b^* = \sigma T^{*4}/\pi$. The solution for the ODA has already been found in Chapter 13, Example 13.5, as

$$G^*(\tau) = C - 3q^*\tau = 4\pi S^*; \quad C = 4\sigma T_1^4 - 2q^*.$$

In this particular case, since S^* is linear, equations (14.75), (14.78), and (14.79) all give the same result; and it is actually a little simpler to evaluate equation (14.78) rather than use equation (14.79). From Fig. 14-8, with $J_i = \sigma T_i^4/\pi$, and replacing τ_s (optical distance from point under consideration to point at wall) by τ/μ (for $0 < \mu = \cos\theta < 1$, or intensities traveling upward) and $-(\tau_L - \tau)/\mu$ (for $-1 < \mu < 0$, or intensities traveling downward) gives

$$
\begin{aligned}
I(\tau, \mu) &= \frac{\sigma T_1^4}{\pi} e^{-\tau/\mu} + S^*(1 - e^{-\tau/\mu}) - \mu \frac{dS^*}{d\tau}\left[1 - \left(1 + \frac{\tau}{\mu}\right)e^{-\tau/\mu}\right] \\
&= S^* + \left[\frac{\sigma T_1^4}{\pi} - \frac{\sigma T_1^4}{\pi} + \frac{q^*}{2\pi}\left(1 + \frac{3}{2}\tau\right)\right]e^{-\tau/\mu} \\
&\qquad + \frac{3}{4\pi}q^*\left[\mu - (\mu + \tau)e^{-\tau/\mu}\right], \quad 0 < \mu < 1; \\
&= \frac{\sigma T_2^4}{\pi} e^{(\tau_L - \tau)/\mu} + S^*\left(1 - e^{(\tau_L - \tau)/\mu}\right) \\
&\qquad - \mu \frac{dS^*}{d\tau}\left[1 - \left(1 - \frac{\tau_L - \tau}{\mu}\right)e^{(\tau_L - \tau)/\mu}\right] \\
&= S^* + \left[\frac{\sigma T_2^4}{\pi} - \frac{\sigma T_1^4}{\pi} + \frac{q^*}{2\pi}\left(1 + \frac{3}{2}\tau\right)\right]e^{(\tau_L - \tau)/\mu} \\
&\qquad + \frac{3}{4\pi}q^*\left[\mu - (\mu - \tau_L + \tau)e^{(\tau_L - \tau)/\mu}\right], \quad -1 < \mu < 0.
\end{aligned}
$$

The heat flux is then determined as

$$
\begin{aligned}
\Psi(\tau) &= \frac{q(\tau)}{\sigma(T_1^4 - T_2^4)} = 2\pi \int_{-1}^{1} I(\tau, \mu)\,\mu\,d\mu \Big/ \sigma(T_1^4 - T_2^4) \\
&= 2E_3(\tau_L - \tau) + \Psi^*\left(1 + \tfrac{3}{2}\tau\right)[E_3(\tau) - E_3(\tau_L - \tau)] \\
&\quad + \tfrac{3}{2}\Psi^*\left[\tfrac{2}{3} - E_4(\tau) - \tau E_3(\tau) - E_4(\tau_L - \tau) - (\tau_L - \tau)E_3(\tau_L - \tau)\right],
\end{aligned}
$$

which, with $\Psi^* = q^*/\sigma(T_1^4 - T_2^4) = 1/\left(1 + \tfrac{3}{4}\tau_L\right)$, reduces to

$$
\Psi(\tau) = \frac{1 + E_3(\tau) + E_3(\tau_L - \tau) - \tfrac{3}{2}[E_4(\tau) + E_4(\tau_L - \tau)]}{1 + \tfrac{3}{4}\tau_L}.
$$

This result is also included in Fig. 14-9 and compared with other solution methods. The accuracy of the IDA is very good, and of the same order as the MDA, but the most severe weakness of the IDA is displayed in Fig. 14-9. In this method radiative equilibrium is not satisfied exactly, i.e., the approximate heat flux fluctuates slightly around the exact (constant) value.

Example 14.8.   Repeat Example 14.6 using the IDA.

Solution

Since there is no scattering, we have $S^* = S = \sigma T_m^4/\pi$ (i.e., the source term is given, and there is no need to go through the P_1-solution). Further, since $T_m = \text{const}$, equation (14.79) is exact and not an approximation, leading to

$$
\begin{aligned}
I &= \frac{\sigma T_w^4}{\pi} e^{-\tau/\mu} + \frac{\sigma T_m^4}{\pi}\left(1 - e^{-\tau/\mu}\right), & 0 < \mu < 1, \\
&= \frac{\sigma T_w^4}{\pi} e^{(\tau_L - \tau)/\mu} + \frac{\sigma T_m^4}{\pi}\left(1 - e^{(\tau_L - \tau)/\mu}\right), & -1 < \mu < 0,
\end{aligned}
$$

that is, the *exact* intensity distribution within the medium. Therefore, for this simple problem, the IDA yields the exact result

$$\Psi(\tau) = \frac{q(\tau)}{\sigma(T_w^4 - T_m^4)} = 2\left[E_3(\tau) - E_3(\tau_L - \tau)\right],$$

as shown in Fig. 14-6.

Some two-dimensional examples have been given for MDA by [27, 29, 31] and for IDA by [32–36], proving their excellent accuracy under all conditions (even when the P_1-approximation fails).

References

1. Jeans, J. H.: "The Equations of Radiative Transfer of Energy," *Monthly Notices Royal Astronomical Society*, vol. 78, pp. 28–36, 1917.
2. Kourganoff, V.: *Basic Methods in Transfer Problems*, Dover Publications, New York, 1963.
3. Davison, B.: *Neutron Transport Theory*, Oxford University Press, London, 1958.
4. Murray, R. L.: *Nuclear Reactor Physics*, Prentice Hall, Englewood Cliffs, NJ, 1957.
5. Sommerfeld, A.: *Partial Differential Equations of Physics*, Academic Press, New York, 1964.
6. Krook, M.: "On the Solution of the Equation of Transfer, I," *Astrophysical Journal*, vol. 122, pp. 488–497, 1955.
7. Cheng, P.: "Study of the Flow of a Radiating Gas by a Differential Approximation," Ph.D. thesis, Stanford University, Stanford, CA, 1965.
8. Cheng, P.: "Dynamics of a Radiating Gas with Application to Flow Over a Wavy Wall," *AIAA Journal*, vol. 4, no. 2, pp. 238–245, 1966.
9. Ou, S.-C. S., and K. N. Liou: "Generalization of the Spherical Harmonic Method to Radiative Transfer in Multi-Dimensional Space," *Journal of Quantitative Spectroscopy and Radiative Transfer*, vol. 28, no. 4, pp. 271–288, 1982.
10. Condiff, D.: "Anisotropic Scattering in Three Dimensional Differential Approximation of Radiation Heat Transfer," in *Fundamentals and Applications of Radiation Heat Transfer*, vol. HTD–72, ASME, pp. 19–29, 1987.
11. Brenner, H.: "The Stokes Resistance of a Slightly Deformed Sphere — II Intrinsic Resistance Operators for an Arbitrary Initial Flow," *Chemical Engineering and Science*, vol. 22, p. 375, 1967.
12. Abramowitz, M., and I. A. Stegun (eds.): *Handbook of Mathematical Functions*, Dover Publications, New York, 1965.
13. Mark, J. C.: "The Spherical Harmonics Method, Part I," Technical Report Atomic Energy Report No. MT 92, National Research Council of Canada, 1944.
14. Mark, J. C.: "The Spherical Harmonics Method, Part II," Technical Report Atomic Energy Report No. MT 97, National Research Council of Canada, 1945.
15. Marshak, R. E.: "Note on the Spherical Harmonics Method as Applied to the Milne Problem for a Sphere," *Phys. Rev.*, vol. 71, pp. 443–446, 1947.
16. Pellaud, B.: "Numerical Comparison of Different Types of Vacuum Boundary Conditions for the P_N Approximation," *Trans. Am. Nucl Soc.*, vol. 9, pp. 434–435, 1966.
17. Schmidt, E., and E. M. Gelbard: "A Double P_N Method for Spheres and Cylinders," *Trans. Am. Nucl Soc.*, vol. 9, pp. 432–433, 1966.
18. MacRobert, T. M.: *Spherical Harmonics*, 3rd ed., Pergamon Press, New York, 1967.
19. Modest, M. F.: "Photon-Gas Formulation of the Differential Approximation in Radiative Transfer," *Letters in Heat and Mass Transfer*, vol. 3, pp. 111–116, 1976.
20. Pipes, L. A., and L. R. Harvill: *Applied Mathematics for Engineers and Physicists*, McGraw-Hill, New York, 1970.
21. Bayazitoğlu, Y., and J. Higenyi: "The Higher-Order Differential Equations of Radiative Transfer: P_3 Approximation," *AIAA Journal*, vol. 17, pp. 424–431, 1979.
22. Kofink, W.: "Complete Spherical Harmonics Solution of the Boltzmann Equation for Neutron Transport in Homogeneous Media with Cylindrical Geometry," *Nuclear Science and Engineering*, vol. 6, pp. 473–486, 1959.

23. Mengüç, M. P., and R. Viskanta: "Radiative Transfer in Three-Dimensional Rectangular Enclosures Containing Inhomogeneous, Anisotropically Scattering Media," *Journal of Quantitative Spectroscopy and Radiative Transfer*, vol. 33, no. 6, pp. 533–549, 1985.

24. Mengüç, M. P., and R. Viskanta: "Radiative Transfer in Axisymmetric, Finite Cylindrical Enclosures," *ASME Journal of Heat Transfer*, vol. 108, pp. 271–276, 1986.

25. Olfe, D. B.: "A Modification of the Differential Approximation for Radiative Transfer," *AIAA Journal*, vol. 5, no. 4, pp. 638–643, 1967.

26. Olfe, D. B.: "Application of a Modified Differential Approximation to Radiative Transfer in a Gray Medium Between Concentric Sphere and Cylinders," *Journal of Quantitative Spectroscopy and Radiative Transfer*, vol. 8, pp. 899–907, 1968.

27. Olfe, D. B.: "Radiative Equilibrium of a Gray Medium Bounded by Nonisothermal Walls," *Progress in Astronautics and Aeronautics*, vol. 23, pp. 295–317, 1970.

28. Olfe, D. B.: "Radiative Equilibrium of a Gray Medium in a Rectangular Enclosure," *Journal of Quantitative Spectroscopy and Radiative Transfer*, vol. 13, pp. 881–895, 1973.

29. Glatt, L., and D. B. Olfe: "Radiative Equilibrium of a Gray Medium in a Rectangular Enclosure," *Journal of Quantitative Spectroscopy and Radiative Transfer*, vol. 13, pp. 881–895, 1973.

30. Wu, C.-Y., W. H. Sutton, and T. J. Love: "Successive Improvement of the Modified Differential Approximation in Radiative Heat Transfer," *Journal of Thermophysics and Heat Transfer*, vol. 1, no. 4, pp. 296–300, 1987.

31. Modest, M. F.: "The Modified Differential Approximation for Radiative Transfer in General Three-Dimensional Media," *Journal of Thermophysics and Heat Transfer*, vol. 3, no. 3, pp. 283–288, 1989.

32. Modest, M. F.: "Two-Dimensional Radiative Equilibrium of a Gray Medium in a Plane Layer Bounded by Gray Non-Isothermal Walls," *ASME Journal of Heat Transfer*, vol. 96C, pp. 483–488, 1974.

33. Modest, M. F.: "Radiative Equilibrium in a Rectangular Enclosure Bounded by Gray Non-Isothermal Walls," *Journal of Quantitative Spectroscopy and Radiative Transfer*, vol. 15, pp. 445–461, 1975.

34. Modest, M. F., and D. Stevens: "Two Dimensional Radiative Equilibrium of a Gray Medium Between Concentric Cylinders," *Journal of Quantitative Spectroscopy and Radiative Transfer*, vol. 19, pp. 353–365, 1978.

35. Modest, M. F.: "The Improved Differential Approximation for Radiative Transfer in General Three-Dimensional Media," in *Heat Transfer Phenomena in Radiation, Combustion and Fires*, vol. HTD–106, ASME, pp. 213–220, 1989.

36. Modest, M. F.: "The Improved Differential Approximation for Radiative Transfer in Multi-Dimensional Media," *ASME Journal of Heat Transfer*, vol. 112, pp. 819–821, 1990.

Problems

14.1 Consider a gray medium at radiative equilibrium contained within a long black cylinder with a surface temperature of $T(r = R, z) = T_w(z)$. Find the relevant boundary conditions for the P_1-approximation directly from equation (14.22), i.e., in a manner similar to the development in Example 14.1.

14.2 Consider a gray, isotropically scattering medium at radiative equilibrium contained between large, isothermal, gray plates at temperatures T_1 and T_2, and emissivities ϵ_1 and ϵ_2, respectively. Determine the radiative heat flux between the plates using the P_1-approximation. Compare the result with the exact answer from Table 12.1.

14.3 Consider a large, isothermal (temperature T_w), gray and diffuse (emissivity ϵ) wall adjacent to a semi-infinite gray absorbing/emitting and linear-anisotropically scattering medium. The medium is isothermal (temperature T_m). Determine the radiative heat flux as a function of distance away from the plate using the P_1-approximation with (*i*) Marshak's, (*ii*) Mark's boundary conditions.

14.4 Consider parallel, gray-diffuse plates that are isothermal at temperatures T_1 and T_2, and with emissivities ϵ_1 and ϵ_2, respectively. The plates are separated by a gray, linear-anisotropically scattering medium of thickness L, which is at radiative equilibrium.

Using the P_1-approximation, determine the temperature distribution within, and the heat flux through, the medium. Compare the heat flux with the exact answer given by Table 12.1 (for isotropic scattering, and optical thicknesses of $\tau_L = \beta L = 0$, 0.1, 0.5, 1, 2, and 5). Show that the radiative heat flux can be obtained from the expression given in Example 14.3, by letting $R_2 = R_1 + L \to \infty$.

14.5 Black spherical particles of $100 \, \mu$m radius are suspended between two cold and black parallel plates 1 m apart. The particles produce heat at a rate of $\pi/10$ W/particle, which must be removed by thermal radiation. The number of particles between the plates is given by

$$N_T(z) = N_0 + \Delta N z / L, \quad 0 < z < L; \qquad N_0 = \Delta N = 212 \text{ particles/cm}^3.$$

(a) Determine the local absorption coefficient and the local heat production rate; introduce an optical coordinate and determine the optical thickness of the entire gap.

(b) Assuming the P_1-approximation to be valid, what are the relevant equations and boundary conditions governing the heat transfer?

(c) What are the heat flux rates at the top and bottom surfaces? What is the entire amount of energy released by the particles? What is the maximum particle temperature?

14.6 Consider parallel, black plates spaced 1 m apart, at constant temperatures $T_1 = 1000$ K and $T_2 = 300$ K, respectively, separated by a gray, nonscattering medium at radiative equilibrium. The absorption coefficient of the medium depends on its temperature according to a power law, $\kappa = \kappa_0 (T/T_0)^n$ ($\kappa_0 = 1 \, \text{m}^{-1}$, $T_0 = 300$ K, and n is an arbitrary, positive constant).

(a) Using the P_1-approximation, outline how to determine the radiative heat flux through, and temperature field within, the medium (i.e., the result may contain unsolved implicit relationships).

(b) Write a small computer program to quantify the results for $n = 0$, 0.5, 1, and 4. Compare with results obtained for a constant κ [evaluated at an average temperature $T_{av} = 0.5 \times (300 + 1000) = 650$ K].

14.7 Two infinitely long concentric cylinders of radii R_1 and R_2 with emissivities ϵ_1 and ϵ_2 both have the same constant surface temperature T_w. The medium between the cylinders has a constant absorption coefficient κ and does not scatter; uniform heat generation $\dot{Q}'''$ takes place inside the medium. Determine the temperature distribution in the medium and heat heat fluxes at the wall if radiation is the only means of heat transfer by using the P_1-approximation.

14.8 An infinite, black, isothermal plate bounds a semi-infinite space filled with black spheres. At any given distance z away from the plate the particle number density is identical, namely, $N_T = 6.3662 \times 10^8 \text{m}^{-3}$. However, the radius of the suspended spheres diminishes monotonically away from the surface as

$$a = a_o e^{-z/L}; \qquad a_o = 10^{-4} \text{ m}, \qquad L = 1 \text{ m}.$$

(a) Determine the absorption coefficient as a function of z (you may make the large-particle assumption).

(b) Determine the optical coordinate as a function of z. What is the total optical thickness of the semi-infinite space?

(c) Assuming that radiative equilibrium prevails and that the differential approximation is valid, set up the boundary conditions.

(d) Solve for heat flux and temperature distribution (as a function of z).

14.9 Consider two parallel black plates both at 1000 K, which are 2 m apart. The medium between the plates emits and absorbs (but does not scatter) with an absorption coefficient of $\kappa = 0.05236\,\text{cm}^{-1}$ (gray medium). Heat is generated by the medium according to the formula

$$\dot{Q}''' = C\sigma T^4, \qquad C = 6.958 \times 10^{-4}\,\text{cm}^{-1},$$

where T is the local temperature of the medium between the plates. Assuming that radiation is the only important mode of heat transfer, determine the heat flux to the plates.

14.10 A furnace burning pulverized coal may be approximated by a gray cylinder at radiative equilibrium with uniform heat generation $\dot{Q}''' = 0.266\,\text{W/cm}^3$, bounded by a cold black wall. The gray and constant absorption and scattering coefficients are, respectively, $0.16\,\text{cm}^{-1}$ and $0.04\,\text{cm}^{-1}$, while the furnace radius is $R = 0.5\,\text{m}$. Scattering may be assumed to be isotropic. Using the P_1-approximation,

(a) set up the relevant equations and their boundary conditions;

(b) calculate the total heat loss from the furnace (per unit length);

(c) calculate the radial temperature distribution; what are centerline and adjacent-to-wall temperatures, respectively?

(d) qualitatively, keeping the extinction coefficient constant, what is the effect of varying the scattering coefficient on (i) heat transfer rates, (ii) temperature levels?

14.11 Estimate the radial temperature distribution in the sun. You may make the following assumptions:

(i) The sun is a sphere of radius R;

(ii) As a result of high temperatures in the sun the absorption and scattering coefficients may be approximated to be constant, i.e., $\kappa_\nu, \beta_\nu = \text{const} \neq f(\nu, T, r)$ (free-free transitions!);

(iii) Due to high temperatures, radiation is the only mode of heat transfer;

(iv) The fusion process may be approximated by assuming that a small sphere at the center of the sun releases heat uniformly corresponding to the total heat loss of the sun (i.e., assume the sun to be concentric spheres with a certain heat flux at the inner boundary $r = r_i$).

(a) Relate the heat production to the effective sun temperature $T_{\text{eff}} = 5762\,\text{K}$.

(b) Would you expect the sun to be optically thin, intermediate, or thick? Why? What are the prevailing boundary conditions?

(c) Find an expression for the temperature distribution (for $r > r_i$).

(d) What is the surface temperature of the sun?

14.12 Repeat Problem 14.11 but replace assumption (iv) by the following: The fusion process may be approximated by assuming that the sun releases heat uniformly throughout its volume corresponding to the total heat loss of the sun.

14.13 Consider a sphere of very hot dissociated gas of radius 5 cm. The gas may be approximated as a gray, linear-anisotropically scattering medium with $\kappa = 0.1\,\text{cm}^{-1}$, $\sigma_s = 0.2\,\text{cm}^{-1}$, $A_1 = 1$. The gas is suspended magnetically in vacuum within a large cold container and is initially at a uniform temperature $T_g = 10,000\,\text{K}$. Using the P_1-approximation and neglecting conduction and convection, specify the total heat loss per unit time from the entire sphere at time $t = 0$. Outline the solution procedure for times $t > 0$.

Hint: Solve the governing equation by introducing a new dependent variable $g(\tau) = \tau(4\pi I_b - G)$.

14.14 A spherical test bomb of 1 m radius is coated with a nonreflective material and cooled. Inside the sphere is nitrogen mixed with spherical particles at a rate of 10^8 particles/m^3. The particles have a radius of $300\,\mu\text{m}$, are diffuse-gray with $\epsilon = 0.5$, and generate heat at a rate of $150\,\text{W}/\text{cm}^3$ of particle volume. Using radiative properties determined in Problem 10.8 [with the two phase functions from Problem 10.8], determine the temperature distribution inside the bomb, using the differential approximation for both cases (*i*) and (*ii*). In particular, what is the gas temperature at the center and at the wall? How much do the two methods differ from one another?

14.15 Consider a gray, isotropically-scattering medium at radiative equilibrium contained between large, isothermal, gray plates at temperatures T_1 and T_2, and emissivities ϵ_1 and ϵ_2, respectively. Determine the radiative heat flux between the plates using the P_3-approximation. Compare the results with the answer from Problem 14.2.

14.16 Do Problem 14.3 using the P_3-approximation with Marshak's boundary condition.

14.17 A hot gray medium is contained between two concentric black spheres of radius $R_1 = 10\,\text{cm}$ and $R_2 = 20\,\text{cm}$. The surfaces of the spheres are isothermal at $T_1 = 2000\,\text{K}$ and $T_2 = 500\,\text{K}$, respectively. The medium absorbs and emits with $n = 1$, $\kappa = 0.05\,\text{cm}^{-1}$, but does not scatter radiation. Determine the heat flux between the spheres using the modified differential approximation (MDA).

Note: This problem requires the numerical solution of a simple ordinary differential equation.

14.18 Repeat Problem 14.17 for concentric cylinders of the same radii. Compare your result with those of Fig. 14-5.

Note: This problem requires the numerical solution of a simple ordinary differential equation.

14.19 Repeat Problem 14.17 using the ordinary P_1-approximation (ODA) and the improved differential approximation (IDA).

Note: This problem requires the numerical solution of an integral.

14.20 Repeat Problem 14.18 using the ordinary P_1-approximation (ODA) and the improved differential approximation (IDA).

Note: This problem requires the numerical solution of an integral.

14.21 Repeat Problem 14.13 using the improved differential approximation (IDA).

Note: This problem requires the numerical solution of an integral.

CHAPTER
15

THE METHOD OF
DISCRETE
ORDINATES
(S_N-APPROXIMATION)

15.1 INTRODUCTION

Like the spherical harmonics method, the *discrete ordinate method* is a tool to transform the equation of transfer (for a gray medium, or on a spectral basis) into a set of simultaneous partial differential equations. Like the P_N-method, the discrete ordinates or S_N-method may be carried out to any arbitrary order and accuracy. First proposed by Chandrasekhar [1] in his work on stellar and atmospheric radiation, the S_N-method originally received little attention in the heat transfer community. Again like the P_N-method, the discrete ordinates method was first systematically applied to problems in neutron transport theory, notably by Lee [2] and Lathrop [3, 4]. There were some early, unoptimized attempts to apply the method to one-dimensional, planar thermal radiation problems (Love *et al.* [5, 6], Hottel *et al.* [7], Roux and Smith [8, 9]). But only very recently has the discrete ordinates method been applied to, and optimized for, general radiative heat transfer problems, primarily through the pioneering works of Truelove [10–12] and Fiveland [13–16]. At present its popularity is clearly surging, with applications to three-dimensional cylinders [17], inhomogeneous atmospheres [18, 19], inhomogeneous spheres [20], etc., to name but a few.

The discrete ordinates method is based on a discrete representation of the directional variation of the radiative intensity. A solution to the transport problem is found by solving the equation of transfer for a set of discrete directions spanning the total solid angle range of 4π. As such, the discrete ordinates method is simply a finite differencing of the directional dependence of the equation of transfer. Integrals over solid angle are approximated by numerical quadrature (e.g., for the evaluation of the radiative source term, the radiative heat flux, etc.).

In this chapter we shall first develop the set of partial differential equations for the general S_N-method and their boundary conditions. This is followed by a section describing how the method may be applied to one-dimensional plane-parallel media, and another dealing with spherical and cylindrical geometries. Finally, its application to two- and three-dimensional problems will be outlined.

15.2   GENERAL RELATIONS

The general equation of transfer for an absorbing, emitting, and anisotropically scattering medium is, according to equation (8.23),

$$\frac{dI}{ds} = \hat{\mathbf{s}} \cdot \boldsymbol{\nabla} I(\mathbf{r}, \hat{\mathbf{s}}) = \kappa(\mathbf{r}) I_b(\mathbf{r}) - \beta(\mathbf{r}) I(\mathbf{r}, \hat{\mathbf{s}}) + \frac{\sigma_s(\mathbf{r})}{4\pi} \int_{4\pi} I(\mathbf{r}, \hat{\mathbf{s}}') \Phi(\mathbf{r}, \hat{\mathbf{s}}', \hat{\mathbf{s}}) \, d\Omega', \quad (15.1)$$

Equation (15.1) is valid for a gray medium or, on a spectral basis, for a nongray medium, and is subject to the boundary condition

$$I(\mathbf{r}_w, \hat{\mathbf{s}}) = \epsilon(\mathbf{r}_w) I_b(\mathbf{r}_w) + \frac{\rho(\mathbf{r}_w)}{\pi} \int_{\hat{\mathbf{n}} \cdot \hat{\mathbf{s}}' < 0} I(\mathbf{r}_w, \hat{\mathbf{s}}') |\hat{\mathbf{n}} \cdot \hat{\mathbf{s}}| \, d\Omega', \quad (15.2)$$

where we have limited ourselves to an enclosure with opaque, diffusely emitting and diffusely reflecting walls. The extension of equation (15.2) to more general boundary conditions is straightforward.

Discrete Ordinates Equations

In the discrete ordinates method, equation (15.1) is solved for a set of n different directions $\hat{\mathbf{s}}_i$, $i = 1, 2, \ldots, n$, and the integrals over direction are replaced by numerical quadratures, that is,

$$\int_{4\pi} f(\hat{\mathbf{s}}) \, d\Omega \simeq \sum_{i=1}^{n} w_i \, f(\hat{\mathbf{s}}_i), \quad (15.3)$$

where the w_i are the quadrature weights associated with the directions $\hat{\mathbf{s}}_i$. Thus, equation (15.1) is approximated by a set of n equations,

$$\hat{\mathbf{s}}_i \cdot \boldsymbol{\nabla} I(\mathbf{r}, \hat{\mathbf{s}}_i) = \kappa(\mathbf{r}) I_b(\mathbf{r}) - \beta(\mathbf{r}) I(\mathbf{r}, \hat{\mathbf{s}}_i) + \frac{\sigma_s(\mathbf{r})}{4\pi} \sum_{j=1}^{n} w_j \, I(\mathbf{r}, \hat{\mathbf{s}}_j) \Phi(\mathbf{r}, \hat{\mathbf{s}}_i, \hat{\mathbf{s}}_j),$$

$$i = 1, 2, \ldots, n, \quad (15.4)$$

subject to the boundary conditions

$$I(\mathbf{r}_w, \hat{\mathbf{s}}_i) = \epsilon(\mathbf{r}_w) I_b(\mathbf{r}_w) + \frac{\rho(\mathbf{r}_w)}{\pi} \sum_{\hat{\mathbf{n}} \cdot \hat{\mathbf{s}}_j < 0} w_j I(\mathbf{r}_w, \hat{\mathbf{s}}_j) |\hat{\mathbf{n}} \cdot \hat{\mathbf{s}}_j|, \quad \hat{\mathbf{n}} \cdot \hat{\mathbf{s}}_i > 0. \quad (15.5)$$

Each beam traveling in a direction of $\hat{\mathbf{s}}_i$ intersects the enclosure surface twice: Once where the beam emanates from the wall ($\hat{\mathbf{n}} \cdot \hat{\mathbf{s}}_i > 0$), and once where it strikes the wall, to be absorbed or reflected ($\hat{\mathbf{n}} \cdot \hat{\mathbf{s}}_i < 0$). The governing equation is first order, requiring only one boundary condition (for the emanating intensity, $\hat{\mathbf{n}} \cdot \hat{\mathbf{s}}_i > 0$). Equations (15.4) together with their boundary conditions (15.5) constitute a set of n simultaneous, first-order, linear partial differential equations for the unknown $I_i(\mathbf{r}) = I(\mathbf{r}, \hat{\mathbf{s}}_i)$. The solution for the I_i may be found using any standard technique (analytical or numerical). If scattering is present ($\sigma_s \neq 0$), the equations are coupled in such a way that generally an iterative procedure is necessary. Even in the absence of scattering, the temperature field is usually not known, but must be calculated from the intensity field, again making iterations necessary. Only if there is no scattering and if the temperature field is given is the solution to the intensities I_i straightforward (as is the exact solution).

Once the intensities have been determined the desired direction-integrated quantities are readily calculated. The radiative heat flux, inside the medium or at a surface, may be found from its definition, equation (8.47),

$$\mathbf{q}(\mathbf{r}) = \int_{4\pi} I(\mathbf{r}, \hat{\mathbf{s}}) \hat{\mathbf{s}} \, d\Omega \simeq \sum_{i=1}^{n} w_i I_i(\mathbf{r}) \hat{\mathbf{s}}_i. \quad (15.6)$$

The incident radiation G [and, through equation (8.54), the divergence of the radiative heat flux] is similarly determined as

$$G(\mathbf{r}) = \int_{4\pi} I(\mathbf{r}, \hat{\mathbf{s}}) \, d\Omega \simeq \sum_{i=1}^{n} w_i I_i(\mathbf{r}). \quad (15.7)$$

At a surface the heat flux may also be determined from surface energy balances [equations (4.1) and (3.16)] as

$$\begin{aligned} \mathbf{q} \cdot \hat{\mathbf{n}}(\mathbf{r}_w) &= \epsilon(\mathbf{r}_w) [\pi I_b(\mathbf{r}_w) - H(\mathbf{r}_w)] \\ &\simeq \epsilon(\mathbf{r}_w) \Big(\pi I_b(\mathbf{r}_w) - \sum_{\hat{\mathbf{n}} \cdot \hat{\mathbf{s}}_i < 0} w_i I_i(\mathbf{r}_w) |\hat{\mathbf{n}} \cdot \hat{\mathbf{s}}_i| \Big). \end{aligned} \quad (15.8)$$

Equations (15.4) and (15.5) can be written in a somewhat more compact form if one limits the analysis to linear-anisotropic scattering, i.e., to a scattering phase function of

$$\Phi(\mathbf{r}, \hat{\mathbf{s}}, \hat{\mathbf{s}}') = 1 + A_1(\mathbf{r}) \hat{\mathbf{s}}' \cdot \hat{\mathbf{s}}. \quad (15.9)$$

Then, using equations (15.6) and (15.7) and/or equation (12.15),

$$\hat{\mathbf{s}}_i \cdot \boldsymbol{\nabla} I_i + \beta I_i = \kappa I_b + \frac{\sigma_s}{4\pi}(G + A_1 \mathbf{q} \cdot \hat{\mathbf{s}}_i), \quad i = 1, 2, \ldots, n, \tag{15.10}$$

with boundary condition

$$I_i = \frac{J_w}{\pi} = I_{bw} - \frac{1 - \epsilon}{\epsilon \pi} \mathbf{q} \cdot \hat{\mathbf{n}}, \quad \hat{\mathbf{n}} \cdot \hat{\mathbf{s}}_i > 0 \tag{15.11}$$

at the enclosure surface. Equation (15.11) shows that, at a point on the boundary, all I_i are equal, since the walls are diffusely emitting and reflecting. Of course, radiative heat flux and incident radiation are unknowns to be determined from directional intensities from the series in equations (15.6) and (15.7). In most cases the intensities are determined through iteration. Equations (15.10) and (15.11) are convenient forms for such a procedure: For each iteration values of G and $\mathbf{q}$ are estimated, and the n intensities I_i are evaluated. The values for G and $\mathbf{q}$ are then updated, and so on.

In general, we distinguish among three different heat transfer scenarios: (1) the temperature profile of the medium and, therefore, $I_b(\mathbf{r})$ are specified; (2) radiation is competing with conduction and/or convection; (3) radiation is of overriding importance, i.e., radiative equilibrium prevails. In scenario (2) the governing equations are highly nonlinear, and it is common to separate the different modes of heat transfer by estimating a temperature profile, which is then iteratively updated (see Chapter 20). Thus, for the first two scenarios the temperature profile is specified, and the radiative heat transfer rates may be calculated directly from equations (15.10) and (15.11) together with the relations (15.6) and (15.7).

Radiative Equilibrium   If radiative equilibrium prevails, then the temperature profile is unknown and needs to be determined. From equations (8.54) and (8.64),

$$\boldsymbol{\nabla} \cdot \mathbf{q} = \dot{Q}''' = \kappa(4\pi I_b - G), \tag{15.12}$$

where $\dot{Q}'''$ is heat generation (per unit volume) within the medium, if internal energy sources are present. Then equation (15.10) reduces to

$$\hat{\mathbf{s}}_i \cdot \boldsymbol{\nabla} I_i + \beta I_i = \frac{1}{4\pi}\left(\beta G + A_1 \sigma_s \mathbf{q} \cdot \hat{\mathbf{s}}_i + \dot{Q}'''\right), \tag{15.13}$$

and

$$I_b = \frac{1}{4\pi}\left(G + \frac{\dot{Q}'''}{\kappa}\right). \tag{15.14}$$

Selection of Discrete Ordinate Directions

The choice of quadrature scheme is arbitrary, although restrictions on the directions $\hat{\mathbf{s}}_i$ and quadrature weights w_i may arise from the desire to preserve symmetry and to satisfy certain conditions. It is customary to choose sets of directions and weights that are completely symmetric (i.e., sets that are invariant after any rotation of 90°), and that satisfy the zeroth, first, and second moments, or

$$\int_{4\pi} d\Omega = 4\pi = \sum_{i=1}^{n} w_i, \tag{15.15a}$$

$$\int_{4\pi} \hat{s}\, d\Omega = 0 = \sum_{i=1}^{n} w_i\, \hat{s}_i, \tag{15.15b}$$

$$\int_{4\pi} \hat{s}\hat{s}\, d\Omega = \frac{4\pi}{3}\boldsymbol{\delta} = \sum_{i=1}^{n} w_i\, \hat{s}_i\hat{s}_i, \tag{15.15c}$$

where $\boldsymbol{\delta}$ is the unit tensor [cf. equation (14.29)]. Different sets of directions and weights satisfying all these criteria have been tabulated, for example, by Lee [2] and Lathrop and Carlson [21]. Fiveland [15] and Truelove [11] have observed that different sets of ordinates may result in considerably different accuracy. They noted that (*i*) the intensity may have directional discontinuity at a wall, and (*ii*) the important radiative heat fluxes at the walls are evaluated through a first moment of intensity over a half range of 2π [equation (15.8)]. They concluded that the set of ordinates and weights should also satisfy the first moment over a half range, that is,

$$\int_{\hat{\mathbf{n}}\cdot\hat{s}<0} |\hat{\mathbf{n}}\cdot\hat{s}|\, d\Omega = \int_{\hat{\mathbf{n}}\cdot\hat{s}>0} \hat{\mathbf{n}}\cdot\hat{s}\, d\Omega = \pi = \sum_{\hat{\mathbf{n}}\cdot\hat{s}_i>0} w_i\, \hat{\mathbf{n}}\cdot\hat{s}_i. \tag{15.16}$$

While it is impossible to satisfy equation (15.16) for arbitrary orientations of the surface normal, it can be satisfied for the principal orientations, if $\hat{\mathbf{n}} = \hat{\mathbf{i}}, \hat{\mathbf{j}}$, or $\hat{\mathbf{k}}$. Sets of ordinates and weights that satisfy (*i*) the symmetry requirement, (*ii*) the moment equations (15.15), and (*iii*) the half-moment equation (15.16) (for the three principal directions of $\hat{\mathbf{n}}$)[1] have been given by Lathrop and Carlson [21]. The first four sets labeled S_2-, S_4-, S_6-, and S_8-approximation are reproduced in Table 15.1. In the table the ξ_i, η_i, and μ_i are the direction cosines of $\hat{s}_i$, or

$$\hat{s}_i = (\hat{s}_i \cdot \hat{\mathbf{i}})\hat{\mathbf{i}} + (\hat{s}_i \cdot \hat{\mathbf{j}})\hat{\mathbf{j}} + (\hat{s}_i \cdot \hat{\mathbf{k}})\hat{\mathbf{k}} = \xi_i\hat{\mathbf{i}} + \eta_i\hat{\mathbf{j}} + \mu_i\hat{\mathbf{k}}. \tag{15.17}$$

Only positive direction cosines are given in Table 15.1, covering one eighth of the total range of solid angles 4π. To cover the entire 4π any or all of the values of ξ_i, η_i, and μ_i may be positive or negative. Therefore, each row of ordinates contains eight different directions. For example, for the S_2-approximation the different directions are $\hat{s}_1 = 0.577350(\hat{\mathbf{i}} + \hat{\mathbf{j}} + \hat{\mathbf{k}})$, $\hat{s}_2 = 0.577350(\hat{\mathbf{i}} + \hat{\mathbf{j}} - \hat{\mathbf{k}})$, ..., $\hat{s}_8 = -0.577350(\hat{\mathbf{i}}+\hat{\mathbf{j}}+\hat{\mathbf{k}})$. Since the symmetric S_2-approximation does not satisfy the half-moment condition, a nonsymmetric S_2-approximation is also included in Table 15.1, as proposed by Truelove [11]. This approximation satisfies equation (15.16) for two principal directions and should be applied to one- and two-dimensional problems, from which the nonsymmetric term drops out (as seen in Example 15.1 in the following

---

[1] With the exception of the symmetric S_2-approximation.

TABLE 15.1
Discrete ordinates for the S_N-approximation (N =2, 4, 6, 8), from [21].

Order of Approximation	Ordinates			Weights
	ξ	η	μ	w
S_2 (symmetric)	0.5773503	0.5773503	0.5773503	1.5707963
S_2 (nonsymmetric)	0.5000000	0.7071068	0.5000000	1.5707963
S_4	0.2958759	0.2958759	0.9082483	0.5235987
	0.2958759	0.9082483	0.2958759	0.5235987
	0.9082483	0.2958759	0.2958759	0.5235987
S_6	0.1838670	0.1838670	0.9656013	0.1609517
	0.1838670	0.6950514	0.6950514	0.3626469
	0.1838670	0.9656013	0.1838670	0.1609517
	0.6950514	0.1838670	0.6950514	0.3626469
	0.6950514	0.6950514	0.1838670	0.3626469
	0.9656013	0.1838670	0.1838670	0.1609517
S_8	0.1422555	0.1422555	0.9795543	0.1712359
	0.1422555	0.5773503	0.8040087	0.0992284
	0.1422555	0.8040087	0.5773503	0.0992284
	0.1422555	0.9795543	0.1422555	0.1712359
	0.5773503	0.1422555	0.8040087	0.0992284
	0.5773503	0.5773503	0.5773503	0.4617179
	0.5773503	0.8040087	0.1422555	0.0992284
	0.8040087	0.1422555	0.5773503	0.0992284
	0.8040087	0.5773503	0.1422555	0.0992284
	0.9795543	0.1422555	0.1422555	0.1712359

section). The name "S_N-approximation" indicates that N different direction cosines are used for each principal direction. For example, for the S_4-approximation $\xi_i = \pm 0.295876$ and ± 0.908248 (or η_i or μ_i). Altogether there are always $n = N(N + 2)$ different directions to be considered (because of symmetry, many of these may be unnecessary for one- and two-dimensional problems). Several other quadrature schemes can be found in the literature. Carlson [22] proposed a set with equal weights w_i (such as the S_2 and S_4 sets in Table 15.1). Recently, two more quadratures and a good review of the applicability of all discrete ordinate sets have been given by Fiveland [23].

15.3   THE ONE-DIMENSIONAL SLAB

We will first demonstrate how the S_N discrete ordinates method is applied to the simple case of a one-dimensional plane-parallel slab bounded by two diffusely emitting and reflecting isothermal plates. As in previous chapters we shall limit ourselves to linear-anisotropic scattering, although extension to arbitrarily anisotropic scattering is straightforward. We avoid it here to make the steps in the development a little easier to follow. If we choose z as the spatial coordinate between the two plates

$(0 \leq z \leq L)$, and introduce the optical coordinate τ with $d\tau = \beta dz$ $(0 \leq \tau \leq \tau_L)$, equation (15.4) is transformed to

$$\mu_i \frac{dI_i}{d\tau} = (1 - \omega) I_b - I_i + \frac{\omega}{4\pi} \sum_{j=1}^{n} w_j I_j [1 + A_1(\mu_i \mu_j + \xi_i \xi_j + \eta_i \eta_j)],$$

$$i = 1, 2, \ldots, n. \quad (15.18)$$

Since for every ordinate j with a positive value for ξ_j there is another with the same, but negative, value, and since the intensity is the same for both ordinates (because of azimuthal symmetry in the one-dimensional slab), the terms involving ξ_j in equation (15.18) add to zero. The same is true for the terms involving η_j, but not for those with μ_j (since the intensity does depend on polar angle θ, and $\mu = \cos\theta$). However, the terms involving μ_j are repeated several times: Each value of μ (counting positive and negative μ-values separately) shown in one row of Table 15.1 corresponds to four different ordinates (combinations of positive and negative values for ξ and η). In addition, a particular value of μ may occur on more than one line of Table 15.1. If all the quadrature weights corresponding to a single μ-value are added together, equation (15.18) reduces to

$$\mu_i \frac{dI_i}{d\tau} = (1 - \omega) I_b - I_i + \frac{\omega}{4\pi} \sum_{j=1}^{N} w'_j I_j (1 + A_1 \mu_i \mu_j),$$

$$i = 1, 2, \ldots, N. \quad (15.19)$$

where the w'_j are the summed quadrature weights. For example, for $\mu = 0.2958759$ in the S_4-approximation the summed quadrature weight is $w' = 4 \times (0.5235987 + 0.5235987) = 4\pi/3$, and so forth. The ordinates and quadrature weights for the one-dimensional slab are listed in Table 15.2. Equation (15.19) could have been found less painfully by using equation (15.10) instead of (15.4), leading directly to

$$\mu_i \frac{dI_i}{d\tau} + I_i = (1 - \omega) I_b + \frac{\omega}{4\pi}(G + A_1 q \mu_i), \quad i = 1, 2, \ldots, N. \quad (15.20)$$

Before proceeding to the boundary conditions of equation (15.20) we should recognize that, of the N different intensities, half emanate from the wall at $\tau = 0$ (with $\mu_i > 0$), and the other half from the wall at $\tau = \tau_L$ (with $\mu_i < 0$). Following the notation of Chapter 12, we replace the N different I_i by

$$I_1^+, I_2^+, \ldots, I_{N/2}^+ \quad \text{and} \quad I_1^-, I_2^-, \ldots, I_{N/2}^-.$$

Then equation (15.20) may be rewritten as

$$\mu_i \frac{dI_i^+}{d\tau} + I_i^+ = (1 - \omega) I_b + \frac{\omega}{4\pi}(G + A_1 q \mu_i), \quad (15.21a)$$

$$-\mu_i \frac{dI_i^-}{d\tau} + I_i^- = (1 - \omega) I_b + \frac{\omega}{4\pi}(G - A_1 q \mu_i), \quad (15.21b)$$

$$i = 1, 2, \ldots, N/2; \quad \mu_i > 0.$$

TABLE 15.2

Discrete Ordinates for the One-Dimensional S_N-approximation ($N = 2, 4, 6, 8$).

Order of Approximation	Ordinates μ	Weights w'
S_2 (symmetric)	0.5773503	6.2831853
S_2 (nonsymmetric)	0.5000000	6.2831853
S_4	0.2958759	4.1887902
	0.9082483	2.0943951
S_6	0.1838670	2.7382012
	0.6950514	2.9011752
	0.9656013	0.6438068
S_8	0.1422555	2.1637144
	0.5773503	2.6406988
	0.8040087	0.7938272
	0.9795543	0.6849436

With this notation the boundary conditions for equation (15.21) follow from equations (15.5) or (15.11) as

$$\tau = 0 : \quad I_i^+ = J_1/\pi = I_{b1} - \frac{1 - \epsilon_1}{\epsilon_1 \pi} q_1, \tag{15.22a}$$

$$\tau = \tau_L : \quad I_i^- = J_2/\pi = I_{b2} + \frac{1 - \epsilon_2}{\epsilon_2 \pi} q_2, \tag{15.22b}$$

$$i = 1, 2, \ldots, N/2, \quad \mu_i > 0.$$

(For the boundary condition at τ_L the sign switches since $\hat{n}$ points in the direction opposite to z).

Radiative heat flux q and incident radiation G are related to the directional intensities through equations (15.6) and (15.7), or

$$q = \sum_{i=1}^{N/2} w_i' \mu_i (I_i^+ - I_i^-), \tag{15.23a}$$

$$G = \sum_{i=1}^{N/2} w_i' (I_i^+ + I_i^-). \tag{15.23b}$$

At the two surfaces the radiative heat flux is more conveniently evaluated from equation (15.8) as

$$\tau = 0 : \quad q_1 = \quad q(0) = \quad \epsilon_1 \left(E_{b1} - \sum_{i=1}^{N/2} w_i' \mu_i I_i^- \right), \tag{15.24a}$$

$$\tau = \tau_L : \quad q_2 = -q(\tau_L) = -\epsilon_2 \left(E_{b2} - \sum_{i=1}^{N/2} w_i' \mu_i I_i^+ \right). \tag{15.24b}$$

Example 15.1.   Consider two large, parallel, gray-diffuse, and isothermal plates, separated by a distance L. One plate is at temperature T_1 with emissivity ϵ_1, the other is at T_2 with ϵ_2. The medium between the two plates is an absorbing/emitting and linear-anisotropically scattering gas ($n = 1$) with extinction coefficient β and single scattering albedo ω. Assuming that radiative equilibrium prevails, determine the radiative heat flux between the two plates using the S_2-approximation.

Solution

For radiative equilibrium we have $I_b = G/4\pi$ and $q = $ const; equations (15.21) and (15.22) become

$$\mu_1 \frac{dI_1^+}{d\tau} + I_1^+ = \frac{1}{4\pi}(G + A_1 \omega \mu_1 q),$$

$$-\mu_1 \frac{dI_1^-}{d\tau} + I_1^- = \frac{1}{4\pi}(G - A_1 \omega \mu_1 q),$$

$$\tau = 0: \quad I_1^+ = J_1/\pi, \qquad \tau = \beta L = \tau_L: \quad I_1^- = J_2/\pi.$$

For the S_2-approximation we have only a single ordinate direction μ_1 (pointing toward τ_L for I_1^+, and toward 0 for I_1^-), where $\mu_1 = 0.57735$ for the symmetric S_2-approximation, and $\mu_1 = 0.5$ for the nonsymmetric S_2-approximation [which satisfies the half-range moment, equation (15.16)]. For the simple S_2-approximation the simultaneous equations (only two in this case) may be separated. We do this here by eliminating I_1^+ and I_1^- in favor of G and q. From equation (15.23), with $w_i' = 2\pi$,

$$G = 2\pi(I_1^+ + I_1^-),$$

$$q = 2\pi \mu_1 (I_1^+ - I_1^-).$$

Therefore, adding and subtracting the two differential equations and multiplying by 2π, lead to

$$\frac{dq}{d\tau} + G = G, \quad \text{or} \quad \frac{dq}{d\tau} = 0,$$

$$\mu_1 \frac{dG}{d\tau} + \frac{1}{\mu_1} q = A_1 \omega \mu_1 q, \quad \text{or} \quad \frac{dG}{d\tau} = -\left(\frac{1}{\mu_1^2} - A_1 \omega\right) q.$$

The first equation is simply a restatement of radiative equilibrium, while the second may be integrated (since $q = $ const), or

$$G = C - \left(\frac{1}{\mu_1^2} - A_1 \omega\right) q\tau.$$

This relation contains two unknown constants (C and q), which must be determined from the boundary conditions, that is,

$$\tau = 0: \quad I_1^+ = \frac{1}{4\pi}\left(G + \frac{q}{\mu_1}\right) = J_1/\pi,$$

$$\tau = \tau_L: \quad I_1^- = \frac{1}{4\pi}\left(G - \frac{q}{\mu_1}\right) = J_2/\pi,$$

TABLE 15.3

Radiative heat flux through a one-dimensional plane-parallel medium at radiative equilibrium; comparison of S_2- and P_1-approximations.

		$\Psi = q/(J_1 - J_2)$		
τ_L	Exact	S_2 (sym)	S_2 (nonsym)	P_1
0.0	1.0000	1.1547	1.0000	1.0000
0.1	0.9157	1.0627	0.9091	0.9302
0.5	0.7040	0.8058	0.6667	0.7273
1.0	0.5532	0.6188	0.5000	0.5714
5.0	0.2077	0.2166	0.1667	0.2105

or

$$\tau = 0: \quad 4J_1 = G + \frac{q}{\mu_1} = C + \frac{q}{\mu_1},$$

$$\tau = \tau_L: \quad 4J_2 = G - \frac{q}{\mu_1} = C - \left(\frac{1}{\mu_1^2} - A_1\omega\right)q\,\tau_L - \frac{q}{\mu_1}.$$

Subtracting, we obtain,

$$\Psi = \frac{q}{J_1 - J_2} = \frac{2\mu_1}{1 + (1/\mu_1^2 - A_1\omega)\,\mu_1\tau_L/2},$$

from which the radiosities may be eliminated through equation (12.44). For the symmetric S_2-approximation, $\mu_1 = 0.57735 = 1/\sqrt{3}$, and with isotropic scattering, $A_1 = 0$, this expression becomes

$$\Psi_{\text{symmetric}} = \frac{1}{\sqrt{3}/2 + 3\tau_L/4}$$

On the other hand, for the nonsymmetric S_2-approximation ($\mu_1 = 0.5$), also with isotropic scattering,

$$\Psi_{\text{nonsymmetric}} = \frac{1}{1 + \tau_L}.$$

The S_2-approximation is the same as the two-flux method discussed in Section 13.3, and the nonsymmetric S_2-method is nothing but the Schuster-Schwarzschild approximation. Results from the two S_2-approximations are compared in Table 15.3 with those from the P_1-approximation and the exact solution. It is seen that the accuracy of the S_2-method is roughly equivalent to the one of the P_1-approximation. The nonsymmetric S_2-approximation is superior to the symmetric one, since the symmetric S_2 does not satisfy the half-moment condition, equation (15.16), and causes substantial errors in the optically thin limit.

As a second example for the one-dimensional discrete ordinates method we shall repeat Example 14.4, which was originally designed to demonstrate the use of the P_3-approximation.

Example 15.2.   Consider an isothermal medium at temperature T, confined between two large, parallel black plates that are isothermal at the (same) temperature T_w. The medium is gray and absorbs and emits, but does not scatter. Determine an expression for the heat transfer rates within the medium using the S_2 and S_4 discrete ordinates approximations.

Solution

For this particularly simple case equations (15.21) reduce to

$$\mu_i \frac{dI_i^+}{d\tau} + I_i = I_b,$$

$$-\mu_i \frac{dI_i^-}{d\tau} + I_i = I_b.$$

Since $I_b =$ const, these equations may be integrated right away, leading to

$$I_i^+ = I_b + C^+ e^{-\tau/\mu_i},$$
$$I_i^- = I_b + C^- e^{\tau/\mu_i}.$$

The integration constants C^+ and C^- may be found from boundary conditions (15.22) as

$$\tau = 0: \quad I_i^+ = I_{bw} = I_b + C^+, \quad \text{or} \quad C^+ = I_{bw} - I_b;$$
$$\tau = \tau_L: \quad I_i^- = I_{bw} = I_b + C^- e^{\tau_L/\mu_i}, \quad \text{or} \quad C^- = (I_{bw} - I_b)\, e^{-\tau_L/\mu_i}.$$

Thus,

$$I_i^+ = I_b + (I_{bw} - I_b)e^{-\tau/\mu_i},$$
$$I_i^- = I_b + (I_{bw} - I_b)e^{-(\tau_L-\tau)/\mu_i}.$$

The radiative heat flux follows then from equation (15.23) as

$$q = \sum_{i=1}^{N/2} w_i'\, \mu_i (I_{bw} - I_b)\left(e^{-\tau/\mu_i} - e^{-(\tau_L-\tau)/\mu_i}\right),$$

or, in nondimensional form,

$$\Psi = \frac{q}{n^2 \sigma (T_w^4 - T^4)} = \frac{1}{\pi} \sum_{i=1}^{N/2} w_i'\, \mu_i \left(e^{-\tau/\mu_i} - e^{-(\tau_L-\tau)/\mu_i}\right).$$

For the nonsymmetric S_2-approximation we have $w_1' = 2\pi$ and $\mu_1 = 0.5$, or

$$\Psi_{S_2} = e^{-2\tau} - e^{-2(\tau_L-\tau)}.$$

For the S_4-approximation, $w_1' = 4\pi/3$, $w_2' = 2\pi/3$, $\mu_1 = 0.2958759$, $\mu_2 = 0.9082483$, and $\sum w_i' \mu_i = \pi$, so that

$$\Psi_{S_4} = 0.3945012\left(e^{-\tau/0.2958759} - e^{-(\tau_L-\tau)/0.2958759}\right)$$
$$+ 0.6054088\left(e^{-\tau/0.9082483} - e^{-(\tau_L-\tau)/0.9082483}\right).$$

The results should be compared with those of Examples 14.2 and 14.4 for the P_1 and P_3-approximations. Note that the S_N-method goes to the correct optically thick limit ($\tau_L \to \infty$)

at the wall, i.e., $\Psi \to 1$ [if the half moment of equation (15.16) is satisfied]. The P_N-approximations, on the other hand, overpredict the optically thick limit for this particular example.

It should be emphasized that this last example—dealing with a nonscattering, isothermal medium—is particularly well suited for the discrete ordinate method. One should not expect that, for a general problem, the S_4-method is easier to apply than the P_3-approximation.

A number of researchers have solved more complicated one-dimensional problems by the discrete ordinates method. Fiveland [15] considered the identical case as presented in this section, but allowed for arbitrarily anisotropic scattering. Solving the system of equations by a finite difference method, he noted that higher-order S_N-methods demand a smaller numerical step Δr, in order to obtain a stable solution. Kumar and coworkers [24] not only allowed arbitrarily anisotropic scattering, but also considered boundaries with specular reflectivities as well as boundaries with collimated irradiation (as discussed in Chapter 16). To solve the set of simultaneous first-order differential equations they employed a subroutine from the IMSL software library [25], which is available on many computers. Stamnes and colleagues [19] investigated the same problem as Kumar and coworkers but also allowed for variable radiative properties and a general bidirectional reflection function at the surfaces. They decoupled the set of simultaneous equations using methods of linear algebra and found exact analytical solutions in terms of eigenvalues and eigenvectors that, in turn, were determined using the EISPACK software library [26].

15.4 ONE-DIMENSIONAL CONCENTRIC SPHERES AND CYLINDERS

To apply the discrete ordinates method to spherical or cylindrical geometries, and to take advantage of the symmetries in a one-dimensional problem, are considerably more difficult for concentric spheres and cylinders than for a plane-parallel slab. The reason is that the local direction cosines change while traveling along a straight line of sight through such enclosures.

Concentric Spheres

Consider two concentric spheres of radius R_1 and R_2, respectively. The inner sphere surface has an emissivity ϵ_1 and is kept isothermal at temperature T_1, while the outer sphere is at temperature T_2 with emissivity ϵ_2. If the temperature within the medium is a function of radius only, then the equation of transfer is given by equation (12.66),

$$\mu \frac{\partial I}{\partial r} + \frac{1 - \mu^2}{r} \frac{\partial I}{\partial \mu} + \beta I = \beta S, \tag{15.25a}$$

or, alternatively,

$$\frac{\mu}{r^2} \frac{\partial}{\partial r} (r^2 I) + \frac{1}{r} \frac{\partial}{\partial \mu} \left[(1 - \mu^2) I \right] + \beta I = \beta S, \tag{15.25b}$$

where $\mu = \cos\theta$ is the cosine of the polar angle, measured from the radial direction (see Fig. 12-5). S is the radiative source function,

$$S(r, \mu) = (1 - \omega) I_b + \frac{\omega}{2} \int_{-1}^{1} I(r, \mu') \Phi(\mu, \mu') \, d\mu', \qquad (15.26a)$$

or

$$S(r, \mu) = (1 - \omega) I_b + \frac{\omega}{4\pi} (G + A_1 q \mu), \qquad (15.26b)$$

if the scattering is limited to the linear-anisotropic case. The additional difficulty lies in the fact that equation (15.25) contains a derivative over direction cosine, μ, that is to be discretized in the discrete ordinates method. Applying the S_N-method to equation (15.25), we obtain

$$\frac{\mu_i}{r^2} \frac{d}{dr} (r^2 I_i) + \frac{1}{r} \left\{ \frac{\partial}{\partial \mu} \left[(1 - \mu^2) I \right] \right\}_{\mu = \mu_i} + \beta I_i = \beta S_i, \quad i = 1, 2, \ldots, N, \qquad (15.27)$$

where S_i is readily determined from equation (15.26) (and is independent of ordinate direction unless the medium scatters anisotropically). Equation (15.27) is only applied to the N principal ordinates since, similar to the slab, there is no azimuthal dependence. Since the direction vector μ is discretized, its derivative must be approximated by finite differences. We may write

$$\left\{ \frac{\partial}{\partial \mu} \left[(1 - \mu^2) I \right] \right\}_{\mu = \mu_i} \simeq \frac{\alpha_{i+1/2} I_{i+1/2} - \alpha_{i-1/2} I_{i-1/2}}{w_i'}, \qquad (15.28)$$

which is a central difference with the $I_{i \pm 1/2}$ evaluated at the boundaries between two ordinates, as shown in Fig. 15-1. Since the differences between any two sequential μ_i are nonuniform, the geometrical coefficients α are nonconstant and need to be determined. The values of α depend only on the differencing scheme and, therefore, are independent of intensity and may be determined by examining a particularly simple intensity field. For example, if $I_{b1} = I_{b2} = I_b = \text{const}$, then $I = I_b = \text{const}$ and

$$\alpha_{i+1/2} - \alpha_{i-1/2} = w_i' \left[\frac{\partial}{\partial \mu} (1 - \mu^2) \right]_{\mu = \mu_i} = -2 w_i' \mu_i, \quad i = 1, 2, \ldots, N. \qquad (15.29)$$

This expression may be used as a recursion formula for $\alpha_{i+1/2}$, if a value for $\alpha_{1/2}$ can be determined. That value is found by noting that $I_{1/2}$ is evaluated at $\mu = -1$ (Fig. 15-1), where $(1 - \mu^2) I = 0$ and, therefore, $\alpha_{1/2} = 0$. Similarly, $I_{N+1/2}$ is evaluated at $\mu = +1$ and also $\alpha_{N+1/2} = 0$. The finite-difference scheme of equations (15.28) and (15.29) satisfies the relation [4]

$$\int_{-1}^{+1} \frac{\partial}{\partial \mu} \left[(1 - \mu^2) I \right] d\mu = (1 - \mu^2) I \Big|_{-1}^{+1} = 0$$

$$= \sum_{i=1}^{N} w_i' \left\{ \frac{\partial}{\partial \mu} \left[(1 - \mu^2) I \right] \right\}_{\mu = \mu_i} = \sum_{i=1}^{N} (\alpha_{i+1/2} I_{i+1/2} - \alpha_{i-1/2} I_{i-1/2})$$

$$= \alpha_{3/2} I_{3/2} - \alpha_{1/2} I_{1/2} + \alpha_{5/2} I_{5/2} - \alpha_{3/2} I_{3/2} + - \cdots \alpha_{N+1/2} I_{N+1/2} - \alpha_{N-1/2} I_{N-1/2}$$

$$= 0.$$

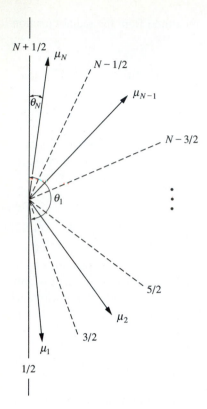

FIGURE 15-1
Directional discretization and discrete ordinate values for one-dimensional problems.

Finally, the intensities at the node boundaries, $I_{i\pm 1/2}$, need to be expressed in terms of node center values, I_i. We shall use here simple, linear averaging, i.e., $I_{i+1/2} \simeq \frac{1}{2}(I_i + I_{i+1})$. Equation (15.27) may now be rewritten as

$$\frac{\mu_i}{r^2} \frac{d}{dr}(r^2 I_i) + \frac{\alpha_{i+1/2}I_{i+1} + (\alpha_{i+1/2} - \alpha_{i-1/2})I_i - \alpha_{i-1/2}I_{i-1}}{2rw_i'} + \beta I_i = \beta S_i,$$

or, carrying out the differentiation and using equation (15.29),

$$\mu_i \frac{dI_i}{dr} + \frac{\mu_i}{r} I_i + \frac{\alpha_{i+1/2}I_{i+1} - \alpha_{i-1/2}I_{i-1}}{2rw_i'} + \beta I_i = \beta S_i, \qquad (15.30a)$$

$$\alpha_{i+1/2} = \alpha_{i-1/2} - 2w_i' \mu_i, \quad \alpha_{1/2} = \alpha_{N+1/2} = 0, \qquad (15.30b)$$

$$i = 1, 2, \ldots, N.$$

Equations (15.30) constitute a set of N simultaneous differential equations in the N unknown intensities I_i, subject to the boundary conditions [cf. equation (15.22)]

$$r = R_1 : \quad I_i = J_1/\pi = I_{b1} - \frac{1 - \epsilon_1}{\epsilon_1 \pi} q_1, \ i = \frac{N}{2} + 1, \frac{N}{2} + 2, \ldots, N, \ (\mu_i > 0),$$

$$(15.31a)$$

$$r = R_2 : \quad I_i = J_2/\pi = I_{b2} + \frac{1-\epsilon_2}{\epsilon_2 \pi} q_2, \quad i = 1, 2, \ldots, \frac{N}{2}, \quad (\mu_i < 0). \quad (15.31b)$$

As for the one-dimensional slab the radiative heat flux and incident radiation are evaluated [cf. equations (15.23) and (15.24)] from

$$G(r) = \sum_{i=1}^{N} w_i' I_i(r), \tag{15.32a}$$

$$q(r) = \sum_{i=1}^{N} w_i' \mu_i I_i(r), \tag{15.32b}$$

and

$$q(R_1) = q_1 = \epsilon_1 \left(E_{b1} + \sum_{\substack{i=1 \\ (\mu_i < 0)}}^{N/2} w_i' \mu_i I_i \right), \tag{15.32c}$$

$$-q(R_2) = q_2 = \epsilon_2 \left(E_{b2} - \sum_{\substack{i=N/2+1 \\ (\mu_i > 0)}}^{N} w_i' \mu_i I_i \right). \tag{15.32d}$$

Example 15.3.   Consider a nonscattering medium at radiative equilibrium that is contained between two isothermal, gray spheres. The absorption coefficient of the medium may be assumed to be gray and constant. Using the S_2-approximation determine the radiative heat flux between the two concentric spheres.

Solution

From equation (15.30) we find, with $N = 2$, that $\alpha_{1/2} = \alpha_{5/2} = 0$, $\alpha_{3/2} = -2w_1' \mu_1 = 2w_2' \mu_2 = 4\pi\mu$ (since $\mu_2 = -\mu_1 > 0$; we keep $\mu = \mu_2$ as a nonnumerical value to allow comparison between the symmetric and nonsymmetric S_2-approximations). Thus, with $\beta = \kappa$ and $S_i(r) = I_b(r) = G/4\pi$ [for a nonscattering medium at radiative equilibrium, equation (15.13)],

$$i = 1 : \quad -\mu \frac{dI_1}{d\tau} - \frac{\mu}{\tau} I_1 + \frac{\mu}{\tau} I_2 + I_1 = \frac{G}{4\pi} = \frac{1}{2}(I_1 + I_2),$$

$$-\mu \frac{dI_1}{d\tau} - \left(\frac{\mu}{\tau} - \frac{1}{2}\right)(I_1 - I_2) = 0,$$

$$i = 2 : \quad \mu \frac{dI_2}{d\tau} + \frac{\mu}{\tau} I_2 - \frac{\mu}{\tau} I_1 + I_2 = \frac{1}{2}(I_1 + I_2),$$

$$\mu \frac{dI_2}{d\tau} - \left(\frac{\mu}{\tau} + \frac{1}{2}\right)(I_1 - I_2) = 0.$$

While addition of the two equations simply leads to a restatement of radiative equilibrium (as in Example 15.1), subtracting them (and multiplying by $w_i' = 2\pi$) leads to

$$-\mu \frac{d}{d\tau}[2\pi(I_1 + I_2)] + 2\pi(I_1 - I_2) = 0,$$

or

$$\frac{dG}{d\tau} = -\frac{q}{\mu^2} = -\frac{\tau^2 q}{\mu^2} \frac{1}{\tau^2}.$$

Since for a medium at radiative equilibrium between concentric spheres $Q = 4\pi r^2 q = $ const and, therefore, $\tau^2 q = $ const, the incident radiation may be found by integration,

$$G(\tau) = \frac{\tau^2 q}{\mu^2} \frac{1}{\tau} + C,$$

where the two constants $(\tau^2 q)$ and C are still unknown and must be determined from the boundary conditions, equations (15.31):

$$I_2(\tau_1) = J_1/\pi, \quad I_1(\tau_2) = J_2/\pi.$$

Using the definitions for q and G, equations (15.32),

$$q = 2\pi\mu(I_2 - I_1) \quad \text{and} \quad G = 2\pi(I_2 + I_1),$$

or

$$I_1 = \frac{1}{4\pi}\left(G - \frac{q}{\mu}\right), \quad I_2 = \frac{1}{4\pi}\left(G + \frac{q}{\mu}\right),$$

the boundary conditions may be restated in terms of q and G as

$$\tau = \tau_1: \quad 4J_1 = G + \frac{q_1}{\mu} = \frac{\tau_1 q_1}{\mu^2} + C + \frac{q_1}{\mu} = \frac{\tau^2 q}{\mu^2}\left(\frac{1}{\tau_1} + \frac{\mu}{\tau_1^2}\right) + C,$$

$$\tau = \tau_2: \quad 4J_2 = G - \frac{q_2}{\mu} = \frac{\tau_2 q_2}{\mu^2} + C - \frac{q_2}{\mu} = \frac{\tau^2 q}{\mu^2}\left(\frac{1}{\tau_2} - \frac{\mu}{\tau_2^2}\right) + C.$$

Subtracting the second boundary condition from the first we obtain

$$\Psi = \frac{\tau^2}{\tau_1^2} \frac{q}{J_1 - J_2} = \frac{1}{\frac{1}{4\mu}\left(1 + \frac{\tau_1^2}{\tau_2^2}\right) + \frac{\tau_1}{4\mu^2}\left(1 - \frac{\tau_1}{\tau_2}\right)}.$$

For the symmetric S_2-approximation, with $\mu = 1/\sqrt{3}$, this equation becomes

$$\Psi_{\text{symmetric}} = \frac{1}{\frac{\sqrt{3}}{4}\left(1 + \frac{\tau_1^2}{\tau_2^2}\right) + \frac{3\tau_1}{4}\left(1 - \frac{\tau_1}{\tau_2}\right)},$$

and for the nonsymmetric approximation with $\mu = 0.5$,

$$\Psi_{\text{nonsymmetric}} = \frac{1}{\frac{1}{2}\left(1 + \frac{\tau_1^2}{\tau_2^2}\right) + \tau_1\left(1 - \frac{\tau_1}{\tau_2}\right)}.$$

The accuracy of the S_2-approximation is very similar to that of the P_1-approximation, for which

$$\Psi_{P_1} = \frac{1}{\frac{1}{2}\left(1 + \frac{\tau_1^2}{\tau_2^2}\right) + \frac{3\tau_1}{4}\left(1 - \frac{\tau_1}{\tau_2}\right)}.$$

Note that the method is very accurate for large τ_1 (large optical thickness) but breaks down for optically thin conditions ($\kappa \to 0$), in particular for small ratios of radii, R_1/R_2. In the limit ($\kappa \to 0$, $R_1/R_2 \to 0$) we find $\Psi_{P_1} = \Psi_{S_2,\text{nonsym}} \to 2$, while the correct limit should go to $\Psi_{\text{exact}} \to 1$.

Numerical solutions to equations (15.30), allowing for anisotropic scattering, variable properties and external irradiation, have been reported by Tsai and colleagues [20] using the S_8 discrete ordinates method with the equal-weight ordinates of Fiveland [15]. The same method was used by Jones and Bayazitoğlu [27] to determine the combined effects of conduction and radiation through a spherical shell.

Concentric Cylinders

The analysis for two concentric cylinders follows along similar lines. Again we consider an absorbing, emitting, and scattering medium contained between two isothermal cylinders with radii R_1 (temperature T_1, diffuse emissivity ϵ_1) and R_2 (temperature T_2, emissivity ϵ_2), respectively. For this case the equation of transfer is given by equation (12.85),

$$\sin\theta\,\cos\psi\frac{\partial I}{\partial r} - \frac{\sin\theta\,\sin\psi}{r}\frac{\partial I}{\partial \psi} + \beta I = \beta S, \qquad (15.33)$$

where polar angle θ is measured from the z-axis, and the azimuthal angle ψ is measured from the local radial direction (cf. Fig. 12-6). S is the radiative source function and has been given by equation (15.26). Introducing the direction cosines $\xi = \hat{s}\cdot\hat{e}_z = \cos\theta$, $\mu = \hat{s}\cdot\hat{e}_r = \sin\theta\,\cos\psi$, and $\eta = \hat{s}\cdot\hat{e}_{\psi_c} = \sin\theta\,\sin\psi$, we may rewrite equation (15.33) as

$$\frac{\mu}{r}\frac{\partial}{\partial r}(r\,I) - \frac{1}{r}\frac{\partial}{\partial \psi}(\eta\,I) + \beta I = \beta S. \qquad (15.34)$$

For a one-dimensional cylindrical medium the symmetry conditions are not as straightforward as for slabs and spheres. Here we have

$$I(r, \theta, \psi) = I(r, \pi - \theta, \psi) = I(r, \theta, -\psi). \qquad (15.35)$$

Therefore, the intensity is the same for positive and negative values of ξ, as well as for positive and negative values of η. Thus, we only need to consider positive values for ξ_i and η_i from Table 15.1, leading to $N_c = N(N+2)/4$ different ordinates for the S_N-approximation, with quadrature weights $w_i'' = 4w_i$. Equation (15.34) may then be written in discrete ordinates form as

$$\frac{\mu_i}{r}\frac{d}{dr}(r\,I_i) - \frac{1}{r}\left\{\frac{\partial}{\partial\psi}(\eta I)\right\}_{\psi=\psi_i} + \beta I_i = \beta S_i, \quad i = 1, 2, \ldots, N_c. \qquad (15.36)$$

As for the concentric spheres case the term in braces is approximated as

$$\left\{\frac{\partial}{\partial\psi}(\eta I)\right\}_{\psi=\psi_i} \simeq \frac{\alpha_{i+1/2}I_{i+1/2} - \alpha_{i-1/2}I_{i-1/2}}{w_i''}, \quad i = 1, 2, \ldots, N_i, \ \xi_i \text{ fixed.} \qquad (15.37)$$

In this relation the subscript $i + 1/2$ implies "toward the next higher value of ψ_i, keeping ξ_i constant." The value of N_i depends on the value of ξ_i. For example, for the S_4-approximation we have from Table 15.1 $N_i = 4$ for $\xi_i = 0.2958759$ (four different values for μ_i, two positive and two negative) and $N_i = 2$ for $\xi_i = 0.9082483$. In the case of concentric cylinders the recursion formula for α, by letting $I = S = \text{const}$ in equation (15.34), is obtained as

$$\alpha_{i+1/2} - \alpha_{i-1/2} = w_i'' \left.\frac{\partial \eta}{\partial \psi}\right|_{\psi = \psi_i} = w_i'' \mu_i, \quad i = 1, 2, \ldots, N_i, \ \xi_i \text{ fixed.} \qquad (15.38)$$

Again, $\alpha_{1/2} = 0$ since at that location $\psi_{1/2} = 0$ and, therefore, $\eta = 0$. Similarly, $\alpha_{N_i+1/2} = 0$ since $\psi_{N_i+1/2} = \pi$ and $\eta = 0$. Finally, using linear averaging for the half-node intensities leads to

$$\mu_i \frac{dI_i}{dr} + \frac{\mu_i}{2r} I_i - \frac{\alpha_{i+1/2} I_{i+1} - \alpha_{i-1/2} I_{i-1}}{2 r w_i''} + \beta I_i = \beta S_i, \quad i = 1, 2, \ldots, N_c, \quad (15.39a)$$

$$\alpha_{i+1/2} = \alpha_{i-1/2} + w_i'' \mu_i, \quad \alpha_{1/2} = \alpha_{N+1/2} = 0, \ i = 1, 2, \ldots N_i, \ \xi_i \text{ fixed.} \quad (15.39b)$$

Equation (15.39) is the set of equations for concentric cylinders, for the $N_c = N(N + 2)/4$ unknown directional intensities I_i, and is equivalent to the set for concentric spheres, equation (15.30). The boundary conditions for cylinders and spheres are basically identical [equations (15.31)], except for some renumbering, as are the expressions for incident intensity and radiative heat flux [equations (15.32)], that is,

$$r = R_1 : \quad I_i = \frac{J_1}{\pi} = I_{b1} - \frac{1 - \epsilon_1}{\epsilon_1 \pi} q_1, \ i = \frac{N_c}{2} + 1, \ \frac{N_c}{2} + 2, \ldots, N_c, \ (\mu_i > 0), \quad (15.40a)$$

$$r = R_2 : \quad I_i = \frac{J_2}{\pi} = I_{b2} + \frac{1 - \epsilon_2}{\epsilon_2 \pi} q_2, \ i = 1, 2, \ldots, \frac{N_c}{2}, \ (\mu_i < 0), \qquad (15.40b)$$

$$G(r) = \sum_{i=1}^{N_c} w_i'' I_i(r), \qquad (15.40c)$$

$$q(r) = \sum_{i=1}^{N_c} w_i'' \mu_i I_i(r), \qquad (15.40d)$$

and

$$q(R_1) = q_1 = \epsilon_1 \left(E_{b1} + \sum_{\substack{i=1 \\ (\mu_i < 0)}}^{N_c/2} w_i'' \mu_i I_i \right), \qquad (15.40e)$$

$$-q(R_2) = q_2 = \epsilon_2 \left(E_{b2} - \sum_{\substack{i=N_c/2+1 \\ (\mu_i > 0)}}^{N_c} w_i'' \mu_i I_i \right). \qquad (15.40f)$$

15.5 MULTIDIMENSIONAL PROBLEMS

While the discrete ordinates method is readily extended to multidimensional configurations, the method results in a set of simultaneous first-order partial differential
equations that generally must be solved numerically. As for one-dimensional geometries, the equation of transfer is slightly different whether a Cartesian, cylindrical,
or spherical coordinate system is employed. We shall first describe the method for
Cartesian coordinate systems, followed by a brief description of the differences for
cylindrical and spherical geometries.

Enclosures Described by Cartesian Coordinates

For Cartesian coordinates equation (15.4) becomes, using equation (15.17),

$$\xi_i \frac{\partial I_i}{\partial x} + \eta_i \frac{\partial I_i}{\partial y} + \mu_i \frac{\partial I_i}{\partial z} + \beta I_i = \beta S_i, \quad i = 1, 2, \ldots, n, \tag{15.41}$$

where S_i is again shorthand for the radiative source function

$$S_i = (1 - \omega) I_b + \frac{\omega}{4\pi} \sum_{j=1}^{n} w_j \Phi_{ij} I_j, \quad i = 1, 2, \ldots, n. \tag{15.42}$$

Equation (15.41) is subject to the boundary conditions in equation (15.5) along each
surface. For example, for a surface parallel to the y-z-plane, with $\hat{n} = \hat{i}$ and $\hat{n} \cdot \hat{s}_j = \hat{s}_j \cdot \hat{i} = \xi_j$, we have for all i with $\xi_i > 0$ ($n/2$ boundary conditions)

$$I_i = J_w/\pi = \epsilon_w I_{bw} + \frac{1 - \epsilon_w}{\pi} \sum_{\substack{\xi_j < 0 \\ (n/2 \text{ values})}} w_j I_j |\xi_j|. \tag{15.43}$$

Although the numerical solution to equation (15.41) may be found through standard
finite differences, the first order of the equations necessitates backward differencing
with large truncation errors. Consequently, it is more common to employ the finite-
volumes approach of Carlson and Lathrop [4] described below.

Two-Dimensional Problems

For clarity, we shall develop the method here for a two-dimensional geometry (i.e.,
for $\partial I/\partial z \equiv 0$). For such a problem the intensity is the same for positive and negative values of μ_i. Thus, we need to consider only positive values of μ_i (and double
the quadrature weights w_i). A general volume element is shown in Fig. 15-2. The
volume element has four face areas A_W and A_E (in the x-direction), and A_S and A_N
(in the y-direction). In a simple rectangular enclosure one would have (per unit length
in the z-direction) $A_W = A_E = \Delta y$, $A_S = A_N = \Delta x$, and $V = \Delta x \Delta y$. However,

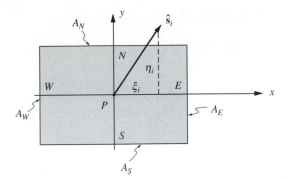

FIGURE 15-2
A general two-dimensional control volume.

the finite-volume approach allows for irregularly shaped elements, which are useful in odd-shaped enclosures, as long as the volume elements are "rectangles" described by curvilinear, rectangular coordinates (such as the volume element $r\Delta r\,\Delta\psi_c$ for a two-dimensional cylindrical coordinate system). The finite-volume formulation of equation (15.41) is obtained by integrating it over a volume element. For example, the term $\partial I_i/\partial x$ transforms to

$$\int_V \frac{\partial I_i}{\partial x}\,dV = \int_{A_E} I_i\,dA_E - \int_{A_W} I_i\,dA_W = I_{Ei}\,A_E - I_{Wi}\,A_W, \qquad (15.44)$$

where I_{Ei} and I_{Wi} are average values of I_i over the faces of A_E and A_W, respectively. Operating similarly on the other terms changes equation (15.41) to

$$\xi_i(A_E\,I_{Ei} - A_W\,I_{Wi}) + \eta_i(A_N\,I_{Ni} - A_S\,I_{Si}) = -\beta\,V I_{pi} + \beta\,V S_{pi}, \qquad (15.45)$$

where I_{pi} and S_{pi} are volume averages. The number of unknowns in equation (15.45) may be reduced by relating cell-edge intensities to the volume-averaged intensity. Most often a linear relationship is chosen, i.e.,

$$I_{pi} = \gamma\,I_{Ni} + (1-\gamma)\,I_{Si} = \gamma\,I_{Ei} + (1-\gamma)\,I_{Wi}, \qquad (15.46)$$

in which γ is a constant $0 \le \gamma \le 1$, and the scheme is known as "weighted diamond differencing" as proposed by Carlson and Lathrop [4]. To date most investigators have employed a symmetric scheme, i.e., $\gamma = \frac{1}{2}$.

For any point on a surface, boundary conditions are given for all directions pointing *away* from the surface. Therefore, the numerical solution of equation (15.41) customarily proceeds as follows: First, the surface radiosities, J_w, and internal radiative source terms, S_i, are estimated (usually by neglecting surface irradiation and volume in-scattering during the first iteration). Next, the lower left corner (corresponding to minimum values of x and y, as indicated in Fig. 15-3) is chosen as a starting point. From that point all outgoing directions lie in the first quadrant (i.e., both direction cosines ξ_i and η_i are positive). The West and South faces of the control volume in that corner are part of the enclosure surface and, therefore, their intensities

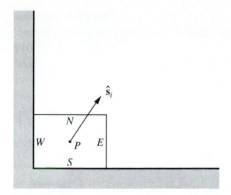

FIGURE 15-3
Enclosure corner, used as starting point for discrete ordinates calculations.

I_{Wi} and I_{Si} are known from the boundary conditions. For any discrete ordinate i, the volume-averaged intensity, I_{pi}, of the corner control volume can be calculated by eliminating I_{Ei} and I_{Ni} from equation (15.45) with the help of equation (15.46). Thus,

$$\gamma(A_E I_{Ei} - A_W I_{Wi}) = A_E I_{pi} - [(1 - \gamma)A_E + \gamma A_W] I_{Wi},$$

etc., and

$$I_{pi} = \frac{\beta V \gamma S_{pi} + \xi_i A_{EW} I_{Wi} + \eta_i A_{NS} I_{Si}}{\beta V \gamma + \xi_i A_E + \eta_i A_N}, \tag{15.47}$$

where

$$A_{EW} = (1 - \gamma) A_E + \gamma A_W, \tag{15.48a}$$
$$A_{NS} = (1 - \gamma) A_N + \gamma A_S, \tag{15.48b}$$

are averaged face areas. Once I_{pi} has been calculated, I_{Ei} and I_{Ni} are readily determined from equation (15.46), and are equal to the West and South intensities for the adjacent control volumes (for increasing x and y); thus, one by one, the first-quadrant intensities can be calculated for all finite volumes in the enclosure. The procedure is then repeated three times, starting from the remaining three corners of the enclosure, covering the remaining three quadrants of directions. For example, for $\xi_i < 0$, the intensity at the East face of the control volume is known, and the West face intensity must be eliminated from equation (15.45), to be determined after I_{pi} has been found. Thus, we may rewrite equations (15.45) and (15.46) for general (positive or negative) values of ξ_i and η_i as

$$|\xi_i| (A_{x_e} I_{x_e i} - A_{x_i} I_{x_i i}) + |\eta_i| (A_{y_e} I_{y_e i} - A_{y_i} I_{y_i i}) = -\beta V I_{pi} + \beta V S_i, \tag{15.49}$$

$$I_{pi} = \gamma I_{x_e i} + (1 - \gamma) I_{x_i i} = \gamma I_{y_e i} - (1 - \gamma) I_{y_i i}, \tag{15.50}$$

where A_{x_i} is the x-direction face area where the beam enters ($= A_W$ for $\xi_i > 0$, and $= A_E$ for $\xi_i < 0$), A_{x_e} is the x-direction face area through which the beam exits ($= A_E$ for $\xi_i > 0$, and $= A_W$ for $\xi_i < 0$), $I_{y_i i}$ and $I_{y_e i}$ are the corresponding y-direction face intensities, and so on. Then equation (15.47) may be generalized to

$$I_{pi} = \frac{\beta V \gamma S_{pi} + |\xi_i| A_x I_{x_i i} + |\eta_i| A_y I_{y_i i}}{\beta V \gamma + |\xi_i| A_{x_e} + |\eta_i| A_{y_e}}, \tag{15.51}$$

where

$$A_x = (1 - \gamma) A_{x_e} + \gamma A_{x_i}, \tag{15.52a}$$
$$A_y = (1 - \gamma) A_{y_e} + \gamma A_{y_i}. \tag{15.52b}$$

Once all directional intensities for each finite volume have been calculated, the values for the wall radiosities and the radiative source term may be updated, and the procedure is repeated until convergence criteria are met. And finally, internal values of incident radiation and radiative heat flux are determined from equations (15.6) and (15.7), while heat fluxes at the walls may be calculated from equations (15.8) or (15.11).

Carlson and Lathrop [4] indicate that, when the cell-averaged intensity predicted from equation (15.51) is used with equation (15.50) to extrapolate the face intensities, these face intensities may become negative, which is physically impossible. While they simply suggest setting negative intensities to zero and continuing computations, this may lead to oscillations and instability. Fiveland [16] showed that such negative intensities may be minimized (but not totally avoided) if finite volume dimensions are kept within

$$\Delta x < \frac{|\xi_i|_{\min}}{\beta(1 - \gamma)}, \quad \Delta y < \frac{|\eta_i|_{\min}}{\beta(1 - \gamma)}. \tag{15.53}$$

Therefore, higher-order S_N-approximations (with their smaller minimum value for ξ_i and η_i), as well as optically thick media (large β), require finer volumetric meshes.

Example 15.4.   A gray, absorbing/emitting (but not scattering) medium is contained within a square enclosure of side lengths L. The medium is at radiative equilibrium and has a constant absorption coefficient such that $\kappa L = 1$. The top and both side walls are at zero temperature, while the bottom wall is isothermal at temperature T_w (with constant blackbody intensity I_{bw}); all four surfaces are black. Calculate the local heat loss from the bottom surface using the discrete ordinates method.

Solution

For the illustrative purposes of this example we shall limit ourselves to the simple non-symmetric S_2-approximation, with the crude nodal system indicated in Fig. 15-4. For the nonsymmetric S_2-approximation (without dependence in the z-direction) we have to consider four discrete ordinates whose direction vectors (projected into the x-y plane) are $\hat{s}_i = \xi_i \hat{i} + \eta_i \hat{j} = \pm 0.5(\hat{i} \pm \hat{j})$, as given by Table 15.1. The quadrature weight for each direction is, after doubling because of the two-dimensionality, $w_i = \pi$. For radiative

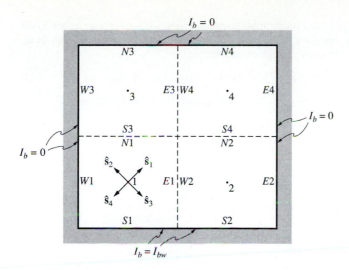

FIGURE 15-4

Square enclosure for Example 15.4.

equilibrium in a nonscattering medium the source function is, from equations (15.12) and (15.42), $S = I_b = G/4\pi$ and is not a function of direction. We choose $\gamma = 0.5$ and, since all nodal surface areas are $A = L/2$, all $|\xi_i| = |\eta_i| = 0.5$, and $\beta V = \kappa(L/2)^2 = 0.25\,\kappa L^2 = 0.25\,L$, equation (15.51) becomes

$$I_{pi} = \frac{\frac{1}{8}S_p + \frac{1}{4}I_{x_i i} + \frac{1}{4}I_{y_i i}}{\frac{1}{8} + \frac{1}{4} + \frac{1}{4}} = \frac{1}{5}(S_p + 2I_{x_i i} + 2I_{y_i i}).$$

We start in the lower left corner with all directions for which $\xi_i > 0$ and $\eta_i > 0$ (i.e., a single direction for the S_2-approximation). For this direction $x_i = $ West and $y_i = $ South. To distinguish among the different nodes we attach the node number after the W, etc. For example, $I_{W2,1}$ is the intensity at the West face of volume element 2, pointing into the direction of $\hat{s}_1$.

$i = 1$ $\left[\hat{s}_1 = 0.5(\hat{i} + \hat{j})\right]$: For all nodes

$$I_{pj,1} = \tfrac{1}{5}(S_{pj} + 2I_{Wj,1} + 2I_{Sj,1}),$$
$$I_{Ej,1} = 2I_{pj,1} - I_{Wj,1},$$
$$I_{Nj,1} = 2I_{pj,1} - I_{Sj,1}, \quad j = 1,2,3,4.$$

Starting at Element 1 we have $I_{W1,1} = 0$, $I_{S1,1} = I_{bw}$, and

$$I_{p1,1} = \tfrac{1}{5}(S_{p1} + 2I_{bw}),$$
$$I_{E1,1} = 2I_{p1,1} = I_{W2,1}, \quad I_{N1,1} = 2I_{p1,1} - I_{bw} = I_{S3,1};$$
$$I_{p2,1} = \tfrac{1}{5}(S_{p2} + 2I_{W2,1} + 2I_{S2,1}) = \tfrac{1}{5}(S_{p2} + 4I_{p1,1} + 2I_{bw}),$$
$$I_{N2,1} = 2I_{p2,1} - I_{bw} = I_{S4,1};$$
$$I_{p3,1} = \tfrac{1}{5}(S_{p3} + 2I_{S3,1}) = \tfrac{1}{5}(S_{p3} + 4I_{p1,1} - 2I_{bw}),$$
$$I_{E3,1} = 2I_{p3,1} = I_{W4,1};$$
$$I_{p4,1} = \tfrac{1}{5}(S_{p4} + 2I_{W4,1} + 2I_{S4,1})$$
$$= \tfrac{1}{5}(S_{p4} + 4I_{p3,1} + 4I_{p2,1} - 2I_{bw});$$

$i = 2 \left[\hat{\mathbf{s}}_2 = 0.5(-\hat{\mathbf{i}} + \hat{\mathbf{j}}) \right]$: In a problem without symmetry we would start in the lower right corner, scanning again over all elements. However, in this problem we can determine the intensities right away through symmetry, as

$$ I_{p1,2} = I_{p2,1}, \ I_{p2,2} = I_{p1,1}, \ I_{p3,2} = I_{p4,1}, \ I_{p4,2} = I_{p3,1}. $$

$i = 3 \left[\hat{\mathbf{s}}_3 = 0.5(\hat{\mathbf{i}} - \hat{\mathbf{j}}) \right]$: Starting in the upper left corner, we have, for all nodes,

$$ I_{pj,3} = \tfrac{1}{5}(S_{pj} + 2I_{Wj,3} + 2I_{Nj,3}), $$
$$ I_{Ej,3} = 2I_{pj,3} - I_{Wj,3}, $$
$$ I_{Sj,3} = 2I_{pj,3} - I_{Nj,3}. $$

Starting at Element 3 with $I_{W3,3} = I_{N3,3} = 0$, we find

$$ I_{p3,3} = \tfrac{1}{5}S_{p3}, $$
$$ I_{S3,3} = 2I_{p3,3} = I_{N1,3}, \ I_{E3,3} = 2I_{p3,3} = I_{W4,3}; $$

$$ I_{p4,3} = \tfrac{1}{5}(S_{p4} + 2I_{W4,3}) = \tfrac{1}{5}(S_{p4} + 4I_{p3,3}), $$
$$ I_{S4,3} = 2I_{p4,3} = I_{N2,3}; $$

$$ I_{p1,3} = \tfrac{1}{5}(S_{p1} + 2I_{N1,3}) = \tfrac{1}{5}(S_{p1} + 4I_{p3,3}), $$
$$ I_{E1,3} = 2I_{p1,3} = I_{W2,3}; $$

$$ I_{p2,3} = \tfrac{1}{5}(S_{p2} + 2I_{W2,3} + 2I_{N2,3}) = \tfrac{1}{5}(S_{p2} + 4I_{p1,3} + 4I_{p4,3}). $$

Also

$$ I_{S1,3} = 2I_{p1,3} - I_{N1,3} = 2(I_{p1,3} - I_{p3,3}), $$
$$ I_{S2,3} = 2I_{p2,3} - I_{N2,3} = 2(I_{p2,3} - I_{p4,3}), $$

which will be needed later for the calculation of wall heat fluxes from equation (15.8).

$i = 4 \left[\hat{\mathbf{s}}_4 = -0.5(\hat{\mathbf{i}} + \hat{\mathbf{j}}) \right]$: Again, by symmetry it follows immediately that

$$ I_{p1,4} = I_{p2,3}, \ I_{p2,4} = I_{p1,3}, \ I_{p3,4} = I_{p4,3}, \ I_{p4,4} = I_{p3,3}, $$

and also

$$ I_{S1,4} = I_{S2,3}, \ I_{S2,4} = I_{S1,3}. $$

Summarizing, we have

$$ I_{p1,1} = I_{p2,2} = \tfrac{1}{5}(S_{p1} + 2I_{bw}), $$
$$ I_{p2,1} = I_{p1,2} = \tfrac{1}{5}(S_{p2} + 4I_{p1,1} + 2I_{bw}) $$
$$ I_{p3,1} = I_{p4,2} = \tfrac{1}{5}(S_{p3} + 4I_{p1,1} - 2I_{bw}) $$
$$ I_{p4,1} = I_{p3,2} = \tfrac{1}{5}(S_{p4} + 4I_{p3,1} + 4I_{p2,1} - 2I_{bw}), $$

$$ I_{p1,3} = I_{p2,4} = \tfrac{1}{5}(S_{p1} + 4I_{p3,3}), $$
$$ I_{p2,3} = I_{p1,4} = \tfrac{1}{5}(S_{p2} + 4I_{p1,3} + 4I_{p4,3}) $$
$$ I_{p3,3} = I_{p4,4} = \tfrac{1}{5}S_{p3}, $$
$$ I_{p4,3} = I_{p3,4} = \tfrac{1}{5}(S_{p4} + 4I_{p3,3}), $$

$$ I_{S1,3} = I_{S2,4} = 2(I_{p1,3} - I_{p3,3}), $$
$$ I_{S2,3} = I_{S1,4} = 2(I_{p2,3} - I_{p4,3}). $$

TABLE 15.4

Nodal intensities of Example 15.4 as a function of iteration, normalized by I_{bw}.

Iteration	$I_{p1,1}$	$I_{p2,1}$	$I_{p3,1}$	$I_{p4,1}$	$I_{p1,3}$	$I_{p2,3}$	$I_{p3,3}$	$I_{p4,3}$	S_{p1}	S_{p3}
1	0.4000	0.7200	0.0000*	0.1760	0.0000	0.0000	0.0000	0.0000	0.2800	0.0440
2	0.4560	0.8208	0.0000*	0.2654	0.0630	0.1191	0.0088	0.0158	0.3647	0.0725
3	0.4729	0.8513	0.0000*	0.2955	0.0846	0.1615	0.0145	0.0261	0.3926	0.0840
≥9	0.4815	0.8667	0.0037	0.3148	0.0963	0.1852	0.0185	0.0333	0.4074	0.0926

*negative values set to zero

The source functions are readily evaluated from equation (15.7) and symmetry as

$$S_{p1} = S_{p2} = \tfrac{1}{4}(I_{p1,1} + I_{p1,2} + I_{p1,3} + I_{p1,4}),$$
$$S_{p3} = S_{p4} = \tfrac{1}{4}(I_{p3,1} + I_{p3,2} + I_{p3,3} + I_{p3,4}).$$

Since the equations are linear, one could substitute the relations for the S_{pj} into the above equations and solve for the unknown $I_{pj,i}$ by matrix inversion. However, in general one would have many more, and much more complicated, equations, which are best solved by iteration. We start by setting all $S_{pj} = 0$, finding values for the $I_{pj,i}$, updating the S_{pj}, reevaluating the $I_{pj,i}$, and so on, until convergence has been reached. The changing values of the intensity (normalized with I_{bw}) as function of iteration are given in Table 15.4. Values accurate to $\simeq 5\%$ are reached after three iterations, and fully converged values (to four significant digits) are obtained after nine iterations. The converged intensities are used to determine the net radiative heat flux from the bottom wall at $x = L/4$ and $x = 3L/4$. From equation (15.8) we have

$$q(x = 0.25\,L) = q(x = 0.75\,L) = \pi I_{bw} - \sum_{i=3}^{4} w_i I_{S1,i} |\eta_i|$$

$$= \pi I_{bw} - \frac{\pi}{2}(I_{S1,3} + I_{S1,4}),$$

or

$$\Psi = \frac{q_{0.25L}}{E_{bw}} = \frac{q_{0.75L}}{E_{bw}} = 1 - \frac{I_{p1,3} - I_{p3,3} + I_{p2,3} - I_{p4,3}}{I_{bw}} = 0.7704.$$

The result is shown in Fig. 15-5 along with exact results reported by Razzaque and coworkers [28], and with S_2- and S_4-calculations of Truelove [11] (for a much finer mesh). Truelove's results demonstrate the importance of good ordinate sets, at least for low-order approximations: The S_2' and S_4' results were obtained with sets that do not obey the half-moment condition of equation (15.16) (as used by Fiveland [14] in a first investigation of rectangular enclosures), while the S_2 and S_4 results were obtained with the sets given in Table 15.1 (using the nonsymmetric ordinates for S_2).

In his early calculations Fiveland [14] applied the S_2-, S_4-, and S_6-approximations to purely scattering rectangular media ($\omega = 1$), and to isothermal, nonscattering media bounded by cold black walls. Truelove [11] repeated some of those results to demonstrate the importance of good ordinate sets, and gave some new results for radiative equilibrium in a square enclosure. Jamaluddin and Smith [29] applied the S_4-approximation to a rectangular, nonscattering enclosure with known temperature

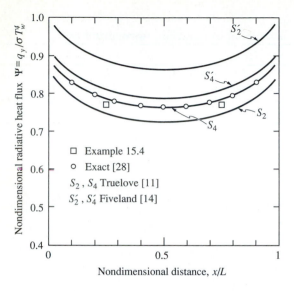

profile. Kim and Lee investigated the effects of strongly anisotropic scattering, using high-order approximations (up to S_{16}) [30], and the effects of collimated irradiation [31]. Finally, combined conduction and radiation in a linear-anisotropically scattering rectangular enclosure has been studied by Baek and Kim [32].

Three-Dimensional Problems

The method can be extended immediately to three-dimensional geometries by giving the control volume Front and Back surfaces, A_F and A_B, and rewriting equation (15.51) as

$$I_{pi} = \frac{\beta V \gamma S_{pi} + |\xi_i| A_x I_{x_i i} + |\eta_i| A_y I_{y_i i} + |\mu_i| A_z I_{z_i i}}{\beta V \gamma + |\xi_i| A_{x_e} + |\eta_i| A_{y_e} + |\mu_i| A_{z_e}}, \tag{15.54}$$

where

$$A_x = (1 - \gamma) A_{x_e} + \gamma A_{x_i}, \tag{15.55a}$$
$$A_y = (1 - \gamma) A_{y_e} + \gamma A_{y_i}, \tag{15.55b}$$
$$A_z = (1 - \gamma) A_{z_e} + \gamma A_{z_i}, \tag{15.55c}$$

and the sub-subscript i again denotes the face where the beam enters, and e where it exits, as explained in the context of equation (15.51). A three-dimensional Cartesian enclosure has eight corners, from each of which $\frac{1}{8}N(N + 2)$ directions must be traced (covering one *octant* of directions), for a total of $N(N + 2)$ ordinates. Some such calculations have been performed by Jamaluddin and Smith [33] (nonscattering medium with prescribed temperature), and by Fiveland [16] and Truelove [12] (both studying the idealized furnace of Mengüç and Viskanta [34], considering a linear-anisotropically scattering medium with internal heat generation at radiative equilibrium).

Multidimensional Cylindrical and Spherical Geometries

Few investigations have dealt with the application of the discrete ordinates method to two- and three-dimensional cylindrical enclosures, and none appears to have investigated multidimensional spherical geometries. The method was first applied to a two-dimensional axisymmetric enclosure by Fiveland [13], who calculated radiative heat flux rates for a cylindrical furnace with known temperature profile. A very similar problem was treated by Jamaluddin and Smith [33] who, a little later, also addressed the case of a three-dimensional cylindrical furnace [17].

The governing equation for multidimensional cylindrical enclosures is readily found by extending equation (15.33) to include changes of azimuthal angle ψ_c (for position within the cylinder) and axial location z along a beam traveling in the direction of $\hat{s}$. From Fig. 12-6 and the development in Section 12.7 it follows that

$$
\begin{aligned}
\frac{dI}{ds} &= \frac{\partial I}{\partial r}\frac{dr}{ds} + \frac{\partial I}{\partial \psi_c}\frac{d\psi_c}{ds} + \frac{\partial I}{\partial z}\frac{dz}{ds} + \frac{\partial I}{\partial \psi}\frac{d\psi}{ds} \\
&= \sin\theta\cos\psi\,\frac{\partial I}{\partial r} + \frac{\sin\theta\,\sin\psi}{r}\frac{\partial I}{\partial \psi_c} + \cos\theta\frac{\partial I}{\partial z} - \frac{\sin\theta\,\sin\psi}{r}\frac{\partial I}{\partial \psi} \\
&= \xi\frac{\partial I}{\partial r} + \frac{\eta}{r}\frac{\partial I}{\partial \psi_c} + \mu\frac{\partial I}{\partial z} - \frac{\eta}{r}\frac{\partial I}{\partial \psi}.
\end{aligned}
$$

Therefore, similar to the one-dimensional case, we have

$$
\frac{\xi}{r}\frac{\partial}{\partial r}(rI) + \frac{\eta}{r}\frac{\partial I}{\partial \psi_c} + \mu\frac{\partial I}{\partial z} - \frac{1}{r}\frac{\partial}{\partial \psi}(\eta I) = \beta S - \beta I. \tag{15.56}
$$

This equation may be solved by the discrete ordinates method as described in this section, with the additional task that the ψ-derivative must be finite-differenced as outlined in the previous section, equation (15.37).

To apply the discrete ordinates method to three-dimensional spherical geometries, the equation of transfer must be written in spherical coordinates by generalizing equation (12.61) or (15.25). In this case, a beam traveling through the medium not only passes through changing position coordinates (r, θ_s, ψ_s), but also sees changes of both polar angle θ and azimuthal angle ψ. This problem apparently has not yet been studied by any researcher, probably for want of important applications.

References

1. Chandrasekhar, S.: *Radiative Transfer*, Dover Publications, 1960.
2. Lee, C. E.: "The Discrete S_N Approximation to Transport Theory," Technical Information Series Report LA2595, Lawrence Livermore Laboratory, 1962.
3. Lathrop, K. D.: "Use of Discrete-Ordinate Methods for Solution of Photon Transport Problems," *Nuclear Science and Engineering*, vol. 24, pp. 381–388, 1966.
4. Carlson, B. G., and K. D. Lathrop: "Transport Theory—The Method of Discrete Ordinates," in *Computing Methods in Reactor Physics*, eds. H. Greenspan, C. N. Kelber, and D. Okrent, Gordon & Breach, New York, 1968.

5. Love, T. J., and R. J. Grosh: "Radiative Heat Transfer in Absorbing, Emitting, and Scattering Media," *ASME Journal of Heat Transfer*, vol. 87, pp. 161–166, 1965.

6. Hsia, H. M., and T. J. Love: "Radiative Heat Transfer Between Parallel Plates Separated by a Nonisothermal Medium with Anisotropic Scattering," *ASME Journal of Heat Transfer*, vol. 89, pp. 197–204, 1967.

7. Hottel, H. C., A. F. Sarofim, L. B. Evans, and I. A. Vasalos: "Radiative Transfer in Anisotropically Scattering Media: Allowance for Fresnel Reflection at the Boundaries," *ASME Journal of Heat Transfer*, vol. 90, pp. 56–62, 1968.

8. Roux, J. A., D. C. Todd, and A. M. Smith: "Eigenvalues and Eigenvectors for Solutions to the Radiative Transport Equation," *AIAA Journal*, vol. 10, no. 7, pp. 973–976, 1972.

9. Roux, J. A., and A. M. Smith: "Radiative Transport Analysis for Plane Geometry with Isotropic Scattering and Arbitrary Temperature," *AIAA Journal*, vol. 12, no. 9, pp. 1273–1277, 1974.

10. Hyde, D. J., and J. S. Truelove: "The Discrete Ordinates Approximation for Multidimensional Radiant Heat Transfer in Furnaces," Technical Report VKAEA Report No. AERE-R 8502, Thermodynamics Division, AERE Harwell, Oxfordshire, February 1977.

11. Truelove, J. S.: "Discrete-Ordinate Solutions of the Radiation Transport Equation," *ASME Journal of Heat Transfer*, vol. 109, no. 4, pp. 1048–1051, 1987.

12. Truelove, J. S.: "Three-Dimensional Radiation in Absorbing-Emitting-Scattering Media Using the Discrete-Ordinates Approximation," *Journal of Quantitative Spectroscopy and Radiative Transfer*, vol. 39, no. 1, pp. 27–31, 1988.

13. Fiveland, W. A.: "A Discrete Ordinates Method for Predicting Radiative Heat Transfer in Axisymmetric Enclosures," ASME Paper 82-HT-20, 1982.

14. Fiveland, W. A.: "Discrete Ordinates Solutions of the Radiative Transport Equation for Rectangular Enclosures," *ASME Journal of Heat Transfer*, vol. 106, pp. 699–706, 1984.

15. Fiveland, W. A.: "Discrete Ordinate Methods for Radiative Heat Transfer in Isotropically and Anisotropically Scattering Media," *ASME Journal of Heat Transfer*, vol. 109, pp. 809–812, 1987.

16. Fiveland, W. A.: "Three-Dimensional Radiative Heat-Transfer Solutions by the Discrete-Ordinates Method," *Journal of Thermophysics and Heat Transfer*, vol. 2, no. 4, pp. 309–316, Oct 1988.

17. Jamaluddin, A. S., and P. J. Smith: "Discrete-Ordinates Solution of Radiative Transfer Equation in Non-Axisymmetric Cylindrical Enclosures," in *Proceedings of the 1988 National Heat Transfer Conference*, vol. HTD–96, ASME, pp. 227–232, 1988.

18. Stamnes, K., and P. Conklin: "A New Multi-Layer Discrete Ordinate Approach to Radiative Transfer in Vertically Inhomogeneous Atmospheres," *Journal of Quantitative Spectroscopy and Radiative Transfer*, vol. 31, no. 3, pp. 273–282, 1984.

19. Stamnes, K., S.-C. Tsay, W. J. Wiscombe, and K. Jayaweera: "Numerically Stable Algorithm for Discrete-Ordinate-Method Radiative Transfer in Multiple Scattering and Emitting Layered Media," *Applied Optics*, vol. 27, no. 12, pp. 2502–2509, 1988.

20. Tsai, J. R., M. N. Özişik, and F. Santarelli: "Radiation in Spherical Symmetry with Anisotropic Scattering and Variable Properties," *Journal of Quantitative Spectroscopy and Radiative Transfer*, vol. 42, no. 3, pp. 187–199, 1989.

21. Lathrop, K. D., and B. G. Carlson: "Discrete-Ordinates Angular Quadrature of the Neutron Transport Equation," Technical Information Series Report LASL-3186, Los Alamos Scientific Laboratory, 1965.

22. Carlson, B. G.: "Tables of Equal Weight Quadrature over the Unit Sphere," Technical Information Series Report LA-4737, Los Alamos Scientific Laboratory, 1971.

23. Fiveland, W. A.: "The Selection of Discrete Ordinate Quadrature Sets for Anisotropic Scattering," in *Fundamentals of Radiation Heat Transfer*, vol. HTD–160, ASME, pp. 89–96, 1991.

24. Kumar, S., A. Majumdar, and C. L. Tien: "The Differential-Discrete-Ordinate Method for Solutions of the Equation of Radiative Transfer," *ASME Journal of Heat Transfer*, vol. 112, no. 2, pp. 424–429, 1990.

25. *IMSL Math/Library*, 1st ed., IMSL, Houston, TX, 1989.

26. Cowell, W. R. (ed.): *Sources and Developments of Mathematical Software*, Prentice Hall, Englewood Cliffs, NJ, 1980.

27. Jones, P. D., and Y. Bayazitoğlu: "Combined Radiation and Conduction from a Sphere in a Participating Medium," in *Proceedings of the Ninth International Heat Transfer Conference*, vol. 6, Washington, D.C., Hemisphere, pp. 397–402, 1990.

28. Razzaque, M. M., D. E. Klein, and J. R. Howell: "Finite Element Solution of Radiative Heat Transfer in a Two-Dimensional Rectangular Enclosure with Gray Participating Media," *ASME Journal of Heat Transfer*, vol. 105, pp. 933–936, 1983.

29. Jamaluddin, A. S., and P. J. Smith: "Predicting Radiative Transfer in Rectangular Enclosures Using the Discrete Ordinates Method," *Combustion Science and Technology*, vol. 59, pp. 321–340, 1988.

30. Kim, T. K., and H. S. Lee: "Effect of Anisotropic Scattering on Radiative Heat Transfer in Two-Dimensional Rectangular Enclosures," *International Journal of Heat and Mass Transfer*, vol. 31, no. 8, pp. 1711–1721, 1988.

31. Kim, T. K., and H. S. Lee: "Radiative Transfer in Two-Dimensional Anisotropic Scattering Media with Collimated Incidence," *Journal of Quantitative Spectroscopy and Radiative Transfer*, vol. 42, pp. 225–238, 1989.

32. Baek, S. W., and T. Y. Kim: "The Conductive and Radiative Heat Transfer in Rectangular Enclosure Using the Discrete Ordinates Method," in *Proceedings of the Ninth International Heat Transfer Conference*, vol. 6, Washington, D.C., Hemisphere, pp. 433–438, 1990.

33. Jamaluddin, A. S., and P. J. Smith: "Predicting Radiative Transfer in Axisymmetric Cylindrical Enclosures Using the Discrete Ordinates Method," *Combustion Science and Technology*, vol. 62, pp. 173–186, 1988.

34. Mengüç, M. P., and R. Viskanta: "Radiative Transfer in Three-Dimensional Rectangular Enclosures Containing Inhomogeneous, Anisotropically Scattering Media," *Journal of Quantitative Spectroscopy and Radiative Transfer*, vol. 33, no. 6, pp. 533–549, 1985.

Problems

15.1 Consider a gray, isothermal and isotropically-scattering medium contained between large, isothermal, gray plates at temperatures T_1 and T_2, and emissivities ϵ_1 and ϵ_2, respectively. Determine the radiative flux between the plates using the S_2-approximation.

15.2 Consider a large, isothermal (temperature T_w), gray and diffuse (emissivity ϵ) wall adjacent to a semi-infinite gray absorbing/emitting and linear-anisotropically scattering medium. The medium is isothermal (temperature T_m). Determine the radiative flux as a function of distance away from the plate using the S_2-approximation.

15.3 Black spherical particles of $100\,\mu$m radius are suspended between two cold and black parallel plates 1 m apart. The particles produce heat at a rate of $\pi/10$ W/particle, which must be removed by thermal radiation. The number of particles between the plates is given by

$$N_T(z) = N_0 + \Delta N z/L, \quad 0 < z < L; \qquad N_0 = \Delta N = 212 \text{ particles/cm}^3.$$

(a) Determine the local absorption coefficient and the local heat production rate; introduce an optical coordinate and determine the optical thickness of the entire gap.

(b) If the S_2-approximation is to be employed, what are the relevant equations and boundary conditions governing the heat transfer?

(c) What are the heat flux rates at the top and bottom surfaces? What is the entire amount of energy released by the particles? What is the maximum particle temperature?

15.4 Two infinitely long, concentric cylinders of radii R_1 and R_2 with emissivities ϵ_1 and ϵ_2 have the same constant surface temperature T_w. The medium between the cylinders has a constant absorption coefficient κ and does not scatter; uniform heat generation $\dot{Q}'''$ takes place inside the medium. Determine the temperature distribution in the medium and heat fluxes at the wall if radiation is the only means of heat transfer, using the S_2-approximation.

15.5 An infinite, black, isothermal plate bounds a semi-infinite space filled with black spheres. At any given distance z away from the plate the particle number density is identical, namely, $N_T = 6.3662 \times 10^8 \, \text{m}^{-3}$. However, the radius of the suspended spheres diminishes monotonically away from the surface as

$$a = a_o e^{-z/L}; \qquad a_o = 10^{-4} \, \text{m}, \qquad L = 1 \, \text{m}$$

(a) Determine the absorption coefficient as a function of z (you may make the large-particle assumption).

(b) Determine the optical coordinate as a function of z. What is the total optical thickness of the semi-infinite space?

(c) Assuming that radiative equilibrium prevails and using the S_2-approximation, set up the boundary conditions and solve for heat flux and temperature distribution (as a function of z).

15.6 Consider two parallel black plates, both at 1000 K, that are 2 m apart. The medium between the plates emits and absorbs (but does not scatter) with an absorption coefficient of $\kappa = 0.05236 \, \text{cm}^{-1}$ (gray medium). Heat is generated by the medium according to the formula

$$\dot{Q}''' = C \sigma T^4, \qquad C = 6.958 \times 10^{-4} \text{cm}^{-1},$$

where T is the local temperature of the medium between the plates. Assuming that radiation is the only important mode of heat transfer, determine the heat flux to the plates using the (symmetric) S_2-approximation.

15.7 A furnace burning pulverized coal may be approximated by a gray cylinder at radiative equilibrium with uniform heat generation $\dot{Q}''' = 0.266 \, \text{W/cm}^3$, bounded by a cold black wall. The gray and constant absorption and scattering coefficients are, respectively, $0.16 \, \text{cm}^{-1}$ and $0.04 \, \text{cm}^{-1}$, while the furnace radius is $R = 0.5 \, \text{m}$. Scattering may be assumed to be isotropic. Using the S_2-approximation,

(a) set up the relevant equations and their boundary conditions;

(b) calculate the total heat loss from the furnace (per unit length);

(c) calculate the radial temperature distribution; what are centerline and the adjacent-to-wall temperatures?

(d) qualitatively, if the extinction coefficient is kept constant, what is the effect of varying the scattering coefficient on (i) heat transfer rates, (ii) temperature levels?

15.8 Estimate the radial temperature distribution in the sun. You may make the following assumptions:

(i) The sun is a sphere of radius R;

(ii) As a result of high temperatures in the sun, the absorption and scattering coefficients may be approximated to be constant, i.e. $\kappa_\nu, \beta_\nu = \text{const} \neq f(\nu, T, r)$ (free-free transitions!);

(iii) As a result of high temperatures, radiation is the only mode of heat transfer;

(iv) The fusion process may be approximated by assuming that a small sphere at the center of the sun releases heat uniformly corresponding to the total heat loss of the sun (i.e., assume the sun to be concentric spheres with a certain flux at the inner boundary $r = r_i$).

(a) Relate the heat production to the effective sun temperature $T_{\text{eff}} = 5762 \, \text{K}$.

(b) Would you expect the sun to be optically thin, intermediate, or thick? Why? What are the prevailing boundary conditions?

(c) Find an expression for the temperature distribution (for $r > r_i$) using the S_2-approximation.

(d) What is the surface temperature of the sun?

15.9 Repeat Problem 15.8 but replace assumption (iv) by the following: The fusion process may be approximated by assuming that the sun releases heat uniformly throughout its volume corresponding to the total heat loss of the sun.

15.10 Consider a sphere of very hot dissociated gas of radius 5 cm. The gas may be approximated as a gray, linear-anisotropically scattering medium with $\kappa = 0.1\,\text{cm}^{-1}$, $\sigma_s = 0.2\,\text{cm}^{-1}$, $A_1 = 1$. The gas is suspended magnetically in vacuum within a large cold container and is initially at a uniform temperature $T_g = 10{,}000\,\text{K}$. Using the S_2-approximation and neglecting conduction and convection, specify the total heat loss per unit time from the entire sphere at time $t = 0$. Outline the solution procedure for times $t > 0$.

Hint: Solve the governing equation by introducing a new dependent variable $g(\tau) = \tau(4\pi I_b - G)$.

15.11 Consider a gray, isothermal, isotropically-scattering medium contained between large, cold, black plates. Determine the local radiative heat flux using the S_4-method. To this purpose, set up the analytical solution using the method of successive approximations, i.e., guess a radiative source function, $S(\tau)$, which is to be improved by successive iterations. Carry out one successive approximation.

15.12 Reconsider Problem 15.11 for a similar medium at radiative equilibrium contained between isothermal black plates at temperatures T_1 and T_2, respectively.

15.13 A hot gray medium is contained between two concentric black spheres of radius $R_1 = 10\,\text{cm}$ and $R_2 = 20\,\text{cm}$. The surfaces of the spheres are isothermal at $T_1 = 2000\,\text{K}$ and $T_2 = 500\,\text{K}$, respectively. The medium absorbs, emits with $n = 1$, $\kappa = 0.05\,\text{cm}^{-1}$, but does not scatter radiation. Determine the heat flux between the spheres using the S_4-approximation. Compare your results with those of Table 12.2.

Note: This problem requires the numerical solution of four simultaneous simple ordinary differential equations.

CHAPTER
16

THE TREATMENT
OF COLLIMATED
IRRADIATION

16.1 INTRODUCTION

In recent years, there has been increasing interest in the analysis of radiative transfer in multidimensional absorbing, emitting, and scattering media with collimated irradiation. By collimated irradiation we mean external radiation that penetrates from the outside into a participating medium (as opposed to emission from a bounding surface), with all light waves being parallel to one another (or approximately so). Typical examples include solar radiation through the atmosphere and into the ocean, laser irradiation of particles or liquids, and so on. In all these problems the intensity incident on a surface dA at location $\mathbf{r}_w$ on the bounding surface of the medium, as shown in Fig. 16-1, may be written as

$$
\begin{aligned}
I_{ow}(\mathbf{r}_w, \hat{\mathbf{s}}) &= q_o(\mathbf{r}_w)\delta\,[\hat{\mathbf{s}} - \hat{\mathbf{s}}_o(\mathbf{r}_w)] \\
&= q_o(\mathbf{r}_w)\delta\,[\mu - \mu_o(\mathbf{r}_w)]\,\delta[\psi - \psi_o(\mathbf{r}_w)],
\end{aligned}
\tag{16.1}
$$

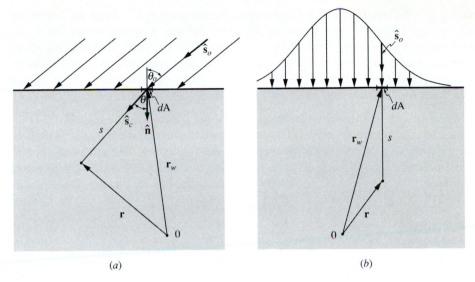

FIGURE 16-1
Collimated irradiation impinging on an arbitrary surface. (a) Solar irradiation, (b) laser irradiation.

where δ is the Dirac-delta function, defined as

$$\delta(x) = \begin{cases} 0, & |x| > 0, \\ \lim\limits_{\epsilon \to 0} \dfrac{1}{\epsilon}, & |x| < \epsilon, \end{cases} \tag{16.2a}$$

$$\int_{4\pi} f(\hat{\mathbf{s}})\,\delta(\hat{\mathbf{s}} - \hat{\mathbf{s}}_o)\,d\Omega = \int_0^{2\pi} \int_{-1}^{+1} f(\mu, \psi)\,\delta(\mu - \mu_o)\delta(\psi - \psi_o)\,d\mu\,d\psi$$

$$= f(\mu_o, \psi_o), \tag{16.2b}$$

and

$$\hat{\mathbf{s}}_o = \cos\theta_o\,\hat{\mathbf{n}} + \sin\theta_o(\cos\psi_o\,\hat{\mathbf{t}}_1 + \sin\psi_o\,\hat{\mathbf{t}}_2), \qquad \mu_o = \cos\theta_o, \tag{16.3}$$

is the direction from which the collimated radiation impinges onto the medium (with $\hat{\mathbf{n}}$ the surface normal pointing into the medium and $\hat{\mathbf{t}}_1$ and $\hat{\mathbf{t}}_2$ two orthogonal unit vectors lying on the boundary surface). Equation (16.1) implies that the incident intensity is zero for all directions except for $\hat{\mathbf{s}}_o$, where it is infinitely large. The total heat flux within the collimated irradiation is determined from

$$\mathbf{q}_o = \int_{4\pi} I_{ow}(\hat{\mathbf{s}})\,\hat{\mathbf{s}}\,d\Omega = q_o \int_{4\pi} \hat{\mathbf{s}}\,\delta(\hat{\mathbf{s}} - \hat{\mathbf{s}}_o)\,d\Omega = q_o\hat{\mathbf{s}}_o, \tag{16.4}$$

that is, q_o is the total radiative heat flux of the collimated irradiation through a surface normal to the rays. The component penetrating into the medium is then

$$\mathbf{q}_c = [1 - \rho(\mathbf{r}_w, \hat{\mathbf{s}}_o)]\,q_o\hat{\mathbf{s}}_c, \tag{16.5}$$

where ρ is the reflectivity of the interface in the direction of $\hat{s}_o$. Since the irradiation penetrating into the medium may be refracted, the unit direction vector inside the medium is denoted as $\hat{s}_c$, which may be different from $\hat{s}_o$. As indicated in the above expressions the magnitude of the irradiation q_o, as well as the direction of irradiation, $\hat{s}_o$, may vary over the surface of the enclosure, while the reflectivity of the surface may vary with position and direction.

In a strictly mathematical sense equation (16.1) introduces nothing new: Collimated irradiation could simply be treated as "strongly directional emission." However, the discontinuity of intensity with direction causes problems with analytical as well as numerical solution techniques, thus warranting a separate approach for this type of problem.

Most earlier works on collimated radiation dealt with solar radiation and other atmospheric or astrophysical applications. They are, therefore, generally limited to one-dimensional cases with uniform irradiation of a planar medium. For this simple case, some exact and approximate solutions have been given by Irvine [1], who used the Henyey-Greenstein phase function, a scattering phase function that adequately approximates the anisotropic scattering behavior of a large number of media [2], as given by equation (10.82). The identical problem for Rayleigh scattering was treated by Kubo [3] without, however, reporting any results. Armaly and El-Baz [4] found some approximate solutions for isotropic scattering in a finite-thickness slab using the common kernel approximation. Their application was in the area of solar collectors. A similar problem was treated by Houf and Incropera [5], who investigated different approximate techniques for solar irradiation of aqueous media.

Only recently, with the advent of the laser as a research and manufacturing tool, has nonsolar collimated radiation received some research attention. Smith [6] investigated the case of a uniform strip of collimated radiation incident on a semi-infinite medium. The resulting two-dimensional integral equation was reduced to one-dimensional form using Fourier transforms. Hunt [7] investigated the effect of a cylindrical collimated beam impinging upon a finite layer. A solution was found for the basic case of Bessel-function varying intensity using Green's functions. The first ones to apply this theory to laser radiation appear to be Beckett and coworkers [8], who investigated numerically the effect of a cylindrical beam with Gaussian variation penetrating through a finite layer. They showed how a diagnostic laser beam can be used to deduce radiative properties of an optically thick slab, such as single-scattering albedo, extinction, and absorption coefficients. Finally, a number of papers by Crosbie and coworkers [9–12] dealt with exact solutions to the general two-dimensional problem of collimated radiation impinging onto an absorbing-scattering layer. First, they treated collimated strip sources irradiating a semi-infinite body [9]; later, they discussed cylindrical beams falling on a semi-infinite body [10, 11]. Collimated irradiation onto a rectangular medium was investigated by Crosbie and Schrenker [12] for isotropic scattering, while Kim and Lee [13] demonstrated the accuracy of the high-order discrete ordinates method by applying it to the same problem with anisotropic scattering. The exact solutions may be used as bench marks for evaluation of approximate methods and may be necessary in cases where the requirement for highly accurate results justifies going through the trouble usually associated with these methods. In the area of heat transfer, however, approximate solutions often result in acceptable predicitions for most practical situations.

In this chapter we shall describe how problems involving radiative transfer in an absorbing, emitting, and anisotropically scattering medium of arbitrary geometry exposed to arbitrary collimated irradiation[1] are dealt with by separating the collimated radiation (as it travels through the medium) from the rest of the radiation field. The problem is thus reduced to one without collimated irradiation, but with modified radiation source term (now including a source due to the scattered part of the collimated irradiation). We shall see that it is possible to incorporate collimated irradiation readily into well-known approximate methods such as the P_1-approximation.

16.2 REDUCTION OF THE PROBLEM

The equation of transfer for an absorbing, emitting, and anisotropically scattering medium is given by equation (8.23) as

$$\hat{\mathbf{s}} \cdot \nabla I(\mathbf{r}, \hat{\mathbf{s}}) = \kappa I_b(\mathbf{r}) - \beta I(\mathbf{r}, \hat{\mathbf{s}}) + \frac{\sigma_s}{4\pi} \int_{4\pi} I(\mathbf{r}, \hat{\mathbf{s}}') \Phi(\hat{\mathbf{s}}, \hat{\mathbf{s}}')\, d\Omega'. \tag{16.6}$$

As usual, the lack of a spectral subscript implies that we deal either with spectral intensity or with a gray medium. We shall limit ourselves here to media with diffusely emitting and reflecting boundaries. Then the boundary condition for equation (16.6) is, for any location $\mathbf{r}_w$ on the surface,

$$I(\mathbf{r}_w, \hat{\mathbf{s}}) = [1 - \rho(\mathbf{r}_w)] I_{ow}(\mathbf{r}_w, \hat{\mathbf{s}}) + \epsilon(\mathbf{r}_w) I_{bw}(\mathbf{r}_w)$$
$$+ \frac{\rho(\mathbf{r}_w)}{\pi} \int_{\hat{\mathbf{n}} \cdot \hat{\mathbf{s}}' < 0} I(\mathbf{r}_w, \hat{\mathbf{s}}') |\hat{\mathbf{n}} \cdot \hat{\mathbf{s}}'|\, d\Omega'. \tag{16.7}$$

Here the first term on the right-hand side represents penetration of collimated radiation, the second term describes emission from the surrounding medium, and the last term is due to diffuse reflection at the interface. This distribution is shown schematically in Fig. 16-2a. Since we assume here diffuse emission and reflection and are looking at spectral relations or a gray medium, we also have $\epsilon = 1 - \rho$.

We now separate the intensity within the medium into two parts: (*i*) the remnant of the collimated beam after partial extinction, by absorption and scattering, along its path, and (*ii*) a fairly diffuse part, which is the result of emission from the boundaries,

---

[1]We shall limit our discussion to unpolarized irradiation, even though laser sources are always polarized (circularly or linearly). Polarization together with specular reflections may result in a number of interesting effects, for example, in the area of laser processing of materials [14, 15].

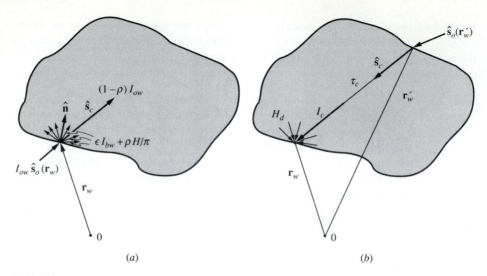

FIGURE 16-2
Radiative intensity at the surface of an enclosure with collimated irradiation: (*a*) Outgoing intensity, (*b*) Incoming intensity.

emission from within the medium, and the radiation scattered away from the collimated irradiation. Thus, we set

$$I(\mathbf{r}, \hat{\mathbf{s}}) = I_c(\mathbf{r}, \hat{\mathbf{s}}) + I_d(\mathbf{r}, \hat{\mathbf{s}}), \tag{16.8}$$

where the collimated remnant of the irradiation obeys the equation of transfer

$$\hat{\mathbf{s}} \cdot \nabla I_c(\mathbf{r}, \hat{\mathbf{s}}) = -\beta I_c(\mathbf{r}, \hat{\mathbf{s}}), \tag{16.9}$$

subject to the boundary condition

$$I_c(\mathbf{r}_w, \hat{\mathbf{s}}) = [1 - \rho(\mathbf{r}_w)] \, q_o(\mathbf{r}_w) \, \delta \, [\hat{\mathbf{s}} - \hat{\mathbf{s}}_c(\mathbf{r}_w)] \,. \tag{16.10}$$

Equations (16.9) and (16.10) are readily solved as[2]

$$I_c(\mathbf{r}, \hat{\mathbf{s}}) = [1 - \rho(\mathbf{r}_w)] \, q_o(\mathbf{r}_w) \, \delta \, [\hat{\mathbf{s}} - \hat{\mathbf{s}}_c(\mathbf{r}_w)] \, e^{-\tau_c}, \tag{16.11}$$

where $\tau_c = \int_0^s \beta \, ds'$ and $s = |\mathbf{r} - \mathbf{r}_w|$ as indicated in Fig. 16-2*b*. Substituting equations (16.8) and (16.9) into equation (16.6) gives the equation of transfer for the noncollimated radiation as

$$\frac{1}{\beta} \hat{\mathbf{s}} \cdot \nabla I_d(\mathbf{r}, \hat{\mathbf{s}}) = \hat{\mathbf{s}} \cdot \nabla_\tau I_d(\mathbf{r}, \hat{\mathbf{s}}) = -I_d(\mathbf{r}, \hat{\mathbf{s}}) + \frac{\omega}{4\pi} \int_{4\pi} I_d(\mathbf{r}, \hat{\mathbf{s}}') \Phi(\hat{\mathbf{s}}, \hat{\mathbf{s}}') \, d\Omega'$$
$$+ (1 - \omega) I_b(\mathbf{r}) + \omega S_c(\mathbf{r}, \hat{\mathbf{s}}), \tag{16.12}$$

---

[2]For simplicity of notation, equation (16.11) and the following development assume collimated radiation from a single direction; if multiple collimated sources are present, I_c is found by summation.

where the abbreviation

$$S_c(\mathbf{r}, \hat{\mathbf{s}}) \equiv \frac{1}{4\pi} \int_{4\pi} I_c(\mathbf{r}, \hat{\mathbf{s}}') \Phi(\hat{\mathbf{s}}, \hat{\mathbf{s}}') \, d\Omega'$$

$$= \frac{1}{4\pi} [1 - \rho(\mathbf{r}_w)] \, q_o(\mathbf{r}_w) \, e^{-\tau_c} \Phi(\hat{\mathbf{s}}, \hat{\mathbf{s}}_c) \qquad (16.13)$$

has been introduced and ∇_τ again implies that the gradient is to be taken with respect to nondimensional optical coordinates. Thus, S_c is a radiative source term resulting from radiation scattered away from the collimated beam; it behaves similarly to I_b, although this "emission" may not be isotropic (in the case of anisotropic scattering). Similarly, substituting equations (16.8) and (16.10) into equation (16.7) gives the boundary condition for equation (16.12) as

$$I_d(\mathbf{r}_w, \hat{\mathbf{s}}) = \epsilon I_{bw}(\mathbf{r}_w) + \frac{\rho(\mathbf{r}_w)}{\pi} \left[H_c(\mathbf{r}_w) + \int_{\hat{\mathbf{n}} \cdot \hat{\mathbf{s}}' < 0} I_d(\mathbf{r}_w, \hat{\mathbf{s}}') |\hat{\mathbf{n}} \cdot \hat{\mathbf{s}}'| \, d\Omega' \right], \qquad (16.14)$$

where

$$H_c(\mathbf{r}_w) \equiv \int_{\hat{\mathbf{n}} \cdot \hat{\mathbf{s}}' < 0} I_c(\mathbf{r}_w, \hat{\mathbf{s}}') |\hat{\mathbf{n}} \cdot \hat{\mathbf{s}}'| \, d\Omega'$$

$$= [1 - \rho(\mathbf{r}'_w)] \, q_o(\mathbf{r}'_w) |\hat{\mathbf{n}} \cdot \hat{\mathbf{s}}'_c| e^{-\tau_c} \qquad (16.15)$$

is a surface irradiation term due to the collimated beam, and its diffuse reflection results in an additional surface source similar to I_{bw}. In this expression $\mathbf{r}'_w$ is the location at which the collimated beam enters the medium with a direction of $\hat{\mathbf{s}}'_c$, and $\mathbf{r}_w$ is the next point on the enclosure surface that the beam hits after traversing through the medium, as shown in Fig. 16-2b.

Inspection of equations (16.12) and (16.14) shows that, for isotropic scattering, the intensity field for I_d is readily determined from standard methods, after replacing I_b within the medium by $I_b + (\sigma_s/\kappa)S_c$, and I_{bw} at the enclosure surface by $I_{bw} + (\rho/\epsilon)H_c/\pi$. In the case of anisotropic scattering the emission term S_c becomes direction-dependent, which may necessitate slight changes in the solution procedure.

Example 16.1.   Consider a plane-parallel slab of an absorbing and isotropically scattering medium as shown in Fig. 16-3. The medium is gray (with absorption coefficient κ, scattering coefficient σ_s, and a refractive index of $n = 1$), cold (i.e., essentially nonemitting) and of constant thickness L. At the top ($z = 0$) the layer is bounded by a nonparticipating gas ($n = 1$) and is exposed to solar radiation impinging at θ_o off-normal. At the bottom of the layer ($z = L$) the medium is bounded by a cold black surface. Determine radiative heat flux (and its divergence) as functions of depth.

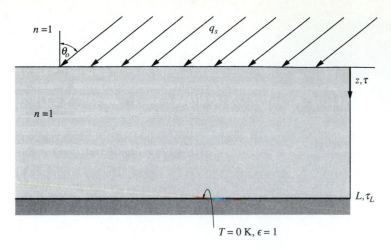

FIGURE 16-3
Geometry for Example 16.1.

Solution

Since both media have the same index of refraction the interface reflectivity is $\rho = 0$ and $\hat{s}_c = \hat{s}_o$; from equation (16.11) we find immediately

$$I_c(\tau, \hat{s}) = q_s e^{-\tau/\mu_o} \delta(\hat{s} - \hat{s}_o),$$

as well as

$$S_c(\tau) = \frac{G_c}{4\pi}(\tau) = \frac{q_s}{4\pi} e^{-\tau/\mu_o},$$

$$\mathbf{q}_c(\tau) = q_s e^{-\tau/\mu_o} \hat{s}_o.$$

The heat flux vector due to collimated irradiation has two components, one in the direction of τ, the other in a direction normal to it (in the plane formed by $\hat{s}_o$ and the surface normal). Thus, the overall problem is two-dimensional. However, inspection of the source term in equation (16.12) shows that the source is isotropic, as are the boundary conditions for equation (16.12). Therefore, I_d can depend only on distance perpendicular to the surfaces, τ, and on polar angle. Thus,

$$\mu \frac{dI_d}{d\tau} + I_d = \omega \left[\frac{1}{4\pi} \int_{4\pi} I_d \, d\Omega' + S_c \right] = \frac{\omega}{4\pi} \left[G_d(\tau) + G_c(\tau) \right].$$

The boundary conditions are

$$\tau = 0: \qquad I(0, \hat{s}) = q_s \delta(\hat{s} - \hat{s}_o), \qquad 0 \le \theta < \frac{\pi}{2},$$

$$\tau = \tau_L: \qquad I(\tau_L, \hat{s}) = 0, \qquad \frac{\pi}{2} < \theta \le \pi,$$

or, after subtracting the collimated component,

$$\tau = 0: \qquad I_d(0, \mu) = 0, \qquad 0 < \mu \le 1,$$

$$\tau = \tau_L: \qquad I_d(\tau, \mu) = 0, \qquad -1 \le \mu < 0.$$

Therefore, with $S = (\omega/4\pi)(G_d + G_c)$, the solution for I_d is, from equation (12.21),

$$G_d(\tau) = 2\pi \int_0^{\tau_L} \frac{\omega}{4\pi}(G_d + G_c)(\tau')E_1(|\tau - \tau'|)\, d\tau'.$$

In nondimensional form, with $\Phi(\tau) = (G_d + G_c)/q_s$, this expression becomes

$$\Phi(\tau) - e^{-\tau/\mu_o} = \frac{\omega}{2} \int_0^{\tau_L} \Phi(\tau')E_1(|\tau - \tau'|)\, d\tau'.$$

This integral equation must be solved numerically in the same fashion as equation (12.41). Once the function $\Phi(\tau)$ has been determined, the diffuse component of the heat flux is found from equation (12.22) as

$$\mathbf{q}_d(\tau) = q_d(\tau)\,\hat{\mathbf{k}} = 2\pi \int_{-1}^{1} I_d\, \mu\, d\mu\, \hat{\mathbf{k}}$$

$$q_d(\tau) = 2\pi \left\{ \int_0^{\tau} \frac{\omega}{4\pi}(G_d + G_c)(\tau')E_2(\tau - \tau')\, d\tau' \right.$$

$$\left. - \int_{\tau}^{\tau_L} \frac{\omega}{4\pi}(G_d + G_c)(\tau')E_2(\tau' - \tau)\, d\tau' \right\}.$$

Finally, the total heat flux in the τ-direction, on a nondimensional basis, is

$$\Psi = \frac{\mathbf{q}_c \cdot \hat{\mathbf{k}} + q_d}{q_s} = \mu_o e^{-\tau/\mu_o} + \frac{\omega}{2}\left[\int_0^{\tau} \Phi(\tau')E_2(\tau - \tau')\, d\tau' - \int_{\tau}^{\tau_L} \Phi(\tau')E_2(\tau' - \tau)\, d\tau' \right].$$

The divergence of the radiative heat flux is found from equation (8.54) as

$$\nabla_\tau \cdot \mathbf{q} = (1 - \omega)(4\pi I_b - G) = -(1 - \omega)(G_c + G_d),$$

or in nondimensional form

$$\frac{1}{q_s}\nabla_\tau \cdot \mathbf{q} = -(1 - \omega)\Phi.$$

Some results for a purely scattering medium ($\omega = 1$) are shown in Fig. 16-4. For that case we find $\nabla \cdot \mathbf{q} = 0$ and, since the tangential component of $\mathbf{q}_c$ does not depend on tangential direction, $\mathbf{q} \cdot \hat{\mathbf{k}} = \text{const}$ and $\Psi = \text{const}$. Thus, evaluating the heat flux at $\tau = 0$, we get

$$\Psi(\omega = 1) = \mu_o - \frac{1}{2}\int_0^{\tau_L} \Phi(\tau')E_2(\tau')\, d\tau'.$$

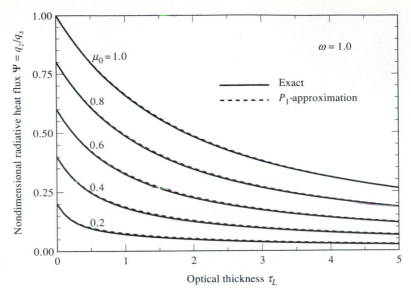

FIGURE 16-4
Nondimensional radiative heat flux in a purely scattering layer with collimated irradiation.

16.3   THE P_1-APPROXIMATION WITH COLLIMATED IRRADIATION

As for problems without collimated irradiation, the exact or approximate solution to equation (16.12), together with its boundary condition (16.14), may be found using a variety of different methods. As an illustration we will show here how the P_1-approximation may be applied to problems with collimated irradiation, following the development of Modest and Tabanfar [16]. The P_1 or *differential approximation* is simple to use (requiring only the solution of an elliptic differential equation) and powerful (applicable to multidimensional geometries, as well as to anisotropic scattering). Unfortunately, the fact that the P_1-approximation is accurate only for smoothly varying (with direction) intensity fields makes the method particularly unsuitable for problems with collimated irradiation. However, once the collimated intensity, I_c, has been removed from the intensity field, the P_1-approximation may be expected to give accurate solutions to equations (16.12) and (16.14) for many situations. To apply the method, we assume again that the remnant intensity can deviate only slightly from isotropic conditions or, from equation (14.30),

$$I_d(\mathbf{r}, \hat{\mathbf{s}}) \simeq \frac{1}{4\pi} [G_d(\mathbf{r}) + 3\mathbf{q}_d(\mathbf{r}) \cdot \hat{\mathbf{s}}].\tag{16.16}$$

As in Section 14.4 we shall limit ourselves to linear-anisotropic scattering,

$$\Phi(\hat{\mathbf{s}}, \hat{\mathbf{s}}') = 1 + A_1 \hat{\mathbf{s}} \cdot \hat{\mathbf{s}}',\tag{16.17}$$

so that

$$
\begin{aligned}
S_c &= \frac{1}{4\pi} \int_{4\pi} I_c(\hat{\mathbf{s}}')(1 + A_1\hat{\mathbf{s}} \cdot \hat{\mathbf{s}}')\,d\Omega' \\
&= \frac{1}{4\pi}(G_c + A_1\mathbf{q}_c \cdot \hat{\mathbf{s}}) = \frac{1}{4\pi}[1 - \rho(\mathbf{r}_w)]q(\mathbf{r}_w)e^{-\tau_c}(1 + A_1\hat{\mathbf{s}} \cdot \hat{\mathbf{s}}_c).
\end{aligned}
\qquad (16.18)
$$

Now, integrating equation (16.12) over all directions (zeroth moment), we obtain

$$
\nabla_\tau \cdot \mathbf{q}_d = (1 - \omega)(4\pi I_b - G_d) + \omega G_c.
\qquad (16.19)
$$

Similarly, integrating equation (16.12) after multiplication with $\hat{\mathbf{s}}$ and invoking equation (16.16), we obtain

$$
\frac{1}{3}\nabla_\tau G_d = -\left(1 - \frac{A_1\omega}{3}\right)\mathbf{q}_d + \frac{A_1\omega}{3}\mathbf{q}_c.
\qquad (16.20)
$$

As for the standard P_1-approximation, the necessary boundary conditions for this set of equations are found by demanding continuity of heat flux normal to the surface at the boundary, that is,

$$
\mathbf{q}_d \cdot \hat{\mathbf{n}}(\mathbf{r}_w) = \int_{4\pi} I_d(\mathbf{r}_w, \hat{\mathbf{s}})\hat{\mathbf{n}} \cdot \hat{\mathbf{s}}\,d\Omega,
\qquad (16.21)
$$

with $I_d(\mathbf{r}_w, \hat{\mathbf{s}})$ from equation (16.16) for incoming directions ($\hat{\mathbf{n}} \cdot \hat{\mathbf{s}} < 0$), and from equation (16.14) for outgoing directions ($\hat{\mathbf{n}} \cdot \hat{\mathbf{s}} > 0$). Calculating first the diffuse irradiation leads to

$$
\begin{aligned}
-H_d(\mathbf{r}_w) &= \int_{\hat{\mathbf{n}}\cdot\hat{\mathbf{s}}<0} I_d(\mathbf{r}_w, \hat{\mathbf{s}})\hat{\mathbf{n}} \cdot \hat{\mathbf{s}}\,d\Omega = \frac{1}{4\pi} \int_0^{2\pi} \int_{\pi/2}^{\pi} (G_d + 3\mathbf{q}_d \cdot \hat{\mathbf{s}})\hat{\mathbf{n}} \cdot \hat{\mathbf{s}}\,d\Omega \\
&= \frac{1}{2} \int_{\pi/2}^{\pi} (G_d + 3\mathbf{q}_d \cdot \hat{\mathbf{n}} \cos\theta) \cos\theta \sin\theta\,d\theta = -\frac{G_d}{4} + \frac{\mathbf{q}_d \cdot \hat{\mathbf{n}}}{2}.
\end{aligned}
\qquad (16.22)
$$

Thus,

$$
\mathbf{q}_d \cdot \hat{\mathbf{n}} = \epsilon\pi I_{bw} + \rho(H_c + H_d) - H_d,
$$

or, after substituting equation (16.22),

$$
\mathbf{r} = \mathbf{r}_w: \qquad 2\mathbf{q}_d \cdot \hat{\mathbf{n}} = \frac{\epsilon(4\pi I_{bw} - G_d) + 4(1 - \epsilon)H_c}{2 - \epsilon}.
\qquad (16.23)
$$

The derivation of equations (16.19) and (16.20) is very similar to the development of the standard P_1-approximation, which has been given in some greater detail in Section 14.4.

Example 16.2. Find the solution to the previous example using the P_1-approximation.

Solution

As for the exact solution we find

$$G_c = q_s e^{-\tau/\mu_o}, \qquad \mathbf{q}_c = G_c \hat{\mathbf{s}}_o,$$

and we realize again that G_d and $\mathbf{q}_d = q_d \hat{\mathbf{k}}$ depend on τ (optical distance perpendicular to the layer) only. Thus, from equations (16.19) and (16.20) and their boundary conditions (16.23), we find that

$$\frac{dq_d}{d\tau} = -(1 - \omega)G_d + \omega G_c,$$

$$\frac{dG_d}{d\tau} = -3q_d,$$

$$\tau = 0: \qquad 2q_d = -G_d,$$

$$\tau = \tau_L: \qquad -2q_d = -G_d.$$

Since the solution procedure for this equation is different for $\omega = 1$ (as opposed to $\omega < 1$), and since we would like to compare the present results with exact ones shown in Fig. 16-4, we shall limit the rest of our discussion to $\omega = 1$. Then

$$\frac{dq_d}{d\tau} = q_s e^{-\tau/\mu_o}, \quad \text{or} \quad q_d = -\mu_o q_s e^{-\tau/\mu_o} + C_1,$$

$$\frac{dG_d}{d\tau} = -3q_d, \quad \text{or} \quad G_d = -3\mu_o^2 q_s e^{-\tau/\mu_o} - 3C_1\tau + C_2.$$

It follows from the boundary conditions that

$$\tau = 0: \qquad 2C_1 + C_2 = (2 + 3\mu_o)\mu_o q_s,$$

$$\tau = \tau_L: \quad (2 + 3\tau_L)C_1 - C_2 = (2 - 3\mu_o)\mu_o q_s e^{-\tau/\mu_o},$$

or

$$C_1 = \frac{2 + 3\mu_o + (2 - 3\mu_o)e^{-\tau_L/\mu_o}}{4 + 3\tau_L}\mu_o q_s,$$

$$q_d = \left[\frac{2 + 3\mu_o + (2 - 3\mu_o)e^{-\tau_L/\mu_o}}{4 + 3\tau_L} - e^{-\tau/\mu_o}\right]\mu_o q_s.$$

Finally,

$$\Psi = \frac{\mathbf{q}_c \cdot \hat{\mathbf{k}} + q_d}{q_s} = \frac{2 + 3\mu_o + (2 - 3\mu_o)e^{-\tau_L/\mu_o}}{4 + 3\tau_L}\mu_o,$$

which, as discussed in the previous example, is constant across the layer. This nondimensional heat flux is compared with the exact result in Fig. 16-4. It is seen that the P_1-approximation gives good accuracy for all cases shown in that figure.

Some two-dimensional examples for the P_1-approximation with collimated irradiation have been given by Modest and Tabanfar [16], comparing with exact results by Crosbie and Koewing [17] and Crosbie and Dougherty [10]. The accuracy of the P_1-approximation was found to be excellent in most cases, since it is generally applied to an "emitting" medium with cold boundaries. As expected, the accuracy of the P_1-approximation decreases if sharp gradients of the radiative source occur within the medium (e.g., a source resulting from scattering of a highly focused, penetrating laser beam).

References

1. Irvine, W. M.: "Multiple Scattering by Large Particles II. Optically Thick Layers," *The Astrophysical Journal*, vol. 152, pp. 823–834, 1968.
2. Van de Hulst, H. C.: *Light Scattering by Small Particles*, John Wiley & Sons, New York, 1957 (also Dover Publications, New York, 1981).
3. Kubo, S.: "Effects of Anisotropic Scattering on Steady One-Dimensional Radiative Heat Transfer Through an Absorbing-Emitting Medium," *J. Phys. Soc. Japan*, vol. 41, no. 3, pp. 894–898, 1976.
4. Armaly, B. F., and H. S. El-Baz: "Radiative Transfer Through an Isotropically Scattering Finite Medium: Approximate Solution," *AIAA Journal*, vol. 15, no. 8, pp. 1180–1185, 1976.
5. Houf, W. G., and F. P. Incropera: "An Assessment of Techniques for Predicting Radiation Transfer in Aqueous Media," *Journal of Quantitative Spectroscopy and Radiative Transfer*, vol. 23, pp. 101–115, 1980.
6. Smith, M. G.: "The Transport Equation with Plane Symmetry and Isotropic Scattering," *Proc. Camb. Phil. Soc.*, vol. 60, p. 909, 1964.
7. Hunt, G. E.: "The Transport Equation of Radiative Transfer with Axial Symmetry," *SIAM J. Appl. Math.*, vol. 16, no. 1, pp. 228–237, 1968.
8. Beckett, P., P. J. Foster, V. Huston, and R. L. Moss: "Radiative Transfer for a Cylindrical Beam Scattered Isotropically," *Journal of Quantitative Spectroscopy and Radiative Transfer*, vol. 14, pp. 1115–1125, 1974.
9. Crosbie, A. L., and T. L. Linsenbardt: "Two-Dimensional Isotropic Scattering in a Semi-Infinite Medium," *Journal of Quantitative Spectroscopy and Radiative Transfer*, vol. 19, pp. 257–284, 1978.
10. Crosbie, A. L., and R. L. Dougherty: "Two-Dimensional Isotropic Scattering in a Semi-Infinite Cylindrical Medium," *Journal of Quantitative Spectroscopy and Radiative Transfer*, vol. 20, pp. 151–173, 1978.
11. Crosbie, A. L., and R. L. Dougherty: "Two-Dimensional Linearly Anisotropic Scattering in a Semi-Infinite Cylindrical Medium Exposed to a Laser Beam," *Journal of Quantitative Spectroscopy and Radiative Transfer*, vol. 28, no. 3, pp. 233–263, 1982.
12. Crosbie, A. L., and R. G. Schrenker: "Multiple Scattering in a Two-Dimensional Rectangular Medium Exposed to Collimated Radiation," *Journal of Quantitative Spectroscopy and Radiative Transfer*, vol. 33, no. 2, pp. 101–125, 1985.
13. Kim, T. K., and H. S. Lee: "Radiative Transfer in Two-Dimensional Anisotropic Scattering Media with Collimated Incidence," *Journal of Quantitative Spectroscopy and Radiative Transfer*, vol. 42, pp. 225–238, 1989.
14. Duley, W. W.: *Laser Processing and Analysis of Materials*, Plenum Press, New York, 1983.
15. Bang, S. Y., and M. F. Modest: "Evaporative Scribing with a Moving CW Laser - Effects of Multiple Reflections and Beam Polarization," in *Proceedings of ICALEO '91, Laser Materials Processing*, vol. 74, San Jose, CA, pp. 288–304, 1992.
16. Modest, M. F., and S. Tabanfar: "A Multi-Dimensional Differential Approximation for Absorbing/Emitting Anistropically Scattering Media with Collimated Irradiation," *Journal of Quantitative Spectroscopy and Radiative Transfer*, vol. 29, pp. 339–351, 1983.
17. Crosbie, A. L., and J. W. Koewing: "Two-Dimensional Radiative Heat Transfer in a Planar Layer Bounded by Nonisothermal Walls," *AIAA Journal*, vol. 17, no. 2, pp. 196–203, 1979.

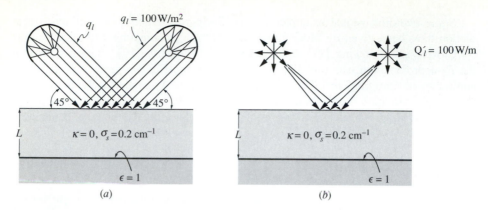

FIGURE 16-5
Geometry for (a) Problem 16.2, (b) Problem 16.8.

Problems

16.1 A semi-infinite, gray, isotropically scattering medium, originally at zero temperature, is subjected to collimated irradiation with a constant flux q_o normal to its nonreflecting surface. Set up the integral relationships governing steady-state temperature and radiative heat flux within the medium, assuming radiative equilibrium.

16.2 In a greenhouse a layer of water (thickness $L = 5\,\text{cm}$) is resting on top of a black substrate. The water is loaded with growing organisms that scatter light isotropically but do not absorb ($\sigma_s = 0.2\,\text{cm}^{-1}$). The water layer is illuminated by two long growth-enhancing lights, fitted with reflector shields that make the light essentially parallel (Fig. 16-5a), each light delivering a heat flux of $q_l = 100\,\text{W/m}^2$ (per unit area normal to the light rays). Using the exact method, calculate energy generated within the water and the radiative heat flux absorbed by the black surface in the zone between the lights, where the heat transfer is essentially one-dimensional. Emission from the water and substrate are negligible.
Hint: Use Figs. 3-16 and 16-4.

16.3 Reconsider the medium described in Example 16.1. Rather than being bounded by a cold black surface at the bottom, the layer is now exposed to the nonparticipating gas as well as to solar irradiation (using mirrors) on both sides. Determine radiative heat flux and its divergence within the layer in terms of the function $\Phi(\tau_L, \omega, \mu_o, \tau)$ given in Example 16.1.

16.4 Solve Problem 16.1 using the P_1-approximation.

16.5 The starship Enterprise is hitting a Klingon cruiser with its phaser gun. The armament of the cruiser is a partially reflecting material that, after some irradiation, partly evaporates, forming a protective gas layer above the surface. Assuming that the surface is at evaporation temperature T_{ev} and has an emissivity ϵ, the gas has an absorption coefficient κ_g and a thickness L, determine the fraction of the heat flux that hits the Klingon ship. Under these conditions you may assume the effects of conduction and convection to be negligible (but not reradiation from the gas). Use the P_1-approximation.

16.6 Reconsider Problem 16.5. After further irradiation, the surface material starts to disintegrate, spewing particulate material into the gas layer. If we make the assumption that

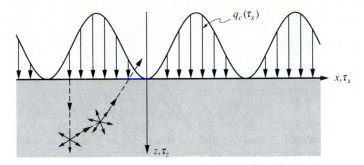

FIGURE 16-6
Geometry for Problem 16.7.

the debris has an absorption coefficient κ_p and (isotropic) scattering coefficient σ_{sp}, how does this modify the surface irradiation?

16.7 Consider a semi-infinite gray medium with a nonreflecting surface. The medium is cold, absorbs (absorption coefficient κ) and scatters isotropically (scattering coefficient σ_s). Collimated radiation obeying the relation

$$\mathbf{q}_c = q_o(1 - \cos\alpha\tau_x)\hat{\mathbf{k}}$$

shines normally onto the medium as shown in Fig. 16-6. Determine the reflectivity of the medium (i.e., the fraction of the irradiation leaving the interface in the opposite direction), using the P_1-approximation.
Hint: To solve the two-dimensional governing equation, set $G_d(\tau_x, \tau_z) = G_1(\tau_z) + G_2(\tau_z)\cos\alpha\tau_x$.
This problem is a special case of solutions given by Crosbie and Koewing [17] (exact) and Modest and Tabanfar [16] (P_1).

16.8 Reconsider Problem 16.2 for lights not fitted with reflector shields as depicted in Fig. 16-5b. Assuming that the figure shows only two of many equally-spaced lights (i.e., using symmetry), set up the solution for the radiative heat flux, using the P_1-approximation. Each light outputs a total of 100 W per meter length. Since this is a two-dimensional problem it will be sufficient to reduce the problem to the solution of a two-dimensional partial differential equation with stated boundary conditions.

CHAPTER
17

THE TREATMENT
OF NONGRAY
EXTINCTION
COEFFICIENTS

17.1 INTRODUCTION

In the preceding chapters on solution methods for the radiative equation of transfer within participating media we have exclusively dealt with gray media, that is, media whose radiative properties (absorption coefficient κ, scattering coefficient σ_s, and phase function Φ, as well as emissivity of boundary surfaces ϵ) do not vary across the electromagnetic spectrum. We noted that most relationships also hold true for a nongray medium on a spectral basis (i.e., as long as the simplification of radiative equilibrium is not invoked). While the assumption of gray surfaces made in the net radiation method of Chapters 5 (diffusely reflecting surfaces) and 6 (partly specular reflecting surfaces) is often a good one over the relatively small relevant part of the spectrum, this is nearly never the case for participating media. Molecular gases below dissociation temperatures (as discussed in detail in Chapter 9) absorb and emit over a multitude of very narrow spectral lines, which may overlap and form vibration-rotation bands. The result is an absorption coefficient that oscillates wildly within each band, and is zero between bands. Similarly, the discussion of radiative properties of suspended particles (Chapter 10) has shown that their absorption and scattering properties may also oscillate strongly across the spectrum (see, for example, Fig. 10-3). However, if particles of varying sizes are present, as is usually the case, the spectral oscillations tend to be damped out so that the assumption of a gray medium becomes a reasonable one. Like molecular gases, semitransparent solids and liquids often display strong absorption bands in the infrared due to photon-phonon coupling, with weak absorption coefficients between bands. Therefore, we conclude that the simplification of a gray participating medium is, except for particle suspensions with variable sizes, a poor assumption that may lead to very significant errors in the analysis. It behooves the engineer to realize that accurate solutions to the equation of transfer (such as exact solutions in two or three dimensions), as opposed to simple approximate ones (such as the P_1-approximation in one or two dimensions), may be meaningless unless the spectral variation of radiation properties is taken into account.

Unfortunately, consideration of spectral variations of radiation properties tends to increase considerably the difficulty of an already extremely difficult problem, or at least make their numerical solution many times more computer-time intensive. All solution methods discussed thus far, whether exact or approximate, are poorly suited for the consideration of nongray properties. In general, radiative heat flux, divergence of heat flux, and/or incident radiation must be evaluated for many, many spectral locations, followed by numerical quadrature of the spectral results. This process will always involve the guessing of a temperature field, followed by an iterative procedure. This statement is true even for the case of radiative equilibrium, since the condition $\nabla \cdot \mathbf{q} = 0$ holds only for total heat flux, but not for spectral heat flux: While it is true that each volume element must emit as much radiative energy as it absorbs, the re-emission of energy must not occur at the same wavelength; the wavelengths of absorption are determined by the local absorption coefficient and by the wavelengths of the incoming radiation (and, thus, depend on the temperature of the *surrounding* medium), while the wavelengths of emission are determined by the local absorption coefficient and the *local* temperature.

Relatively few papers in the literature have dealt with radiation from nongray media—perhaps because the problem quickly becomes very involved even for simple situations, perhaps because nongray media do not lend themselves very well to the presentation of results (rather than showing results systematically for nondimensional parameters, one can only present data for a distinct medium at distinct conditions). Wang [1, 2] considered the one-dimensional slab using an exponential kernel approximation as outlined by Sparrow and Cess [3]. An exact solution for an optically thin slab at radiative equilibrium has been given by Cess, Mighdoll and Tiwari [4]. Cess and Tiwari [5] found the temperature distribution in a slab with internal heat source, using the exponential kernel approximation together with constant absorption coefficient and linearized emissive power. Yuen and Rasky [6] and Yuen [7] used the P_1-approximation and two different functions for the molecular-gas absorption coefficient (bypassing the band-absorptance approach) to calculate the heat flux through a slab at radiative equilibrium. A number of investigators have obtained exact solutions to the relatively simple problems of one-dimensional isothermal media. Edwards and Balakrishnan [8] found exact expressions for the heat flux in an isothermal gas slab and gave results for the high-pressure limit (strong spectral line overlap). Modest [9] considered the same problem but included absorbing/emitting nongray particles as well as nonisothermal temperature fields in his analysis. Edwards [10] gave an expression for the heat loss from an isothermal gas sphere; while Crosbie and Khalil [11] considered an isothermal spherical layer, employing a number of approximate absorption coefficients. Finally, heat loss from an isothermal gas cylinder has been discussed by Wassel and Edwards [12].

The complexity and time consumption of nongray property treatment may be decreased considerably if one considers limiting situations, or if some simple approximations are made for the spectral dependence of the absorption and/or scattering coefficients. In the following we shall first consider the simplest method, known as the *mean beam length method,* in which the entire participating medium is assumed to be a single isothermal zone that exchanges heat with finite surface areas, much like the net radiation method of Chapter 5. Next the *picket fence* or *box model* is discussed, in which the absorption coefficient is assumed to attain a finite number of values that remain constant over finite wavenumber regions. The following section deals much more rigorously with the band nature of molecular gases, but is more or less limited to one-dimensional, plane-parallel, nonscattering media confined between black plates. Finally, the weighted-sum-of-gray-gases model is presented, which is very simple, accurate and powerful, although it is also limited to nonscattering media within black-walled enclosures.

17.2   THE MEAN BEAM LENGTH METHOD

The idea of a *mean beam length* was first advanced by Hottel [13] for the determination of radiative heat fluxes from an isothermal volume of hot combustion gases to cold black furnace walls. With some difficulty the method may be extended to include the effects of hot and gray walls. We include here only a brief discussion of the method, primarily for historical reasons and since the notion of a mean beam length is sometimes employed by other methods (see, for example, the picket-fence model in the next section). A somewhat more detailed account has been given by Hottel [13] and Hottel and Sarofim [14]. Today, with the availability of fast digital computers the method is somewhat outdated and is commonly replaced by the related zonal method, discussed in detail in Chapter 18, which allows not only for hot and gray walls, but also for a number of isothermal subvolumes within the enclosure.

Definition of Mean Beam Lengths

Consider a hot, isothermal, nonscattering gas volume radiating toward a black area element dA on its surface, as shown in Fig. 17-1a. The spectral heat flux arriving at and absorbed by dA from a volume element is equal to the spectral emission by dV

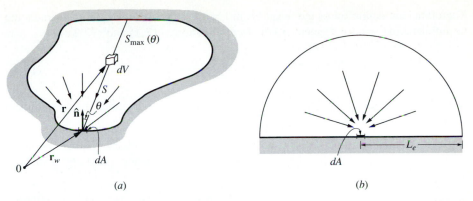

FIGURE 17-1
Isothermal gas volume radiating to surface element; (a) arbitrary gas volume, (b) equivalent hemisphere radiating to center of its base.

into all (4π) directions $\times$ the fraction intercepted by dA $\times$ the fraction transmitted along the path from dV to dA. Thus, from equation (8.49) the spectral heat flux arriving at dA from all volume elements may be written as

$$q_\eta dA(\mathbf{r}_w) = \int_V (4\pi\kappa_\eta I_{b\eta}dV) \times \left(\frac{dA\cos\theta}{4\pi S^2}\right) \times e^{-\kappa_\eta S},$$

or

$$q_\eta(\mathbf{r}_w) = I_{b\eta} \int_V e^{-\kappa_\eta S} \frac{\kappa_\eta \cos\theta\, dV}{S^2}, \tag{17.1}$$

where we have chosen wavenumber η as the spectral variable. We notice that, in general, the heat flux from a hot gas volume arriving at a surface element dA is proportional to the blackbody intensity $I_{b\eta}$ and a factor which depends on the spectral absorption coefficient as well as on the geometry of the medium. While the integral factor will not be trivial to evaluate for most geometries, it is readily determined for a hemispherical volume radiating to the center of its base, dA, as shown in Fig. 17-1b. For this case $S = r$ and $dV = r^2 \sin\theta\, dr\, d\theta\, d\psi$, leading to

$$q_\eta = I_{b\eta} \int_{\psi=0}^{2\pi} \int_{\theta=0}^{\pi/2} \int_{r=0}^{R} e^{-\kappa_\eta r} \kappa_\eta \cos\theta \sin\theta\, dr\, d\theta\, d\psi$$

$$= \pi I_{b\eta}(1 - e^{-\kappa_\eta R}) = \pi I_{b\eta}\epsilon_\eta, \tag{17.2}$$

where we have employed the definition for the spectral emissivity of an isothermal layer, equation (8.15). It is clear that the radiative heat fluxes arriving at dA, either from an arbitrary volume V or from a hemisphere of radius R, may be made equal if an appropriate value for the radius of the hemisphere is chosen. Thus, as far as the spectral, hemispherical irradiation onto dA is concerned, there is no difference whether the emission originated from an arbitrary volume or from an equivalent hemisphere of the correct radius $R = L_e$, where L_e is known as the mean beam length.

Therefore, the definition of the mean beam length for an arbitrary volume irradiating an infinitesimal surface element dA is, from equations (17.1) and (17.2),

$$\frac{q_\eta}{\pi I_{b\eta}} = 1 - e^{-\kappa_\eta L_e} = \frac{1}{\pi} \int_V e^{-\kappa_\eta S} \frac{\kappa_\eta \cos \theta \, dV}{S^2}. \tag{17.3}$$

Note that the magnitude of the mean beam length depends on absorption coefficient as well as on geometry.

It is also common to define a mean beam length for an arbitrary volume irradiating a finite surface, by replacing the local heat flux q_η in equation (17.1) by a surface-averaged value, or

$$\frac{q_{\eta,av}}{\pi I_{b\eta}} = 1 - e^{-\kappa_\eta L_e} = \frac{1}{A} \int_A \int_V e^{-\kappa_\eta S} \frac{\kappa_\eta \cos \theta \, dV}{S^2} dA. \tag{17.4}$$

Example 17.1.   Determine the mean beam length for an isothermal gas layer of thickness L radiating to (a) an infinitesimal surface element, (b) an entire bounding surface.

Solution

The mean beam length may be evaluated by first finding from equation (12.38) the radiative heat flux hitting a surface element, or by integrating equation (17.3) directly. Choosing the latter for illustrative purposes, we express V in terms of a cylindrical coordinate system with its origin at dA. Thus, $dV = 2\pi r \, dr \, dz$, $S = \sqrt{r^2 + z^2}$, and $\cos \theta = z/S$, leading to

$$\frac{q_\eta}{E_{b\eta}} = \frac{1}{\pi} \int_{z=0}^{L} \int_{r=0}^{\infty} e^{-\kappa_\eta S} \frac{\kappa_\eta z \, 2\pi r \, dr \, dz}{S^3}.$$

By replacing the integration variable r by S, this expression becomes, with $r \, dr = S \, dS$,

$$\frac{q_\eta}{E_{b\eta}} = 2 \int_{z=0}^{L} \int_{S=z}^{\infty} e^{-\kappa_\eta S} \frac{\kappa_\eta z \, dS \, dz}{S^2} = 2\kappa_\eta \int_{z=0}^{L} E_2(\kappa_\eta z) dz,$$

where the definition for the exponential integral has been employed [see equation (12.31) or Appendix E]. Integrating, we obtain

$$\frac{q_\eta}{E_{b\eta}} = -2E_3(\kappa_\eta z) \Big|_0^L = 1 - 2E_3(\kappa_\eta L),$$

which, of course, would also have followed immediately from equation (12.38), if only emission from the medium had been considered ($T_1 = T_2 = 0$).

Thus, from equation (17.3), the mean beam length from the gas layer to a surface element dA is

$$L_e = \frac{1}{\kappa_\eta} \ln \frac{1}{2E_3(\kappa_\eta L)}.$$

The mean beam length for the entire surface, equation (17.4), is the same since, for this one-dimensional problem, local and average heat flux are identical.

Mean Beam Lengths
for Optically Thin Media

Equations (17.3) and (17.4) are generally not trivial to evaluate and, for nongray media, the integrations need to be carried out for different absorption coefficients if total rather than spectral heat fluxes are desired (as is usually the case). However, the relationships become much simpler if optically thin media are considered, i.e., if $\kappa_\eta L \ll 1$, where L is a characteristic dimension of the medium. If we expand the exponents in equations (17.3) and (17.4), and drop terms of order κ_η^2 and higher, we find the mean beam length for an optically thin volume radiating to a point on its surface, L_0, is

$$1 - (1 - \kappa_\eta L_0) = \frac{1}{\pi} \int_V 1 \times \frac{\kappa_\eta \cos\theta \, dV}{S^2},$$

or

$$L_0 = \frac{1}{\pi} \int_V \frac{\cos\theta \, dV}{S^2}. \tag{17.5}$$

If we express the volume in terms of a spherical coordinate system centered at dA, with S as the radius, we may write $dV = S^2 dS \sin\theta \, d\theta \, d\psi = S^2 dS \, d\Omega$, and equation (17.5) becomes

$$L_0 = \frac{1}{\pi} \int_{\psi=0}^{2\pi} \int_{\theta=0}^{\pi/2} \int_{S=0}^{S_{\max}(\theta,\psi)} \cos\theta \sin\theta \, dS \, d\theta \, d\psi$$

$$= \frac{1}{\pi} \int_{2\pi} S_{\max}(\hat{\mathbf{s}}) \cos\theta \, d\Omega. \tag{17.6}$$

Similarly, from equation (17.4), the mean beam length for an optically thin volume radiating to a finite surface is

$$L_0 = \frac{1}{\pi A} \int_A \int_{2\pi} S_{\max}(\mathbf{r}_w, \hat{\mathbf{s}}) \cos\theta \, d\Omega \, dA. \tag{17.7}$$

By employing physical arguments, one finds that the solution of equation (17.7) is trivial for the case that A is the entire area bounding the volume V: The total emission from the entire volume is, from equation (8.49), $4\pi\kappa_\eta I_{b\eta} V$. Since, for an optically thin medium, no self-absorption occurs, all of this energy must be absorbed by the (black) bounding surface. Therefore, the average heat flux onto the surface is

$$q_\eta = 4\pi\kappa_\eta I_{b\eta} V / A$$

and, from equation (17.3) (with $\kappa_\eta L_0 \ll 1$),

$$\frac{q_\eta}{\pi I_{b\eta}} = \kappa_\eta L_0 = 4\kappa_\eta V / A$$

or

$$L_0 = 4\frac{V}{A}. \tag{17.8}$$

The mean beam lengths for optically thin media, L_0, are often called *geometric mean beam lengths*, based on the work by Dunkle [15].

Example 17.2. Determine the mean beam lengths of Example 17.1 for an optically thin gas layer.

Solution

From the last example we have

$$\frac{q_\eta}{E_{b\eta}} = 1 - e^{-\kappa_\eta L_e} = 1 - 2E_3(\kappa_\eta L),$$

which, for $\kappa_\eta L \ll 1$, becomes

$$\frac{q_\eta}{E_{b\eta}} = 1 - 1 + \kappa_\eta L_0 = 1 - 2(\tfrac{1}{2} - \kappa_\eta L) = 2\kappa_\eta L,$$

or

$$L_0 = 2L.$$

The mean beam length for the entire surface is, of course, again the same. This could also have been found immediately from equation (17.8) as

$$L_0 = 4\frac{V}{A} = 4\frac{A_{\text{plate}} \times L}{2A_{\text{plate}}} = 2L.$$

Obviously, equation (17.8) is trivial to evaluate for any geometry, but even equations (17.7) (mean beam length to a part of the bounding surface) and (17.5) (mean beam length to a point on the bounding surface) are readily integrated for many configurations.

The geometric mean beam lengths between a gas volume and a bounding surface for a number of configurations, as collected by Hottel and Sarofim [14], have been summarized in Table 17.1.

Spectrally Averaged Mean Beam Lengths

The spectral heat flux, generated by emission from an isothermal volume, that is absorbed by an element of the black bounding surface (or the average heat flux onto a finite area) is given by equation (17.2) as

$$q_\eta = \epsilon_\eta(L_e)\pi I_{b\eta} = (1 - e^{-\kappa_\eta L_e})\pi I_{b\eta},$$

where the mean beam length L_e depends on the spectral absorption coefficient as well as the geometry of the volume. However, Hottel noticed that the spectral heat flux q_η is not very sensitive to the spectral fluctuations of L_e, and that replacing the spectrally varying L_e by an *average mean beam length* L_m (independent of κ_η) predicts spectral heat fluxes with acceptable accuracy. This fact is demonstrated in Fig. 17-2, which shows the ratio of exact and approximate spectral heat fluxes, that is,

$$\frac{q_\eta[\kappa_\eta, L_e = L_e(\kappa_\eta)]}{q_\eta(\kappa_\eta, L_e = L_m = \text{const})} = \frac{1 - e^{-\kappa_\eta L_e}}{1 - e^{-\kappa_\eta L_m}}. \tag{17.9}$$

TABLE 17.1

Mean beam lengths for radiation from a gas volume to a surface on its boundary

Geometry of gas volume	Characterizing dimension L	Geometric mean beam length L_0/L	Average mean beam length L_m/L	L_m/L_0
Sphere radiating to its surface	Diameter, $L = D$	0.67	0.65	0.97
Infinite circular cylinder to bounding surface	Diameter, $L = D$	1.00	0.94	0.94
Semi-infinite circular cylinder to:	Diameter, $L = D$			
Element at center of base		1.00	0.90	0.90
Entire base		0.81	0.65	0.80
Circular cylinder (height/diameter =1) to:	Diameter, $L = D$			
Element at center of base		0.76	0.71	0.92
Entire surface		0.67	0.60	0.90
Circular cylinder (height/diameter =2) to:	Diameter, $L = D$			
Plane base		0.73	0.60	0.82
Concave surface		0.82	0.76	0.93
Entire surface		0.80	0.73	0.91
Circular cylinder (height/diameter =0.5) to:	Diameter, $L = D$			
Plane base		0.48	0.43	0.90
Concave surface		0.53	0.46	0.88
Entire surface		0.50	0.45	0.90
Infinite semicircular cylinder to center of plane rectangular face	Radius, $L = R$		1.26	
Infinite slab to its surface	Slab thickness, L	2.00	1.76	0.88
Cube to a face	Edge L	0.67	0.6	0.90
Rectangular $1 \times 1 \times 4$ parallelepipeds:	Shortest edge, L			
To 1×4 face		0.90	0.82	0.91
To 1×1 face		0.86	0.71	0.83
To all faces		0.89	0.81	0.91

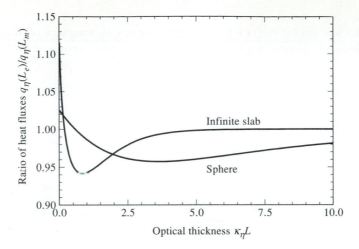

FIGURE 17-2
Ratio of spectral heat flux from isothermal gas volume to surface, with that evaluated using the average mean beam length.

Two different geometries have been considered in Fig. 17-2, namely, an infinite slab radiating to a point on its boundary (or to an entire face), as given by Example 17.1, and a spherical volume radiating to a point on its surface (or to its entire surface). Inspection of Fig. 17-2 shows that the error in the evaluation of the spectral heat flux, if the average mean beam length is used, is never more than ~5% (if a suitable L_m is chosen). This statement may be generalized to other geometries. Values for the average mean beam lengths have also been included in Table 17.1. Inspection of the ratio between average and optically thin mean beam lengths, L_m/L_0, shows that their value is generally in the vicinity of 0.9. Therefore, a value of

$$L_m \simeq 0.9 L_0 = 3.6 \frac{V}{A} \tag{17.10}$$

is recommended for geometries for which values for L_e are not available.

Besides saving computational effort for the evaluation of L_e, employing an average value L_m in equation (17.9) has the tremendous advantage that it allows the straightforward spectral integration of equation (17.9), resulting in a total heat flux of

$$q = \int_0^\infty q_\eta \, d\eta = \int_0^\infty \left(1 - e^{-\kappa_\eta L_m} \right) \pi I_{b\eta}(T) \, d\eta$$

$$= \epsilon(L_m, T) \pi I_b(T) = \epsilon(L_m, T) n^2 \sigma T^4, \tag{17.11}$$

where $\epsilon(L_m, T)$ is the total emissivity of an isothermal gas layer of thickness L_m.

Example 17.3.   Combustion products at $p = 5$ bar, $T = 1000$ K, consisting of 70% N_2, 10% CO_2, and 20% H_2O are contained within a spherical container of radius $R = 75$ cm. Assuming that the container wall is cold and black, estimate the radiative heat flux to the wall.

Solution

The radiative heat flux to the container walls is readily found from equation (17.11), once the total emissivity for the gas mixture has been determined for an average mean beam length of $L_m = 0.65D \simeq 100$ cm, as indicated by Table 17.1. Total emissivities for carbon dioxide–steam mixtures have been discussed in Chapter 9, and the total emissivity for this particular gas mixture for a 100 cm thick layer has already been evaluated in Example 9.10 (surprise!) as $\epsilon = 0.593$. Therefore,

$$q = 0.593 \times 5.670 \times 10^{-12} \times 1000^4 \text{ W/cm}^2 = 3.36 \text{ W/cm}^2.$$

Example 17.4.   A 1 m thick isothermal layer of pure CO_2 at a pressure of 1 bar and a temperature of 1700 K is confined between two parallel, cold, black plates. Estimate the total heat loss from the gas using the mean beam length approach.

Solution

Again, the heat flux to the walls is readily determined from equation (17.11) if the total emissivity for the mean beam length is known, which in this case is $L_m = 1.76L = 176$ cm. The total emissivity for the CO_2 may be determined from Fig. 9-17 or equation (9.95). Using Fig. 9-17, we find ϵ_0 (1 bar, 1 bar, 1700 K) $\simeq 0.14$. The correction factor ϵ/ϵ_0 is determined from equation (9.96) with the correlation constants given in Table 9.5, which leads to $\epsilon/\epsilon_0 \simeq 1.00$ and, therefore, $\epsilon \simeq 0.14$. Substituting this value into equation (17.11), we obtain

$$-q(0) = q(L) = 0.14 \times 5.670 \times 10^{-12} \ 1700^4 \text{ W/cm}^2 = 6.63 \text{ W/cm}^2,$$

and the total heat lost from both sides is $2 \times 6.63 = 13.26$ W/cm^2.

17.3   SEMIGRAY APPROXIMATIONS

It is common practice in engineering to treat properties as "constants," that is, as being independent of one or more dependent variables, primarily to linearize the problem. For example, in heat conduction it is generally assumed that the thermal conductivity is independent of temperature. Very accurate results can be obtained with such an analysis if (*i*) the temperature variation of the material's conductivity is not too strong, and (*ii*) an appropriate, constant "effective conductivity" can be found. It is tempting to use such simplifying assumptions in the calculation of radiative heat fluxes, in particular as far as spectral variations are concerned. A number of researchers, such as Viskanta [16], Finkleman and coworkers [17–19], and Traugott [20, 21], have introduced several different "effective" absorption coefficients and incorporated them into "semigray" schemes, all with limited success.

Consider the volume of a participating medium at a uniform temperature T. If the medium is optically thin (i.e., it emits but does not absorb any of the emitted radiation), the total heat loss from the volume is, according to equation (8.49),

$$Q = 4V \int_0^\infty \kappa_\eta E_{b\eta} \, d\eta, \tag{17.12}$$

or, with the definition of the Planck-mean absorption coefficient, equation (9.100),

$$Q = 4V \kappa_P n^2 \sigma T^4. \tag{17.13}$$

This expression is equivalent to taking the direction-integrated equation of transfer (or conservation of radiative energy), equation (8.55), and integrating it over the entire volume after dropping the self-absorption term. Therefore, for optically thin media, it is reasonable to set

$$\nabla \cdot \mathbf{q} = \int_0^\infty \kappa_\eta (4\pi I_{b\eta} - G_\eta) \, d\eta \simeq \kappa_P (4\pi I_b - G). \tag{17.14}$$

On the other hand, for optically thick media radiative heat flux obeys the diffusion limit or, from equation (13.19) for an isotropically scattering medium,

$$\mathbf{q}_\eta \simeq -\frac{1}{3\beta_\eta} \nabla E_{b\eta}; \tag{17.15}$$

and, using the definition of the Rosseland-mean extinction coefficient, equation (9.104),

$$\mathbf{q} = -\int_0^\infty \frac{1}{3\beta_\eta} \nabla E_{b\eta} \, d\eta = -\frac{1}{3\beta_R} \nabla E_b. \tag{17.16}$$

Apparently, to make accurate calculations using a gray model, the effective absorption coefficient must be close to the Planck-mean for optically thin situations and close to the Rosseland-mean for optically thick cases. A simple (i.e., not dependent on geometry and, through it, optical thickness) average value should only be expected to give accurate results if Planck-mean and Rosseland-mean are of similar value (while in real life they frequently are orders of magnitude apart, in particular for molecular gases).

Replacing E_b in equation (17.16) by incident radiation G gives, together with equation (17.14), a semigray P_1-approximation,

$$\nabla \cdot \mathbf{q} = \kappa_P (4\pi I_b - G), \tag{17.17a}$$
$$\nabla G = -3\beta_R \mathbf{q}. \tag{17.17b}$$

Eliminating $\mathbf{q}$ leads to a single equation for G,

$$\nabla^2 G - 3\beta_R \kappa_P (G - 4\pi I_b) = 0, \tag{17.18}$$

where we have assumed κ_P and β_R to be constant (spatially) for simplicity. Thus, comparing equation (17.18) with (14.38) (setting $A_1 = 0$ for isotropic scattering) leads to an effective absorption coefficient of

$$\kappa_{\text{eff}} \simeq \sqrt{\kappa_P \kappa_R}, \tag{17.19}$$

which is quite commonly employed in gray analyses. However, equation (17.19) should only be used with great caution, (*i*) since accurate answers can be expected only if κ_P / κ_R is close to unity, and (*ii*) since use of the Rosseland-mean absorption coefficient is problematic for pure molecular gases. In the second instance the diffusion limit applies only to optically thick situations, while all molecular gases have transparent regions over large parts of the spectrum.

Consider a gas-particulate mixture, whose absorption coefficient may be written as

$$\kappa_\eta = \kappa_{p\eta} + \kappa_{g\eta} = \kappa_{p\eta} + \sum_{n=1}^{N} \kappa_{n\eta}. \tag{17.20}$$

Here $\kappa_{p\eta}$ is the spectral absorption coefficient of the particles and $\kappa_{g\eta}$ that of the gas, which is composed of N individual vibration-rotation bands, each with its own spectral absorption coefficient $\kappa_{n\eta}$ (for a qualitative picture see Fig. 17-5 in the following section). We shall also assume that the bands are relatively narrow, do not overlap, and may be described by the wide band model of Chapter 9. Then the Planck-mean absorption coefficient may be evaluated as

$$\kappa_P = \frac{1}{E_b} \int_0^\infty \kappa_\eta E_{b\eta} \, d\eta = \kappa_{p,P} + \sum_{n=1}^{N} \frac{1}{E_b} \int_{\text{band } n} \kappa_{n\eta} E_{b\eta} \, d\eta$$

$$\simeq \kappa_{p,P} + \sum_{n=1}^{N} \frac{E_{b\eta_n}}{E_b} \int_{\text{band } n} \kappa_{n\eta} \, d\eta = \kappa_{p,P} + \sum_{n=1}^{N} \frac{E_{b\eta_n}}{E_b} \alpha_n, \tag{17.21}$$

where α_n is the band strength parameter and $E_{b\eta_n}$ is the spectral, blackbody emissive power at the band center, both for band n. The Rosseland-mean absorption coefficient may be evaluated similarly as

$$\frac{1}{\kappa_R} = \int_0^\infty \frac{1}{\kappa_\eta} \frac{dE_{b\eta}}{dE_b} d\eta = \frac{1}{\kappa_{p,R}} - \int_0^\infty \left(\frac{1}{\kappa_{p\eta}} - \frac{1}{\kappa_\eta} \right) \frac{dE_{b\eta}}{dE_b} d\eta$$

$$\simeq \frac{1}{\kappa_{p,R}} - \sum_{n=1}^{N} \left(\frac{dE_{b\eta}}{dE_b} \right)_{\eta_n} \int_{\text{band } n} \left(\frac{1}{\kappa_{p\eta_n}} - \frac{1}{\kappa_{p\eta_n} + \kappa_{n\eta}} \right) d\eta, \tag{17.22}$$

where $\kappa_{p\eta_n}$ is a constant average value, assuming that $\kappa_{p\eta}$ does not vary greatly across each band. We shall also assume that, inside the integral, $\kappa_{n\eta}$ may be replaced by the narrow band average, $(S/d)_\eta$, for which the wide band model stipulates [cf. equation (9.66)]

$$\left(\frac{S}{d} \right)_\eta \simeq \frac{\alpha_n}{\omega_n} e^{-t|\eta - \eta_n|/\omega_n}, \tag{17.23}$$

where ω_n is the band width parameter and $t = 1$ for a band with head, and $t = 2$ for a symmetric band. Substituting this expression into equation (17.22) leads to

$$
\int_{\text{band } n} \left(\frac{1}{\kappa_{p\eta_n}} - \frac{1}{\kappa_{p\eta_n} + \kappa_{n\eta}} \right) d\eta = \frac{1}{\kappa_{p\eta_n}} \int_0^\infty \frac{(S/d)_\eta}{\kappa_{p\eta_n} + (S/d)_\eta} d\eta
$$

$$
= \frac{\alpha_n}{\kappa_{p\eta_n}} \int_0^\infty \frac{e^{-x} dx}{\kappa_{p\eta_n} + (\alpha_n/\omega_n)e^{-x}}
$$

$$
= \frac{\omega_n}{\kappa_{p\eta_n}} \ln\left(1 + \frac{\alpha_n}{\omega_n \kappa_{p\eta_n}} \right), \qquad (17.24)
$$

regardless of the value of t (cf. the development of Example 9.3). Thus,

$$
\frac{1}{\kappa_R} = \frac{1}{\kappa_{p,R}} - \sum_{n=1}^N \frac{\omega_n}{\kappa_{p\eta_n}} \left(\frac{dE_{b\eta}}{dE_b} \right)_{\eta_n} \ln\left(1 + \frac{\alpha_n}{\omega_n \kappa_{p\eta_n}} \right). \qquad (17.25)
$$

It is evident from equation (17.25) that for a pure molecular gas, $\kappa_R \to 0$. This statement is also true if the assumption of a narrow band is relaxed: It is readily observed that $1/\kappa_{n\eta}$ tends toward infinity faster than $E_{b\eta}$ tends to zero for both $\eta \to 0$ and $\eta \to \infty$. Clearly, for a pure molecular gas $\kappa_P/\kappa_R \to \infty$, so that (*i*) κ_R and equation (17.19) are not suitable for the determination of κ_{eff}, and (*ii*) accurate predictions should not be expected from the semigray approach.

Example 17.5.   A molecular gas is confined between two parallel, black plates, spaced 1 m apart, which are kept isothermal at $T_1 = 1200\,\text{K}$ and $T_2 = 800\,\text{K}$, respectively. The (hypothetical) gas has a single vibration-rotation band in the infrared, with an average absorption coefficient of

$$
\left(\frac{S}{d} \right)_\eta = \frac{\alpha}{\omega} e^{-2|\eta - \eta_0|/\omega}, \qquad \eta_0 = 3000\,\text{cm}^{-1}, \qquad \omega = 200\,\text{cm}^{-1},
$$

and an overlap parameter of β (see the discussion of narrow band and wide band models in Chapter 9). Assuming convection and conduction to be negligible, estimate the radiative heat flux between the two plates using the semigray model. Carry out the analysis for variable values of (α/ω) and β. Repeat the calculations for the same gas mixed with nonscattering particles whose absorption coefficient is $\kappa_p = 0.1\,\text{m}^{-1}$ (gray).

Solution

To make an "equivalent" gray analysis, a suitable gray absorption coefficient must be found. Since for a pure molecular gas the Rosseland-mean is inappropriate, and for want of any better value, we choose the Planck-mean absorption coefficient, which leads to

$$
\tau_P = \kappa_P L = \alpha \frac{E_{b\eta_0}}{\sigma T^4} L = \tau_L \frac{\omega E_{b\eta_0}(T)}{\sigma T^4}, \qquad \tau_L = \left(\frac{\alpha}{\omega} \right) L.
$$

Consequently, τ_P depends on the local temperature of the gas, even if $(\alpha/\omega) = \text{const}$. To simplify the analysis we use a constant Planck-mean absorption coefficient evaluated at some average temperature, say $T_{av} = 1000\,\text{K}$. Thus, $\eta_0/T_{av} = 3\,\text{cm}^{-1}/\text{K}$ and, from Appendix C,

$$\tau_P = \frac{200 \times 1.36576 \times 10^{-8}}{5.670 \times 10^{-8} \times 1000} \times \tau_L = 0.0482\,\tau_L.$$

For a gray medium the radiative heat flux between the two plates is determined from Example 13.5 for the P_1-approximation as

$$\Psi_{\text{gray}} = \frac{1}{1 + \frac{3}{4}\tau_P} = \frac{1}{1 + 0.0362\,\tau_L}.$$

If a particle background is present, it is better to utilize κ_{eff} from equation (17.19). Thus, with $\tau_p = \kappa_p L = 0.1 \times 1 = 0.1$,

$$\tau_P = 0.1 + 0.0482\,\tau_L,$$

$$\frac{1}{\tau_R} = \frac{1}{\tau_p} - \frac{1}{\tau_p}\left(\omega\frac{dE_{b\eta}}{dE_b}\right)_{\eta_n}\ln\left(1 + \frac{\tau_L}{\tau_p}\right).$$

From equation (1.14) it follows that

$$\omega\frac{dE_{b\eta}}{dE_b} = \frac{\omega}{4\sigma T^3}\frac{dE_{b\eta}}{dT} = \frac{\omega}{4\sigma T^3}\frac{d}{dT}\left[\frac{C_1\eta^3}{\exp(C_2\eta/T) - 1}\right]$$

$$= \frac{\omega C_1\eta^3 \exp(C_2\eta/T)\,C_2\eta/T^2}{4\sigma T^3\,[\exp(C_2\eta/T) - 1]^2}$$

$$= \frac{1}{4}\frac{\omega E_{b\eta}}{E_b}\frac{(C_2\eta/T)\exp(C_2\eta/T)}{\exp(C_2\eta/T) - 1},$$

$$\left(\omega\frac{dE_{b\eta}}{dE_b}\right)_{\eta_n} = \frac{1}{4} \times 0.0482 \times \frac{4.3164\,e^{4.3164}}{e^{4.3164} - 1} = 0.0527,$$

$$\tau_R = \frac{0.1}{1 - 0.0527\,\ln(1 + 10\,\tau_L)}.$$

τ_P and τ_R may be calculated for any τ_L, and the heat flux becomes

$$\Psi = \frac{1}{1 + \frac{3}{4}\sqrt{\tau_P\,\tau_R}}.$$

Representative results are shown in Fig. 17-3, together with results obtained with the Monte Carlo method [22]. Clearly, for a pure molecular gas the semigray approximation fails miserably, since the Planck-mean is much too large to be a good effective absorption coefficient for optically thick bands. With a particle background the method performs considerably better, with κ_P/κ_R ranging in value between 1 (for $\tau_L = 0$) and 31 ($\tau_L = 100$). The semigray approach cannot account for spectral windows, nor for line structure (line overlap parameter β): If there is little line overlap (small β) radiation can travel unimpeded through "mini-windows" between strong spectral lines. For a gray gas the heat flux must always tend to zero for optically thick gases.

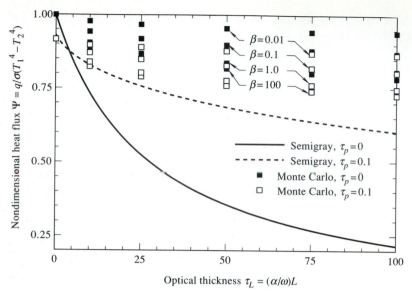

FIGURE 17-3
Nondimensional radiative heat flux for radiative equilibrium across a molecular gas-particulate layer bounded by parallel, black walls, calculated by the semigray method.

17.4   THE STEPWISE-GRAY MODEL (PICKET FENCE)

Another simple way to incorporate the effects of absorption-emission bands of molecular gases in radiative heat transfer calculations is to approximate the band absorptances through the box model described in Section 9.6. In this model the spectral absorption coefficient for a molecular gas with N vibration-rotation bands is approximated (see Fig. 17-4) as

$$\kappa_\eta \simeq \sum_{n=1}^{N} \kappa_n \left[H(\eta - \eta_n + \tfrac{1}{2}\Delta\eta_n) - H(\eta - \eta_n - \tfrac{1}{2}\Delta\eta_n) \right], \tag{17.26}$$

where η_n is the wavenumber at the band center, $\Delta\eta_n$ is the band width and κ_n is the absorption coefficient of the nth band (assumed constant for each band). Finally, the function $H(x)$ is *Heaviside's unit step function*,

$$H(x) = \begin{cases} 0, & x < 0, \\ 1, & x > 0. \end{cases} \tag{17.27}$$

If the molecular gas is accompanied by absorbing and/or scattering particles (e.g., soot or ash particles), the absorption coefficient of equation (17.26) must be augmented

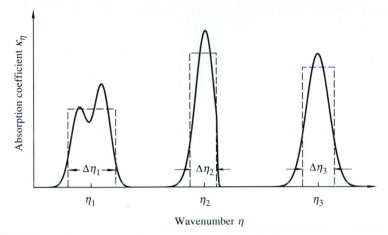

FIGURE 17-4
The picket fence model: Box model approximation of molecular gas bands.

by the extinction coefficient of the background (see Fig. 17-5):

$$\beta_\eta = \beta_{p\eta} + \kappa_{g\eta} = \sum_{m=1}^{M} \beta_m \left[H(\eta - \eta'_{m-1}) - H(\eta - \eta'_m) \right],$$ (17.28a)

where

$$\eta'_m = \begin{cases} 0, & m = 0 \\ \eta_m + \frac{1}{2}\Delta\eta_m, & 1 < m < M, \\ \infty, & m = M. \end{cases}$$ (17.28b)

Here we have broken up the spectrum into N gas bands and $M - N$ "spectral windows." Equation (17.28) may also be employed in the modeling of semitransparent media. If the background material can be approximated as gray, equation (17.28) reduces to

$$\beta_\eta = \beta_p + \sum_{n=1}^{N} \kappa_n \left[H(\eta - \eta_n + \frac{1}{2}\Delta\eta_n) - H(\eta - \eta_n - \frac{1}{2}\Delta\eta_n) \right].$$ (17.29)

A number of researchers have used various forms of the picket fence model to solve nongray radiation problems. Originally proposed by Chandrasekhar [23], the model has primarily been applied to one-dimensional plane media at radiative equilibrium, for example by Siewert and Zweifel [24], Kung and Sibulkin [25], and Reith, Siewert and Özişik [26]. Greif [27] applied the method to combined conduction/radiation in a plane layer. Modest [28] showed that for gases with a single band strength (see Fig. 17-6) the picket fence approach can be incorporated into the P_1-approximation, making multidimensional calculations possible. This was extended to the general picket fence model by Modest and Sikka [22], and a consistent method for the determination of box model parameters was given.

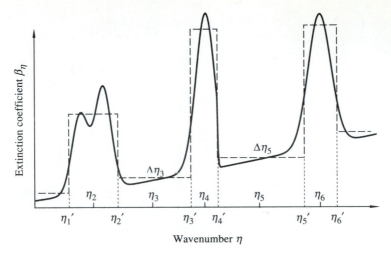

FIGURE 17-5
The picket fence model: Molecular gases mixed with suspended particles.

Comparing the picket fence model with the mean beam length method, we see that the mean beam length method can model the spectral variations of the absorption coefficient very well, but it is limited to isothermal, black-walled enclosures with nonscattering media. The picket fence model, on the other hand, can handle non-isothermal, scattering media bounded by nonblack walls, while its spectral modeling is rather crude.

How well the picket fence model predicts radiative heat fluxes (or their divergence) for nongray media largely depends on how well "optimum" box parameters are determined for a given medium. To find appropriate values for these parameters, one must realize that the exact integral relationships that govern radiative heat transfer in a participating medium [see, for example, equation (8.30)] contain the spectral absorption coefficient in the form of the spectral emissivity (or its derivative)

$$\epsilon_\eta = 1 - e^{-\kappa_\eta X}. \tag{17.30}$$

Here $X = L$ if the linear absorption coefficient is used, or L multiplied by the partial density or pressure of the absorbing gas if either mass or pressure absorption coefficient is used; and L is the geometric path length over which absorption/emission is being considered. Spectrally integrated heat fluxes (or their divergence), therefore, depend strongly on the total band absorptances of the medium,

$$A(X) = \int_{\text{band}} \epsilon_\eta \, d\eta = \int_{\text{band}} (1 - e^{-\kappa_\eta X}) \, d\eta. \tag{17.31}$$

Thus, the aim of the box model must be to approximate the total band absorptance as well as possible for all possible conditions. For a gas without particle background, as shown in Fig. 17-4, we have for band n,

$$A_n(X) \simeq \Delta\eta_n(1 - e^{-\kappa_n X}), \tag{17.32}$$

where $\Delta\eta_n$ and κ_n are two parameters that may have arbitrary dependence on all gas conditions (pressure, temperature, line overlap, etc.), *except* path length X. Consequently, $A_n(X)$ as calculated by equation (17.32) ranges in value between 0 and $\Delta\eta_n$ and can coincide with the exact value of $A_n(X)$ (as discussed in Chapter 9) for precisely two values of X. It is this restriction that limits the accuracy of the picket fence model, since the "exact" value of the total band absorptance increases as $\ln(\kappa_\eta X)$ for optically thick conditions (cf. Table 9.2). Modest and Sikka [22] found that best results are obtained by choosing $\Delta\eta_n$ and κ_n in such a way that equation (17.32) predicts the correct band absorptance for optically thin situations (X small) and for a characteristic length X_m based on the mean beam length L_m (as listed in Table 17.1). In the optically thin limit we have

$$X \ll X_m : \quad A_n(x) = \int_{\text{band}} \kappa_\eta X \, d\eta = \alpha_n X = \kappa_n \Delta\eta_n X , \tag{17.33}$$

where α_n is the integrated absorption coefficient, or

$$\alpha_n \simeq \kappa_n \Delta\eta_n . \tag{17.34}$$

At the mean beam length we have

$$X = X_m : \quad A_n(X_m) = \Delta\eta_n(1 - e^{-\kappa_n X_m}), \tag{17.35}$$

where $A_n(X_m)$ must be evaluated from any appropriate wide band model. Equations (17.34) and (17.35) constitute a set of two equations for the unknowns κ_n and $\Delta\eta_n$, which are readily solved, especially if $\kappa_n X_m \gg 1$ (which will be the case for most important bands).

Radiative Equilibrium in a Plane Gas Layer with Single Band Strength

As a first simple case we consider radiative equilibrium in a one-dimensional plane-parallel layer of molecular gases confined between two isothermal, gray-diffuse plates. The medium does not scatter and the absorption coefficient obeys equation (17.26). As a further simplification we assume that all bands are of equal strength, as shown in Fig. 17-6, that is,

$$\kappa_1 = \kappa_2 = \cdots = \kappa_N = \overline{\kappa}, \tag{17.36}$$

and that the band width $\Delta\eta_n$ does not vary with location or temperature. For this simple case the spectral values for incident radiation, G_η, and radiative heat flux, q_η, are readily found from equations (12.34) and (12.35) as

$$G_\eta(\tau_\eta) = 2\Big\{ J_{1\eta}E_2(\tau_\eta) + J_{2\eta}E_2(\tau_{L\eta} - \tau_\eta)$$
$$+ \int_0^{\tau_L} E_{b\eta}(\tau_\eta')E_1(|\tau_\eta - \tau_\eta'|) \, d\tau_\eta' \Big\}, \tag{17.37}$$

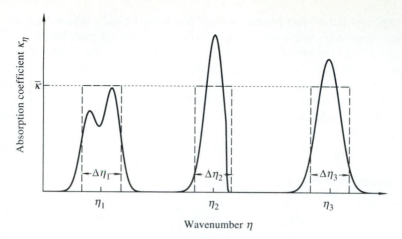

FIGURE 17-6
The picket fence model: Box models for molecular gases with single band strength.

$$q_\eta(\tau_\eta) = 2\left\{ J_{1\eta}E_3(\tau_\eta) - J_{2\eta}E_3(\tau_{L\eta} - \tau_\eta) \right.$$

$$\left. + \int_0^{\tau_\eta} E_{b\eta}(\tau_\eta')E_2(\tau_\eta - \tau_\eta')\,d\tau_\eta' - \int_{\tau_\eta}^{\tau_{L\eta}} E_{b\eta}(\tau_\eta')E_2(\tau_\eta' - \tau_\eta)\,d\tau_\eta' \right\}, \quad (17.38)$$

where

$$\tau_\eta = \int_0^z \overline{\kappa}\,dz \sum_{n=1}^N \left[H(\eta - \eta_n + \tfrac{1}{2}\Delta\eta_n) - H(\eta - \eta_n - \tfrac{1}{2}\Delta\eta_n) \right]$$

$$= \begin{cases} \overline{\tau}, & \text{within bands,} \\ 0, & \text{within windows.} \end{cases} \quad (17.39)$$

Integrating equations (17.37) and (17.38) over all *bands* (excluding windows) results in

$$G_B(\overline{\tau}) = 2\left\{ J_{B1}E_2(\overline{\tau}) + J_{B2}E_2(\overline{\tau}_L - \overline{\tau}) \right.$$

$$\left. + \int_0^{\overline{\tau}_L} E_B(\overline{\tau}')E_1(|\overline{\tau} - \overline{\tau}'|)\,d\overline{\tau}' \right\}, \quad (17.40)$$

$$q_B(\overline{\tau}) = 2\left\{ J_{B1}E_3(\overline{\tau}) - J_{B2}E_3(\overline{\tau}_L - \overline{\tau}) \right.$$

$$\left. + \int_0^{\overline{\tau}} E_B(\overline{\tau}')E_2(\overline{\tau} - \overline{\tau}')\,d\overline{\tau}' - \int_{\overline{\tau}}^{\overline{\tau}_L} E_B(\overline{\tau}')E_2(\overline{\tau}' - \overline{\tau})\,d\overline{\tau}' \right\}, \quad (17.41)$$

where the subscript B denotes a quantity integrated over all bands, for example,

$$G_B = \sum_{n=1}^{N} \int_{\eta_n - \frac{1}{2}\Delta\eta_n}^{\eta + \frac{1}{2}\Delta\eta_n} G_\eta \, d\eta. \tag{17.42}$$

On the other hand, integrating the equation of transfer over the *entire spectrum* (including windows) gives, from equation (8.55),

$$\nabla \cdot \mathbf{q} = \frac{dq}{dz} = \int_0^\infty \kappa_\eta (4E_{b\eta} - G_\eta) \, d\eta = \overline{\kappa}(4E_B - G_B) = 0, \tag{17.43}$$

where the zero is due to the fact that radiative equilibrium prevails. Since integrating the equation of transfer over the bands only (excluding windows) would have resulted in the identical right-hand side, we conclude that $dq_B/dz = 0$ or $q_B = \text{const}$, and $G_B = 4E_B$. Thus, equations (17.40) and (17.41) are identical to the gray case, equations (12.34) and (12.35), after replacing total values by band-integrated values, or

$$\Phi_B = \frac{E_B(\overline{\tau}) - J_{B2}}{J_{B1} - J_{B2}} = \frac{1}{2}\left[E_2(\overline{\tau}) + \int_0^{\overline{\tau}_L} \Phi_B(\overline{\tau}')E_1(|\overline{\tau} - \overline{\tau}'|)\, d\overline{\tau}'\right], \tag{17.44}$$

$$\Psi_B = \frac{q_B}{J_{B1} - J_{B2}} = 1 - 2\int_0^{\overline{\tau}_L} \Phi_B(\overline{\tau}')E_2(\overline{\tau}')\, d\overline{\tau}'. \tag{17.45}$$

The total heat flux between the plates is then determined by adding to this the heat flux over the spectral windows

$$q_W = \int_{\text{windows}} (J_{1\eta} - J_{2\eta})\, d\eta = J_1 - J_2 - (J_{B1} - J_{B2}), \tag{17.46}$$

or

$$\Psi = \frac{q}{J_1 - J_2} = \frac{q_W + q_B}{J_1 - J_2} = 1 - \frac{J_{B1} - J_{B2}}{J_1 - J_2}(1 - \Psi_B). \tag{17.47}$$

We note that, for a gray medium, Ψ varies between 1 (vacuum) and 0 (opaque medium), while the minimum heat flux for a nongray medium is

$$\Psi_{\text{min}} = 1 - \frac{J_{B1} - J_{B2}}{J_1 - J_2}, \tag{17.48}$$

since radiation will travel unimpeded from surface to surface over spectral windows, even if the bands are opaque.

Example 17.6.   Pure CO_2 at 1 bar $= 100\,\text{kPa}$ pressure is confined between two parallel, black plates, spaced 1 m apart, which are kept isothermal at $T_1 = 1000\,\text{K}$ and $T_2 = 2000\,\text{K}$, respectively. Assuming conduction and convection to be negligible, estimate the radiative heat flux between the two plates.

Solution

From Table 9.3 we find that carbon dioxide has three important bands in the infrared: At $3660 \, \text{cm}^{-1}$ ($2.7 \, \mu\text{m}$), $2410 \, \text{cm}^{-1}$ ($4.3 \, \mu\text{m}$), and $667 \, \text{cm}^{-1}$ ($15 \, \mu\text{m}$). Since the walls are at 1000 and 2000 K, respectively, the most important wavelength regime will be between approximately $1 \, \mu\text{m}$ and $4 \, \mu\text{m}$ (Wien's displacement law, Chapter 1). At first glance it would appear that the $2.7 \, \mu\text{m}$ band is the most important. However, the $4.3 \, \mu\text{m}$ band is approximately 20 times stronger and, therefore, should be modeled most accurately. Before we can employ the picket fence model we must find suitable box model parameters for these bands. We shall do this by comparing the band absorptances of the box model with those of the (more accurate) exponential wide band model [22], by applying equations (17.34) and (17.35). Since we should like to use a single $\overline{\kappa}$ for all bands we cannot use all of these conditions, and shall apply equation (17.35) only for the (most important) $4.3 \, \mu\text{m}$ band. Thus,

$$\overline{\kappa} \Delta \eta_i = \alpha_i, \quad \text{for all three bands,}$$

$$\Delta \eta_{4.3} = A_{4.3}(X_m), \quad \text{for the } 4.3 \, \mu\text{m band.}$$

In the last relation we have assumed $\overline{\kappa} X_m \gg 1$, which needs to be verified.

To simplify the analysis further we shall calculate the box model parameters at a single temperature, say $T_{av}^4 = (T_1^4 + T_2^4)/2$ or $T_{av} \simeq 1700 \, \text{K}$. With a gas constant of $R = 0.18892 \, \text{kJ/kg K}$ for CO_2 [29] and the data in Table 9.3, and Fig. 9-13 we get

$$\rho_{CO_2} = p/RT = 100 \, \text{kPa}/(0.18892 \times 1700 \, \text{kJ/kg}) = 311.4 \, \text{g/m}^3,$$

$2.7 \, \mu\text{m}$ band:

$$\alpha \rho_{CO_2} = \alpha_0 (\alpha/\alpha_0)\rho = 4.0 \times 1.61 \times 311.4 = 2005 \, \text{cm}^{-1}/\text{m},$$

$$\omega = 23.5 \times \sqrt{17} = 96.9 \, \text{cm}^{-1},$$

$4.3 \, \mu\text{m}$ band:

$$\alpha \rho_{CO_2} = 110.0 \times 1 \times 311.4 = 34{,}254 \, \text{cm}^{-1}/\text{m},$$

$$\omega = 11.2 \times \sqrt{17} = 46.2 \, \text{cm}^{-1},$$

$$\gamma = \gamma_0(\gamma/\gamma_0) = 0.24723 \times 24.7 = 6.14,$$

$$P_e = 1.3^{0.8} = 1.234, \quad \beta = 6.14 \times 1.234 = 7.57,$$

$15 \, \mu\text{m}$ band:

$$\alpha \rho_{CO_2} = 19.0 \times 1 \times 311.4 = 5917 \, \text{cm}^{-1}/\text{m},$$

$$\omega = 12.7 \times \sqrt{17} = 52.4 \, \text{cm}^{-1}.$$

With an average mean beam length of $L_m = 1.76 \times 1 \, \text{m} = 1.76 \, \text{m}$, we obtain for the important $4.3 \, \mu\text{m}$ band from Table 9.3

$$\Delta \eta_{4.3} = A_{4.3} = \omega_{4.3} \left[\ln \frac{\alpha_{4.3} \rho_{CO_2} L_m}{\omega_{4.3}} + 1 \right]$$

$$= 46.2 \, \text{cm}^{-1} \left[\ln \frac{34{,}254 \times 1.76}{46.2} + 1 \right] = 377.6 \, \text{cm}^{-1},$$

and

$$\overline{\kappa} = (\alpha \rho_{CO_2}/\Delta \eta)_{4.3} = 34{,}254/377.6 = 90.7 \, \text{m}^{-1} = 0.907 \, \text{cm}^{-1}.$$

Noting that $\bar{\kappa}$ is a *linear* absorption coefficient, and multiplying with L_m, we find $\bar{\kappa}L_m = 0.907 \times 176 = 160 \gg 1$, so neglecting the exponential in equation (17.35) was indeed justified.

The widths of the other two bands follow from the same relationship as

$$\Delta\eta_{2.7} = \alpha_{2.7}\rho_{CO_2}/\bar{\kappa} = 2005/90.7 = 22.1\,\text{cm}^{-1},$$
$$\Delta\eta_{15} = \alpha_{15}\rho_{CO_2}/\bar{\kappa} = 5917/90.7 = 65.2\,\text{cm}^{-1}.$$

We are now in a position to calculate the nondimensional heat flux between the plates from equation (17.47) with $\bar{\tau} = \bar{\kappa}L = 90.7$. $\Psi_B(\bar{\tau})$ obeys the same equation as $\Psi_b(\tau)$ in Chapter 12 and, thus, may be evaluated from Table 12.1. Since the bands are essentially opaque we find $\Psi_B = 0.015 \ll 1$, and

$$\Psi = 1 - 0.985\frac{E_{B1} - E_{B2}}{E_{b1} - E_{b2}} \simeq 1 - \frac{0.985}{E_{b1} - E_{b2}}\sum_{n=1}^{3}(E_{b\eta_n,1} - E_{b\eta_n,2})\Delta\eta_n,$$

where radiosity is replaced by emissive power for the black-walled enclosure, and we assume that the bands are narrow (to justify evaluation of E_B as value at band center $\times$ band width). To look up values for spectral emissive power in Appendix C, we write with $\eta_{2.7} = 3660\,\text{cm}^{-1}$, $\eta_{4.3} \simeq 10^4/4.3 = 2326\,\text{cm}^{-1}$ (since the 4.3 μm band is a band with head, it is better to evaluate $E_{b\eta}$ near the center of the band), and $\eta_{15} = 667\,\text{cm}^{-1}$,

$$\Psi = 1 - \frac{0.985}{\sigma(T_1^4 - T_2^4)}\sum_{n=1}^{3}\left[\frac{E_{b\eta}(\eta_n/T_1)}{T_1^3}T_1^3 - \frac{E_{b\eta}(\eta_n/T_2)}{T_2^3}T_2^3\right]\Delta\eta_n$$

$$= 1 - \frac{0.985}{5.670 \times 10^{-8}(1000^4 - 2000^4)}$$
$$\times \Big[\big(0.9523 \times 10^{-8+9} - 1.7747 \times 2^3 \times 10\big) \times 22.1$$
$$+ \big(1.7157 \times 10 - 1.3589 \times 2^3 \times 10\big) \times 377.6$$
$$+ \big(0.6885 \times 10 - 0.2251 \times 2^3 \times 10\big) \times 65.2\Big]$$
$$= 0.956.$$

Thus, the heat flux is reduced only by 4.7%, as compared with the no-gas case, to

$$q = 0.956 \times 5.670 \times 10^{-8}(2000^4 - 1000^4) = 813,000\,\text{W/m}^2 = 81.3\,\text{W/cm}^2.$$

Note that, since all bands are essentially opaque, the (somewhat arbitrary) choice for band widths is of extreme importance in this model.

The P_1-Approximation for Radiative Equilibrium in a Gas with Single Band Strength

The method discussed in the previous section enables us to calculate heat transfer rates for a nongray medium at radiative equilibrium in a simple way. Unfortunately, the method is limited to one-dimensional plane-parallel media. Following the treatment of Modest [28], we shall now show that the same picket-fence absorption coefficient may also be applied to the P_1 or differential approximation, making solutions for arbitrary multidimensional geometries possible.

The governing equations for the P_1-approximation, on a spectral basis and for a nonscattering medium, are, from Section 14.4

$$\nabla \cdot \mathbf{q}_\eta = \kappa_\eta(4E_{b\eta} - G_\eta),$$ (17.49)

$$\nabla G_\eta = -3\kappa_\eta \mathbf{q}_\eta,$$ (17.50)

subject to the boundary condition

$$2\mathbf{q}_\eta \cdot \hat{\mathbf{n}} = 4J_{w\eta} - G_\eta.$$ (17.51)

Assuming again that the absorption coefficient may be approximated by equation (17.36), we integrate equation (17.49) over the entire spectrum, and over all bands only, resulting in

$$\nabla \cdot \mathbf{q} = \nabla \cdot \mathbf{q}_B = \overline{\kappa}(4E_B - G_B) = 0,$$ (17.52)

where the zero is again due to the fact that radiative equilibrium is assumed. Now, integrating equations (17.50) and (17.51) over all bands gives

$$\nabla G_B = -3\overline{\kappa}\mathbf{q}_B,$$ (17.53)

$$2\mathbf{q}_B \cdot \hat{\mathbf{n}} = 4J_{Bw} - G_B.$$ (17.54)

The heat flux may be eliminated from these equations, leading to a single elliptic equation in G_B as

$$\nabla_{\overline{\tau}}^2 G_B = 0,$$ (17.55)

with boundary condition

$$-\tfrac{2}{3}\hat{\mathbf{n}} \cdot \nabla_{\overline{\tau}} G_B + G_B = 4J_{Bw},$$ (17.56)

where the subscript $\overline{\tau}$ indicates that the gradients are with respect to optical coordinates $d\overline{\tau} = \overline{\kappa}ds$. Once the band-integrated incident radiation, G_B, has been determined, the heat flux for the gas bands follows as

$$\mathbf{q}_B = -\frac{1}{3}\nabla_{\overline{\tau}} G_B.$$ (17.57)

Finally, to calculate total heat transfer rates, the heat fluxes through the optical windows, $\mathbf{q}_W$, must be determined independently through standard methods (Chapters 5 and 6), as indicated in the previous section for one-dimensional, plane-parallel media, equation (17.46).

Example 17.7.   Repeat Example 17.6 using the differential approximation.

Solution

Since we use the same box model as in the previous example to approximate the absorption coefficient, we shall again use $\overline{\kappa} = 0.907\,\text{cm}^{-1}$, $\Delta\eta_{2.7} = 22.1\,\text{cm}^{-1}$, $\Delta\eta_{4.3} = 377.6\,\text{cm}^{-1}$, and $\Delta\eta_{15} = 65.2\,\text{cm}^{-1}$. Equations (17.52) through (17.54) or equations (17.55) and (17.56) are identical to the general P_1-approximation for gray media (except for the added subscript B). Thus, for radiative equilibrium in a one-dimensional plane-parallel medium

$$\Psi_B = \frac{1}{1 + \frac{3}{4}\overline{\tau}} = 0.015.$$

The heat flux over the spectral windows is, of course, the same as calculated for the previous example, as is the expression for total heat flux, equation (17.47). We conclude that the heat flux evaluation using the P_1-approximation gives the identical result as the exact[1] method.

The P_1-Approximation for the General Picket Fence Model

Both the exact method for one-dimensional plane-parallel media and the P_1-approximation for general geometries are readily extended to the general picket fence model in which the extinction coefficient is approximated by equation (17.28), although we shall limit our discussion to the P_1-approximation, following the development of Modest and Sikka [22]. Further limiting ourselves to an absorbing, emitting, and linear anisotropically scattering medium, the governing equations, on a spectral basis, for the P_1-approximation are again equations (17.49) through (17.51), except that equation (17.50) needs to be augmented by a scattering term, or

$$\nabla G_\eta = -[3\kappa_\eta + (3 - A_{1\eta})\sigma_{s\eta}]\,\mathbf{q}_\eta. \tag{17.58}$$

Although not necessary we shall, for simplicity, assume that the absorption coefficient obeys equation (17.29), i.e., we have a gas with N vibration-rotation bands and a gray particle background. Integrating equations (17.49), (17.58), and (17.51) over all wavenumbers outside the N bands (windows), we obtain

$$\nabla \cdot \mathbf{q}_W = \kappa_p (4E_W - G_W), \tag{17.59}$$

$$\nabla G_W = -[3\kappa_p + (3 - A_1)\sigma_s]\,\mathbf{q}_W, \tag{17.60}$$

$$2\mathbf{q}_W \cdot \hat{\mathbf{n}} = 4J_{Ww} - G_W. \tag{17.61}$$

Similarly, for each gas band we get

$$\nabla \cdot \mathbf{q}_n = (\kappa_p + \kappa_n)(4E_n - G_n), \tag{17.62}$$

$$\nabla G_n = -[3(\kappa_p + \kappa_n) + (3 - A_1)\sigma_s]\,\mathbf{q}_n, \tag{17.63}$$

$$2\mathbf{q}_n \cdot \hat{\mathbf{n}} = 4J_{nw} - G_n, \qquad n = 1, 2, \cdots, N. \tag{17.64}$$

---

[1] Exact calculation of radiative heat flux based on very approximate expressions for the spectral variation of the absorption coefficient.

Here the subscript W indicates integration over the entire spectrum outside gas bands, and subscript n indicates spectral integration over band n. If the temperature field is known, equations (17.59) through (17.64) constitute a set of $N + 1$ elliptic partial differential equations in the unknowns G_W and G_n ($n = 1, 2, \cdots, N$). If the temperature field is not known (e.g., if radiative equilibrium prevails), a temperature distribution must be assumed and must be determined through iteration. Total quantities are found by adding windows and bands, or

$$E_b = E_W + \sum_{n=1}^{N} E_n, \tag{17.65}$$

$$q = q_W + \sum_{n=1}^{N} q_n, \tag{17.66}$$

$$G = G_W + \sum_{n=1}^{N} G_n. \tag{17.67}$$

Example 17.8.   Consider a 1 m thick isothermal layer of pure CO_2 at a pressure of 100 kPa and a temperature of 1700 K. The gas contains a certain amount of (nonscattering) soot that adds a gray absorption coefficient of $\kappa_p = 1\,m^{-1}$ to that of the gas. The gas is confined between two parallel, cold, black plates. Determine the total heat loss from the gas-soot mixture.

Solution

From Example 17.6 we already know that CO_2 has three important bands in the infrared, at 2.7 μm, 4.3 μm and 15 μm. In that example, we did our calculations for a temperature of 1700 K and a pressure of 100 kPa; thus, those results are directly applicable here. However, for added accuracy, we would like to calculate individual κ_n for each band [by applying equation (17.35)], rather than choosing a single $\bar{\kappa}$. We obtain:

2.7 μm band:

$$\gamma = \gamma_0(\gamma/\gamma_0) = 0.1334 \times 34.6 = 4.616,$$

$$P_e = 1.3^{0.65} = 1.186, \ \beta = 4.616 \times 1.186 = 5.47,$$

$$\Delta\eta_{2.7} = \omega\left[\ln\left(\frac{\alpha\rho_{CO_2}}{\omega}L_m\right)+1\right] = 96.9\left[\ln\frac{2005 \times 1.76}{96.9}+1\right] = 445.3\,cm^{-1},$$

$$\kappa_{2.7} = \alpha\rho_{CO_2}/\Delta\eta = 2005/445.3 = 4.5\,m^{-1} = 0.045\,cm^{-1};$$

4.3 μm band:

$$\Delta\eta_{4.3} = 377.6\,cm^{-1},$$

$$\kappa_{4.3} = 0.907\,cm^{-1};$$

15 μm band:

$$\gamma = 0.06157 \times 24.6 = 1.515,$$

$$P_e = 1.3^{0.7} = 1.202, \ \beta = 1.515 \times 1.202 = 1.82$$

$$\Delta\eta_{15} = 52.4\left[\ln\frac{5917 \times 1.76}{52.4}+1\right] = 329.7\,cm^{-1},$$

$$\kappa_{15} = 5917/329.7 = 17.9\,m^{-1} = 0.179\,cm^{-1}$$

The solution to the present problem for a gray medium has already been given in Example 14.2 as

$$-q(0) = q(L) = \frac{2E_b \sinh \frac{1}{2}\sqrt{3}\tau_L}{\sinh \frac{1}{2}\sqrt{3}\tau_L + \frac{1}{2}\sqrt{3}\cosh \frac{1}{2}\sqrt{3}\tau_L}.$$

Thus, we may write for the spectral windows

$$q_W(L) = \frac{2E_W \sinh \frac{1}{2}\sqrt{3}\tau_{pL}}{\sinh \frac{1}{2}\sqrt{3}\tau_{pL} + \frac{1}{2}\sqrt{3}\cosh \frac{1}{2}\sqrt{3}\tau_{pL}},$$

and, similarly, for each band,

$$q_n(L) = \frac{2E_n \sinh \frac{1}{2}\sqrt{3}(\tau_{pL}+\tau_{nL})}{\sinh \frac{1}{2}\sqrt{3}(\tau_{pL}+\tau_{nL}) + \frac{1}{2}\sqrt{3}\cosh \frac{1}{2}\sqrt{3}(\tau_{pL}+\tau_{nL})}.$$

Assuming again that the gas bands are relatively narrow, so that

$$E_n \simeq E_{b\eta_n}\Delta\eta_n = \frac{E_{b\eta_n}(\eta_n/T)}{T^3}T^3\Delta\eta_n,$$

we obtain from Appendix C,

$$
\begin{aligned}
E_{2.7} &= E_{b\eta}\left(\tfrac{3660}{1700}\mathrm{cm}^{-1}/\mathrm{K}\right)\Delta\eta_{2.7} \\
&= 1.7649 \times 10^{-8} \times 1700^3 \frac{\mathrm{W}}{\mathrm{m^2cm^{-1}}} \times 445.3\,\mathrm{cm}^{-1} = 38{,}612\,\mathrm{W/m^2},
\end{aligned}
$$

$$
\begin{aligned}
E_{4.3} &= E_{b\eta}\left(\tfrac{2326}{1700}\mathrm{cm}^{-1}/\mathrm{K}\right)\Delta\eta_{4.3} \\
&= 1.5548 \times 10^{-8} \times 1700^3 \frac{\mathrm{W}}{\mathrm{m^2cm^{-1}}} \times 377.6\,\mathrm{cm}^{-1} = 28{,}844\,\mathrm{W/m^2},
\end{aligned}
$$

$$
\begin{aligned}
E_{15} &= E_{b\eta}\left(\tfrac{667}{1700}\mathrm{cm}^{-1}/\mathrm{K}\right)\Delta\eta_{15} \\
&= 0.2979 \times 10^{-8} \times 1700^3 \frac{\mathrm{W}}{\mathrm{m^2cm^{-1}}} \times 329.7\,\mathrm{cm}^{-1} = 4825\,\mathrm{W/m^2},
\end{aligned}
$$

and

$$
\begin{aligned}
E_W = E_b - \sum_{n=1}^{3} E_n &= 5.670 \times 10^{-8} \times 1700^4 - 38{,}612 - 28{,}844 - 4825 \\
&= 401{,}283\,\mathrm{W/m^2}.
\end{aligned}
$$

Finally:

$$
\begin{aligned}
q_W(L) &= \frac{2 \times 401{,}283 \times \sinh \frac{1}{2}\sqrt{3}}{\sinh \frac{1}{2}\sqrt{3} + \frac{1}{2}\sqrt{3}\cosh \frac{1}{2}\sqrt{3}} \\
&= 0.8935 \times 401{,}283 = 358{,}546\,\mathrm{W/m^2},
\end{aligned}
$$

$$q_{2.7}(L) = \frac{2 \times 38{,}612 \times \sinh \frac{1}{2}\sqrt{3}(5.5)}{\sinh 2.75\sqrt{3} + \frac{1}{2}\sqrt{3}\cosh 2.75\sqrt{3}}$$

$$= 1.072 \times 38{,}612 = 41{,}392\ \mathrm{W/m^2},$$

$$q_{4.3}(L) = \frac{2 \times 28{,}844 \times \sinh \frac{1}{2}\sqrt{3}(91.7)}{\sinh 45.85\sqrt{3} + \frac{1}{2}\sqrt{3}\cosh 45.85\sqrt{3}}$$

$$= 1.072 \times 28{,}844 = 30{,}915\ \mathrm{W/m^2},$$

$$q_{15}(L) = \frac{2 \times 4825 \times \sinh \frac{1}{2}\sqrt{3}(18.9)}{\sinh 9.45\sqrt{3} + \frac{1}{2}\sqrt{3}\cosh 9.45\sqrt{3}}$$

$$= 1.072 \times 4825 = 5171\ \mathrm{W/m^2}.$$

The total heat lost from both sides is

$$q = 2\left(q_w + \sum_{n=1}^{3} q_n\right)$$

$$= 2 \times (358{,}546 + 41{,}392 + 30{,}915 + 5171) = 87.20\ \mathrm{W/cm^2}.$$

We note that all three gas bands are essentially opaque (with the characteristic 7% overprediction for the P_1-approximation with temperature discontinuity). This heat flux should be compared with the maximum possible,

$$q_{\max} = 2\sigma T^4 = 2 \times 5.670 \times 10^{-12} \times 1700^4 = 94.71\ \mathrm{W/cm^2}.$$

The importance of considering spectral variations for the evaluation of radiative heat fluxes from molecular gases (or gas-particulate mixtures) is perhaps best understood by comparing a nongray analysis with an "equivalent" gray model. While we shall consider here only the simple case of radiative equilibrium in a one-dimensional slab, similar conclusions can be drawn for radiation heat loss from hot (multidimensional) gas bodies.

Example 17.9.   Repeat Example 17.5, using the picket fence model.

Solution

The functional relationship of $(S/d)_\eta$ is the same as the one used in the exponential wide band model in Chapter 9, so Table 9.2 may be used to determine the total band absorptance. Thus, from equations (17.34) and (17.35),

$$\alpha = \int_{\mathrm{band}} \kappa_\eta\, d\eta = \int_{\mathrm{band}} \left(\frac{S}{d}\right)_\eta d\eta = \bar{\kappa}\Delta\eta_e \quad \text{or} \quad \tau_L = \frac{\Delta\eta_e}{\omega}\bar{\tau}_L,$$

$$\frac{A(X_m)}{\omega} = A^*(\tau_m,\,\beta) = \frac{\Delta\eta_e}{\omega}\left(1 - e^{-\bar{\tau}_m}\right),$$

TABLE 17.2
Picket fence model parameters for Example 17.5

τ_L	β	A^*	$\bar{\tau}_L$	$\Delta\eta_e/\omega$
1	0.01	0.2553	3.9125	0.2556
	0.10	0.7390	1.1849	0.8439
	≥ 1.00	1.5653	0.1360	7.3516
10	0.01	0.8290	12.0620	0.8290
	0.10	2.4653	4.0530	2.4673
	≥ 1.00	3.8679	2.5567	3.9114
100	0.01	2.5553	39.1341	2.5553
	0.10	4.7679	20.9736	4.7679
	≥ 1.00	6.1705	16.2062	6.1705

where

$$\tau_L = \frac{\alpha}{\omega}L, \quad \tau_m = \frac{\alpha}{\omega}L_m = 1.76\,\tau_L, \quad \bar{\tau} = \bar{\kappa}L, \quad \bar{\tau}_m = \bar{\kappa}L_m = 1.76\,\bar{\tau}.$$

With the aid of Table 9.2 one can readily determine the unknowns $\bar{\tau}_L$ and $(\Delta\eta_e/\omega)$, as shown in Table 17.2 for a few representative values. The radiative heat flux between the two plates may be calculated as in Examples 17.6 and 17.7, leading to [cf. equation (17.47)]

$$\Psi = \frac{q}{\sigma(T_1^4 - T_2^4)} = 1 - \frac{E_{B1} - E_{B2}}{E_{b1} - E_{b2}}(1 - \Psi_B)$$

$$\simeq 1 - \frac{\Delta\eta_e}{\omega} \times \frac{\omega\,[E_{b\eta_0}(T_1) - E_{b\eta_0}(T_2)]}{\sigma(T_1^4 - T_2^4)} \times \left(1 - \frac{1}{1 + \frac{3}{4}\bar{\tau}_L}\right).$$

The temperature-dependent constant is evaluated from Appendix C, with $\eta_0/T_1 = 3000/1200 = 2.5\,\text{cm}^{-1}/\text{K}$ and $\eta_0/T_2 = 3000/800 = 3.75\,\text{cm}^{-1}/\text{K}$, as

$$\frac{\omega\,[E_{b\eta_0}(T_1) - E_{b\eta_0}(T_2)]}{\sigma(T_1^4 - T_2^4)}$$

$$= \frac{200 \times (1.64683 \times 10^{-8} \times 1200^3 - 0.89873 \times 10^{-8} \times 800^3)}{5.670 \times 10^{-8}(1200^4 - 800^4)} = 0.0506$$

Therefore,

$$\Psi \simeq 1 - \frac{0.0506\Delta\eta_e/\omega}{1 + 4/3\bar{\tau}_L}.$$

Representative results are shown in Fig. 17-7, together with exact results found by the Monte Carlo method [22, 30], results from the weighted-sum-of-gray-gases method (discussed later in this chapter) and the semigray approach. It is seen that the picket fence model gives surprisingly good answers, overpredicting heat fluxes by only a few percent.[2]

---

[2]In this figure the value of ω in the evaluation of τ_m and A^* has been decreased by 20% (i.e., for *wide band* calculations), as suggested by Edwards [31], in order to compare with results from the other methods, which use $\omega = 200\,\text{cm}^{-1}$ for *spectral* calculations; see also footnote 9 in the discussion of the exponential wide band model in Chapter 9.

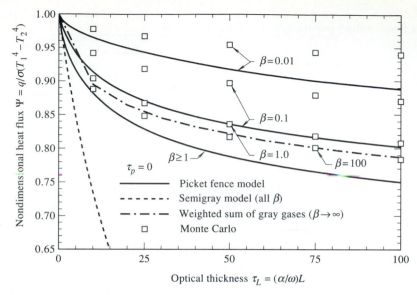

FIGURE 17-7
Nondimensional radiative heat flux for radiative equilibrium across a molecular gas layer bounded by parallel, black walls, as calculated by various methods.

The heat flux is seen to increase with decreasing line overlap parameter β, since radiation can travel from plate to plate through the "mini-windows" between optically thick spectral lines.

17.5 GENERAL FORMULATION FOR NONSCATTERING MEDIA

While the stepwise gray model is very convenient, it is unfortunately not necessarily very accurate. We have already seen that the rather arbitrary choice for the band width can introduce serious errors. In addition, there are situations in which even the most careful choice for the band width leads to unacceptable results. Consider, for example, flow of an absorbing/emitting gas inside a tube. Let the gas temperature be equal to the wall temperature at the wall (no slip) and hotter inside. The gas will emit and absorb radiation over the spectral regions of its vibration-rotation bands, and there will be no net radiative heat flux over the spectral regions of the windows. If the radius of the tube is sufficiently large and the box model is employed, such that $\overline{\kappa}R \gg 1$, then the spectral heat flux for the bands will also vanish as a result of the diffusion limit (since there is no temperature discontinuity). Thus, the stepwise gray model predicts a zero total radiative heat flux for this case, which is clearly not realistic. The reason for this error is that the box model cannot take into account the effects of the exponentially decaying band wings of a vibration-rotation band: No matter how optically thick the band center is, there will always be a portion of the band wings that has an intermediate optical thickness and, thus, contributes strongly to the radiative heat flux.

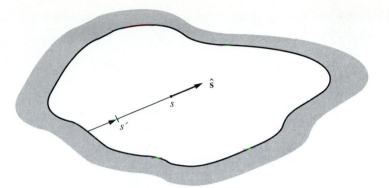

FIGURE 17-8
Spectral intensity within an arbitrary black-walled enclosure.

In this section we shall develop the general formulation for spectrally integrated intensities, incident radiation, and radiative heat fluxes for an absorbing/emitting (but not scattering) medium confined within a black-walled enclosure. The spectral absorption coefficient of the medium may have arbitrary functional form (although we shall look also at the important special cases of molecular gases and particulate suspensions), and the geometry may be arbitrary and multidimensional. Although this formulation is not so general as one would like (not allowing for nonblack surfaces and/or a scattering medium), it has a number of important applications, most notably heat transfer within combustion chambers, where the medium consists of combustion gases and (nonscattering) soot, and where the walls are soot covered (and nearly black).

The equation of transfer for the radiative intensity at a wavenumber η and along a path s is, for a nonscattering medium, from equation (8.23),

$$\frac{dI_\eta}{ds} = \kappa_\eta(I_{b\eta} - I_\eta), \tag{17.68}$$

with the formal solution, equation (8.31),

$$I_\eta(s) = I_{bw\eta} \exp\left(-\int_0^s \kappa_\eta ds'\right) + \int_0^s I_{b\eta}(s') \exp\left(-\int_s^{s'} \kappa_\eta ds''\right)\kappa_\eta(s')\, ds', \tag{17.69}$$

where $I_{bw\eta} = I_{b\eta}(T_w)$ is the intensity emitted into the medium from the (black) wall at $s = 0$, as shown in Fig. 17-8. Integrating this expression over the entire spectrum, we obtain the total intensity as

$$I(s) = \int_0^\infty I_\eta\, d\eta = \int_0^\infty I_{bw\eta} \exp\left(-\int_0^s \kappa_\eta ds'\right)d\eta$$
$$+ \int_0^s \int_0^\infty I_{b\eta}(s') \exp\left(-\int_{s'}^s \kappa_\eta ds''\right)\kappa_\eta(s')\, d\eta\, ds'. \tag{17.70}$$

In Sections 8.3 and 8.4 we defined the spectral absorptivity and emissivity of a participating medium as

$$\alpha_\eta(0 \to s) = \epsilon_\eta(0 \to s) = 1 - \exp\left(-\int_0^s \kappa_\eta ds'\right). \tag{17.71}$$

For a constant absorption coefficient the absorptivity depends on the thickness of the gas layer as well as the (constant) absorption coefficient. If κ_η is not constant but varies spatially and/or with temperature, the absorptivity depends on the variation of κ_η along the entire path, here denoted by the argument $0 \to s$. Substituting equation (17.71) into (17.70), and using

$$\frac{\partial \alpha_\eta}{\partial s'}(s' \to s) = \frac{\partial}{\partial s'}\left[1 - \exp\left(-\int_{s'}^s \kappa_\eta ds''\right)\right] = -\kappa_\eta(s') \exp\left(-\int_{s'}^s \kappa_\eta ds''\right) \tag{17.72}$$

where Leibnitz' rule was used for the differentiation of an integral,[3] we get

$$I(s) = \int_0^\infty I_{bw\eta} [1 - \alpha_\eta(0 \to s)] \, d\eta - \int_0^s \int_0^\infty I_{b\eta}(s') \frac{\partial \alpha_\eta}{\partial s'}(s' \to s) \, d\eta \, ds'. \tag{17.73}$$

With the definition of the *total absorptivity* as

$$\alpha(T, s' \to s) = \frac{1}{I_b(T)} \int_0^\infty \alpha_\eta(s' \to s) I_{b\eta}(T) \, d\eta$$

$$= \frac{1}{I_b(T)} \int_0^\infty \left[1 - \exp\left(-\int_{s'}^s \kappa_\eta ds''\right)\right] I_{b\eta}(T) \, d\eta, \tag{17.74}$$

we may consolidate equation (17.73) as[4]

$$I(s) = [1 - \alpha(T_w, 0 \to s)] I_{bw} - \int_0^s \frac{\partial \alpha}{\partial s'}[T(s'), s' \to s] I_b(s') \, ds'. \tag{17.75}$$

To determine total heat flux, the total intensity is, after multiplication with the unit vector $\hat{s}$, integrated over all directions, or

$$\mathbf{q} = \int_0^\infty \mathbf{q}_\eta \, d\eta = \int_{4\pi} I \hat{s} \, d\Omega. \tag{17.76}$$

To evaluate the divergence of the radiative heat flux (for a known temperature field) or the temperature field (for radiative equilibrium), the equation governing conservation

---

[3] Leibnitz' rule was first introduced in Chapter 5, equation (5.1).
[4] Note that, in concurrence with the definition of total absorptivity, the derivative in $\partial \alpha/\partial s'$ is only with respect to the path $s' \to s$, and *not* with respect to the s' in the temperature $T(s')$.

of radiative energy must be integrated over all wavenumbers or, from equation (8.55),

$$\nabla \cdot \mathbf{q} = \nabla \cdot \int_0^\infty \mathbf{q}_\eta \, d\eta = \int_0^\infty \kappa_\eta \left(4\pi I_{b\eta} - \int_{4\pi} I_\eta d\Omega\right) d\eta$$

$$= 4\pi\kappa_P I_b - \int_{4\pi} \left(\int_0^\infty \kappa_\eta I_\eta d\eta\right) d\Omega, \tag{17.77}$$

where κ_P is the Planck-mean absorption coefficient first defined in equation (9.100). Thus, multiplying equation (17.69) by $\kappa_\eta(s)$ and integrating, we get

$$\int_0^\infty \kappa_\eta(s) I_\eta(s) \, d\eta = \int_0^\infty I_{bw\eta}\kappa_\eta(s) \exp\left(-\int_0^s \kappa_\eta ds''\right) d\eta$$

$$+ \int_0^s \int_0^\infty I_{b\eta}(s') \exp\left(-\int_{s'}^s \kappa_\eta ds''\right)\kappa_\eta(s')\kappa_\eta(s) \, d\eta \, ds'. \tag{17.78}$$

Differentiating equation (17.72) with respect to s, again using Leibnitz' rule, leads to

$$\frac{\partial^2 \alpha_\eta}{\partial s \partial s'}(s' \to s) = \kappa_\eta(s')\kappa_\eta(s) \exp\left(-\int_{s'}^s \kappa_\eta ds''\right). \tag{17.79}$$

Therefore,

$$\int_0^\infty \kappa_\eta I_\eta d\eta = \int_0^\infty I_{bw\eta}\frac{\partial \alpha_\eta}{\partial s}(0 \to s) \, d\eta + \int_0^s \int_0^\infty I_{b\eta}(s')\frac{\partial^2 \alpha_\eta}{\partial s \partial s'}(s' \to s) \, d\eta \, ds'$$

$$= \frac{\partial \alpha}{\partial s}(T_w, 0 \to s)I_{bw} + \int_0^s \frac{\partial^2 \alpha}{\partial s \partial s'}[T(s'), s' \to s] I_b(s') \, ds'. \tag{17.80}$$

The functional form of $\alpha(T, s' \to s)$ depends, of course, on the local properties of the participating medium. We shall briefly discuss the special cases of pure gas and gas-particulate mixtures.

Pure Molecular Gas

If the medium is a molecular gas with N vibration-rotation bands, the absorption coefficient may be stated as

$$\kappa_\eta = \sum_{n=1}^N \kappa_{n\eta}. \tag{17.81}$$

We shall now assume that each band is fairly narrow, i.e., that the blackbody intensity does not vary appreciably over each band, and that the bands do not overlap.[5] Then

---

[5]These assumptions are usually very good except in the limit of extreme optical thickness, which tends to widen bands. Even at lesser optical thickness some bands of important gases overlap, most notably the 2.7 μm bands of CO_2 and H_2O (see Chapter 9).

the total absorptivity may be evaluated as

$$\alpha(T, s' \to s) \simeq \sum_{n=1}^{N} \frac{\omega_n I_{b\eta_n}}{I_b}(T) \frac{1}{\omega_n} \int_0^\infty \left[1 - \exp\left(-\int_{s'}^s \kappa_{n\eta} ds''\right)\right] d\eta$$

$$= \sum_{n=1}^{N} \frac{\omega_n I_{b\eta_n}}{I_b}(T) A_n^*(s' \to s), \qquad (17.82)$$

where $I_{b\eta_n}$ is the Planck function at the center (or head) of band n, ω_n is the band width parameter, and A_n^* is the nondimensional band absorptance, as discussed in detail in Chapter 9. A solution to equations (17.76) and (17.77) for a one-dimensional water vapor mixture contained between black plates has recently been given by Kim and coworkers [32], using narrow band as well as wide band correlations together with the discrete ordinates method of Chapter 15.

Molecular Gas with Suspended Particles

If the molecular gas contains a suspension of nonscattering particles, the absorption coefficient may be written as

$$\kappa_\eta = \kappa_{p\eta} + \kappa_{g\eta} = \kappa_{p\eta} + \sum_{n=1}^{N} \kappa_{n\eta}. \qquad (17.83)$$

We shall now assume that, not only are the gas bands narrow and nonoverlapping, but also that the absorption coefficient of the particles, $\kappa_{p\eta}$, does not vary appreciably over each band. Then

$$\alpha(T, s' \to s) = \frac{1}{I_b(T)} \int_0^\infty \left[1 - \exp\left(-\int_{s'}^s \kappa_{p\eta} ds''\right)\exp\left(-\int_{s'}^s \kappa_{g\eta} ds''\right)\right] I_{b\eta}(T) d\eta$$

$$= \frac{1}{I_b(T)} \int_0^\infty \left[1 - \exp\left(-\int_{s'}^s \kappa_{p\eta} ds''\right)\right] I_{b\eta}(T) d\eta$$

$$+ \frac{1}{I_b(T)} \int_0^\infty \exp\left(-\int_{s'}^s \kappa_{p\eta} ds''\right)\left[1 - \exp\left(-\int_{s'}^s \kappa_{g\eta} ds''\right)\right] I_{b\eta}(T) d\eta. \quad (17.84)$$

The first term in equation (17.84) is simply the absorptivity for the particle background (without molecular gas). In the second spectral integral we note that the integrand is nonzero only when $\kappa_{g\eta}$ is nonzero, i.e., over the vibration-rotation bands of the gas. Since we assume the gas bands to be spectrally narrow, the particle attenuation term may be evaluated at the band center and taken outside the spectral integral. Then

$$\alpha(T, s' \to s) \simeq \alpha_p(T, s' \to s)$$

$$+ \sum_{n=1}^{N} \frac{\omega_n I_{b\eta_n}}{I_b}(T) \exp\left(-\int_{s'}^s \kappa_{p\eta_n} ds''\right) A_n^*(s' \to s). \qquad (17.85)$$

Radiative Equilibrium

In the case of radiative equilibrium equation (17.80) is used to determine the temperature field. Substituting the result into equation (17.77), with $\nabla \cdot \mathbf{q} = 0$, yields

$$4\pi\kappa_P I_b = \int_{4\pi}\left[\frac{\partial\alpha}{\partial s}(T_w, 0\rightarrow s)I_{bw} + \int_0^s \frac{\partial^2\alpha}{\partial s\partial s'}[T(s'), s'\rightarrow s]I_b(s')\,ds'\right]d\Omega. \quad (17.86)$$

This is a single (but rather complicated) integral equation for the unknown temperature. Once the temperature field is known, the heat flux follows from equations (17.76) and (17.75) as

$$\mathbf{q} = \int_{4\pi}\left[[1 - \alpha(T_w, 0\rightarrow s)]I_{bw} - \int_0^s \frac{\partial\alpha}{\partial s'}[T(s'), s'\rightarrow s]\,I_b(s')\,ds'\right]\hat{s}\,d\Omega. \quad (17.87)$$

In equations (17.86) and (17.87) the absorptivities and their derivatives are obtained from equation (17.82) (pure molecular gas) or (17.85) (gas-particulate mixture).

Medium with Known Temperature Field

If the temperature field is known, the local heat flux may be determined from equation (17.87), while the divergence of the heat flux follows from equations (17.77) and (17.80) as

$$\nabla \cdot \mathbf{q} = 4\pi\kappa_P I_b$$
$$- \int_{4\pi}\left[\frac{\partial\alpha}{\partial s}(T_w, 0\rightarrow s)I_{bw} + \int_0^s \frac{\partial^2\alpha}{\partial s\partial s'}[T(s'), s'\rightarrow s]I_b(s')\,ds'\right]d\Omega. \quad (17.88)$$

Again, in equation (17.88) the absorptivities and their derivatives are obtained from equations (17.82) (pure molecular gas) or (17.85) (gas-particulate mixture).

The general relationships developed in this section will be employed over the following two sections to carry out heat transfer calculations. First we shall look in some detail at radiative transfer within an isothermal, one-dimensional nongray medium. Then follows a section describing how the concept of a weighted-sum-of-gray-gases can be applied to the general nongray medium problem.

17.6 THE WIDE BAND MODEL FOR ISOTHERMAL MEDIA

In this section we shall develop some of the simpler expressions governing heat transfer in a isothermal, nongray plane-parallel medium at temperature T_m, bounded by two black surfaces, both at temperature T_w. We shall also present very simple results for isothermal gases of spherical and cylindrical shape.

For an isothermal medium, if we assume that the absorption coefficient is a function of temperature only (and is, therefore, constant throughout the medium), the definition of the total absorptivity, equation (17.74), simplifies to

$$\alpha(T, s) = \frac{1}{I_b(T)} \int_0^\infty \left(1 - e^{-\kappa_\eta s}\right) I_{b\eta}(T) \, d\eta, \tag{17.89}$$

that is, the total absorptivity no longer depends on the entire path, but only on the path length itself. A similar statement holds true for the nondimensional band absorptance of molecular gases, which now depends only on the optical path length (as well as a line overlap parameter), as discussed in Chapter 9, or

$$A_n^*(s' \to s) = A^*[\kappa_n(s - s'), \beta_n], \tag{17.90}$$

where $\kappa_n = \rho_a \alpha_n / \omega_n$ is the gas absorption coefficient at the band center or band head,[6] and β_n is the overlap parameter. In addition, for an isothermal medium, the integral in equation (17.75) may be evaluated so that

$$I(s) = [1 - \alpha(T_w, s)] I_{bw} + \alpha(T_m, s) I_{bm}, \tag{17.91}$$

where $I_{bm} = I_b(T_m)$ and T_m is the (constant) temperature of the medium.

The Isothermal Gas Slab

For a slab $0 < z < L$ the intensity as given by equation (17.91) may emanate from either of the two bounding surfaces. Splitting intensity into I^+ and I^- ("positive" and "negative" directions), as was done in Section 12.2, leads to $s = z/\mu$ for "positive" directions $(0 < \mu < 1)$ and to $s = -(L - z)/\mu$ for "negative" directions $(-1 < \mu < 0)$, as shown in Fig. 12-1. Substituting into equation (17.91) results in

$$I^+(z) = [1 - \alpha(T_w, z/\mu)] I_{bw} + \alpha(T_m, z/\mu) I_{bm},$$
$$0 < \mu < 1, \tag{17.92a}$$
$$I^-(z) = [1 - \alpha(T_w, -(L-z)/\mu)] I_{bw} + \alpha(T_m, -(L-z)/\mu) I_{bm},$$
$$-1 < \mu < 0. \tag{17.92b}$$

The local heat flux is then determined from

$$\begin{aligned}
q(z) &= 2\pi \left[\int_{-1}^0 I^- \mu \, d\mu + \int_0^{+1} I^+ \mu \, d\mu \right] \\
&= 2\pi \left\{ \int_0^1 \left[\alpha\left(T_m, \frac{z}{\mu}\right) - \alpha\left(T_m, \frac{L-z}{\mu}\right) \right] \mu \, d\mu \, I_b(T_m) \right. \\
&\quad \left. - \int_0^1 \left[\alpha\left(T_w, \frac{z}{\mu}\right) - \alpha\left(T_w, \frac{L-z}{\mu}\right) \right] \mu \, d\mu \, I_b(T_w) \right\}. \tag{17.93}
\end{aligned}$$

---

[6]Note that, in wide band correlations, the band strength parameter α is usually based on a mass absorption coefficient and must, therefore, be corrected by multiplying with the partial density of the absorbing gas, ρ_a.

If the medium is a molecular gas, equation (17.82) is substituted into this expression. Using emissive powers in favor of intensities, $E_{b\eta} = \pi I_{b\eta}$, then leads to

$$q(z) = \sum_{n=1}^{N} \omega_n [E_{b\eta_n}(T_m) - E_{b\eta_n}(T_w)][A_s(\kappa_n z, \beta_n) - A_s(\kappa_n(L-z), \beta_n)], \quad (17.94)$$

where A_s, termed *slab band absorptance* by Edwards and Balakrishnan [8], is defined as

$$A_s(\tau, \beta) = 2 \int_0^1 A^* \left(\frac{\tau}{\mu}, \beta \right) \mu \, d\mu. \quad (17.95)$$

The slab band absorptance may be evaluated explicitly for a number of band absorptance correlations, such as the Edwards and Menard correlation of Table 9.2 and the high-pressure limit given by equation (9.81), and must be evaluated numerically for more involved correlations such as the one by Felske and Tien, equation (9.82). We shall consider here only the high-pressure limit ($\beta \to \infty$),

$$A^*(\tau, \infty) = E_1(\tau) + \ln \tau + \gamma_E, \quad (17.96)$$

or

$$A_s(\tau, \infty) = 2 \int_0^1 \left[E_1 \left(\frac{\tau}{\mu} \right) + \ln \frac{\tau}{\mu} + \gamma_E \right] \mu \, d\mu. \quad (17.97)$$

Integrating by parts we obtain, with $E_1'(x) = -E_0(x) = e^{-x}/x$,

$$A_s(\tau, \infty) = \mu^2 \left[E_1 \left(\frac{\tau}{\mu} \right) + \ln \frac{\tau}{\mu} + \gamma_E \right] \Big|_0^1 - \int_0^1 \mu^2 \left(\frac{\tau}{\mu^2} \frac{e^{-\tau/\mu}}{\tau/\mu} - \frac{1}{\mu} \right) d\mu,$$

or

$$A_s(\tau, \infty) = E_1(\tau) + \ln \tau + \gamma_E + \tfrac{1}{2} - E_3(\tau). \quad (17.98)$$

Figure 17-9 shows a plot of the slab band absorptance together with the equivalent function for isothermal spheres and cylinders.

Example 17.10. Consider the gas layer of Example 17.8, but without the suspended particles. Calculate the radiative transfer from the gas to the walls using the exponential wide band model.

Solution

The line overlap parameter β has already been calculated for all three bands in Examples 17.6 ($\beta_{4.3} = 7.57$) and 17.8 ($\beta_{2.7} = 5.47$, $\beta_{15} = 1.82$), and is substantially larger than unity for all bands. Thus, the high-pressure limit, equation (17.97), may be used. Applying equation (17.94) at $z = L$ for the three important CO_2 bands, we obtain [with $T_w = 0$ and $A_s(0, \beta) = 0$]

$$q_n(L) = \omega_n E_{b\eta_n}(T_m) A_s \left(\frac{\alpha_n \rho_{CO_2}}{\omega_n} L, \beta \right).$$

All required terms, have already been calculated in the previous example, and we find the following:[7]

---

[7]Note that all three bands are optically so thick that the exponential integrals essentially vanish.

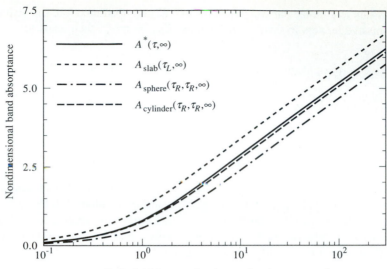

Optical thickness at band center/head τ, τ_L, $\tau_D = 2\tau_R$

FIGURE 17-9
Isothermal gas band absorptances for slabs, spheres and cylinders.

2.7 μm band:
$$\alpha_{2.7}\rho_{CO_2}L/\omega_{2.7} = 2005 \times 1/96.9 = 20.69$$
$$A_{s2.7}(20.69,\infty) = E_1(20.69)+\ln 20.69+0.5772+0.5-E_3(20.69) = 4.107$$
$$q_{2.7}(L) = 1.7649\times 10^{-8}\times 1700^3 \times 96.9\times 4.107 = 34,508\ \text{W/m}^2$$

4.3 μm band:
$$\alpha_{4.3}\rho_{CO_2}L/\omega_{4.3} = 34,254 \times 1/46.2 = 741.43$$
$$A_{s4.3}(741.43,\infty) = E_1(741.43)+\ln 741.43+0.5772+0.5-E_3(741.43) = 7.686$$
$$q_{4.3}(L) = 1.5548\times 10^{-8}\times 1700^3 \times 46.2\times 7.686 = 27,125\ \text{W/m}^2$$

15 μm band:
$$\alpha_{15}\rho_{CO_2}L/\omega_{15} = 5917 \times 1/52.4 = 112.92$$
$$A_{s15}(112.92,\infty) = E_1(112.92)+\ln 112.92+0.5772+0.5-E_3(112.92) = 5.804$$
$$q_{15}(L) = 0.2979\times 10^{-8}\times 1700^3 \times 52.4\times 5.804 = 4805\ \text{W/m}^2.$$

Finally, the total heat loss from the gas is

$$q = 2 \times [34,508 + 27,125 + 4805] = 13.29\ \text{W/cm}^2.$$

Comparing with results from Example 17.8 we find that the present values are $\sim$15% lower for each band. Since the gas bands are essentially opaque, the particle absorption coefficient has no effect on the q_n (as opposed to q_W). Roughly one half of the discrepancy is due to the fact that we used the P_1-approximation in the previous example. Therefore, we conclude that the results from the crude stepwise-gray model differ by only $\sim$10% from the much more sophisticated exponential wide band model.

Comparison with Example 17.2 shows that the present results agree excellently with those of the mean beam length method.

The Isothermal Slab with Gases and Particles

We shall now consider a slab $0 < z < L$ consisting of a mixture of molecular gases and particles, both at constant temperature T_m, and bounded by two isothermal black walls, both at T_w. The relations developed in the previous paragraph up to and including equation (17.93) hold. We shall limit ourselves here to the simple case of gray particles ($\kappa_{p\eta} = \kappa_p = \text{const}$), while the original reference [9] dealt with the somewhat more general case of a nongray particle absorption coefficient (as well as several prescribed temperature fields). For the gas-particulate mixture the absorptivity is then evaluated from equation (17.85) as

$$\alpha(T, s) = 1 - e^{-\kappa_p s} + \sum_{n=1}^{N} \frac{\omega_n I_{b\eta_n}}{I_b}(T) e^{-\kappa_p s} A^*(\kappa_n s, \beta_n). \tag{17.99}$$

Substituting this expression into equation (17.93) gives

$$
\begin{aligned}
q(z) &= 2 \int_0^1 \left[e^{-\kappa_p(L-z)/\mu} - e^{-\kappa_p z/\mu} \right] \mu \, d\mu \, [E_b(T_m) - E_b(T_w)] \\
&\quad + 2 \sum_{n=1}^{N} \omega_n [E_{b\eta_n}(T_m) - E_{b\eta_n}(T_w)] \left[\int_0^1 e^{-\kappa_p z/\mu} A^*(\kappa_n z/\mu, \beta_n) \mu \, d\mu \right. \\
&\qquad \left. - \int_0^1 e^{-\kappa_p(L-z)/\mu} A^*\!\left(\kappa_n(L-z)/\mu, \beta_n\right) \mu \, d\mu \right] \\
&= 2 [E_b(T_m) - E_b(T_w)] [E_3(\kappa_p(L-z)) - E_3(\kappa_p z)] \\
&\quad + \sum_{n=1}^{N} \omega_n [E_{b\eta_n}(T_m) - E_{b\eta_n}(T_w)] \\
&\qquad \times \left[A_{pg}(\kappa_p z, \kappa_n z, \beta_n) - A_{pg}\!\left(\kappa_p(L-z), \kappa_n(L-z), \beta_n\right) \right], \tag{17.100}
\end{aligned}
$$

where A_{pg} is a slab band absorptance modified for the presence of particles,

$$A_{pg}(\tau_p, \tau_g, \beta) = 2 \int_0^1 e^{-\tau_p/\mu} A^*\!\left(\frac{\tau_g}{\mu}, \beta\right) \mu \, d\mu. \tag{17.101}$$

We shall again limit ourselves to the high-pressure limit ($\beta \to \infty$). The slab band absorptance may then be evaluated by using the integral form for the band absorptance [9, 33]

$$A^*(\tau, \infty) = E_1(\tau) + \ln \tau + \gamma_E = \int_0^1 (1 - e^{-\tau u}) \frac{du}{u}. \tag{17.102}$$

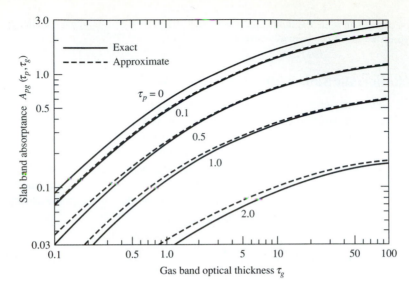

FIGURE 17-10
Slab band absorptance for gas-particulate mixtures.

Therefore,

$$
\begin{aligned}
A_{pg}(\tau_p, \tau_g, \infty) &= 2\int_{\mu=0}^{1} e^{-\tau_p/\mu} \int_{u=0}^{1} (1 - e^{-u\tau_g/\mu}) \frac{du}{u}\, \mu\, d\mu \\
&= 2\int_{u=0}^{1}\int_{\mu=0}^{1}\left[e^{-\tau_p/\mu} - e^{-(\tau_p+u\tau_g)/\mu} \right] \mu\, d\mu\, \frac{du}{u} \\
&= 2\int_{0}^{1} [E_3(\tau_p) - E_3(\tau_p + u\,\tau_g)]\, \frac{du}{u}.
\end{aligned} \tag{17.103}
$$

Equation (17.103) cannot be evaluated explicitly, but an approximation has been given by Modest [9] as

$$
A_{pg}(\tau_p, \tau_g, \infty) \simeq 2E_3(\tau_p)\left[E_1(\tau_g) + \ln \tau_g + \gamma_E + \tfrac{1}{2} \right] - E_3(\tau_p + \tau_g). \tag{17.104}
$$

As shown in Fig. 17-10 the approximation gives very accurate results except for appreciable τ_p accompanied by small values for τ_g (i.e., situations when gas radiation and, therefore, the accuracy of A_{pg} will be unimportant).

Example 17.11. Repeat Example 17.8 using the exponential wide band model.

Solution

Setting $z = L$ and $T_w = 0$ in equation (17.100), we obtain

$$
q(L) = \sigma T_m^4 [1 - 2E_3(\kappa_p L)] + \sum_{n=1}^{3} \omega_n E_{b\eta_n}(T_m) A_{pg}(\kappa_p L, \kappa_n L, \beta_n).
$$

Again, assuming the high-pressure limit to hold, following the calculations in the previous example with $\kappa_p L = 1$, and using the approximation of equation (17.104), we have the following:

2.7 μm band:

$$A_{pg2.7}(1,\ 20.69,\ \infty)\ =\ 2E_3(1)[4.107] - 0 = 2 \times 0.10969 \times 4.107 = 0.901,$$

$$q_{2.7}\ =\ 1.7649 \times 10^{-8} \times 1700^3 \times 96.9 \times 0.901 = 7570\ \text{W/m}^2$$

4.3 μm band:

$$A_{pg4.3}(1,\ 741.43,\ \infty)\ =\ 2E_3(1)[7.686] - 0 = 1.686,$$

$$q_{4.3}\ =\ 1.5548 \times 10^{-8} \times 1700^3 \times 46.2 \times 1.686 = 5950\ \text{W/m}^2$$

15 μm band:

$$A_{pg15}(1,\ 112.92,\ \infty)\ =\ 2E_3(1)[5.804] - 0 = 1.273,$$

$$q_{15}\ =\ 0.2979 \times 10^{-8} \times 1700^3 \times 52.4 \times 1.273 = 1054\ \text{W/m}^2$$

Particles:

$$q_p\ =\ 5.670 \times 10^{-8} \times 1700^4[1 - 2E_3(1)] = 369{,}671\ \text{W/m}^2.$$

And the total heat lost from the slab follows as

$$q\ =\ 2 \times (7570 + 5950 + 1054 + 369{,}671) = 76.45\ \text{W/cm}^2.$$

Comparing this with the result from Example 17.8, we again find the former to overpredict the heat flux by ~15% (half of it due to the P_1-approximation). In the stepwise-gray approximation heat fluxes are broken up into window heat flux and total (gas and particles) band heat fluxes. For the exponential wide band model the total particle heat flux is calculated (for the entire spectrum), and the q_n give only the gas contributions.

The Isothermal Gas Sphere

We shall now consider an isothermal sphere $0 < r < R$ at temperature T_m, bounded by an isothermal black surface at T_w. Inspecting Fig. 17-11a, we find that a beam at r in a direction of $\hat{s}$ has traveled through the medium for a distance of $S = r\mu + \sqrt{R^2 - r^2(1 - \mu^2)}$, where $\mu = \cos\theta$, and that this distance is independent of azimuthal angle. Thus, from equation (17.91) we find

$$I(r, \mu)\ =\ \left[1 - \alpha\!\left(T_w,\ r\mu + \sqrt{R^2 - r^2(1 - \mu^2)}\right)\right] I_b(T_w)$$

$$+\ \alpha\!\left(T_m,\ r\mu + \sqrt{R^2 - r^2(1 - \mu^2)}\right) I_b(T_m). \qquad (17.105)$$

Substituting equation (17.82) into (17.105) leads to

$$I(r, \mu) = I_b(T_w) + \sum_{n=1}^{N} \omega_n\, [I_{b\eta_n}(T_m) - I_{b\eta_n}(T_w)]$$

$$\times A^*\!\left(\kappa_n\left[r\mu + \sqrt{R^2 - r^2(1 - \mu^2)}\right],\ \beta_n\right). \qquad (17.106)$$

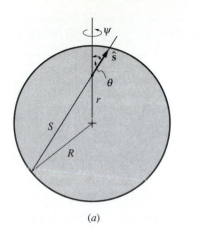 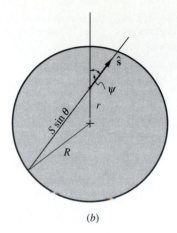

(a)   (b)

FIGURE 17-11
Intensity field in (a) a sphere and (b) a cylinder.

Therefore, the local heat flux may be evaluated as

$$q(r) = 2\pi \int_{-1}^{+1} I(r, \mu) \mu \, d\mu$$

$$= \sum_{n=1}^{N} \omega_n [E_{b\eta_n}(T_m) - E_{b\eta_n}(T_w)] A_{\text{sphere}}(\kappa_n r, \kappa_n R, \beta_n), \qquad (17.107)$$

where

$$A_{\text{sphere}}(\tau_r, \tau_R, \beta) = 2 \int_{-1}^{+1} A^*\left(\tau_r \mu + \sqrt{\tau_R^2 - \tau_r^2(1-\mu^2)}, \beta\right) \mu \, d\mu. \qquad (17.108)$$

In general, the *sphere band absorptance* must be evaluated numerically. To calculate total heat loss from the sphere (i.e., the heat flux at $r = R$), integration may be carried out for a number of band absorptance correlations, in particular the high-pressure limit ($\beta \to \infty$). By doing so we obtain (noting that, for $\mu < 0$, the argument of A^* vanishes)

$$A_{\text{sphere}}(\tau_R, \tau_R, \infty) = 2 \int_0^1 A^*(\tau_D \mu, \infty) \mu \, d\mu$$

$$= 2 \int_0^1 [E_1(\tau_D \mu) + \ln(\tau_D \mu) + \gamma_E] \mu \, d\mu$$

$$= \mu^2 [E_1(\tau_D \mu) + \ln(\tau_D \mu) + \gamma_E] \Big|_0^1 - \int_0^1 \mu^2 \frac{1 - e^{-\tau_D \mu}}{\mu} d\mu,$$

or

$$A_{\text{sphere}}(\tau_R, \tau_R, \infty) = E_1(\tau_D) + \ln \tau_D + \gamma_E - \tfrac{1}{2}$$
$$+ \frac{1}{\tau_D^2}\left[1 - (1 + \tau_D)e^{-\tau_D}\right], \qquad (17.109)$$

where $\tau_D = 2\tau_R = 2\kappa_n R$ is the optical thickness at the band center (or head) based on diameter. Equation (17.109) is also included in Fig. 17-9. Note that, although $A_{\text{slab}} > A^*$ (since the average path cutting through a slab is larger than L), $A_{\text{sphere}} < A^*$ (since the average path cutting through a sphere is less than D).

Example 17.12. Consider an isothermal spherical gas body with diameter D and at temperature T_m, with a single vibration-rotation band in the infrared. The spectral location η_0, and band width ω is such that, at this temperature, $\omega E_{b\eta_0}(T_m)/\sigma T_m^4 = 0.05$. The gas is contained in a cold, nonreflecting enclosure, and is at such high pressure that spectral lines are strongly overlapped. Determine the (nondimensional) total radiative heat loss from the sphere as function of band strength α and band width ω.

Solution

Since the surroundings are cold we assume $T_w \simeq 0$; and the heat loss from the surface of the sphere is, from equations (17.107) and (17.109),

$$q(R) = \omega E_{b\eta_0}(T_m)A_{\text{sphere}}(\kappa R, \kappa R, \infty)$$

or

$$\Psi = \frac{q(R)}{\sigma T_m^4} = \frac{\omega E_{b\eta_0}(T_m)}{\sigma T_m^4} A_{\text{sphere}}(\tau_R, \tau_R, \infty).$$

The nondimensional heat loss is shown in Fig. 17-12 and compared with results from the following section on the *weighted-sum-of-gray-gases model*.

The Isothermal Gas Cylinder

Finally, we briefly discuss the case of an isothermal cylindrical gas volume, $0 < r < R$, at temperature T_m, bounded by an isothermal black wall at T_w. The case is very similar to that of the isothermal sphere. However, because of the more limited symmetry for the one-dimensional cylinder, it is preferable to place the directional coordinate system as shown in Fig. 17-11b: The polar angle θ is measured from the axis of the cylinder, and the azimuthal angle is measured in the plane normal to it as shown in the figure. Let S be the distance that a beam, coming from a point on the cylinder wall, has traveled before passing point r in the direction of $\hat{s}$. The distance indicated in Fig. 17-11b is, therefore, the projection of S into the plane of the paper, $S \sin \theta$, and

$$S = \left(r \cos \psi + \sqrt{R^2 - r^2 \sin^2 \psi}\right)\Big/ \sin \theta. \qquad (17.110)$$

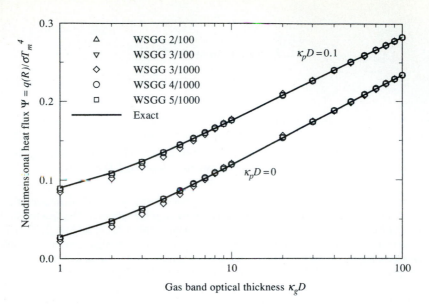

FIGURE 17-12
Total heat loss from an isothermal sphere.

From equation (17.91),

$$I(r,\theta,\psi) = \left[1 - \alpha\left(T_w, (r\cos\psi + \sqrt{R^2 - r^2\sin^2\psi})/\sin\theta\right)\right] I_b(T_w)$$

$$+ \alpha\left(T_m, (r\cos\psi + \sqrt{R^2 - r^2\sin^2\psi})/\sin\theta\right) I_b(T_m).$$

Again, substituting equation (17.82) into this expression leads to

$$I(r,\theta,\psi) = I_b(T_w) + \sum_{n=1}^{N} \omega_n\,[I_{b\eta_n}(T_m) - I_{b\eta_n}(T_w)]$$

$$\times A^*\left(\kappa_n(r\cos\psi + \sqrt{R^2 - r^2\sin^2\psi})/\sin\theta,\ \beta_n\right), \quad (17.111)$$

and the radial heat flux may be calculated as

$$q(r) = \int_{\psi=0}^{2\pi}\int_{\theta=0}^{\pi/2} I(r,\theta,\psi)\hat{\mathbf{s}}\cdot\hat{\mathbf{e}}_r\,\sin\theta\,d\theta\,d\psi$$

$$= 4\int_{0}^{\pi}\int_{0}^{\pi/2} I(r,\theta,\psi)\sin^2\theta\,d\theta\,\cos\psi\,d\psi. \quad (17.112)$$

Here we have made use of the symmetry in both θ and ψ; and the radial component of the direction unit vector is $s_r = \hat{\mathbf{s}}\cdot\hat{\mathbf{e}}_r = \sin\theta\,\cos\psi$. This expression leads to

$$q(r) = \sum_{n=1}^{N} \omega_n\,[E_{b\eta_n}(T_m) - E_{b\eta_n}(T_w)]\,A_{\text{cylinder}}(\kappa_n r,\ \kappa_n R,\ \beta_n), \quad (17.113)$$

where we have again replaced I_b by E_b/π and the *cylinder band absorptance* is defined as

$$A_{\text{cylinder}}(\tau_r, \tau_R, \beta) = \frac{4}{\pi} \int_0^\pi \int_0^{\pi/2} A^* \left((\tau_r \cos\psi + \sqrt{\tau_R^2 - \tau_r^2 \sin^2\psi}) / \sin\theta, \beta \right)$$
$$\times \sin^2\theta \, d\theta \cos\psi \, d\psi. \quad (17.114)$$

The cylinder band absorptance must always be evaluated numerically. The case of $r = R$ and $\beta \to \infty$ is also included in Fig. 17-9. Note that $A_{\text{sphere}} < A_{\text{cylinder}} < A_{\text{slab}}$ as is to be expected from the length of the average path that a beam traverses through these three bodies.

17.7 THE WEIGHTED-SUM-OF-GRAY-GASES MODEL

The concept of a weighted-sum-of-gray-gases approach was first presented by Hottel [14] within the framework of the zonal method, which is described in detail in Chapter 18. Modest [30] has demonstrated that this approach can be applied to the directional equation of transfer, equation (17.68), and, therefore, to any solution method for the equation of transfer (exact, P_N-approximation, discrete ordinates method, etc.). In this method the nongray gas is replaced by a number of gray gases, for which the heat transfer rates are calculated independently. The total heat flux is then found by adding the heat fluxes of the gray gases after multiplication with certain weight factors.

As a starting point consider equation (17.75). For mathematical simplicity we shall limit ourselves here to a constant (or averaged) absorption coefficient, while variable properties were considered in Modest [30]. Thus, we have for the spectrally integrated intensity:

$$I(s) = [1 - \alpha(T_w, s)]I_{bw} - \int_0^s \frac{\partial \alpha}{\partial s'}[T(s'), s - s']I_b(s') \, ds', \quad (17.115)$$

where the total absorptivity α may be obtained from equation (17.82) (for gases), from equation (17.85) (for gas-particulate mixtures), or from charts or correlations such as the ones by Leckner [34] for steam and carbon dioxide.

From the definition of the absorptivity, equation (17.89), it follows that, for a gray medium with $\kappa_\eta = \kappa = \text{const}$,

$$\alpha(T, s) = 1 - e^{-\kappa s}. \quad (17.116)$$

We shall now assume that the absorptivity of equation (17.116) may be approximated by a *weighted sum of gray gases*, or

$$\alpha(T, s) \simeq \sum_{k=0}^K a_k(T)\left(1 - e^{-\kappa_k s}\right). \quad (17.117)$$

For mathematical simplicity we have chosen the gray-gas absorption coefficients κ_k to be constants, while the weight factors a_k may be functions of temperature. Neither a_k nor κ_k are allowed to depend on pathlength s. Depending on the material, the quality of the fit, and the accuracy desired, a K of 2 or 3 usually gives results of satisfactory accuracy [14]. Since, for an infinitely thick medium, the absorptivity approaches unity, we find

$$\sum_{k=0}^{K} a_k(T) = 1. \tag{17.118}$$

Still, for a molecular gas with its "spectral windows" it would take very large path lengths indeed for the absorptivity to be close to unity. For this reason equation (17.117) starts with $k = 0$ (with an implied $\kappa_0 = 0$), to allow for spectral windows.

Substituting equation (17.117) into equation (17.115) and using

$$1 - \alpha(T, s) = \sum_{k=0}^{K} a_k(T) e^{-\kappa_k s},$$

$$\frac{\partial \alpha}{\partial s'}(T, s - s') = \frac{\partial}{\partial s'} \sum_{k=0}^{K} a_k \left[1 - e^{-\kappa_k(s-s')} \right] = -\sum_{k=0}^{K} a_k \kappa_k e^{-\kappa_k(s-s')},$$

leads to

$$
\begin{aligned}
I(s) &= \sum_{k=0}^{K} a_k(T_w) e^{-\kappa_k s} I_{bw} + \int_0^s \sum_{k=0}^{K} a_k[T(s')] \kappa_k e^{-\kappa_k(s-s')} I_b(s') ds' \\
&= \sum_{k=0}^{K} \left\{ [a_k I_b](T_w) e^{-\kappa_k s} + \int_0^s [a_k I_b](s') e^{-\kappa_k(s-s')} \kappa_k ds' \right\}.
\end{aligned}
\tag{17.119}
$$

Setting

$$I(s) = \sum_{k=0}^{K} I_k(s), \tag{17.120}$$

and comparing equations (17.119) and (17.69) we find that I_k satisfies the equation of transfer

$$\frac{dI_k}{ds} = \kappa_k([a_k I_b] - I_k), \tag{17.121}$$

subject to the boundary condition

$$s = 0: \qquad I_k = [a_k I_b](T_w). \tag{17.122}$$

This expression is, of course, the equation of transfer for a gray gas with constant absorption coefficient κ_k, but with blackbody intensity I_b (for medium as well as surfaces) replaced by a weighted intensity $a_k I_b$. Thus, if the temperature field is known (or assumed), the intensity field (or simply the heat fluxes) must be determined for $k = 0, 1, \cdots, K$, using any standard solution method. The results are then added to give the total intensity (or radiative heat flux). Note that, as for the stepwise-gray approximation, it will always be necessary to know or assume a temperature profile: For radiative equilibrium the condition $\nabla \cdot \mathbf{q} = 0$ applies to the total heat flux only and, in general $\nabla \cdot \mathbf{q}_k \neq 0$.

Absorptivity Curve Fit for a Simple Gas

In general, the curve fit of the total absorptivity of the medium should be tailored to the medium at hand, and depends on composition, pressure levels, temperature levels, number of molecular gas bands, and so on. Only if the fit is optimized will one be able to achieve acceptable accuracy with a weighted sum of two or three gray gases. Unfortunately, the curve fit indicated in equation (17.117) is a nonlinear one, and is further complicated by the fact that the a_k may be functions of temperature, pressure, composition, and so forth. As a result of these difficulties the curve fitting effort may become more involved than the heat transfer calculations themselves! Some weighted-gray-gas absorptivity fits for important gases have been reported in the literature for use with the zonal method (Chapter 18), e.g., by Smith and coworkers [35] for water vapor–carbon dioxide mixtures and by Farag and Allam [36] for carbon dioxide. It would be desirable to have a "cookbook" formula available, even at the expense of numerical efficiency.

We shall limit ourselves here to the case of a gas-particulate mixture with gray particulates and spatially-independent absorption coefficients. The total absorptivity of such a medium is described by equation (17.99). Suppose the band absorptance $A^*(\tau, \beta)$ can be approximated by a weighted sum of gray gases, that is,

$$A^*(\tau, \beta) \simeq \sum_{l=1}^{L} c_l(\beta)\left[1 - e^{-b_l(\beta)\tau}\right], \qquad (17.123)$$

where the $b_l(\beta)$ and $c_l(\beta)$ are functions to be determined. If we substitute this expression into equation (17.99), and use the abbreviations

$$\psi_n = \frac{\omega_n I_{b\eta_n}}{I_b}(T), \quad b_{ln} = b_l(\beta_n), \quad c_{ln} = c_l(\beta_n),$$

we find

$$\alpha(T, s) \simeq 1 - e^{-\kappa_p s} + \sum_{n=1}^{N} \psi_n e^{-\kappa_p s} \sum_{l=1}^{L} c_{ln}\left(1 - e^{-b_{ln}\kappa_n s}\right)$$

$$= \left[1 - \sum_{n=1}^{N}\sum_{l=1}^{L} c_{ln}\psi_n(T)\right]\left(1 - e^{-\kappa_p s}\right)$$

$$+ \sum_{n=1}^{N}\sum_{l=1}^{L} c_{ln}\psi_n(T)\left[1 - e^{-(\kappa_p + b_{ln}\kappa_n)s}\right]. \qquad (17.124)$$

TABLE 17.3
Weighted-sum-of-gray-gases parameters for band absorptance at high pressures,
$A^*(\tau, \infty)$.

l	$L = 2$		$L = 2$		$L = 3$	
	c_l	b_l	c_l	b_l	c_l	b_l
1	2.997	0.2385×10^{-1}	3.308	0.3272×10^{-2}	2.741	0.1491×10^{-1}
2	2.416	0.2992	4.192	0.9466×10^{-1}	1.656	0.1101
3					2.746	0.2409
4					1.397	0.5267
RMS[†]	0.0285/100		0.0875/1000		0.0189/100	

	$L = 3$		$L = 4$		$L = 5$	
l	c_l	b_l	c_l	b_l	c_l	b_l
1	2.891	0.2046×10^{-2}	2.748	0.1522×10^{-2}	2.682	0.1216×10^{-2}
2	2.195	0.2092×10^{-1}	1.693	0.1144×10^{-1}	1.479	0.7984×10^{-2}
3	2.746	0.2409	1.769	0.6255×10^{-1}	1.361	0.3212×10^{-1}
4			1.868	0.4057	1.401	0.1269
5					1.357	0.5291
RMS[†]	0.0189/1000		0.0042/1000		0.0010/1000	

[†]RMS = a/b implies a root-mean-square error of a over the range of the fit
between $\tau = 0$ and $\tau = b$.

This equation is clearly of the same type as the desired form, equation (17.117). A number of curve fits for equation (17.123) for the high-pressure limit ($\beta \rightarrow \infty$) are given in Table 17.3 [30]. In these approximations the root-mean-square error, integrated between $\tau = 0$ and $\tau = 100$ or 1000, has been minimized. Inspection of equation (17.124) reveals that the nongray medium is represented by a weighted sum of $N \times L + 1$ gray gases.

Example 17.13.   Repeat Example 17.12 using the weighted-sum-of-gray-gases approach. In addition, determine the heat lost from the sphere if the gas is mixed with gray particulates, such that the absorption coefficient for the particles is $\kappa_p = 0.1/D$.

Solution

For the case of $N = 1$ and $\psi_n = 0.05$ equation (17.124) reduces to

$$\alpha(T_m, s) = \left(1 - 0.05\sum_{l=1}^{L} c_l\right)\left(1 - e^{-\kappa_p s}\right) + 0.05\sum l = 1^L c_l\left[1 - e^{-(\kappa_p + b_l \kappa_g)s}\right],$$

$$= \sum_{k=0}^{L} a_k\left(1 - e^{-\kappa_k s}\right),$$

where

$$a_k = \begin{cases} 1 - 0.05 \sum_{l=1}^{L} c_l, & k = 0, \\ 0.05 c_k, & k \geq 1, \end{cases}$$

$$\kappa_k = \begin{cases} \kappa_p, & k = 0, \\ \kappa_p + b_k \kappa_g, & k \geq 1, \end{cases}$$

and κ_g is the absorption coefficient of the gas at the band center (or head). Thus, the problem must be solved for L gray gases with absorption coefficients $(b_k \kappa_g)$ (gas only), or for $L + 1$ gray gases with absorption coefficients κ_p and $(\kappa_p + b_k \kappa_g)$ (gas and particulates).

The radiative intensity leaving an isothermal, gray sphere at $r = R$ is, from equation (17.105) (after setting $T_w = 0$),

$$I(R, \mu) = \alpha(T_m, D\mu) I_{bm} = \left(1 - e^{-\kappa D\mu}\right) I_{bm}.$$

The total heat flux leaving the surface of the sphere is found by integrating the normal component of the intensity over all directions, or

$$q(R) = \int_{2\pi} I |\hat{s} \cdot \hat{n}| d\Omega = 2\pi \int_0^1 I(R, \mu) \mu \, d\mu$$

$$= 2\pi I_{bm} \int_0^1 (1 - e^{-\kappa D\mu}) \mu \, d\mu,$$

$$= \pi I_{bm} \left\{ 1 - \frac{2}{(\kappa D)^2} \left[1 - (1 + \kappa D) e^{-\kappa D} \right] \right\}.$$

Applying this result to the weighted-sum-of-gray-gases model, i.e., replacing I_{bm} by $a_k I_{bm}$ and κ by κ_k, we find for the total nondimensional heat flux

$$\Psi = \frac{q(R)}{\sigma T_m^4} = \frac{\sum_{k=0}^{L} q_k(R)}{\sigma T_m^4}$$

$$= \left(1 - 0.05 \sum_{l=1}^{L} c_l\right) \left\{ 1 - \frac{2}{(\kappa_p D)^2} \left[1 - (1 + \kappa_p D) e^{-\kappa_p D} \right] \right\}$$

$$+ 0.05 \sum_{k=1}^{L} c_k \left\{ 1 - \frac{2}{(\kappa_k D)^2} \left[1 - (1 + \kappa_k D) e^{-\kappa_k D} \right] \right\},$$

where $\kappa_k = \kappa_p + b_k \kappa_g$. A plot of this function (for $\kappa_p = 0$ and $\kappa_p = 0.1/D$) is included in Fig. 17-12 for different curve-fit qualities, L.

Example 17.14.   Reconsider the isothermal medium of Example 17.13 (with and without particles). Assuming the medium is confined between two parallel cold and black plates a distance L apart, calculate the radiative heat flux within the slab, using the weighted-sum-of-gray-gases approach together with the P_1-approximation.

Solution

The P_1-approximation for an isothermal medium with absorption coefficient κ_k and a Planck function of $[a_k I_b]$, bounded by cold and black plates, is

$$\frac{dq_k}{dz} = \kappa_k\left(4\pi[a_k I_b] - G_k\right),$$

$$\frac{dG_k}{dz} = -3\kappa_k q_k,$$

$$z = 0: \quad 2q_k + G_k = 0,$$
$$z = L: \quad -2q_k + G_k = 0.$$

The answer to this simple set of equations follows immediately, as a special case of Example 14.2, as

$$q_k(z) = a_k E_{bm}\frac{2\sinh\sqrt{3}\kappa_k(z-L/2)}{\sinh\frac{1}{2}\sqrt{3}\kappa_k L + \frac{1}{2}\sqrt{3}\cosh\frac{1}{2}\sqrt{3}\kappa_k L}.$$

Since the medium is identical to the one of Example 17.13, their total absorptivities and, thus, the values for the correlation coefficients a_k and κ_k are identical as well. The total nondimensional heat flux follows as

$$\Psi = \frac{q}{\sigma T_m^4} = \frac{\sum_{k=0}^{L} q_k}{\sigma T_m^4} = \sum_{k=0}^{L}\frac{2a_k\sinh\sqrt{3}\kappa_k(z-L/2)}{\sinh\frac{1}{2}\sqrt{3}\kappa_k L + \frac{1}{2}\sqrt{3}\cosh\frac{1}{2}\sqrt{3}\kappa_k L},$$

where

$$k = 0: \qquad a_0 = 1 - 0.05\sum_{l=1}^{L} c_l, \qquad \kappa_0 = \kappa_p,$$

$$k \geq 1: \qquad a_k = 0.05 c_l, \qquad \kappa_k = \kappa_p + b_k\kappa_g.$$

The result is shown in Fig. 17-13 for $\kappa_p L = 0$ and $\kappa_p L = 0.1$, for a number of absorptivity curve fits from Table 17.3. For the slab more accurate curve fits are required than for a sphere to achieve similar accuracy (here as compared to the spectral integration of the P_1-approximation). This observation is due to the fact that, for a sphere, no path through the medium is larger than D while, for a slab, infinitely long paths exist at grazing angles to the surfaces. Thus, the absorptivity fit must be accurate over much larger optical ranges.

References

1. Wang, L. S.: "The Role of Emissivities in Radiative Transport Calculations," *Journal of Quantitative Spectroscopy and Radiative Transfer*, vol. 8, p. 1233, 1968.
2. Wang, L. S.: "An Integral Equation of Radiative Equilibrium in Infrared Radiating Gases," *Journal of Quantitative Spectroscopy and Radiative Transfer*, vol. 8, p. 851, 1968.
3. Sparrow, E. M., and R. D. Cess: *Radiation Heat Transfer*, Hemisphere, New York, 1978.
4. Cess, R. D., P. Mighdoll, and S. N. Tiwari: "Infrared Radiative Heat Transfer in Non-gray Gases," *International Journal of Heat and Mass Transfer*, vol. 10, pp. 1521–1532, 1967.
5. Cess, R. D., and S. N. Tiwari: "The Large Path Length Limit for Infrared Gaseous Radiation," *Applied Scientific Research*, vol. 19, p. 439, 1968.

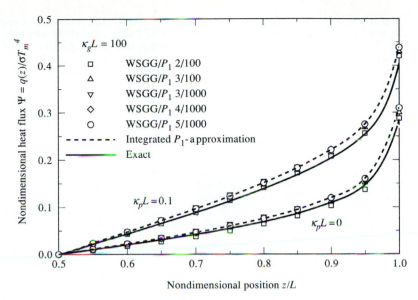

FIGURE 17-13
Local radiative heat fluxes in an isothermal gas-particulate slab with single gas band.

6. Yuen, W. W., and D. J. Rasky: "Application of the P-1 Approximation of Radiative Heat Transfer in a Non-Gray Medium," *ASME Journal of Heat Transfer*, vol. 103, pp. 182–183, 1981.

7. Yuen, W. W.: "Radiative Heat Transfer Characteristics of a Soot/Gas Mixture," in *Proceedings of the Seventh International Heat Transfer Conference*, Washington, D.C., Hemisphere, pp. 577–582, 1982.

8. Edwards, D. K., and A. Balakrishnan: "Slab Band Absorptance for Molecular Gas Radiation," *Journal of Quantitative Spectroscopy and Radiative Transfer*, vol. 12, pp. 1379–1387, 1972.

9. Modest, M. F.: "Radiative Heat Transfer in a Plane-Layer Mixture of Non-Gray Particulates and Molecular Gases," *Journal of Quantitative Spectroscopy and Radiative Transfer*, vol. 26, pp. 523–533, 1981.

10. Edwards, D. K.: "Molecular Gas Band Radiation," in *Advances in Heat Transfer*, vol. 12, Academic Press, New York, pp. 115–193, 1976.

11. Crosbie, A. L., and H. K. Khalil: "Effect of Line or Band Shape on the Radiative Flux of an Isothermal Spherical Layer," *Journal of Quantitative Spectroscopy and Radiative Transfer*, vol. 13, pp. 977–993, 1973.

12. Wassel, A. T., and D. K. Edwards: "Molecular Gas Band Radiation in Cylinders," *ASME Journal of Heat Transfer*, vol. 96, pp. 21–26, 1974.

13. Hottel, H. C.: "Radiant Heat Transmission," in *Heat Transmission*, ed. W. H. McAdams, 3d ed., ch. 4, McGraw-Hill, New York, 1954.

14. Hottel, H. C., and A. F. Sarofim: *Radiative Transfer*, McGraw-Hill, New York, 1967.

15. Dunkle, R. V.: "Geometric Mean Beam Lengths for Radiant Heat Transfer Calculations," *ASME Journal of Heat Transfer*, vol. 86, no. 1, pp. 75–80, 1964.

16. Viskanta, R.: "Concerning the Definitions of the Mean Absorption Coefficient," *International Journal of Heat and Mass Transfer*, vol. 7, no. 9, pp. 1047–1049, 1964.

17. Finkleman, D., and K. Y. Chien: "Semigrey Radiative Transfer," *AIAA Journal*, vol. 6, no. 4, pp. 755–758, 1968.

18. Finkleman, D.: "Numerical Studies in Semigray Radiative Transfer," *AIAA Journal*, vol. 7, pp. 1602–1605, 1969.

19. Finkleman, D.: "A Note on Boundary Conditions for Use with the Differential Approximation to Radiative Transfer," *International Journal of Heat and Mass Transfer*, vol. 12, pp. 653–656, 1969.

20. Traugott, S. C.: "Radiative Heat-flux Potential for a Nongrey Gas," *AIAA Journal*, vol. 4, no. 3, pp. 541–542, 1966.

21. Traugott, S. C.: "On Grey Absorption Coefficients in Radiative Transfer," *Journal of Quantitative Spectroscopy and Radiative Transfer*, vol. 8, pp. 971–999, 1968.

22. Modest, M. F., and K. K. Sikka: "The Application of the Stepwise-Gray P-1 Approximation to Molecular Gas-Particulate Mixtures," in *Fundamentals of Radiation Heat Transfer*, vol. HTD–160, ASME, pp. 97–103, July 1991 (to appear in JQSRT).

23. Chandrasekhar, S.: "The Radiative Equilibrium of the Outer Layers of a Star with Special Reference to the Blanketing Effects of the Reversing Layer," *Monthly Notices Royal Astron. Soc.*, vol. 96, pp. 21–42, 1935.

24. Siewert, C. E., and P. F. Zweifel: "An Exact Solution of Equations of Radiative Transfer for Local Thermodynamic Equilibrium in the Non-Gray Case: Picket Fence Approximation," *Ann. Phys. (N.Y.)*, vol. 36, pp. 61–85, 1966.

25. Kung, S. C., and M. Sibulkin: "Radiative Transfer in a Nongray Gas Between Parallel Walls," *Journal of Quantitative Spectroscopy and Radiative Transfer*, vol. 9, pp. 1447–1461, 1969.

26. Reith, R. J., C. E. Siewert, and M. N. Özişik: "Non-Grey Radiative Heat Transfer in Conservative Plane-Parallel Media with Reflecting Boundaries," *Journal of Quantitative Spectroscopy and Radiative Transfer*, vol. 11, pp. 1441–1462, 1971.

27. Greif, R.: "Energy Transfer by Radiation and Conduction with Variable Gas Properties," *International Journal of Heat and Mass Transfer*, vol. 7, pp. 891–900, 1964.

28. Modest, M. F.: "A Simple Differential Approximation for Radiative Transfer in Non-Gray Gases," *ASME Journal of Heat Transfer*, vol. 101, pp. 735–736, 1979.

29. Van Wylen, G. H., and R. E. Sonntag: *Fundamentals of Classical Thermodynamics*, John Wiley & Sons, New York, 1985.

30. Modest, M. F.: "The Weighted-Sum-of-Gray-Gases Model for Arbitrary Solution Methods in Radiative Transfer," *ASME Journal of Heat Transfer*, vol. 113, no. 3, pp. 650–656, 1991.

31. Edwards, D. K.: "Molecular Gas Band Radiation," in *Advances in Heat Transfer*, vol. 12, Academic Press, New York, pp. 115–193, 1976.

32. Kim, T. K., J. A. Menart, and H. S. Lee: "Nongray Radiative Gas Analyses Using the S-N Discrete Ordinates Method," *ASME Journal of Heat Transfer*, vol. 113, no. 4, pp. 946–952, 1991.

33. Abramowitz, M., and I. A. Stegun (eds.): *Handbook of Mathematical Functions*, Dover Publications, New York, 1965.

34. Leckner, B.: "Spectral and Total Emissivity of Water Vapor and Carbon Dioxide," *Combustion and Flame*, vol. 19, pp. 33–48, 1972.

35. Smith, T. F., Z. F. Shen, and J. N. Friedman: "Evaluation of Coefficients For the Weighted Sum of Gray Gases Model," *ASME Journal of Heat Transfer*, vol. 104, pp. 602–608, 1982.

36. Farag, I. H., and T. A. Allam: "Gray-Gas Approximation of Carbon Dioxide Standard Emissivity," *ASME Journal of Heat Transfer*, vol. 103, pp. 403–405, 1981.

Problems

17.1 Two parallel, infinite, black plates at constant temperatures T_1 and T_2 are separated by a nongray medium of geometrical thickness $d = 10$ cm that is at radiative equilibrium. The absorption characteristics of the medium are such that they can be approximated by

$$\kappa_\lambda = \begin{cases} \overline{\kappa} = 1\,\text{cm}^{-1} & 3\,\mu\text{m} < \lambda < 7\,\mu\text{m}, \\ 0 & \text{elsewhere.} \end{cases}$$

Calculate the nondimensional heat flux, $q/\sigma(T_1^4 - T_2^4)$, for a number of T_2 ($T_2 = 500$ K,

750 K, 1000 K, 1500 K, and 2000 K) and $T_1 = 300$ K by

(a) the differential approximation, using a gray gas with Planck-mean absorption coefficient κ_P,

(b) the nongray differential approximation.

For the evaluation of κ_P you may use $T_m = (T_1 + T_2)/2$. Plot, compare, and discuss your results.

17.2 A cold-walled cylindrical furnace of 1 m radius contains pure CO_2 that is isothermal at 1700 K and at a pressure of p atm. Using the (i) gray and (ii) nongray differential approximation with single bandstrength $\bar{\kappa}$, determine the nondimensional wall heat flux $\Psi = q_w/\sigma T^4$ as a function of pressure. Plot Ψ vs. p (actual calculations for $p = 0.001, 0.01, 0.1$, and 1.0 should suffice; for simplification, you may assume that band width is not a function of p).

17.3 Repeat Problem 17.2, adding steam at 0.1 atm partial pressure to the medium. You may assume that only the 2.7 and 6.3 μm bands are of importance.

17.4 An infinitely long cylinder of radius $R = 10$ cm is bounded by a cold black wall. Inside the cylinder there is uniform heat generation of $\dot{Q}''' = 38,136$ W/m^3. Estimate wall heat fluxes and temperature distributions using the differential approximation if

(a) the medium has a band at $\lambda = 4$ μm of width $\Delta\lambda = 1$ μm; across the band it has a constant absorption coefficient such that $\bar{\kappa}R = 100$.

(b) the medium is gray with an "appropriately" chosen κ_P, say by evaluating κ_P at the volume-averaged temperature T_{av}, that is

$$T_{av}^4 = \frac{1}{V}\int_V T^4 dV .$$

17.5 Consider a sphere of very hot molecular gas of radius 50 cm. The gas has a single vibration-rotation band at $\eta_0 = 3000$ cm^{-1}, is suspended magnetically in a vacuum within a large cold container and is initially at a uniform temperature $T_g = 3000$ K. For this gas $(\rho_a\alpha)(T) = 500$ cm^{-2}, $\omega(T) = 100\sqrt{T/100\,\text{K}}$ cm^{-1} and $\beta \gg 1$. These properties imply that the absorption coefficient may be determined from

$$\kappa_\eta = \kappa_0 e^{-2|\eta-\eta_0|/\omega}, \qquad \kappa_0 = \frac{\rho_a\alpha}{\omega}$$

and the band absorptance from

$$A(s) = \omega A^* = \omega[E_1(\kappa_0 s) + \ln(\kappa_0 s) + \gamma_E], \qquad \gamma_E = 0.577216.$$

Using the stepwise-gray model together with the P_1-approximation and neglecting conduction and convection, specify the total heat loss per unit time from the entire sphere at time $t = 0$. Outline the solution procedure for times $t > 0$.
Hint: Solve the governing equation by introducing a new dependent variable $g(\tau) = \tau(4\pi I_b - G)$.

17.6 Repeat Problem 17.5 using the exact integral relations together with the exponential wide band model.

17.7 Repeat Problem 17.5 using the weighted-sum-of-gray gases approach together with the P_1-approximation.

17.8 Repeat Problem 17.5 for varying line overlap β, say $\beta = 0.01$, 0.1, 1, and 10. Plot heat loss at $t = 0$ vs. β.

Hint: Use Table 9.2 or some other correlation for the band absorptance.

17.9 Repeat Problem 17.8 using the exact integral relations together with the exponential wide band model.

CHAPTER
18

THE ZONAL
METHOD

18.1 INTRODUCTION

The zonal method for the determination of radiative heat transfer rates within an absorbing, emitting, and isotropically scattering medium is an extension of the net radiation method developed in Chapter 5 for surface exchange (i.e., for enclosures without a participating medium). In this method the enclosure is subdivided into a finite number of isothermal volume and surface area zones. An energy balance is then performed for the radiative exchange between any two zones, employing precalculated "exchange areas." This process leads to a set of simultaneous equations for the unknown temperatures or heat fluxes. The method was first developed by Hottel and Cohen [1] for an absorbing, emitting, nonscattering gray gas with constant absorption coefficient. Hottel and Sarofim [2] extended it to deal with nonconstant and nongray absorption coefficients as well as with isotropically scattering media. The discussion by Hottel and Sarofim [2] is extensive but is limited to three-dimensional problems (i.e., only volume zones finite in all three dimensions are discussed). The method was extended by Walther and coworkers [3] to allow for linear variation of emissive power through a one-dimensional slab zone and by Einstein to deal with two-dimensional zones in Cartesian [4] and cylindrical [5] coordinate systems. A slight variation of the method has been given by Larsen and Howell [6], who expressed the energy balances in terms of "exchange factors," which are physically measurable quantities. Such measurements were carried out by Liu and Howell [7].

639

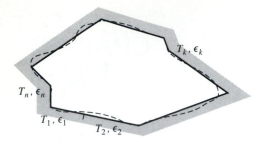

FIGURE 18-1
General enclosure broken into N isothermal surface zones.

18.2 SURFACE EXCHANGE — NO PARTICIPATING MEDIUM

We shall start by rederiving the relations for radiative exchange between surfaces in the zonal form, since doing so clearly demonstrates the similarities between the standard net radiation method and the zonal method. As in the net radiation approach we break up the surface of the enclosure into N isothermal subsurfaces, or zones, as shown in Fig. 18-1.

Black Surfaces — Direct Exchange Areas

If the enclosure consists of only black surfaces then the net exchange of radiative energy between any two surfaces is, from equation (5.9),

$$Q_{i \leftrightarrow j} = -Q_{j \leftrightarrow i} = \overline{s_i s_j}(E_{bi} - E_{bj}), \qquad i, j = 1, 2, \ldots, N, \qquad (18.1)$$

where $Q_{i \leftrightarrow j}$ is the net energy exchange between zones i and j (positive if zone i is losing heat as a result of the exchange) and the

$$\overline{s_i s_j} = \overline{s_j s_i} = A_i F_{i \to j} = A_j F_{j \to i} = \int_{A_i} \int_{A_j} \frac{\cos \theta_i \cos \theta_j}{\pi S_{ij}^2} dA_j \, dA_i \qquad (18.2)$$

are known as *direct exchange areas*. Unlike the view factors, $F_{i \to j}$, the $\overline{s_i s_j}$ are not nondimensional but have the dimensions of area. While this is a minor inconvenience the formulation of equation (18.2) has the advantage that the principle of reciprocity is more easily applied (eliminating a common source of error). Although the use of *direct exchange factors* (reducing to view factors in the absence of a participating medium) would be just as acceptable, use of exchange areas is accepted practice. Summing equation (18.1) over all zones yields the net heat flux at zone i as

$$Q_i = A_i q_i = \sum_{j=1}^{N} \overline{s_i s_j}(E_{bi} - E_{bj}) = A_i E_{bi} - \sum_{j=1}^{N} \overline{s_i s_j} E_{bj}, \quad i = 1, 2, \ldots, N. \quad (18.3)$$

Since the total heat flux leaving surface i is $A_i E_{bi}$, we find that

$$\sum_{j=1}^{N} \overline{s_i s_j} = A_i, \qquad (18.4)$$

which is equivalent to equation (4.17), or $\sum_{j=1}^{N} F_{i \to j} = 1$.

Gray Diffuse Surfaces — Total Exchange Areas

If the surfaces are not black but are partially reflective, then the energy exchange between zones is not only by direct travel, but also may include contributions due to single and multiple reflections from any number of surface zones. If the reflective behavior of the surfaces is gray and diffuse, then the reflected radiation leaving a surface cannot be distinguished from emitted radiation and equation (18.1) still holds after replacing emissive power E_b by radiosity J (as described in detail in Chapter 5). Thus,

$$Q_{i \leftrightarrow j} = -Q_{j \leftrightarrow i} = \overline{s_i s_j} \, (J_i - J_j), \quad i, j = 1, 2, \ldots, N, \tag{18.5}$$

and, performing an energy balance on A_i,

$$Q_i = A_i q_i = J_i - H_i = J_i - \sum_{j=1}^{N} \overline{s_i s_j} \, J_j, \quad i = 1, 2, \ldots, N. \tag{18.6}$$

Thus, similar to the net radiation method, we must now assume that the surface zones are small enough so that their *radiosities* do not vary appreciably across them.

We may eliminate the radiosities from equation (18.6) by using equation (5.26),

$$q_i = \frac{\epsilon_i}{1 - \epsilon_i} (E_{bi} - J_i), \quad i = 1, 2, \ldots, N. \tag{18.7}$$

This leads to

$$\sum_{j=1}^{N} \left(\frac{A_j \delta_{ij}}{\epsilon_j} - \frac{1 - \epsilon_j}{\epsilon_j} \, \overline{s_i s_j} \right) q_j = \sum_{j=1}^{N} (A_j \delta_{ij} - \overline{s_i s_j}) E_{bj}, \quad i = 1, 2, \ldots, N, \tag{18.8}$$

which is equivalent to equation (5.37). Thus, if all temperatures are known, the unknown wall heat fluxes may be determined by matrix inversion. If the zonal temperatures are not known, but must be determined iteratively by considering conduction and/or convection as well, then one matrix inversion must be performed for every iteration. To avoid such unnecessary matrix inversions (since the exchange areas do not depend on temperature), Hottel and Cohen [1] introduced the concept of *total exchange areas*, defined by

$$Q_{i \leftrightarrow j} = -Q_{j \leftrightarrow i} = \overline{S_i S_j} \, (E_{bi} - E_{bj}), \quad i, j = 1, 2, \ldots, N, \tag{18.9}$$

which include energy exchange by direct travel as well as by paths with one or more surface reflections. This concept implies that reciprocity holds also for total exchange areas, that is,

$$\overline{S_i S_j} = \overline{S_j S_i}, \quad \text{all } i, j. \tag{18.10}$$

Once the $\overline{S_i S_j}$ have been determined, the heat flux for each zone is found immediately and without matrix inversion from

$$Q_i = A_i q_i = \sum_{j=1}^{N} \overline{S_i S_j} (E_{bi} - E_{bj})$$

$$= \epsilon_i A_i E_{bi} - \sum_{j=1}^{N} \overline{S_i S_j} E_{bj}, \quad i = 1, 2, \ldots, N, \qquad (18.11)$$

where the last part of equation (18.11) follows from the fact that total emission from zone i is $\epsilon_i A_i E_{bi}$. We conclude that, for a nonparticipating medium,

$$\sum_{j=1}^{N} \overline{S_i S_j} = \epsilon_i A_i, \quad i = 1, 2, \ldots, N. \qquad (18.12)$$

There are many ways of expressing the total exchange areas $\overline{S_i S_j}$ in terms of direct exchange areas $\overline{s_i s_j}$. Hottel and Cohen [1] and Hottel and Sarofim [2] achieved this by setting the emissive power of all zones to zero except for zone k, for which the emissive power is set to unity. From equation (18.11) it follows then that, for this case,

$$Q_i = -\overline{S_i S_k} = {}_k J_i - \sum_{j=1}^{N} \overline{s_i s_j} {}_k J_j, \quad i = 1, 2, \ldots, N, \qquad (18.13)$$

where the presubscript k for the J_i implies that these artificial radiosities are for a single emitting zone with unit emissive power. The Q_i can be eliminated using equation (18.7), resulting in N simultaneous equations for the unknown ${}_k J_i$ ($i = 1, 2, \ldots, N$, k fixed). After their determination the $\overline{S_i S_k}$ may be determined from equation (18.13). Utilizing a different approach, Noble [8] cast the governing equations in matrix form and used elegant matrix manipulation to evaluate the total exchange areas. We shall follow here an approach similar to Noble's but shall use a somewhat more conventional notation to accomodate the reader who is not very familiar with matrix manipulation.

Making a simple heat balance for zone i,

$$Q_i = \epsilon_i A_i E_{bi} - \epsilon_i A_i H_i, \qquad (18.14)$$

(stating that the net heat flux leaving zone i is the difference between emitted and absorbed irradiation) we find from equation (18.6) that the direct exchange areas are related to the absorbed irradiation, $\epsilon_i A_i H_i$, where H_i is total irradiation per unit area on zone i, or

$$A_i H_i = \sum_{j=1}^{N} \overline{s_i s_j} J_j, \quad i = 1, 2, \ldots, N. \qquad (18.15)$$

We now eliminate the radiosities from equation (18.15) using its definition, equation (5.18),

$$J_j = \epsilon_j E_{bj} + \rho_j H_j, \qquad (18.16)$$

which leads to

$$\sum_{j=1}^{N}\left(\frac{\delta_{ij}}{\epsilon_j} - \frac{\rho_j \overline{s_i s_j}}{\epsilon_j A_j}\right)\epsilon_j A_j H_j = \sum_{j=1}^{N}\overline{s_i s_j}\,\epsilon_j E_{bj}, \quad i = 1, 2, \ldots, N. \tag{18.17}$$

Equation (18.17) is a set of N linear equations in the N unknown H_j. The general solution to this equation is most easily found by casting it into matrix form. We define two $N \times N$ matrices (or *second rank tensors*),

$$\mathbf{T} = \begin{pmatrix} \dfrac{1}{\epsilon_1} - \dfrac{\rho_1 \overline{s_1 s_1}}{\epsilon_1 A_1} & -\dfrac{\rho_2 \overline{s_1 s_2}}{\epsilon_2 A_2} & \cdots & -\dfrac{\rho_N \overline{s_1 s_N}}{\epsilon_N A_N} \\[2.5ex] -\dfrac{\rho_1 \overline{s_2 s_1}}{\epsilon_1 A_1} & \dfrac{1}{\epsilon_2} - \dfrac{\rho_2 \overline{s_2 s_2}}{\epsilon_2 A_2} & \cdots & -\dfrac{\rho_N \overline{s_2 s_N}}{\epsilon_N A_N} \\[2.5ex] \vdots & \vdots & \ddots & \vdots \\[2.5ex] -\dfrac{\rho_1 \overline{s_N s_1}}{\epsilon_1 A_1} & -\dfrac{\rho_2 \overline{s_N s_2}}{\epsilon_2 A_2} & \cdots & \dfrac{1}{\epsilon_N} - \dfrac{\rho_N \overline{s_N s_N}}{\epsilon_N A_N} \end{pmatrix} \tag{18.18}$$

and

$$\mathbf{S} = \begin{pmatrix} \overline{s_1 s_1}\,\epsilon_1 & \overline{s_1 s_2}\,\epsilon_2 & \cdots & \overline{s_1 s_N}\,\epsilon_N \\[1.5ex] \overline{s_2 s_1}\,\epsilon_1 & \overline{s_2 s_2}\,\epsilon_2 & \cdots & \overline{s_2 s_N}\,\epsilon_N \\[1.5ex] \vdots & \vdots & \ddots & \vdots \\[1.5ex] \overline{s_N s_1}\,\epsilon_1 & \overline{s_N s_2}\,\epsilon_2 & \cdots & \overline{s_N s_N}\,\epsilon_N \end{pmatrix}, \tag{18.19}$$

as well as two vectors $\mathbf{h}$ and $\mathbf{e_b}$,[1]

$$\mathbf{h} = \begin{pmatrix} \epsilon_1 A_1 H_1 \\ \epsilon_2 A_2 H_2 \\ \vdots \\ \epsilon_N A_N H_N \end{pmatrix}, \quad \mathbf{e_b} = \begin{pmatrix} E_{b1} \\ E_{b2} \\ \vdots \\ E_{bN} \end{pmatrix}. \tag{18.20}$$

Equation (18.17) may then be rewritten[2] as

$$\mathbf{T} \cdot \mathbf{h} = \mathbf{S} \cdot \mathbf{e_b}. \tag{18.21}$$

---

[1] For easy readability of matrix manipulations we shall follow here the convention that a two-dimensional matrix is denoted by a bold capitalized letter, while a vector is written as a bold lowercase letter.

[2] Again, for easy readability of matrix manipulations we adopt the convention that a dot product denotes summation over the closest indices on both sides of the dot, e.g., $\mathbf{T} \cdot \mathbf{h} = \left[\sum_j T_{ij} h_j\right]$, $\mathbf{D} \cdot \mathbf{T} \cdot \mathbf{h} = \left[\sum_i \sum_j D_{ki} T_{ij} h_j\right]$, etc. Thus, a tensor dotted with a vector gives a vector, a tensor dotted with a tensor results in another tensor, and so forth.

If we form the dot product of equation (18.21) with the *inverse* of **T**, or **T**$^{-1}$, we get

$$\mathbf{T}^{-1}{\cdot}\mathbf{T}\cdot\mathbf{h} = \mathbf{h} = \mathbf{T}^{-1}\cdot\mathbf{S}\cdot\mathbf{e_b}, \tag{18.22}$$

where we have made use of the fact that $\mathbf{T}^{-1}\cdot\mathbf{T} = \boldsymbol{\delta}$, where $\boldsymbol{\delta}$ is the *unit tensor* with elements δ_{ij}, and $\boldsymbol{\delta}\cdot\mathbf{h} = \left[\sum_j \delta_{ij}h_j\right] = [h_i] = \mathbf{h}$. Thus, we may write, in series notation,

$$\epsilon_i A_i H_i = \sum_{k=1}^{N}\sum_{j=1}^{N}(T^{-1})_{ik}S_{kj}E_{bj}. \tag{18.23}$$

By comparing this expression with equations (18.11) and (18.14) we find the *total exchange areas* as

$$\overline{S_iS_j} = \sum_{k=1}^{N}(T^{-1})_{ik}S_{kj}, \tag{18.24}$$

or, in matrix notation

$$\overline{\mathbf{SS}} = \mathbf{T}^{-1}\cdot\mathbf{S}. \tag{18.25}$$

We note in passing that, for every black zone, an entire column of **T** has elements that are zero (with the exception of the diagonal term), simplifying the inversion of **T**.

Example 18.1. Evaluate the radiative heat flux between two infinitely long concentric cylinders separated by a nonparticipating medium. Both cylinders are isothermal and are covered with a gray, diffusely emitting and reflecting material. The inner cylinder is of radius R_1 with temperature T_1 and emissivity ϵ_1. The outer cylinder has the corresponding values of R_2, T_2, and ϵ_2. Use the zonal method and employ the concept of total exchange areas.

Solution

Letting the inner cylinder be Zone 1 and the outer cylinder Zone 2 we have

$$\overline{s_1s_2} = \overline{s_2s_1} = A_1F_{1\to2} = A_1, \quad \overline{s_1s_1} = 0, \quad \overline{s_2s_2} = A_2F_{2\to2} = A_2 - A_1.$$

Therefore, we get

$$\mathbf{T} = \begin{pmatrix} \dfrac{1}{\epsilon_1} & -\dfrac{\rho_2}{\epsilon_2}\dfrac{A_1}{A_2} \\[2ex] -\dfrac{\rho_1}{\epsilon_1} & 1 + \dfrac{\rho_2}{\epsilon_2}\dfrac{A_1}{A_2} \end{pmatrix},$$

$$\mathbf{S} = \begin{pmatrix} 0 & A_1\epsilon_2 \\[2ex] A_1\epsilon_1 & (A_2 - A_1)\epsilon_2 \end{pmatrix}.$$

The inverse of **T** follows immediately as

$$\mathbf{T}^{-1} = \frac{1}{T}\begin{pmatrix} 1 + \dfrac{\rho_2}{\epsilon_2}\dfrac{A_1}{A_2} & \dfrac{\rho_2}{\epsilon_2}\dfrac{A_1}{A_2} \\[3mm] \dfrac{\rho_1}{\epsilon_1} & \dfrac{1}{\epsilon_1} \end{pmatrix},$$

where

$$T = |\mathbf{T}| = \frac{1}{\epsilon_1} + \frac{\rho_2}{\epsilon_2}\frac{A_1}{A_2}.$$

The total exchange areas are calculated from

$$\overline{S_i S_j} = \sum_{k=1}^{2} (T^{-1})_{ik} S_{kj},$$

or

$$\overline{S_1 S_1} = T_{11}^{-1} S_{11} + T_{12}^{-1} S_{21} = \frac{1}{T}\frac{\rho_2}{\epsilon_2}\frac{A_1}{A_2} A_1 \epsilon_1,$$

$$\overline{S_1 S_2} = T_{11}^{-1} S_{12} + T_{12}^{-1} S_{22} = \overline{S_2 S_1} = \frac{1}{T}\left[\left(1 + \frac{\rho_2}{\epsilon_2}\frac{A_1}{A_2}\right) A_1 \epsilon_2 + \frac{\rho_2}{\epsilon_2}\frac{A_1}{A_2}(A_2 - A_1)\epsilon_2 \right] = \frac{A_1}{T},$$

$$\overline{S_2 S_2} = T_{21}^{-1} S_{12} + T_{22}^{-1} S_{22} = \frac{1}{T}\left[\frac{\rho_1}{\epsilon_1} A_1 \epsilon_2 + \frac{\epsilon_2}{\epsilon_1}(A_2 - A_1) \right].$$

The net heat flux at Surface 1 follows as

$$Q_1 = A_1 q_1 = \overline{S_1 S_2}(E_{b1} - E_{b2})$$

or

$$\Psi_1 = \frac{q_1}{\sigma(T_1^4 - T_2^4)} = \frac{1}{T} = \frac{1}{\dfrac{1}{\epsilon_1} + \dfrac{A_1}{A_2}\left(\dfrac{1}{\epsilon_2} - 1\right)}.$$

This is, of course, somewhat of a hard way to arrive at the well-known result. The method will save considerable computer time, however, if many zones and iterative determination of the temperature field are involved.

18.3 RADIATIVE EXCHANGE IN GRAY ABSORBING/EMITTING MEDIA

We shall now consider radiative transfer through a gray absorbing, emitting, but nonscattering medium with constant absorption coefficient. The medium is confined by an enclosure with gray, diffusely emitting and reflecting surfaces. Again, the surfaces are broken into N isothermal zones (with weakly varying radiosities), while the medium is broken into K isothermal volume zones. There will now be radiative transfer between surface zones and other surface zones, from surface zones to volume zones and vice versa, and from volume zones to volume zones.

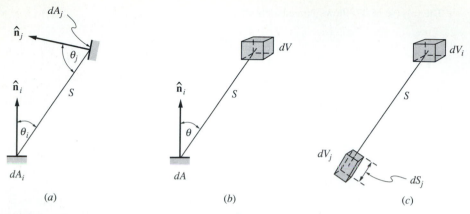

FIGURE 18-2
Radiative exchange between (a) two differential surface elements, (b) differential volume and surface elements, (c) two differential volume elements.

Direct Exchange Areas

Surface–Surface Exchange

The direct exchange area between two surface zones is defined, as for pure surface exchange, by

$$Q_{i \to j} = \overline{s_i s_j}\, J_i,$$

where $Q_{i \to j}$ is the total heat flux coming from zone i that travels directly (without reflections) to zone j. The exchange areas are the same as in the last section, except that only a fraction of the energy leaving i toward j will arrive at j (while the rest will be absorbed by the medium). Thus, similar to the development of standard view factors developed in Chapter 4, we have for the heat flux leaving zone i arriving at j
=intensity leaving dA_i into the direction of dA_j × area normal to ray × solid angle subtended by dA_j as seen from dA_i × fraction transmitted, or

$$\left(\frac{J_1}{\pi} \right) \times (dA_i \cos \theta_i) \times \left(\frac{dA_j \cos \theta_j}{S^2} \right) \times e^{-\kappa S},$$

as shown in Fig. 18-2a. The *surface-to-surface direct exchange area* may, therefore, be determined from

$$\overline{s_i s_j} = \int_{A_i} \int_{A_j} e^{-\kappa S}\, \frac{\cos \theta_i \cos \theta_j}{\pi S^2}\, dA_j\, dA_i. \qquad (18.26)$$

Equation (18.26) is identical to equation (18.2) except for the transmission factor $e^{-\kappa S}$.

Volume–Surface Exchange

The volume-to-surface direct exchange area is defined similarly through

$$Q_{i \to j} = \overline{g_i s_j} \, E_{bi}.$$

Inspecting Fig. 18-2b, we find that the heat flux emitted from volume zone i arriving at surface zone j = energy emitted from dV_i into all 4π directions as given by equation (8.49) × fraction leaving toward dA_j × fraction transmitted, or

$$(4\kappa E_{bi} \, dV_i) \times \left(\frac{dA_j \cos \theta_j}{4\pi S^2} \right) \times e^{-\kappa S},$$

leading to a *volume-to-surface direct exchange area* of

$$\overline{g_i s_j} = \int_{V_i} \int_{A_j} e^{-\kappa S} \frac{\cos \theta_j}{\pi S^2} \kappa \, dA_j \, dV_i. \tag{18.27}$$

Volume–Volume Exchange

The volume-to-volume direct exchange area is related to heat flux leaving one zone and absorbed by another after direct travel (without reflections) through

$$Q_{i \to j} = \overline{g_i g_j} \, E_{bi}.$$

The magnitude of the direct exchange area is most easily derived by orienting the receiving volume, $dV_j = dA_j \times dS_j$, as shown in Fig. 18-2c: The heat flux from i intercepted by j is equal to energy emitted from dV_i into all 4π directions × fraction leaving toward dA_j × fraction transmitted × fraction absorbed over dS_j, or

$$(4\kappa E_{bi} \, dV_i) \times \left(\frac{dA_j}{4\pi S^2} \right) \times (e^{-\kappa S}) \times (\kappa \, dS_j),$$

leading to a *volume-to-volume direct exchange area* of

$$\overline{g_i g_j} = \int_{V_i} \int_{V_j} e^{-\kappa S} \frac{\kappa^2}{\pi S^2} \, dV_j \, dV_i. \tag{18.28}$$

Making the same arguments in the opposite direction, or by inspecting equations (18.26) through (18.28), we find that reciprocity for direct exchange areas holds, that is,

$$\overline{s_i s_j} = \overline{s_j s_i}, \quad \overline{g_i s_j} = \overline{s_j g_i}, \quad \overline{g_i g_j} = \overline{g_j g_i}. \tag{18.29}$$

Making an energy balance on surface zone i, we have

$$
\begin{aligned}
Q_{si} &= A_i q_i = A_i (J_i - H_{si}) = A_i \epsilon_i (E_{bsi} - H_{si}), \\
&= \sum_{j=1}^{N} \overline{s_j s_i} \, (J_i - J_j) + \sum_{k=1}^{K} \overline{g_k s_i} \, (J_i - E_{bgk}) \\
&= \epsilon_i \left(A_i E_{bsi} - \sum_{j=1}^{N} \overline{s_j s_i} \, J_j - \sum_{k=1}^{K} \overline{g_k s_i} \, E_{bgk} \right), \quad i = 1, 2, \ldots, N, \tag{18.30}
\end{aligned}
$$

where we have added the subscripts s and g to distinguish between emissive powers and irradiation of surface and volume zones, respectively.

The first version of equation (18.30) uses the concept that the net heat flux at zone i is equal to the sum of the net exchange between any two zones, and the second states that net heat flux is equal to total emission minus the absorbed fraction of total irradiation. Realizing that the exchange areas do not depend on temperatures and that, for an isothermal enclosure, all heat fluxes go to zero (as well as $J_j = E_{bj}$), we get the relationship

$$\sum_{j=1}^{N} \overline{s_j s_i} + \sum_{k=1}^{K} \overline{g_k s_i} = A_i, \quad i = 1, 2, \ldots, N. \tag{18.31}$$

Similarly, making an energy balance over a volume zone, we find[3]

$$Q_{gi} = \kappa V_i(4E_{bgi} - G_i) = \sum_{j=1}^{N} \overline{s_j g_i}(E_{bgi} - J_j) + \sum_{k=1}^{K} \overline{g_k g_i}(E_{bgi} - E_{bgk})$$

$$= 4\kappa V_i E_{bgi} - \sum_{j=1}^{N} \overline{s_j g_i} J_j - \sum_{k=1}^{K} \overline{g_k g_i} E_{bgk}, \quad i = 1, 2, \ldots, K. \tag{18.32}$$

Comparing, or looking at an isothermal enclosure, we find

$$\sum_{j=1}^{N} \overline{s_j g_i} + \sum_{k=1}^{K} \overline{g_k g_i} = 4\kappa V_i, \quad i = 1, 2, \ldots, K. \tag{18.33}$$

For a volume zone the Q_{gi} represents the net radiative source within volume V_i and is, therefore,

$$Q_{gi} = \int_{V_i} \nabla \cdot \mathbf{q} \, dV_i. \tag{18.34}$$

If the Q_{si} for surface zones are eliminated through equation (18.7), and if the Q_{gi} for volume zones are known (by assuming radiative equilibrium) or the gas zone emissive powers are known (through connection with an overall energy equation), then equations (18.30) and (18.32) form a system of $N + K$ equations in the unknowns J_i ($i = 1, 2, \ldots, N$) and Q_{gk} or E_{bgk} ($k = 1, 2, \ldots, K$).

Total Exchange Areas

Again, we would like to eliminate the radiosities, J_j, from equations (18.30) and (18.32) to avoid repeated matrix inversions in case the zonal temperatures must be

---

[3] Here $\kappa V_i G_i = \int_{V_i} \int_{4\pi} \kappa I \, d\Omega \, dV_i$ is the total incident radiation absorbed by V_i.

determined by iteration. The total exchange areas are defined by

$$Q_{si} = \epsilon_i A_i E_{bsi} - \sum_{j=1}^{N} \overline{S_i S_j} E_{bsj} - \sum_{k=1}^{K} \overline{S_i G_k} E_{bgk}, \qquad i = 1, 2, \ldots, N, \quad (18.35)$$

$$Q_{gi} = 4\kappa V_i E_{bgi} - \sum_{j=1}^{N} \overline{G_i S_j} E_{bsj} - \sum_{k=1}^{K} \overline{G_i G_k} E_{bgk}, \qquad i = 1, 2, \ldots, K. \quad (18.36)$$

The law of reciprocity follows again from the definition of total exchange areas, that is,

$$\overline{S_j S_i} = \overline{S_i S_j}, \qquad \overline{G_k S_i} = \overline{S_i G_k}, \qquad \overline{G_k G_i} = \overline{G_i G_k}, \qquad (18.37)$$

and, setting all emissive powers equal in equations (18.35) and (18.36) (resulting in zero heat fluxes everywhere) gives the relations

$$\sum_{j=1}^{N} \overline{S_i S_j} + \sum_{k=1}^{K} \overline{S_i G_k} = \epsilon_i A_i, \qquad i = 1, 2, \ldots, N, \qquad (18.38)$$

$$\sum_{j=1}^{N} \overline{G_i S_j} + \sum_{k=1}^{K} \overline{G_i G_k} = 4\kappa V_i, \qquad i = 1, 2, \ldots, K. \qquad (18.39)$$

To determine the total exchange factors, we extract the surface irradiation from equation (18.30), giving

$$A_i H_{si} = \sum_{j=1}^{N} \overline{s_i s_j} J_j + \sum_{k=1}^{N} \overline{s_i g_k} E_{bgk}, \qquad i = 1, 2, \ldots, N, \qquad (18.40)$$

or, after using equation (18.16),

$$\sum_{j=1}^{N} \left(\frac{\delta_{ij}}{\epsilon_j} - \frac{\rho_j \, \overline{s_i s_j}}{\epsilon_j A_j} \right) \epsilon_j A_j H_{sj} = \sum_{j=1}^{N} \overline{s_i s_j} \, \epsilon_j E_{bsj} + \sum_{k=1}^{K} \overline{s_i g_k} E_{bgk},$$

$$i = 1, 2, \ldots, N. \qquad (18.41)$$

In vector notation we may write[4]

$$\mathbf{T} \cdot \mathbf{h_s} = \mathbf{S} \cdot \mathbf{e_{bs}} + \overline{\mathbf{sg}} \cdot \mathbf{e_{bg}}, \qquad (18.42)$$

where $\mathbf{T}$, $\mathbf{h_s}$, $\mathbf{S}$, $\mathbf{e_{bs}}$ are defined as in the previous section (although the direct exchange areas in them are different because of the transmission factor), while the other two

---

[4]We augment our vector notation rules here somewhat by stating that a bolded and barred two-letter name may be used for a matrix (besides a bold capital letter).

are defined as

$$
\overline{sg} = \begin{pmatrix} \overline{s_1 g_1} & \overline{s_1 g_2} & \cdots & \overline{s_1 g_K} \\ \overline{s_2 g_1} & \overline{s_2 g_2} & \cdots & \overline{s_2 g_K} \\ \vdots & \vdots & \ddots & \vdots \\ \overline{s_N g_1} & \overline{s_N g_2} & \cdots & \overline{s_N g_K} \end{pmatrix}, \quad \mathbf{e_{bg}} = \begin{pmatrix} E_{bg1} \\ E_{bg2} \\ \vdots \\ E_{bgK} \end{pmatrix}. \tag{18.43}
$$

Note that $\mathbf{T}$ and $\mathbf{S}$ are $N \times N$ matrices and $\mathbf{h_s}$ and $\mathbf{e_{bs}}$ are N-vectors, while $\overline{sg}$ is a $N \times K$ matrix and $\mathbf{e_{bg}}$ is a K-vector. Inverting $\mathbf{T}$ gives

$$
\mathbf{h_s} = \mathbf{T}^{-1} \cdot \mathbf{S} \cdot \mathbf{e_{bs}} + \mathbf{T}^{-1} \cdot \overline{sg} \cdot \mathbf{e_{bg}}, \tag{18.44}
$$

and, after comparing with equation (18.35), the total exchange areas can be identified as

$$
\overline{SS} = \mathbf{T}^{-1} \cdot \mathbf{S}, \quad \text{or} \quad \overline{S_i S_j} = \sum_{l=1}^{N} (T^{-1})_{il} S_{lj}, \tag{18.45}
$$

$$
\overline{SG} = \mathbf{T}^{-1} \cdot \overline{sg}, \quad \text{or} \quad \overline{S_i G_k} = \sum_{l=1}^{N} (T^{-1})_{il} \overline{s_l g_k}. \tag{18.46}
$$

A similar procedure for the volume zones gives, from equation (18.32),

$$
\begin{aligned}
\kappa V_i G_i &= \sum_{j=1}^{N} \overline{g_i s_j} J_j + \sum_{k=1}^{K} \overline{g_i g_k} E_{bgk}, \\
&= \sum_{j=1}^{N} \overline{g_i s_j} (\epsilon_j E_{bsj} + \rho_j H_{sj}) + \sum_{k=1}^{K} \overline{g_i g_k} E_{bgk}, \quad i = 1, 2, \ldots, K. \tag{18.47}
\end{aligned}
$$

Remembering that $h_{sj} = \epsilon_j A_j H_{sj}$ and using equations (18.44) and (18.45), we find that equation (18.47) is transformed to

$$
\begin{aligned}
\kappa V_i G_i &= \sum_{j=1}^{N} \overline{g_i s_j} \epsilon_j E_{bsj} + \sum_{j=1}^{N} \frac{\overline{g_i s_j}}{A_j} \frac{\rho_j}{\epsilon_j} \sum_{l=1}^{N} \overline{S_j S_l} E_{bsl} \\
&+ \sum_{j=1}^{N} \frac{\overline{g_i s_j}}{A_j} \frac{\rho_j}{\epsilon_j} \sum_{k=1}^{K} \overline{S_j G_k} E_{bgk} + \sum_{k=1}^{K} \overline{g_i g_k} E_{bgk}, \quad i = 1, 2, \ldots, K. \tag{18.48}
\end{aligned}
$$

Switching the dummy counters j and l in the second term on the right-hand side and rearranging, we have

$$
\begin{aligned}
\kappa V_i G_i &= \sum_{j=1}^{N} \left(\overline{g_i s_j} \epsilon_j + \sum_{l=1}^{N} \frac{\overline{g_i s_l}}{A_l} \frac{\rho_l}{\epsilon_l} \overline{S_l S_j} \right) E_{bsj} \\
&+ \sum_{k=1}^{K} \left(\overline{g_i g_k} + \sum_{l=1}^{N} \frac{\overline{g_i s_l}}{A_l} \frac{\rho_l}{\epsilon_l} \overline{S_l G_k} \right) E_{bgk}, \quad i = 1, 2, \ldots, K. \tag{18.49}
\end{aligned}
$$

By comparison with equation (18.36), we find

$$\overline{G_i S_j} = \overline{g_i s_j}\,\epsilon_j + \sum_{l=1}^{N} \frac{\overline{g_i s_l}}{A_l}\frac{\rho_l}{\epsilon_l}\,\overline{S_l S_j}, \quad i = 1, 2, \ldots, K, \tag{18.50}$$

$$\overline{G_i G_k} = \overline{g_i g_k} + \sum_{l=1}^{N} \frac{\overline{g_i s_l}}{A_l}\frac{\rho_l}{\epsilon_l}\,\overline{S_l G_k}, \quad i = 1, 2, \ldots, K, \tag{18.51}$$

or, in matrix notation

$$\mathbf{GS} = \mathbf{R} + \mathbf{Q}\cdot\overline{\mathbf{SS}}, \tag{18.52}$$

$$\mathbf{GG} = \overline{\mathbf{gg}} + \mathbf{Q}\cdot\overline{\mathbf{SG}}, \tag{18.53}$$

where

$$\mathbf{R} = \begin{pmatrix} \overline{g_1 s_1}\,\epsilon_1 & \overline{g_1 s_2}\,\epsilon_2 & \cdots & \overline{g_1 s_N}\,\epsilon_N \\[1ex] \overline{g_2 s_1}\,\epsilon_1 & \overline{g_2 s_2}\,\epsilon_2 & \cdots & \overline{g_2 s_N}\,\epsilon_N \\[1ex] \vdots & \vdots & \ddots & \vdots \\[1ex] \overline{g_K s_1}\,\epsilon_1 & \overline{g_K s_2}\,\epsilon_2 & \cdots & \overline{g_K s_N}\,\epsilon_N \end{pmatrix}, \tag{18.54}$$

$$\mathbf{Q} = \begin{pmatrix} \dfrac{\overline{g_1 s_1}}{A_1}\dfrac{\rho_1}{\epsilon_1} & \dfrac{\overline{g_1 s_2}}{A_2}\dfrac{\rho_2}{\epsilon_2} & \cdots & \dfrac{\overline{g_1 s_N}}{A_N}\dfrac{\rho_N}{\epsilon_N} \\[2ex] \dfrac{\overline{g_2 s_1}}{A_1}\dfrac{\rho_1}{\epsilon_1} & \dfrac{\overline{g_2 s_2}}{A_2}\dfrac{\rho_2}{\epsilon_2} & \cdots & \dfrac{\overline{g_2 s_N}}{A_N}\dfrac{\rho_N}{\epsilon_N} \\[2ex] \vdots & \vdots & \ddots & \vdots \\[2ex] \dfrac{\overline{g_K s_1}}{A_1}\dfrac{\rho_1}{\epsilon_1} & \dfrac{\overline{g_K s_2}}{A_2}\dfrac{\rho_2}{\epsilon_2} & \cdots & \dfrac{\overline{g_K s_N}}{A_N}\dfrac{\rho_N}{\epsilon_N} \end{pmatrix}. \tag{18.55}$$

Because of reciprocity $\overline{G_k S_j} = \overline{S_j G_k}$ or $\mathbf{GS} = \mathbf{SG}^T$ (where the superscript T denotes the *transpose* of a matrix) and

$$\mathbf{R} + \mathbf{Q}\cdot\overline{\mathbf{SS}} = \overline{\mathbf{gs}}\cdot(\mathbf{T}^{-1})^T. \tag{18.56}$$

Obviously, the $\overline{G_k S_j}$ may be evaluated using either of equations (18.46) and (18.50) (or both, using one as a checking mechanism).

Example 18.2.   Consider a gray, absorbing-emitting (but not scattering) isothermal medium confined between two parallel, isothermal, gray and diffuse plates. The temperature of the medium is T_m, its absorption coefficient is κ, and the distance between the plates is L. Both plates have the same temperature, T_w, and emissivity, ϵ. Determine the net radiative heat flux at the plates.

Solution

We begin by evaluating the direct exchange areas. Since the problem is one-dimensional with infinitely large plates, all direct and total exchange areas will become infinitely large. Thus, we shall evaluate all exchange areas per unit area, using a tilde instead of a bar in the notation. From equation (18.26)

$$\widetilde{s_1 s_2} = \lim_{A_1 \to \infty} \frac{1}{A_1} \int_{A_1} \int_{A_2} e^{-\kappa S} \frac{\cos \theta_1 \cos \theta_2}{\pi S^2} dA_2 \, dA_1,$$

$$= \int_{A_2} e^{-\kappa S} \frac{\cos \theta_1 \cos \theta_2}{\pi S^2} dA_2,$$

since the exchange factor between any point on Surface 1 and all of Surface 2 is the same everywhere. Thus, we get, with $\cos \theta_1 = \cos \theta_2 = L/S$, $S^2 = r^2 + L^2$ and $\tau_L = \kappa L$,

$$\widetilde{s_1 s_2} = \int_0^\infty e^{-\kappa S} \frac{L^2}{\pi S^4} 2\pi r \, dr = 2 \int_1^\infty e^{-\tau_L \gamma} \frac{d\gamma}{\gamma^3} = 2E_3(\tau_L),$$

where $\gamma = S/L = \sqrt{(r/L)^2 + 1}$. We also have $\widetilde{s_1 s_1} = \widetilde{s_2 s_2} = 0$ and $\widetilde{s_2 s_1} = \widetilde{s_1 s_2}$. From equation (18.31), it follows immediately that

$$\widetilde{g s_1} = 1 - \widetilde{s_1 s_2} = \widetilde{g s_2} = \widetilde{g s} = \widetilde{s g},$$

and, from equation (18.33),

$$\widetilde{g g} = 4\tau_L - 2\widetilde{s g}.$$

To determine the total exchange factors, we first establish the relevant matrices,

$$\mathbf{T} = \begin{pmatrix} \dfrac{1}{\epsilon} & -\dfrac{\rho}{\epsilon} \widetilde{s_1 s_2} \\ -\dfrac{\rho}{\epsilon} \widetilde{s_1 s_2} & \dfrac{1}{\epsilon} \end{pmatrix}, \quad T = \frac{1}{\epsilon^2}\left[1 - \rho^2 \left(\widetilde{s_1 s_2}\right)^2\right],$$

$$\mathbf{T}^{-1} = \frac{1}{T}\begin{pmatrix} \dfrac{1}{\epsilon} & \dfrac{\rho}{\epsilon}\widetilde{s_1 s_2} \\ \dfrac{\rho}{\epsilon}\widetilde{s_1 s_2} & \dfrac{1}{\epsilon} \end{pmatrix}, \quad \mathbf{S} = \begin{pmatrix} 0 & \epsilon\,\widetilde{s_1 s_2} \\ \epsilon\,\widetilde{s_1 s_2} & 0 \end{pmatrix},$$

where $\mathbf{S}$ is also evaluated per unit area. Now, from equations (18.45) and (18.46)

$$\widetilde{S_1 S_1} = \frac{1}{T}\rho\left(\widetilde{s_1 s_2}\right)^2 = \widetilde{S_2 S_2}, \quad \widetilde{S_1 S_2} = \frac{1}{T}\widetilde{s_1 s_2} = \widetilde{S_2 S_1},$$

$$\widetilde{S_1 G} = \frac{1}{T}\left(\frac{1}{\epsilon} + \frac{\rho}{\epsilon}\widetilde{s_1 s_2}\right)\widetilde{s g} = \widetilde{S_2 G} = \widetilde{S G} = \widetilde{G S}.$$

The last result may be verified by comparing with the result from equation (18.50). Finally with

$$\mathbf{Q} = \left(\frac{\rho}{\epsilon}\widetilde{s g} \quad \frac{\rho}{\epsilon}\widetilde{s g}\right),$$

$$\widetilde{G G} = \widetilde{g g} + 2\frac{\rho}{\epsilon}\widetilde{s g}\,\widetilde{S G} = \widetilde{g g} + \frac{2\rho}{\epsilon^2 T}\left(1 + \rho\widetilde{S_1 S_2}\right)\left(\widetilde{s g}\right)^2.$$

These results are readily (and should be) verified by substituting them into equations (18.38) and (18.39).

The net heat flux to a plate is calculated from equation (18.35) as

$$q_w = \left(\epsilon - \widetilde{S_1 S_1} - \widetilde{S_1 S_2} \right) E_{bw} - \widetilde{SG} \, E_{bm} = \widetilde{SG} \, (E_{bw} - E_{bm}),$$

or

$$\Psi = \frac{q_w}{\sigma \, (T_w^4 - T_m^4)} = \widetilde{SG} = \frac{\epsilon \, [1 - 2E_3(\tau_L)]}{1 - 2\rho E_3(\tau_L)},$$

which is, of course, the exact result.

A number of researchers have applied the zonal method to gray media without scattering. For example, Einstein calculated combined convective/radiative heat fluxes for two-dimensional duct flow for a gray gas in a black-walled duct [4, 5] and Modest investigated radiative equilibrium in a rectangular enclosure with a single nonblack wall [9].

18.4 RADIATIVE EXCHANGE IN GRAY MEDIA WITH ISOTROPIC SCATTERING

We shall now extend the zonal method to the most general case to which it can be applied by including the effects of isotropic scattering and spatially varying absorption and scattering coefficients. As before, we break up the surfaces into N isothermal zones (with weakly varying radiosities) and the medium into K isothermal volume zones (which, as we shall see, should have weakly varying radiative source terms). We shall rederive the expressions for direct exchange areas by making allowance for property variations as well as for isotropic scattering.

Direct Exchange Areas

Surface–Surface Exchange

The direct exchange area between two surface zones applies now to total heat flux coming from zone i traveling to j directly without reflections and without being scattered. Therefore, equation (18.26) must be rewritten as

$$\overline{s_i s_j} = \int_{A_i} \int_{A_j} e^{-\beta_{ij} S} \, \frac{\cos \theta_i \cos \theta_j}{\pi S^2} \, dA_j \, dA_i, \tag{18.57}$$

where

$$\beta_{ij} = \frac{1}{S} \int_0^S \beta \, dS \tag{18.58}$$

is the *average extinction coefficient* between zones i and j. In general, of course, the value for β_{ij} would be different for any two zones.

Volume–Surface Exchange

Radiative energy leaves an infinitesimal volume not only in the form of isotropic emission ($= 4\pi\kappa_i I_{bgi} = 4\kappa_i E_{bgi}$) but also as isotropic out-scattering ($= \sigma_{si} G_i$, where σ_s is the scattering coefficient).[5]

Defining a *volume zone radiosity* as

$$J_g = (1 - \omega)E_{bg} + \frac{\omega}{4}G, \tag{18.59}$$

we may express the direct exchange of energy between volume and surface zones as

$$Q_{i\leftrightarrow j} = \overline{g_i s_j}(J_{gi} - J_{sj}), \tag{18.60}$$

and equation (18.27) becomes

$$\overline{g_i s_j} = \int_{V_i}\int_{A_j} e^{-\beta_{ij}S} \frac{\cos\theta_j}{\pi S^2}\beta_i\, dA_j\, dV_i. \tag{18.61}$$

Therefore, with this definition we demand that we may assume the volume zone radiosity, J_g, to be constant throughout the zone.

Volume–Volume Exchange

By the same reasoning we need to redefine the volume-to-volume direct exchange areas as

$$\overline{g_i g_j} = \int_{V_i}\int_{V_j} e^{-\beta_{ij}S} \frac{\beta_i\beta_j}{\pi S^2}\, dV_j\, dV_i, \tag{18.62}$$

where the β_i is due the source term over dV_i, and the β_j is associated with the extinction in dV_j.

Rewriting the energy balance for surface zone i, equation (18.30), we get

$$\begin{aligned}
Q_{si} &= A_i q_i = A_i(J_{si} - H_{si}) = A_i \epsilon_i(E_{bsi} - H_{si}) \\
&= \sum_{j=1}^{N} \overline{s_i s_j}(J_{si} - J_{sj}) + \sum_{k=1}^{K} \overline{s_i g_k}(J_{si} - J_{gk}) \\
&= \epsilon_i\left(A_i E_{bsi} - \sum_{j=1}^{N} \overline{s_i s_j} J_{sj} - \sum_{k=1}^{K} \overline{s_i g_k} J_{gk} \right), \quad i = 1, 2, \ldots, N. \tag{18.63}
\end{aligned}$$

---

[5]See also the development for the radiative source function for isotropic scattering, $S = \omega I_b + (1 - \omega)G/4\pi$, in equation (8.27).

Similarly, for a volume zone, equation (18.32) is rewritten as

$$Q_{gi} = \kappa_i V_i (4E_{bgi} - G_i) = \beta_i V_i (4J_{gi} - G_i)$$

$$= \sum_{j=1}^{N} \overline{g_i s_j} (J_{gi} - J_{sj}) + \sum_{k=1}^{K} \overline{g_i g_k} (J_{gi} - J_{gk})$$

$$= 4\kappa_i V_i E_{bgi} - (1 - \omega_i) \left(\sum_{j=1}^{N} \overline{g_i s_j} J_{sj} + \sum_{k=1}^{K} \overline{g_i g_k} J_{gk} \right), \quad i = 1, 2, \ldots, K . \quad (18.64)$$

Examining an isothermal enclosure, we find that equations (18.31) and (18.33) continue to hold, if the absorption coefficient is replaced by the extinction coefficient everywhere, and if the extinction coefficient (and, therefore, the direct exchange areas) does not depend on temperature:

$$\sum_{j=1}^{N} \overline{s_j s_i} + \sum_{k=1}^{K} \overline{g_k s_i} = A_i, \quad i = 1, 2, \ldots, N, \quad (18.65)$$

$$\sum_{j=1}^{N} \overline{s_j g_i} + \sum_{k=1}^{K} \overline{g_k g_i} = 4\beta_i V_i, \quad i = 1, 2, \ldots, K . \quad (18.66)$$

Eliminating the radiosities from equations (18.63) and (18.64) through equations (18.16) and (18.59), we obtain, in a similar fashion as for equations (18.41) and (18.47),

$$\sum_{j=1}^{N} \left(\frac{\delta_{ij}}{\epsilon_j} - \frac{\rho_j \overline{s_i s_j}}{\epsilon_j A_j} \right) h_{sj} - \sum_{k=1}^{K} \frac{\overline{s_i g_k} \omega_k}{4\kappa_k V_k} h_{gk}$$

$$= \sum_{j=1}^{N} \overline{s_i s_j} \epsilon_j E_{bsj} + \sum_{k=1}^{K} \overline{s_i g_k} (1 - \omega_k) E_{bgk}, \quad i = 1, 2, \ldots, N, \quad (18.67)$$

$$- \sum_{j=1}^{N} \frac{\overline{g_i s_j} \rho_j}{\epsilon_j A_j} h_{sj} + \sum_{k=1}^{K} \left(\frac{\delta_{ik}}{1 - \omega_k} - \frac{\overline{g_i g_k} \omega_k}{4\kappa_k V_k} \right) h_{gk}$$

$$= \sum_{j=1}^{N} \overline{g_i s_j} \epsilon_j E_{bsj} + \sum_{k=1}^{K} \overline{g_i g_k} (1 - \omega_k) E_{bgk}, \quad i = 1, 2, \ldots, K, \quad (18.68)$$

where $h_{gk} = \kappa_k V_k G_k$. Equations (18.67) and (18.68) are a system of $N + K$ equations in the $N + K$ unknown h_{sj} and h_{gk}.[6] A solution may be found by standard techniques.

Total Exchange Areas

If the temperature field is to be determined iteratively, determination of *total exchange areas* is again desirable so long as radiative properties and, therefore, direct and total

---

[6]This is valid if the medium temperature field is "known." In the case of radiative equilibrium we have, from equation (18.64), $G_k = 4E_{bk}$ or $h_{gk} = 4\kappa_k V_k E_{bgk}$ as the unknowns for the volume zones, and the second series on the right-hand side of equation (18.68) is eliminated.

exchange areas do not vary with temperature (although they may vary with location). We may rewrite equations (18.67) and (18.68) in matrix notation as

$$\mathbf{T} \cdot \mathbf{h_s} - \mathbf{U} \cdot \mathbf{h_g} = \mathbf{S} \cdot \mathbf{e_{bs}} + \mathbf{V} \cdot \mathbf{e_{bg}}, \tag{18.69}$$

$$-\mathbf{Q} \cdot \mathbf{h_s} + \mathbf{W} \cdot \mathbf{h_g} = \mathbf{R} \cdot \mathbf{e_{bs}} + \mathbf{X} \cdot \mathbf{e_{bg}}, \tag{18.70}$$

where $\mathbf{T}, \mathbf{S}, \mathbf{Q}, \mathbf{R}, \mathbf{e_{bs}}, \mathbf{e_{bg}}$, and $\mathbf{h_s}$ have been defined in the previous section and

$$\mathbf{U} = \begin{pmatrix} \dfrac{\overline{s_1 g_1}\,\omega_1}{4\kappa_1 V_1} & \dfrac{\overline{s_1 g_2}\,\omega_2}{4\kappa_2 V_2} & \cdots & \dfrac{\overline{s_1 g_K}\,\omega_K}{4\kappa_K V_K} \\[2ex] \dfrac{\overline{s_2 g_1}\,\omega_1}{4\kappa_1 V_1} & \dfrac{\overline{s_2 g_2}\,\omega_2}{4\kappa_2 V_2} & \cdots & \dfrac{\overline{s_2 g_K}\,\omega_K}{4\kappa_K V_K} \\[2ex] \vdots & \vdots & \ddots & \vdots \\[2ex] \dfrac{\overline{s_N g_1}\,\omega_1}{4\kappa_1 V_1} & \dfrac{\overline{s_N g_2}\,\omega_2}{4\kappa_2 V_2} & \cdots & \dfrac{\overline{s_N g_K}\,\omega_K}{4\kappa_K V_K} \end{pmatrix}, \tag{18.71}$$

$$\mathbf{W} = \begin{pmatrix} \dfrac{1}{1-\omega_1} - \dfrac{\overline{g_1 g_1}\,\omega_1}{4\kappa_1 V_1} & -\dfrac{\overline{g_1 g_2}\,\omega_2}{4\kappa_2 V_2} & \cdots & -\dfrac{\overline{g_1 g_K}\,\omega_K}{4\kappa_K V_K} \\[2ex] -\dfrac{\overline{g_2 g_1}\,\omega_1}{4\kappa_1 V_1} & \dfrac{1}{1-\omega_2} - \dfrac{\overline{g_2 g_2}\,\omega_2}{4\kappa_2 V_2} & \cdots & -\dfrac{\overline{g_2 g_K}\,\omega_K}{4\kappa_K V_K} \\[2ex] \vdots & \vdots & \ddots & \vdots \\[2ex] -\dfrac{\overline{g_K g_1}\,\omega_1}{4\kappa_1 V_1} & -\dfrac{\overline{g_K g_2}\,\omega_2}{4\kappa_2 V_2} & \cdots & \dfrac{1}{1-\omega_K} - \dfrac{\overline{g_K g_K}\,\omega_K}{4\kappa_K V_K} \end{pmatrix}, \tag{18.72}$$

$$\mathbf{V} = \begin{pmatrix} \overline{s_1 g_1}(1-\omega_1) & \overline{s_1 g_2}(1-\omega_2) & \cdots & \overline{s_1 g_K}(1-\omega_K) \\[2ex] \overline{s_2 g_1}(1-\omega_1) & \overline{s_2 g_2}(1-\omega_2) & \cdots & \overline{s_2 g_K}(1-\omega_K) \\[2ex] \vdots & \vdots & \ddots & \vdots \\[2ex] \overline{s_N g_1}(1-\omega_1) & \overline{s_N g_2}(1-\omega_2) & \cdots & \overline{s_N g_K}(1-\omega_K) \end{pmatrix}, \tag{18.73}$$

$$\mathbf{X} = \begin{pmatrix} \overline{g_1 g_1}(1-\omega_1) & \overline{g_1 g_2}(1-\omega_2) & \cdots & \overline{g_1 g_K}(1-\omega_K) \\[2ex] \overline{g_2 g_1}(1-\omega_1) & \overline{g_2 g_2}(1-\omega_2) & \cdots & \overline{g_2 g_K}(1-\omega_K) \\[2ex] \vdots & \vdots & \ddots & \vdots \\[2ex] \overline{g_K g_1}(1-\omega_1) & \overline{g_K g_2}(1-\omega_2) & \cdots & \overline{g_K g_K}(1-\omega_K) \end{pmatrix}, \tag{18.74}$$

$$\mathbf{h_g} = \begin{pmatrix} \kappa_1 V_1 G_1 \\ \kappa_2 V_2 G_2 \\ \vdots \\ \kappa_K V_K G_K \end{pmatrix}. \tag{18.75}$$

Inverting $\mathbf{W}$, we can solve equation (18.70) for $\mathbf{h_g}$ as

$$\mathbf{h_g} = \mathbf{W}^{-1} \cdot \mathbf{Q} \cdot \mathbf{h_s} + \mathbf{W}^{-1} \cdot \mathbf{R} \cdot \mathbf{e_{bs}} + \mathbf{W}^{-1} \cdot \mathbf{X} \cdot \mathbf{e_{bg}}, \tag{18.76}$$

and substituting this expression into equation (18.69), we find

$$\mathbf{h_s} = \mathbf{P}^{-1} \cdot \mathbf{C} \cdot \mathbf{e_{bs}} + \mathbf{P}^{-1} \cdot \mathbf{D} \cdot \mathbf{e_{bg}}, \tag{18.77}$$

$$\mathbf{h_g} = \mathbf{W}^{-1} \cdot (\mathbf{Q} \cdot \mathbf{P}^{-1} \cdot \mathbf{C} + \mathbf{R}) \cdot \mathbf{e_{bs}} + \mathbf{W}^{-1} \cdot (\mathbf{Q} \cdot \mathbf{P}^{-1} \cdot \mathbf{D} + \mathbf{X}) \cdot \mathbf{e_{bg}}, \tag{18.78}$$

where

$$\mathbf{C} = \mathbf{S} + \mathbf{U} \cdot \mathbf{W}^{-1} \cdot \mathbf{R}, \qquad \mathbf{D} = \mathbf{V} + \mathbf{U} \cdot \mathbf{W}^{-1} \cdot \mathbf{X}, \tag{18.79a}$$

$$\mathbf{P} = \mathbf{T} - \mathbf{U} \cdot \mathbf{W}^{-1} \cdot \mathbf{Q}. \tag{18.79b}$$

In terms of total exchange areas, equations (18.63) and (18.64) are rewritten as

$$Q_{si} = \epsilon_i A_i E_{bsi} - \sum_{j=1}^{N} \overline{S_i S_j} E_{bsj} - \sum_{k=1}^{K} \overline{S_i G_k} E_{bgk}, \quad i = 1, 2, \ldots, N, \tag{18.80}$$

$$Q_{gi} = 4\kappa_i V_i E_{bgi} - \sum_{j=1}^{N} \overline{G_i S_j} E_{bsj} - \sum_{k=1}^{K} \overline{G_i G_k} E_{bgk}, \quad i = 1, 2, \ldots, K, \tag{18.81}$$

which are the same as equations (18.35) and (18.36) (but with modified total exchange areas).

Thus, by comparison we find

$$\overline{\mathbf{SS}} = \mathbf{P}^{-1} \cdot \mathbf{C} \quad \text{or} \quad \overline{S_i S_j} = \sum_{l=1}^{N} P_{il}^{-1} C_{lj}, \tag{18.82}$$

$$\overline{\mathbf{SG}} = \mathbf{P}^{-1} \cdot \mathbf{D} \quad \text{or} \quad \overline{S_i G_k} = \sum_{l=1}^{N} P_{il}^{-1} D_{lk}, \tag{18.83}$$

$$\overline{\mathbf{GS}} = \mathbf{W}^{-1} \cdot (\mathbf{Q} \cdot \mathbf{P}^{-1} \cdot \mathbf{C} + \mathbf{R}) = \overline{\mathbf{SG}}^T,$$

$$\text{or} \quad \overline{G_k S_i} = \sum_{l=1}^{K} W_{kl}^{-1} \left(\sum_{m=1}^{N} \sum_{n=1}^{N} Q_{lm} P_{mn}^{-1} C_{ni} + R_{li} \right), \tag{18.84}$$

$$\overline{\mathbf{GG}} = \mathbf{W}^{-1} \cdot (\mathbf{Q} \cdot \mathbf{P}^{-1} \cdot \mathbf{D} + \mathbf{X}),$$

$$\text{or} \quad \overline{G_k G_i} = \sum_{l=1}^{K} W_{kl}^{-1} \left(\sum_{m=1}^{N} \sum_{n=1}^{N} Q_{lm} P_{mn}^{-1} D_{ni} + X_{li} \right). \tag{18.85}$$

Setting all emissive powers equal in equations (18.80) and (18.81) shows that equations (18.38) and (18.39) still hold, *without* replacing absorption coefficient κ by extinction coefficient β on the right-hand side of equation (18.39),

$$\sum_{j=1}^{N} \overline{S_i S_j} + \sum_{k=1}^{K} \overline{S_i G_k} = \epsilon_i A_i, \qquad i = 1, 2, \ldots, N, \tag{18.86}$$

$$\sum_{j=1}^{N} \overline{G_i S_j} + \sum_{k=1}^{K} \overline{G_i G_k} = 4\kappa_i V_i, \qquad i = 1, 2, \ldots, K. \tag{18.87}$$

Temperature Dependent Properties

If the absorption and extinction coefficients depend on temperature, then all direct exchange areas also depend on temperature. Thus, if the temperature field is found iteratively, then the direct exchange areas must be recalculated after every iteration. This requirement, in turn, eliminates the advantages of using total exchange areas. Therefore, for these problems one would, in general, solve equations (18.67) and (18.68) (or some variation) directly. There is, however, one important case for which a set of total exchange areas can be calculated beforehand (and which do not enter the iteration).

For this we shall make the following two assumptions: (*i*) The medium does not scatter, and (*ii*) the temperature dependence of the average absorption coefficient between two zones, κ_{ij}, may be neglected (or is only updated after a given number of iterations for the temperature field. With these assumptions we may write

$$\overline{s_i s_j} = (\overline{s_i s_j})_{\text{av}}, \quad \overline{s_i g_j} = \overline{g_j s_i} = (\overline{s_i g_j})_{\text{av}} \frac{\kappa_j}{\kappa_{\text{av}}}, \quad \overline{g_i g_j} = (\overline{g_i g_j})_{\text{av}} \frac{\kappa_i \kappa_j}{\kappa_{\text{av}}^2}, \tag{18.88}$$

where $(\overline{s_i s_j})_{\text{av}}$, etc., are direct exchange areas evaluated from equations (18.57), (18.61), and (18.62), with β_i and β_j (or κ_i and κ_j for a nonscattering medium) replaced by an average absorption coefficient κ_{av}. The assumption of a temperature independent κ_{ij} thus implies that the average direct exchange areas do not depend on temperature. Substituting equation (18.88) into (18.45), and realizing that neither **T** nor **S** depends on temperature, we find

$$\overline{S_i S_j} = \sum_{l=1}^{N} (T^{-1})_{il} S_{lj} = (\overline{S_i S_j})_{\text{av}}, \tag{18.89}$$

that is, the $\overline{S_i S_j}$ do not depend on temperature either. From equation (18.46)

$$\overline{S_i G_k} = \sum_{l=1}^{N} (T^{-1})_{il} (\overline{s_l g_k})_{\text{av}} \frac{\kappa_k}{\kappa_{\text{av}}} = (\overline{S_i G_k})_{\text{av}} \frac{\kappa_k}{\kappa_{\text{av}}}, \tag{18.90}$$

that is, the $\overline{(SG)}_{av}$ may be calculated once and for all, and the actual $\overline{SG}$ are found by a simple multiplication. Similarly, from equation (18.51),

$$\overline{G_i G_k} = \left[\overline{(g_i g_k)}_{av} + \sum_{l=1}^{N} \frac{\overline{(g_i s_l)}_{av}}{A_l} \frac{\rho_l}{\epsilon_l} \overline{(S_l G_k)}_{av} \right] \frac{\kappa_i \kappa_k}{\kappa_{av}^2} = \overline{(G_i G_k)}_{av} \frac{\kappa_i \kappa_k}{\kappa_{av}^2}. \qquad (18.91)$$

Equations (18.35) and (18.36) may then be rewritten in terms of average total exchange areas.

Example 18.3. Repeat Example 18.2 for a medium that not only emits and absorbs, but that also scatters isotropically.

Solution

We shall still limit ourselves to a single gas zone even though this will no longer render the analysis exact: For that, the radiative source term, $4(1 - \omega)E_b + \omega G$, must be constant across the volume zone.

It follows immediately that the direct exchange areas remain unchanged (except that $\tau_L = \beta L$ is now based on the extinction coefficient). From equations (18.80) and (18.86) it follows that the nondimensional heat flux at the surface is still evaluated as

$$\Psi = \frac{q_w}{\sigma(T_w^4 - T_m^4)} = \widetilde{SG},$$

and that only the evaluation of $\widetilde{SG}$ from the direct exchange areas is different from the previous example. While the tensors **T** and **Q** remain unchanged, we still need to evaluate **U**, **V**, **W**, and **X** for the evaluation of $\widetilde{SG}$. Thus,

$$\mathbf{U} = \frac{\widetilde{sg}\,\omega}{4\tau_L(1 - \omega)} \begin{pmatrix} 1 \\ 1 \end{pmatrix}; \qquad \mathbf{V} = \widetilde{sg}(1 - \omega)\begin{pmatrix} 1 \\ 1 \end{pmatrix};$$

$$\mathbf{W} = \left(\frac{4\tau_L - \omega\,\widetilde{gg}}{4\tau_L(1 - \omega)} \right); \qquad \mathbf{X} = \left(\widetilde{gg}(1 - \omega) \right),$$

where **W** and **X** are single-term 1×1 matrices. Thus,

$$\mathbf{W}^{-1} = (1/W) \quad \text{and} \quad D_{ij} = V_{ij} + \sum_{k=1}^{1} \sum_{l=1}^{1} U_{ik} W_{kl}^{-1} X_{lj},$$

or

$$\mathbf{D} = \left[\widetilde{sg}(1 - \omega) + \frac{\widetilde{sg}\,\omega}{4\tau_L(1 - \omega)} \times \frac{4\tau_L(1 - \omega)}{4\tau_L - \omega\,\widetilde{gg}} \times \widetilde{gg}\,(1 - \omega) \right] \begin{pmatrix} 1 \\ 1 \end{pmatrix}$$

$$= \frac{4\tau_L(1 - \omega)\widetilde{sg}}{4\tau_L - \omega\,\widetilde{gg}} \begin{pmatrix} 1 \\ 1 \end{pmatrix}.$$

Similarly,

$$P_{ij} = T_{ij} - \frac{U_{i1}Q_{1j}}{W} = T_{ij} - \frac{\widetilde{sg}\,\omega}{4\tau_L(1 - \omega)} \times \frac{4\tau_L(1 - \omega)}{4\tau_L - \omega\,\widetilde{gg}} \times \frac{\rho}{\epsilon}\widetilde{sg}$$

$$= T_{ij} - \frac{\rho}{\epsilon} \frac{\omega\,(\widetilde{sg})^2}{4\tau_L - \omega\,\widetilde{gg}},$$

or

$$
\mathbf{P} = \begin{pmatrix}
\dfrac{1}{\epsilon} - \dfrac{\rho}{\epsilon}\dfrac{\omega(\overline{sg})^2}{4\tau_L - \omega\,\overline{gg}} & -\dfrac{\rho}{\epsilon}\left[\overline{s_1 s_2} + \dfrac{\omega(\overline{sg})^2}{4\tau_L - \omega\,\overline{gg}}\right] \\[3mm]
-\dfrac{\rho}{\epsilon}\left[\overline{s_1 s_2} + \dfrac{\omega(\overline{sg})^2}{4\tau_L - \omega\,\overline{gg}}\right] & \dfrac{1}{\epsilon} - \dfrac{\rho}{\epsilon}\dfrac{\omega(\overline{sg})^2}{4\tau_L - \omega\,\overline{gg}}
\end{pmatrix}
$$

$$
\mathbf{P}^{-1} = \dfrac{1}{P}\begin{pmatrix}
\dfrac{1}{\epsilon} - \dfrac{\rho}{\epsilon}\dfrac{\omega(\overline{sg})^2}{4\tau_L - \omega\,\overline{gg}} & \dfrac{\rho}{\epsilon}\left[\overline{s_1 s_2} + \dfrac{\omega(\overline{sg})^2}{4\tau_L - \omega\,\overline{gg}}\right] \\[3mm]
\dfrac{\rho}{\epsilon}\left[\overline{s_1 s_2} + \dfrac{\omega(\overline{sg})^2}{4\tau_L - \omega\,\overline{gg}}\right] & \dfrac{1}{\epsilon} - \dfrac{\rho}{\epsilon}\dfrac{\omega(\overline{sg})^2}{4\tau_L - \omega\,\overline{gg}}
\end{pmatrix}
$$

and

$$
P = P_{11}^2 - P_{12}^2 = \dfrac{1}{\epsilon^2}\left[1 - 2\rho\dfrac{\omega(\overline{sg})^2}{4\tau_L - \omega\,\overline{gg}} - \rho^2(\overline{s_1 s_2})^2 - 2\rho^2\overline{s_1 s_2}\dfrac{\omega(\overline{sg})^2}{4\tau_L - \omega\,\overline{gg}}\right]
$$

$$
= \dfrac{1}{\epsilon^2}\left[1 - (\rho\overline{s_1 s_2})^2 - 2\rho(1 + \rho\overline{s_1 s_2})\dfrac{\omega(\overline{sg})^2}{4\tau_L - \omega\,\overline{gg}}\right].
$$

Finally,

$$
\overline{SG} = \mathbf{P}^{-1}\cdot\mathbf{D} = \begin{pmatrix}
P_{11}^{-1}D_{11} + P_{12}^{-1}D_{21} \\[2mm]
P_{21}^{-1}D_{11} + P_{22}^{-1}D_{21}
\end{pmatrix}
$$

$$
= \epsilon\,\dfrac{\left(1 + \rho\overline{s_1 s_2}\right)4\tau_L(1 - \omega)\overline{sg}/\left(4\tau_L - \omega\,\overline{gg}\right)}{1 - \left(\rho\overline{s_1 s_2}\right)^2 - 2\rho\left(1 + \rho\overline{s_1 s_2}\right)\dfrac{\omega(\overline{sg})^2}{4\tau_L - \omega\,\overline{gg}}}\begin{pmatrix}1\\1\end{pmatrix}
$$

$$
= \dfrac{4\tau_L(1 - \omega)\epsilon\overline{sg}}{\left(1 - \rho\overline{s_1 s_2}\right)\left(4\tau_L - \omega\,\overline{gg}\right) - 2\rho\omega(\overline{sg})^2}\begin{pmatrix}1\\1\end{pmatrix}.
$$

Substituting for $\overline{s_1 s_2}$, $\overline{sg}$ and $\overline{gg}$, which have already been evaluated in Example 18.2 (although τ_L is now based on extinction coefficient), this reduces to

$$
\Psi = \dfrac{q_w}{\sigma(T_1^4 - T_2^4)} = \overline{SG} = \dfrac{\epsilon[1 - 2E_3(\tau_L)]}{1 - 2\rho\,E_3(\tau_L) + \dfrac{2\epsilon\omega}{4\tau_L(1 - \omega)}[1 - 2E_3(\tau_L)]}.
$$

The zonal method has been applied to gray, isotropically scattering media by a number of researchers. For example, Naraghi and Kassemi [10] treated the same problem of a rectangular medium at radiative equilibrium as Modest [9], but included isotropic scattering.

A different approach than described here for the evaluation of total exchange areas was taken by Vercammen and Froment [11], who used the Monte Carlo method with subsequent data smoothing (to eliminate statistical scatter).[7] A similar approach was also taken by Naraghi and Chung [12]. Naraghi and Chung also presented what they called a "unified matrix approach" by combining equations (18.69) and (18.70) into a single matrix equation,

$$\begin{pmatrix} \mathbf{T} & -\mathbf{U} \\ -\mathbf{Q} & \mathbf{W} \end{pmatrix} \cdot \begin{pmatrix} \mathbf{h_s} \\ \mathbf{h_g} \end{pmatrix} = \begin{pmatrix} \mathbf{S} & \mathbf{V} \\ \mathbf{R} & \mathbf{X} \end{pmatrix} \cdot \begin{pmatrix} \mathbf{e_{bs}} \\ \mathbf{e_{bg}} \end{pmatrix}, \tag{18.92}$$

which may be inverted to give

$$\begin{pmatrix} \overline{\mathbf{SS}} & \overline{\mathbf{SG}} \\ \overline{\mathbf{GS}} & \overline{\mathbf{GG}} \end{pmatrix} = \begin{pmatrix} \mathbf{T} & -\mathbf{U} \\ -\mathbf{Q} & \mathbf{W} \end{pmatrix}^{-1} \cdot \begin{pmatrix} \mathbf{S} & \mathbf{V} \\ \mathbf{R} & \mathbf{X} \end{pmatrix}. \tag{18.93}$$

It is not clear whether this approach will result in increased computer efficiency, since here a single $(N + K) \times (N + K)$ matrix must be inverted while, in the standard method, two matrices must be inverted (one $N \times N$ and the other $K \times K$).

18.5 RADIATIVE EXCHANGE THROUGH A NONGRAY MEDIUM

In the foregoing discussion we have assumed that the medium and the surfaces are gray, although all equations in this chapter up to this point are equally valid, on a spectral basis, for a nongray enclosure with a nongray medium. Unfortunately, the zonal method—like many other solution methods for radiative transfer problems—is not well suited for nongray media. However, Hottel and coworkers [1, 2, 13] have demonstrated that the zonal method may be applied to a nongray medium in one very special, but very important, case: An absorbing, emitting (but not scattering) medium confined in a black-walled enclosure. This important situation is often encountered in furnace applications where mixtures of combustion gases and soot are confined by furnace refractories blackened by soot.

In this method, known as the *weighted-sum-of-gray-gases* method, the medium is assumed to consist of different fractions of gray gases with different (but gray) absorption coefficients.[8]

Surface–Surface Exchange

From equation (18.26) it follows that the radiative heat flux leaving dA_i and arriving at dA_j is

$$Q_{i \to j} = \int_0^\infty \int_{A_i} \int_{A_j} e^{-\kappa_\nu S} \frac{\cos\theta_i \cos\theta_j}{\pi S^2} dA_j \, dA_i \, J_{\nu i} \, d\nu, \tag{18.94}$$

where, for simplicity, it is assumed that the spectral absorption coefficient, κ_ν, does not depend on temperature or location. For a black enclosure $J_\nu = E_{b\nu}$, and

---

[7] For a detailed description of the Monte Carlo method, see Chapter 19.
[8] See also the more detailed discussion in Section 17.7.

equation (18.94) may be integrated over the entire spectrum by using the definition for *medium emissivity*, equation (9.89),

$$\epsilon_m(T, S) = \frac{1}{\sigma T^4} \int_0^\infty (1 - e^{-\kappa_\nu S}) E_{b\nu}(T) \, d\nu, \tag{18.95}$$

leading to

$$Q_{i \to j} = \int_{A_i} \int_{A_j} [1 - \epsilon_m(T_i, S)] \frac{\cos\theta_i \cos\theta_j}{\pi S^2} dA_j \, dA_i \, E_{bi} = \overrightarrow{s_i s_j} E_{bi}. \tag{18.96}$$

Similarly, we have

$$Q_{j \to i} = \int_{A_i} \int_{A_j} [1 - \epsilon_m(T_j, S)] \frac{\cos\theta_i \cos\theta_j}{\pi S^2} dA_j \, dA_i \, E_{bj} = \overleftarrow{s_i s_j} F_{bj}. \tag{18.97}$$

Both exchange areas now have arrows, rather than bars, since they are no longer equal, but depend on the temperature of the emitting surface, or

$$Q_{i \leftrightarrow j} = \overrightarrow{s_i s_j} E_{bi} - \overleftarrow{s_i s_j} E_{bj}. \tag{18.98}$$

Hottel [13] has shown that the emissivity of the medium may be written in the form

$$\epsilon_m(T, S) \simeq \sum_{l=0}^L a_l(T)\left(1 - e^{-\kappa_l S}\right), \quad \sum_{l=0}^L a_l = 1, \tag{18.99}$$

where $l = 0$ with $\kappa_0 = 0$ has been included for the case in which the medium consists entirely of molecular gases with large windows between bands. Experience has shown that $L = 3$ usually gives sufficiently accurate results, and even $L = 1$ (gray plus clear gas) often suffices [2]. Using this "weighted-sum-of-gray-gases" we have

$$Q_{i \leftrightarrow j} \simeq \sum_{l=0}^L (\overline{s_i s_j})_l \, [a_l(T_i) E_{bi} - a_l(T_j) E_{bj}], \tag{18.100}$$

where $(\overline{s_i s_j})_l$ is $\overline{s_i s_j}$ evaluated for an absorption coefficient κ_l. Thus, the solution for the nongray medium is identical to that for a gray medium, except that emissive power is replaced by a weighted emissive power, and the calculations have to be carried out for a number of gray gases.

Surface–Volume Exchange

Similarly, from equation (18.27) it follows that, for a black enclosure,

$$\begin{aligned}
Q_{i \to j} &= \int_0^\infty \int_{V_i} \int_{A_j} \frac{\cos\theta_j}{\pi S^2} \kappa_\nu e^{-\kappa_\nu S} dA_j \, dV_i \, E_{b\nu i} \, d\nu \\
&= \int_{V_i} \int_{A_j} \frac{\cos\theta_j}{\pi S^2} \frac{d}{dS} \left[\int_0^\infty (1 - e^{-\kappa_\nu S}) E_{b\nu i} \, d\nu \right] dA_j \, dV_i \\
&= \int_{V_i} \int_{A_j} \frac{d\epsilon_m}{dS}(T_i, S) \frac{\cos\theta_j}{\pi S^2} dA_j \, dV_i \, E_{bi} = \overrightarrow{g_i s_j} E_{bi}.
\end{aligned} \tag{18.101}$$

Note that E_{bi} can be pulled inside the operator d/dS, since it is assumed constant over V_i. Therefore, using equation (18.99), we find

$$Q_{i \leftrightarrow j} = \overrightarrow{g_i s_j} E_{bi} - \overleftarrow{g_i s_j} E_{bj}$$

$$\simeq \sum_{l=1}^{L} (\overrightarrow{g_i s_j})_l [a_l(T_i) E_{bi} - a_l(T_j) E_{bj}], \qquad (18.102)$$

where the summation starts with $l = 1$ since there is no emission within the medium for $\kappa_0 = 0$.

Volume–Volume Exchange

By the same reasoning equation (18.28) becomes[9]

$$Q_{i \to j} = \int_0^\infty \int_{V_i} \int_{V_j} \frac{1}{\pi S^2} \kappa_\nu^2 e^{-\kappa_\nu S} \, dV_j \, dV_i \, E_{b\nu i} \, d\nu$$

$$= -\int_{V_i} \int_{V_j} \frac{d^2 \epsilon_m}{dS^2}(T_i, S) \frac{1}{\pi S^2} dV_j \, dV_i \, E_{bi} = \overrightarrow{g_i g_j} E_{bi}, \qquad (18.103)$$

and

$$Q_{i \leftrightarrow j} = \overrightarrow{g_i g_j} E_{bi} - \overleftarrow{g_i g_j} E_{bj}$$

$$\simeq \sum_{l=1}^{L} (\overrightarrow{g_i g_j})_l [a_l(T_i) E_{bi} - a_l(T_j) E_{bj}]. \qquad (18.104)$$

Therefore, the net heat flow between any two surface or volume zones is calculated by determining the heat fluxes for a number of enclosures filled with a gray gas and with weighted emissive powers, followed by adding these heat fluxes.

Note that the approximation of equation (18.99) is not necessary: The directional direct exchange areas may be evaluated in terms of the total emissivity $\epsilon_m(T, S)$, leading to more accurate results. The approximation is made for convenience, reducing the problem to the simpler gray case with somewhat less involved evaluations of the direct exchange areas. The major advantage of the weighted-sum-of-gray-gases method is that the direct exchange areas $(\overline{s_i s_j})_l$, $(\overline{s_i g_k})_l$, and $(\overline{g_i g_k})_l$ do not depend on temperature and, therefore, (i) may be looked up in tables and graphs such as the ones given by [2] (this fact was of major importance during the method's early days, when fast computers were not yet available), and (ii) do not have to be recalculated, if the temperature field is found by iteration.

While equations (18.100) or (18.98), (18.102), and (18.104) are strictly valid only for black enclosures without scattering, a case can be made to extend this to reflecting

---

[9]In the light of property variations, equation (18.62), a more correct representation of $\kappa^2 e^{-\kappa S}$ would be $d^2 \epsilon_m / dS_{i \to j} \, dS_{j \to i}$, where $S = S_{i \to j} = -S_{j \to i}$, i.e., measured from different points of origin.

surfaces and to scattering media: The difference in direct and total exchange areas is due to the average total vs. direct distance traveled between zones i and j. Since S does not appear directly in these equations[10] we may assume that, approximately, they are also valid for total exchange areas,

$$\overrightarrow{S_iS_j}E_{bi} - \overleftarrow{S_iS_j}E_{bj} \simeq \sum_{l=0}^{L} \left(\overleftrightarrow{S_iS_j}\right)_l [a_l(T_i)E_{bi} - a_l(T_j)E_{bj}], \tag{18.105}$$

$$\overrightarrow{G_iS_j}E_{bi} - \overleftarrow{G_iS_j}E_{bj} \simeq \sum_{l=1}^{L} (\overleftrightarrow{G_iS_j})_l [a_l(T_i)E_{bi} - a_l(T_j)E_{bj}], \tag{18.106}$$

$$\overrightarrow{G_iG_j}E_{bi} - \overleftarrow{G_iG_j}E_{bj} \simeq \sum_{l=1}^{L} (\overleftrightarrow{G_iG_j})_l [a_l(T_i)E_{bi} - a_l(T_j)E_{bj}], \tag{18.107}$$

and equations (18.80) and (18.81) may be used to determine net heat fluxes.

Considerable work has appeared in the literature dealing with the nongray zonal method. Nelson [14] pointed out that the use of equation (18.99) is not necessary, except for numerical efficiency for iterative calculations. He also proposed an approximation to apply the zonal method to reflecting boundaries. Correlations of the type of equation (18.99) have been determined for CO_2–H_2O gas mixtures by Smith and coworkers [15]. They later applied the weighted-sum-of-gray-gases model to combined radiative and convective heat transfer in a pipe [16]. A similar problem was addressed by Sistino [17].

18.6 DETERMINATION OF DIRECT EXCHANGE AREAS

The use of equations (18.63) and (18.64) together with the definition of direct exchange areas, or equations (18.80) and (18.81) together with the definition of total exchange areas, demands that the variation of the radiosity must be neglected over each surface zone, and that the variation of the radiation source term must be neglected over each volume zone. Hottel and Sarofim [2] state that, if 5% accuracy in the calculation of heat fluxes is to be assured, the zone sizes should be such that no zone has an optical depth (extinction coefficient × largest dimension of zone over which radiosity or source varies) exceeding the value of 0.4.

Accuracy of the zone method may be improved in two ways: (*i*) Increasing the number of zones, and (*ii*) increasing the accuracy of the numerical quadrature used to determine individual direct exchange areas. Increasing the number of zones allows the use of simple efficient techniques for the evaluation of exchange areas, but re-

---

[10]The total average distance between i and j, including reflections and scattering events, does of course to some degree depend on absorption coefficient.

quires the time-consuming inversion of a large matrix. Modest [9] found that, for sufficiently many zones, all exchange areas but one may be calculated by evaluating the integrand between zonal centers, multiplied by the applicable zonal areas and/or volumes. The error made for the closest zones is then offset by applying equations (18.65) and (18.66) for the last one ($\overline{g_i g_i}$ for volume-volume areas, common-face $\overline{g_i s_j}$ for volume-surface areas, and common-boundary $\overline{s_i s_j}$ for a corner zone). Using relatively few zones with accurately evaluated exchange areas, on the other hand, requires only the inversion of a small matrix. But the numerical quadrature for the evaluation of exchange areas becomes time-consuming since many internal nodal points will be required. Still, if the temperature field is to be determined by iteration, considerable computer time may be saved since the evaluation of exchange areas does not enter the iteration. Accuracy for the zonal method with relatively few zones may be increased by allowing the emissive power to vary across a zone. Walther and coworkers [3] have shown for one-dimensional plane-parallel slab zones that a linear emissive power variation across the slab can be incorporated in a rigorous fashion. For two-dimensional rectangular volume zones Einstein [4, 5] included linear emissive power variations in a semirigorous manner (for two-dimensional Cartesian and cylindrical geometries).

References

1. Hottel, H. C., and E. S. Cohen: "Radiant Heat Exchange in a Gas-Filled Enclosure: Allowance for Nonuniformity of Gas Temperature," *AIChE Journal*, vol. 4, pp. 3–14, 1958.
2. Hottel, H. C., and A. F. Sarofim: *Radiative Transfer*, McGraw-Hill, New York, 1967.
3. Walther, V. A., J. Dörr, and E. Eller: "Mathematische Berechnung der Temperaturverteilung in der Glasschmelze mit Berücksichtigung von Wärmeleitung und Wärmestrahlung," *Glastechnishe Berichte*, vol. 26, pp. 133–140, 1953.
4. Einstein, T. H.: "Radiant Heat Transfer to Absorbing Gases Enclosed Between Parallel Flat Plates with Flow and Conduction," *NASA TR R-154*, 1963.
5. Einstein, T. H.: "Radiant Heat Transfer to Absorbing Gases Enclosed in a Circular Pipe with Conduction, Gas Flow, and Internal Heat Generation," *NASA TR R-156*, 1963.
6. Larsen, M. E., and J. R. Howell: "The Exchange Factor Method: An Alternative Zonal Formulation of Radiating Enclosure Analysis," *ASME Journal of Heat Transfer*, vol. 108, no. 1, pp. 936–942, 1985.
7. Liu, H. P., and J. R. Howell: "Measurement of Radiation Exchange Factors," *ASME Journal of Heat Transfer*, vol. 109, no. 2, pp. 470–477, 1987.
8. Noble, J. J.: "The Zone Method: Explicit Matrix Relations for Total Exchange Areas," *International Journal of Heat and Mass Transfer*, vol. 18, no. 2, pp. 261–269, 1975.
9. Modest, M. F.: "Radiative Equilibrium in a Rectangular Enclosure Bounded by Gray Non-Isothermal Walls," *Journal of Quantitative Spectroscopy and Radiative Transfer*, vol. 15, pp. 445–461, 1975.
10. Naraghi, M. H. N., and M. Kassemi: "Radiative Transfer in Rectangular Enclosures: A Discretized Exchange Factor Solution," in *Proceedings of the 1988 National Heat Transfer Conference*, vol. HTD–96, ASME, pp. 259–268, 1988.
11. Vercammen, H. A. J., and G. F. Froment: "An Improved Zone Method Using Monte Carlo Techniques for the Simulation of Radiation in Industrial Furnaces," *International Journal of Heat and Mass Transfer*, vol. 23, pp. 329–337, 1980.
12. Naraghi, M. H. N., and B. T. F. Chung: "A Unified Matrix Formulation for the Zone Method: A Stochastic Approach," *International Journal of Heat and Mass Transfer*, vol. 28, no. 2, pp. 245–251, 1985.

13. Hottel, H. C.: "Radiant Heat Transmission," in *Heat Transmission*, ed. W. H. McAdams, 3d ed., ch. 4, McGraw-Hill, New York, 1954.

14. Nelson, D. A.: "Radiation Heat Transfer in Molecular-Gas Filled Enclosures," ASME paper no. 77-HT-16, 1977.

15. Smith, T. F., Z. F. Shen, and J. N. Friedman: "Evaluation of Coefficients For the Weighted Sum of Gray Gases Model," *ASME Journal of Heat Transfer*, vol. 104, pp. 602–608, 1982.

16. Smith, T. F., Z. F. Shen, and A. M. Al-Turki: "Radiative and Conductive Transfer in a Cylindrical Enclosure for a Real Gas," *ASME Journal of Heat Transfer*, vol. 107, pp. 482–485, 1985.

17. Sistino, A. J.: "Mean Beam Length and the Zone Method (without and with Scattering) For a Cylindrical Enclosure," ASME paper no. 82-HT-3, 1982.

Problems

18.1 For a one-dimensional problem of a gas medium confined between two parallel plates, determine the direct exchange factor $\overline{g_i g_j}$ (i.e., per unit area), where medium zone i is an infinitely large slab of thickness Δz (located between $z_i \leq z \leq z_i + \Delta z$), and j is another medium zone (located at $z_j \leq z \leq z_j + \Delta z$).

18.2 For a two-dimensional problem determine the direct exchange factor $\overline{g_i g_j}'$ (i.e., per unit length) between two infinitely long volume zones of constant rectangular cross section, as shown in Fig. 18-3a (the zones are infinitely long in the direction normal to the figure). You may assume that the zones are sufficiently far apart so that $S \simeq \text{const}$ across the cross section of the zones. In a similar fashion determine $\overline{g_i s_l}'$ and $\overline{s_m s_l}'$.

18.3 Consider a two-dimensional medium confined in a cylinder (with variation in the radial and axial directions). Determine the direct exchange area $\overline{g_i g_j}$ between two toroidal volume zones of rectangular cross section, as shown in Fig. 18-3b.

18.4 For Problem 18.1 determine the total exchange area $\widetilde{G_i G_j}$ (per unit area) for the case that the bottom surface ($z = 0$) is a diffuse reflector (emissivity ϵ), while the top surface ($z = L$) is black.

18.5 Do Example 18.2 for a molecular gas with one vibration-rotation band in the infrared at wavenumber η_0, assuming that the total emissivity of the gas may be approximated as

$$\epsilon(T, S) = \frac{\omega E_{b\eta 0}}{E_b}(T)\left(1 - e^{-\kappa_0 S}\right),$$

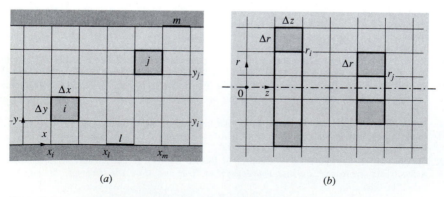

(a)   (b)

FIGURE 18-3
Two-dimensional zones for Cartesian and cylindrical systems.

where ω is the band width parameter, and $E_{b\eta 0}$ is the spectral emissive power at the band center.

The following problems are considerably more involved and require the development of a small computer program.

18.6 Consider a rectangular enclosure of width $W = 1\,\text{m}$ and height $H = 20\,\text{cm}$ (and infinite length, without any dependence on that direction). The bottom surface has a temperature of $T_1 = 1000\,\text{K}$ and emissivity $\epsilon_1 = 0.8$. The other three walls are at $T_2 = T_3 = T_4 = 400\,\text{K}$, $\epsilon_2 = \epsilon_3 = \epsilon_4 = 0.5$. There is no participating medium in the enclosure. Break up the enclosure surfaces into zones of $4\,\text{cm}$ width, and calculate the local heat transfer rates by the zonal method with total exchange areas.

18.7 Do Problem 18.6 for an enclosure filled with an absorbing/emitting gray medium with absorption coefficient $\kappa = 5\,\text{m}^{-1}$. The enclosure is in radiative equilibrium. Considering $4\,\text{cm} \times 4\,\text{cm}$ volume zones, calculate surface heat fluxes for the case of black walls.

18.8 Do Problem 18.7 for an isothermal medium at $T_m = 400\,\text{K}$.

18.9 Do Problem 18.7 for a gray-walled enclosure with emissivity values taken from Problem 18.6.

18.10 Do Problem 18.9 for a medium that also scatters isotropically (with $\sigma_s = 3\,\text{m}^{-1}$).

18.11 Consider a one-dimensional problem such as Problem 18.1. The plates are a distance $L = 1\,\text{m}$ apart, the bottom plate is at $T_1 = 1000\,\text{K}$ with emissivity $\epsilon_1 = 0.5$, the top plate is at $T_2 = 300\,\text{K}$ and $\epsilon_2 = 0.5$. The medium between the plates is isothermal at $T_m = 1000\,\text{K}$ and has an absorption coefficient of

$$\kappa_\eta = \kappa_0 e^{-2|\eta - \eta_0|/\omega}, \qquad \kappa_0 = 10\,\text{cm}^{-1}, \qquad \eta_0 = 3000\,\text{cm}^{-1}, \qquad \omega = 200\,\text{cm}^{-1},$$

(that is, it has a single molecular gas band modeled by the exponential wide band approximation). Calculate the heat gain of the top plate using

(a) spectral direct exchange areas,

(b) the weighted-sum-of-gray-gases approach.

18.12 Repeat Problem 18.11 for a medium at radiative equilibrium.

18.13 Consider a sphere of very hot dissociated gas of radius $5\,\text{cm}$. The gas may be approximated as a gray, isotropically scattering medium with $\kappa = 0.1\,\text{cm}^{-1}$, $\sigma_s = 0.2\,\text{cm}^{-1}$. The gas is suspended magnetically in vacuum within a large cold container and is initially at a uniform temperature $T_g = 10{,}000\,\text{K}$. Using the zonal method and neglecting conduction and convection, specify the total heat loss per unit time from the entire sphere at $t = 0$. Outline the solution for times $t > 0$.

18.14 Consider a sphere of nonscattering gas, initially at uniform $T = 3000\,\text{K}$, with a single vibration-rotation band at $\eta_0 = 3000\,\text{cm}^{-1}$. The gas, with a radius of $20\,\text{cm}$, is suspended magnetically in vacuum within a large cold container. For this gas it is known that bandstrength $\rho_a \alpha(T) = 500\,\text{cm}^{-2}$, bandwidth $\omega = 100\,\sqrt{T/100\,\text{K}}\,\text{cm}^{-1}$, and line overlap $\beta \gg 1$. These properties imply that the absorption coefficient may be determined from

$$\kappa_\eta = \kappa_0 e^{-2|\eta - \eta_0|/\omega}, \qquad \kappa_0 = \frac{\rho_a \alpha}{\omega};$$

and the band absorptance from

$$A(s) = \omega A^* = \omega \left[E_1(\kappa_0 s) + \ln(\kappa_0 s) + \gamma_E \right].$$

Neglecting conduction and convection, specify the total heat loss per unit time from the entire sphere at time $t = 0$, using the zonal method. Outline the solution procedure for times $t > 0$.

CHAPTER 19

THE MONTE CARLO METHOD FOR THERMAL RADIATION

19.1 INTRODUCTION

Very few exact, closed-form solutions to thermal radiation problems exist, even in the absence of a participating medium. Under most circumstances the solution has to be found by numerical means. For most engineers, who are used to dealing with partial differential equations, this implies use of *finite difference* and *finite element* techniques. These methods are, of course, applicable to thermal radiation problems whenever a solution method is chosen that transforms the governing equations into sets of partial differential equations (e.g., the P_N-method or the method of discrete ordinates). In general, however, radiative transfer is governed by integral equations. Still, similar techniques to finite differences and finite elements may be used by employing *numerical quadrature* for the evaluation of integrals.

With these techniques the solutions to relatively simple problems are readily found. However, if the geometry is involved, if radiative properties vary with direction, and/or if (anisotropic) scattering is of importance, then a solution by conventional numerical techniques may quickly become extremely involved if not impossible.

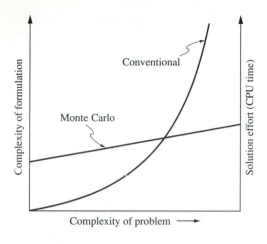

FIGURE 19-1
Comparison of Monte Carlo and conventional
solution techniques.

Many mathematical problems may also be solved by statistical methods, through sampling techniques, to any degree of accuracy. For example, consider predicting the outcome of the next presidential elections. Establishing a mathematical model that would predict voter turnout and voting behavior is, of course, impossible, let alone finding the analytical solution to such a model. However, if an appropriate sampling technique is chosen, the outcome can be predicted by conducting a poll. The accuracy of its prediction depends primarily on the sample size, i.e., how many people have been polled. Solving mathematical problems statistically always involves the use of random numbers, which may be picked, e.g., by placing a ball into a spinning roulette wheel. For this reason these sampling methods are called *Monte Carlo methods* (named after the principality of Monte Carlo in the south of France, famous for its casino). There is no single scheme to which the name *Monte Carlo* applies. Rather, any method of solving a mathematical problem with an appropriate statistical sampling technique is commonly referred to as a Monte Carlo method.

Problems in thermal radiation are particularly well suited to solution by a Monte Carlo technique, since energy travels in discrete parcels (photons) over (usually) relatively long distances along a (usually) straight path before interaction with matter. Thus, solving a thermal radiation problem by Monte Carlo implies tracing the history of a statistically meaningful random sample of photons from their points of emission to their points of absorption. The advantage of the Monte Carlo method is that even the most complicated problem may be solved with relative ease, as schematically indicated in Fig. 19-1. For a trivial problem, setting up the appropriate photon sampling technique alone may require more effort than finding the analytical solution. As the complexity of the problem increases, however, the complexity of formulation and the solution effort increase much more rapidly for conventional techniques. For problems beyond a certain complexity, the Monte Carlo solution will be preferable. Unfortunately, there is no way to determine a priori precisely where this crossover point in complexity lies. The disadvantage of Monte Carlo methods is that, as statistical methods, they are subject to statistical error (very similar to the unavoidable error associated with experimental measurements).

The name and the systematic development of Monte Carlo methods dates from about 1944 [1], although some crude mathematical sampling techniques were used off and on during previous centuries. Their first use as a research tool stems from the attempt to model neutron diffusion in fission material, for the development of the atomic bomb during World War II. The method was first applied to thermal radiation problems in the early 1960s by Fleck [2, 3] and Howell and Perlmutter [4–6].

For a thorough understanding of Monte Carlo methods, a good background in statistical methods is necessary, which goes beyond the scope of this book. In this chapter the method as applied to thermal radiation is outlined, and statistical considerations are presented in an intuitive way rather than in a rigorous mathematical fashion. For a more detailed description, the reader may want to consult the books by Hammersley and Handscomb [1], Cashwell and Everett [7], and Schreider [8], or the monographs by Kahn [9], Brown [10], Halton [11] and Haji-Sheikh [12]. A first monograph dealing specifically with Monte Carlo methods as applied to thermal radiation has been given by Howell [13]. Another, very recent one by Walters and Buckius [14] emphasizes the treatment of scattering.

Probability Distributions

When a political poll is conducted, people are not selected at random from a telephone directory. Rather, people are randomly selected from different groups according to *probability distributions*, to ensure that representative numbers of barbers, housewives, doctors, smokers, gun owners, bald people, heat transfer engineers, etc. are included in the poll. Similarly, in order to follow the history of radiative energy bundles in a statistically meaningful way, the points, directions and wavelengths of emission, reflective behavior, etc. must be chosen according to probability distributions.

As an example, consider the total radiative heat flux being emitted from a surface, i.e., the total emissive power,

$$E = \int_0^\infty E_\lambda \, d\lambda = \int_0^\infty \epsilon_\lambda E_{b\lambda} \, d\lambda. \tag{19.1}$$

Between the wavelengths of λ and $\lambda + d\lambda$ the emitted heat flux is $E_\lambda \, d\lambda = \epsilon_\lambda E_{b\lambda} \, d\lambda$, and the fraction of energy emitted over this wavelength range is

$$P(\lambda) \, d\lambda = \frac{E_\lambda \, d\lambda}{\int_0^\infty E_\lambda \, d\lambda} = \frac{E_\lambda}{E} \, d\lambda. \tag{19.2}$$

We may think of all the photons leaving the surface as belonging to a set of N energy bundles of equal energy (each consisting of many photons of a single wavelength).

Then each bundle carries the amount of energy (E/N) with it, and the probability that any particular bundle has a wavelength between λ and $\lambda + d\lambda$ is given by the *probability density function* $P(\lambda)$. The fraction of energy emitted over all wavelengths between 0 and λ is then

$$R(\lambda) = \int_0^\lambda P(\lambda) \, d\lambda = \frac{\int_0^\lambda E_\lambda \, d\lambda}{\int_0^\infty E_\lambda \, d\lambda}. \tag{19.3}$$

It is immediately obvious that $R(\lambda)$ is also the probability that any given energy bundle has a wavelength between 0 and λ, and it is known as the *cumulative distribution function*. The probability that a bundle has a wavelength between 0 and ∞ is, of course, $R(\lambda \rightarrow \infty) = 1$, a certainty. Equation (19.3) implies that if we want to simulate emission from a surface with N energy bundles of equal energy, then the fraction $R(\lambda)$ of these bundles must have wavelengths smaller than λ.

Now consider a pool of random numbers equally distributed between the values 0 and 1. Since they are equally distributed, this implies that a fraction R of these random numbers have values less than R itself. Let us now pick a single random number, say R_0. Inverting equation (19.3), we find $\lambda(R_0)$, i.e., the wavelength corresponding to a cumulative distribution function of value R_0, and we assign this wavelength to one energy bundle. If we repeat this process many times, then the fraction R_0 of all energy bundles will have wavelengths below $\lambda(R_0)$, since the fraction R_0 of all our random numbers will be below this value. Thus, in order to model correctly the spectral variation of surface emission, using N bundles of equal energy, their wavelengths may be determined by picking N random numbers between 0 and 1, and inverting equation (19.3).

Random Numbers

If we throw a ball onto a spinning roulette wheel, the ball will eventually settle on any one of the wheel's numbers (between 0 and 36). If we let the roulette wheel decide on another number again and again, we will obtain a *set of random numbers* between 0 and 36 (or between 0 and 1, if we divide each number by 36). Unless the croupier throws in the ball and spins the wheel in a regular (nonrandom) fashion,[1] any number may be chosen each time with equal probability, regardless of what numbers have been picked previously. However, if sufficiently many numbers are picked, we may expect that roughly half (i.e., 18/37) of all the picked numbers will be between 0 and 17, for example.

During the course of a Monte Carlo simulation, generally somewhere between 10^5 and 10^7 random numbers need to be drawn, and they need to be drawn very rapidly. Obviously, spinning a roulette wheel would be impractical. One solution to

---

[1]This is, of course, the reason casinos tend to employ a number of croupiers, each of whom works only for a very short period each day.

this problem is to store a (externally determined) set of random numbers. However, such a table would require a prohibitive amount of computer storage, unless it were a relatively small table that would be used repeatedly (thus destroying the true randomness of the set). The only practical answer is to generate the random numbers within the computer itself. This appears to be a contradiction, since a digital computer is the incarnation of logic (nonrandomness). Substantial research has been carried out on how to generate sets of sufficiently random numbers using what are called *pseudorandom number generators*. A number of such generators exist that, after making the choice of a starting point, generate a new pseudorandom number from the previous one. The randomness of such a set of numbers depends on the quality of the generator as well as the choice of the starting point and should be tested by different "randomness tests." For a more detailed discussion of pseudorandom number generators, the reader is referred to Hammersley and Handscomb [1], Schreider [8], or Taussky and Todd [15].

Accuracy Considerations

Since Monte Carlo methods are statistical methods, the results, when plotted against number of samples, will generally fluctuate randomly around the correct answer. If a set of truly random numbers is used for the sampling, then these fluctuations will decrease as the number of samples increases. Let the answer obtained from the Monte Carlo method after tracing N energy bundles be $S(N)$, and the exact solution obtained after sampling infinitely many energy bundles $S(\infty)$. For some simple problems it is possible to calculate directly the probability that the obtained answer, $S(N)$, differs by less than a certain amount from the correct answer, $S(\infty)$.

Even if it were possible to directly calculate the confidence level for more complicated situations, this would not take into account the pseudorandomness of the computer-generated random number set. That this effect can be rather substantial is seen from Fig. 19-2, which depicts the Monte Carlo evaluation of the view factor between two parallel black plates [16]. Both sets of data use the same computer code and the same random number generator (on a UNIVAC 1110). If a starting value of 1 is used, the results are still fairly inaccurate after 5000 bundles; if a starting value of 12,345 is used (this number gave the fastest-converging results of the ones tested for the random number generator used here), good convergence is achieved after only 4000 bundles. Obviously, careful investigation of the random number generator can increase convergence and accuracy and thus decrease computer time considerably. Randomness tests performed on sets of generated numbers showed that a starting value of 12,345 performs well in all tests and indeed results in a "better" set of random numbers than the starting value of 1 for the random number generator employed by Modest and Poon [16].

For radiative heat transfer calculations the most straightforward way of estimating the error associated with the sampling result $S(N)$ is to break up the result into a

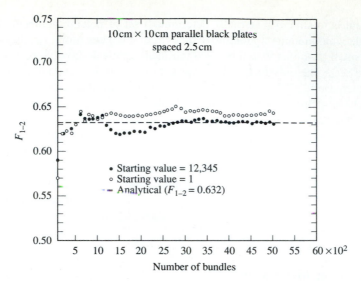

FIGURE 19-2
Convergence of Monte Carlo method for different sets of random numbers.

number of I subsamples $S(N_i)$, such that

$$N = N_1 + N_2 + \ldots + N_I = \sum_{i=1}^{I} N_i, \tag{19.4}$$

$$S(N) = \frac{1}{N}\left(N_1 S(N_1) + \ldots + N_I S(N_I)\right) = \frac{1}{N}\sum_{i=1}^{I} N_i S(N_i). \tag{19.5}$$

Normally, each subsample would include identical amounts of bundles, leading to

$$N_i = N/I; \qquad i = 1, 2, \ldots, I, \tag{19.6}$$

$$S(N) = \frac{1}{I}\sum_{i=1}^{I} S(N_i). \tag{19.7}$$

The I subsamples may then be treated as if they were independent experimental measurements of the same quantity. We may then calculate the *variance* or *adjusted mean square deviation of the mean*

$$\sigma_m^2 = \frac{1}{I(I-1)}\sum_{i=1}^{I} [S(N_i) - S(N)]^2. \tag{19.8}$$

The *central limit theorem* states that the mean $S(N)$ of I measurements $S(N_i)$ follows a Gaussian distribution, whatever the distribution of the individual measurements. This implies that we can say with 68.3% confidence that the correct answer $S(\infty)$ lies within the limits of $S(N) \pm \sigma_m$, with 95.5% confidence within $S(N) \pm 2\sigma_m$, or with

99% confidence within $S(N) \pm 2.58\sigma_m$. Details on statistical analysis of errors may be found in any standard book on experimentation, for example, the one by Barford [17].

19.2 HEAT TRANSFER RELATIONS FOR RADIATIVE EXCHANGE BETWEEN SURFACES

In the absence of a participating medium and assuming a refractive index of unity, the radiative heat flux leaving or going into a certain surface, using the Monte Carlo technique, is governed by the following basic equation:

$$q(\mathbf{r}) = \epsilon(\mathbf{r})\sigma T^4(\mathbf{r}) - \int_A \epsilon(\mathbf{r}')\,\sigma T^4(\mathbf{r}')\frac{d\mathcal{F}_{dA'\to dA}}{dA}\,dA', \qquad (19.9)$$

where

$$
\begin{array}{lll}
q(\mathbf{r}) & = & \text{local surface heat flux at location } \mathbf{r}, \\
T(\mathbf{r}) & = & \text{surface temperature at location } \mathbf{r}, \\
\epsilon(\mathbf{r}) & = & \text{total hemispherical emissivity of the surface at } \mathbf{r}, \\
A & = & \text{surface area of the enclosure, and} \\
d\mathcal{F}_{dA'\to dA} & = & \text{generalized radiation exchange factor between} \\
& & \text{surface elements } dA' \text{ and } dA.
\end{array}
$$

In equation (19.9) the first term on the right-hand side describes the emission from the surface, and the integrand of the second term is the fraction of energy, originally emitted from the surface at $\mathbf{r}'$, which eventually gets absorbed at location $\mathbf{r}$. Therefore, the definition for the generalized exchange factor must be:

$$
\begin{array}{lll}
d\mathcal{F}_{dA'\to dA} & \equiv & \text{fraction of the total energy emitted by } dA' \text{ that is} \qquad (19.10) \\
& & \text{absorbed by } dA, \text{ either directly or after any} \\
& & \text{number and type of reflections.}
\end{array}
$$

This definition appears to be the most compatible one for solution by ray-tracing techniques and is therefore usually employed for calculations by the Monte Carlo method. Figure 19-3 shows a schematic of an arbitrary enclosure with energy bundles emitted at dA' and absorbed at dA.

If the enclosure is not closed, i.e., has openings into space, some artificial closing surfaces must be introduced. For example, an opening directed into outer space without irradiation from the sun or earth can be replaced by a black surface at a temperature of $0\,\mathrm{K}$. If the opening is irradiated by the sun, it is replaced by a nonreflecting surface with zero emissivity for all angles but the solar angle, etc.

The enclosure surface is now divided into J subsurfaces, and equation (19.9) reduces to

$$Q_i = \int_{A_i} q_i\,dA_i = \epsilon_i\sigma T_i^4 A_i - \sum_{j=1}^{J}\epsilon_j\sigma T_j^4 A_j\,\mathcal{F}_{j\to i} \;-\; q_{\text{ext}}A_s\mathcal{F}_{s\to i},$$

$$1 \le i \le J, \qquad (19.11)$$

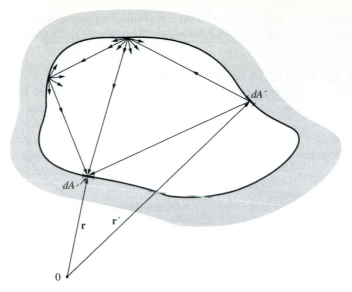

FIGURE 19-3
Possible energy bundle paths in an arbitrary enclosure.

where

q_{ext} = external energy entering through any opening in the enclosure,
A_s = area of the opening irradiated from external sources,

and the ϵ_j and T_j are suitable average values for each subsurface, i.e.,

$$\epsilon_j \sigma T_j^4 = \frac{1}{A_j} \int_{A_j} \epsilon \sigma T^4 dA. \tag{19.12}$$

Although heat flow rates Q_i can be calculated directly by the Monte Carlo method, it is of advantage to instead determine the exchange factors: Although the Q_is depend on all surface temperatures in the enclosure, the $\mathcal{F}_{i \to j}$s either do not (gray surfaces) or depend only on the temperature of the emitting surface (nongray surfaces), provided that surface reflectivities (and absorptivities) are independent of temperature (as they are to a very good degree of accuracy).

A large statistical sample of energy bundles N_i is emitted from surface A_i, each of them carrying the amount of radiative energy

$$\Delta E_i = \epsilon_i \sigma T_i^4 A_i / N_i. \tag{19.13}$$

If N_{ij} of these bundles become absorbed by surface A_j either after direct travel or after any number of reflections, the exchange factor may be calculated from

$$\mathcal{F}_{i \to j} = \lim_{N_i \to \infty} \left(\frac{N_{ij}}{N_i} \right) \simeq \left(\frac{N_{ij}}{N_i} \right)_{N_i \gg 1}. \tag{19.14}$$

19.3 RANDOM NUMBER RELATIONS FOR SURFACE EXCHANGE

In order to calculate the exchange factor by tracing the history of a large number of energy bundles, we need to know how to pick statistically meaningful energy bundles as explained in Section 19.1: For each emitted bundle we need to determine a point of emission, a direction of emission, and a wavelength of emission. Upon impact of the bundle onto another point of the enclosure surface, we need to decide whether the bundle is reflected and, if so, into what direction.

Points of Emission

Similar to equation (19.1) we may write for the total emission from a surface A_j:

$$E_j = \int_{A_j} \epsilon \sigma T^4 \, dA. \tag{19.15}$$

Since integration over an area is a double integral, we may rewrite this equation, without loss of generality, as

$$E_j = \int_{x=0}^{X} \int_{y=0}^{Y} \epsilon \sigma T^4 \, dy \, dx = \int_0^X E_j' \, dx, \tag{19.16}$$

where

$$E_j'(x) = \int_0^Y \epsilon \sigma T^4 \, dy. \tag{19.17}$$

Thus, we may apply equation (19.3) and find

$$R_x = \frac{1}{E_j} \int_0^x E_j' \, dx. \tag{19.18}$$

This relationship may be inverted to find the x-location of the emission point as a function of a random number R_x:

$$x = x(R_x). \tag{19.19}$$

Once the x-location has been determined, equation (19.3) may also be applied to equation (19.17), leading to an expression for the y-location of emission:

$$R_y = \frac{1}{E_j'(x)} \int_0^y \epsilon \sigma T^4 \, dy, \tag{19.20}$$

and

$$y = y(R_y, x). \tag{19.21}$$

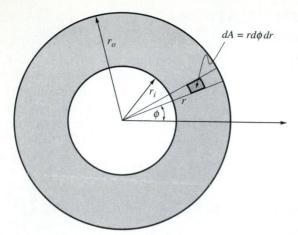

FIGURE 19-4
Geometry for Example 19.1.

Note that the choice for the y-location depends not only on the random number R_y, but also on the location of x.

If the emissive power may be separated in x and y, i.e., if

$$E = \epsilon \sigma T^4 = E_x(x)E_y(y),\tag{19.22}$$

then equation (19.18) reduces to

$$R_x = \int_0^x E_x(x)dx \Big/ \int_0^X E_x(x)dx,\tag{19.23}$$

and equation (19.20) simplifies to

$$R_y = \int_0^y E_y(y)dy \Big/ \int_0^Y E_y(y)dy,\tag{19.24}$$

that is, choices for x- and y-locations become independent of one another. In the simplest case of an isothermal surface with constant emissivity, these relations reduce to

$$x = R_x X, \quad y = R_y Y.\tag{19.25}$$

Example 19.1.   Given a ring surface element on the bottom of a black isothermal cylinder with inner radius $r_i = 10\,\text{cm}$ and outer radius $r_o = 20\,\text{cm}$, as indicated in Fig. 19-4, calculate the location of emission for a pair of random numbers $R_r = 0.5$ and $R_\phi = 0.25$.

Solution

We find

$$E = \int_A E_b dA = E_b \int_0^{2\pi} \int_{r_i}^{r_o} r\,dr\,d\phi.$$

Since this expression is separable in r and ϕ, this leads to

$$R_\phi = \int_0^\phi d\phi \bigg/ \int_0^{2\pi} d\phi = \frac{\phi}{2\pi}, \quad \text{or} \quad \phi = 2\pi R_\phi,$$

and

$$R_r = \int_{r_i}^r r\,dr \bigg/ \int_{r_i}^{r_o} r\,dr = \frac{r^2 - r_i^2}{r_o^2 - r_i^2},$$

or

$$r = \sqrt{r_i^2 + (r_o^2 - r_i^2)R_r}.$$

Therefore, $\phi = 2\pi \times 0.25 = \pi/2$ and $r = \sqrt{100 + (400 - 100)0.5} = 15.8\,\text{cm}$. While, as expected for a random number of 0.25, the emission point angle is 90° away from the $\phi = 0$ axis, the r-location does not fall onto the midpoint. This is because the cylindrical ring has more surface area at larger radii, resulting in larger total emission. This implies that more energy bundles must be emitted from the outer part of the ring.

Wavelengths of Emission

Once an emission location has been chosen, the wavelength of the emitted bundle needs to be determined (unless all surfaces in the enclosure are gray; in that case the wavelength of the bundle does not enter the calculations, and its determination may be omitted). The process of finding the wavelength has already been outlined in Section 19.1, leading to equation (19.3), i.e.,

$$R_\lambda = \frac{1}{\epsilon\sigma T^4} \int_0^\lambda \epsilon_\lambda E_{b\lambda}\,d\lambda, \tag{19.26}$$

and, after inversion,

$$\lambda = \lambda(R_\lambda, x, y). \tag{19.27}$$

We note that the choice of wavelength, in general, depends on the choice for the emission location (x, y), unless the surface is isothermal with constant emissivity. If the surface is black or gray, equation (19.26) reduces to the simple case of

$$R_\lambda = \frac{1}{\sigma T^4} \int_0^\lambda E_{b\lambda}\,d\lambda = f(\lambda T). \tag{19.28}$$

Directions of Emission

The spectral emissive power (for a given position and wavelength) is

$$E_\lambda = \int_{2\pi} \epsilon_\lambda' I_{b\lambda} \cos\theta\, d\Omega = \frac{1}{\pi} E_{b\lambda} \int_0^{2\pi} \int_0^{\pi/2} \epsilon_\lambda' \cos\theta \sin\theta\, d\theta\, d\psi. \tag{19.29}$$

As we did for choosing the (two-dimensional) point of emission, we write

$$R_\psi = \frac{E_{b\lambda}}{\pi E_\lambda} \int_0^\psi \int_0^{\pi/2} \epsilon_\lambda' \cos\theta \sin\theta \, d\theta \, d\psi$$

$$= \frac{1}{\pi} \int_0^\psi \int_0^{\pi/2} \frac{\epsilon_\lambda'}{\epsilon_\lambda} \cos\theta \sin\theta \, d\theta \, d\psi, \tag{19.30}$$

or

$$\psi = \psi(R_\psi, x, y, \lambda). \tag{19.31}$$

We note from equation (19.30) that ψ does not usually depend on emission location, unless the emissivity changes across the surface. However, ψ does depend on the chosen wavelength, unless spectral and directional dependence of the emissivity are separable. Once the azimuthal angle ψ is found, the polar angle θ is determined from

$$R_\theta = \int_0^\theta \epsilon_\lambda' \cos\theta \sin\theta \, d\theta \Big/ \int_0^{\pi/2} \epsilon_\lambda' \cos\theta \sin\theta \, d\theta, \tag{19.32}$$

or

$$\theta = \theta(R_\theta, x, y, \lambda, \psi). \tag{19.33}$$

Most surfaces tend to be isotropic so that the directional emissivity does not depend on azimuthal angle ψ. In that case $\epsilon_\lambda = 2\int_0^{\pi/2} \epsilon_\lambda' \cos\theta \sin\theta \, d\theta$, and equation (19.30) reduces to

$$R_\psi = \frac{\psi}{2\pi}, \quad \text{or} \quad \psi = 2\pi R_\psi, \tag{19.34}$$

and the choice of polar angle becomes independent of azimuthal angle. For a diffuse emitter, equation (19.32) simplifies to

$$R_\theta = \sin^2\theta, \quad \text{or} \quad \theta = \sin^{-1}\sqrt{R_\theta}. \tag{19.35}$$

Absorption and Reflection

When radiative energy impinges on a surface, the fraction α_λ' will be absorbed, which may depend on the wavelength of irradiation, the direction of the incoming rays, and, perhaps, the local temperature. Of many incoming bundles the fraction α_λ' will therefore be absorbed while the rest, $1 - \alpha_\lambda'$, will be reflected. This can clearly be simulated by picking a random number, R_α, and comparing it with α_λ': If $R_\alpha \le \alpha_\lambda'$, the bundle is absorbed, while if $R_\alpha > \alpha_\lambda'$, it is reflected.

The direction of reflection depends on the bidirectional reflection function of the material. The fraction of energy reflected into all possible directions is equal to the directional-hemispherical spectral reflectivity, or

$$
\rho_\lambda^{\prime\ominus}(\lambda, \theta_i, \psi_i) = \int_{2\pi} \rho_\lambda^{\prime\prime}(\lambda, \theta_i, \psi_i, \theta_r, \psi_r) \cos \theta_r \, d\Omega_r
$$

$$
= \int_0^{2\pi} \int_0^{\pi/2} \rho_\lambda^{\prime\prime}(\lambda, \theta_i, \psi_i, \theta_r, \psi_r) \cos \theta_r \, \sin \theta_r \, d\theta_r \, d\psi_r. \tag{19.36}
$$

As before, the direction of reflection may then be determined from

$$
R_{\psi_r} = \frac{1}{\rho_\lambda^{\prime\ominus}} \int_0^{\psi_r} \int_0^{\pi/2} \rho_\lambda^{\prime\prime}(\lambda, \theta_i, \psi_i, \theta_r, \psi_r) \cos \theta_r \, \sin \theta_r \, d\theta_r \, d\psi_r, \tag{19.37}
$$

and

$$
R_{\theta_r} = \frac{\displaystyle\int_0^{\theta_r} \rho_\lambda^{\prime\prime}(\lambda, \theta_i, \psi_i, \theta_r, \psi_r) \cos \theta_r \, \sin \theta_r \, d\theta_r}{\displaystyle\int_0^{\pi/2} \rho_\lambda^{\prime\prime}(\lambda, \theta_i, \psi_i, \theta_r, \psi_r) \cos \theta_r \, \sin \theta_r \, d\theta_r}. \tag{19.38}
$$

If the surface is a diffuse reflector, i.e., $\rho_\lambda^{\prime\prime}(\lambda, \theta_i, \psi_i, \theta_r, \psi_r) = \rho_\lambda^{\prime\prime}(\lambda) = \rho_\lambda^{\prime\ominus}(\lambda)/\pi$, then equations (19.37) and (19.38) reduce to

$$
R_{\psi_r} = \frac{\psi_r}{2\pi}, \quad \text{or} \quad \psi_r = 2\pi R_{\psi_r}, \tag{19.39}
$$

and

$$
R_{\theta_r} = \sin^2 \theta_r, \quad \text{or} \quad \theta_r = \sin^{-1} \sqrt{R_{\theta_r}}, \tag{19.40}
$$

which are the same as for diffuse emission. For a purely specular reflector, the reflection direction follows from the law of optics as

$$
\psi_r = \psi_i + \pi, \quad \theta_r = \theta_i, \tag{19.41}
$$

that is, no random numbers are needed.[2]

19.4 SURFACE DESCRIPTION

When Monte Carlo simulations are applied to very simple configurations such as flat plates, e.g., Toor and Viskanta [18], the surface description, bundle intersection points, intersection angles, reflection angles, etc. are relatively obvious and straightforward. If more complicated surfaces are considered, such as the second-order polynomial description by Weiner and coworkers [19] or the arbitrary-order polymial description by Modest and Poon [16] and Modest [20], a systematic way to describe surfaces is preferable. It appears most logical to describe surfaces in vectorial form,

---

[2]Mathematically, equation (19.41) may also be obtained from equations (19.37) and (19.38) by replacing $\rho_\lambda^{\prime\prime}$ by an appropriate Dirac-delta function.

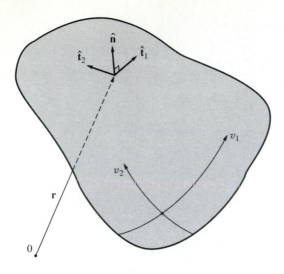

FIGURE 19-5
Surface description in terms of a position vector.

as indicated in Fig. 19-5,

$$\mathbf{r} = \sum_{i=1}^{3} x_i(v_1, v_2)\,\hat{\mathbf{e}}_i, \quad v_{1\min} \leq v_1 \leq v_{1\max}, \quad v_{2\min}(v_1) \leq v_{2\max}(v_1), \qquad (19.42)$$

that is, $\mathbf{r}$ is the vector pointing from the origin to a point on the surface, v_1 and v_2 are two surface parameters, the x_i are the (x, y, z) coordinates of the surface point, and the $\hat{\mathbf{e}}_i$ are unit vectors $(\hat{\mathbf{i}}, \hat{\mathbf{j}}, \hat{\mathbf{k}})$ into the x, y, z directions, respectively. We may define two unit tangents to the surface at any point as

$$\hat{\mathbf{t}}_1 = \frac{\partial \mathbf{r}}{\partial v_1} \bigg/ \left| \frac{\partial \mathbf{r}}{\partial v_1} \right|, \quad \hat{\mathbf{t}}_2 = \frac{\partial \mathbf{r}}{\partial v_2} \bigg/ \left| \frac{\partial \mathbf{r}}{\partial v_2} \right|. \qquad (19.43)$$

While it is usually a good idea to choose the surface parameters v_1 and v_2 perpendicular to one another (making $\hat{\mathbf{t}}_1$ and $\hat{\mathbf{t}}_2$ perpendicular to each other), this is not necessary. In either case, one can evaluate the unit surface normal as

$$\hat{\mathbf{n}} = \frac{\hat{\mathbf{t}}_1 \times \hat{\mathbf{t}}_2}{|\hat{\mathbf{t}}_1 \times \hat{\mathbf{t}}_2|}, \qquad (19.44)$$

where it has been assumed that v_1 and v_2 have been ordered such that $\hat{\mathbf{n}}$ is the *outward* surface normal.

19.5 RAY TRACING

Points of emission may be found by establishing a relationship such as equation (19.15) for the general vectorial surface description given by equation (19.42). The infinitesimal area element on the surface may be described by

$$dA = \left| \frac{\partial \mathbf{r}}{\partial v_1} \times \frac{\partial \mathbf{r}}{\partial v_2} \right| dv_1\, dv_2 = |\hat{\mathbf{t}}_1 \times \hat{\mathbf{t}}_2| \left| \frac{\partial \mathbf{r}}{\partial v_1} \right| \left| \frac{\partial \mathbf{r}}{\partial v_2} \right| dv_1\, dv_2. \qquad (19.45)$$

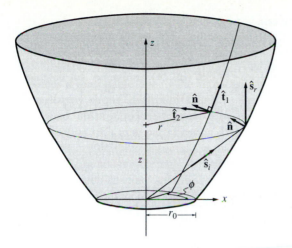

FIGURE 19-6
Rocket nozzle diffuser geometry for Example 19.2.

Thus, if we replace x by v_1 and y by v_2, emission points (v_1, v_2) are readily found from equations (19.18) and (19.20).

Example 19.2.   Consider the axisymmetric rocket nozzle diffuser shown in Fig. 19-6. Assuming that the diffuser is gray and isothermal, establish the appropriate random number relationships for the determination of emission points.

Solution

The diffuser surface is described by the formula

$$z = a(r^2 - r_0^2), \quad 0 \le z \le L, \quad r_0 \le r \le r_L, \quad a = \frac{1}{2r_0},$$

where L is the length of the diffuser and r_0 and r_L are its radius at $z = 0$ and L, respectively. In vectorial form, we may write

$$\mathbf{r} = r \cos\phi\hat{\mathbf{i}} + r \sin\phi\hat{\mathbf{j}} + a(r^2 - r_0^2)\hat{\mathbf{k}},$$

where ϕ is the azimuthal angle in the x-y-plane, measured from the x-axis. This suggests the choice $v_1 = r$ and $v_2 = \phi$. The two surface tangents are now readily calculated as

$$\hat{\mathbf{t}}_1 = \frac{\cos\phi\hat{\mathbf{i}} + \sin\phi\hat{\mathbf{j}} + 2ar\hat{\mathbf{k}}}{\sqrt{1 + 4a^2r^2}},$$

$$\hat{\mathbf{t}}_2 = -\sin\phi\hat{\mathbf{i}} + \cos\phi\hat{\mathbf{j}}.$$

It is seen that $\hat{\mathbf{t}}_1 \cdot \hat{\mathbf{t}}_2 = 0$, i.e., the tangents are perpendicular to one another. The surface normal is then

$$\hat{\mathbf{n}} = \hat{\mathbf{t}}_1 \times \hat{\mathbf{t}}_2 = \frac{1}{\sqrt{1 + 4a^2r^2}} \begin{pmatrix} \hat{\mathbf{i}} & \hat{\mathbf{j}} & \hat{\mathbf{k}} \\ \cos\phi & \sin\phi & 2ar \\ -\sin\phi & \cos\phi & 0 \end{pmatrix}$$

$$= \frac{-2ar(\cos\phi\hat{\mathbf{i}} + \sin\phi\hat{\mathbf{j}}) + \hat{\mathbf{k}}}{\sqrt{1 + 4a^2r^2}},$$

and, finally, an infinitesimal surface area is

$$dA = |\cos\phi\hat{\mathbf{i}} + \sin\phi\hat{\mathbf{j}} + 2ar\hat{\mathbf{k}}|| - r\sin\phi\hat{\mathbf{i}} + r\cos\phi\hat{\mathbf{j}}|dr\,d\phi$$
$$= \sqrt{1 + 4a^2r^2}\,r\,dr\,d\phi.$$

Since there is no dependence on azimuthal angle ϕ in either dA or the emissive power, we find immediately

$$R_\phi = \frac{\phi}{2\pi}, \text{ or } \phi = 2\pi R_\phi,$$

and for the radial position parameter

$$R_r = \frac{\int_{r_0}^{r}\sqrt{1+4a^2r^2}\,r\,dr}{\int_{r_0}^{r_L}\sqrt{1+4a^2r^2}\,r\,dr} = \frac{(1+4a^2r^2)^{3/2}\Big|_{r_0}^{r}}{(1+4a^2r^2)^{3/2}\Big|_{r_0}^{r_L}}$$

$$= \frac{(1+4a^2r^2)^{3/2} - (1+4a^2r_0^2)^{3/2}}{(1+4a^2r_L^2)^{3/2} - (1+4a^2r_0^2)^{3/2}}.$$

The above expression is readily solved to give an explicit expression for $r = r(R_r)$.

Once a point of emission has been found, a wavelength and a direction are calculated from equations (19.26), (19.30), and (19.32). As shown in Fig. 19-7, the direction may be specified as a *unit direction vector* with polar angle θ measured from the surface normal, and azimuthal angle ψ measured from $\hat{\mathbf{t}}_1$, leading to

$$\hat{\mathbf{s}} = \frac{\sin\theta}{\sin\alpha}[\sin(\alpha - \psi)\hat{\mathbf{t}}_1 + \sin\psi\hat{\mathbf{t}}_2] + \cos\theta\hat{\mathbf{n}}, \tag{19.46}$$

and

$$\sin\alpha = \left|\hat{\mathbf{t}}_1 \times \hat{\mathbf{t}}_2\right|, \tag{19.47}$$

where α is the angle between $\hat{\mathbf{t}}_1$ and $\hat{\mathbf{t}}_2$. If $\hat{\mathbf{t}}_1$ and $\hat{\mathbf{t}}_2$ are perpendicular ($\alpha = \pi/2$), equation (19.46) reduces to

$$\hat{\mathbf{s}} = \sin\theta[\cos\psi\hat{\mathbf{t}}_1 + \sin\psi\hat{\mathbf{t}}_2] + \cos\theta\hat{\mathbf{n}}. \tag{19.48}$$

As also indicated in Fig. 19-7, the intersection point of an energy bundle emitted at location $\mathbf{r}_e$, traveling into the direction $\hat{\mathbf{s}}$, with a surface described in vectorial form may be determined as

$$\mathbf{r}_e + D\hat{\mathbf{s}} = \mathbf{r}, \tag{19.49}$$

where $\mathbf{r}$ is the vector describing the intersection point, and D is the distance traveled by the energy bundle. Equation (19.49) may be written in terms of its x, y, z components

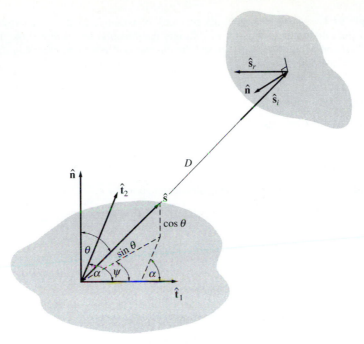

FIGURE 19-7
Vector description of emission direction and point of impact.

and solved for D by forming the dot products with unit vectors $\hat{\mathbf{i}}$, $\hat{\mathbf{j}}$, and $\hat{\mathbf{k}}$:

$$D = \frac{x(v_1, v_2) - x_e}{\hat{\mathbf{s}} \cdot \hat{\mathbf{i}}} = \frac{y(v_1, v_2) - y_e}{\hat{\mathbf{s}} \cdot \hat{\mathbf{j}}} = \frac{z(v_1, v_2) - z_e}{\hat{\mathbf{s}} \cdot \hat{\mathbf{k}}}. \tag{19.50}$$

Equation (19.50) is a set of three equations in the three unknowns v_1, v_2, and D: First v_1 and v_2 are calculated, and it is determined whether the intersection occurs within the confines of the surface under scrutiny. If so, and if more than one intersection is a possibility (in the presence of convex surfaces, protruding corners, etc.), then the path length D is also determined; if more than one intersection is found, the correct one is the one after the shortest positive path.

If the bundle is reflected, and if reflection is nonspecular, a reflection direction is chosen from equations (19.37) and (19.38). This direction is then expressed in vector form using equation (19.46). If the surface is a specular reflector, the direction of reflection is determined from equation (19.41), or in vector form as

$$\hat{\mathbf{s}}_r = \hat{\mathbf{s}}_i + 2|\hat{\mathbf{s}}_i \cdot \hat{\mathbf{n}}|\hat{\mathbf{n}}. \tag{19.51}$$

Once the intersection point and the direction of reflection have been determined, a new intersection may be found from equation (19.50), etc., until the bundle is absorbed.

Example 19.3.   Consider again the geometry of Example 19.2. An energy bundle is emitted from the origin ($x = y = z = 0$) into the direction $\hat{s} = 0.8\hat{i} + 0.6\hat{k}$. Determine the intersection point on the diffuser and the direction of reflection, assuming the diffuser to be a specular reflector.

Solution

With $r_e = 0$ and equations (19.46) and (19.50), we find

$$D = \frac{r \cos \phi}{0.8} = \frac{r \sin \phi}{0} = \frac{a(r^2 - r_0^2)}{0.6}.$$

Obviously, $\phi = 0,$[3] and solving the quadratic equation for r,

$$r^2 - r_0^2 = \frac{3}{4}\frac{r}{a} = \tfrac{3}{2}rr_0, \text{ or } r = 2r_0 \text{ and } z = \frac{1}{2r_0}(4r_0^2 - r_0^2) = \tfrac{3}{2}r_0.$$

At that location we form the unit vectors, as given in Example 19.2,

$$\hat{t}_1 = \frac{1}{\sqrt{5}}(\hat{i} + 2\hat{k}), \ \hat{t}_2 = \hat{j}, \text{ and } \hat{n} = \frac{1}{\sqrt{5}}(-2\hat{i} + \hat{k}).$$

Therefore, the direction of reflection is determined from equation (19.51) as

$$\hat{s}_r = 0.8\hat{i} + 0.6\hat{k} + 2\left|\frac{-2 \times 0.8 + 0.6}{\sqrt{5}}\right|\frac{-2\hat{i} + \hat{k}}{\sqrt{5}} = \hat{k},$$

as is easily verified from Fig. 19-6.

19.6   HEAT TRANSFER RELATIONS FOR PARTICIPATING MEDIA

If the enclosure is filled with an absorbing, emitting, and/or scattering medium, equations (19.9) through (19.14) for the evaluation of surface heat fluxes must be augmented by a term to account for emission from within the medium, and the definition for the generalized radiation exchange factor must be altered to allow for absorption and/or scattering. Again assuming a refractive index of unity, from equation (8.49) the total emission per unit volume is $4\kappa_P \sigma T^4$ and, therefore,

$$q(\mathbf{r}) = \epsilon(\mathbf{r})\sigma T^4(\mathbf{r}) - \int_A \epsilon(\mathbf{r}')\sigma T^4(\mathbf{r}')\frac{d\mathcal{F}_{dA' \to dA}}{dA} dA',$$

$$- \int_V 4\kappa_P(\mathbf{r}'')\sigma T^4(\mathbf{r}'')\frac{d\mathcal{F}_{dV'' \to dA}}{dA} dV'', \qquad (19.52)$$

---

[3] In computer calculations care must be taken here and elsewhere to avoid division by zero.

where

$\kappa_P(\mathbf{r}'')$ =   local Planck-mean absorption coefficient of the medium at $\mathbf{r}''$,

$d\mathcal{F}_{dV'' \to dA}$ =   generalized radiation exchange factor between volume elements dV'' and surface element dA.

Equation (19.10) still applies to all exchange factors, including $d\mathcal{F}_{dV'' \to dA}$, with the added stipulation that energy bundles may be attenuated by absorption and/or redirected by scattering.

A similar equation is needed to describe the net amount of radiative energy deposited (or withdrawn) per unit volume of the medium, i.e., the divergence of the radiative heat flux. From equations (8.54) and (9.100), it follows that

$$\nabla \cdot \mathbf{q} = 4\kappa_P \sigma T^4 - \int_0^\infty \kappa_\lambda G_\lambda \, d\lambda, \tag{19.53}$$

where G_λ is the spectral incident radiation. The first term in equation (19.53) describes emission from within the volume, and the second term gives the absorbed fraction per unit volume of all radiation incident on the element. For Monte Carlo calculations this term may be replaced by an expression similar to the one in equation (19.52), i.e.,

$$\nabla \cdot \mathbf{q} = 4\kappa_P \sigma T^4 - \int_A \epsilon(\mathbf{r}') \sigma T^4(\mathbf{r}') \frac{d\mathcal{F}_{dA' \to dV}}{dV} \, dA'$$
$$- \int_V 4\kappa_P(\mathbf{r}'') \sigma T^4(\mathbf{r}'') \frac{d\mathcal{F}_{dV'' \to dV}}{dV} \, dV''. \tag{19.54}$$

For numerical calculations it is again necessary to break up the enclosure into a number J finite subsurfaces and K finite subvolumes, transforming equations (19.52) and (19.54) to

$$Q_i = \epsilon_i \sigma T_i^4 A_i - \sum_{j=1}^J \epsilon_j \sigma T_j^4 A_j \mathcal{F}_{j \to i} - q_{\text{ext}} A_s \mathcal{F}_{s \to i}$$
$$- \sum_{k=1}^K 4\kappa_{Pk} \sigma T_k^4 V_k \mathcal{F}_{k \to i}, \qquad i = 1, 2, \ldots, J, \tag{19.55}$$

$$\int_{V_l} \nabla \cdot \mathbf{q} \, dV = 4\kappa_{Pl} \sigma T_l^4 V_l - \sum_{j=1}^J \epsilon_j \sigma T_j^4 A_j \mathcal{F}_{j \to l} - q_{\text{ext}} A_s \mathcal{F}_{s \to l}$$
$$- \sum_{k=1}^K 4\kappa_{Pk} \sigma T_k^4 V_k \mathcal{F}_{k \to l}, \qquad l = 1, 2, \ldots, K, \tag{19.56}$$

where the T_k are suitably defined average temperatures within the medium

$$\kappa_{Pk} \sigma T_k^4 = \frac{1}{V_k} \int_{V_k} \kappa_P \sigma T^4 \, dV. \tag{19.57}$$

Using this formulation, all generalized exchange factors may then be evaluated through equation (19.14).

19.7   RANDOM NUMBER RELATIONS FOR PARTICIPATING MEDIA

Besides the random number relations established in the last section we need to find additional expressions for emission from within the volume, for absorption by the medium, and for scattering.

Points of Emission within Medium

The total emission from a subvolume V_k is given by

$$E_k = \int_{V_k} 4\kappa_P \sigma T^4 dV .$$
(19.58)

Using Cartesian coordinates with $dV = dx\, dy\, dz$, equation (19.58) may be rewritten as

$$E_k = \int_0^X \left(\int_0^Y \int_0^Z 4\kappa_P \sigma T^4 dz\, dy \right) dx = \int_0^X E_k'(x)dx .$$
(19.59)

Then, following the development of equations (19.15) through (19.20), points of emission may be related to random numbers through

$$R_x = \frac{1}{E_k} \int_0^x E_k'\, dx = \frac{\int_0^x \int_0^Y \int_0^Z \kappa_P \sigma T^4\, dz\, dy\, dx}{\int_0^X \int_0^Y \int_0^Z \kappa_P \sigma T^4\, dz\, dy\, dx} ,$$
(19.60)

$$R_y = \int_0^y \int_0^Z \kappa_P \sigma T^4\, dz\, dy \left/ \int_0^Y \int_0^Z \kappa_P \sigma T^4\, dz\, dy \right. ,$$
(19.61)

$$R_z = \int_0^z \kappa_P \sigma T^4 dz \left/ \int_0^Z \kappa_P \sigma T^4 dz \right. ,$$
(19.62)

or

$$x = x(R_x), \quad y = y(R_y, x), \quad z = z(R_z, x, y).$$
(19.63)

Again, choices for x, y, and z become independent of another if the emission term is separable, e.g., for an isothermal medium with uniform absorption coefficient.

Wavelengths for Emission from within Medium

As for surface emission, the choice of emission wavelength, in general, depends on emission location (x, y, z), unless the volume is isothermal with constant absorption coefficient. From equation (8.49) and the definition of the Planck-mean absorption coefficient it follows immediately that

$$R_\lambda = \frac{\pi}{\kappa_P \sigma T^4} \int_0^\lambda \kappa_\lambda I_{b\lambda} \, d\lambda, \tag{19.64}$$

and, after inversion,

$$\lambda = \lambda(R_\lambda, x, y, z). \tag{19.65}$$

Directions for Emission from within Medium

Under local thermodynamic equilibrium conditions emission within a participating medium is isotropic, i.e., all possible directions are equally likely for the emission of a photon. All possible directions from a point within the medium are contained within the solid angle of $4\pi = \int_0^{2\pi} \int_0^\pi \sin\theta \, d\theta \, d\psi$, where polar angle θ and azimuthal angle ψ are measured from arbitrary reference axes. Thus, since the integrand is separable,

$$R_\psi = \frac{\psi}{2\pi}, \quad \text{or } \psi = 2\pi R_\psi, \tag{19.66}$$

$$R_\theta = \frac{1}{2} \int_0^\theta \sin\theta \, d\theta = \frac{1}{2}(1 - \cos\theta), \tag{19.67}$$

or

$$\theta = \cos^{-1}(1 - 2R_\theta). \tag{19.68}$$

If the polar angle is measured from the z-axis and ψ from the x-axis, then the unit direction vector for emission may be expressed as

$$\hat{\mathbf{s}} = \sin\theta(\cos\psi\hat{\mathbf{i}} + \sin\psi\hat{\mathbf{j}}) + \cos\theta\hat{\mathbf{k}}. \tag{19.69}$$

Absorption within Medium

When radiative energy travels through a participating medium, the energy is attenuated by absorption and scattered. Equation (8.8) gives the absorptivity for a photon path of length S as

$$\alpha_\lambda = 1 - \exp\left(-\int_0^S \kappa_\lambda ds\right). \tag{19.70}$$

Therefore, the fraction of energy penetrating through a layer of thickness S is

$$R_\kappa = \exp\left(-\int_0^S \kappa_\lambda ds\right). \tag{19.71}$$

This implies that, if the total radiative energy is divided into bundles of equal energy content, the fraction R_κ will be transmitted over a distance S or farther. Thus, we may relate the distance that any one bundle travels before absorption to a random number by inverting equation (19.71).

This inversion is readily obtained if the absorption coefficient does not vary throughout the medium (κ_λ =const). Under these conditions

$$S = \frac{1}{\kappa_\lambda} \ln \frac{1}{R_\kappa}, \tag{19.72}$$

and the bundle is allowed to travel a total distance S through the medium before being absorbed (unless it is absorbed by a surface before traveling this far).

If the absorption coefficient is not uniform (because of temperature dependence or because of a nonisotropic medium), inversion of equation (19.71) is considerably more difficult. Usually, the optical path is evaluated by breaking the volume up into K subvolumes with constant absorption coefficient. Then

$$\int_0^s \kappa_\lambda ds \simeq \sum_l \kappa_{\lambda_l} s_l, \tag{19.73}$$

where the summation is over those subvolumes (l) through which the bundle has traveled, and s_l is the geometric distance the bundle travels through these elements. As long as

$$\int_0^s \kappa_\lambda ds < \int_0^S \kappa_\lambda ds = \ln \frac{1}{R_\kappa}, \tag{19.74}$$

the bundle is not absorbed and is allowed to travel on.

Scattering within Medium

Attenuation by scattering obeys the same relationships as for absorption, with the absorption coefficient replaced by the scattering coefficient. Thus,

$$S = \frac{1}{\sigma_{s\lambda}} \ln \frac{1}{R_\sigma} \tag{19.75}$$

is the distance a bundle travels in a medium with uniform scattering coefficient before being scattered, or

$$\int_0^s \sigma_{s\lambda} ds < \int_0^S \sigma_{s\lambda} ds = \ln \frac{1}{R_\sigma} \tag{19.76}$$

for a medium with variable scattering coefficient.

Once a photon bundle is scattered, it will travel on into a new direction. The probability that the scattered bundle will travel within a cone of solid angle $d\Omega'$ around the direction $\hat{s}'$, after originally traveling in the direction $\hat{s}$ (cf. Fig. 19-8), is

$$P(\hat{s}')d\Omega' = \Phi(\hat{s} \cdot \hat{s}')d\Omega',$$

where Φ is the scattering phase function. Therefore, we may establish polar and azimuthal angles for scattering as

$$R_\psi = \int_0^{\psi'} \int_0^\pi \Phi(\hat{s} \cdot \hat{s}') \sin\theta' \, d\theta' \, d\psi' \Big/ \int_0^{2\pi} \int_0^\pi \Phi(\hat{s} \cdot \hat{s}') \sin\theta' \, d\theta' \, d\psi', \qquad (19.77)$$

and

$$R_\theta = \int_0^{\theta'} \Phi(\hat{s} \cdot \hat{s}') \sin\theta' \, d\theta' \Big/ \int_0^\pi \Phi(\hat{s} \cdot \hat{s}') \sin\theta' \, d\theta'. \qquad (19.78)$$

For linear anisotropic scattering, from equation (10.88),

$$\Phi(\hat{s} \cdot \hat{s}') = 1 + A_1 \hat{s} \cdot \hat{s}' = 1 + A_1 \cos\theta', \qquad (19.79)$$

where it is assumed that the polar angle θ' is measured from an axis pointing into the $\hat{s}$-direction, and the azimuthal angle ψ' is measured in a plane normal to $\hat{s}$. Equations (19.77) and (19.78), then, reduce for linear anisotropic scattering to

$$R_\psi = \frac{\psi'}{2\pi}, \quad \text{or } \psi' = 2\pi R_\psi, \qquad (19.80)$$

$$R_\theta = \frac{1}{2}\left(1 - \cos\theta' + \frac{A_1}{2}\sin^2\theta'\right). \qquad (19.81)$$

For isotropic scattering ($A_1 = 0$) these relations are identical to those for (by nature isotropic) emission, equations (19.66) and (19.68).

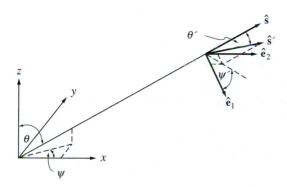

FIGURE 19-8
Local coordinate system for scattering direction.

The new direction vector, $\hat{s}'$, must then be found by introducing a local coordinate system at the point of scattering, with $\hat{s}$ pointing into its z-direction (i.e., from where the polar angle θ' is measured), as shown in Fig. 19-8. The local x-direction (from where ψ' is measured) and y-direction are given by

$$\hat{e}_1 = \mathbf{a} \times \hat{s}/|\mathbf{a} \times \hat{s}|, \quad \hat{e}_2 = \hat{s} \times \hat{e}_1, \tag{19.82}$$

where $\mathbf{a}$ is any arbitrary vector. The first of equations (19.82) ensures that the local x-axis is perpendicular to $\hat{s}$, and the second makes the coordinate system right-handed. Similar to equation (19.69), the new direction vector may now be expressed as

$$\hat{s}' = \sin \theta' (\cos \psi' \hat{e}_1 + \sin \psi' \hat{e}_2) + \cos \theta' \hat{s}. \tag{19.83}$$

If scattering is isotropic the scattering direction does not depend on the original path $\hat{s}$ (all directions are equally likely). In that case, the choice of a local coordinate is totally arbitrary, and equation (19.69) may be used directly.

Example 19.4. Consider again the geometry of Example 19.2. The medium within the diffuser is gray with absorption and scattering coefficients of $\kappa r_0 = 1$ and $\sigma_s r_0 = 2$, respectively, and an anisotropy factor of $A_1 = 1$ (strong forward scattering). How far will the energy bundle of Example 19.3 travel before being absorbed and/or scattered, if random numbers $R_\kappa = 0.200$ and $R_\sigma = 0.082$ are drawn? If scattering occurs, determine the energy bundle's new direction after the scattering event, for $R_\psi = 0.25$ and $R_\theta = 0.13$.

Solution

From equations (19.72) and (19.75) $S_\kappa = r_0 \ln(1/0.20) = 1.61 r_0$, and $S_\sigma = (r_0/2) \times \ln(1/0.082) = 1.25 r_0$. From Example 19.3 we know that the bundle must travel a distance of $D = (2r_0) \cos 0/0.8 = 2.5 r_0$ before hitting the diffuser. Since $S_\sigma < S_\kappa < D$, this implies that the bundle will scatter before hitting the diffuser, after which it will travel another distance of $S_\kappa - S_\sigma = 0.36 r_0$ before being absorbed (over which distance it may be scattered again or hit a diffuser wall). The location at which the scattering occurs is, from equation (19.49),

$$x = x_e + S_\sigma \hat{s} \cdot \hat{i} = 1.25 r_0 \times 0.8 = 1.0 r_0, \quad y = 0, \quad z = 1.25 r_0 \times 0.6 = 0.75 r_0.$$

From equations (19.80) and (19.81) we find $\psi' = 2\pi \times 0.25 = \pi/2$ and $\cos \theta' = 2\sqrt{1 - R_\theta} - 1 = 2\sqrt{1 - 0.13} - 1 = 0.8655$, or $\theta' = 30°$. Here the polar angle θ' is measured from the direction of $\hat{s} = 0.8\hat{i} + 0.6\hat{k}$ and ψ' in the plane normal to it. At the scattering point we may introduce a local coordinate system with, say, $\mathbf{a} = \hat{j}$, or

$$\hat{e}_1 = \hat{j} \times \hat{s}/|\hat{j} \times \hat{s}| = 0.6\hat{i} - 0.8\hat{k}, \quad \hat{e}_2 = \hat{s} \times \hat{e}_1 = \hat{j},$$

and, from equation (19.83),

$$\hat{s}' = \tfrac{1}{2}(0 + 1 \times \hat{j}) + \tfrac{1}{2}\sqrt{3}(0.8\hat{i} - 0.6\hat{k}) = 0.4\sqrt{3}\hat{i} + 0.5\hat{j} - 0.3\sqrt{3}\hat{k}.$$

Treatment of Spectral Line Structure Effects

If the participating medium contains an absorbing/emitting molecular gas, the gas will have a number of vibration-rotation bands, which in turn consist of hundreds of overlapping spectral lines (cf. the discussion on gas properties in Chapter 9). If line overlap is not very pronounced (small line overlap parameter β), the absorption coefficient becomes a strongly gyrating function of wavelength (cf. Fig. 9-7), making the use of equation (19.64) (emission wavelength) and equation (19.71) (absorption location) impractical: (*i*) Many digits of accuracy would be required in the evaluation of λ to ascertain whether emission occurs near a line center (with large κ_λ) or between lines (small κ_λ), (*ii*) accurate knowledge of the spectral gyration of κ_λ is rarely known. To simplify the analysis one may employ the narrow band model described in Chapter 9. First, the absorption coefficient is split into two components,

$$\kappa_\lambda = \kappa_{p\lambda} + \kappa_{g\lambda}, \tag{19.84}$$

where $\kappa_{p\lambda}$ is the (spectrally smooth) absorption coefficient of other participating material (such as particles or ions), and $\kappa_{g\lambda}$ is the rapidly varying gas absorption coefficient. Taking a narrow band average over the Planck function-weighted absorption coefficient leads to

$$\int_0^\lambda \kappa_{g\lambda} I_{b\lambda}\, d\lambda = \int_0^\lambda \left(\frac{1}{\delta\lambda} \int_{\delta\lambda} \kappa_{g\lambda} I_{b\lambda}\, d\lambda'\right) d\lambda \simeq \int_0^\lambda \overline{\kappa}_{g\lambda} I_{b\lambda}\, d\lambda,$$

where $\overline{\kappa}_{g\lambda} = (S/d)_\lambda$ is the narrow band average of the gas absorption coefficient.[4] The wavelength of emission is determined with equation (19.64) from

$$R_\lambda = \frac{\pi}{(\kappa_{pP} + \kappa_{gP})\sigma T^4} \int_0^\lambda (\kappa_{p\lambda} + \overline{\kappa}_{g\lambda}) I_{b\lambda}\, d\lambda, \tag{19.85}$$

and again, after inversion,

$$\lambda = \lambda(R_\lambda, x, y, z). \tag{19.86}$$

Application of the narrow band model to find the location of absorption within the participating medium is somewhat more complicated. The random number relations are different for photon bundles emitted from a surface (with spectrally smooth emissivity ϵ_λ,) as opposed to bundles emitted from within the medium (with strongly varying absorption coefficient κ_λ). Bundles emitted from a wall are equally likely to have wavelengths close to the center of a line or the gap between two lines, causing them to travel a certain distance before absorption. Bundles emitted from within the medium are likely to have wavelengths for which κ_λ is large [as easily seen by looking at equations (8.49) or (19.58) on a spectral basis], making them much more

---

[4]Due to the dearth of letters in the alphabet, care must be taken to distinguish the line strength parameter $S_\lambda = \int_{\text{line}} \kappa_\lambda d\lambda$ in the locally averaged absorption coefficient from the geometric distance S.

likely to be absorbed near the point of emission. We will limit our discussion here to the case of a spatially constant absorption coefficient, i.e., $\kappa_\lambda = \kappa_\lambda(\lambda)$. The more general case of a spatially varying (i.e., temperature- and/or concentration-dependent) absorption coefficient may be found in the original paper of Modest [21].

The amount of energy emitted by a surface element dA over a wavelength range $d\lambda$ into a pencil of rays $d\Omega$ is

$$\epsilon_\lambda I_{b\lambda} \, dA_p \, d\lambda \, d\Omega,$$

where $dA_p = dA|\hat{n} \cdot \hat{s}|$ is the projected area normal to the pencil of ray. Of this, the amount

$$\epsilon_\lambda I_{b\lambda} \, dA_p \, d\lambda \, d\Omega \, e^{-\kappa_\lambda S}$$

penetrates a distance S into the medium. Taking a narrow band average of both expressions leaves the first one untouched while the second becomes

$$\epsilon_\lambda I_{b\lambda} \, dA_p \, d\lambda \, d\Omega \frac{1}{\delta\lambda} \int_{\delta\lambda} e^{-\kappa_\lambda S} \, d\lambda = \epsilon_\lambda I_{b\lambda} \, dA_p \, d\lambda \, d\Omega \, (1 - \overline{\alpha}_\lambda),$$

where $\overline{\alpha}_\lambda$ is the narrow band average of the spectral absorptivity. The ratio of the two expressions gives the *fraction of energy* traveling a distance S. Thus, using the statistical model summarized in Table 9.1, we find

$$R_\kappa = 1 - \overline{\alpha}_\lambda \simeq \exp\left(-\kappa_{p\lambda} S - \frac{\tau}{\sqrt{1 + \tau/\beta}}\right), \tag{19.87}$$

where $\tau = \overline{\kappa}_{g\lambda} S$ and β is the line overlap parameter. In the high-pressure limit (strong line overlap with $\beta \to \infty$) equation (19.87) reduces to equation (19.71). Explicit inversion of equation (19.87) is not possible (unless $\kappa_{p\lambda} = 0$).

If emission is from a volume element, we have for a volume dV, a wavelength range $d\lambda$, and a pencil of rays $d\Omega$, the total emitted energy

$$\kappa_\lambda I_{b\lambda} \, dV \, d\lambda \, d\Omega,$$

of which the amount

$$\kappa_\lambda I_{b\lambda} \, dV \, d\lambda \, d\Omega \, e^{-\kappa_\lambda S}$$

is transmitted over a distance of S. Taking the narrow band average of both expressions and dividing the second by the first gives the transmitted fraction as

$$\begin{aligned}
R_\kappa &= \frac{1}{\delta\lambda} \int_{\delta\lambda} \kappa_\lambda e^{-\kappa_\lambda S} \, d\lambda \Big/ \frac{1}{\delta\lambda} \int_{\delta\lambda} \kappa_\lambda \, d\lambda \\
&= -\frac{1}{\kappa_{p\lambda} + \overline{\kappa}_{g\lambda}} \frac{d}{dS}\left(\frac{1}{\delta\lambda} \int_{\delta\lambda} e^{-\kappa_\lambda S} \, d\lambda\right) = \frac{1}{\kappa_{p\lambda} + \overline{\kappa}_{g\lambda}} \frac{d\overline{\alpha}_\lambda}{dS} \\
&= \frac{\kappa_{p\lambda} + \dfrac{\overline{\kappa}_{g\lambda}}{\sqrt{1 + \tau/\beta}} \dfrac{\beta + \tau/2}{\beta + \tau}}{\kappa_{p\lambda} + \overline{\kappa}_{g\lambda}} \exp\left(-\kappa_{p\lambda} S - \frac{\tau}{\sqrt{1 + \tau/\beta}}\right). \tag{19.88}
\end{aligned}$$

Again, equation (19.88) reduces to equation (19.71) for $\beta \to \infty$.

All other random number relations, since they do not involve the spectral absorption coefficient, are unaffected by spectral line effects.

Example 19.5. Consider a photon bundle traveling through a molecular gas. The wavelength of the bundle is such that $\overline{\kappa}_{g\lambda} = 1\,\text{cm}^{-1}$ and $\beta = 0.1$. Drawing a random number of $R_\kappa = 0.200$, how far will the bundle travel before absorption, if it was emitted (a) by a gray wall, (b) from within the gas?

Solution

(a) If the bundle originates from a wall, we have from equation (19.87)

$$R_\kappa = 0.200 = \exp\left(-\frac{\tau}{\sqrt{1 + 10\tau}}\right).$$

By trial and error (or solution of a quadratic equation), it follows that $\tau = 25.9$ and $S = \tau/\overline{\kappa}_{g\lambda} = 25.9\,\text{cm}$.

(b) For medium emission, equation (19.88) is applicable, and

$$R_\kappa = 0.200 = \frac{1}{\sqrt{1+10\tau}}\frac{1 + 5\tau}{1 + 10\tau}e^{-\tau/\sqrt{1+10\tau}},$$

or $\tau \simeq 0.48$ and $S = 0.48\,\text{cm}$.

Therefore, as expected, the bundle travels *much* farther if emitted from a wall. For comparison, in a gray medium the bundle would have traveled

$$S = \frac{1}{\kappa}\ln\frac{1}{R_\kappa} = \frac{1}{1\,\text{cm}^{-1}}\ln\frac{1}{0.200} = 1.61\,\text{cm}$$

for both cases.

Some Monte Carlo results for gas-particulate mixtures with line structure effects are shown in Chapter 17, in Figs. 17-3 and 17-7.

19.8 OVERALL ENERGY CONSERVATION

The temperature field within the medium is determined from overall conservation of energy, as given by equation (8.61). In the absence of conduction and convection, i.e., if radiative equilibrium prevails, this equation reduces to the simple form of $\nabla \cdot \mathbf{q}_R = 0$, where $\mathbf{q}_R$ is the radiative heat flux. Whether an analytical technique or a Monte Carlo method is used, the solution is simplest for a gray medium at radiative equilibrium, followed by the case of radiative equilibrium in a nongray medium and, finally, the gray and nongray medium in the presence of conduction and/or convection.

Gray Medium at Radiative Equilibrium

Radiative equilibrium implies that anywhere within the medium the material absorbs precisely as much radiative energy as it emits. Therefore, for every photon bundle absorbed at location **r**, another photon bundle of the same strength must be emitted at the same location. The direction of the new photon is determined from equations (19.66) and (19.68). We note that these relations are identical to those for isotropic scattering, equations (19.80) and (19.81), since emission is always isotropic. The wavelength of the newly emitted energy bundle may be determined from equation (19.64) and depends on the local temperature. However, if the medium and the walls are gray, then the wavelength of the bundle is irrelevant (indeed, does not have to be determined). Thus, if absorption and scattering coefficients are independent of temperature, knowledge of the temperature field is not required to find the solution: Energy bundles are emitted from the bounding walls (according to their temperatures) and are followed until they are absorbed by a wall (after perhaps numerous scattering and absorption–re-emission events inside the medium). Numerically, the process is identical to a purely scattering medium, with the extinction coefficient $\beta = \kappa + \sigma_s$ replaced by an effective scattering coefficient $\sigma_s' = \beta$. The temperature field inside the medium is determined by keeping track of the total re-emitted energy from a control volume V_i:

$$Q_{\text{abs},i} = \sum_{j=1}^{N_i} Q_{ij} = Q_{\text{em},i} = 4\sigma\kappa_i T_i^4 V_i, \tag{19.89}$$

or

$$T_i = \left(\sum_{j=1}^{N_i} Q_{ij}/4\sigma\kappa_i V_i\right)^{1/4}, \tag{19.90}$$

where the Q_{ij} are the amounts of energy carried by the N_i photon bundles that have been absorbed within V_i (after emission from a wall and, possibly, re-emission from within the medium).

This solution is limited to the case of constant properties, since absorption and scattering locations depend on local values of absorption and scattering coefficients. If these properties depend on temperature, a temperature field must be guessed to determine them, and an iteration becomes necessary.

Nongray Medium at Radiative Equilibrium

If the medium is nongray, the wavelength of each re-emitted bundle must be determined from equation (19.64), requiring knowledge of the temperature field. Therefore, the solution becomes an iterative process: First a temperature field is guessed, and employing this guess, the solution proceeds similar to the one described above for a gray medium, after which local temperatures are recalculated from equation (19.90), etc., until the solution converges.

There is another way to obtain a solution. Based on the guess of the temperature field we "know" how much energy is emitted from each subvolume. We may therefore separate the emission and absorption processes: Photon bundles are emitted not only by the walls, but also by the medium, and they are then traced until they are absorbed by either wall or medium (i.e., there is no re-emission in this method). This leads to different values for $Q_{abs,i}$ and $Q_{em,i}$ in equation (19.89), which may be used to update the temperature field. This method of solution is usually inferior since emission depends very strongly on the (unknown) temperature field, while nongray behavior is only implicitly influenced by the temperature.

Coupling with Conduction and/or Convection

If conduction and/or convection are of importance the radiation problem must be solved simultaneously with overall conservation of energy, equation (8.61). Since the energy equation is usually solved by conventional numerical methods (although a Monte Carlo solution is, in principle, possible; see, e.g., Haji-Sheikh [12]), an iteration in the temperature field is necessary: Similar to radiative equilibrium in a nongray medium, a temperature field is guessed and used to solve the radiation problem, leading to volume emission rates, $Q_{em,i}$, and absorption rates, $Q_{abs,i}$, for each subvolume. The net radiative source is then

$$(\nabla \cdot \mathbf{q}_R)_i = \frac{1}{V_i} (Q_{em,i} - Q_{abs,i}), \tag{19.91}$$

which is substituted into the solution for equation (8.61) to predict an updated temperature field.

19.9   EFFICIENCY CONSIDERATIONS

The accuracy of results for generalized radiation exchange factors, wall heat fluxes, or net radiation sources within the medium, as characterized by the standard deviation, equation (19.8), is determined by the statistical scatter of the results. The scatter may be expected to be inversely proportional to the number of bundles absorbed by a subsurface or subvolume. This number of bundles, on the other hand, is directly proportional to both total number of bundles and size of subsurface/subvolume. Thus, in order to achieve good spatial resolution (small element sizes), very large numbers of bundles—often several million—must be emitted and traced. Consequently, even with the availability of today's fast digital computers, it is imperative that the Monte Carlo implementation and its ray tracings be as numerically efficient as possible, if many hours of CPU time for each computer run are to be avoided.

Inversion of Random Number Relations

Many of the random number relationships governing emission location, wavelength, direction, etc., cannot be inverted explicitly. For example, to determine the wavelength of emission, even for a simple black surface, for a given random number R_λ requires the solution of the transcendental equation (19.26),

$$R_\lambda = \frac{1}{\sigma T^4} \int_0^\lambda E_{b\lambda} d\lambda = f(\lambda T). \tag{19.92}$$

In principle, this requires guessing a λ, calculating R_λ, etc. until the correct wavelength is found; this would then be repeated for each emitted photon bundle. It would be much more efficient to invert equation (19.92) once and for all before the first energy bundle is traced as

$$\lambda T = f^{-1}(R_\lambda). \tag{19.93}$$

This is done by first calculating $R_{\lambda,j}$ corresponding to a $(\lambda T)_j$ for a sufficient number of points $j = 0, 1, \cdots, J$. These data points may then be used to obtain a polynomial description

$$\lambda T = A + BR_\lambda + CR_\lambda^2 + \cdots, \tag{19.94}$$

as proposed by Howell [13]. With the math libraries available today on most digital computers it would, however, be preferable to invert equation (19.93) using a (cubic) spline.

Even more efficient is the method employed by Modest and Poon [16] and Modest [20], who used a cubic spline to determine values of $(\lambda T)_j$ for $(J + 1)$ equally spaced random numbers

$$(\lambda T)_j = f^{-1}\left(R_\lambda = \frac{j}{J}\right), \quad j = 0, 1, 2, \ldots, J. \tag{19.95}$$

If, for example, a random number $R_\lambda = 0.6789$ is picked, it is immediately known that (λT) lies between $(\lambda T)_m$ and $(\lambda T)_{m+1}$, where m is the largest integer less than $J \times R_\lambda$ ($=67$ if $J = 100$). The actual value for (λT) may then be found by (linear) interpolation.

The quantity to be determined may depend on more than a single random number. For example, to fix an emission location in a three-dimensional medium with nonseparable emissive power requires the determination of

$$x = x(R_x), \quad y = y(R_y, x), \quad z = z(R_z, x, y). \tag{19.96}$$

That is, first the x-location is chosen, requiring the interpolation between and storage of J data points $x_j(R_j)$; next the y-location is determined, requiring a double interpolation and storage of a $J \times K$ array for $y_{jk}(R_k, x_j)$; and finally z is found from a triple interpolation from a $J \times K \times L$ array for $z_{jkl}(R_l, x_j, y_k)$. This may lead to excessive computer storage requirements if J, K, L are chosen too large: If $J = K = L = 100$, an array with one million numbers needs to be stored for the determination of z-locations alone! The problem may be alleviated by choosing a better interpolation scheme together with smaller values for J, K, L (for example, a choice of $J = K = L = 40$ reduces storage requirements to 64,000 numbers).

Energy Partitioning

In the general Monte Carlo method, a ray of fixed energy content is traced until it is absorbed. In the absence of a participating medium, the decision whether the bundle is absorbed or reflected is made after every impact on a surface. Thus, on the average it will take $1/\alpha$ tracings until the bundle is absorbed. Therefore, it takes $1/\alpha$ tracings to add one statistical sample to the calculation of one of the $\mathscr{F}_{i \to j}$s. If the configuration has openings, a number of bundles may be reflected a few times before they escape into space without adding a statistical sample to any of the $\mathscr{F}_{i \to j}$s. Thus, the ordinary Monte Carlo method becomes extremely inefficient for open configurations and/or highly reflective surfaces. The former problem may be alleviated by partitioning the energy of emitted bundles. This was first applied by Sparrow and coworkers [22, 23], who, before determining a direction of emission, split the energy of the bundle into two parts: The part leaving the enclosure through the opening (equal to the view factor from the emission point to the opening) and the rest (which will strike a surface). A direction is then determined, limited to those that make the bundle hit an enclosure surface. The procedure is repeated after every reflection. This method guarantees that each bundle will contribute to the statistical sample for exchange factor evaluation. A somewhat more general and more easily implemented energy partitioning scheme was applied by Modest and Poon [16, 20]: Rather than drawing a random number R_α to decide whether a bundle is (fully) absorbed or not, they partition the energy of a bundle at each reflection into the fraction α, which is absorbed, and the fraction $\rho = 1 - \alpha$, which is reflected. The bundle is then traced until it either leaves the enclosure or until its energy is depleted (below a certain fraction of original energy content). This method adds to the statistical sample of a $\mathscr{F}_{i \to j}$ after *every* tracing and thus leads to vastly faster convergence for highly reflective surfaces.

Energy partitioning is equally successfully applied to problems in participating media: In optically thick media, bundles emitted in the interior rarely travel far enough (before absorption) to hit a bounding surface, although it is usually the surface heat fluxes that are of primary interest. Modest [24] extended the partitioning concept by depleting the energy content of a bundle along its path: No random number is drawn to determine the distance traveled until absorption, equation (19.72); rather, the energy content is depleted due to gradual absorption. The depleted amount is added to the absorption rates of the subvolumes through which the bundle travels. Again, the bundle is traced until it leaves the enclosure or until its energy is depleted. The method results in tremendous computer time savings for optically thick media. However, this type of energy partitioning is limited to nonscattering media with known (or iterated) temperature field (i.e., it cannot be applied to the standard method for radiative equilibrium, where photon bundles are absorbed and re-emitted at selected locations).

Other Efficiency Improvements

There are many other ways to make a particular Monte Carlo simulation compu- tationally more efficient; these are often connected to the particular geometry under scrutiny. For example, large amounts of computer time may be wasted because it is not immediately known which of the many subsurfaces the traveling bundle will hit. In general, an intersection between every surface and the bundle must be calculated. Only then can it be determined whether this intersection is legitimate, i.e., whether it occurs within the bounds of the surface. Often the overall enclosure can be broken up into a (relatively small) number of basic surfaces (dictated by geometry), which in turn are broken up into a number of smaller, isothermal subsurfaces. Furthermore, a bundle emitted or reflected from some subsurface may not be able to hit some basic surface by any path. In other cases, if a possible point of impact on some surface has been determined, it may not be necessary to check the remaining surfaces, etc. There are no fixed rules for the computational structure of a Monte Carlo code. In these applications the proverbial "dash of ingenuity" can go a long way in making a computation efficient.

References

1. Hammersley, J. M., and D. C. Handscomb: *Monte Carlo Methods*, John Wiley & Sons, New York, 1964.
2. Fleck, J. A.: "The Calculation of Nonlinear Radiation Transport by a Monte Carlo Method," Technical Report UCRL-7838, Lawrence Radiation Laboratory, 1961.
3. Fleck, J. A.: "The Calculation of Nonlinear Radiation Transport by a Monte Carlo Method: Statistical Physics," *Methods in Computational Physics*, vol. 1, pp. 43–65, 1961.
4. Howell, J. R., and M. Perlmutter: "Monte Carlo Solution of Thermal Transfer through Radiant Media between Gray Walls," *ASME Journal of Heat Transfer*, vol. 86, no. 1, pp. 116–122, 1964.
5. Howell, J. R., and M. Perlmutter: "Monte Carlo Solution of Thermal Transfer in a Nongrey Nonisothermal Gas with Temperature Dependent Properties," *AIChE Journal*, vol. 10, no. 4, pp. 562–567, 1964.
6. Perlmutter, M., and J. R. Howell: "Radiant Transfer through a Gray Gas Between Concentric Cylinders Using Monte Carlo," *ASME Journal of Heat Transfer*, vol. 86, no. 2, pp. 169–179, 1964.
7. Cashwell, E. D., and C. J. Everett: *A Practical Manual on the Monte Carlo Method for Random Walk Problems*, Pergamon Press, New York, 1959.
8. Schreider, Y. A.: *Method of Statistical Testing – Monte Carlo Method*, Elsevier, New York, 1964.
9. Kahn, H.: "Applications of Monte Carlo," *Report for Rand Corp.*, vol. Rept. No. RM-1237-AEC (AEC No. AECU-3259), 1956.
10. Brown, G. W.: "Monte Carlo Methods," in *Modern Mathematics for the Engineer*, McGraw-Hill, New York, pp. 279–307, 1956.
11. Halton, J. H.: "A Retrospective and Prospective Survey of the Monte Carlo Method," *SIAM Rev.*, vol. 12, no. 1, pp. 1–63, 1970.
12. Haji-Sheikh, A.: "Monte Carlo Methods," in *Handbook of Numerical Heat Transfer*, John Wiley & Sons, New York, pp. 673–722, 1988.
13. Howell, J. R.: "Application of Monte Carlo to Heat Transfer Problems," in *Advances in Heat Transfer*, eds. J. P. Hartnett and T. F. Irvine, vol. 5, Academic Press, New York, 1968.
14. Walters, D. V., and R. O. Buckius: "Monte Carlo Methods for Radiative Heat Transfer in Scattering Media," in *Annual Review of Heat Transfer*, vol. 5, Hemisphere, New York, 1992.
15. Taussky, O., and J. Todd: "Generating and Testing of Pseudo-Random Numbers," in *Symposium on Monte Carlo Methods*, John Wiley & Sons, New York, pp. 15–28, 1956.
16. Modest, M. F., and S. C. Poon: "Determination of Three-Dimensional Radiative Exchange Factors for the Space Shuttle by Monte Carlo," ASME paper no. 77-HT-49, 1977.
17. Barford, N. C.: *Experimental Measurements: Precision, Error and Truth*, Addison-Wesley, London, 1967.

18. Toor, J. S., and R. Viskanta: "A Numerical Experiment of Radiant Heat Exchange by the Monte Carlo Method," *International Journal of Heat and Mass Transfer*, vol. 11, no. 5, pp. 883–887, 1968.

19. Weiner, M. M., J. W. Tindall, and L. M. Candell: "Radiative Interchange Factors by Monte Carlo," ASME paper no. 65-WA/HT-51, 1965.

20. Modest, M. F.: "Determination of Radiative Exchange Factors for Three Dimensional Geometries with Nonideal Surface Properties," *Numerical Heat Transfer*, vol. 1, pp. 403–416, 1978.

21. Modest, M. F.: "The Monte Carlo Method Applied to Gases with Spectral Line Structure," *Numerical Heat Transfer*, 1992 (in print).

22. Heinisch, R. P., E. M. Sparrow, and N. Shamsundar: "Radiant Emission from Baffled Conical Cavities," *Journal of the Optical Society of America*, vol. 63, no. 2, pp. 152–158, 1973.

23. Shamsundar, N., E. M. Sparrow, and R. P. Heinisch: "Monte Carlo Solutions — Effect of Energy Partitioning and Number of Rays," *International Journal of Heat and Mass Transfer*, vol. 16, pp. 690–694, 1973.

24. Modest, M. F.: "Radiative Heat Transfer Fluxes Through the Exit of GE Combustor Transition Piece," Technical Report (private communication to General Electric Co.), 1980.

Problems

Because of the nature of the Monte Carlo technique, each of the following problems requires the development of a small computer code. However, all problem solutions can be outlined by giving relevant relations, equations, and a detailed flow chart.

19.1 Consider two concentric parallel disks of radius R, spaced a distance H apart. Both plates are isothermal (at T_1 and T_2, respectively), are gray diffuse emitters with emissivity ϵ, and are gray reflectors with diffuse reflectivity component ρ^d and purely specular component ρ^s. Write a computer code that calculates the generalized exchange factor $\mathscr{F}_{1\to2}$ and, taking advantage of the fact that $\mathscr{F}_{1\to2} = \mathscr{F}_{2\to1}$, calculate the total heat loss from each plate. Compare with the analytical solution treating each surface as a single node.

19.2 Repeat Problem 19.1, but calculate heat fluxes directly, i.e., without first calculating exchange factors.

19.3 Determine the view factor for Configuration 39 of Appendix D, for $h = w = l$. Compare with exact results.

19.4 Repeat Problem 5.17 for $T_1 = T_2 = 1000\,\text{K}$, $\epsilon_1 = \epsilon_2 = 0.5$. Use the Monte Carlo method, employing the energy partitioning of Sparrow and coworkers [22, 23].

19.5 Repeat Problem 5.18; Compare with the exact solutions for several values of ϵ.

19.6 Repeat Problem 6.1, using the Monte Carlo method. Compare with the solution from Chapter 6 for a few values of D/L and ϵ, and $T_1 = 1000\,\text{K}$, $T_2 = 2000\,\text{K}$. How can the problem be done by emitting bundles from only one surface?

19.7 Repeat Problem 6.7 using the Monte Carlo method.

19.8 Repeat Example 7.3 using the Monte Carlo method.

19.9 Repeat Example 7.4 using the Monte Carlo method.

19.10 Repeat Problem 7.9 using the Monte Carlo method.

19.11 Consider radiative equilibrium in a plane-parallel medium between two isothermal, diffusely emitting and reflecting gray plates ($T_1 = 300\,\text{K}$, $\epsilon_1 = 0.5$, $T_2 = 2000\,\text{K}$, $\epsilon_2 = 0.8$) spaced $L = 1\,\text{m}$ apart. The medium has constant absorption and scattering coefficients ($\kappa = 0.01\,\text{cm}^{-1}$, $\sigma_s = 0.04\,\text{cm}^{-1}$), and scattering is linear-anisotropic with

$A_1 = 0.5$. Calculate the radiative heat flux and the temperature distribution within the medium by the Monte Carlo method. Compare with results from the P_1-approximation.

19.12 Consider an isothermal plane-parallel slab ($T = 1000\,\text{K}$) between two cold, gray, diffuse surfaces ($\epsilon = 0.5$). The medium absorbs and emits but does not scatter. Prepare a standard Monte Carlo solution to obtain the radiative heat loss from the medium for optical thickness $\kappa L = 0.2, 1, 5, 10$. Compare with the exact solution.

19.13 Repeat Problem 19.12 using energy partitioning. Compare the efficiency of the two methods.

19.14 A molecular gas is confined between two parallel, black plates, spaced 1 m apart, that are kept isothermal at $T_1 = 1200\,\text{K}$ and $T_2 = 800\,\text{K}$, respectively. The (hypothetical) gas has a single vibration-rotation band in the infrared, with an average absorption coefficient of

$$\overline{\kappa}_{g\eta} = \left(\frac{S}{d}\right)_\eta = \frac{\alpha}{\omega}e^{-2|\eta-\eta_0|/\omega}, \quad \eta_0 = 3000\,\text{cm}^{-1}, \quad \omega = 200\,\text{cm}^{-1}$$

and a line overlap parameter of β (see the discussion of narrow band and wide band models in Chapter 9). Assuming convection and conduction to be negligible, determine the radiative heat flux between the two plates, using the Monte Carlo method. Carry out the analysis for variable values of (α/ω) and β, and plot nondimensional radiative heat flux vs. $(S/d)_0 L$ with β as a parameter.

19.15 Consider a sphere of very hot molecular gas of radius 50 cm. The gas has a single vibration-rotation band at $\eta_0 = 3000\,\text{cm}^{-1}$, is suspended magnetically in a vacuum within a large, cold container, and is initially at a uniform temperature $T_g = 3000\,\text{K}$. For this gas, $\rho_a\alpha(T) = 500\,\text{cm}^{-2}$, $\omega = 100\sqrt{T/100\,\text{K}}\,\text{cm}^{-1}$, $\beta \gg 1$. This implies that the absorption coefficient may be determined from

$$\kappa_\eta = \kappa_0 e^{-2|\eta-\eta_0|/\omega}, \quad \kappa_0 = \frac{\rho_a\alpha}{\omega}$$

and the band absorptance from

$$A(s) = \omega A^* = \omega[E_1(\kappa_0 s) + \ln(\kappa_0 s) + \gamma_E].$$

Find the total heat loss from the sphere and its temperature distribution by the Monte Carlo method (including $t > 0$).

19.16 Consider a sphere of very hot dissociated gas of radius 5 cm. The gas may be approximated as a gray, isotropically scattering medium with $\kappa = 0.1\,\text{cm}^{-1}$, $\sigma_s = 0.2\,\text{cm}^{-1}$. The gas is suspended magnetically in a vacuum within a large, cold container and is initially at a uniform temperature $T_g = 10,000\,\text{K}$. Using the Monte Carlo method and neglecting conduction and convection, specify the total heat loss per unit time from the entire sphere at $t = 0$. Outline the solution for times $t > 0$.

CHAPTER 20

RADIATION COMBINED WITH CONDUCTION AND CONVECTION

20.1 INTRODUCTION

Up to this point we have considered only the analysis of radiative heat transfer without interaction with other modes of heat transfer; i.e., we have limited ourselves to cases of radiative equilibrium and cases of specified temperature fields. In practical systems, of course, it is nearly always the case that radiation occurs in conjunction with conduction and/or convection, and two or three heat transfer modes must be accounted for simultaneously. The interaction may be quite simple, or it may be rather involved. For example, heat loss from an isothermal surface of known temperature, adjacent to a radiatively nonparticipating medium, may occur by convection as well as radiation; however, convective and radiative heat fluxes are independent of one another, can be calculated independently, and may simply be added. If boundary conditions are more complex (i.e., surface temperatures are not specified) but the medium is radiatively nonparticipating, then radiation enters the remaining conduction/convection problem as a nonlinear boundary condition. If the medium is radiatively participating, then overall conservation of energy, equation (8.61), needs to be solved, which always leads to a nonlinear integro-differential equation.

Many important applications of interactions between radiation and other modes of heat transfer have been reported in the literature. Discussion of all of these could easily, by itself, fill a book as voluminous as this one. We will therefore limit ourselves here to the discussion of a few very basic cases (*i*) to show the basic trends of how the different modes of heat transfer interact with one another, and (*ii*) to outline some of the numerical schemes that have been used to solve such problems. We will begin with a short section on combined radiation and conduction and/or convection in radiatively nonparticipating media. This is followed by two sections dealing with combined radiation and conduction in participating media, the latter one including change-of-phase effects. Combined radiation and convection is treated in three subsequent sections, the first two dealing with external and internal flows, respectively. Since, despite the high speed with which thermal radiation reaches equilibrium, turbulent convection combined with radiation often requires some special consideration, it is dealt with in a separate section. In all cases involving participating media, we will limit our theoretical developments to a simple plane-parallel geometry with a gray medium, since our aim is to investigate only the general trends of the interaction among the different modes of heat transfer. At the end of each section a short description of more advanced problems is given, as well as a list of references.

20.2 SURFACE RADIATION WITH CONDUCTION AND/OR CONVECTION

In a number of important applications, a heat transfer analysis needs to be performed for a radiatively nonparticipating medium, or for an opaque substance surrounded by a nonparticipating medium. In such cases, thermal radiation enters the considerations only as a nonlinear boundary condition at opaque surfaces. If the medium adjacent to opaque surfaces is nonmoving (e.g., vacuum or transparent solid), heat transfer is by conduction with a radiation boundary condition. If the adjacent medium is a flowing (transparent) gas or liquid, heat transfer is by both conduction and convection, with a radiation boundary condition.

Conduction and Surface Radiation—Fins

The vast majority of combined conduction-surface radiation applications involve heat transfer through vacuum, e.g., heat loss from space vehicles or vacuum insulations. As a single example we will discuss here the performance of a simple rectangular-fin radiator used to reject heat from a spacecraft.

Consider a tube with a set of radial fins, as schematically shown in Fig. 20-1. In order to facilitate the analysis, we will make the following assumptions:

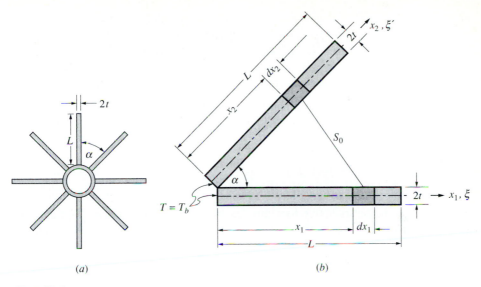

FIGURE 20-1
Schematic of a space radiator tube with longitudinal fins.

1. The thickness of each fin, $2t$, is much less than its length in the radial direction, L, which in turn is much less than the fin extent in the direction of the tube axis. This implies that heat conduction within the fin may be calculated by assuming that the fin temperature is a function of radial distance, x, only.

2. End losses from the fin tips (by convection and radiation) are negligible, i.e., $\partial T_i / \partial x_i (L) = 0$.

3. The thermal conductivity of the fin material, k, is constant.

4. The base temperatures of all fins are the same, i.e., $T_1(0) = T_2(0) = T_b$, and the fin arrangement is symmetrical, i.e., $T_1(x_1) = T_2(x_2 = x_1)$, etc.

5. The surfaces are coated with an opaque, gray, diffusely emitting and reflecting material of uniform emissivity ϵ.

6. There is no external irradiation falling into the fin cavities ($H_o = 0$, $T_\infty = 0$).

The first three assumptions are standard simplifications made for the analysis of thin fins (see, e.g., Holman [1]), and the other three have been made to make the radiation part of the problem more tractable. Performing an energy balance on an infinitesimal volume element (of unit length in the axial direction) $dV = 2t\,dx$, one finds:

conduction going in at x across cross-sectional area $(2t)$

$\qquad = $ conduction going out at $x + dx$

$\qquad\qquad + $ net radiative loss from top and bottom surfaces $(2dx)$

or

$$-2tk \left.\frac{dT}{dx}\right|_x = -2tk \left.\frac{dT}{dx}\right|_{x+dx} + 2\, q_R\, dx.$$

Expanding the outgoing conduction term into a truncated Taylor series,

$$\left.\frac{dT}{dx}\right|_{x+dx} = \left.\frac{dT}{dx}\right|_x + dx \left.\frac{d^2T}{dx^2}\right|_x + \cdots,$$

then leads to

$$\frac{d^2T}{dx^2} = \frac{1}{tk}\, q_R. \tag{20.1}$$

Here $q_R(x)$ is the net radiative heat flux leaving a surface element of the fin, which may be determined in terms of surface radiosity, J, from equations (5.24) and (5.25) as[1]

$$q_R(x_1) = J(x_1) - \int_{x_2=0}^{L} J(x_2)\, dF_{d1-d2}, \tag{20.2}$$

$$J(x_1) = \epsilon\sigma T^4(x_1) + (1-\epsilon)\int_{x_2=0}^{L} J(x_2)\, dF_{d1-d2}. \tag{20.3}$$

The expression for radiative heat flux may be simplified by eliminating the integral, equation (5.26),

$$q_R(x_1) = \frac{\epsilon}{1-\epsilon}\left[\sigma T_1^4(x_1) - J_1(x_1)\right]. \tag{20.4}$$

The view factor between two infinitely long strips may be found from Appendix D, Configuration 5, or from Example 4.1 as

$$F_{d1-d2} = \frac{x_1 \sin^2\alpha\, x_2\, dx_2}{2\, S_0^3} = \frac{\sin^2\alpha\, x_1\, x_2\, dx_2}{2(x_1^2 - 2x_1x_2 \cos\alpha + x_2^2)^{3/2}}. \tag{20.5}$$

Equation (20.1) requires two boundary conditions, namely,

$$T(x=0) = T_b, \qquad \frac{dT}{dx}(x=L) = 0. \tag{20.6}$$

Before we attempt a numerical solution, it is a good idea to summarize the mathematical problem in terms of nondimensional variables and parameters,

$$\theta(\xi) = \frac{T(x)}{T_b}, \qquad \mathcal{J}(\xi) = \frac{J(x)}{\sigma T_b^4}, \qquad N_c = \frac{kt}{\sigma T_b^3 L^2}, \qquad \xi = \frac{x}{L}, \tag{20.7}$$

---

[1] For the radiative exchange it is advantageous to attach subscripts 1 and 2 to the x-coordinates to distinguish contributions from different plates, even though $T(x)$, $J(x)$, $q_R(x)$, etc., are the same along each of the fins.

where θ and $\mathcal{J}$ are nondimensional temperature and radiosity, and N_c is usually called the *conduction-to-radiation parameter*. With these definitions,

$$\frac{d^2\theta}{d\xi^2} = \frac{1}{N_c}\frac{\epsilon}{1-\epsilon}\left[\theta^4(\xi) - \mathcal{J}(\xi)\right], \tag{20.8a}$$

$$\mathcal{J}(\xi) = \epsilon\,\theta^4(\xi) + (1-\epsilon)\int_{\xi'=0}^{1} \mathcal{J}(\xi')\,K(\xi,\xi')\,d\xi', \tag{20.8b}$$

$$K(\xi,\xi') = \frac{1}{2}\sin^2\alpha\,\frac{\xi\xi'}{(\xi^2 - 2\xi\xi'\cos\alpha + \xi'^2)^{3/2}}, \tag{20.8c}$$

subject to

$$\theta(\xi = 0) = 1, \quad \frac{d\theta}{d\xi}(\xi = 1) = 0. \tag{20.8d}$$

As for convection-cooled fins, a *fin efficiency*, η_f, is defined, comparing the heat loss from the actual fin to that of an ideal fin (a black fin, which is isothermal at T_b). The total heat loss from an ideal fin ($\epsilon = 1$, $J = \sigma T_b^4$) is readily determined from equation (20.2) and Appendix D, Configuration 34, as

$$Q_{\text{ideal}} = 2L\,q_{R,\text{ideal}} = 2L\,\sigma T_b^4(1 - F_{1-2}) = 2L\sin\frac{\alpha}{2}\sigma T_b^4, \tag{20.9}$$

while the actual heat loss follows from Fourier's law applied to the base, or by integrating over the length of the fin, as

$$Q_{\text{actual}} = -2tk\left.\frac{dT}{dx}\right|_{x=0} = 2\int_0^L q_R(x)\,dx. \tag{20.10}$$

Thus,

$$\eta_f = \frac{Q_{\text{actual}}}{Q_{\text{ideal}}} = -\frac{N_c}{\sin\frac{\alpha}{2}}\left.\frac{d\theta}{d\xi}\right|_0 = \frac{1}{\sin\frac{\alpha}{2}}\frac{\epsilon}{1-\epsilon}\int_0^1 (\theta^4 - \mathcal{J})\,d\xi, \tag{20.11}$$

where the last expression is obtained by integrating equation (20.8a) across the fin.

The set of equations (20.8) is readily solved by a host of different methods, including the net radiation method [finite-differencing equation (20.8b) into finite-width isothermal strips, to which equation (5.34) can be applied] or any of the solution methods for Fredholm equations discussed in Section 5.5. Because of the nonlinear nature of the equations it is always advisable to employ the method of successive approximations, i.e., a temperature field is guessed, a radiosity distribution is calculated, an updated temperature field is determined by solving the differential equation (for a known right-hand side), etc.

Sample results for the efficiency, as obtained by Sparrow and coworkers [2], are shown in Fig. 20-2. The variation of the fin efficiency is similar to that for a

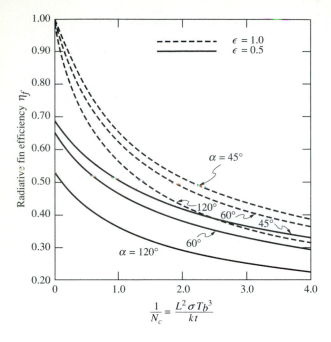

$$\frac{1}{N_c} = \frac{L^2 \sigma T_b^3}{kt}$$

FIGURE 20-2
Radiative fin efficiency for longitudinal plate fins [2].

convectively cooled fin (with the heat transfer coefficient replaced by a "radiative heat transfer coefficient," $h_R = 4\epsilon\sigma T_b^3$). Maximum efficiency is obtained for $N_c \rightarrow \infty$, i.e., when conduction dominates and the fin is essentially isothermal. For $\epsilon < 1$ the efficiency is limited to values $\eta_f < 1$ since a black configuration will always lose more heat. It is also observed that the fin efficiency (but not the actual heat lost) increases as the opening angle α decreases: For small opening angles irradiation from adjacent fins reduces the net radiative heat loss by a large fraction, but not as much as for the "ideal" fin (with irradiation from adjacent fins, which are black and at T_b).

Many other studies discussing the interaction of surface radiation and one-dimensional conduction may be found in the literature. For example, Hering [3] and Tien [4] considered the fins of Fig. 20-1 with specularly reflecting surfaces, and Sparrow and coworkers [2] investigated the influence of external irradiation. Fins connecting parallel tubes were studied by Bartas and Sellers [5], Sparrow and coworkers [6, 7], and Lieblein [8]. Single annular fins (i.e., annular disks attached to the outside of tubes) were studied by Chambers and Sommers [9] (rectangular cross section), Keller and Holdredge [10] (variable cross-section), and Mackay [11] (with external irradiation), while Sparrow and colleagues [12] investigated the interaction between adjacent fins. Various other publications have appeared dealing with different geometries, surface properties (including nongrayness effects), irradiation conditions, etc. A partial listing is given with [13–31].

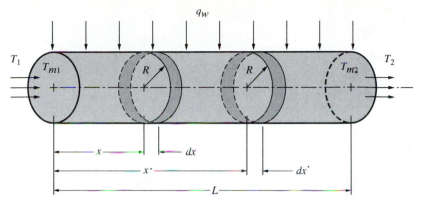

FIGURE 20-3
Forced convection and radiation of a transparent medium flowing through a circular tube, subject to constant wall heat flux.

Convection and Surface Radiation — Tube Flow

As in the case of pure convection heat transfer, it is common to distinguish between external flow and internal flow applications. If the flowing medium is air or some other relatively inert gas, the assumption of a transparent, or radiatively nonpartici-pating, medium is often justified. As an example we will consider here the case of a transparent gas flowing through a cylindrical tube of diameter $D = 2R$ and length L, which is heated uniformly at a rate of q_w (per unit surface area). As schemati-cally shown in Fig. 20-3, the fluid enters the tube at $x = 0$ with a mean, or bulk, temperature T_{m1}. Over the length of the tube the supplied heat flux q_w is dissipated from the inner surface by convection (to the fluid) and radiation (to the openings and to other parts of the tube wall), while the outer surface of the tube is insulated. The two open ends of the tube are exposed to radiation environments at temperatures T_1 and T_2, respectively. The inner surface of the tube is assumed to be gray, diffusely emitting and diffusely reflecting, with a uniform emissivity ϵ. Finally, for a simplified analysis, we will assume that the convective heat transfer coefficient, h, between tube wall and fluid is constant, independent of the radiative heat transfer, and known.

With these simplifications an energy balance on a control volume $dV = \pi R^2 \times dx$ yields:

enthalpy flux in at x +   convective flux in over dx =   enthalpy flux out at $x + dx$,

or

$$\dot{m} c_p T_m(x) + h\,[T_w(x) - T_m(x)]\,2\pi R dx = \dot{m}\,c_p\,T_m(x + dx)$$

$$= \dot{m} c_p \left[T_m(x) + \frac{dT_m}{dx}(x)\,dx \right],$$

or

$$\frac{dT_m}{dx} = \frac{2h}{\rho c_p u_m R} [T_w(x) - T_m(x)], \tag{20.12}$$

where axial conduction has been neglected, and the mass flow rate has been expressed in terms of mean velocity as $\dot{m} = \rho u_m \pi R^2$. Equation (20.12) is a single equation for the unknown wall and bulk temperatures $T_w(x)$ and $T_m(x)$ and is subject to the inlet condition

$$T_m(x = 0) = T_{m1}. \tag{20.13}$$

An energy balance for the tube surface states that the prescribed heat flux q_w is dissipated by conduction and radiation or, applying equation (5.26) for the radiative heat flux,

$$q_w = h[T_w(x) - T_m(x)] + \frac{\epsilon}{1 - \epsilon} [\sigma T_w^4(x) - J(x)]. \tag{20.14}$$

The radiosity $J(x)$ is found from equation (5.24) as

$$J(x) = \epsilon \sigma T_w^4(x) + (1 - \epsilon) \left\{ \sigma T_1^4 F_{dx-1} + \sigma T_2^4 F_{dx-2} + \int_0^L J(x') dF_{dx-dx'} \right\}, \tag{20.15}$$

where F_{dx-1} is the view factor from the circular strip of width dx at x to the opening at $x = 0$, F_{dx-2} is the one to the opening at $x = L$, and $dF_{dx-dx'}$ is the view factor between two circular strips located at x and x', as indicated in Fig. 20-3. All view factors are readily determined from Appendix D, Configurations 9 and 31, and will not be repeated here. Equations (20.12), (20.14), and (20.15) are a set of three simultaneous equations in the unknown $T_w(x)$, $T_m(x)$ and $J(x)$, which must be solved numerically. Before we attempt such a solution, it is best to recast the equations in nondimensional form. Defining the following variables and parameters,

$$\xi = \frac{x}{D}, \quad \theta(\xi) = \left(\frac{\sigma T^4}{q_w} \right)^{1/4}, \quad \mathcal{J}(\xi) = \frac{J}{q_w}, \tag{20.16a}$$

$$\text{St} = \frac{h}{\rho c_p u_m D}, \quad H = \frac{h}{q_w} \left(\frac{q_w}{\sigma} \right)^{1/4}, \tag{20.16b}$$

transforms equations (20.12) through (20.15) to

$$\frac{d\theta_m}{d\xi} = 4 \text{St} [\theta_w(\xi) - \theta_m(\xi)], \quad \theta_m(\xi = 0) = \theta_{m1}, \tag{20.17}$$

$$1 = H[\theta_w(\xi) - \theta_m(\xi)] + \frac{\epsilon}{1 - \epsilon} [\theta_w^4(\xi) - \mathcal{J}(\xi)], \tag{20.18}$$

$$\mathcal{J}(\xi) = \epsilon \theta_w^4(\xi) + (1 - \epsilon) \left\{ \theta_1^4 F_{d\xi-1} + \theta_2^4 F_{d\xi-2} + \int_0^{L/D} \mathcal{J}(\xi') dF_{d\xi-d\xi'} \right\}. \tag{20.19}$$

Equation (20.18) becomes indeterminate for $\epsilon = 1$. For the case of a black tube $\mathcal{J} = \theta_w^4$, and equations (20.18) and (20.19) may be combined as

$$1 = H\left[\theta_w(\xi) - \theta_m(\xi)\right] + \theta_w^4(\xi) - \theta_1^4 F_{d\xi-1}$$
$$- \theta_2^4 F_{d\xi-2} - \int_0^{L/D} \theta_w^4(\xi')\, dF_{d\xi-d\xi'}. \tag{20.20}$$

Example 20.1. A transparent gas flows through a black tube subject to a constant heat flux. The convective heat transfer coefficient is known to be constant such that Stanton numbers and the nondimensional heat transfer coefficient are evaluated as St = 2.5×10^{-3} and $H = 0.8$. The environmental temperatures at both ends are equal to the local gas temperatures, i.e., $\theta_1 = \theta_{m1}$ and $\theta_2 = \theta_{m2} = \theta_m(\xi = L/D)$, and the nondimensional inlet temperature is given as $\theta_{m1} = 1.5$. Determine the (nondimensional) wall temperature variation as a function of relative tube length, L/D, using the numerical quadrature approach of Example 5.10.

Solution

Since the tube wall is black we have only two simultaneous equations, (20.17) and (20.20), in the two unknowns θ_m and θ_w. However, the equations are nonlinear; therefore, an iterative procedure is necessary. For simplicity, we will adopt a simple backward finite-difference approach for the solution of equation (20.17), and the numerical quadrature scheme of equation (5.50) for the integral in equation (20.20). Evaluating temperatures at $N + 1$ nodal points $\xi_i = i\Delta\xi$, $(i = 0, 1, \ldots, N)$ where $\Delta\xi = L/(ND)$, this implies

$$\left(\frac{d\theta_m}{d\xi}\right)_{\xi_i} \simeq \frac{\theta_m(\xi_i) - \theta_m(\xi_{i-1})}{\Delta\xi}, \quad i = 1, 2, \ldots, N,$$

$$\int_0^{L/D} \theta_w^4(\xi') \frac{dF_{d\xi-d\xi'}}{d\xi'}\, d\xi' \simeq \frac{L}{D} \sum_{j=0}^{N} c_j\, \theta_w^4(\xi_j)\, K(\xi_i, \xi_j), \quad i = 0, 1, \ldots, N,$$

where the c_j are quadrature weights and, from Configuration 9 in Appendix D,[2]

$$K(\xi_i, \xi_j) = 1 - \frac{X_{ij}(2X_{ij}^2 + 3)}{2(X_{ij}^2 + 1)}; \quad X_{ij} = |\xi_i - \xi_j|.$$

Similarly, the two view factors to the openings are evaluated from Configuration 31 in Appendix D as

$$F_{d\xi_i-k} = \frac{X_{ij}^2 + \frac{1}{2}}{\sqrt{X_{ij}^2 + 1}} - X_{ij},$$

where

$$j = 0 \quad \text{if} \quad k = 1 \quad \text{(opening at } \xi = \xi_0 = 0),$$
$$j = N \quad \text{if} \quad k = 2 \quad \text{(opening at } \xi = \xi_N = L/D).$$

---

[2]Note that $K(\xi, \xi')$ has a sharp peak at $\xi' = \xi$. Therefore, and also in light of the truncation error in the finite-differencing of $d\theta_m/d\xi$, it is best to limit the quadrature scheme to Simpson's rule [32].

To solve for the unknown $\theta_m(\xi_i)$ and $\theta_w(\xi_i)$, we adopt the following iterative procedure:

1. A wall temperature is guessed for all wall nodes, say,

$$\theta_w(\xi_i) = \theta_1, \quad i = 0, 1, \ldots, N.$$

2. A temperature difference is calculated from equation (20.20), i.e.,

$$\phi_i = H\left[\theta_w(\xi_i) - \theta_m(\xi_i)\right] = 1 - \theta_w^4(\xi_i) + \theta_1^4 F_{d\xi_i-1} + \theta_2^4 F_{d\xi_i-2}$$
$$+ \frac{L}{D}\sum_{j=0}^{N} c_j\,\theta_w^4(\xi_j)\,K(\xi_i,\xi_j).$$

3. The gas bulk temperature is calculated from equation (20.17) as

$$\theta_m(\xi_i) = \theta_m(\xi_{i-1}) + \frac{4\,\mathrm{St}\,\Delta\xi}{H}\phi_i; \quad \theta_m(\xi_0) = \theta_1.$$

4. An updated value for the wall temperatures is then determined from the definition for ϕ_i, that is,

$$\theta_w^{\mathrm{new}}(\xi_i) = \omega\left[\theta_m(\xi_i) + \frac{1}{H}\phi_i\right] + (1-\omega)\,\theta_w^{\mathrm{old}}(\xi_i),$$

where ω is known as the *relaxation parameter*. The iteration scheme is called underrelaxed if $\omega < 1$, and overrelaxed if $\omega > 1$. If ω is chosen too large, the iteration will become unstable and not converge at all. A good or optimal value for the relaxation parameter must usually be found by trial and error. Detailed discussions on relaxation may be found in standard numerical analysis texts such as [33, 34]. Some representative results are shown in Fig. 20-4 for several values of L/D. Because of the strong nonlinearity of the problem, and the crude numerical scheme employed here, large numbers of nodes are necessary to achieve good accuracy ($N \simeq 40L/D$), together with strong underrelaxation ($\omega < 0.02$).

For the case of pure convection ($\epsilon = 0$, or $\phi_i \equiv 1$) the tube wall temperature rises linearly with axial distance, since constant wall heat flux implies a linear increase in bulk temperature and, therefore, (assuming a constant heat transfer coefficient) in surface temperature. This is not the case if radiation is present, in particular for short tubes (small L/D). Near both ends of the tube, much of the radiative energy leaves through the openings, causing a distinct drop in surface temperature. For long tubes ($L/D > 50$) the surface temperature rises almost linearly over the central parts of the tube, although the temperature stays below the convection-only case: Due to the higher temperatures downstream, some net radiative heat flux travels upstream, making overall heat transfer a little more efficient. It should be noted here that the assumption of a constant heat transfer coefficient is not particularly realistic, since it implies a fully developed thermal profile. It is well known that for pure convection $h \to \infty$ at the inlet and, thus, $\theta_w(\xi = 0) = 1$ [1]. Near the inlet of a tube the actual temperature distribution for pure convection is very similar to the one depicted in Fig. 20-3, which is driven by radiation losses. Although for pure convection a fully developed thermal profile and constant h are eventually reached (at $L/D > 20$ for turbulent flow), in the presence of radiation a constant heat transfer coefficient is *never* reached (because the radiation term makes the governing equations nonlinear).

A number of researchers have investigated combined convection and radiation

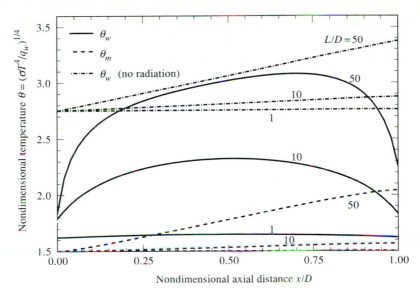

FIGURE 20-4
Axial surface temperature development for combined convection and surface radiation in a black tube subjected to constant wall heat flux.

for a transparent flowing medium. Flow through circular tubes was considered by Siegel and coworkers [35–37] for a number of situations, but always assuming a constant and known heat transfer coefficient. Dussan and Irvine [38] and Chen [39] calculated the local convection rate by solving the two-dimensional energy equation for the flowing medium, but they made severe simplifications in the evaluation of radiative heat fluxes. The most general tube flow analysis has been carried out by Thorsen and Kanchanagom [40, 41]. Similar problems for parallel-plate channel flow were investigated by Keshock and Siegel [42] (for a constant heat transfer coefficient) and Lin and Thorsen [43] (for two-dimensional convection calculations). Combined radiation and forced convection of external flow across a flat plate has been addressed by Cess [44, 45], Sparrow and Lin [46], and Sohal and Howell [47].

20.3 COMBINED RADIATION AND CONDUCTION

Throughout the remainder of this chapter we will deal with the interaction of radiation with conduction and/or convection within an absorbing, emitting, and scattering medium. We start in this section by discussing the interaction between radiation and conduction in a stationary, radiatively participating medium. Since we are primarily interested in general trends and in evaluation methods, we will limit ourselves here to the relatively simple example of steady-state heat transfer through a one-dimensional, absorbing-emitting (but not scattering) gray medium, confined between two parallel, isothermal, gray, diffusely emitting and reflecting plates.

The energy equation for simultaneous conduction and radiation in a participating medium is, from equation (8.61),

$$\rho c_v \frac{\partial T}{\partial t} = \nabla \cdot (k \nabla T) + \dot{Q}''' - \nabla \cdot \mathbf{q}_R \tag{20.21}$$

For a one-dimensional, planar medium at steady state and without internal heat generation, this reduces to equation (8.62), or

$$\frac{d}{dz}\left(k \frac{dT}{dz} - q_R\right) = 0, \tag{20.22}$$

subject to the boundary conditions

$$z = 0: \quad T(0) = T_1, \tag{20.23a}$$
$$z = L: \quad T(L) = T_2. \tag{20.23b}$$

The radiative heat flux, or its divergence

$$\frac{dq_R}{dz} = \int_0^\infty \kappa_\eta (4\pi I_{b\eta} - G_\eta)\, d\eta, \tag{20.24}$$

may be obtained by any of the methods discussed in the preceding chapters.

For simplicity, we will assume that all properties are constant (i.e., thermal conductivity k, absorption coefficient κ, and refractive index n). Note that the assumption of a semitransparent medium implies that the absorptive index (also denoted by the letter k)—although directly related to the absorption coefficient—is negligible[3] in the evaluation of the blackbody intensity, i.e., $I_b = n^2 \sigma T^4/\pi$. Introducing the nondimensional variables and parameters

$$\xi = \frac{z}{L}, \quad \theta = \frac{T}{T_1}, \quad \Psi_R = \frac{q_R}{n^2 \sigma T_1^4}, \quad g = \frac{G}{4n^2 \sigma T^4};$$
$$\tau_L = \kappa L, \quad \theta_L = \frac{T_2}{T_1}, \quad N = \frac{k\kappa}{4\sigma T_1^3},$$

reduces equations (20.22) through (20.24) to

$$\frac{d^2\theta}{d\tau^2} = \frac{1}{4N}\frac{d\Psi_R}{d\tau}, \tag{20.25}$$
$$\frac{d\Psi_R}{d\tau} = 4(\theta^4 - g), \tag{20.26}$$
$$\theta(0) = 1, \quad \theta(\tau_L) = \theta_L. \tag{20.27}$$

Here τ_L is the optical thickness of the medium, and N is known as the *conduction-to-radiation parameter*. For optically thick slabs ($\tau_L \gg 1$) N gives a good estimate of the

---

[3] See the discussion on the value of k in Section 3.5.

relative importance of conductive and radiative heat fluxes: From equations (13.17) and (13.18),

$$\tau_L \gg 1: \quad \frac{q_C}{q_R} = \frac{-k\,\partial T/\partial z}{-k_R\,\partial T/\partial z} = \frac{k}{k_R} = \frac{3}{4}\frac{k\kappa}{4n^2\sigma T^4},$$

which gives the ratio of heat fluxes in terms of a *local* temperature. The situation is a little more complicated for optically thin situations ($\tau_L \ll 1$), for which the temperature field of the entire enclosure must be considered. For example, for an optically thin slab bounded by two black walls at T_1 and T_2, respectively, from equation (13.7),

$$\tau_L \ll 1: \quad q_R \simeq n^2\sigma(T_1^4 - T_2^4) = 4n^2\sigma T_{\rm av}^3(T_1 - T_2),$$

$$q_C = -k\frac{\partial T}{\partial z} \simeq k(T_1 - T_2)/L,$$

$$\frac{q_C}{q_R} \simeq \frac{k/L}{4n^2\sigma T_{\rm av}^3} = \frac{1}{\tau_L}\frac{k\kappa}{4n^2\sigma T_{\rm av}^3}.$$

If, in an optically thin slab, emission from within the slab (rather than from its boundaries) dominates the radiative heat flux, then q_R becomes proportional to κ [cf. equation (8.49)], and

$$\tau_L \ll 1 \text{ (emission dominated)} : \quad \frac{q_C}{q_R} = \mathbb{O}\!\left(\frac{N}{\tau_L^2}\right).$$

As representative examples for combined radiation and conduction in a slab, we will discuss here solutions for the radiative heat flux using the exact integral formulation (as presented in Chapter 12) and the differential or P_1-approximation (described in Sections 13.4 and 14.4). Similarly, equation (20.22) may be solved by a variety of numerical techniques. Since the equation is nonlinear (because of the T^4-dependence for the radiative heat flux), analytical solutions are not possible, and numerical schemes require an iterative solution. For illustrative purposes we will limit ourselves here to a finite-difference solution of equations (20.22) and (20.23).

Exact Formulation

The exact formulation for incident radiation G and radiative heat flux q_R for a one-dimensional slab with specified temperature distribution has been given by equations (12.54) and (12.55). For a nonscattering medium the radiative source term reduces to $S(\tau) = I_b(\tau) = n^2\sigma T^4(\tau)/\pi$ [as given by equation (12.56)], and the radiative heat flux, as given by equation (12.55), becomes, in nondimensional form,

$$\Psi_R(\tau) = 2\left\{ \mathcal{I}_1 E_3(\tau) - \mathcal{I}_2 E_3(\tau_L - \tau) + \int_0^\tau \theta^4(\tau')\,E_2(\tau - \tau')\,d\tau' \right.$$
$$\left. - \int_\tau^{\tau_L} \theta^4(\tau')\,E_2(\tau' - \tau)\,d\tau' \right\}, \qquad (20.28)$$

where we have introduced the nondimensional radiosities $\mathcal{J}_i = J_i/n^2\sigma T_1^4$. Equation (20.28) may be integrated by parts, using the recursion relations of Appendix E, leading to

$$\Psi_R(\tau) = 2\left\{(\mathcal{J}_1 - 1)E_3(\tau) - (\mathcal{J}_2 - \theta_L^4)E_3(\tau_L - \tau)\right.$$

$$\left. - \int_0^\tau \frac{d\theta^4}{d\tau'}(\tau')E_3(\tau - \tau')\,d\tau' - \int_\tau^{\tau_L} \frac{d\theta^4}{d\tau'}(\tau')E_3(\tau' - \tau)\,d\tau'\right\}, \qquad (20.29)$$

and, using Leibnitz' rule [48], as given by equation (3.101),

$$\frac{d\Psi_R}{d\tau} = 2\left\{(1 - \mathcal{J}_1)E_2(\tau) + (\theta_L^4 - \mathcal{J}_2)E_2(\tau_L - \tau)\right.$$

$$\left. + \int_0^\tau \frac{d\theta^4}{d\tau'}(\tau')E_2(\tau - \tau')\,d\tau' - \int_\tau^{\tau_L} \frac{d\theta^4}{d\tau'}(\tau')E_2(\tau' - \tau)\,d\tau'\right\}. \qquad (20.30)$$

Equation (20.30) must be solved simultaneously with equation (20.25) and its boundary conditions (20.27). For nonblack surfaces two additional relations are required for the determination of the radiosities $\mathcal{J}_1$ and $\mathcal{J}_2$. These may be obtained by applying equation (20.29) (evaluation of the radiative heat flux in terms of radiosities and medium temperature) at the two boundaries, eliminating the radiative heat flux through equation (12.44) (relating heat flux to radiosity and surface temperature). For the illustrative purposes of our present discussion, we will limit ourselves to black surfaces, i.e., $\mathcal{J}_1 = 1$ and $\mathcal{J}_2 = \theta_L^4$, and

$$\frac{d\Psi_R}{d\tau} = 2\left\{\int_0^\tau \frac{d\theta^4}{d\tau'}(\tau')E_2(\tau - \tau')\,d\tau' - \int_\tau^{\tau_L} \frac{d\theta^4}{d\tau'}(\tau')E_2(\tau' - \tau)\,d\tau'\right\}. \qquad (20.31)$$

For this simple case, substitution of equation (20.31) into (20.25) gives a single nonlinear integro-differential equation for the unknown temperature, θ. Once the temperature field has been determined, the total heat flux follows as

$$q = -k\frac{dT}{dz} + q_R = \text{const},$$

or, in nondimensional form,

$$\Psi = \frac{q}{n^2\sigma T_1^4} = -4N\frac{d\theta}{d\tau} + \Psi_R = \text{const}. \qquad (20.32)$$

Example 20.2.   An absorbing-emitting medium is contained between two large, parallel, isothermal, black plates at temperatures T_1 and $T_2 = 0.5\,T_1$, respectively. Determine the steady-state temperature distribution within the medium and the total heat flux between the two plates, if heat is transferred by conduction and radiation. Discuss the influence of the conduction-to-radiation parameter, N, and of the optical thickness of the layer, τ_L.

Solution

The numerical solution to the governing equation may be found in a number of ways. We will employ here $J + 1$ equally spaced nodes $\tau = 0, \Delta\tau, 2\Delta\tau, \ldots, J\Delta\tau = \tau_L$ with nodal temperatures θ_i, $(i = 0, 1, 2, \ldots, J)$ and simple finite-differencing for the conduction term,

$$\frac{d^2\theta}{d\tau^2} \simeq \frac{\theta_{i+1} - 2\theta_i + \theta_{i-1}}{\Delta\tau^2} + \mathbb{O}(\Delta\tau^2),$$

with a truncation error of order $\Delta\tau^2$. The divergence of the radiative heat flux, equation (20.31), will be calculated by approximating the emissive power, θ^4, by a spline function, followed by analytical evaluation of the piecewise integrals. In order to obtain the same truncation error as for the conduction term, $\mathbb{O}(\Delta\tau^2)$, the prediction of $d\theta^4/d\tau'$ must be accurate to $\mathbb{O}(\Delta\tau)$ [since the piecewise integration decreases the truncation error by $\mathbb{O}(\Delta\tau)$]. Thus, for the emissive power a linear spline is sufficient, or

$$\theta^4(\tau) = \theta_i^4 + B_i(\tau - \tau_i) + \mathbb{O}(\Delta\tau^2) = \frac{\theta_i^4(\tau_{i+1} - \tau) + \theta_{i+1}^4(\tau - \tau_i)}{\Delta\tau} + \mathbb{O}(\Delta\tau^2),$$

$$\frac{d\theta^4}{d\tau}(\tau) = \frac{\theta_{i+1}^4 - \theta_i^4}{\Delta\tau} + \mathbb{O}(\Delta\tau),$$

$$\tau_i < \tau < \tau_{i+1}, \quad i = 0, 1, 2, \ldots, J - 1.$$

Substituting this into equation (20.31) leads to

$$\left(\frac{d\Psi_R}{d\tau}\right)_i \simeq 2\sum_{j=1}^{i} \frac{\theta_j^4 - \theta_{j-1}^4}{\Delta\tau} \int_{\tau_{j-1}}^{\tau_j} E_2(\tau_i - \tau') \, d\tau'$$

$$-2\sum_{j=i+1}^{J} \frac{\theta_j^4 - \theta_{j-1}^4}{\Delta\tau} \int_{\tau_{j-1}}^{\tau_j} E_2(\tau' - \tau_i) \, d\tau'$$

$$= \frac{2}{\Delta\tau} \sum_{j=1}^{i} \left(\theta_j^4 - \theta_{j-1}^4\right)[E_3(\tau_i - \tau_j) - E_3(\tau_i - \tau_{j-1})]$$

$$+ \frac{2}{\Delta\tau} \sum_{j=i+1}^{J} \left(\theta_j^4 - \theta_{j-1}^4\right)[E_3(\tau_j - \tau_i) - E_3(\tau_{j-1} - \tau_i)]$$

$$= \frac{2}{\Delta\tau} \sum_{j=1}^{J} \left(\theta_j^4 - \theta_{j-1}^4\right)[E_3(|i - j|\Delta\tau) - E_3(|i + 1 - j|\Delta\tau)].$$

Equating both sides of equation (20.25), we find

$$\theta_{i-1} - 2\theta_i + \theta_{i+1} = \frac{\Delta\tau}{2N} \sum_{j=1}^{J}(\theta_j^4 - \theta_{j-1}^4)[E_3(|i - j|\Delta\tau) - E_3(|i + 1 - j|\Delta\tau)],$$

$$i = 1, 2, \ldots, J - 1,$$

$$\theta_0 = 1, \quad \theta_J = \theta_L.$$

If N is relatively large $(N > 0.1)$, heat transfer is dominated by conduction, and the solution proceeds as follows:

1. A temperature profile is guessed (e.g., the linear profile for pure conduction), and the $(d\Psi_R/d\tau)_i$ are calculated based on these temperatures.

2. A new temperature profile is determined by inverting the simple tridiagonal matrix for θ.

3. The temperature profile is iterated on, using underrelaxation as necessary (as discussed in the previous example).

If N is small, radiation dominates, and the process should be reversed:

1. A temperature profile is guessed, the conduction contribution is calculated, and an emissive power field is determined by inverting the full matrix for the θ_i^4 on the right-hand side.

2. A new temperature profile is deduced from the emissive powers, etc.

Once the temperature profile is known, the total heat flux follows from equations (20.29) and (20.32) as

$$
\begin{aligned}
\Psi_i \;\simeq\; & -\frac{2N}{\Delta\tau}(\theta_{i+1}-\theta_{i-1}) - \frac{2}{\Delta\tau}\left\{\sum_{j=1}^{i}\left(\theta_j^4-\theta_{j-1}^4\right)\int_{\tau_{j-1}}^{\tau_j} E_3(\tau_i-\tau')\,d\tau'\right. \\
& \left. +\sum_{j=i+1}^{J}\left(\theta_j^4-\theta_{j-1}^4\right)\int_{\tau_{j-1}}^{\tau_j} E_3(\tau'-\tau_i)\,d\tau'\right\} \\
=\; & -\frac{2N}{\Delta\tau}(\theta_{i+1}-\theta_{i-1}) - \frac{2}{\Delta\tau}\left\{\sum_{j=1}^{i}\left(\theta_j^4-\theta_{j-1}^4\right)[E_4(\tau_i-\tau_j)-E_4(\tau_i-\tau_{j-1})]\right. \\
& \left. -\sum_{j=i+1}^{J}\left(\theta_j^4-\theta_{j-1}^4\right)[E_4(\tau_j-\tau_i)-E_4(\tau_{j-1}-\tau_i)]\right\},
\end{aligned}
$$

$$i = 1, 2, \ldots, J-1.$$

This value for the nondimensional heat flux should be the same for all nodes.

Representative results are shown in Figs. 20-5 and 20-6. Figure 20-5 shows the nondimensional temperature variation within the slab for an intermediate optical thickness of $\tau_L = 1$, calculated by two different methods: By the integral formulation of the present example, and by the P_1-approximation. For $N = 0$ there is no conduction, and the temperature profile is discontinuous at the walls, as first indicated in Fig. 12-3. For very small values of N the temperature profile remains similar except near the walls, where the medium temperature must rapidly approach the surface temperatures. As N increases, the influence of conduction increases, and the temperature profile rapidly becomes linear. For optically thin situations (not shown) the effect is even more pronounced: Larger temperature jumps at the wall for $N = 0$ and an already near-linear temperature profile for $N = 0.01$. This behavior may be explained by noting that—for small τ_L—little emission and absorption takes place inside the medium; radiative heat flux travels directly from surface to surface.

Representative nondimensional heat fluxes are shown in Fig. 20-6 and are compared with approximate methods, which will be discussed a little later. Since the optical thickness of a slab acts as a radiative barrier between two surfaces at different temperatures, the net heat flux increases with decreasing τ_L. That $q/n^2\sigma T_1^4$ increases with increasing N may be interpreted in two opposite ways: If the increase of N is due to an increase in thermal conductivity k, then the conductive and total heat fluxes increase. However, if the increase

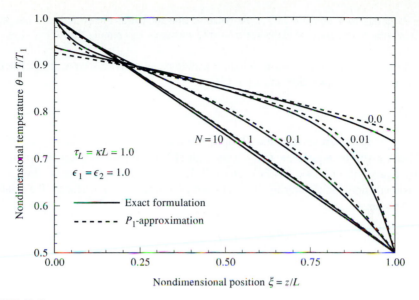

FIGURE 20-5

Nondimensional temperature distribution for combined radiation and conduction across a gray slab of optical thickness $\tau_L = 1$, bounded by black plates with a temperature ratio of $\theta_L = T_2/T_1 = 0.5$.

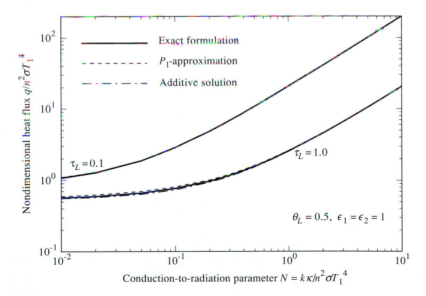

FIGURE 20-6

Nondimensional total heat flux for combined radiation and conduction across a gray slab, bounded by black plates with a temperature ratio of $\theta_L = T_2/T_1 = 0.5$.

in N is due to a decrease in T_1, the radiative and total heat fluxes *decrease* due to the decreasing temperature levels (since $q/n^2\sigma T_1^4$ increases less rapidly than N).

Simple combined conduction-radiation problems such as this were first treated by Viskanta and Grosh [49, 50] and Lick [51].

P_1-Approximation

The governing equations for the P_1-approximation and their boundary conditions have been given by equations (13.42) through (13.44) for the one-dimensional slab, and by equations (14.34), (14.36), and (14.46) for general geometries. For a one-dimensional, gray, nonscattering slab between two gray-diffuse surfaces, the relations may be summarized as

$$\frac{dq}{d\tau} = 4\pi I_b - G, \tag{20.33}$$

$$\frac{dG}{d\tau} = -3q, \tag{20.34}$$

$$\tau = 0: \qquad 2q = 4J_1 - G = \frac{\epsilon_1}{2 - \epsilon_1}(4\pi I_{b1} - G), \tag{20.35a}$$

$$\tau = \tau_L: \qquad -2q = 4J_2 - G = \frac{\epsilon_2}{2 - \epsilon_2}(4\pi I_{b2} - G), \tag{20.35b}$$

or, in nondimensional form (as given at the beginning of this section),

$$\frac{d\Psi_R}{d\tau} = 4(\theta^4 - g), \tag{20.36}$$

$$\frac{dg}{d\tau} = -\frac{3}{4}\Psi_R, \tag{20.37}$$

$$\tau = 0: \qquad \Psi_R = 2(\mathcal{J}_1 - g) = \frac{2\epsilon_1}{2 - \epsilon_1}(1 - g), \tag{20.38a}$$

$$\tau = \tau_L: \qquad -\Psi_R = 2(\mathcal{J}_2 - g) = \frac{2\epsilon_2}{2 - \epsilon_2}(\theta_L^4 - g). \tag{20.38b}$$

The radiative heat flux, Ψ_R, may be eliminated from equations (20.36) through (20.38), leading to

$$\frac{d^2g}{d\tau^2} + 3(\theta^4 - g) = 0, \tag{20.39}$$

$$\tau = 0: \qquad \frac{dg}{d\tau} + \frac{3}{2}\frac{\epsilon_1}{2 - \epsilon_1}(1 - g) = 0, \tag{20.40a}$$

$$\tau = \tau_L: \qquad \frac{dg}{d\tau} - \frac{3}{2}\frac{\epsilon_2}{2 - \epsilon_2}(\theta_L^4 - g) = 0. \tag{20.40b}$$

This second-order differential equation for the incident radiation is connected to the overall energy equation by combining equations (20.25) and (20.26), or

$$\frac{d^2\theta}{d\tau^2} = \frac{1}{N}(\theta^4 - g),\tag{20.41}$$

with its boundary condition (20.27). A solution is obtained by guessing a temperature field, followed by the determination of the incident radiation field from equations (20.39) and (20.40). This, in turn, is used to find an updated temperature field from equations (20.41) and (20.27). Using suitable underrelaxation (generally necessary because of the nonlinearity of the problem), an iteration is performed until converged temperature and incident radiation fields have been obtained. At that point the net heat flux may be calculated from equation (20.32) after evaluation of the radiative heat flux from equation (20.37), or

$$\Psi_R = -\frac{4}{3}\frac{d g}{d\tau}.\tag{20.42}$$

Example 20.3.   Repeat the previous example, employing the P_1-approximation.

Solution

We will use a simple finite-difference method for the solution of overall energy as well as the P_1-approximation. As before, we will break up the optical thickness τ_L into $J + 1$ equally spaced nodes: $i = 0, 1, \ldots, J$ with $\tau_i = i\Delta\tau$ and $\Delta\tau = \tau_L/N$. Thus, equation (20.41) becomes

$$\theta_{i-1} - 2\theta_i + \theta_{i+1} = \varphi_i = \frac{\Delta\tau^2}{N}(\theta_i^4 - g_i), \qquad i = 1, 2, \ldots, J - 1,$$

$$\theta_0 = 1, \qquad \theta_J = \theta_L.$$

Similarly, equation (20.39) transforms to

$$g_{i-1} - (2 + 3\Delta\tau^2)\, g_i + g_{i+1} = -3\Delta\tau^2\theta_i^4, \qquad i = 1, 2, \ldots, J - 1.$$

Two more relations are needed at the two walls. The two boundary conditions for g are of the *third kind*, i.e., they contain both the dependent variable and its normal derivative. In order to retain the overall truncation error of $\mathbb{O}(\Delta\tau^2)$ for all relations, and to retain the tridiagonal nature of the finite-difference equations, it is best to use the *method of artifical nodes* [34]. In this method hypothetical nodes outside the medium (i.e., inside the walls) are introduced on each side, as indicated in the sketch of Fig. 20-7, and equations (20.39) and (20.40) are finite-differenced at the walls as if the boundary nodes were well inside the medium. Thus, with

$$\left(\frac{d g}{d\tau}\right)_0 \simeq \frac{g_1 - g_{-1}}{2\Delta\tau},$$

$$g_{-1} - (2 + 3\Delta\tau^2)\, g_0 + g_1 = -3\Delta\tau^2,$$

$$-g_{-1} \qquad - 3\Delta\tau\, g_0 + g_1 = -3\Delta\tau.$$

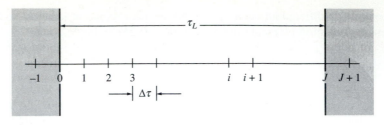

FIGURE 20-7
Nodal system for a one-dimensional slab, with artificial nodes "-1" and "$J+1$" inside the walls.

Adding,

$$- [2 + 3\Delta\tau(1 + \Delta\tau)]\, g_0 + 2g_1 = -3\Delta\tau(1 + \Delta\tau).$$

Similarly, at the other boundary,

$$g_{N-1} - (2 + 3\Delta\tau^2)\, g_N + g_{N+1} = -3\Delta\tau^2\theta_L^4,$$

$$-g_{N-1} \qquad + 3\Delta\tau\, g_N + g_{N+1} = 3\Delta\tau\theta_L^4,$$

and, after subtracting,

$$2g_{N-1} - [2 + 3\Delta\tau(1 + \Delta\tau)]\, g_N = -3\Delta\tau(1 + \Delta\tau)\,\theta_L^4.$$

Therefore, we have two simultaneous tridiagonal systems for the unknown θ_i and g_i. These systems are readily solved by guessing a distribution for the φ_i (say, $\varphi_i = 0$) and inverting the tridiagonal matrix for θ_i. With this the right-hand side for the g_i can be calculated, and the tridiagonal matrix for g_i can be inverted. At this point new values for φ_i may be determined, etc. Once the iteration has converged, the net heat flux is obtained from

$$\Psi_i = \frac{2N}{\Delta\tau}(\theta_{i-1} - \theta_{i+1}) + \frac{2}{3\Delta\tau}(g_{i-1} - g_{i+1}).$$

Some sample results are included in Figs. 20-5 and 20-6 for comparison with the exact results. It is observed that the accuracy of the temperature profile is as expected from the differential approximation (cf. Chapters 13 and 14). Also as expected, the accuracy improves with increasing N, i.e., when conduction dominates more and more over radiation. Similar observations hold true for the evaluation of net heat fluxes.

Wang and Tien [52] apparently were the first ones to employ the P_1- or differential approximation for combined radiation and conduction.

Additive Solutions

Since the evaluation of simultaneous heat transfer by conduction and radiation is rather cumbersome, it is tempting to treat each mode of energy transfer separately (as if the other one weren't there), followed by adding the two resulting heat fluxes. This simple method gives the correct heat flux for the two limiting situations (when only one mode of heat transfer is present); the question is, how accurate the method is for intermediate situations.

The energy flux by pure steady-state conduction through a one-dimensional slab of thickness L is given by

$$q_C = k\frac{T_1 - T_2}{L}, \tag{20.43}$$

while the radiative heat flux for a gray, nonscattering medium at radiative equilibrium, confined between two isothermal black plates is, from Example 13.5,

$$q_R = \frac{n^2\sigma(T_1^4 - T_2^4)}{1 + \frac{3}{4}\tau_L}, \tag{20.44}$$

where we have used the result obtained from the differential approximation, in order to make a closed-form expression possible. Adding these two heat fluxes yields the approximate net heat flux, which, in nondimensional form, may be written as

$$\Psi = \frac{q}{n^2\sigma T_1^4} \simeq \frac{4N}{\tau_L}(1 - \theta_L) + \frac{1 - \theta_L^4}{1 + \frac{3}{4}\tau_L}, \tag{20.45}$$

which is also included in Fig. 20-7. It is observed that the additive solution is surprisingly accurate. Einstein [53] and Cess [45] have shown that the method is within 10% of exact results for black plates, although somewhat larger errors are observed for strongly reflecting surfaces. Howell [54] has demonstrated the relative accuracy of the method for concentric cylinders. Since the method has no physical foundation, it is impossible to predict its accuracy for general geometries. In addition, the method cannot be used to predict the temperature field, since pure conduction and pure radiation each predict their own—conflicting—profiles.

Since the early 1960s numerous articles on combined conduction-radiation problems have appeared in the literature. Most of the early papers dealt with very simple one-dimensional problems [49–52, 55–60]. A number of investigations dealt with the effects of scattering in a one-dimensional slab [61–73]; others considered spectral/nongray effects in varying degrees of sophistication [74–82]. The effects of external irradiation on the combined-mode heat transfer in a one-dimensional slab have been discussed in various investigations [79, 83–88] and the influence of transient conduction in others [67, 87–102]. Various numerical schemes for the solution of the governing nonlinear integro-differential equation have been employed, such as collocation with B-spline trial functions [103], collocation with Chebyshev polynomials [71], Galerkin methods [66, 67] and finite-element methods [67]. In addition to the "exact" integral expressions, a number of different approximate methods were used to evaluate the radiative heat flux, such as the two-flux method [62, 92, 104], the diffusion method [87], the exponential kernel approximation [51], the P_1-approximation [52, 68, 89], the zonal method [82], the Monte Carlo method [70, 105, 106], and others. The few available experimental measurements of conduction-radiation interaction demonstrate the validity of theoretical models for glass [107, 108], aerogel [109], glass particles [71], fiberglass [69], porous media [110], and gases [111].

While the vast majority of investigations have dealt with the interaction in a one-dimensional slab, a few papers have considered other geometries, such as one-dimensional spheres [54, 112–115], one-dimensional cylinders [116–121], and rectangles and other two- and three-dimensional configurations [70, 122–130].

20.4 MELTING AND SOLIDIFICATION WITH INTERNAL RADIATION

Melting and solidification of materials is of importance in many applications and has been studied for over a century. Until the 1950s attention had been focused exclusively on melting and solidification of opaque materials, i.e., situations where the influence of internal radiative heat transfer may be neglected. Early investigations into the effects of radiation have assumed that, as in the case of opaque bodies, there is a distinct interface between liquid and solid zones [131–141], even though meteorologists had already realized that internal melting may occur within ice (e.g., [142, 143]). Only recently has it been postulated by Chan and coworkers [144] that there exists a two-phase zone between the pure liquid and pure solid zones, as shown schematically in Fig. 20-8. The existence of such a two-phase layer in the presence of an internal radiation field may be explained as follows. Consider the melting of a semi-infinite solid, which is initially isothermal at its melting temperature T_m. A constant radiative heat flux is supplied to the face of the solid, as indicated in Fig. 20-8. If the material is opaque, the incident heat flux is absorbed by a thin surface layer

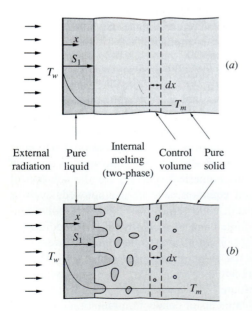

FIGURE 20-8
Melting zones within a semi-infinite body: (a) Opaque medium, (b) semitransparent medium.

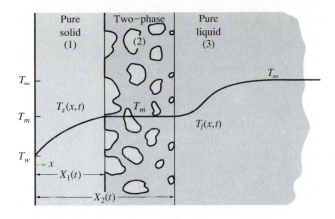

FIGURE 20-9
Solidification of a semitransparent liquid at T_∞, subjected to a
cold boundary T_w ($T_w < T_m <$
T_∞).

at $x = 0$, and heat transfer inside the medium is by conduction alone. Melting then
proceeds with a distinct interface as indicated in Fig. 20-8a, and as described in many
papers and textbooks, e.g., [145]. If the material is semitransparent the external
radiation penetrates deep into the solid, and some of the energy is absorbed internally,
say, in the strip dx. This absorbed energy cannot be conducted away (the solid is
isothermal at T_m), nor can it raise the sensible heat without first melting the solid
within the layer. Since the amount of energy absorbed over a short period of time
cannot be sufficient to melt all of the material within the layer dx instantaneously,
only gradual—and, therefore, partial—melting can be expected. As the amount of
absorbed energy decreases for increasing distance away from the surface, the melt
fraction will decrease along with it. For the more general case, if there is solid at
temperatures below the melting point, absorbed radiative energy will be used first
to raise the sensible heat of the material, resulting in a purely solid zone. Similar
conclusions about the existence of a two-phase zone or "mushy zone" can be reached
by replacing the external heat flux by a hot surface (with its surface emission), or by
considering solidification rather than melting.

For the illustrative purposes of the present section, we will limit our consideration
to a semi-infinite body, which is originally liquid and isothermal at temperature T_∞
($T_\infty > T_m$, the melting temperature of the medium). For times $t > 0$ the temperature
of the face at $x = 0$ is changed to, and kept at, a temperature T_w, which is lower than
the melting/solidification temperature T_m. This results in a three-layer system with a
qualitative temperature distribution as shown in Fig. 20-9. To keep the analysis simple,
we will further assume that liquid and solid have identical and constant properties
($k_l = k_s = k$, $\kappa_l = \kappa_s = \kappa$, etc.), that the medium does not scatter, and that
the face is black ($\epsilon_w = 1$). Consideration of variable properties, different boundary
conditions, different geometry, and/or melting instead of freezing is straightforward
(but very tedious) and will not be discussed here. In the following pages we will set up
the relevant energy equations governing the three zones, and the boundary conditions
that they require, following the development of Chan and coworkers [144].

Pure Solid Region. If, at $t = 0$, the temperature of the face is lowered instantaneously to $T_w < T_m$, this requires the instantaneous formation of an (infinitesimally thin) layer of pure solid, which will grow with time. The governing equation for the temperature within the solid zone follows from equation (8.61) as

$$\rho c \frac{\partial T}{\partial t} = k \frac{\partial^2 T}{\partial x^2} - \frac{dq_R}{dx},$$

(20.46)

which—assuming for now the location of the solid-mushy zone interface $X_1(t)$ to be known—requires an initial condition and two boundary conditions, that is,

$$t = 0 : \qquad T(x, 0) = T_\infty,$$

(20.47a)

$$x = 0 : \qquad T(0, t) = T_w,$$

(20.47b)

$$x = X_1(t) : \qquad T(X_1, t) = T_m.$$

(20.47c)

We defer, for the moment, the evaluation of the radiative heat flux since this is done in the same way for all three zones.

Two-Phase Region (Mushy Zone). In the presence of a two-phase region, at least a part of the solidification takes place over a finite *volume* (rather than only at a distinct interface). Since during solidification the medium releases heat in the amount of L J/kg (where L is the *heat of fusion*), this gives rise to a volumetric heat source in the amount of

$$\dot{Q}''' = L \dot{m}_s''' = L \rho_s \dot{V}_s''' = L \rho_s \frac{\partial f_s}{\partial t},$$

(20.48)

where $\dot{m}_s'''$ and $\dot{V}_s'''$ are the mass and volume of solid formed per unit time and volume, respectively, ρ_s is the density of the pure solid, and f_s is the local solid fraction. Thus, with this heat source the energy equation (8.61) becomes

$$\rho c \frac{\partial T}{\partial t} = k \frac{\partial^2 T}{\partial x^2} - \frac{dq_R}{dx} + \rho L \frac{\partial f_s}{\partial t},$$

(20.49)

where we have omitted the subscript s from ρ_s in the heat source term, since we assume that $\rho_s = \rho_l = \rho = $ const. Since everywhere within the two-phase zone liquid and solid coexist and are assumed to be in local thermodynamic equilibrium, this implies that the temperature in the mushy zone is uniformly at the melting point, and there can be no sensible heat change ($\partial T / \partial t = 0$) and no conduction ($\partial^2 T / \partial x^2 = 0$). Thus, the energy equation simply becomes a relationship for the determination of the solid fraction, or

$$\frac{\partial f_s}{\partial t} = \frac{1}{\rho L} \frac{dq_R}{dx},$$

(20.50)

subject to the initial condition

$$t = 0 : \qquad f_s(x, 0) = 0.$$

(20.51)

Pure Liquid Region. The energy equation for the pure liquid region is identical to the one for the solid, but with different boundary conditions since the zone extends from $x = X_2(t)$ to $x \to \infty$:

$$\rho c \frac{\partial T}{\partial t} = k \frac{\partial^2 T}{\partial x^2} - \frac{dq_R}{dx}, \tag{20.52}$$

$$t = 0: \qquad T(x, 0) = T_\infty, \tag{20.53a}$$

$$x = X_2(t): \qquad T(X_2, t) = T_m, \tag{20.53b}$$

$$x \to \infty: \qquad T(\infty, t) = T_\infty. \tag{20.53c}$$

Radiative Heat Flux. The radiative heat flux within a semitransparent, semi-infinite medium bounded by a black wall, as well as its divergence, are readily found from equations (12.55) through (12.57):

$$q_R(\tau) = 2 \left[E_{bw} E_3(\tau) + \int_0^\tau E_b(\tau') E_2(\tau - \tau') \, d\tau \right.$$

$$\left. - \int_\tau^\infty E_b(\tau') E_2(\tau' - \tau) \, d\tau' \right], \tag{20.54}$$

$$\frac{dq_R}{d\tau}(\tau) = 4E_b(\tau) - 2 \left[E_{bw} E_2(\tau) + \int_0^\infty E_b(\tau') E_1(|\tau - \tau'|) \, d\tau' \right], \tag{20.55}$$

where $\tau = \kappa x$ is the usual optical coordinate, and we assume here that the absorption coefficient is constant and the same for both liquid and solid. Note that $q_R(\tau)$ and $dq_R/d\tau$ are continuous everywhere, including interfaces[4] (which is not true for the divergence of the conductive heat flux, as we will see from the interface conditions below).

Interface Conditions. Finally, we need two conditions for the determination of the location of the two interfaces between solid and mushy zones, $X_1(t)$, and between mushy zone and pure liquid, $X_2(t)$. These are obtained by performing energy balances over infinitesimal volumes adjacent to the interface, as depicted in Fig. 20-10. Consider a volume of thickness dX_1 at the solid-mushy zone interface, dX_1 being the thickness that becomes purely solid over a time period dt. An energy balance gives:

energy conducted in at $X_1(t)$ + energy radiated in at $X_1(t)$

$$+ \text{energy released during } dt$$

= energy conducted out at $X_1(t + dt)$ + energy radiated out at $X_1(t + dt)$,

or

$$-k \frac{\partial T}{\partial x}\bigg|_{X_1 - 0} dt + q_R(X_1) \, dt + \rho L(1 - f_s) \, dX_1$$

$$= -k \frac{\partial T}{\partial x}\bigg|_{X_1 + dX_1 + 0} dt + q_R(X_1 + dX_1) \, dt, \tag{20.56}$$

---

[4] This is also true for variable/different absorption coefficients, for which equations (20.54) and (20.55) continue to hold with $\tau = \int_0^x \kappa(x) \, dx$.

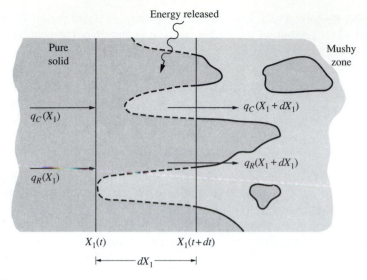

FIGURE 20-10
Energy balance at the moving interface between solid and mushy zones.

where the subscripts ± 0 imply locations on the left of the interface (-0), i.e., in the solid, and on the right of the interface $(+0)$, i.e., in the mushy zone. The heat release term contains the factor $(1 - f_s)$ because the fraction f_s is already solid. Noting that $T = T_m = \text{const}$ inside the mushy zone, it follows that $\partial T / \partial x |_{X_1 + dX_1 + 0} = 0$. The radiative heat flux, on the other hand, is continuous and cancels out from the interface condition once dt and dX_1 are shrunk to zero, and equation (20.56) becomes simply

$$x = X_1(t): \quad -k \left. \frac{\partial T}{\partial x}\right|_{X_1 - 0} + \rho L (1 - f_s) \frac{dX_1}{dt} = 0, \tag{20.57}$$

subject to

$$t = 0: \qquad X_1(0) = 0. \tag{20.58}$$

Note that there does not appear to be any requirement of $f_s \rightarrow 1$ at the interface (smooth transition from mushy zone to pure solid).

Similar to equation (20.56) we find for the mushy zone–liquid interface

$$-k \left. \frac{\partial T}{\partial x}\right|_{X_2 - 0} dt + q_R(X_2)\, dt + \rho L\, f_s\, dX_2$$

$$= -k \left. \frac{\partial T}{\partial x}\right|_{X_2 + dX_2 + 0} dt + q_R(X_2 + dX_2)\, dt, \tag{20.59}$$

where $(1 - f_s)$ is replaced by f_s since the fraction f_s solidifies from pure liquid. Upon shrinking dt and dX_2, the q_R cancel again, and the conduction term within the mushy zone vanishes, or

$$\rho L f_s \frac{dX_2}{dt} = -k \left. \frac{\partial T}{\partial x} \right|_{X_2+0}. \tag{20.60}$$

Now, in order for freezing to occur, we must have $dX_2/dt > 0$ and $\partial T/\partial x \geq 0$. Since the solid fraction must be nonnegative, this implies that the left-hand side of equation (20.60) should be positive and the right-hand side should be negative. This apparent contradiction can be overcome only if both sides of equation (20.60) are identically equal to zero, or,

$$x = X_2(t): \qquad f_s(X_2, t) = 0, \qquad \frac{\partial T}{\partial x}(X_2, t) = 0. \tag{20.61}$$

This implies that there is no distinct interface between mushy zone and liquid: Temperature, heat flux, and solid fraction are continuous across this "interface." Mathematically, one distinguishes between mushy zone and pure liquid, since in the mushy zone f_s is the unknown variable ($T = T_m$ is known), and in the liquid zone the temperature is unknown ($f_s = 0$ is known). The location of interface X_2 is found implicitly by evaluating $f_s(x, t)$ and determining the location where $f_s = 0$.

In summary, in order to predict the solidification of a semitransparent solid, it is necessary to simultaneously solve equations (20.46) and (20.47) (solid), equations (20.50) and (20.51) (mushy zone), equations (20.52) and (20.53) (liquid), together with the interface conditions, equations (20.57) and (20.61). Note that—for an opaque medium—the radiative source within the medium vanishes ($q_R = 0$) and, from equations (20.50) and (20.51), $f_s(x, t) = 0$; that is, the mushy zone shrinks to a point, collapsing the two interfaces as expected for pure conduction. This system of equations is nonlinear, even in the absence of radiation, making exact analytical solutions impossible to find. Chan and coworkers [144] have presented approximate results for a few simple situations. For example, Fig. 20-11 shows the development of the solid and mushy zones for the case of a liquid that is initially uniform at melting temperature.

Example 20.4.   Consider a large (i.e., semi-infinite) block of clear ice exposed to solar radiation on one of its faces. The ice is initially at a uniform 0°C, i.e., at its melting temperature. Heat transfer from the surfaces of the ice (except the solar irradiation) may be neglected, as may the radiative emission from within the ice. Determine the development of the mushy zone for small times. Indicate how the movement of the liquid-mushy zone interface may be calculated.

Solution

Since the side walls are insulated, the problem is one-dimensional; and since the block is "very large," we may assume that it is essentially a semi-infinite body with solar irradiation on its (otherwise insulated) left face at $x = 0$. Since, in this example, we consider the melting of a solid, the order of zones is reversed, i.e., we have pure liquid for $0 \leq x \leq X_1$,

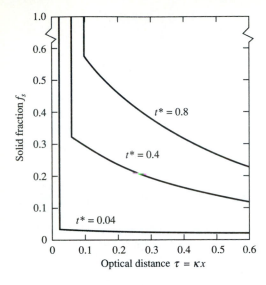

FIGURE 20-11
Solidification of a semi-infinite, semitransparent medium initially isothermal at melting temperature: Development of solid and mushy zones; $\theta_w = T_w/T_m = 0.9$, $N = k\kappa/4n^2\sigma T_m^3 = 0.75$, Ste $= L/cT_m = 500$, $t^* = 2\kappa n^2\sigma(T_m^4 - T_w^4)t/\rho L$.

the mushy zone for $X_1 < x < X_2$, and pure solid for $x > X_2$. In the present example $X_2 \to \infty$, since the ice is everywhere at the melting point. Also, since the face temperature is not increased abruptly, there is no instantaneous formation of a pure liquid layer and $X_1 = 0$ for some time $t > 0$. The solar irradiation is not absorbed by the surface but penetrates into the ice, causing a local radiative heat flux—if emission from and scattering by the ice is neglected—of

$$q_R(x) = q_{\text{sol}} e^{-\kappa x},$$

where q_{sol} is the strength of solar irradiation penetrating into the ice (after losing some of its strength due to reflection at the interface at $x = 0$) (see Chapter 16).

The purely liquid zone is essentially described by equations (20.46) and (20.47):

$$\rho c \frac{\partial T}{\partial t} = k \frac{\partial^2 T}{\partial x^2} + q_{\text{sol}} \kappa e^{-\kappa x},$$

$$t = t_0: \qquad T(x, t_0) = T_m,$$

$$x = 0: \qquad \frac{\partial T}{\partial x}(x, t) = 0,$$

$$x = X_1(t): \qquad T(X_1, t) = T_m,$$

where t_0 is the time at which a purely liquid zone starts to exist, and the boundary condition at $x = 0$ has been replaced to reflect the lack of heat transfer at the surface.

The heat generation term of equation (20.48) becomes a sink, and, while the expression is correct as is, it appears more logical to work with a liquid fraction, $f_l = 1 - fs$, in the case of melting. Thus, equation (20.50) becomes

$$\frac{\partial f_l}{\partial t} = -\frac{1}{\rho L} \frac{dq_R}{dx} = \frac{q_{\text{sol}} \kappa}{\rho L} e^{-\kappa x},$$

$$t = 0: \qquad f_l(0) = 0.$$

Finally, the interface equation at $x = X_1(t)$ must be rewritten as

$$-k \left.\frac{\partial T}{\partial x}\right|_{X_1-0} = \rho L(1-f_l)\frac{dX_1}{dt},$$

$$t = t_0: \quad X_1 = 0,$$

where f_s has been replaced by f_l, L by $-L$ (since melting *requires* heat rather than *releasing* it). Since $\partial T/\partial x = 0$ at $x = 0$, no liquid layer can grow until $f_l = 1$ at $x = 0$. After this has taken place (at time $t = t_0$) the temperature may rise at $x = 0$, and $\partial T/\partial x$ becomes negative at $x = X_1 - 0$; therefore, $f_l(X_1)$ must diminish again, and $dX_1/dt > 0$.

For times $t < t_0$, the equation for the mushy zone is readily solved, leading to

$$f_l(x, t) = \frac{q_{sol}\kappa t}{\rho L}e^{-\kappa x}, \quad 0 = X_1(t) < x < \infty.$$

From this relationship it follows that a purely liquid zone starts at

$$t_0 = \frac{\rho L}{q_{sol}\kappa},$$

that is, when $f_l = 1$ at $x = 0$. For times larger than t_0, the relation for the liquid fraction, $f_l(x, t)$, within the mushy zone continues to hold, but only for $x \geq X_1 > 0$. The temperature profile within the liquid zone and the location of its interface must be determined by simultaneously solving the conduction and interface equations (with known values of f_l).

Since the original postulation by Chan and coworkers [144], the notion of a mushy zone has found widespread acceptance among other researchers [101, 146–148].

20.5 COMBINED RADIATION AND CONVECTION IN BOUNDARY LAYERS

In this section we will briefly discuss how at high temperatures the presence of thermal radiation affects the temperature distribution in a thermal boundary layer and, therefore, the heat transfer rate to or from a wall. Again, since we are mainly interested in the basic nature of interaction between convective and radiative heat transfer, we will limit ourselves to a single simple case, laminar flow over a flat plate.

Consider steady, laminar flow of a viscous, compressible, absorbing/emitting (but not scattering) gray fluid over an isothermal gray-diffuse plate, as illustrated in Fig. 20-12. Making the standard boundary layer assumptions [149], conservation of mass, momentum and energy follow as

$$\frac{\partial}{\partial x}(\rho u) + \frac{\partial}{\partial y}(\rho v) = 0, \tag{20.62}$$

$$\rho\left(u\frac{\partial u}{\partial x} + v\frac{\partial u}{\partial y}\right) = \frac{\partial}{\partial y}\left(\mu\frac{\partial u}{\partial y}\right) - \frac{dp}{dx}, \tag{20.63}$$

$$\rho c_p\left(u\frac{\partial T}{\partial x} + v\frac{\partial T}{\partial y}\right) = \frac{\partial}{\partial y}\left(k\frac{\partial T}{\partial y}\right) - \frac{\partial q_R}{\partial y} + \mu\left(\frac{\partial u}{\partial y}\right)^2, \tag{20.64}$$

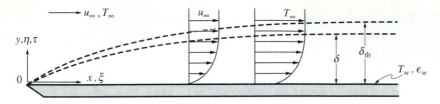

FIGURE 20-12
Laminar flow of an absorbing/emitting fluid over an isothermal gray-diffuse plate.

subject to the boundary conditions

$$x = 0: \quad u(0, y) = u_\infty, \quad T(0, y) = T_\infty; \tag{20.65a}$$

$$y = 0: \quad u(x, 0) = v(x, 0) = 0, \quad T(x, 0) = T_w; \tag{20.65b}$$

$$y \to \infty: \quad u(x, \infty) = u_\infty, \quad T(x, \infty) = T_\infty. \tag{20.65c}$$

Equations (20.63) and (20.64) incorporate the standard boundary layer assumptions of $\partial u/\partial y \gg \partial u/\partial x$ and $\partial T/\partial y \gg \partial T/\partial x$ (momentum and heat transfer rates across the boundary layer are much larger than along the plate, which is dominated by convection), as well as the simplified dissipation function $(\partial u/\partial y)^2$. Similarly, one may drop the x-wise radiation term in favor of the radiative heat flux across the boundary layer. This is readily justified by using the diffusion approximation to get an order-of-magnitude estimate for the radiative heat flux: From equation (13.20), $q_R = -k_R \nabla T$, and—since $\partial T/\partial y \gg \partial T/\partial x$—radiation along the plate may be neglected as compared to radiation across the boundary layer. Therefore, assuming that the radiative heat flux is one-dimensional, q_R may be approximated from equation (12.55) (with $\tau = \int_0^y \kappa dy$ and $\tau_L \to \infty$) as[5]

$$q_R(x, y) = 2J_w(x) E_3(\tau)$$
$$+ 2 \int_0^\tau E_b(x, \tau') E_2(\tau - \tau') \, d\tau' - 2 \int_\tau^\infty E_b(x, \tau') E_2(\tau' - \tau) \, d\tau', \tag{20.66}$$

and

$$\frac{1}{\kappa} \frac{\partial q_R}{\partial y}(x, y) = \frac{\partial q_R}{\partial \tau} = 4E_b(x, \tau) - 2J_w E_2(\tau)$$
$$- 2 \int_0^\infty E_b(x, \tau') E_1(|\tau - \tau'|) \, d\tau'. \tag{20.67}$$

Alternatively, the radiative heat flux may be evaluated from any of the approximate methods discussed in Chapter 13.

It should be remembered that photons carry momentum, thus causing radiation pressure and radiation stress (cf. Section 1.8), and that a control volume stores ra-

---

[5]Equations (20.66) and (20.67) are approximate since they assume that the local value of E_b is independent of x.

diative energy [cf. equation (8.22) and Section 8.8]. However, these effects are generally negligible except at extremely high temperatures ($> 50,000\,\text{K}$ at 1 atm pressure) [150, 151] and will not be included here.

To improve the clarity of development, we will make the additional assumptions of constant fluid properties (ρ, c_p, μ, k, κ = const), slow flow (negligible dissipation term), a black plate [$\epsilon_w = 1$, or $J_w = E_b(T_w) = E_{bw}$], and constant free stream values (u_∞, T_∞ = const). Then equations (20.62) through (20.64) and (20.67) reduce to

$$\frac{\partial u}{\partial x} + \frac{\partial v}{\partial y} = 0, \tag{20.68}$$

$$u \frac{\partial u}{\partial x} + v \frac{\partial u}{\partial y} = \nu \frac{\partial^2 u}{\partial y^2}, \tag{20.69}$$

$$u \frac{\partial T}{\partial x} + v \frac{\partial T}{\partial y} = \alpha \frac{\partial^2 T}{\partial y^2} - \frac{1}{\rho c_p} \frac{\partial q_R}{\partial y}, \tag{20.70}$$

$$\frac{\partial q_R}{\partial y} = 2\kappa \left[2E_b(x, \tau) - E_{bw} E_2(\tau) - \int_0^\infty E_b(x, \tau') E_1(|\tau - \tau'|)\, d\tau' \right], \tag{20.71}$$

subject to boundary conditions (20.65). Here $\nu = \mu/\rho$ is the kinematic viscosity, and $\alpha = k/\rho c_p$ is the thermal diffusivity. Introducing the stream function ψ as

$$u = \frac{\partial \psi}{\partial y}, \quad v = -\frac{\partial \psi}{\partial x}, \tag{20.72}$$

eliminates the continuity equation and transforms the momentum and energy equations to

$$\frac{\partial \psi}{\partial y} \frac{\partial^2 \psi}{\partial x \partial y} - \frac{\partial \psi}{\partial x} \frac{\partial^2 \psi}{\partial y^2} = \nu \frac{\partial^3 \psi}{\partial y^3}, \tag{20.73}$$

$$\frac{\partial \psi}{\partial y} \frac{\partial T}{\partial x} - \frac{\partial \psi}{\partial x} \frac{\partial T}{\partial y} = \alpha \frac{\partial^2 T}{\partial y^2} - \frac{1}{\rho c_p} \frac{\partial q_R}{\partial y}. \tag{20.74}$$

Making the standard[6] coordinate transformation from x and y to the nondimensional ξ and η, where

$$\xi = \frac{4n^2 \sigma T_\infty^3 \kappa x}{\rho c_p u_\infty}, \quad \eta = \left(\frac{u_\infty}{\nu x} \right)^{1/2} y, \tag{20.75}$$

and introducing new nondimensional dependent variables

$$f = \frac{\psi}{(\nu u_\infty x)^{1/2}}, \quad \theta = \frac{T}{T_\infty}, \quad \Psi_R = \frac{q_R}{n^2 \sigma T_\infty^4} \tag{20.76}$$

---

[6]Except for the nondimensionalization factor for ξ.

reduces momentum and energy equations to

$$\frac{d^3 f}{d\eta^3} + \frac{1}{2} f \frac{d^2 f}{d\eta^2} = 0, \tag{20.77}$$

$$\frac{1}{\mathrm{Pr}} \frac{\partial^2 \theta}{\partial \eta^2} + \frac{1}{2} f \frac{\partial \theta}{\partial \eta} = \frac{df}{d\eta} \xi \frac{\partial \theta}{\partial \xi} + \frac{1}{4} \left(\frac{\xi}{N \, \mathrm{Pr}} \right)^{1/2} \frac{\partial \Psi_R}{\partial \eta}. \tag{20.78}$$

In this equation $\mathrm{Pr} = \nu/\alpha = \mu c_p/k$ is the *Prandtl number* of the fluid, and N is the conduction-to-radiation parameter previously introduced as

$$N \equiv \frac{k\kappa}{4n^2 \sigma T_\infty^3}. \tag{20.79}$$

Sometimes, a convection-to-radiation parameter, or *Boltzmann number,* is also introduced, which is defined as

$$\mathrm{Bo} \equiv \frac{\rho c_p u_\infty}{n^2 \sigma T_\infty^3} = 4 \left(\frac{N \, \mathrm{Re}_x \, \mathrm{Pr}}{\xi} \right)^{1/2}, \tag{20.80}$$

where $\mathrm{Re}_x = u_\infty x/\nu$ is the local *Reynolds number.* Very similar to the conduction-to-radiation parameter N, the Boltzmann number gives a qualitative measure of the relative magnitudes of convective and radiative heat fluxes.

Equation (20.77) contains no ξ-derivative since η turns out to be a similarity variable; i.e., no term in the equation (except the ξ-derivative) contains ξ, and the boundary conditions for f do not depend on ξ, collapsing to

$$\eta = 0 : \qquad f = \frac{df}{d\eta} = 0, \tag{20.81a}$$

$$\eta \to \infty : \qquad \frac{df}{d\eta} = 1. \tag{20.81b}$$

Thus, equation (20.77) is an ordinary differential equation for the unknown f, which is a function of the similarity variable η alone. Equation (20.77) and its solution was first given by Blasius and is well documented in fluid mechanics texts, such as [152]. The energy equation (20.78) is a partial differential equation for the unknown θ, subject to the boundary conditions

$$\eta = 0 : \qquad \theta = \frac{T_w}{T_\infty} = \theta_w, \tag{20.82a}$$

$$\eta \to \infty : \qquad \theta = 1, \tag{20.82b}$$

$$\xi = 0 : \qquad \theta = 1. \tag{20.82c}$$

Since the boundary conditions at $x = 0$ correspond to both $\xi = 0$ and $\eta \to \infty$, equation (20.78) can also reduce to a similarity solution, but only if $\Psi_R \propto \xi^{-1/2}$. This is *not* the case if Ψ_R is evaluated from equation (20.71) or most approximate

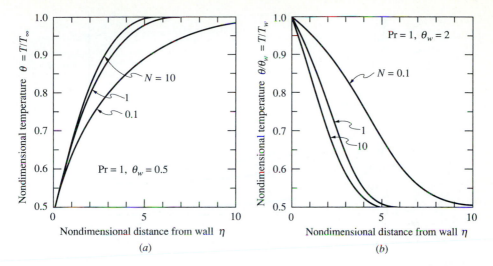

FIGURE 20-13
Similarity profiles for nondimensional temperature profiles across an optically thick laminar boundary over a flat plate; Pr $=1$, (a) $\theta_w = T_w/T_\infty = 0.5$, (b) $\theta_w = 2$.

methods discussed in Chapter 13. However, if the thermal boundary layer is optically very thick, so that the diffusion approximation becomes applicable, one finds from equation (13.20)

$$\Psi_R = -\frac{4}{3\kappa}\frac{\partial\theta^4}{\partial y} = -\frac{4}{3(N\,\mathrm{Pr}\,\xi)^{1/2}}\frac{\partial\theta^4}{\partial\eta}. \qquad (20.83)$$

This expression is substituted into equation (20.78), resulting in the ordinary differential equation

$$\frac{1}{\mathrm{Pr}}\frac{d^2\theta}{d\eta^2} + \frac{1}{2}f\frac{d\theta}{d\eta} = -\frac{1}{3N\,\mathrm{Pr}}\frac{d^2\theta^4}{d\eta^2}, \qquad (20.84)$$

since then θ is a function of the similarity variable η only. The interaction of radiation and convection in an optically thick laminar boundary layer of a gray gas was first investigated by Viskanta and Grosh [153] and others [154–157]. Figure 20-13 shows the similarity profile for the nondimensional temperature, as obtained using the diffusion approximation [153], for a number of different values for the conduction-to-radiation parameter N. For $N = 10$ the temperature profile was found to be within 2% of the pure convection case (which numerically corresponds to $N \to \infty$). When radiation is present, the thermal boundary layer was always found to thicken, which may be explained by the fact that radiation provides an *additional* means to diffuse energy. Even for strong radiation (large T_∞) the thickening of the thermal boundary layer may be limited if the fluid is optically thick (large κ). However, if the absorption coefficient is small (optically thin fluid), the thickening of the thermal boundary may

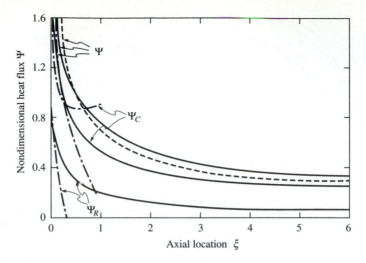

FIGURE 20-14
Comparison of conductive, radiative, and total heat fluxes for a laminar boundary layer over a flat plate:
— · — optically thin solution from [45]; — — — optically thick solution from [153]; and — — — exact
solution from [158]; $N = 0.1$, Pr $=1.0$, $\theta_w = 0.1$.

become so large as to invalidate the basic boundary assumptions (i.e., the neglect of
conduction and radiation in the x-direction).

Figure 20-14 shows nondimensional radiative, conductive, and total surface heat
fluxes along the plate for a representative case as evaluated by three different methods.
The radiative heat flux is evaluated according to the definition in equation (20.76),
and the conductive heat flux is defined as

$$\Psi_C = -k \left.\frac{\partial T}{\partial y}\right|_{y=0} \Big/ n^2 \sigma T_\infty^4 = -4 \left(\frac{N}{\Pr\xi}\right)^{1/2} \left.\frac{\partial \theta}{\partial \eta}\right|_{\eta=0}, \tag{20.85}$$

and

$$\Psi = \Psi_C + \Psi_R. \tag{20.86}$$

The "exact" results are a numerical solution of equation (20.78) with the radiation term evaluated from equation (20.71), as obtained by Zamuraev [158] (and
reported by Viskanta [159]). In the optically thick solution Ψ_R is evaluated from
equation (20.83) as

$$\Psi_R = -\frac{4}{3(N\Pr\xi)^{1/2}} \left.\frac{\partial \theta^4}{\partial \eta}\right|_{\eta=0}, \tag{20.87}$$

and displays a simple $\xi^{-1/2}$ dependence. The optically thin solution has been taken
from Cess [45, 160], who postulated a two-region temperature field consisting of a
very thin conventional thermal boundary layer (in which radiation is neglected in favor of conduction) and an outer region with slowly changing temperature (in which

conduction is neglected). As seen from Fig. 20-14, the diffusion approximation predicts the wall heat flux accurately over the entire length of the plate, while the optically thin approximation fails a short distance away from the leading edge (apparently since downstream the boundary layer grows too thick to neglect radiation and/or the outer layer becomes too nonisothermal to neglect conduction). Other early optically thin models have been reported by Smith and Hassan [161] and Tabaczynski and Kennedy [162]; Pai and Tsao [163] used the exponential kernel approach, and Oliver and McFadden [164] solved the "exact" relations, equation (20.71), by the method of successive approximations, stopping after three iterations. Dissipation effects [160, 165–167] as well as hypersonic conditions [166, 168–170] have been considered by a number of investigators. The influences of scattering [171, 172], nongray radiation properties [173, 174], external irradiation [175, 176], as well as turbulent boundary layers [177–179] have also been addressed.

The effects of radiation are often even more important when combined with free convection rather than forced convection. The radiation effects on a vertical free-convection boundary layer have been modeled by Cess [180] for the optically thin case and by Arpaci [181] for the optically thin and thick cases, while Cheng and Özişik [182] and Desrayaud and Lauriat [183] considered scattering effects. Webb and Viskanta [184] investigated the effects of external irradiation, and Webb [185] verified the model with experiment. Thermal stability of horizontal layers with radiation has also found some attention [186–189]. Combined radiation and free convection within enclosed spaces has also found considerable interest, particularly square cavities [190–194] and parallel vertical plates [195, 196]. In addition, a vertical annulus [197] and cubical cavities [198] have been studied.

20.6 COMBINED RADIATION AND CONVECTION IN DUCT FLOW

Forced convective heat transfer in circular and noncircular ducts, for laminar and turbulent flow, has been thoroughly studied for situations in which radiative heat transfer may be neglected. The case of a transparent medium with radiating boundaries has been briefly discussed in Section 20.2. In this section we will examine the interaction of radiation and convection for a radiatively participating medium flowing through a duct. In the spirit of the previous sections we will again limit our theoretical development to one particularly simple case, namely hydrodynamically developed laminar flow of an incompressible, constant-property fluid through a parallel-plate channel. This is commonly referred to as *Poiseuille flow*. We will assume that the fluid is gray, absorbing, and emitting (but not scattering), and that the plates are gray and diffuse, a distance L apart, and isothermal, as indicated in Fig. 20-15. The fully developed velocity distribution for Poiseuille flow follows readily from equations (20.68) and (20.69), setting $u = u(y)$, as

$$u = 6u_m \frac{y}{L}\left(1 - \frac{y}{L}\right), \quad v = 0, \tag{20.88}$$

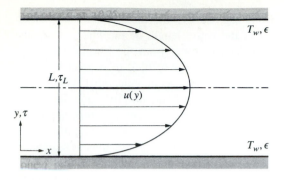

FIGURE 20-15
Thermally developing Poiseuille flow of a gray, absorbing, and emitting fluid between gray-diffuse plates.

where u_m is the mean velocity across the duct. Thus, the energy equation (20.70) reduces to

$$u(y)\frac{\partial T}{\partial x} = \alpha\frac{\partial^2 T}{\partial y^2} - \frac{1}{\rho c_p}\frac{\partial q_R}{\partial y}, \tag{20.89}$$

if again we limit ourselves to the case in which conduction and radiation in the flow direction (along x) are negligible as compared to their transverse values (along y). This is generally a good assumption for channel locations that are a few plate spacings L removed from the inlet [199]. Equation (20.89) is subject to the boundary conditions

$$x = 0: \quad T = T_i, \tag{20.90a}$$
$$y = 0, L: \quad T = T_w, \tag{20.90b}$$

and the radiative heat flux may be obtained from equation (12.55) as[7]

$$q_R(x, y) = 2J_w(x)\left[E_3(\tau) - E_3(\tau_L - \tau)\right] + 2\int_0^\tau E_b(x, \tau')E_2(\tau - \tau')\,d\tau'$$
$$-2\int_\tau^{\tau_L} E_b(x, \tau')E_2(\tau' - \tau)\,d\tau'. \tag{20.91}$$

The radiative heat flux could, of course, instead be evaluated by any of the approximate methods discussed in Chapter 13.

Introducing similar nondimensional variables and parameters as in the previous section,

$$\theta = \frac{T}{T_w}, \quad \Psi_R = \frac{q_R}{n^2\sigma T_w^4}, \tag{20.92a}$$

$$\xi = \frac{x}{L\,\mathrm{Re}_m\,\mathrm{Pr}} = \frac{x}{L}\bigg/\frac{u_m L}{\nu}\frac{\nu}{\alpha}, \quad \eta = \frac{y}{L}, \quad \tau = \kappa y, \tag{20.92b}$$

$$N = \frac{k\kappa}{4n^2\sigma T_w^3}, \quad \tau_L = \kappa L, \tag{20.92c}$$

---

[7] Again using an approximate, i.e., one-dimensional, solution by neglecting the x-wise variation of emissive power in the evaluation of q_R.

transforms equations (20.88) through (20.91) to

$$6\eta(1-\eta)\frac{\partial\theta}{\partial\xi} = \frac{\partial^2\theta}{\partial\eta^2} - \frac{\tau_L}{4N}\frac{d\Psi_R}{d\eta}, \tag{20.93}$$

$$\xi = 0: \quad \theta = T_i/T_w = \theta_i, \quad \eta = 0, 1: \quad \theta = 1, \tag{20.94}$$

$$\Psi_R = 2\left[E_3(\tau) - E_3(\tau_L - \tau)\right.$$
$$\left. + \int_0^\tau \theta^4(\xi, \tau')E_2(\tau - \tau')\,d\tau' - \int_\tau^{\tau_L} \theta^4(\xi, \tau')E_2(\tau' - \tau)\,d\tau'\right], \tag{20.95}$$

where, for simplicity, we have limited ourselves to black channel walls.

Equation (20.93) and its boundary conditions must be solved simultaneously with equation (20.95), making it a nonlinear integro-differential system. Equation (20.93) is a parabolic differential equation allowing a straightforward numerical solution technique, marching forward from $\xi = 0$. While, in principle, an explicit numerical solution is possible if small enough steps in ξ are taken, in practice implicit methods are employed. Because of the nonlinearity, this requires guessing the temperature field for the next ξ-location (as function of η), followed by an iterative procedure until convergence criteria are met. This scheme is then repeated for all downstream locations. The Poiseuille flow problem described here was first solved by Kurosaki [200] and, a little later, with results reported for higher temperatures (smaller N), by Echigo and coworkers [201]. Figure 20-16 shows the axial development of the local Nusselt number for the case of $\theta_i = 0.5$ (cold fluid, hot wall), with the Nusselt number defined as

$$\text{Nu}_x(\xi) = \frac{q_w L}{k[T_w - T_m(\xi)]}, \tag{20.96}$$

where $q_w = q_C + q_R$ is total heat flux per unit area at the wall, by radiation and conduction. In terms of nondimensional quantities, the local Nusselt number becomes

$$\text{Nu}_x(\xi) = \frac{4}{1 - \theta_m(\xi)}\left[-\frac{\partial\theta}{\partial\eta} + \frac{\tau_L}{4N}\Psi_R\right]_{\eta=0}. \tag{20.97}$$

It is apparent from Fig. 20-16 that—due to the nonlinear radiative contribution—no fully developed temperature profile, and consequently no asymptotic Nusselt number, develop. Rather, for the heated wall case ($T_w > T_i$), the Nusselt number goes through a minimum at a certain downstream location, behind which it tends to increase again. The location of the maximum moves toward the inlet with increasing importance of radiation. This phenomenon may be explained as follows: Downstream from the inlet the convective heat flux always decreases more rapidly than the temperature difference, $T_w - T_m(\xi)$, causing a steady decrease in the convective contribution to the Nusselt number; the fractional radiative heat flux, on the other hand, increases

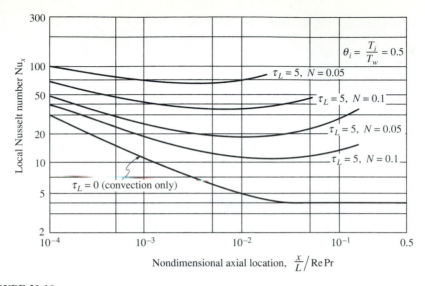

FIGURE 20-16
Influence of optical thickness and conduction-radiation parameter on Nusselt number development in Poiseuille flow—heated wall.

monotonically with x, leading to the observed behavior. Qualitatively, this Nusselt number development is the same for all channel flows with heated walls (regardless of geometry, turbulent flow, presence of scattering, nongrayness, etc.). The heat transfer behavior is somewhat different if a hot fluid enters a cold-walled duct ($T_w < T_i$). This is shown in Fig. 20-17 for turbulent tube flow of a gas seeded with small particles, from Azad and Modest [202]. The Nusselt number always decreases monotonically, somewhat similar to the pure convection case, and eventually appears to reach an asymptotic value. However, as seen from Fig. 20-18, in the presence of thermal radiation this "fully developed" case is never reached until the bulk temperature is essentially equal to the wall temperature (note that, for pure convection, the bulk temperature has changed only by $\approx 20\%$ of maximum when fully developed conditions have been reached). Therefore, it may be concluded that *no thermally fully developed conditions can exist* for forced convection in duct flow combined with appreciable thermal radiation (i.e., radiative heat fluxes too large to be approximated by a linear expression in temperature). This fact was not realized by a number of early investigations on the subject, which employed "thermally developed" conditions to obtain relatively simple results [203–210]. Figures 20-17 and 20-18 also demonstrate how temperature level and optical thickness influence the variation of Nusselt number and bulk temperature. Reducing N/τ_R (which does not depend on absorption coefficient and, for a given medium and tube radius, implies raising temperatures) for a constant optical thickness results in increased heat transfer rates due entirely to an increase in radiative heat flux. The radiative heat flux goes through a maximum at an intermediate optical thickness, $\tau_R \approx 1$ (for constant N/τ_R). This is readily explained

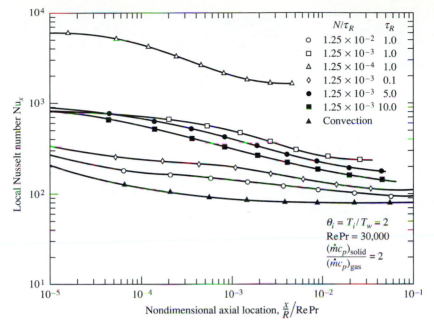

FIGURE 20-17
Influence of optical thickness and conduction-radiation parameter on Nusselt number development for gas-particulate flow through a tube—cooled wall.

by examining the optical limits. In the optically thin limit the medium does not emit or absorb any radiation, resulting in purely convective heat transfer. On the other hand, in the optically thick limit any emitted radiation is promptly absorbed again in the immediate vicinity of the emission point, again reducing radiative heat flux to zero.

A simple one-dimensional temperature profile does exist in the case of *Couette flow* (two infinite parallel plates moving at different velocities), since the entire problem becomes one-dimensional. The analysis for this case reduces to the same equations arrived at in the previous section for combined conduction and radiation, which have been solved numerically by Goulard and Goulard [211] and Viskanta and Grosh [212].

As indicated earlier, the Poiseuille flow problem was originally investigated by Kurosaki [200], using the exact integral relations for the radiative heat flux. The problem had been addressed a little earlier by Timofeyev and coworkers [213], using the two-flux method. The case of slug flow between parallel plates, with rigorous modeling of the radiative heat flux, has been treated by Lii and Özişik [214]. The influence of scattering on Poiseuille flow has been discussed by a number of investigators [215–218]. Yener and coworkers [219, 220] examined the same problem for turbulent flow conditions, while Echigo and Hasegawa [221] addressed a laminar, scattering gas-particulate mixture. All of these publications neglected axial radiation.

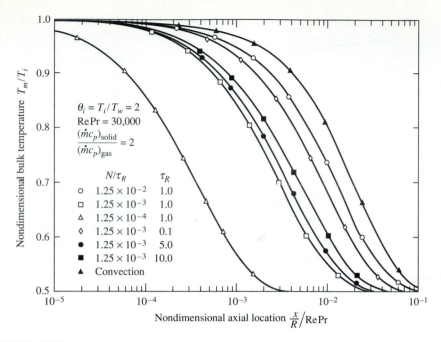

FIGURE 20-18
Influence of optical thickness and conduction-radiation parameter on bulk temperature development for gas-particulate flow through a tube—cooled wall.

Two-dimensional radiation for Poiseuille flow has been studied by Einstein [53] (nonscattering fluid) and Kassemi and Chung [222] (isotropically scattering fluid), using the zonal method, and by Kim and Lee [223] (anisotropically scattering fluid), using the discrete ordinates method.

Combined convection and radiation in thermally developing tube flow appears to have been investigated first by Einstein [224], deSoto [199], and Echigo and coworkers [225], considering two-dimensional (axial and radial) radiation. Landram and colleagues [226] considered optically thin radiation in a turbulent pipe. Pearce and Emery [227] investigated laminar flow (considering also the effects of a nongray gas), and Habib and Greif [228] studied a similar case for turbulent flow, all using a one-dimensional (radial) radiation term. Gau and Chi [229] also investigated nongray effects, using a one-dimensional P_1-approximation for laminar and turbulent tube flows, while Smith and coworkers [230] used the two-dimensional zonal method and weighted-sum-of-gray-gases approach for a similar problem.

Gas-particulate suspension flows were first addressed by Echigo and colleagues [231, 232] for laminar and turbulent flow of nonscattering media, respectively. Anisotropic scattering in tube suspension flows has been treated by Modest and coworkers for gray [202, 233] and nongray [234] carrier gases. Nongray effects have also been studied by Al-Turki and Smith [235], using the zonal method.

20.7 INTERACTION OF RADIATION WITH TURBULENCE

During the development of the equation of transfer in Chapter 8, we noted that heat transfer due to thermal radiation is essentially instantaneous, depending on the temporal temperature distribution as well as the temporal concentration field of the absorbing, emitting, and/or scattering medium. During turbulent flow the temperature field and, for mixtures, the concentration field undergo rapid and irregular local oscillations (but slow compared with the response time of thermal radiation). The governing equations, such as equations (20.62) through (20.64), are then rewritten in terms of time-averaged quantities (denoted by an overbar), e.g.,

$$\bar{\rho}(x, y) = \frac{1}{\delta t} \int_{\delta t} \rho(x, y, t) \, dt, \tag{20.98}$$

where δt is the (small) time interval used for averaging. Commonly, the so-called Favre averaging (or mass-weighted averaging), denoted by a tilde, is also employed, that is,

$$\tilde{\phi} = \overline{\rho \phi} \Big/ \bar{\rho}, \tag{20.99}$$

where ϕ is the quantity to be averaged. This leads to

$$\frac{\partial}{\partial x} (\bar{\rho} \tilde{u}) + \frac{\partial}{\partial y} (\bar{\rho} \tilde{v}) = 0 \tag{20.100}$$

$$\bar{\rho} \left(\tilde{u} \frac{\partial \tilde{u}}{\partial x} + \tilde{v} \frac{\partial \tilde{u}}{\partial y} \right) = \frac{\partial}{\partial y} \left[(\bar{\mu} + \mu_t) \frac{\partial \tilde{u}}{\partial y} \right] - \frac{d \bar{p}}{dx}, \tag{20.101}$$

$$\overline{\rho c_p} \left(\tilde{u} \frac{\partial \tilde{T}}{\partial x} + \tilde{v} \frac{\partial \tilde{T}}{\partial y} \right) = \frac{\partial}{\partial y} \left[(\bar{k} + k_t) \frac{\partial \tilde{T}}{\partial y} \right] - \frac{\partial \overline{q_R}}{\partial y}, \tag{20.102}$$

where μ_t and k_t are turbulent viscosity and conductivity, respectively. The divergence of the radiative heat flux in equation (20.102) is a time-averaged value. For example, for a boundary layer flow of a gray, constant-property fluid over a black isothermal plate, from equation (20.71),

$$\frac{\partial \overline{q_R}}{\partial y} = 2\kappa \left[2 \overline{E_b(x, \tau)} - E_{bw} E_2(\tau) - \int_0^\infty \overline{E_b(x, \tau')} E_1(|\tau - \tau'|) \, d\tau' \right], \tag{20.103}$$

where $\overline{E_b}$ is a time-averaged emissive power,

$$\overline{E_b(x, \tau)} = \frac{1}{\delta t} \int_{\delta t} n^2 \sigma T^4(x, \tau, t) \, dt. \tag{20.104}$$

Since $\overline{E_b}$ (and, in the case of nongray and/or variable properties, more complicated time-averaged quantities) is not available from standard turbulence models, the nonlinear relationship between radiation and fluctuating temperature and concentration fields has generally been ignored. However, it has become clear during recent years that the common practice of evaluating time-averaged values for emissive power and radiative properties at time-averaged values for temperature and concentrations, e.g., $\overline{E_b(x,\tau)} \simeq E_b[\tilde{T}(x,\tau)]$, can very seriously underpredict the emission from a hot medium. This was demonstrated experimentally by Gore and Faeth [236] for turbulent diffusion flames, and theoretically by Gore and coworkers [237] and Song and Viskanta [238]. The stochastic approach commonly used in turbulence modeling provides time averages of dependent variables, such as $\tilde{T}$, as well as second-order correlations, such as $\tilde{T'^2}$, where T' is temperature fluctuation around the mean, or

$$T(x,y,t) = \tilde{T}(x,y) + T'(x,y,t), \quad \tilde{T'} = 0. \tag{20.105}$$

With this knowledge quantities such as $E_b(T)$ may be approximated by a truncated Taylor series, leading to

$$E_b(T) \simeq E_b(\tilde{T}) + T'\frac{dE_b}{d\tilde{T}} + \frac{1}{2}T'^2\frac{d^2E_b}{d\tilde{T}^2} + \cdots,$$

and

$$\overline{E_b(T)} \simeq E_b(\tilde{T}) + \frac{1}{2}\frac{d^2E_b}{d\tilde{T}^2}\tilde{T'^2}. \tag{20.106}$$

Substitution of equation (20.106) into (20.103), then provides an expression for the divergence of the radiative heat flux that accounts for the radiation-turbulence interaction in an approximate fashion.

A detailed discussion of radiation-turbulence interaction is beyond the scope of this book, since it requires thorough understanding of turbulence modeling; the reader is referred to the pertinent literature [236–245]. An example of the interaction effects is shown in Fig. 20-19, which shows the radiative heat flux to the wall of a channel with a combusting jet entering at $x = 0$ [238]. If the jet is preheated to 1000 K before entering the channel, the effects of radiation-turbulence interaction appear to be minimal. However, if the jet enters at room temperature, underprediction of flame emission causes a very severe underprediction of heat fluxes in the colder sections of the channel.

To date, the interaction between radiation and turbulence is still poorly understood. Sample calculations seem to indicate that, while unimportant in many applications, the interaction tends to dominate radiative heat fluxes when large, turbulent flames are present.

References

1. Holman, J. P.: *Heat Transfer*, 7th ed., McGraw-Hill Book Company, New York, 1990.
2. Sparrow, E. M., E. R. G. Eckert, and T. F. Irvine: "The Effectiveness of Radiating Fins with Mutual Irradiation," *Journal of the Aerospace Sciences*, no. 28, pp. 763–772, 1961.

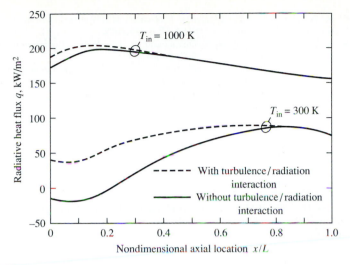

FIGURE 20-19
Comparison of the total heat flux distributions along the wall of a parallel channel with a natural gas–air flame, with and without turbulence–radiation interaction; from Song and Viskanta [238].

3. Hering, R. G.: "Radiative Heat Exchange Between Conducting Plates with Specular Reflection," *ASME Journal of Heat Transfer*, vol. C88, pp. 29–36, 1966.

4. Tien, C. L.: "Approximate Solutions of Radiative Exchange Between Conducting Plates with Specular Reflection," *ASME Journal of Heat Transfer*, vol. 89C, pp. 119–120, 1967.

5. Bartas, J. G., and W. H. Sellers: "Radiation Fin Effectiveness," *ASME Journal of Heat Transfer*, vol. 82C, pp. 73–75, 1960.

6. Sparrow, E. M., and E. R. G. Eckert: "Radiant Interaction Between Fins and Base Surfaces," *ASME Journal of Heat Transfer*, vol. C84, pp. 12–18, 1962.

7. Sparrow, E. M., V. K. Jonsson, and W. J. Minkowycz: "Heat Transfer from Fin-Tube Radiators Including Longitudinal Heat Conduction and Radiant Interchange Between Longitudinally Non-Isothermal Finite Surfaces," *NASA TN D-2077*, 1963.

8. Lieblein, S.: "Analysis of Temperature Distribution and Radiant Heat Transfer Along a Rectangular Fin," *NASA TN D-196*, 1959.

9. Chambers, R. L., and E. V. Sommers: "Radiation Fin Efficiency for One-Dimensional Heat Flow in a Circular Fin," *ASME Journal of Heat Transfer*, vol. 81C, no. 4, pp. 327–329, 1959.

10. Keller, H. H., and E. S. Holdredge: "Radiation Heat Transfer for Annular Fins of Trapezoid Profile," *ASME Journal of Heat Transfer*, vol. 92, no. 6, pp. 113–116, 1970.

11. Mackay, D. B.: *Design of Space Powerplants*, Prentice-Hall, Englewood Cliffs, NJ, 1963.

12. Sparrow, E. M., G. B. Miller, and V. K. Jonsson: "Radiating Effectiveness of Annular-Finned Space Radiators Including Mutual Irradiation between Radiator Elements," *Journal of Aerospace Sciences*, vol. 29, pp. 1291–1299, 1962.

13. Abarbanel, S. S.: "Time Dependent Temperature Distribution in Radiating Solids," *J. Math Phys.*, vol. 39, no. 4, pp. 246–257, 1960.

14. Eckert, E. R. G., T. F. Irvine, and E. M. Sparrow: "Analytical Formulation for Radiating Fins with Mutual Irradiation," *American Rocket Society Journal*, vol. 30, pp. 644–646, 1960.

15. Nilson, E. N., and R. Curry: "The Minimum Weight Straight Fin of Triangular Profile Radiating to Space," *Journal of the Aerospace Sciences*, vol. 27, p. 146, 1960.

16. Hickman, R. S.: "Transient Response and Steady-State Temperature Distribution in a Heated, Radiating, Circular Plate," Technical Report 32-169, California Institute of Technology, Jet Propulsion Laboratory, 1961.

17. Heaslet, M. A., and H. Lomax: "Numerical Predictions of Radiative Interchange Between Conducting Fins with Mutual Irradiations," *NASA TR R-116*, 1961.

18. Nichols, L. D.: "Surface-Temperature Distribution on Thin-Walled Bodies Subjected to Solar Radiation in Interplanetary Space," *NASA TN D-584*, 1961.

19. Schreiber, L. H., R. P. Mitchell, G. D. Gillespie, and T. M. Olcott: "Techniques for Optimization of a Finned-Tube Radiator," *ASME Paper No. 61-SA-44*, June 1961.

20. Olmstead, W. E., and S. Raynor: "Solar Heating of a Rotating Spherical Space Vehicle," *International Journal of Heat and Mass Transfer*, vol. 5, pp. 1165–1177, 1962.

21. Wilkins, J. E.: "Minimum-Mass Thin Fins and Constant Temperature Gradients," *J. Soc. Ind. Appl. Math*, vol. 10, no. 1, pp. 62–73, 1962.

22. Hrycak, P.: "Influence of Conduction on Spacecraft Skin Temperatures," *AIAA Journal*, vol. 1, pp. 2619–2621, 1963.

23. Karlekar, B. V., and B. T. Chao: "Mass Minimization of Radiating Trapezoidal Fins with Negligible Base Cylinder Interaction," *International Journal of Heat and Mass Transfer*, vol. 6, pp. 33–48, 1963.

24. Stockman, N. O., and J. L. Kramer: "Effect of Variable Thermal Properties on One-Dimensional Heat Transfer in Radiating Fins," *NASA TN D-1878*, 1963.

25. Kotan, K., and O. A. Arnas: "On the Optimization of the Design Parameters of Parabolic Radiating Fins," *ASME Paper No. 65-HT-42*, August 1965.

26. Mueller, H. F., and N. D. Malmuth: "Temperature Distribution in Radiating Heat Shields by the Method of Singular Perturbations," *International Journal of Heat and Mass Transfer*, vol. 8, pp. 915–920, 1965.

27. Russell, L. D., and A. J. Chapman: "Analytical Solution of the 'Known-Heat-Load' Space Radiator Problem," *Journal of Spacecraft and Rockets*, vol. 4, no. 3, pp. 311–315, 1967.

28. Frost, W., and A. H. Eraslan: "An Iterative Method for Determining the Heat Transfer from a Fin with Radiative Interaction Between the Base and Adjacent Fin Surfaces," *AIAA Paper No. 68-772*, June 1968.

29. Donovan, R. C., and W. M. Rohrer: "Radiative Conducting Fins on a Plane Wall, Including Mutual Irradiation," *ASME Paper No. 69-WA/HT-22*, November 1969.

30. Schnurr, N. M., A. B. Shapiro, and M. A. Townsend: "Optimization of Radiating Fin Arrays with Respect to Weight," *ASME Journal of Heat Transfer*, vol. 98, no. 4, pp. 643–648, 1976.

31. Eslinger, R., and B. Chung: "Periodic Heat Transfer in Radiating and Convecting Fins or Fin Arrays," *AIAA Journal*, vol. 17, no. 10, pp. 1134–1140, 1979.

32. Fröberg, C. E.: *Introduction to Numerical Analysis*, Addison-Wesley, Reading, Massachusetts, 1969.

33. Hornbeck, R. W.: *Numerical Methods*, Quantum Publishers, Inc., New York, 1975.

34. Ferziger, J. H.: *Numerical Methods for Engineering Application*, John Wiley & Sons, New York, 1981.

35. Siegel, R., and M. Perlmutter: "Convective and Radiant Heat Transfer for Flow of a Transparent Gas in a Tube with Gray Wall," *International Journal of Heat and Mass Transfer*, vol. 5, pp. 639–660, 1962.

36. Perlmutter, M., and R. Siegel: "Heat Transfer by Combined Forced Convection and Thermal Radiation in a Heated Tube," *ASME Journal of Heat Transfer*, vol. C84, pp. 301–311, 1962.

37. Siegel, R., and E. G. Keshock: "Wall Temperature in a Tube with Forced Convection, Internal Radiation Exchange and Axial Wall Conduction," *NASA TN D-2116*, 1964.

38. Dussan, B. I., and T. F. Irvine: "Laminar Heat Transfer in a Round Tube with Radiating Flux at the Outer Wall," in *Proceedings of the Third International Heat Transfer Conference*, vol. 5, Washington, D.C., Hemisphere, pp. 184–189, 1966.

39. Chen, J. C.: "Laminar Heat Transfer in a Tube with Nonlinear Radiant Heat-Flux Boundary Conditions," *International Journal of Heat and Mass Transfer*, vol. 9, pp. 433–440, 1966.

40. Thorsen, R. S.: "Heat Transfer in a Tube with Forced Convection, Internal Radiation Exchange, Axial Wall Heat Conduction and Arbitrary Wall Heat Generation," *International Journal of Heat and Mass Transfer*, vol. 12, pp. 1182–1187, 1969.

41. Thorsen, R. S., and D. Kanchanagom: "The Influence of Internal Radiation Exchange, Arbitrary Wall Heat Generation and Wall Heat Conduction on Heat Transfer in Laminar and Turbulent Flows,"

in *Proceedings of the Fourth International Heat Transfer Conference*, vol. 3, New York, Elsevier, pp. 1–10, 1970.

42. Keshock, E. G., and R. Siegel: "Combined Radiation and Convection in a Asymmetrically Heated Parallel Plate Flow Channel," *ASME Journal of Heat Transfer*, vol. 86C, pp. 341–350, 1964.

43. Lin, S. T., and R. S. Thorsen: "Combined Forced Convection and Radiation Heat Transfer in Asymmetrically Heated Parallel Plates," in *Proceedings of the Heat Transfer and Fluid Mechanics Institute*, Stanford University Press, pp. 32–44, 1970.

44. Cess, R. D.: "The Effect of Radiation Upon Forced-Convection Heat Transfer," *Applied Scientific Research Part A*, vol. 10, pp. 430–438, 1962.

45. Cess, R. D.: "The Interaction of Thermal Radiation with Conduction and Convection Heat Transfer," in *Advances in Heat Transfer*, vol. 1, Academic Press, New York, pp. 1–50, 1964.

46. Sparrow, E. M., and S. H. Lin: "Boundary Layers with Prescribed Heat Flux–Application to Simultaneous Convection and Radiation," *International Journal of Heat and Mass Transfer*, vol. 8, pp. 437–448, 1965.

47. Sohal, M., and J. R. Howell: "Determination of Plate Temperature in Case of Combined Conduction, Convection and Radiation Heat Exchange," *International Journal of Heat and Mass Transfer*, vol. 16, pp. 2055–2066, 1973.

48. Wylie, C. R.: *Advanced Engineering Mathematics*, 5th ed., McGraw-Hill, New York, 1982.

49. Viskanta, R., and R. J. Grosh: "Effect of Surface Emissivity on Heat Transfer by Simultaneous Conduction and Radiation," *International Journal of Heat and Mass Transfer*, vol. 5, pp. 729–734, 1962.

50. Viskanta, R., and R. J. Grosh: "Heat Transfer by Simultaneous Conduction and Radiation in an Absorbing Medium," *ASME Journal of Heat Transfer*, vol. 84, pp. 63–72, 1963.

51. Lick, W.: "Energy Transfer by Radiation and Conduction," in *Proceedings of the Heat Transfer and Fluid Mechanics Institute*, Palo Alto, California, Stanford University Press, pp. 14–26, 1963.

52. Wang, L. S., and C. L. Tien: "Study of the Interaction Between Radiation and Conduction by a Differential Method," in *Proceedings of the Third International Heat Transfer Conference*, vol. 5, Washington, D.C., Hemisphere, pp. 190–199, 1966.

53. Einstein, T. H.: "Radiant Heat Transfer to Absorbing Gases Enclosed Between Parallel Flat Plates with Flow and Conduction," *NASA TR R-154*, 1963.

54. Howell, J. R.: "Determination of Combined Conduction and Radiation of Heat through Absorbing Media by the Exchange Factor Approximation," *Chemical Engineering Progress Symposium Series*, vol. 61, no. 59, pp. 162–171, 1965.

55. Wang, L. S., and C. L. Tien: "A Study of Various Limits in Radiation Heat-Transfer Problems," *International Journal of Heat and Mass Transfer*, vol. 10, pp. 1327–1338, 1967.

56. Chang, Y.-P.: "A Potential Treatment of Energy Transfer by Conduction, Radiation, and Convection," *AIAA Journal*, vol. 5, pp. 1024–1026, 1967.

57. Timmons, D. H., and J. O. Mingle: "Simultaneous Radiation and Conduction with Specular Reflection," *AIAA Paper No. 68-28*, January 1968.

58. Tien, C. L., P. S. Jagannathan, and B. F. Armaly: "Analysis of Lateral Conduction and Radiation Along Two Parallel Long Plates," *AIAA Journal*, vol. 7, pp. 1806–1808, 1969.

59. Tarshis, L. A., S. O'Hara, and R. Viskanta: "Heat Transfer by Simultaneous Conduction and Radiation for Two Absorbing Media in Intimate Contact," *International Journal of Heat and Mass Transfer*, vol. 12, pp. 333–347, 1969.

60. Anderson, E. E., R. Viskanta, and W. H. Stevenson: "Heat Transfer through Semitransparent Solids," *ASME Journal of Heat Transfer*, vol. 95, no. 2, pp. 179–186, 1973.

61. Viskanta, R.: "Heat Transfer by Conduction and Radiation in Absorbing and Scattering Materials," *ASME Journal of Heat Transfer*, vol. 87, pp. 143–150, 1965.

62. Bergquam, J. B., and R. A. Seban: "Heat Transfer by Conduction and Radiation in Absorbing and Scattering Materials," *ASME Journal of Heat Transfer*, vol. 93, pp. 236–238, 1971.

63. Dayan, A., and C. L. Tien: "Heat Transfer in a Gray Planar Medium with Linear Anisotropic Scattering," *ASME Journal of Heat Transfer*, vol. 97, pp. 391–396, 1975.

64. Roux, J. A., and A. M. Smith: "Combined Conductive and Radiative Heat Transfer in an Absorbing Scattering Infinite Slab," *ASME Journal of Heat Transfer*, vol. 100, no. 1, pp. 98–104, 1978.

65. Yuen, W. W., and L. W. Wang: "Heat Transfer by Conduction and Radiation in a One-Dimensional Absorbing, Emitting and Anisotropically Scattering Medium," *ASME Journal of Heat Transfer*, vol. 102, pp. 303–307, 1980.

66. Wu, S. T., R. E. Ferguson, and L. L. Altigilbers: "Application of Finite-Element Techniques to the Interaction of Conduction and Radiation in a Participating Medium," in *Heat Transfer and Thermal Control, Progress in Aeronautics and Astronautics*, vol. 78, AIAA, New York, pp. 61–69, 1981.

67. Fernandes, R., J. Francis, and J. N. Reddy: "A Finite-Element Approach to Combined Conductive and Radiative Heat Transfer in a Planar Medium," in *Heat Transfer and Thermal Control, Progress in Aeronautics and Astronautics*, vol. 78, New York, AIAA, pp. 92–109, 1981.

68. Ratzell III, A. C., and J. R. Howell: "Heat Transfer by Conduction and Radiation in One-Dimensional Planar Medium Using Differential Approximation," *ASME Journal of Heat Transfer*, vol. 104, pp. 388–391, 1982.

69. Houston, R. I., and S. A. Korpela: "Heat Transfer through Fiberglass Insulation," in *Proceedings of the Seventh International Heat Transfer Conference, Munich*, vol. 2, Washington, D.C., Hemisphere, pp. 499–504, 1982.

70. Mishkin, M., and G. J. Kowalski: "Application of Monte Carlo Techniques to the Steady State Radiative and Conductive Heat Transfer through a Participating Medium," *ASME Paper No. 83-WA/HT-27*, 1983.

71. Kamiuto, K., and M. Iwamoto: "Combined Conductive and Radiative Heat Transfer Through a Glass Particle Layer," in *Proceedings of the Second ASME/JSME Conference*, vol. 4, pp. 77–84, 1987.

72. Ho, C. H., and M. N. Özişik: "Simultaneous Conduction and Radiation in a Two-Layer Planar Medium," *Journal of Thermophysics and Heat Transfer*, vol. 1, no. 2, pp. 154–161, 1987.

73. Ho, C. H., and M. N. Özişik: "Combined Conduction and Radiation in a Two-Layer Planar Medium with Flux Boundary Condition," *Numerical Heat Transfer*, vol. 11, no. 3, p. 321, 1987.

74. Greif, R.: "Energy Transfer by Radiation and Conduction with Variable Gas Properties," *International Journal of Heat and Mass Transfer*, vol. 7, pp. 891–900, 1964.

75. Lick, W.: "Transient Energy Transfer by Radiation and Conduction," *International Journal of Heat and Mass Transfer*, vol. 8, pp. 119–127, 1965.

76. Echigo, R., S. Hasegawa, and Y. Miyazaki: "Composite Heat Transfer with Thermal Radiation in Nongray Medium: Part I: Interaction of Radiation with Conduction," *International Journal of Heat and Mass Transfer*, vol. 14, pp. 2001–2015, 1971.

77. Crosbie, A. L., and R. Viskanta: "Interaction of Heat Transfer by Conduction and Radiation in a Nongray Planar Medium," *Wärme- und Stoffübertragung*, vol. 4, pp. 205–212, 1971.

78. Anderson, E. E., and R. Viskanta: "Spectral and Boundary Effects on Coupled Conduction-Radiation Heat Transfer through Semitransparent Solids," *Wärme- und Stoffübertragung*, vol. 1, pp. 14–24, 1973.

79. Viskanta, R., and D. M. Kim: "Heat Transfer Through Irradiated Semi-Transparent Layer at High Temperature," *ASME Journal of Heat Transfer*, vol. 102, pp. 388–390, 1980.

80. Vasilev, M. G., and V. S. Yuferev: "Radiative-Conductive Heat Transfer in a Thin Semitransparent Plate in the Guide-Wave Approximation for a Temperature and Frequency-Dependent Absorption Coefficient," *J. Appl. Mech. Techn. Phys.*, vol. 22, pp. 80–85, 1981.

81. Ramerth, D., and R. Viskanta: "Heat Transfer Through Coal-Ash Deposits," *ASME Paper No. 82-HT-11*, 1982.

82. Smith, T. F., A. M. Al-Turki, K. H. Byun, and T. K. Kim: "Radiative and Conductive Transfer for a Gas/Soot Mixture Between Diffuse Parallel Plates," *Journal of Thermophysics and Heat Transfer*, vol. 1, no. 1, pp. 50–55, 1987.

83. Wendlandt, B. C. H.: "Temperature in an Irradiated Thermally Conducting Medium," *Journal of Physics D: Applied Physics*, vol. 6, pp. 657–660, 1973.

84. Gilpin, R. R., R. B. Roberton, and B. Singh: "Radiative Heating in Ice," *ASME Journal of Heat Transfer*, vol. 99, pp. 227–232, 1977.

85. Viskanta, R., and E. D. Hirleman: "Combined Conduction-Radiation Heat Transfer Through an Irradiated Semitransparent Plate," *ASME Journal of Heat Transfer*, vol. 100, pp. 169–172, 1978.

86. Zakhidov, R. A., S. Y. Bogomolov, D. A. Kirgizbaev, and S. I. Klychev: "Determination of the Temperature Field of Semitransparent Materials when they are Heated with Optical Radiant," *Applied Solar Energy*, vol. 23, no. 6, p. 36, 1987.
87. Kowalski, G. J.: "Transient Response of An Optically Thick Medium Exposed to Short Pulses of Laser Radiation," in *Fundamentals and Applications in Radiation Heat Transfer*, vol. HTD–72, ASME, pp. 67–74, 1987.
88. Heping, T., B. Maestre, and M. Lallemand: "Transient and Steady-State Combined Heat Transfer in Semi-Transparent Materials Subjected to a Pulse or a Step Irradiation," *ASME Journal of Heat Transfer*, vol. 113, no. 1, pp. 166–173, 1991.
89. Hazzak, A. S., and J. V. Beck: "Unsteady Combined Conduction-Radiation Energy Transfer Using a Rigorous Differential Method," *International Journal of Heat and Mass Transfer*, vol. 13, pp. 517–522, 1970.
90. Lii, C. C., and M. N. Özişik: "Transient Radiation and Conduction in an Absorbing, Emitting, Scattering Slab with Reflective Boundaries," *International Journal of Heat and Mass Transfer*, vol. 15, pp. 1175–1179, 1972.
91. Weston, K. C., and J. L. Hauth: "Unsteady, Combined Radiation and Conduction in an Absorbing, Scattering, and Emitting Medium," *ASME Journal of Heat Transfer*, vol. 95, pp. 357–364, 1973.
92. Matthews, L. K., R. Viskanta, and F. P. Incropera: "Combined Conduction and Radiation Heat Transfer in Porous Materials Heated by Intense Solar Radiation," *Solar Energy*, vol. 107, no. 1, p. 29, 1985.
93. Tong, T. W., D. L. McElroy, and D. W. Yarbrough: "Transient Conduction and Radiation Heat Transfer in Porous Thermal Insulations," *Journal of Thermal Insulation*, vol. 9, pp. 13–29, July 1985.
94. Barker, C., and W. H. Sutton: "The Transient Radiation and Conduction Heat Transfer in a Gray Participating Medium with Semi-Transparent Boundaries," in *Radiation Heat Transfer*, vol. HTD–49, ASME, pp. 25–36, 1985.
95. Sutton, W. H.: "A Short Time Solution for Coupled Conduction and Radiation in a Participating Slab Geometry," *ASME Journal of Heat Transfer*, vol. 108, no. 2, pp. 465–466, 1986.
96. Burka, A. L.: "Nonsteady Combined Heat Transfer, Taking Account of Scattering Anisotropy," *High Temperature*, vol. 25, no. 1, p. 99, 1987.
97. Crosbie, A. L., and M. Pattabongse: "Radiative Ignition in a Planar Medium," *Journal of Quantitative Spectroscopy and Radiative Transfer*, vol. 37, no. 2, p. 193, 1987.
98. Glass, D. E., M. N. Özişik, and D. S. McRae: "Hyperbolic Heat Conduction with Radiation in an Absorbing and Emitting Medium," *Numerical Heat Transfer*, vol. 12, no. 3, p. 321, 1987.
99. Rish III, J. W., and J. A. Roux: "Heat Transfer Analysis of Fiberglass Insulations with and without Foil Radiant Barriers," *Journal of Thermophysics and Heat Transfer*, vol. 1, no. 1, pp. 43–49, 1987.
100. Rubtsov, N. A., and E. P. Golova: "Taking into Account Scattering in the Study of Nonstationary Radiative Conductive Heat Transfer in Multilayered Media," *High Temperature*, vol. 25, no. 4, p. 559, 1987.
101. Webb, B. W., and R. Viskanta: "Crystallographic Effects During Radiative Melting of Semi-transparent Materials," *Journal of Thermophysics and Heat Transfer*, vol. 1, no. 4, pp. 313–320, 1987.
102. Tsai, J.-H., and J.-D. Lin: "Transient Combined Conduction and Radiation with Anisotropic Scattering," *Journal of Thermophysics and Heat Transfer*, vol. 4, no. 1, pp. 92–97, 1990.
103. Chawla, T. C., and S. H. Chan: "Solution of Radiation-Conduction Problems with Collocation Method Using B-Splines as Approximating Functions," *International Journal of Heat and Mass Transfer*, vol. 22, no. 12, pp. 1657–1667, 1979.
104. Campo, A., and A. Tremante: "Two-Flux Model Applied to Combined Conduction-Radiation in a Gray Planar Medium," *Wärme- und Stoffübertragung*, vol. 21, no. 4, p. 221, 1987.
105. Kholodov, N. M., Z. H. Flom, and P. S. Koltun: "Calculation of Radiative Conductive Transfer in a Semitransparent Plate by the Monte Carlo Method," *Journal of Engineering Physics*, vol. 42, pp. 333–338, 1982.

106. Abed, A. A., and J. F. Sacadura: "A Monte Carlo-Finite Difference Method for Coupled Radiation-Conduction Heat Transfer in Semi-Transparent Media," *ASME Journal of Heat Transfer*, vol. 105, no. 4, p. 931, 1983.

107. Nishimura, M., M. Hasatani, and S. Sugiyama: "Simultaneous Heat Transfer by Radiation and Conduction: High-Temperature One-Dimensional Heat Transfer in Molten Glass," *Intern. Chem. Eng.*, vol. 8, pp. 739–745, 1968.

108. Eryou, N. D., and L. R. Glicksman: "An Experimental and Analytical Study of Radiative and Conductive Heat Transfer in Molten Glass," *ASME Journal of Heat Transfer*, vol. 94, no. 2, pp. 224–230, 1972.

109. Scheuerpflug, P., R. Caps, D. Buettner, and J. Fricke: "Apparent Thermal Conductivity of Evacuated, SiO_2 Aerogel Tiles Under Variation of Radiative Boundary Conditions," *International Journal of Heat and Mass Transfer*, vol. 28, no. 12, pp. 2299–2306, 1985.

110. Kamiuto, K., Y. Miyoshi, I. Kinoshita, and S. Hasegawa: "Combined Conductive and Radiative Heat Transfer in an Optically Thick Porous Body (Case of Coorierite Porous Bodies)," *Bulletin of Japan Society of Mechanical Engineers*, vol. 27, p. 1136, 1984.

111. Schimmel, W. P., J. L. Novotny, and F. A. Olsofka: "Interferometric Study of Radiation-Conduction Interaction," in *Proceedings of the Fourth International Heat Transfer Conference*, New York, Elsevier, September 1970.

112. Viskanta, R., and R. L. Merriam: "Heat Transfer by Combined Conduction and Radiation Between Concentric Spheres Separated by Radiating Medium," *ASME Journal of Heat Transfer*, vol. 90, pp. 248–256, 1968.

113. Tsai, J. R., and M. N. Özişik: "Transient Combined Conduction and Radiation in an Absorbing, Emitting, and Isotropically Scattering Solid Sphere," *Journal of Quantitative Spectroscopy and Radiative Transfer*, vol. 38, no. 4, pp. 243–251, 1987.

114. Jones, P. D., and Y. Bayazitoğlu: "Combined Radiation and Conduction from a Sphere in a Participating Medium," in *Proceedings of the Ninth International Heat Transfer Conference*, Washington, D.C., Hemisphere, pp. 397–402, 1990.

115. Thynell, S. T.: "Interaction of Conduction and Radiation in Anisotropically Scattering, Spherical Media," *Journal of Thermophysics and Heat Transfer*, vol. 4, no. 3, pp. 299–304, 1990.

116. Gordoninejad, F., and J. Francis: "A Finite Difference Solution to Transient Combined Conductive and Radiative Heat Transfer in an Annular Medium," *ASME Journal of Heat Transfer*, vol. 106, no. 4, pp. 888–891, 1984.

117. Kim, T. K., and T. F. Smith: "Radiative and Conductive Transfer for a Real Gas Cylinder Enclosure with Gray Walls," *International Journal of Heat and Mass Transfer*, vol. 28, no. 12, pp. 2269–2277, 1985.

118. Youssef, A. M., and J. A. Harris: "The P-1 Solution to the Conduction/Radiation Problem in a Cylindrical Medium with a Constant Heat Source," in *Proceedings of the 1988 National Heat Transfer Conference*, vol. HTD–96, ASME, pp. 187–192, 1988.

119. Tsai, J. R., and M. N. Özişik: "Transient Combined Conduction and Radiation in an Absorbing, Emitting, and Isotropically Scattering Solid Cylinder," *Journal of Applied Physics*, vol. 64, no. 8, p. 3820, 1988.

120. Harris, J. A.: "Solution of the Conduction/Radiation Problem with Linear-Anisotropic Scattering in an Annular Medium by the Spherical Harmonics Method," *ASME Journal of Heat Transfer*, vol. 111, no. 1, pp. 194–196, 1989.

121. Pandey, D. K.: "Combined Conduction and Radiation Heat Transfer in Concentric Cylindrical Media," *Journal of Thermophysics and Heat Transfer*, vol. 3, no. 1, pp. 75–82, 1989.

122. Amlin, D. W., and S. A. Korpela: "Influence of Thermal Radiation on the Temperature Distribution in Semi-Transparent Solid," *ASME Journal of Heat Transfer*, vol. 102, pp. 76–80, 1980.

123. Ratzell III, A. C., and J. R. Howell: "Two-Dimensional Energy Transfer in Radiatively Participating Media with Conduction by the P-N Approximation," in *Proceedings of the Seventh International Heat Transfer Conference, Munich*, vol. 2, Washington, D.C., Hemisphere, pp. 535–540, 1982.

124. Shih, T.-M., and Y. N. Chen: "A Discretized-Intensity Method Proposed for Two-Dimensional Systems Enclosing Radiative and Conductive Media," *Numerical Heat Transfer*, vol. 6, pp. 117–134, 1983.

125. Razzaque, M. M., J. R. Howell, and D. E. Klein: "Coupled Radiative and Conductive Heat Transfer in a Two-Dimensional Enclosure with Gray Participating Media Using Finite Elements," in *Proceedings of the First JSME/ASME Joint Thermal Conference, Honolulu*, vol. 4, pp. 41–48, 1983.

126. Yücel, A., and M. L. Williams: "Heat Transfer by Combined Conduction and Radiation in Axisymmetric Enclosures," *Journal of Thermophysics and Heat Transfer*, vol. 1, no. 4, pp. 301–306, 1987.

127. Yücel, A., and M. L. Williams: "Interaction of Conduction and Radiation in Cylindrical Geometry without Azimuthal Symmetry," in *Proceedings of the 1988 National Heat Transfer Conference*, vol. HTD–96, ASME, pp. 281–288, 1988.

128. Ho, C. H., and M. N. Özişik: "Combined Conduction and Radiation in a Two-Dimensional Rectangular Enclosure," *Numerical Heat Transfer*, vol. 13, no. 2, p. 229, 1988.

129. Yuen, W. W., and E. E. Takara: "Analysis of Combined Conductive-Radiative Heat Transfer in a Two-Dimensional Rectangular Enclosure with a Gray Medium," *ASME Journal of Heat Transfer*, vol. 110, no. 2, pp. 468–474, 1988.

130. Baek, S. W., and T. Y. Kim: "The Conductive and Radiative Heat Transfer in Rectangular Enclosure Using the Discrete Ordinates Method," in *Proceedings of the Ninth International Heat Transfer Conference*, Washington, D.C., Hemisphere, pp. 433–438, 1990.

131. Habib, I. S.: "Solidification of Semi-Transparent Materials by Conduction and Radiation," *International Journal of Heat and Mass Transfer*, vol. 14, pp. 2161–2164, 1971.

132. Habib, I. S.: "Solidification of a Semi-Transparent Cylindrical Medium by Conduction and Radiation," *ASME Journal of Heat Transfer*, vol. 95, pp. 37–41, 1973.

133. Abrams, M., and R. Viskanta: "The Effects of Radiative Heat Transfer upon the Melting and Solidification of Semi-Transparent Crystals," *ASME Journal of Heat Transfer*, vol. 96, pp. 184–190, 1974.

134. Viskanta, R., and E. E. Anderson: "Heat Transfer in Semi-Transparent Solids," in *Advances in Heat Transfer*, vol. 11, Academic Press, New York, pp. 317–441, 1975.

135. Cho, C., and M. N. Özişik: "Effects of Radiation on Melting of Semi-Infinite Medium," in *Proceedings of the Sixth International Heat Transfer Conference*, vol. 3, Washington, D.C., Hemisphere, pp. 373–378, 1978.

136. Seki, N., M. Sugawara, and S. Fukusako: "Radiative Melting of Horizontal Clear Ice Layer," *Wärme- und Stoffübertragung*, vol. 11, pp. 207–216, 1978.

137. Seki, N., M. Sugawara, and S. Fukusako: "Radiative Melting of Ice Layer Adhering to a Vertical Surface," *Wärme- und Stoffübertragung*, vol. 12, pp. 137–144, 1979.

138. Seki, N., M. Sugawara, and S. Fukusako: "Back Melting of a Horizontal Cloudy Ice Layer with Radiative Heating," *ASME Journal of Heat Transfer*, vol. 101, pp. 90–95, 1979.

139. Diaz, L. A., and R. Viskanta: "Melting of a Semitransparent Material by Irradiation from an External Radiation Source," in *Spacecraft Radiative Heat Transfer and Temperature Control*, vol. 83, New York, AIAA, pp. 38–60, 1982.

140. Diaz, L. A., and R. Viskanta: "Radiation Induced Melting of a Semitransparent Phase Change Material," *AIAA Paper No. 82-0848*, 1982.

141. Viskanta, R., and X. Wu: "Effect of Radiation on the Melting of Glass Batch," *Glastechnische Berichte*, vol. 56, pp. 138–147, 1983.

142. Dorsey, N. E.: *Properties of Ordinary Water Substance*, Hafner, New York, p. 404, 1963.

143. Knight, C. A.: *The Freezing of Supercooled Liquids*, Van Nostrand, Princeton, p. 125, 1967.

144. Chan, S. H., D. H. Cho, and G. Kocamustafaogullari: "Melting and Solidification with Internal Radiative Transfer – A Generalized Phase Change Model," *International Journal of Heat and Mass Transfer*, vol. 26, no. 4, pp. 621–633, 1983.

145. Carslaw, H. S., and J. C. Jaeger: *Conduction of Heat in Solids*, 2nd ed., Oxford University Press, 1959.

146. Oruma, F. O., M. N. Özişik, and M. A. Boles: "Effects of Anisotropic Scattering on Melting and Solidification of a Semi-Infinite, Semi-Transparent Medium," *International Journal of Heat and Mass Transfer*, vol. 28, no. 2, pp. 441–449, 1985.

147. Burka, A. L., N. A. Rubtsov, and N. A. Savvinova: "Nonsteady-State Radiant-Conductive Heat Exchange in a Semitransparent Medium with Phase Transition," *Journal of Applied Mechanics and Technical Physics*, vol. 28, no. 1, p. 91, 1987.

148. Chan, S. H., and K. Y. Hsu: "The Mushy Zone in a Phase Change Model of a Semitransparent Material with Internal Radiative Transfer," *ASME Journal of Heat Transfer*, vol. 110, pp. 260–264, February 1988.

149. Kays, W. M., and M. E. Crawford: *Convective Heat and Mass Transfer*, McGraw-Hill, 1980.

150. Pai, S. I.: "Inviscid Flow of Radiation Gas Dynamics," *J. Math Phys. Sci.*, vol. 39, pp. 361–370, 1969.

151. Özişik, M. N.: *Radiative Transfer and Interactions With Conduction and Convection*, John Wiley & Sons, New York, 1973.

152. Schlichting, H.: *Boundary Layer Theory*, 7th ed., McGraw-Hill, New York, 1979.

153. Viskanta, R., and R. J. Grosh: "Boundary Layer in Thermal Radiation Absorbing and Emitting Media," *International Journal of Heat and Mass Transfer*, vol. 5, pp. 795–806, 1962.

154. Rumynskii, A. N.: "Boundary Layers in Radiating and Absorbing Media," *American Rocket Society Journal*, vol. 32, pp. 1135–1138, 1962.

155. Goulard, R.: "The Transition from Black Body to Rosseland Formulations in Optically Thick Flows," *International Journal of Heat and Mass Transfer*, vol. 7, pp. 1145–1146, 1964.

156. Novotny, J. L., and K. T. Yang: "The Interaction of Thermal Radiation in Optically Thick Boundary Layers," *ASME Paper No. 67-HT-9*, 1967.

157. Pai, S. I., and A. P. Scaglione: "Unsteady Laminar Boundary Layers of an Infinite Plate in an Optically Thick Radiating Gas," *Applied Scientific Research*, vol. 22, pp. 97–112, 1970.

158. Zamuraev, V. P.: *Zh. Prikl. Mekhan. i Tekhn. Fiz.*, vol. 3, p. 73, 1964.

159. Viskanta, R.: "Radiation Transfer and Interaction of Convection with Radiation Heat Transfer," in *Advances in Heat Transfer*, vol. 3, Academic Press, New York, pp. 175–251, 1966.

160. Cess, R. D.: "Radiation Effects upon Boundary Layer Flow of an Absorbing Gas," *ASME Journal of Heat Transfer*, vol. 86C, pp. 469–475, 1964.

161. Smith, A. M., and H. A. Hassan: "Nongray Radiation Effects on the Boundary Layer at Low Eckert Numbers," *ASME Paper No. 66-WA/HT-35*, 1966.

162. Tabaczynski, R. J., and L. A. Kennedy: "Thermal Radiation Effects in Laminar Boundary-Layer Flow," *AIAA Journal*, vol. 5, pp. 1893–1894, 1967.

163. Pai, S. I., and C. K. Tsao: "A Uniform Flow of a Radiating Gas over a Flat Plate," *Proceedings of the Third International Heat Transfer Conference*, vol. 5, pp. 129–137, 1966.

164. Oliver, C. C., and P. W. McFadden: "The Interaction of Radiation and Convection in the Laminar Boundary Layer," *ASME Journal of Heat Transfer*, vol. C88, pp. 205–213, 1966.

165. Rumynskii, A. N.: "Boundary Layer with an Opaque Underlayer," *American Rocket Society Journal*, vol. 32, pp. 1139–1140, 1962.

166. Koh, J. C. Y., and C. N. DeSilva: "Interaction Between Radiation and Convection in the Hypersonic Boundary Layer on a Flat Plate," *American Rocket Society Journal*, vol. 32, pp. 739–743, 1962.

167. Taitel, Y., and J. P. Hartnett: "Equilibrium Temperature in a Boundary Layer Flow over a Flat Plate of Absorbing-Emitting Gas," *ASME Paper No. 66-WA/HT-48*, 1966.

168. Robin, M. N., R. I. Souloukhin, and I. B. Yutevich: "Influence of Reflection of Radiation on Radiative-Convective Heat Exchange During Hypersonic Flow over a Blunt Body," *J. Appl. Mech. Tech. Phys.*, vol. 21, pp. 239–245, 1980.

169. Golubkin, V. N.: "On the Asymptotic Theory of the Three-Dimensional Flow of a Hypersonic Stream of Radiating Gas Around a Body," *Appl. Math. Mech.*, vol. 47, p. 493, 1983.

170. Tiwari, S. N., K. Y. Szema, J. N. Moss, and S. V. Subramanian: "Convective and Radiative Heating of a Saturn Entry Probe," *International Journal of Heat and Mass Transfer*, vol. 27, pp. 191–206, 1984.

171. Dombrowski, L. A.: "Radiation-Convection Heat Transfer by an Optically Thick Boundary Layer on a Plate," *High Temperature*, vol. 19, pp. 100–109, 1981.

172. Yücel, A., and Y. Bayazitoğlu: "Radiative Heat Transfer in Absorbing, Emitting and Anisotropically Scattering Boundary Layer," *AIAA Paper No. 83-1504*, 1983.

173. Yücel, A., R. H. Kehtarenavaz, and Y. Bayazitoğlu: "Interaction of Radiation with the Boundary Layer: Nongray Media," *ASME Paper No. 83-HT-33*, 1983.

174. Soufiani, A., and J. Taine: "Application of Statistical Narrow-Band Model to Coupled Radiation and Convection at High Temperature," *International Journal of Heat and Mass Transfer*, vol. 30, no. 3, pp. 437–448, 1987.

175. Fritsch, C. A., R. J. Grosh, and M. W. Wild: "Radiative Heat Transfer Through an Absorbing Boundary Layer," *ASME Journal of Heat Transfer*, vol. 86, no. 4, pp. 296–304, 1966.

176. Houf, W. G., F. P. Incropera, and R. Viskanta: "Thermal Conditions in Irradiated, Slowly Moving Liquid Layers," *ASME Journal of Heat Transfer*, vol. 107, no. 1, pp. 92–98, 1985.

177. Elliott, J. M., R. I. Vachon, D. F. Dyer, and J. R. Dunn: "Application of the Patankar-Spalding Finite Difference Procedure to Turbulent Radiating Boundary Layer Flow," *International Journal of Heat and Mass Transfer*, vol. 14, pp. 667–672, 1971.

178. Goswami, D. Y., and R. I. Vachon: "Turbulent Boundary Layer Flow of Absorbing, Emitting and Axisymmetrically Scattering Gaseous Medium," *AIAA Paper No. 80-1518*, 1980.

179. Naidenov, V. I., and S. A. Shindini: "Interaction of Radiation with Turbulent Fluctuations in a Boundary Layer," *High Temperature*, vol. 19, pp. 106–109, 1981.

180. Cess, R. D.: "The Interaction of Thermal Radiation with Free Convection Heat Transfer," *International Journal of Heat and Mass Transfer*, vol. 9, pp. 1269–1277, 1966.

181. Arpaci, V. S.: "Effect of Thermal Radiation on the Laminar Free Convection from a Heated Vertical Plate," *International Journal of Heat and Mass Transfer*, vol. 11, pp. 871–881, 1968.

182. Cheng, E. H., and M. N. Özişik: "Radiation with Free Convection in an Absorbing, Emitting and Scattering Medium," *International Journal of Heat and Mass Transfer*, vol. 15, pp. 1243–1252, 1972.

183. Desrayaud, G., and G. Lauriat: "Natural Convection of a Radiating Fluid in a Vertical Layer," *ASME Journal of Heat Transfer*, vol. 107, no. 3, pp. 710–712, 1985.

184. Webb, B. W., and R. Viskanta: "Analysis of Radiation-Induced Natural Convection in Rectangular Enclosures," *Journal of Thermophysics and Heat Transfer*, vol. 1, no. 2, pp. 146–153, 1987.

185. Webb, B. W.: "Interaction of Radiation and Free Convection on a Heated Vertical Plate: Experiment and Analysis," *Journal of Thermophysics and Heat Transfer*, vol. 4, no. 1, pp. 117–120, 1990.

186. Arpaci, V. S., Kabiri-Bamoradian, and E. Cesmebasi: "Finite Amplitude Benard Convection of Purely Radiating Fluids," *ASME Paper No. 78-HT-40*, 1978.

187. Epstein, M., F. B. Cheung, T. C. Chawla, and G. M. Hauser: "Effective Thermal Conductivity for Combined Radiation and Free Convection in an Optically Thick Heated Fluid Layer," *ASME Journal of Heat Transfer*, vol. 103, pp. 114–120, 1981.

188. Bakan, S.: "Thermal Stability of Radiating Fluids: The Scattering Problem," *Physics of Fluids*, vol. 27, no. 12, p. 2969, 1984.

189. Yang, W. M.: "Thermal Instability of a Fluid Layer Induced by Radiation," *Numerical Heat Transfer – Part A: Applications*, vol. 17, pp. 365–376, 1990.

190. Chang, L. C., K. T. Yang, and J. R. Lloyd: "Radiation–Natural Convection Interactions in Two-Dimensional Enclosures," *ASME Journal of Heat Transfer*, vol. 105, no. 1, pp. 89–95, 1983.

191. Webb, B. W., and R. Viskanta: "Analysis of Radiation-Induced Melting with Natural Convection in the Melt," in *Fundamentals and Applications of Radiation Heat Transfer*, vol. HTD–72, ASME, pp. 75–82, 1987.

192. Yücel, A., S. Acharya, and M. L. Williams: "Combined Natural Convection and Radiation in a Square Enclosure," in *Proceedings of the 1988 National Heat Transfer Conference*, vol. HTD–96, ASME, pp. 209–218, 1988.

193. Fusegi, T., and B. Farouk: "A Computational and Experimental Study of Natural Convection and Surface/Gas Radiation Interactions in a Square Cavity," *ASME Journal of Heat Transfer*, vol. 112, pp. 802–804, 1990.

194. Tan, Z., and J. R. Howell: "Combined Radiation and Natural Convection in a Two-Dimensional Participating Square Medium," *International Journal of Heat and Mass Transfer*, vol. 34, no. 3, pp. 785–794, 1991.

195. Carpenter, J. R., D. G. Briggs, and V. Sernas: "Combined Radiation and Developing Laminar Free Convection Between Vertical Flat Plates with Asymmetric Heating," *ASME Paper No. 75-HT-19*, 1975.

196. Yamada, Y.: "Combined Radiation and Free Convection Heat Transfer in a Vertical Channel with Arbitrary Wall Emissivities," *International Journal of Heat and Mass Transfer*, vol. 31, no. 2, pp. 429–440, 1988.

197. Campo, A., and U. Lacoa: "Influence of Thermal Radiation on Natural Convection Inside Vertical Annular Enclosures," in *Proceedings of the 1988 National Heat Transfer Conference*, vol. HTD–96, ASME, pp. 219–226, 1988.

198. Fusegi, T., K. Ishii, B. Farouk, and K. Kuwahara: "Three-Dimensional Study of Convection-Radiation Interactions in a Cubical Enclosure Field with a Non-Gray Gas," in *Proceedings of the Ninth International Heat Transfer Conference*, Washington, D.C., Hemisphere, pp. 421–426, 1990.

199. DeSoto, S.: "Coupled Radiation, Conduction and Convection in Entrance Region Flow," *International Journal of Heat and Mass Transfer*, vol. 11, pp. 39–53, 1968.

200. Kurosaki, Y.: "Heat Transfer by Simultaneous Radiation and Convection in an Absorbing and Emitting Medium in a Flow Between Parallel Plates," in *Proceedings of the Fourth International Heat Transfer Conference*, vol. 3, No. R2.5, New York, Elsevier, 1970.

201. Echigo, R., K. Kamiuto, and S. Hasegawa: "Analytical Method on Composite Heat Transfer with Predominant Radiation–Analysis by Integral Equation and Examination on Radiation Slip," in *Proceedings of the Fifth International Heat Transfer Conference*, vol. 1, Japan, JSME, pp. 103–107, 1974.

202. Azad, F. H., and M. F. Modest: "Combined Radiation and Convection in Absorbing, Emitting and Anistropically Scattering Gas-Particulate Tube Flow," *International Journal of Heat and Mass Transfer*, vol. 24, pp. 1681–1698, 1981.

203. Viskanta, R.: "Interaction of Heat Transfer by Conduction, Convection, and Radiation in a Radiating Fluid," *ASME Journal of Heat Transfer*, vol. 85, pp. 318–328, 1963.

204. Viskanta, R.: "Heat Transfer in a Radiating Fluid with Slug Flow in a Parallel Plate Channel," *Applied Scientific Research Part A*, vol. 13, pp. 291–311, 1964.

205. Chen, J. C.: "Simultaneous Radiative and Convective Heat Transfer in an Absorbing, Emitting, and Scattering Medium in Slug Flow Between Parallel Plates," *AIChE Journal*, vol. 10, no. 2, pp. 253–259, 1964.

206. DeSoto, S., and D. K. Edwards: *Radiative Emission and Absorption in Non-Isothermal Nongray Gases in Tubes*, Stanford University Press, Stanford, CA, pp. 358–372, 1965.

207. Edwards, D. K., and A. Balakrishnan: "Nongray Radiative Transfer in a Turbulent Gas Layer," *International Journal of Heat and Mass Transfer*, vol. 16, pp. 1003–1015, 1973.

208. Edwards, D. K., and A. Balakrishnan: "Self Absorption of Radiation in Turbulent Molecular Gases," *Combustion and Flame*, vol. 20, pp. 401–417, 1973.

209. Balakrishnan, A., and D. K. Edwards: "Established Laminar and Turbulent Channel Flow of a Radiating Molecular Gas," in *Proceedings of the Fifth International Heat Transfer Conference*, vol. 1, Japan, JSME, pp. 93–97, 1974.

210. Wassel, A. T., and D. K. Edwards: "Molecular Radiation in a Laminar or Turbulent Pipe Flow," *ASME Journal of Heat Transfer*, vol. 98, pp. 101–107, 1976.

211. Goulard, R., and M. Goulard: *Energy Transfer in the Couette Flow of a Radiant and Chemically Reacting Gas*, Stanford University Press, Stanford, CA, pp. 126–139, 1959.

212. Viskanta, R., and R. J. Grosh: "Temperature Distribution in Couette Flow with Radiation," *American Rocket Society Journal*, vol. 31, pp. 839–840, 1961.

213. Timofeyev, V. N., F. R. Shklyar, V. M. Malkin, and K. H. Berland: "Combined Heat Transfer in an Absorbing Stream Moving in a Flat Channel. Parts I, II, and III," *Heat Transfer, Soviet Research*, vol. 1, no. 6, pp. 57–93, November 1969.

214. Lii, C. C., and M. N. Özişik: "Heat Transfer in an Absorbing, Emitting and Scattering Slug Flow Between Parallel Plates," *ASME Journal of Heat Transfer*, vol. 95C, pp. 538–540, 1973.

215. Bergquam, J. B., and N. S. Wang: "Heat Transfer by Convection and Radiation in an Absorbing, Scattering Medium Flowing Between Parallel Plates," *ASME Paper No. 76-HT-50*, 1976.

216. Chawla, T. C., and S. H. Chan: "Spline Collocation Solution of Combined Radiation-Convection in Thermally Developing Flows with Scattering," *Numerical Heat Transfer*, vol. 3, pp. 47–76, 1980.

217. Chawla, T. C., and S. H. Chan: "Combined Radiation and Convection in Thermally Developing Poiseuille Flow with Scattering," *ASME Journal of Heat Transfer*, vol. 102, pp. 297–302, 1980.

218. Mengüç, M. P., Y. Yener, and M. N. Özişik: "Interaction of Radiation in Thermally Developing Laminar Flow in a Parallel Plate Channel," *ASME Paper No. 83-HT-35*, 1983.

219. Yener, Y., B. Shahidi-Zandi, and M. N. Özişik: "Simultaneous Radiation and Forced Convection in Thermally Developing Turbulent Flow Through a Parallel Plate Channel," *ASME Paper No. 84-WA/HT-15*, 1984.

220. Yener, Y., and M. N. Özişik: "Simultaneous Radiation and Forced Convection in Thermally Developing Turbulent Flow Through a Parallel Plate Channel," *ASME Journal of Heat Transfer*, vol. 108, no. 4, pp. 985–987, 1986.

221. Echigo, R., and S. Hasegawa: "Radiative Heat Transfer by Flowing Multiphase Medium—Part I. An Analysis on Heat Transfer of Laminar Flow Between Parallel Flat Plates," *International Journal of Heat and Mass Transfer*, vol. 15, pp. 2519–2534, 1972.

222. Kassemi, M., and B. T. F. Chung: "Two-Dimensional Convection and Radiation with Scattering from a Poiseuille Flow," *Journal of Thermophysics and Heat Transfer*, vol. 4, no. 1, pp. 98–105, 1990.

223. Kim, T. K., and H. O. Lee: "Two-Dimensional Anisotropic Scattering Radiation in a Thermally Developing Poiseuille Flow," *Journal of Thermophysics and Heat Transfer*, vol. 4, no. 3, pp. 292–298, 1990.

224. Einstein, T. H.: "Radiant Heat Transfer to Absorbing Gases Enclosed in a Circular Pipe with Conduction, Gas Flow, and Internal Heat Generation," *NASA TR R-156*, 1963.

225. Echigo, R., S. Hasegawa, and K. Kamiuto: "Composite Heat Transfer in a Pipe with Thermal Radiation of Two-Dimensional Propagation," *International Journal of Heat and Mass Transfer*, vol. 18, pp. 1149–1159, 1975.

226. Landram, C. S., R. Greif, and I. S. Habib: "Heat Transfer in Turbulent Pipe Flow with Optically Thin Radiation," *ASME Journal of Heat Transfer*, vol. 91, pp. 330–336, 1969.

227. Pearce, B. E., and A. F. Emery: "Heat Transfer by Thermal Radiation and Laminar Forced Convection to an Absorbing Fluid in the Entry Region of a Pipe," *ASME Journal of Heat Transfer*, vol. 92, pp. 221–230, 1970.

228. Habib, I. S., and R. Greif: "Heat Transfer to a Flowing Non-Gray Radiating Gas: An Experimental and Theoretical Study," *International Journal of Heat and Mass Transfer*, vol. 13, pp. 1571–1582, 1970.

229. Gau, C., and D. C. Chi: "A Simple Numerical Study of Combined Radiation and Convection Heat Transfer in the Entry Region of a Circular Pipe Flow," in *Proceedings of the Second ASME/JSME Conference*, vol. 3, pp. 635–643, 1987.

230. Smith, T. F., Z. F. Shen, and A. M. Al-Turki: "Radiative and Convective Transfer in a Cylindrical Enclosure for a Real Gas," *ASME Journal of Heat Transfer*, vol. 107, no. 2, p. 482, 1985.

231. Echigo, R., S. Hasegawa, and H. Tamehiro: "Radiative Heat Transfer by Flowing Multiphase Medium—Part II. An Analysis on Heat Transfer of Laminar Flow in an Entrance Region of Circular Tube," *International Journal of Heat and Mass Transfer*, vol. 15, pp. 2595–2610, 1972.

232. Tamehiro, H., R. Echigo, and S. Hasegawa: "Radiative Heat Transfer by Flowing Multiphase Medium– Part III. An Analysis on Heat Transfer of Turbulent Flow in a Circular Tube," *International Journal of Heat and Mass Transfer*, vol. 16, pp. 1199–1213, 1973.

233. Modest, M. F., B. R. Meyer, and F. H. Azad: "Combined Convection and Radiation in Tube Flow of an Absorbing, Emitting and Anisotropically Scattering Gas-Particulate Suspension," *ASME Paper No. 80-HT-27*, 1980.

234. Tabanfar, S., and M. F. Modest: "Combined Radiation and Convection in Tube Flow with Non-Gray Gases and Particulates," *ASME Journal of Heat Transfer*, vol. 109, pp. 478–484, 1987.

235. Al-Turki, A. M., and T. F. Smith: "Radiative and Convective Transfer in a Cylindrical Enclosure for a Gas/Soot Mixture," *ASME Journal of Heat Transfer*, vol. 109, no. 1, p. 259, 1987.

236. Gore, J. P., and G. M. Faeth: "Structure and Spectral Radiation Properties of Turbulent Ethylene/Air Diffusion Flames," in *Proceedings of the Twenty-First Symposium on Combustion*, pp. 1521–1531, 1986.

237. Gore, J. P., S. M. Jeng, and G. M. Faeth: "Spectral and Total Radiation Properties of Turbulent Carbon Monoxide/Air Diffusion Flames," *AIAA Journal*, vol. 25, no. 2, pp. 339–345, 1987.

238. Song, B., and R. Viskanta: "Interaction of Radiation with Turbulence: Application to a Combustion System," *Journal of Thermophysics and Heat Transfer*, vol. 1, no. 1, pp. 56–62, 1987.

239. Cox, G.: "On Radiant Heat Transfer from Turbulent Flames," *Combustion Science and Technology*, vol. 17, pp. 75–78, 1977.

240. Kabashnikov, V. P., and G. I. Kmit: "Influence of Turbulent Fluctuations of Thermal Radiation," *Journal of Applied Spectroscopy*, vol. 31, pp. 963–967, 1979.

241. Jeng, S. M., M. C. Lai, and G. M. Faeth: "Nonluminous Radiation in Turbulent Buoyant Axisymmetric Flames," *Combustion Science and Technology*, vol. 40, pp. 41–53, 1984.

242. Faeth, G. M.: "Heat and Mass Transfer in Flames," in *Proceedings of the Eighth International Heat Transfer Conference*, Washington, D.C., Hemisphere, pp. 151–160, 1986.

243. Chan, S. H., and C. F. Chern: "A Method for Turbulence-Radiation Interaction Analysis in Multiphase Liquid Metal Diffusion Flames," in *Proceedings of the Ninth International Heat Transfer Conference*, vol. 4, Washington, D.C., Hemisphere, pp. 143–148, 1990.

244. Soufiani, A., P. Mignon, and J. Taine: "Radiation-Turbulence Interaction in Channel Flows of Infrared Active Gases," in *Proceedings of the Ninth International Heat Transfer Conference*, vol. 6, Washington, D.C., Hemisphere, pp. 403–408, 1990.

245. Kounalakis, M. E., Y. R. Sivathanu, and G. M. Faeth: "Infrared Radiation Statistics of Nonluminous Turbulent Diffusion Flames," *ASME Journal of Heat Transfer*, vol. 113, no. 2, pp. 437–445, 1991.

246. Vader, D. T., R. Viskanta, and F. P. Incropera: "Design and Testing of a High-Temperature Emissometer for Porous and Particulate Dielectrics," *Rev. Sci. Instrum.*, vol. 57, no. 1, pp. 87–93, 1986.

247. Sikka, K. K.: "High Temperature Normal Spectral Emittance of Silicon Carbide Based Materials," M.S. thesis, The Pennsylvania State University, University Park, PA, 1991.

Problems

20.1 A satellite shaped like a sphere ($R = 1$ m) has a gray-diffuse surface coating with $\epsilon_s = 0.3$ and is fitted with a long, thin, cylindrical antenna, as shown in the adjacent sketch. The antenna is a specular reflector with $\epsilon_a = 0.1$, $k_a = 100$ W/m K, and $d = 1$ cm. Satellite and antenna are exposed to solar radiation of strength $q_{sol} = 1300$ W/m^2 from a direction normal to the antenna. Assuming that the satellite produces heat at a rate of 4 kW and—due to a high-conductivity shell—is essentially isothermal, determine the equilibrium temperature distribution along the antenna. (Hint: Use the fact that $d \ll R$ not only for conduction calculations, but also for the calculation of view factors.)

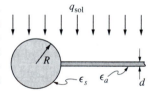

20.2 In the emissometer of Vader and coworkers [246] and Sikka [247], the sample is kept inside a long silicon carbide tube that in turn, is inside a furnace, as shown in the sketch. The furnace is heated with a number of SiC heating elements, providing a uniform flux over a 45 cm length as shown. Assume that there is no heat loss through the refractory brick or the bottom of the furnace, that the inside heat transfer coefficient for free convection (with air at 600°C) is 10 W/m^{2}K, that the silicon carbide tube is gray-diffuse ($\epsilon = 0.9$, $k = 100$ W/m K), and that the sample temperature is equal to the SiC tube temperature at the same height. What must be the steady-state power load on the furnace to maintain a sample temperature of 1000°C? In this configuration a detector receiving radiation from a small center spot of the sample is supposedly getting the same amount as from

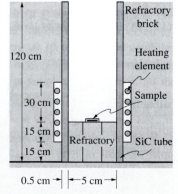

a blackbody at 1000°C (cf. Table 5.1). What is the actual emissivity sensed by the detector, i.e., what systematic error is caused by this near-blackbody, if the sample is gray and diffuse with $\epsilon_s = 0.5$?

20.3 A long, thin, cylindrical needle ($L \gg D$) is attached perpendicularly to a large, isothermal base plate at $T = T_b = $ const. The base plate is gray and diffuse ($\epsilon_b = \alpha_b$), while the needle is nongray and diffuse ($\epsilon \neq \alpha$). The needle exchanges heat by convection and radiation with a large, isothermal environment at T_∞.

    (a) Neglecting heat losses from the free tip of the needle, formulate the problem for the calculation of needle temperature distribution, total heat loss, and fin efficiency.

    (b) Implement the solution numerically for $L = 1$ m, $D = 1$ cm, $k = 10$ W/m K, $h = 40$ W/m²K, $\epsilon = 0.8$, $\alpha = 0.4$, $\epsilon_b = 0.8$, $T_b = 1000$ K, $T_\infty = 300$ K.

20.4 A thermocouple with a 0.5 mm diameter bead is used to measure the local temperature of a hot, radiatively nonparticipating gas flowing through an isothermal, gray-diffuse tube ($T_w = 300$ K, $\epsilon_w = 0.8$). The thermocouple is a diffuse emitter/specular reflector with $\epsilon_b = 0.5$, and the heat transfer coefficient between bead and gas is 30 W/m²K.

    (a) Determine the thermocouple error as function of gas temperature (i.e., $|T_b - T_g|$ vs. T_g).

    (b) In order to reduce the error, a radiation shield in the form of a thin, stainless-steel cylinder ($\epsilon = 0.1$, $R = 2$ mm, $L = 20$ mm) is placed over the thermocouple. This also reduces the heat transfer coefficient between bead and gas to 15 W/m²K, which is equal to the heat transfer coefficient on the inside of the shield. On the outside of the cylinder the heat transfer coefficient is 30 W/m²K. Determine error vs. gas temperature for this case.

To simplify the problem, you may make the following assumptions: (*i*) the leads of the thermocouple may be neglected, (*ii*) the shield is very long as far as the radiation analysis is concerned; (*iii*) the shield reflects diffusely.

20.5 Repeat Problem 5.19 for the case in which a radiatively nonparticipating, stationary gas ($k = 0.04$ W/m K) is filling the 1 cm–thick gap between surface and shield.

20.6 A vat of molten glass is heated from below by a gray, diffuse surface with $T = 1800$ K and $\epsilon = 0.8$. The glass layer is 1 m thick, and its top is exposed to free convection and radiation with an ambient space at 1000 K (heat transfer coefficient for free convection = 5 W/m²K). Neglecting convection within the melt, estimate the temperature distribution within the glass, using radiative properties of glass as given in Figs. 1-18 and 3-16. What is the total heat loss from the bottom surface?

20.7 Estimate the total heat flux for Problem 20.6, as well as the glass–air interface temperature, by using the additive solution method.

20.8 A glass sphere ($D = 4$ cm) initially at uniform temperature $T_i = 300$ K is placed into a furnace, whose walls and inert gas are at a uniform $T_g = T_w = 1500$ K. Assuming the glass to be gray and nonscattering ($\kappa = 1$ cm^{-1}, $n = 1.5$, $k = 1.5$ W/m K) and a sphere/furnace gas heat transfer coefficient of 10 W/m²K, determine the sphere's temperature distribution as function of time.

20.9 A 1 cm–thick quartz window (assumed gray with $\kappa = 1$ cm^{-1} and $n = 1.5$) forms the barrier between a furnace and the ambient, resulting in face temperatures of $T_1 = 800$ K and $T_2 = 400$ K. Estimate the conductive, radiative, and total heat fluxes passing through the window ($k = 1.5$ W/m K).

20.10 Repeat Problem 5.19 for the case in which a gray, isotropically-scattering, stationary gas ($\kappa = 2\,\mathrm{cm}^{-1}$, $k = 0.04\,\mathrm{W/m\,K}$) is filling the 1 cm–thick gap between surface and shield.

20.11 A sheet of ice 20 cm thick is lying on top of black soil. Initially, ice and soil are at $-10°\mathrm{C}$ when the sun begins to shine, hitting the top of the ice with a strength of $800\,\mathrm{W/m^2}$ (normal to the rays), at an off-normal angle of $30°$. Assume the ground to be insulated, ice and water to have constant and equal properties (k, ρ, c_p), a gray absorption coefficient (for solar light) of $\kappa \simeq 1\,\mathrm{cm}^{-1}$, and a gray reflectivity of 0.02. Neglecting emission from and scattering by the ice, as well as convection losses/gains at the surface, determine the transient temperature distribution within the ice/water until the time when all ice has melted.

20.12 Consider a gray medium separating an axle from its bearing. The gap is so narrow that the movement between axle and bearing may be approximated by Couette flow (two infinite parallel plates, one stationary and the other moving at constant velocity U). The movement is so rapid that viscous dissipation must be considered [$\Phi = (\partial u/\partial y)^2$, where $u = u(y)$ is the velocity at a distance y from the lower, stationary plate]. The medium is gray and nonscattering with a constant absorption coefficient, and both surfaces are isothermal (at different temperatures) and gray-diffuse. Set up the necessary equations and boundary conditions to calculate the net heat transfer rates on the two surfaces.

20.13 Consider a solar water heater as shown in the adjacent sketch. A 5 mm–thick layer of water is flowing down a black, insulated plate as shown while exposed to sunshine. The water is seeded with a fine powder that gives it a gray absorption coefficient of $\kappa = 5\,\mathrm{cm}^{-1}$. The top of the water layer loses heat by free convection ($h = 10\,\mathrm{W/m^2K}$) to the ambient at $T_{amb} = 300\,\mathrm{K}$. At the top of the collector ($x = 0$) the water enters at a uniform temperature of $T_0 = 300\,\mathrm{K}$. The velocity profile may be considered fully developed everywhere. Determine the cumulative collected solar energy as a function of x.

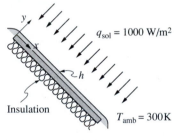

20.14 Consider a gas-particulate mixture flowing through an isothermal tube ($\epsilon_w = 1$, $T_w = 400\,\mathrm{K}$). The gas is radiatively nonparticipating and has constant velocity u across the tube cross section such that $\mathrm{Pe} = \mathrm{Re}\,\mathrm{Pr} = uD/\alpha = 30{,}000$. The particles are very small, gray, and uniformly distributed such that $\kappa_p R = 5$ (no scattering) and $(\dot{m}c_p)_{particles}/(\dot{m}c_p)_{gas} = 2$. The particles are so small that they are essentially at the same temperature as the gas surrounding them. Using the diffusion approximation for the radiative heat transfer, set up the relevant equations and boundary conditions for the calculation of local bulk temperature and local total heat flux. Obtain a numerical solution (after neglecting axial conduction and radiation), and compare with Figs. 20-17 and 20-18.

APPENDIX

A

CONSTANTS AND CONVERSION FACTORS

TABLE A.1

Physical constants

Speed of light in vacuum	c_0	$=2.9979 \times 10^8$ m/s
First Planck function constant	C_1	$=3.7419 \times 10^{-6}$ W m^2 $= 2\pi h c_0^2$
Second Planck function constant	C_2	$=14{,}388 \ \mu$m K $=h c_0/k$
Wien's constant	C_3	$=2897.8 \ \mu$m K
Electron charge	e	$=1.6022 \times 10^{-19}$ C
Planck's constant	h	$=6.6262 \times 10^{-34}$ J s
Modified Planck's constant	$\hbar$	$=1.0546 \times 10^{-34}$ J s $= h/2\pi$
Boltzmann's constant	k	$=1.3806 \times 10^{-23}$ J/K
Electron rest mass	m_e	$=9.1096 \times 10^{-31}$ kg
Neutron rest mass	m_n	$=1.6749 \times 10^{-27}$ kg
Proton rest mass	m_p	$=1.6726 \times 10^{-27}$ kg
Avogadro's number	N_A	$=6.0222 \times 10^{23}$ molecules/mol
Solar constant (at mean R_{SE})	q_{sol}	$=1353$ W/m^2
Radius of earth (mean)	R_{earth}	$=6.371 \times 10^6$ m
Radius of solar disk	R_{sun}	$=6.9598 \times 10^8$ m
Earth-sun distance (mean)	R_{SE}	$=1.4960 \times 10^{11}$ m
Universal gas constant	$\mathcal{R}$	$=8.3143$ J/mol K
Effective surface temperature of sun	T_{sun}	$=5762$ K
Molar volume of ideal gas	$\mathcal{V}_{\text{mol}}$	$=22.4136 \ \ell/$mol
		$=22.4136$ m^3/kmol
Electrical permittivity of vacuum	ϵ_0	$=8.8542 \times 10^{-12}$ C^2/N m^2
Magnetic permeability of vacuum	μ_0	$=4\pi \times 10^{-7}$ N s^2/C^2
Stefan-Boltzmann constant	σ	$=5.6696 \times 10^{-8}$ W/m^{2}K^4

TABLE A.2

Conversion factors

Acceleration	$1\,\text{m/s}^2$	$=4.2520\times10^7\,\text{ft/h}^2$
Area	$1\,\text{m}^2$	$=1550.0\,\text{in}^2$
		$=10.764\,\text{ft}^2$
Diffusivity	$1\,\text{m}^2/\text{s}$	$=3.875\times10^4\,\text{ft}^2/\text{h}$
Energy	$1\,\text{J}$	$=9.4787\times10^{-4}\,\text{Btu}$
	$1\,\text{eV} = 1.6022\times10^{-19}\text{J}$	$=1.5187\times10^{-22}\,\text{Btu}$
Force	$1\,\text{N}$	$=0.22481\,\text{lb}_f$
Heat transfer rate	$1\,\text{W}$	$=3.4123\,\text{Btu/h}$
Heat flux	$1\,\text{W/m}^2$	$=0.3171\,\text{Dtu/h ft}^2$
Heat generation rate	$1\,\text{W/m}^3$	$=0.09665\,\text{Btu/h ft}^3$
Heat transfer coefficient	$1\,\text{W/m}^2\text{K}$	$=0.17612\,\text{Btu/h ft}^2\,°\text{F}$
Intensity	$1\,\text{W/m}^2\,\text{sr}$	$=0.3171\,\text{Btu/h ft}^2\,\text{sr}$
Kinematic viscosity	$1\,\text{m}^2/\text{s}$	$=3.875\times10^4\,\text{ft}^2/\text{h}$
Latent heat	$1\,\text{J/kg}$	$=4.2995\times10^{-4}\,\text{Btu/lb}_m$
Length	$1\,\text{m}$	$=39.370\,\text{in}$
		$=3.2808\,\text{ft}$
	$1\,\text{km}$	$=0.62137\,\text{mi}$
Mass	$1\,\text{kg}$	$=2.2046\,\text{lb}_m$
Mass density	$1\,\text{kg/m}^3$	$=0.062428\,\text{lb}_m/\text{ft}^3$
Mass flow rate	$1\,\text{kg/s}$	$=7936.6\,\text{lb}_m/\text{h}$
Power	$1\,\text{W}$	$=3.4123\,\text{Btu/h}$
Pressure and stress	$1\,\text{Pa} = 1\,\text{N/m}^2$	$=1.4504\times10^{-4}\,\text{lb}_f/\text{in}^2$
	$1.0133\times10^5\,\text{N/m}^2$	$=1$ standard atmosphere
Specific heat	$1\,\text{J/kg K}$	$=2.3886\times10^{-4}\,\text{Btu/lb}_m\,°\text{F}$
Temperature	K	$=(5/9)°\text{R}$
		$=(5/9)(°\text{F} + 459.67)$
		$=°\text{C} + 273.15$
Temperature difference	$1\,\text{K}$	$=1°\text{C}$
		$=(9/5)°\text{R} = (9/5)°\text{F}$
Thermal conductivity	$1\,\text{W/m K}$	$=0.57782\,\text{Btu/h ft}\,°\text{F}$
Thermal resistance	$1\,\text{K/W}$	$=0.52750\,°\text{F h/Btu}$
Velocity and speed	$1\,\text{m/s}$	$=3.2808\,\text{ft/s}$
		$=2.2364\,\text{mph}$
Viscosity (dynamic)	$1\,\text{Ns/m}^2 = 1\,\text{kg/s m}$	$=2419.1\,\text{lb}_m/\text{ft h}$
Volume	$1\,\text{m}^3$	$=6.1023\times10^4\,\text{in}^3$
		$=35.314\,\text{ft}^3$
Volume flow rate	$1\,\text{m}^3/\text{s}$	$=1.2713\times10^5\,\text{ft}^3/\text{h}$
		$=2.1189\times10^3\,\text{ft}^3/\text{min}$

TABLE A.3
Conversion factors for spectral variables

Wavelength to energy	$a\ \mu m = a \times 10^3\ nm$	$\cong 1.240/a\ eV$
to frequency	$a\ \mu m = a \times 10^4\ \text{Å}$	$\cong 2.9979 \times 10^{14}/a\ Hz$
to wavenumber	$a\ \mu m$	$\cong 10^4/a\ cm^{-1}$
Energy to frequency	$a\ eV$	$\cong 2.418 \times 10^{14} a\ Hz$
to wavelength	$a\ eV$	$\cong 1.240/a\ \mu m$
to wavenumber	$a\ eV$	$\cong 8.066 \times 10^3 a\ cm^{-1}$
Wavenumber to energy	$a\ cm^{-1}$	$\cong 1.240 \times 10^{-4}\ a\ eV$
to frequency	$a\ cm^{-1}$	$\cong 2.9979 \times 10^{10}\ a\ Hz$
to wavelength	$a\ cm^{-1}$	$\cong 10^4/a\ \mu m$
Frequency to energy	$a\ Hz$	$\cong 4.136 \times 10^{-15}\ a\ eV$
to wavelength	$a\ Hz$	$\cong 2.9979 \times 10^{14}/a\ \mu m$
to wavenumber	$a\ Hz$	$\cong 3.336 \times 10^{-11}\ a\ cm^{-1}$

TABLES FOR RADIATIVE PROPERTIES OF OPAQUE SURFACES

In this appendix, tables of total normal emissivities, as well as a number of total normal solar absorptivities, are given. The data have been collected from several surveys [1–8] that, in turn, have assembled their data from a multitude of references dating back all the way into the 1920s. As seen from the tables, there can sometimes be considerable differences in total emissivity for ostensibly the same material, as reported by different researchers. While these discrepancies are partially due to varying accuracy, the primary reason is, as outlined in Chapter 3, the fact that surface layers, surface roughness, oxidation, etc., strongly affect the emissivity of materials. Therefore, it should be realized that the total normal emissivity or absorptivity of a given surface may, in actuality, differ considerably from these reported values.

In estimating the total hemispherical emissivity from total normal data, one should keep in mind that:

1. Materials with high emissivity tend to behave like dielectrics, resulting in a hemispherical emissivity that is 3% to 5% larger than the normal one (cf. Fig. 3-19).

2. Materials with low emissivity tend to behave like metals, resulting in hemispherical emissivities that may be up to 25% larger than normal ones (cf. Fig. 3-9).

762

References

1. Edwards, D. K., A. F. Mills, and V. E. Denny: *Transfer Processes*, 2nd ed., Hemisphere/McGraw-Hill, New York, 1979.
2. Hottel, H. C.: "Radiant Heat Transmission," in *Heat Transmission*, ed. W. H. McAdams, 3d ed., ch. 4, McGraw-Hill, New York, 1954.
3. Hottel, H. C., and A. F. Sarofim: *Radiative Transfer*, McGraw-Hill, New York, 1967.
4. Gubareff, G. G., J. E. Janssen, and R. H. Torborg: "Thermal Radiation Properties Survey," 2d ed., Honeywell Research Center, Minneapolis, MI, 1960.
5. Wood, W. D., H. W. Deem, and C. F. Lucks: *Thermal Radiative Properties*, Plenum Publishing Company, New York, 1964.
6. Touloukian, Y. S., and D. P. DeWitt (eds.): *Thermal Radiative Properties: Metallic Elements and Alloys*, vol. 7 of *Thermophysical Properties of Matter*, Plenum Press, New York, 1970.
7. Touloukian, Y. S., and D. P. DeWitt (eds.): *Thermal Radiative Properties: Nonmetallic Solids*, vol. 8 of *Thermophysical Properties of Matter*, Plenum Press, New York, 1972.
8. Svet, D. I.: *Thermal Radiation: Metals, Semiconductors, Ceramics, Partly Transparent Bodies, and Films*, Plenum Publishing Company, New York, 1965.

TABLE B.1

Total emissivity and solar absorptivity of selected surfaces (as compiled by Edwards et al. [1])

	Temperature [°C]	Total normal emissivity	Extraterrestrial solar absorptivity
Alumina, flame-sprayed	−25	0.80	0.28
Aluminum foil			
As received	20	0.04	
Bright dipped	20	0.025	0.10
Aluminum, vacuum-deposited on			
duPont mylar	20	0.025	0.10
Aluminum alloy 6061, as received	20	0.03	0.37
Aluminum alloy 75S-T6, weathered			
20,000 hr on a DC6 aircraft	65	0.16	0.54
Aluminum, hard-anodized, 6061-T6	−25	0.84	0.92
Aluminum, soft-anodized, *Reflectal*			
aluminum alloy	−25	0.79	0.23
Aluminum, 7075-T6, sandblasted with			
60 mesh silicon carbide grit	20	0.30	0.55
Aluminized silicone resin paint	95	0.20	0.27
Dow Corning XP-310	425	0.22	
Beryllium	150	0.18	0.77
	370	0.21	
	600	0.30	
Beryllium, anodized	150	0.90	
	370	0.88	
	600	0.82	
Black paint			
Parson's optical black	−25	0.95	0.975
Black silicone, high-heat			
National Lead Co. 46H47	−25 to 750	0.93	0.94
Black epoxy paint, Cat-a-lac			
Finch Paint and Chem. Co. 463-1-8	−25	0.89	0.95
Black enamel paint, Rinshed-Mason	95	0.81	
Heated 1000 h at 375°C in air	425	0.80	
Chromium plate	95	0.12	
Heated 50 h at 600°C	400	0.15	
	35	0.15	0.78
Copper, electroplated	20	0.03	0.47
Black-oxidized in Ebonol C	35	0.16	0.91

TABLE B.1

Total emissivity and solar absorptivity of selected surfaces (as compiled by Edwards et al. [1]) (cont'd.)

	Temperature [°C]	Total normal emissivity	Extraterrestrial solar absorptivity
Glass, second surface mirror			
Aluminized	−25	0.83	0.13
Silvered	−25	0.83	0.13
Gold, coated on stainless steel	95	0.09	
Heated in air at 540°C	400	0.14	
Coated on 3M tape Y9814	20	0.025	0.21
Graphite, crushed on sodium silicate	−25	0.91	0.96
Inconel X			
Oxidized 4 h at 1000°C	−25	0.71	0.90
Oxidized 10 h at 700°C	95	0.81	
	425	0.79	
Magnesium-thorium alloy	95	0.07	
	260	0.06	
Magnesium, Dow 7 coating	370	0.36	
Mylar, duPont film, aluminized on second surface			
0.0625 mm thick	20	0.37	0.17
0.025 mm thick	20	0.63	0.17
0.075 mm thick	20	0.81	0.24
Nickel, electroplated	20	0.03	0.22
Nickel, Tabor solar absorber, electro-oxidized nickel on copper			
110-30	35	0.05	0.85
125-30	35	0.11	0.85
Platinum-coated stainless steel	95	0.13	
	400	0.15	
Annealed in air 300 h at 375°C	95	0.11	
	425	0.13	
Silica, Corning Glass 7940M sintered, powdered, fused silica	35	0.84	0.08
Silica, second surface mirror			
Aluminized	20	0.83	0.14
Silvered	20	0.83	0.07
Silicon solar cell, boron-doped, no coverglass	35	0.32	0.94

TABLE B.1

Total emissivity and solar absorptivity of selected surfaces (as compiled by Edwards et al. [1]) (cont'd.)

	Temperature [°C]	Total normal emissivity	Extraterrestrial solar absorptivity
Silver			
Plated on nickel on stainless steel	95	0.06	
	400	0.08	
Heated 300 h at 375°C	95	0.11	
	425	0.13	
Silver Chromatone paint	20	0.24	0.20
Stainless steel			
Type 312, heated 300 h at 260°C	95	0.27	
	425	0.32	
Type 301 with Armco black oxide	−25	0.75	0.89
Type 410, heated to 700°C in air	35	0.13	0.76
Type 303, sandblasted heavily with 80 mesh aluminum oxide grit	95	0.42	0.68
Titanium			
75A	95	0.10	
	425	0.19	
75A, oxidized 300 h at 450°C	35	0.21	0.80
	425	0.25	
C-110M, oxidized 100 h at 425°C in air	35	0.16	0.52
C-110M, oxidized 300 h at 450°C in air	35	0.20	0.77
Evaporated 80 to 100 μm thick and oxidized 3 h at 400°C in air	35	0.14	0.75
Anodized	−25	0.73	0.51
White acrylic resin paint	95	0.92	
Sherwin-Williams M49WC8-CA-10144	200	0.87	
White epoxy paint, Cat-a-lac Finch Paint and Chemical Co. 483-1-8	−25	0.88	0.25
White potassium zirconium silicate inorganic spacecraft coating	20	0.89	0.13
Zinc, blackened by Tabor Solar collector electrochemical treatment 120-20	35	0.12	0.89

TABLE B.2
Total normal emissivity of various surfaces

	Temperature[a] [°C]	Total normal emissivity[a]
A. Metals and their oxides		
Aluminum		
Highly polished plate, 98.3% pure	225–575	0.039–0.057
Commercial sheet	100	0.09
Rough polish	100	0.18
Rough plate	40	0.055–0.07
Oxidized at 600°C	200–600	0.11–0.19
Heavily oxidized	95–500	0.20–0.31
Aluminum oxide	275–500	0.63–0.42
	500–825	0.42–0.26
Al-surfaced roofing	40	0.216
Aluminum alloys[b]		
Alloy 75 ST: A, B_1, C	25	0.11, 0.10, 0.08
Alloy 75 ST: A[c]	230–480	0.22–0.16
Alloy 75 ST: B_1[c]	230–425	0.20–0.18
Alloy 75 ST: C[c]	230–500	0.22–0.15
Alloy 24 ST: A, B_1, C	25	0.09
Alloy 24 ST: A[c]	230–485	0.17–0.15
Alloy 24 ST: B_1[c]	230–505	0.20–0.16
Alloy 24 ST: C[c]	230–460	0.16–0.13
Calorized surfaces, heated at 600°C		
Copper	200–600	0.18–1.19
Steel	200–600	0.52–0.57
Antimony, polished	35–260	0.28–0.31
Bismuth, bright	75	0.34
Brass		
Highly polished		
73.2% Cu, 26.7% Zn	245–355	0.028–0.031
62.4% Cu, 36.8% Zn, 0.4% Pb, 0.3% Al	255–375	0.033–0.037
82.9% Cu, 17.0% Zn	275	0.030
Polished	100	0.06
	40–315	0.10
Rolled plate, natural surface	22	0.06
Rolled plate, rubbed with coarse emery	22	0.20
Dull plate	50–350	0.22
Oxidized by heating at 600°C	200–600	0.61–0.59
Chromium (see nickel alloys for Ni-Cr steels):		
Polished	40–1100	0.08–0.36
	100	0.075

TABLE B.2
Total normal emissivity of various surfaces (cont'd.)

	Temperature[a] [°C]	Total normal emissivity[a]
Copper		
Carefully polished electrolytic copper	80	0.018
Polished	115	0.023
	100	0.052
Commercial emeried, polished, pits remaining	19	0.030
Commercial, scraped shiny, not mirror-like	22	0.072
Plate heated long time, covered with		
thick oxide layer	25	0.78
Plate heated at 600°C	200–600	0.57
Cuprous oxide	800–1100	0.66–0.54
Molten copper	1075–1275	0.16–0.13
Dow metal[b]		
A, B$_1$, C	25	0.15, 0.15, 0.12
A[c]	230–400	0.24–0.20
B$_1$[c]	230–425	0.16
C[c]	230–405	0.21–0.18
Gold, pure, highly polished	225–625	0.018–0.035
Inconel[b]		
Types X and B: surface A, B$_2$, C	25	0.19–0.21
Type X: surface A[c]	230–880	0.55–0.78
Type X: surface B$_2$[c]	230–855	0.60–0.75
Type X: surface C[c]	230–900	0.62–0.73
Type B: surface A[c]	230–880	0.35–0.55
Type B: surface B$_2$[c]	230–950	0.32–0.51
Type B: surface C[c]	230–1000	0.35–0.40
Iron and steel (not including stainless)		
Metallic surfaces (or very thin oxide layer)		
Electrolytic iron, highly polished	175–225	0.052–0.064
Steel, polished	100	0.066
Iron, polished	425–1025	0.14–0.38
Iron, roughly polished	100	0.17
Iron, freshly emeried	20	0.24
Cast iron, polished	200	0.21
Cast iron, newly turned	22	0.44
Cast iron, turned and heated	880–990	0.60–0.70
Wrought iron, highly polished	40–250	0.28
Polished steel casting	770–1035	0.52–0.56

TABLE B.2
Total normal emissivity of various surfaces (cont'd.)

	Temperature[a] [°C]	Total normal emissivity[a]
Ground sheet steel	935–1100	0.55–0.61
Smooth sheet iron	900–1040	0.55–0.60
Mild steel[b]: A, B$_2$, C	25	0.12, 0.15, 0.10
Mild steel[b]: A[c]	230–1065	0.20–0.32
Mild steel[b]: B$_2$[c]	230–1050	0.34–0.35
Mild steel[b]: C[c]	230–1065	0.27–0.31
Oxidized surfaces		
Iron plate, pickled, then rusted red	20	0.61
Iron plate, completely rusted	20	0.69
Iron, dark gray surface	100	0.31
Rolled sheet steel	21	0.66
Oxidized iron	100	0.74
Cast iron, oxidized at 600°C	200–600	0.64–0.78
Steel, oxidized at 600°C	200–600	0.79
Smooth, oxidized electrolytic iron	125–525	0.78–0.82
Iron oxide	500–1200	0.85–0.89
Rough ingot iron	925–1115	0.87–0.95
Sheet steel		
Strong, rough oxide layer	25	0.80
Dense, shiny oxide layer	25	0.82
Cast plate, smooth	23	0.80
Cast plate, rough	23	0.82
Cast iron, rough, strongly oxidized	40–250	0.95
Wrought iron, dull oxidized	20–360	0.94
Steel plate, rough	40–370	0.94–0.97
Molten surfaces		
Cast iron	1300–1400	0.29
Mild steel	1600–1800	0.28
Steel, several different kinds with 0.25–		
1.2% C (slightly oxidized surface)	1560–1710	0.27–0.39
Steel	1500–1650	0.42–0.53
	1520–1650	0.43–0.40
Pure iron	1515–1770	0.42–0.45
Armco iron	1520–1690	0.40–0.41
Lead		
Pure (99.96%), unoxidized	125–225	0.057–0.075
Gray oxidized	25	0.28
Oxidized at 150°C	200	0.63

TABLE B.2
Total normal emissivity of various surfaces (cont'd.)

	Temperature[a] [°C]	Total normal emissivity[a]
Magnesium		
Magnesium oxide	275–825	0.55–0.20
	900–1705	0.20
Magnesium, polished	35–260	0.07–0.13
Mercury	0–100	0.09–0.12
Molybdenum		
Filament	725–2595	0.096–0.202
Massive, polished	100	0.071
Polished	35–260	0.05–0.08
	540–1370	0.10–0.18
	2750	0.29
Monel metal[b]		
Oxidized at 600°C	200–600	0.41–0.46
K Monel 5700: A, B$_2$, C	25	0.23, 0.17, 0.14
K Monel 5700: A[c]	230–875	0.46–0.65
K Monel 5700: B$_2$[c]	230–955	0.54–0.77
K Monel 5700: C[c]	230–975	0.35–0.53
Nickel		
Electroplated, polished	23	0.045
Technically pure (98.9% Ni, + Mn), polished	225–375	0.07–0.087
Polished	100	0.072
Electroplated, not polished	20	0.11
Wire	185–1005	0.096–0.186
Plate, oxidized by heating at 600°C	200–600	0.37–0.48
Nickel oxide	650–1255	0.59–0.86
Nickel alloys		
Chromnickel	50–1035	0.64–0.76
Copper-nickel, polished	100	0.059
Nichrome wire, bright	50–1000	0.65–0.79
Nichrome wire, oxidized	50–500	0.95–0.98
Nickel-silver, polished	100	0.135
Nickelin (18–32% Ni; 55–68% Cu; 20% Zn), gray oxidized	20	0.262
Type ACI-HW (60% Ni; 12% Cr), smooth, black, firm adhesive oxide coat from service	270–560	0.89–0.82
Platinum		
Pure, polished plate	225–625	0.054–0.104
Strip	925–1625	0.12–0.17

TABLE B.2
Total normal emissivity of various surfaces (cont'd.)

	Temperature[a] [°C]	Total normal emissivity[a]
Filament	27–1225	0.036–0.192
Wire	225–1375	0.073–0.182
Silver		
Polished, pure	225–625	0.020–0.032
Polished	40–370	0.022–0.031
	100	0.052
Stainless steel[b]		
Polished	100	0.074
Type 301: A, B_2, C	25	0.21, 0.27, 0.16
Type 301: A[c]	230–950	0.57–0.55
Type 301: B_2[c]	230–940	0.54–0.63
Type 301: C[c]	230–900	0.51–0.70
Type 316: A, B_2, C	25	0.28, 0.28, 0.17
Type 316: A[c]	230–870	0.57–0.66
Type 316: B_2[c]	230–1050	0.52–0.50
Type 316: C[c]	230–1050	0.26–0.31
Type 347: A, B_2, C	25	0.39, 0.35, 0.17
Type 347: A[c]	230–900	0.52–0.65
Type 347: B_2[c]	230–875	0.51–0.65
Type 347: C[c]	230–900	0.49–0.64
Type 304: (8% Cr; 18% Ni)		
Light silvery, rough, brown after heating	215–490	0.44–0.36
After 42 h heating at 525°C	215–525	0.62–0.73
Type 310 (25% Cr; 20% Ni), brown,		
splotched, oxidized from furnace service	215–525	0.90–0.97
Allegheny metal no. 4, polished	100	0.13
Allegheny alloy no. 66, polished	100	0.11
Tantalum filament	1340–3000	0.19–0.31
Thorium oxide	275–500	0.58–0.36
	500–825	0.36–0.21
Tin		
Bright tinned iron	25	0.043, 0.064
Bright	50	0.06
Commercial tin-plated sheet iron	100	0.07, 0.08
Tungsten		
Filament, aged	27–3300	0.032–0.35
Filament	3300	0.39
Polished coat	100	0.066

TABLE B.2
Total normal emissivity of various surfaces (cont'd.)

	Temperature[a] [°C]	Total normal emissivity[a]
Zinc		
Commercial 99.1% pure, polished	225–325	0.045–0.053
Oxidized by heating at 400°C	400	0.11
Galvanized sheet iron, fairly bright	27	0.23
Galvanized sheet iron, gray oxidized	25	0.28
Zinc, galvanized sheet	100	0.21
B. Refractories, building materials, paints, and miscellaneous		
Alumina (99.5–85% Al_2O_3; 0–12% SiO_2; 0–1% Fe_2O_3)		
Effect of mean grain size	1010–1565	
10 μm		0.30–0.18
50 μm		0.39–0.28
100 μm		0.50–0.40
Alumina on Inconel	540–1100	0.65–0.45
Alumina-silica (showing effect of Fe)	1010–1565	
80–58% Al_2O_3; 16–38% SiO_2; 0.4% Fe_2O_3		0.61–0.43
36–26% Al_2O_3; 50–60% SiO_2; 1.7% Fe_2O_3		0.73–0.62
61% Al_2O_3; 35% SiO_2; 2.9% Fe_2O_3		0.78–0.68
Asbestos		
Board	23	0.96
Paper	35–370	0.93–0.94
Brick		
Red, rough, but no gross irregularities	20	0.93
Grog brick, glazed	1100	0.75
Building	1000	0.45
Fireclay	1000	0.75
White refractory	1100	0.29
Carbon		
Filament	1040–1405	0.526
Rough plate	100–320	0.77
	320–500	0.77–0.72
Graphitized	100–320	0.76–0.75
	320–500	0.75–0.71
Candle soot	95–270	0.952
Lampblack-waterglass coating	100–275	0.96–0.95
Thin layer on iron plate	20	0.927
Thick coat	20	0.967
Lampblack, 0.075 mm or thicker	40–370	0.945
Lampblack, rough deposit	100–500	0.84–0.78
Lampblack, other blacks	50–1000	0.96
Graphite, pressed, filed surface	250–510	0.98

TABLE B.2

Total normal emissivity of various surfaces (cont'd.)

	Temperature[a] [°C]	Total normal emissivity[a]
Carborundum (87% SiC; density 2.3 g/cm^3)	1010–1400	0.92–0.81
Concrete tiles	1000	0.63
Concrete, rough	38	0.94
Enamel, white fused, on iron	20	0.90
Glass		
Smooth	20	0.94
Pyrex, lead, and soda	260–540	0.95–0.85
Gypsum, 5 mm thick on smooth or		
blackened plate	20	0.903
Ice		
Smooth	0	0.966
Rough crystals	0	0.985
Magnesite refractory brick	1000	0.38
Marble, light gray, polished	20	0.93
Paints, lacquers, varnishes		
White enamel varnish on rough iron plate	72	0.906
Black shiny lacquer, sprayed on iron	25	0.875
Black shiny shellac on tinned iron sheet	20	0.821
Black matte shellac	75–145	0.91
Black or white lacquer	35–95	0.80–0.95
Flat black lacquer	35–95	0.96–0.98
Oil paints, 16 different, all colors	100	0.92–0.96
Aluminum paints and lacquers		
10% Al, 22% lacquer body, on rough or		
smooth surface	100	0.52
Other Al paints, varying age and Al content	100	0.27–0.67
Al lacquer, varnish binder, on rough plate	20	0.39
Al paint, after heating at 325°C	150–315	0.35
Lacquer coatings, 0.025–0.37 mm thick on		
aluminum alloys	35–150	0.87–0.97
Clear silicone vehicle coatings, 0.025–0.375 mm		
On mild steel	260	0.66
On stainless steels, 316, 301, 347	260	0.68, 0.75, 0.75
On Dow metal	260	0.74
On Al alloys 24 ST, 75 ST	260	0.77, 0.82
Aluminum paint with silicone vehicle,		
two coats on Inconel	260	0.29

TABLE B.2
Total normal emissivity of various surfaces (cont'd.)

	Temperature[a] [°C]	Total normal emissivity[a]
Paper		
White	35	0.95
Thin, pasted on tinned or blackened plate	20	0.92, 0.94
Roofing	20	0.91
Plaster, rough lime	10–88	0.91
Porcelain, glazed	20	0.92
Quartz		
Rough, fused	20	0.93
Glass, 1.98 mm thick	280–840	0.90–0.41
Glass, 6.88 mm thick	280–840	0.93–0.47
Opaque	280–840	0.92–0.68
Rubber		
Hard, glossy plate	23	0.94
Soft, gray, rough (reclaimed)	25	0.86
Sandstone	35–260	0.83–0.90
Serpentine, polished	23	0.90
Silica (98% SiO_2; Fe-free), effect of grain size	1010–1565	
10 μm		0.42–0.33
70–600 μm		0.62–0.46
Silicon carbide	150–650	0.83–0.96
Slate	35	0.67–0.80
Soot, candle	90–260	0.95
Water	0–100	0.95–0.963
Wood		
Sawdust	35	0.75
Oak, planed	20	0.90
Beech	70	0.94
Zirconium silicate	240–500	0.92–0.80
	500–830	0.80–0.52

[a] When temperatures and emissivities appear in pairs separated by dashes, they correspond, i.e., linear interpolation is permissible.

[b] Identification of surface treatment: A, cleaned with tolune, then methanol; B_1, cleaned with soap and water, tolune, and methanol in succession; B_2, cleaned with abrasive soap and water, tolune, and methanol; C, polished on buffing wheel to mirror surface, cleaned with soap and water.

[c] Results after repeated heating and cooling.

BLACKBODY
EMISSIVE POWER
TABLE

$n\lambda T$ [μm K]	η/nT [cm^{-1}/K]	$E_{b\lambda}/n^3T^5$ [W/m$^2\mu$m K^5]	$E_{b\eta}/nT^3$ [W/m^{2}cm^{-1}K^3]	$f(n\lambda T)$
1000	10.0000	0.02110×10^{-11}	0.00211×10^{-8}	0.00032
1100	9.0909	0.04846	0.00586	0.00091
1200	8.3333	0.09329	0.01343	0.00213
1300	7.6923	0.15724	0.02657	0.00432
1400	7.1429	0.23932	0.04691	0.00779
1500	6.6667	0.33631	0.07567	0.01285
1600	6.2500	0.44359	0.11356	0.01972
1700	5.8824	0.55603	0.16069	0.02853
1800	5.5556	0.66872	0.21666	0.03934
1900	5.2632	0.77736	0.28063	0.05210
2000	5.0000	0.87858	0.35143	0.06672
2100	4.7619	0.96994	0.42774	0.08305
2200	4.5455	1.04990	0.50815	0.10088
2300	4.3478	1.11768	0.59125	0.12002
2400	4.1667	1.17314	0.67573	0.14025
2500	4.0000	1.21659	0.76037	0.16135
2600	3.8462	1.24868	0.84411	0.18311
2700	3.7037	1.27029	0.92604	0.20535
2800	3.5714	1.28242	1.00542	0.22788
2900	3.4483	1.28612	1.08162	0.25055

$n\lambda T$ [μm K]	η/nT [cm^{-1}/K]	$E_{b\lambda}/n^3T^5$ [W/m$^2\mu$m K^5]	$E_{b\eta}/nT^3$ [W/m^{2}cm^{-1}K^3]	$f(n\lambda T)$
3000	3.3333	1.28245×10^{-11}	1.15420×10^{-8}	0.27322
3100	3.2258	1.27242	1.22280	0.29576
3200	3.1250	1.25702	1.28719	0.31809
3300	3.0303	1.23711	1.34722	0.34009
3400	2.9412	1.21352	1.40283	0.36172
3500	2.8571	1.18695	1.45402	0.38290
3600	2.7778	1.15806	1.50084	0.40359
3700	2.7027	1.12739	1.54340	0.42375
3800	2.6316	1.09544	1.58181	0.44336
3900	2.5641	1.06261	1.61623	0.46240
4000	2.5000	1.02927	1.64683	0.48085
4100	2.4390	0.99571	1.67380	0.49872
4200	2.3810	0.96220	1.69731	0.51599
4300	2.3256	0.92892	1.71758	0.53267
4400	2.2727	0.89607	1.73478	0.54877
4500	2.2222	0.86376	1.74912	0.56429
4600	2.1739	0.83212	1.76078	0.57925
4700	2.1277	0.80124	1.76994	0.59366
4800	2.0833	0.77117	1.77678	0.60753
4900	2.0408	0.74197	1.78146	0.62088
5000	2.0000	0.71366	1.78416	0.63372
5100	1.9608	0.68628	1.78502	0.64606
5200	1.9231	0.65983	1.78419	0.65794
5300	1.8868	0.63432	1.78181	0.66935
5400	1.8519	0.60974	1.77800	0.68033
5500	1.8182	0.58608	1.77288	0.69087
5600	1.7857	0.56332	1.76658	0.70101
5700	1.7544	0.54146	1.75919	0.71076
5800	1.7241	0.52046	1.75081	0.72012
5900	1.6949	0.50030	1.74154	0.72913
6000	1.6667	0.48096	1.73147	0.73778
6100	1.6393	0.46242	1.72066	0.74610
6200	1.6129	0.44464	1.70921	0.75410
6300	1.5873	0.42760	1.69716	0.76180
6400	1.5625	0.41128	1.68460	0.76920
6500	1.5385	0.39564	1.67157	0.77631
6600	1.5152	0.38066	1.65814	0.78316
6700	1.4925	0.36631	1.64435	0.78975
6800	1.4706	0.35256	1.63024	0.79609
6900	1.4493	0.33940	1.61587	0.80219

$n\lambda T$ [μm K]	η/nT [cm^{-1}/K]	$E_{b\lambda}/n^3T^5$ [W/m^2 μm K^5]	$E_{b\eta}/nT^3$ [W/m^2 cm^{-1}K^3]	$f(n\lambda T)$
7000	1.4286	0.32679×10^{-11}	1.60127×10^{-8}	0.80807
7100	1.4085	0.31471	1.58648	0.81373
7200	1.3889	0.30315	1.57152	0.81918
7300	1.3699	0.29207	1.55644	0.82443
7400	1.3514	0.28146	1.54126	0.82949
7500	1.3333	0.27129	1.52600	0.83436
7600	1.3158	0.26155	1.51069	0.83906
7700	1.2987	0.25221	1.49535	0.84359
7800	1.2821	0.24326	1.48000	0.84796
7900	1.2658	0.23468	1.46465	0.85218
8000	1.2500	0.22646	1.44933	0.85625
8100	1.2346	0.21857	1.43405	0.86017
8200	1.2195	0.21101	1.41882	0.86396
8300	1.2048	0.20375	1.40366	0.86762
8400	1.1905	0.19679	1.38857	0.87115
8500	1.1765	0.19011	1.37357	0.87456
8600	1.1628	0.18370	1.35866	0.87786
8700	1.1494	0.17755	1.34385	0.88105
8800	1.1364	0.17164	1.32916	0.88413
8900	1.1236	0.16596	1.31458	0.88711
9000	1.1111	0.16051	1.30013	0.88999
9100	1.0989	0.15527	1.28580	0.89277
9200	1.0870	0.15024	1.27161	0.89547
9300	1.0753	0.14540	1.25755	0.89807
9400	1.0638	0.14075	1.24363	0.90060
9500	1.0526	0.13627	1.22985	0.90304
9600	1.0417	0.13197	1.21622	0.90541
9700	1.0309	0.12783	1.20274	0.90770
9800	1.0204	0.12384	1.18941	0.90992
9900	1.0101	0.12001	1.17622	0.91207
10,000	1.0000	0.11632	1.16319	0.91415
10,200	0.9804	0.10934	1.13759	0.91813
10,400	0.9615	0.10287	1.11260	0.92188
10,600	0.9434	0.09685	1.08822	0.92540
10,800	0.9259	0.09126	1.06446	0.92872
11,000	0.9091	0.08606	1.04130	0.93184
11,200	0.8929	0.08121	1.01874	0.93479
11,400	0.8772	0.07670	0.99677	0.93758
11,600	0.8621	0.07249	0.97538	0.94021
11,800	0.8475	0.06856	0.95456	0.94270

$n\lambda T$ [μm K]	η/nT [cm^{-1}/K]	$E_{b\lambda}/n^3T^5$ [W/m^2μm K^5]	$E_{b\eta}/nT^3$ [W/m^{2}cm^{-1}K^3]	$f(n\lambda T)$
12,000	0.8333	0.06488×10^{-11}	0.93430×10^{-8}	0.94505
12,200	0.8197	0.06145	0.91458	0.94728
12,400	0.8065	0.05823	0.89540	0.94939
12,600	0.7937	0.05522	0.87674	0.95139
12,800	0.7813	0.05240	0.85858	0.95329
13,000	0.7692	0.04976	0.84092	0.95509
13,200	0.7576	0.04728	0.82374	0.95680
13,400	0.7463	0.04494	0.80702	0.95843
13,600	0.7353	0.04275	0.79076	0.95998
13,800	0.7246	0.04069	0.77493	0.96145
14,000	0.7143	0.03875	0.75954	0.96285
14,200	0.7042	0.03693	0.74456	0.96418
14,400	0.6944	0.03520	0.72998	0.96546
14,600	0.6849	0.03358	0.71579	0.96667
14,800	0.6757	0.03205	0.70198	0.96783
15,000	0.6667	0.03060	0.68853	0.96893
15,200	0.6579	0.02923	0.67544	0.96999
15,400	0.6494	0.02794	0.66270	0.97100
15,600	0.6410	0.02672	0.65029	0.97196
15,800	0.6329	0.02556	0.63820	0.97288
16,000	0.6250	0.02447	0.62643	0.97377
16,200	0.6173	0.02343	0.61496	0.97461
16,400	0.6098	0.02245	0.60379	0.97542
16,600	0.6024	0.02152	0.59290	0.97620
16,800	0.5952	0.02063	0.58228	0.97694
17,000	0.5882	0.01979	0.57194	0.97765
17,200	0.5814	0.01899	0.56186	0.97834
17,400	0.5747	0.01823	0.55202	0.97899
17,600	0.5682	0.01751	0.54243	0.97962
17,800	0.5618	0.01682	0.53308	0.98023
18,000	0.5556	0.01617	0.52396	0.98081
18,200	0.5495	0.01555	0.51506	0.98137
18,400	0.5435	0.01496	0.50638	0.98191
18,600	0.5376	0.01439	0.49790	0.98243
18,800	0.5319	0.01385	0.48963	0.98293

$n\lambda T$ [μm K]	η/nT [cm^{-1}/K]	$E_{b\lambda}/n^3T^5$ [W/m^2 μm K^5]	$E_{b\eta}/nT^3$ [W/m^{2}cm^{-1}K^3]	$f(n\lambda T)$
19,000	0.5263	0.01334×10^{-11}	0.48155×10^{-8}	0.98340
19,200	0.5208	0.01285	0.47367	0.98387
19,400	0.5155	0.01238	0.46597	0.98431
19,600	0.5102	0.01193	0.45845	0.98474
19,800	0.5051	0.01151	0.45111	0.98515
20,000	0.5000	0.01110	0.44393	0.98555
21,000	0.4762	0.00931	0.41043	0.98735
22,000	0.4545	0.00786	0.38049	0.98886
23,000	0.4348	0.00669	0.35364	0.99014
24,000	0.4167	0.00572	0.32948	0.99123
25,000	0.4000	0.00492	0.30767	0.99217
26,000	0.3846	0.00426	0.28792	0.99297
27,000	0.3704	0.00370	0.26999	0.99367
28,000	0.3571	0.00324	0.25366	0.99429
29,000	0.3448	0.00284	0.23875	0.99482
30,000	0.3333	0.00250	0.22510	0.99529
31,000	0.3226	0.00221	0.21258	0.99571
32,000	0.3125	0.00196	0.20106	0.99607
33,000	0.3030	0.00175	0.19045	0.99640
34,000	0.2941	0.00156	0.18065	0.99669
35,000	0.2857	0.00140	0.17158	0.99695
36,000	0.2778	0.00126	0.16317	0.99719
37,000	0.2703	0.00113	0.15536	0.99740
38,000	0.2632	0.00103	0.14810	0.99759
39,000	0.2564	0.00093	0.14132	0.99776
40,000	0.2500	0.00084	0.13501	0.99792
41,000	0.2439	0.00077	0.12910	0.99806
42,000	0.2381	0.00070	0.12357	0.99819
43,000	0.2326	0.00064	0.11839	0.99831
44,000	0.2273	0.00059	0.11352	0.99842
45,000	0.2222	0.00054	0.10895	0.99851
46,000	0.2174	0.00049	0.10464	0.99861
47,000	0.2128	0.00046	0.10059	0.99869
48,000	0.2083	0.00042	0.09677	0.99877
49,000	0.2041	0.00039	0.09315	0.99884
50,000	0.2000	0.00036	0.08974	0.99890

D

VIEW FACTOR CATALOGUE

In this appendix a small number of view factor relations and figures are presented. A much larger collection from a variety of references has been compiled by Howell [1], from which the present list has been extracted. For three important configurations, with relatively complicated formulae for their evaluation, the view factors are also given in graphical form. Finally, a list is given of papers and monographs that either deal with evaluation methods for view factors, or present results for specified configurations (ordered by date of publication). No attempt of completeness has been made.

Note: In all expressions in which inverse trigonometric functions appear, the principal value is to be taken; i.e., for any argument ξ,

$$-\frac{\pi}{2} \leq \sin^{-1}\xi \leq +\frac{\pi}{2}; \quad 0 \leq \cos^{-1}\xi \leq \pi; \quad -\frac{\pi}{2} \leq \tan^{-1}\xi \leq +\frac{\pi}{2}.$$

1

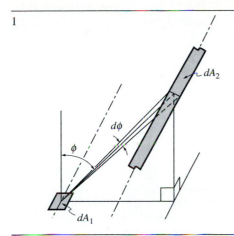

Differential strip element of any length z to infinitely long strip of differential width on parallel line; plane containing element does not intercept strip

$$dF_{d1-d2} = \frac{\cos\phi\,d\phi}{2}$$

2

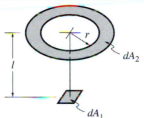

Differential planar element to differential coaxial ring parallel to the element

$$R = r/l$$

$$dF_{d1-d2} = \frac{2R}{(1 + R^2)^2}\,dR$$

3

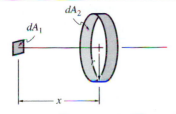

Differential planar element on and normal to ring axis to inside of differential ring

$$X = x/r$$

$$dF_{d1-d2} = \frac{2X}{(X^2 + 1)^2}\,dX$$

4

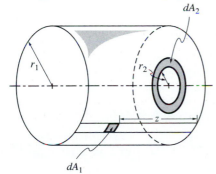

Element on surface of right-circular cylinder to coaxial differential ring on cylinder base, $r_2 < r_1$

$$Z = z/r_1, \quad R = r_2/r_1$$
$$X = 1 + Z^2 + R^2$$

$$dF_{d1-d2} = \frac{2Z(X - 2R^2)R\,dR}{(X^2 - 4R^2)^{3/2}}$$

5

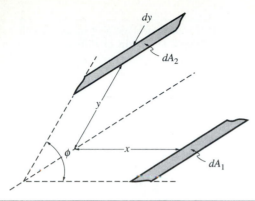

Parallel differential strip elements in intersecting planes

$$Y = y/x$$

$$dF_{d1-d2} = \frac{Y \sin^2 \phi \, dY}{2(1 + Y^2 - 2Y \cos \phi)^{3/2}}$$

6

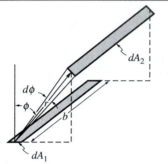

Strip of finite length b and of differential width, to differential strip of same length on parallel generating line

$$B = b/r$$

$$dF_{d1-d2} = \tan^{-1} B \, \frac{\cos \phi}{\pi} \, d\phi$$

7

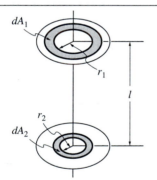

Differential ring element to ring element on coaxial disk

$$R = r_2/r_1, \quad L = l/r_1$$

$$dF_{d1-d2} = \frac{2RL^2[L^2 + R^2 + 1]dR}{[(L^2 + R^2 + 1)^2 - 4R^2]^{3/2}}$$

8

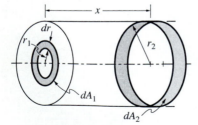

Ring element on base to circumferential ring element on interior of right-circular cylinder

$$X = x/r_2, \quad R = r_1/r_2$$

$$dF_{d1-d2} = \frac{2X(X^2 + R^2 - 1)dX}{[(X^2 + R^2 + 1)^2 - 4R^2]^{3/2}}$$

9

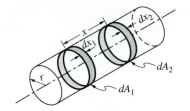

Two ring elements on the interior of right-circular cylinder

$$X = x/2r$$

$$dF_{d1-d2} = \left[1 - \frac{X(2X^2 + 3)}{2(X^2 + 1)^{3/2}}\right]dX_2$$

10

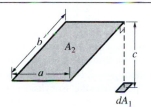

Differential planar element to finite parallel rectangle; normal to element passes through corner of rectangle

$$A = a/c, \quad B = b/c$$

$$F_{d1-2} = \frac{1}{2\pi}\left\{\frac{A}{\sqrt{1+A^2}}\tan^{-1}\frac{B}{\sqrt{1+A^2}} + \frac{B}{\sqrt{1+B^2}}\tan^{-1}\frac{A}{\sqrt{1+B^2}}\right\}$$

11

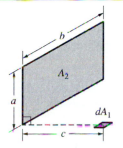

Differential planar element to rectangle in plane 90° to plane of element

$$X = a/b, \quad Y = c/b$$

$$F_{d1-2} = \frac{1}{2\pi}\left(\tan^{-1}\frac{1}{Y}\right.$$
$$\left. - \frac{Y}{\sqrt{X^2 + Y^2}}\tan^{-1}\frac{1}{\sqrt{X^2 + Y^2}}\right)$$

12

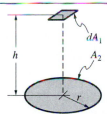

Differential planar element to circular disk in plane parallel to element; normal to element passes through center of disk

$$H = h/r$$

$$F_{d1-2} = \frac{1}{H^2 + 1}$$

13

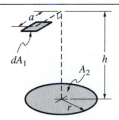

Differential planar element to circular disk in plane parallel to element

$$H = h/a, \quad R = r/a$$
$$Z = 1 + H^2 + R^2$$

$$F_{d1-2} = \frac{1}{2}\left[1 - \frac{Z - 2R^2}{\sqrt{Z^2 - 4R^2}}\right]$$

14

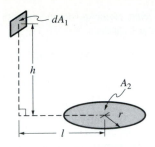

Differential planar element to circular disk; planes containing element and disk intersect at $90°$; $l \geq r$

$$H = h/l, \quad R = r/l$$
$$Z = 1 + H^2 + R^2$$

$$F_{d1-2} = \frac{H}{2}\left[\frac{Z}{\sqrt{Z^2 - 4R^2}} - 1\right]$$

15

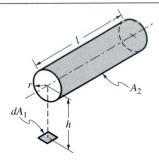

Differential planar element to right-circular cylinder of finite length and radius; normal to element passes through one end of cylinder and is perpendicular to cylinder axis

$$L = l/r, \quad H = h/r$$
$$X = (1 + H)^2 + L^2$$
$$Y = (1 - H)^2 + L^2$$

$$F_{d1-2} = \frac{L}{\pi H}\left[\frac{1}{L}\tan^{-1}\frac{L}{\sqrt{H^2-1}} + \frac{X-2H}{\sqrt{XY}}\tan^{-1}\sqrt{\frac{X(H-1)}{Y(H+1)}} - \tan^{-1}\sqrt{\frac{H-1}{H+1}}\right]$$

16

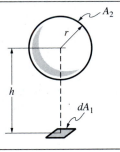

Differential planar element to sphere; normal to center of element passes through center of sphere

$$F_{d1-2} = \left(\frac{r}{h}\right)^2$$

17

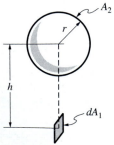

Differential planar element to sphere; tangent to element passes through center of sphere

$$H = h/r$$

$$F_{d1-2} = \frac{1}{\pi}\left[\tan^{-1}\frac{1}{\sqrt{H^2-1}} - \frac{\sqrt{H^2-1}}{H^2}\right]$$

18

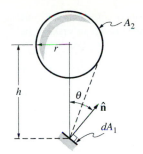

Differential planar element to sphere; element plane does not intersect sphere

$$\theta \le \cos^{-1} \frac{r}{h}$$

$$F_{d1-2} = \left(\frac{r}{h}\right)^2 \cos \theta$$

19

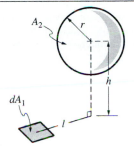

Differential planar element to sphere

$$L = l/r, \quad H = h/r$$

$$H \ge 1: \quad F_{d1-2} = \frac{H}{(L^2 + H^2)^{3/2}}$$

$$-1 < H < 1: \quad F_{d1-2} = \frac{1}{\pi} \left\{ \frac{H}{(L^2 + H^2)^{3/2}} \cos^{-1} \frac{-H}{L\sqrt{L^2 + H^2 - 1}} \right.$$
$$\left. - \frac{\sqrt{(L^2 + H^2 - 1)(1 - H^2)}}{L^2 + H^2} - \sin^{-1} \frac{\sqrt{H^2 + L^2 - 1}}{L^2} + \frac{\pi}{2} \right\}$$

20

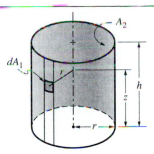

Differential element on longitudinal strip inside cylinder to inside cylinder surface

$$Z = z/2R, \quad H = h/2R$$

$$F_{d1-2} = 1 + H - \frac{Z^2 + \frac{1}{2}}{\sqrt{Z^2 + 1}} - \frac{(H - Z)^2 + \frac{1}{2}}{\sqrt{(H - Z)^2 + 1}}$$

21

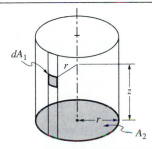

Differential element on longitudinal strip on inside of right-circular cylinder to base of cylinder

$$Z = z/r$$

$$F_{d1-2} = \frac{Z^2 + 2}{2\sqrt{Z^2 + 4}} - \frac{Z}{2}$$

22

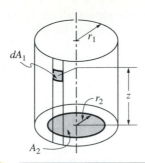

Differential element on surface of right-circular cylinder to disk on base of cylinder, $r_2 < r_1$ (see Configuration 13)

$$Z = z/r_1, \quad R = r_2/r_1$$

$$X = 1 + Z^2 + R^2$$

$$F_{d1-2} = \frac{Z}{2} \left\{ \frac{X}{\sqrt{X^2 - 4R^2}} - 1 \right\}$$

23

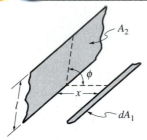

Infinite differential strip to parallel infinite plane of finite width; plane and plane containing strip intersect at arbitrary angle ϕ

$$X = x/l$$

$$F_{d1-2} = \frac{1}{2} + \frac{\cos \phi - X}{2 \sqrt{1 + X^2 - 2X \cos \phi}}$$

24

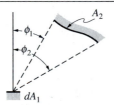

Differential strip element of any length to an infinitely long strip of finite width; cross section of A_2 is arbitrary (but does not vary perpendicular to the paper); plane of dA_1 does not intersect A_2

$$F_{d1-2} = \tfrac{1}{2}(\sin \phi_2 - \sin \phi_1)$$

25

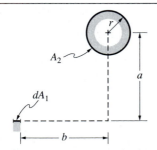

Differential strip element of any length to infinitely long parallel cylinder; $r < a$

$$A = a/r, \quad B = b/r$$

$$F_{d1-2} = \frac{A}{A^2 + B^2}$$

26

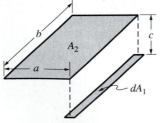

Differential strip element to rectangle in plane parallel to strip; strip is opposite one edge of rectangle

$$X = a/c, \quad Y = b/c$$

$$F_{d1-2} = \frac{1}{\pi Y} \left(\sqrt{1 + Y^2} \tan^{-1} \frac{X}{\sqrt{1 + Y^2}} - \tan^{-1} X + \frac{XY}{\sqrt{1 + X^2}} \tan^{-1} \frac{Y}{\sqrt{1 + X^2}} \right)$$

27

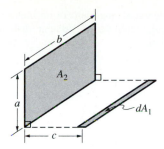

Differential strip element to rectangle in plane 90° to plane of strip

$$X = a/b, \quad Y = c/b$$

$$F_{d1-2} = \frac{1}{\pi}\left[\tan^{-1}\frac{1}{Y} + \frac{Y}{2}\ln\frac{Y^2(X^2+Y^2+1)}{(Y^2+1)(X^2+Y^2)} - \frac{Y}{\sqrt{X^2+Y^2}}\tan^{-1}\frac{1}{\sqrt{X^2+Y^2}}\right]$$

28

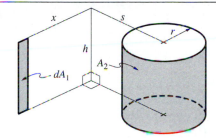

Differential strip element to exterior of right-circular cylinder of finite length; strip and cylinder are parallel and of equal length; plane containing strip does not intersect cylinder

$$S = s/r, \quad X = x/r, \quad H = h/r$$

$$A = H^2 + S^2 + X^2 - 1$$

$$B = H^2 - S^2 - X^2 + 1$$

$$F_{d1-2} = \frac{S}{S^2+X^2}\left(1 - \frac{1}{\pi}\left[\cos^{-1}\frac{B}{A} - \frac{\sqrt{A^2+4H^2}}{2H}\cos^{-1}\frac{B}{A\sqrt{S^2+X^2}}\right.\right.$$

$$\left.\left. + \frac{B}{2H}\sin^{-1}\frac{1}{\sqrt{S^2+X^2}}\right]\right) - \frac{A}{4H}$$

29

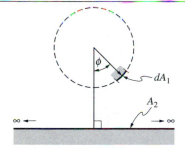

Differential strip element of any length on exterior of cylinder to plane of infinite length and width

$$F_{d1-2} = \tfrac{1}{2}(1 + \cos\phi)$$

30

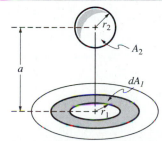

Differential ring element on surface of disk to coaxial sphere

$$R_1 = r_1/a, \quad R_2 = r_2/a$$

$$F_{d1-2} = \frac{R_2^2}{(1+R_1^2)^{3/2}}$$

31

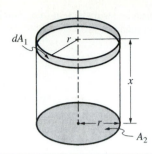
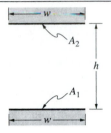

Differential ring element on interior of right-circular cylinder to circular disk at end of cylinder

$$X = x/2r$$

$$F_{d1-2} = \frac{X^2 + \frac{1}{2}}{\sqrt{X^2 + 1}} - X$$

32

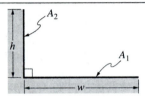

Two infinitely long, directly opposed parallel plates of the same finite width

$$H = h/w$$

$$F_{1-2} = F_{2-1} = \sqrt{1 + H^2} - H$$

33

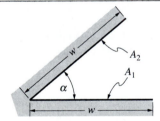

Two infinitely long plates of unequal widths h and w, having one common edge, and at an angle of $90°$ to each other

$$H = h/w$$

$$F_{1-2} = \frac{1}{2}\left(1 + H - \sqrt{1 + H^2}\right)$$

34

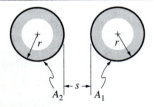

Two infinitely long plates of equal finite width w, having one common edge, forming a wedge-like groove with opening angle α

$$F_{1-2} = F_{2-1} = 1 - \sin\frac{\alpha}{2}$$

35

Infinitely long parallel cylinders of the same diameter

$$X = 1 + \frac{s}{2r}$$

$$F_{1-2} = \frac{1}{\pi}\left(\sin^{-1}\frac{1}{X} + \sqrt{X^2 - 1} - X\right)$$

36

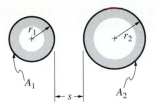

Two infinite parallel cylinders of different radius

$$R = r_2/r_1, \quad S = s/r_1,$$
$$C = 1 + R + S$$

$$F_{1-2} = \frac{1}{2\pi} \left\{ \pi + \sqrt{C^2 - (R+1)^2} - \sqrt{C^2 - (R-1)^2} \right. $$
$$\left. + (R-1)\cos^{-1}\frac{R-1}{C} - (R+1)\cos^{-1}\frac{R+1}{C} \right\}$$

37

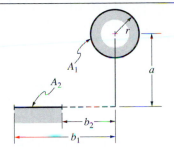

Exterior of infinitely long cylinder to unsymmetrically placed, infinitely long parallel rectangle; $r \le a$

$$B_1 = b_1/a, \quad B_2 = b_2/a$$

$$F_{1-2} = \frac{1}{2\pi}\left(\tan^{-1}B_1 - \tan^{-1}B_2\right)$$

38

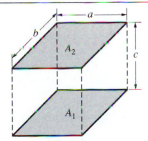

Identical, parallel, directly opposed rectangles

$$X = a/c, \quad Y = b/c$$

$$F_{1-2} = \frac{2}{\pi XY}\left\{ \ln\left[\frac{(1+X^2)(1+Y^2)}{1+X^2+Y^2}\right]^{1/2} + X\sqrt{1+Y^2}\,\tan^{-1}\frac{X}{\sqrt{1+Y^2}} \right.$$
$$\left. + Y\sqrt{1+X^2}\,\tan^{-1}\frac{Y}{\sqrt{1+X^2}} - X\tan^{-1}X - Y\tan^{-1}Y \right\}$$

39

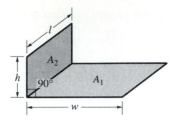

Two finite rectangles of same length, having one common edge, and at an angle of 90° to each other

$$H = h/l, \quad W = w/l$$

$$F_{1-2} = \frac{1}{\pi W}\left(W\tan^{-1}\frac{1}{W} + H\tan^{-1}\frac{1}{H} - \sqrt{H^2+W^2}\,\tan^{-1}\frac{1}{\sqrt{H^2+W^2}} \right.$$

$$\left. +\frac{1}{4}\ln\left\{ \frac{(1+W^2)(1+H^2)}{1+W^2+H^2}\left[\frac{W^2(1+W^2+H^2)}{(1+W^2)(W^2+H^2)}\right]^{W^2}\left[\frac{H^2(1+H^2+W^2)}{(1+H^2)(H^2+W^2)}\right]^{H^2}\right\} \right)$$

40

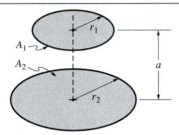

Disk to parallel coaxial disk of unequal radius

$$R_1 = r_1/a, \quad R_2 = r_2/a$$

$$X = 1 + \frac{1+R_2^2}{R_1^2}$$

$$F_{1-2} = \frac{1}{2}\left\{ X - \sqrt{X^2 - 4\left(\frac{R_2}{R_1}\right)^2} \right\}$$

41

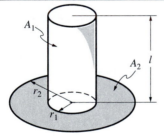

Outer surface of cylinder to annular disk at end of cylinder

$$R = r_1/r_2, \quad L = l/r_2,$$

$$A = L^2 + R^2 - 1,$$

$$B = L^2 - R^2 + 1$$

$$F_{1-2} = \frac{A}{8RL} + \frac{1}{2\pi}\left[\cos^{-1}\frac{A}{B} - \frac{1}{2L}\sqrt{\frac{(A+1)^2}{R^2} - 4}\,\cos^{-1}\frac{AR}{B} - \frac{A}{2RL}\sin^{-1}R\right]$$

42

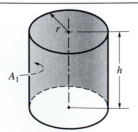

Inside surface of right-circular cylinder to itself

$$H = h/2r$$

$$F_{1-1} = 1 + H - \sqrt{1+H^2}$$

43

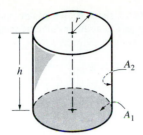

Base of right-circular cylinder to inside surface of cylinder

$$H = h/2r$$

$$F_{1-2} = 2H \left[\sqrt{1 + H^2} - H \right]$$

44

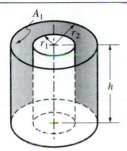

Interior of finite-length, right-circular coaxial cylinder to itself

$$R = r_2/r_1, \quad H = h/r_1$$

$$F_{1-1} = 1 - \frac{1}{R} - \frac{H^2 + 4R^2 - H}{4R} + \frac{1}{\pi} \left(\frac{2}{R} \tan^{-1} \frac{2\sqrt{R^2 - 1}}{H} \right.$$

$$\left. - \frac{H}{2R} \left\{ \frac{\sqrt{4R^2 + H^2}}{H} \sin^{-1} \frac{H^2 + 4(R^2 - 1) - 2H^2/R^2}{H^2 + 4(R^2 - 1)} - \sin^{-1} \frac{R^2 - 2}{R^2} \right\} \right)$$

45

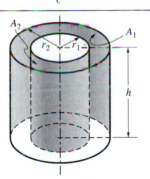

Interior of outer right-circular cylinder of finite length to exterior of inner right-circular coaxial cylinder

$$R = r_2/r_1, \quad H = h/r_1$$

$$F_{1-2} = \frac{1}{R} \left(1 - \frac{H^2 + R^2 - 1}{4H} - \frac{1}{\pi} \left\{ \cos^{-1} \frac{H^2 - R^2 + 1}{H^2 + R^2 - 1} \right. \right.$$

$$\left. \left. - \frac{\sqrt{(H^2 + R^2 + 1)^2 - 2R^2}}{2H} \cos^{-1} \frac{H^2 - R^2 + 1}{R(H^2 + R^2 - 1)} - \frac{H^2 - R^2 + 1}{2H} \sin^{-1} \frac{1}{R} \right\} \right)$$

46

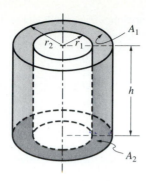

Interior of outer right-circular cylinder of finite length to annular end enclosing space between coaxial cylinders

$$H = h/r_2, \quad R = r_1/r_2,$$
$$X = \sqrt{1 - R^2},$$
$$Y = \frac{R(1 - R^2 - H^2)}{1 - R^2 + H^2}$$

$$F_{1-2} = \frac{1}{\pi} \left\{ R \left(\tan^{-1} \frac{X}{H} - \tan^{-1} \frac{2X}{H} \right) + \frac{H}{4} \left[\sin^{-1}(2R^2 - 1) - \sin^{-1} R \right] \right.$$
$$+ \frac{X^2}{4H} \left(\frac{\pi}{2} + \sin^{-1} R \right) - \frac{\sqrt{(1 + R^2 + H^2)^2 - 4R^2}}{4H} \left(\frac{\pi}{2} + \sin^{-1} Y \right)$$
$$\left. + \frac{\sqrt{4 + H^2}}{4} \left[\frac{\pi}{2} + \sin^{-1} \left(1 - \frac{2R^2 H^2}{4X^2 + H^2} \right) \right] \right\}$$

47

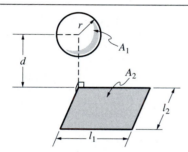

Sphere to rectangle, $r < d$

$$D_1 = d/l_1, \quad D_2 = d/l_2$$
$$F_{1-2} = \frac{1}{4\pi} \tan^{-1} \sqrt{\frac{1}{D_1^2 + D_2^2 + D_1^2 D_2^2}}$$

48

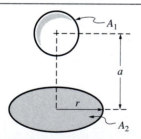

Sphere to coaxial disk

$$R = r/a$$
$$F_{1-2} = \frac{1}{2} \left[1 - \frac{1}{\sqrt{1 + R^2}} \right]$$

49

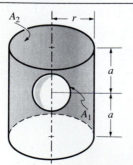

Sphere to interior surface of coaxial right-circular cylinder; sphere within ends of cylinder

$$R = r/a$$
$$F_{1-2} = \frac{1}{\sqrt{1 + R^2}}$$

50

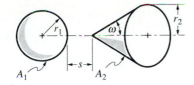

Sphere to coaxial cone

$$S = s/r_1, \quad R = r_2/r_1$$

for $\omega \geq \sin^{-1} \dfrac{1}{S+1}$:

$$F_{1-2} = \frac{1}{2}\left[1 - \frac{1+S+R\cot\phi}{\sqrt{(1+S+R\cot\phi)^2+R^2}} \right]$$

51

Infinite plane to row of cylinders

$$F_{1-2} = \frac{1}{\pi}\left(\cos^{-1}\frac{D}{s} + \frac{s}{D} - \sqrt{\left(\frac{s}{D}\right)^2 - 1} \right)$$

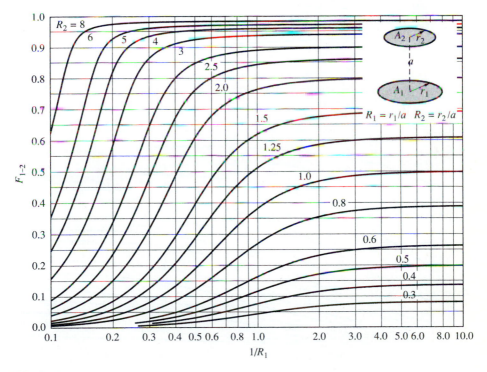

FIGURE D-1
View factor between parallel, coaxial disks of unequal radius (Configuration 40).

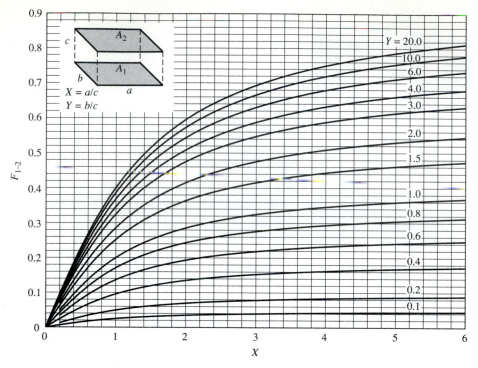

FIGURE D-2
View factor between identical, parallel, directly opposed rectangles (Configuration 38).

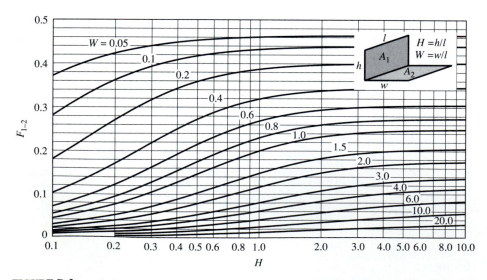

FIGURE D-3
View factor between perpendicular rectangles with common edge (Configuration 39).

References

1. Howell, J. R.: *A Catalog of Radiation Configuration Factors*, McGraw Hill, New York, 1982.
2. Hamilton, D. C., and W. R. Morgan: "Radiant Interchange Configuration Factors," *NASA TN 2836*, 1952.
3. Hottel, H. C.: "Radiant Heat Transmission," in *Heat Transmission*, ed. W. H. McAdams, 3d ed., ch. 4, McGraw-Hill, New York, 1954.
4. Leuenberger, H., and R. A. Pearson: "Compilation of Radiant Shape Factors for Cylindrical Assemblies," ASME paper no. 56-A-144, 1956.
5. Jakob, M.: *Heat Transfer*, vol. 2, John Wiley & Sons, New York, 1957.
6. Usiskin, C. M., and R. Siegel: "Thermal Radiation from a Cylindrical Enclosure with Specified Wall Heat Flux," *ASME Journal of Heat Transfer*, vol. 82C, pp. 369–374, 1960.
7. Cunningham, F. G.: "Power Input to a Small Flat Plate from a Diffusely Radiating Sphere with Application to Earth Satellites," *NASA TN D-710*, 1961.
8. Moon, P.: *Scientific Basis of Illuminating Engineering*, Dover Publications, New York, 1961 (originally published by McGraw-Hill, New York, 1936).
9. Nichols, L. D.: "Surface-Temperature Distribution on Thin-Walled Bodies Subjected to Solar Radiation in Interplanetary Space," *NASA TN D-584*, 1961.
10. Robbins, W. H.: "An Analysis of Thermal Radiation Heat Transfer in a Nuclear-Rocket Nozzle," *NASA TN D-586*, 1961.
11. Buschman, A. J., and C. M. Pittman: "Configuration Factors for Exchange of Radiant Energy between Axisymmetrical Sections of Cylinders, Cones, and Hemispheres and Their Bases," *NASA TN D-944*, 1961.
12. Sparrow, E. M., and J. L. Gregg: "Radiant Interchange Between Circular Disks Having Arbitrarily Different Temperatures," *ASME Journal of Heat Transfer*, vol. 83C, pp. 494–502, 1961.
13. Stephens, C. W., and A. M. Haire: "Internal Design Considerations for Cavity-Type Solar Absorbers," *American Rocket Society Journal*, vol. 31, no. 7, pp. 896–901, 1961.
14. Goetze, D., and C. B. Grosch: "Earth-Emitted Infrared Radiation Incident Upon a Satellite," *Journal of Aerospace Science*, vol. 29, no. 11, pp. 521–524, 1962.
15. Robbins, W. H., and C. A. Todd: "Analysis, Feasibility, and Wall-Temperature Distribution of a Radiation-Cooled Nuclear-Rocket Nozzle," *NASA TN D-878*, 1962.
16. Sparrow, E. M., and E. R. G. Eckert: "Radiant Interaction between Fin and Base Surfaces," *ASME Journal of Heat Transfer*, vol. 84C, no. 1, pp. 12–18, 1962.
17. Sparrow, E. M., L. U. Albers, and E. R. G. Eckert: "Thermal Radiation Characteristics of Cylindrical Enclosures," *ASME Journal of Heat Transfer*, vol. 84C, pp. 73–81, 1962.
18. Sparrow, E. M., and V. K. Jonsson: "Absorption and Emission Characteristics of Diffuse Spherical Enclosures," *NASA TN D-1289*, 1962.
19. Sparrow, E. M., and V. K. Jonsson: "Absorption and Emission Characteristics of Diffuse Spherical Enclosures," *ASME Journal of Heat Transfer*, vol. 84, pp. 188–189, 1962.
20. Sparrow, E. M., G. B. Miller, and V. K. Jonsson: "Radiative Effectiveness of Annular-Finned Space Radiators, Including Mutual Irradiation between Radiator Elements," *Journal of the Aerospace Sciences*, vol. 29, no. 11, pp. 1291–1299, 1962.
21. Sparrow, E. M., and V. K. Jonsson: "Radiant Emission Characteristics of Diffuse Conical Cavities," *Journal of the Optical Society of America*, vol. 53, pp. 816–821, 1963.
22. Sparrow, E. M., and V. K. Jonsson: "Thermal Radiation Absorption in Rectangular-Groove Cavities," *ASME Journal of Applied Mechanics*, vol. E30, pp. 237–234, 1963.
23. Wiebelt, J. A., and S. Y. Ruo: "Radiant-Interchange Configuration Factors for Finite Right Circular Cylinders to Rectangular Planes," *International Journal of Heat and Mass Transfer*, vol. 6, no. 2, pp. 143–146, 1963.
24. Morizumi, S. J.: "Analytical Determination of Shape Factors from a Surface Element to an Axisymmetric Surface," *AIAA Journal*, vol. 2, no. 11, pp. 2028–2030, 1964.
25. Sotos, C. J., and N. O. Stockman: "Radiant Interchange View Factors and Limits of Visibility for Differential Cylindrical Surfaces with Parallel Generating Lines," *NASA TN D-2556*, 1964.
26. Watts, R. G.: "Radiant Heat Transfer to Earth Satellites," *ASME Journal of Heat Transfer*, vol. 87, no. 3, pp. 369–373, 1965.

27. Bien, D. D.: "Configuration Factors for Thermal Radiation from Isothermal Inner Walls of Cones and Cylinders," *ASME Journal of Heat Transfer*, vol. 3, no. 1, pp. 155–156, 1966.
28. Bobco, R. P.: "Radiation from Conical Surfaces with Nonuniform Radiosity," *AIAA Journal*, vol. 4, no. 3, pp. 544–546, 1966.
29. Feingold, A.: "Radiant-Interchange Configuration Factors between Various Selected Plane Surfaces," *Proceedings of the Royal Society London*, vol. 292, no. 1428, pp. 51–60, 1966.
30. Kezios, S. P., and W. Wulff: "Radiative Heat Transfer through Openings of Variable Cross Sections," *Third International Heat Transfer Conference, AIChE*, vol. 5, pp. 207–218, 1966.
31. Wiebelt, J. A.: *Engineering Radiation Heat Transfer*, Holt, Rinehart & Winston, New York, 1966.
32. Hottel, H. C., and A. F. Sarofim: *Radiative Transfer*, McGraw-Hill, New York, 1967.
33. Hsu, C. J.: "Shape Factor Equations for Radiant Heat Transfer between Two Arbitrary Sizes of Rectangular Planes," *Canadian Journal of Chemical Engineering*, vol. 45, no. 1, pp. 58–60, 1967.
34. Liebert, C. H., and R. R. Hibbard: "Theoretical Temperatures of Thin-Film Solar Cells in Earth Orbit," *NASA TN D-4331*, 1968.
35. Campbell, J. P., and D. G. McConnell: "Radiant Interchange Configuration Factors for Spherical and Conical Surfaces to Spheres," *NASA TN D-4457*, 1968.
36. Grier, N. T.: "Tabulations of Configuration Factors Between and Two Spheres and Their Parts," *NASA SP-3050*, 1969.
37. Grier, N. T., and R. D. Sommers: "View Factors for Toroids and Their Parts," *NASA TN D-5006*, 1969.
38. Sommers, R. D., and N. T. Grier: "Radiation View Factors for a Toroid: Comparison of Eckert's Technique and Direct Computation," *ASME Journal of Heat Transfer*, vol. 91, no. 3, pp. 459–461, 1969.
39. Wakao, N., K. Kato, and N. Furuya: "View Factor Between Two Hemispheres in Contact and Radiation Heat-Transfer Coefficient in Packed Beds," *International Journal of Heat and Mass Transfer*, vol. 12, pp. 118–120, 1969.
40. Feingold, A., and K. G. Gupta: "New Analytical Approach to the Evaluation of Configuration Factors in Radiation from Spheres and Infinitely Long Cylinders," *ASME Journal of Heat Transfer*, vol. 92, no. 1, pp. 69–76, 1970.
41. Rein, R. G., C. M. Sliepcevich, and J. R. Welker: "Radiation View Factors for Tilted Cylinders," *Journal of Fire and Flammability*, vol. 1, pp. 140–153, 1970.
42. Crawford, M.: "Configuration Factor Between Two Unequal, Parallel, Coaxial Squares," ASME paper no. 72-WT/HT-16, November 1972.
43. Chung, B. T. F., and P. S. Sumitra: "Radiation Shape Factors from Plane Point Sources," *ASME Journal of Heat Transfer*, vol. 94, pp. 328–330, 1972.
44. Ballance, J. O., and J. Donovan: "Radiation Configuration Factors for Annular Rings and Hemispherical Sectors," *ASME Journal of Heat Transfer*, vol. 95, no. 2, pp. 275–276, 1973.
45. Reid, R. L., and J. S. Tennant: "Annular Ring View Factors," *AIAA Journal*, vol. 11, no. 10, pp. 1446–1448, 1973.
46. Holchendler, J., and W. F. Laverty: "Configuration Factors for Radiant Heat Exchange in Cavities Bounded at the Ends by Parallel Disks and Having Conical Centerbodies," *ASME Journal of Heat Transfer*, vol. 96, no. 2, pp. 254–257, 1974.
47. Rea, S. N.: "Rapid Method for Determining Concentric Cylinder Radiation View Factors," *AIAA Journal*, vol. 13, no. 8, pp. 1122–1123, 1975.
48. Juul, N. H.: "Diffuse Radiation Configuration View Factors Between Two Spheres and Their Limits," *Letters in Heat and Mass Transfer*, vol. 3, no. 3, pp. 205–211, 1976.
49. Juul, N. H.: "Investigation of Approximate Methods for Calculation of the Diffuse Radiation Configuration View Factors Between Two Spheres," *Letters in Heat and Mass Transfer*, vol. 3, pp. 513–522, 1976.
50. Minning, C. P.: "Calculation of Shape Factors between Parallel Ring Sectors Sharing a Common Centerline," *AIAA Journal*, vol. 14, no. 6, pp. 813–815, 1976.
51. Minning, C. P.: "Calculation of Shape Factors between Rings and Inverted Cones Sharing a Common Axis," *ASME Journal of Heat Transfer*, vol. 99, no. 3, pp. 492–494, 1977.
52. Feingold, A.: "A New Look at Radiation Configuration Factors Between Disks," *ASME Journal of Heat Transfer*, vol. 100, no. 4, pp. 742–744, 1978.

53. Sparrow, E. M., and R. D. Cess: *Radiation Heat Transfer*, Hemisphere, New York, 1978.
54. Chekhovskii, I. R., V. V. Sirotkin, V. C. Yu, and V. A. Chebanov: "Determination of Radiative View Factors for Rectangles of Different Sizes," *High Temperature*, July 1979.
55. Juul, N. H.: "Diffuse Radiation View Factors from Differential Plane Sources to Spheres," *ASME Journal of Heat Transfer*, vol. 101, no. 3, pp. 558–560, 1979.
56. Minning, C. P.: "Shape Factors Between Coaxial Annular Disks Separated by a Solid Cylinder," *AIAA Journal*, vol. 17, no. 3, pp. 318–320, 1979.
57. Modest, M. F.: "Solar Flux Incident on an Orbiting Surface after Reflection from a Planet," *AIAA Journal*, vol. 18, no. 6, pp. 727–730, 1980.
58. Gross, U., K. Spindler, and E. Hahne: "Shape Factor Equations for Radiation Heat Transfer Between Plane Rectangular Surfaces of Arbitrary Position and Size with Parallel Boundaries," *Letters in Heat and Mass Transfer*, vol. 8, p. 219, 1981.
59. Chung, T. J., and J. Y. Kim: "Radiation View Factors by Finite Elements," *ASME Journal of Heat Transfer*, vol. 104, p. 792, 1982.
60. Chung, B. T. F., and M. H. N. Naraghi: "A Simpler Formulation for Radiative View Factors From Spheres to a Class of Axisymmetric Bodies," *ASME Journal of Heat Transfer*, vol. 104, p. 201, 1982.
61. Juul, N. H.: "View Factors in Radiation Between Two Parallel Oriented Cylinders," *ASME Journal of Heat Transfer*, vol. 104, p. 235, 1982.
62. Naraghi, M. H. N., and B. T. F. Chung: "Radiation Configuration Factors Between Disks and a Class of Axisymmetric Bodies," *ASME Journal of Heat Transfer*, vol. 104, p. 426, 1982.
63. Mahbod, B., and R. L. Adams: "Radiation View Factors Between Axisymmetric Subsurfaces within a Cylinder with Spherical Centerbody," *ASME Journal of Heat Transfer*, vol. 106, no. 1, p. 244, 1984.
64. Eichberger, J. I.: "Calculation of Geometric Configuration Factors in an Enclosure Whose Boundary Is Given by an Arbitrary Polygon in the Plane," *Wärme- und Stoffübertragung*, vol. 19, no. 4, p. 269, 1985.
65. Mathiak, F. U.: "Calculation of Form-Factors for Plane Areas with Polygonal Boundaries," *Wärme- und Stoffübertragung*, vol. 19, no. 4, p. 273, 1985.
66. Shapiro, A. B.: "Computer Implementation, Accuracy and Timing of Radiation View Factor Algorithms," *ASME Journal of Heat Transfer*, vol. 107, no. 3, pp. 730–732, 1985.
67. Maxwell, G. M., M. J. Bailey, and V. W. Goldschmidt: "Calculations of the Radiation Configuration Factor Using Ray Casting," *Computer Aided Design*, vol. 18, no. 7, p. 371, 1986.
68. Stefanizzi, P.: "Reliability of the Monte Carlo Method in Black Body View Factor Determination," *Termotecnica*, vol. 40, no. 6, p. 29, 1986.
69. Eddy, T. L., and G. E. Nielsson: "Radiation Shape Factors for Channels with Varying Cross Sections," *ASME Journal of Heat Transfer*, vol. 110, no. 1, p. 264, 1988.
70. Frankel, J. I., and T. P. Wang: "Radiative Exchange Between Gray Fins Using a Coupled Integral Equation Formulation," *Journal of Thermophysics and Heat Transfer*, vol. 2, no. 4, pp. 296–302, Oct 1988.
71. Modest, M. F.: "Radiative Shape Factors Between Differential Ring Elements on Concentric Axisymmetric Bodies," *Journal of Thermophysics and Heat Transfer*, vol. 2, no. 1, pp. 86–88, 1988.
72. Mel'man, M. M., and G. G. Trayanov: "View Factors in a System of Parallel Contacting Cylinders," *Journal of Engineering Physics*, vol. 54, no. 4, p. 401, 1988.
73. Naraghi, M. H. N.: "Radiation View Factors from Differential Plane Sources to Disks—A General Formulation," *Journal of Thermophysics and Heat Transfer*, vol. 2, no. 3, pp. 271–274, 1988.
74. Sabet, M., and B. T. F. Chung: "Radiation View Factors from a Sphere to Nonintersecting Planar Surfaces," *Journal of Thermophysics and Heat Transfer*, vol. 2, no. 3, pp. 286–288, 1988.
75. Naraghi, M. H. N.: "Radiative View Factors from Spherical Segments to Planar Surfaces," *Journal of Thermophysics and Heat Transfer*, vol. 2, no. 4, pp. 373–375, Oct 1988.
76. Chung, B. T. F., and M. M. Kermani: "Radiation View Factors from a Finite Rectangular Plate," *ASME Journal of Heat Transfer*, vol. 111, no. 4, p. 1115, 1989.
77. van Leersum, J.: "A Method for Determining a Consistent Set of Radiation View Factors from a Set Generated by a Nonexact Method," *International Journal of Heat and Fluid Flow*, vol. 10, no. 1, p. 83, 1989.

78. Emery, A. F., O. Johansson, M. Lobo, and A. Abrous: "A Comparative Study of Methods for Computing the Diffuse Radiation Viewfactors for Complex Structures," *ASME Journal of Heat Transfer*, vol. 113, no. 2, pp. 413–422, 1991.
79. Rushmeier, H. E., D. R. Baum, and D. E. Hall: "Accelerating the Hemi-Cube Algorithm for Calculating Radiation Form Factors," *ASME Journal of Heat Transfer*, vol. 113, no. 4, pp. 1044–1047, 1991.
80. Sika, J.: "Evaluation of Direct-Exchange Areas for a Cylindrical Enclosure," *ASME Journal of Heat Transfer*, vol. 113, no. 4, pp. 1040–1043, 1991.

EXPONENTIAL INTEGRAL FUNCTIONS

The exponential integral functions $E_n(x)$ and their derivatives occur frequently in radiative heat transfer calculations; therefore, a summary of their properties as well as a brief tabulation are given here. More detailed discussions of their properties may be found in the books by Chandrasekhar [1] and Kourganoff [2], or in mathematical handbooks such as [3]. Detailed tabulations are given in [3], and formulae for their numerical evaluation are listed in [3, 4].

The *exponential integral of order n* is defined as

$$E_n(x) = \int_1^\infty e^{-xt} \frac{dt}{t^n}, \quad n = 0, 1, 2, \ldots, \tag{E.1}$$

or, setting $\mu = 1/t$,

$$E_n(x) = \int_0^1 e^{-x/\mu} \mu^{n-2} d\mu, \quad n = 0, 1, 2, \ldots. \tag{E.2}$$

Differentiating equation (E.1), a first recurrence relationship is found as

$$\frac{dE_n}{dx}(x) = -E_{n-1}(x), \quad n = 1, 2, \ldots, \tag{E.3}$$

where

$$E_0(x) = \int_1^\infty e^{-xt}\,dt = \frac{e^{-x}}{x}. \tag{E.4}$$

A second recurrence is found by integrating equation (E.3), or

$$\int_x^\infty E_n(x)\,dx = E_{n+1}(x), \quad n = 0, 1, 2, \ldots. \tag{E.5}$$

An algebraic recurrence between consecutive orders may be obtained by integrating equation (E.1) by parts, or

$$E_{n+1}(x) = \frac{1}{n}\left[e^{-x} - xE_n(x)\right], \quad n = 1, 2, 3, \ldots. \tag{E.6}$$

The integral of equation (E.1) may be solved in a general series expansion as [3]

$$E_n(x) = \frac{(-x)^{n-1}}{(n-1)!}(\psi_n - \ln x) + \sum_{\substack{m=0 \\ m \neq n-1}}^\infty \frac{(-x)^m}{m!(n-1-m)}, \quad n = 1, 2, 3, \ldots, \tag{E.7a}$$

where

$$\psi_n = \begin{cases} -\gamma_E, & n = 1, \\ -\gamma_E + \displaystyle\sum_{m=1}^{n-1} \frac{1}{m}, & n \geq 2, \end{cases} \tag{E.7b}$$

and

$$\gamma_E = \int_1^\infty \left(1 - e^{-t}\right)\frac{dt}{t} = 0.577216\ldots \tag{E.7c}$$

is known as *Euler's constant*. Substituting values for n, one obtains

$$E_1(x) = -(\gamma_E + \ln x) + x - \frac{x^2}{2!2} + \frac{x^3}{3!3} - \frac{x^4}{4!4} + - \ldots, \tag{E.8}$$

$$E_2(x) = 1 + x(\gamma_E - 1 + \ln x) - \frac{x^2}{2!1} + \frac{x^3}{3!2} - \frac{x^4}{4!3} + - \ldots, \tag{E.9}$$

$$E_3(x) = \frac{1}{2} - x + \frac{x^2}{2}\left(-\gamma_E + \frac{3}{2} - \ln x\right) + \frac{x^3}{3!1} - \frac{x^4}{4!2} + - \ldots. \tag{E.10}$$

A function related to E_1, which often occurs in radiation calculations, is

$$\mathrm{Ein}(x) = \int_0^1 \left(1 - e^{-xt}\right)\frac{dt}{t} = \int_0^\infty \left(1 - e^{-xe^{-\xi}}\right)d\xi$$

$$= E_1(x) + \ln x + \gamma_E = x - \frac{x^2}{2!2} + \frac{x^3}{3!3} - + \ldots. \tag{E.11}$$

For vanishing values of x it follows from equation (E.7), or directly integrating equation (E.1),

$$E_n(0) = \begin{cases} +\infty, & n = 1, \\ \dfrac{1}{n-1}, & n \geq 2, \end{cases} \tag{E.12a}$$

$$\text{Ein}(0) = 0. \tag{E.12b}$$

For large values of x, the asymptotic expansion for the exponential integrals is given by [3]

$$E_n(x) = \frac{e^{-x}}{x}\left[1 - \frac{n}{x} + \frac{n(n+1)}{x^2} - \frac{n(n+1)(n+2)}{x^3} + - \ldots\right], \quad n = 0, 1, 2, \ldots$$

$$\tag{E.13}$$

To estimate the relative magnitude of different orders of exponential integrals, the following inequalities are sometimes handy [3]:

$$\frac{n-1}{n}E_n(x) < E_{n+1}(x) < E_n(x), \quad n = 1, 2, 3, \ldots, \tag{E.14}$$

$$\frac{1}{x+n} < e^x E_n(x) \leq \frac{1}{x+n-1}, \quad n = 1, 2, 3, \ldots. \tag{E.15}$$

References

1. Chandrasekhar, S.: *Radiative Transfer*, Dover Publications, New York, 1960 (originally published by Oxford University Press, London, 1950).
2. Kourganoff, V.: *Basic Methods in Transfer Problems*, Dover Publications, New York, 1963.
3. Abramowitz, M., and I. A. Stegun (eds.): *Handbook of Mathematical Functions*, Dover Publications, New York, 1965.
4. Breig, W. F., and A. L. Crosbie: "Numerical Computation of a Generalized Exponential Integral Function," *Math Comp.*, vol. 28, no. 126, pp. 575–579, 1974.

TABLE E.1
Values of exponential integral functions

x	Ein	E_1	E_2	E_3	E_4
0.00	0.000000	∞	1.000000	0.500000	0.333333
0.01	0.009975	4.037929	0.949671	0.490277	0.328382
0.02	0.019900	3.354707	0.913105	0.480968	0.323526
0.03	0.029776	2.959118	0.881672	0.471998	0.318762
0.04	0.039603	2.681263	0.853539	0.463324	0.314085
0.05	0.049382	2.467898	0.827835	0.454919	0.309494
0.06	0.059112	2.295307	0.804046	0.446761	0.304986
0.07	0.068794	2.150838	0.781835	0.438833	0.300559
0.08	0.078428	2.026941	0.760961	0.431120	0.296209
0.09	0.088015	1.918744	0.741244	0.423610	0.291935
0.10	0.097554	1.822924	0.722545	0.416291	0.287736
0.15	0.144557	1.464461	0.641039	0.382276	0.267789
0.20	0.190428	1.222650	0.574201	0.351945	0.249447
0.25	0.235204	1.044283	0.517730	0.324684	0.232543
0.30	0.278920	0.905677	0.469115	0.300042	0.216935
0.35	0.321609	0.794215	0.426713	0.277669	0.202501
0.40	0.363305	0.702380	0.389368	0.257286	0.189135
0.45	0.404039	0.625331	0.356229	0.238663	0.176743
0.50	0.443842	0.559773	0.326644	0.221604	0.165243
0.60	0.520769	0.454379	0.276184	0.191551	0.144627
0.70	0.594310	0.373769	0.234947	0.166061	0.126781
0.80	0.664669	0.310597	0.200852	0.144324	0.111290
0.90	0.732039	0.260184	0.172404	0.125703	0.097812
1.00	0.796600	0.219384	0.148496	0.109692	0.086062
1.10	0.858517	0.185991	0.128281	0.095881	0.075801
1.20	0.917946	0.158408	0.111104	0.083935	0.066824
1.30	0.975031	0.135451	0.096446	0.073576	0.058961
1.40	1.029907	0.116219	0.083890	0.064576	0.052064
1.50	1.082700	0.100020	0.073101	0.056739	0.046007
1.60	1.133528	0.086308	0.063803	0.049906	0.040682
1.70	1.182499	0.074655	0.055771	0.043937	0.035997
1.80	1.229716	0.064713	0.048815	0.038716	0.031870
1.90	1.275274	0.056204	0.042780	0.034143	0.028232
2.00	1.319263	0.048901	0.037534	0.030133	0.025023
2.10	1.361767	0.042614	0.032966	0.026614	0.022189
2.20	1.402864	0.037191	0.028983	0.023521	0.019686
2.30	1.442627	0.032502	0.025504	0.020800	0.017473
2.40	1.481125	0.028440	0.022461	0.018405	0.015515
2.50	1.518421	0.024915	0.019798	0.016295	0.013782
3.00	1.688876	0.013048	0.010642	0.008931	0.007665
3.50	1.836949	0.006970	0.005802	0.004945	0.004296
4.00	1.967289	0.003779	0.003198	0.002761	0.002423
4.50	2.083366	0.002073	0.001779	0.001552	0.001374
5.00	2.187802	0.001148	0.000996	0.000878	0.000783

ACKNOWLEDGMENTS

Figs. 1-1, 1-6, 1-7, 1-9: Incropera, F. P., and DeWitt, D. P. *Fundamentals of Heat and Mass Transfer,* 3rd ed. Copyright © 1990 John Wiley & Sons, Inc., New York. Reprinted with permission of John Wiley & Sons, Inc.

Fig. 1-3: Courtesy of NASA.

Fig. 1-14: White, F. M. *Heat Transfer,* © 1984 by Addison-Wesley Publishing Co. Reprinted by permission of Addison-Wesley Publishing Co., Inc., Reading, MA.

Fig. 1-16: Edwards, D. K. "Radiation Interchange in a Nongray Enclosure Containing an Isothermal CO_2–N_2 Gas Mixture," *ASME Journal of Heat Transfer,* vol. 84C, 1962, pp. 1–11.

Fig. 1-17: Meyorott, R. E., Sokoloff, J., and Nicholls, R. W. "Absorption Coefficients of Air," in *Air Force Cambridge Research Center,* 1960, Geophysics Research Paper No. 68.

Fig. 1-18: Neuroth, N. Der Einfluss der Temperatur auf die Spektrale Absorption von Gläsern im Ultraroten, I (Effect of Temperature on Spectral Absorption of Glasses in the Infrared), *Glastech Ber.* 25, 242–249, 1952.

Figs. 2-2, 3-2, 3-4, 20-1*a*, *b*, 20-3, 20-12, 20-15: Özişik, M. N. *Radiative Transfer and Interactions with Conduction and Convection,* John Wiley & Sons, Inc., New York, 1973. Reprinted by permission of the author.

Figs. 2-10, 3-6, 4-21, 9-7, 10-12, 10-13: Siegel, R., and Howell, J. R. *Thermal Radiation Heat Transfer,* Hemisphere Publishing Corp., New York, 2nd ed., 1981. Reprinted by permission.

Figs. 2-12, 2-14, 10-9, 10-16: Bohren, C. F., and Hoffman, D. R. *Absorption and Scattering of Light by Small Particles.* Copyright © 1983 John Wiley & Sons, Inc., New York. Reprinted with permission of John Wiley & Sons, Inc.

Fig. 3-1*a, b*: Schmidt, E., and Eckert, E. R. G. "Über die Richtungsverteilung der Wärmestrahlung von Oberflächen," *Forschung auf dem Gebiete des Ingenieurwesens,* vol. 17, 1935, p. 175.

Fig. 3-5: Torrance, K. E., and Sparrow, E. M. "Biangular Reflectance of an Electric Nonconductor as a Function of Wavelength and Surface Roughness," *ASME Journal of Heat Transfer,* vol. 87, 1965, pp. 283–292.

Figs. 3-8, 3-11: Parker, W. J., and Abbott, G. L. "Theoretical and Experimental Studies of the Total Emittance of Metals." In *Symposium on Thermal Radiation of Solids,* NASA SP-55, 1965, pp. 11–28.

Fig. 3-10: Dunkle, R. V. "Emissivity and Inter-Reflection Relationships for Infinite Parallel Specular Surfaces." In *Symposium on Thermal Radiation of Solids,* NASA SP-55, 1965, pp. 39–44.

Fig. 3-13: From: Spitzer, W. G., Kleinman, D. A., Frosch, C. J., and Walsh, D. J. "Optical Properties of Silicon Carbide." In O'Connor, J. R., and Smittens, J., eds., *Silicon Carbide—A High Temperature Semi-Conductor, Proceedings of the 1959 Conference on Silicon Carbide, Boston, MA.* Pergamon Press, 1960, pp. 347–365.

Figs. 3-14, 3-20: Jasperse, J. R., Kahan, A., Plendl, J. N., and Mitra, S. S. "Temperature Dependence of Infrared Dispersion in Ionic Crystals LiF and MgO," *Physical Review,* vol. 140, no. 2, 1966, pp. 526–542.

Fig. 3-15: Touloukian, Y. S., and DeWitt, D. P., eds. "Thermal Radiative Properties: Nonmetallic Solids." In *Thermophysical Properties of Matter,* vol. 8 (1972). Plenum Press Publishing Corp., New York.

Fig. 3-16: *American Institute of Physics Handbook,* 3rd ed. Ch. 6, McGraw-Hill Publishing Co., New York, © 1972. Reprinted with permission of McGraw-Hill Publishing Co., Inc.

Fig. 3-17: Brandenberg, W. M. "The Reflectivity of Solids at Grazing Angles." In *Measurement of Thermal Radiation of Solids,* 1963, pp. 75–82, NASA.

Fig. 3-21: Riethof, T. R., and DeSantis, V. J. "Techniques of Measuring Normal Spectral Emissivity of Conductive Refractory Compounds at High Temperatures." In *Measurement of Thermal Radiation Properties of Solids,* NASA SP-31, 1963, pp. 565–584.

Fig. 3-23: Torrance, K. E., and Sparrow, E. M. "Theory for Off-Specular Reflection from Roughened Surfaces," *Journal of the Optical Society of America,* vol. 57, no. 9, 1967, pp. 1105–1114.

Figs. 3-24: Sparrow, E. M., and Cess, R. D. *Radiation Heat Transfer,* Hemisphere Publishing Corp., New York, 1978. Reprinted by permission.

Fig. 3-25: Pezdirtz, G. F., and Jewell, R. A. "A Study of the Photodegradation of Selected Thermal Control Surfaces." In *Symposium on Thermal Radiation of Solids,* NASA SP-55, 1965, pp. 433–441.

Fig. 3-29: Fan, J. C. C., and Bachner, F. J. "Transparent Heat Mirrors for Solar-Energy Applications," *Applied Optics,* vol. 15, no. 4, pp. 1012–1017, 1976.

Fig. 3-31: Duffie, J. A., and Beckman, W. A. *Solar Energy Thermal Processes.* Copyright © 1974 John Wiley & Sons, Inc., New York. Reprinted with permission of John Wiley & Sons, Inc.

Fig. 3-32, 3-46b: From: Edwards, D. K., Nelson, K. E., Roddick, R. D., and Gier, J. T. "Basic Studies on the Use and Control of Solar Energy." Technical Report, The University of California, 1960, Report No. 60-93.

Fig. 3-34: Trombe, F., Foex, M., and Lephat, V. M. "Research on Selective Surfaces for Air Conditioning Dwellings," *Proceedings of the UN Conference on New Sources of Energy,* vol. 4, 1964, pp. 625–638.

Fig. 3-35: Brandenberg, W. M., and Clausen, O. W. "The Directional Spectral Emittance of Surfaces Between 200 and 600 C," in *Symposium on Thermal Radiation of Solids,* ed. S. Katzoff, NASA SP-55, 1965, pp. 313–319.

Figs. 3-36, 3-37, 3-38: Courtesy of the Oriel Product Catalog.

Fig. 3-40: Funai, A. I. "A Multichamber Calorimeter for High-Temperature Emittance Studies." In *Measurement of Thermal Radiation Properties of Solids,* NASA SP-31, 1963, pp. 317–327.

Figs. 3-41, 3-42: Fussell, W. B., and Stair, F. "Preliminary Studies Toward the Determination of Spectral Absorption Coefficients of Homogeneous Dielectric Material in the Infrared at Elevated Temperatures." In *Symposium on Thermal Radiation of Solids,* NASA SP-55, 1965, pp. 287–292.

Fig. 3-44: Birkebak, R. C., and Eckert, E. R. G. "Effect of Roughness of Metal on Angular Distribution of Monochromatic Related Radiation," *ASME Journal of Heat Transfer,* vol. 87, pp. 85–94. Reprinted with permission of ASME.

Fig. 3-45: Dunkle, R. V. "Spectral Reflection Measurements," in *First Symposium—Surface Effects on Spacecraft Materials,* John Wiley & Sons, Inc., New York, pp. 117–137, 1960.

Fig. 3-47: Touloukian, Y. S., and DeWitt, D. P., eds. *Thermal Radiative Properties: Metallic Elements and Alloys,* vol. 7 of *Thermophysical Properties of Matter,* Plenum Press Publishing Corp., New York, 1970.

Fig. 6-1: Sarofim, A. F., and Hottel, H. C. "Radiation Exchange Among Non-Lambert Surfaces," *ASME Journal of Heat Transfer,* vol. 88C, 1966, pp. 37–44.

Fig. 6-11: Reprinted with permission from *Solar Energy,* vol. 7, no. 3, Hollands, K. G. T. "Directional Selectivity Emittance and Absorptance Properties of Vee Corrugated Specular Surfaces," 1963, pp. 108–116.

Fig. 6-18: Toor, J. S. "Radiant Heat Transfer Analysis Among Surfaces Having Direction Dependent Properties by the Monte Carlo Method," M.S. Thesis, Purdue University, 1967.

Figs. 9-5, 9-8, 9-18, 9-19: Tien, C. L. *Thermal Radiation Properties of Gases,* vol. 5 of *Advances in Heat Transfer,* Academic Press, 1968, pp. 253–324.

Fig. 9-9: Plass, G. N., "Models for Spectral Band Absorption," *Journal of the Optical Society of America,* vol. 48, no. 10, 1958, pp. 690–703.

Figs. 9-16, 9-17: Reprinted by permission of Elsevier Science Publishing Co., Inc. from "Spectral and Total Emissivity of Water Vapor and Carbon Dioxide" by B. Leckner, *Combustion and Flame,* vol. 19, pp. 33–48. © 1972 by the Combustion Institute.

Figs. 9-20, 9-21: Tien, C. L., and Giedt, W. H. "Experimental Determination of Infrared Absorption of High-Temperature Gases," *Advances in Thermophysical Properties at Extreme Temperatures and Pressures,* ASME, 1965, pp. 167–173. Reprinted by permission of ASME.

Fig. 9-22: Bevans, J. T., Dunkle, R. V., Edwards, D. K., Gier, J. T., Levenson, L. L., and Oppenheim, A. K. "Apparatus for the Determination of the Band Absorption of Gases at Elevated Pressures and Temperatures," *Journal of the Optical Society of America,* vol. 50, 1960, pp. 130–136.

Fig. 9-23: Reprinted with permission from *Journal of Quantitative Spectroscopy and Radiative Transfer,* vol. 12, Tien, C. L., Modest, M. F. and McCreight, C. R., "Infrared Radiation Properties of Nitrous Oxide," 1972, Pergamon Press.

Fig. 10-2: Tien, C. L., and Drolen, B. L. "Thermal Radiation in Particulate Media with Dependent and Independent Scattering." In *Annual Review of Numerical Fluid Mechanics and Heat Transfer,* vol. 1, Hemisphere Publishing Corp., New York, 1987, pp. 1–32.

Figs. 10-3, 10-11: Vande Hulst, H. C. *Light Scattering by Small Particles,* Dover Publications, 1982.

Figs. 10-4, 10-5: Kattawar, G. W., and Plass, G. N. "Electromagnetic Scattering from Absorbing Spheres," *Applied Optics,* vol. 6, no. 8, 1967, pp. 1377–1383.

Figs. 10-7, 10-8, 12-4: Modest, M. F., and Azad, F. H. "The Influence and Treatment of Mie-Anisotropic Scattering in Radiative Heat Transfer," *ASME Journal of Heat Transfer,* vol. 102, 1980, pp. 92–98. Reprinted by permission of ASME.

Figs. 10-17, 10-21: Hottel, H. C., Sarofim, A. F., Vasalos, I. A., and Dalzell, W. H. "Multiple Scatter: Comparison of Theory with Experiment," *ASME Journal of Heat Transfer,* vol. 92, 1970, pp. 285–291. Reprinted by permission of ASME.

Figs. 10-18, 10-19, 10-20: Smart, C., Jacobsen, R., Kerker, M., Kratohvil, P., and Matijevic, E. "Experimental Study of Multiple Light Scattering," *Journal of the Optical Society of America,* vol. 55, no. 8, 1965, pp. 947–955.

Fig. 10-22: Reprinted with permission from *Journal of Quantitative Spectroscopy and Radiative Transfer,* vol. 33, no. 4, Crosbie, A. L., and Davidson, G. W. "Dirac-Delta Function Approximations to the Scattering Phase Function," 1985, Pergamon Press.

Fig. 10-24: Buckius, R. O., and Hwang, D. C. "Radiation Properties for Polydispersions: Application to Coal," *ASME Journal of Heat Transfer,* vol. 102, 1980, pp. 99–103. Reprinted by permission of ASME.

Fig. 10-25: Millikan, R. C. "Optical Properties of Soot," *Journal of the Optical Society of America,* vol. 51, 1961, pp. 698–699.

Fig. 10-26: Felske, J. D., Charalmpopoulos, T. T., and Hura, H. S. "Determination of Refractive Indices of Soot Particles from the Reflectivities of Compressed Soot Particles," *Combustion Science and Technology,* vol. 37, pp. 263–284, 1984.

Fig. 10-27: Reprinted with permission from *Journal of Quantitative Spectroscopy and Radiative Transfer,* vol. 8, Lee, S. C., and Tien, C. L., "Effect of Soot Shape on Soot Radiation," 1983, Pergamon Press.

Fig. 10-28: Lee, S. C., and Tien, C. L. "Optical Constants of Soot in Hydrocarbon Flames," in *Eighteenth Symposium (International) on Combustion,* 1980, pp. 1159–1166.

Figs. 11-1, 11-2: Smakula, A. "Synthetic Crystals and Polarizing Materials," *Optica Acta,* vol. 9, 1962, pp. 205–222.

Fig. 11-3: Boyd, I. W., Binnie, J. I., Wilson, B., and Colles, M. J., "Absorption of Infrared Radiation in Silicon," *Journal of Applied Physics,* vol. 55, no. 8, 1984, pp. 3061–3063.

Fig. 11-5: Barker, A. J. "The Effect of Melting on the Multiphonon Infrared Absorption Spectra of KBr, NaCL, and LiF," *Journal of Physics C: Solid State Physics,* vol. 5, 1972, pp. 2276–2282. Reprinted with permission of The Institute of Physics.

Fig. 11-6: Myers, V. H., Ono, A., and DeWitt, D. P. "A Method for Measuring Optical Properties of Semitransparent Materials at High Temperatures," *AIAA Journal,* vol. 24, no. 2, 1986, pp. 321–326.

Figs. 11-7a, b: Ebert, J. L., and Self, S. A., "The Optical Properties of Molten Coal Slag." In *Heat Transfer Phenomena in Radiation, Combustion and Fires,* vol. HTD-106, ASME, 1989, pp. 123–126. Reprinted by permission of ASME.

Figs. 13-2, 14-4: Reprinted with permission from *Journal of Quantitative Spectroscopy and Radiative Transfer,* vol. 23, Modest, M. F., and Azad, F. H. "The Differential Approximation for Radiative Transfer in an Emitting, Absorbing and Anisotropically Scattering Medium," 1980, Pergamon Press.

Fig. 14-7: Modest, M. F. "The Improved Differential Approximation for Radiative Transfer in Multi-Dimensional Media," *ASME Journal of Heat Transfer,* vol. 112, 1990, pp. 819–821. Reprinted with permission of ASME.

Fig 14-9: Modest, M. F., "The Improved Differential Approximation for Radiative Transfer in General Three-Dimensional Media." In *Heat Transfer Phenomena in Radiation, Combustion and Fires,* vol. HTD-106, ASME, 1989, pp. 213–220. Reprinted with permission of ASME.

Fig. 15-5: Abstracted from: Truelove, J. S. "Discrete-ordinate Solutions of the Radiation Transport Equation," *ASME Journal of Heat Transfer,* vol. 109, no. 4, 1987, pp. 1048–1051. Reprinted with permission of ASME.

Fig. 18-2: Hottel, H. C., and Sarofim, A. F. *Radiative Transfer,* McGraw-Hill Publishing Co., New York, 1967. Reprinted by permission of McGraw-Hill, Inc.

Fig. 19-2: Modest, M. F., and Poon, S. C. "Determination of Three-Dimensional Radiative Exchange Factors for the Space Shuttle by Monte Carlo," ASME paper no. 77-HT-49, 1977. Reprinted with permission of ASME.

Fig. 20-2: Sparrow, E. M., Eckert, E. R. G., and Irvine, T. P. "The Effectiveness of Radiating Fins with Mutual Irradiation," *Journal of the Aerospace Sciences,* no. 28, pp. 763–772, 1961.

Figs. 20-8, 20-9, 20-11: Reprinted with permission from *International Journal of Heat and Mass Transfer,* vol. 26, no. 4, Chan, S. H., Cho, D. H., and Kocamustafaogullari, G. "Melting and Solidification with Internal Radiative Transfer—A Generalized Phase Change Model," 1983, Pergamon Press.

Figs. 20-13a, b: Reprinted with permission from *International Journal of Heat and Mass Transfer,* vol. 5, Viskanta, R., and Grosh, R. J. "Effect of Surface Emissivity on Heat Transfer by Simultaneous Conduction and Radiation," 1962, Pergamon Press.

Fig. 20-14: Viskanta, R. "Radiation Transfer and Interaction of Convection with Radiation Heat Transfer." In *Advances in Heat Transfer,* vol. 3, Academic Press, Inc., New York, 1966, pp. 175–251.

Fig. 20-16: Kurosaki, Y. "Heat Transfer by Simultaneous Radiation and Convection in an Absorbing and Emitting Medium in a Flow Between Parallel Plates." In *Proceedings of the Fourth International Heat Transfer Conference,* vol. 3, no. R2.5, 1970.

Figs. 20-17, 20-18: Reprinted with permission from *International Journal of Heat and Mass Transfer,* vol. 24, Azad, F. H. and Modest, M. F. "Combined Radiation and Convection in Absorbing, Emitting and Anistropically Scattering Gas-Particulate Tube Flow," 1981, Pergamon Press.

Fig. 20-19: Song, B., and Viskanta, R. "Interaction of Radiation with Turbulence: Application to a Combustion System," *Journal of Thermophysics and Heat Transfer,* vol. 1, no. 1, 1987, pp. 56–62.

Unnumbered figures in Appendix D: Howell, J. R. *A Catalog of Radiation Configuration Factors,* McGraw-Hill, New York, 1982. Reprinted with permission.

AUTHOR INDEX

Bevans, J. T., 150, 281, 287, 293, 373, 374, 378
Bhattacharyya, A., 440, 447
Bhushan, B., 147
Bien, D. D., 796
Binnie, J. I., 447
Birkebak, R. C., 140, 150, 235, 268
Birmingham, B. W., 269
Blackwell, T. M., 147
Blokh, A. G., 434
Blomberg, M., 440, 441, 447
Bobco, R. P., 280, 292, 796
Bogomolov, S. Y., 749
Bohren, C. F., 46, 49, 73, 146, 386, 397, 433,
        442, 445, 448
Boles, M. A., 751
Boyd, I. W., 440, 441, 447
Boynton, F. P., 378
Bradar, F. J., 150
Brandenberg, W. M., 98, 128, 146
Branstetter, J. R., 277, 281, 292
Breene, R. G., 377
Breig, W. F., 801
Brenner, H., 504, 536
Brewster, M. Q., 413, 433
Briggs, D. G., 753
Brosmer, M. A., 377, 378
Brown, G. W., 671, 700
Bryant, F. D., 433
Buckius, R. O., 422, 424, 428, 429, 434, 435,
        671, 700
Buckley, H., 214, 226
Buettner, D., 750
Burak, L. D., 434
Burch, D. E., 377
Burka, A. L., 749, 752
Burkhard, D. G., 255, 269
Burrell, G. J., 435
Buschman, A. J., 795
Butler, C. P., 149
Byun, K. H., 748

Campbell, J. P., 796
Campo, A., 749, 754
Candell, L. M., 701
Caps, R., 447, 750
Carlson, B. G., 545, 546, 559, 560, 562, 567,
        568
Carpenter, J. R., 753
Carslaw, H. S., 751
Cartigny, J. D., 413, 433
Case, K. M., 478
Cashwell, E. D., 671, 700
Catton, I., 145
Cengel, Y. A., 479
Cesmebasi, E., 753

Cess, R. D., 33, 162, 188, 255, 268, 319, 360,
        371, 378, 587, 634, 713, 723, 736,
        737, 747, 752, 753, 797
Chambers, R. L., 708, 745
Chan, C. S., 494
Chan, S. H., 360, 377, 378, 479, 500, 724,
        725, 729, 731, 749, 751, 752, 754,
        756
Chandrasekhar, S., 46, 73, 323, 376, 476, 478,
        541, 567, 601, 636, 799, 801
Chang, L. C., 753
Chang, Y.-P., 747
Chao, B. T., 746
Chapman, A. J., 746
Charalmpopoulos, T. T., 430, 435, 436
Chawla, T. C., 479, 480, 749, 753, 754
Chebanov, V. A., 797
Chekhovskii, I. R., 797
Chen, J. C., 713, 746, 754
Chen, S. H. P., 149
Chen, W., 479
Chen, Y. N., 319, 494, 500, 750
Cheng, E. H., 737, 753
Cheng, P., 504, 520, 536
Chern, C. F., 756
Cheung, F. B., 753
Chi, D. C., 742, 755
Chien, K. Y., 635
Chin, J. H., 494, 500
Chippett, S., 430, 436
Cho, C., 751
Cho, D. H., 751
Choi, H. Y., 319
Chou, Y. S., 320
Chu, C. M., 318, 319, 388, 433
Chu, K. H., 377
Chung, B., 746
Chung, B. T. F., 661, 665, 742, 755, 797, 798
Chung, T. J., 797
Churchill, S. W., 318, 319, 388, 433, 494,
        500
Clark, G. C., 388, 433
Clark, J. A., 146
Clausen, O. W., 98, 128, 146
Clement, D., 150
Cogley, A. C., 474, 478
Cohen, E. S., 320, 376, 639, 641, 642, 665
Colles, M. J., 447
Condiff, D., 479, 504, 536
Conklin, P., 568
Courant, R., 214, 226
Cowell, W. R., 568
Cox, G., 756
Crawford, M., 797
Crawford, M. E., 319, 752